HANDBOOK
OF SMALL
ELECTRIC MOTORS

HANDBOOK OF SMALL ELECTRIC MOTORS

William H. Yeadon, P.E. Editor in Chief

Alan W. Yeadon, P.E. Associate Editor

Yeadon Energy Systems, Inc
Yeadon Engineering Services, P.C.

McGraw-Hill

New York   Chicago   San Francisco   Lisbon   London
Madrid   Mexico City   Milan   New Delhi
San Juan   Seoul   Singapore
Sydney   Toronto

Library of Congress Cataloging-in-Publication Data

Handbook of small electric motors / William H. Yeadon, editor in chief, Alan W. Yeadon, associate editor.
  p. cm.
  ISBN 0-07-072332-X
  1. Electric motors, Fractional horsepower.   I. Yeadon, William H.   II. Yeadon, Alan W.

TK2537 .H34   2001
621.46—dc21   00-048974

McGraw-Hill
A Division of The McGraw-Hill Companies

Copyright © 2001 by The McGraw-Hill Companies, Inc. All rights reserved. Printed in the United States of America. Except as permitted under the United States Copyright Act of 1976, no part of this publication may be reproduced or distributed in any form or by any means, or stored in a data base or retrieval system, without the prior written permission of the publisher.

1 2 3 4 5 6 7 8 9 0   DOC/DOC   0 7 6 5 4 3 2 1

ISBN 0-07-072332-X

The sponsoring editor for this book was Scott Grillo and the production supervisor was Sherri Souffrance. It was set in Times Roman by North Market Street Graphics.

Printed and bound by R. R. Donnelley & Sons Company.

McGraw-Hill books are available at special quantity discounts to use as premiums and sales promotions, or for use in corporate training programs. For more information, please write to the Director of Special Sales, Professional Publishing, McGraw-Hill, Two Penn Plaza, New York, NY 10121-2298. Or contact your local bookstore.

This handbook is intended to be used as a reference for information regarding the design and manufacture of electric motors. It is not intended to encourage or discourage any motor type, design, or process. Some of the configurations or processes described herein may be patented. It is the responsibility of the user of this information to determine if any infringement may occur as a result thereof.

> Information contained in this work has been obtained by The McGraw-Hill Companies, Inc. ("McGraw-Hill") from sources believed to be reliable. However, neither McGraw-Hill nor its authors guarantee the accuracy or completeness of any information published herein, and neither McGraw-Hill nor its authors shall be responsible for any errors, omissions, or damages arising out of use of this information. This work is published with the understanding that McGraw-Hill and its authors are supplying information but are not attempting to render engineering or other professional services. If such services are required, the assistance of an appropriate professional should be sought.

This handbook is dedicated to my wife, Luci Yeadon, who took most of the photographs for it.

William H. Yeadon
Editor in Chief

CONTENTS

Contributors xi
Preface xiii
Acknowledgments xv

Chapter 1. Basic Magnetics   1.1

1.1. Units / *1.1*
1.2. Definition of Terms / *1.6*
1.3. Estimating the Permeance of Probable Flux Paths / *1.21*
1.4. Electromechanical Forces and Torques / *1.32*
1.5. Magnetic Materials / *1.43*
1.6. Losses / *1.52*
1.7. Magnetic Moment (or Magnetic Dipole Moment) / *1.58*
1.8. Magnetic Field for a Current Loop / *1.65*
1.9. Helmholtz Coil / *1.67*
1.10. Coil Design / *1.67*
1.11. Reluctance Actuator Static and Dynamic (Motion) Analysis / *1.78*
1.12. Moving-Coil Actuator Static and Dynamic (Motion) Analysis / *1.83*
1.13. Electromagnetic Forces / *1.89*
1.14. Energy Approach (Energy-Coenergy) / *1.91*

Chapter 2. Materials   2.1

2.1. Magnetic Materials / *2.1*
2.2. Lamination Steel Specifications / *2.4*
2.3. Lamination Annealing / *2.6*
2.4. Core Loss / *2.46*
2.5. Pressed Soft Magnetic Material for Motor Applications / *2.51*
2.6. Powder Metallurgy / *2.59*
2.7. Magnetic Test Methods / *2.71*
2.8. Characteristics of Permanent Magnets / *2.80*
2.9. Insulation / *2.163*
2.10. Magnet Wire / *2.176*
2.11. Lead Wire and Terminations / *2.189*

Chapter 3. Mechanics and Manufacturing Methods   3.1

3.1. Motor Manufacturing Process Flow / *3.1*
3.2. End Frame Manufacturing / *3.4*
3.3. Housing Materials and Manufacturing Processes / *3.10*
3.4. Shaft Materials and Machining / *3.12*
3.5. Shaft Hardening / *3.14*
3.6. Rotor Assembly / *3.15*

- 3.7. Wound Stator Assembly Processing / *3.21*
- 3.8. Armature Manufacturing and Assembly / *3.22*
- 3.9. Assembly, Testing, Painting, and Packing / *3.23*
- 3.10. Magnetic Cores / *3.25*
- 3.11. Bearing Systems for Small Electric Motors / *3.46*
- 3.12. Sleeve Bearings / *3.72*
- 3.13. Process Control in Commutator Fusing / *3.79*
- 3.14. Armature and Rotor Balancing / *3.87*
- 3.15. Brush Holders for Small Motors / *3.99*
- 3.16. Varnish Impregnation / *3.103*
- 3.17. Adhesives / *3.109*
- 3.18. Magnetizers, Magnetizing Fixtures, and Test Equipment / *3.124*
- 3.19. Capacitive-Discharge Magnetizing / *3.138*

Chapter 4. Direct-Current Motors   4.1

- 4.1. Theory of DC Motors / *4.1*
- 4.2. Lamination, Field, and Housing Geometry / *4.46*
- 4.3. Commutation / *4.84*
- 4.4. PMDC Motor Performance / *4.96*
- 4.5. Series DC and Universal AC Performance / *4.116*
- 4.6. Shunt-Connected DC Motor Performance / *4.130*
- 4.7. Compound-Wound DC Motor Calculations / *4.134*
- 4.8. DC Motor Windings / *4.138*
- 4.9. Automatic Armature Winding Pioneering Theory and Practice / *4.140*

Chapter 5. Electronically Commutated Motors   5.1

- 5.1. Brushless Direct-Current (BLDC) Motors / *5.1*
- 5.2. Step Motors / *5.47*
- 5.3. Switched-Reluctance Motors / *5.99*

Chapter 6. Alternating-Current Induction Motors   6.1

- 6.1. Introduction / *6.1*
- 6.2. Theory of Single-Phase Induction Motor Operation / *6.5*
- 6.3. Three-Phase Induction Motor Dynamic Equations and Steady-State Equivalent Circuit / *6.38*
- 6.4. Single-Phase and Polyphase Induction Motor Performance Calculations / *6.45*

Chapter 7. Synchronous Machines   7.1

- 7.1. Induction Synchronous Motors / *7.1*
- 7.2. Hysteresis Synchronous Motors / *7.5*
- 7.3. Permanent-Magnet Synchronous Motors / *7.9*
- 7.4. Performance Calculation and Analysis / *7.16*

Chapter 8. Application of Motors   8.1

- 8.1. Motor Application Requirements / *8.1*
- 8.2. Velocity Profiles / *8.4*

8.3. Current Density / *8.9*
8.4. Thermal Analysis for a PMDC Motor / *8.10*
8.5. Summary of Motor Characteristics and Typical Applications / *8.32*
8.6. Electromagnetic Interference (EMI) / *8.32*
8.7. Electromagnetic Fields and Radiation / *8.35*
8.8. Controlling EMI / *8.38*

Chapter 9. Testing   9.1

9.1. Speed-Torque Curves / *9.1*
9.2. AC Motor Thermal Tests / *9.5*
9.3. DC Motor Testing / *9.9*
9.4. Motor Spectral Analysis / *9.15*
9.5. Resonance Control in Small Motors / *9.27*
9.6. Fatigue and Lubrication Tests / *9.36*
9.7. Qualification Tests for Adhesives and Plastic Assemblies / *9.42*
9.8. Trends in Test Automation / *9.43*

Chapter 10. Drives and Controls   10.1

10.1. Measurement Systems Terminology / *10.1*
10.2. Environmental Standards / *10.4*
10.3. Feedback Elements / *10.9*
10.4. Comparisons Between the Various Technologies / *10.42*
10.5. Future Trends in Sensor Technology / *10.48*
10.6. Selection of Short-Circuit Protection and Control for Design E Motors / *10.51*
10.7. Switched-Reluctance Motor Controls / *10.65*
10.8. Basic Stepping-Motor Control Circuits / *10.70*
10.9. Current Limiting for Stepping Motors / *10.80*
10.10. Microstepping / *10.93*
10.11. Brushless DC Motor Drive Schemes / *10.97*
10.12. Motor Drive Electronic Commutation Patterns / *10.119*
10.13. Performance Characteristics of BLDC Motors / *10.122*

References   R.1

Bibliography   B.1

Index   I.1
About the Contributors   C.1
About the Editors   E.1

CONTRIBUTORS

Larry C. Anderson *American Hoffman Corporation* (Sec. 3.14)
John S. Bank *Phoenix Electric Manufacturing Company* (Sec. 3.15)
Warren C. Brown *Link Engineering Company* (Sec. 3.10.6)
Joseph H. Bularzik *Magnetics International, Inc.* (Sec. 2.5)
Peter Caine *Oven Systems, Inc.* (Sec. 3.16)
David Carpenter *Vector Fields, Ltd.* (contributed the finite-element plots in Chap. 4)
John Cocco *Loctite Corporation* (Sec. 3.17)
Philip Dolan *Oberg Industries* (Sec. 3.10.5)
Birch L. DeVault *Cutler-Hammer* (Sec. 10.6)
Brad Frustaglio *Yeadon Energy Systems, Inc.* (Sec. 6.4)
Francis Hanejko *Hoeganaes Corporation* (Sec. 2.6)
Duane C. Hanselman *University of Maine* (Secs. 5.1.4 and 10.11)
Daniel P. Heckenkamp *Cutler-Hammer* (Sec. 10.6)
Leon Jackson *LDJ Electronics* (Sec. 3.19)
Dan Jones *Incremotion Associates* (Secs. 5.1.3, 10.12, and 10.13)
Douglas W. Jones *University of Iowa* (Secs. 5.2.10 and 10.8 to 10.10)
Mark A. Juds *Eaton Corporation* (Secs. 1.1 to 1.12)
Robert R. Judd *Judd Consulting Associates* (Secs. 2.2 and 2.3)
Ramani Kalpathi *Dana Corporation* (Sec. 10.7)
John Kauffman *Phelps Dodge Magnet Wire Company* (Sec. 2.10)
Todd L. King *Eaton Corporation* (Sec. 10.6)
H. R. Kokal *Magnetics International, Inc.* (Sec. 2.5)
Robert F. Krause *Magnetics International, Inc.* (Sec. 2.5)
Barry Landers *Electro-Craft Motion Control* (Chap. 9)
Roger O. LaValley *Magnetic Instrumentation, Inc.* (Sec. 3.18)
Bill Lawrence *Oven Systems, Inc.* (Sec. 3.16)
Andrew E. Miller *Software and motor designer* (Secs. 4.5, 6.4.3, and 6.4.4)
Stanley D. Payne *Windamatics Systems, Inc.* (Sec. 3.10.4)
Derrick Peterman *LDJ Electronics* (Sec. 3.19)

Curtis Rebizant   *Integrated Engineering Software* (contributed the boundary element plots in Figs. 5.58 to 5.61)

Earl F. Richards   *University of Missouri* (Secs. 1.14, 4.1, 6.2, 6.3, and 8.4 to 8.8)

Robert M. Setbacken   *Renco Encoders, Inc.* (Secs. 10.1 to 10.5)

Karl H. Schultz   *Schultz Associates* (Secs. 3.1 to 3.9)

Joseph J. Stupak Jr.   *Oersted Technology Corporation* (Sec. 2.8)

Chris A. Swenski   *Yeadon Energy Systems, Inc.* (Secs. 3.6.5, 6.4, and 7.4)

Harry J. Walters   *Oberg Industries* (Sec. 3.10.5)

Alan W. Yeadon   *Yeadon Engineering Services, PC* (Secs. 3.10, 3.11, and 4.2 to 4.5)

Luci Yeadon   *Luci's Photography* (contributed most of the photographs in this handbook)

William H. Yeadon   *Yeadon Engineering Services, PC* (Secs. 1.13, 2.1, 2.9, 2.11, 3.10 to 3.12, 4.3, 4.6 to 4.8, 5.1 to 5.3, 6.1, 6.4, 7.1 to 7.4, and 8.1 to 8.3)

PREFACE

When I was first approached about writing this handbook, it seemed like a fairly straightforward task. There was information available from a variety of sources. There were many capable people in the field who could contribute, and there was historical data from many sources.

The intent of this book is to cover the operating theory, practical design approaches, and manufacturing methods for the most common motors now in use. We have tried to meet this intent by including as much information as possible. The universe of motor information is huge, although most of the information is old. It became apparent that much information would have to be left out. We tried to include the basics along with that information we felt was most necessary and useful to those designing, manufacturing, and using motors. This is not a design course or a highly theoretical text but a place where people can go to get practical answers.

We are setting up a section at our Web site, www.yeadoninc.com, for people to comment about the book and tell us of improvements we can make and things they would like to have added to the next edition.

ACKNOWLEDGMENTS

Over the course of my career I have had the privilege to meet many of the giants of this industry. Many I have met through my association with the Small Motors and Motion Association (SMMA) and others through business relationships. Included among them are Dr. Cyril G. Veinott, Professor Philip H. Trickey, Dr. Ben Kuo, Dr. Duane Hanselman, and those authors who have contributed to this handbook.

There is, however, one person of whom I must make special mention. He is Dr. Earl Richards, Professor Emeritus of the University of Missouri at Rolla. This man never ceases to amaze me. He is always willing to help out selflessly with projects of this type. I have taught many motor design courses with him. When a student asks questions of him, he can start at the lowest level of understanding necessary and develop in a very understandable way a logical and reasonable answer to the question. His ability to communicate and teach is truly amazing. He has been very helpful in the preparation of this book.

I also need to acknowledge the dedication of my secretary, Kristina Wodzinski. Without her tireless effort this work would not have been completed.

CHAPTER 1
BASIC MAGNETICS

Chapter Contributors

Mark A. Juds
Earl F. Richards
William H. Yeadon

Electric motors convert electrical energy into mechanical energy by utilizing the properties of electromagnetic energy conversion. The different types of motors operate in different ways and have different methods of calculating the performance, but all utilize some arrangement of magnetic fields. Understanding the concepts of electromagnetics and the systems of units that are employed is essential to understanding electric motor operation. The first part of this chapter covers the concepts and units and shows how forces are developed. Nonlinearity of magnetic materials and uses of magnetic materials are explained. Energy and coenergy concepts are used to explain forces, motion, and activation. Finally, this chapter explains how motor torque is developed using these concepts.

1.1 UNITS*

Although the rationalized mks system of units [Système International (SI) units] is used in this discussion, there are also at least four other systems of units—cgs, esu, emu, and gaussian] that are used when describing electromagnetic phenomena. These systems are briefly described as follows.

MKS. Meter, kilogram, second.
CGS. Centimeter, gram, second.

---

* Sections 1.1 to 1.12 contributed by Mark A. Juds, Eaton Corporation.

ESU. CGS with $e_0 = 1$, based on Coulomb's law for electric poles.
EMU. CGS with $\mu_0 = 1$, based on Coulomb's law for magnetic poles.
Gaussian. CGS with electric quantities in esu and magnetic quantities in emu. The factor 4π appears in Maxwell's equation.
Rationalized mks. $\mu_0 = 4\pi \times 10^{-7}$ H/m, based on the force between two wires.
Rationalized cgs. $\mu_0 = 1$, based on Coulomb's law for magnetic poles. The factor 4π appears in Coulomb's law.

The rationalized mks and the rationalized cgs systems of units are the most widely used. These systems are defined in more detail in the following subsections.

1.1.1 The MKS System of Units

The rationalized mks system of units (also called SI units) uses the magnetic units tesla (T) and amps per meter (A/m), for flux density B and magnetizing force H, respectively. In this system, the flux density B is defined first (before H is defined), and is based on the force between two current-carrying wires. A distinction is made between B and H in empty (free) space, and the treatment of magnetization is based on amperian currents (equivalent surface currents).

Total or normal flux density B, T

$$B = \mu_0 (H + M) \tag{1.1}$$

Intrinsic flux density B_i, T

$$B_i = \mu_0 M \tag{1.2}$$

Permeability of free space μ_0, (T·m)/A

$$\mu_0 = 4\pi \times 10^{-7} \tag{1.3}$$

Magnetization M, A/m

$$M = \frac{B_I}{\mu_0} = \frac{m}{V} \tag{1.4}$$

Magnetic moment M, J/T

$$m = MV = pl \tag{1.5}$$

Magnetic pole strength p, A·m

$$p = \frac{m}{l} \tag{1.6}$$

1.1.2 The CGS System of Units

The rationalized cgs system of units uses the magnetic units of gauss (G) and oersted (Oe) for flux density B and magnetizing force H, respectively. In this system, the magnetizing force, or field intensity, H is defined first (before B is defined), and is based on the force between two magnetic poles. No distinction is made between B and H in empty (free) space, and the treatment of magnetization is based on magnetic poles. The unit *emu* is equivalent to an erg per oersted and is understood to mean the electromagnetic unit of magnetic moment.

Total or normal flux density B, G

$$B = H + 4\pi M \tag{1.7}$$

Intrinsic flux density B_i, G

$$B_i = 4\pi M \tag{1.8}$$

Permeability of free space μ_0, G/Oe

$$\mu_0 = 1 \tag{1.9}$$

Magnetization M, emu/cm^3

$$M = \frac{B_i}{4\pi} = \frac{m}{v} \tag{1.10}$$

Magnetic moment m, emu

$$m = MV = pl \tag{1.11}$$

Magnetic pole strength p, emu/cm

$$p = \frac{m}{l} \tag{1.12}$$

The magnetization or magnetic polarization M is sometimes represented by the symbols I or J, and the intrinsic flux density B_i is then represented as $4\pi M$, $4\pi I$, or $4\pi J$.

1.1.3 Unit Conversions

Magnetic Flux ϕ

1.0 Wb = 10^8 line
= 10^8 maxwell
= 1.0 V·s
= 1.0 H·A
= 1.0 T·m^2

Magnetic Flux Density B

1.0 T = 1.0 Wb/m^2
= 10^8 line/m^2
= 10^8 maxwell/m^2
= 10^4 G
= 10^9 γ

$1.0 \text{ G} = 1.0 \text{ line/cm}^2$
$= 10^{-4} \text{ T}$
$= 10^5 \text{ }\gamma$
$= 6.4516 \text{ line/in}^2$

Magnetomotive or Magnetizing Force *NI*

$1.0 \text{ A·turn} = 0.4 \text{ }\pi \text{ Gb}$
$= 0.4 \text{ }\pi \text{ Oe·cm}$

Magnetic Field Intensity *H*

$1.0 \text{ (A·turn)/m} = 4\pi \times 10^{-3} \text{ Oe}$
$= 4\pi \times 10^{-3} \text{ Gb/cm}$
$= 0.0254 \text{ (A·turn)/in}$
$1.0 \text{ Oe} = 79.5775 \text{ (A·turn)/m}$
$= 1.0 \text{ Gb/cm}$
$= 2.02127 \text{ (A·turn)/in}$

Permeability μ

$1.0 \text{ (T·m)/(A·turn)} = 10^7/4\pi \text{ G/Oe}$
$= 1.0 \text{ Wb/(A·turn·m)}$
$= 1.0 \text{ H/m}$
$= 1.0 \text{ N/(amp·turn)}^2$
$1.0 \text{ G/Oe} = 4\pi \times 10^{-7} \text{ H/m}$

Inductance *L*

$1.0 \text{ H} = 1.0 \text{ (V·s·turn)/A}$
$= 1.0 \text{ (Wb·turn)/A}$
$= 10^8 \text{ (line·turn)/A}$

Energy *W*

$1.0 \text{ J} = 1.0 \text{ W·s}$
$= 1.0 \text{ V·A·s}$
$= 1.0 \text{ Wb·A·turn}$
$= 1.0 \text{ N·m}$
$= 10^8 \text{ line·A·turn}$
$= 10^7 \text{ erg}$

Energy Density *w*

$1.0 \text{ MG·Oe} = 7.958 \text{ kJ/m}^3$
$= 7958 \text{ (T·A·turn)/m}$
$1.0 \text{ (T·A·turn)/m}^3 = 1.0 \text{ J/m}^3$

Force F

$1.0\text{ N} = 1.0\text{ J/m}$
$\phantom{1.0\text{ N}} = 0.2248\text{ lb}$
$1.0\text{ lb} = 4.448\text{ N}$

Magnetic Moment m

$1.0\text{ emu} = 1.0\text{ erg/Oe}$
$\phantom{1.0\text{ emu}} = 1.0\text{ erg/G}$
$\phantom{1.0\text{ emu}} = 10.0\text{ A·cm}^2$
$\phantom{1.0\text{ emu}} = 10^{-3}\text{ A·m}^2$
$\phantom{1.0\text{ emu}} = 10^{-3}\text{ J/T}$
$\phantom{1.0\text{ emu}} = 4\pi\text{ G·cm}^3$
$\phantom{1.0\text{ emu}} = 4\pi \times 10^{-10}\text{ Wb·m}$
$\phantom{1.0\text{ emu}} = 10^{-7}\text{ (N·m)/Oe}$

Magnetic Moment of the Electron Spin $\beta = eh/4\pi m_e$

$1.0\text{ Bohr magneton} = \beta = 9.274 \times 10^{-24}\text{ J/T}$
$\phantom{1.0\text{ Bohr magneton}} = \beta = 9.274 \times 10^{-24}\text{ A·m}^2$
$\phantom{1.0\text{ Bohr magneton}} = \beta = 9.274 \times 10^{-21}\text{ erg/G}$

Constants

Permeability of free space	$\mu_0 = 1.0\text{ G/Oe}$
	$\mu_0 = 4\pi \times 10^{-7}\text{ (T·m)/(A·turn)}$
	$\mu_0 = 4\pi \times 10^{-7}\text{ H/m}$
Electron charge	$e = 1.602177 \times 10^{-19}\text{ C}$
Electron mass	$m_e = 9.109390 \times 10^{-31}\text{ kg}$
Plank's constant	$h = 6.6262 \times 10^{-34}\text{ J·s}$
Velocity of light	$c = 2.997925 \times 10^8\text{ m/s}$
Pi	$\pi = 3.1415926536$
Acceleration of gravity	$g = 9.807\text{ m/s}^2$
	$= 32.174\text{ ft/s}^2$
	$= 386.1\text{ in/s}^2$

Miscellaneous

Length l	1.0 m	$= 39.37\text{ in}$
		$= 1.094\text{ yd}$
	1.0 in	$= 25.4\text{ mm}$
Time t	1.0 day	$= 24\text{ h}$
	1.0 min	$= 60\text{ s}$
Velocity v	1.0 m/s	$= 3.6\text{ km/h}$

$$= 3.281 \text{ ft/s}$$
Acceleration a $1.0 \text{ m/s}^2 = 3.281 \text{ ft/s}^2$
$$= 39.37 \text{ in/s}^2$$

1.2 DEFINITION OF TERMS

The Greeks discovered in 600 B.C. that certain metallic rocks found in the district of Magnesia in Thessaly would attract or repel similar rocks and would also attract iron. This material was called Magnes for the district of Magnesia, and is a naturally magnetic form of magnetite (Fe_3O_4), more commonly known as lodestone. *Lodestone* means "way stone," in reference to its use in compasses for guiding sailors on their way.

A bar-shaped permanent magnet suspended on a frictionless pivot (like a compass needle) will align with the earth's magnetic field. The end of the bar magnet that points to the earth's geographic north is designated as the *north magnetic pole* and the opposite end is designated as the *south magnetic pole*. If any tiny compass needles are placed around the bar magnet, they will line up to reveal the magnetic field shape of the bar magnet. Connecting lines along the direction of the compass needles show that the magnetic field lines emerge from one pole of the bar magnet and enter the opposite pole of the bar magnet. These magnetic field lines do not stop or end, but pass through the magnet to form closed curves or loops.

By convention, the magnetic field lines emerge from the north magnetic pole and enter through the south magnetic pole. Two permanent magnets will attract or repel each other in an effort to minimize the length of the magnetic field lines, which is why like poles repel and opposite poles attract. Therefore, since the north magnetic pole of a bar magnet points to the earth's geographic north, the earth's geographic north pole has a south magnetic polarity.

Hans Oersted discovered in 1820 that a compass needle is deflected by an electric current, and for the first time showed that electricity and magnetism are related. The magnetic field around a current-carrying wire can be examined by placing many tiny compass needles on a plane perpendicular to the axis of the wire. This shows that the magnetic field lines around a wire can be envisioned as circles centered on the wire and lying in planes perpendicular to the wire.

The direction of the magnetic field around a wire can be determined by using the *right-hand rule*, as follows (see Fig. 1.1). The thumb of your right hand is pointed in the direction of the current, where current is defined as the flow of positive charge from + to −. The fingers of your right hand curl around the wire to point in the direction of the magnetic field. If the current is defined as the flow of negative charge from − to +, then the left-hand rule must be used.

To summarize:

1. The north magnetic pole of a bar magnet will point to the earth's geographic north.
2. Magnetic field lines emerge from the north magnetic pole of a permanent magnet.
3. Magnetic field lines encircle a current-carrying wire.
4. Magnetic field lines do not stop or end, but form closed curves or loops that always encircle a current-carrying wire and/or pass through a permanent magnet.
5. The right-hand rule is used with current flowing from positive to negative and with the magnetic field lines emerging from the north magnetic pole.

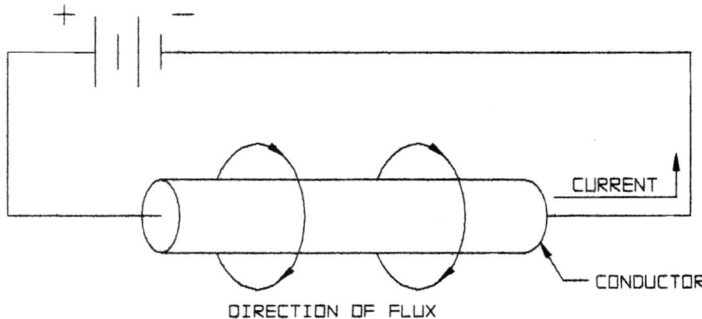

FIGURE 1.1 Direction of flux. *(Courtesy of Eaton Corporation.)*

1.2.1 System Performance

ϕ = magnetic flux
NI = magnetomotive or magnetizing force
$\mathcal{R}$ = reluctance
$\mathcal{P}$ = permeance
λ = flux linkage

Figure 1.2 shows a magnetic circuit based on an electrical analogy. In general, a coil with N turns of wire and I amperes (amps) provides the magnetomotive force NI that pushes the magnetic flux ϕ through a region (or a material) with a cross sectional area a and a magnetic flux path length l. In the electrical analogy, a voltage V provides the electromotive force that pushes an electrical current I through a region. The amount of magnetomotive force required per unit of magnetic flux is called reluctance $\mathcal{R}$. The amount of voltage required per amp is called resistance R.

$$\mathcal{R} = \frac{NI}{\phi} \tag{1.13}$$

$$NI = \phi\mathcal{R} \tag{1.14}$$

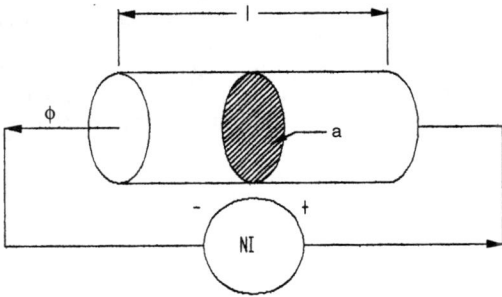

FIGURE 1.2 Magnetic circuit with electrical analogy. *(Courtesy of Eaton Corporation.)*

Electrical Analogy

$$R = \frac{V}{I} \tag{1.15}$$

$$V = IR \quad \text{(Ohm's law)} \tag{1.16}$$

$$\mathcal{P} = \frac{\phi}{NI} = \frac{1}{\mathcal{R}} \tag{1.17}$$

$$\phi = NI\,\mathcal{P} \tag{1.18}$$

$$\lambda = N\phi \tag{1.19}$$

$$V = \frac{d\lambda}{dt} = \frac{d(N\phi)}{dt} \quad \text{(Faraday's law)} \tag{1.20}$$

Visualization of Flux Linkage λ**.** Figure 1.3 shows 10 magnetic flux lines passing through 4 turns of wire in a coil. Each turn of the coil is linked to the 10 magnetic flux lines, like links in a chain. Therefore, the total flux linkage λ is obtained by multiplying the turns N by the magnetic flux ϕ, as defined in Eq. (1.19). In this case, the magnetic flux linkage $\lambda = 40$ line turns, where the units of N are turns and the units of magnetic flux ϕ are lines.

1.2.2 Material Properties

B = magnetic flux density
H = magnetic field intensity
μ = magnetic permeability

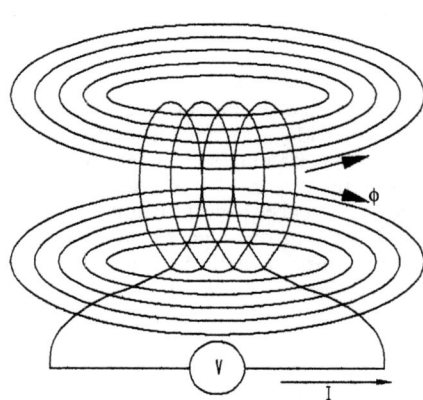

FIGURE 1.3 Flux linkage visualization. *(Courtesy of Eaton Corporation.)*

The magnetic flux density B is defined as the magnetic flux per unit of cross-sectional area a. The magnetic field intensity H is defined as the magnetomotive force per unit of magnetic flux path l. The magnetic permeability μ of the material is defined as the ratio between the magnetic flux density B and the magnetic field intensity. The permeability can also be obtained graphically from the magnetization curve shown in Fig. 1.4. In the electrical analogy, the current density J, the electric field intensity E, and the electrical conductivity σ are defined using ratios of similar physical parameters.

$$B = \frac{\phi}{a} \tag{1.21}$$

$$H = \frac{NI}{l} \tag{1.22}$$

$$\mu = \frac{B}{H} \tag{1.23}$$

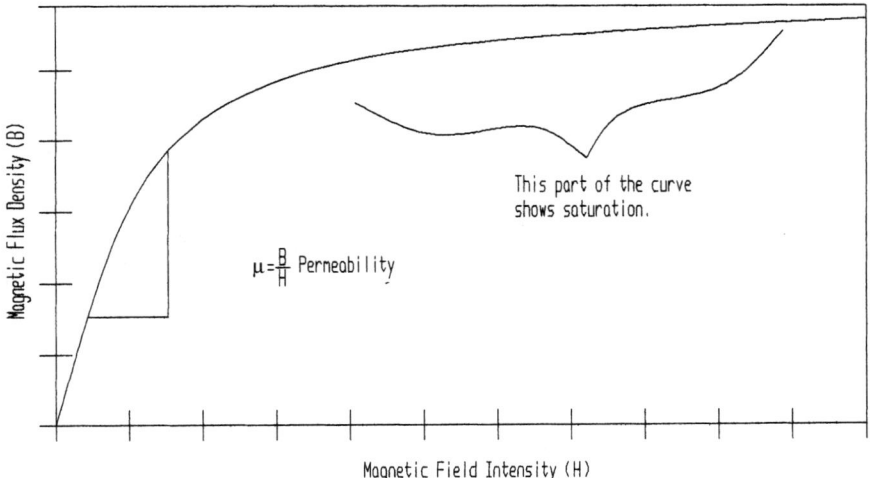

FIGURE 1.4 Magnetization *B-H* curve showing permeability μ. *(Courtesy of Eaton Corporation.)*

Electrical Analogy

$$J = \frac{I}{a} \tag{1.24}$$

$$E = \frac{V}{l} \tag{1.25}$$

$$\sigma = \frac{J}{E} \tag{1.26}$$

Permeability of free space, H/m or (T·m)/A:

$$\mu_0 = 4\pi \times 10^{-7} \tag{1.27}$$

Relative permeability:

$$\mu_r = \frac{\mu}{\mu_0} \tag{1.28}$$

where μ is the permeability of a material at any given point.

1.2.3 System Properties

$\mathcal{R}$ = reluctance
$\mathcal{P}$ = permeance
L = inductance

The reluctance $\mathcal{R}$, as defined in equation (1.17), can be written in a form based on the material properties and the geometry (μ, a, and l), as shown in Eq. (1.29).

$$\mathcal{R} = \frac{NI}{\phi} = \frac{Hl}{Ba} = \frac{l}{\mu a} \qquad (1.29)$$

The same can be done for the permeance $\mathcal{P}$, as shown in equation [1-30].

$$\mathcal{P} = \frac{1}{\mathcal{R}} = \frac{\phi}{NI} = \frac{\mu a}{l} \qquad (1.30)$$

The inductance L is defined as the magnetic flux linkage λ per amp I, as shown in Eq. (1.31). The inductance can be written in a form based on the material properties and the geometry (μ, a, and L), also shown in Eq. (1.31). The B–H curve can be easily changed into a λ–I curve, as shown in Fig. 1.5, and the inductance can then be obtained graphically.

$$L = \frac{\lambda}{I} = \frac{N\phi}{I} = N^2 \frac{\phi}{NI} = N^2 \mathcal{P} \qquad (1.31)$$

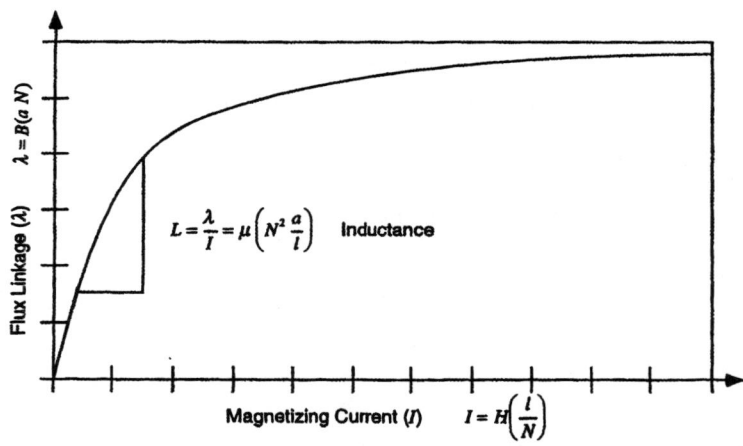

FIGURE 1.5 Magnetization λ-I curve showing inductance L. (*Courtesy of Eaton Corporation.*)

1.2.4 Energy

W_e = input electric energy
W_f = stored magnetic field energy
W_m = output mechanical energy
W_{co} = magnetic field coenergy

Energy Balance. All systems obey the first law of thermodynamics, which states that energy is conserved. This means that energy is neither created nor destroyed. Therefore, an energy balance can be written for a general system stating that the change in energy input to the system ΔW_e equals the change in energy stored in the system ΔW_f plus the change in energy output from the system ΔW_m. This energy balance is illustrated in the following equation.

$$\Delta W_e = \Delta W_f + \Delta W_m \qquad (1.32)$$

Input Electric Energy W_e**.** The input electrical energy can be calculated by integrating the coil voltage and current over time, as follows.

$$W_e = \int_0^t VI\,dt \tag{1.33}$$

Substituting Faraday's law, Eq. (1.20), into Eq. (1.33) shows that the input electrical energy is equal to the product of the coil magnetizing current I and the flux linkage λ.

$$W_e = \int_0^t I\,\frac{d\lambda}{dt}\,dt = \int_0^\lambda I\,d\lambda \tag{1.34}$$

$$W_e = I\lambda \tag{1.35}$$

Stored Magnetic Field Energy. As can be seen in Fig. 1.5, the flux linkage λ is a function of the magnetizing current I and depends on the material properties or the magnetization curve. The stored magnet field energy is calculated by integrating Eq. (1.34) over the magnetization curve. By inspection of Eq. (1.34), the area of integration lies above the magnetization curve, as shown in Fig. 1.6.

The stored magnetic field energy can be calculated for linear materials by substituting Eq. (1.31) into Eq. (1.34) as follows. Linear materials are characterized by a constant value of inductance L or permeability μ.

$$W_f = \int_0^\lambda I\,d\lambda = \int_0^\lambda \frac{\lambda}{L}\,d\lambda \tag{1.36}$$

$$W_f = \frac{1}{2}\frac{\lambda^2}{L} = \frac{1}{2}I\lambda = \frac{1}{2}LI^2 = \frac{1}{2}NI\phi = \frac{1}{2}(NI)^2\mathcal{P} = \frac{1}{2}\phi^2\mathcal{R}$$

where

$$L \text{ and } \mu \text{ are constant} \tag{1.37}$$

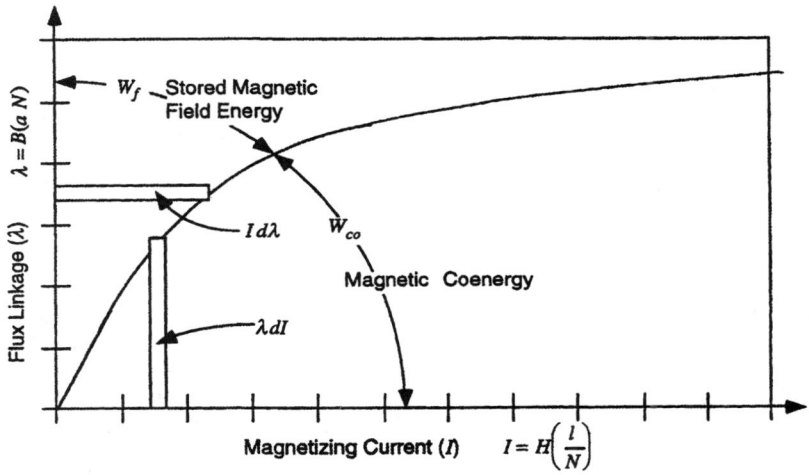

FIGURE 1.6 Stored magnetic field energy and magnetic coenergy. *(Courtesy of Eaton Corporation.)*

Output Mechanical Energy W_m. The change in mechanical energy ΔW_m is equal to the product of the mechanical force F and the distance over which it acts ΔX.

$$\Delta W_m = F \Delta X \tag{1.38}$$

In the limit as $\Delta X \to 0$,

$$dW_m = F \, dX \tag{1.39}$$

When a mechanical system includes a spring gradient k,

$$F = kX \tag{1.40}$$

Substituting Eq. (1.40) into Eq. (1.38) gives Eq. (1.41); integrating gives Eq. (1.42), the energy to change for a mechanical system with a linear spring.

$$\Delta W_m = \int_0^x F \, dX = \int_0^x kX \, dX \tag{1.41}$$

$$\Delta W_m = \frac{1}{2} kX^2 = \frac{1}{2} FX \tag{1.42}$$

Magnetic Coenergy W_{co}. The stored magnetic field energy W_f is derived from Eq. (1.34), and is represented by the area above the magnetization curve as previously described and as shown in Fig. 1.7. The magnetic coenergy W_{co} is represented by the area under the curve, and can be derived by starting with Eq. (1.34), as follows.

$$\Delta W_f = \int_0^\lambda I \, d\lambda \tag{1.43}$$

$$\Delta W_{co} = I\lambda - \Delta W_f = I\lambda - \int_0^\lambda I \, d\lambda = \int_0^I \lambda \, dI \tag{1.44}$$

The magnetic coenergy can be calculated for linear materials by substituting Eq. (1.31) into Eq. (1.44) as follows. Linear materials are characterized by a constant value of inductance L or permeability μ:

$$\Delta W_{co} = \int_0^I LI \, dI = \frac{1}{2} LI^2 = \frac{1}{2} NI\phi \tag{1.45}$$

where L and μ are constant.

A comparison of Eqs. (1.37) and (1.45) shows that the stored magnetic field energy W_f and the magnetic coenergy W_{co} are equal if the magnetic materials have linear properties:

$$\Delta W_f = \Delta W_{co} \tag{1.46}$$

where L and μ are constant.

Electromechanical Energy Conservation (Mechanical Force, Torque). The mechanical forces and torques produced by electromagnetic actuators are derived using the energy balance from Eq. (1.32). A graphical representation for the electromechanical energy conversion is shown in Fig. 1.7 to help in visualizing the derivation. Figure 1.7 shows two operating states for an electromagnetic actuator. State 1 is characterized by coil current I_1, flux linkage λ_1, and flux path length l_1. State 2 is characterized by coil current I_2, flux linkage λ_2, and flux path length l_2. The change in flux path length represents mechanical motion, which implies that there is a change in the mechanical energy. The change in the magnetization curve from flux path length l_1 to l_2 reflects a change in the inductance, as described in Eqs. (1.30) and (1.31).

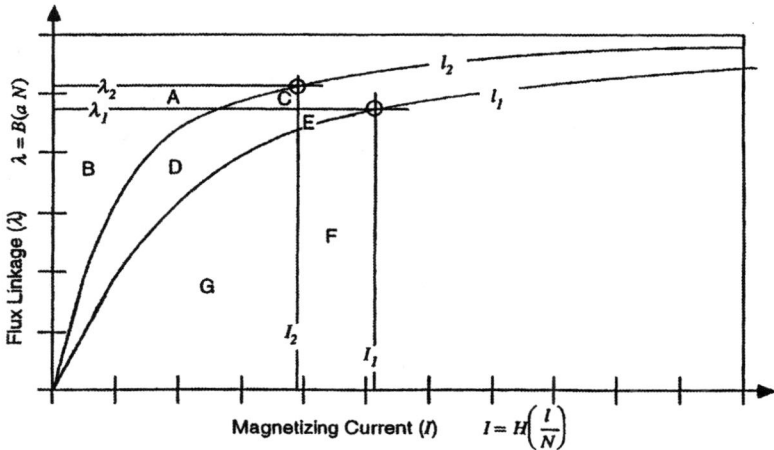

FIGURE 1.7 Graphical visualization of electromechanical energy conversion. *(Courtesy of Eaton Corporation.)*

The change in electric energy and the change in stored magnetic field energy are defined as follows.

$$\Delta W_e = W_{e2} - W_{e1} \tag{1.47}$$

$$\Delta W_f = W_{f2} - W_{f1} \tag{1.48}$$

The electric energy and the stored magnetic field energy for states 1 and 2 can be obtained by using the applicable regions above and below the magnetization curve designated as A, B, C, D, E, F, and G in Fig. 1.7.

Electric energy:

$$W_{e1} = I_1 \lambda_1 = B + D + E + F + G \tag{1.49}$$

$$W_{e2} = I_2 \lambda_2 = A + B + C + D + G \tag{1.50}$$

$$\Delta W_e = (A + B + C + D + G) - (B + D + E + F + G) \tag{1.51}$$

$$\Delta W_e = A + C - E - F \tag{1.52}$$

Stored magnetic field energy:

$$W_{f1} = \int_0^{\lambda_1} I\, d\lambda = B + D + E \tag{1.53}$$

$$W_{f2} = \int_0^{\lambda_2} I\, d\lambda = A + B \tag{1.54}$$

$$\Delta W_f = (A + B) - (B + D + E) \tag{1.55}$$

$$\Delta W_f = A - D - E \tag{1.56}$$

The resulting change in the mechanical energy is obtained by rewriting the energy balance, Eq. (1.32), and substituting the results from Eqs. (1.52) and (1.56), as follows.

$$\Delta W_m = \Delta W_e - \Delta W_f \qquad (1.57)$$

$$\Delta W_m = (A + C - E - F) - (A - D - E) \qquad (1.58)$$

$$\Delta W_m = D + C - F \qquad (1.59)$$

The magnetic coenergy for states 1 and 2 can also be obtained by using the applicable regions below the magnetization curve from Fig. 1.7, as follows.

$$W_{co1} = \int_0^{I_1} \lambda \, dI = G + F \qquad (1.60)$$

$$W_{co2} = \int_0^{I_2} \lambda \, dI = D + C + G \qquad (1.61)$$

$$W_{co} = (D + C - F) - (G + F) \qquad (1.62)$$

$$\Delta W_{co} = D + C - F \qquad (1.63)$$

$$\Delta W_m = W_{co} \qquad (1.64)$$

A comparison of the results from Eqs. (1.59) and (1.63) shows that the change in mechanical energy is equal to the change in magnetic coenergy.

$$\Delta W_m = W_{co} \qquad (1.65)$$

The mechanical force can be calculated as follows, by substituting Eq. (1.38) into Eq. (1.65).

$$F \Delta X = \Delta W_{co} \qquad (1.66)$$

$$F = \frac{\Delta W_{co}}{\Delta X} \qquad (1.67)$$

$$F = \frac{dW_{co}}{dX} \quad \text{in the limit as } \Delta X \to 0 \qquad (1.68)$$

The mechanical torque can be calculated using Eq. (1.66) and the radius r that relates the force F, the torque T, the linear displacement X, and the angular displacement θ.

$$F = \frac{T}{r} \qquad (1.69)$$

$$\Delta X = r \, \Delta \theta \qquad (1.70)$$

$$F \Delta X = T \Delta \theta = \Delta W_{co} \qquad (1.71)$$

$$T = \frac{\Delta W_{co}}{\Delta \theta} \qquad (1.72)$$

$$T = \frac{dW_{co}}{d\theta} \quad \text{in the limit as } \Delta X \to 0 \qquad (1.73)$$

When the energy equations are applied to an air gap where L and μ are constant, the coenergy is equal to the stored field energy ($\Delta W_f = \Delta W_{co}$), and Eq. (1.37) can be substituted into Eqs. (1.68) and (1.73) as follows:

$$F = \frac{d}{dX}\left[\frac{1}{2}(NI)^2 \mathcal{P}\right] = \frac{d}{dX}\left(\frac{1}{2}\phi^2 \mathcal{R}\right) \qquad (1.74)$$

$$T = \frac{d}{d\theta}\left[\frac{1}{2}(NI)^2 \mathcal{P}\right] = \frac{d}{d\theta}\left(\frac{1}{2}\phi^2 \mathcal{R}\right) \tag{1.75}$$

for constant L, μ.

1.2.5 Application of the Force and Energy Equations to an Actuator

The purpose of this section is to show how the energy and force equations can be applied to an actuator to determine the armature force. The reluctance actuator shown in Fig. 1.8a will be used for this discussion. The saturable iron regions of the actuator include the armature, which moves in the x direction, two stationary poles, and a coil core. The magnetic flux generally remains in the iron regions; however, it must cross air gap 1 and air gap 2 to reach the armature. Some of the magnetic flux finds alternative air paths which bypass the armature; these flux paths are called *leakage flux paths*. The air flux path shape and the reluctance equations are derived from the reluctance definition $\mathcal{R} = 1/\mu a$ in Sec. 1.3.

The first step in constructing an equivalent reluctance model is to identify each iron flux path and each air flux path for the actuator. The iron flux paths include the core, pole 1, pole 2, and the armature, and the reluctances are designated as $\mathcal{R}_{core}$, $\mathcal{R}_{P1}$, $\mathcal{R}_{P2}$, and $\mathcal{R}_{arm}$, respectively. The air flux paths include gap 1, gap 2, leakage 1, and leakage 2, and the reluctances are designated as $\mathcal{R}_{g1}$, $\mathcal{R}_{g2}$, $\mathcal{R}_{L1}$, and $\mathcal{R}_{L2}$, respectively. By observing the expected paths of the magnetic flux (Fig. 1.8a), an equivalent reluctance network can be assembled (Fig. 1.8b), in which each reluctance is modeled as an equivalent electrical resistor. The coil is shown as an amp-turn NI source in series with the core reluctance (Fig. 1.8b), and is modeled as an equivalent voltage source directed toward the left (from − to +), according to the right-hand rule. The magnetic flux ϕ is modeled as an equivalent electric current.

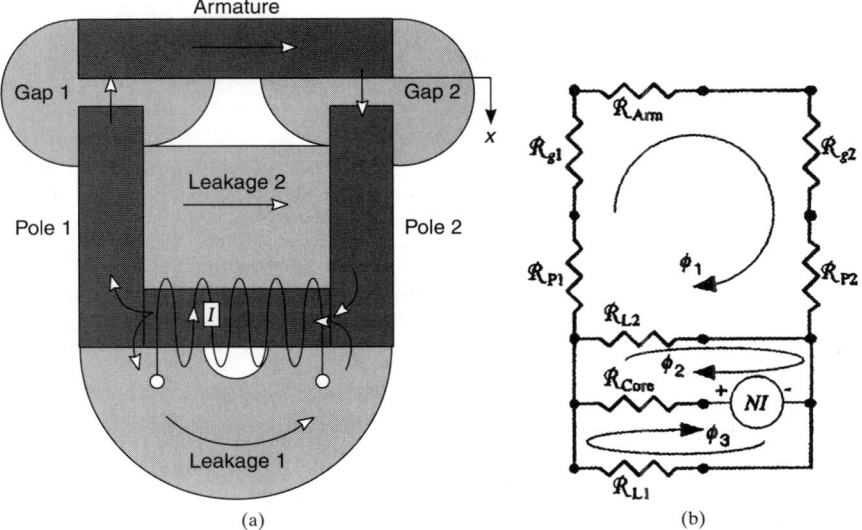

FIGURE 1.8 (a) Actuator iron and air flux paths, and (b) equivalent reluctance network. *(Courtesy of Eaton Corporation.)*

Three magnetic flux loops (ϕ_1, ϕ_2, ϕ_3) can be defined in the equivalent reluctance network (Fig. 1.8b). Three loop equations can be written in which all of the amp-turn NI drops around each flux loop must sum to zero (magnetic Kirchoff's law), as follows.

Flux loop ϕ_1

$$\sum NI = 0 = \phi_1(\mathcal{R}_{P1} + \mathcal{R}_{P2} + \mathcal{R}_{g1} + \mathcal{R}_{g2} + \mathcal{R}_{arm} + \mathcal{R}_{L2}) - \phi_2(\mathcal{R}_{L2}) \tag{1.76}$$

Flux loop ϕ_2

$$\sum NI = 0 = -\phi_1(\mathcal{R}_{L2}) + \phi_2(\mathcal{R}_{L2} + \mathcal{R}_{core}) - \phi_3(\mathcal{R}_{core}) - NI \tag{1.77}$$

Flux loop ϕ_3

$$\sum NI = 0 = -\phi_2(\mathcal{R}_{core}) + \phi_3(\mathcal{R}_{L1} + \mathcal{R}_{core}) + NI \tag{1.78}$$

The only unknowns in these equations are the magnetic fluxes (ϕ_1, ϕ_2, ϕ_3), which can be determined by simultaneously solving the loop equations. The resulting core flux and the leakage flux in the second leakage path can be determined as follows:

Magnetic flux in the core

$$\phi_{core} = \phi_2 - \phi_3 \tag{1.79}$$

Magnetic flux in the second leakage path

$$\phi_{L2} = \phi_2 - \phi_1 \tag{1.80}$$

The amp-turn NI drop across each one of the reluctances $\mathcal{R}$ can be determined as follows.

Amp-turn drop across iron armature

$$NI_{arm} = \phi_1 \mathcal{R}_{arm} \tag{1.81}$$

Amp-turn drop across iron pole 1

$$NI_{P1} = \phi_1 \mathcal{R}_{P1} \tag{1.82}$$

Amp-turn drop across iron pole 2

$$NI_{P2} = \phi_1 \mathcal{R}_{P2} \tag{1.83}$$

Amp-turn drop across iron core

$$NI_{core} = \phi_{core} \mathcal{R}_{core} \tag{1.84}$$

Amp-turn drop across air gap 1

$$NI_{g1} = \phi_1 \mathcal{R}_{g1} \tag{1.85}$$

Amp-turn drop across air gap 2

$$NI_{g2} = \phi_1 \mathcal{R}_{g2} \tag{1.86}$$

Amp-turn drop across air leakage 1

$$NI_{L1} = \phi_3 \mathcal{R}_{L1} \tag{1.87}$$

Amp-turn drop across air leakage 2

$$NI_{L2} = \phi_{L2} \mathcal{R}_{L2} \tag{1.88}$$

The magnetic field intensity H and the magnetic flux density B for the nonlinear iron regions can be determined as follows.

Iron armature

$$H_{arm} = \frac{NI_{arm}}{l_{arm}} \qquad B_{arm} = \frac{\phi_1}{a_{arm}} \tag{1.89}$$

Iron pole 1

$$H_{P1} = \frac{NI_{P1}}{l_{P1}} \qquad B_{P1} = \frac{\phi_1}{a_{P1}} \tag{1.90}$$

Iron pole 2

$$H_{P2} = \frac{NI_{P2}}{l_{P2}} \qquad B_{P2} = \frac{\phi_1}{a_{P2}} \tag{1.91}$$

Iron core

$$H_{core} = \frac{NI_{core}}{l_{core}} \qquad B_{core} = \frac{\phi_{core}}{a_{core}} \tag{1.92}$$

This completes the solution for the state of the magnetic field in the actuator. The x-direction force on the armature can now be determined by calculating the change in the magnetic coenergy as a function of armature displacement in the x-direction. The magnetic coenergy can be calculated for the entire actuator or for just the working air gaps.

Magnetic Coenergy Applied to the Entire Actuator. In general, the magnetic coenergy is calculated from Eq. (1.44), which requires knowledge of the nonlinear magnetic properties in terms of the flux linkage λ and current I. However, the magnetic properties for ferromagnetic materials are normally published in terms of the magnetic flux density B and magnetic field intensity H, as B-H curves. By substitution of the definitions for flux linkage $\lambda = N\phi$, magnetic flux density $\phi = Ba$, and magnetic field intensity $NI = Hl$ into Eq. (1.44), the magnetic coenergy can be calculated from the B-H property characteristics, as follows:

$$W_{co} = \int_0^I \lambda \, dI = \int_0^{NI} \phi \, dNI = al \int_0^H B \, dH \tag{1.93}$$

Iron is a ferromagnetic material and by definition it has nonlinear magnetic properties. Therefore, the magnetic coenergy in each of the iron reluctances must be calculated by integrating the area under the B-H curve, as follows. The total magnetic coenergy in the iron is the summation of the iron coenergies.

Magnetic coenergy in the armature

$$W_{co,arm} = a_{arm} l_{arm} \int_0^{H_{arm}} B dH \qquad (1.94)$$

Magnetic coenergy in pole 1

$$W_{co,P1} = a_{P1} l_{P1} \int_0^{H_{P1}} B dH \qquad (1.95)$$

Magnetic coenergy in pole 2

$$W_{co,P2} = a_{P2} l_{P2} \int_0^{H_{P2}} B dH \qquad (1.96)$$

Magnetic coenergy in the core

$$W_{co,core} = a_{core} l_{core} \int_0^{H_{core}} B dH \qquad (1.97)$$

Total magnetic coenergy in iron

$$W_{co,iron} = W_{co,apm} + W_{co,P1} + W_{co,P2} + W_{co,core} \qquad (1.98)$$

Air is not a ferromagnetic material, and by definition it has linear or constant magnetic properties. Therefore, the magnetic coenergy in each of the air reluctances is identical to the stored magnetic field energy ($\Delta W_f = \Delta W_{co}$). The magnetic coenergy in each of the air reluctances can be calculated from Eq. (1.37), as follows. The total magnetic coenergy in the air is the summation of the air coenergies.

Magnetic coenergy in air gap 1

$$W_{co,g1} = \frac{1}{2} N I_{g1} \phi_1 \qquad (1.99)$$

Magnetic coenergy in air gap 2

$$W_{co,g2} = \frac{1}{2} N I_{g2} \phi_1 \qquad (1.100)$$

Magnetic coenergy in leakage 1

$$W_{co,L1} = \frac{1}{2} N I_{L1} \phi_3 \qquad (1.101)$$

Magnetic coenergy in leakage 2

$$W_{co,L2} = \frac{1}{2} N I_{L2} \phi_{L2} \qquad (1.102)$$

Total magnetic coenergy in air

$$W_{co,air} = W_{co,g1} + W_{co,g2} + W_{co,L1} + W_{co,L2} \qquad (1.103)$$

The total magnetic coenergy for the entire actuator is the summation of the iron and the air coenergies, as follows.

$$W_{co,tot} = W_{co,iron} + W_{co,air} \qquad (1.104)$$

BASIC MAGNETICS 1.19

The armature force F_{arm} can be obtained by calculating the total actuator magnetic coenergy $W_{co,tot}$ at each of two armature positions, x_1 and x_2. The resulting armature force is the average force over the armature position change $x = x_1$ to x_2, and in the direction of the armature position change.

Total actuator magnetic coenergy at $x = x_1$

$$W_{co1} = W_{co,tot}|_{x1} \tag{1.105}$$

Total actuator magnetic coenergy at $x = x_2$

$$W_{co2} = W_{co,tot}|_{x2} \tag{1.106}$$

Average armature force over $x = x_1$ to x_2

$$F_{arm} = \frac{\Delta W_{co}}{\Delta x} = \frac{W_{co2} - W_{co1}}{x_2 - x_1} \tag{1.107}$$

Magnetic Coenergy Applied to the Working Air Gaps. Since the armature force is produced across the working air gaps (gap 1 and gap 2), the armature force can be determined by considering the coenergy change in the working gaps alone. The total magnetic coenergy in the working air gaps is the summation of the air gap coenergies from Eqs. (1.99) and (1.100), as follows:

Magnetic coenergy of air gap 1

$$W_{co,g1} = \frac{1}{2} NI_{g1}\phi_1 = \frac{1}{2} NI_{g1}^2 \mathcal{P}_{g1} \tag{1.108}$$

Magnetic coenergy of air gap 2

$$W_{co,g2} = \frac{1}{2} NI_{g2}\phi_1 = \frac{1}{2} NI_{g2}^2 \mathcal{P}_{g2} \tag{1.109}$$

Total working air gap coenergy

$$W_{co,gap} = W_{co,g1} + W_{co,g2} = \frac{1}{2} NI_{g1}^2 \mathcal{P}_{g1} + \frac{1}{2} NI_{g2}^2 \mathcal{P}_2 \tag{1.110}$$

The armature force F_{arm} can be obtained by calculating the total working air gap magnetic coenergy $W_{co,gap}$ at each of two armature positions x_1 and x_2. The resulting armature force is the average force over the armature position change $x = x_1$ to x_2, and in the direction of the armature position change.

Coenergy change

$$\Delta W_{co} = W_{co,gap}|_{x2} - W_{co,gap}|_{x1} \tag{1.111}$$

Coenergy change

$$\Delta W_{co} = \frac{1}{2}\left[NI_{g1}^2 \mathcal{P}_{g1} + NI_{g2}^2 \mathcal{P}_{g2}\right]_{x2} - \frac{1}{2}\left[NI_{g1}^2 \mathcal{P}_{g1} + NI_{g2}^2 \mathcal{P}_{g2}\right]_{x1} \tag{1.112}$$

Average force

$$F_{\text{arm}} = \frac{\Delta W_{\text{co}}}{\Delta x} = \frac{\lfloor NI_{g1}^2 \mathcal{P}_{g1} + NI_{g2}^2 \mathcal{P}_{g2} \rfloor_{x2} - \lfloor NI_{g1}^2 \mathcal{P}_{g1} + NI_{g2}^2 \mathcal{P}_{g2} \rfloor_{x1}}{2(x_2 - x_1)} \quad (1.113)$$

1.2.6 Summary of Magnetic Terminology

The analysis and design of electromagnetic devices can be accomplished by using the relations presented in Secs. 1.2.1 to 1.2.4. The key equations from these sections are listed here.

System Performance

Magnetic Ohm's law

$$NI = \phi \mathcal{R} \quad (1.14)$$

$$\phi = NI \mathcal{P} \quad (1.18)$$

Flux linkage

$$\lambda = N\phi \quad (1.19)$$

Faraday's law

$$V = \frac{d\lambda}{dt} = \frac{d(N\phi)}{dt} \quad (1.20)$$

System Properties

Reluctance

$$\mathcal{R} = \frac{NI}{\phi} = \frac{1}{\mu a} \quad (1.29)$$

Permeance

$$\mathcal{P} = \frac{1}{\mathcal{R}} = \mu \frac{a}{l} \quad (1.30)$$

Inductance

$$L = \frac{\lambda}{I} = \frac{N\phi}{I} = N^2 P \quad (1.31)$$

Material Properties

Magnetic flux density

$$B = \frac{\phi}{a} \quad (1.21)$$

Magnetic field intensity

$$H = \frac{NI}{l} \quad (1.22)$$

BASIC MAGNETICS

Permeability

$$\mu = \frac{B}{H} \quad (1.23)$$

Permeability of free space, (T·m)/A

$$\mu_0 = 4\pi \times 10^{-7} \quad (1.27)$$

Relative permeability

$$\mu_r = \frac{\mu}{\mu_0} \quad (1.28)$$

Energy

Magnetic field energy

$$W_f = \int_0^\lambda I\, d\lambda \quad (1.36)$$

$$W_f = \frac{1}{2} I\lambda = \frac{1}{2} LI^2 = \frac{1}{2} NI\phi = \frac{1}{2}(NI)^2 \mathcal{P} = \frac{1}{2}\phi^2 \mathcal{R} \quad \text{for constant } L, \mu \quad (1.37)$$

Magnetic coenergy

$$W_{co} = \int_0^I \lambda\, dI \quad (1.44)$$

$$W_{co} = W_f \quad \text{for constant } L, \mu \quad (1.46)$$

Mechanical energy

$$\Delta W_m = F \Delta X \quad (1.38)$$

Force or Torque

Force

$$F = \frac{dW_{co}}{dX} \quad (1.68)$$

$$F = \frac{d}{dX}\left[\frac{1}{2}(NI)^2 \mathcal{P}\right] = \frac{d}{dX}\left(\frac{1}{2}\phi^2 \mathcal{R}\right) \quad \text{for constant } L, \mu \quad (1.74)$$

Torque

$$T = \frac{dW_{co}}{d\theta} \quad (1.73)$$

$$T = \frac{d}{d\theta}\left[\frac{1}{2}(NI)^2 \mathcal{P}\right] = \frac{d}{d\theta}\left(\frac{1}{2}\phi^2 \mathcal{R}\right) \quad \text{for constant } L, \mu \quad (1.75)$$

1.3 ESTIMATING THE PERMEANCE OF PROBABLE FLUX PATHS

Defining the permeance of the steel parts is very simple because the field is generally confined to the steel. Therefore, the flux path is very well defined because it has the same geometry as the steel parts.

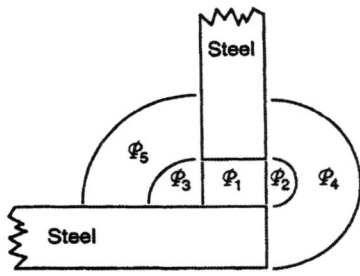

FIGURE 1.9 Air gap permeance paths. *(Courtesy of Eaton Corporation.)*

The flux path through air is complex. In general, the magnetic flux in the air is perpendicular to the steel surfaces and spreads out into a wide area. As an example, Fig. 1.9 shows five of the flux paths for a typical air gap between two pieces of steel. The total permeance of the air gap is equal to the sum of the permeances for the parallel flux paths. The permeance of each path can be calculated based on the dimensions shown in Fig. 1.10 and on Eq. (1.30), as follows. Path $\mathcal{P}_1$ is the direct face-to-face flux path. Paths $\mathcal{P}_2, \mathcal{P}_3, \mathcal{P}_4$, and $\mathcal{P}_5$ are generally identified as fringing paths. H. C. Roters (1941) recommends that the value for dimension h, as shown in Fig. 1.10, should be equal to 90 percent of the smaller thickness of the two steel parts, $h = 0.9t$. However, it is easier to remember the slightly larger value of $h = 1.0t$, and there is no significant loss in accuracy. Two examples of magnetic flux lines are shown in Fig. 1.11 (iron filings on a U-shaped magnet) and Fig. 1.12 (finite element result flux-line plot). In these examples it is easy to see the general flux path shapes shown in Fig. 1.9.

Path $\mathcal{P}_1$. The direct face-to-face air gap flux path has the same geometry as the perpendicular interface region between the two steel parts.

$$a_1 = tw \tag{1.114}$$

$$l_1 = g \tag{1.115}$$

$$\mathcal{P}_1 = \mu_0 \frac{a_1}{l_1} = \mu_0 \frac{tw}{g} \tag{1.116}$$

Path $\mathcal{P}_2$ (Half Cylinder). The cross-sectional area of this flux path varies along the length of the path. Therefore, Eq. (1.30) is modified as follows, where v_2 is the vol-

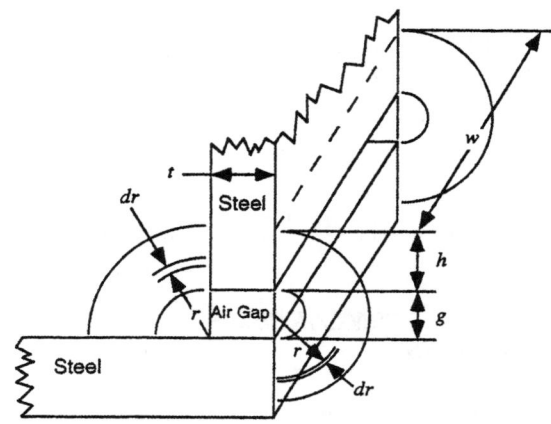

FIGURE 1.10 Air gap and steel part dimensions. *(Courtesy of Eaton Corporation.)*

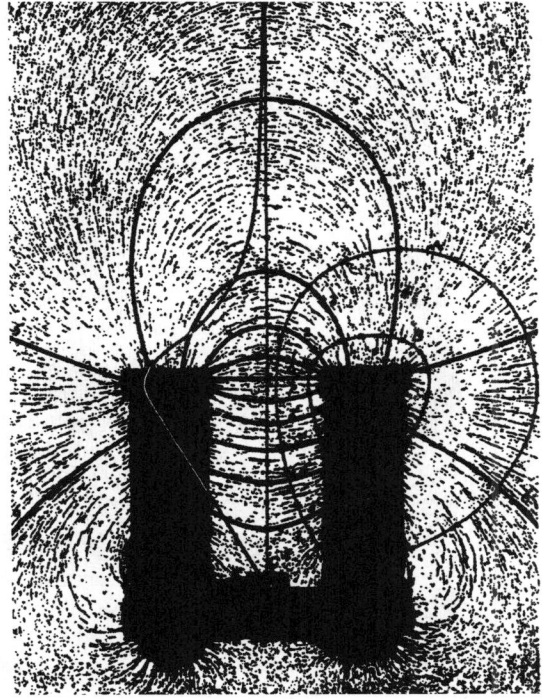

FIGURE 1.11 Magnetic flux lines illustrated by iron filings on U-shaped magnet. *(Courtesy of H.C. Roters and Eaton Corporation.)*

ume of flux path $\mathcal{P}_2$. Roters (1941) uses a graphical approximation to the mean path length, resulting in a permeance with a value of $\mathcal{P}_2 = 0.26\,\mu_0 w$, which is slightly larger than that shown here.

$$\mathcal{P}_2 = \mu_0 \frac{a_2}{l_2} \frac{l_2}{l_2} = \mu_0 \frac{v_2}{l_2^2} \tag{1.117}$$

$$r_2 = \frac{g}{2} \tag{1.118}$$

$$v_2 = \frac{w}{2} \pi r_2^2 = \frac{\pi}{8} w g^2 \tag{1.119}$$

$$l_2 = \frac{1}{2}(g + \pi r_2) = 1.285 g \quad \text{(average of inner and outer paths)} \tag{1.120}$$

$$\mathcal{P}_2 = \mu_0 \frac{v_2}{l_2^2} = 0.24 \mu_0 w \tag{1.121}$$

Path $\mathcal{P}_3$ (Quarter Cylinder). This flux path is very similar to flux path $\mathcal{P}_2$, and the calculation method is identical. Roters (1941) uses a graphical approximation to the

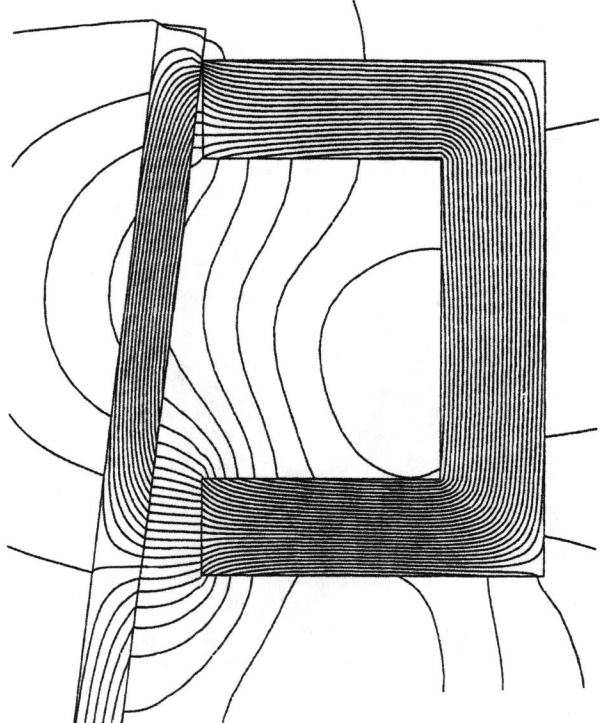

FIGURE 1.12 Magnetic flux lines illustrated by a finite element solution flux-line plot on a U-shaped magnet. *(Courtesy of the Ansoft/DMAS finite element program and Eaton Corporation.)*

mean path length, resulting in a permeance with a value of $\mathcal{P}_3 = 0.52\ \mu_0 w$, which is slightly larger than that shown here.

$$r_3 = g \tag{1.122}$$

$$l_3 = \frac{1}{2}\left(g + \frac{\pi}{2} r_3\right) = 1.285g \qquad \text{(average of inner and outer paths)} \tag{1.123}$$

$$v_3 = \frac{w}{4}\pi r_3^2 = \frac{\pi}{4} wg^2 \tag{1.124}$$

$$\mathcal{P}_3 = \mu_0 \frac{v_3}{l_3^2} = 0.48\mu_0 w \tag{1.125}$$

Path $\mathcal{P}_4$ (Half Cylindrical Shell). The cross-sectional area of this flux path is constant. However, the magnetic flux path length increases as the radius r increases. Therefore, Eq. (1.30) is written in differential form and the permeance is calculated

BASIC MAGNETICS

by integrating over the radius as follows. Roters (1941) uses the same procedure and shows the same results.

$$d\mathcal{P}_4 = \mu_0 \frac{da_4}{l_4} \tag{1.126}$$

$$\frac{g}{2} < r_4 < \frac{g}{2} + h \tag{1.127}$$

$$da_4 = w\, dr_4 \tag{1.128}$$

$$l_4 = \pi r_4 \tag{1.129}$$

$$\mathcal{P}_4 = \mu_0 \frac{w}{\pi} \int_{g/2}^{g/2+h} \frac{1}{r_4}\, dr_4 = \mu_0 \frac{w}{\pi} \ln\left(\frac{g/2+h}{g/2}\right) \tag{1.130}$$

$$\mathcal{P}_4 = \mu_0 \frac{w}{\pi} \ln\left(1 + 2\frac{h}{g}\right) \tag{1.131}$$

Path $\mathcal{P}_5$ (Quarter Cylindrical Shell). This flux path is very similar to flux path $\mathcal{P}_4$, and the calculation method is identical. Roters (1941) uses the same procedure and shows the same result.

$$d\mathcal{P}_5 = \mu_0 \frac{da_5}{l_5} \tag{1.132}$$

$$g < r_5 < g + h \tag{1.133}$$

$$da_5 = w\, dr_5 \tag{1.134}$$

$$l_5 = \frac{\pi}{2} r_5 \tag{1.135}$$

$$\mathcal{P}_5 = 2\mu_0 \frac{w}{\pi} \int_{g}^{g+h} \frac{1}{r_5}\, dr_5 = 2\mu_0 \frac{w}{\pi} \ln\left(\frac{g+h}{g}\right) \tag{1.136}$$

$$\mathcal{P}_5 = 2\mu_0 \frac{w}{\pi} \ln\left(1 + \frac{h}{g}\right) \tag{1.137}$$

Paths $\mathcal{P}_6$ and $\mathcal{P}_7$. There are additional flux paths that extend into and out of the page in Fig. 1.9. Based on the geometry shown in Fig. 1.10, there is a half-cylinder flux path ($\mathcal{P}_6$) extending out of the page over path $\mathcal{P}_1$ and into the page behind path $\mathcal{P}_1$, which is identical in shape to path $\mathcal{P}_2$. There is also a half cylindrical shell flux path ($\mathcal{P}_7$) extending out of the page over path $\mathcal{P}_6$ and into the page behind path $\mathcal{P}_6$, which is identical in shape to path $\mathcal{P}_4$. The values for flux paths $\mathcal{P}_6$ and $\mathcal{P}_7$ can be obtained from Eqs. (1.121) and (1.131) by using the proper dimensions for the flux paths from Fig. 1.10, as follows.

$$\mathcal{P}_6 = 0.24\mu_0 t \tag{1.138}$$

$$\mathcal{P}_7 = \mu_0 \frac{t}{\pi} \ln\left(1 + 2\frac{h}{g}\right) \tag{1.139}$$

Path $\mathcal{P}_8$ (Spherical Octant). There are also spherical flux paths on the corners, as shown in Fig. 1.13. The permeance of these flux paths can be estimated using the same technique demonstrated for evaluating flux paths $\mathcal{P}_2$ through $\mathcal{P}_5$, as follows.

The cross-sectional area of flux path $\mathcal{P}_8$ varies along the path length. Therefore, Eq. (1.30) is modified as follows, where v_8 is the volume of flux path $\mathcal{P}_8$. Roters (1941) uses a graphical approximation to the mean flux path length, resulting in a permeance with a value of $\mathcal{P}_8 = 0.308\ \mu_0 g$, which is slightly smaller than that shown here.

$$\mathcal{P}_8 = \mu_0 \frac{a_8}{l_8} = \mu_0 \frac{a_8}{l_8} \frac{l_8}{l_8} = \mu_0 \frac{v_8}{l_8^2} \tag{1.140}$$

$$r_8 = g \tag{1.141}$$

$$v_8 = \frac{1}{8} \frac{4}{3} \pi r_8^3 = 0.5236 g^3 \tag{1.142}$$

$$l_8 = \frac{1}{2}\left(r_8 + \frac{\pi}{2} r_8\right) = 1.285 g \quad \text{(average of inner and outer paths)} \tag{1.143}$$

$$\mathcal{P}_8 = \mu_0 \frac{v_8}{l_8^2} = 0.317 \mu_0 g \tag{1.144}$$

Path $\mathcal{P}_9$ (Spherical Shell Octant). The cross-sectional area of this flux path varies along the path length, and the magnetic flux path length increases as the radius

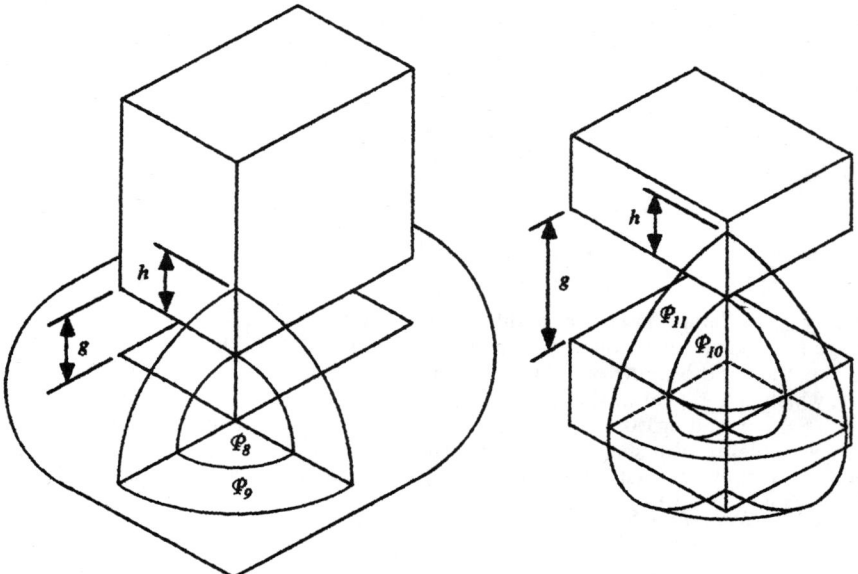

FIGURE 1.13 Corner flux paths in the shape of spherical octants and quadrants. *(Courtesy of Eaton Corporation.)*

increases. Therefore, Eq. (1.30) is written in differential form, and the permeance is calculated by integrating over the radius as follows, where v_9 is the volume of flux path $\mathcal{P}_9$. Roters (1941) uses a graphical approximation to both the mean path length and the mean path area, resulting in a permeance with a value of $\mathcal{P}_9 = 0.50\mu_0 h$, which is slightly smaller than that shown here.

$$\mathcal{P}_9 = \mu_0 \frac{a_9}{l_9} \frac{l_9}{l_9} = \mu_0 \frac{v_9}{l_9^2} \tag{1.145}$$

$$g < r_9 < g + h \tag{1.146}$$

$$l_9 = \frac{\pi}{2} r_9 \tag{1.147}$$

$$dv_9 = \frac{1}{8} 4\pi r_9^2 \, dr_9 \tag{1.148}$$

$$d\mathcal{P}_9 = \mu_0 \frac{dv_9}{l_9^2} = \mu_0 2 \frac{\pi r_9^2 \, dr_9}{\pi^2 r_9^2} \tag{1.149}$$

$$\mathcal{P}_9 = \mu_0 \frac{2}{\pi} \int_g^{g+h} dr_9 \tag{1.150}$$

$$\mathcal{P}_9 = 0.64 \mu_0 h \tag{1.151}$$

Path $\mathcal{P}_{10}$ (Spherical Quadrant). The cross-sectional area of this flux path varies along the path length. Therefore, Eq. (1.30) is modified as follows, where v_{10} is the volume of flux path $\mathcal{P}_{10}$. Roters (1941) uses a graphical approximation to the mean path length, resulting in a permeance with a value of $\mathcal{P}_{10} = 0.077 \, \mu_0 g$, which is slightly smaller than that shown here.

$$\mathcal{P}_{10} = \mu_0 \frac{a_{10}}{l_{10}} = \mu_0 \frac{a_{10}}{l_{10}} \frac{l_{10}}{l_{10}} = \mu_0 \frac{v_{10}}{l_{10}^2} \tag{1.152}$$

$$r_{10} = \frac{g}{2} \tag{1.153}$$

$$v_{10} = \frac{1}{4} \frac{4}{3} \pi r_{10}^3 = 0.1309 g^3 \tag{1.154}$$

$$l_{10} = \frac{1}{2}(2r_{10} + \pi r_{10}) = 1.285g \quad \text{(average of inner and outer paths)} \tag{1.155}$$

$$\mathcal{P}_{10} = \mu_0 \frac{v_{10}}{l_{10}^2} = 0.079 \mu_0 g \tag{1.156}$$

Path $\mathcal{P}_{11}$ (Spherical Shell Quadrant). The cross-sectional area of this flux path varies along the path length, and the magnetic flux path length increases as the radius increases. Therefore, Eq. (1.30) is written in differential form, and the permeance is calculated by integrating over the radius as follows, where v_9 is the volume of flux path $\mathcal{P}_9$. Roters (1941) uses a graphical approximation to both the mean path length and the mean path area, resulting in a permeance with a value of $\mathcal{P}_9 = 0.25$ $\mu_0 h$, which is slightly smaller than that shown here.

$$\mathcal{P}_{11} = \mu_0 \frac{a_{11}}{l_{11}} \frac{l_{11}}{l_{11}} = \mu_0 \frac{v_{11}}{l_{11}^2} \tag{1.157}$$

$$\frac{g}{2} < r_{11} < \frac{g}{2} + h \tag{1.158}$$

$$l_{11} = \pi r_{11} \tag{1.159}$$

$$dv_{11} = \frac{1}{4} 4\pi r_{11}^2 \, dr_{11} \tag{1.160}$$

$$d\mathcal{P}_{11} = \mu_0 \frac{dv_{11}}{l_{11}^2} = \mu_0 \frac{\pi r_{11}^2 \, dr_{11}}{\pi^2 r_{11}^2} \tag{1.161}$$

$$\mathcal{P}_{11} = \mu_0 \frac{1}{\pi} \int_{g/2}^{g/2 + h} dr_{11} \tag{1.162}$$

$$\mathcal{P}_{11} = 0.32 \mu_0 h \tag{1.163}$$

Total Permeance. The total permeance of the air gap is equal to the sum of the individual parallel flux paths, as follows:

$$\mathcal{P}_{\text{total}} = \mathcal{P}_1 + \mathcal{P}_2 + \mathcal{P}_3 + \mathcal{P}_4 + \mathcal{P}_5 + 2(\mathcal{P}_6 + \mathcal{P}_7) + 4(\mathcal{P}_8 + \mathcal{P}_9 + \mathcal{P}_{10} + \mathcal{P}_{11}) \tag{1.164}$$

1.3.1 Summary of Flux Path Permeance Equations

All of the flux path permeances are based on Eq. (1.30). The relationships shown here are for the special case of 90° and 180° angles between steel surfaces. However, the techniques that are demonstrated here can be applied to any geometry. The final forms of the flux path permeances for Figs. 1.9, 1.10, and 1.14 follow. All of the magnetic flux paths represent fringing regions except for the direct face-to-face flux path ($\mathcal{P}_1$).

Direct face-to-face flux path (Figs. 1.9 and 1.10)

$$\mathcal{P}_1 = \mu_0 \frac{wt}{g} \tag{1.116}$$

Half cylinder (Figs. 1.9 and 1.10)

$$\mathcal{P}_2 = 0.24 \mu_0 w \tag{1.121}$$

Quarter cylinder (Figs. 1.9 and 1.10)

$$\mathcal{P}_3 = 0.48 \mu_0 w \tag{1.125}$$

Half cylindrical shell (Figs. 1.9 and 1.10)

$$\mathcal{P}_4 = \mu_0 \frac{w}{\pi} \ln\left(1 + 2\frac{h}{g}\right) \tag{1.131}$$

Quarter cylindrical shell (Figs. 1.9 and 1.10)

$$\mathcal{P}_5 = 2\mu_0 \frac{w}{\pi} \ln\left(1 + \frac{h}{g}\right) \tag{1.137}$$

Spherical octant (Fig. 1.13)

$$\mathcal{P}_8 = 0.317\mu_0 g \tag{1.144}$$

Spherical shell octant (Fig. 1.13)

$$\mathcal{P}_9 = 0.64\mu_0 h \tag{1.151}$$

Spherical quadrant (Fig. 1.13)

$$\mathcal{P}_{10} = 0.079\mu_0 g \tag{1.156}$$

Spherical shell quadrant (Fig. 1.13)

$$\mathcal{P}_{11} = 0.32\mu_0 h \tag{1.163}$$

Thickness of the shells (Fig. 1.10)

$$h = t \tag{1.165}$$

1.3.2 Leakage Flux Paths

The magnetic flux paths shown in Figs. 1.9, 1.10, and 1.13 (the direct face-to-face flux path $\mathcal{P}_1$ and the fringing flux paths $\mathcal{P}_2, \mathcal{P}_3, \mathcal{P}_4,$ and $\mathcal{P}_5$) are based on the air gap geometry. These flux paths carry the magnetic flux across the working gaps g from the magnet poles to the armature, as shown for the permanent-magnet reluctance actuator in Fig. 1.14. Also shown are the leakage flux paths $\mathcal{P}_{L1}, \mathcal{P}_{L2},$ and $\mathcal{P}_{L3}$, which carry the magnetic flux between the magnet poles, and prevent some of the magnetic flux from reaching the armature and the working gaps g. The effect of each flux path is described here.

Direct Face-to-face Flux Path ($\mathcal{P}_1$). This flux path is the highest-efficiency producer of the force on the armature. It also produces the majority of the force on the armature. The total magnetic flux through this path is limited by the saturation flux for the materials in the magnet poles and the armature.

Fringing Flux Paths ($\mathcal{P}_2, \mathcal{P}_3, \mathcal{P}_4,$ and $\mathcal{P}_5$). The fringing flux paths increase the system permeance, increase the total magnetic flux, and produce a lower-efficiency force on the armature than does the direct flux path. Initial magnetic performance estimates commonly use only the direct face-to-face flux path and ignore the fringing flux paths, in order to make the first calculations very easy and fast. If the steel in the magnet poles and armature is not saturated, then the fringing flux paths increase the total magnetic flux and increase the armature force. If the steel in the magnet poles and armature is saturated, then adding the fringing flux paths does not change the total magnetic flux, and since the fringing flux paths take magnetic flux away from the direct flux path, the armature force is decreased.

Leakage Flux Paths ($\mathcal{P}_{L1}, \mathcal{P}_{L2},$ and $\mathcal{P}_{L3}$). These flux paths take magnetic flux away from both the direct flux path and the fringing flux paths, and produce no force on the

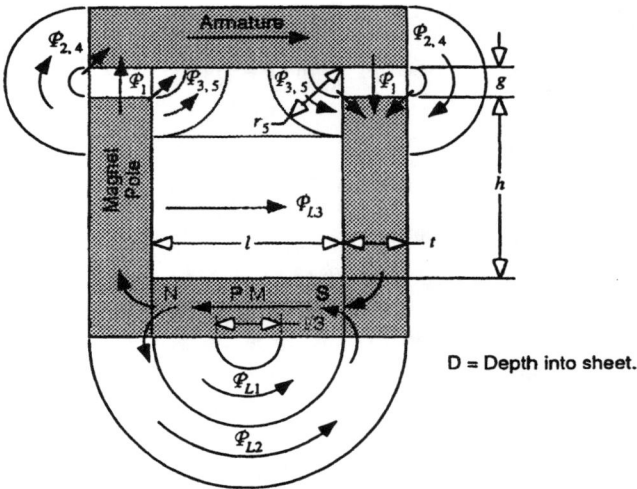

FIGURE 1.14 Two-dimensional air flux paths around a permanent magnet reluctance actuator. $\mathcal{P}_{L1}$, $\mathcal{P}_{L2}$, and $\mathcal{P}_{L3}$ are leakage flux paths. $\mathcal{P}_2$, $\mathcal{P}_3$, $\mathcal{P}_4$, and $\mathcal{P}_5$ are fringing flux paths (also shown in Fig. 1.9). *(Courtesy of Eaton Corporation.)*

armature. Also, the magnetic flux carried by the leakage flux paths must be carried by a portion of the magnet poles. This causes the magnet pole material to reach the saturation flux limit sooner than expected. Therefore, the leakage flux paths cause a large decrease in the armature force. When a permanent magnet is placed near the working gap, the armature force is increased, because some of the leakage flux becomes fringing flux. This essentially minimizes the leakage flux. Conversely, the leakage flux paths are useful in permanent-magnet systems as a means of protecting the permanent magnet from large demagnetizing fields. In this case, some of the demagnetizing flux bypasses the permanent magnet through the leakage flux paths.

The leakage flux paths can be evaluated by using the following procedures.

Path $\mathcal{P}_{L1}$ (Half Cylindrical Shell). This flux path is identical in shape to path $\mathcal{P}4$. The diameter of the internal half cylindrical shell d_1 is equal to 33 percent of the permanent-magnet length. The radius of the external half cylindrical shell r_1 is equal to 50% of the permanent-magnet length. These permeance relationships are shown here, based on Eqs. (1.126) through (1.131). This flux path is valid only for alnico and earlier permanent magnets, which have effective poles at 70 percent of the magnet length. Ferrite and rare-earth permanent magnets have effective poles at 95 percent of the magnet length; therefore, no magnetic flux is generated in this path and the permeance is zero.

$$d_1 = \frac{1}{3} \tag{1.166}$$

$$r_1 = \frac{1}{2} \tag{1.167}$$

$$\frac{1}{6} < r < \frac{1}{2} \tag{1.168}$$

$$\mathcal{P}_{L1} = \mu_0 \frac{D}{\pi} \int_{1/6}^{1/2} \frac{dr}{r} = \mu_0 \frac{D}{\pi} \ln\left(\frac{6}{2}\right) \tag{1.169}$$

For alnico and earlier magnets:

$$\mathcal{P}_{L1} = 0.350\mu_0 D \tag{1.170}$$

For ferrite and rare-earth magnets:

$$\mathcal{P}_{L1} = 0 \tag{1.171}$$

Path $\mathcal{P}_{L2}$ (Half Cylindrical Shell). This flux path is also identical in shape to path $\mathcal{P}_4$. The dimensions can be obtained by inspection from Fig. 1.14, and the permeance relationships shown here are based on Eqs. (1.126) through (1.131).

$$\frac{1}{2} < r < \frac{1}{2} + t \tag{1.172}$$

$$\mathcal{P}_{L2} = \mu_0 \frac{D}{\pi} \int_{1/2}^{1/2+t} \frac{dr}{r} \tag{1.173}$$

$$\mathcal{P}_{L2} = \mu_0 \frac{D}{\pi} \ln\left(1 + 2\frac{t}{l}\right) \tag{1.174}$$

Path $\mathcal{P}_{L3}$ (Direct Pole-to-Pole). This flux path is identical in shape to path $\mathcal{P}_1$. The lower edge of the path is located at the permanent magnet. The upper edge of the path is located at a distance equal to radius r_5 below the armature. Radius r_5 is defined by equating the length of a flux line along the radius to 50 percent of the length of a flux line in the leakage flux path $\mathcal{P}_{L3}$. These permeance relationships are shown here, based on Eqs. (1.114) through (1.116):

$$r_5 = \frac{1}{\pi} \tag{1.175}$$

$$a = D(h + g - r_5) = D\left(h + g - \frac{l}{\pi}\right) \tag{1.176}$$

$$\mathcal{P}_{L3} = \mu_0 \frac{a}{l} = \mu_0 D \left(\frac{h+g}{l} - \frac{1}{\pi}\right) \tag{1.177}$$

for a permanent magnet.

If the permanent magnet is replaced with a coil, then this path has only half of this permeance. The magnetic flux at the bottom of the flux path encircles zero coil amp-turns, and the magnetic flux at the top of the flux path encircles all of the coil amp-turns. Therefore, a coil actuator contains only 50 percent of the magnetic flux in path $\mathcal{P}_{L3}$ as compared to a permanent-magnet actuator. This corresponds to a 50 percent permeance value for the coil actuator.

For a coil:

$$\mathcal{P}_{L3} = \frac{1}{2} \mu_0 \frac{a}{l} = \mu_0 \frac{D}{2} \left(\frac{h+g}{l} - \frac{1}{\pi}\right) \tag{1.178}$$

Total Permeance and Equivalent Reluctance Circuit. The total reluctance of the armature air gaps ($\mathcal{P}_1$, $\mathcal{P}_2$, $\mathcal{P}_3$, $\mathcal{P}_4$, and $\mathcal{P}_5$) and the leakage flux paths ($\mathcal{P}_{L1}$, $\mathcal{P}_{L2}$, and $\mathcal{P}_{L3}$), as seen from the permanent magnet, can be calculated by drawing the equivalent reluctance network, as shown in Fig. 1.15. The permeance of each armature air gap is the sum of the air gap permeances.

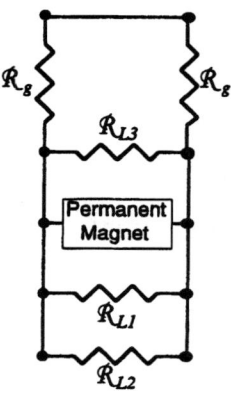

FIGURE 1.15 Equivalent reluctance network for the permanent magnet leakage shown in Fig. 1.14. *(Courtesy of Eaton Corporation.)*

Armature air gap permeance

$$\mathcal{P}_g = \mathcal{P}_1 + \mathcal{P}_2 + \mathcal{P}_3 + \mathcal{P}_4 + \mathcal{P}_5 \quad (1.179)$$

Armature air gap reluctance

$$\mathcal{R}_g = \frac{1}{\mathcal{P}_g} \quad (1.180)$$

The reluctances of the leakage flux paths are listed here as the inverse of the permeance values.

$$\mathcal{R}_{L1} = \frac{1}{\mathcal{P}_{L1}} \quad (1.181)$$

$$\mathcal{R}_{L2} = \frac{1}{\mathcal{P}_{L2}} \quad (1.182)$$

$$\mathcal{R}_{L3} = \frac{1}{\mathcal{P}_{L3}} \quad (1.183)$$

The total reluctance of the circuit shown in Fig. 1.15 as seen from the permanent magnet can be written as follows:

$$\mathcal{R}_T = \frac{1}{1/2\mathcal{R}_g + 1/\mathcal{R}_{L1} + 1/\mathcal{R}_{L2} + 1/\mathcal{R}_{L3}} \quad (1.184)$$

The total reluctance can be obtained by substituting Eqs. (1.179) through (1.183) into Eq. (1.184).

Total circuit reluctance

$$\frac{1}{\mathcal{P}_T} = \frac{1}{\mathcal{P}_g/2 + \mathcal{P}_{L1} + \mathcal{P}_{L2} + \mathcal{P}_{L3}} \quad (1.185)$$

Total circuit permeance

$$\mathcal{P}_T = \frac{1}{2}(\mathcal{P}_1 + \mathcal{P}_2 + \mathcal{P}_3 + \mathcal{P}_4 + \mathcal{P}_5) + (\mathcal{P}_{L1} + \mathcal{P}_{L2} + \mathcal{P}_{L3}) \quad (1.186)$$

1.4 ELECTROMECHANICAL FORCES AND TORQUES

The mechanical force and torque equations, Eqs. (1.68) and (1.73), are general equations which can be applied to an entire system or individually to a specific air gap.

When these equations are applied to a nonlinear system, the coenergy W_{co} must be obtained by integrating each region over the local magnetization curve:

$$F = \frac{1}{2} \frac{d}{dX} [(NI)^2 \mathcal{P}] = \frac{1}{2} \frac{d}{dX} (\phi^2 \mathcal{R}) \qquad (1.74)$$

$$T = \frac{1}{2} \frac{d}{d\theta} [(NI)^2 \mathcal{P}] = \frac{1}{2} \frac{d}{d\theta} (\phi^2 \mathcal{R}) \qquad (1.75)$$

for constant L, μ.

Equations (1.74) and (1.75) can be applied to a specific air gap to obtain the mechanical force or torque. The magnetic medium in the air gap is linear. Therefore, the coenergy W_{co} and the stored field energy W_f are equal and can be obtained from Eq. (1.37), where NI is the magnetizing force in the air gap, ϕ is the magnetic flux passing across the air gap, and $\mathcal{P}$ is the air gap permeance.

The total magnetizing force provided by the coil NI_{coil} is used to magnetize both the steel part and the air gaps. In general, the magnetizing force in the steel parts is small, and the resulting magnetizing force in the air gap NI is nearly equal to the coil magnetizing force ($NI \approx NI_{coil}$). However, when the steel parts become saturated, the magnetizing force in the steel parts becomes large and the magnetizing force in the air gap becomes small ($NI \ll NI_{coil}$). The saturation of the steel parts produces a magnetic flux limiting condition in which the second form of Eqs. (1.74) and (1.75) can be used effectively.

1.4.1 General Air Gap Linear Equations

An actuator or motor generally has one or two magnetizing sources. Some examples are listed here.

- A *reluctance motor* has one stationary coil and a moving armature that changes the reluctance $\mathcal{R}$ or the permeance $\mathcal{P}$ of the air gap as it moves. The mechanical force is generated by the change in reluctance.
- A *flux-transfer reluctance motor* has two stationary coils and a moving armature that changes the reluctance $\mathcal{R}$ or the permeance $\mathcal{P}$ of the air gap as it moves. The mechanical force is generated by the change in reluctance. Usually one of the stationary coils provides a constant magnetizing force to produce a bias magnetic field which is modified by the second coil. The coil that produces the constant magnetizing force can be a permanent magnet.
- A *moving coil motor* has one moving coil that changes the number of turns which contribute to the total magnetizing force in the motor. There may be a second coil (stationary) that provides a constant magnetizing force to increase the air gap stored magnetic field energy and the resulting force or torque. The second coil can be replaced by a permanent magnet. The permanent magnet and coil can be interchanged to produce a *moving magnet motor*. The mechanical force is generated by the change in the number of turns and is also called the *Lorentz force*.

The magnetizing force NI in Eqs. (1.74) and (1.75) is expanded here to show the contribution from a second coil $N_a I_a$ (either stationary or on a moving armature) and the contribution from a bias field coil $N_f I_f$ (stationary).

$$NI = N_a I_a + N_f I_f \qquad (1.187)$$

$$(NI)^2 = (N_aI_a)^2 + 2(N_aI_aN_fI_f) + (N_fI_f)^2 \tag{1.188}$$

Multiplying by the permeance gives the energy expression needed in Eqs. (1.74) and (1.75). Also, the definition of the self-inductance of the moving armature coil L_a, the self-inductance of the bias field coil L_f, and the mutual inductance of both coils L_{af} can be substituted into Eq. (1.188) as follows:

$$(NI)^2\mathcal{P} = (N_aI_a)^2\mathcal{P} + 2(N_aI_aN_fI_f)\mathcal{P} + (N_fI_f)^2\mathcal{P} \tag{1.189}$$

$$L_a = N_a^2\mathcal{P} \tag{1.190}$$

$$L_f = N_f^2\mathcal{P} \tag{1.191}$$

$$L_{af} = N_aN_f\mathcal{P} \tag{1.192}$$

$$(NI)^2\mathcal{P} + I_a^2L_a + 2I_aI_fL_{af} + I_f^2L_f \tag{1.193}$$

Substitution of Eq. (1.193) into Eqs. (1.74) and (1.75) gives the following motor force and torque equations. It is assumed that the currents I_a and I_f are constant with respect to the armature positions X and θ.

$$F = \frac{1}{2} I_a^2 \frac{dL_a}{dX} + I_aI_f \frac{dL_{af}}{dX} + \frac{1}{2} I_f^2 \frac{dL_f}{d} \tag{1.194}$$

$$T = \frac{1}{2} I_a^2 \frac{dL_a}{d\theta} + I_aI_f \frac{dL_{af}}{d\theta} + \frac{1}{2} I_f^2 \frac{dL_f}{d\theta} \tag{1.195}$$

Differentiation of each inductance term can be carried out in Eq. (1.194) as follows. It is assumed that field coil turns N_f are constant with respect to the armature positions X and θ.

$$\frac{1}{2} I_a^2 \frac{dL_a}{dX} = \frac{1}{2} (N_aI_a)^2 \frac{d\mathcal{P}}{dX} + N_aI_a^2\mathcal{P} \frac{dN_a}{dX} \tag{1.196}$$

$$I_aI_f \frac{dL_{af}}{dX} = \frac{1}{2} N_aN_fI_aI_f \frac{d\mathcal{P}}{dX} + N_fI_fI_a\mathcal{P} \frac{dN_a}{dX} \tag{1.197}$$

$$\frac{1}{2} I_f^2 \frac{dL_f}{dX} = \frac{1}{2} (N_fI_f)^2 \frac{d\mathcal{P}}{d} \tag{1.198}$$

The terms in Eqs. (1.196), (1.197), and (1.198) can be combined, or Eq. (1.198) can be differentiated directly to produce the following force and torque equations.

$$F = \frac{1}{2} (NI)^2 \frac{d\mathcal{P}}{dX} + N_aI_a^2\mathcal{P} \frac{dN_a}{dX} + I_aN_fI_f\mathcal{P} \frac{dN_a}{dX} \tag{1.199}$$

$$T = \frac{1}{2} (NI)^2 \frac{d\mathcal{P}}{d\theta} + N_aI_a^2\mathcal{P} \frac{dN_a}{d\theta} + I_aN_fI_f\mathcal{P} \frac{dN_a}{d\theta} \tag{1.200}$$

The magnetic flux produced by the armature coil ϕ_a and the bias field coil ϕ_f are defined and substituted into Eqs. (1.199) and (1.200) as follows.

$$\phi_a = N_aI_a\mathcal{P} \tag{1.201}$$

$$\phi_f = N_f I_f \mathcal{P} \tag{1.202}$$

$$F = \frac{1}{2}(NI)^2 d\frac{\mathcal{P}}{dX} + I_a \phi_a \frac{dN_a}{dX} + I_a \phi_f \frac{dN_a}{dX} \tag{1.203}$$

$$T = \frac{1}{2}(NI)^2 d\frac{\mathcal{P}}{d\theta} + I_a \phi_a \frac{dN_a}{d\theta} + I_a \phi_f \frac{dN_a}{d\theta} \tag{1.204}$$

The first term in Eqs. (1.203) and (1.204) represents the reluctance force produced by a change in air gap permeance as a function of the armature position. The second term represents the Lorentz force on the moving coil due to the interaction with the magnetic flux produced by the armature coil (this is typically called the *armature reaction force*). The third term represents the Lorentz force on the moving coil due to the interaction with the magnetic flux produced by the bias field coil.

1.4.2 Reluctance Actuators with Small Air Gaps and Saturation

When an air gap is very small, the permeance $\mathcal{P}_1$ becomes very large in comparison to the fringing flux paths $\mathcal{P}_2$ through $\mathcal{P}_{10}$. Therefore, the total permeance $\mathcal{P}_{total}$ can be closely approximated by the direct face-to-face permeance $\mathcal{P}_1$, Eq. (1.30).

Reluctance Normal Force. The force for the actuator in Fig. 1.16 can be written as follows, based on Eqs. (1.203) and (1.30). The motion of the armature X is defined to be in the direction to close the air gap, and the initial length of the air gap is defined to be g_0. Also, $N_a = I_a = 0$ because there is only one coil. The area of the pole face is $a = T_P D$.

$$l = g = g_0 - X \tag{1.205}$$

$$\mathcal{P} = \mu_0 \frac{a}{g_0 - X} \quad \text{for small } g \tag{1.206}$$

This is the permeance of the flux path between one pole and the armature. The total permeance of the flux path includes two air gap permeances in series. Therefore, the total permeance $\mathcal{P}_T$ is equal to one-half of the permeance for one air gap:

$$\mathcal{P}_T = \frac{1}{2}\mathcal{P} = \frac{1}{2}\mu_0 \frac{a}{g_0 - X} \tag{1.207}$$

$$\frac{d\mathcal{P}_T}{dX} = \frac{1}{2}\mu_0 \frac{a}{(g_0 - X)^2} = \frac{1}{2}\mu_0 \frac{a}{g^2} \tag{1.208}$$

$$F_T = \frac{1}{2}(NI)^2 \frac{d\mathcal{P}_T}{dX} = \frac{1}{4}(NI)^2 \mu_0 \frac{a}{g^2} \tag{1.209}$$

for small g and limited NI.

All of the variables in Eq. (1.209) are constant except the armature position g. Therefore, the armature reluctance force is proportional to the inverse square of the armature position (or to the size of the working air gap). If the steel parts become saturated in some armature positions, then the air gap magnetizing force is reduced. Under this condition the air gap magnetizing force is a function of the armature position.

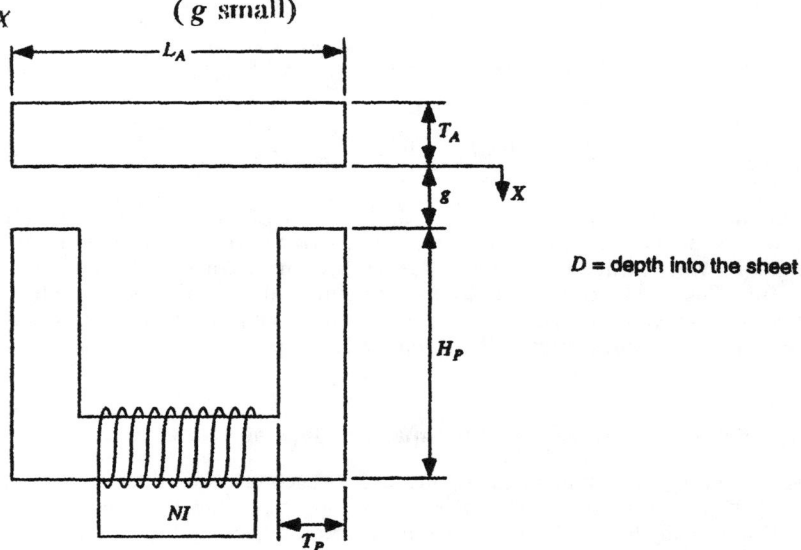

FIGURE 1.16 Actuator with reluctance force produced normal to the armature bottom surface, in the direction of motion X. *(Courtesy of Eaton Corporation.)*

When the steel parts become saturated the total magnetic flux ϕ in the system reaches a maximum limit, as shown in Fig. 1.4. Equation (1.209) can be written as follows by substituting Eqs. (1.18), (1.21), and (1.30):

$$F_T = \frac{1}{4}(NI)^2 \mu_0 \frac{a}{g^2} = \frac{1}{2}\left(\frac{\phi}{\mathcal{P}_T}\right)^2 \frac{\mathcal{P}_T}{g} = \frac{\phi^2}{2\mathcal{P}_T g} \quad (1.210)$$

for small g

$$F_T = \frac{\phi^2}{\mu_0 a} = \frac{B^2 a}{\mu_0} \quad (1.211)$$

for small g and limited ϕ and B.

All of the variables in Eq. (1.211) are constant. Therefore, the armature reluctance force is constant regardless of the armature position when the system is saturated. Due to symmetry, the force on each side of the armature or on each pole of the magnet frame is half of the total force, as shown here:

$$F = \frac{B^2 a}{2\mu_0} \quad (1.212)$$

for a single pole, where g is small.

Maximum Possible Reluctance Normal Force. The reluctance normal force in Eq. (1.212) can be divided by the pole area a to obtain the normal magnetic pressure p on the pole, as shown in Eq. (1.213). The maximum normal magnetic pressure is dependent only on the saturation magnetic flux density (for small gaps). Steel typically saturates at 1.60 T. Therefore, the maximum possible normal magnetic pressure for this steel is 150 lb/in²:

$$p = \frac{F}{a} = \frac{B^2}{2\mu_0} = 150 \text{ lb/in}^2 \qquad (1.213)$$

for a single pole, where g is small.

This relationship can be converted into the following simple design strategy. If 150 lb of magnetic force is required, then at least 1.0 in² of steel is needed. Conversely, if we are limited to 1.0 in² of steel, then the maximum possible magnetic force will be 150 lb.

Reluctance Tangential Force. The force for the actuator in Fig. 1.17 can be written as follows, based on Eqs. (1.203) and (1.30). The motion of the armature x is defined to be in the direction to insert the armature into the stator cup. Also, $N_a = I_a = 0$ because there is only one coil:

$$l = C \qquad (1.214)$$

$$a = \pi D x \qquad (1.215)$$

$$\mathcal{P} = \mu_0 \frac{\pi D x}{C} \qquad (1.216)$$

$$\frac{d\mathcal{P}}{dx} = \mu_0 \frac{\pi D}{C} \qquad (1.217)$$

$$F = \frac{1}{2}(NI)^2 \mu_0 \frac{\pi D}{C} \qquad (1.218)$$

where C is small and NI is limited.

All of the variables in Eq. (1.218) are constant. Therefore, the armature reluctance force is constant regardless of the armature position. If the steel parts become saturated in some armature positions, then the air gap magnetizing force is reduced. Under this condition, the air gap magnetizing force is a function of the armature position, and a new air gap magnetizing force must be calculated at each position.

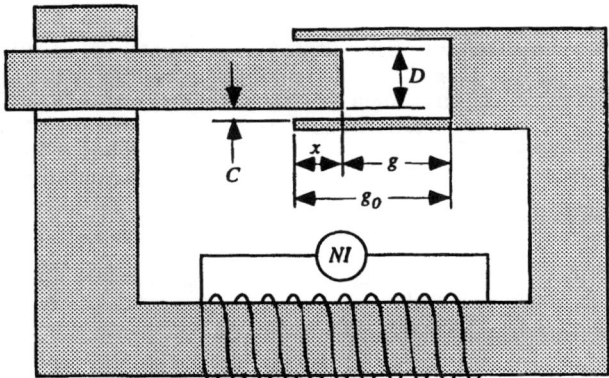

FIGURE 1.17 Actuator with reluctance force produced tangential to the armature side surface, in the direction of motion x. *(Courtesy of Eaton Corporation.)*

When the steel parts become saturated, the total magnetic flux ϕ in the system reaches a maximum limit, as shown in Fig. 1.4. Equation (1.218) can be rewritten as follows by substituting Eqs. (1.18) and (1.30):

$$F = \frac{1}{2}\left(\frac{\phi}{\mathcal{P}}\right)^2 \mu_0 \frac{\pi D}{C} = \frac{1}{2}\left(\frac{\phi C}{\mu_0 \pi D x}\right)^2 \mu_0 \frac{\pi D}{C} \tag{1.219}$$

$$F = \frac{\phi^2 C}{2\mu_0 \pi D x^2} = \frac{\phi^2 C}{2\mu_0 \pi D (g_0 - g)^2} \tag{1.220}$$

where C is small and ϕ is limited.

All of the variables in Eq. (1.220) are constant except for the armature position x. Therefore, the armature reluctance force is inversely proportional to the square of the armature position when the system is saturated. Equation (1.221) is the flux density B limited form of Eq. (1.220), which was obtained by substituting Eqs. (1.21) and (1.215) into Eq. (1.220). Equation (1.221) shows that the force is independent of the armature position x as long as the air gap flux density B is constant. The force is constant, and the air gap flux density is constant if the actuator steel parts are not saturated:

$$F = \frac{B^2}{2\mu_0} \pi D C \tag{1.221}$$

where C is small and B is limited.

Pole Shaping. An example of the performance variations that are possible by shaping the pole faces can be observed by comparing Eqs. (1.209) and (1.218) and Figs. 1.16 and 1.17. The normal force on the pole faces shown in Fig. 1.16 and Eq. (1.218) is inversely proportional to the square of the armature position, $F \propto 1/g^2$. The tangential force F on the pole faces shown in Fig. 1.17 and Eq. (1.209) is constant regardless of the armature position. A force characteristic between these two limiting conditions can be achieved by using a cup-cone configuration. Fig. 1.16 is similar to the limiting configuration of a flat-face cone, and Fig. 1.17 is similar to the limiting configuration of a straight-sided cup. Therefore, it is possible to shape the force characteristic as a function of the armature position by modifying the pole face, as shown in Fig. 1.18.

Reluctance Tangential Torque. The torque for the actuator in Fig. 1.19 can be written as follows, based on Eqs. (1.204) and (1.30). The motion of the armature θ is defined to be in the direction to align the armature vertically with the stator poles. Also, $N_a = I_a = 0$ because there is only one coil:

$$l = g \tag{1.222}$$

$$a = \theta r D \tag{1.223}$$

$$\mathcal{P} = \mu_0 \frac{a}{l} = \mu_0 \frac{\theta r D}{g} \quad \text{for small } g \tag{1.224}$$

$$\mathcal{P}_T = \frac{1}{2}\mathcal{P} = \mu_0 \frac{\theta r D}{2g} \tag{1.225}$$

$$\frac{d\mathcal{P}_T}{d\theta} = \mu_0 \frac{r D}{2g} \tag{1.226}$$

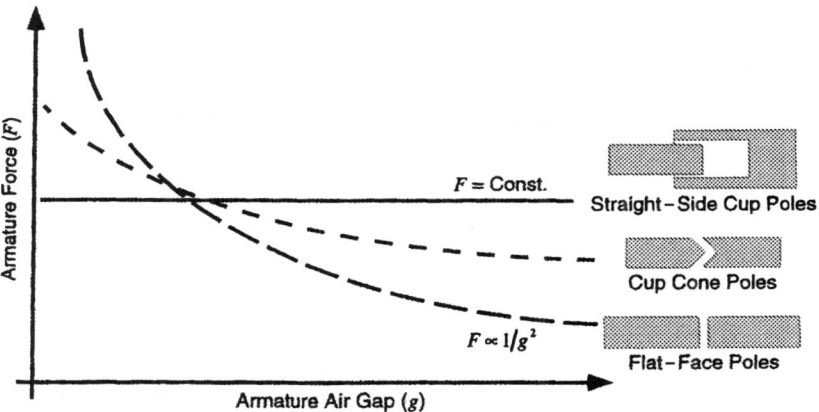

FIGURE 1.18 Force curve performance variations due to modification of the pole shape. *(Courtesy of Eaton Corporation.)*

$$T_T = \frac{1}{2}\left[(NI)^2 \frac{d\mathcal{P}_T}{d\theta}\right] = \frac{1}{4}(NI)^2 \mu_0 \frac{rD}{g} \qquad (1.227)$$

where g is small and NI is limited.

All of the variables in Eq. (1.227) are constant. Therefore, the armature reluctance torque is constant regardless of the armature position. If the steel parts become saturated in some armature positions, then the air gap magnetizing force is reduced.

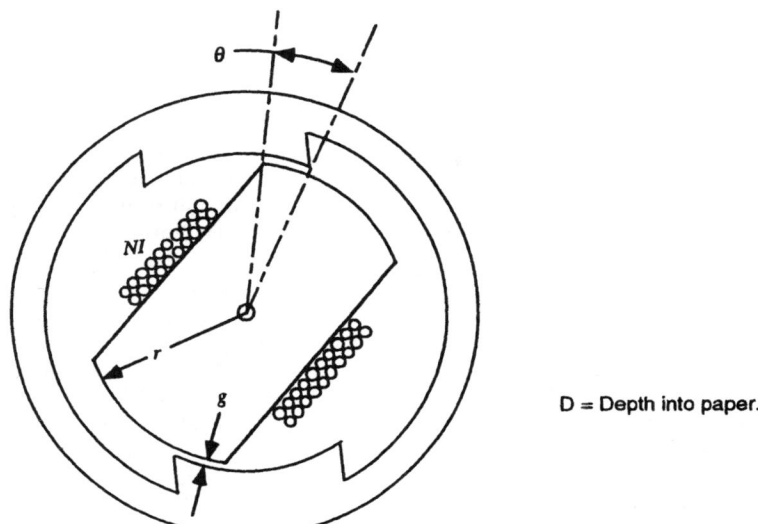

FIGURE 1.19 Actuator with reluctance torque produced tangential to the armature end surface, in the direction of motion θ. *(Courtesy of Eaton Corporation.)*

Under this condition the air gap magnetizing force is a function of the armature position, and a new air gap magnetizing force must be calculated at each position.

When the steel parts become saturated, the total magnetic flux ϕ in the system reaches a maximum limit, as shown in Fig. 1.4. Equation (1.227) can be rewritten as follows by substituting Eqs. (1.18) and (1.30):

$$T_T = \frac{1}{4}(NI)^2 \mu_0 \frac{rD}{g} = \left(\frac{\phi}{\mathcal{P}_T}\right)^2 \frac{\mathcal{P}_T}{2\theta} = \frac{\phi^2}{2\theta\mathcal{P}_T} \tag{1.228}$$

$$T_T = \frac{1}{\theta^2} \frac{\phi^2 g}{\mu_0 rD} \tag{1.229}$$

where g is small and ϕ is limited.

All of the variables in Eq. (1.229) are constant except for the armature position θ. Therefore, the armature reluctance torque is inversely proportional to the square of the armature position when the system is saturated. Again, as shown in the previous section, the torque produced at each end of the armature is one-half of the total torque:

$$T = \frac{1}{\theta_2} \frac{\phi^2 g}{2\mu_0 rD} \tag{1.230}$$

where g is small and ϕ is limited.

Equation (1.231) is the flux density B limited form of Eq. (1.230), which was obtained by substituting Eqs. (1.21) and (1.223) into Eq. (1.230). Equation (1.231) shows that the torque is independent of the rotation angle θ as long as the air gap flux density B is constant. The torque is constant, and the air gap flux density is constant if the actuator steel parts are not saturated:

$$T = \frac{B^2}{2\mu_0} grD \tag{1.231}$$

where g is small and B is limited.

Moving-Coil Actuator Lorentz Force. The force for the moving-coil actuator in Fig. 1.20 can be written as follows, based on Eqs. (1.203) and (1.30). The motion of the armature X is defined to be in the direction to bring more of the moving coil into the magnetic circuit. Also, $\mathcal{P}$ is constant, because the size of the air gap does not change as the position of the moving coil changes. Therefore, the rate of change of the permeance with respect to the armature position is zero.

$$d\frac{\mathcal{P}}{dX} = 0 \tag{1.232}$$

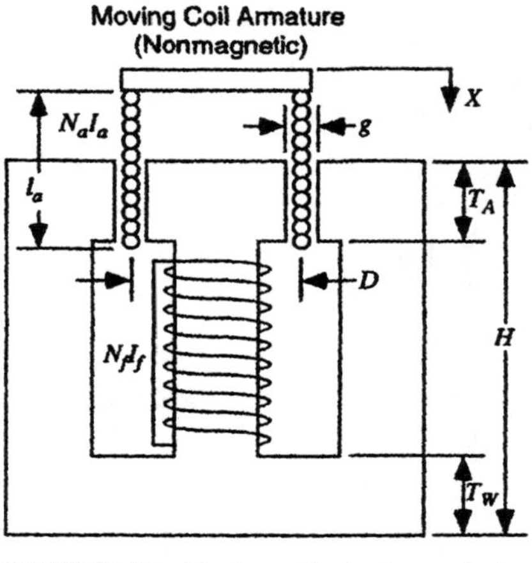

FIGURE 1.20 Moving-coil actuator producing a Lorentz force in the direction of motion X. *(Courtesy of Eaton Corporation.)*

The length of wire on the moving coil that is in the air gap permeance path can be calculated as follows. The permeance is based on the direct face-to-face flux path with the assumption of a small gap.

$$l_{\text{wire}} = \pi D N_a \frac{T_A}{l_a} \qquad \text{for small } g \tag{1.233}$$

The rate of change of the turns in the moving coil with respect to the moving-coil position is calculated by dividing the number of turns in the moving coil by the length of the moving coil, as follows.

$$\frac{dN_a}{dX} = \frac{N_a}{l_a} = \frac{l_{\text{wire}}}{\pi D T_A} \tag{1.234}$$

The total magnetic flux across the moving coil in the small air gap is equal to the magnetic flux from the bias field ϕ_a plus the magnetic flux from the moving coil ϕ_f. The total magnetic flux density B across the moving coil in the small air gap can also be calculated as shown here.

$$\phi = \phi_a + \phi_f = B\pi D T_A \qquad \text{for small } g \tag{1.235}$$

Substituting Eqs. (1.232), (1.233), (1.234), and (1.235) into Eq. (1.203) gives the following moving-coil force characteristic, known as the *Lorentz force*.

$$F = I_a(\phi_a + \phi_f)\frac{dN_a}{dX} \tag{1.236}$$

$$F = I_a \phi \frac{dN_a}{dX} = I_a(B\pi D T_A)\left(\frac{l_{\text{wire}}}{\pi D T_A}\right) \tag{1.237}$$

$$F = I_a B l_{\text{wire}} \qquad \text{for small } g \tag{1.238}$$

Reluctance Normal Torque. The torque for the actuator shown in Fig. 1.21 can be written as follows, based on Eqs. (1.204) and (1.30). The motion of the armature θ is defined to be in the direction to open the air gap g. Since the air gap thickness varies as a function of the radius from the pivot point, the equations are written in differential form.

$$\theta = \sin^{-1}\left(\frac{g}{R_2}\right) \cong \frac{g}{R_2} \tag{1.239}$$

$$l = r\theta \tag{1.240}$$

$$da = T\,dr \tag{1.241}$$

$$d\mathcal{P} = \mu_0 \frac{da}{l} = \mu_0 \frac{T\,dr}{r\theta} \tag{1.242}$$

The permeance of the flux path between each pole and the armature is obtained by integrating over the entire pole face.

$$\mathcal{P} = \mu_0 \frac{T}{\theta} \int_{R1}^{R2} \frac{dr}{r} = \mu_0 \frac{T}{\theta} \ln\left(\frac{R_2}{R_1}\right) \tag{1.243}$$

$$\mathcal{P}_T = \frac{1}{2}\mathcal{P} = \frac{1}{2}\mu_0 \frac{T}{\theta} \ln\left(\frac{R_2}{R_1}\right) \tag{1.244}$$

Differentiating the permeance with respect to the armature position and substituting the result into Eq. (1.204) gives the total torque on the armature.

$$\frac{d\mathcal{P}_T}{d\theta} = -\frac{1}{2}\mu_0 \frac{T}{\theta^2} \ln\left(\frac{R_2}{R_1}\right) \tag{1.245}$$

$$T_T = \frac{1}{2}(NI)^2 \frac{d\mathcal{P}_T}{d\theta} = -\frac{1}{4}(NI)^2 \mu_0 \frac{T}{\theta^2} \ln\left(\frac{R_2}{R_1}\right) \tag{1.246}$$

The location of the force centroid on the armature can be obtained by dividing the total armature torque, Eq. (1.246), by the total force on the armature. The total armature force is calculated by integrating the differential force over the radial length of the armature, as follows. The first step is to convert the permeance equation Eq. (1.242) into linear coordinates.

$$X = r\theta \tag{1.247}$$

$$d\mathcal{P}_T = \frac{1}{2} d\mathcal{P} = \frac{1}{2}\mu_0 \frac{T\,dr}{X} \tag{1.248}$$

$$dF_T = \frac{1}{2}(NI)^2 \frac{d}{dX}(d\mathcal{P}_T) \tag{1.249}$$

$$\frac{d}{dX}(d\mathcal{P}_T) = -\frac{1}{2}\mu_0 \frac{T\,dr}{X^2} = -\frac{1}{2}\mu_0 \frac{T\,dr}{r^2\theta^2} \tag{1.250}$$

$$dF_T = -\frac{1}{4}(NI)^2 \mu_0 \frac{T\,dr}{r^2\theta^2} \tag{1.251}$$

Integrating Eq. (1.251) gives the total armature force, as follows.

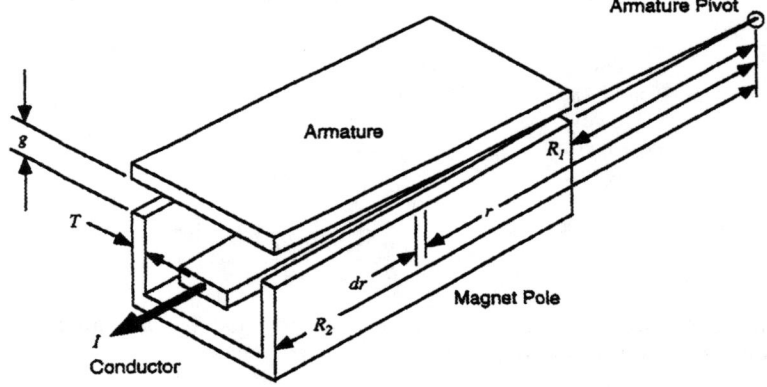

FIGURE 1.21 Actuator with reluctance torque produced normal to the armature bottom surface, in the direction of motion θ. *(Courtesy of Eaton Corporation.)*

$$F_T = -\frac{1}{4}(NI)^2\mu_0 \frac{T}{\theta^2} \int_{R1}^{R2} \frac{dr}{r^2} \qquad (1.252)$$

$$F_T = -\frac{1}{4}(NI)^2\mu_0 \frac{T}{\theta^2} \left(\frac{R_2 - R_1}{R_2 R_1} \right) \qquad (1.253)$$

Dividing the torque equation Eq. (1.246) by the force equation Eq. (1.253) gives the position of the force centroid.

$$R_{\text{cent}} = \frac{T_T}{F_T} = \left(\frac{R_2 R_1}{R_2 - R_1} \right) \ln\left(\frac{R_2}{R_1} \right) \qquad (1.254)$$

1.5 MAGNETIC MATERIALS

Magnetic materials are generally divided into two categories, soft materials and hard materials. The soft magnetic materials are easy to magnetize and demagnetize; they have high permeability and low losses. The hard magnetic materials are difficult to magnetize and demagnetize. Therefore, the hard magnetic materials are generally referred to as *permanent magnets*.

Permanent magnets were initially made from quench-hardened steels. Some references suggest that good permanent-magnet steel was available from China as early as 500 A.D. These permanent-magnet steels, which were initially made from plain carbon steel, steadily progressed to highly alloyed cobalt steels with a carbon content as high as 1.2 percent by 1920. These permanent-magnet steels are mechanically very hard materials, with Brinell hardness values as high as 690.

High-permeability low-loss materials are made from very low carbon annealed steels. These low-alloy annealed steels are mechanically very soft materials, with Brinell hardness values as low as 130.

The permanently magnetic materials became known as *hard magnetic materials*, and the high-permeability low-loss materials became known as *soft magnetic materials*, with the terms *soft* and *hard* referring to the mechanical hardness of the material. The following sections describe the material properties for both the soft and hard materials.

The electrons in the atoms of materials circulate around the atoms in orbits, creating an *orbital* magnetic field. The electrons also spin on their own axes, producing a *spin* magnetic field. In most materials, there are other electrons which cancel these magnetic fields. However, magnetic materials such as iron, nickel, and cobalt have lone electrons which contribute a net magnetic field to each atom. Groups of atoms with the same magnetic field direction are called *domains*. Also, groups of atoms that form into one continuous crystal structure are called *grains*. In general, there are approximately 10^{15} atoms in a domain, 10^6 domains in a grain, and 10^2 grains in a cubic centimeter.

During magnetization of some materials, the domain boundaries (or walls) move, so that the aligned domains become larger and the misaligned domains become smaller. If the magnetic material contains alloying elements, such as in the case of the quench-hardened permanent-magnet steels, nonmagnetic carbide inclusions are formed in the grain boundaries. The motion of a domain boundary through a nonmagnetic inclusion requires more energy, and this makes it more difficult to magnetize and demagnetize the permanent-magnet steels.

In other materials, the domain walls are pinned at the grain boundaries by carbide or other intermetallic compounds. If the energy required to move the domain wall is not exceeded during magnetization, the domain wall will remain pinned, and the entire domain will rotate to align with the magnetizing field.

1.5.1 Soft Magnetic Materials

Soft magnetic materials typically have a very narrow magnetic hysteresis loop to minimize losses and to maximize the permeability μ_{max}, as shown in Fig. 1.22. Equation (1.37) and Fig. 1.6 show that the area within the hysteresis loop represents energy. When ac coils and motors drive the magnetic material around the hysteresis loop, energy is lost in the form of heat. Therefore, minimizing the width of the hysteresis loop (coercive force H_c) will also minimize the hysteresis loss. The maximum slope of the hysteresis loop μ_{max} is approximately equal to the ratio of the remnant flux density B_r to the coercive force H_c. Therefore, minimizing the width of the hysteresis loop (coercive force H_c) will also maximize the permeability μ_{max}.

The magnetic permeability μ_{max} reflects the ability of a material to carry magnetic flux. As mentioned, steel materials are composed of uniform magnetic regions called domains. When a magnetic field intensity is applied to a steel material, the magnetic domains rotate to align with the magnetic field. As the magnetic field intensity is increased, additional domains rotate into alignment with the magnetic field. Residual stress adds energy to the crystal structure and prevents the domains from rotating. The result is an increase in coercive force.

Small coercive force values are typically obtained in steels by minimizing the carbon content and by eliminating the residual strain with a full anneal. The properties of various soft magnetic materials are listed in Table 1.1.

1.5.2 Hard Magnetic Materials (Permanent Magnets)

Hard magnetic materials (permanent magnets) typically have a very wide magnetic hysteresis loop to maximize the operating magnetic field energy, as shown in Fig. 1.23. Permanent magnets are magnetized in quadrants I and III and are used in magnetic devices in quadrants II and IV. Since quadrants II and IV are identical, the properties and performance characteristics described here are based on quadrant II.

Both the normal and intrinsic hysteresis loops are shown in Fig. 1.23. Most permanent-magnet specification sheets show both curves, and both curves are required to fully determine the permanent-magnet performance at different temperatures. However, in general, only the normal curve is required to determine the performance of the permanent magnet and the system at constant temperature.

The intrinsic curve represents the added magnetic flux that the permanent-magnet material produces. The normal curve represents the total measurable or usable magnetic flux which is carried in combination by the air (free space) and by the permanent-magnet material. For example, imagine that a coil is placed in air with a flux meter located at one end of the coil axis. The magnetic flux measured by the meter is the flux carried by the air, and is called the *air flux*. When a magnetic material is placed in the center of the coil, the magnetic flux increases. The amount of the increase is called the *intrinsic flux*, and the total of the air flux plus the intrinsic flux is called the *normal flux*. The total or normal flux is the new flux-meter reading.

The quadrant II operating region generally lies on the normal curve between the remnant flux density B_r and the coercive force H_c. As the external system acts to

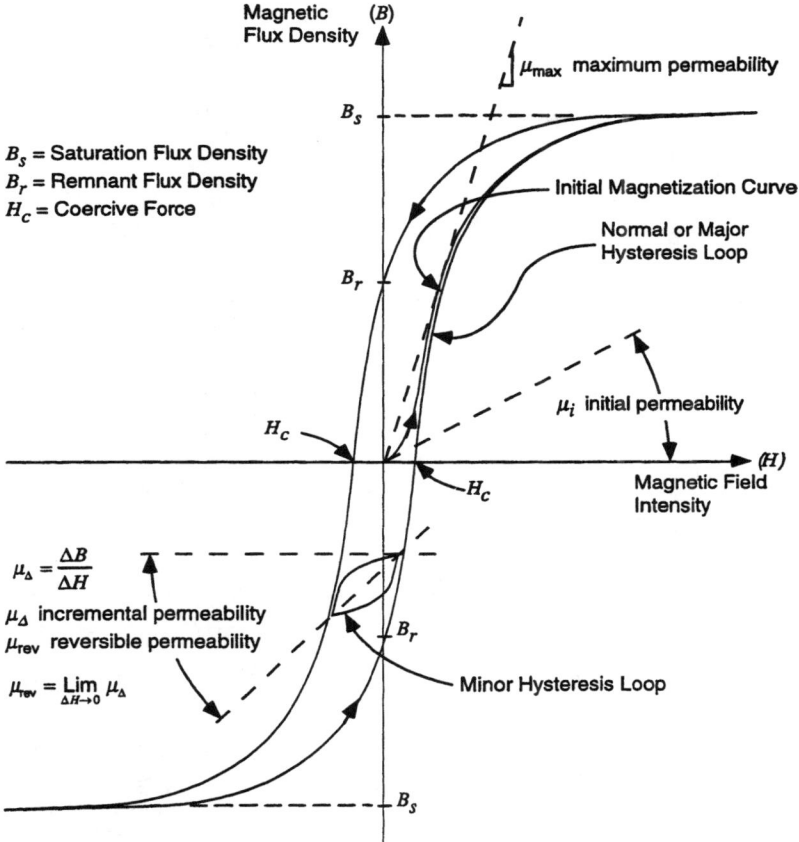

FIGURE 1.22 Magnetic hysteresis loop and characteristic parameters. *(Courtesy of Eaton Corporation.)*

demagnetize the permanent magnet, the operating point will move from B_r toward H_c. As the external system acts to magnetize the permanent magnet, the operating point will move from H_c toward B_r. As long as the operating point remains on the linear slope of the normal curve between the B_r and H_c points, the magnetizing and demagnetizing cycles will be reversible. However, if the demagnetizing field becomes large enough to move the operating point beyond the linear region toward the H_c point, the subsequent magnetizing cycle will follow a different return path.

For example, if the demagnetizing field becomes large enough to move the operating point beyond the coercive force H_c point to the recoil point, the ensuing magnetizing cycle will follow the recoil line at a slope of μ_{rev} rather than return up the nonlinear demagnetizing path toward the H_c point. The magnetizing and demagnetizing cycles will then be reversible along the linear recoil line.

Permanent magnets are almost always designed to operate on the linear region between the remnant flux density B_r and the coercive force H_c. The position of the operating point can be identified by the pair of coordinates (B_d, H_d), the demagne-

TABLE 1.1 Soft Magnetic Material Properties

	Basic material property and cost data								Energy density force per area $B_{sat}^2/2\mu_0$		Cost per unit		Energy CD_2/E_3, $/J	Energy per unit cost E_3/CD_2, J/$
	Flux density B_{sat} ($H=100$ Oe), T	Relative perm. u_r ($B = B_{sat}/2$)	Coercive force H_c ($B = 10$ kG), Oe	Electrical conduct. s, 1/Ωm	Curie temp. T_C, °C	Density		Cost C, $/lb*	E_2, kN/m²	E_3, lb/in²	Volume CD_2, $/in³			
Material						D_1, g/cc	D_2, lb/in³							
Copper	—	1	—	58.00E+6	—	8.949	0.323	2.80	—	—	0.90	—	—	—
Carbon steel														
1008/1010	1.84	3.29E+3	1.20	5.00E+6	760	7.84	0.283	0.35	1347	195	0.10	0.004	222.87	
1018	1.78	2.54E+3	1.30	5.00E+6	760	7.84	0.283	0.44	1261	183	0.12	0.006	165.91	
1020	1.78	2.54E+3	1.30	5.00E+6	760	7.84	0.283	—	1261	183	—	—	—	
1030	1.74	1.16E+3	2.40	5.00E+6	760	7.84	0.283	—	1205	175	—	—	—	
CH135 (0.005% C)	1.78	4.45E+3	0.61	5.00E+6	760	7.84	0.283	—	1261	183	—	—	—	
Ferrite—Phillips														
3C81	0.50	2.70E+3	0.19	1.0	210	4.80	0.173	—	99	14	—	—	—	
3E2A	0.41	5.00E+3	0.06	2.0	170	4.80	0.173	—	67	10	—	—	—	
4C4	0.38	125	3.10	10.0E−6	350	4.50	0.163	—	57	8	—	—	—	
Carpenter														
430F stainless	1.42	1.8E+3	2.00	1.67E+6	671	7.75	0.280	—	802	116	—	—	—	
Core iron 0.06% C	1.70	2.7E+3	1.40	10.0E+6	760	7.86	0.284	—	1150	167	—	—	—	
1.0% Si core iron A	1.80	4.5E+3	0.90	4.00E+6	810	7.75	0.280	—	1289	187	—	—	—	
2.5% Si core iron B	1.80	5.0E+3	0.70	2.50E+6	799	7.65	0.276	—	1289	187	—	—	—	
4.0% Si core iron C	1.65	4.0E+3	0.60	1.72E+6	788	7.60	0.275	—	1083	157	—	—	—	
High perm 49	1.50	50.0E+3	0.07	2.08E+6	475	8.25	0.298	8.25	895	130	2.46	0.168	5.97	
HyMu 80	0.73	200.E+3	0.02	1.72E+6	460	8.75	0.316	13.99	212	31	4.42	1.272	0.79	
Hyperco-50	2.28	8.00E+3	0.60	2.50E+6	940	8.11	0.293	47.09	2068	300	13.80	0.407	2.46	
Armco														
M-15, 2.7% Si	1.76	8.00E+3	0.35	1.96E+6	800	7.70	0.278	—	1232	179	—	—	—	
M-19, 2.7% Si	1.72	8.19E+3	0.36	2.00E+6	800	7.70	0.278	—	1177	171	—	—	—	
M-22, 2.0% Si	1.72	7.82E+3	0.39	2.00E+6	800	7.75	0.280	0.66	1177	171	0.18	0.010	104.39	
USS M-27, 2.0% Si	1.71	5.03E+3	0.48	2.27E+6	800	7.75	0.280	0.59	1163	169	0.17	0.009	115.42	
M-36, 2.0% Si	1.75	7.81E+3	0.50	2.27E+6	800	7.75	0.280	—	1219	177	—	—	—	
M-43, 1.6% Si	1.76	7.40E+3	0.55	1.60E+6	800	7.80	0.282	0.45	1232	179	0.13	0.006	159.28	
M-45, 1.6% Si	1.77	7.08E+3	0.81	2.94E+6	800	7.80	0.282	—	1247	181	—	—	—	
Metglas—Allied														
2605CO	1.80	400E+3	0.05	813E+3	415	7.56	0.273	53.98	1289	187	14.74	0.698	1.43	
2714A	0.57	1.00E+6	0.01	704E+3	205	7.59	0.274	104.33	129	19	28.61	13.504	0.07	

* Approximate cost for comparison purposes only.

tizing flux density and field intensity. The product $B_d H_d$ is called the *energy product* and represents the energy supplied to the system from the permanent magnet. The maximum energy product BH_{max} is a relative measure of the strength of a permanent magnet and is always listed on the material specification sheet.

Typical Properties of Permanent Magnets. Some typical properties of common permanent-magnet materials are listed in Table 1.2, and the quadrant II normal curves are plotted in Fig. 1.24. There are obvious design and performance tradeoffs involved in selecting a permanent-magnet material, including the shape of the normal curve, the energy product, and the operating temperature.

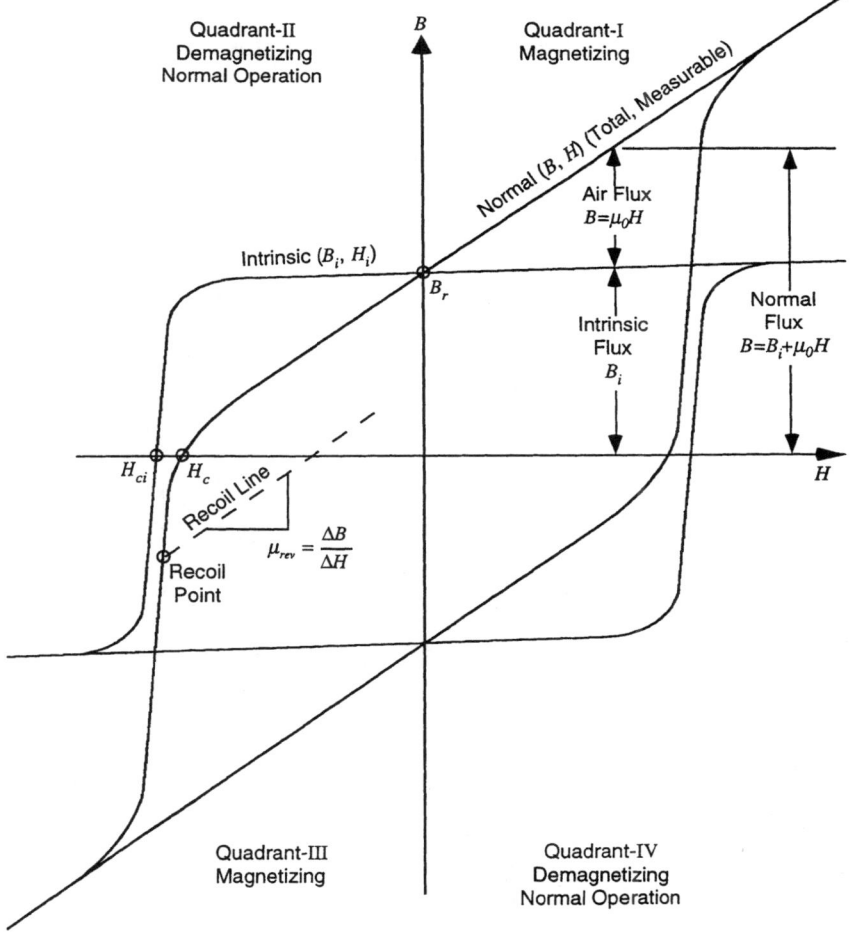

FIGURE 1.23 Hard magnetic material hysteresis loop and characteristic parameters. *(Courtesy of Eaton Corporation.)*

Load Line. The permanent-magnet operating point must be determined before the size and the material of the permanent magnet can be finalized. The operating point is the intersection of the permanent-magnet recoil line (in this case the normal curve) and the system load lines, as shown in Fig. 1.25. Since the load line passes through the operating point (B_d, H_d) and the origin $(0, 0)$, the slope of the load line μ_L can be defined as follows.

$$\mu_L = \frac{B_d}{H_d} \tag{1.255}$$

The system load line can be determined by using Eq. (1.18), which states that the system magnetic flux is equal to the product of the system magnetizing force and the system permeance $\mathcal{P}_{sys}$. In a system with a permanent magnet and no coil (see Fig. 1.26), the system magnetizing force is proportional to the permanent-magnet field intensity H_d, and the system magnetic flux is proportional to the permanent-magnet flux density B_d, as described in Eqs. (1.21) and (1.22).

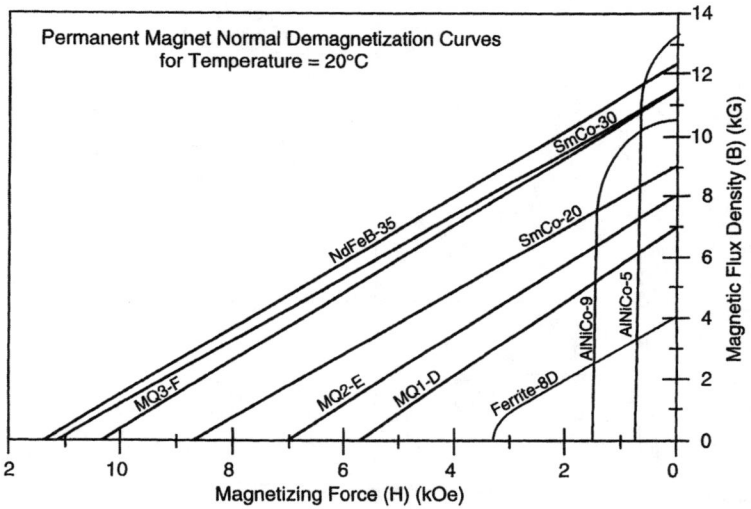

FIGURE 1.24 Normal demagnetization curves for several common materials. *(Courtesy of Eaton Corporation.)*

$$\phi_{PM} = NI_{PM}\mathcal{P}_{sys} \tag{1.256}$$

$$B_d a_{PM} = H_d l_{PM} \mathcal{P}_{sys} \tag{1.257}$$

Solving Eq. (1.257) for the load-line slope results in the following definition of the load line. In some references, this slope is also called a *permeance coefficient*.

$$\mu_L = \frac{B_d}{H_d} = \mathcal{P}_{sys} \frac{l_{PM}}{a_{PM}} \tag{1.258}$$

TABLE 1.2 Hard Magnetic Material (Permanent-Magnet) Properties

Material	Material property and cost data									Energy per unit volume		Cost per unit volume	Energy	Energy per unit cost
	Residual flux density B_r, kG	Coercive force H_C, kOe	Energy product BH_{max}, Mg·Oe	Curie temp. T_C, °C	Operating temp. T_{max}, °C	Density		Cost C, $/lb*	E_2, BH_{max} kJ/m³	E_3, BH_{max} J/in³	Volume CD_2, $/in³	CD_2/E_3, $/J	E_3/CD_2, J/$
						D_1, g/cc	D_2, lb/in³						
Ceramic													
Ceramic 1	2.20	3.25	1.05	450	300	4.99	0.180	2	8.4	0.14	0.36	2.63	0.380
Ceramic 5	3.95	2.45	3.50	450	300	4.99	0.180	3	27.9	0.46	0.54	1.18	0.845
Ceramic 8	3.90	3.25	3.50	450	300	4.99	0.180	4	27.9	0.46	0.72	1.58	0.634
Ceramic 10	4.20	3.05	4.20	450	300	4.99	0.180	8	33.4	0.55	1.44	2.63	0.380
Alnico													
Alnico 5	12.50	0.64	5.5	860	540	7.30	0.263	40	43.8	0.72	10.54	14.70	0.068
Alnico 5-7	13.50	0.74	7.5	860	540	7.30	0.263	55	59.7	0.98	14.49	14.82	0.067
Alnico 8	8.20	1.65	5.3	860	550	7.30	0.263	48	42.2	0.69	12.65	18.30	0.055
Alnico 8 HC	7.20	1.90	5.0	860	550	7.30	0.263	48	39.8	0.65	12.65	19.40	0.052
Alnico 9	10.60	1.50	10.5	860	540	7.30	0.263	100	83.6	1.37	26.35	19.24	0.052
Magnequench													
MQ1-C 9H (NdFeB)	6.30	5.60	9	470	125	6.10	0.220	90	79.6	1.30	19.49	14.95	0.067
MQ2-E 15 (NdFeB)	8.25	7.20	15	335	180	7.60	0.274	—	119.4	1.96	—	—	—
MQ3-F 42 (NdFeB)	13.10	12.30	42	370	180	7.60	0.274	—	334.2	5.48	—	—	—
Samarium cobalt													
SmCo 18	8.60	7.20	18	775	250	8.30	0.300	110	143.2	2.35	32.95	14.04	0.071
SmCo 22	9.85	8.75	22	820	250	8.30	0.300	120	175.1	2.87	35.95	12.53	0.080
SmCo 26 HS	10.60	9.80	27	820	380	8.30	0.300	150	214.9	3.52	44.94	12.76	0.078
SmCo 28	10.70	10.30	28	820	350	8.30	0.300	150	222.8	3.65	44.94	12.31	0.081
SmCo 32	11.60	9.50	32	820	350	8.30	0.300	160	254.6	4.17	47.93	11.49	0.087
Neodymium iron boron													
NdFeB-24	10.00	9.60	24	310	210	7.45	0.269	80	191.0	3.13	21.51	6.87	0.145
NdFeB-24 UH	9.80	7.50	24	300	80	7.45	0.269	60	191.0	3.13	16.13	5.16	0.194
NdFeB-28	10.80	10.10	28	310	150	7.45	0.269	90	222.8	3.65	24.20	6.63	0.151
NdFeB-28 UH	10.90	10.40	28	310	190	7.45	0.269	80	222.8	3.65	21.51	5.89	0.170
NdFeB-32 SH	11.60	11.10	32	310	180	7.45	0.269	90	254.6	4.17	24.20	5.80	0.172
NdFeB-38	12.55	11.70	38	365	130	7.45	0.269	70	302.4	4.96	18.82	3.80	0.263
NdFeB-42 H	13.30	12.70	42	310	120	7.45	0.269	80	334.2	5.48	21.51	3.93	0.255
NdFeB-48	14.10	12.90	48	310	80	7.45	0.269	80	382.0	6.26	21.51	3.44	0.291

* Approximate cost for comparison purposes only.

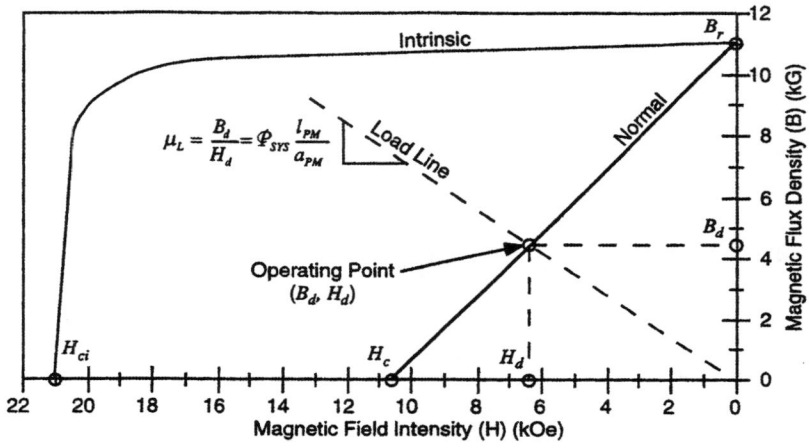

FIGURE 1.25 Quadrant II demagnetization curves for NdFeB at 20°C, showing the operating point and the load line with no coil. *(Courtesy of Eaton Corporation.)*

When a coil is added to the system, as shown in Fig. 1.26, Eq. (1.256) is modified as follows to include the coil magnetizing force. In this case, it is assumed that the coil magnetizing force is negative, or propagates in a direction that would demagnetize the permanent magnet.

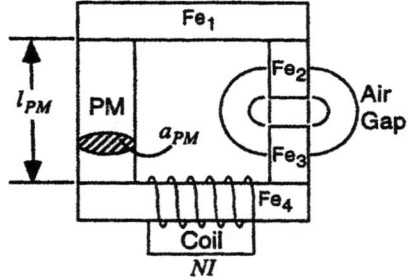

FIGURE 1.26 Sketch of a permanent magnet in a simple system with a coil. *(Courtesy of Eaton Corporation.)*

$$\phi_{PM} = (NI_{PM} - NI)\mathcal{P}_{sys} \qquad (1.259)$$

$$B_d a_{PM} = (H_d l_{PM} - NI)\mathcal{P}_{sys} \qquad (1.260)$$

$$B_d = \left(H_d - \frac{NI}{l_{PM}}\right)\mathcal{P}_{sys}\frac{l_{PM}}{a_{PM}} \qquad (1.261)$$

$$\mu_L = \frac{B_d}{H_d - NI/l_{PM}} = \mathcal{P}_{sys}\frac{l_{PM}}{a_{PM}} \qquad (1.262)$$

This load-line equation shows that the slope of the load line is independent of the coil. However, the denominator of Eq. (1.262) indicates that the load line passes through both the operating point (B_d, H_d) and a point on the H axis $(0, H_{coil})$, as shown in Fig. 1.27. The slope between these two points can be written as follows and equated with the result from Eq. (1.262).

$$\mu_L = \frac{B_d}{H_d - H_{coil}} = \frac{B_d}{H_d - NI/l_{PM}} \qquad (1.263)$$

$$H_{coil} = \frac{NI}{l_{PM}} \qquad (1.264)$$

This shows that the H axis intercept is the magnetic field intensity of the coil as seen across the permanent magnet. This makes sense because Fig. 1.27 is a plot of the permanent-magnet material properties. Also, the slope of the load line represents the permeance of the system as seen across the permanent magnet.

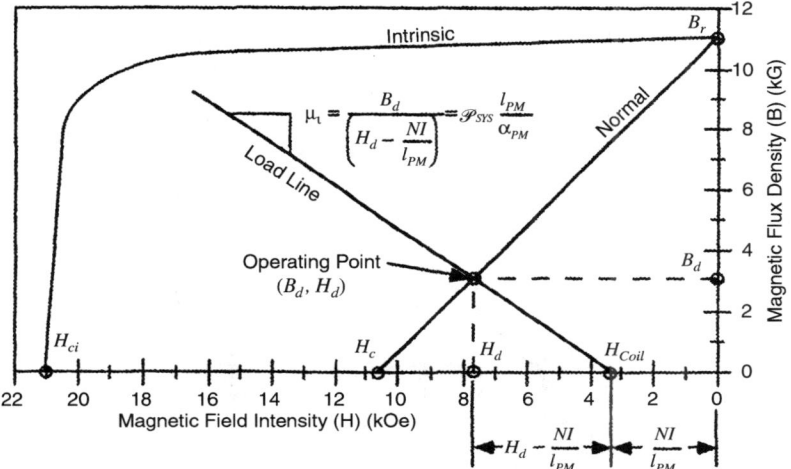

FIGURE 1.27 Quadrant II demagnetization curves for NdFeB at 20°C, showing the operating point and the load line with a demagnetizing coil. *(Courtesy of Eaton Corporation.)*

Permanent-Magnet Reluctance Model. Magnetic systems can be modeled using electric circuit analogies, as discussed in Secs. 1.2.1 and 1.2.2. In general, from Eq. (1.14), a coil is analogous to a voltage source, and the reluctance of each material and air gap in the system is analogous to a resistor. A permanent magnet can be modeled as a voltage source with a series resistance as follows.

1. The recoil line is the path of the operating point. The slope of the recoil line μ_{rev} defines the magnetic flux density change as a function of the magnetic field intensity change. Therefore, the permanent-magnet reluctance can be defined from Eq. (1.29) as follows.

$$\mathcal{R}_{PM} = \frac{l_{PM}}{\mu_{rev} a_{PM}} \qquad (1.265)$$

2. Extending the recoil line (beyond the normal and intrinsic curves if needed) to intersect with the H axis gives the effective coercive force of the recoil line H_R, as shown in Fig. 1.28. The permanent-magnet magnetizing force provided at zero magnetic flux is then defined from Eq. (1.22) as follows.

$$NI_{PM} = H_R l_{PM} \qquad (1.266)$$

The reluctance model for the permanent magnet is included in Fig. 1.29 with the entire reluctance model for the system from Fig. 1.26. The iron reluctance values $\mathcal{R}_{Fe1}$, $\mathcal{R}_{Fe2}$, $\mathcal{R}_{Fe3}$, and $\mathcal{R}_{Fe4}$ can be easily calculated from Eq. (1.29) when the geometry and the soft magnetic properties are known. The reluctance of the air gap $\mathcal{R}_{air}$ is simply the inverse of the air gap permeance. The air gap permeance can be easily calculated from the equations in Sec. 1.2.5 when the geometry of the air gap is known. The load line for the system shown in Fig. 1.28 can be calculated from the reluctances in Fig. 1.29 as follows.

$$\mathcal{R}_{sys} = \mathcal{R}_{Fe1} + \mathcal{R}_{Fe2} + \mathcal{R}_{Fe3} + \mathcal{R}_{Fe4} + \mathcal{R}_{air} \qquad (1.267)$$

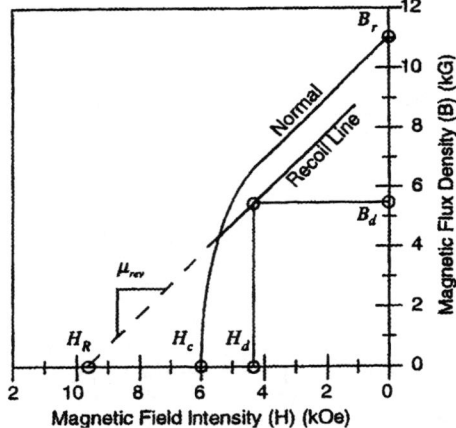

FIGURE 1.28 Example of extending the recoil interest with the H axis. *(Courtesy of Eaton Corporation.)*

$$\mathcal{P}_{sys} = \frac{1}{\mathfrak{R}_{sys}} \tag{1.268}$$

$$\mu_L = \mathcal{P}_{sys} \frac{l_{PM}}{a_{PM}} \tag{1.269}$$

$$H_{coil} = \frac{NI}{l_{PM}} \tag{1.270}$$

The inductance of the coil in Fig. 1.26 can also be calculated based on the system permeance $\mathcal{P}_{sys}$ and Eq. (1.31), as follows.

$$L = N^2 \mathcal{P}_{sys} \tag{1.271}$$

1.6 LOSSES

Power losses generally take the form of hysteresis loss, eddy current loss, coil electrical resistance loss, and mechanical friction. The losses that are discussed here are the losses in the magnetic material, called core loss P_C, which includes the hysteresis loss P_H and the eddy current loss P_E. Data from U.S. Steel for the typical 60-Hz core loss in steel motor-lamination materials (M19 through M45) indicates that approximately 67 percent of the core loss is due to hysteresis effects and 33 percent of the core loss is due to eddy current effects. The published core loss properties of magnetic materials show only the total core loss. Therefore, the hysteresis loss and the eddy current loss can be approximated as follows. The units of core loss are in power per unit mass, for example, watt per kilogram.

Total core loss P_C, W/kg

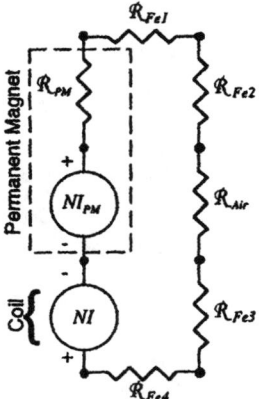

FIGURE 1.29 Permanent magnet reluctance model for the system in Fig. 1.26. *(Courtesy of Eaton Corporation.)*

$$P_C = P_H + P_E \quad (1.272)$$

Hysteresis core loss P_H, W/kg

$$P_H = \frac{2}{3} P_C \quad (1.273)$$

Eddy current core loss P_E, W/kg

$$P_E = \frac{1}{3} P_C \quad (1.274)$$

1.6.1 Hysteresis Power Loss

The hysteresis power loss is caused by forcing the magnetic material around the hysteresis loop. Each time the magnetic domains change direction, external energy is required. As shown in Sec. 1.2.4, the area inside the hysteresis loop represents energy (Fig. 1.22). Each time the magnetic material is forced around the hysteresis loop, energy equal to the area inside the hysteresis loop is lost. Therefore, the hysteresis power loss is proportional to the product of the hysteresis loop area and the cyclic frequency, as shown here. The outer hysteresis loop area w_H, J/m³, can be approximated as a rectangle having a height of $2B_S$ and a width of $2H_C$.

$$w_H = 4B_S H_C \quad (1.275)$$

Dividing the hysteresis loop area by the magnetic material density ρ and multiplying it by the cyclic frequency f gives the outer loop hysteresis core loss P_H, W/kg, as follows.

$$P_H = 4B_S H_C f \frac{1}{\rho} \quad (1.276)$$

The area of a minor hysteresis loop can be approximated by substituting the peak flux density B_P in place of the saturation flux density B_S, and by assuming that the minor loop width remains constant.

$$P_H = 4B_P H_C f \frac{1}{\rho} \quad (1.277)$$

It should be noted here that the hysteresis core loss is directly proportional to the excitation frequency.

1.6.2 Eddy Current Power Loss

Eddy currents are induced around a changing magnetic field as defined by the induced voltage from Faraday's law and the conductivity of the material. An example of a changing magnetic flux density B inducing an eddy current density J_e in a magnetic steel material is shown in Fig. 1.30. At low frequencies, the magnetic flux density is uniform over the steel cross section $T \times W$. However, at high frequencies

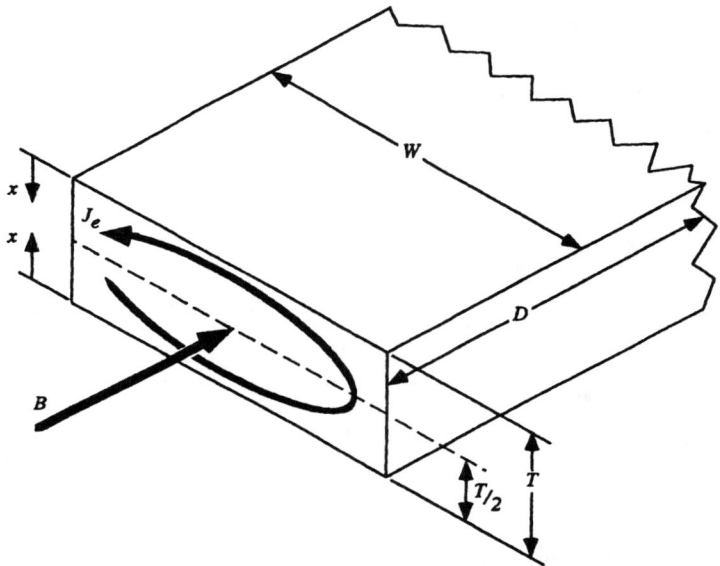

FIGURE 1.30 Magnetic flux density B in a steel lamination induces eddy current density J_e and produces an opposing magnetic flux according to Lenz's law. *(Courtesy of Eaton Corporation.)*

the magnetic flux moves toward a thin layer near the outer surface of the magnetic material called the *skin*. This is called the *skin effect*. The skin thickness is defined as follows, where μ is the magnetic permeability, f is the cyclic frequency, ω is the radian frequency, and σ is the electrical conductivity.

$$s = \frac{1}{\sqrt{\pi f \mu \sigma}} \qquad s = \sqrt{\frac{2}{\omega \mu \rho}} \tag{1.278}$$

Low Frequencies, No Skin Effect (T < 0.5s). The voltage induced over the length of the eddy current path is defined by Faraday's law as follows. Under the condition of $T < 0.5s$, it can be assumed that the magnetic flux density is uniform over the cross-sectional area of the lamination.

$$V_e = E_e l_e = \frac{d\lambda}{db} = \frac{d(N\phi)}{dt} = \frac{d(NBa)}{dt} \tag{1.279}$$

$$E_e l_e = Na \frac{dB}{dt} \tag{1.280}$$

$$E_e = N \frac{a}{l_e} \frac{dB}{dt} \tag{1.281}$$

The number of turns made by the eddy current around the magnetic flux is one. The magnetic flux density is applied over an area a starting at the center of the lamination, and the eddy current path length l_e is the distance around the magnetic flux density area. Also, it is assumed that the lamination thickness T is small compared to the width W, and the flux density is a sinusoidal function of time as follows.

$$N = 1 \tag{1.282}$$

$$a = 2W\left(\frac{T}{2} - x\right) \tag{1.283}$$

$$y = \frac{T}{2} - x \tag{1.284}$$

$$a = 2yW \tag{1.285}$$

$$l_e = 2W \quad \text{where } T \ll W \tag{1.286}$$

$$B = B_P \sin(\omega t) \tag{1.287}$$

Substituting Eqs. (1.282) through (1.287) into Eq. (1.281) gives the following expression for the electric field intensity in the path of the eddy current.

$$E_e = yB_P\omega \cos(\omega t) \tag{1.288}$$

The eddy current density is then obtained by multiplying the electric field intensity by the conductivity of the material in the path of the eddy current. The root mean square (RMS) value for the eddy current density is then obtained by dividing by the square root of 2, as follows.

$$J_e = E_e\sigma = y\sigma B_P\omega \cos(\omega t) \tag{1.289}$$

$$J_{\text{RMS}} = \frac{1}{\sqrt{2}} y\sigma b_P\omega \tag{1.290}$$

The power loss per unit volume due to the eddy current density is equal to the square of the eddy current density RMS value divided by the conductivity. The total eddy current power loss can be obtained by integrating over the entire volume. The eddy current power loss per unit mass (or the eddy current core loss) is obtained by dividing by the total mass of the eddy current path $m = \rho v$, as follows.

$$\int_0^{P_E} dP_E = \frac{1}{\rho v} \int_0^v J_{\text{RMS}}^2 \frac{1}{\sigma} \, dv \quad \text{W/kg} \tag{1.291}$$

The differential volume of the eddy current path and the total volume are defined from Fig. 1.30 as follows.

$$dv = l_e D \, dy \tag{1.292}$$

$$v = WDT \tag{1.293}$$

Substituting Eqs. (1.290), (1.292), and (1.293) into Eq. (1.291) gives the following expression for the eddy current core loss:

$$P_E = \frac{2WD}{\sigma\rho WDT} \int_0^{T/2} \left(\frac{1}{\sqrt{2}} y\sigma B_P\omega\right)^2 dy \tag{1.294}$$

$$P_E = \frac{\sigma B_p^2 \omega^2}{\rho T} \int_0^{T/2} y^2 \, dy \tag{1.295}$$

$$P_E = \frac{\sigma B_p^2 \omega^2}{\rho T} \left(\frac{y^3}{3}\right)_0^{T/2} \tag{1.296}$$

$$P_E = \frac{\sigma B_p^2 \omega^2 T^2}{24\rho} \quad \text{W/kg} \tag{1.297}$$

where $T < 0.5s$.

It should be noted here that the eddy current core loss is directly proportional to the square of the excitation frequency, the square of the flux density, and the square of the lamination thickness.

High Frequencies, Large Skin Effect (T > 5.0s). Under the condition of $T > 5.0s$, the magnetic flux density is not uniform over the cross-sectional area of the lamination. According to Bozorth (1993), it can be assumed that magnetic flux density varies exponentially over the cross-sectional area as follows.

$$B = B_p e^{-x/s} \sin(\omega t) \tag{1.298}$$

Substituting Eq. (1.298) into Eq. (1.281) and following the process previously used for Eqs. (1.288) through (1.297) gives the following general expression for the eddy current core loss.

$$P_E = \frac{\sigma B_p^2 \omega^2}{\rho T} \int_0^{T/2} (y e^{-x/s})^2 \, dy \tag{1.299}$$

$$P_E = \frac{\sigma B_p^2 \omega^2 T^2}{24\rho} \left[3\left(\frac{s}{T}\right) - 6\left(\frac{s}{T}\right)^2 - 6\left(\frac{s}{T}\right)^3 (1 - e^{-T/s})\right] \quad \text{W/kg} \tag{1.300}$$

The eddy current core loss equation Eq. (1.300) is asymptotic to the following two limiting equations.

$$P_E = \frac{\sigma B_p^2 \omega^2 T^2}{24\rho} \quad \text{for } T < 0.5s \tag{1.301}$$

$$P_E = \frac{\sigma B_p^2 \omega^2 T^2}{24\rho} 3\left(\frac{s}{T}\right) \quad \text{for } T > 5.0s \tag{1.302}$$

As can be seen from Eq. (1.298), the magnetic flux is not completely confined to the depth of one skin thickness. However, Steinmetz defined a depth of penetration d which is described in Roters (1941) and Bozorth as the required surface layer thickness that will contain all of the magnetic flux at a uniform magnetic flux density equal to the magnetic flux density at the outside surface. The depth of penetration is shown in Eq. (1.303).

$$d = \frac{1}{\sqrt{2\pi f \mu \sigma}} \qquad d = \frac{1}{\sqrt{\omega \mu \sigma}} \tag{1.303}$$

Comparing the skin depth s in Eq. (1.278) to the depth of penetration d in Eq. (1.303) gives the following relationship.

$$s = d\sqrt{2} \tag{1.304}$$

$$d = \frac{s}{\sqrt{2}} \tag{1.305}$$

The depth of penetration d can be used to determine the total effective magnetic flux cross-sectional area and the peak magnetic flux density at the surface. This provides the capability to consider the effects due to saturation on performance, such as determining the limitations on peak force and peak inductance.

1.6.3 Reflected Core Loss Resistance

The core loss of an electromagnetic device can be modeled as a reflected resistance in the coil or as a wider hysteresis loop. The reflected core loss resistance R_C in the coil can be calculated by considering the total power loss in the core $P_C \rho v$ and the power loss of the coil $I^2 R$ as follows, where P_C is the core loss, W/kg; ρ is the density, kg/m^3; and v is the volume, m^3.

$$P = I^2 R + P_C \rho v \tag{1.306}$$

$$P_C \rho v = I^2 R_C \tag{1.307}$$

$$R_C = \frac{P_C \rho v}{I^2} \tag{1.308}$$

1.6.4 Imaginary Permeability

As described in Sec. 1.6.1, the hysteresis loop area represents an energy loss. Also, for nonsaturating conditions, the hysteresis loop can be modeled as a rotating vector system based on Eq. (1.23) as follows.

$$B = \mu H = (\mu_R + j\mu_i) H \tag{1.309}$$

The system inductance and impedance can be written based on equations [1-30], [1-31] and [1-308].

Inductance

$$L = N^2 \mu \frac{a}{l} = N^2 (\mu_R + j\mu_i) \frac{a}{l} \tag{1.310}$$

Impedance

$$Z = R + j\omega L \tag{1.311}$$

$$Z = \left(R - \omega N^2 \mu_i \frac{a}{l} \right) + j\left(\omega N^2 \mu_R \frac{a}{l} \right) \tag{1.312}$$

The first term of Eq. (1.312) represents the resistive impedance, and the second term represents the reactive impedance. Therefore, only the first term contributes to the power loss, as shown in Eq. (1.313).

$$P = I^2 \left(R - \omega N^2 \mu_i \frac{a}{l} \right) \tag{1.313}$$

The first term of Eq. (1.313) represents the power loss in the coil resistance, and the second term represents the power loss in the core. Therefore, the second term can be equated with the power loss in the core from Eq. (1.306), and the imaginary permeability can be determined, as follows.

$$P_c \rho v = -I^2 \omega N^2 \mu_i \frac{a}{l} \tag{1.314}$$

$$\mu_i = -\frac{P_c \rho v l}{N^2 I^2 \omega a} \tag{1.315}$$

Equation (1.22) can be written for NI, Eq. (1.23) can be written for H, and volume of the core is simply the product of the magnetic flux path length and the magnetic flux cross-sectional area. Substitution of these relations into Eq. (1.315) gives the imaginary permeability as a function of the excitation (B, ω) and the material properties (P_c, ρ, μ).

$$NI = Hl \tag{1.316}$$

$$H = \frac{B}{\mu} \tag{1.317}$$

Volume of the core

$$v = al \tag{1.318}$$

Imaginary permeability, H/m

$$\mu_i = -\mu^2 \frac{P_c \rho}{B^2 \omega} \tag{1.319}$$

Imaginary relative permeability

$$\mu_{ri} = \mu_r^2 \left(\frac{\mu_0 P_c \rho}{B^2 \omega} \right) \tag{1.320}$$

1.7 MAGNETIC MOMENT (OR MAGNETIC DIPOLE MOMENT)

This section describes the magnetic moment and its use in determining the properties of magnetic materials. The magnetic moment (which is based on a current loop) can be used to model the field around a permanent magnet. Correlations for approximately the magnetic field distribution for a magnetic moment and for a current loop are also presented in this section.

The properties of a magnetic material are described based on magnetization curves, which can be presented in a number of different ways, as listed here. These properties and their relationship to the atomic magnetic moment (Bohr magneton) are explained in this section.

- Total or normal flux density B versus magnetizing force H

- Intrinsic flux density B_i, $\mu_0 M$, $4\pi M$, $4\pi I$, or $4\pi J$ versus magnetizing force H
- Magnetic moment m versus magnetizing force H
- Magnetic moment per unit volume m/V versus magnetizing force H
- Magnetic polarization M, I, or J versus magnetizing force H

1.7.1 Magnetic Moment for a Current Loop

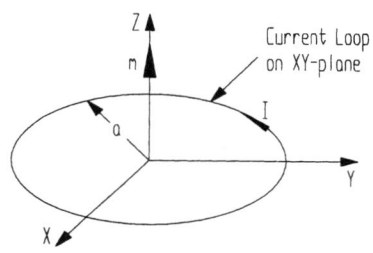

FIGURE 1.31 Magnetic moment for a current loop. *(Courtesy of Eaton Corporation.)*

The magnitude of the magnetic moment m for a current loop is equal to the loop area πa_2 times the current I (see Fig. 1.31). The direction of the magnetic moment m is in the direction of the thumb as the fingers of the right hand follow the current I.

$$m = \pi a^2 I \quad \text{A·m}^2 \text{ or J/T} \quad (1.321)$$

1.7.2 Magnetic Moment for a Magnetic Material

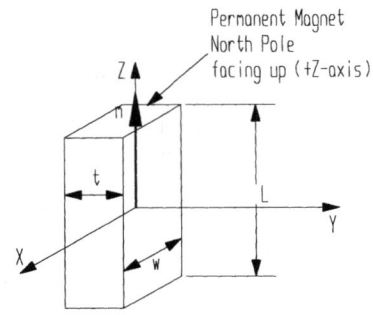

FIGURE 1.32 Magnetic moment for a permanent magnet. *(Courtesy of Eaton Corporation.)*

The magnetic moment m for a permanent magnet is equal to the magnetization M times the volume V, with a direction perpendicular to the north pole face (see Fig. 1.32). The magnetization or magnetic polarization M of a permanent magnet is equal to the operating point intrinsic flux density B_{di} (see Fig. 1.33) divided by the permeability of free space $\mu_0 = 4\pi \times 10^{-7}$ (T·m)/A or H/m.

$$m = MV = \left(\frac{B_{di}}{\mu_0}\right) wtl \quad \text{A·m}^2 \text{ or J/T} \quad (1.322)$$

Approximate Magnetic Moment for a Permanent Magnet. High-energy permanent magnets (such as NdFeB and SmCo) typically have a very small recoil permeability μ_R, so that $\mu_R \approx \mu_0$. Therefore, the magnetic polarization M and the magnetic moment m can be approximated as follows.

Deriving from Fig. 1.33:

$$M = \frac{m}{V} = \frac{B_{di}}{\mu_0} = \frac{\mu_R}{\mu_0} H_c - \left(\frac{\mu_R}{\mu_0} - 1\right) H_d \quad \text{A/m} \quad (1.323)$$

Rewriting Eq. (1.323), where $\mu_R \approx \mu_0$:

$$M = \frac{m}{V} = \frac{B_{di}}{\mu_0} = H_c \quad \text{A/m} \quad (1.324)$$

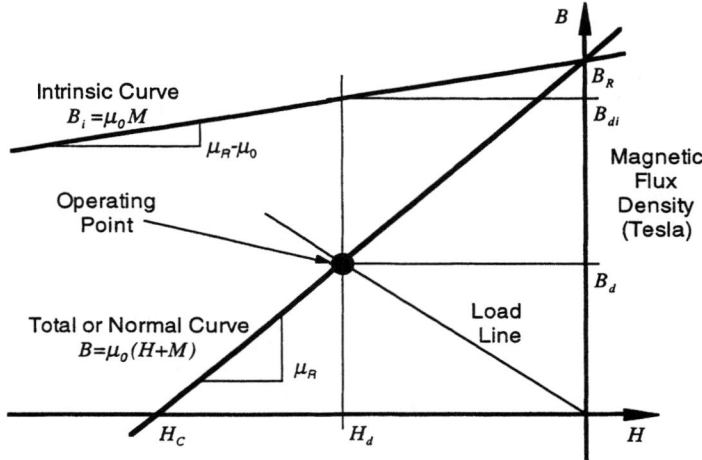

FIGURE 1.33 Quadrant II permanent magnet demagnetizing curve. *(Courtesy of Eaton Corporation.)*

Combining Eqs. (1.322) and (1.324), where $\mu_R \approx \mu_0$:

$$m \approx H_c(wtl) \quad \text{A·m}^2 \tag{1.325}$$

Permanent Magnet Current Loop Model. A high-energy permanent magnet can be modeled as a current loop by equating the magnetic moments as follows, where H_c is coercive force. This is 99.99 percent accurate for short magnets with high coercive force (NdFeB or SmCo), and 70 percent accurate for long magnets with low coercive force (alnico).

Combining Eqs. (1.321) and (1.325), where $\mu_R \approx \mu_0$:

$$\pi a^2 I \approx H_c(wtl) \quad \text{A·m}^2 \tag{1.326}$$

Equivalent loop area (for $\mu_R \approx \mu_0$)

$$\pi a^2 = wt \quad \text{m}^2 \tag{1.327}$$

Equivalent loop current (for $\mu_R \approx \mu_0$)

$$I \approx H_c l \quad \text{A} \tag{1.328}$$

Rewriting Eq. (1.327) for equivalent loop radius:

$$a = \sqrt{\frac{wt}{\pi}} \quad \text{m} \tag{1.329}$$

Permanent Magnet Pole Strength and Dipole Model. The magnetic dipole moment m can also be written as the product of the magnetic pole strength p and the distance l between the poles. The magnetic pole strength p of a high-energy permanent magnet can be approximated by the product of the coercive force H_c and the pole face area wt.

Magnetic moment

$$m = pl \quad \text{A·m}^2 \tag{1.330}$$

Combining Eqs. (1.325) and (1.330), where $\mu_R \approx \mu_0$:

$$pl \approx H_c wt \quad \text{A·m}^2 \tag{1.331}$$

Dividing Eq. (1.331) by 1, where $\mu_R \approx \mu_0$:

$$p \approx H_c wt \quad \text{A·m} \tag{1.332}$$

1.7.3 Torque on a Magnetic Moment in a Uniform Field

Torque T is produced on a magnetic dipole moment m by a uniform magnetic field B (see Fig. 1.34). The torque is in the direction to align the magnetic moment with the direction of the uniform magnetic field.

Torque on magnetic moment **T**, N·m

$$\mathbf{T} = \mathbf{m} \times \mathbf{B} \tag{1.333}$$

Magnetic moment **m**, A · m²

$$\mathbf{m} = 3s^2 I \hat{z} \tag{1.334}$$

Uniform magnetic field **B**, T

$$\mathbf{B} = -B_y \hat{y} + B_z \hat{z} \tag{1.335}$$

Combining Eqs. (1.333), (1.334), and (1.335):

$$\mathbf{T} = 3s^2 I B_y \hat{x} = m B_y \hat{x} \quad \text{N·m} \tag{1.336}$$

This torque can also be derived from the Lorentz forces on the current loop, as follows.

Mechanical torque **T**, N·m

$$\mathbf{T} = \mathbf{R} \times \mathbf{F} = 2\mathbf{R}_R \times \mathbf{F}_R \tag{1.337}$$

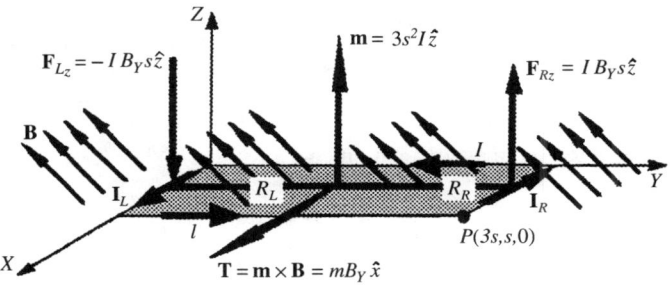

FIGURE 1.34 Torque on a magnetic moment in a uniform magnetic field. *(Courtesy of Eaton Corporation.)*

Radius vector to Lorentz force $\mathbf{R}_R$, m

$$\mathbf{R}_R = -\mathbf{R}_L = \frac{3}{2} s\hat{y} \quad (1.338)$$

Current vector $\mathbf{I}_R$, A

$$\mathbf{I}_R = -\mathbf{I}_L = -I\hat{x} \quad (1.339)$$

Lorentz force $\mathbf{F}_R$, N

$$\mathbf{F}_R = [\mathbf{I}_R \times \mathbf{B}]s \quad (1.340)$$

Combining Eqs. (1.335), (1.339), and (1.340):

$$\mathbf{F}_R = IB_Z s\hat{y} + IB_y s\hat{z} \quad \text{N} \quad (1.341)$$

Combining Eqs. (1.337), (1.338), and (1.341):

$$\mathbf{T} = 2\left[\frac{3}{2} s\hat{y}\right](1B_Z s\hat{y} + IB_y s\hat{z}) \quad \text{N·m} \quad (1.342)$$

Combining terms in Eq. (1.342):

$$\mathbf{T} = 3s^2 IB_y \hat{x} = mB_y \hat{x} \quad \text{N·m} \quad (1.343)$$

1.7.4 Magnetic Moment in Atoms (Bohr Magneton)

The magnetic moment in various types of materials is a result of the following factors.

- *Electron orbit.* An electron in an orbit around a nucleus is analogous to a small current loop, in which the current is opposite to the direction of electron travel. This factor is significant only for diamagnetic and paramagnetic materials, where it is the same order of magnitude as the electron spin magnetic moment. The magnetic properties of most materials (diamagnetic, paramagnetic, and antiferromagnetic) are so weak that they are commonly considered to be non-magnetic.
- *Electron spin.* The electron cannot be accurately modeled as a small current loop. However, relativistic quantum theory predicts a value for the spin magnetic moment (or Bohr magneton β) as shown following in Eq. (1.344). In an atom with many electrons, only the spin of electrons in shells which are not completely filled contribute to the magnetic moment. This factor is at least an order of magnitude larger than the electron orbit magnetic moment for ferromagnetic, antiferromagnetic, and superparamagnetic materials.

Bohr magneton (spin magnetic moment)

$$\beta = \frac{he}{4\pi m_e} = 9.274 \times 10^{-24} \text{ J/T} \quad (1.344)$$

Planck's constant $\quad h = 6.262 \times 10^{-34}$ J·s

Charge of an electron $\quad e = 1.6022 \times 10^{-19}$ C

Mass of an electron $m_e = 9.1094 \times 10^{-31}$ kg

- *Nuclear spin.* This factor is insignificant relative to the overall magnetic properties of materials. However, it is the basis for nuclear magnetic resonance imaging (MRI).
- *Exchange force.* The exchange force is an interaction force (or coupling) between the spins of neighboring electrons. This is a quantum effect related to the indistinguishability of electrons, so that nothing changes if the two electrons change places. The exchange force can be positive or negative, and in some materials the net spins of neighboring atoms are strongly coupled.

Chromium and manganese (in which each atom is strongly magnetic) have a strong negative exchange coupling, which forces the electron spins of neighboring atoms to be in opposite directions and results in antiferromagnetic (very weak) magnetic properties. Iron, cobalt, and nickel have unbalanced electron spins (so that each atom is strongly magnetic) and have a strong positive exchange coupling. Therefore, the spins of neighboring atoms point in the same direction and produce a large macroscopic magnetization. This large-scale atomic cooperation is called *ferromagnetism.*

1.7.5 Intrinsic Saturation Flux Density

The theoretical intrinsic saturation flux density for iron, nickel, and cobalt can be calculated using Eq. (1.345) as follows. Typically, the intrinsic saturation flux density B_i is measured and the number of Bohr magnetons per atom n_0 is calculated.

The ferromagnetic properties of materials disappear when the temperature becomes high enough. This temperature is called the *Curie temperature T_c* or the *Curie point.* Iron ($T_c = 770°C$), nickel ($T_c = 358°C$), and cobalt ($T_c = 1130°C$) are the only ferromagnetic materials that have Curie points above room temperature. Some rare earth metals like gadolinium ($T_c = 16°C$), dysprosium ($T_c = -168°C$), and holmium are ferromagnetic, but their Curie points are below room temperature.

Theoretical intrinsic saturation flux density B_i, T

$$B_i = \mu_0 M = \mu_0 \beta N_0 \frac{n_0 d}{A} C_V \tag{1.345}$$

Permeability of free space	$\mu_0 = 4\pi \times 10^{-7}$ H/m
Bohr magneton (spin magnetic moment)	$\beta = 9.27 \times 10^{-24}$ J/T
Avogadro's number	$N_0 = 6.025 \times 10^{23}$ atom/mol
Volume units conversion factor	$C_V = 1 \times 10^6$ cm^3/m^3

Iron

Number of Bohr magnetons	$n_0 = 2.218$ magneton/atom
Density	$d = 7.874$ g/cm^3
Atomic weight	$A = 55.85$ g/mol
Intrinsic saturation flux	$B_i = 2.195$ T

Nickel

Number of Bohr magnetons	$n_0 = 0.604$ magneton/atom
Density	$d = 8.90$ g/cm^3
Atomic weight	$A = 58.69$ g/mole
Intrinsic saturation flux	$B_i = 0.643$ T

Cobalt

Number of Bohr magnetons	$n_0 = 1.715$ magneton/atom
Density	$d = 8.84$ g/cm^3
Atomic weight	$A = 58.94$ g/mol
Intrinsic Saturation flux	$B_i = 1.804$ T

1.7.6 Magnetic Far Field for a Magnetic Dipole Moment

The magnetic field is well defined for points far from a magnetic dipole moment (far field, $R \gg a$; see Fig. 1.35). This means that the distance R must be large compared to the size of the magnetic dipole radius a or length l, or the size of the magnetic dipole a or l must be small compared to the distance R:

$$\mathbf{B}_x = \frac{\mu_0}{4\pi} \frac{m}{R^3} \left(3\frac{xz}{R^2}\right)\hat{x} \quad \text{T} \tag{1.346}$$

$$\mathbf{B}_y = \frac{\mu_0}{4\pi} \frac{m}{R^3} \left(3\frac{yz}{R^2}\right)\hat{y} \quad \text{T} \tag{1.347}$$

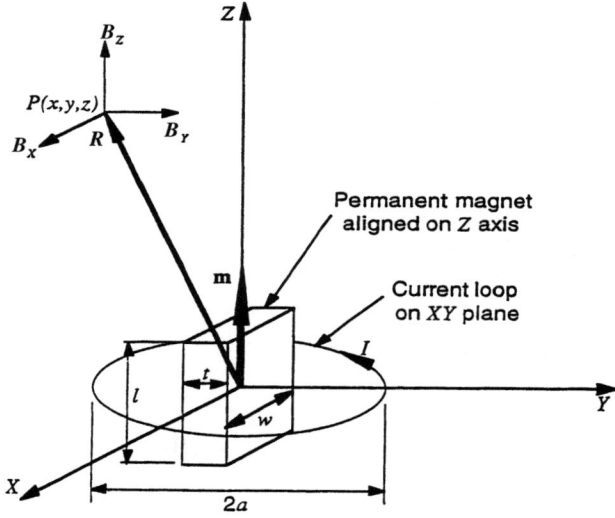

FIGURE 1.35 Geometric configuration for a magnetic dipole moment. *(Courtesy of Eaton Corporation.)*

$$\mathbf{B}_Z = \frac{\mu_0}{4\pi} \frac{m}{R^3}\left(3\frac{z^2}{R^2} - 1\right)\hat{z}F \quad \text{T} \tag{1.348}$$

where $R \gg a$ and

$$F = \frac{1}{[1+(a/R)^2]^{3/2}} \tag{1.349}$$

is the B_z correction factor for $R \approx a$.

1.8 MAGNETIC FIELD FOR A CURRENT LOOP

The magnetic field produced by a moving charge can be calculated by the Biot-Savart law. Integrating the Biot-Savart law over a current path gives the total magnetic field produced by the entire current path. This technique can be used to determine the magnetic field produced by a current loop, such as a single-turn wire loop or a magnetic moment. Further integration of the magnetic field over the area of the wire loop gives the inductance.

1.8.1 Biot-Savart Law

The Biot-Savart law gives the exact field distribution (in the absence of magnetic materials) for both the near and far fields, for any current path, or for any current path segment. The Biot-Savart law as applied to a circular wire loop in the XY plane is shown in Fig. 1.36 (Note that Eqs. (1.346) to (1.348) may be used to approximate the far field, $R \gg a$).

$$d\mathbf{B} = \frac{\mu_0 I}{4\pi} \frac{d\mathbf{l} \times \mathbf{r}}{r^3} \quad \text{T} \tag{1.350}$$

1.8.2 Axial Field (B_z).

The Biot-Savart equation Eq. (1.350) can be easily integrated to obtain the axial (Z axis) magnetic field B_z ($x=0, y=0, R=z$).

Magnetic field on the Z axis (exact)

$$B_Z = \frac{\mu_0 I}{4\pi r^3}\left(\int_0^{2\pi a} r\, dl\right)\left(\frac{a}{r}\right) \quad \text{T} \tag{1.351}$$

Combining terms, $r = \sqrt{a^2 + z^2}$, $P = (0, 0, z)$

$$B_Z = \frac{\mu_0 a^2 I}{2(a^2+z^2)^{3/2}} \quad \text{T} \tag{1.352}$$

Magnetic field B_Z, $P = (0,0,0)$, $R = 0$

$$|B_Z|_{Z=0} = \frac{\mu_0 I}{2a} \quad \text{T} \tag{1.353}$$

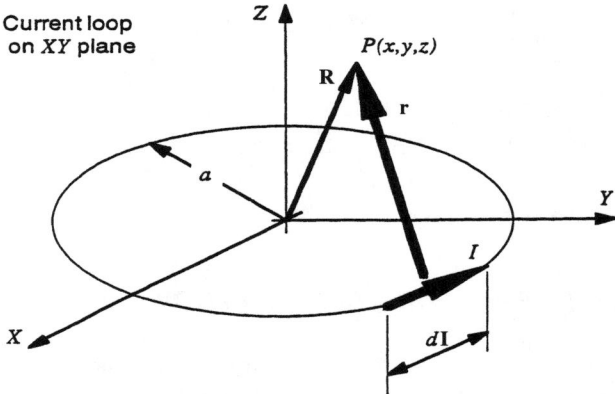

FIGURE 1.36 Geometric configuration for the Biot-Sauart law. *(Courtesy of Eaton Corporation.)*

1.8.3 Approximate Axial Field B_z, Based on a Parallel Straight Wire Assumption

The magnetic field on the XY plane ($z = 0$, $R \leq a$) can be very roughly approximated on the field between two parallel straight wires as follows.

$$B_Z \approx \frac{\mu_0 I}{2\pi r} + \frac{\mu_0 I}{2\pi(2a-r)} = \frac{\mu_0 I}{2\pi r}\left(\frac{2a}{2a-r}\right) \quad \text{T} \tag{1.354}$$

Substituting $r = a - R$

$$B_Z \approx \frac{\mu_0 I}{2\pi(a-R)}\left(\frac{2a}{a+r}\right) \quad \text{T} \tag{1.355}$$

Combining terms, $P = (x, y, 0)$

$$B_Z \approx \frac{\mu_0 a I}{\pi(a^2 - R^2)} \quad \text{T} \tag{1.356}$$

The final result is 36 percent lower than Eq. (1.353); $P = (0, 0, 0)$, $R = 0$.

$$B_Z|_{Z=0} \approx \frac{\mu_0 I}{\pi a} \quad \text{T} \tag{1.357}$$

1.8.4 Inductance for a Single-Turn Wire Loop

The following approximation for the inductance of a single-turn wire loop is very accurate for small wires ($a/r_0 > 10$), from Plonsey and Collin (1961) (r_0 = wire radius). The classical solution (which involves elliptic integrals) is required when the wire is large.

$$L = \mu_0 a \left[\ln\left(8\frac{a}{r_0}\right) - 1.75\right] \quad \text{H} \tag{1.358}$$

where $a/r_0 > 10$

1.8.5 Approximate Inductance Based on a Parallel Straight Wire Assumption

The inductance of a single-turn wire loop can be very roughly approximated (based on the field between two parallel straight wires) by integrating Eq. (1.356) over the area of the loop and dividing by the current, as follows, where r_0 = wire radius:

$$L = \frac{\lambda}{I} = \frac{\phi}{I} \approx \frac{1}{I} \int_0^{a-r_0} B_Z 2\pi R \, dR = 2\mu_0 a \int_0^{a-r_0} \frac{R}{a^2 - R^2} \, dR \quad (1.359)$$

$$L \approx \mu_0 a \left[2 \ln\left(\frac{a}{r_0}\right) - \ln\left(2\frac{a}{r_0} - 1\right) \right] \text{ H} \quad (1.360)$$

Error < 8.6 percent for $a/r_0 > 3$.

The error in Eq. (1.360) relative to the Plonsey and Collin approximation Eq. (1.358) ranges from +5.3 percent at $a/r_0 = 3$ to –8.6 percent at $a/r_0 = 13$, and falls off to –4.5 percent at $a/r_0 = 1000$. The + indicates that Eq. (1.360) gives a larger value than Eq. (1.358), and the – indicates that Eq. (1.360) gives a smaller value than Eq. (1.358).

1.9 HELMHOLTZ COIL

The Helmholtz coil (Fig. 1.37) is a pair of identical coils with a mean coil axial separation equal to the mean coil radius a, where the resulting axial magnetic field B_Z is uniform within 10 percent inside a sphere of radius $0.1a$ located midway between the two coils, $z = a/2$. The best performance is obtained when both the width w and the thickness t of the winding cross section are less than 20 percent of the mean radius. The first, second, and third derivatives of the axial field $d^n B_Z/dz^n$ ($n = 1, 2, 3$) are all equal to zero at $z = a/2$.

The axial field B_Z at the midpoint z between the two coils ($z = a/2$) can be obtained by doubling the field from Eq. (1.352), where NI = the amp-turns in each coil.

Multiplying Eq. (1.352) times 2, evaluated at $z = a/2$:

$$B_Z = 2 \frac{\mu_0 a^2 NI}{2[a^2 + (a/2)^2]^{3/2}} \quad \text{T} \quad (1.361)$$

Rewriting Eq. (1.361) for $z = a/2$:

$$B_Z = 0.8^{3/2} \frac{\mu_0 NI}{a} \quad \text{T} \quad (1.362)$$

Maximum allowable coil winding cross section:

$$t, w \leq 0.2a \quad (1.363)$$

1.10 COIL DESIGN

AC and dc coils need to be designed differently to function properly. Some of the common terms relating to both types of coils are defined here. The electrical conductivity σ of copper is defined at a temperature of 20°C (σ_{20}). Copper also has a temperature coefficient of resistance α. As a coil gets hot, its electrical resistance increases. The electrical conductivity at temperatures other then 20°C can be calculated as follows.

Copper conductivity at 20°C

$$\sigma_{20} = 5.8 \times 10^7 \ \Omega\text{m}^{-1} \tag{1.364}$$

Copper temperature coefficient, for $0 < T < 100$°C

$$\alpha_T = \frac{100\%}{k_{\text{Cu}} + T} \quad \%/°\text{C} \tag{1.365}$$

Copper reference temperature constant

$$k_{\text{Cu}} = 234°\text{C} \tag{1.366}$$

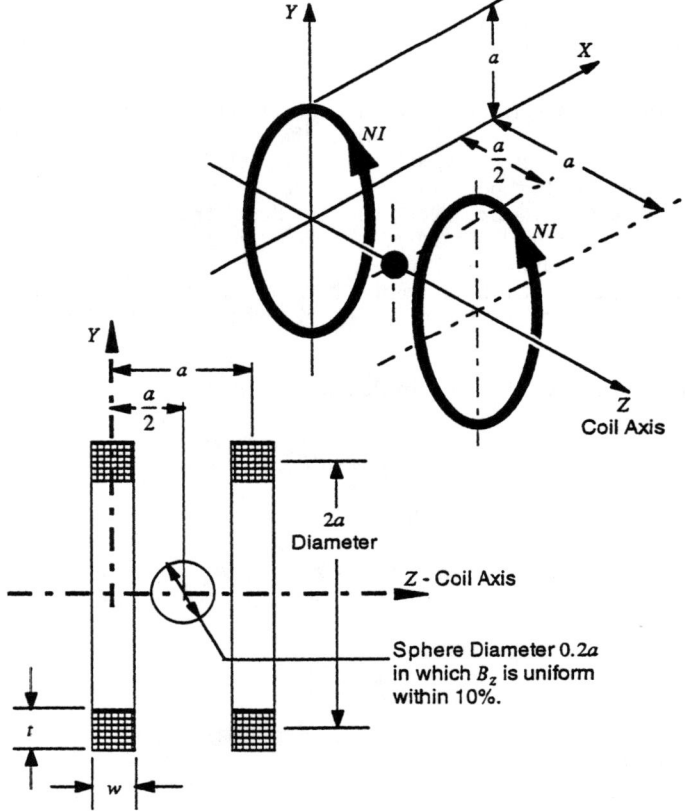

FIGURE 1.37 Helmholtz coil configuration: two identical coils in the *XY* plane, with radius *a*, and with an axial separation *a* equal to the radius. *(Courtesy of Eaton Corporation.)*

Copper temperature coefficient at $T = 20°C$

$$\alpha_{20} = \frac{100\%}{234.5 + 20} = 0.393 \ \%/°C \tag{1.367}$$

Copper conductivity at T, °C

$$\sigma_T = \frac{\sigma_{20}}{1 + (\alpha_{20}/100)(T - 20)} \ \Omega m^{-1} \tag{1.368}$$

The bare copper wire diameter d_B can be used to determine the resistance of a coil. The bare wire diameter is a function of the American Wire Gauge (AWG) number as follows, where the bare wire diameter d_0 for AWG = 0 is 0.00826 m = 0.325 in.

$$d_B = d_0(1.123^{-AWG}) \qquad d_0 = 0.00826 \ m \tag{1.369}$$

The total wire diameter includes the thickness of the electrical insulation layer (usually a varnish coating) and is used to determine the coil size or the number of turns that will fit on a bobbin. The wire insulation thickness is specified by the number of insulation layers that are built up on the wire, such as single build and double build. The total insulated wire diameter can be calculated based on a curve fit of the tables from MWS Wire Industries (1985), as follows. These equations give the total insulated wire diameter in units of meters and are accurate to within ±4 percent over the wire size range of AWG 4 to 60.

Single build

$$d = (0.00820)(0.8931)^{AWG} \tag{1.370}$$

Double (or heavy) build

$$d = (0.00804)(0.8956)^{AWG} \tag{1.371}$$

Triple build

$$d = (0.00794)(0.8980)^{AWG} \tag{1.372}$$

Quadruple build

$$d = (0.00792)(0.8997)^{AWG} \tag{1.373}$$

The typical winding density n for a given wire can be calculated based on a square lay, assuming each wire uses a square region equal to its diameter. Coils are usually wound with much less precision, so a winding density factor C is included to account for voids caused by loose windings or terminations. The winding density factor typically varies from 0.70 up to 0.95, and has units of turns per unit cross-sectional area.

$$n = C \frac{1}{d^2} \tag{1.374}$$

The maximum winding density can be achieved if the wires form a triangular or a hexagonal cross-section pattern. The maximum winding density n_{max} and the corresponding maximum winding density factor C_{max} are shown here.

$$n_{max} = \frac{2}{\sqrt{3}} \frac{1}{d^2} \tag{1.375}$$

$$C_{max} = 1.157 \tag{1.376}$$

The geometry of the bobbin is shown in Fig. 1.38. The bobbin length l, outside diameter D_0, and inside diameter D_I are used to determine the mean turn length l_M and the winding cross-sectional area A_W.

Mean turn length

$$l_M = \pi \frac{D_0 + D_I}{2} \tag{1.377}$$

Winding cross-sectional area

$$A_W = l \frac{D_0 - D_I}{2} \tag{1.378}$$

1.10.1 DC Coil

The typical requirement in a dc coil design is to determine the turns N, the wire size AWG, the temperature rise ΔT, the coil resistance R, and the inductance L, given the required amp-turns NI. The required amp-turns can be obtained from Eq. (1.17) by knowing the required magnetic flux ϕ and the system reluctance $\mathcal{R}_{sys}$.

The first thing that should be done is to estimate the maximum temperature of the coil, or the maximum allowable temperature unit. Then calculate the copper conductivity at the elevated temperature. Use of the elevated temperature conductivity will guarantee that the coil will provide the required amp-turns at the maximum temperature. The resistance of the coil can be determined by dividing the length of the wire by the conductivity of the wire and by the cross-sectional area of the wire $a_B = \pi d_B^2/4$, as follows.

$$R = \frac{Nl_M}{\sigma_T a_B} = 4 \frac{Nl_M}{\sigma_T \pi d_B^2} \tag{1.379}$$

The electrical current in the coil can be calculated by solving Ohm's law for the current. Multiplying by the number of turns in the winding gives the amp-turns produced by the coil.

$$I = \frac{V}{R} \tag{1.380}$$

$$NI = N \frac{V}{R} \tag{1.381}$$

where I = amperes
V = volts
R = ohms

Substitution of Eq. (1.379) into Eq. (1.381) gives the following equation, which can be solved for the bare wire diameter. Note that in Eq. (1.382) the coil NI is independent of the number of turns N. This is true as long as any change in the number of turns does not significantly change the mean turn length l_M.

$$NI = \frac{\pi}{4} \frac{V \sigma_T d_B^2}{l_M} \tag{1.382}$$

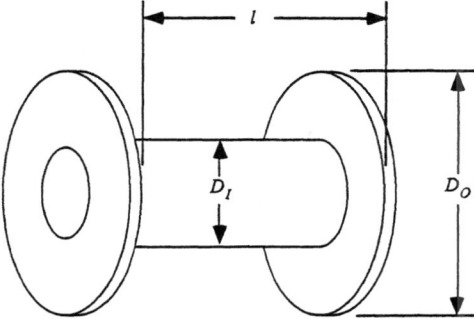

FIGURE 1.38 Cylindrical bobbin dimensions for coil winding. *(Courtesy of Eaton Corporation.)*

$$d_B \geq \sqrt{\frac{4l_M NI}{\pi \sigma_T V}} \qquad (1.383)$$

The bare wire diameter must be larger than or equal to the expression on the right of Eq. (1.383) to guarantee that the minimum coil NI is at least as large as the required NI. Equation (1.369) can be written to solve for the AWG wire size, as follows. The AWG sizes are typically in integer increments; therefore, only the rounded-down integer value of the expression is useful in this case. AWG wire sizes are available in half-gauge increments, but in small quantities they are 2 to 3 times more expensive. Note that the AWG number is dimensionless and that both d_0 and d_B must have the same units:

$$\text{AWG} = \text{int}\left[\frac{\ln(d_0/d_B)}{\ln(1.123)}\right] \qquad (1.384)$$

where $d_0 = 0.00826$ m $= 0.325$ in.

Now that the AWG wire size is determined, the actual bare wire diameter can be obtained from Eq. (1.369), and the total wire diameter can be obtained from Eqs. (1.370) through (1.373). The total number of turns that will fit on the bobbin can be calculated by using Eqs. (1.374) and (1.378).

$$N = nA_W = \frac{CA_W}{d^2} \qquad (1.385)$$

The coil resistance can be obtained from Eq. (1.379), the coil inductance can be obtained from Eq. (1.31), and the coil current can be obtained from Ohm's law. The power dissipation at the elevated temperature can be calculated as follows.

$$Q = \frac{V^2}{R} \quad \text{W} \qquad (1.386)$$

The steady-state temperature rise of the coil can be determined as shown here, where h is the combined heat transfer coefficient for convection and radiation, A is the heat transfer surface area of the coil, and ΔT is temperature rise above the ambient temperature.

$$Q = hA\Delta T \qquad (1.387)$$

$$h = 0.007 \text{ W}/(\text{in}^2 \cdot {}^\circ\text{C}) \tag{1.388}$$

$$A = \pi(D_0 + D_I)l + 2\pi\left(\frac{D_0^2 - D_I^2}{4}\right) \tag{1.389}$$

$$\Delta T = \frac{Q}{hA} \tag{1.390}$$

The maximum temperature of the coil T_{max} is equal to the ambient temperature T_A plus the temperature rise ΔT.

$$T_{max} = T_A + \Delta T \tag{1.391}$$

1.10.2 AC Coil

The typical requirement in an ac coil design is to determine the turns N, the wire size AWG, the temperature rise ΔT, the coil resistance R, and the inductance L, given the required magnetic flux ϕ. The performance of an ac coil is defined by Ohm's law and Faraday's law.

$$V = IR + \frac{d\lambda}{dt} \tag{1.392}$$

Equations (1.18) and (1.19) can be substituted into Eq. (1.392) in place of the coil current I and the flux linkages λ.

$$V = \phi \frac{R}{N\mathcal{P}} + N \frac{d\phi}{dt} \tag{1.393}$$

If the magnetic flux is assumed to be sinusoidal, the voltage and the magnetic flux magnitudes can be related as follows.

$$\phi = \phi_P \sin(\omega t) \tag{1.394}$$

$$V = \phi_P \frac{R}{N\mathcal{P}} \sin(\omega t) + N\phi_P \omega \cos(\omega t) \tag{1.395}$$

$$V_P = \phi_P \sqrt{\left(\frac{R}{N\mathcal{P}}\right)^2 + (N\omega)^2} \tag{1.396}$$

In general, the resistive impedance of an ac coil is far less than the reactive impedance. Therefore, Eq. (1.396) can be simplified as follows to solve for the number of turns N.

$$\frac{R}{N\mathcal{P}} \ll N\omega \tag{1.397}$$

$$V_P = N\phi_P\omega \tag{1.398}$$

$$N = \frac{V_P}{\phi_P \omega} \tag{1.399}$$

Equation (1.385) can be written to solve for the maximum allowable total wire diameter based on the number of turns and the winding cross-sectional area.

BASIC MAGNETICS

$$d = \sqrt{\frac{CA_W}{N}} \quad (1.400)$$

Equations (1.369) through (1.373) were curve-fit to a single expression for the total wire diameter. These equations can be written to solve for the AWG wire size, as was done in Eq. (1.384), and as shown here for single-build insulation, Eq. (1.401). The AWG wire sizes are in integer increments; therefore, only the rounded-up integer value of the expression is useful in this case. Note that this equation uses the wire diameter in units of meters.

$$\text{AWG} = \text{int}\left(\frac{\ln(d/0.00820)}{\ln(0.8931)}\right) \quad (1.401)$$

With the AWG wire size determined, the actual bare wire diameter can be obtained from Eq. (1.369), the total wire diameter can be obtained from Eqs. (1.370) through (1.373), and the coil resistance can be obtained from Eq. (1.379).

With the coil resistance defined, the actual peak magnetic flux ϕ_A can be determined from Eq. (1.402), for the condition when the coil resistive impedance is significant.

$$\phi_A = \frac{V_P}{\sqrt{(R/N\mathcal{P})^2 + (N\omega)^2}} \quad (1.402)$$

If the actual magnetic flux is less than the required magnetic flux ($\phi_A < \phi_P$), then the number of turns N must be reduced, and the calculation procedure should return to Eq. (1.400) to obtain a new wire diameter based on the new number of turns. This procedure should continue until the actual peak magnetic flux is approximately equal to the required peak magnetic flux ($\phi_A \approx \phi_P$), within an acceptable tolerance.

If $\phi_A \approx \phi_P$, then the procedure should continue as follows. The coil inductance L can be obtained from Eq. (1.31), and the coil impedance Z and the coil current I can be determined as follows.

$$L = N^2 \mathcal{P} \quad (1.403)$$

$$Z = \sqrt{R^2 + (L\omega)^2} \quad (1.404)$$

$$I = \frac{V}{Z} \quad (1.405)$$

The power dissipation at the elevated temperature can be calculated as follows, and the coil temperature rise can be obtained from Eqs. (1.387) through (1.391).

$$Q = I^2 R \quad (1.406)$$

1.10.3 Copper Area Ratio Approximation

The wire diameter equations Eqs. (1.370) to (1.373) can be simplified based on the estimated copper area ratio K_A. The copper area ratio is defined in Eq. (1.407) as follows, where A_C is the copper cross-sectional area and A_W is the total winding cross-sectional area from Eq. (1.378).

$$K_A = \frac{A_C}{A_W} \quad (1.407)$$

With a winding density factor C of 0.70, the following values of the copper area ratio K_A can be used. A winding density factor of 0.70 is conservative, because most coils can be wound with a higher winding density. Therefore, a coil designed with this approximation will be manufacturable.

Copper area ratio K_A, for C = 0.70	Valid wire gauge range	Number of installation layers
0.50 ± 0.04	10 < AWG < 38	Single build
0.45 ± 0.07	10 < AWG < 38	Double build
0.40 ± 0.10	10 < AWG < 38	Triple build
0.35 ± 0.13	10 < AWG < 38	Quadruple build

The copper cross-sectional area can be obtained by rewriting Eq. (1.407) and including the effect of different values for the winding density factor, as follows.

$$A_C = K_A A_W \left(\frac{C}{0.7} \right) \tag{1.408}$$

The copper cross-sectional area can also be calculated from the number of turns N and the bare wire diameter d_B, as follows.

$$A_C = N \frac{\pi}{4} d_B^2 \tag{1.409}$$

Equations (1.408) and (1.409) can be combined to determine the number of turns or to determine the bare wire diameter. In the case of a dc coil design, the number of turns N can be determined from Eq. (1.410) without needing to use the set of diameter equations Eqs. (1.370) to (1.373) and Eq. (1.385). In the case of an ac coil design, the bare wire diameter d_B can be determined from Eq. (1.411) without needing to use equation (1.400), the set of diameter equations Eqs. (1.370) to (1.373), and Eq. (1.401).

Approximate number of turns for dc coils

$$N = \text{int} \left[\frac{C A_W}{d_B^2} (1.82 K_A) \right] \tag{1.410}$$

Approximate maximum bare wire diameter for ac coils

$$d_B \leq \sqrt{\frac{C A_W}{N} (1.82 K_A)} \tag{1.411}$$

1.10.4 Coil Design Procedures Using the Copper Area Ratio Approximation

The equations given previously in Sec. 1.10 for the coil geometry, dc coil performance, ac coil performance, and copper area ratio are listed in the following two subsections. These general design procedures for dc and ac coils will determine the wire size, the number of turns, and the coil performance from the known coil voltage, bobbin geometry, and required magnetic performance NI or ϕ.

DC Coil Design Procedure

1. Given:
   Coil dc voltage V

Required amp-turns NI
Expected maximum coil temperature T
Ambient temperature T_A
Copper area ratio K_A
Winding density factor C
Axial bobbin winding length l
Inside bobbin winding diameter D_I
Outside bobbin winding diameter D_O

2. Calculate:
Copper conductivity at maximum temperature σ_T

$$\sigma_T = \frac{\sigma_{20}}{1 + (\alpha_{20}/100)(T - 20°C)} \quad \Omega m^{-1} \tag{1.368}$$

Coil mean turn length l_M

$$l_M = \pi \frac{D_O + D_I}{2} \tag{1.377}$$

Bobbin winding cross-sectional area A_W

$$A_W = l \frac{D_O - D_I}{2} \tag{1.378}$$

Coil heat transfer surface area A

$$A = \pi(D_O + D_I)l + 2\pi \left(\frac{D_O^2 - D_I^2}{4} \right) \tag{1.389}$$

3. Calculate wire size, turns, and performance:
Minimum bare wire diameter d_B

$$d_B \geq \sqrt{\frac{4l_M NI}{\pi \sigma_T V}} \tag{1.383}$$

AWG wire size, where $d_0 = 0.00826$ m

$$AWG = \text{int} \left[\frac{\ln (d_0/d_B)}{\ln (1.123)} \right] \tag{1.384}$$

Actual bare wire diameter D_B

$$d_B = d_0(1.123^{-AWG}) \tag{1.369}$$

Number of turns N

$$N = \text{int} \left[\frac{CA_W}{d_B^2} (1.82K_A) \right] \tag{1.410}$$

Resistance R, Ω

$$R = 4 \frac{Nl_M}{\sigma_T \pi d_B^2} \tag{1.379}$$

Permeance $\mathcal{P}$ (See Sec. 1.3.)

Inductance L, H

$$L = N^2 \mathcal{P} \qquad (1.31)$$

Power dissipation Q, W

$$Q = \frac{V^2}{R} \qquad (1.386)$$

Temperature rise ΔT, °C, where $h = 0.007$ W/(in$^2 \cdot$°C)

$$\Delta T = \frac{Q}{hA} \qquad (1.390)$$

Maximum coil temperature T_{max}, °C (iterate T and T_{max})

$$T_{max} = T_A + \Delta T \qquad (1.391)$$

Minimum coil current I, A, at temperature T

$$I = \frac{V}{R}$$

AC Coil Design Procedure

1. Given:
   Coil peak voltage V_P, frequency ω
   Required peak magnetic flux ϕ_P
   Expected maximum coil temperature T
   Ambient temperature T_A
   Copper area ratio K_A
   Winding density factor C
   Axial bobbin winding length l
   Inside bobbin winding diameter D_I
   Outside bobbin winding diameter D_O

2. Calculate conductivity and geometry:
   Copper conductivity at maximum temperature σ_T

$$\sigma_T = \frac{\sigma_{20}}{1 + (\alpha_{20}/100)(T - 20°C)} \quad \Omega\text{m}^{-1} \qquad (1.368)$$

Coil mean turn length l_M

$$l_M = \pi \frac{D_O + D_I}{2} \qquad (1.377)$$

Bobbin winding cross-sectional area A_W

$$A_W = l \frac{D_O - D_I}{2} \qquad (1.378)$$

Coil heat transfer surface area A

$$A = \pi(D_O + D_I)l + 2\pi \left(\frac{D_O^2 - D_I^2}{4} \right) \qquad (1.389)$$

3. Calculate wire size, turns, and performance:

BASIC MAGNETICS

Number of turns N

$$N = \frac{V_P}{\phi_P \omega} \qquad (1.399)$$

Minimum bare wire diameter d_B

$$d_B \leq \sqrt{\frac{CA_W}{N}(1.82K_A)} \qquad (1.411)$$

AWG wire size, where $d_0 = 0.00826$ m

$$\text{AWG} = \text{int}\left[\frac{\ln(d_0/d_B)}{\ln(1.123)}\right] \qquad (1.384)$$

Actual bare wire diameter d_B

$$d_B = d_0(1.123^{-\text{AWG}}) \qquad (1.369)$$

Resistance R, Ω

$$R = 4\frac{Nl_M}{\sigma_T \pi d_B^2} \qquad (1.379)$$

Permeance $\mathcal{P}$. (See Sec. 1.3.)
Actual magnetic flux ϕ_A (iterate N and ϕ_A)

$$\phi_A = \frac{V_P}{\sqrt{(R/N\mathcal{P})^2 + (N\omega)^2}} \qquad (1.402)$$

Inductance L, H

$$L = N^2\mathcal{P} \qquad (1.31)$$

Coil impedance Z, Ω

$$Z = \sqrt{R^2 + (L\omega)^2} \qquad (1.404)$$

Minimum coil current I, A, at temperature T

$$I = \frac{V}{Z} \qquad (1.405)$$

Power dissipation Q, W

$$Q = I^2 R \qquad (1.406)$$

Temperature rise ΔT, °C, where $h = 0.007$ W/(in$^2 \cdot$ °C)

$$\Delta T = \frac{Q}{hA} \qquad (1.390)$$

Maximum coil temperature ($T_{\max}$, °C), (iterate T and $T_{\max}$)

$$T_{\max} = T_A + \Delta T \qquad (1.391)$$

1.11 RELUCTANCE ACTUATOR STATIC AND DYNAMIC (MOTION) ANALYSIS

As shown in Figs. 1.2 and 1.29, the electrical analogy of the magnetomotive force supplied by a coil is an electrical voltage source, and the electrical analog for the reluctance of the steel part and the air gaps is an electrical resistor. Analysis of the reluctance actuator shown in Fig. 1.39 is based on reluctance circuits that correspond to the electrical analogy. Section 1.11.1 shows the static dc steady-state analysis and Sec. 1.11.2 shows the dynamic analysis with time-varying coil voltage and armature motion.

1.11.1 Static Analysis (DC, Steady State)

The reluctance circuit for the system in Fig. 1.39 can be solved by writing the equation for each flux loop, based on Eq. (1.14). In general, the magnetomotive forces around each loop are summed to zero, $\sum NI = 0$. This results in two flux-loop equations with two unknowns. The circuit elements and the loop equations for both of the flux loops are listed here.

$\mathcal{R}_A$ = reluctance of the steel armature
$\mathcal{R}_g$ = reluctance of the air gap between the armature and the magnet pole
$\mathcal{R}_L$ = reluctance of the leakage air path between the vertical legs of the magnet pole
$\mathcal{R}_H$ = reluctance of a vertical leg of the steel magnet pole
$\mathcal{R}_W$ = reluctance of the bottom of the steel magnet pole

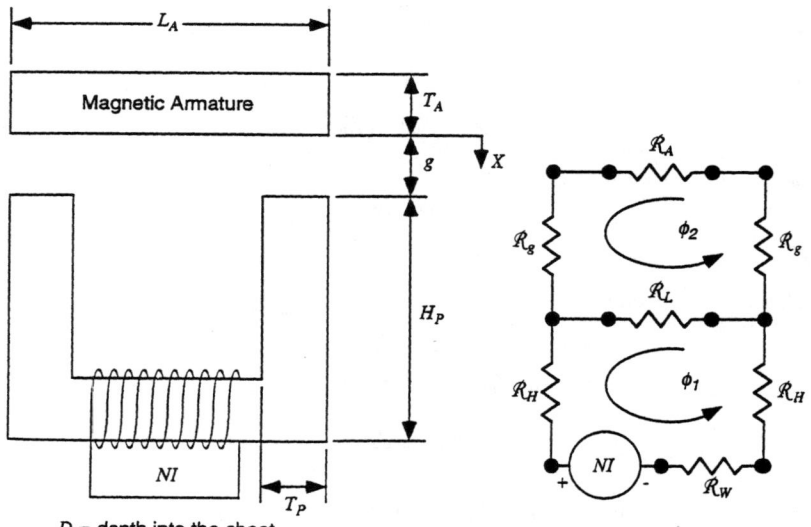

FIGURE 1.39 (a) Example of a reluctance actuator in which only one coil is used, and (b) the corresponding reluctancy circuit. *(Courtesy of Eaton Corporation.)*

Loop ϕ_1

$$NI - \phi_1(2\mathcal{R}_H + \mathcal{R}_W + \mathcal{R}_L) + \phi_2(\mathcal{R}_L) = 0 \qquad (1.412)$$

Loop ϕ_2

$$-\phi_2(2\mathcal{R}_g + \mathcal{R}_A + \mathcal{R}_L) + \phi_1(\mathcal{R}_L) = 0 \qquad (1.413)$$

The reluctances for the actuator in Fig. 1.39 can be easily calculated from Eqs. (1.29) and (1.116) through (1.163), and the coil current I can be calculated by dividing the dc source voltage V by the coil resistance R. These equations can then be solved for the loop fluxes as follows.

$$\phi_2 = \phi_1 \frac{\mathcal{R}_L}{(2\mathcal{R}_g + \mathcal{R}_A + \mathcal{R}_L)} \qquad (1.414)$$

$$\phi_2 = C\phi_1 \qquad (1.415)$$

Leakage factor

$$C = \frac{\mathcal{R}_L}{(2\mathcal{R}_g + \mathcal{R}_A + \mathcal{R}_L)} \qquad (1.416)$$

$$NI = \phi_1 \left[2\mathcal{R}_H + \mathcal{R}_W + \mathcal{R}_L - \frac{\mathcal{R}_L^2}{(2\mathcal{R}_g + \mathcal{R}_A + \mathcal{R}_L)} \right] = \phi_1 \mathcal{R}_{sys} \qquad (1.417)$$

$$\mathcal{R}_{sys} = \left[2\mathcal{R}_H + \mathcal{R}_W + \mathcal{R}_L - \frac{\mathcal{R}_L^2}{(2\mathcal{R}_g + \mathcal{R}_A + \mathcal{R}_L)} \right] \qquad (1.418)$$

Flux-loop solution

$$\phi_1 = \frac{NI}{\mathcal{R}_{sys}} = NI\mathcal{P}_{sys} \qquad (1.419)$$

DC steady-state solution

$$\phi_1 = N \frac{V}{R} \mathcal{P}_{sys} \qquad (1.420)$$

The system reluctance defined in Eq. (1.418) can be used to calculate the inductance of the coil from Eq. (1.31), and if a permanent magnet replaces the coil, a load line can be calculated from Eq. (1.269) as follows.

$$L = N^2 \mathcal{P}_{sys} \qquad (1.421)$$

$$\mu_L = \mathcal{P}_{sys} \frac{l_{PM}}{a_{PM}} \qquad (1.422)$$

The force on the armature can be calculated from Eq. (1.74) by determining the magnetizing force across the air gap as follows. The system in Fig. 1.39 does not have an armature coil; therefore, $N_a = 0$.

$$(NI)_g = \phi_2 \mathcal{R}_g \qquad (1.423)$$

$$F = 2\left[\frac{1}{2}(NI)_g^2 d\frac{\mathcal{P}_g}{dx}\right] \qquad (1.424)$$

1.11.2 Dynamic Analysis (AC and Motion)

The force in Eq. (1.424) is valid for steady-state dc current with no armature motion. If a transient solution is required, or if the source voltage is a function of time (such as ac voltage), or if the armature is allowed to move, then the coil voltage must be defined using both Ohm's law and Faraday's law. Equation (1.417) is also used as the definition of magnetic flux for this system.

$$V = IR + \frac{d\lambda_1}{dt} = IR + \frac{d(N\phi_1)}{dt} = IR + \frac{d(N^2 I \mathcal{P}_{sys})}{dt} \qquad (1.425)$$

$$V = IR + N^2 \mathcal{P}_{sys}\frac{dI}{dt} + N^2 I\frac{d\mathcal{P}_{sys}}{dt} + 2NI\mathcal{P}_{sys}\frac{dN}{dt} \qquad (1.426)$$

The time derivatives of the coil turns N and the system permeance $\mathcal{P}_{sys}$ in Eq. (1.424) can be expanded as a function of the armature position X and the armature velocity v by using the chain rule as follows.

$$v = \frac{dX}{dt} \qquad (1.427)$$

$$\frac{dN}{dt} = \frac{dN}{dX}\frac{dX}{dt} = v\frac{dN}{dX} \qquad (1.428)$$

$$\frac{d\mathcal{P}_{sys}}{dt} = \frac{d\mathcal{P}_{sys}}{dX}\frac{dX}{dt} = v\frac{d\mathcal{P}_{sys}}{dX} \qquad (1.429)$$

Substitution of Eqs. (1.428) and (1.429) into Eq. (1.426) gives the general form of the coil equation.

$$V = IR + N^2 \mathcal{P}_{sys}\frac{dI}{dt} + N^2 Iv\frac{d\mathcal{P}_{sys}}{dX} + 2NI\mathcal{P}_{sys}v\frac{dN}{dX} \qquad (1.430)$$

The second term can be written using the inductance definition, Eq. (1.31), and the fourth term can be written using the permeance definition, Eq. (1.419), to provide a more familiar appearance, as follows.

$$V = IR + L\frac{dI}{dt} + N^2 Iv\frac{d\mathcal{P}_{sys}}{dX} + 2\phi_1 v\frac{dN}{dX} \qquad (1.431)$$

The first term represents the resistance voltage drop, the second term represents the inductive voltage drop, the third term represents the voltage drop produced by the armature velocity and by changing reluctance, and the fourth term represents the voltage drop produced by the armature velocity and by the change in the number of turns that link the magnetic flux. Both the third term and the fourth term contain the armature velocity and are usually referred to as

BASIC MAGNETICS

the *back electromotive force* (emf). The coil in the system shown in Fig. 1.42 (see Sec. 1.12) has a constant number of turns linking the flux. Therefore, the fourth term can be ignored. The resulting voltage equation can be written as follows.

$$V = IR + N^2 Iv \frac{d\mathcal{P}_{sys}}{dX} + N^2 \mathcal{P}_{sys} \frac{dI}{dt} \tag{1.432}$$

Dynamic solution for I

$$V = I\left(R + N^2 v \frac{d\mathcal{P}_{sys}}{dX}\right) + N^2 \mathcal{P}_{sys} \frac{dI}{dt} \tag{1.433}$$

Equation (1.433) can be solved for the current I, and the magnetic flux ϕ_1 can be obtained by solving Eq. (1.419) based on the coil current. Equation (1.433) describes the time-varying coil current I as a function of the known variables (the voltage, the system permeance, the number of turns in the coil, the armature velocity, and the coil resistance). Equation (1.433) also shows that the coil current is a function of the armature velocity.

An alternative solution method for this system is to substitute Eq. (1.419) into Eq. (1.425), and then rearrange the terms as follows to solve for the magnetic flux.

$$V = \phi_1 \frac{R}{N\mathcal{P}_{sys}} + \frac{d(N\phi_1)}{dt} \tag{1.434}$$

$$V = \phi_1 \frac{R}{N\mathcal{P}_{sys}} + N \frac{d\phi_1}{dt} + \phi_1 \frac{dN}{dt} \tag{1.435}$$

$$\frac{dN}{dt} = 0 \tag{1.436}$$

Dynamic solution for ϕ_1

$$V = \phi_1 \frac{R}{N\mathcal{P}_{sys}} + N \frac{d\phi_1}{dt} \tag{1.437}$$

DC steady-state solution

$$\phi_1 = N \frac{V}{R} \mathcal{P}_{sys} \tag{1.438}$$

Equation (1.437) can be solved for the magnetic flux ϕ_1, and the coil current can be obtained by solving Eq. (1.419) based on the magnetic flux. Equation (1.437) describes the time-varying magnetic flux ϕ_1 as a function of the known variables (the coil voltage, the system permeance, the number of turns in the coil, and the coil resistance), and it shows that the magnetic flux is not a function of the armature velocity. Equation (1.438) shows that the dc steady-state solution is identical to Eq. (1.420).

1.11.3 Mechanical Dynamics

The armature velocity and position can be determined by solving the dynamic system shown in Figs. 1.40 and 1.41. Where F_{mag} is the magnetic force from Eq. (1.424),

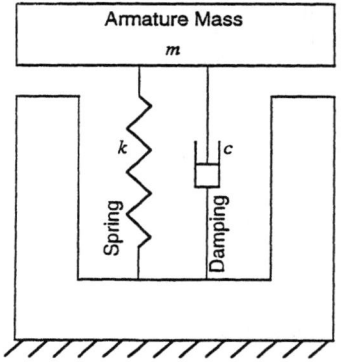

FIGURE 1.40 Reluctance actuator mass, spring, and damping dynamic system. *(Courtesy of Eaton Corporation.)*

F_g is the force due to gravity, F_I is the inertia force in opposition to the acceleration a, F_D is the damping force in opposition to the velocity v, F_s is the spring force in opposition to the displacement or position X, and F_0 is the initial spring force.

Force balance

$$\sum F = 0 = F_I + F_D + F_S - F_g - F_{mag} \quad (1.439)$$

Inertia force

$$F_I = ma = m\frac{d^2X}{dt^2} \quad (1.440)$$

Damping force

$$F_D = cv = c\frac{dX}{dt} \quad (1.441)$$

Spring force

$$F_S = kX + F_0 \quad (1.442)$$

Gravity force

$$F_g = mg \quad (1.443)$$

Force balance

$$m\frac{d^2X}{dt^2} + c\frac{dX}{dt} + kX = F_{mag} + mg - F_0 \quad (1.444)$$

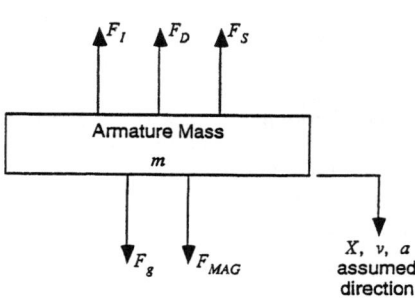

FIGURE 1.41 Reluctance actuator armature free-body diagram. *(Courtesy of Eaton Corporation.)*

The performance of this actuator can be calculated by solving the differential equation for the coil current I, Eq. (1.433), or the differential equation for the magnetic flux ϕ_1, Eq. (1.437), and the differential equation for motion, Eq. (1.444). Equations (1.419), (1.415), and (1.424) are needed to solve for the other variables, ϕ_1 or I, ϕ_2, and F_{mag}. Equations (1.433) and (1.437) are first-order nonlinear differential equations, and the system permeance and the armature velocity are functions of the armature position and the armature force. Therefore, this set of equations must be solved with an iterative finite-difference technique, such as fourth-order Runge-Kutta.

1.12 MOVING-COIL ACTUATOR STATIC AND DYNAMIC (MOTION) ANALYSIS

Analysis of the moving-coil actuator shown in Fig. 1.42 is based on reluctance circuits that correspond to the electrical analogy. Section 1.12.1 shows the static dc steady-state analysis and Sec. 1.12.2 shows the dynamic analysis with time-varying coil voltage and armature motion.

1.12.1 Static Analysis (DC, Steady State)

The reluctance circuit for the system in Fig. 1.42 can be solved by writing the equation for each flux loop, based on Eq. (1.14). In general, the magnetomotive forces around each loop are summed to zero, $\sum NI = 0$. This results in two flux-loop equations with two unknowns. The circuit elements and the loop equations for both of the flux loops are listed here.

$\mathcal{R}_g$ = reluctance of the air gap between the center pole and the outer pole

$\mathcal{R}_L$ = reluctance of the leakage air path between the center pole and the outer pole

$\mathcal{R}_C$ = reluctance of the center steel magnet pole

$\mathcal{R}_H$ = reluctance of the outer steel magnet pole

$\mathcal{R}_W$ = reluctance of the bottom of the steel magnet pole

$N_a I_a$ = magnetomotive force of the moving armature coil

$N_f I_f$ = magnetomotive force of the stationary field coil

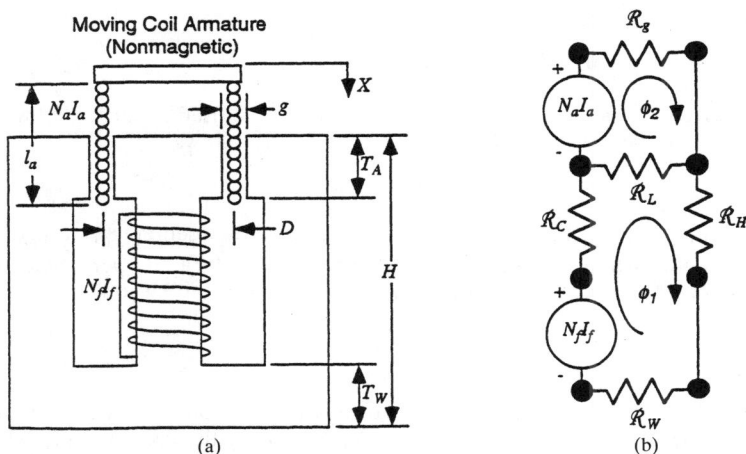

FIGURE 1.42 (*a*) A moving-coil actuator in which the field coil is stationary and the armature coil moves. This shape of this actuator is cylindrical. (*b*) The corresponding reluctance circuit. *(Courtesy of Eaton Corporation.)*

Loop ϕ_2

$$N_a I_a - \phi_2(\mathcal{R}_g + \mathcal{R}_L) + \phi_1(\mathcal{R}_L) = 0 \tag{1.445}$$

Loop ϕ_1

$$N_f I_f - \phi_1(\mathcal{R}_H + \mathcal{R}_C + \mathcal{R}_W + \mathcal{R}_L) + \phi_2(\mathcal{R}_L) = 0 \tag{1.446}$$

Both of the coils and the magnet pole are cylindrical. Therefore, the outer legs of the magnet pole are one piece of steel in the shape of a tube. The reluctance circuit in Fig. 1.42 is shown with the coils on the center pole and only one reluctance value $\mathcal{R}_H$ for the outer tubular pole. Equations (1.445) and (1.446) can be solved for the second loop magnetic flux ϕ_2 as follows.

$$\phi_1 = \frac{\phi_2 \mathcal{R}_L + N_f I_f}{\mathcal{R}_H + \mathcal{R}_C + \mathcal{R}_W + \mathcal{R}_L} \tag{1.447}$$

$$N_a I_a = \phi_2(\mathcal{R}_g + \mathcal{R}_L) - \frac{\phi_2 \mathcal{R}_L^2 + N_f I_f \mathcal{R}_L}{\mathcal{R}_H + \mathcal{R}_C + \mathcal{R}_W + \mathcal{R}_L} \tag{1.448}$$

$$N_a I_a = \phi_2 \left(\mathcal{R}_g + \mathcal{R}_L - \frac{\mathcal{R}_L^2}{\mathcal{R}_H + \mathcal{R}_C + \mathcal{R}_W + \mathcal{R}_L} \right) - N_f I_f \left(\frac{\mathcal{R}_L}{\mathcal{R}_H + \mathcal{R}_C + \mathcal{R}_W + \mathcal{R}_L} \right) \tag{1.449}$$

$$\mathcal{R}_{sys} = \mathcal{R}_g + \mathcal{R}_L - \frac{\mathcal{R}_L^2}{\mathcal{R}_H + \mathcal{R}_C + \mathcal{R}_W + \mathcal{R}_L} \qquad \mathcal{P}_{sys} = \frac{1}{\mathcal{R}_{sys}} \tag{1.450}$$

Leakage factor

$$C = \frac{\mathcal{R}_L}{\mathcal{R}_H + \mathcal{R}_C + \mathcal{R}_W + \mathcal{R}_L} \tag{1.451}$$

$$N_a I_a = \phi_2 \mathcal{R}_{sys} - N_f I_f C \tag{1.452}$$

Flux-loop solution

$$\phi_2 = (N_a I_a + N_f I_f C) \mathcal{P}_{sys} \tag{1.453}$$

The system reluctance defined in Eq. (1.461) can be used to calculate the inductance of the moving armature coil from Eq. (1.31), as follows.

$$L_a = N_a^2 \mathcal{P}_{sys} \tag{1.454}$$

The force on the armature can be calculated from Eq. (1.203). The system in Fig. 1.42 does not have a moving magnetic armature; therefore, the permeance of the working gap $\mathcal{P}_g$ is constant. Also, I_a, I_f, and N_f are assumed to be constant.

$$F = I_a \phi_a \frac{dN_a}{dX} + I_a \phi_f \frac{dN_a}{dX} \tag{1.455}$$

The rate of change of the moving coil turns with respect to the armature position is simply the number of turns in the coil N_a divided by the coil length l_a. Also, the total length of the wire in the air gap l_{wire} is equal to the product of the wire path circumference and the number of turns in the gap, as shown following. At this point it

BASIC MAGNETICS 1.85

is assumed that all of the air gap flux exists in the direct face-to-face interface (no fringing flux). This assumption is valid for small air gaps, where $g \ll \sqrt{\pi D T_A}$.

$$\frac{dN_a}{dX} = \frac{N_a}{l_a} \tag{1.456}$$

$$l_{\text{wire}} = \pi D N_a \frac{T_a}{l_a} \tag{1.457}$$

These equations can be combined to give the following result.

$$\frac{dN_a}{dX} = \frac{N_a}{l_a} = \frac{l_{\text{wire}}}{\pi D T_a} \tag{1.458}$$

The armature flux ϕ_a and the field coil flux ϕ_f can be obtained from Eq. (1.453) by alternately setting the current in each coil to zero. The flux densities can then be obtained by dividing by the cross-sectional area of the gap.

$$\phi = (N_a I_a)\mathcal{P}_{\text{sys}} \qquad B_a = \frac{\phi_a}{\pi D T_a} \tag{1.459}$$

$$\phi_f = (N_f I_f C)\mathcal{P}_{\text{sys}} \qquad B_f = \frac{\phi_f}{\pi D T_a} \tag{1.460}$$

Combining Eqs. (1.458), (1.459), and (1.460) with Eq. (1.455) gives the following equation for the force on the moving armature coil. Equation (1.461) is also called the *Lorentz force*.

$$F = I_a (B_a + B_f) l_{\text{wire}} \qquad \text{for small } g \tag{1.461}$$

1.12.2 Dynamic Analysis (AC and Motion)

The force in Eq. (1.461) is valid for steady-state dc current with no armature motion and for small air gaps, where $g \ll \sqrt{\pi D T_A}$.

If a transient solution is required, or if the armature is allowed to move, then the current in the moving coil must be defined with both Ohm's law and Faraday's law, Eq. (1.16) and (1.20), as follows. In this derivation, the field coil is assumed to have a constant current and a constant number of turns. The permeance of the working air gap is also assumed to be constant.

$$V_a = I_a R_a + \frac{d\lambda_a}{dt} = I_a R_a + \frac{d(N_a \phi_2)}{dt} \tag{1.462}$$

$$\phi_2 = (N_a I_a + N_f I_f C)\mathcal{P}_{\text{sys}} \tag{1.463}$$

$$V_a = I_a R_a + \frac{d[(N_a^2 I_a + N_a N_f I_f C)\mathcal{P}_{\text{sys}}]}{dt} \tag{1.464}$$

$$V_a = I_a R_a + N_a^2 \mathcal{P}_{\text{sys}} \frac{dI_a}{dt} + C N_f I_f \mathcal{P}_{\text{sys}} \frac{dN_a}{dt} + 2 N_a I_a \mathcal{P}_{\text{sys}} \frac{dN_a}{dt} \tag{1.465}$$

Substitution of Eqs. (1.428), (1.457), (1.459), and (1.460) into Eq. (1.465) gives the following result for the voltage on the moving armature coil.

$$V_a = I_a R_a + L_a \frac{dI_a}{dt} + \phi_f v \frac{dN_a}{dX} + 2\phi_a v \frac{dN_a}{dX} \quad (1.466)$$

The first term represents the resistance voltage drop, the second term represents the inductive voltage drop, the third term represents the voltage drop produced by the velocity of the armature coil as it passes through the field produced by the field coil, and the fourth term represents the voltage drop produced by the velocity of the armature coil as it passes though the field produced by itself (this is an armature reaction voltage). Both the third term and the fourth term contain the armature velocity and are usually referred to as the *back emf*. Equation (1.465) can be written as follows.

Dynamic solution for I_a

$$V_a - CN_f I_f \mathcal{P}_{sys} v \frac{dN_a}{dX} = I_a \left(R_a + 2N_a \mathcal{P}_{sys} v \frac{dN_a}{dX} \right) + N_a^2 \mathcal{P}_{sys} \frac{dI_a}{dt} \quad (1.467)$$

Equation (1.467) can be solved for the current I_a, and the magnetic flux ϕ_2 can be obtained by solving Eq. (1.453) based on the coil current. Equation (1.467) describes the time-varying coil current I as a function of the known variables (the coil voltage, the system permeance, the number of turns in each coil, the current in the bias field coil, the armature velocity, and the coil resistance). Equation (1.467) also shows that the coil current is a function of the armature velocity.

An alternative solution method for this system is to solve Eq. (1.452) for the coil current I_a, substitute the result into Eq. (1.462), and then rearrange the terms as follows to solve for the magnetic flux.

$$I_a = \frac{\phi_2}{N_a \mathcal{P}_{sys}} - \frac{N_f I_f C}{N_a} \quad (1.468)$$

$$V_a = \left(\frac{\phi_2}{N_a \mathcal{P}_{sys}} - \frac{N_f I_f C}{N_a} \right) R_a + \frac{d(N_a \phi_2)}{dt} \quad (1.469)$$

$$V_a = \phi_2 \frac{R_a}{N_a P_{sys}} - \frac{R_a N_f I_f C}{N_a} + N_a \frac{d\phi_2}{dt} + \phi_2 \frac{dN_a}{dt} \quad (1.470)$$

$$V_a = \phi_2 \frac{R_a}{N_a P_{sys}} - \frac{R_a N_f I_f C}{N_a} + N_a \frac{d\phi_2}{dt} + \phi_2 v \frac{dN_a}{dX} \quad (1.471)$$

$$V_a + \frac{R_a N_f I_f C}{N_a} = \phi_2 \left(\frac{R_a}{N_a \mathcal{P}_{sys}} + v \frac{dN_a}{dX} \right) + N_a \frac{d\phi_2}{dt} \quad (1.472)$$

Equation (1.472) describes the magnetic flux ϕ_2 as a function of the known variables (the coil voltage, the system permeance, the number of turns in each coil, the current in the bias field coil, the armature velocity, and the coil resistance). Equation (1.472) also shows that the magnetic flux is a function of the armature velocity.

The performance of the system shown in Fig. 1.42 can be calculated by solving the differential equation for the coil current I_4, Eq. (1.467), or the differential equation for the magnetic flux ϕ_2, Eq. (1.472). In either case, Eqs. (1.452), (1.447), (1.461),

(1.440), (1.441), and (1.442) are needed to solve for the other variables, ϕ_2 or I_a, ϕ_1, F_{mag}, a, v, and X.

The armature velocity and position can be calculated by solving the mechanical dynamic equations listed earlier in Eqs. (1.439) to (1.442). Equations (1.467) and (1.472) are first-order nonlinear differential equations. The number of turns on the moving armature coil that link the air gap flux and the armature velocity are functions of the armature position and the armature force. Therefore, the best way to solve this equation is to use an iterative finite-difference technique, such as fourth-order Runge-Kutta.

Figure 1.43 shows a block diagram for the control of this moving-coil actuator. In general, a reference position X_{ref} is input to the control and compared to the measured position X_m. The difference between the reference and measured positions is converted to a coil voltage V_a by the transfer function $G(s)$, and the actual positon X is converted into the measured position by the transfer function $H_m(s)$. K_D is the mechanical damping constant of the actuator, and represents friction, bearings, oil viscosity, and air movement (wind resistance). The other two constants, K_F and K_{emf}, are obtained directly from Eqs. (1.455) and (1.466), as follows.

Average turns per unit length over the entire coil

$$\frac{dN_a}{dX} = \frac{N_a}{l_a} \tag{1.473}$$

Rewriting Eq. (1.455)

$$F = I_a(\phi_a + \phi_f)\frac{N_a}{l_a} \tag{1.474}$$

Force per amp proportionality constant

$$K_F = (\phi_a + \phi_f)\frac{N_a}{l_a} \tag{1.475}$$

Block diagram form for Eq. (1.474)

$$F = I_a K_F \tag{1.476}$$

Rewriting Eq. (1.466)

$$V_a = I_a R_a + L_a \frac{dI_a}{dt} + v(\phi_f + 2\phi_a)\frac{N_a}{l_a} \tag{1.477}$$

Volts per velocity proportionality constant

$$K_{emf} = (\phi_f + 2\phi_a)\frac{N_a}{l_a} \tag{1.478}$$

Block diagram form for Eq. (1.477)

$$V_a = I_a R_a + L_a \frac{dI_a}{dt} + v K_{emf} \tag{1.479}$$

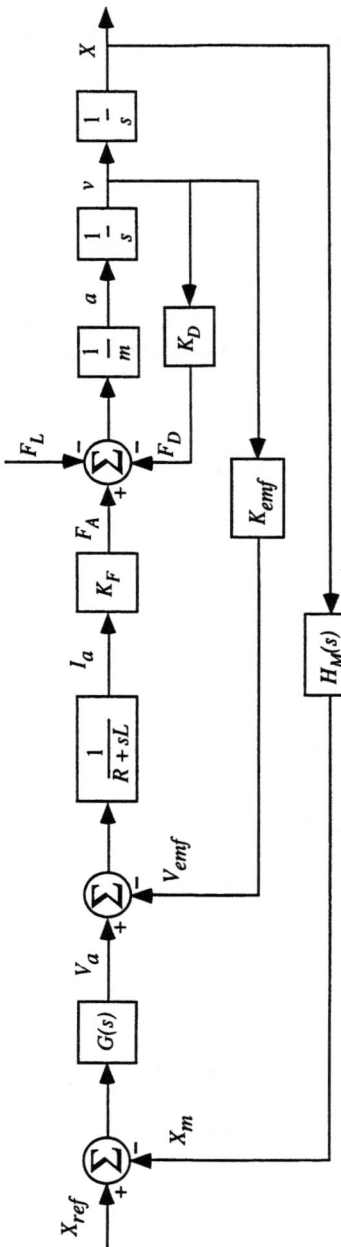

FIGURE 1.43 Block diagram for moving-coil actuator. *(Courtesy of Eaton Corporation.)*

1.13 ELECTROMAGNETIC FORCES*

A force is exerted on a conductor if it is carrying current and is placed in a magnetic field. Fig. 1.44 shows the direction of the force when a current I flows through the conductor in the direction shown. The force F on the conductor is as follows.

$$F = BIl \qquad (1.480)$$

where B = flux density
l = length of conductor being linked by flux
I = current in conductor

Torque T may be produced electromagnetically if current-carrying conductors are arranged such that they may pivot on an axis that is centered in a magnetic field, as shown in Fig. 1.45.

---

* Section contributed by William H. Yeadon, Yeadon Engineering Services.

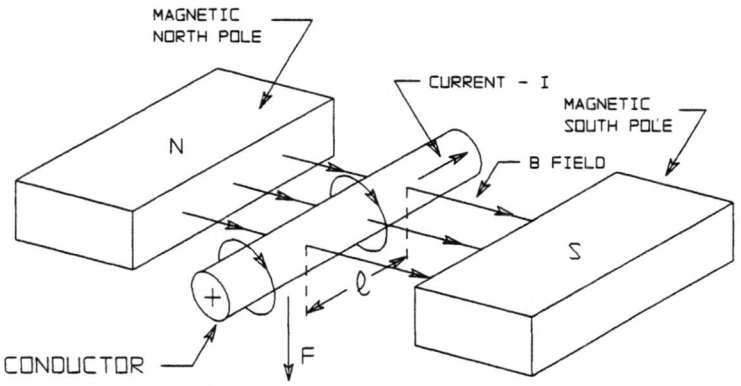

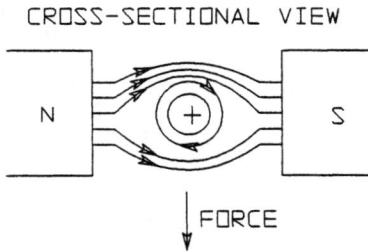

FIGURE 1.44 Force on a conductor. *(Courtesy of Yeadon Engineering Services, P.C.)*

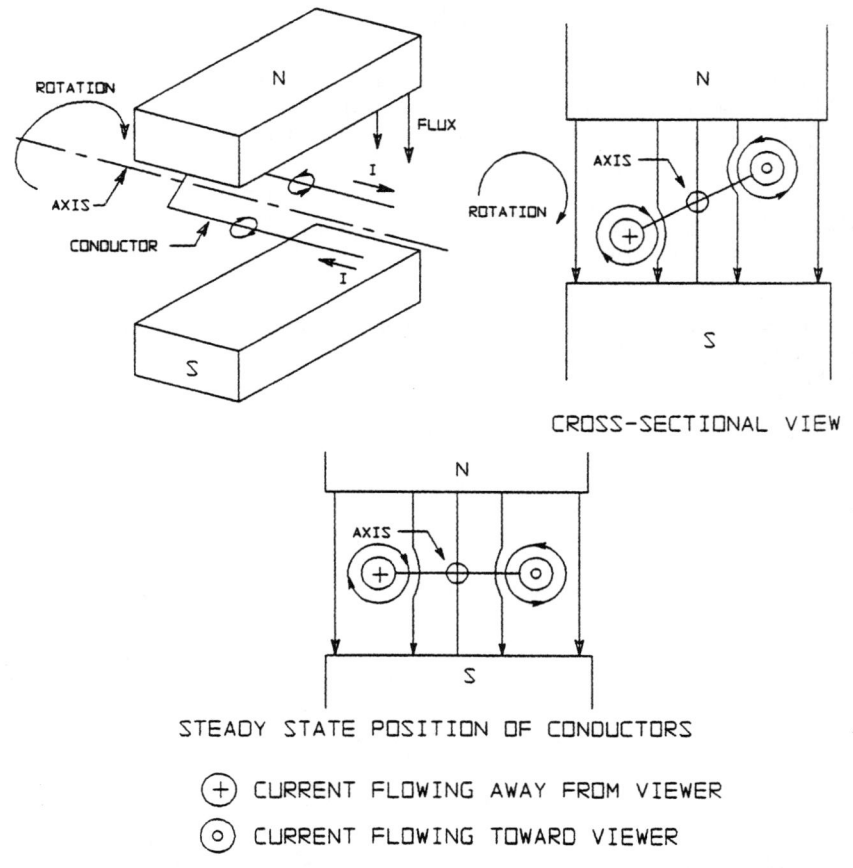

FIGURE 1.45 Electromagnetic torque. *(Courtesy of Yeadon Engineering Services, P.C.)*

The flux in the gap between the magnetic north and south poles will interact with the flux produced by the current in the conductors. If the current is in the direction indicated in Fig. 1.45, the conductors will tend to rotate in the direction shown.

The following equation is used to determine the torque produced.

$$T = \left(K \frac{N\phi P}{a} \right) I \tag{1.481}$$

where N = number of current carrying conductors
 ϕ = flux per pole
 P = number of poles
 a = number of parallel current paths
 I = current in the conductors
 K = constant

The bottom drawing in Fig. 1.45 shows the resting position of the conductors. If the direction of the current is switched as the conductors reach this position, the conductors will continue to rotate on the axis in the same direction. Commutation is the process of switching the directions of the currents to allow for continuous rotation.

In motors, the conductors are contained by magnetic steel teeth. The motor field sets up flux through these teeth. The current in the conductors causes the field to distort, setting up a net torque. The steel increases the amount of flux available to produce torque by lowering the circuit reluctance.

An electric motor is a device for converting electrical power to mechanical power (usually rotational).

1.14 ENERGY APPROACH (ENERGY-COENERGY)*

Another approach to determining forces and torque is through energy concepts. The principle of conservation is the basis for the application of the energy approach.

This section considers the idea of changes in stored field energy. As an introduction to the concept, consider a magnet core, as shown in Fig. 1.46, around which a winding has been placed. This is a single-energy-source system since only one input is involved and no mechanical movement occurs.

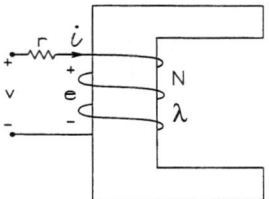

FIGURE 1.46 Magnetic core single-energy-source system. *(Courtesy of Earl F. Richards.)*

Around the loop, Kirchoff's voltage law suggests Eq. (1.482):

$$v = ir + e \tag{1.482}$$

where e can be expressed by Faraday's law:

$$e = \frac{d\lambda}{dt} = N\frac{d\phi}{dt} \tag{1.483}$$

where ϕ = effective flux
λ = coil flux linkages
N = number of turns in coil
r = coil resistance

Multiplying Eq. (1.482) by i, we obtain the following power expression:

$$p = vi = i^2 r + ei \quad \text{W} \tag{1.484}$$

where vi is the instantaneous electrical power input. Integrating the energy expression yields the following:

---

* Section contributed by Earl F. Richards.

$$\int_0^T vi\, dt = \int_0^T i^2 r\, dt + \int_0^T ei\, dt \quad \text{J} \tag{1.485}$$

Electrical energy input = electrical energy loss + electrical field energy.
The electrical field energy at time T is as follows.

$$W_\phi = \int_0^T ei\, dt \tag{1.486}$$

Then

$$W_\phi = \int_0^T \frac{d\lambda}{dt} i\, dt = \int_0^{\lambda_T} i\, d\lambda \tag{1.487}$$

where λ_T is the flux linkage at time T, as illustrated in Fig. 1.47.
Since $\lambda = N\phi$ and $d\lambda = N\, d\phi$, then

$$W_\phi = \int_0^{\phi_T} iN\, d\phi = \int_0^{\phi_T} \mathcal{F}\, d\phi \tag{1.488}$$

where ϕ_T is the flux at time T, as illustrated in Fig. 1.48.

The area W'_ϕ is called the *coenergy* and is useful in developing an expression for electromechanical forces or torque. It can be found as follows.

$$W'_\phi = \int_0^{I_T} \lambda\, di = \int_0^{\mathcal{F}_T} \phi\, d\mathcal{F} \tag{1.489}$$

It is obvious that

$$W_\phi + W'_\phi = \lambda_T I_T = \phi_T \mathcal{F}_T \tag{1.490}$$

If the system is linear without saturation, it is obvious that $W_\phi = W'_\phi$.

The slopes of the characteristic are as follows:

$$\frac{\phi}{\mathcal{F}} = \frac{1}{\mathcal{R}} \quad \text{and} \quad \frac{\lambda}{i} = L \tag{1.491}$$

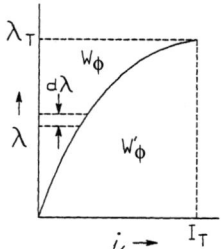

FIGURE 1.47 Flux linkage versus current. *(Courtesy of Earl F. Richards.)*

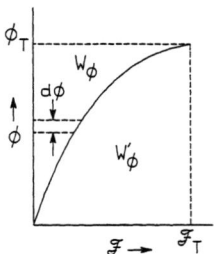

FIGURE 1.48 Flux versus magnetizing force. *(Courtesy of Earl F. Richards.)*

where $\mathcal{R}$ and L are the reluctance and inductance of the system, respectively.

Suppose we now add an armature to Fig. 1.46 and redraw the coil so that an air gap exists as shown in Fig. 1.49.

Note that under this assumption mechanical motion is involved in the coil current i and the position of the armature x, that is, $\lambda = \lambda(i, x)$.

Since mechanical motion is involved, an additional equation involving Newton's law of motion is required to describe the dynamics of the system. Before we pursue this analysis, it is very informative to look at a graphical analysis which will give insight to the energy approach.

As previously shown, for the case in which the flux linkage is a function of the coil current i, only a two-dimensional plot is drawn (Fig. 1.46). However, because such a plot requires three dimensions, if we use a two-dimensional plot we now have a choice of considering constant current or constant flux linkage, giving a graphical approach to assist in understanding the energy principles. In either case, our interest is in looking at changes in energy for an energy balance.

Let us first assume we will excite the coil shown in Fig. 1.49 with a constant-current source of magnitude i. Refer to Fig. 1.50 and note that five distinct areas are shown. Before the armature movement, the stored field energy area in Fig. 1.50 was C + D. If the armature now moves from x_0 to $x_0 + \Delta x$, the air gap decreases and a differential change in flux and flux linkage occurs. The additional energy supplied by the electrical circuit is as follows:

$$\Delta W_{el} + I\Delta\lambda + A + B + E \tag{1.492}$$

The differential energy balance equation is as follows:

$$\Delta W_{el} = \Delta W_\phi + \Delta W_M \tag{1.493}$$

where ΔW_M is the mechanical energy output and W_ϕ is the additional energy stored in the magnetic field.

But

$$\Delta W_M = F\Delta x \tag{1.494}$$

The fields of energy before and after the movement Δx are as follows:

Energy before movement

$$\Delta W_{\phi i} = C + D \tag{1.495}$$

Energy after movement

$$\Delta W_{\phi f} = A + C \tag{1.496}$$

The increase in stored energy is then

$$\Delta W_\phi = \Delta W_{\phi f} - \Delta W_{\phi i} = (A + C) - (C + D) = A - D \tag{1.497}$$

$$\Delta W_M = \Delta W_{el} - \Delta W_f = (A + B + E) - (A + D) = B + D + E \tag{1.498}$$

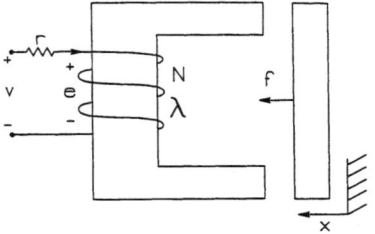

FIGURE 1.49 Electromagnetic actuator. *(Courtesy of Earl F. Richards.)*

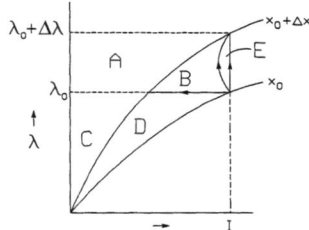

FIGURE 1.50 Energy areas before and after movement. *(Courtesy of Earl F. Richards.)*

The average force on the armature is as follows:

$$f\Delta x = \Delta W_M = B + D + E \tag{1.499}$$

The force here is in a direction to shorten the air gap. If the current restraint is relieved and the $\lambda - i$ trajectory could follow the indicated curved path from x_0 to $x_0 + \Delta x$, we then have the following.

$$\Delta W_{el} = A + B \tag{1.500}$$

$$\Delta W_\phi = (A + C) - (C + D) = A - D \tag{1.501}$$

Then

$$\Delta W_M = \Delta W_{el} - \Delta W_\phi = (A + B) - (A - D) = B + D \tag{1.502}$$

Note that here ΔW_{el} and ΔW_M are reduced by an equal amount, namely area E. This means that the average force f is smaller than when the current was constant. If constant flux linkage is assumed, we have the following.

$$\Delta W_{el} = 0 \quad \text{because} \quad \frac{d\lambda}{dt} = 0 \tag{1.503}$$

Then

$$\Delta W_{\phi f} - \Delta W_{\phi i} = C - D(C + D) = -D \tag{1.504}$$

$$\Delta W_M = \Delta W_{el} - \Delta W_\phi = 0 - D = D \tag{1.505}$$

Then

$$f = \frac{\Delta W_M}{\Delta X} = \frac{D}{\Delta X} \tag{1.506}$$

Note that the average force is reduced from the constant-current case. All of the mechanical energy is obtained from a reduction in the field energy since the electrical input was zero.

Let us now return to the original problem and see how we can derive the mathematical expressions for the force or torque.

Let us again consider Fig. 1.49. The differential energy balance equation is as follows.

$$\Delta W_{el} = \Delta W_\phi + \Delta W_M \tag{1.507}$$

$$= \Delta W_\phi + f\Delta x \tag{1.508}$$

or

$$dW_{el} = dW_\phi + dW_M = dW_\phi + f\,dx \tag{1.509}$$

where

$$dW_{el} = i\,d\lambda \tag{1.510}$$

We have a choice in selection of independent variables. There are λ, i, and x. Our choice must be as follows.

BASIC MAGNETICS

$$\lambda = \lambda(i, x) \quad \text{or} \quad i = i(\lambda, x) \tag{1.511}$$

The usual test is to take i and x as independent variables, although the other choice can be made.

With i and x as independent variables

$$W_\phi = W_\phi(i, x) \qquad \lambda = \lambda(i, x) \tag{1.512}$$

Then

$$dW_\phi = \frac{\partial W_\phi}{\partial i} di + \frac{\partial W_\phi}{\partial x} dx \tag{1.513}$$

$$d\lambda = \frac{\partial \lambda}{\partial i} di + \frac{\partial \lambda}{\partial x} dx \tag{1.514}$$

Substituting into the differential energy balance

$$i\left(\frac{\partial \lambda}{\partial i} di \frac{\partial \lambda}{\partial x} dx\right) = \frac{\partial W_\phi}{\partial i} di + \frac{\partial W_\phi}{\partial x} dx + f\, dx \tag{1.515}$$

$$f\, dx = \left(i \frac{\partial \lambda}{\partial i} - \frac{\partial W_\phi}{\partial i}\right) di + \left(i \frac{\partial \lambda}{\partial x} - \frac{\partial W_\phi}{\partial x}\right) dx \tag{1.516}$$

From which, since dx and di are independent

$$f\, dx = \left(i \frac{\partial \lambda}{\partial i} - \frac{\partial W_\phi}{\partial i}\right) di \quad \text{and} \quad \left(i \frac{\partial \lambda}{\partial x} - \frac{\partial W_\phi}{\partial x}\right) dx = 0 \tag{1.517}$$

$$f = i \frac{\partial \lambda}{\partial x} - \frac{\partial W_\phi}{\partial x} \tag{1.518}$$

Since we are interested in the force only, this solution will be used.

At this point it is appropriate to introduce coenergy W'_ϕ. Recall that

$$W_\phi + W'_\phi = \lambda i \tag{1.519}$$

$$W_\phi = \lambda i - W'_\phi \tag{1.520}$$

Taking the differential with respect to x

$$\frac{\partial W_\phi}{\partial x} = i \frac{\partial \lambda}{\partial x} - \frac{\partial W'_\phi}{\partial x} \tag{1.521}$$

$$f = i \frac{\partial \lambda}{\partial x} + \frac{\partial W'_\phi}{\partial x} - i \frac{\partial \lambda}{\partial x} = \frac{\partial W'_\phi}{\partial x} \tag{1.522}$$

Substituting into the force equation Eq. (1.524) gives an important result that can be expanded to include not only translational systems but also rotational systems and multiexcited systems of P magnetic poles.

Summarizing without proof, these are as follows.

$$f = \frac{\partial W'_\phi}{\partial x} (i_1, \cdots i_2, i_n, x) \tag{1.523}$$

$$T = \frac{P}{2} \frac{\partial W'_\phi}{\partial \theta_r} (i_1, i_2, \cdots i_n, x, \theta_r) \tag{1.524}$$

where x = linear displacement
 T = torque
 P = number of poles
 n = number of excitation windings having currents
 f = force
 θ_r = angular position of rotor with respect to a reference

CHAPTER 2
MATERIALS

Chapter Contributors

Allegheny-Teledyne
Joseph H. Bularzik
Francis Hanejko
Robert R. Judd
Harold R. Kokal
Robert F. Krause
Phelps Dodge Company
Joseph J. Stupak
United States Steel Corporation
William H. Yeadon

The purpose of this chapter is to assist in the selection of materials used in electric motors. Material choices are largely a function of the motor's application. All materials commonly used in electric motors are covered in this chapter, including lamination steel, magnets, wire, and insulation.

2.1 MAGNETIC MATERIALS*

2.1.1 Steel Selection

Steel is used in most electric motors as the primary flux-carrying member. It is used in stator cores, rotor cores, armature assemblies, field assemblies, housings, and shafting. It may be solid, laminated, or in powdered iron forms. Magnetic properties vary with the type being used. This section will cover the magnetic and mechanical properties of these steels.

By way of review from Chap. 1: A rectangular block of magnetic material is wound with a coil of wire, as in Fig. 2.1. If the coil of wire in Fig. 2.1 gradually has its current increased from zero, a magnetizing force $\mathcal{F}$ will be produced. The block of steel will be subjected to a magnetic field intensity H.

---

*Section contributed by William H. Yeadon, Yeadon Engineering Services, PC.

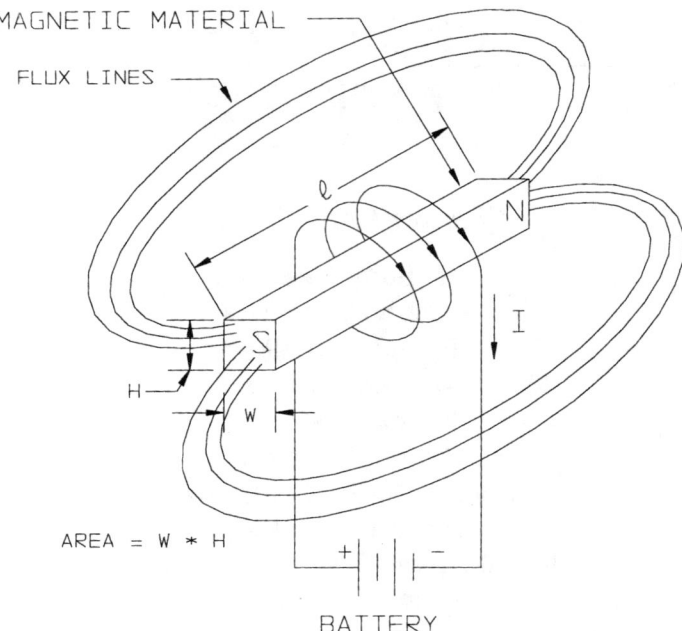

FIGURE 2.1 Magnetization of materials.

This field intensity is proportional to the current times the number of turns of wire per inch of magnetic material being magnetized:

$$H = \frac{\mathscr{F}}{\ell} \quad (A \cdot \text{turns})/\text{in or A/m} \tag{2.1}$$

As H is increased, there is a flux established in the block of material. Since the area of the block is known, the flux density is:

$$B = \frac{\phi}{\text{area}} \quad \text{lines/in}^2, \text{W/m}^2, \text{ or T} \tag{2.2}$$

As the current increases, the flux density B is increased along the virgin magnetization curve shown in Fig. 2.2. Eventually B will be increased only as if the steel were air. This is called the *saturation point* of the material. As the applied field is decreased, the flux density B is decreased, but at zero H some B_r (residual flux density) still exists. To drive B to zero it is necessary to drive H negative and hold it at this value. If H is driven negative so that it is numerically equal to $+H$, the hysteresis loop shown in Fig. 2.2 would exist. The H required to overcome B_r results in losses in magnetic circuits where the flux is continually reversed. These losses are commonly referred to as *hysteresis losses.*

Since in most electric motors the material is alternately magnetized and demagnetized, a changing field exists.

Steinmetz defines *hysteresis power loss* as:

$$P_{\text{hys}} = \sigma_h f B^{1.6} \quad \text{W/lb} \tag{2.3}$$

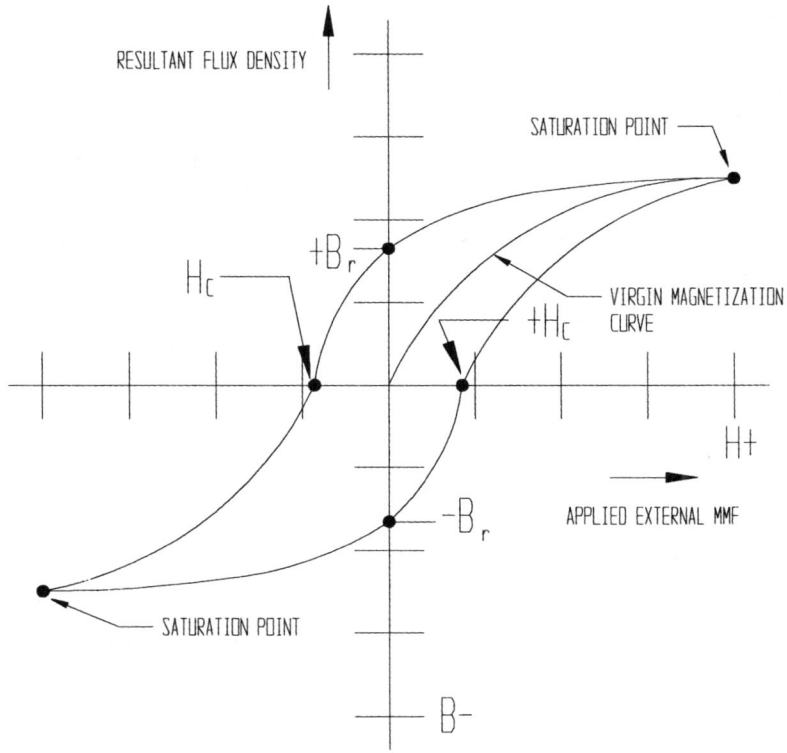

FIGURE 2.2 Hysteresis curves of magnetic material.

where B = flux density
f = frequency, Hz = (number of poles × r/min) ÷ 120
σ_h = constant based on the quality of the iron and its volume and density

Richter predicts hysteresis power loss as:

$$P_{\text{hys}} = \sigma_h \frac{f}{60} \left(\frac{B}{64{,}500} \right)^2 \quad \text{W/lb} \tag{2.4}$$

In addition, a changing magnetic field induces voltages in conductors moving relative to the field. If a completed electrical path exists, currents will be set up in the conductor, limited only by the resistance of the conductor material. These currents are referred to as *eddy currents* and they cause unwanted power losses. In the case of electric motors, eddy current losses in the cores become significant.

Stator cores are laminated to reduce eddy current losses. Richter determines eddy power losses as:

$$P_{\text{eddy}} = \sigma_e \left(\Delta \frac{f}{60} \frac{B}{64{,}500} \right)^2 \quad \text{W/lb} \tag{2.5}$$

where σ_e = constant based on the quality of the iron, containing an element of resistivity of the material and the material density
Δ = thickness of the laminations
f = frequency
B = flux density

These losses (hysteresis and eddy) are added together and called *core losses*.

In practice, iron losses are derived from curves supplied by the steel manufacturers. Units are in watts per pound or watts per kilogram of steel.

Hysteresis losses are reduced by improving the grade of steel and by annealing the laminations. Annealing the laminations changes the grain structure of the steel to allow for easy magnetization. Eddy current losses are reduced by using thinner laminations and increasing the resistivity of the steel. Adding silicon to steel reduces eddy losses but increases die wear during punching because silicon increases steel hardness.

As a general rule, as the grade number increases, the induction level increases and the core loss increases, but cost goes down. For example, see Table 2.1.

TABLE 2.1 Comparison of Steel Grades

Material type	Flux density B @ 100 Oe	W/lb @ 18 kG	Relative cost/lb
M-19	17.5 kG	3.0	Higher
M-50	17.8 kG	4.4	Medium
Low-carbon CRS	18.5 kG	6.0	Lower

2.2 LAMINATION STEEL SPECIFICATIONS*

All motor designs must eventually be brought to production to achieve their final goal. Most motor producers want a minimum of two steel suppliers for a given lamination type. This means that someone has to find more than one steel sheet supplier that can provide the same magnetic quality and punchability. U.S. domestic suppliers do not make this a simple task. They typically have an in-house name for their steel grades that is little help in inferring magnetic quality. The old American Iron and Steel Institute (AISI) electrical steel M series is an example. AISI abandoned this series as an industry standard in 1983 when they published their last *Electrical Steels* steel products manual. However, the grade designation still exists in the Armco and WCI product lines and in older Temple steel material specifications, but all three specifications having the same M number may not have the same magnetic characteristics.

The American Society for Testing and Materials (ASTM) has attempted to unify steel specifications by means of a universal naming system that is published in ASTM specification A664. The result is a mixed-unit alphanumeric string, such as 47S200, where the first two numbers are the sheet thickness in millimeters times 100, the next letter is a steel-grade annealing treatment and testing procedure designation, and the next three numbers are the core loss in watts per pound divided by 100. If the core loss is given in watts per kilogram instead of watts per pound, an "M" is appended to the string to indicate a metric core loss measurement. Because of the

---

*Sections 2.2 and 2.3 contributed by Robert R. Judd, Judd Consulting.

mixed units, this effort is not intellectually pleasing, but no one can deny the overall need for it.

Many foreign manufacturers and standardization bodies have recognized the need for meaningful electrical sheet specifications and have adopted a specification name similar to that of the ASTM effort. All specifications have much more magnetic property detail than can be conveyed in an identifying name. For instance, no permeability or magnetization curve shape is indicated in the steel name, but some indications of minimum permeability or minimum induction at a designated magnetizing field will be given in the specification detail.

The punchability of steel sheet with identical magnetic quality from two different suppliers is rarely the same. This forces the press shop to have a set of dies for each steel supplier of a given part. This can raise the costs of keeping several steel suppliers for one part. Also, the subtleties of producing flat, round laminations from a large sheet usually involve a trial-and-error procedure for the die shop. This means that parts for which there are multiple steel suppliers are a multiple headache for the die shop.

The information in Table 2.1 illustrates the M-grade (motor grade) steels categorization system.

Magnetic properties are given in a variety of units. The conversion chart in Table 2.2 is provided for convenience.

Laminated cores are normally considered because of the necessity of reducing the core losses which occur at high switching frequencies. There are, however, some applications where low cost is a higher priority than efficiency. In these cases powdered metal cores may be considered. Their induction levels are similar to those of annealed sheet steel, but the core losses may be four to five times greater. There are some recent advances in powdered iron that make them suitable for these applications. They are discussed in a later section.

The following figures show magnetic property curves of several materials. Note that many of the scales are in different units.

The new Temple product description was created to simplify material selection. Each description incorporates the gauge, material family, and maximum core loss into a concise, six-character label. The first two characters in the new description indicate the thickness of the material, for example, 29 for 0.014 in thick. The third character is a letter which indicates the material family, such as "G" for grain ori-

TABLE 2.2 Electromagnetic Unit Systems

Quantity	Symbol	MKS	CGS	English
Flux density	B	Teslas (Webers/m^2) $1\,T = 6.452 \times 10^4$ lines/in^2 $1\,T = 10^4\,G$	Gauss $1\,G = 6.452$ lines/in^2	Kilolines/in^2 1 kline/in^2 = 0.155 kG 1 kline/in^2 = 1.55×10^{-2} T
Magnetic field intensity	H	Amps/m (A·T/m) 1 A·T/m = 0.01257 Oe	Oersted 1 Oe = 2.021 A·T/in	Amps/in (A·T/in) 1 A·T/in = 0.4947 Oe
Magnetic flux	ϕ	Webers 1 Wb = 10^8 maxwells	Maxwells 1 maxwell = 1 line	Kilolines 1 kline = 10^{-5} Wb
Reluctance	$\mathcal{R}$	Henries/m	Henries/cm	H/in
Permeability of free space	μ_0	$4\pi \times 10^{-7}$ Wb/(A·T/m)	1 maxwell/(Gb/cm)	3.19 lines/(A·T·in)

ented, "N" for nonoriented, and "T" for Tempcor. The last three characters define the material's maximum core loss. The inclusion of the maximum core loss in the product description eliminates the need to cross-index the M grade with a core loss chart. To illustrate the system, 26N174 is the description of 26-gauge, nonoriented silicon steel with a maximum core loss of 1.74 W/lb.

Figures 2.3 through 2.33 show typical properties of magnetic motor steels, courtesy of Temple Steel Company.

Allegheny-Teledyne company also produces alloy steels with varying properties for motor applications. Figures 2.34 through 2.39 show typical properties of nickel-iron alloys and steels, courtesy of Allegheny-Teledyne Company.

Figures 2.40 through 2.59 show typical properties of nonoriented silicon steels.

2.3 LAMINATION ANNEALING

The type of annealing to be discussed here is the final annealing of laminations punched from semiprocessed electrical sheets. Other types of annealing that enhance the quality of laminations are the stress relief annealing of laminations punched from fully processed electrical sheet and the annealing of hot band coils before cold rolling. Stress relief annealing is done to flatten laminations and to recrystallize the crystals damaged during punching. This damage extends from the punched edge to a distance from the edge equal to the sheet thickness, and it severely degrades the magnetic quality of the affected volume. In a small motor, this

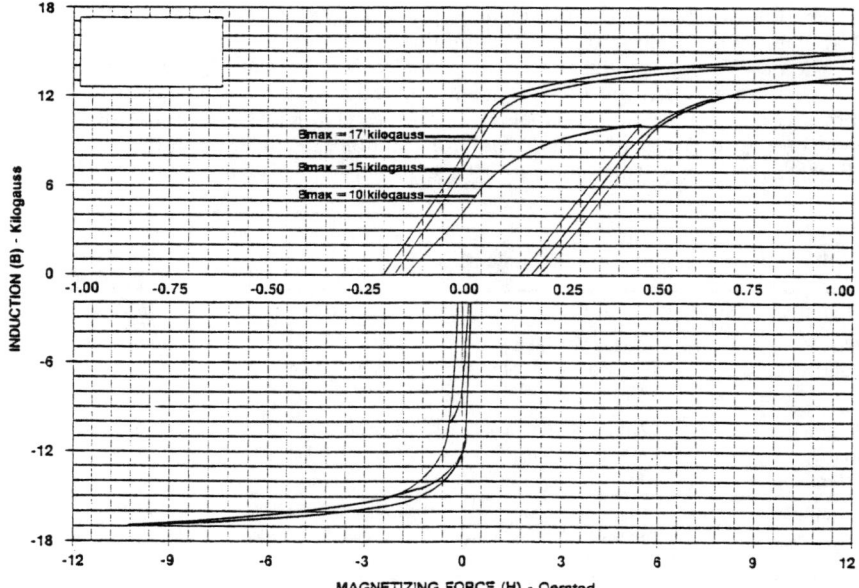

FIGURE 2.3 B-H magnetization loops for 29G066 75–25% (29 06). Values based on ASTM 596 and A773; 75 percent parallel grain and 25 percent cross grain after annealing.

MATERIALS

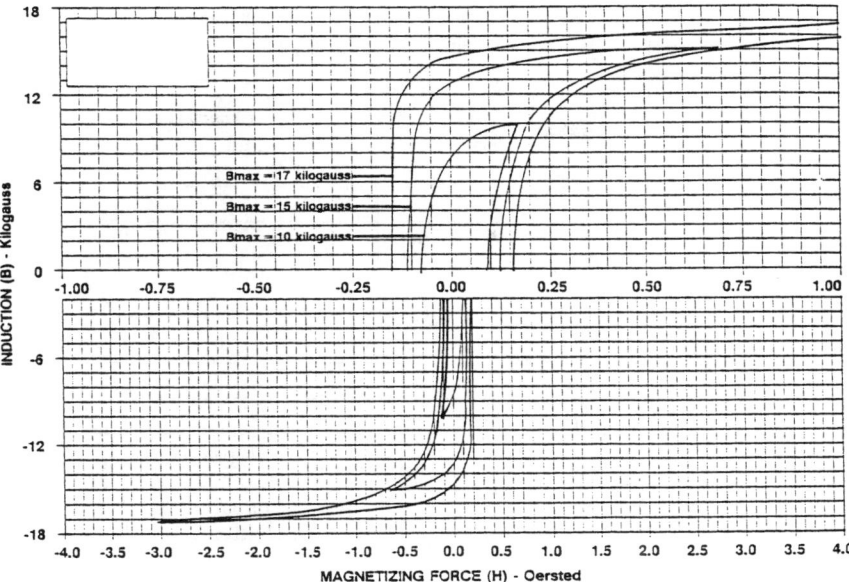

FIGURE 2.4 *B-H* magnetization loops for 29G066 100% (29 06). Values based on ASTM 596 and A773; 100 percent parallel grain after annealing.

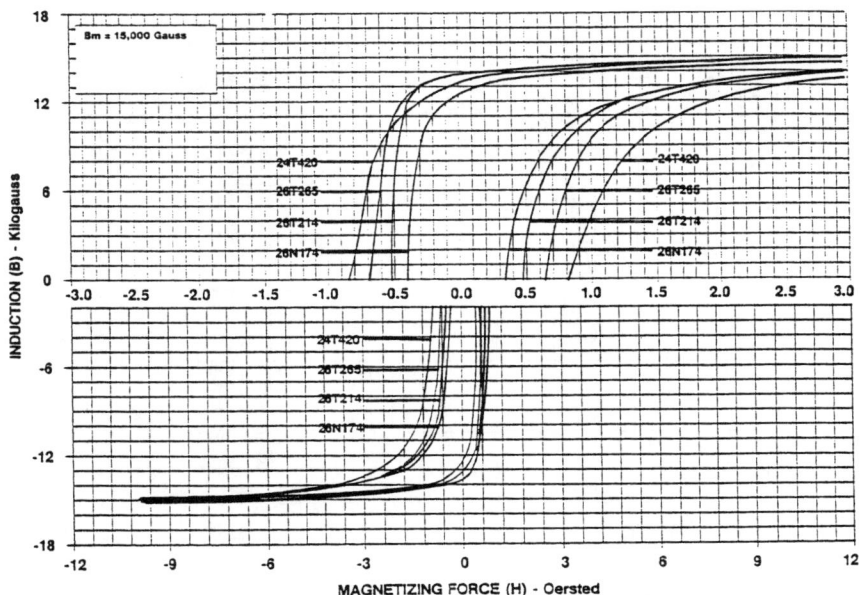

FIGURE 2.5 *B-H* magnetization loops for 26N174, 26T214, 26T265, and 24T240. Typical values based on ASTM 596 and A773; half parallel and half cross grain after annealing.

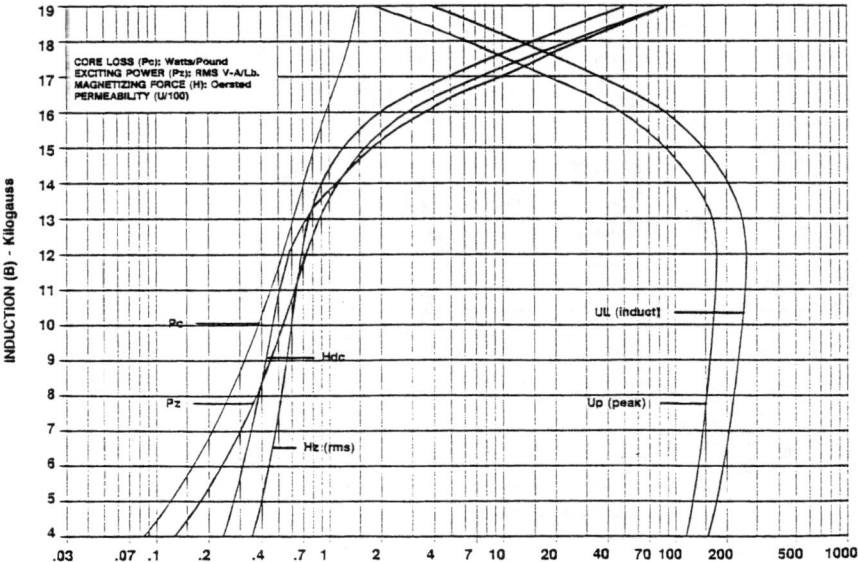

FIGURE 2.6 29G066 75–25% (0.99 W/lb maximum 29 06). Typical values based on Epstein samples; 75 percent parallel grain and 25 percent cross grain at 60 Hz after annealing.

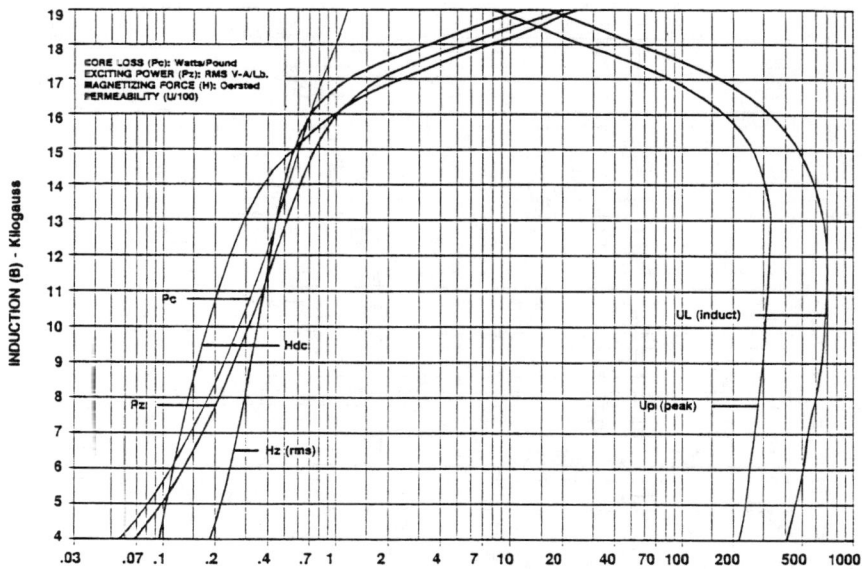

FIGURE 2.7 29G066 100% (0.66 W/lb maximum 29 06). Typical values based on Epstein samples; 100 percent parallel grain at 60 Hz after annealing.

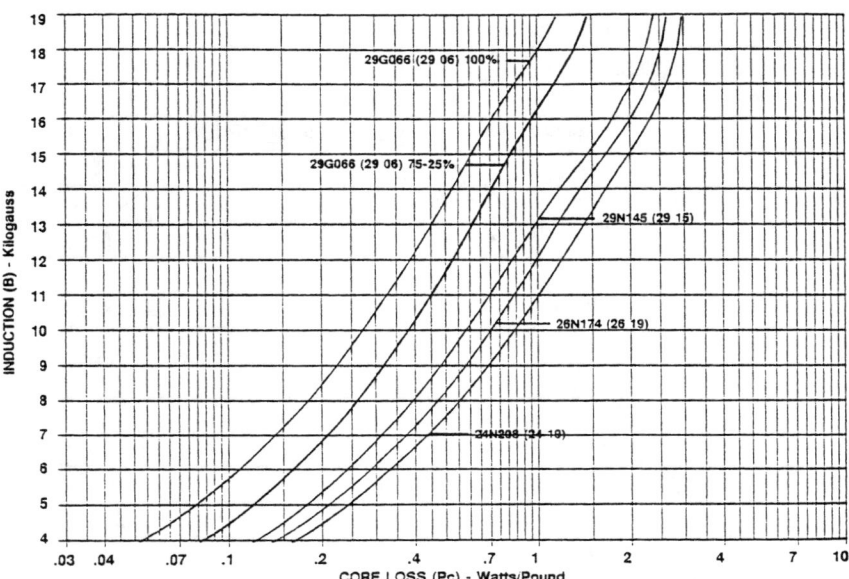

FIGURE 2.8 29G066 (29 06), 29N145 (29 15), 26N174 (26 19), and 24N208 (24 19). Typical core loss values, W/lb, based on Epstein samples (ASTM A343); half parallel and half cross grain (except where noted) at 60 Hz after annealing.

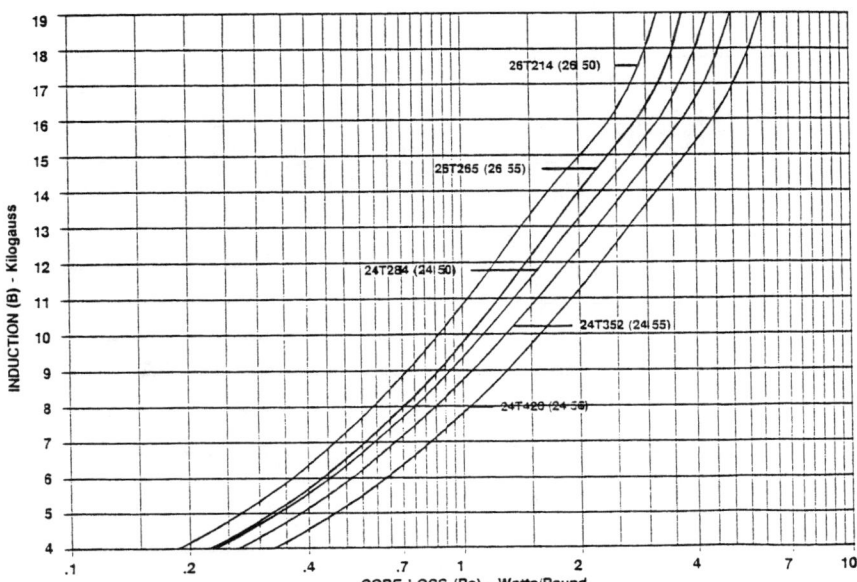

FIGURE 2.9 26T214 (26 50), 26T265 (26 55), 24T284 (24 50), 24T352 (24 55), and 24T420 (24 56). Typical core loss values, W/lb, based on Epstein samples (ASTM A343); half parallel and half cross grain at 60 Hz after annealing.

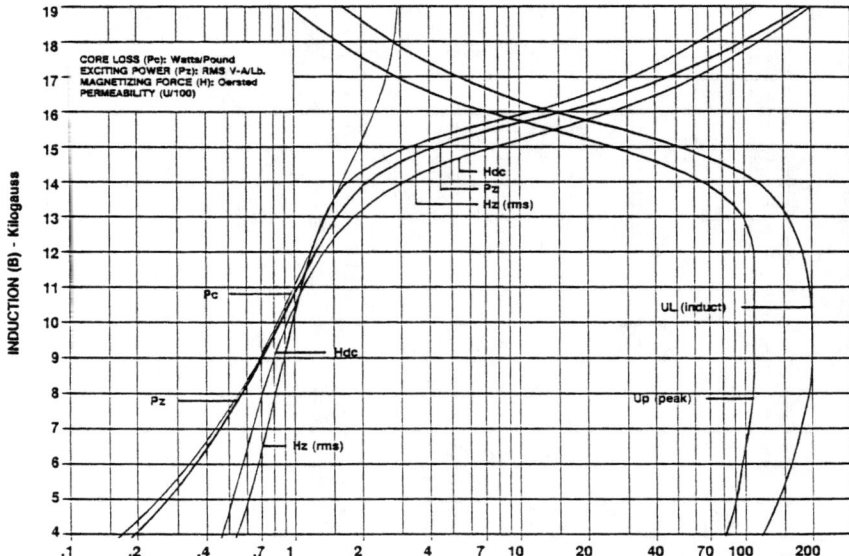

FIGURE 2.10 24N208 (2.08 W/lb maximum 24 19). Typical magnetization curves based on Epstein samples; half parallel and half cross grain at 60 Hz after annealing.

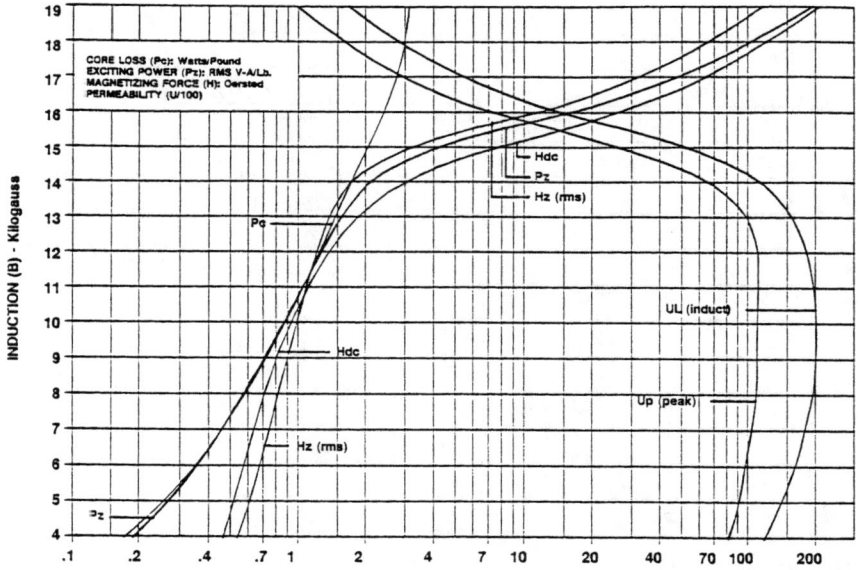

FIGURE 2.11 24N218 (2.18 W/lb maximum 24 22). Typical magnetization curves based on Epstein samples; half parallel and half cross grain at 60 Hz after annealing.

MATERIALS

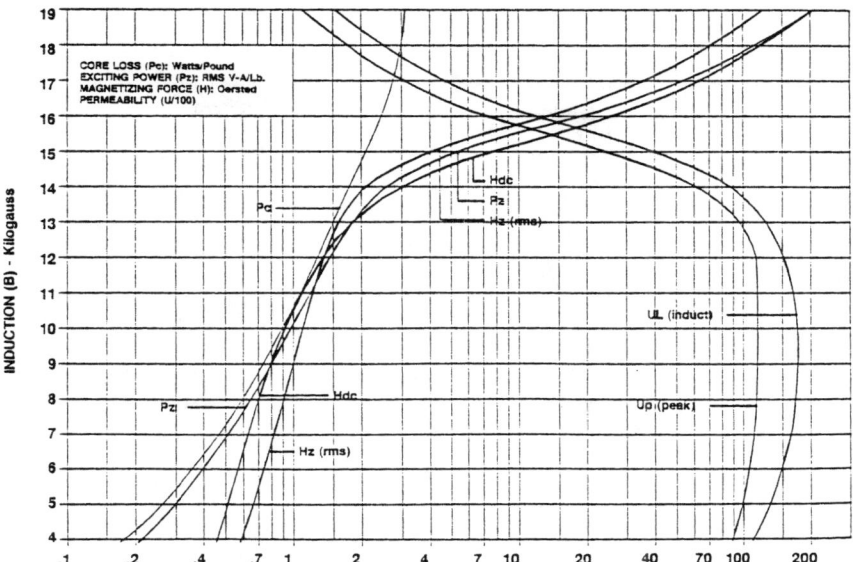

FIGURE 2.12 24N225 (2.25 W/lb maximum 24 27). Typical magnetization curves based on Epstein samples; half parallel and half cross grain at 60 Hz after annealing.

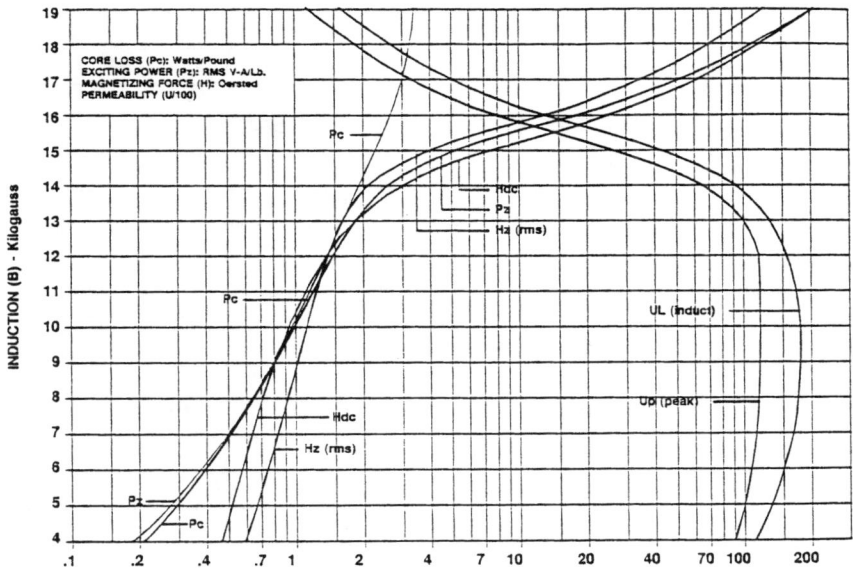

FIGURE 2.13 24N240 (2.40 W/lb maximum 24 36). Typical magnetization curves based on Epstein samples; half parallel and half cross grain at 60 Hz after annealing.

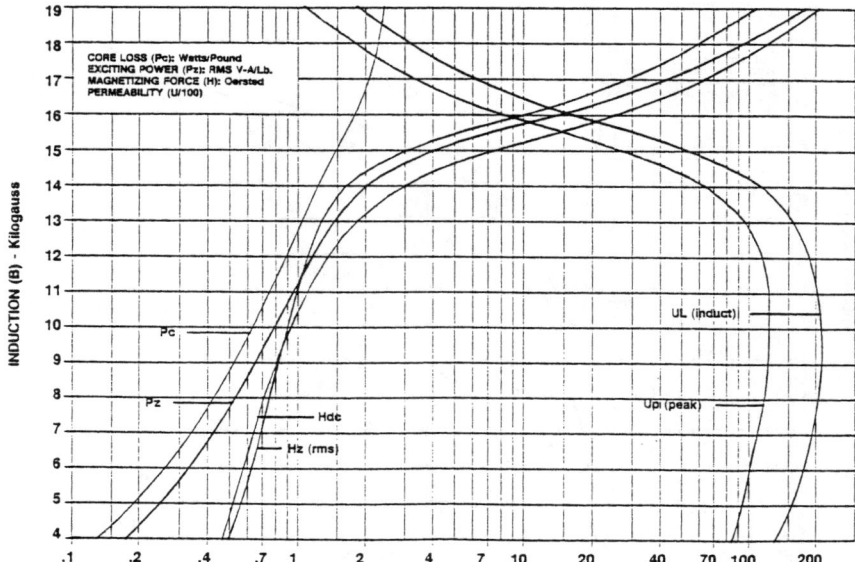

FIGURE 2.14 26N158 (1.58 W/lb maximum 26 14). Typical magnetization curves based on Epstein samples; half parallel and half cross grain at 60 Hz after annealing.

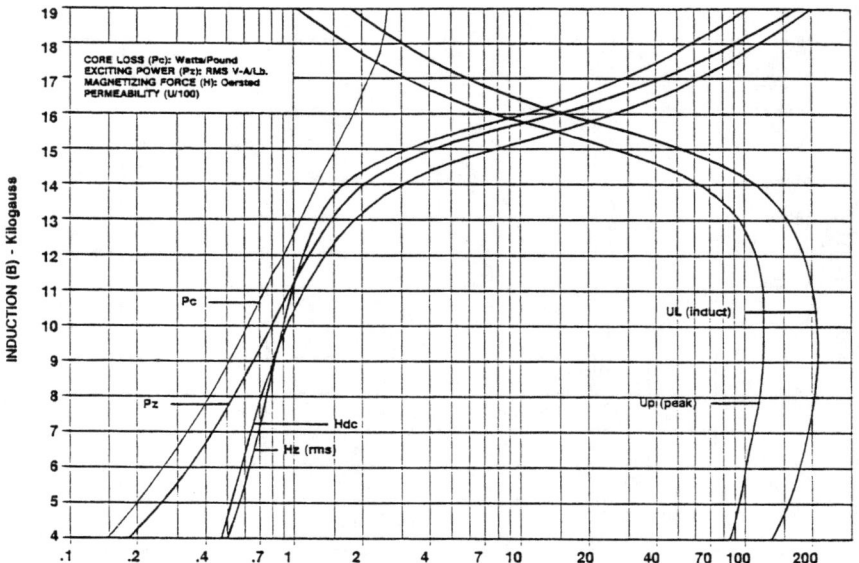

FIGURE 2.15 26N174 (1.74 W/lb maximum 26 19). Typical magnetization curves based on Epstein samples; half parallel and half cross grain at 60 Hz after annealing.

MATERIALS

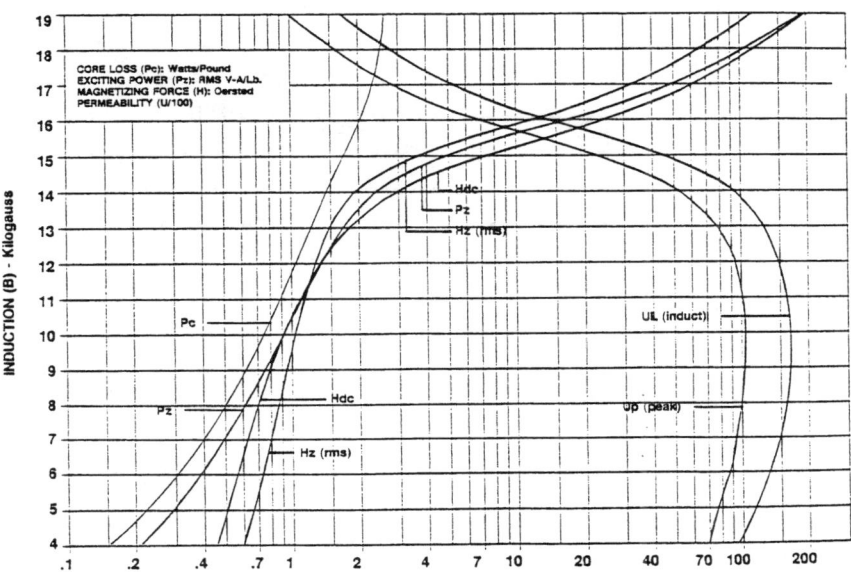

FIGURE 2.16 26N185 (1.85 W/lb maximum 26 22). Typical magnetization curves based on Epstein samples; half parallel and half cross grain at 60 Hz after annealing.

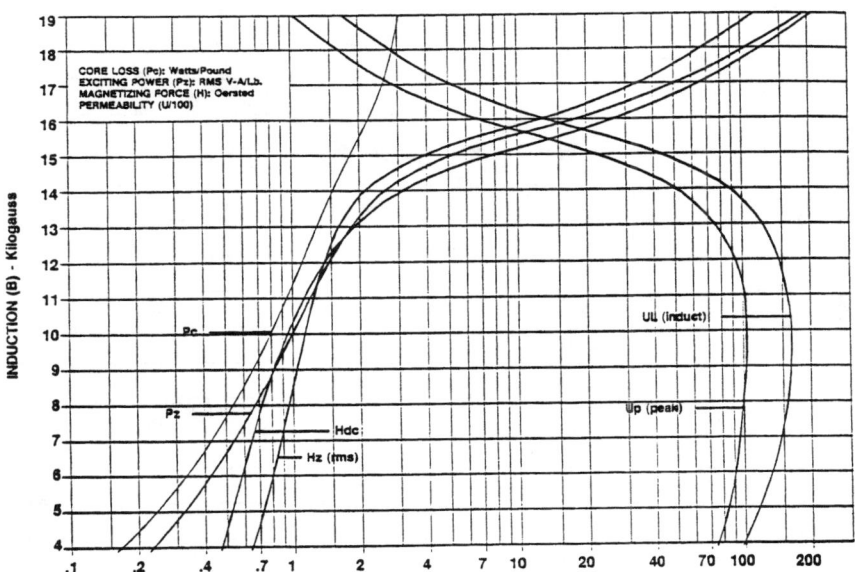

FIGURE 2.17 26N190 (1.90 W/lb maximum 26 27). Typical magnetization curves based on Epstein samples; half parallel and half cross grain at 60 Hz after annealing.

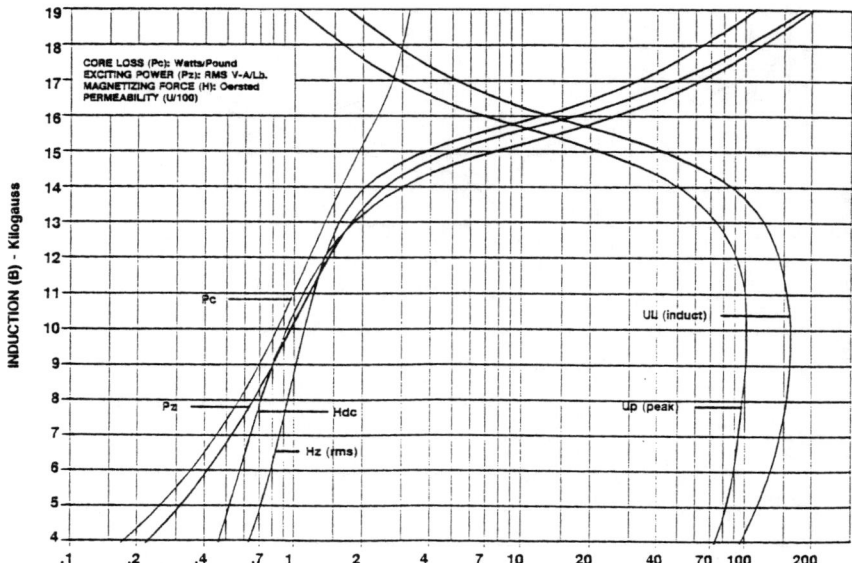

FIGURE 2.18 26N205 (2.05 W/lb maximum 26 36). Typical magnetization curves based on Epstein samples; half parallel and half cross grain at 60 Hz after annealing.

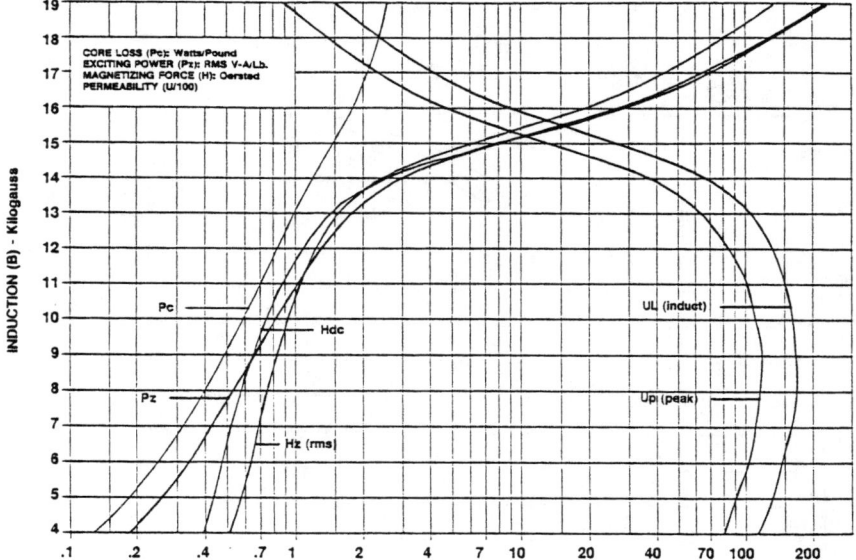

FIGURE 2.19 29N145 (1.45 W/lb maximum 29 15). Typical magnetization curves based on Epstein samples; half parallel and half cross grain at 60 Hz after annealing.

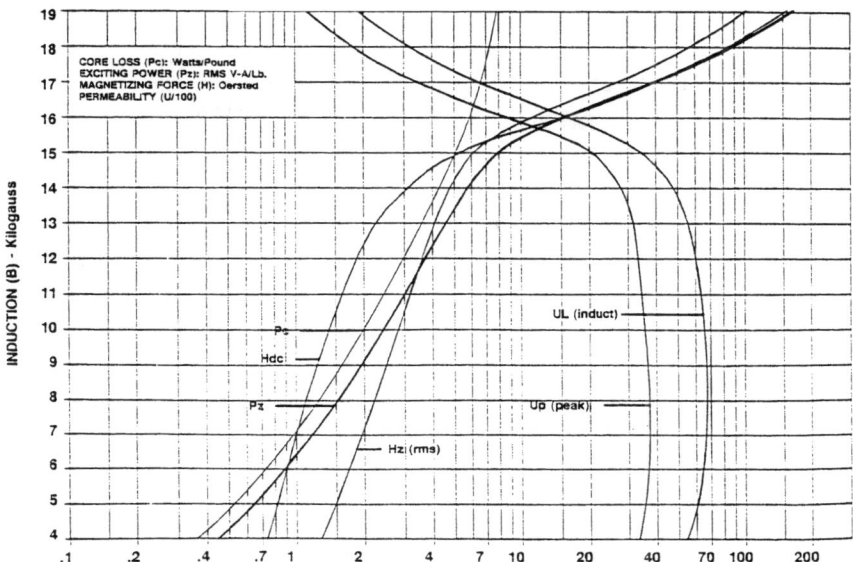

FIGURE 2.20 22T600 (6.00 W/lb maximum 22 56). Typical magnetization curves based on Epstein samples; half parallel and half cross grain at 60 Hz after annealing.

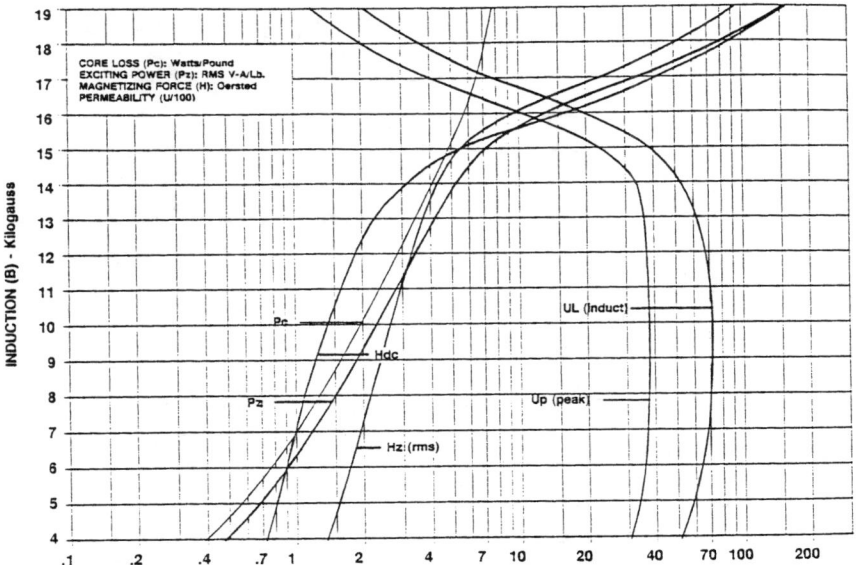

FIGURE 2.21 23T500 (5.00 W/lb maximum 23 56). Typical magnetization curves based on Epstein samples; half parallel and half cross grain at 60 Hz after annealing.

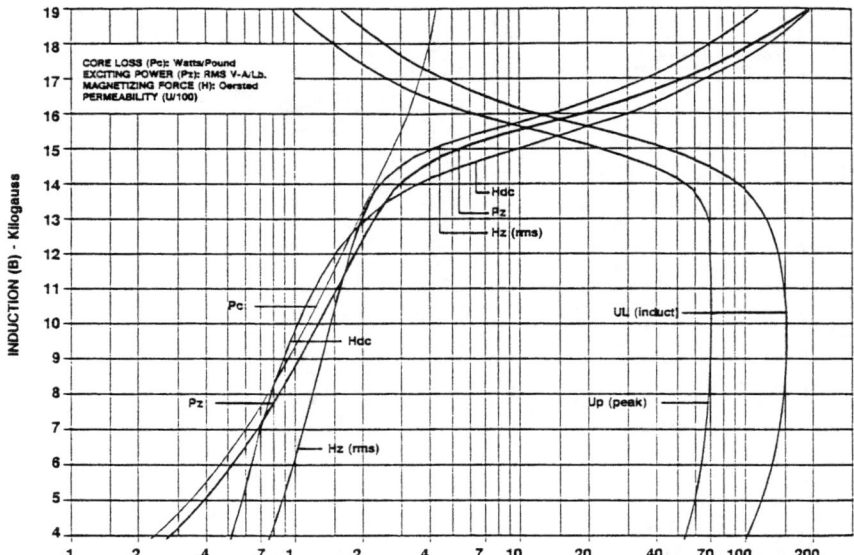

FIGURE 2.22 24T284 (2.84 W/lb maximum 24 50). Typical magnetization curves based on Epstein samples; half parallel and half cross grain at 60 Hz after annealing.

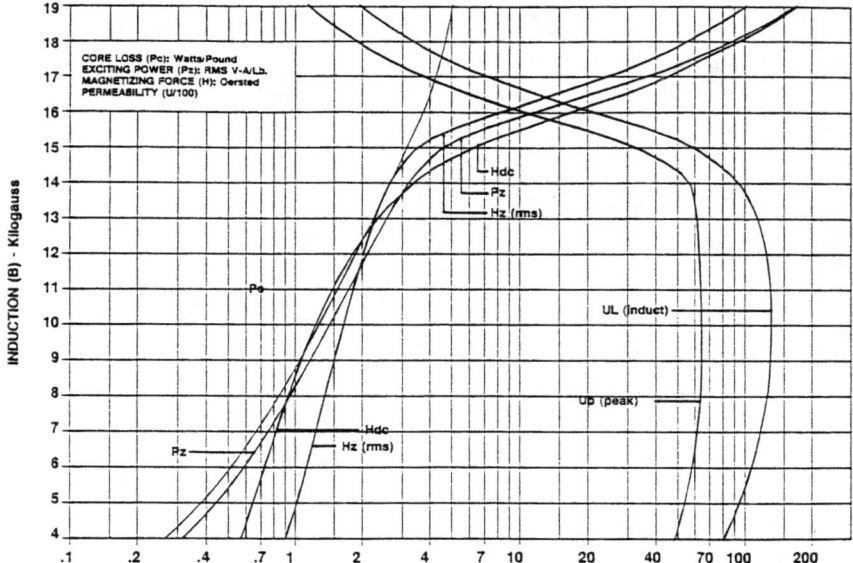

FIGURE 2.23 24T352 (3.52 W/lb maximum 24 55). Typical magnetization curves based on Epstein samples; half parallel and half cross grain at 60 Hz after annealing.

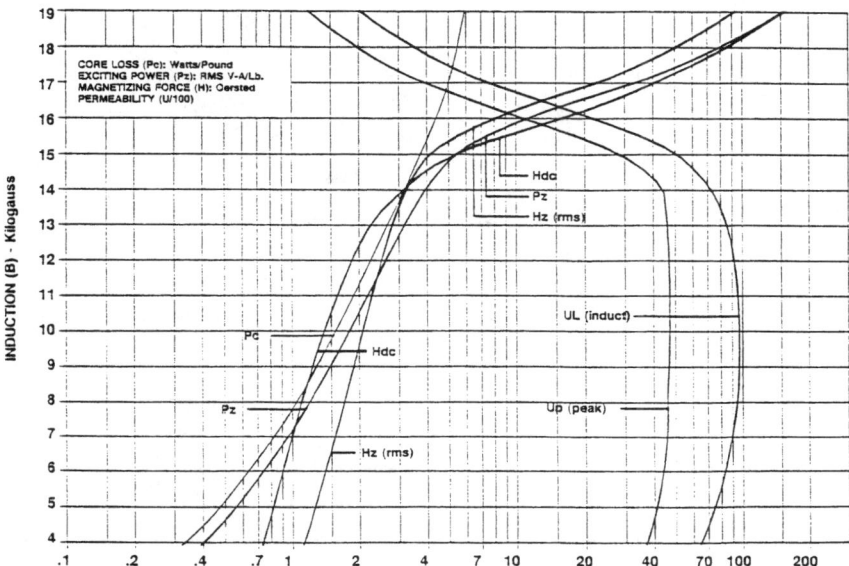

FIGURE 2.24 24T420 (4.20 W/lb maximum 24 56). Typical magnetization curves based on Epstein samples; half parallel and half cross grain at 60 Hz after annealing.

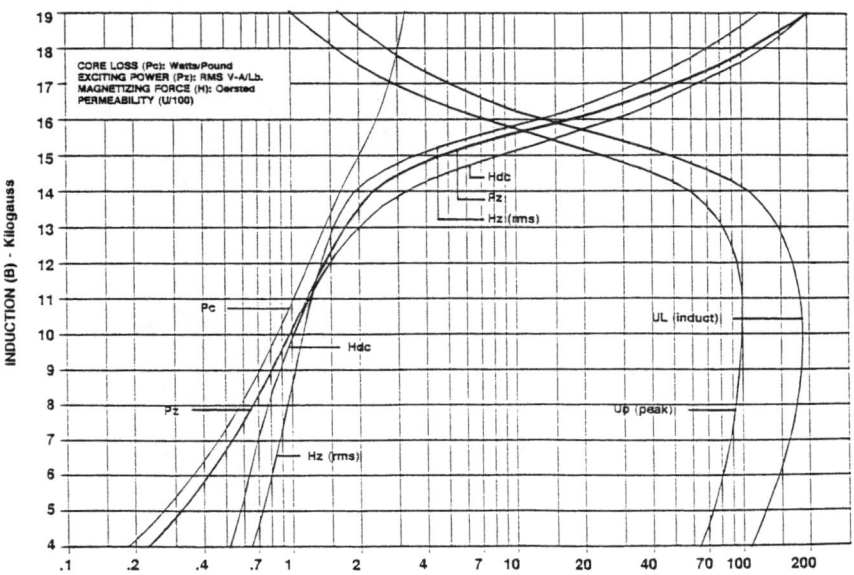

FIGURE 2.25 26T214 (2.14 W/lb maximum 26 50). Typical magnetization curves based on Epstein samples; half parallel and half cross grain at 60 Hz after annealing.

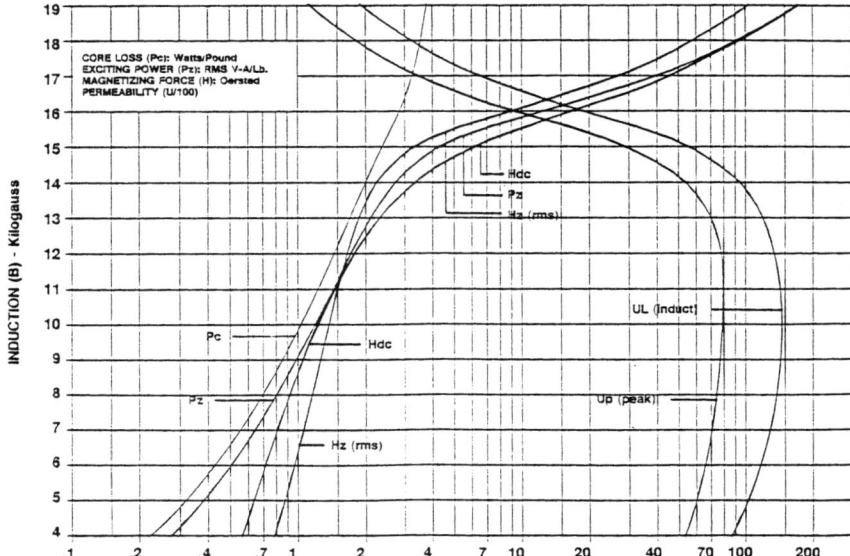

FIGURE 2.26 26T265 (2.65 W/lb maximum 26 55). Typical magnetization curves based on Epstein samples; half parallel and half cross grain at 60 Hz after annealing.

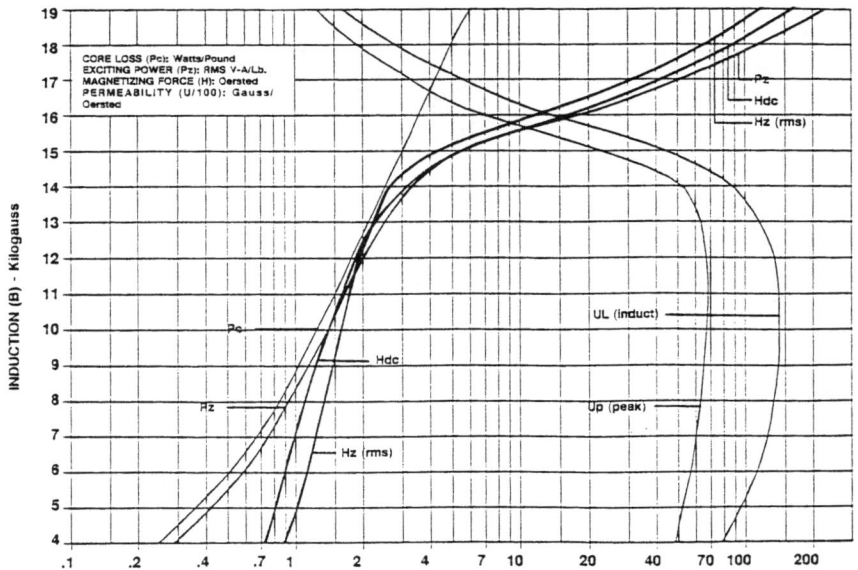

FIGURE 2.27 26T330 (3.30 W/lb maximum 26 56). Typical magnetization curves based on Epstein samples; half parallel and half cross grain at 60 Hz after annealing.

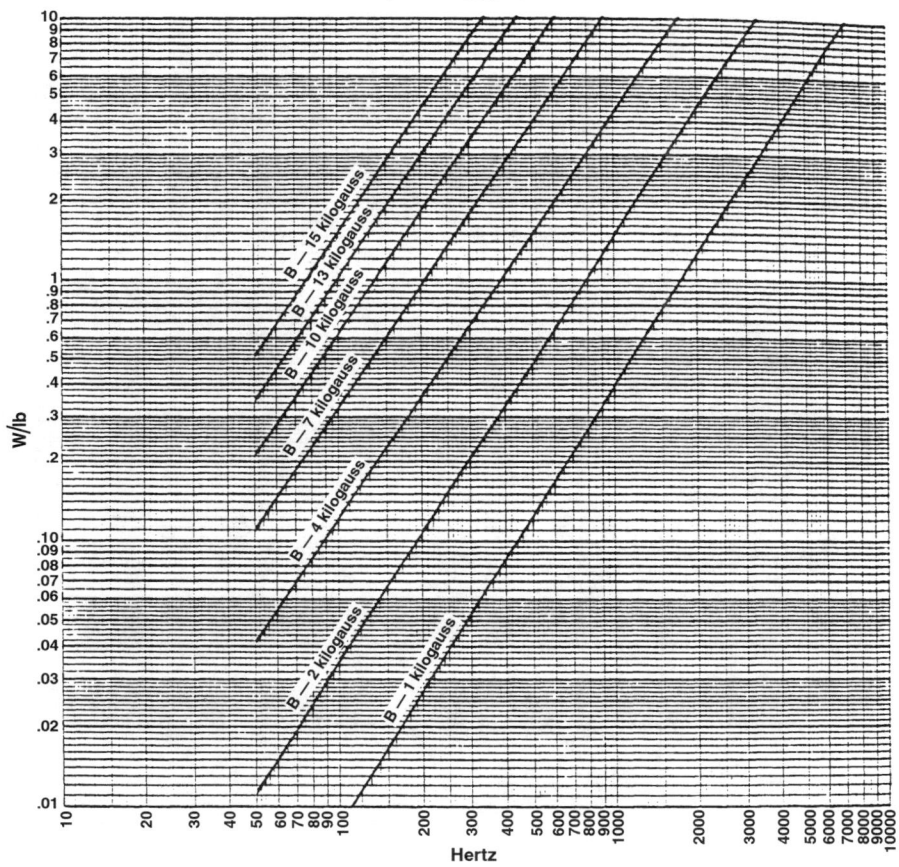

FIGURE 2.28 Core loss versus frequency for 29G066 (29 06).

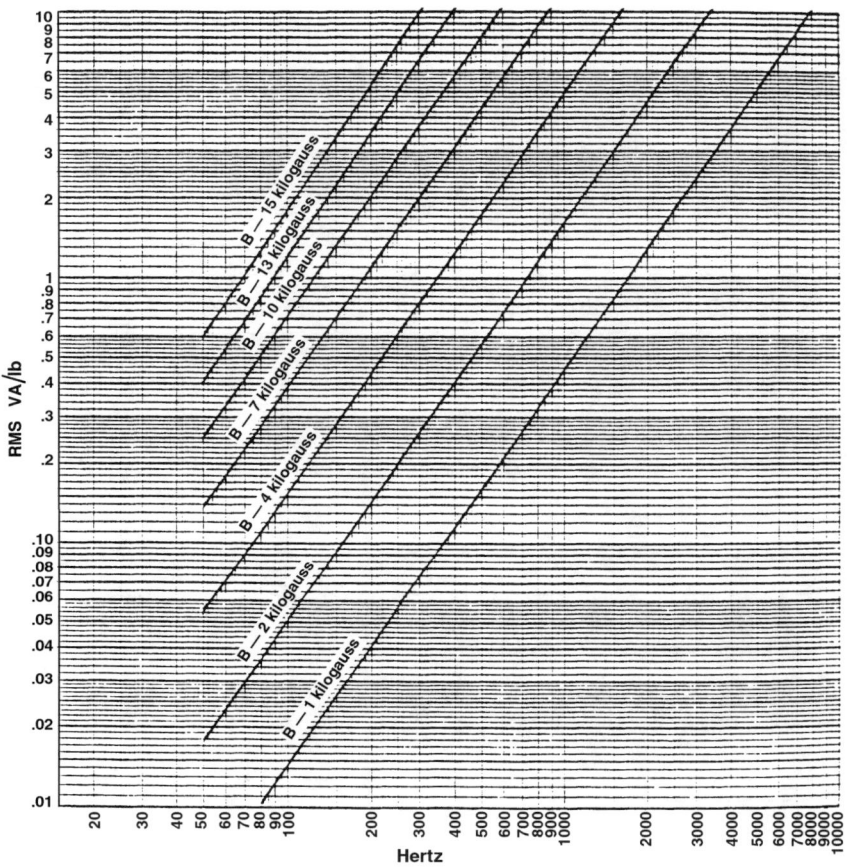

FIGURE 2.29 Exciting power versus frequency for 29G066 (29 06); 100 percent parallel grain.

MATERIALS

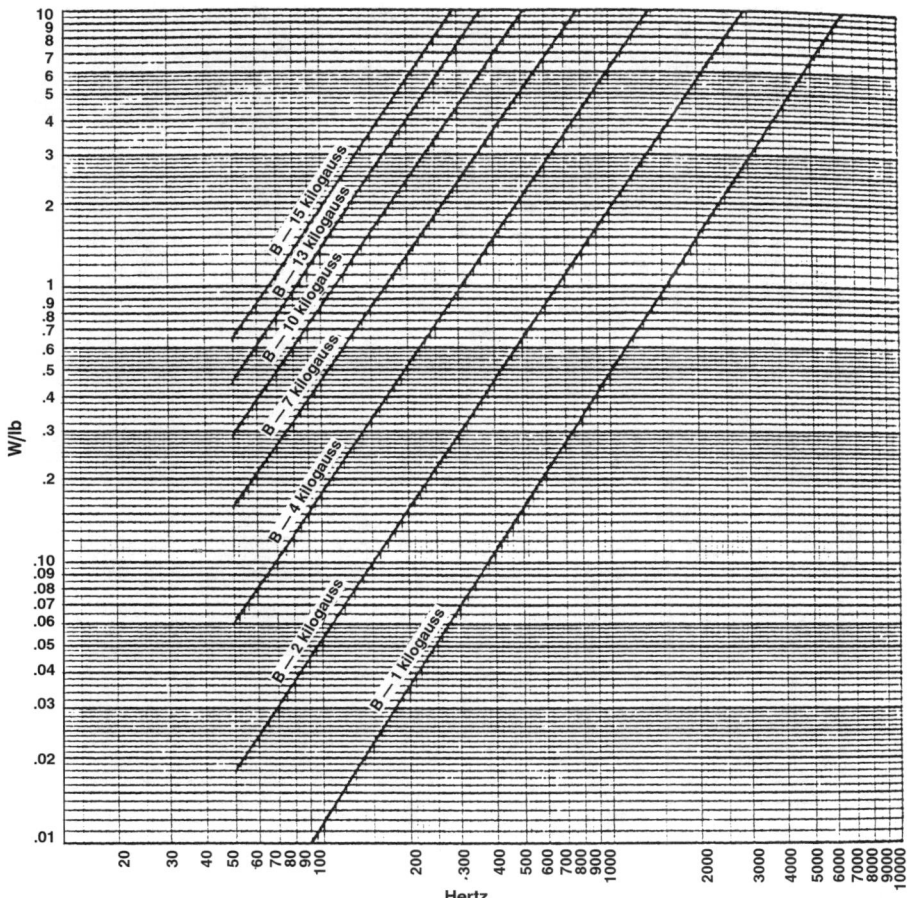

FIGURE 2.30 Core loss versus frequency for 29G066 (29 06); 75 percent parallel grain and 25 percent cross grain.

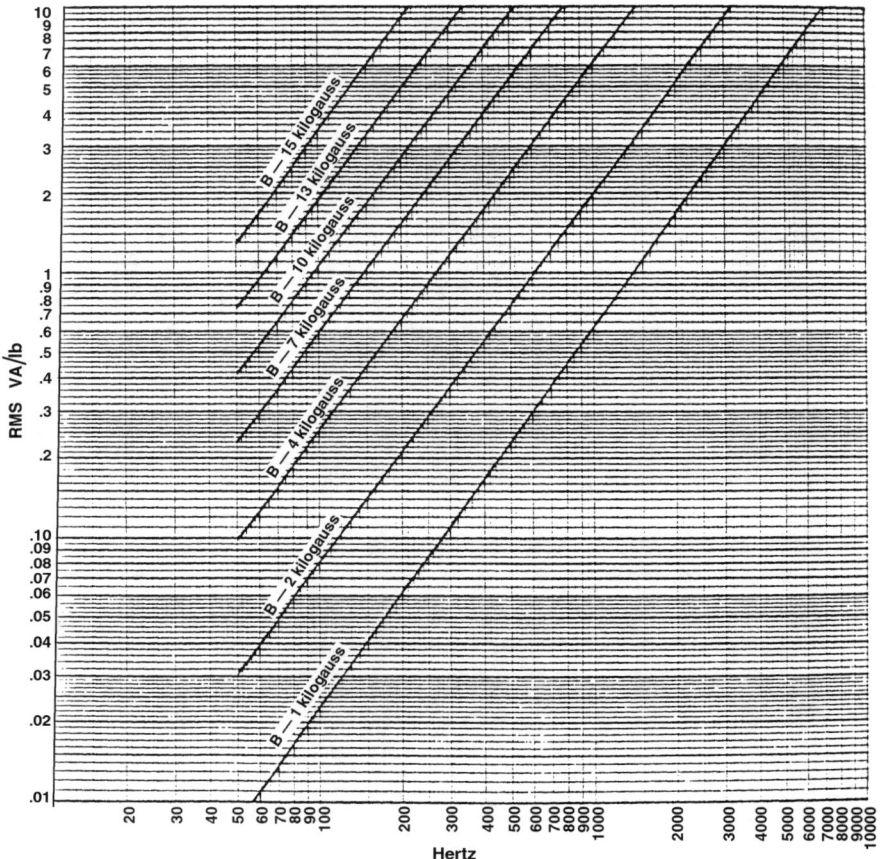

FIGURE 2.31 Exciting power versus frequency for 29G066 (29 06); 75 percent parallel grain and 25 percent cross grain.

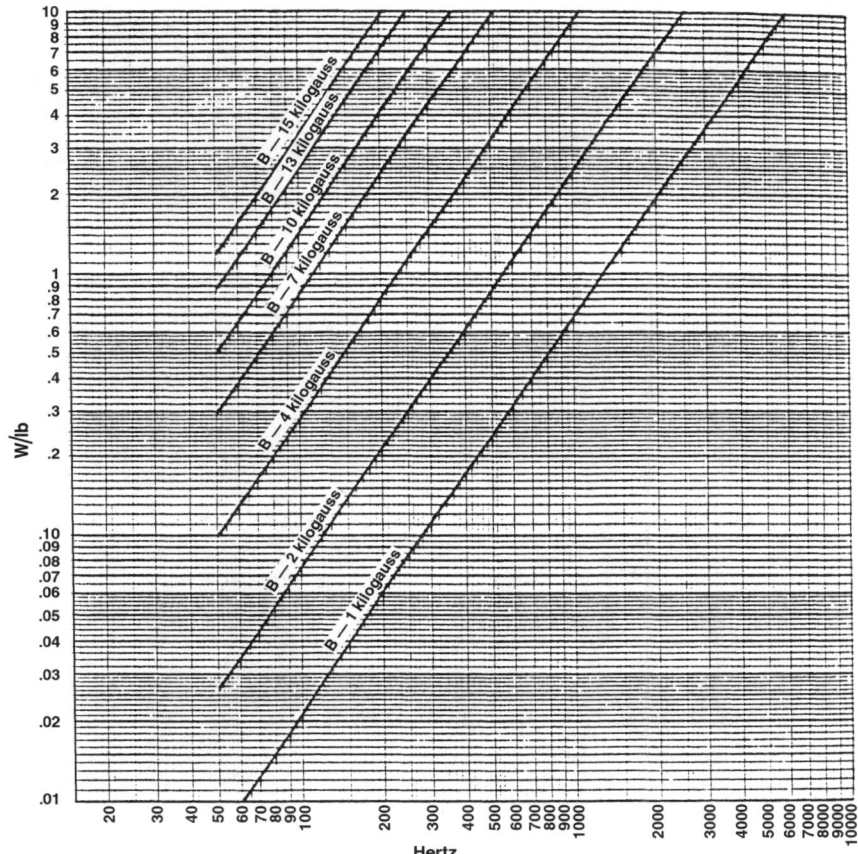

FIGURE 2.32 Core loss versus frequency for 26N174 (26 19); half parallel grain and half cross grain.

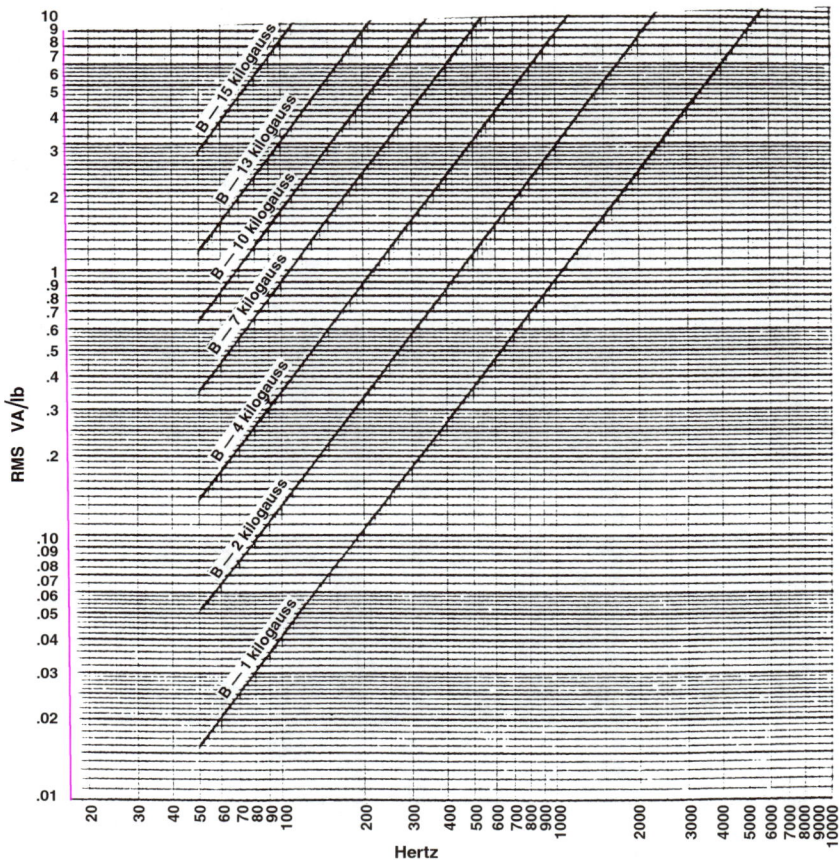

FIGURE 2.33 Exciting power versus frequency for 29G066 (29 06); half parallel grain and half cross grain.

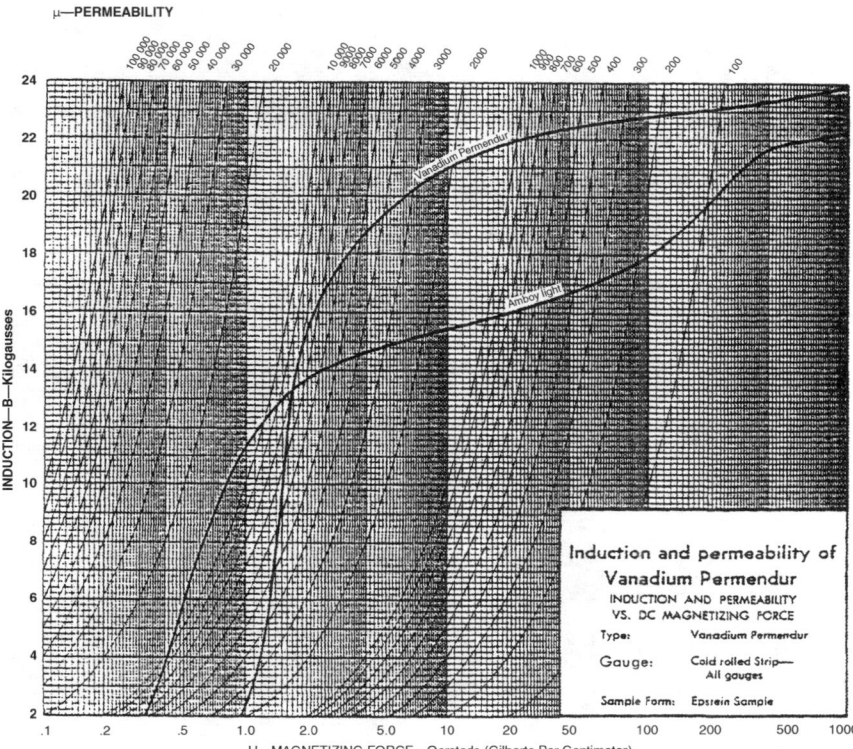

FIGURE 2.34 Induction and permeability of vanadium permendur.

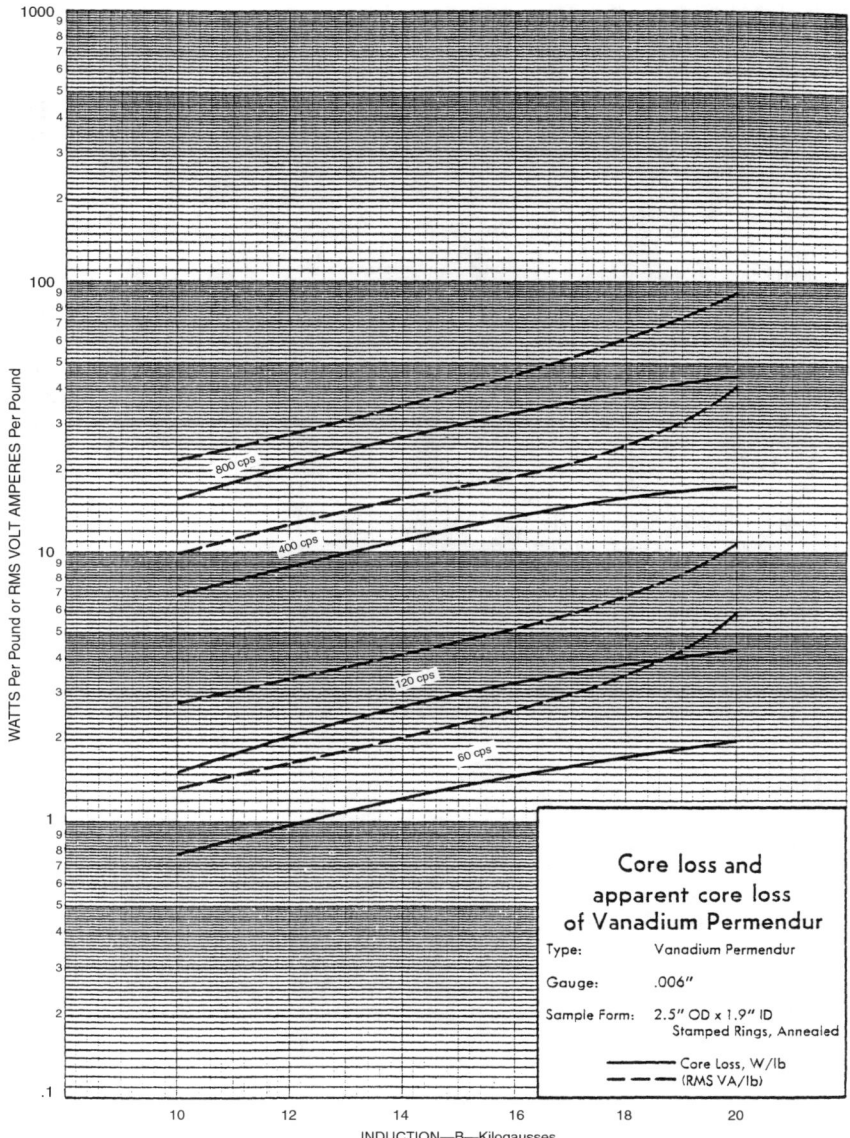

FIGURE 2.35 Core loss and apparent core loss of 0.006-in vanadium permendur.

MATERIALS

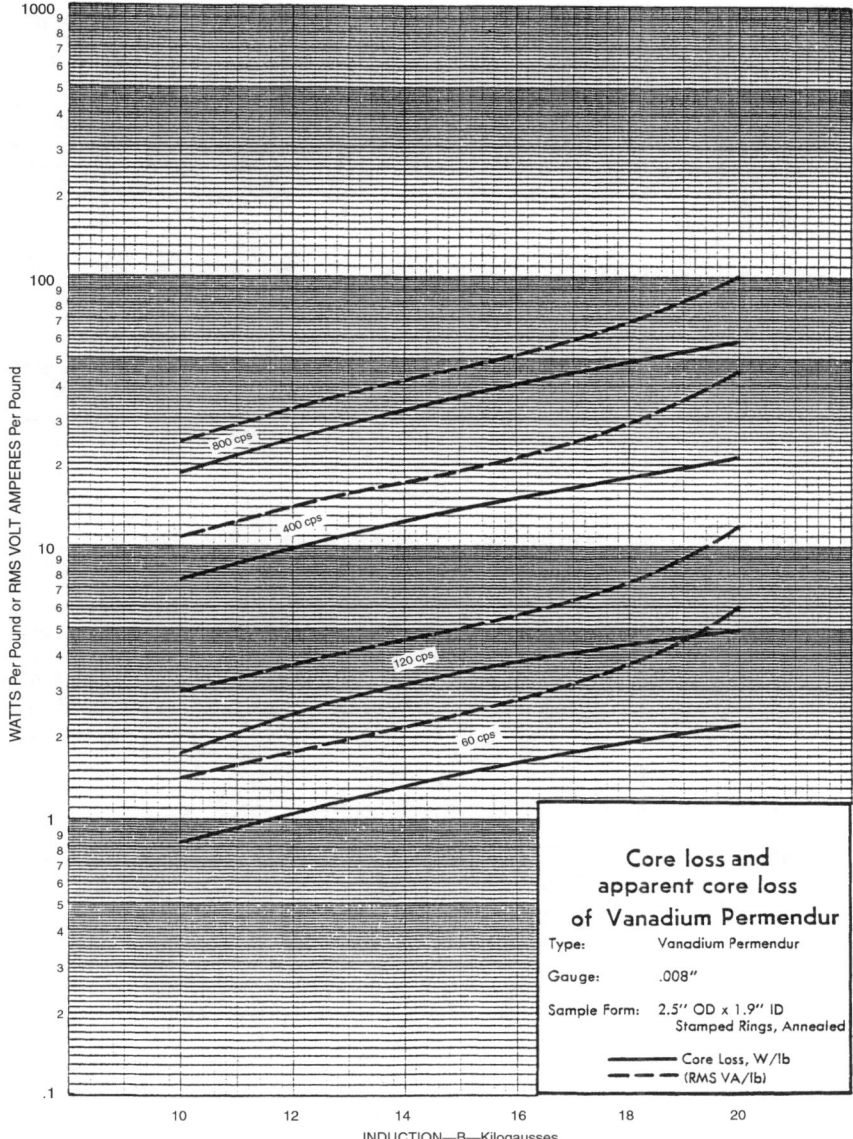

FIGURE 2.36 Core loss and apparent core loss of 0.008-in vanadium permendur.

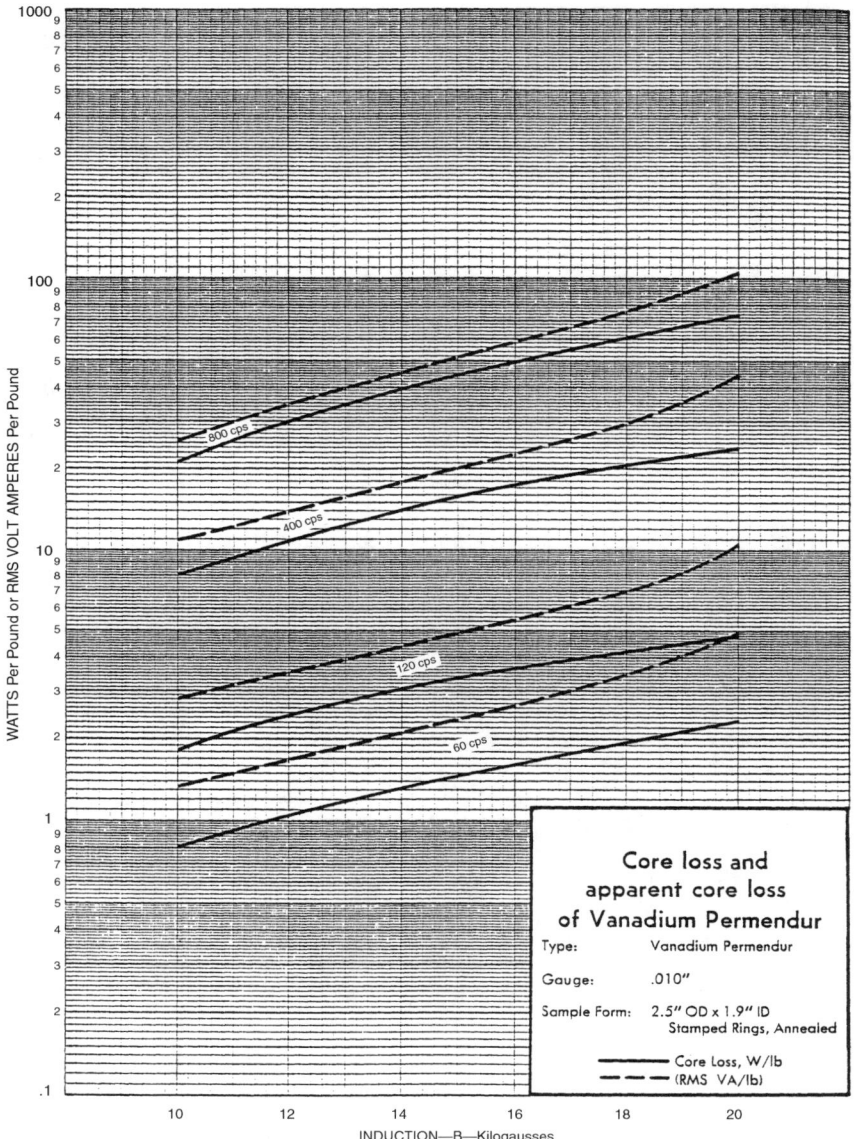

FIGURE 2.37 Core loss and apparent core loss of 0.010-in vanadium permendur.

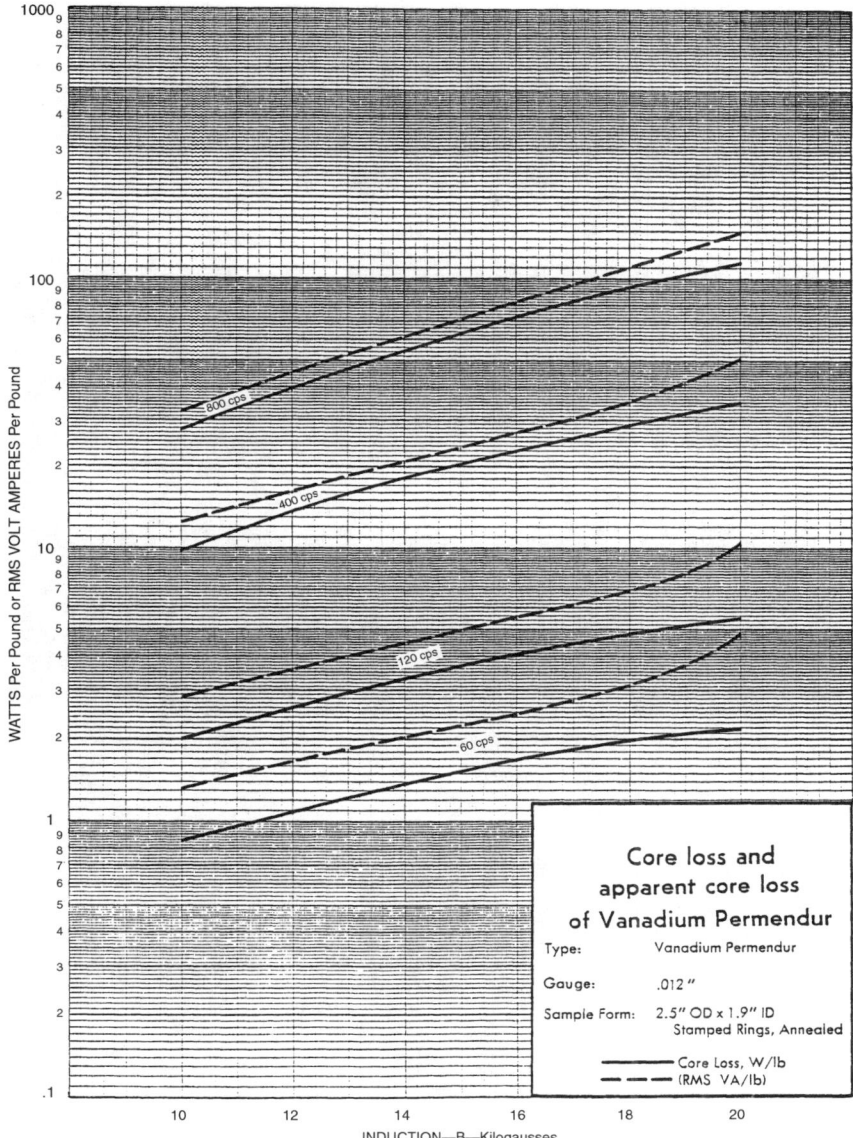

FIGURE 2.38 Core loss and apparent core loss of 0.012-in vanadium permendur.

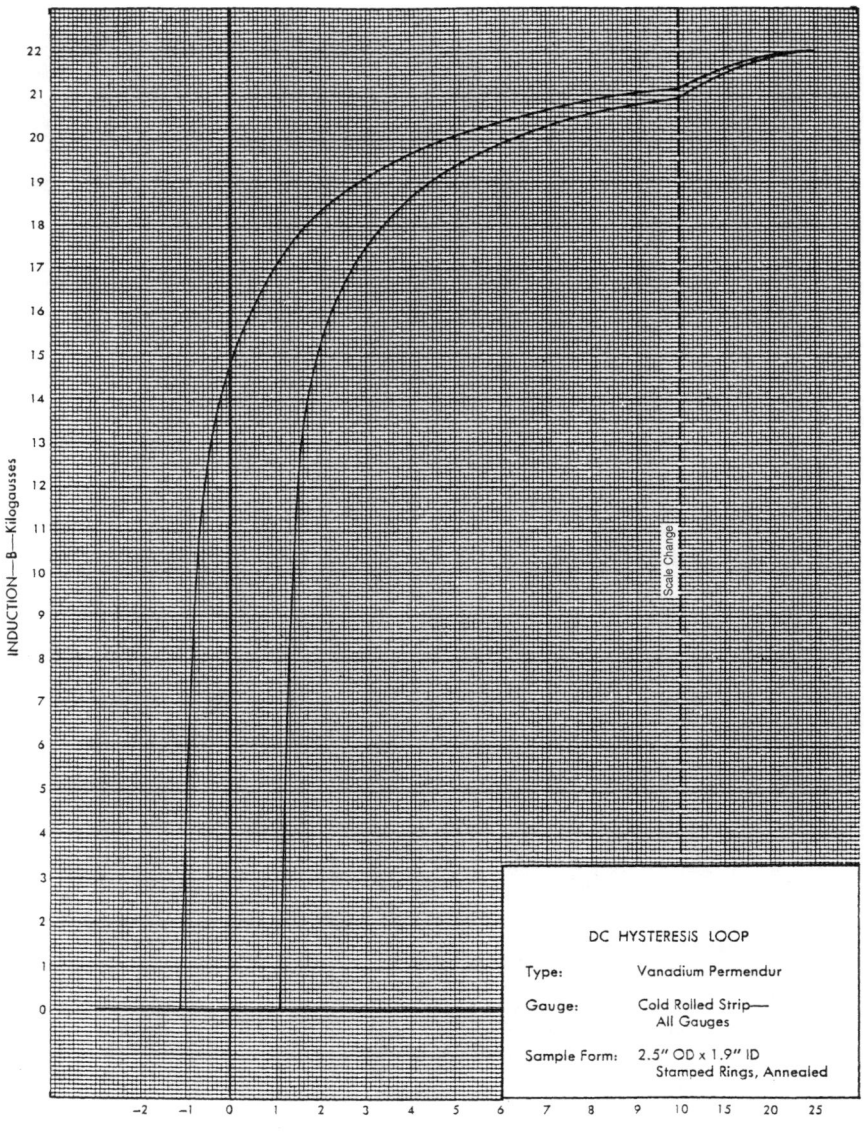

FIGURE 2.39 DC hysteresis loop for vanadium permendur.

MATERIALS

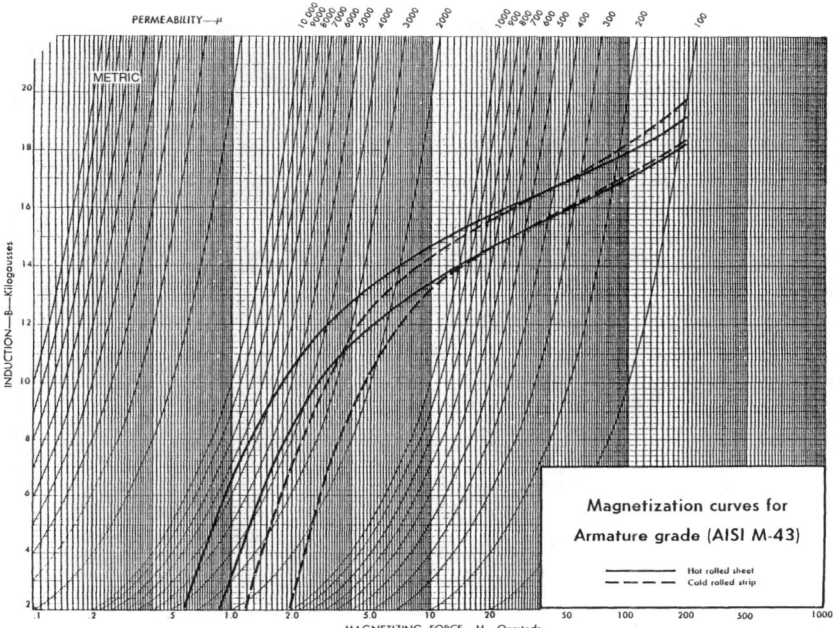

FIGURE 2.40 Magnetization curves for armature grade (AISI M-43), metric units.

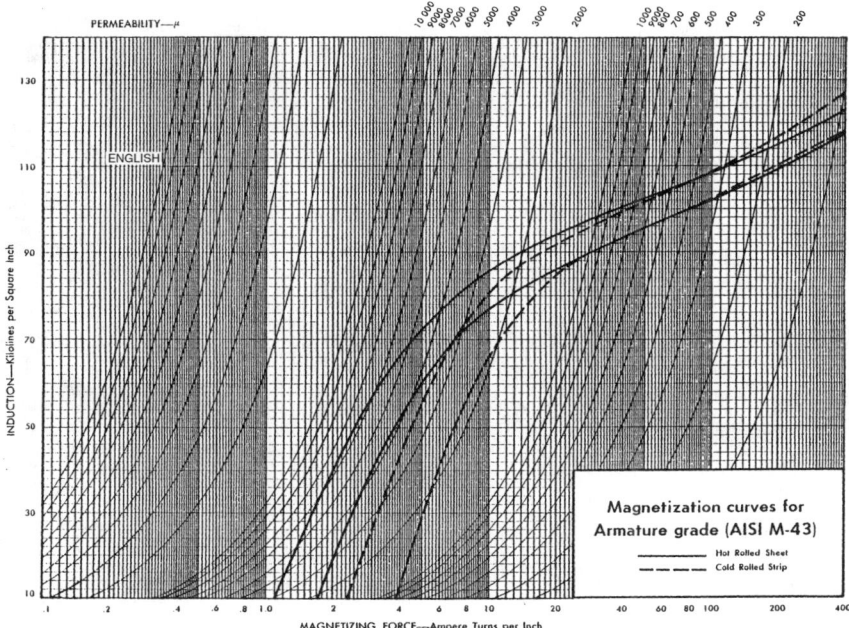

FIGURE 2.41 Magnetization curves for armature grade (AISI M-43), English units.

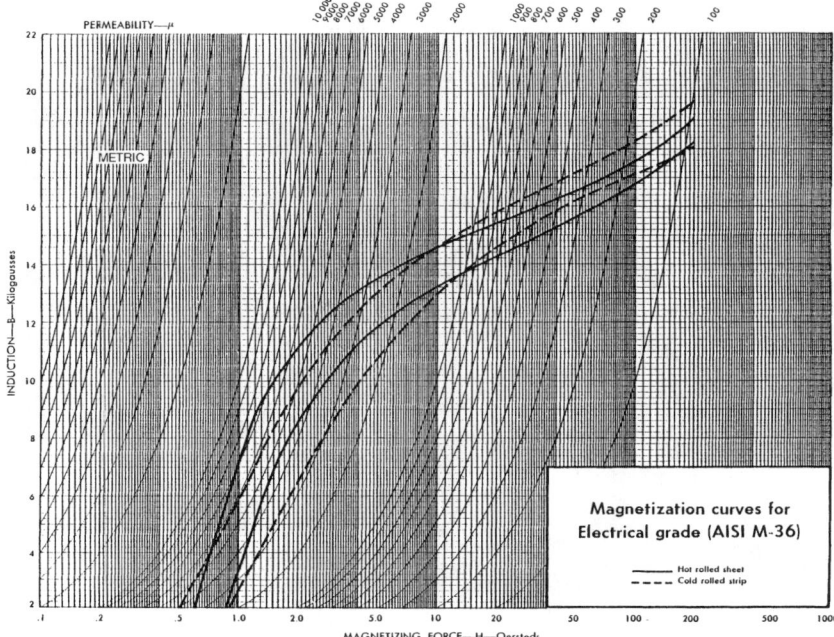

FIGURE 2.42 Magnetization curves for electrical grade (AISI M-36), metric units.

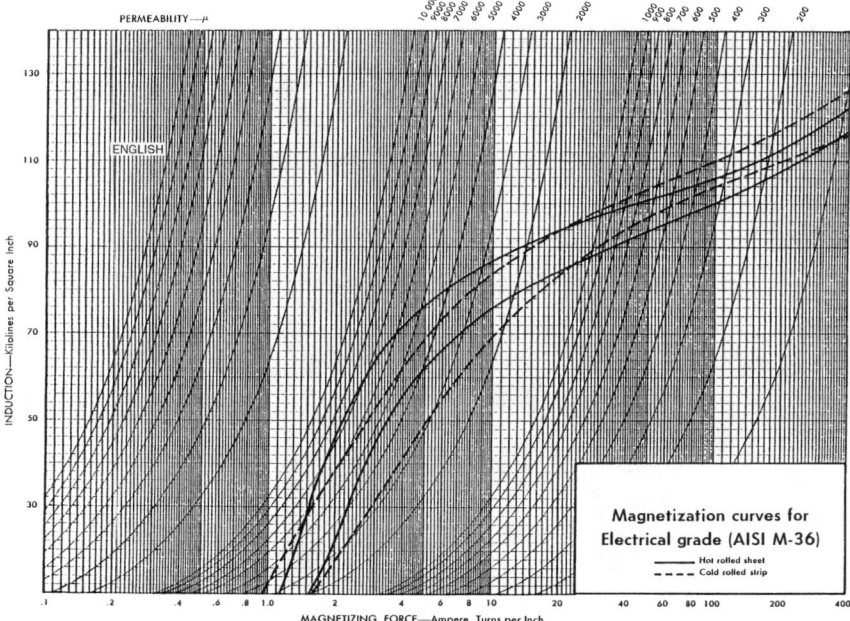

FIGURE 2.43 Magnetization curves for electrical grade (AISI M-36), English units.

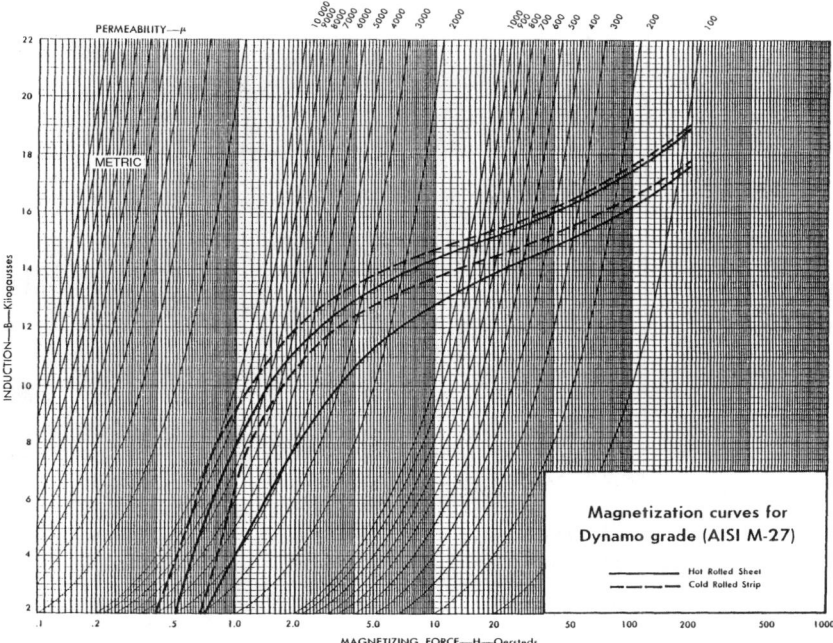

FIGURE 2.44 Magnetization curves for dynamo grade (AISI M-27), metric units.

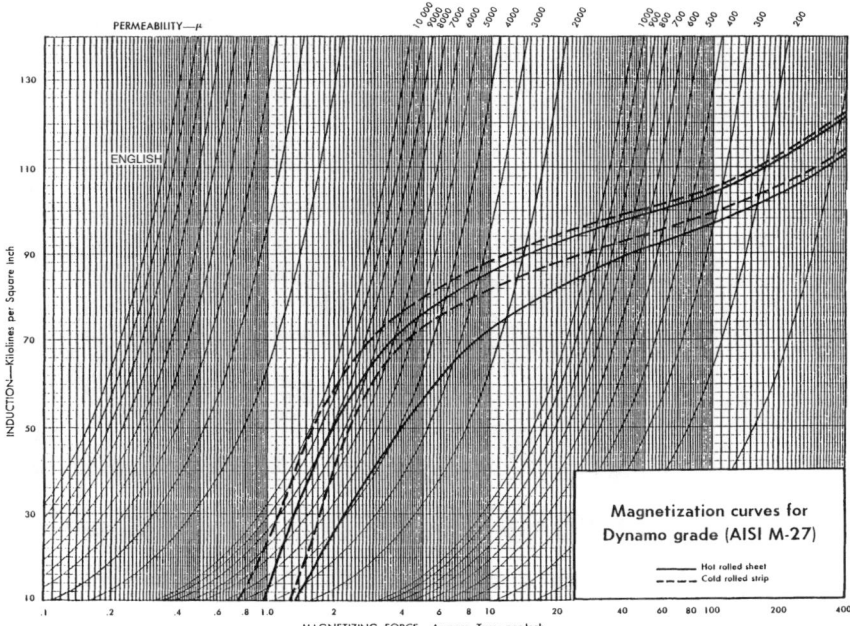

FIGURE 2.45 Magnetization curves for dynamo grade (AISI M-27), English units.

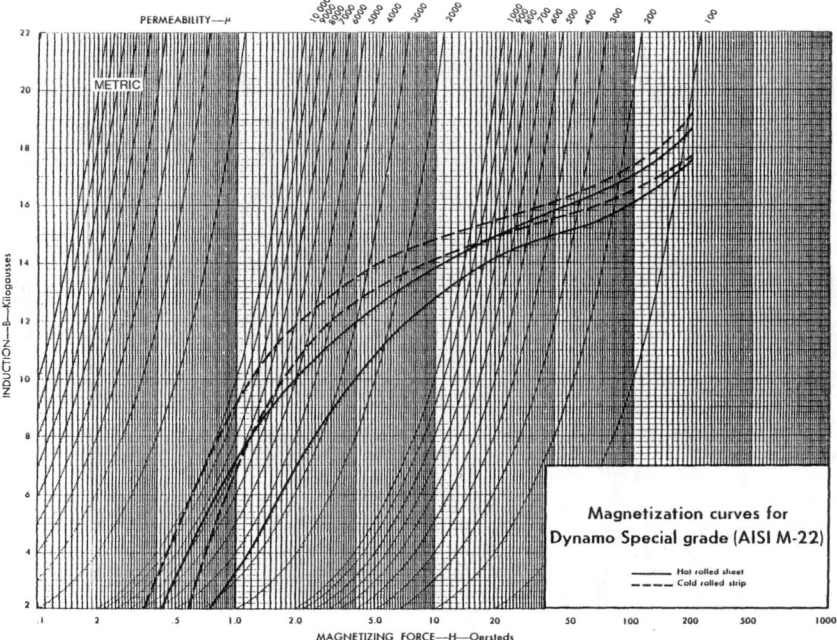

FIGURE 2.46 Magnetization curves for dynamo special grade (AISI M-22), metric units.

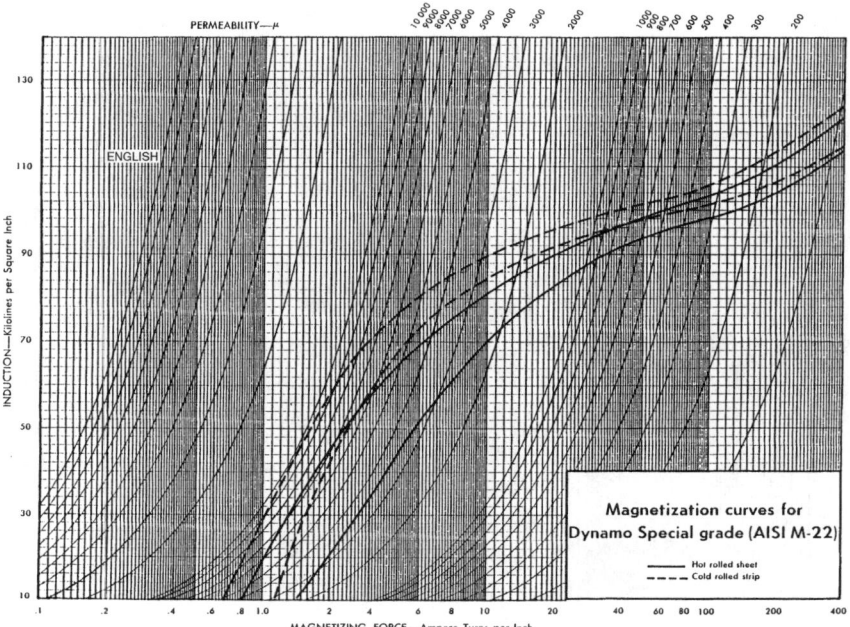

FIGURE 2.47 Magnetization curves for dynamo special grade (AISI M-22), English units.

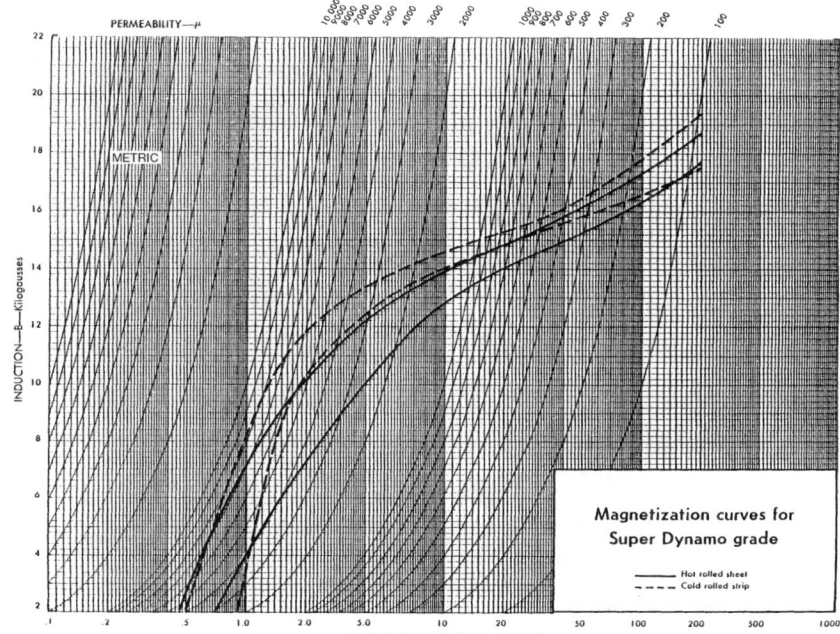

FIGURE 2.48 Magnetization curves for super dynamo grade, metric units.

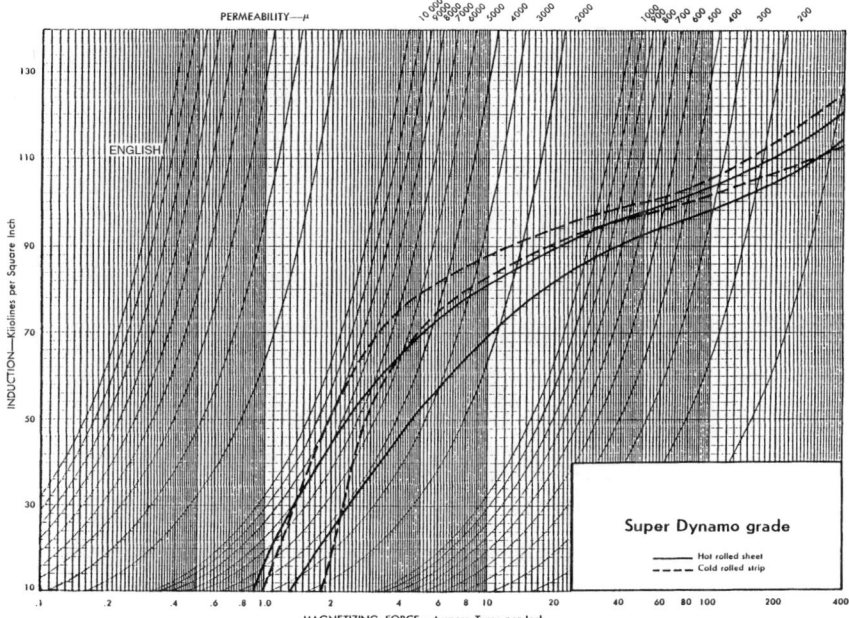

FIGURE 2.49 Magnetization curves for super dynamo grade, English units.

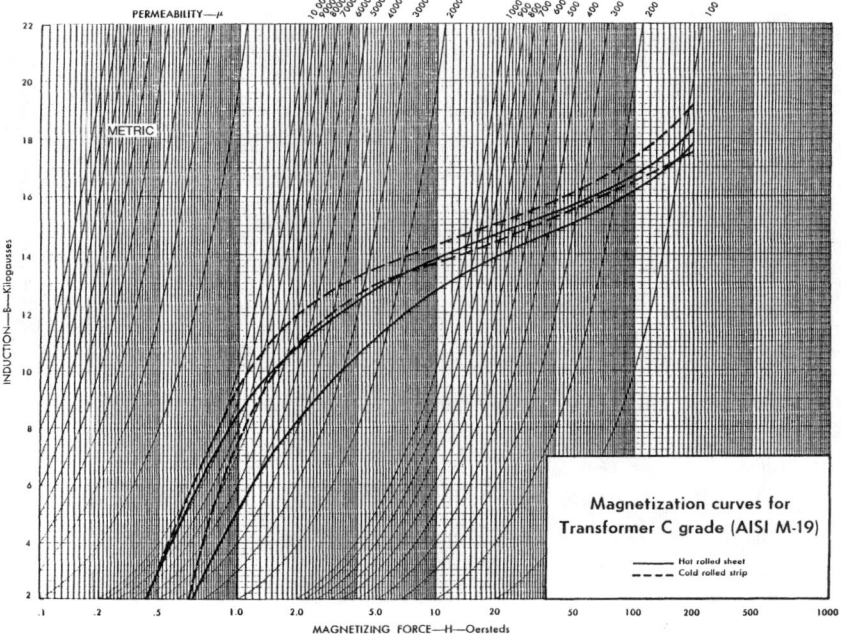

FIGURE 2.50 Magnetization curves for transformer C grade (AISI M-19), metric units.

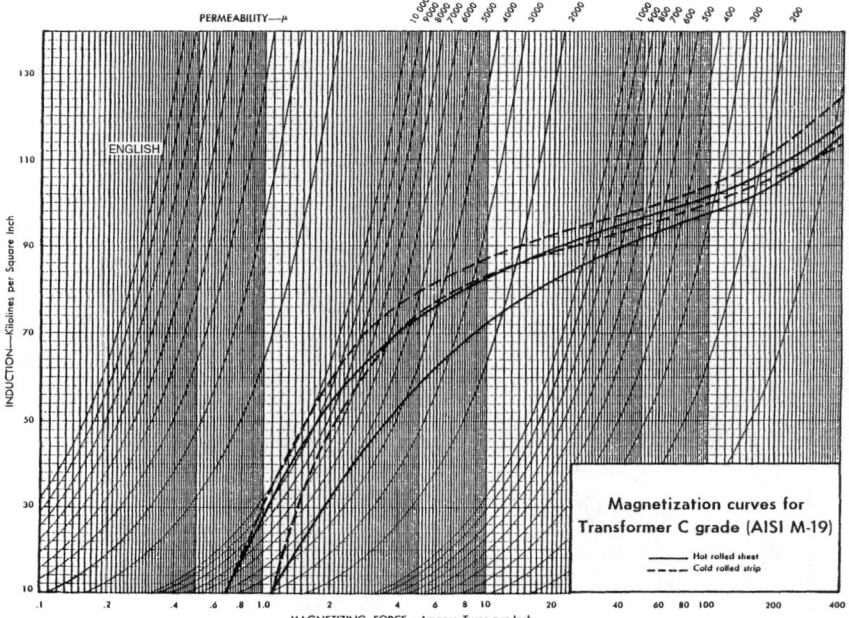

FIGURE 2.51 Magnetization curves for transformer C grade (AISI M-19), English units.

MATERIALS 2.37

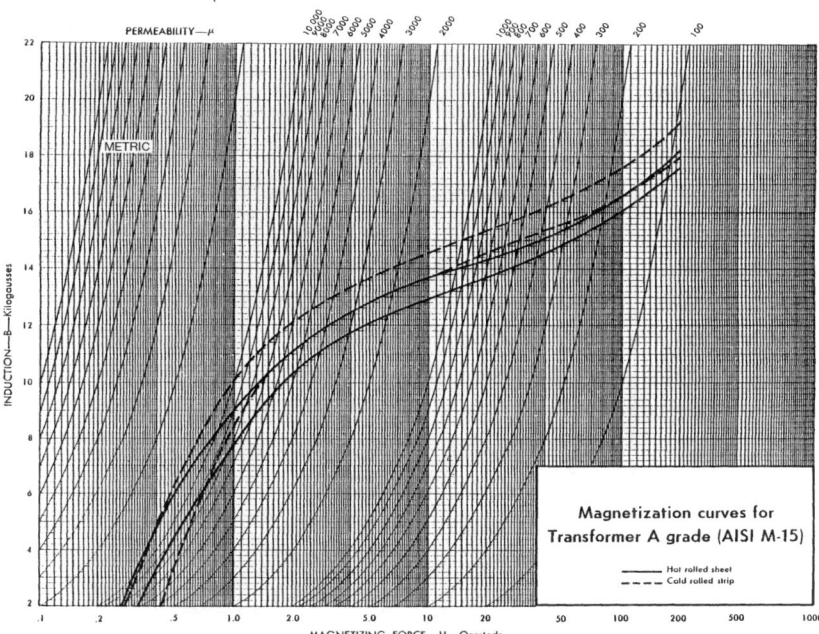

FIGURE 2.52 Magnetization curves for transformer A grade (AISI M-15), metric units.

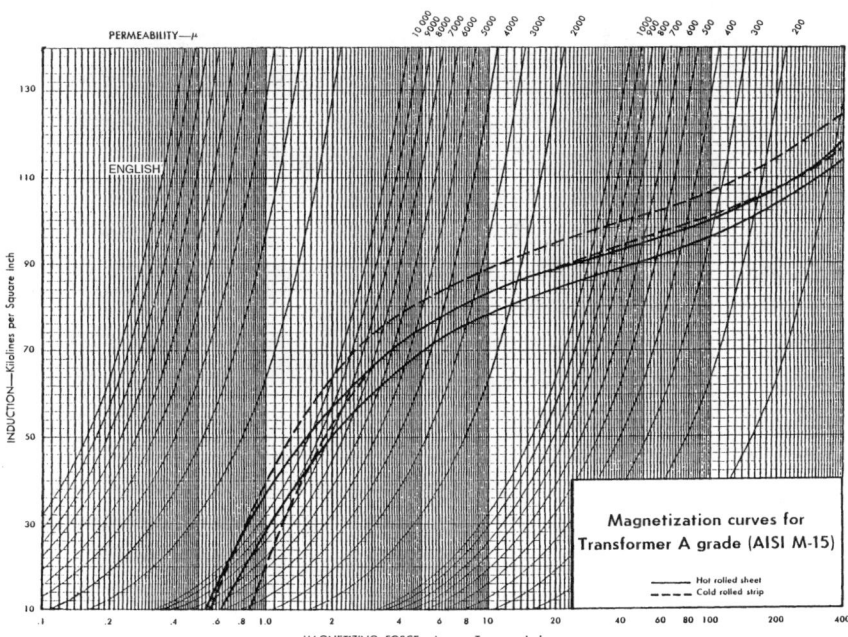

FIGURE 2.53 Magnetization curves for transformer A grade (AISI M-15), English units.

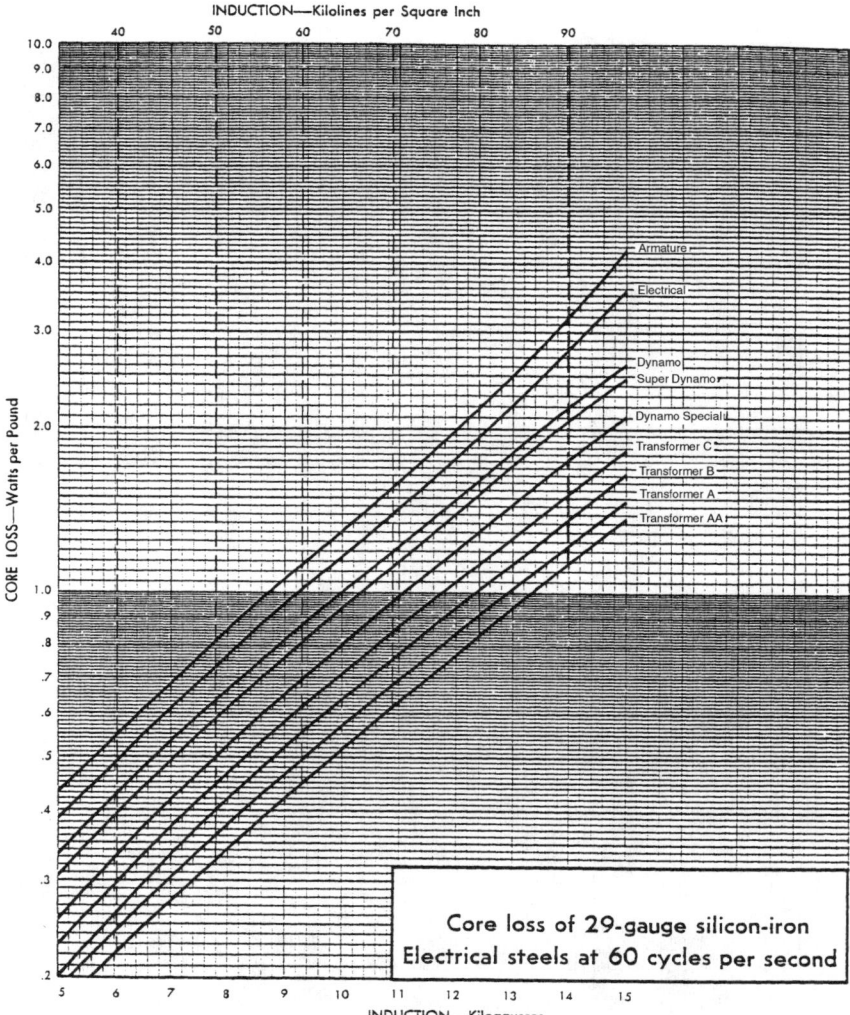

FIGURE 2.54 Core loss of 29-gauge silicon-iron electrical steels at 60 cps.

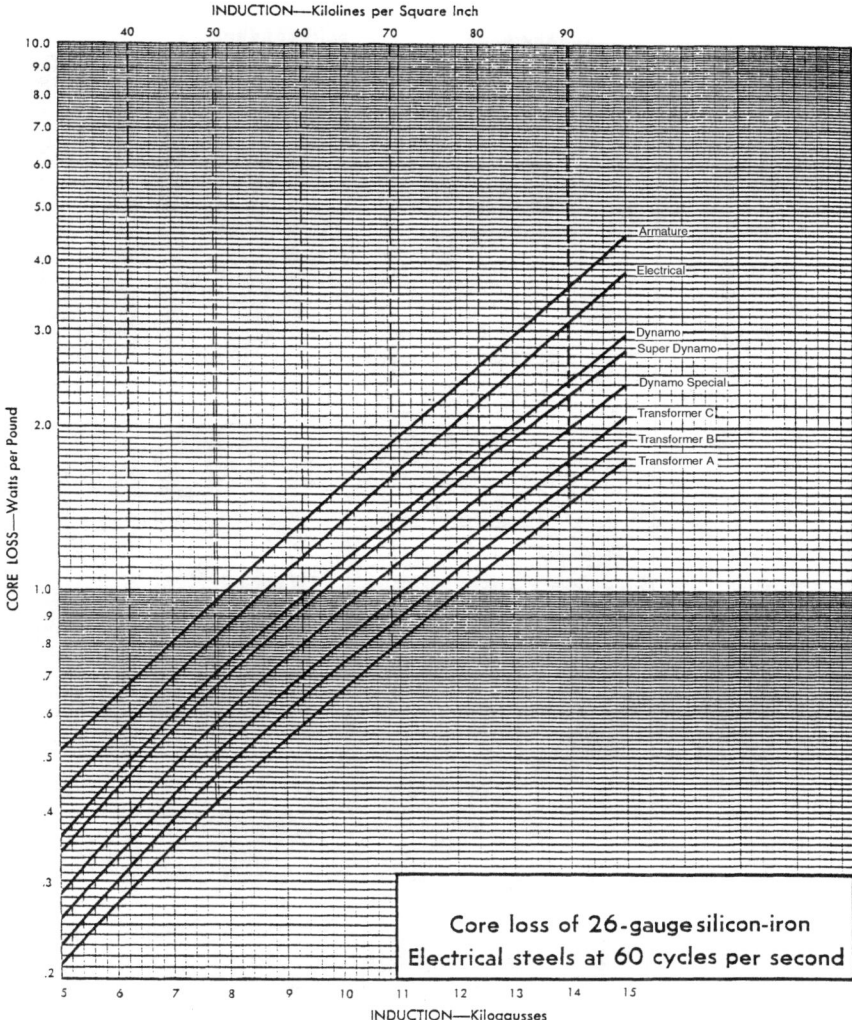

FIGURE 2.55 Core loss of 26-gauge silicon-iron electrical steels at 60 cps.

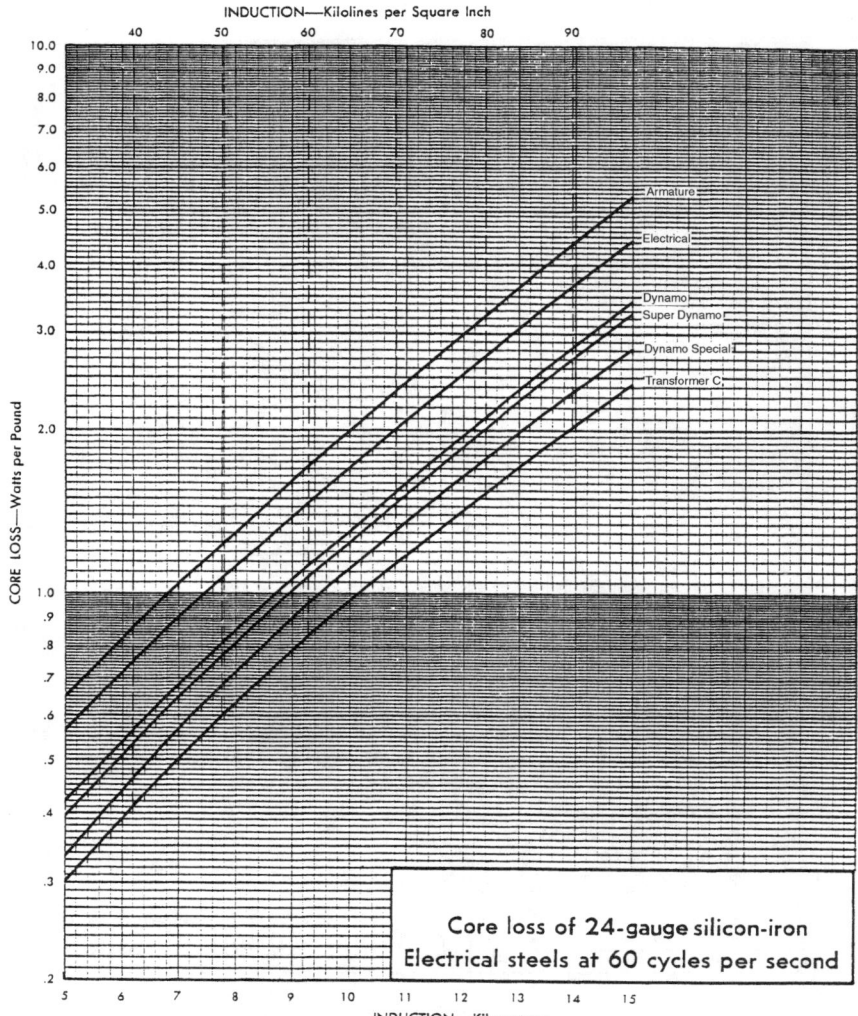

FIGURE 2.56 Core loss of 24-gauge silicon-iron electrical steels at 60 cps.

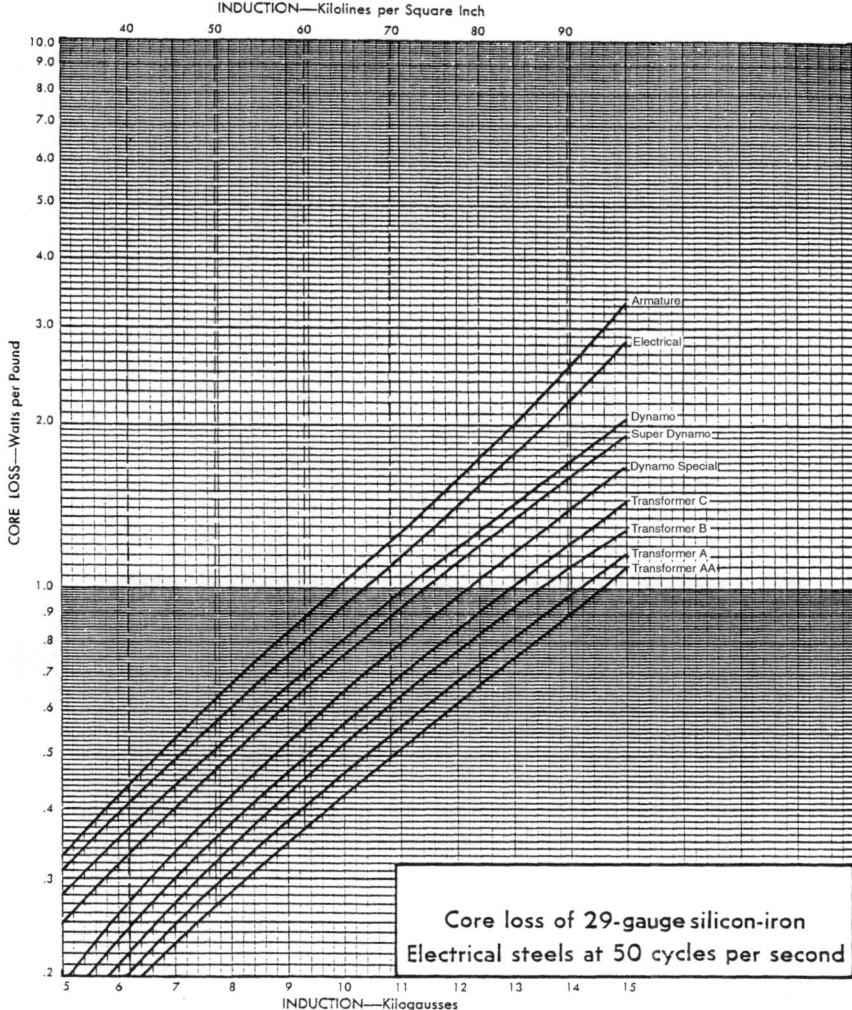

FIGURE 2.57 Core loss of 29-gauge silicon-iron electrical steels at 50 cps.

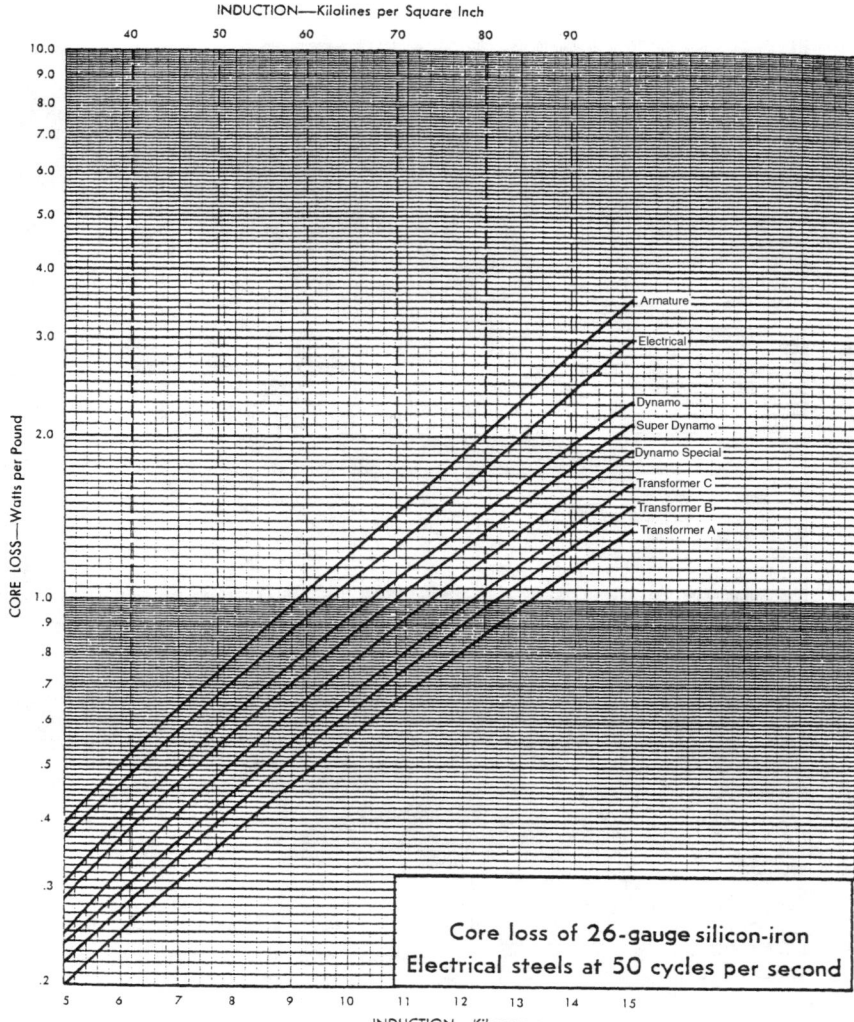

FIGURE 2.58 Core loss of 26-gauge silicon-iron electrical steels at 50 cps.

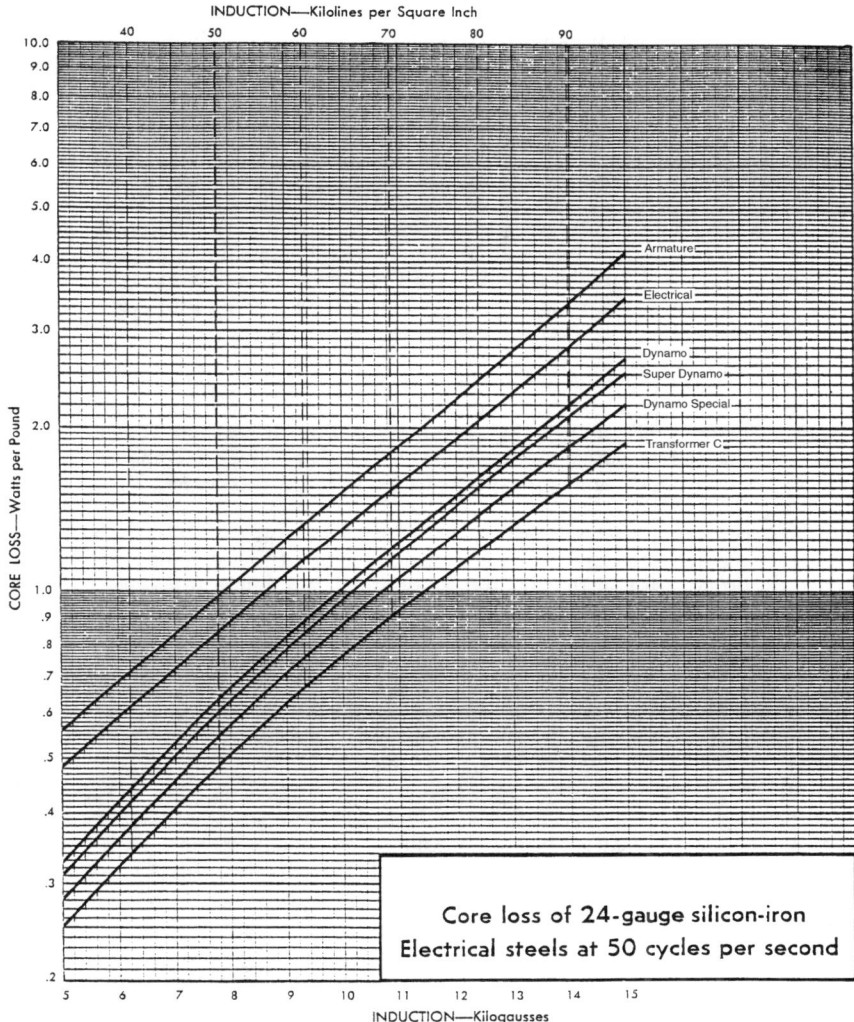

FIGURE 2.59 Core loss of 24-gauge silicon-iron electrical steels at 50 cps.

can be an appreciable percentage of the lamination teeth cross section. Because the teeth carry a very high flux density, punching damage can severely reduce small motor efficiency. The annealing of hot band coils is done in the producing steel mill on high-quality lamination sheet, primarily to enhance permeability.

2.3.1 Stator Lamination Annealing

Semiprocessed lamination sheet is received from the producing mill in the heavily temper-rolled condition. This condition enhances the punchability of the sheet and provides energy for the metallurgical process of grain growth that takes place during the annealing treatment. Annealing of the laminations is done for several reasons. Among them are the following.

Cleaning. Punched laminations carry some of the punching lubricant on their surfaces. This can be a water-based or a petroleum-based lubricant. It must be removed before the laminations enter the high-temperature zone of the annealing furnace to avoid sticking and carburization problems. This is done by preheating the laminations in an air or open-flame atmosphere to 260 to 427°C (500 to 800°F).

Carbon Control. Carbon in solution in steel can form iron carbides during mill processing, annealing, and electromagnetic device service. These carbides have several effects on properties—all detrimental. They affect metallurgical processing in the producing mill, degrading permeability and, to some extent, core loss. They pin grain boundaries during annealing, slowing grain growth. They pin magnetic domain walls in devices, inhibiting magnetization and thus increasing core losses and magnetizing current. If the carbides precipitate during device use, the process is called *aging*.

Because of these problems, the amount of carbon is kept as low as is practical during mill processing. The best lamination steels are produced to carbon contents of less than 50 ppm. Steels of lesser quality can be produced with up to 600-ppm carbon, but in the United States, 400 ppm is presently a practical upper limit. Laminated cores cannot run efficiently with these high carbon contents, so the carbon is removed by decarburization during annealing. The annealing atmosphere contains water vapor and carbon dioxide, which react with carbon in the steel to form carbon monoxide. The carbon monoxide is removed as a gas from the furnace. This process works well for low-alloy steels, but for steels with appreciable amounts of silicon and aluminum, the same water vapor and carbon dioxide provide oxygen that diffuses into the steel, forming subsurface silicates and aluminates. These subsurface oxides impede magnetic domain wall motion, lowering permeability and raising core loss.

Grain Growth. The grain diameter that minimizes losses in laminations driven at common power frequencies is 80 to 180 μm. As the driving frequency increases, this diameter will decrease. Presently, the temper-rolling percentage and the annealing time and temperature are designed to achieve grain diameters of 80 to 180 μm.

Coating. Laminations punched from semiprocessed steel are uncoated, while those punched from fully processed sheet are typically coated at the steel mill with a *core plate* coating. This coating insulates laminations from each other to reduce interlamination eddy currents, protects the steel from rust, reduces contact between laminations from burrs, and reduces die wear by acting as a lubricant during stamping. The semiprocessed steel laminations are also improved by a coating, but economics precludes coating them at the steel mill. Instead, they are coated at the end of the annealing treatment when the laminations are cooling from 566°C to about 260°C (1050°F to about 500°F). The moisture content of the annealing atmosphere is controlled to form a surface oxide coating of magnetite. This oxide of iron is very adherent and has a reasonably high insulating value. Therefore, it can be used for the same purposes as the relatively expensive core plate coating on the fully processed steel laminations. This magnetite coating is referred to as a *blue coating* or *bluing* because blue to blue-gray is its predominant color.

2.3.2 Rotor Lamination Annealing

Sometimes rotors are annealed with the stators, but often they are only given a rotor blue anneal. This is similar to the end of the stator anneal, mentioned previously. The rotors are heated to about 371°C (700°F) in a steam-containing atmosphere to form a magnetite oxide on their surface. They are then die-cast with aluminum to form conductor bars and end rings. The magnetite oxide prevents adherence of the aluminum to the steel laminations and thereby reduces rotor losses.

2.3.3 Annealing Furnaces and Cycles

The actual annealing can be done in a batch or a continuous annealing furnace. Both are used commercially, but most large production shops prefer the continuous furnace because it fits the scheduling of the other operations in the shop. A typical annealing furnace is a multichambered roller hearth furnace. The first chamber is a burn-off oven. It heats the laminations to less than 427°C (800°F) and evaporates the punching lubricant. Usually the transfer between the burn-off and the high-heat chamber is in a hooded open space that removes smoke that may escape from the burn-off oven. The steel enters the high-heat chamber through a flame curtain or a purgeable vestibule, or both. The laminations are heated to the soak temperature in the high-heat chamber, held at that temperature if necessary, and then slowly cooled to bluing temperature [less than 566°C (1050°F)]. Bluing can take place in the same furnace chamber that houses the high-heat zone or in a separate bluing chamber. The need for adequate moisture in the bluing zone is the controlling factor that will determine whether a separate chamber is necessary. If a steam blue is chosen, a purgeable transfer vestibule between the high-heat chamber and the bluing chamber is recommended.

The soak temperature and time in the high-heat chamber are functions of the steel composition and the amount of temper mill extension in the punched sheet. For low-alloy sheet with 0.02 percent carbon (200 ppm) and about 5 percent temper mill extension, a typical high-temperature annealing cycle is 788°C (1450°F) for 2 h in an atmosphere containing a hydrogen-to-water ratio of two to three. If the atmosphere is an EXO gas type with at least 5 percent carbon dioxide, the soak time can be reduced to about 1 h.

Steels with 1.3 to 2 percent silicon plus aluminum should be annealed at a higher temperature because of the sluggish grain growth at 788°C (1450°F). A typical soak temperature for these steels is 816 to 843°C (1500 to 1550°F). These higher-alloy steels will have the best magnetic quality if they are purchased as ultra-low-carbon (<50 ppm carbon) sheet and are annealed in an atmosphere with a hydrogen-to-water ratio greater than 20.

Very high alloy electrical sheet with alloy contents in excess of 2 percent should be annealed at temperatures around 899°C (1650°F). When the carbon content of the steel is less than 50 ppm, there is no need to hold the laminations at high temperature in the high-heat chamber. The soak time previously mentioned is for decarburization. The other metallurgical process that takes place at high temperature is grain growth. Grain growth is more sensitive to temperature than to time at that temperature, so the annealing cycle for ultra-low-carbon steels can be shortened by minimizing the time at a particular temperature.

2.3.4 Methods of Eliminating Annealing

Fully processed lamination steels are available as an alternative to annealing. Silicon and aluminum additions increase the resistivity of steel and are substitution elements that don't affect the regularity of the crystal structure. They are used for electrical steels to improve core loss.

2.4 CORE LOSS*

2.4.1 Determination of Hysteresis and Eddy Current Coefficient and Hysteresis Exponent Flux Density and Frequency Variable

In using the classical equations to calculate hysteresis and eddy current losses, it is necessary to know the values of the coefficients and hysteresis exponent for the gauge and grade involved. These can be obtained by simultaneous solution of three equations, with the flux density and frequency as variables.

The hysteresis loss equation is:

$$P_h = \left(\frac{W}{D10^7} \right) f_\eta B^x \, \text{W} \qquad (2.6)$$

and the eddy loss equation is:

$$P_e = \left(\frac{1.6545 d^2 W}{D10^{16}} \right) f^2 B^2 \lambda \, \text{W} \qquad (2.7)$$

with the total loss:

$$P = P_h + P_e \, \text{W} \qquad (2.8)$$

where W = weight of core, g
 f = frequency, Hz
 d = thickness of core lamination, cm
 B = flux density, G
 D = density of core material
 η = hysteresis loss coefficient
 λ = eddy loss coefficient
 x = hysteresis loss exponent for B

---

*Section contributed by United States Steel Corporation.

Let
$$A = \frac{W}{D10^7} \quad \text{and} \quad C = \frac{1.645 d^2 W}{D 10^{16}} \qquad (2.9)$$

Thus, for two frequencies f_1 and f_2 and two flux densities B_1 and B_2, these three equations can be set up:

$$P_1 = A_\eta f_1 B_1^x + C f_1^2 B_1^2 \lambda \qquad (2.10)$$

$$P_2 = A_\eta f_2 B_2^x + C f_2^2 B_2^2 \lambda \qquad (2.11)$$

$$P_3 = A_\eta f_1 B_2^x + C f_1^2 B_2^2 \lambda \qquad (2.12)$$

Solving Eqs. (2.4) through (2.6) simultaneously with $a = f_2/f_1$,

$$x = \frac{\log\left[\dfrac{B_2^2(P_2 - a^2 P_3)}{B_2^2 a P_1(1-a) + B_1^2(P_2 - a P_3)}\right]}{\log(B_2/B_1)} \qquad (2.13)$$

Note: The flux density in Eq. (2.7) can be stated in kilogauss, to reduce the size of the number.

$$\lambda = \frac{P_2 B_1^x - a P_1 B_2^x}{C f_2^2 / a (a B_2^2 B_1^x - B_1^2 B_2^x)} \qquad (2.14)$$

$$\eta = \frac{P_3 - f_1^2 \lambda C B_2^2}{f_1 A B_2^x} \qquad (2.15)$$

For an example, choosing 26-gauge USS Dynamo grade, the core losses are as follows, as obtained from a standard Epstein sample:

$B_2 = 10{,}000$ G $\quad f_2 = 60$ Hz $\quad P_2 = 19.48$ W
$B_2 = 10{,}000$ G $\quad f_1 = 30$ Hz $\quad P_3 = 8.30$ W
$B_1 = 4000$ G $\quad f_1 = 30$ Hz $\quad P_1 = 1.80$ W

with other values as:

$W = 10{,}000$ g
$d = 0.4675$ cm
$a = 2.0$
$D = 7.5$

$$A = \frac{1.333}{10^4} \qquad C = \frac{4.78}{10^{10}} \qquad (2.16)$$

Substituting these values in Eqs. (2.7) through (2.9),

$$x = 1.61 \qquad \lambda = 3.365 * 10^4 \qquad \eta = 6.225 * 10^{-4}$$

Substituting these values in Eqs. (2.1) through (2.3),

$$P = \frac{8.3}{10^8} fB^{1.61} + \frac{1.61}{10^{11}} f^2 B^2 \text{ W} \qquad (2.17)$$

for a 22-lb standard Epstein sample.

Equation (2.10) checked the observed values over the flux density range of about 2000 to 12,000 G with a precision of 3 percent, and between 4000 and 10,000 G at 1 percent in the frequency range of 25 to 70 cycles.

These equations are limited in application to the flux range of about 2000 to 12,000 G. Above 12,000 G the hysteresis exponent increases in value, and it also changes below about 2000 G. It is suggested that the constants for these equations be determined for a range of flux density and frequency not exceeding the region for which calculations are desired—for instance, 8000 to 11,000 G, 50 and 60 Hz.

Note: This separation method and the equations apply only to essentially nonoriented material of uniform properties.

Hysteresis and Eddy Current Loss Separation Equations, Constant Flux Density and Variable Frequency. When it is desirable to obtain information concerning the hysteresis and eddy current loss components of the total core loss for one or more frequencies, the data required are the core losses at two frequencies for a given flux density.

Let P_1 = total core loss, W, at frequency f_1, Hz/s
P_2 = total core loss, W, at frequency f_2, Hz/s
(If f_2 is greater in value than f_1, P_2 will be greater in value than P_1.)
P_e = eddy current loss, W
P_h = hysteresis loss, W
a = ratio of f_2 to f_1
K_1 = eddy current loss constant
K_2 = hysteresis loss constant

The two equations used are:

$$P_1 = K_1 f_1^2 + K_2 f_1 \text{ W} \tag{2.18}$$

$$P_2 = K_1 f_2^2 + K_2 f_2 \text{ W} \tag{2.19}$$

From Eqs. (2.11) and (2.12),

$$K_1 = \frac{a(P_2 - aP_1)}{f_2^2(a-1)} \tag{2.20}$$

$$K_2 = \frac{a^2 P_1 - P_2}{f_2(a-1)} \tag{2.21}$$

The eddy loss, in watts, for frequency f_1 is:

$$P_e = K_1 f_1^2 = \frac{P_2 - aP_1}{a(a-1)} \text{ W} \tag{2.22}$$

The eddy loss, in watts, for frequency f_2 is:

$$P_e = K_1 f_2^2 = \frac{a(P_2 - aP_1)}{a-1} \text{ W} \tag{2.23}$$

The hysteresis loss, in watts, for frequency f_1 is:

$$P_h = K_2 f_1 = \frac{a^2 P_1 - P_2}{a(a-1)} \text{ W} \tag{2.24}$$

The hysteresis loss, in watts, for frequency f_2 is:

$$P_h = K_2 F_2 = \frac{a^2 P_1 - P_2}{(a-1)} \text{ W} \tag{2.25}$$

To calculate the total loss P_T at frequencies (f_n) between and in the vicinity of f_1 and f_2,

$$P_T = P_e + P_h = K_1 f_n^2 + K_2 f_n \text{ W} \tag{2.26}$$

with K_1 and K_2 values obtained from Eqs. (2.13) and (2.14).

As 60 Hz is the principal frequency in the United States, Table 2.3 provides values of the ratio a, with f_2 at 60 cycles and f_1 at other, lower frequencies.

TABLE 2.3 Variation of a with Frequency

	For $f_2 = 60$ Hz/s		
f_1, Hz	a	a^2	$(a-1)$
50	1.200	1.440	0.200
45	1.333	1.777	0.333
40	1.500	2.250	0.500
35	1.715	2.941	0.715
30	2.000	4.000	1.000
25	2.400	5.760	1.400

In obtaining separation data, it is suggested that the two measuring frequencies should not be separated more than necessary for the range desired. If the separation is desired at 60 cycles, the measuring frequencies 60 and either 50 or 70 Hz are recommended.

For example, taking 24-gauge USS Dynamo, calculate the hysteresis and eddy loss at 50 and 60 Hz at 10,000 G. The measured data at 10,000 G are as follows:

At 50 Hz (f_1), the total loss $P_1 = 0.812$ W/lb.
At 60 Hz (f_2), the total loss $P_2 = 1.039$ W/lb.
From Table 2.3, $a = 1.2$.

From Eq. (2.15), the eddy loss P_e at 50 Hz (f_1) can be obtained as follows:

$$P_e = \frac{P_2 - aP_1}{a(a-1)} = 4.1666 P_2 - 5.0 P_1 = 0.269 \text{ W/lb} \tag{2.27}$$

Similarly, using Eqs. (2.23) through (2.25), the loss components for 50 and 60 cycles can be calculated and listed as in Table 2.4 for 10,000 G.

Separation of losses can also be calculated graphically, and the procedure is well known, being a graphical expression of Eq. (2.18) or (2.19), which is used after dividing by the frequency. The calculation method is more flexible and probably more definite than the graphical method.

Note: This separation method and the equations apply only to essentially nondirectional material of uniform properties.

TABLE 2.4 Variation of Core Loss with Frequency

	50 Hz	60 Hz
Eddy loss:	0.269 W/lb	0.388 W/lb
Hysteresis loss:	0.543 W/lb	0.651 W/lb
Total loss:	0.812 W/lb	1.039 W/lb

2.4.2 Calculation of Core Losses at Frequencies Intermediate to Two Frequencies for a Given Flux Density

At times it is necessary to know the core loss at some frequency lying between core loss data available at two frequencies f_1 and f_2, such as 25 and 60 Hz, at some flux density, as 60 Hz can be considered as the major power frequency and 25 Hz at one time was the minor frequency found in the United States. To calculate the core loss at some frequency intermediate to f_1 and f_2 cycles, the following equation can be used:

$$P_\eta = \frac{f_\eta}{f_2(a-1)} \left(\frac{f_\eta(aP_2 - a^2 P_1)}{f_2} + a^2 P_1 - P_2 \right) \tag{2.28}$$

where P_η = calculated core loss at frequency f_η and flux density B
P_2 = core loss at frequency f_2 and flux density B
P_1 = core loss at frequency f_1 and flux density B
f_η = intermediate frequency (f_1 is less than f_η is less than f_2), in cycles per second
a = ratio of f_2 to f_1

Note: The core losses P_1, P_2, and P_η can be stated in watts, watts per pound, and so on, provided all three are stated in the same unit.

In the application of Eq. (2.19), suppose the core loss is required at 40 Hz at 8000 G for 29-gauge USS Transformer 58 grade, having data available at 25 and 60 Hz at 8000 G. Thus,

$f_1 = 25$ Hz $\quad\quad P_1 = 0.123$ W/lb
$f_\eta = 40$ Hz $\quad\quad P_2 = 0.368$ W/lb
$f_2 = 60$ Hz $\quad\quad B = 8000$ G
$a = 60/25 = 2.4 \quad\quad a^2 = 5.760$

and Eq. (2.28) becomes:

$$P_\eta = \frac{40}{60(1.4)} \left[\frac{40(0.8832 - 0.7075)}{60} + 0.7075 - 0.368 \right] \tag{2.29}$$

$P = 0.2175$ W/lb

The calculated losses at intermediate frequencies are well within ±2 percent of the measured value, for the enclosing frequencies of 25 and 60 Hz.

As the ratio a of the enclosing frequencies f_1 and f_2 approaches 1, the precision increases in calculating the losses.

2.5 PRESSED SOFT MAGNETIC MATERIAL FOR MOTOR APPLICATIONS*

2.5.1 Abstract

Pressed powdered iron has been used in motor applications for many years. The use has been limited to mechanical parts or to magnetic circuit components where the flux is not changing. Attempts to use these parts as primary alternating flux carriers resulted in high core loss in turn resulting from eddy currents, which usually made them unsuitable. Some companies have developed methods of manufacturing these kinds of parts that keep eddy current losses within reason. Although the electrical and mechanical characteristics are not identical to those of lamination steels, they are comparable. The following section describes this type of material.

A new soft magnetic material has been developed for ac and dc motor applications. The motor components are made by a powder metallurgical process, in contrast to the traditional method of stacking punched laminations. This new material not only offers many manufacturing advantages but exhibits a core loss that is comparable to that of many grades of motor lamination steels at 60 Hz. Because of its low eddy current loss, the material has excellent high-frequency magnetic properties, making it an ideal material for brushless dc motors operating in the high-frequency range.

2.5.2 Introduction

All electromagnetic devices that operate at power frequencies (50 or 60 Hz) use lamination steel sheet, which, for most devices (motors, small transformers, ballast transformers, etc.) is punched, annealed, and stacked by the manufacturer to form the magnetic core of the device. This manufacturing process has some major shortcomings. The most obvious is that a significant amount of scrap is generated in the lamination punching operation, sometimes exceeding 40 percent. Often the device manufacturer is driven to reduce scrap at the expense of optimizing the electrical efficiency. Also, the freedom to design the most efficient motors is limited by the two-dimensional constraints of stacking individual lamination sheets.

In addition to design and scrap considerations, one must look at market trends to see which devices are on the horizon and how the use of lamination steels will affect their economic and electrical efficiency. One area of growth is variable-speed permanent magnet motors. These devices use a permanent magnet rotor; that is, the rotor is solid rather than made from a stack of electrical steel laminations. The net effect is that now, since there is no use for the hole punched from the motor stator, the lamination scrap rate is dramatically increased. This has the effect of significantly increasing the cost of making the motor stator.

Finally, variable-speed motors place different demands on the lamination steel. The speed of permanent magnet motors is controlled by changing the frequency in the motor stator windings; frequencies as high as 1200 Hz are currently being used.

---

*Section contributed by Robert F. Krause, Joseph H. Bularzik, and Harold R. Kokal, Magnetics International.

As the frequency increases, the core loss of the steel increases dramatically. This can be understood by examining the nature of the core loss.

The core loss P_c of an electrical sheet product can be considered to be composed of two components: the hysteresis loss P_h and the eddy current loss P_e (see Eq. [2.20]).

$$P_c = P_h + P_e \tag{2.30}$$

whereas the hysteresis loss increases linearly with frequency, the eddy current loss increases with the square of the magnetizing frequency (see Eq. [2.21]).

$$P_e = \frac{kt^2 B^2 f^2}{6\rho} \tag{2.31}$$

where t = sheet thickness
B = magnetic induction
f = frequency
ρ = resistivity of the steel

Thus, the higher the magnetizing frequency, the more important is the eddy current component of the total core loss of the steel. (Note that the physical properties of the steel that affect the eddy current loss are the sheet thickness and the steel resistivity.)

The resultant high core losses not only reduce motor efficiency but limit its operating range. These high core losses will ultimately have to be addressed by the motor designer as well as the steel manufacturer. On the steel side, as is obvious from Eq. (2.21), thinner gauges and higher alloy contents are needed to meet the growing demand for high-speed motors. Both the thinner gauge and the higher-alloy trends will increase the cost required to produce these steel grades. In addition, silicon added to the steel to increase the resistivity also decreases the magnetic saturation of the steel. This decrease in magnetic saturation negatively impacts many motor designs.

The Accucore concept (Bularzik, Krause, and Kokal, 1998, pp. 38–42) is directed toward addressing the aforementioned problems—a reduction in scrap, a greater electrical efficiency of the finished electromagnetic device, and almost limitless design freedom. An Accucore pressed magnetic component doesn't have the same design constraints that are inherent in a stack of laminations. A powder metal part is scrapless and is, therefore, free of design constraints based upon simply reducing scrap. Furthermore, three degrees of freedom are available to the designer rather than the two degrees imposed by simply stacking steel laminations on top of one another. The zero scrap and increased degree of design freedom will allow for optimization of the magnetic circuit around an improved electrical efficiency for the device.

Another advantage of a powder metallurgically pressed component is related to the frequency dependence of the core loss. It is possible to make a magnetic component that exhibits an almost linear dependence of core loss with magnetizing frequency, as opposed to the almost squared dependence of core loss on frequency of most lamination steel grades. Since variable-speed permanent magnet motors operate at frequencies up to 1200 Hz, higher efficiency in these devices is possible.

2.5.3 Sample Preparation and Testing

Pressed components and test rings are formed from acicular metal particles by traditional powder metallurgical forming methods (Kokal, Bularzik, and Krause, 1997, pp. 61–75). Discrete particles are made by a proprietary manufacturing process that allows control of the particles' physical parameters. Particle chemistry is controlled to achieve desired magnetic properties. The particles are annealed and coated with a metalloorganic substance to fully insulate the particles from each other. Particles are lubricated with a commonly used lubricant such as acrawax or a stearate salt. To produce components, the lubricated particles are fed to a standard powder metallurgy press feed system to form commercial parts. Finished parts are heated in air to remove the lubricant. No further treatment is needed prior to use.

To evaluate the magnetic properties of the particles and for quality control, toroidal rings are made and tested to ASTM standards. Test rings are prepared by hand filling the cavity in a ring-shaped die and pressing it in a laboratory press. Rings are evaluated with a Donart ac tester capable of testing at 50, 60, 100, and 400 Hz. The dc properties are evaluated on a Walker AMH-20 at near dc conditions.

2.5.4 AC Magnetic Properties

The 60- and 400-Hz properties of a pressed compact are compared to Epstein pack tests of a typical cold-rolled motor lamination (CRML) steel and an M-19-grade silicon steel, shown in Figs. 2.60 and 2.61, respectively.

At the typical ASTM-designated test induction of 15 kG and 60 Hz, the core loss of the pressed test core is significantly poorer than that of the M-19 and slightly poorer than that of the CRML steel (see Fig. 2.60). However, in the high-induction range of greater than 18 kG, where many motors operate, the pressed core exhibits a lower core loss than the CRML. At 400 Hz (Fig. 2.61), the loss behavior of the pressed core is significantly better than that of the CRML and very similar to that of the M-19 over the entire induction range. The core loss of the pressed core exhibits a lower dependence with the frequency than the M-19 silicon steel. The difference in the core loss as a function of frequency is thought to be related to the different loss components that constitute the core loss.

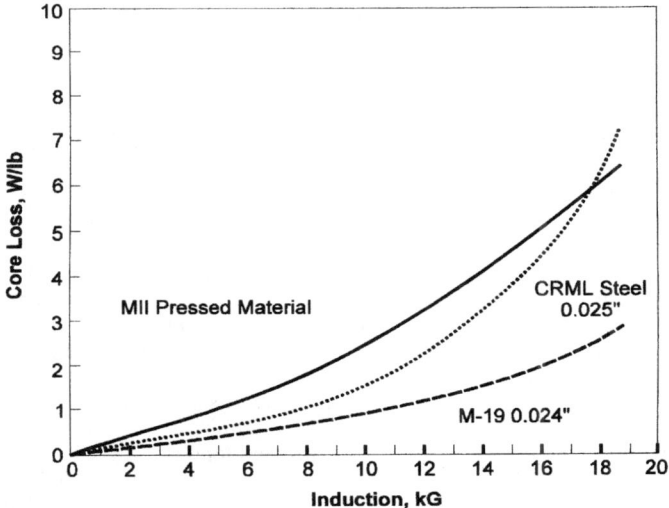

FIGURE 2.60 Core loss at 60 Hz.

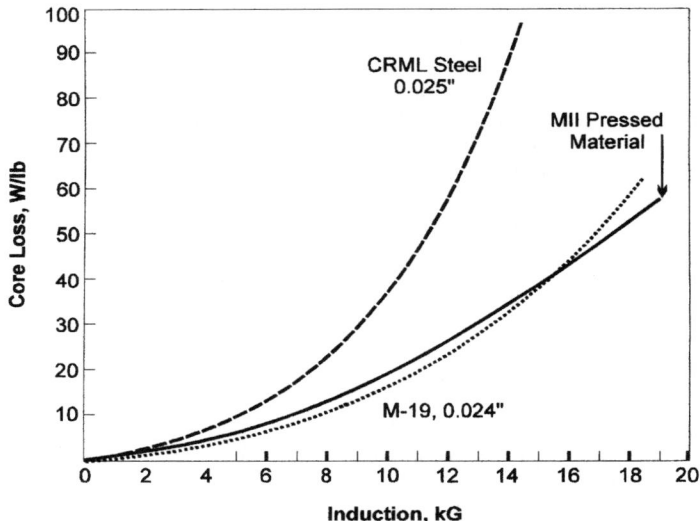

FIGURE 2.61 Core loss at 400 Hz.

To understand the loss mechanism more completely, one can measure the losses as a function of frequency and express them in a classical manner, as shown in Eq. (2.32).

$$P_t = P_h f + P_e f^2 \qquad (2.32)$$

where P_t is the total loss, P_h the hysteresis loss, and P_e the eddy current loss (Cullity, 1972, p. 502). Dividing by frequency, one obtains Eq. (2.33):

$$\frac{P_t}{f} = P_h + P_e f \qquad (2.33)$$

Note that the loss per cycle P_t/f is equal to a constant P_h and a frequency-dependent term (see Fig. 2.62).

Note that for the pressed core only about 5 percent of the total core loss at 60 Hz is due to eddy current loss. The remaining 95 percent is hysteresis loss. A typical lamination steel will exhibit a 50/50 eddy current/hysteresis ratio at 60 Hz.

Figure 2.63 shows the total core loss of the pressed material at 60 Hz, separated into hysteresis loss and eddy current loss components (the eddy current loss comprises the classic eddy current loss and anomalous loss.) Over the entire induction range, the eddy current losses are a very small part of the total losses.

The relatively small eddy current losses keep the total losses low at higher frequencies. Figure 2.64 shows the comparison of the loss per cycle, measured at 15 kG, of the pressed material to a CRML steel and a nonoriented silicon steel, M-19. In both comparisons the hysteresis loss of the pressed material is greater, but the eddy current loss is less. At low frequencies the pressed material has the highest core loss. At around 60 Hz the pressed material losses are comparable to the CRML steel losses and are much lower at higher frequencies. The M-19 steel has the lowest core

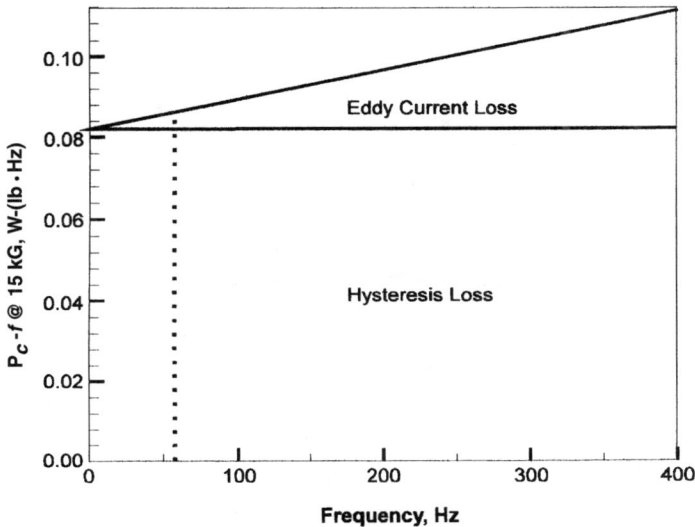

FIGURE 2.62 Eddy current and hysteresis loss versus frequency, Hz.

loss at the lower frequencies. Yet at around 400 Hz the core loss for the pressed material is comparable to that of the M-19 steel. The pressed material has the lowest loss at frequencies greater than 400 Hz.

The lower total core loss for the pressed material is obviously due to the lower eddy current losses. Figures 2.65 and 2.66 show the percentage of eddy current loss for each of the materials at 60 and 400 Hz, respectively. At both frequencies the per-

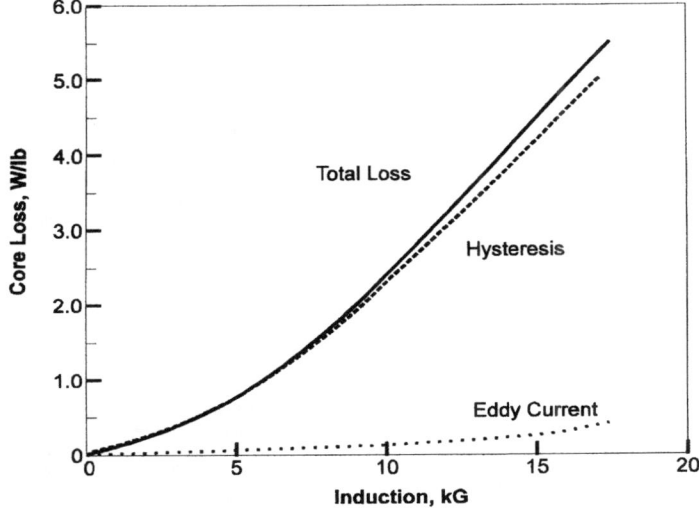

FIGURE 2.63 Core loss components at 60 Hz.

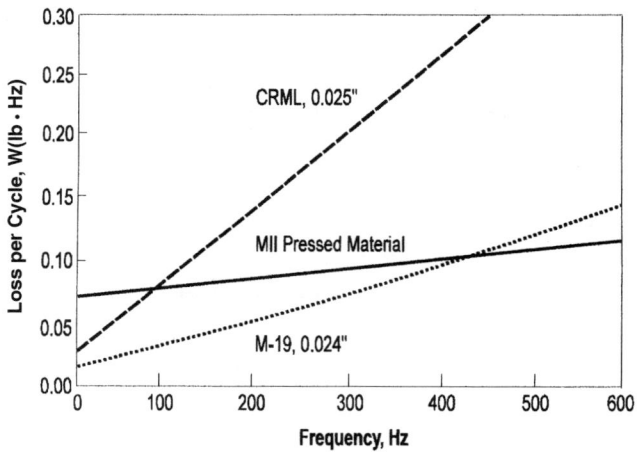

FIGURE 2.64 Comparison core loss versus frequency.

centage of the total core loss due to eddy current losses is the lowest for the pressed material. The low eddy current loss for this pressed material makes it a good candidate for motors.

2.5.5 Applicability in Motors

In most motor design where optimum efficiency is not the major criterion, the induction level in many portions of the motor stator is significantly greater than 18 kG.

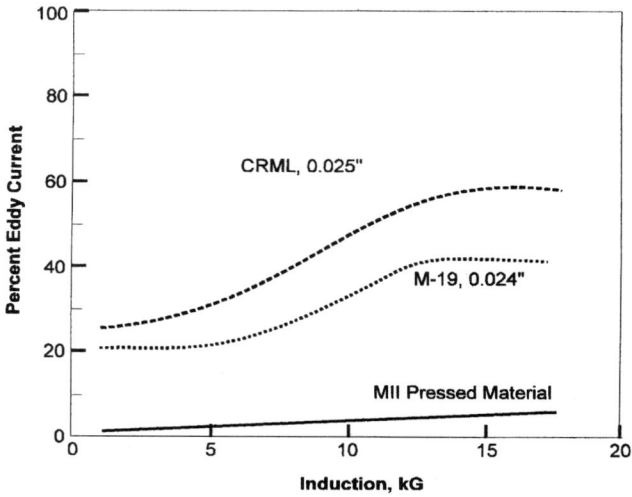

FIGURE 2.65 Percent eddy current loss at 60 Hz.

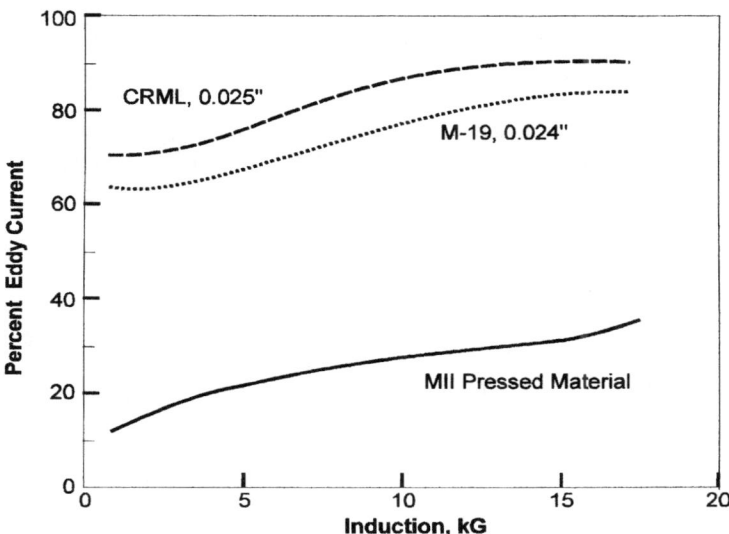

FIGURE 2.66 Percent eddy current loss at 40 Hz.

Additionally, the waveform of the induction in most motors is rich in harmonics. Thus, the core loss in the motor is not the core loss guaranteed by the steel manufacturer at 15 kG and under sinusoidal flux conditions, but is higher due to the higher operating induction and harmonics in the induction waveform.

In the case of electronically controlled motors, the imposed voltage is not normally sinusoidal but is a square wave, as shown in Fig. 2.67. As a result, the flux density is very nonsinusoidal.

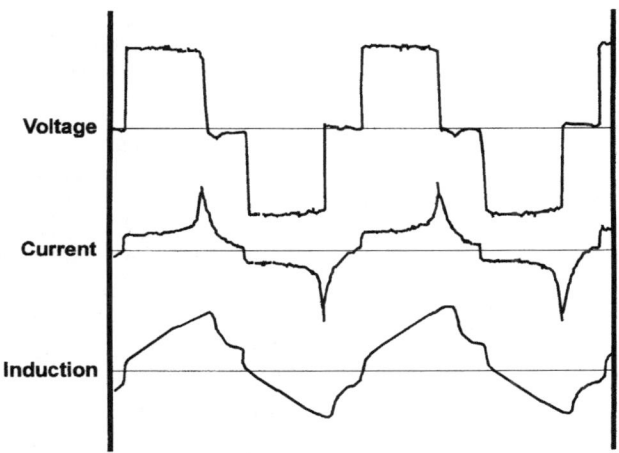

FIGURE 2.67 Electronically commutated motor waveform.

A Fourier analysis of this nonsinusoidal waveform is shown in Fig. 2.68. The core loss under these magnetizing conditions could be expressed as shown in Eq. (2.34):

$$P_c = P_c\,(f_0 B_0) + P_c\,(n_1 f_0 B_1) + P_c(n_2 f_0 B_2) + \dots \tag{2.34}$$

where the total core loss P_c will be higher than the core loss measured under sinusoidal flux conditions. One might anticipate that a material exhibiting a low eddy current loss component might enhance the performance of these types of motors.

FIGURE 2.68 Fourier analysis of waveform.

2.5.6 Motor Results

Two example four-pole brushless stator cores were made from pressed particles and built into motors. One motor operated in the 400-Hz range, while the second operated in the 700-Hz range. Both motors were compared to identical motors in which the stator cores were made from punched 24-mil-thick M-19 silicon steel laminations. The data are presented in Table 2.5.

TABLE 2.5 Performance of Pressed Core Compared to Standard Laminated and Stacked Core

	Torque, oz · in.	Current, A	Speed, r/min	Efficiency, %	Temp. rise, C°
(a) Low-speed, four-pole brushless motor					
Standard	4.0	4.6	11,401	67.2	14.0
Pressed	4.1	4.4	11,662	69.4	12.6
(b) High-speed, four-pole brushless motor					
Standard	9.4	7.1	15,746	64.5	49.4
Pressed	9.4	6.6	16,881	74.5	27.8

In the case of the motor operating in the 400-Hz frequency range, Table 2.5a, the torque was held constant at approximately 4.0 oz · in, and the performance was measured. The pressed core performed statistically better than the laminated core. This is attributed to the fact that, although the core loss of the pressed core is comparable to the M-19 silicon steel in the 400-Hz frequency range (Fig. 2.64), the loss associated with the higher-frequency harmonics imposed on the core by the power supply benefits from the lower eddy current losses exhibited by the pressed core (Fig. 2.66).

In the case of the motor operating in the 700-Hz frequency range, the torque was held constant at approximately 9.4 oz · in, and the performance measured. In this case the pressed core outperformed the standard M-19 silicon steel laminated core (Table 2.5b). It had a lower current draw, ran at a higher speed, and had a lower temperature rise. Overall, the efficiency of the pressed core was greater by 10 percentage points. This motor performance is obviously the result of the much lower core losses of the pressed material in the 700-Hz frequency range (Fig. 2.64).

2.5.7 Conclusions

These are new cases where the pressed metal stator cores yield equivalent or superior results to those shown by the normal laminated steel curves. This is particularly true in high-frequency applications.

2.6 POWDER METALLURGY*

2.6.1 Introduction

Powder metallurgy (P/M) is a mass production manufacturing technology utilizing metal powders with compaction and sintering technology that enables the production of net or near net shape parts. Advantages of the P/M process include a high level of material utilization (little or no scrap), flexibility of design, and high production rates. The consequence of these advantages is parts with lower final unit cost relative to other manufacturing techniques.

This section will review the P/M process, including part production techniques, structural P/M material, and the magnetic properties of iron-based P/M materials for both dc and ac applications.

2.6.2 P/M Process

Figure 2.69 is a schematic of the P/M part manufacturing process. The iron P/M parts production process begins with the manufacture of iron powder. Volume production of iron powder is accomplished in one of two ways: (1) the solid-state reduction of iron ore or mill scale and (2) the melting and water atomization of iron and low-alloy powders. The solid-state reduction processes, although the oldest powder manufacturing techniques, are declining in usage because of limited alloying potential

---

*Section contributed by Francis Hanejko, Hoeganes Corporation.

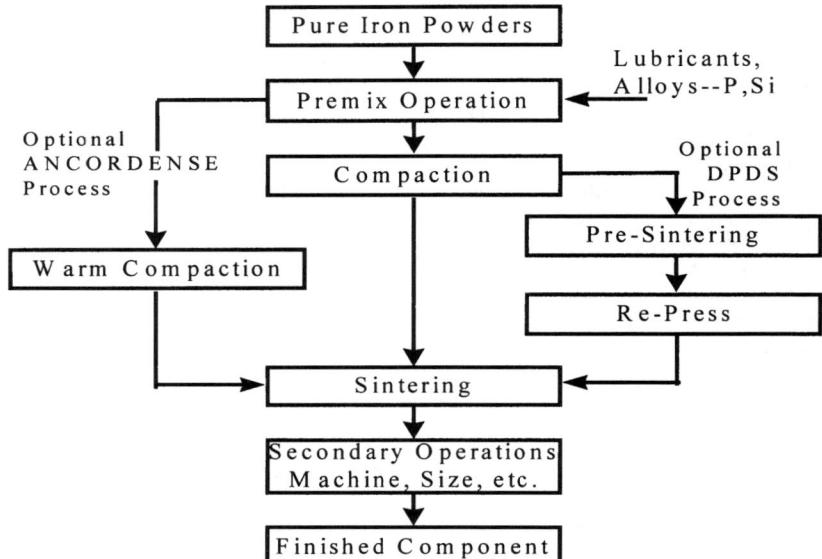

FIGURE 2.69 Schematic of powder metallurgy processing.

and the inherent disadvantages of reduced material compressibility. These solid-state reduced materials are well suited for lower-density applications, such as self-lubricating bearings and components requiring a high degree of strength in the as-pressed condition.

Powders produced by the melting and atomizing process use conventional steelmaking technology to produce liquid metal that is then reduced to iron powder by impinging high-pressure water jets on a stream of the liquid metal. These water-atomized powders offer the advantages of greater compressibility, greater purity (through liquid metal refining techniques), and the ability to prealloy the powders for enhanced P/M part performance. An excellent review of the various powder manufacturing techniques is presented in volume 7 of the ASM *Metals Handbook* (1998).

The objective of the powder-making process is to produce a raw material that offers a high level of compressibility with sufficient strength in the as-pressed condition to facilitate part production. As such, the iron powder is annealed to lower the carbon level to the lowest possible level and metallurgically soften the iron, yielding reduced compressive strength to give enhanced green density. After the powder manufacturing step, the iron powder is premixed with additives such as graphite, nickel, copper, and pressing lubricants. The graphite, copper, and nickel are added to enhance the mechanical strength of the P/M part after sintering. Pressing lubricants are added to the powder blends to facilitate ejection of the P/M parts from the die after the compaction step.

After the premixing stage, the powder is ready for compaction. This step involves the filling of a die cavity with the premixed iron powder and then applying pressure to the powder to consolidate the loose powder into a green compact. By varying the compacting pressure, the green density of the P/M part can be varied; typically, higher green densities result in higher levels of mechanical and magnetic properties.

The range of compacting pressures is usually 20 to 50 t/in^2 of part surface area. The green compact is then transferred to a sintering furnace which operates at a temperature below the melting point of the iron with a protective atmosphere. A typical sintering temperature is 1120°C (2050°F) with sintering atmospheres consisting of blends of nitrogen and hydrogen. The sintering step removes the lubricant from the powder compact, causes metallurgical bonding of the powder particles, and alloys the various premix additives. At the completion of the sintering stage, many P/M parts are complete; that is, no further manufacturing steps are required.

A recent variation on the standard compaction process is warm compaction of iron powder. In this process, a specifically designed powder blend and compaction tooling are heated to approximately 138 to 149°C (280 to 300°F). This elevated temperature compaction results in higher green and sintered densities with a corresponding increase in the magnetic and physical properties of the P/M part. Warm compaction is ideally suited for high-performance applications where restriking or double pressing of the part is not practical.

The greatest percentage of P/M parts are used for structural applications requiring mechanical strength, fatigue endurance, surface wear, and so on. The Metal Powder Industries Federation has established material codes and material property standards for the range of iron P/M alloys used commercially. The interested reader is referred to MPIF Standard 35 for the range of P/M alloys and mechanical properties available by means of the powder process in both the as-sintered and the heat-treated conditions.

Standard 35 is intended to help potential application engineers in the design of new structural parts. However, it is somewhat limited in information concerning the magnetic performance of P/M materials. ASTM has published material standards for magnetic alloys.

2.6.3 Magnetic Properties of P/M Materials

DC Applications/DC Magnetic Performance. DC magnetic components are produced by the conventional press and sintered P/M process. Magnetic properties of a P/M part can be varied by the raw material, the compaction, and the sintering conditions utilized. The range of alloys available varies from high-purity irons to alloys of nickel and iron; Table 2.6 summarizes the most common ferromagnetic powders used and the resulting magnetic properties. Note that the alloys of iron phosphorus and iron silicon are not prealloyed powders; rather, they are produced using a high-purity iron powder which is then premixed with either an iron-phosphorus intermetallic or an intermetallic of iron silicon (either 31 percent or 50 percent silicon). This premixing approach preserves the high level of compressibility of the iron powder and allows for part-specific compositions utilizing various premix additives.

TABLE 2.6 Typical Properties of Sintered P/M Materials

Alloy system	Density range, g/cm^3	Maximum permeability	Hc, Oe	Induction at 15 Oe, kG	Resistivity μΩ·cm
Iron	6.8–7.2	1800–4000	1.5–2.5	10–13	10
Iron-phosphorus	6.7–7.4	2500–6000	1.2–2.0	10–14	30
Iron-silicon	6.8–7.5	4000–10000	0.3–1.0	8–11	60
400 series	5.9–7.2	500–2000	1.5–3.0	5–10	50
50Ni/50Fe	7.2–7.6	5000–15000	0.2–0.5	9–14	45

The material families presented in Table 2.6 represent the current commercial iron P/M alloys. High-purity iron powders are used in applications where the key magnetic characteristic is saturation induction. Saturation induction is a function of the part density; that is, higher part densities give higher saturation levels, regardless of the material specified. Figure 2.70 is a graph showing the maximum saturation induction and the induction at 15 Oe for both a high-purity iron (trade name Ancorsteel 1000B) and an iron alloyed with 0.45 percent phosphorus (trade name Ancorsteel 45P). The iron-phosphorus alloy has higher induction levels at 15 Oe. However, the saturation induction is unaffected by this alloy addition. The saturation induction of the iron and iron-phosphorus alloys is a linear function of the density.

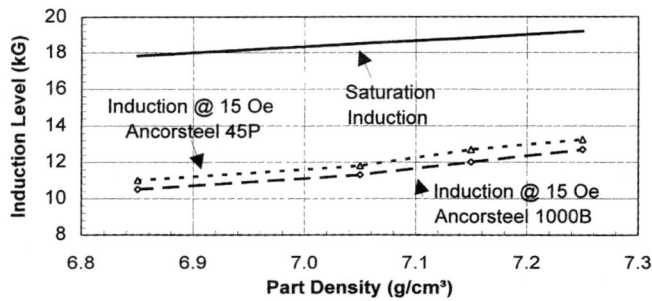

FIGURE 2.70 Induction for Ancorsteel 45P and Ancorsteel 1000B.

Alloys of iron and phosphorus are used in applications requiring higher dc permeability without going to the expense of using a silicon-containing steel. Iron-phosphorus alloys can be sintered at 1120°C (2050°F) using either pure hydrogen or hydrogen-nitrogen blends as the sintering atmosphere. Alloys containing silicon require sintering temperatures of 1260°C (2300°F) to completely homogenize the silicon, thus giving the high level of magnetic properties. This elevated-temperature sintering, although becoming more commonplace, does increase the part cost. Iron-silicon alloys are used in applications where the low coercive force coupled with the high levels of permeability and resistivity are necessary.

The 400 series stainless steels are used in applications requiring magnetic performance coupled with corrosion resistance. Antilock brake wheel sensors represent the largest application for the 400 series stainless steels. The unique configuration of the wheel sensors combined with 100,000-mile durability make the 400 series stainless steels the logical choice. Prealloys of 50Ni/50Fe are the highest-cost P/M alloy and are used in the most demanding applications. Typically, this material is used in flux return paths for computer printing devices where the response time and low coercive forces are critical.

Processing Considerations for Sintered P/M Magnetic Materials. P/M processing is unique in that the magnetic properties can be tailored via part density and sintering conditions to meet the specific part requirements. As mentioned earlier, density has a significant effect on the part performance. Higher-density P/M parts exhibit increased permeability and saturation induction without any degradation of the coercive force. Techniques to increase the part density include double press/double sinter, warm compaction, or restriking a fully sintered part.

Increasing the density by either double press/double sinter or warm compaction processing results in sintered densities approaching 7.4 to 7.5 g/cm^3. At these density levels, the permeability and saturation induction approach the values achieved for fully dense wrought steels. Table 2.7 shows a comparison of a low-carbon wrought steel with pure iron and phosphorus irons pressed to 7.3 to 7.35 g/cm^3. The wrought steel was evaluated in the as-forged condition. Performance of the P/M materials is comparable in both permeability and saturation induction to the wrought AISI 1008 at 15 Oe. The P/M materials are superior in terms of lower coercive force values. Interestingly, the mechanical properties of the warm compacted Ancorsteel 45P are similar to the low-carbon steel forging. Thus, the P/M alternative produces a part that gives equivalent magnetic properties along with comparable mechanical properties.

TABLE 2.7 Comparison of Ancordense Processed Ancorsteel 1000B, Ancorsteel 45P, and AISI 1008

Property	AISI 1008	Ancorsteel 1000B @ 7.3 g/cm^3	Ancorsteel 45P @ 7.35 g/cm^3
Maximum permeability	1900	2700	2700
Induction @ 15 Oe, kG	14.4	15.0	15.1
Hc, Oe	3.00	2.10	1.90
Yield strength, lb/in^2 (MPa)	42,000 (285)	21,000 (145)	42,000 (285)
Tensile strength, lb/in^2 (MPa)	56,000 (385)	32,800 (225)	59,400 (405)
Elongation, %	37	13.7	12

Repressing or restriking a sintered P/M part has a deleterious effect on the magnetic properties. Table 2.8 shows the effect of repressing on the magnetic properties with and without an annealing step. The decrease in magnetic performance resulting from the restriking operation is eliminated when the part is annealed. An excellent review of the effects of secondary processing on the magnetic performance of P/M parts has been given by Frayman, Ryan, and Ryan (1996, pp. 25–37). Very often during the initial development of a P/M part, simple slugs are pressed and sintered. These slugs are then machined in the final part geometry. The machining step introduces stress within the part, thus lowering the overall magnetic performance. Annealing will eliminate the induced stress and fully restore the magnetic properties of the sintered part. Annealing temperatures for the iron, iron-phosphorus, and iron-silicon alloys are in the range of 843 to 899°C (1550 to 1650°F), with cooling rates not to exceed 11°C/min (20°F/min). A protective atmosphere of either pure hydrogen or a hydrogen-nitrogen atmosphere is recommended. Annealing of the 50Ni/50Fe is done at 982 to 1038°C (1800 to 1900°F) in a hydrogen-containing atmosphere, followed by slow cooling at a rate not to exceed 5.5°C/min (10°F/min).

The structure-sensitive properties of magnetic materials are the permeability, coercive force, and residual induction. With P/M magnetic parts, the permeability and residual induction are affected by the density of the component. However, it has been shown that the coercive force is not density sensitive. Permeability is affected by both the density and the microstructure of the final part. Several key parameters that influence the structure-sensitive properties in magnetic materials are grain size, pore size and morphology, and material purity (in particular, residual interstitial elements such as carbon and nitrogen). Figure 2.71 presents the permeability of

TABLE 2.8 Effect of Repressing on the Magnetic Properties of Ancorsteel 45P at 6.8 g/cm^3

Condition	Maximum permeability	Hc at 15 Oe, in Oe	Induction at 15 Oe, in kG
As sintered	2260	1.98	11.0
Sintered and sized	1160	2.69	9.8
Sintered—sized and annealed	2270	2.22	11.2

Ancorsteel 45P sintered at 1120°C (2050°F) and 1260°C (2300°F). Sintering at the higher temperature results in a larger grain size, greater pore rounding, and, consequently, higher permeability.

Figure 2.72 illustrates the effect of both sintering temperature and elevated nitrogen levels on the coercive force and permeability. The data illustrate that sintering at an elevated temperature results in a significantly lower coercive force and greater permeability. However, the presence of increased nitrogen levels results in a degradation of the magnetic properties. Carbon has a similar effect to nitrogen on the magnetic performance; that is, the higher the carbon content, the lower the magnetic response. High nitrogen and carbon contents are the result of improper sintering and/or improper atmosphere selection. Care is necessary to make certain that proper lubricant burnout is effected and that a clean furnace is utilized to ensure a low sintered carbon content. Nitrogen content can be minimized by utilizing pure hydrogen in the sintering atmosphere.

Iron P/M alloys containing phosphorus and silicon are made by premixing a high-purity iron powder with an iron-phosphorus or iron-silicon intermetallic. The most common iron-phosphorus intermetallic is Fe$_3$P, which contains approximately 16 percent by weight of phosphorus. This premix additive has a melting temperature of ~1066°C (1950°F); as such, sintering and complete homogenization of iron-phosphorus alloys can be accomplished at 1121°C (2050°F). Higher-temperature

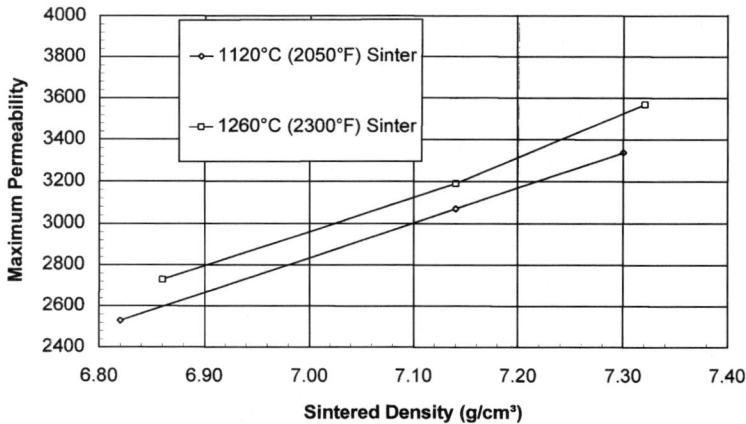

FIGURE 2.71 Permeability of Ancorsteel 45P sintered at 1120°C (2050°F) and 1260°C (2300°F).

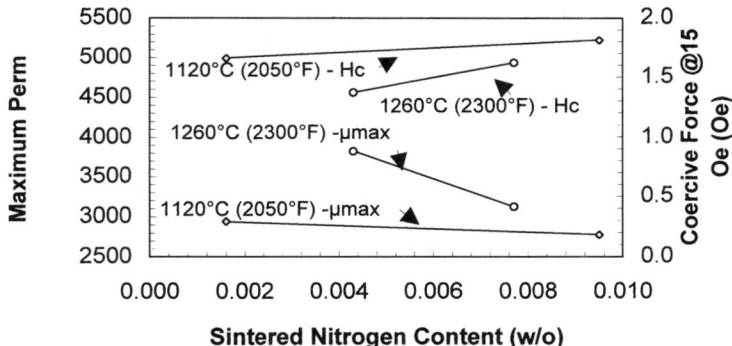

FIGURE 2.72 Effects of sintering temperature and elevated nitrogen levels on permeability and coercive force.

sintering will result in greater pore rounding and grain growth, thus enhancing the structure-sensitive properties of the iron-phosphorus material. Unlike the iron-phosphorus intermetallic, the iron-silicon intermetallic has a melting point in excess of 1371°C (2500°F); sintering at 1121°C (2050°F) will not completely homogenize the silicon within the iron matrix. To achieve complete homogeneity, sintering at 1260°C (2300°F) is necessary. Once properly sintered, the magnetic properties shown in Table 2.6 are achieved.

Sintering of the 400 series stainless steels and 50Ni/50Fe can be done at either 1121 or 1260°C (2050°F or 2300°F) with the higher temperature yielding a higher level of magnetic performance through enhanced densification and grain growth. A word of caution on the 400 series stainless steels concerns the need to use a 100 percent hydrogen atmosphere. If a nitrogen-containing atmosphere is used, the affinity of the chromium for nitrogen will result in high nitrogen levels within the sintered part. These high nitrogen levels severely degrade the magnetic performance.

P/M Materials for AC Magnetic Applications. The rationale for using laminated steel assemblies for ac magnetic devices is well understood—specifically, the reduction of the eddy current losses resulting from the alternating magnetic field. A fully sintered P/M material is not suitable for ac applications because the inherent thickness of the P/M part will result in large eddy current losses even at low ac frequencies. Attempts at making thin laminations via the P/M process were not successful because the thinnest part practical via P/M is 0.060 in thick. Additionally, the cost of pressing and sintering laminations exceeds the cost of laminations made by the conventional stamping method.

Iron powder has been used in the manufacture of switch-mode power supplies, light dimmers, and loading coils for high-frequency applications. These applications utilize iron powder that has been treated with an electrically insulating surface coating and subsequently premixed with a polymeric binder and pressing lubricant. This so-called dust core technology utilizes the inherent fine particle size of the powder material to minimize the eddy current losses at higher frequencies. Limitations in compaction capacity and part densities focused this technology into generally smaller parts with generally higher operating frequencies.

Iron Powder–Polymer Composites. To extend the range of the dust core technology, experimental work was directed to develop new powder-coating technologies and new compaction techniques that enabled higher part densities with higher saturation induction levels and higher permeabilities. If a suitable coating technology could be developed, powder compacts potentially could displace laminations in low-frequency applications such as electric motors and transformers.

In the mid-1990s a new family of iron powder materials was introduced that was intended for medium- to high-frequency ac applications. These materials combined polymer processing and iron powder metallurgy to produce an iron powder–polymer composite. The basic processing steps of these materials is illustrated in Fig. 2.73. The process starts with a high-purity iron powder that is coated with a polymer via fluid bed processing. The suitable polymer is dissolved in a solvent and is subsequently sprayed onto the iron powder. After the coating step, the coated iron powder is compacted in heated tooling. Once compacted, the part is suitable for use in ac applications; no sintering is required. The polymer coating, typically less than 1 percent by weight, both electrically insulates the iron particles and adds mechanical strength to the as-pressed component. An optional particle oxide coating can be applied for applications requiring greater interparticle resistivity.

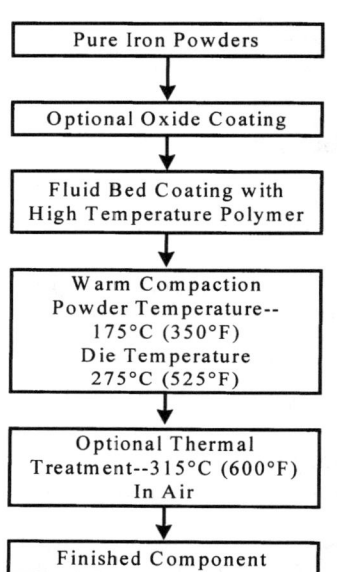

FIGURE 2.73 Processing steps for iron powder–polymer composites.

Iron powder–polymer composites are designed to be used in the as-pressed condition. As such, the strength of the green part must be sufficient to withstand the stresses associated with the winding or final assembly of the component. The strength of the insulated iron is approximately 15,000 lb/in^2 (100 MPa) in the as-pressed condition. After an optional low-temperature heat treatment [316°C (600°F) for 1 h], the strength increases to nearly 35,000 lb/in^2 (240 MPa). Comparable strength of the standard P/M material in the green or as-compacted condition is approximately 3000 lb/in^2 (20 MPa).

Summary of DC Magnetic Performance of Iron Powder–Polymer Composites. Three distinct coated iron powders are currently available. Table 2.9 shows the composition and the dc magnetic properties of these three materials. Although three grades are presented, the manufacturing process for these powders is flexible; thus, powders can be customized to meet specific application requirements.

The effect of frequency on the permeability of these three materials is shown in Fig. 2.74. The Ancorsteel SC120 material provides the highest degree of permeabil-

TABLE 2.9 Magnetic Characteristics of Ancorsteel

Ancorsteel material	Polymer coating (w/o)	Oxide coating	Density at 50 t/in^2 g/cm^3	Initial permeability	Maximum permeability	H_c, Oe	Induction at 40 Oe, kG
SC120	0.60	No	7.45	120	425	4.7	11.1
SC100	0.75	No	7.40	100	400	4.8	10.9
TC80	0.75	Yes	7.20	80	210	4.7	7.7

ity at the low-frequency levels, whereas the TC80 material gives the best high-frequency performance. The increasing eddy currents as the frequency increases result in a decrease in effective permeability and the roll-off observed. The SC120 material is designed to provide the highest performance at lower frequency levels by controlling the particle size and utilizing just a single polymer coating. The TC80 material uses a finer particle size distribution, the oxide coating, and a higher level of polymer coating, minimizing the eddy current losses. The performance of the SC100 material lies between those of the other two grades.

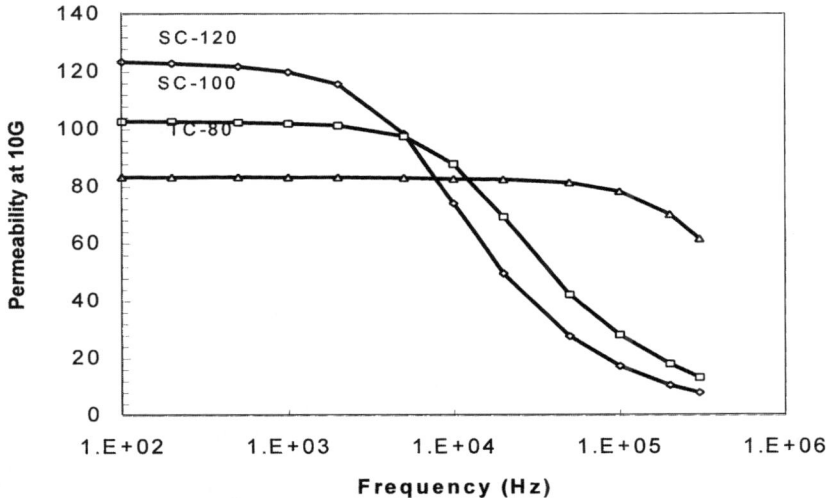

FIGURE 2.74 Effect of operating frequency on the 10-G permeability of the iron powder–polymer composites.

Figure 2.75 shows the effect of frequency on total core losses for the TC80 material compared with a lamination steel. The lamination steel is a nonoriented 3 w/o silicon iron rolled to a thickness of 0.2 mm (0.007 in). At low frequency levels, where the core losses are dominated by hysteresis losses, the laminated material shows lower levels of losses than the coated iron powder material. This limits the low-frequency performance of plastic-coated iron compacts, as the hysteresis losses are greater than those of laminated steels. However, the reduced eddy current loss inherent in the plastic-coated iron results in lower levels of losses and thus higher efficiency at the higher-frequency range where the total core losses are dominated by the eddy current losses.

Low-Core-Loss Insulated Powders. The widespread usage of iron powder–polymer composites remains limited because of the high hysteresis loss affecting high core losses at lower frequencies (<200 Hz). High hysteresis losses result from the cold-working imparted to the iron during compaction. Iron powder–polymer composites are used in the as-compacted condition; the cold-working during compaction affects the structure-sensitive magnetic properties (in particular, the permeability and coercive force). For pure iron, the coercive force of a fully annealed material toroid at induction level of 12,000 G is approximately 2.0 Oe. Even minor

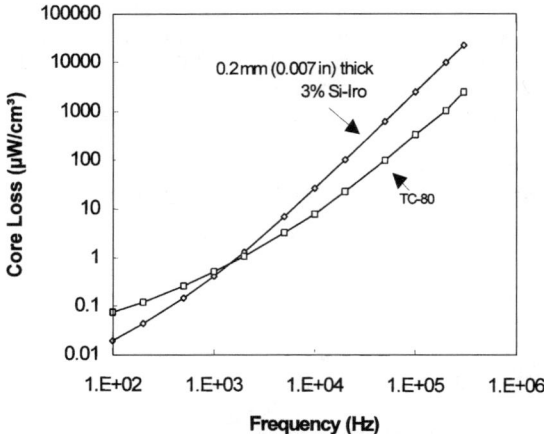

FIGURE 2.75 Frequency on total core losses for TC80 material compared with a lamination steel.

amounts of cold work raise the coercive force to approximately 4.5 Oe. This increase in coercive force raises the hysteresis losses, thus increasing the overall losses at low frequencies.

Thus, for the nonsintered iron powder composites to gain greater acceptance, this low-frequency deficiency must be overcome. To investigate the effects of powder compaction on the magnetic properties, a pure iron powder was surface-coated with 0.03 w/o phosphoric acid and then premixed with 0.75 w/o zinc stearate. Magnetic toroids were compacted over a range of pressures from 10 t/in² (137 MPa) to 50 t/in² (685 MPa). The as-compacted toroids were then cured at 177°C (350°F) and subsequently tested for their magnetic properties.

Table 2.10 presents the effects of compaction pressure on the dc permeability and coercive force of the pressed and cured iron toroid. Compaction pressures as low as 10 t/in² (135 MPa) increase the coercive force to approximately 3.5 Oe, whereas compaction at 50 t/in² (685 MPa) raises the coercive force to approximately 4.5 Oe. Magnetic data for pure iron compacts sintered at 1120°C (2050°F) gave a coercive force of approximately 2 Oe. Thus, even minor amounts of cold-working result in a significant increase in the coercive force, and, consequently, a significant increase in the hysteresis loss.

TABLE 2.10 Effects of Compaction Pressure on the Magnetic Properties of Iron Powder at 40 Oe

Compacting pressure, t/in²	Density, g/cm³	Hc, Oe	Maximum permeability	Induction, kG
10	5.70	3.3	97	3.3
20	6.47	4.1	179	5.9
30	6.92	4.3	225	7.4
40	7.14	4.4	245	8.2
50	7.26	4.4	245	8.3

Testing was done to determine the minimum annealing temperature to reduce the coercive force. Experimental testing determined that a 650°C (1200°F) annealing cycle is adequate to raise the dc permeability and lower the dc coercive force, while also yielding reduced ac core losses. However, conventional iron powder–polymer composite materials cannot withstand this temperature without degradation of the polymeric material. This effort suggests that a new type of insulating material is needed that is compatible with conventional P/M techniques and allows for a modified magnetic annealing at a minimum temperature of 650°C (1200°F).

A proprietary compound was developed that met these criteria. This coating material is compatible with the iron powder and completely wets the powder surface. Once coated, the powder is compacted at 150°C (300°F) using lubricated dies. Annealing of the compacts is accomplished at 650°C (1200°F) in a nitrogen atmosphere for a minimum time of 1 h.

Figure 2.76 presents the dc and 60-Hz ac hysteresis curves of this new material. Both the ac and dc curves exhibit low coercive force and a low hysteresis loop area. The nontraditional look of the *B-H* curves of Fig. 2.76 is a result of the insulating material acting as a distributed air gap within the part, thus shearing the *B-H* curve. Note the similarity of the dc and ac curves, indicating that the losses are primarily hysteresis losses with minimal eddy current losses.

Comparisons of this new insulated material were made relative to cold-rolled motor laminations and an M-19 silicon steel, both at 60 Hz and 200 Hz. Figure 2.77 shows the dc hysteresis curves of the annealable iron powder composite and the M-50 steel lamination material. The wrought laminated materials exhibit significantly greater dc permeability and dc saturation. The reason for the reduced dc performance of the insulated powder is the presence of the powder coating.

Figure 2.78 presents the core loss as a function of the induction at 60 Hz for the annealable insulated powder material compared to M-50 and M-19 steel laminations. The annealable insulated powder material exhibits a total core loss at 60 Hz and 10,000 G of approximately 3 W/lb. This value is considerably lower than the value for the M-50 lamination steel and higher than that for the M-19 lamination steel. The total core loss at 200 Hz for the annealable insulated powder material is presented in Fig. 2.79. At this higher frequency, the annealable insulated powder has lower total core loss relative to the M-50 but is still higher than that of the M-19

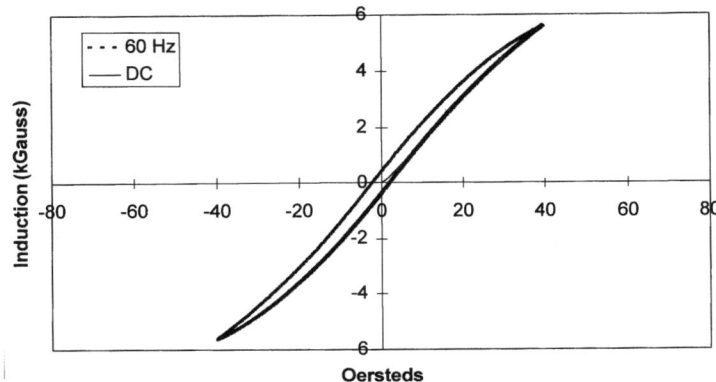

FIGURE 2.76 DC and 60-Hz ac hysteresis curves.

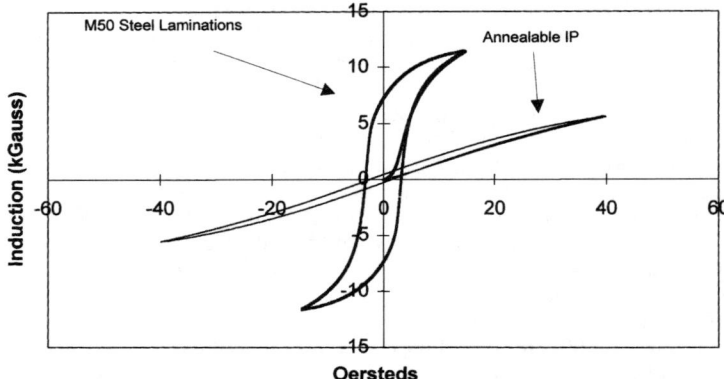

FIGURE 2.77 DC hysteresis curves of annealing iron powder composite and M-50 lamination material.

material. These data indicate that the annealable insulated powder material can successfully replace components made from the M-50-type materials but is not a replacement for the M-19 at the two frequencies examined thus far.

Limitations of this annealable material are the low green strength of the annealable insulated powder ($\approx$10,000 lb/in^2 TRS) and the low dc permeability. The low permeability is a direct consequence of the highly insulating coating. Efforts are under way to enhance the permeability of these materials without degrading the overall magnetic performance. However, this low-core-loss material is designed for powder applications where low total core losses are necessary and the consequence of the low permeability can be minimized by component design.

2.6.4 Summary

Powder metal processing is a mass production metal-forming process capable of producing a wide range of magnetic components for both dc and ac applications. For

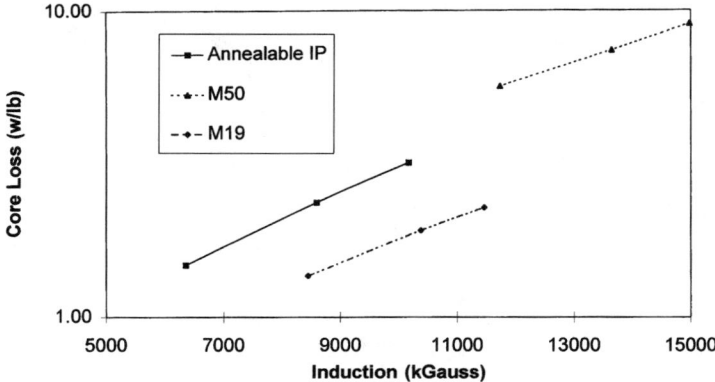

FIGURE 2.78 Core loss at 60 Hz.

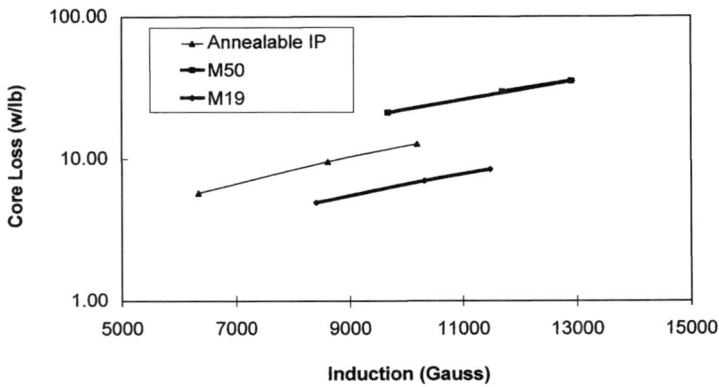

FIGURE 2.79 Core loss at 200 Hz.

dc applications, the designer has the choice of a number of material systems, each with unique properties that can be customized to meet specific part requirements. Care must be taken in the manufacture of the sintered P/M parts to minimize the carbon and nitrogen pickup. By varying the alloying, sintered density, and processing, the resulting magnetic properties cover a broad range of applications.

For ac applications, the P/M industry has developed two families of powders that can be utilized in alternating magnetic fields. The choice between the iron powder–polymer composites and a powder that can be given a low-temperature annealing is based on the part function and ultimate part requirements.

2.7 MAGNETIC TEST METHODS*

The characteristics of the magnetic core materials discussed in this chapter must be determined by certain specified procedures, which are usually in accordance with the standard test methods of the ASTM. The data obtained from such measurements can be used in a comparison of the electrical and magnetic properties and as a guide in the selection of the best material for a particular design. The test methods utilized can be classified in two general groups: those using direct current and those using alternating current as a source of power.

2.7.1 Direct-Current Tests

All direct-current test data are obtained by ballistic test methods. Ring samples, or an appropriate test sample in one of several different permeameters, were used with the basic circuit illustrated in Fig. 2.80 or in a modification of this circuit.

---

*Section courtesy of Allegheny-Teledyne.

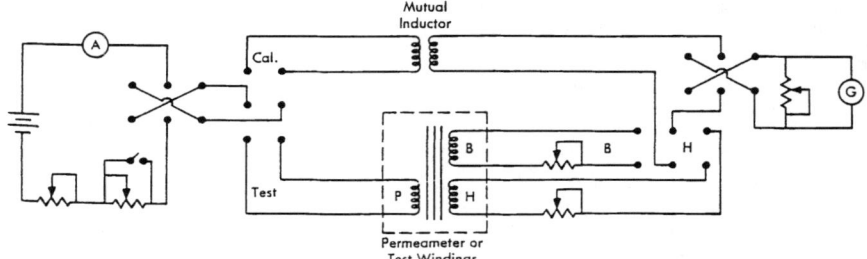

FIGURE 2.80 Basic circuit diagram for demagnetic testing.

For a more detailed discussion of this circuit and its adaptation to various test methods, see ASTM A-34 on standard methods of test for normal induction and hysteresis of magnetic materials. The test methods described by ASTM cover most of the generally accepted test methods likely to be used for obtaining direct-current magnetic properties.

Test Samples. In general, the form of the test sample will be determined by the type of material to be tested, the type of test desired, and the availability of the material from which the test sample will be made. The following is a partial list of sample forms which might be used in obtaining direct-current test data on soft magnetic materials.

- Stamped or machined rings or links
- Wound tape toroid or other form
- Epstein strips
- Split strips
- Small strips or bars
- Bars and rods
- Cut cores and laminations

Tape-wound cores and Epstein strips are not practical for heavy-gauge materials. If the material being tested has no pronounced directional magnetic properties, the ring or toroidal winding is the preferred sample form. For nonoriented materials having some directional properties, an Epstein sample cut one-half in the direction of best properties or in the direction of rolling and one-half across this direction is the preferred form. With oriented materials, the wound toroidal core or Epstein strips cut parallel to the rolling direction are preferred over other types of specimens. Where the size and shape of material to be tested are limited, it may be necessary, with some sacrifice in accuracy, to use other forms of test specimens such as bars, rods, small strips, cut cores, laminations, or other special shapes. For reliable test results, permeameters require properly selected samples of correct size and shape and are limited in ranges of magnetizing force, permeability, and induction. Permeameter tests generally are not as reliable as those made on ring or toroidal specimens of proper dimensions. In most cases, manufactured parts are of such size and shape that they cannot be adapted to fit into standard test equipment and therefore cannot be tested accurately for material characteristic properties.

When testing ring or toroidal specimens, first the B pickup coil, then the magnetizing winding are wound almost directly on the specimen. Care must be taken to avoid strain due to pressure of insulation or winding. Frequently, a close-fitting, rigid insulating case is used to avoid these strains. Magnetizing forces are calculated from the number of turns in the magnetizing coil N, the magnetizing current I, and the mean length of magnetic path in the test specimen ℓ_m, using the following relation:

$$H = \frac{0.4\pi NI}{\ell_m} \text{ Oe} \qquad (2.35)$$

The corresponding induction is measured with the aid of the B coil and the ballistic circuit, the ballistic galvanometer having previously been calibrated to become direct reading in terms of gauss per millimeter of deflection. This calibration is obtained by calculating a current which must be reversed in the primary of the standard mutual inductor to give the desired number of millimeters of deflection for any chosen induction in gauss. The number of B coil pickup turns N, the cross-sectional area in square centimeters of test sample A, the value of the mutual inductance in henries L_m, and the induction in gauss B, at the chosen value of deflection, are used in the following relation to determine the value of the calibrating current:

$$I = \frac{BNA}{L_m \times 10^8} \text{ A} \qquad (2.36)$$

The galvanometer series resistance is adjusted to give, on current reversal, a deflection equal to that chosen to represent the value of B used in the formula.

If accurate machined test samples are available, the cross-sectional area is calculated from physical measurements. For all other samples of uniform cross section, it must be calculated from the weight in grams, the density δ in grams per cubic centimeter, and the mean length ℓ_m in centimeters, as follows:

$$A = \frac{\text{weight}}{\delta \ell_m} \text{ cm}^2 \qquad (2.37)$$

Normal induction curves and hysteresis loops are run in accordance with ASTM designation A-34. In all cases, care must be taken to properly demagnetize the test specimen prior to the measurement of magnetic properties. A drift in low-induction characteristics may be observed subsequent to demagnetization. For best results, a 24-hour storage period in a magnetically shielded container should precede the test run. In practice, however, a reasonable time is allowed to elapse, and the tests are then performed.

To realize fully the benefits of the previous magnetization, the normal induction curve must be obtained by taking a regular series of test points beginning with the lowest value of induction and proceeding upward toward saturation. In addition, the sample must be in a uniform cyclic condition for each of these test values.

Direct-current demagnetization is achieved by first magnetizing to a high induction, then by a long series of slow reversals the current is reduced in small increments to zero. When alternating currents must be used, the lowest available power frequency is chosen, and demagnetization is obtained by first magnetizing to high induction, then slowly reducing the applied field to zero. It is most complete after heat treatment above the Curie point. This condition may not be stable, however, and the first curve obtained cannot always be repeated, even after careful demagnetization using one of the other methods.

Permanent magnets are generally tested in the same manner as described here, with the exception that suitable test specimens are normally of solid bar or rod form and are usually run in a saturation or high H permeameters such as described in Bureau of Standards RP548 and RP1242.

Direct-current tests on all classes of magnetic materials may be run over a wide range of temperatures in the same manner as that used at room temperature. Proper care must be taken to maintain the test specimen at the desired constant temperature for each test run. In many cases, this requires auxiliary equipment of a special nature not generally available for normal test work.

2.7.2 Alternating-Current Tests

Many magnetic materials have widespread use at commercial power frequencies. For this reason, the 60-cycle properties of most magnetic materials have been generated.

Epstein Frame. Core loss and permeability testing at 60 cycles per second are fairly well standardized over a wide range of magnetic inductions for both the 50-cm butt joint and the 25-cm standard double lap joint Epstein frames. Air gap and strain effects have been reduced in the 25-cm double lap joint test frame (employing a 28-cm test sample, thereby permitting more dependable permeability measurements). The basic circuit used for 60-cycle core loss and permeability measurements in this type of testing is illustrated by Fig. 2.81. A detailed description of the test method is given in ASTM designation A-343.

Because of the large sample size required by the Epstein frame and the desirability of testing certain highly oriented materials and materials of extremely high permeability, a test sample in the form of a wound toroid or ring carrying its own test windings is frequently substituted for the Epstein test frame in the circuit illustrated in Fig. 2.81. The test is made in the same manner as with the standard Epstein frame.

When making ac core loss and permeability measurements, it is customary to maintain sinusoidal voltages. ASTM standard test methods assume this condition and, for core loss measurements, apply corrections when the form factor departs from 1.11 by more than 1 percent. In this test method, when true root-mean-square-reading instruments are used, it becomes important to know the waveform of the voltage or current being measured.

At inductions above the point of maximum permeability of the normal magnetization curve, ac permeabilities may be determined by the method previously

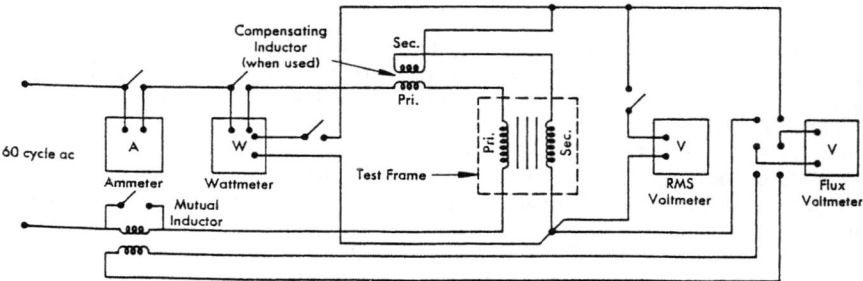

FIGURE 2.81 Basic circuit diagram for 60-cycle ac core loss and permeability measurements.

described. Note that this method assumes that the ratio of the magnetizing component of the current to the total current is nearly unity. Due to the presence of loss components in the current, which lower this ratio, permeabilities obtained by this method may not be the same as those obtained from a direct-current ballistic test.

The magnetizing force for individual test points, in oersteds, is calculated from the following:

$$H = \frac{0.4\pi N I_m}{\ell_m} \quad (2.38)$$

where N = number of magnetizing winding turns
I_m = peak amperes
ℓ_m = mean length of magnetic path, cm
H = magnetizing force for the given test point, Oe

For the standard 25-cm double lap joint test frame, this formula reduces to

$$H = 10 I_m \quad (2.39)$$

For all samples made from sheet or strip materials, the specimen cross-sectional area is calculated from the weight, density, and length. For Epstein samples, the length used to calculate the area is four times the sample length. For other sample forms of uniform cross section, the length is the mean length of the magnetic path.

Induction is calculated from the measured voltage using an average, or root-mean-square volts on a sinusoidal waveform. The formula becomes

$$B_m = \frac{E \times 10^8}{4.44 f N A} \quad (2.40)$$

This test method is usable for core loss and volt-ampere measurements from moderately low inductions up to those approaching saturation. In the upper region, exciting currents become large, and instrumentation and other problems are magnified.

Incremental core loss and ac permeability tests can be made using the preceding test method. Where the operating inductions permit, these tests are usually made with a bridge or electronic instruments.

Owen Bridge. The standard Owen bridge test frame for Epstein samples has 100- and 1000-turn windings, but other sample forms, with appropriate windings, may also be used. This bridge circuit is illustrated in Fig. 2.82 and described in ASTM designation A-343.

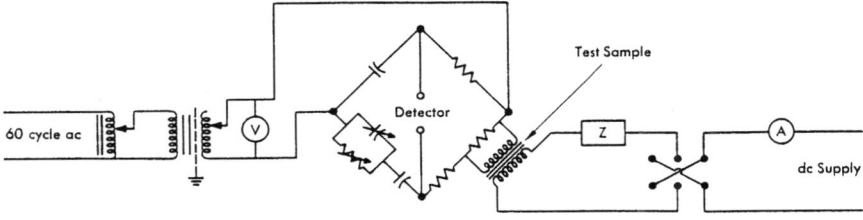

FIGURE 2.82 Circuit diagrams for Owen bridge test method.

Hay and Maxwell Bridges. Hay and Maxwell bridges are also adaptable to these measurements. The bridge methods are the most widely used means for obtaining low-induction properties, but methods using direct reading meters or electronic instruments are also popular. Alternating-current potentiometers may be used, but they are not readily available.

The modified Hay bridge is rapidly gaining popularity for bridge-type measurements. It has been adopted by ASTM and appears in the A-343 standards. This method may be used with Epstein-type test frames as well as with other sample forms. Its circuit diagram is illustrated in Fig. 2.83.

Permeability Measurements. Permeability measurements over very wide ranges of inductions and frequencies may be made using electronic instruments and the circuit diagram of Fig. 2.84. These meters will withstand large overloads and may be calibrated to read directly in terms of magnetizing force and induction.

Frequently, test methods which are designed especially for quality control purposes have sufficient accuracy for other tests and at the same time are fast and convenient to use. The direct impedance substitution method for determining low-induction ac permeability is one type. It uses a more versatile arrangement of the simplified circuit and is shown in Fig. 2.85.

When using this test method, it is desirable to keep the resistance of the test coil and the inductance of the decade resistors as low as possible. The core materials under test have relatively high permeability, but the coil resistance and core losses still produce in-phase exciting current components. The type of core, gauge, size, and other pertinent facts are always known; therefore, for comparative test purposes, it

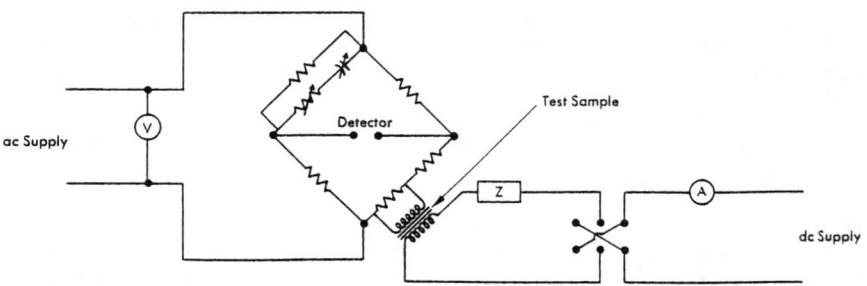

FIGURE 2.83 Circuit diagram of modified Hay bridge.

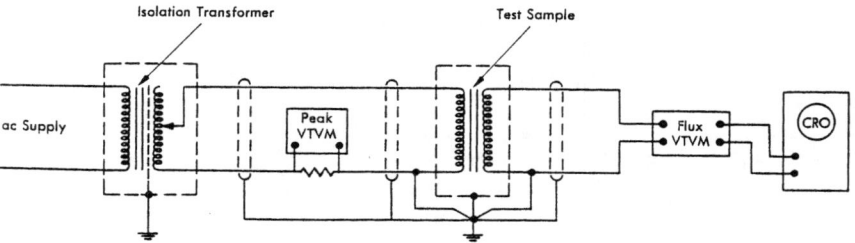

FIGURE 2.84 Circuit diagram for ac permeability.

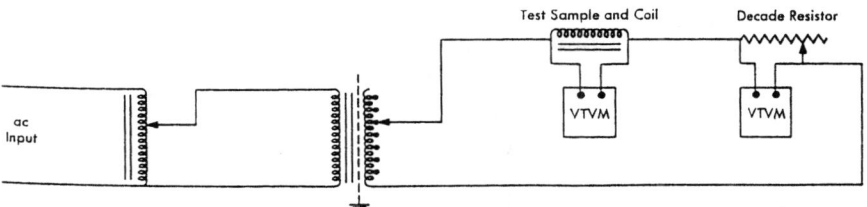

FIGURE 2.85 Circuit diagram for ac permeability measurement by direct impedance substitution.

is reasonably accurate to assume that the voltage drop across the test winding is entirely reactive. Under these conditions, the formulas

$$L = \frac{4\pi N^2 \mu A}{10^9 \ell_m} H \qquad X_L = 2\pi f L \qquad (2.41)$$

are combined and developed to create the following working equation:

$$\mu = \frac{\text{constant}}{E_R} \qquad (2.42)$$

where L = inductance, H
 N = test coil turns
 μ = effective permeability
 A = cross section, cm^2
 ℓ_m = mean length of magnetic path, cm
 f = frequency
 X_L = inductive reactance
 E_R = voltage drop across the series decade resistor

AC Hysteresis Loop Tracer. The dynamic hysteresis loops are also of value in design applications employing many of the newer magnetic materials. These loops are most conveniently obtained with the aid of a suitable oscilloscope with wideband dc amplifiers, using the test circuit illustrated in Fig. 2.86.

In obtaining this type of data, care must be taken to ensure that none of the harmonics present in either the voltage or current waveforms are attenuated by the amplifiers and that the phase shift in the amplifiers and the integrator circuit is held within certain limits. In this circuit, R_1 should be as low as possible and R_2 should have a value at least 10 times the capacitive reactance of condenser C at the test frequency.

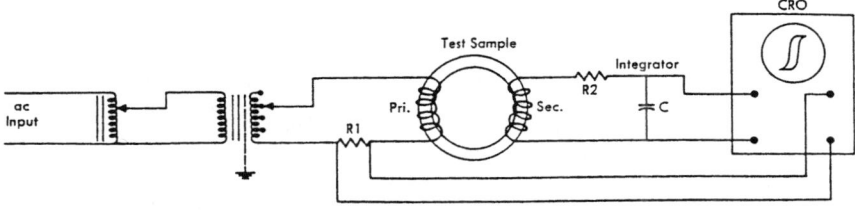

FIGURE 2.86 Schematic diagram for ac hysteresis loop tracer.

Dynamic hysteresis loops are of interest under two conditions of excitation. The condition of most general interest is one in which the sinusoidal flux in the core is maintained at all times and the exciting current is allowed to distort to the nonsinusoidal form required to maintain this flux. The other condition of general interest is that in which the sinusoidal exciting current is maintained and the core flux is permitted to distort as required to maintain sinusoidal excitation.

Under conditions of sinusoidal core flux, no harmonics are present in the voltage wave being integrated, and the R-C integrator illustrated in the circuit in Fig. 2.86 is adequate, provided its phase shift is within required limits. In the case of the core with sinusoidal exciting current, however, substantial percentages of harmonics may be present in the integrated voltage, and a simple R-C integrator may no longer suffice to give a reliable presentation of the dynamic hysteresis loop.

Constant-Current Flux Resetting. Another type of test, which is becoming very popular as a means of evaluating core materials for magnetic amplifiers and saturable reactor applications, is the constant-current flux resetting test. This method uses the basic circuit of Fig. 2.87, which employs a half wave of excitation to drive the core into saturation. A constant value of direct current is used as a means of resetting the core flux during the interval between the half waves of exciting current. An integrating voltmeter is normally used to measure the change in peak induction as a function of the dc resetting current, with a given constant value of half-wave excitation. Under these conditions, this function is a type of magnetization curve similar to the control characteristic curve of a magnetic amplifier.

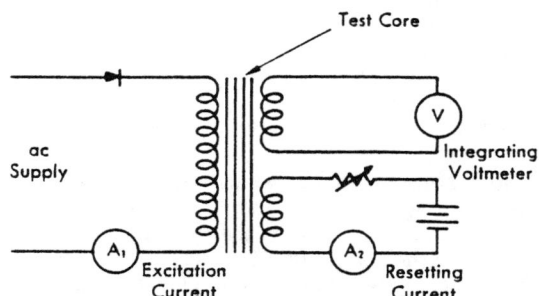

FIGURE 2.87 Constant-current flux resetting test circuit.

Low-Induction Tests. Low-induction tests are usually made with bridge equipment or with ac potentiometers, mentioned previously (see ASTM A-343). Very useful information over a broad range of inductions, extending to extremely low inductions, may be obtained with the circuit of Fig. 2.84.

Tests below 20 G may be made at 60 Hz, provided adequate isolation with electrostatic and magnetic shielding is incorporated into the test equipment. Filters may also be used if necessary to eliminate interference. These measurements normally require amplifiers and electronic equipment, which are supplied from 60-Hz power sources. Without the isolation and shielding, 60-Hz pickup and hum are likely to lead to erroneous test values at this frequency. For this reason, it may be desirable to select test frequencies which are not a multiple of the power frequency; 100 Hz is a commonly used low-level test frequency for such measurements. An audio oscillator may be used directly as a power source at low levels of induction.

Core Loss and AC Permeability Tests at Audio and Ultrasonic Frequencies. The methods of testing for 60-Hz core loss and ac permeability as previously described can be expanded for measurements at higher frequencies. As test frequencies go up, many additional problems are introduced, notably instrumentation and power supply. Because of capacity and stray field effects, improved techniques must be employed to obtain dependable test results.

The circuit diagram of the 60-Hz test in Fig. 2.81 is usually modified to the form shown in Fig. 2.88.

Test samples may be either Epstein strips, tape-wound cores, stamped rings, laminations, or special shapes having uniform closed magnetic paths. Restrictions on geometrical shape must be observed for accurate testing. For ring or toroidal shapes, a ratio of mean diameter to radial width of magnetic path of 10 to 1 or greater is desirable.

Special test frames are prepared for Epstein strips and usually have primary and secondary winding in the range of 24 to 240 turns. For the lower-frequency range, a standard 60-Hz, 700-turn test frame may be used for convenience. Watch for resonant effects, which are likely to appear in this frame. In all test frames where double lap joints are used, the vertical dimension of the frame must be kept as small as possible to avoid unnecessary calculations required to correct for air flux in the pickup coil. If care is used to minimize resistance, interturn or interlayer capacitance, stray pickup, and so on, compensating mutual inductors may be used with these test frames.

For the lower frequencies, direct indicating meters of good quality are now available. Because of frequency limitations on these meters, electronic instruments must be used over most of the frequency range.

High-quality power sources are essential. They must be capable of maintaining sinusoidal flux for all inductions at which tests will be made.

Calculations for this method are essentially the same as for the 60-Hz core loss test method described previously.

When incremental permeability measurements are to be made, it will be necessary to provide a third winding to supply the dc field. The ac blocking impedance used in the dc circuit must be designed to function effectively for any harmonics present as well as at the fundamental test frequency.

Oscilloscope measurements are frequently made at audio and ultrasonic frequencies, particularly on hysteresis loops, peak exciting current, and peak inductions or voltage. When hysteresis loops are being examined under sinusoidal flux conditions, the integrator and amplifiers used should have negligible phase shift, and the amplifiers must be capable of passing all harmonics produced in the exciting current

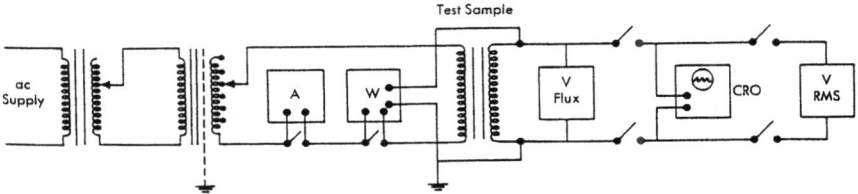

FIGURE 2.88 Circuit diagram for ac core loss and permeability measurements at audio and ultrasonic frequencies.

without attenuation. When hysteresis loops under sinusoidal exciting currents are being examined, the integrators must also be capable of integration over a wide range of frequencies, particularly when the fundamental frequency involved is rather high.

2.8 CHARACTERISTICS OF PERMANENT MAGNETS*

2.8.1 The Meaning of Magnetic North and South

In order to avoid confusion, it is important to have a single, consistent convention for the meaning of magnetic north and south. Of course, it has been known for thousands of years that like poles repel each other and opposite poles attract. It was also known for a long time before the nature of magnets was understood that a magnet suspended from a thread or allowed to float on a block of wood or in a ceramic cup would tend to rotate until one of its two poles would point to the earth's geographic north pole, and the other would turn toward the south. It was not understood, however, that the earth itself was a giant magnet. The pole of the magnet which turned toward the earth's north pole was called the *north-seeking pole,* or simply the north magnetic pole of the magnet. This is the ancient and present meaning of a magnetic north pole. Since opposite poles attract, however, it can be seen that the earth's geographic north pole must be a magnetic south pole! Many have had difficulty with this convention, but it has been so well established over hundreds or thousands of years that it is impossible to change now. If uncertainty exists about which polarity a magnet has, it is not difficult to repeat the old experiment. Hang the magnet on a thread (but not one which has a great deal of twist, as a torque will be exerted on the magnet due to the twist and its weight) or float it in a nonmetallic cup, away from any steel object (such as a steel basin or belt buckle), and if the direction of geographic north is known, the polarity will follow.

***Flux Density* B *and Coercivity* H.** Two important properties of permanent magnet materials are the *flux density* (also called the *magnetic induction*) B and the *coercivity* (also somewhat ambiguously referred to as the *magnetic field strength*) H. These two quantities are related, exist at every point in the magnet and its surroundings, and in general vary from one position to another. They are *vectors*—that is to say, each has a scalar (i.e., a number) value attached to it and also a direction. In free space the two have the same direction at a given point and are related by a simple constant called the *permeability of free space* μ_0, but within a magnet the relationship is more complicated. In some materials the two do not even have the same direction. These two quantities are fundamentally different, the flux density playing a similar role in magnetic circuits as current (per-unit area) in electrical circuits, and the coercivity of magnetic circuits resembling the electrical voltage (per-unit length).

$B = \mu_0 H$ in free space or, practically, in air, plastic, etc.

In most materials which are not more magnetic than space, including air, organic substances such as plastic and wood, and most metals, however, the magnetic permeability is almost indistinguishable from that of free space.

It is frequently convenient to specify permeability not in absolute units but in relationship to that of free space. This is defined as the relative permeability μ_r:

---

*Section contributed by Joseph J. Stupak, Oersted Technologies.

$B = \mu_r \mu_0 H$ in isotropic, magnetically permeable materials

The B-H Curve of Magnetic Materials.   Suppose an unmagnetized sample of magnetic material were placed in a volume of space in which H could be varied at will. This could be done, for example, inside a very long coil of electrically conductive wire, through which the electric current may be varied. In practice a short coil must be used, and a yoke of conductive material (steel) is used to prevent changes in the field at the coil ends, in a device called a *permeameter*. The flux density B in the magnet may be measured with a hall sensor (the construction of which is described later) or by other means.

In Fig. 2.89, H is initially small, and B may be linearly related to it by a nearly constant ratio (the initial permeability). If, from this region, H were reduced to zero, B would also go to zero; the part is not magnetized. Beyond some point, however, as H increases, the ratio of B to H rises much more sharply. If, now, from the steep curve H were returned to zero again, B would drop to some positive value above zero. It is said that the part is partially magnetized. As H is further increased, the ratio of B to H decreases again, eventually to the same ratio as in free space. The part of the curve traced from the origin to here is different from that of subsequent parts of the curve, and is called the *virgin curve*. If H is then reduced to zero in the magnet (which would require a small but nonzero current in the coil to compensate for small losses in the yoke), B reaches a value known as the *remanance B*. Higher values of H than that reached at which $B/H = \mu_0$ do not increase B_r, and the magnet is said to be fully magnetized or saturated.

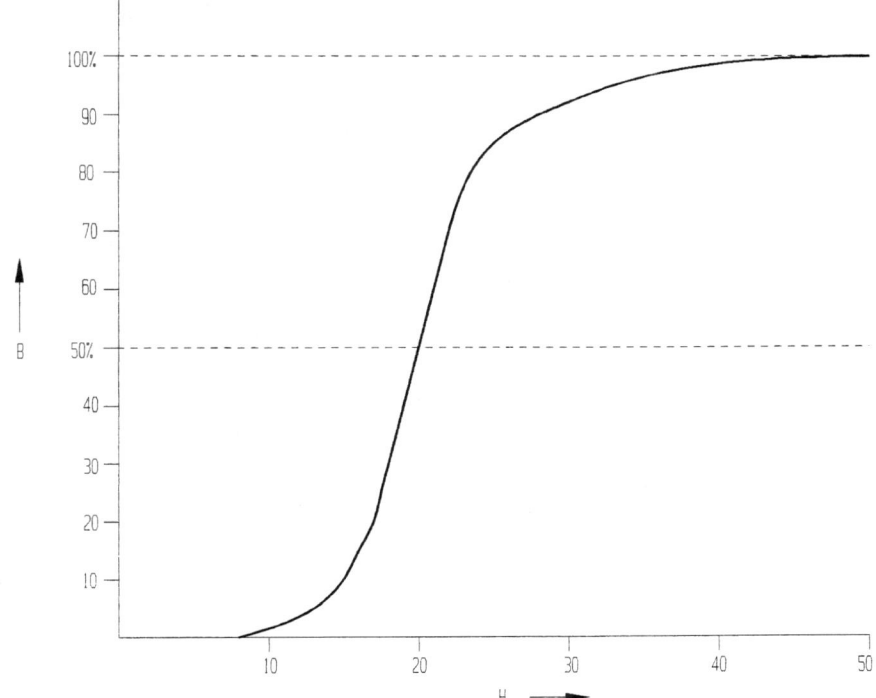

FIGURE 2.89   Initial magnetization curve, percent B_r versus H_s, for a neodymium-iron sample.

If, next, the coercive force H is increased in the negative direction, B moves downward toward the left, as shown in Fig. 2.90. The slope of the curve in this region is called the *recoil permeability*. This is the normal region of use of the magnet (along the line of recoil permeability, through and on both sides of B_r). If, then, H is made even more negative, a point is eventually reached at which the curve begins to slope steeply downward. This location is called the *knee of the curve*. If the operating point—that is to say, the point representing B and H—is moved down around the knee, and if H is reduced to zero, B does not again retrace the path up and around the curve, but instead moves inward at the slope of the recoil permeability to some value on the B axis less than B_r, and the magnet is partially demagnetized. This line is said to be on a *minor loop*, whereas the outer line is called the *major loop*. If H is increased still more, B is eventually driven to zero. The value of H at which the curve crosses the H axis is labeled H_c, the subscript standing for *coercive force*. Continuing to increase H negatively, B drops rapidly in the negative direction below the curve, until the slope again levels out to the value of free space. If H is now returned to zero, B returns to the B axis at $-B_r$. If H is again increased in the positive direction (please note that "positive" for this discussion is an arbitrarily chosen direction, along the magnet axis), the curve traces out the same curve in the fourth quadrant as it did in the second, reflected about the B and H axes. Further increases in H bring the curve back to the line previously traced in this quadrant, closing the major loop. If H is repeatedly cycled between sufficiently large positive and negative values, the curve is traced repeatedly around the major loop.

The curve described here is called the *normal curve* and is the one used for magnetics design. Another form of the curve, called the *intrinsic curve*, is used for materials studies and represents the part of the field produced by the magnet material

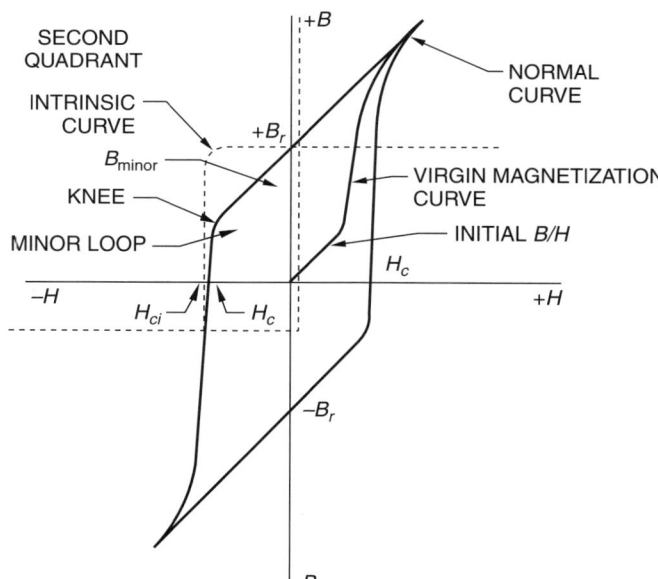

FIGURE 2.90 Typical four-quadrant magnet B-H curve.

alone. This curve has the free-space value of B, which would result from the applied coercive field H subtracted from it at every point. It passes through B_r and $-B_r$, since H is zero there, but has a higher B in the second quadrant and a lower one in the first quadrant than the normal curve. This curve is often presented along with the normal curve on the same graph.

Since the major loop in the first quadrant simply follows the recoil permeability slope and the curve is symmetric about the B and H axes, all the information needed for magnetics design is contained in the second quadrant alone. For this reason, manufacturers normally publish only this part of the curve, with a positive B vertical axis and a negative H horizontal axis. The shape of the B-H curve is a function of temperature, and B_r, H_c, and the location of the knee may all vary significantly. Some materials are little affected by normal ambient changes, but others may be so changed that magnets which operate well at room temperature are partially demagnetized by operation at higher or lower temperatures.

The ratio of B to H in a magnet is a function of its permeability and its shape. This relationship is a straight line passing through the origin (where B and H equal zero). The location where the load line crosses the B-H curve of the material establishes a point representing the actual value of B and H in the magnet. If this point is below the curve knee, the part is partially demagnetized. In general, this occurs if the part thickness, in the direction of flux, is too small compared to its dimensions at a right angle to the direction of flux. A famous calculation by Evershed (1920) gives an approximate but very good estimate of the slope of the load line, as a function of cross-sectional shape, length, and so on, for a magnet in free space or air. The load line is moved horizontally left or right by additional coercive forces caused by electric currents through a coil, the flux of which links the magnet.

Some materials, such as Ceramic 7 in Fig. 2.91, have a B-H curve with a knee which occurs below the H axis—that is, only at negative values of B—in the second quadrant. Since bar and plate magnet shapes cannot produce reversed flux by shape alone, these materials are immune from shape demagnetization. Ceramic 7 is less powerful than Ceramic 8, as shown in Fig. 2.92, but the shape of this B-H curve makes Ceramic 7 more useful in some applications.

The Maximum Energy Product. If the units of B and H are multiplied together, they are found to be equivalent to energy. It can, in fact, be shown that the magnetic energy stored in a region of space is:

$$U = \int \tfrac{1}{2} BH \, dV \qquad (2.43)$$

Some magnetic materials are capable of producing a large amount of flux, but have only a limited coercivity H. Others have a lower flux density, but higher H. A magnet of either type can be used to produce magnetic flux in an air gap of a magnetic circuit, but the shapes of the magnets will be different for different materials. The product $(B \times H)$, without the factor ½, is called the *maximum energy product*. On the B-H graph of a particular material, values of (BH) may be chosen, and for each choice of B there is a corresponding value of H. When these lines are plotted, they are found to be hyperboles, symmetric about the 45° line from the origin. Some of these curves do not touch the B-H curve, and others cross and then recross the curve. Only one curve just touches the B-H curve at a single, tangent point. The value of this curve is called the *BH product*, or maximum energy product, and in the United States is usually expressed today in mega-gauss-oersteds (MGOe).

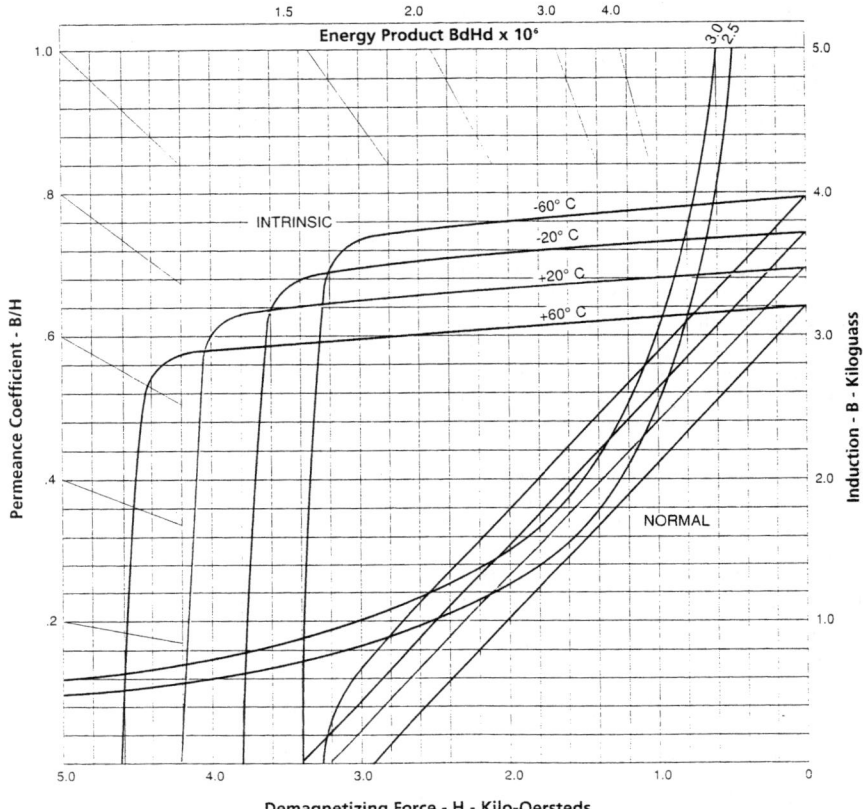

FIGURE 2.91 B-H curve for ceramic 7. *(Courtesy of Arnold Engineering.)*

Other Characteristics of Magnetic Materials. Certain other characteristics of magnet materials are also of considerable interest to magnetics designers and users. When magnetic flux changes with time in a region of space, an electric field is set up around the region (Faraday's induction law). Almost any useful application of magnets results in such changes, with the flux density B varying in the magnet. If the magnet is electrically conductive, eddy currents will be set up inside the magnets themselves, with the effect of slowing down the rate of change of flux and also producing heating in the magnets. These induced currents may be surprisingly large, and on occasion the heating in the magnets may become great enough to damage them, bonding agents, or their surroundings. A way to reduce this effect is to laminate the magnets, with the surface planes between laminations lying parallel to the direction of magnetic flux, as shown in Fig. 2.93. Ceramic magnets have extremely high resistivity, but neodymium-iron and samarium-cobalt in solid, sintered form, are good conductors. The bonded and molded products are less so. Alnico and cunife are also relatively good conductors and may cause problems in this way.

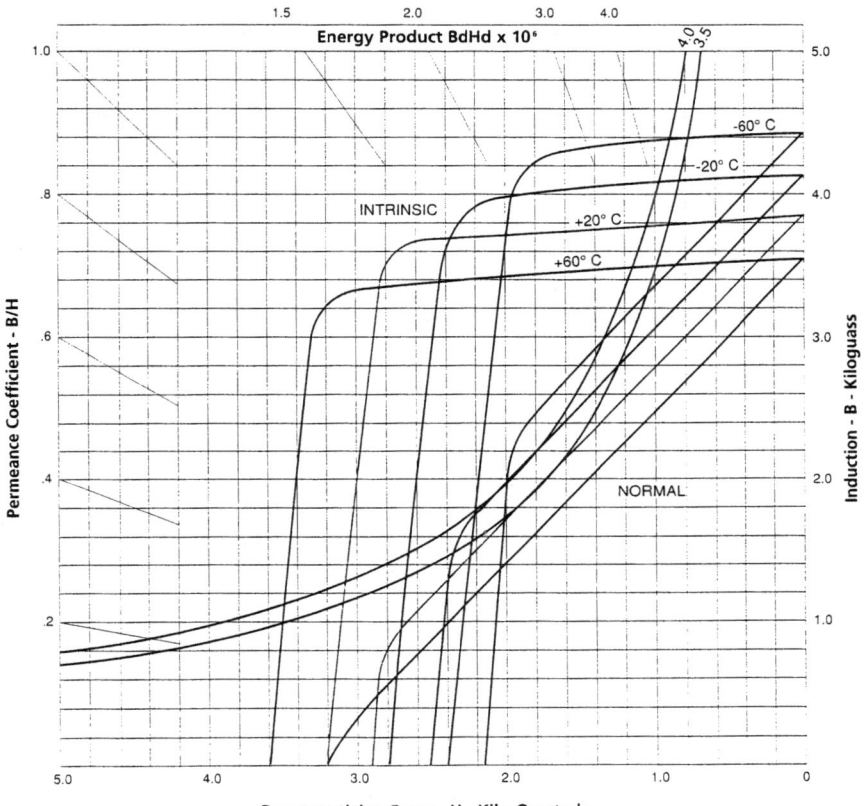

FIGURE 2.92 B-H curve for ceramic 8. *(Courtesy of Arnold Engineering.)*

Many magnetic materials are brittle and may shatter in shipment or use unless protected. Edge chipping in these materials is a common problem.

Magnetic materials often show variations of strength with temperature, becoming more or less powerful with temperature rise. These effects are reversible to some degree; that is to say, when the magnets are restored to their original temperature, their original magnetic characteristics return. There may also be irreversible effects, such as a slow permanent degradation of magnetic properties, when operated for a long time at elevated temperatures. If a magnet is heated to a temperature usually referred to as the *Curie temperature,* however, the magnet will abruptly lose all magnetism and revert to its original, unmagnetized or virgin state. After cooling to ambient temperature, the magnet is unharmed and may be remagnetized as before. Some magnet materials are routinely demagnetized in this way (including neodymium-iron).

Magnets are exposed to mechanical forces caused by magnetism and by tangential forces exerted on rotating bodies, impacts, and temperature expansion differentials between the magnet and the materials to which it is bonded. Magnet material strength and thermal expansion are, therefore, sometimes important.

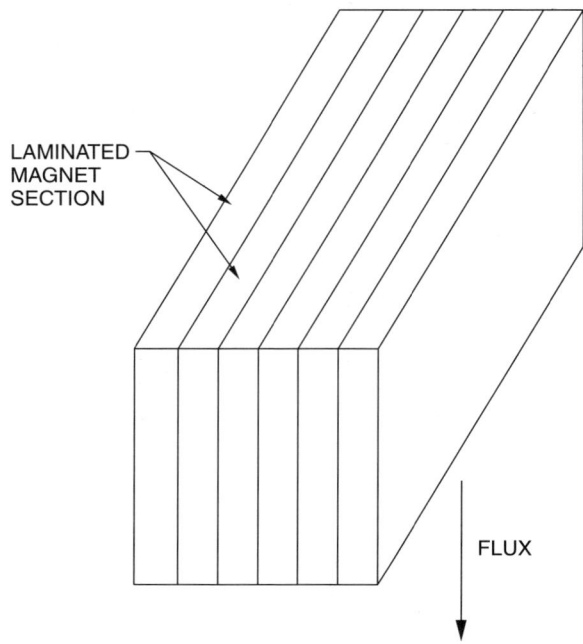

FIGURE 2.93 Laminated magnets.

Magnetic Coatings. Some magnetic materials—neodymium-iron in particular—are very chemically active and will oxidize (rust) if not protected. Considerable progress has been made recently in reducing the chemical reactivity of neodymium-iron, and the products of the companies which have been leading this research are much less reactive. To suppress oxidization, and sometimes to help resist chipping, and so forth, many magnets are coated. A good magnet coat should be conformal; that is to say, the thickness of the coat should not be greatly reduced at edges and corners, but should be approximately uniform everywhere. If the coat is for chemical resistance, it should not have pinholes, tiny holes through which oxygen may enter. It should have good peel resistance, so that the magnet, when bonded to a substrate, will not separate at the magnet-coating surface (instead of at the bond-substrate surface). After meeting these requirements, the coating should also be as thin as possible, so as to introduce as little additional air gap as possible to the magnetic circuit. For some applications—notably motors and actuators for use in computer peripheral hard disk drives—the coating must not degas—that is, give off substances which could coat other nearby objects and interfere with their operation.

Among coating methods in common use today, the least expensive is probably nickel plating. The most expensive, and probably the best, is an aluminum coat sputtered on in a vacuum, the surface of which is then given a chromate conversion (a chemical film). Another method is E-coat, which is a very thin epoxy coat applied from a liquid bath by electrostatic means. An excellent proprietary organic coat meeting all the aforementioned requirements, with low degasing characteristics and which is inexpensive as well, is also available.

Bonding Materials for Magnets. Adhesives used to bond magnets into place need to produce thin, strong bond joints and must be able to fill any voids resulting from mismatches in shape between magnets and substrates. The bond should be thin, to avoid causing additional coercive force loss as a result of the increased effective air gap. The bond thickness should repeat well from one part to another. However, it must also be flexible enough to accommodate the differences of expansion between the magnet and the substrate. If this need is not addressed, the magnets may break free due to thermal stresses, shock, or magnetostriction during use. The bonding agent must, of course, adhere well to the substrate material (usually steel or silicon-iron). It must also lend itself well to high-volume production for most uses and should not be toxic or present other dangers to production personnel.

Table 2.11 presents some characteristics of common magnet materials. "Common" is a name by which these materials are generally known throughout the industry. The *MMPA brief designation* is a method developed by the Magnetic Material Producers Association as a means of generalizing the magnet performance included under the name, irrespective of the type of material it is. The first number is the maximum energy product. The second number is the intrinsic coercive force. For instance, Ceramic 8 could be called 3.5/3.1.

2.8.2 Characteristics of Available Materials

Some older materials, such as lodex, cunico, and vicalloy, were once on the market but are now obsolete. A rare earth material similar to neodymium-iron called praesodymium-iron was previously available but is no longer in production. Cunife, another old material, is rare but is available from a small number of suppliers. The names of some of these older materials follow a pattern in use many years ago, in which the name is put together from the chemical symbols of the components. Thus, cunife is copper, nickel, and iron (chemical symbol Fe). Cunico was copper, nickel, and cobalt, and alnico is aluminum, nickel, and cobalt. The major materials available today are the ceramics (strontium and barium ferrites), various forms of alnico, two forms of samarium-cobalt (Sm, Co_5, and Sm_2Co17, called "one-five" and "two-seventeen," respectively), and neodymium-iron. These materials, in addition, may be in solid (often sintered) form, bonded, or molded.

The material from which most magnets is formed consists of tiny flakes or powder granules, each of which is essentially a single magnetic domain. These flakes have a preferred direction of magnetization and may be magnetized in either direction along that line, north-south or south-north. Solid magnets are made by placing the powder into a mold, pressing the part to shape, then removing the part and sintering it in a furnace at high temperature until the grain boundaries fuse together. Bonded magnets are produced by coating individual magnetic particles with a binder such as nylon, placing the powder in a mold, and then pressing it at a high temperature and pressure into a finished shape. Injection-molded materials are made by injecting a thermoplastic-containing magnetic powder into a mold at high pressure and temperature, then allowing it to cool, and ejecting the part. These parts contain less magnetic material and are therefore less powerful than bonded parts, and they are in turn less powerful than those made of solid magnetic material. Magnets of these types may be made much stronger if the magnetic particles are oriented by a strong magnetic field as the part is being formed. If the particles are distributed in random directions, the material is called *isotropic* or *nonoriented*. Such magnets may be magnetized in any direction. If all the particles are rotated so that their axes are all or substantially parallel, the part is called *anisotropic* or *oriented*. The energy product for an oriented material of the same particle density may be as much as

TABLE 2.11 Properties of Magnetic Materials

Material name	MMPA brief designation	Max. energy product	B_r, G	H_c, Oe	H_{ci}	Relative recoil permeability	Density, lb/in^3	Mechanical state
Cunife		1	4800	4400		2.200	0.311	Ductile
Alnico								
Alnico 1	1.4/0.48	1.4	7100	450	480	7.000	0.249	Brittle
Alnico 2	1.7/0.58	1.65	7750	580	580	6.400	0.256	Brittle
Alnico 3	1.35/0.50	1.35	6400	560	500	7.000	0.249	Brittle
Alnico 4		1.45	6000	660		4.500	0.253	Brittle
Alnico 5	5.5/0.64	5.5	12700	640	640	2.200	0.265	Brittle
Alnico 5-7	7.5/0.74	7.5	13400	740	740	1.900	0.365	Brittle
Alnico 5DG	6.5/0.67	6.5	13300	685	670	2.000	0.263	Brittle
Alnico 6	3.9/0.80	3.65	10500	760	800	4.200	0.268	Brittle
Alnico 8	5.3/1.9	6.75	9000	1600	1860	2.000	0.263	Brittle
Alnico 9	9.0/1.5	10.5	10500	1500	1500	1.500	0.264	Brittle
Ceramic (ferrite) 3250								
Ceramic 1	1.0/3.3	1	2300	1850	3250	1.100	0.180	Brittle
Ceramic 5	3.4/2.5	3.6	3950	2400	2500	1.060	0.178	Brittle
Ceramic 7	2.7/4.0	3.3	3950	2400	4000	1.060	0.178	Brittle
Ceramic 8	3.5/3.1	4.3	4300	2500	3050	1.060	0.179	Brittle
Ceramic bonded	0.4/3.5	0.4	2450	2200	3500	1.040	0.134	Flexible
Samarium cobalt								
SmCo$_5$	19/20	24	9500	9000	20000	1.050	0.300	Brittle
Sm$_2$Co$_{17}$	24/16	28	10900	6500	16000	1.050	0.300	Brittle
Sm$_2$Co$_{17}$ bonded	17/12	17	8900	6800	12000	1.050	0.252	Somewhat brittle
Neodymium-iron								
Neo 35	35/14	35	12300	11300	14000	1.090	0.270	Brittle
Neo 40	40/12	40	12900	12100	12000	1.090	0.271	Brittle
Neo 45	45/17	45	13375	13000	17000	1.090	0.271	Brittle
Neodymium bonded	10/	10	7400	6000		1.200	0.230	Somewhat brittle

Material names	Max. suggested use temp, °F	Coercive field to magnetize to 96% B_{sat}, Oe	Tensile strength, ob/m	Electrical resistivity μ, Ω-cm	Curie temp., °F	Composition	Notes
Cunife	770°	3000	123000			Copper, nickel, iron	
Alnico							
Alnico 1	840°	2250	4000		1430°		
Alnico 2	840°	2900	3000	65	1500°	Aluminum, nickel, cobalt	
Alnico 3	840°	2800	12000	60	1900°	Aluminum, nickel, cobalt	
Alnico 4	840°	3300	9000	65	1430°	Aluminum, nickel, cobalt	

TABLE 2.11 Properties of Magnetic Materials (*Continued*)

Material names	Max. suggested use temp, °F	Coercive field to magnetize to 96% B_{sat}, Oe	Tensile strength, ob/m	Electrical resistivity μ, Ω-cm	Curie temp., °F	Composition	Notes
Alnico 5	930°	3200	5400	47	1630°	Aluminum, nickel, cobalt	
Alnico 5-7	930°	5700	5000		1630°	Aluminum, nickel, cobalt	
Alnico 5DG	930°	3500	5200		1630°	Aluminum, nickel, cobalt	
Alnico 6	930°	3800	23000	50	1610°	Aluminum, nickel, cobalt	
Alnico 8	932°	8000	10000	50	1500°	Aluminum, nickel, cobalt	
Alnico 9	932°	7500	7000	50	1500°		
Ceramic (ferrite)							
Ceramic 1	450°	10000			750°		
Ceramic 5	480°	10000	4000		892°		
Ceramic 7	480°	10000	4000		842°		
Ceramic 8	480°	10000	4000		842°		
Ceramic bonded		10000					
Samarium cobalt							
$SmCo_5$	480°	30000	5800	55	1340°	Samarium, cobalt	Bright metal
Sm_2Co_{17}	572°	35000	2800	86	1472°	Samarium, cobalt	Bright metal
Sm_2Co_{17} bonded	302°	20000	2800	0.43×10^6			Grainy surface
Neodymium-iron							
Neo 35	302°	32000	12000	150			
Neo 40	329°	32000	19200	150			
Neo 45	329°	32000	19200	150			
Neodymium bonded	302°	20000		0.43×10^6			

about three times that of nonoriented material, and B_r and H_c for the oriented material will be increased by as much as about 1.7 (that is, the square root of 3) times that of the nonoriented material (these values will be lower if the material is not completely oriented).

Some magnet material is produced as particles molded into rubberlike plastic in flexible sheets—the familiar refrigerator magnets. This material is flexible and ductile, and is easily cut to shape with scissors, knives, or steel-rule dies. Unfortunately, it is not very strong magnetically, presently reaching only about 1.8 MGOe energy product in oriented form for ferrite powder, but it is sometimes used in inexpensive permanent-magnet motors. It is also possible to produce flexible sheet, which is enhanced with neodymium-iron material (or made entirely with that powder, instead of ferrite). It is as difficult to magnetize as solid neodymium, however, and is far weaker.

Cunife. This material, discovered in 1937, has an orange-coppery appearance, and is composed of copper, nickel, and iron. It is available as small-diameter wire and strip only, oriented along the long axis of the part. Although not very powerful and having little resistance to demagnetization, it has the advantage for some applications of being ductile and can be formed to shape without breaking. It is highly electrically conductive. This material is very nearly obsolete, and its future supply could be uncertain.

Alnico. Alnico is made of aluminum, nickel, and cobalt. The cobalt fraction is rather high (about 30 percent), and cobalt has been of such uncertain supply in the past, due to wars and suspected deliberate manipulation, that it is shunned by many as unreliable. It dates from a Japanese discovery in 1932, followed by improvements from many researchers in different countries. Alnico has a high flux density but relatively little coercivity, that is, resistance to demagnetization. Alnico parts used in air, without the aid of permeable (steel) poles to help shunt the flux, must be long and thin to avoid shape demagnetization. The material is hard and brittle. It may have a bright, polished, chrome-like appearance or be a dull, rough gray if unground. Alnico is little affected by temperature changes, compared to some other magnet materials, and its Curie point is very high. There are large numbers of grades, with somewhat differing properties.

Ceramic (Strontium and Barium Ferrite) Magnets. As much as 2500 years ago, miners in an area of what is now Turkey, called Magnesia, noticed bizarre interactions between the ore they were mining and their steel tools. The ore, called *lodestone,* was ferric ferrite, a sort of natural magnet. Our word *magnet* is derived from the name of the region. Two different types of material have been developed from the ore in recent times, called *soft ferrites* and *hard ferrites.* In this case, "soft" refers to the property of being highly magnetically permeable, but not retaining a magnetic field after the exterior coercive force is removed. Such ferrites, made of magnesium, zinc, nickel, and manganese, are useful for the cores of high-frequency transformers. Ferrite material is also useful to absorb microwave and radar energy (in stealth aircraft, in the walls of apartment buildings near radio transmitters, etc.). Magnetically hard ferrites, on the other hand, are made of barium and strontium ferrite and retain their magnetic fields after magnetizing. Ferrite magnets are very inexpensive, in terms of magnetic energy per unit cost. They are brittle and are so hard that they can be cut effectively only with diamond tools. Grade 1 is anisotropic (not oriented), which can be a useful characteristic for some purposes. Grade 5 is readily available and very inexpensive. Grade 7, as noted earlier, has a B-H curve knee which is below the H axis. It has a high level of resistance to demagnetization, and cannot be demagnetized by shape alone (in bar or plate form). Grade 8 in various subgrades is very powerful. The materials most often used for new design of ferrite permanent-magnet motors and actuators are grades 7 and 8. A few manufacturers have elevated the variations of grade 8 to grade 9 or even 10, but these are found to be equivalent to those of others referring to them as variations of grade 8, and these additional designations are not widely accepted.

Samarium-Cobalt. This material uses a high percentage of cobalt (about 30 percent), and has the same limitations in that regard as Alnico. It is less powerful than neodymium-iron and is more expensive (except in certain bonded forms). It is also very difficult to magnetize—more so than neodymium-iron. It is very brittle and is producible only in small pieces, which must be bonded together for larger assemblies. On the other hand, it has a very low variation of field strength with temperature, which is a problem with neodymium-iron, and has a very high Curie point.

Neodymium-Iron (Neodymium-Iron-Boron). This is the latest and most powerful of magnet materials, with available forms on the market of up to 45 MGOe energy product and some limited production of material up to 50 MGOe (that is, more than 10 times stronger than the best ferrite). The price, originally prohibitive for all but the most extreme requirements, has dropped greatly as production has increased, and the cost continues to decline. It is now within reach for many applications. The material has also been improved, with lower chemical reactivity, improved consistency, high recommended temperatures for long-term use, and easier magnetization from the virgin state. Some manufacturers have learned how to make radially oriented neodymium-iron rings, which are very useful in motors and rotary and linear actuators. It is brittle, except in some molded and sheet forms. The material is very chemically active and should be protected from oxidation, even in the newer, low-reactivity forms. This material is so powerful that persons unfamiliar with it are amazed at the strength of even very small magnets, and larger pieces (perhaps 2 in and greater in cross section) can be dangerous to handle. Large magnets are either magnetized in place after assembly, if possible, or handled by remotely controlled equipment and hydraulic rams. Delicate mechanical devices such as watches and instruments may be bent and destroyed, and magnetic-tape and strip records such as credit cards and computer backup memory may be erased yards away.

2.8.3 Magnetizing

Permanent magnets are usually shipped from the manufacturer unmagnetized. There are a number of reasons for this. As noted in the section on neodymium-iron, powerful modern magnets in magnetized form may present hazards to personnel and other equipment. Magnetically permeable dirt is widespread, and once a magnetized magnet is contaminated it is very difficult to clean (the dirt is not wiped off, but slips around a wiping cloth, for example). Shipping presents problems because of the possibility of erasing records, destroying machinery, and affecting instruments and electronics. Shipping of magnetized parts is facing increasing regulation. Magnetized parts exert forces on other parts and surrounding steel, which may break the magnets themselves. For some types of motors and actuators, pole transitions become very important, and the manufacturer may want or need to keep control of the process. Assembly of magnetized parts may be difficult or dangerous, so magnetization after assembly may be required.

In order to magnetize a permanent magnet, the surrounding magnetic field must be raised to or above a limiting field, usually labeled H_s (the subscript s standing for "saturation," although, in fact, the part may not be completely saturated magnetically at that field). Once the field is strong enough, magnetization takes place in an extremely short time, in order to overcome eddy currents induced by the rate of change of flux. These induced currents may occur in the magnet itself, if it is electrically conductive, as well as in the surrounding parts to which it is bonded, or in the magnetizing fixture.

On occasion, some magnets of low coercivity may be magnetized by other means, such as passing them through the field of dc electromagnets or other permanent magnets, but almost all magnets today are magnetized by immersing them in a momentary, pulsed field caused by passing a pulse of electrical current through a conductor of copper wire, producing a magnetic field in the vicinity of the wire for a short time. The wire may be coiled in a number of turns, to reduce the required current, and it may or may not be strengthened and focused by use of permeable pole material such as mild steel or vanadium-permandur. The latter metal is an alloy of iron and cobalt, with about 2 percent vanadium added. Whereas mild steel saturates at about 20,500 G, vanadium-permandur saturates at about 24,000 G, a useful increase for some purposes, in spite of the high cost of this material.

The magnetizing information of the part of the B-H curve may be redrawn in intrinsic form (that is, the coercive field times the permeability of free space is subtracted from the flux density, so that the result represents the contribution to the field caused by the material only), and the flux density component is normalized as percent B_r. The coercive field is then labeled H_S, to indicate that the purpose of the curve is to show the degree of magnetic saturation of the material. The information presented by this curve, as shown for one material in Fig. 2.90, is very useful in controlling the magnetizing process. All manufacturers have this information on their products, but they are often reluctant to give it to users. This is because they also publish a "recommended field to magnetize," a single field value. Manufacturers want this figure to be as small as possible, showing that their material is easy to magnetize by comparison to their competition. In fact, however, the value given is often one which produces substantially less than full magnetization. For neodymium-iron, it is common for this figure to represent about 96 percent of saturation, but the actual percentage could even be substantially less. If the customer is in possession of the entire curve, the sales misstatement becomes clear. The value of B_r used, however, is obtained at an extremely high field, usually on the order of 100,000 G, a field obtainable only for very small samples of simple shape at a low cycle rate in the laboratory. It is not practical to reach such fields in volume production.

The very high energy pulse needed for magnetization is produced by a device usually called a *magnetizer*.

2.8.4 Magnetizer Circuits

Coil and Rectifier. A simple magnetizer circuit, usable only for low coercivity materials such as alnico and cunife, is shown in Fig. 2.94. This circuit is sometimes used today for steel assemblies containing ferrite magnets as well, in which the steel parts help concentrate the flux to increase the field, although the parts, on testing, are found to be less than fully magnetized.

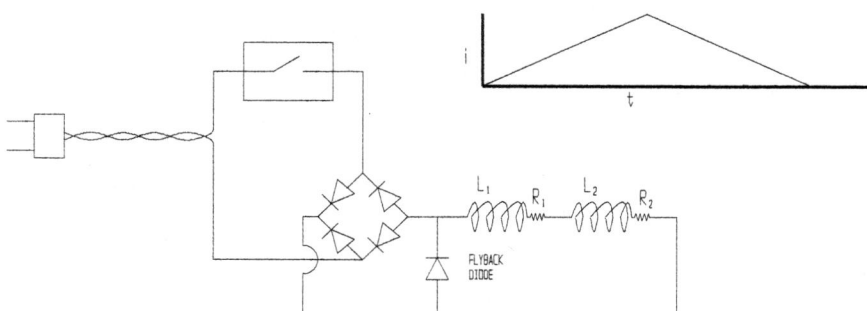

FIGURE 2.94 Long-cycle magnetizer.

In this circuit, line ac power (possibly boosted in voltage by a transformer) controlled by an on-off switch is simply rectified by a diode bridge and passed through two very large coils containing thousands of turns, which are wound around the ends of a steel C frame yoke at the air gap. The huge inductance of the coil limits the current as it builds up, requiring possibly as long as a second to reach rated current. When this current is reached, the power is shut off. The current in the coil cannot be abruptly cut off, or the inductance would produce a voltage high enough to destroy it, in order to complete the circuit and dissipate the stored energy. A flyback diode is therefore added to allow the energy to die out in the coil resistance. Devices of this sort may operate at 460 V ac, and draw up to 60 A or more peak. They are highly inefficient, and produce large amounts of waste heat, which may limit the length of time the device may be operated without a cooling period of hours. The circuitry is simple and inexpensive, but the large coils and heavy steel C frame are costly, heavy, and large. Magnetizers of this type have often been used in the past by the speaker industry, but are falling into disfavor because of performance limitations.

Half-Cycle Machines. It may seem surprising to some, but it is actually possible to draw current of thousands of amps directly from the power lines for a very brief time. A half-cycle machine (see Fig. 2.95) simply turns on the current as the line voltage passes through zero, then turns off again when the current reaches zero on the next half-cycle, about $\frac{1}{120}$ of a second later. As switching is done at the crossings, little power is dissipated in the switch (which may be a silicon-controlled rectifier [SCR] or ignitron tube). Of course, this procedure does great violence to the line and to equipment attached to it. Electromechanical power meters are unable to respond to such a powerful surge in such a short time, so the power use is unrecorded. Such equipment is still in use in the United States in some places, although it may present a hazard to other equipment.

Capacitive Discharge Magnetizers. Most magnetizing is performed by capacitive-discharge machines (as shown in Figs. 2.96 and 2.97), of which several different circuits are in use. Electric power is taken from the line, raised in voltage by a line transformer—or, in equipment of more modern design, a high-frequency "chopper" and small high-frequency transformer—and is then rectified to dc current. This current is used to charge capacitor banks which store the electrical energy until the time comes to release a pulse. The circuit is then discharged at a very high current into the fixture, where part of the energy is converted to a powerful magnetic pulse of brief duration, possibly a few milliseconds or tens of milliseconds long. The current in the

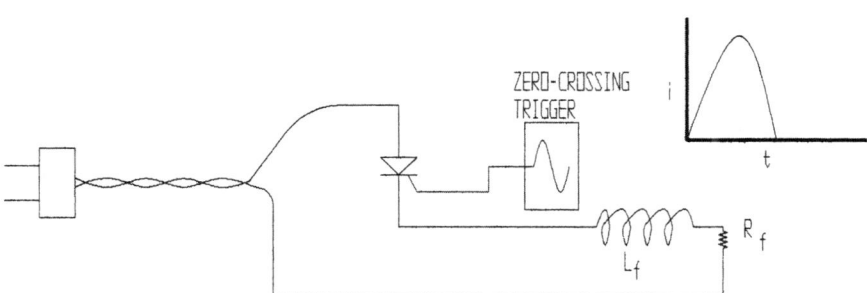

FIGURE 2.95 Half-cycle magnetizer.

fixture, typically thousands or tens of thousands of amps, is then either dissipated in the coil winding through a flyback diode or partially absorbed by the fixture resistance and partially by being returned to the capacitors. After the pulse, if some of the energy was returned to the capacitors (at opposite polarity), it is dissipated by an auxiliary circuit.

In the circuit shown in Fig. 2.96, the capacitors used are polarized and can accept charge in only one direction. This type of capacitor, the aluminum electrolytic capacitor, is relatively inexpensive and small for the amount of energy stored. The negative terminal of this capacitor may not be raised more than a few volts above that of its positive pole, however, without leading to failure. As the maximum operating voltage of the capacitor may be less than that required for the magnetizer, they may be connected in a series for higher voltage. They are then connected also in parallel to increase total capacitance of the bank. The negative terminals of the capacitors are connected directly to ground, and it is wise to add diodes across the banks to ensure that a reversed voltage cannot occur across them after discharge (the result of uneven internal leakage). The control switch may be an SCR or an ignitron. The name of this latter device, a large mercury-filled vacuum tube, should be pronounced "ig-NAI-tron," according to those who manufacture them, but is often mispronounced as "IG-ne-tron" by others who have only encountered the name in printed form. Ignitrons were used before the introduction of high-power SCRs, but, although rugged, they have a high voltage drop across them (up to 200 V, although 30 V is more typical), which varies with temperature, recent use, and age. By comparison, the voltage across an SCR is very small (a volt or two) and stable, resulting in efficient, repeatable performance. Ignitrons perform best if maintained at a temperature higher than ambient (perhaps 95°F). They are often equipped with water-cooling jackets, but their rate of use in magnetizers is usually not high enough to maintain the higher temperature. It may be best, therefore, to use a combination water heater-chiller with such tubes, maintaining their temperature at a level recommended by the manufacturer, to prolong tube life. It was formerly the practice to rebuild ignitron tubes after they began to decline in performance, but because of

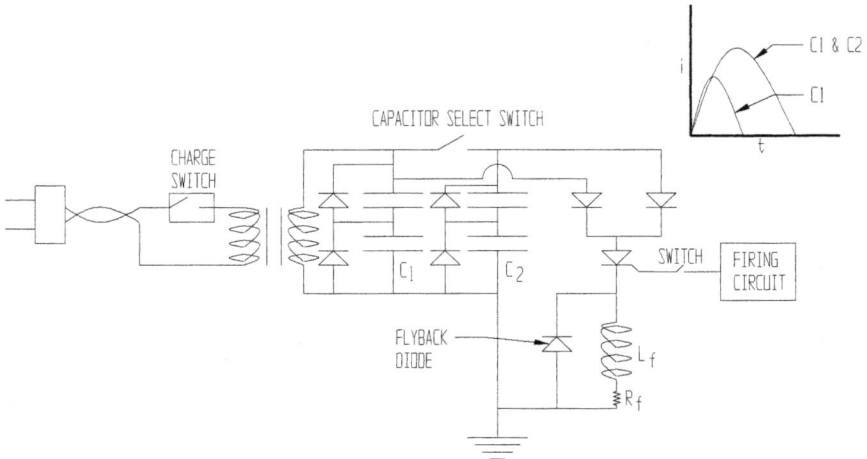

FIGURE 2.96 Unipolar capacitor discharge.

environmental restrictions it is difficult or impossible to find companies willing to perform this service at present in the United States.

The unipolar circuit shown in Fig. 2.96 has another (patented) feature: variable capacitance. This feature is often useful, as it can be shown that no workable fixture design may be possible for use with a particular magnetizer of a given, fixed capacitance, because either the field may be too low at any peak current which will not overheat the fixture, or at the other end of the scale because any voltage high enough to produce a sufficiently large peak current to magnetize the part would destroy the electrical insulation.

A magnetizer circuit using bipolar capacitors is shown in Fig. 2.97. These capacitors may be charged in either direction, so that they have no dedicated positive or negative poles. These capacitors are much more expensive than the unipolar variety and must be charged to a much higher voltage to obtain a reasonable power density. The higher voltage in turn requires more volume within the winding volume to be used for insulation. Safety may also be an issue. On the other hand, since some of the energy of the pulse is returned to the capacitors after the pulse, this design is more efficient than the previous one, resulting in less heating. For many purposes the difference in efficiency is minimal, but for small parts made of powerful material with a large number of poles, this design has advantages. The high voltage leads to fewer turns of larger wire, lowering the circuit inductance, and this, with the lower capacitance, leads to very short pulses, with resultant higher eddy currents. In large parts, this may be a problem.

After the power switch (either an SCR or ignitron) is turned on, the magnetizer-fixture combination behaves (at least in its first-order, linear approximation) as a capacitor, a resistor, and inductance connected in series, with an initial voltage E_c across the capacitor. The voltages across each component are:

Inductance: $\quad E_l = L\, di/dt$ \hfill (2.44)

Resistance: $\quad E = i R$ \hfill (2.45)

Capacitance: $\quad E_c = \left(\dfrac{1}{C}\right) \int i\, dt$ \hfill (2.46)

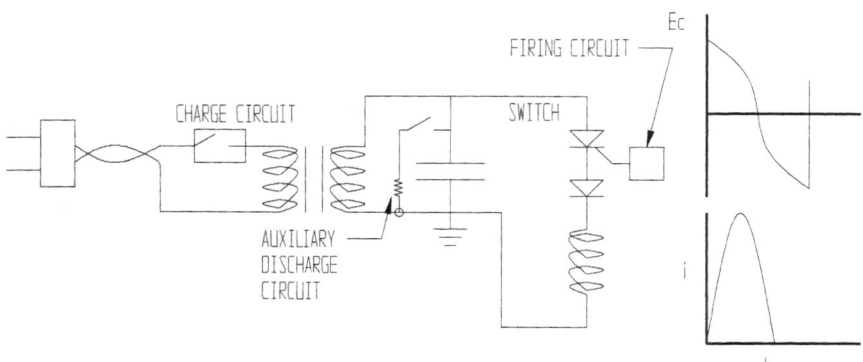

FIGURE 2.97 Bipolar capacitor discharge magnetizer.

We seek a solution for the peak current at any time during the discharge, since the minimum magnetic field will be proportional to this current (eddy currents neglected).

The sum of the voltages around the closed circuit must be equal to zero:

$$E_1 + E_r + E_c = 0 \qquad (2.47)$$

Differentiating and dividing by L:

$$\frac{d^2i}{dt^2} + \left(\frac{R}{L}\right)\frac{di}{dt} + \left(\frac{1}{LC}\right)i = 0 \qquad (2.48)$$

The solution to this equation may be expressed in various ways. The author prefers a solution in terms of voltage E and resistance R, as it seems natural to consider a current, at least for the time-invariant case, as voltage divided by resistance. However, there are other possible ways to write the result, which have sometimes been used by others.

The solution is also simplified if, following a long tradition, we introduce two new variables, the damping constant ρ and the natural frequency ω_0. The damping constant is a dimensionless parameter. The natural frequency, however, is radians per time (seconds) and represents the frequency at which the system would "ring" (oscillate) if no resistance were present.

$$\rho = \frac{R}{2}\left(\frac{C}{L}\right)^{1/2} \qquad (2.49)$$

$$\omega = \left(\frac{1}{LC}\right)^{1/2} \qquad (2.50)$$

With these substitutions, we find that the solution breaks up into three parts, depending on whether the damping constant ρ is less than 1, equal to 1, or greater than 1. These cases are called the *underdamped, critically damped,* and *overdamped* cases, respectively. One other case exists, however, in which $R = 0$. This situation does not occur for real non-superconducting circuits, but is approached for very small resistance. For this limiting case,

$$i_{\text{peak}} = E_0\left(\frac{C}{L}\right)^{1/2} \qquad (2.51)$$

The solution for the damping constant $\rho = 1$ is of relatively simple form:

$$i_{\text{peak}} = \left(\frac{E}{R}\right)\left(\frac{2}{e}\right) = 0.73576\ldots\frac{E}{R} \qquad (2.52)$$

Whenever possible, a fixture is designed to be underdamped ($p < 1$), because this results in more of the available energy being transformed into magnetic energy. The solution is:

$$i_{\text{peak}} = \left(\frac{E}{R}\right)h(\rho) \qquad (2.53)$$

Where h, a function of ρ, is:

$$h = 2\rho\, e^{-g} \qquad (2.54)$$

with

$$g = \left[\frac{\rho}{(1+\rho^2)^{1/2}} \tan^{-1}\left(\frac{(1-\rho^2)^{1/2}}{\rho}\right) \right] \quad (2.55)$$

For the overdamped case ($\rho > 1$), the solution is:

$$i_{peak} = \left(\frac{E}{R}\right) q(\rho) \quad (2.56)$$

where

$$q = 2v \left\{ \exp\left[-v \tan h^{-1}\left(\frac{1}{v}\right)\right] \right\} \sin h \tan h^{-1}\left(\frac{1}{v}\right) \quad (2.57)$$

and

$$v = \frac{\rho}{(\rho^2 - 1)^{1/2}} \quad (2.58)$$

2.8.5 Fixture Structure

A magnetizing fixture is much more than a holder for the magnet while it is being magnetized. Its primary function is to change as much of the electrical energy present in the circuit as possible into magnetic energy for the brief instant needed to magnetize the part, in a field of the required strength, shape, and direction. The major conditions which the fixture must satisfy are:

1. It must produce a field of sufficient strength to completely magnetize the part.
2. The conductor suddenly heats up momentarily, in such a short time (possibly less than a millisecond) that the heat generated does not have time to cross the electrical insulation to the surrounding structure. This short-term temperature rise must not be high enough to destroy the electrical insulation.
3. The fixture as a whole must reject heat fast enough to permit operation at the required cycle rate without overheating. The time constant of this cooling process is much longer than that of (2), from a few seconds to many hours for large steel C frames.
4. The fixture must properly locate the part and restrain it against forces caused by the magnetizing pulse. Both the fixture and the part may expand and contract due to heating and cooling, and also because of magnetostriction, so too tight a fit may lead to broken parts. Also, parts may break due to bending forces induced in them from magnetic effects. Ring magnets for motor rotors, for example, may be magnetized with a number of poles, causing uneven forces. On the other hand, if a ring magnet in an annular space containing a number of evenly spaced magnetizing poles is allowed too much clearance, the ring may move to one side. The result after magnetization is a wider pole spacing on the section of the ring nearest the outer diameter, and a smaller pole width on the side closest to the center. This variation could lead to a noisy, vibrating, and less efficient motor.

It is often advantageous to reduce motor cogging, the tendency for a magnetized rotor to "jump" from one angular position to another as it is rotated, by skewing either the motor lamination stack or the magnet transitions (see Fig. 2.98). Skewing means that the transition lines in the axial direction, from one pole to the next and from one tooth slot to the next, are at an angle between pole and tooth. When one crosses the other during rotation, the forces which would cause this tendency of the rotor to pull forward and back are spread out and superimposed on each other, averaging out. Skewing may be done by twisting the lamination stack, but this presents difficulties in manufacturing, and the assemblies are less repeatable in cosine (because of the stairstep shape caused by lamination thickness). If the magnetizing fixture is skewed, the difficulties are confronted only once, and the resulting product is more repeatable. However, this is more difficult to do than to make a "straight" fixture. Instead of a true spiral shape to the poles, it is much easier and just as effective to allow the pole edge to follow the intersection of a flat plane with a cylinder, and this is almost always done.

2.8.6 Cooling Means

Fixtures must reject considerable heat. If the cycle rate is very low, the fixture may simply be allowed to cool from free convection of the surrounding air. At production rates, however, more effective means must be used. A fan blowing ambient air is several times better than convective cooling, if most of the thermal drop occurs in the boundary layer at the fixture surface. For straight coils, the configuration known as a *Bitter coil* (after its inventor, Dr. Francis Bitter) permits cooling air over both sides of each conductor turn. This arrangement was conceived as a means of cooling very large coils driven by large, steady (dc) currents, and claims made by some manufacturers for this construction for rapidly pulsed magnetizing coils are not always justified by the facts.

Chilled air may be supplied directly from compressed air by a vortex tube (Hilsch tube), a device with no moving parts. Cold air may also be supplied by blowing air through a heat exchanger cooled by conventional refrigerating equipment. A more effective means than air cooling, however, is to use a recirculating fluid. For short-term operation or where plenty of inexpensive water is available, line water is sometimes used once and drained. The fluid may be cooled by returning it to a plenum tank, and then recirculated.

Frequently, the most efficient practice is to cool the fluid using a chiller. Chillers are readily available from a large number of suppliers. The fluid chosen may be

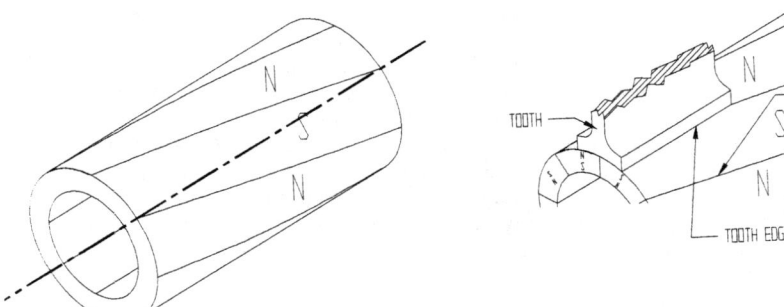

FIGURE 2.98 Ring magnets with skewed poles.

water, which is very effective at removing heat. To avoid a possible safety hazard because of the high voltage present, and the possibility of dangerous voltages caused by rate of change of magnetic flux, a nonelectrically conductive fluid such as oil or Freon may be chosen. For chillers operating at very low temperatures, a fluid with a low freezing point is required. The internal construction of the chiller, if one is used, may limit the available choice of cooling fluid, and the device manual or the manufacturer should be consulted.

2.8.7 Safety Considerations

The levels of voltage and current used in magnetizing are high enough to present a considerable hazard. Voltages above perhaps 60 V and currents of 50 mA or more could potentially kill as a result of heart paralysis: these voltages are far less than those used in magnetizing. Coffee or other conductive liquids should not be allowed on top of or near magnetizing equipment. Flammable liquids, vapors, or gases could be ignited by a spark, as could occur at the electrical connections between the fixture and the magnetizer cables. The magnetizer cable/fixture connections must be protected from accidental contact with the operator, conductive metals, and so on. The output cables may jump during the magnetizing pulse, due to magnetic attraction or repulsion of the cables from each other and nearby electrically or magnetically conductive objects, or even from other regions of the same cable (if bent or looped). A magnetizer should never be operated with damaged or undersized output cables or with insecurely attached cables. A nonconductive switchboard-grade rubber mat on the floor under the operator's feet is an additional useful precaution. It is important that the mat be nonconductive at high voltage, as some rubber mats have filler which render them conductive (for static control in electronic assembly, for example).

Hands or other body parts should not be in or near the fixture during magnetizing. Careless operators will occasionally try to hold a part in a fixture by hand. This is a dangerous practice, and if the operator has a ring on the hand, it could result in a serious burn (as the result of eddy currents heating up the ring) or damage to the hand as the ring is repelled from the field by the magnetic field set up by the eddy currents.

Persons with medical equipment which might be sensitive to strong electric pulses, such as pacemakers, should not be closer than perhaps 5 ft of a magnetizer or fixture in operation.

Magnets will occasionally shatter when magnetized. It is also possible that the fixture, cable, or magnetizer could fail, generating flying debris. The operator should therefore wear eye protection. A foreign object accidentally placed in a magnetizer might be expelled violently. If the object is magnetically permeable, it is attracted toward the center and may pass through it as the pulse ends, flying out the opposite side. If the object is electrically conductive, large eddy currents may be induced in it, which oppose the fixture field, throwing the part outward. Of the two, the latter seems to be more dangerous, and this writer has seen large metal objects embedded in shop ceilings 20 ft or so above a magnetizing fixture (as a result), on two occasions. Operators and others nearby should not be in line of the fixture openings. Of course, magnetizers should not be used as toys, as the results can be extremely dangerous.

Magnetic dirt and magnetically permeable contamination are very common. Some of this contamination is in the form of tiny, nearly invisible metal splinters. These do not normally present much of a hazard, but when they attach themselves to a magnet, the particles may align themselves at right angles to the magnet surface. When the magnets are handled, these may become painful splinters in the hand, which are hard to find and remove. Large magnets may be powerful enough to pinch parts of the hand or even to crush body parts. These forces may occur quite unexpectedly, as a magnet which does not feel dangerous is carried past other magnets or steel parts, tools, or the like. For example, tools or other magnets may even be accelerated toward a magnet being held by an operator from a considerable distance, becoming projectiles.

If possible, the bodies of magnetizers and fixtures should be connected to earth ground, not only through the green ground wire of the electrical service, but by a cable or copper strap able to pass considerable current directly to a well-grounded point such as a water pipe. In case of a short, the very large currents could possibly burn through a light-gauge wire.

2.8.8 Magnet Measurements

Gaussmeters. In 1896, Edwin Herbert Hall discovered in the process of working on his doctoral thesis that if electrical current was passed through a thin strip of gold while it was exposed to a magnetic field, a small but measurable voltage was developed across this strip at right angles to both the direction of current and field, proportional to both. This effect results from the Lorentz force on moving electrons in a magnetic field, which forces them to one side of the strip. They build up a charge there until the charge is just sufficient to counter the effect of the magnetic field. These devices are known today as *Hall-effect sensors*. They are made of semiconductor material, not for the amplifying effects used in transistors, but merely because such materials have a high electrical resistance. The higher resistance forces the electrons in the current stream to move at a higher speed, which increases the resulting voltage. A Hall sensor measures magnetic field strength in a very small region, nearly at a point (a typical sensor might have an active site on the order of 0.030 in across). Only the part of the magnetic vector which is normal (that is, at right angles) to the Hall element is measured, so the sensor must be oriented in that direction. Most gaussmeters on the market today use Hall sensors. A few, however, use some other principle, such as magnetoresistors, which change their resistance in a magnetic field; magnetoresonance, a method used in medical MRI scanners; and, for less accurate devices, mechanical gaussmeters, which use the attraction of two permeable materials for each other, against a spring, in the presence of a magnetic field.

Fluxmeters. As a gaussmeter measures the magnetic flux density nearly at a point, a fluxmeter measures the total magnetic flux across an area (that is, the integral of flux density over an area or, for constant flux density, the flux density times the area). It is more accurate, however, to say that what is measured is not flux, but the change of flux. According to Faraday's law of induction,

$$E = \frac{N \, d\phi}{dt} \tag{2.59}$$

Where E = voltage across a coil of n turns
$\phi = \int B \, dA$ = flux   (2.60)
B = magnetic flux density
A = area
N = number of turns on the coil linked by flux ϕ

Integrating this equation,

$$\phi = \left(\frac{1}{n}\right) \int E\, dt + \phi_0 \qquad (2.61)$$

Where ϕ_0 is an arbitrary constant of integration, normally set to zero to begin the measurement.

That is to say, it is possible to measure an amount of flux passing through an object such as a magnet by placing the object to be measured in a tight-fitting coil, then removing the object, while integrating the voltage across the coil with time. This is the principle of the fluxmeter. Alternately, it is possible to remove the object, rotate it end for end 180°, and reinsert it into the coil. In this case the change of flux is twice that through the object. The sensor for a fluxmeter is just a coil of wire, usually made at the time by the operator. The wire may usually be of small diameter, because very little current flows during the measurement. The larger the number of turns, the larger the signal. However, if the coil resistance becomes relatively high (possibly 50 Ω or more), some fluxmeters with relatively low input resistance may require a correction.

Helmholtz Coil. A Helmholtz coil, in its usual, basic configuration, consists of two similar concentrated coils of small winding cross section compared to coil radius, arranged on a single axis, at a spacing of one coil radius along their common centerline. If electric current is passed through the coils, a very uniform magnetic field is produced in the space between them. If a Helmholtz coil is connected as a sensor to a fluxmeter, then if a bar, plate, or arc magnet is placed in the center with the magnetic axis parallel to the coil axis, and the magnet is then removed (or rotated 180°), the resultant output of the fluxmeter can be shown to be proportional to the magnetic moment of the magnet. The magnetic moment may be defined either as the product of the magnetic flux through the magnet times the pole spacing of the magnet, or as the average axial flux density of the magnet times the magnet volume:

$$M = \phi\, I_2 = B_{av}\, V_m \qquad (2.62)$$

where M = magnetic moment
ϕ = flux through the magnet
I_ρ = pole spacing within the magnet
B_{av} = average flux density in the axial direction, in the magnet
V_m = magnet geometric volume

The combination of a fluxmeter and a Helmholtz coil becomes an accurate, fast, and easy way to determine the strength of a magnet with one measurement. Although originally intended for use with bar or plate magnets, the method can also be used with arc segments (which are used in some permanent-magnet motor rotors).

Magnetic Field Indicating Sheet (Green Paper). This plastic sheet (usually, but not always, green) is often used to inspect magnetic parts, especially for the transitions between magnetic poles (north-south transitioning to south-north, etc.). Unfortunately, a lack of understanding of the way the sheet is constructed often leads to misinterpretation of the results.

Microscopic flakes of nickel are first coated with an oil, in which a plastic material has been dissolved. The plastic then separates out, forming a skin around the oil drop. The flake is then free to rotate within the shell of plastic, which is invisibly small in diameter.

Many layers of these spheres (perhaps 30) are deposited onto a plastic sheet, which forms a support. The support sheet may be on the order of 0.005 in thick, and the layers of spheres may add on the order of 0.002 in to the total thickness.

When no magnetic field is present, the flakes lie flat in the bottom of their spheres, reflecting upward the color of the plastic sheet. When a magnetic field is present in the plane of the sheet, the brightness is intensified. On the other hand, if a magnetic field is present which is normal to the sheet—that is, at approximately right angles into or out of the sheet—the flakes stand on end, aligning with the field. When this occurs, the light reflected off them bounces back and forth until it is absorbed, in a manner similar to the light in a metal tube, and no light is reflected (it is black). It can be seen, then, that the green color means either that no field is present, or that there is a field, in the plane of the sheet. For example, a region in which flux is leaving the sheet at 45° from left to right as it rises will appear to be black when viewed from the right side. The same region viewed from the left side, however, will appear to be green! In order to have a consistent result from the indications of this material, it must be viewed from directly overhead, not from an angle.

The nickel flakes saturate at a relatively small field. This observer noted a change in color at about 10 G and full transition to black at about 100 G for one type of sheet. The transition width for two neodymium-iron magnets side by side in air may be from on the order of +4000 G to –4000 G, but the part of this transition which is indicated by the plastic sheet is much narrower—on the order of $\frac{1}{40}$ as wide. Based on the indications of the plastic sheet, some have thought they were seeing a very narrow transition between magnet poles, to a degree which is physically impossible.

2.8.9 Magnetic Curves

The following are representative of typical commercial magnetic materials (see Tables 2.12–2.71). These curves do not cover all available materials. Figures 2.99 through 2.130 are supplied courtesy of Arnold Engineering Company. Curves in Figs. 2.131 through 2.145 are supplied courtesy of Magnequench.

MATERIALS

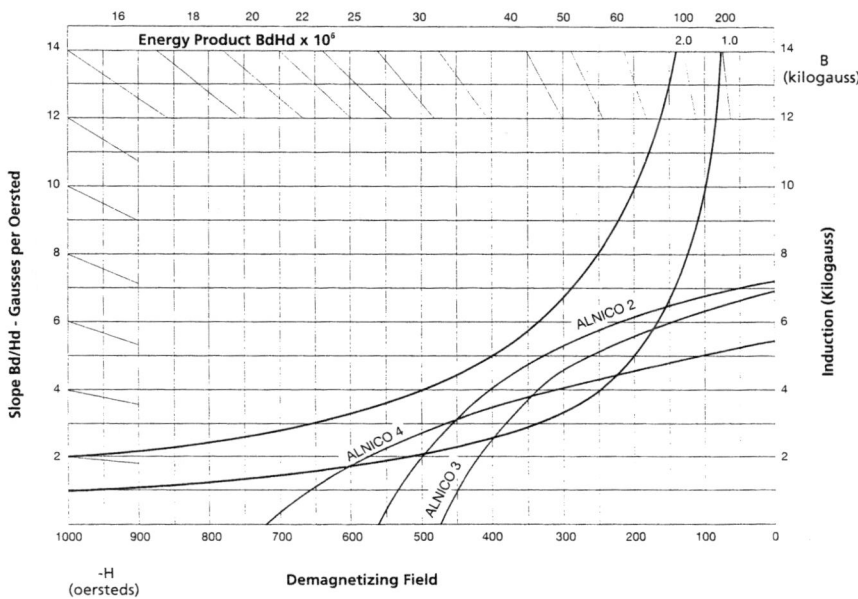

FIGURE 2.99 Typical demagnetization curves for alnico 2, 3, and 4. *(Courtesy of Arnold Engineering Company.)*

TABLE 2.12 Magnetic and Physical Properties (Typical Values)

	Max. energy, product $B_d \times H_d$		Residual induction B_r		Peak magnetic force		Coercive force H_c		Recoil permeability		Permeance coefficient B/H @ (B_dH_d) max.		Induction at max. energy product	
	MGOe	kJ/m³	G	mT	Oe	kA/m	Oe	kA/m	G/Oe	10^{-3} (T·m)/kA	G/Oe	10^{-3} (T·m)/kA	G	mT
Alnico 2	1.60	12.7	7200	720	2000	160	560	45	6.2	7.8	12.0	15.0	4400	440
Alnico 3	1.40	11.1	7000	700	2000	160	475	38	5.1	6.4	14.0	17.5	4450	445
Alnico 4	1.35	10.7	5500	550	3000	240	720	57	4.1	5.2	7.0	9.0	3100	310

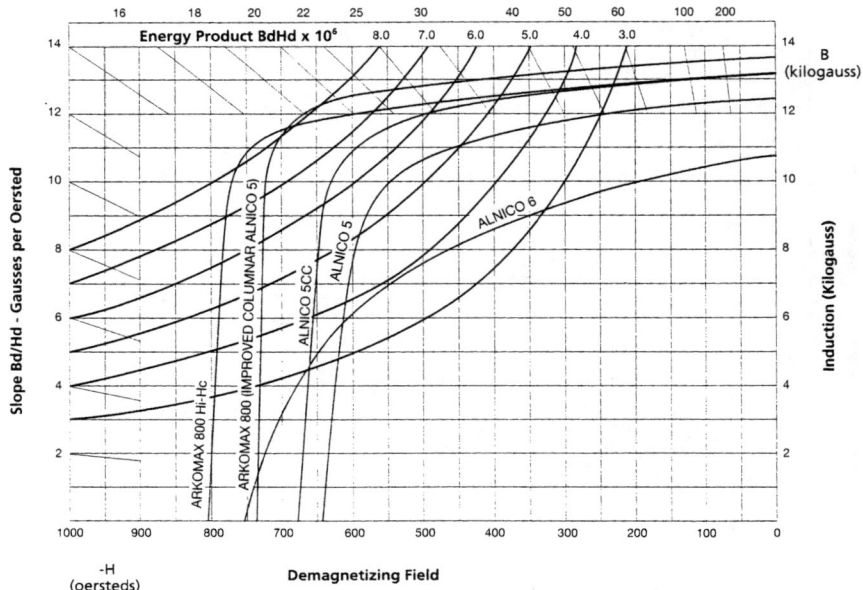

FIGURE 2.100 Typical demagnetization curves for alnico 5, 5cc, and 6 and Arkomax 800 and 800 Hi-Hc. *(Courtesy of Arnold Engineering Company.)*

TABLE 2.13 Magnetic and Physical Properties (Typical Values)

	Max. energy, product $B_d \times H_d$		Residual induction B_r		Peak magnetic force		Coercive force H_c		Recoil permeability		Permeance coefficient B/H @ $(B_d H_d)$ max.		Induction at max. energy product	
										10^{-3} (T·m)/		10^{-3} (T·m)/		
	MGOe	kJ/m³	G	mT	Oe	kA/m	Oe	kA/m	G/Oe	kA	G/Oe	kA	G	mT
Alnico 5	5.50	43.8	12,500	1250	3000	240	640	51	3.7	4.6	19.0	24.0	10,000	1000
Alnico 5cc	6.50	51.7	13,200	1320	3000	240	675	54	2.4	3.0	18.5	23.0	11,000	1100
Alnico 6	3.90	30.0	10,800	1080	3000	240	750	60	5.6	7.0	14.0	17.5	7,400	740
ArKomax® 800	8.10	64.5	13,700	1370	3000	240	740	59	2.0	2.5	18.0	22.5	12,000	1200
ArKomax® H_i-H_c	8.10	64.5	13,200	1320	3000	240	810	64	2.0	2.5	15.5	19.5	11,200	1120

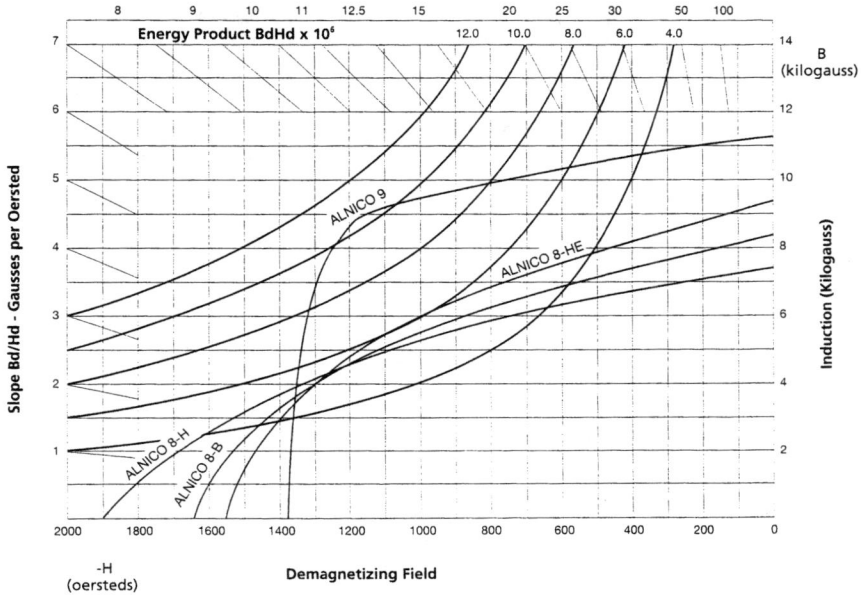

FIGURE 2.101 Typical demagnetization curves for alnico 8B, 8HE, 8H, and 9. *(Courtesy of Arnold Engineering Company.)*

TABLE 2.14 Magnetic and Physical Properties (Typical Values)

	Max. energy product $B_d \times H_d$		Residual induction B_r		Peak magnetic force		Coercive force H_c		Recoil permeability	Permeance coefficient B/H @ (B_dH_d) max.		Induction at max. energy product	
	MGOe	kJ/m³	G	mT	Oe	kA/m	Oe	kA/m	G/Oe	10^{-3} (T·m)/ kA	G/Oe	10^{-3} (T·m)/ kA	G	mT
Anico 8B	5.50	43.8	8,300	830	6000	480	1650	131	2.0	2.5	4.5	5.5	5000	500
Alnico 8HE	6.00	47.7	9,300	930	6000	480	1550	123	2.0	2.5	5.5	7.0	5750	575
Alnico 8H	5.50	43.8	7,400	740	6000	480	1900	151	2.0	2.5	3.5	4.5	4400	440
Alnico 9	10.50	83.6	11,200	1120	6000	480	1375	109	1.3	1.6	7.5	9.5	8900	890

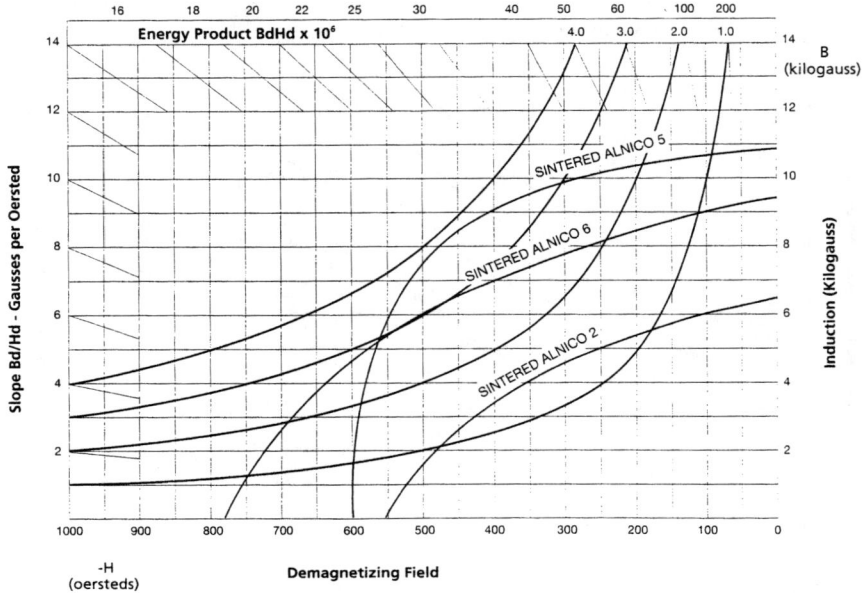

FIGURE 2.102 Typical demagnetization curves for sintered alnico 2, 5, and 6. *(Courtesy of Arnold Engineering Company.)*

TABLE 2.15 Magnetic and Physical Properties (Typical Values)

	Max. energy, product $B_d \times H_d$		Residual induction B_r		Peak magnetic force		Coercive force H_c		Recoil permeability	Permeance coefficient B/H @ $(B_d H_d)$ max.		Induction at max. energy product	
	MGOe	kJ/m³	G	mT	Oe	kA/m	Oe	kA/m	G/Oe	10^{-3} (T·m)/ kA	G/Oe	10^{-3} (T·m)/ kA	G	mT
Sint. Alnico 2	1.40	11.1	6,600	660	2000	160	550	44	5.6	7.0	12.0	15.0	4100	410
Sint. Alnico 5	3.75	29.8	10,800	1080	3000	240	600	48	5.2	6.5	17.0	21.5	8000	800
Sint. Alnico 6	3.00	23.9	9,400	940	3000	240	780	62	5.4	6.8	12.0	15.0	6000	600

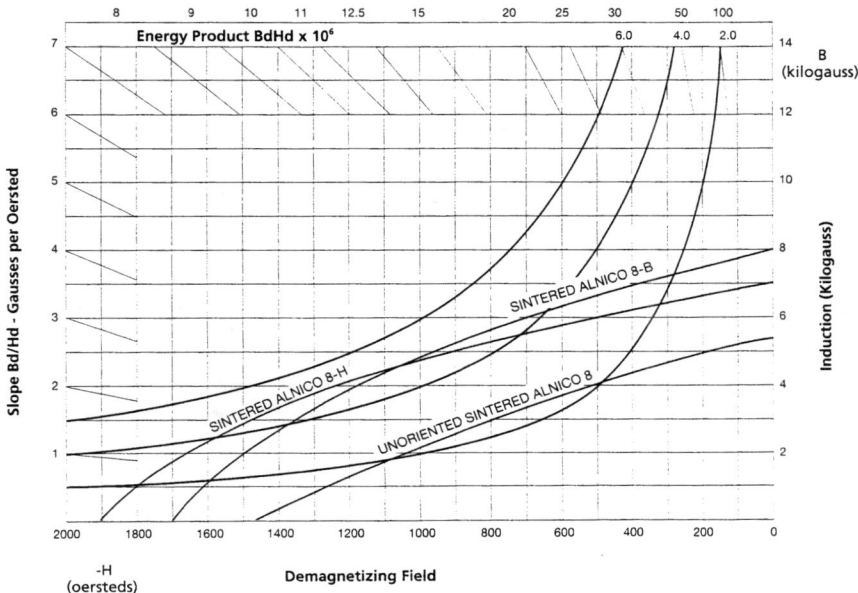

FIGURE 2.103 Typical demagnetization curves for sintered alnico 8B and unoriented sintered alnico 8H. *(Courtesy of Arnold Engineering Company.)*

TABLE 2.16 Magnetic and Physical Properties (Typical Values)

	Max. energy, product $B_d \times H_d$		Residual induction B_r		Peak magnetic force		Coercive force H_c		Recoil permeability		Permeance coefficient B/H @ (B_dH_d) max.		Induction at max. energy product	
	MGOe	kJ/m³	G	mT	Oe	kA/m	Oe	kA/m	G/Oe	10^{-3} (T·m)/ kA	G/Oe	10^{-3} (T·m)/ kA	G	mT
Sint. Alnico 8B	5.00	39.8	8000	800	6000	480	1700	135	1.8	2.3	4.5	5.5	4750	475
Sint. Alnico 8H	5.00	39.8	7000	700	6000	480	1850	147	1.9	2.4	3.4	4.3	4100	410
Unoriented Sint. Alnico 8	2.4	19.1	5500	550	6000	480	1475	117	2.6	3.3	3.8	4.7	3000	300

2.108 CHAPTER TWO

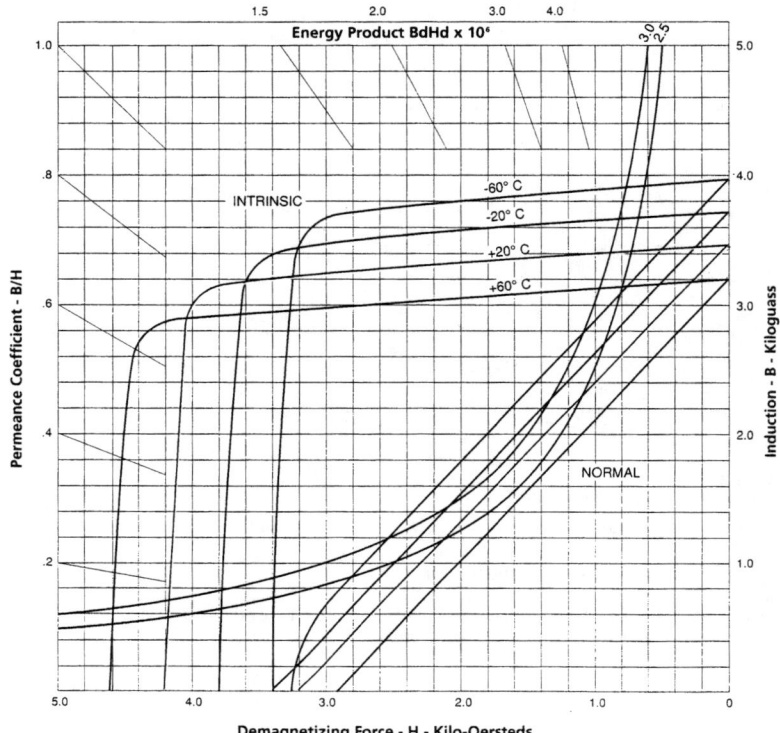

FIGURE 2.104 Normal and intrinsic demagnetization curves for Arnox 7. Typical values; range ±5 percent. *(Courtesy of Arnold Engineering Company.)*

TABLE 2.17 Magnetic and Material Characteristics (Typical Values, 20°C)

Normal peak energy product (B_dH_d) max.		Residual induction B_r (G)		Coercive force H_c (Oe)		Intrinsic coercive force H_{ci} (Oe)		Intrinsic peak energy product (B_eH_d) max.		Density	
MGOe	kJ/m³	B_r	mT	H_c	kA/m	H_{ci}	kA/m	MGOe	kJ/m³	lb/in³	g/cm³
2.7	21.5	3450	345	3200	255	4200	335	12.0	95.5	0.172	4.75

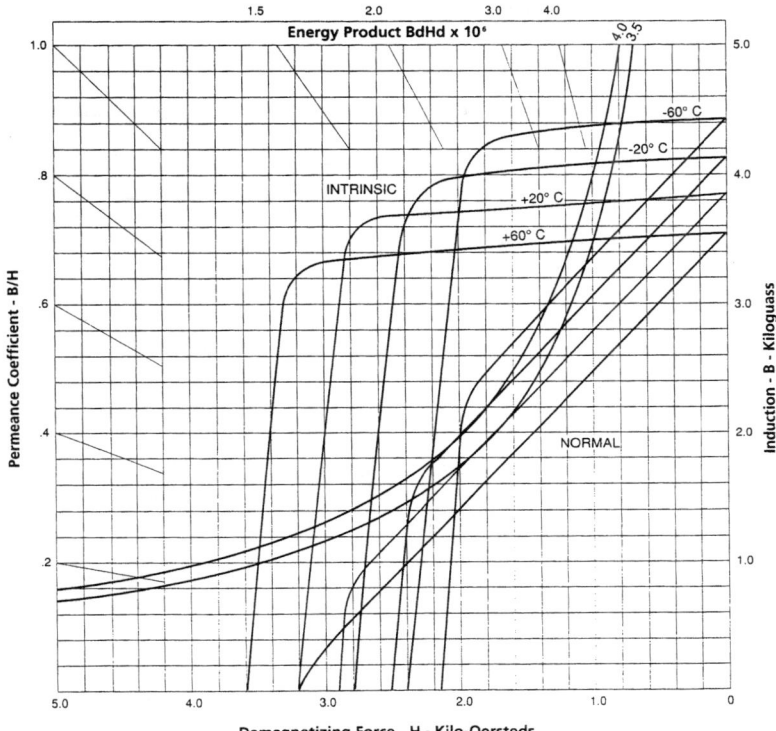

FIGURE 2.105 Normal and intrinsic demagnetization curves for Arnox 8. Typical values; range ±5 percent. *(Courtesy of Arnold Engineering Company.)*

TABLE 2.18 Magnetic and Material Characteristics (Typical Values, 20°C)

Normal peak energy product (B_dH_d) max.		Residual induction B_r (G)		Coercive force H_c (Oe)		Intrinsic coercive force H_{ci} (Oe)		Intrinsic peak energy product (B_eH_d) max.		Density	
MGOe	kJ/m³	B_r	mT	H_c	kA/m	H_{ci}	kA/m	MGOe	kJ/m³	lb/in³	g/cm³
3.5	27.9	3850	385	2900	230	3150	250	10.0	80.0	0.177	4.9

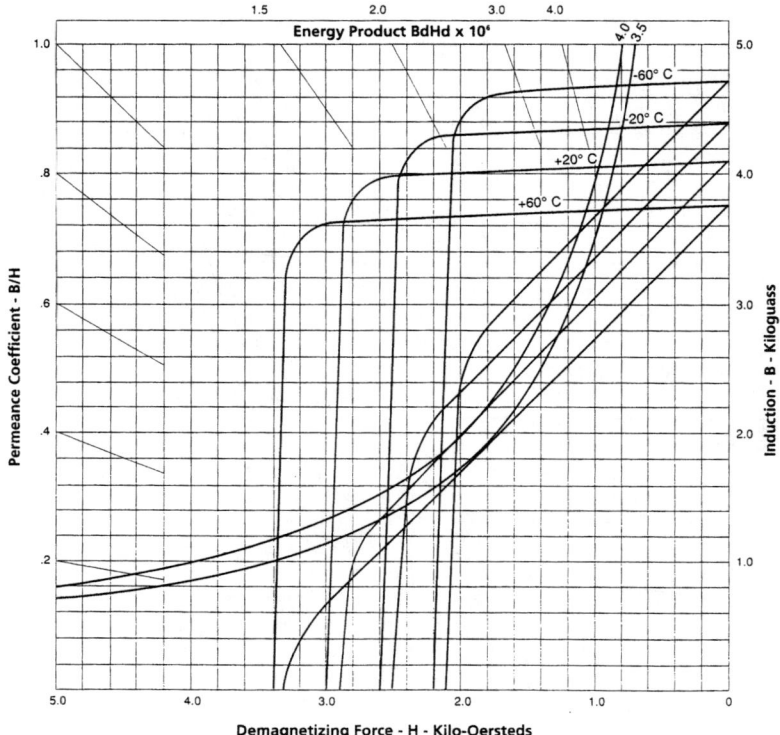

FIGURE 2.106 Normal and intrinsic demagnetization curves for Arnox 8B. Typical values; range ±5 percent. *(Courtesy of Arnold Engineering Company.)*

TABLE 2.19 Magnetic and Material Characteristics (Typical Values, 20°C)

Normal peak energy product (B_dH_d) max.		Residual induction B_r (G)		Coercive force H_c (Oe)		Intrinsic coercive force H_{ci} (Oe)		Intrinsic peak energy product (B_eH_d) max.		Density	
MGOe	kJ/m³	B_r	mT	H_c	kA/m	H_{ci}	kA/m	MGOe	kJ/m³	lb/in³	g/cm³
4.0	31.8	4100	410	2900	230	3000	240	10.5	83.5	0.175	4.85

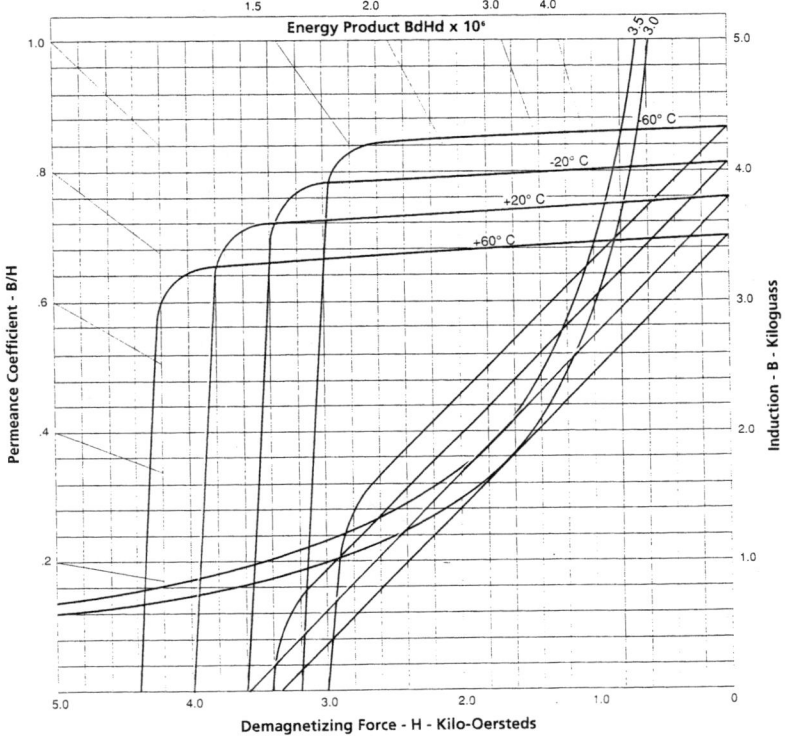

FIGURE 2.107 Normal and intrinsic demagnetization curves for Arnox 8H. Typical values; range ±5 percent. *(Courtesy of Arnold Engineering Company.)*

TABLE 2.20 Magnetic and Material Characteristics (Typical Values, 20°C)

Normal peak energy product (B_dH_d) max.		Residual induction B_r (G)		Coercive force H_c (Oe)		Intrinsic coercive force H_{ci} (Oe)		Intrinsic peak energy product (B_eH_d) max.		Density	
MGOe	kJ/m³	B_r	mT	H_c	kA/m	H_{ci}	kA/m	MGOe	kJ/m³	lb/in³	g/cm³
3.4	27.1	3800	380	3600	285	4000	320	13.0	103.0	0.175	4.85

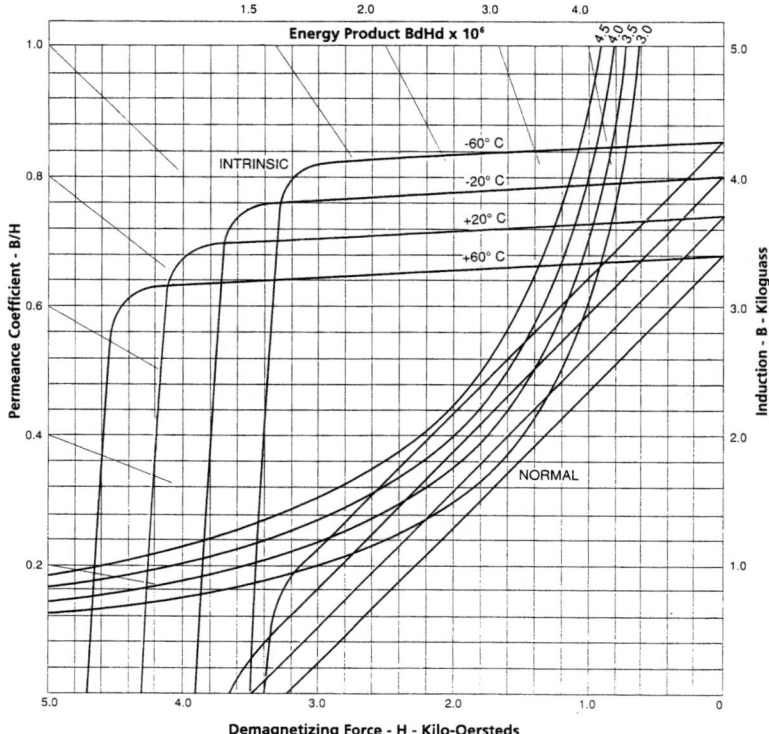

FIGURE 2.108 Normal and intrinsic demagnetization curves for Arnox 9. Typical values; range ±5 percent. *(Courtesy of Arnold Engineering Company.)*

TABLE 2.21 Magnetic and Material Characteristics

Normal peak energy product $(B_d H_d)$ max.		Residual induction B_r (G)		Coercive force H_c (Oe)		Intrinsic coercive force H_{ci} (Oe)		Intrinsic peak energy product $(B_e H_d)$ max.		Density	
MGOe	kJ/m³	B_r	mT	H_c	kA/m	H_{ci}	kA/m	MGOe	kJ/m³	lb/in³	g/cm³
3.25	25.9	3700	370	3500	280	4300	340	13.5	107.0	0.173	4.80

TABLE 2.22 Typical Demagnetization Curve at 23°C (73°F)

Plastiform 2021 features:
Polymer bonded, injection molded, ferrite
Energy product (BH_{max}) 1.5 MGOe
B_rH_x of 6.0 MGOe
Close dimensional and magnetic tolerances
Intricate shapes
Recommended maximum use temperature of 150°C (300°F)

Magnetic			Mechanical		
Property	CGS	SI	Property	CGS	SI
B_r (residual induction)	2450 G	245 mT	Tensile strength	4000 lb/in²	27.5 MPa
H_c (coercive force)	2250 Oe	180 kA/m	Elongation at break	<1%	
H_{ci} (intrinsic coercivity)	4500 Oe	360 kA/m	Density	0.13 lb/in³	3.6 g/cm³
BH_{max} (maximum energy product)	1.5 MGOe	11.9 kJ/m³	Hardness	87 ShoreD	
B_rH_x (intrinsic parameter)	6.0 MGOe	45 kJ/m³	Maximum recommended operating temperature	300°F	150°C
Reversible temperature coefficient on induction, 20–100°C (68–212°F)	−0.10% per °F	−0.18% per °C			
Reversible temperature coefficient of coercivity, 20–100°C (68–212°F)	0.06% per °F	0.10% per °C			
Required magnetizing force	10,000 Oe	790 kA/m			

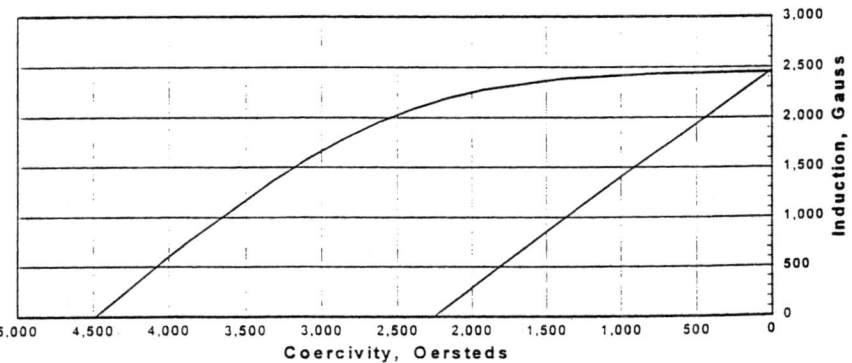

FIGURE 2.109 Typical demagnetization curve at 23°C (73°F) for Plastiform 2021. *(Courtesy of Arnold Engineering Company.)*

TABLE 2.23 Typical Properties at 23°C (73°F)

Plastiform 2023 features:
Polymer bonded, injection molded, ferrite
Energy product (BH_{max}) 1.7 MGOe
B_rH_x of 7.7 MGOe
Close dimensional and magnetic tolerances
Intricate shapes
Recommended maximum use temperature of 150°C (300°F)

Magnetic			Mechanical		
Property	CGS	SI	Property	CGS	SI
B_r (residual induction)	2650 G	265 mT	Tensile strength	4000 lb/in^2	27.5 MPa
H_c (coercive force)	2500 Oe	198 kA/m	Elongation at break	<1%	
H_{ci} (intrinsic coercivity)	4800 Oe	380 kA/m	Density	0.13 lb/in^3	3.6 g/cm^3
BH_{max} (maximum energy product)	1.7 MGOe	13.5 kJ/m^3	Hardness	87 Shore D	
B_rH_x (intrinsic parameter)	7.7 MGOe	61 kJ/m^3	Maximum recommended operating temperature	300°F	150°C
Reversible temperature coefficient on induction, 20–100°C (68–212°F)	−0.10% per °F	−0.18% per °C			
Reversible temperature coefficient of coercivity, 20–100°C (68–212°F)	0.06% per °F	0.10% per °C			
Required magnetizing force	10,000 Oe	790 kA/m			

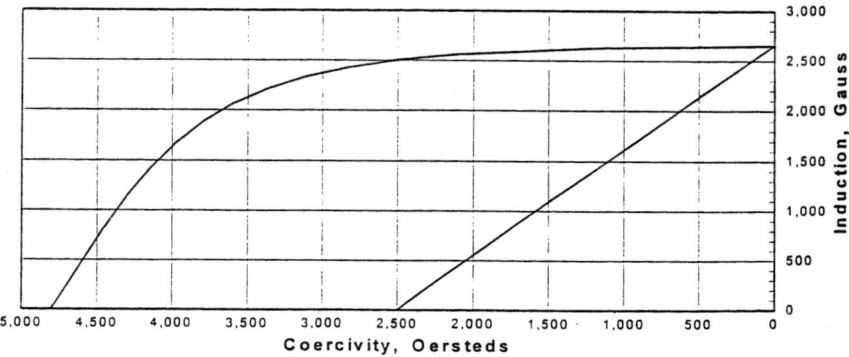

FIGURE 2.110 Typical demagnetization curve at 23°C (73°F) for Plastiform 2023. *(Courtesy of Arnold Engineering Company.)*

TABLE 2.24 Typical Properties at 23°C (73°F)

Plastiform 2120 features:
Polymer bonded, injection molded, anisotropic samarium-cobalt
Energy product (BH_{max}) 8.6 MGOe
B_rH_x of 65 MGOe
Close dimensional and magnetic tolerances
Intricate shapes
Recommended maximum use temperature of 180°C (356°F)

Magnetic			Mechanical		
Property	CGS	SI	Property	CGS	SI
B_r (residual induction)	6200 G	620 mT	Tensile strength	5200 lb/in²	36 MPa
H_c (coercive force)	5350 Oe	425 kA/m	Elongation at break	<1%	
H_{ci} (intrinsic coercivity)	16,500 Oe	1310 kA/m	Density	0.20 lb/in³	5.6 g/cm³
BH_{max} (maximum energy product)	8.6 MGOe	68 kJ/m³	Hardness	90 Shore D	
B_rH_x (intrinsic parameter)	65 MGOe	515 kJ/m³	Maximum recommended operating temperature	356°F	180°C
Reversible temperature coefficient on induction, 20–100°C (68–212°F)	−0.02% per °F	−0.35% per °C			
Reversible temperature coefficient of coercivity, 20–100°C (68–212°F)	−0.1% per °F	−0.20% per °C			
Required magnetizing force	35,000 Oe	2800 kA/m			

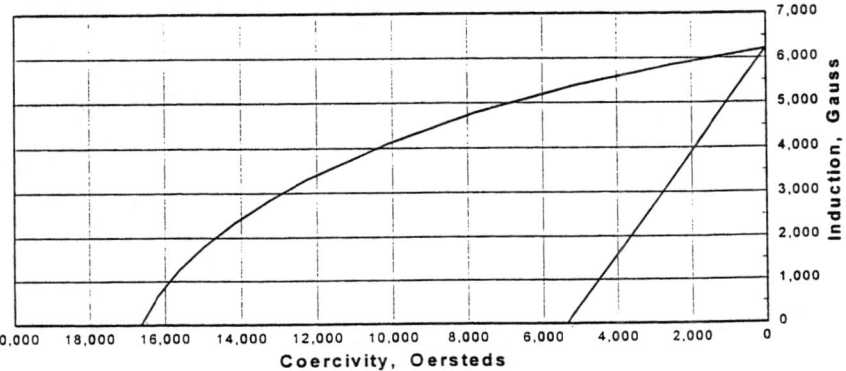

FIGURE 2.111 Typical demagnetization curve at 23°C (73°F) for Plastiform 2102. *(Courtesy of Arnold Engineering Company.)*

TABLE 2.25 Typical Properties at 23°C (73°F)

Plastiform 2201 features:
Polymer bonded, injection molded, isotropic neodymium-iron-boron
Energy product (BH_{max}) 5.0 MGOe
B_rH_x of 65 MGOe
Close dimensional and magnetic tolerances
Intricate shapes
Recommended maximum use temperature of 125°C (260°F)

Magnetic			Mechanical		
Property	CGS	SI	Property	CGS	SI
B_r (residual induction)	4900 G	490 mT	Tensile strength	2500 lb/in²	17.3 MPa
H_c (coercive force)	4100 Oe	330 kA/m	Elongation at break	<2%	
H_{ci} (intrinsic coercivity)	15,000 Oe	1200 kA/m	Density	0.18 lb/in³	5.1 g/cm³
BH_{max} (maximum energy product)	5.0 MGOe	40 kJ/m³	Hardness	85 Shore D	
B_rH_x (intrinsic parameter)	65 MGOe	520 kJ/m³	Maximum recommended operating temperature	260°F	125°C
Reversible temperature coefficient on induction, 20–100°C (68–212°F)	–0.07% per °F	–0.13% per °C			
Reversible temperature coefficient of coercivity, 20–100°C (68–212°F)	–0.22% per °F	–0.040% per °C			
Required magnetizing force	30,000 Oe	2370 kA/m			

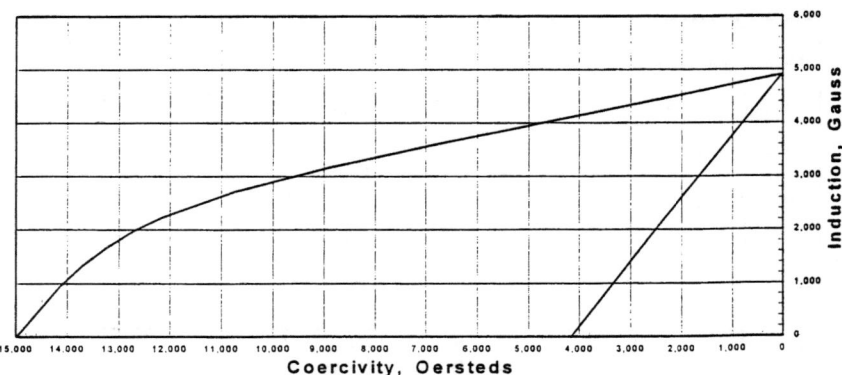

FIGURE 2.112 Typical demagnetization curve at 23°C (73°F) for Plastiform 2201. *(Courtesy of Arnold Engineering Company.)*

TABLE 2.26 Typical Properties at 23°C (73°F)

Plastiform 2202 features:
Polymer bonded, injection molded, isotropic neodymium-iron-boron
Energy product (BH_{max}) 5.0 MGOe
B_rH_x of 20 MGOe
Close dimensional and magnetic tolerances
Intricate shapes
Recommended maximum use temperature of 125°C (260°F)

Magnetic			Mechanical		
Property	CGS	SI	Property	CGS	SI
B_r (residual induction)	5100 G	510 mT	Tensile strength	2500 lb/in²	17.3 MPa
H_c (coercive force)	3800 Oe	300 kA/m	Elongation at break	<2%	
H_{ci} (intrinsic coercivity)	7500 Oe	600 kA/m	Density	0.18 lb/in³	5.1 g/cm³
BH_{max} (maximum energy product)	50. MGOe	40 kJ/m³	Hardness	85 Shore D	
B_rH_x (intrinsic parameter)	20 MGOe	160 kJ/m³	Maximum recommended operating temperature	260°F	125°C
Reversible temperature coefficient on induction, 20–100°C (68–212°F)	−0.06% per °F	−0.105% per °C			
Reversible temperature coefficient of coercivity, 20–100°C (68–212°F)	−0.22% per °F	−0.040% per °C			
Required magnetizing force	30,000 Oe	2370 kA/m			

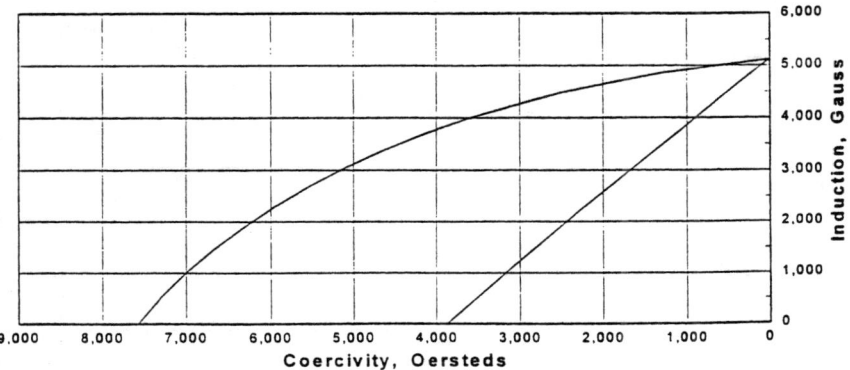

FIGURE 2.113 Typical demagnetization curve at 23°C (73°F) for Plastiform 2202. *(Courtesy of Arnold Engineering Company.)*

TABLE 2.27 Typical Properties at 23°C (73°F)

Plastiform 2203 features:
Polymer bonded, injection molded, isotropic neodymium-iron-boron
Energy product (BH_{max}) 4.8 MGOe
B_rH_x of 14 MGOe
Close dimensional and magnetic tolerances
Intricate shapes
Recommended maximum use temperature of 125°C (260°F)

Magnetic			Mechanical		
Property	CGS	SI	Property	CGS	SI
B_r (residual induction)	5600 G	560 mT	Tensile strength	3900 lb/in²	27 MPa
H_c (coercive force)	3000 Oe	240 kA/m	Elongation at break	<2%	<2%
H_{ci} (intrinsic coercivity)	4300 Oe	340 kA/m	Density	0.18 lb/in³	5.1 g/cm³
BH_{max} (maximum energy product)	4.8 MGOe	38 kJ/m³	Hardness	85 Shore D	85 Shore D
B_rH_x (intrinsic parameter)	14 MGOe	110 kJ/m³	Maximum recommended operating temperature	260°F	125°C
Reversible temperature coefficient on induction, 20–100°C (68–212°F)	−0.06% per °F	−0.1% per °C			
Reversible temperature coefficient of coercivity, 20–100°C (68–212°F)	−0.22% per °F	−0.04% per °C			
Required magnetizing force	25,000 Oe	1900 kA/m			

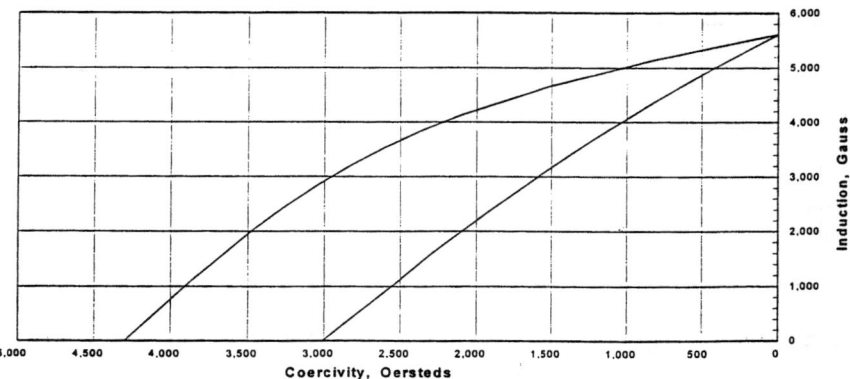

FIGURE 2.114 Typical demagnetization curve at 23°C (73°F) for Plastiform Neolite 2203 isotropic magnets. *(Courtesy of Arnold Engineering Company.)*

TABLE 2.28 Typical Properties at 23°C (73°F)

Plastiform 2204 features:
Polymer bonded, injection molded, isotropic neodymium-iron-boron
Energy product (BH_{max}) 5.2 MGOe
B_rH_x of 42 MGOe
Close dimensional and magnetic tolerances
Intricate shapes
Recommended maximum use temperature of 150°C (300°F)

Magnetic			Mechanical		
Property	CGS	SI	Property	CGS	SI
B_r (residual induction)	5100 G	510 mT	Tensile strength	4700 lb/in^2	32.5 MPa
H_c (coercive force)	4200 Oe	330 kA/m	Elongation at break	<1%	
H_{ci} (intrinsic coercivity)	14,000 Oe	1100 kA/m	Density	0.18 lb/in^3	5.1 g/cm^3
BH_{max} (maximum energy product)	5.2 MGOe	41 kJ/m^3	Hardness	85 Shore D	
B_rH_x (intrinsic parameter)	42 MGOe	335 kJ/m^3	Maximum recommended operating temperature	300°F	150°C
Reversible temperature coefficient on induction, 20–100°C (68–212°F)	−0.07% per °F	−0.13% per °C			
Reversible temperature coefficient of coercivity, 20–100°C (68–212°F)	−0.22% per °F	−0.40% per °C			
Required magnetizing force	25,000 Oe	2370 kA/m			

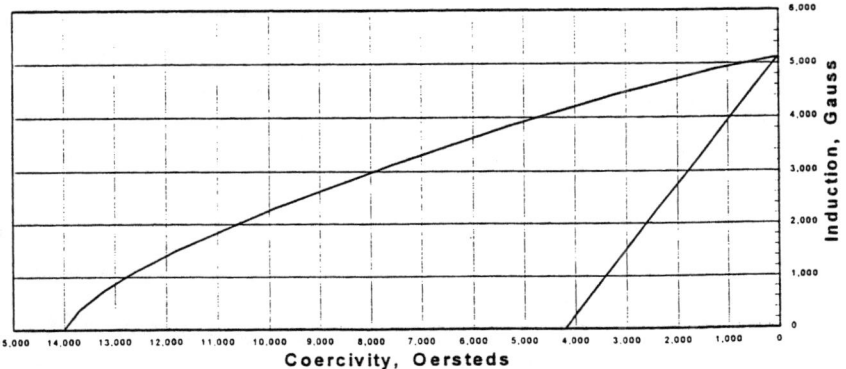

FIGURE 2.115 Typical demagnetization curve at 23°C (73°F) for Plastiform 2204 isotropic magnets. *(Courtesy of Arnold Engineering Company.)*

TABLE 2.29 Typical Properties at 23°C (73°F)

Plastiform 2205 features:
Polymer bonded, injection molded, isotropic neodymium-iron-boron
Energy product (BH_{max}) 5.5 MGOe
B_rH_x of 30 MGOe
Close dimensional and magnetic tolerances
Intricate shapes
Recommended maximum use temperature of 180°C (356°F)

Magnetic			Mechanical		
Property	CGS	SI	Property	CGS	SI
B_r (residual induction)	5180 G	510 mT	Tensile strength	5500 lb/in²	38 MPa
H_c (coercive force)	4250 Oe	330 kA/m	Elongation at break	<1%	
H_{ci} (intrinsic coercivity)	11,000 Oe	1100 kA/m	Density	0.19 lb/in³	5.3 g/cm³
BH_{max} (maximum energy product)	5.5 MGOe	41 kJ/m³	Hardness	90 Shore D	
B_rH_x (intrinsic parameter)	30 MGOe	240 kJ/m³	Maximum recommended operating temperature	356°F	180°C
Reversible temperature coefficient on induction, 20–100°C (68–212°F)	−0.06% per °F	−0.10% per °C			
Reversible temperature coefficient of coercivity, 20–100°C (68–212°F)	−0.22% per °F	−0.40% per °C			
Required magnetizing force	30,000 Oe	2370 kA/m			

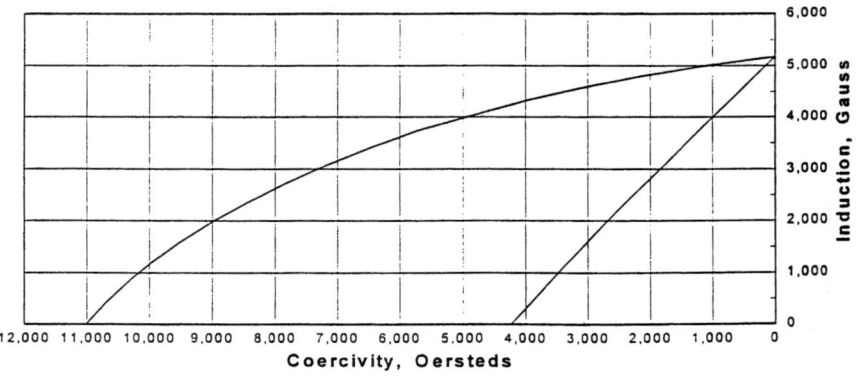

FIGURE 2.116 Typical demagnetization curve at 23°C (73°F) for Plastiform 2205 isotropic magnets. *(Courtesy of Arnold Engineering Company.)*

TABLE 2.30 Typical Properties at 23°C (73°F)

Plastiform 2301 features:
  Injection molded, polymer bonded, neodymium-iron-boron magnet
  Energy product (BH_{max}) 10 MGOe
  BrHx of 45 MGOe
  Close dimensional tolerances
  Intricate shapes
  Recommended maximum use temperature of 100°C (210°F)

Magnetic			Mechanical		
Property	CGS	SI	Property	CGS	SI
B_r (residual induction)	7150 G	710 mT	Tensile strength	2900 lb/in²	20 MPa
H_c (coercive force)	5500 Oe	440 kA/m	Elongation at yield	<4%	
H_{ci} (intrinsic coercivity)	10,000 Oe	800 kA/m	Elongation at break	<5%	
BH_{max} (maximum energy product)	10.0 MGOe	83 kJ/m³	Density	0.17 lb/in³	4.8 g/cm³
$B_r H_x$ (intrinsic parameter)	45 MGOe	360 kJ/m³	Hardness	78 Shore D	
Reversible temperature coefficient on induction, 20–100°C (68–212°F)	−0.06% per °F	−0.1% per °C	Maximum recommended operating temperature	210°F	100°C
Reversible temperature coefficient of coercivity, 20–100°C (68–212°F)	−0.3% per °F	−0.6% per °C			
Required magnetizing force	25,000 Oe	1900 kA/m			

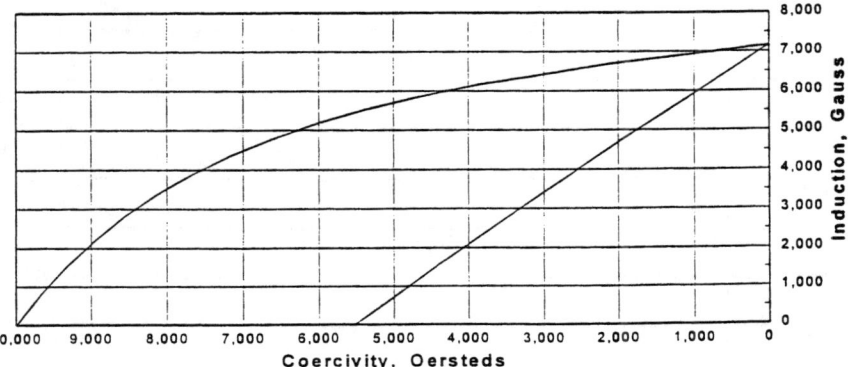

FIGURE 2.117 Typical demagnetization curve at 23°C (73°F) for Plastiform 2301 anisotropic magnets. *(Courtesy of Arnold Engineering Company.)*

TABLE 2.31 Typical Properties at 23°C (73°F)

Plastiform Hybrids 2401, 2402, and 2403 Ferrite-Neo Magnets features:
Polymer bonded, injection molded, oriented ferrite and neodymium-iron-boron blends
Custom blends available
Energy products (BH_{max}) from 1.7 to 4.8 MGOe
B_rH_x of 6 to 9 MGOe
Intricate shapes with close dimensional and magnetic tolerances
Recommended maximum use temperature of 150°C (300°F)
Excellent high- and low-temperature performance

Properties (oriented)	Units	2401	2402	2403
B_r	G	3,000	3,500	4,000
	mT	300	350	400
H_c	Oe	2,400	2,500	2,500
	kA/m	190	200	200
H_{ci}	Oe	3,850	3,800	3,700
	kA/m	310	300	290
BH_{max}	MGOe	2.0	2.6	3.0
	kJ/m³	15.9	20.7	23.9
B_rH_x	MGOe	6.1	6.3	6.5
	kJ/m³	48.5	50.1	51.7
Reversible temperature coefficient	%/°F	−0.08	−0.08	−0.07
of induction B_r	%/°C	−0.14	−0.14	−0.12
Reversible temperature coefficient	%/°F	−0.02	−0.08	−0.14
of coercivity H_{ci}	%/°C	−0.04	−0.14	−0.24
Required magnetizing force for >95% of saturation	kOe	10,000	14,000	17,000
Tensile strength	lb/in²	4,200	5,000	5,000
	MPa	29.0	34.5	34.5
Elongation at break	%	1	1	1
Density	lb/in³	0.14	0.15	0.16
	g/cm³	3.9	4.1	4.4
Hardness	Shore D	88	88	87
Maximum recommended	°F	300	300	300
operating temperature	°C	150	150	150

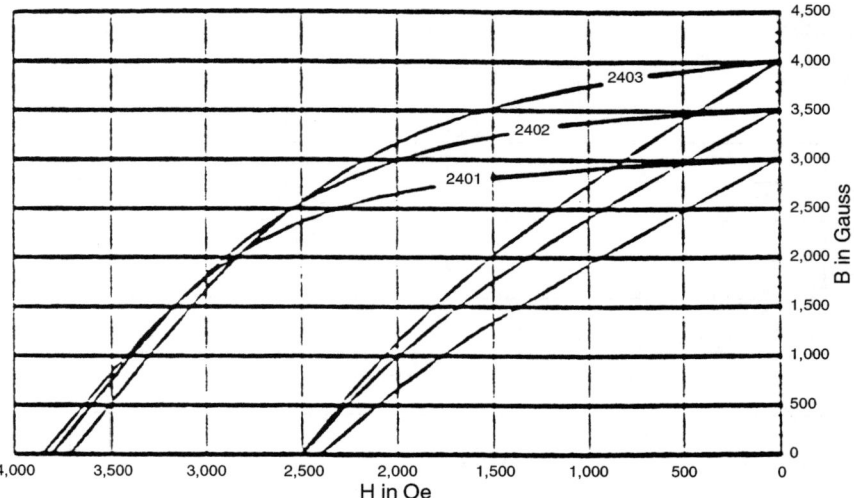

FIGURE 2.118 Quadrant II demagnetization curves for Plastiform hybrid 2401, 2402, and 2403 ferrite-neodymium magnets. *(Courtesy of Arnold Engineering Company.)*

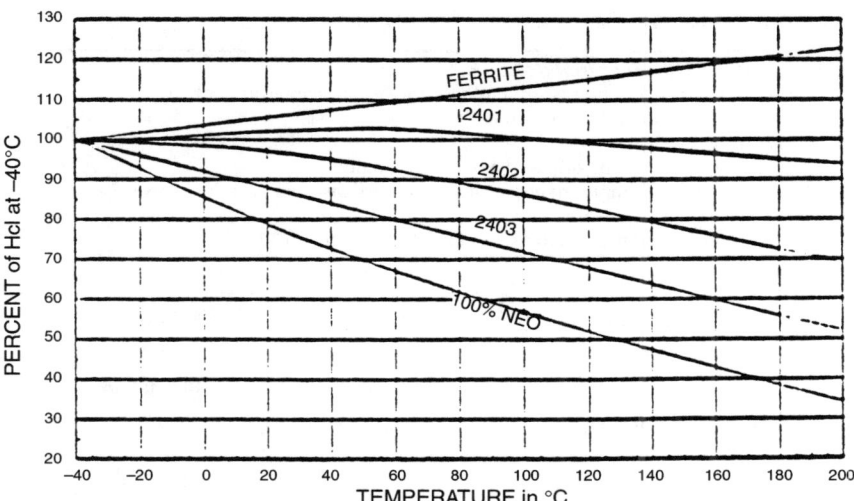

FIGURE 2.119 Coercivity versus temperature for Plastiform hybrid 2401, 2402, and 2403 ferrite-neodymium magnets. *(Courtesy of Arnold Engineering Company.)*

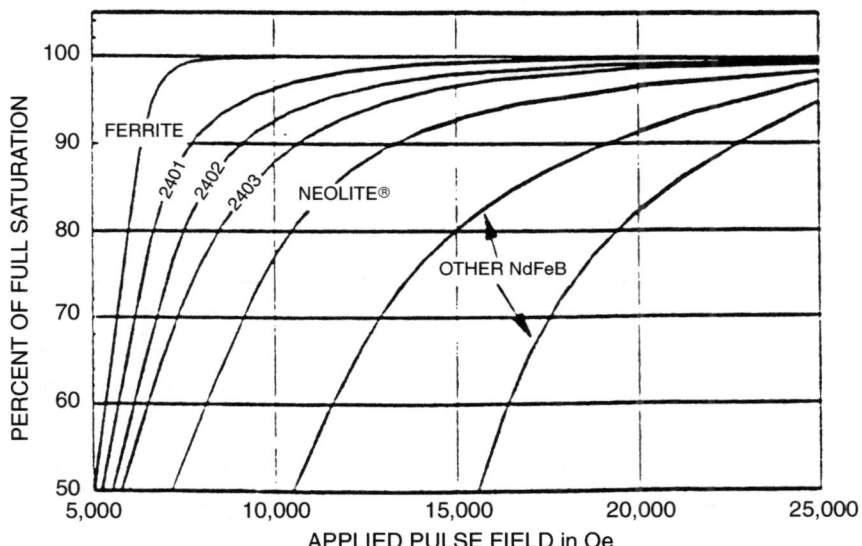

FIGURE 2.120 Saturating field requirements of Plastiform hybrid 2401, 2402, and 2403 ferrite-neodymium magnets. *(Courtesy of Arnold Engineering Company.)*

TABLE 2.32 Typical Properties at 23°C (73°F)

Plastiform 3101 Anisotropic Magnets features:
  Epoxy bonded, compression molded, anisotropic samarium-cobalt
  Energy product (BH_{max}) 9 MGOe
  B_rH_x of 80 MGOe
  Anisotropic, may be oriented axially or radially
  Best suited to simple shapes such as cylinders, discs, and arcs or where
    maximum MGOe is needed
  Recommended maximum use temperature of 150°C (300°F)

Properties	CGS	SI
B_r (residual induction)	6200 G	620 mT
H_c (coercive force)	5700 Oe	450 kA/m
H_{ci} (intrinsic coercivity)	18,000 Oe	1420 kA/m
BH_{max} (maximum energy product)	9 MGOe	70 kJ/m^3
B_rH_x (intrinsic parameter)	80 MGOe	635 kJ/m^3
Reversible temperature coefficient of induction, 20–100°C (68–212°F)	–0.02% per °F	–0.035% per °C
Reversible temperature coefficient of coercivity, 20–100°C (68–212°F)	–0.10% per °F	–0.18% per °C
Required magnetizing force	35,000 Oe	2765 kA/m
Density	0.23 lb/in^3	6.5 g/cm^3
Recommended maximum operating temperature	300°F	150°C

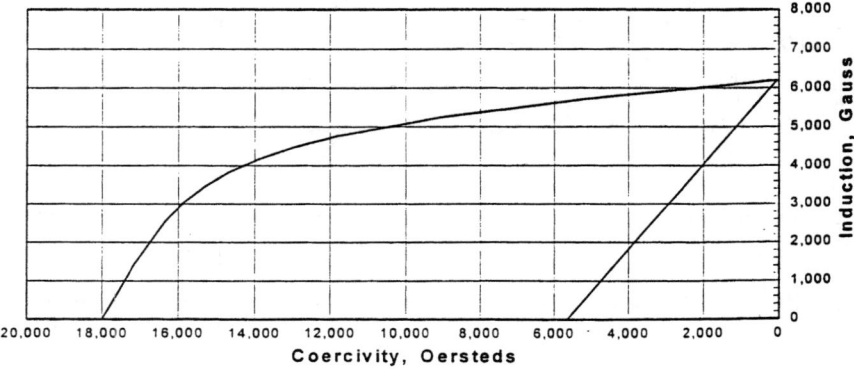

FIGURE 2.121 Typical demagnetization curve at 23°C (73°F) for Plastiform 3101 anisotropic magnets. *(Courtesy of Arnold Engineering Company.)*

TABLE 2.33 Typical Properties at 23°C (73°F)

Plastiform 3201 Isotropic Magnets features:
  Epoxy bonded, compression molded, isotropic neodymium-iron-boron
  Energy product (BH_{max}) 9 MGOe
  B_rH_x of 82 MGOe
  Isotropic, may be magnetized in any direction and pattern
  Best suited to simple shapes such as cylinders, discs, and arcs or where
    maximum MGOe is needed
  Recommended maximum use temperature of 120°C (250°F)

Properties	CGS	SI
B_r (residual induction)	6300 G	620 mT
H_c (coercive force)	5600 Oe	440 kA/m
H_{ci} (intrinsic coercivity)	15,000 Oe	1185 kA/m
BH_{max} (maximum energy product)	9 MGOe	70 kJ/m³
B_rH_x (intrinsic parameter)	82 MGOe	650 kJ/m³
Reversible temperature coefficient of induction, 20–100°C (68–212°F)	–0.07% per °F	–0.13% per °C
Reversible temperature coefficient of coercivity, 20–100°C (68–212°F)	–0.22% per °F	–0.40% per °C
Required magnetizing force	35,000 Oe	2765 kA/m
Density	0.21 lb/in³	5.9 g/cm³
Recommended maximum operating temperature	250°F	120°C

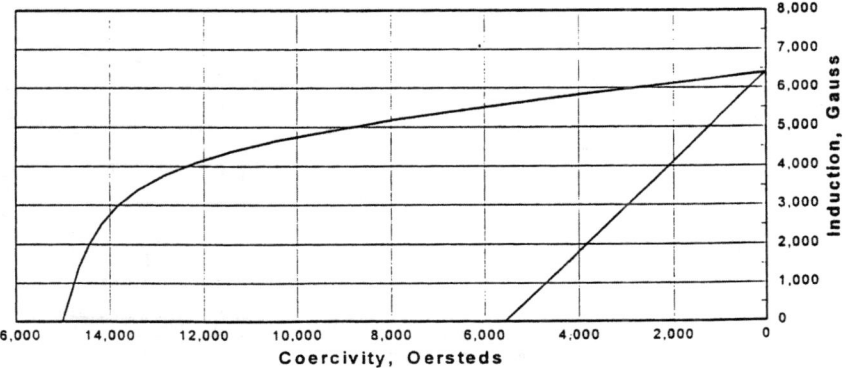

FIGURE 2.122 Typical demagnetization curve at 23°C (73°F) for Plastiform 3201 isotropic magnets. *(Courtesy of Arnold Engineering Company.)*

TABLE 2.34 Typical Properties at 23°C (73°F)

Plastiform 3202 Isotropic Magnets features:
Epoxy bonded, compression molded, isotropic neodymium-iron-boron
Energy product (BH_{max}) 10 MGOe
B_rH_x of 50 MGOe
Isotropic, may be magnetized in any direction and pattern
Best suited to simple shapes such as cylinders, discs, and arcs or where maximum MGOe is needed
Recommended maximum use temperature of 120°C (250°F)

Properties	CGS	SI
B_r (residual induction)	6900 G	690 mT
H_c (coercive force)	5300 Oe	420 kA/m
H_{ci} (intrinsic coercivity)	9500 Oe	750 kA/m
BH_{max} (maximum energy product)	10 MGOe	80 kJ/m³
B_rH_x (intrinsic parameter)	50 MGOe	395 kJ/m³
Reversible temperature coefficient of induction, 20–100°C (68–212°F)	–0.06% per °F	–0.11% per °C
Reversible temperature coefficient of coercivity, 20–100°C (68–212°F)	–0.22% per °F	–0.40% per °C
Required magnetizing force	30,000 Oe	2370 kA/m
Density	0.21 lb/in³	5.95 g/cm³
Recommended maximum operating temperature	250°F	120°C

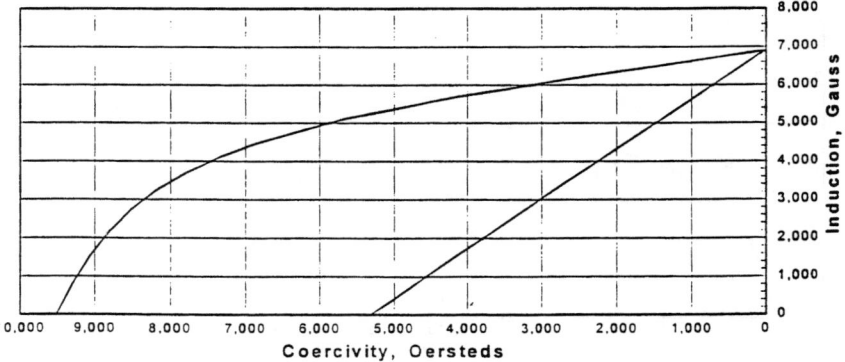

FIGURE 2.123 Typical demagnetization curve at 23°C (73°F) for Plastiform 3202 isotropic magnets. *(Courtesy of Arnold Engineering Company.)*

TABLE 2.35 Typical Properties at 23°C (73°F)

Plastiform 3203 Isotropic Magnets features:
Epoxy bonded, compression molded, isotropic neodymium-iron-boron
Energy product (BH_{max}) 9.5 MGOe
B_rH_x of 48 MGOe
Isotropic, may be magnetized in any direction and pattern
Best suited to simple shapes such as cylinders, discs, and arcs or where maximum MGOe is needed
Recommended maximum use temperature of 140°C (285°F)

Properties	CGS	SI
B_r (residual induction)	6600 G	660 mT
H_c (coercive force)	5600 Oe	445 kA/m
H_{ci} (intrinsic coercivity)	9700 Oe	775 kA/m
BH_{max} (maximum energy product)	9.5 MGOe	76 kJ/m³
B_rH_x (intrinsic parameter)	48 MGOe	380 kJ/m³
Reversible temperature coefficient of induction, 20–100°C (68–212°F)	–0.06% per °F	–0.11% per °C
Reversible temperature coefficient of coercivity, 20–100°C (68–212°F)	–0.22% per °F	–0.40% per °C
Required magnetizing force	30,000 Oe	2370 kA/m
Density	0.208 lb/in³	5.75 g/cm³
Recommended maximum operating temperature	285°F	140°C

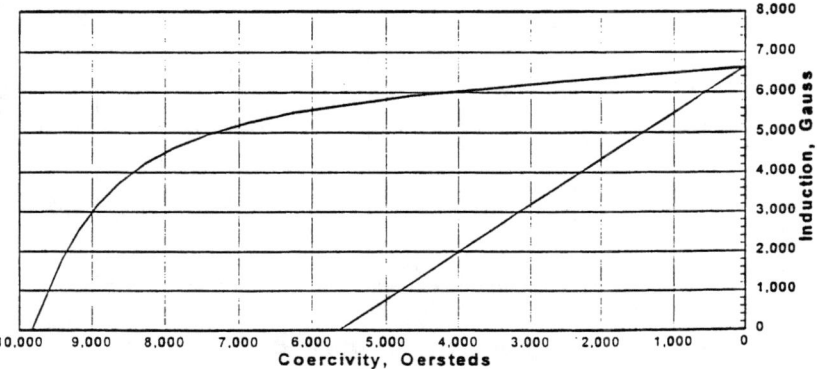

FIGURE 2.124 Typical demagnetization curve at 23°C (73°F) for Plastiform 3203 isotropic magnets. *(Courtesy of Arnold Engineering Company.)*

TABLE 2.36 Solvent Resistance at 23°C (73°F)

Plastiform B-1060 Magnet Material

Solvent	% change from original dimensions*			
	4 h	1 day	14 days	90 days
Gasoline				
Unleaded premium	<0.1	+0.3	+2.9	+3.7
Unleaded regular	<0.1	<0.1	+2.7	+3.6
Motor oil	<0.1	<0.1	<0.1	<0.1
Ethylene glycol	<0.1	<0.1	<0.1	<0.7
Brake fluid	<0.1	<0.1	<0.1	<0.1
Transmission fluid	<0.1	<0.1	<0.1	<0.1
Battery acid	<0.1	<0.1	<0.1	+1.1
Mineral spirits	<0.1	<0.1	+0.4	+2.3
Carburetor cleaner	<0.1	<0.1	†	†
Water (ASTM D-570 saturation)				
Water (ASTM-D-570 27-h immersion)		+0.16		<0.1

Note: Results of a series of full immersion tests on a typical Plastiform B-1060 molded ring showing the percent increase in length over time.
  * All values shown are typical and are not intended for specification purposes. Specification values will be proved upon request.
  † Sample swell >5.0%.

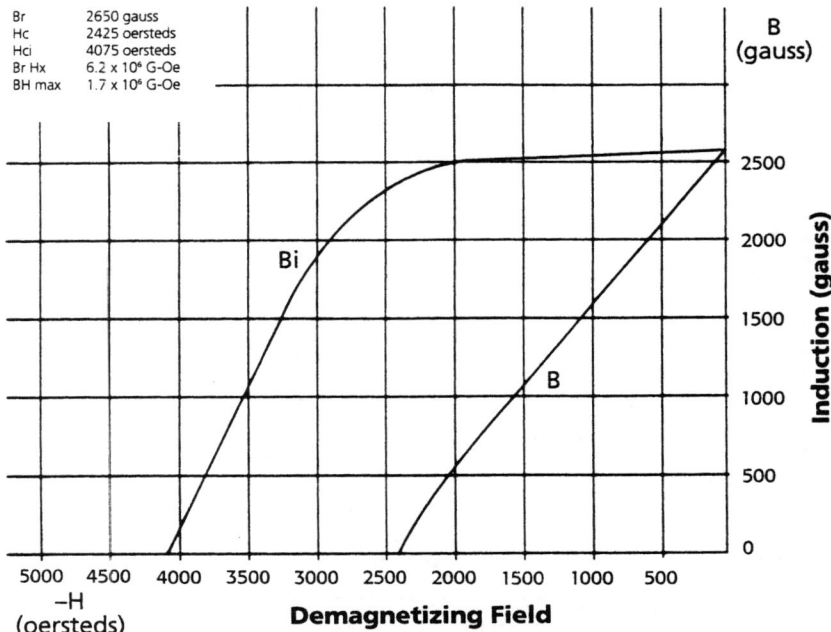

FIGURE 2.125 Typical demagnetization curve at 23°C (73°F) for Plastiform B-1060 magnet material. *(Courtesy of Arnold Engineering Company.)*

TABLE 2.37 Solvent Resistance at 23°C (73°F)

Plastiform B-1061 Magnet Material

Solvent	% change from original dimensions*			
	4 h	1 day	14 days	90 days
Gasoline				
Unleaded premium	<0.1	+0.3	+2.9	+3.7
Unleaded regular	<0.1	<0.1	+2.7	+3.6
Motor oil	<0.1	<0.1	<0.1	<0.1
Ethylene glycol	<0.1	<0.1	<0.1	<0.7
Brake fluid	<0.1	<0.1	<0.1	<0.1
Transmission fluid	<0.1	<0.1	<0.1	<0.1
Battery acid	<0.1	<0.1	<0.1	+1.1
Mineral spirits	<0.1	<0.1	+0.4	+2.3
Carburetor cleaner	<0.1	<0.1	†	†
Water (ASTM D-570 saturation)				
Water (ASTM-D-570 27-h immersion)		+0.16		<0.1

Note: Results of a series of full immersion tests on a typical Plastiform B-1061 molded ring showing the percent increase in length over time.

* All values shown are typical and are not intended for specification purposes. Specification values will be proved upon request.

† Sample swell >5.0%.

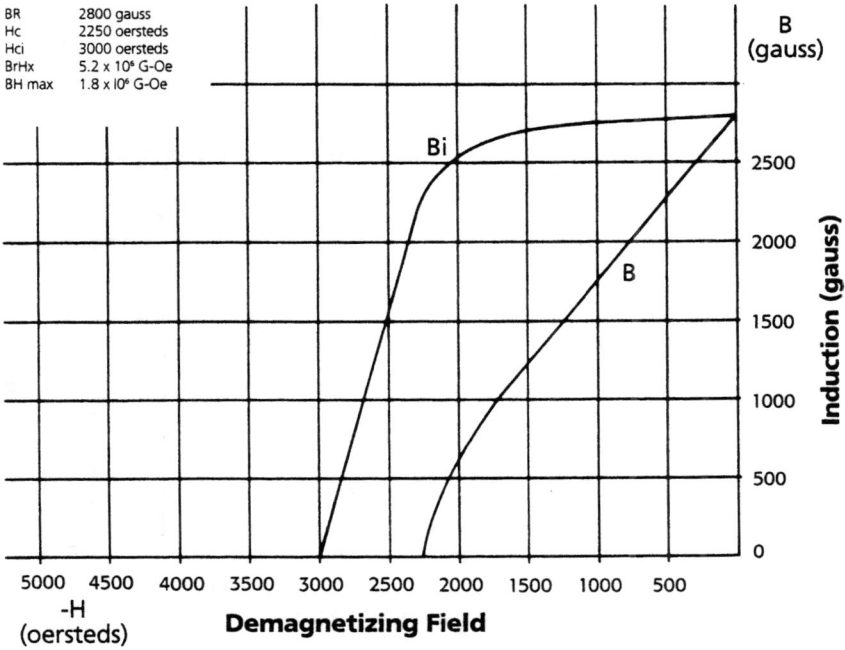

BR 2800 gauss
Hc 2250 oersteds
Hci 3000 oersteds
BrHx 5.2 x 10⁶ G-Oe
BH max 1.8 x 10⁶ G-Oe

FIGURE 2.126 Typical demagnetization curve at 23°C (73°F) for Plastiform B-1061 magnet material. *(Courtesy of Arnold Engineering Company.)*

TABLE 2.38 Solvent Resistance at 23°C (73°F)

Plastiform B-1062 Magnet Material

Solvent	% change from original dimensions*			
	4 h	1 day	14 days	90 days
Gasoline				
Unleaded premium	<0.1	+0.3	+2.9	+3.7
Unleaded regular	<0.1	<0.1	+2.7	+3.6
Motor oil	<0.1	<0.1	<0.1	<0.1
Ethylene glycol	<0.1	<0.1	<0.1	+0.7
Brake fluid	<0.1	<0.1	<0.1	<0.1
Transmission fluid	<0.1	<0.1	<0.1	<0.1
Battery acid	<0.1	<0.1	<0.1	+1.1
Mineral spirits	<0.1	<0.1	+0.4	+2.3
Carburetor cleaner	<0.1	<0.1	†	†
Water (ASTM D-570 saturation)				
Water (ASTM-D-570 27-h immersion)		+0.16		<0.1

Note: Results of a series of full immersion tests on a typical Plastiform B-1061 molded ring showing the percent increase in length over time.
  * All values shown are typical and are not intended for specification purposes. Specification values will be proved upon request.
  † Sample swell >5.0%.

Br   2760 gauss
Hc   2650 oersteds
Hci   5300 oersteds
Br Hx   10 x 10⁶ gauss-oersteds
BH max   1.9 x 10⁶ gauss-oersteds

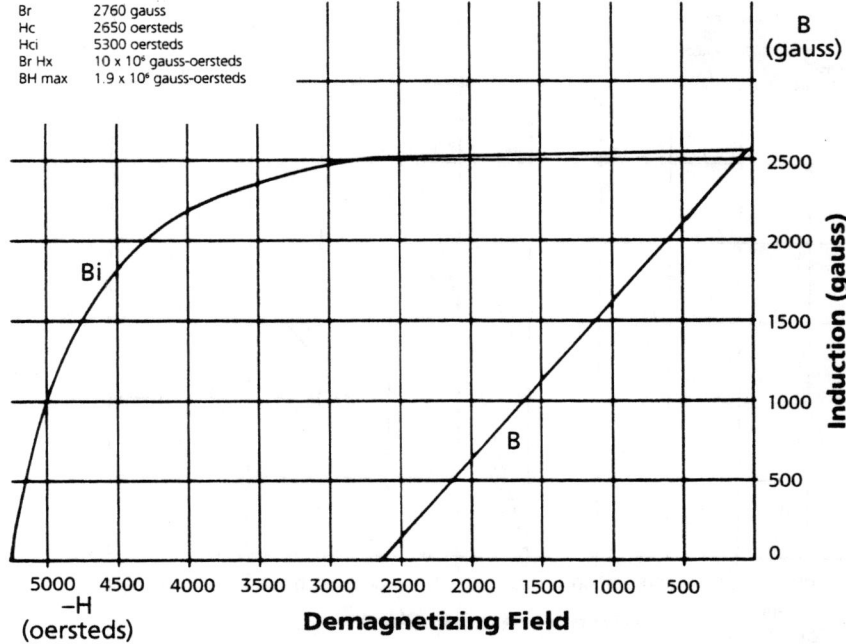

FIGURE 2.127 Typical demagnetization curve at 23°C (73°F) for Plastiform B-1027 magnet material. *(Courtesy of Arnold Engineering Company.)*

TABLE 2.39 Typical Chemical Resistance* (Nitrile Rubber Binder)

Plastiform B-1013, B-1030 Magnet Material Chemical (7 days immersion @ RT)	Performance*
Motor oil	Good
Transmission oil	Good
Hydraulic fluid	Good
Kerosene	Good
JP-4 fuel	Fair
Gasoline	Fair
Heptance	Fair
Antifreeze	Good
Clorox	Good
Turpentine	Good
Water	Good
Detergents	Good
Salt spray	Good
Aromatic hydrocarbons (benzene, toluene, xylene)	Poor
Chlorinated hydrocarbons (carbon tetrachloride, trichloroethylene)	Poor
Ketones	Poor
Alcohols	Fair
Acids, inorganic (HCl, H_2SO_4)	Poor

Note: All values shown are typical and are not intended for specification purposes. Specification values will be provided upon request.

* Good-minor or no effect; up to 10% swell in thickness. Fair-moderate effect; 10–25% swell in thickness. Poor-severe effect; greater than 25% swell in thickness.

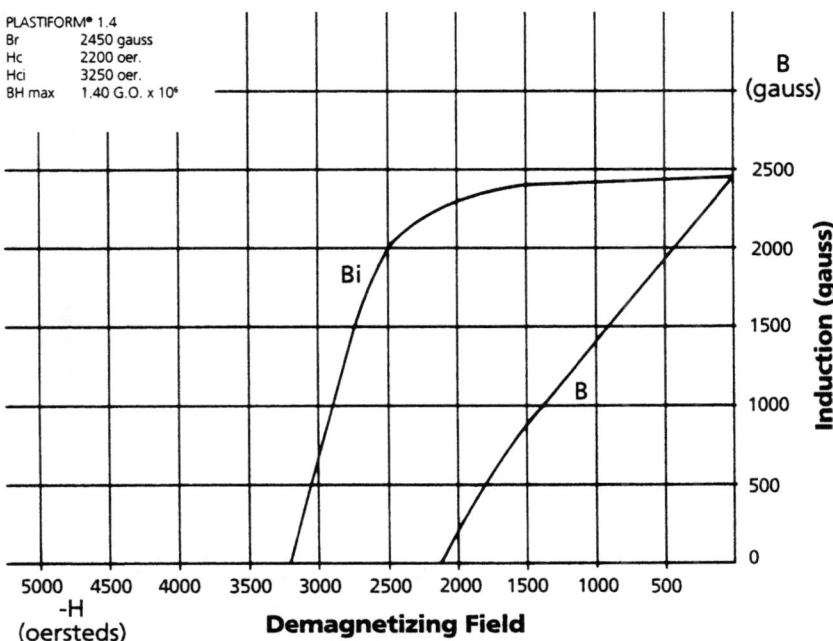

FIGURE 2.128 Typical demagnetization curve at 23°C (73°F) for Plastiform B-1013 and B-1030 magnet material. *(Courtesy of Arnold Engineering Company.)*

TABLE 2.40 Typical Chemical Resistance* (Nitrile Rubber Binder)

Plastiform B-1033 Magnetic Material Chemical (7 days immersion @ RT)	Performance*
Motor oil	Good
Transmission oil	Good
Hydraulic fluid	Good
Kerosene	Good
JP-4 fuel	Fair
Gasoline	Fair
Heptance	Fair
Antifreeze	Good
Clorox	Good
Turpentine	Good
Water	Good
Detergents	Good
Salt spray	Good
Aromatic hydrocarbons (benzene, toluene, xylene)	Poor
Chlorinated hydrocarbons (carbon tetrachloride, trichloroethylene)	Poor
Ketones	Poor
Alcohols	Fair
Acids, inorganic (HCl, H_2SO_4)	Poor

Note: All values shown are typical and are not intended for specification purposes. Specification values will be provided upon request.

* Good-minor or no effect; up to 10% swell in thickness. Fair-moderate effect; 10–25% swell in thickness. Poor-severe effect; greater than 25% swell in thickness.

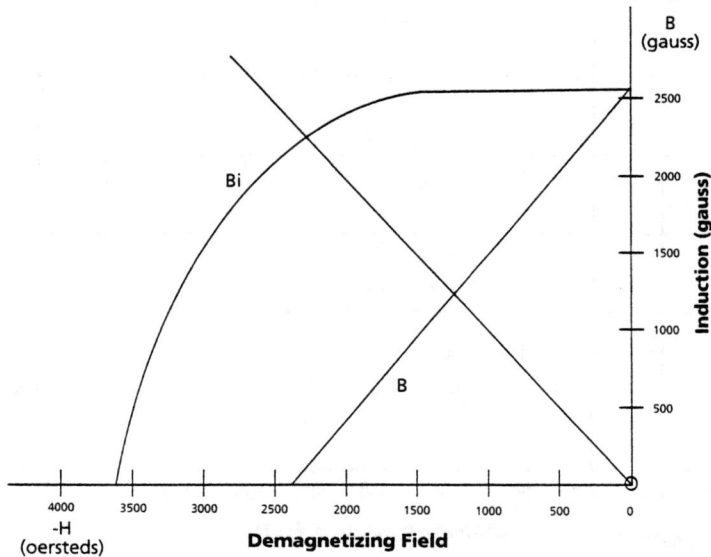

FIGURE 2.129 Typical demagnetization curve at 23°C (73°F) for Plastiform B-1033 magnet material. *(Courtesy of Arnold Engineering Company.)*

TABLE 2.41 Typical Magnetic Properties

Plastiform B-1037 Magnet Material

Properties	Typical values	
	CGS/U.S. units	SI units
Maximum energy product B_dH_d max., at 23°C	1.08 GOe × 10^6	8.6 T(A·t)/m × 10^3
Residual induction* H_c	2150 G	215 mT
Coercive force* H_c, at 23°C (73°F)	1652 Oe	1315 A·t/cm
Coercive force intrinsic* H_{ci}, at 23°C (73°F)	2150 Oe	1710 A·t/cm
Incremental permeability at 23°C (73°F)	1.08 ratio	1.08 ratio
Thermal coefficient of magnetization, 40–120°C (104–248°F)	−0.105% per °F	−0.19% per °C
Thermal coefficient of intrinsic coercive force, 40–120°C (104–248°F)	0.12% per °F	0.22% per °C
Peak magnetizing force required	10,000 Oe	8000 A·t/cm

* Pole-coil hysteresis graph.

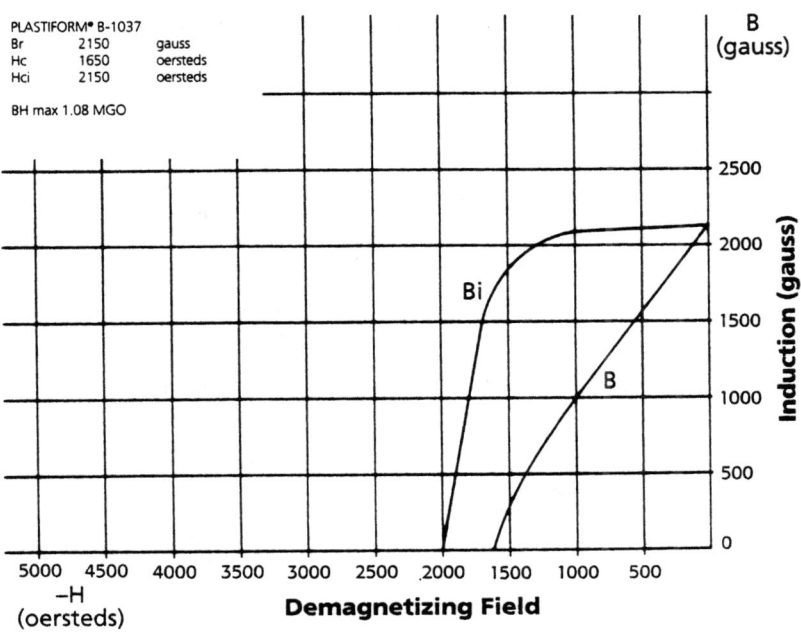

FIGURE 2.130 Typical demagnetization curve at 23°C (73°F) for Plastiform B-1037 magnet material. *(Courtesy of Arnold Engineering Company.)*

TABLE 2.42 Magnetic Characteristics

MQ1-A 9H Magnets		
Residual induction B_r	6.3 kg	0.63 T
Coercive force H_c	5.6 kOe	446 kA/m
Intrinsic coercivity H_{ci}	15 kOe	1194 kA/m
Energy product BH_{max}	9 MGOe	72 kJ/m^3
Recoil permeability μ_r	1.15	
Temperature coefficient of B_r to 100°C (212°F)	−0.13%/°C	
Temperature coefficient of H_{ci} to 100°C (212°F)	−0.40%/°C	
Required magnetizing force (open circuit) H_s	45 kOe	3582 kA/m
Maximum operating temperature*	110°C	

Note: Magnetic properties are typical at room temperature.
* Maximum operating temperature is dependent upon permeance coefficient, coating and, environment.

TABLE 2.43 Physical Properties

MQ1-A 9H Magnets	
Density	6.0 g/cm^3
Coefficient of thermal expansion, 25–200°C (77–392°F)	4.8 μm/m°C
Compressive strength	396 kg/cm^2
Tensile strength	380 kg/cm^2
Young's modulus	8600 kg/cm^2
Poisson ratio	0.2
Transverse rupture strength	320 kg/cm^2
Hardness	30 Rockwell B
Electrical resistivity	14,000 μΩ-cm
Specific heat	0.42 Ws/g°C
Thermal conductivity	0.02 W/cm°C
Curie temperature	305°C

Note: Physical properties are typical at room temperature.
* Maximum operating temperature is dependent upon permeance coefficient, coating, and environment.

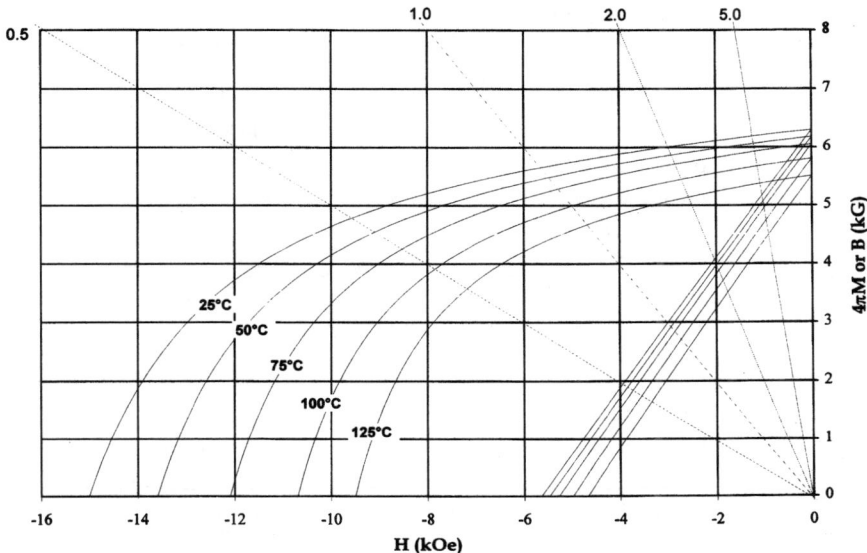

FIGURE 2.131 Demagnetization curves for MQ1-A 9H magnets. *(Courtesy of Magnequench.)*

TABLE 2.44 Magnetic Characteristics

MQ1-B 10 Magnets		
Residual induction B_r	6.9 kG	0.69 T
Coercive force H_c	5.2 kOe	414 kA/m
Intrinsic coercivity H_{ci}	9 kOe	716 kA/m
Energy product BH_{max}	10 MGOe	80 kJ/m^3
Recoil permeability μ_r	1.22	
Temperature coefficient of B_r to 100°C (212°F)	−0.105%/°C	
Temperature coefficient of H_{ci} to 100°C (212°F)	−0.40%/°C	
Required magnetizing force (open circuit) H_s	35 kOe	2786 kA/m
Maximum operating temperature*	110°C	

Note: Magnetic properties are typical at room temperature.
* Maximum operating temperature is dependent upon permeance coefficient, coating, and environment.

TABLE 2.45 Physical Properties

MQ1-B 10 Magnets	
Density	6.0 g/cm^3
Coefficient of thermal expansion, 25–200°C (77–392°F)	4.8 μm/m°C
Compressive strength	396 kg/cm^2
Tensile strength	380 kg/cm^2
Young's modulus	8600 kg/cm^2
Poisson ratio	0.2
Transverse rupture strength	320 kg/cm^2
Hardness	30 Rockwell B
Electrical resistivity	14,000 μΩ-cm
Specific heat	0.42 Ws/g°C
Thermal conductivity	0.02 W/cm°C
Curie temperature	305°C

Note: Physical properties are typical at room temperature.
* Maximum operating temperature is dependent upon permeance coefficient, coating, and environment.

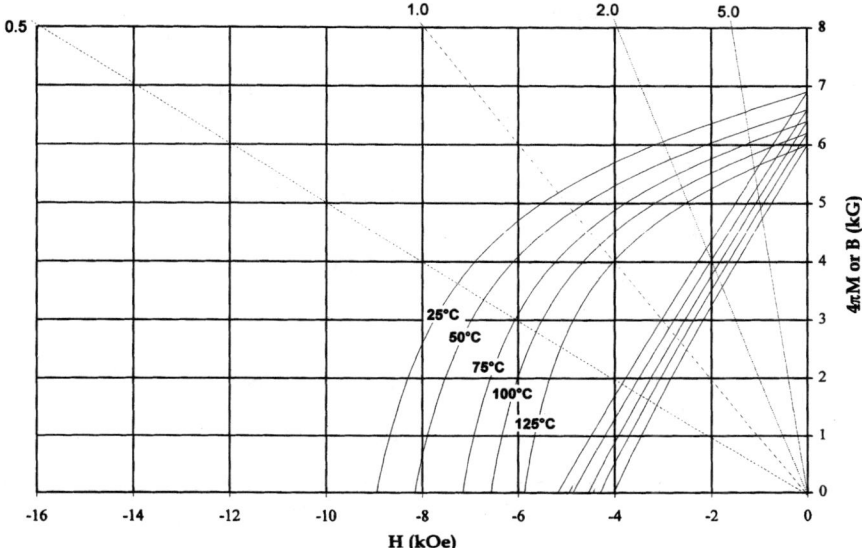

FIGURE 2.132 Demagnetization curves for MQ1-B 10 magnets. *(Courtesy of Magnequench.)*

TABLE 2.46 Magnetic Characteristics

MQ1-C 9H Magnet		
Residual induction B_r	6.3 kG	0.63 T
Coercive force H_c	5.6 kOe	446 kA/m
Intrinsic coercivity H_{ci}	16 kOe	1274 kA/m
Energy product BH_{max}	9 MGOe	72 kJ/m^3
Recoil permeability μ_r	1.15	
Temperature coefficient of B_r to 100°C (212°F)	−0.07%/°C	
Temperature coefficient of H_{ci} to 100°C (212°F)	−0.40%/°C	
Required magnetizing force (open circuit) H_s	45 kOe	3582 kA/m
Maximum operating temperature*	125°C	

Note: Magnetic properties are typical at room temperature.
* Maximum operating temperature is dependent upon permeance coefficient, coating and, environment.

TABLE 2.47 Physical Properties

MQ1-C 9H Magnet	
Density	6.1 g/cm^3
Coefficient of thermal expansion, 25–200°C (77–392°F)	4.8 μm/m°C
Compressive strength	396 kg/cm^2
Tensile strength	380 kg/cm^2
Young's modulus	8600 kg/cm^2
Poisson ratio	0.2
Transverse rupture strength	320 kg/cm^2
Hardness	30 Rockwell B
Electrical resistivity	14,000 μΩ-cm
Specific heat	0.42 Ws/g°C
Thermal conductivity	0.02 W/cm°C
Curie temperature	470°C

Note: Physical properties are typical at room temperature.
* Maximum operating temperature is dependent upon permeance coefficient, coating, and environment.

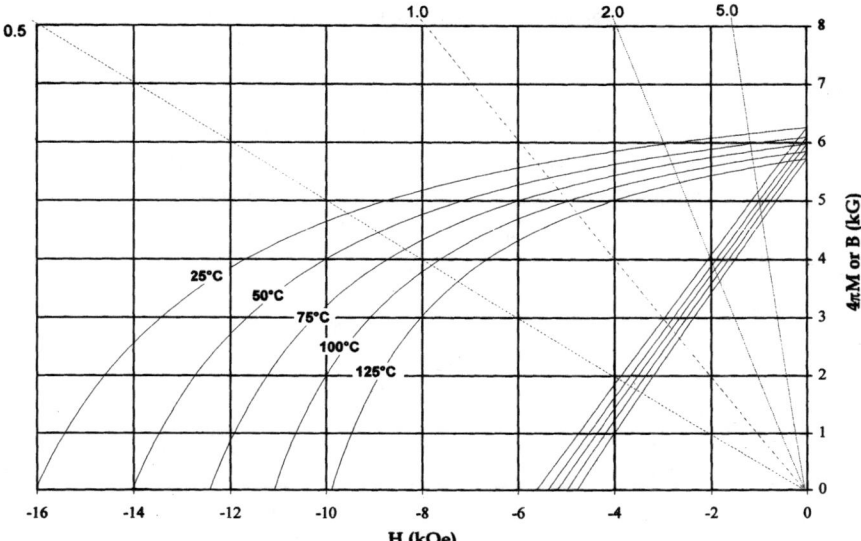

FIGURE 2.133 Demagnetization curves for MQ1-C 9H magnets. *(Courtesy of Magnequench.)*

TABLE 2.48 Magnetic Characteristics

MQ1-D 10 Magnets		
Residual induction B_r	6.8 kG	0.68 T
Coercive force H_c	5.6 kOe	446 kA/m
Intrinsic coercivity H_{ci}	10.2 kOe	812 kA/m
Energy product BH_{max}	10 MGOe	80 kJ/m^3
Recoil permeability μ_r	1.22	
Temperature coefficient of B_r to 100°C (212°F)	−0.07%/°C	
Temperature coefficient of H_{ci} to 100°C (212°F)	−0.40%/°C	
Required magnetizing force (open circuit) H_s	35 kOe	2786 kA/m
Maximum operating temperature*	110°C	

Note: Magnetic properties are typical at room temperature.
* Maximum operating temperature is dependent upon permeance coefficient, coating and, environment.

TABLE 2.49 Physical Properties

MQ1-D 10 Magnets	
Density	6.1 g/cm^3
Coefficient of thermal expansion, 25–200°C (77–392°F)	4.8 μm/m°C
Compressive strength	396 kg/cm^2
Tensile strength	380 kg/cm^2
Young's modulus	8600 kg/cm^2
Poisson ratio	0.2
Transverse rupture strength	320 kg/cm^2
Hardness	30 Rockwell B
Electrical resistivity	14,000 μΩ-cm
Specific heat	0.42 Ws/g°C
Thermal conductivity	0.02 W/cm°C
Curie temperature	470°C

Note: Physical properties are typical at room temperature.
* Maximum operating temperature is dependent upon permeance coefficient, coating, and environment.

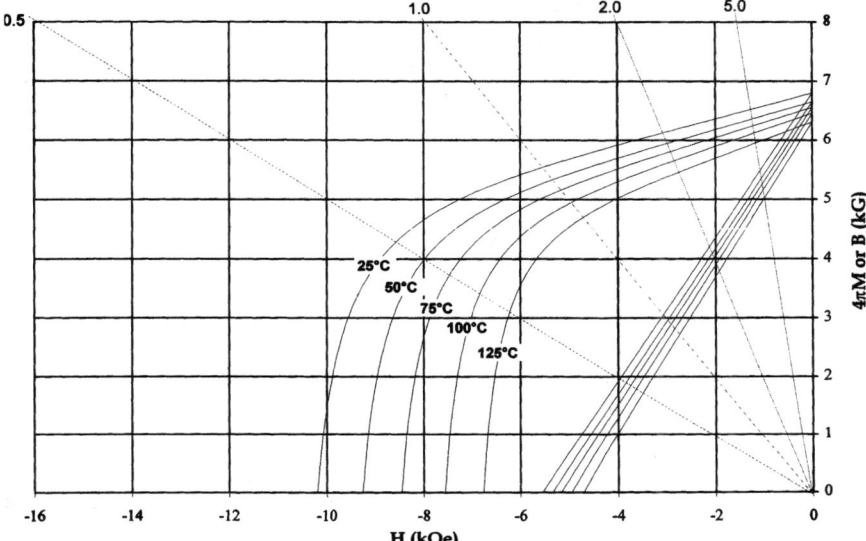

FIGURE 2.134 Demagnetization curves for MQ1-D 10 magnets. *(Courtesy of Magnequench.)*

TABLE 2.50 Magnetic Characteristics

MQ1-O 8 Magnets		
Residual induction B_r	6.1 kG	0.61 T
Coercive force H_c	5.3 kOe	422 kA/m
Intrinsic coercivity H_{ci}	12.5 kOe	955 kA/m
Energy product BH_{max}	8.5 MGOe	68 kJ/m³
Recoil permeability μ_r	1.15	
Temperature coefficient of B_r to 100°C (212°F)	–0.13%/°C	
Temperature coefficient of H_{ci} to 100°C (212°F)	–0.40%/°C	
Required magnetizing force (open circuit) H_s	45 kOe	3582 kA/m
Maximum operating temperature*	140°C	

Note: Magnetic properties are typical at room temperature.
* Maximum operating temperature is dependent upon permeance coefficient, coating and, environment.

TABLE 2.51 Physical Properties

MQ1-O 8 Magnets	
Density	6 g/cm³
Coefficient of thermal expansion, 25–200°C (77–392°F)	4.8 μm/m°C
Compressive strength	396 kg/cm²
Tensile strength	380 kg/cm²
Young's modulus	8600 kg/cm²
Poisson ratio	0.2
Transverse rupture strength	320 kg/cm²
Hardness	30 Rockwell B
Electrical resistivity	14,000 μΩ-cm
Specific heat	0.42 Ws/g°C
Thermal conductivity	0.02 W/cm°C
Curie temperature	305°C

Note: Physical properties are typical at room temperature.
* Maximum operating temperature is dependent upon permeance coefficient, coating, and environment.

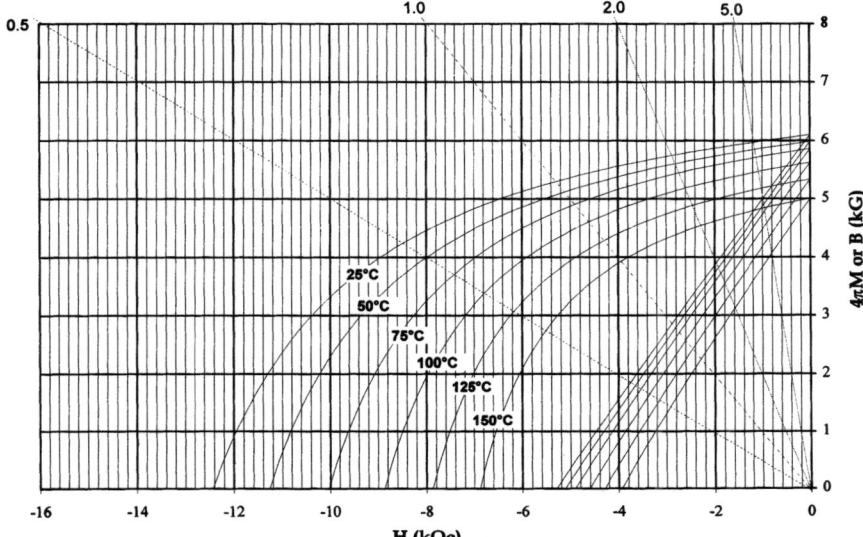

FIGURE 2.135 Demagnetization curves for MQ1-O 8 magnets. *(Courtesy of Magnequench.)*

TABLE 2.52 Magnetic Characteristics

MQ2-F 14H Magnets		
Residual induction $B_r \parallel$	8 kG	0.8 T
Residual induction $B_r \perp$	7.2 kG	0.72 T
Coercive force H_c	7 kOe	557 kA/m
Intrinsic coercivity H_{ci}	>18 kOe	1433 kA/m
Energy product BH_{max}	14 MGOe	111 kJ/m^3
Recoil permeability μ_r	1.14	
Temperature coefficient of B_r to 100°C (212°F)	−0.09%/°C	
Temperature coefficient of H_{ci} to 100°C (212°F)	−0.50%/°C	
Required magnetizing force (open circuit) H_s	45 kOe	3582 kA/m
Maximum operating temperature*	200°C	

Note: Magnetic properties are typical at room temperature.
* Maximum operating temperature is dependent upon permeance coefficient, coating and, environment.

TABLE 2.53 Physical Properties

MQ2-F 14H Magnets	
Density	7.6 g/cm^3
Coefficient of thermal expansion, 25–200°C (77–392°F)	3.5 µm/m°C
Compressive strength	6×10^3 kg/cm^2
Young's modulus	1.42×10^6 kg/cm^2
Poisson ratio	0.27
Transverse rupture strength	1.6×10^3 kg/cm^2
Hardness	60 Rockwell B
Electrical resistivity	143 µΩ-cm
Specific heat	0.42 Ws/g°C
Thermal conductivity	0.07 W/cm°C
Curie temperature	370°C

Note: Physical properties are typical at room temperature.
* Maximum operating temperature is dependent upon permeance coefficient, coating, and environment.

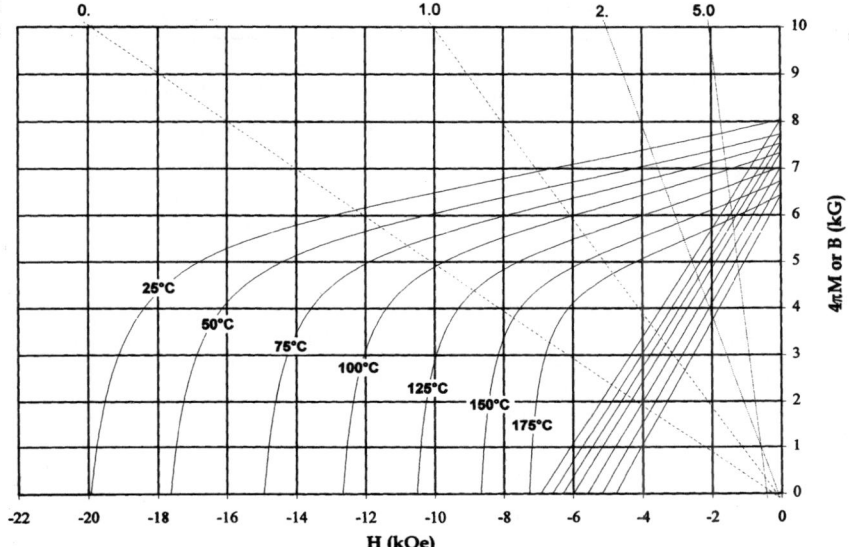

FIGURE 2.136 Demagnetization curves for MQ2-F 14H magnets. *(Courtesy of Magnequench.)*

TABLE 2.54 Magnetic Properties

MQ2-E 15 Magnets		
Residual induction $B_r \parallel$	8.25 kG	0.825 T
Residual induction $B_r \perp$	7.4 kG	0.74 T
Coercive force H_c	7.2 kOe	573 kA/m
Intrinsic coercivity H_{ci}	17.5 kOe	1393 kA/m
Energy product BH_{max}	15 MGOe	119 kJ/m^3
Recoil permeability μ_r	1.14	
Temperature coefficient of B_r to 100°C (212°F)	−0.10%/°C	
Temperature coefficient of H_{ci} to 100°C (212°F)	−0.50%/°C	
Required magnetizing force (open circuit) H_s	45 kOe	3582 kA/m
Maximum operating temperature*	180°C	

Note: Magnetic properties are typical at room temperature.
* Maximum operating temperature is dependent upon permeance coefficient, coating and, environment.

TABLE 2.55 Physical Properties

MQ2-E 15 Magnets	
Density	7.6 g/cm^3
Coefficient of thermal expansion, 25–200°C (77–392°F)	1.24 μm/m°C
Compressive strength	6×10^3 kg/cm^2
Young's modulus	1.42×10^6 kg/cm^2
Poisson ratio	0.27
Transverse rupture strength	1.6×10^3 kg/cm^2
Hardness	60 Rockwell B
Electrical resistivity	143 μΩ-cm
Specific heat	0.42 Ws/g°C
Thermal conductivity	0.07 W/cm°C
Curie temperature	335°C

Note: Physical properties are typical at room temperature.
* Maximum operating temperature is dependent upon permeance coefficient, coating, and environment.

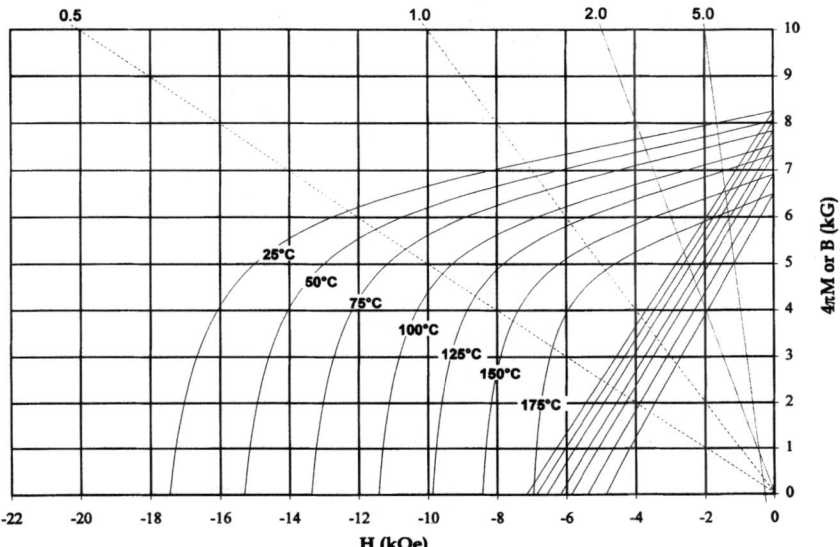

FIGURE 2.137 Demagnetization curves for MQ2-E 15 magnets. *(Courtesy of Magnequench.)*

TABLE 2.56 Magnetic Characteristics

MQ3-E 34 Magnets		
Residual induction $B_r\|$	8.25 kG	0.825 T
Coercive force H_c	7.2 kOe	573 kA/m
Intrinsic coercivity H_{ci}	17.5 kOe	1393 kA/m
Energy product BH_{max}	15 MGOe	119 kJ/m^3
Recoil permeability μ_r	1.14	
Temperature coefficient of B_r to 100°C (212°F)	−0.10%/°C	
Temperature coefficient of H_{ci} to 100°C (212°F)	−0.50%/°C	
Required magnetizing force (open circuit) H_s	45 kOe	3582 kA/m
Maximum operating temperature*	180°C	

Note: Magnetic properties are typical at room temperature.
* Maximum operating temperature is dependent upon permeance coefficient, coating and, environment.

TABLE 2.57 Physical Properties

MQ3-E 34 Magnets	
Density	7.6 g/cm^3
Coefficient of thermal expansion, 25–200°C (77–392°F)	1.24 µm/m°C
Compressive strength	6 × 10^3 kg/cm^2
Young's modulus	1.42 × 10^6 kg/cm^2
Poisson ratio	0.27
Transverse rupture strength	1.6 × 10^3 kg/cm^2
Hardness	60 Rockwell B
Electrical resistivity	143 µΩ-cm
Specific heat	0.42 Ws/g°C
Thermal conductivity	0.07 W/cm°C
Curie temperature	335°C

Note: Physical properties are typical at room temperature.
* Maximum operating temperature is dependent upon permeance coefficient, coating, and environment.

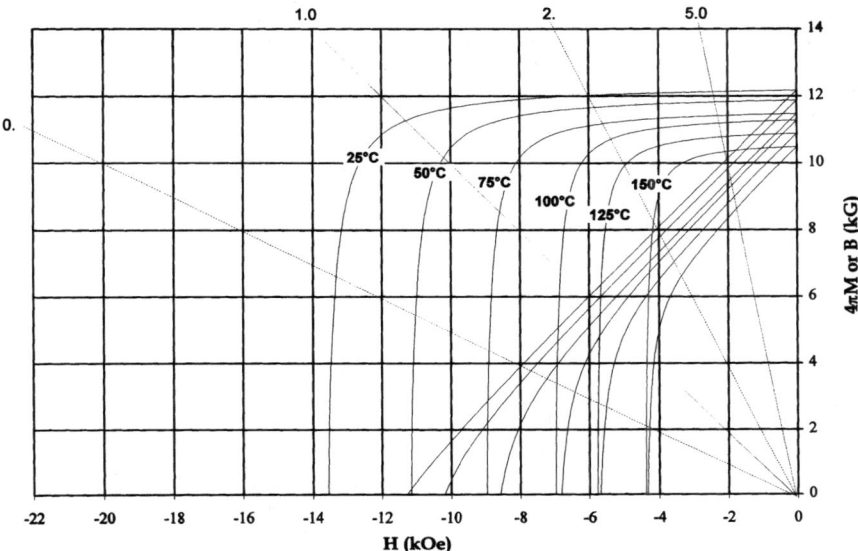

FIGURE 2.138 Demagnetization curves for MQ3-E 34 magnets. *(Courtesy of Magnequench.)*

TABLE 2.58 Magnetic Characteristics

MQ3-E 38 Magnets		
Residual induction $B_r \parallel$	12.8 kG	1.28 T
Coercive force H_c	11.4 kOe	907 kA/m
Intrinsic coercivity H_{ci}	12.5 kOe	995 kA/m
Energy product BH_{max}	38 MGOe	302 kJ/m³
Recoil permeability μ_r	1.09	
Temperature coefficient of B_r to 100°C (212°F)	−0.10%/°C	
Temperature coefficient of H_{ci} to 100°C (212°F)	−0.60%/°C	
Required magnetizing force (open circuit) H_s	35 kOe	2786 kA/m
Maximum operating temperature*	150°C	

Note: Magnetic properties are typical at room temperature.
* Maximum operating temperature is dependent upon permeance coefficient, coating and, environment.

TABLE 2.59 Physical Properties

MQ3-E38 Magnets	
Density	7.6 g/cm³
Coefficient of thermal expansion∥, 25–200°C (77–392°F)	4.3 μm/m°C
Coefficient of thermal expansion ⊥, 25–200°C (77–392°F)	−1 μm/m°C
Compressive strength	5.3×10^3 kg/cm²
Young's modulus	1.6×10^6 kg/cm²
Poisson ratio	0.27
Transverse rupture strength	1.7×10^3 kg/cm²
Hardness	60 Rockwell B
Electrical resistivity	130 μΩ-cm
Specific heat	0.42 Ws/g°C
Thermal conductivity	0.07 W/cm°C
Curie temperature	335°C

Note: Physical properties are typical at room temperature.
* Maximum operating temperature is dependent upon permeance coefficient, coating, and environment.

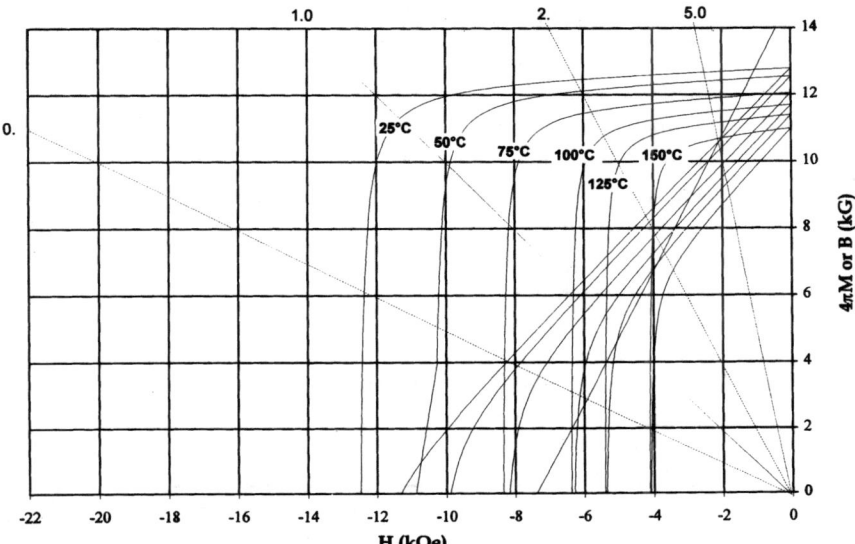

FIGURE 2.139 Demagnetization curves for MQ3-E 38 magnets. *(Courtesy of Magnequench.)*

TABLE 2.60 Magnetic Characteristics

MQ3-F 34H Magnets		
Residual induction $B_r \parallel$	12.8 kG	1.28 T
Coercive force H_c	11.4 kOe	907 kA/m
Intrinsic coercivity H_{ci}	12.5 kOe	995 kA/m
Energy product BH_{max}	38 MGOe	302 kJ/m³
Recoil permeability μ_r	1.09	
Temperature coefficient of B_r to 100°C (212°F)	−0.10%/°C	
Temperature coefficient of H_{ci} to 100°C (212°F)	−0.60%/°C	
Required magnetizing force (open circuit) H_s	35 kOe	2786 kA/m
Maximum operating temperature*	150°C	

Note: Magnetic properties are typical at room temperature.
* Maximum operating temperature is dependent upon permeance coefficient, coating and, environment.

TABLE 2.61 Physical Properties

MQ3-F34H Magnets	
Density	7.6 g/cm³
Coefficient of thermal expansion$\parallel$, 25–200°C (77–392°F)	4.3 µm/m°C
Coefficient of thermal expansion $\perp$, 25–200°C (77–392°F)	−1 µm/m°C
Compressive strength	5.3×10^3 kg/cm²
Young's modulus	1.6×10^6 kg/cm²
Poisson ratio	0.27
Transverse rupture strength	1.7×10^3 kg/cm²
Hardness	60 Rockwell B
Electrical resistivity	130 µΩ-cm
Specific heat	0.42 Ws/g°C
Thermal conductivity	0.07 W/cm°C
Curie temperature	335°C

Note: Physical properties are typical at room temperature.
* Maximum operating temperature is dependent upon permeance coefficient, coating, and environment.

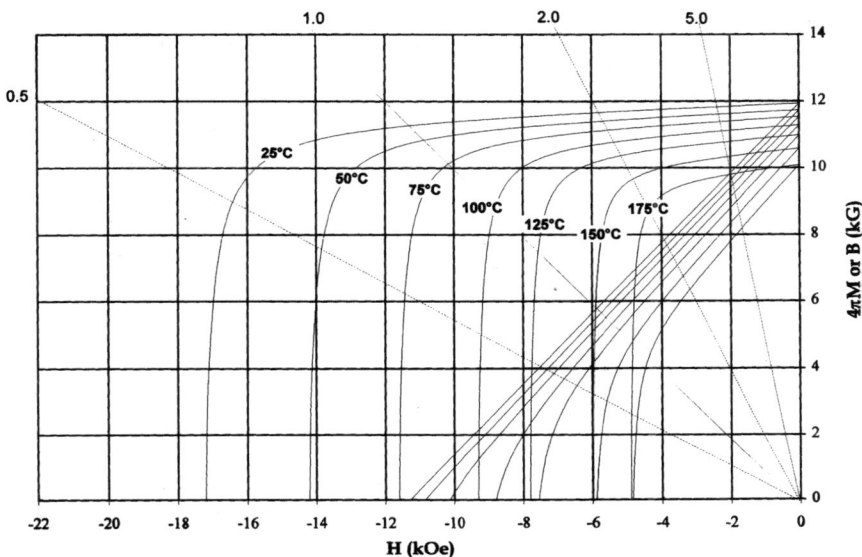

FIGURE 2.140 Demagnetization curves for MQ3-F 34H magnets. *(Courtesy of Magnequench.)*

TABLE 2.62 Magnetic Characteristics

MQ3-F 3H Magnets		
Residual induction B_r	12.5 kG	1.25 T
Coercive force H_c	11.5 kOe	915 kA/m
Intrinsic coercivity H_{ci}	16.5 kOe	1313 kA/m
Energy product BH_{max}	36 MGOe	287 kJ/m³
Recoil permeability μ_r	1.07	
Temperature coefficient of B_r to 100°C (212°F)	−0.09%/°C	
Temperature coefficient of H_{ci} to 100°C (212°F)	−0.60%/°C	
Required magnetizing force (open circuit) H_s	35 kOe	2786 kA/m
Maximum operating temperature*	180°C	

Note: Magnetic properties are typical at room temperature.
* Maximum operating temperature is dependent upon permeance coefficient, coating and, environment.

TABLE 2.63 Physical Properties

MQ3-F 3H Magnets	
Density	7.6 g/cm³
Coefficient of thermal expansion∥, 25–200°C (77–392°F)	6 μm/m°C
Coefficient of thermal expansion ⊥, 25–200°C (77–392°F)	0.6 μm/m°C
Compressive strength	5.3×10^3 kg/cm²
Young's modulus	1.6×10^6 kg/cm²
Poisson ratio	0.27
Transverse rupture strength	1.7×10^3 kg/cm²
Hardness	60 Rockwell B
Electrical resistivity	130 μΩ-cm
Specific heat	0.42 Ws/g°C
Thermal conductivity	0.07 W/cm°C
Curie temperature	370°C

Note: Physical properties are typical at room temperature.
* Maximum operating temperature is dependent upon permeance coefficient, coating, and environment.

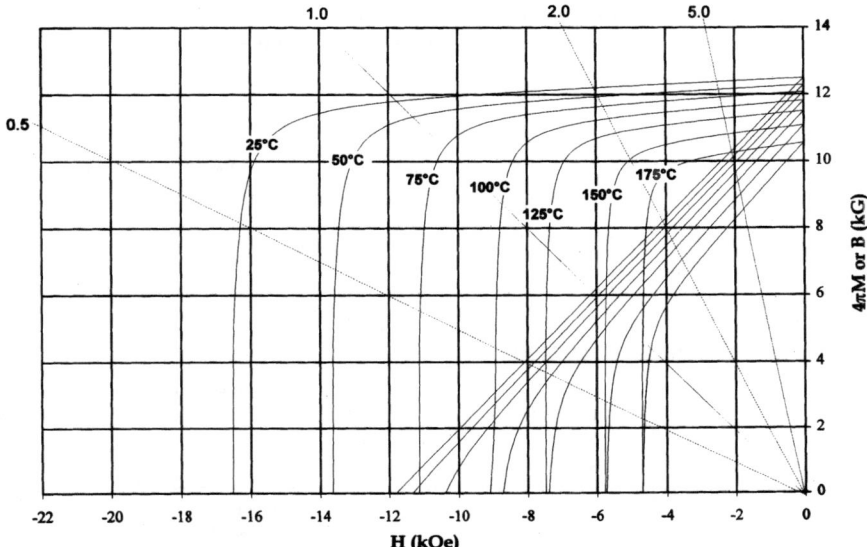

FIGURE 2.141 Demagnetization curves for MQ3-F 36H magnets. *(Courtesy of Magnequench.)*

TABLE 2.64 Magnetic Characteristics

MQ3-G 30SH Magnets		
Residual induction $B_r \parallel$	12.8 kG	1.28 T
Coercive force H_c	11.4 kOe	907 kA/m
Intrinsic coercivity H_{ci}	12.5 kOe	995 kA/m
Energy product BH_{max}	38 MGOe	302 kJ/m³
Recoil permeability μ_r	1.09	
Temperature coefficient of B_r to 100°C (212°F)	−0.10%/°C	
Temperature coefficient of H_{ci} to 100°C (212°F)	−0.60%/°C	
Required magnetizing force (open circuit) H_s	35 kOe	2786 kA/m
Maximum operating temperature*	150°C	

Note: Magnetic properties are typical at room temperature.
* Maximum operating temperature is dependent upon permeance coefficient, coating and, environment.

TABLE 2.65 Physical Properties

MQ3-G 30SH Magnets	
Density	7.6 g/cm³
Coefficient of thermal expansion∥, 25–200°C (77–392°F)	4.3 μm/m°C
Coefficient of thermal expansion ⊥, 25–200°C (77–392°F)	−1 μm/m°C
Compressive strength	5.3×10^3 kg/cm²
Young's modulus	1.6×10^6 kg/cm²
Poisson ratio	0.27
Transverse rupture strength	1.7×10^3 kg/cm²
Hardness	60 Rockwell B
Electrical resistivity	130 μΩ-cm
Specific heat	0.42 Ws/g°C
Thermal conductivity	0.07 W/cm°C
Curie temperature	335°C

Note: Physical properties are typical at room temperature.
* Maximum operating temperature is dependent upon permeance coefficient, coating, and environment.

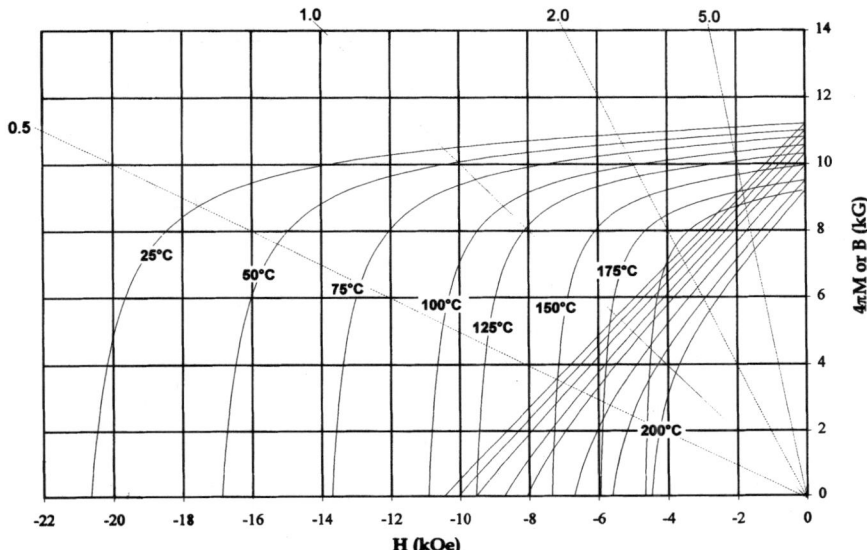

FIGURE 2.142 Demagnetization curves for MQ3-G 30SH magnets. *(Courtesy of Magnequench.)*

TABLE 2.66 Magnetic Characteristics

MQ3-G 32SH Magnets		
Residual induction $B_r \parallel$	13.1 kG	1.31 T
Coercive force H_c	12.3 kOe	979 kA/m
Intrinsic coercivity H_{ci}	16 kOe	1274 kA/m
Energy product BH_{max}	42 MGOe	334 kJ/m^3
Recoil permeability μ_r	1.06	
Temperature coefficient of B_r to 100°C (212°F)	−0.09%/°C	
Temperature coefficient of H_{ci} to 100°C (212°F)	−0.60%/°C	
Required magnetizing force (open circuit) H_s	35 kOe	2786 kA/m
Maximum operating temperature*	180°C	

Note: Magnetic properties are typical at room temperature.
* Maximum operating temperature is dependent upon permeance coefficient, coating and, environment.

TABLE 2.67 Physical Properties

MQ3-G 32SH Magnets	
Density	7.6 g/cm^3
Coefficient of thermal expansion$\parallel$, 25–200°C (77–392°F)	6 μm/m°C
Coefficient of thermal expansion $\perp$, 25–200°C (77–392°F)	0.6 μm/m°C
Compressive strength	5.3×10^3 kg/cm^2
Young's modulus	1.6×10^6 kg/cm^2
Poisson ratio	0.27
Transverse rupture strength	1.7×10^3 kg/cm^2
Hardness	60 Rockwell B
Electrical resistivity	130 μΩ-cm
Specific heat	0.42 Ws/g°C
Thermal conductivity	0.07 W/cm°C
Curie temperature	370°C

Note: Physical properties are typical at room temperature.
* Maximum operating temperature is dependent upon permeance coefficient, coating, and environment.

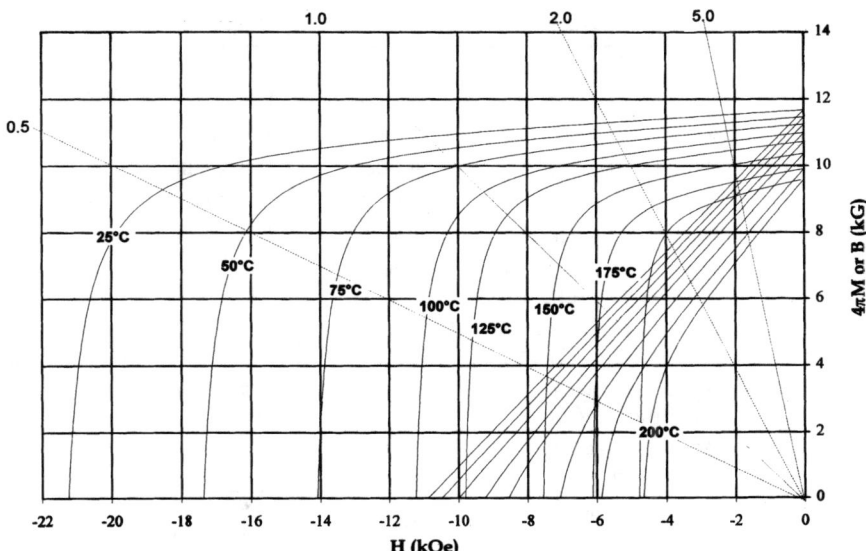

FIGURE 2.143 Demagnetization curves for MQ3-G 32SH magnets. *(Courtesy of Magnequench.)*

TABLE 2.68 Magnetic Characteristics

MQ3-F 42 Magnets		
Residual induction B_r	13.1 kG	1.31 T
Coercive force H_c	12.3 kOe	979 kA/m
Intrinsic coercivity H_{ci}	16 kOe	1274 kA/m
Energy product BH_{max}	42 MGOe	334 kJ/m^3
Recoil permeability μ_r	1.06	
Temperature coefficient of B_r to 100°C (212°F)	−0.09%/°C	
Temperature coefficient of H_{ci} to 100°C (212°F)	−0.60%/°C	
Required magnetizing force (open circuit) H_s	35 kOe	2786 kA/m
Maximum operating temperature*	180°C	

Note: Magnetic properties are typical at room temperature.
* Maximum operating temperature is dependent upon permeance coefficient, coating and, environment.

TABLE 2.69 Physical Properties

MQ3-F 42 Magnets	
Density	7.6 g/cm^3
Coefficient of thermal expansion∥, 25–200°C (77–392°F)	6 μm/m°C
Coefficient of thermal expansion ⊥, 25–200°C (77–392°F)	0.6 μm/m°C
Compressive strength	5.3 × 10^3 kg/cm^2
Young's modulus	1.6 × 10^6 kg/cm^2
Poisson ratio	0.27
Transverse rupture strength	1.7 × 10^3 kg/cm^2
Hardness	60 Rockwell B
Electrical resistivity	130 μΩ-cm
Specific heat	0.42 Ws/g°C
Thermal conductivity	0.07 W/cm°C
Curie temperature	370°C

Note: Physical properties are typical at room temperature.
* Maximum operating temperature is dependent upon permeance coefficient, coating, and environment.

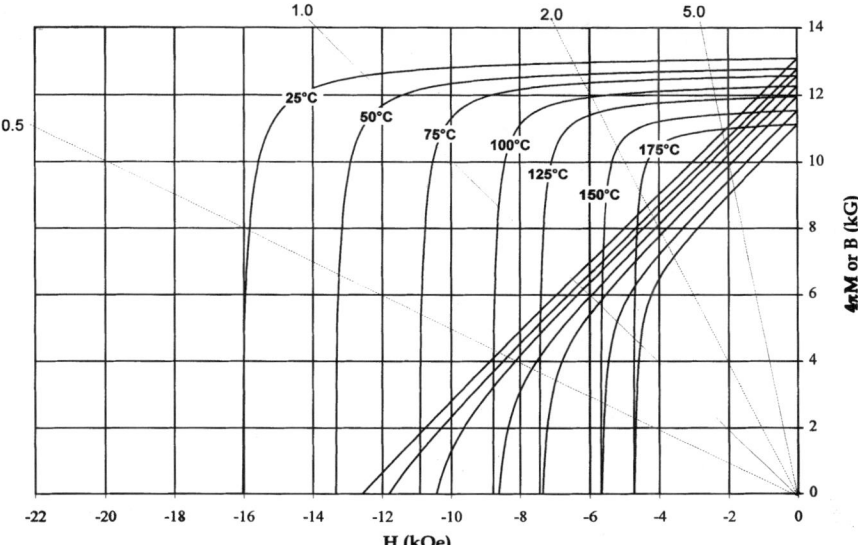

FIGURE 2.144 Demagnetization curves for MQ3-F 42 magnets. *(Courtesy of Magnequench.)*

TABLE 2.70 Magnetic Characteristics

MQ3-F 40 Magnets		
Residual induction $B_r \|$	12.8 kG	1.28 T
Coercive force H_c	11.4 kOe	907 kA/m
Intrinsic coercivity H_{ci}	12.5 kOe	995 kA/m
Energy product BH_{max}	38 MGOe	302 kJ/m^3
Recoil permeability μ_r	1.09	
Temperature coefficient of B_r to 100°C (212°F)	−0.10%/°C	
Temperature coefficient of H_{ci} to 100°C (212°F)	−0.60%/°C	
Required magnetizing force (open circuit) H_s	35 kOe	2786 kA/m
Maximum operating temperature*	150°C	

Note: Magnetic properties are typical at room temperature.
* Maximum operating temperature is dependent upon permeance coefficient, coating and, environment.

TABLE 2.71 Physical Properties

MQ3-F 40 Magnets	
Density	7.6 g/cm^3
Coefficient of thermal expansion$\|$, 25–200°C (77–392°F)	4.3 μm/m°C
Coefficient of thermal expansion $\perp$, 25–200°C (77–392°F)	−1 μm/m°C
Compressive strength	5.3×10^3 kg/cm^2
Young's modulus	1.6×10^6 kg/cm^2
Poisson ratio	0.27
Transverse rupture strength	1.7×10^3 kg/cm^2
Hardness	60 Rockwell B
Electrical resistivity	130 μΩ-cm
Specific heat	0.42 Ws/g°C
Thermal conductivity	0.07 W/cm°C
Curie temperature	335°C

Note: Physical properties are typical at room temperature.
* Maximum operating temperature is dependent upon permeance coefficient, coating, and environment.

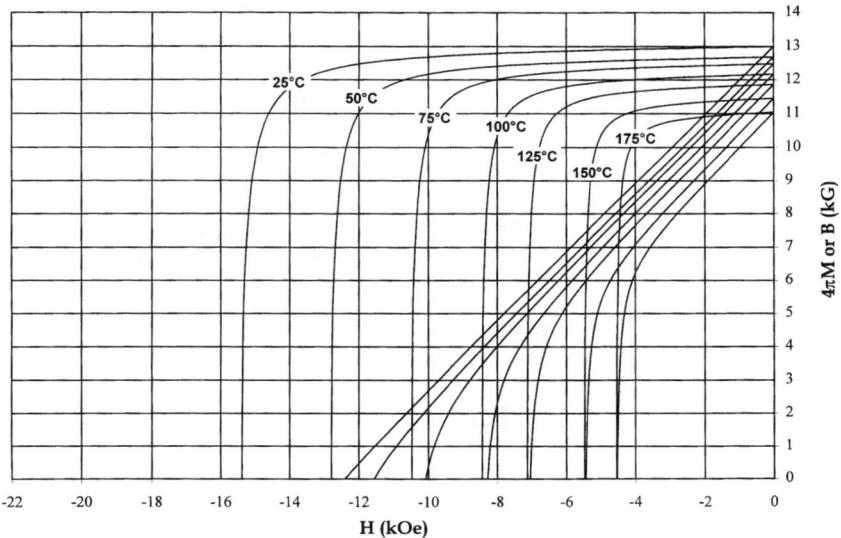

FIGURE 2.145 Demagnetization curves for MQ3-F 40 magnets. *(Courtesy of Magnequench.)*

2.9 INSULATION*

It is necessary to insulate motor laminations from the magnet wire that is used in order to meet safety agency requirements and to prevent the magnet wire from shorting to the laminations (see Fig. 2.146). The material may be an integral epoxy powder coating which covers the slots and ends of the laminations, or it may consist of precut and -formed slot liners and end lamination insulators. (See Figs. 2.147 through 2.151.) The epoxy insulation should be a minimum of 0.010 in thick and be

---

*Section contributed by William H. Yeadon, Yeadon Engineering Services, PC.

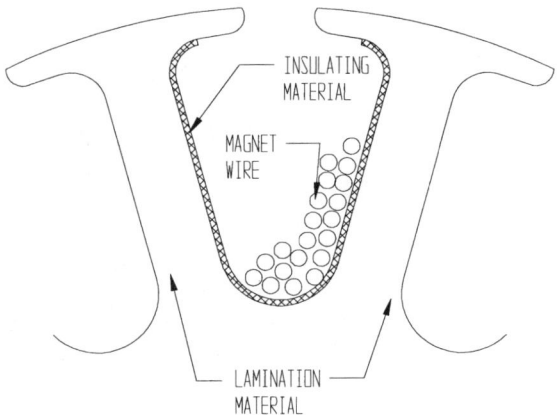

FIGURE 2.146 Copper wire wound in insulated motor slot.

FIGURE 2.147 Armature with integral epoxy powder coating.

free of voids and pinholes. The insulator slot materials should be a minimum of 0.007 in thick and be rated for use at the intended operating temperature of the motor.

Usual insulation classes are:

        Class A = 105°C (221°F)   Class B = 130°C (266°F)
        Class F = 155°C (311°F)   Class H = 180°C (324°F)

The insulators are placed in the slots and ends where necessary. (See Fig. 2.152.) The magnet wire is then wound in place. In the case of universal motors with arma-

FIGURE 2.148 Field with integral epoxy powder coating.

FIGURE 2.149 Preformed slot liners.

FIGURE 2.150 End lamination insulator.

FIGURE 2.151 End lamination insulator.

tures, where the motor operates at high speeds, slot wedges are placed in the slots between the magnet wire and the slot openings. (See Fig. 2.153.) Some armatures have additional varnish coatings to secure the individual strands of wire to keep them from rubbing on each other under high-speed conditions. Varnish must also be rated for the appropriate thermal class. Tables 2.72 and 2.73 list typical properties of some common varnishes used in electrical equipment.

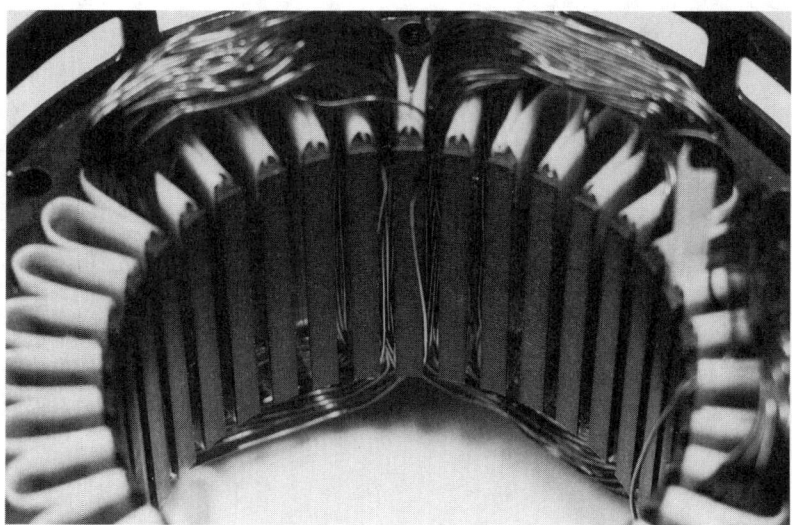

FIGURE 2.152 Partially wound stator and coil assembly.

FIGURE 2.153 Armature assembly showing slot wedges.

The stator must have slot pegs to hold the wire from migrating into the bore and to provide proper dielectric strength. (See Fig. 2.154.)

Safety and regulatory agencies such as Underwriters Laboratories and the Canadian Standards Association set specific requirements for insulation systems and clearances within the motors. They should be contacted for the standard that applies to the particular product being designed. Typical insulating materials are listed in Table 2.74.

FIGURE 2.154 Slot wedges (pegs).

TABLE 2.72 Thermosetting Baking Varnishes

Product	Product features	Appearance	Thinner/reducer	Flash point, °F/°C	Thermal index ASTM D 32521 MW-35, °C	Curing cycle: hours baked (once unit reaches temp.)	Dielectric strength (V/mil) ASTM D 115, dry/wet	Bond strength (lb) STSM D 2519 25°C (77°F)	Bond strength (lb) STSM D 2519 155°C (311°F)	UL-recognized insulation systems 130°C (266°F)	155°C (311°F)	180°C (356°F)	200°C (392°F)	220°C (428°F)
Hi-Therm BC-325	High bond strength, fast cure; gives a marproof film and has low outgassing.	Clear amber	T-100	45°/7°	200	0.5–4 @ 250–325°F 120–165°C	3400/2800	35	4.5				✓	
Hi-Therm BC-340	Flexible, tough film, high bond strength; especially good for form-wound coils; meets MIL-I-24092D.	Clear amber	T-50	81°/27°	195	0.5–4 @ 275–325°F 135–165°C	4100/2900	35	5.0			✓	✓	
Hi-Therm BC-346-A	High-temperature (220°C) varnish; tough, flexible, and QPL and UL recognized; meets MIL-I-24092D. Class 180 for processing at land-based facilities.	Clear amber	T-200-X T-100	54°/12°	215	1–5 @ 135–165°F 275–350°C	4000/2900	28	2.5		✓	✓	✓	✓
Hi-Therm BC-350	High-bond-strength, flexible, polyester varnish; can run at high temperatures; meets MIL-I-24092D.	Clear amber	T-100-X	81°/27°	200	0.5–3 @ 275–350°F 135–175°C	4200/3400	35	5	✓				

2.168

TABLE 2.72 Thermosetting Baking Varnishes (*Continued*)

Product	Product features	Appearance	Thinner/reducer	Flash point, °F/°C	Thermal index ASTM D 32521 MW-35, °C	Curing cycle: hours baked (once unit reaches temp.)	Dielectric strength (V/mil) ASTM D 115, dry/wet	Bond strength (lb) STSM D 2519 25°C (77°F)	Bond strength (lb) STSM D 2519 155°C (311°F)	UL-recognized insulation systems 130°C (266°F)	155°C (311°F)	180°C (356°F)	200°C (392°F)	220°C (428°F)
Dolphon BC-352	Class H epoxy varnish; excellent for hermetic applications; high bond strength; outstanding moisture and chemical resistance.	Clear amber	T-352	81°/27°	180	2–6 @ 325–350°F 165–175°F	4600/3600	55	18				✓	
Hi-Therm BB-353	High-temperature, glossy black version of BC-346-A with excellent appearance.	Glossy black	T-100 T-200-X	50°/10°	200	1–5 @ 275–325°F 135–165°C	2600/2400	28	2.5		✓	✓	✓	✓
Hi-Therm BC-354	High-temperature varnish; excellent flexibility and wetting properties; high-build, silicone alternative. QPL approval for MIL-I-2409D.	Clear amber	T-200-X	81°/27°	215	2–8 @ 275–350°F 135–165°C	4100/3000	28	2.5					
Hi-Therm BC-359	Energy-saving, solderable, fast-cure/low-temperature-cure varnish; flexible, with superior moisture chemical and abrasion resistance. Rule 66 compliant.	Clear amber	T-100	45°/7°	180	0.5–10 @ 230–325°F 110–165°C	3100/3000	28	3.5	✓	✓	✓	✓	

TABLE 2.72 Thermosetting Baking Varnishes (*Continued*)

Product	Product features	Appearance	Thinner/reducer	Flash point, °F/°C	Thermal index ASTM D 32521 MW-35, °C	Curing cycle: hours baked (once unit reaches temp.)	Dielectric strength (V/mil) ASTM D 115, dry/wet	Bond strength (lb) STSM D 2519 25°C (77°F)	Bond strength (lb) STSM D 2519 155°C (311°F)	UL-recognized insulation systems 130°C (266°F)	155°C (311°F)	180°C (356°F)	200°C (392°F)	220°C (428°F)
Hi-Therm BC-359-MS	Specially formulated for motor repair shops, this varnish has all the same features as BC-359 plus requires no pre-thinning.	Clear amber	T-100	45°/7°	180	230–325°F 110–165°C 0.5–10 @	3100/3000	26	3.5	✓	✓	✓	✓	
Hi-Therm BC-363	Fast-curing general-purpose product; UL recognized, high dielectric, high bond strength.	Clear amber	T-200-X	50°/10°	180	230–325°F 110–165°C 0.5–10 @	4500/3500	38	6.0	✓	✓	✓	✓	✓
Aqua-Therm BC-365	Water-based, environmentally friendly, easy-to-use, fast-curing varnish; low viscosity, excellent moisture resistance, flexible.	Clear amber	Water	>200°/>93°	180	275–325°F 135–165°C 1–6 @	1900/1850	30	10	✓	✓	✓	✓	

Source: Courtesy of John C. Dolph Company.

TABLE 2.73 Solventless Resins

Product	Product features	Appearance	Thinner/reducer	Flash point, °F/°C	Thermal index ASTM D 32521 MW-35, °C	Curing cycle: hours baked (once unit reaches temp.)	Dielectric strength (V/mil) ASTM D 115, dry/wet	Bond strength (lb) STSM D 2519 25°C (77°F)	Bond strength (lb) STSM D 2519 155°C (311°F)	UL-recognized insulation systems 130°C (266°F)	155°C (311°F)	180°C (356°F)	200°C (392°F)	220°C (428°F)
Dolphon CC-1105	Solventless high-flash, unsaturated polyester; nonflammable, extra high bond strength, fast cure; resists refrigerants (R123) and requires minimum cleanup. Bake 6 h at 325°F for maximum hermetic resistance.	Clear amber	VSR-3015	329°/165°	205	0.25–1.2 @ 275–350°F 135–175°C	4300/3800	42	20	✓	✓	✓	✓	✓
Dolphon CC-1105-HTC	Solventless, unsaturated polyester resin; high thermal conductivity, high bond strength, good with inverter duty and hermetics applications.	Yellow-orange	VSR-3015	>200°/>93°	205	0.05–8 @ 275–350°F 135–175°C	2500/1200	44	21	✓	✓	✓	✓	✓
Dolphon CC-1115	One-part, high-build VPI epoxy resin; superior electrical properties, excellent moisture and chemical resistance, excellent appearance.	Translucent	None	>200°/>93°		3–9 @ 300–325°F 150–165°C	1400/800	56	7		✓		✓	

TABLE 2.73 Solventless Resins (*Continued*)

Product	Product features	Appearance	Thinner/reducer	Flash point, °F/°C	Thermal index ASTM D 32521 MW-35, °C	Curing cycle: hours baked (once unit reaches temp.)	Dielectric strength (V/mil) ASTM D 115, dry/wet	Bond strength (lb) STSM D 2519 25°C (77°F)	Bond strength (lb) STSM D 2519 155°C (311°F)	UL-recognized insulation systems 130°C (266°F)	155°C (311°F)	180°C (356°F)	200°C (392°F)	220°C (428°F)
Dolphon CC-1118-LV	One-part, high-flash epoxy; excellent electrical properties and moisture resistance; good for VPI applications; approved use on sealed units per MIL-M-17060E; passes Navy submergence tests; resists refrigerants R(123a).	Translucent	None	>200°/>93°		4–10 @ 300–325°F 150–165°C	2200/1900	61	8		✓		✓	
Dolphon CC-1133	Flexible, one-part solventless unsaturated polyester; offers reduced emissions, excellent appearance, and good electrical properties; good for VPI and dip applications; noise reducer and nonbrittle.	Clear amber	VR-3017	142°/61°	180	1–2 @ 300–325°F 150–165°C	3500/2200	35	10	✓	✓	✓	✓	✓
Dolphon CC-1141	Unique epoxy-modified resin; dip and bake or VPI; excellent electrical properties and moisture resistance; good with hermetics.	Translucent amber	None	>200°/93°	205	1–5 @ 300–350°F 150–165°C	3800/3200	35	15	✓	✓	✓	✓	✓

TABLE 2.73 Solventless Resins (*Continued*)

Product	Product features	Appearance	Thinner/reducer	Flash point, °F/°C	Thermal index ASTM D 32521 MW-35, °C	Curing cycle: hours baked (once unit reaches temp.)	Dielectric strength (V/mil) ASTM D 115, dry/wet	Bond strength (lb) STSM D 2519 25°C (77°F)	Bond strength (lb) STSM D 2519 155°C (311°F)	UL-recognized insulation systems 130°C (266°F)	155°C (311°F)	180°C (356°F)	200°C (392°F)	220°C (428°F)
Dolphon CC-1305	Solventless, high-flash unsaturated polyester; nonflammable, flexible, higher build, and low emissions.	Clear amber	VSR-3015	>200°/ >93°	180	1–2 @ 300–325°F 150–165°C	4400/3600	40	12	✓	✓	✓	✓	✓
Dolphon CC-1405	Solventless, high-flash unsaturated polyester; nonflammable, low emissions, superior moisture and chemical resistance.	Clear amber	VSR-3015	>200°/ 93°	180	0.25–3 @ 275–350°F 135–175°C	3800/3000	30	12	✓	✓	✓	✓	✓

Source: Courtesy of John C. Dolph Company.

TABLE 2.74 Insulating Materials

Material		Insulation class	Thickness, in	Dielectric strength, V	
Chemical name	Common trade name				
Polyester film	Mylar	130°C (266°F) B	0.003	9,000	
			0.005	15,000	
			0.0075	19,500	
			0.010	21,000	
			0.14	25,200	
				Flat	Bent
	Dacron-Mylar-Dacron	155°C (311°F) F	0.006	8,400	7,800
			0.009	9,700	8,700
			0.011	12,500	12,500
			0.0135	18,000	16,000
			0.013		
			0.015	12,500	12,500
			0.020	19,000	19,000
	Dacron-Mylar-Dacron	180°C (356°F) H	0.009	8,500	
			0.011	11,000	
			0.035	15,000	
			0.013	8,800	
			0.015	11,500	
			0.020	18,800	
				Flat	Bent
	Nomex-Mylar-Nomex	180°C (356°F)	0.009	14,000	
			0.011	16,000	13,200
			0.0135	24,000	14,900
			0.016		21,000
			0.013	16,000	14,500
			0.015	20,000	19,000
			0.020	22,000	
	Nomex	220°C (428°F)	0.005	3,350	
			0.007	6,150	
			0.010	8,250	
			0.012	10,330	
			0.015	12,400	
			0.020	16,100	
			0.030	21,300	
	Nomex-Kapton-Nomax	220°C (428°F)	0.015	25,000	

Source: Courtesy of DuPont.

TABLE 2.75 Powder Resins Typical Property Data

Temp class	Product no.	Description	Specific gravity	Cut-through resistance	Edge coverage	Impact resistance	Gel time @ 193°C (380°F) hot plate	Dielectric strength	Volume resistivity	Color
B	260 260CG	Spray and fluid bed dip application	1.43	215°C (410°F)	35–45	100 (11.3)	12–16 s	1000 (12–15-mil coating)	10^{15}	Green
B	262	Spray and fluid bed dip applications	1.34	130°C (266°F)	38–48	100 (11.3)	12–16 s	1000 (10-mil coating)	10^{13}	Red
B	263	Spray and fluid bed dip applications in high-temperature cut-through resistance	1.47	290°C (554°F)	40–50	100 (11.3)	8–14 s	1000 (12–15-mil coating)	10^{15}	Green
B	270	Spray and fluid bed dip applications for higher-temperature cut-through and bridging gaps	1.48	250°C (482°F)	35–40	120 (13.8)	12–16 s	1000 (10-mil coating)	10^{13}	Green
B	5555	Cold electrostatic fluid bed, hot venturi spray, or hot fluid bed dip for fractional horsepower motor stators and armatures	1.7	>340°C (644°F)		160 (18.1)	8–12 s	1300 (V/mil)		Green
B	5388	Electrostatic fluid bed process, superior cut-through resistance and well heat, chemical and moisture resistance	1.57	>340°C (644°F)	35 (11.3)	100	25–35 s	1100 (V/mil)		Blue
B	5133	Electrostatic coating for cold as well as heated parts	1.45	160°C (320°F)	15 (13.8)	120		500 (V/mil)	5×10^{14}	Light blue

Source: Courtesy of 3M Company.

Some motors require phase barrier insulation. Usually these are motors rated above 230 V ac. Phase barriers are inserted in the slots, between the coils of different phases sharing the same slot. Additional insulation, sometimes referred to as H insulators, is provided to isolate the end turns of the individual phases. These insulators are generally made of the same material as the slot liners, although the thickness may vary based on dielectric and coil-forming requirements.

Integral epoxy insulation material properties are listed in Table 2.75.

2.10 MAGNET WIRE*

Magnet wire is a key component of all electric motors. It is the current carry and conductor that sets up the required magnetic field that causes rotation and produces torque. Selecting the proper magnet wire for the application requires some basic understanding of magnet wire ratings and sizes. These include dimensions, physical tests, electrical tests, and cure tests. These items are covered in detail in NEMA MW 1000-1997. A brief summary is given here.

2.10.1 AWG Sizes

American Wire Gauge is a standardized geometric progression series of wire sizes used in the United States. Three AWG size reductions equal one-half cross-sectional area equals double feet per pound. Nominal bare diameter = $(0.0050)*(1.1229322)^{(36 - \text{AWG number})}$. Wire dimensions and terminology are shown in Figs. 2.155 through 2.160. Wire properties are shown in Tables 2.76 through 2.79.

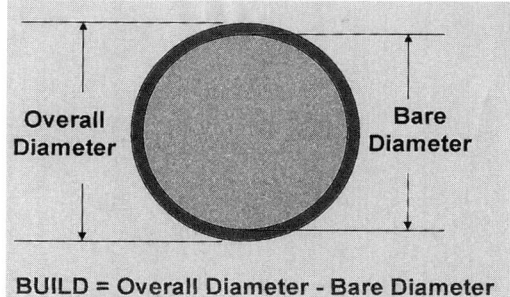

FIGURE 2.155 Wire and insulation dimensions. Build = overall diameter − bare diameter.

---

*Section courtesy of Phelps Dodge Company.

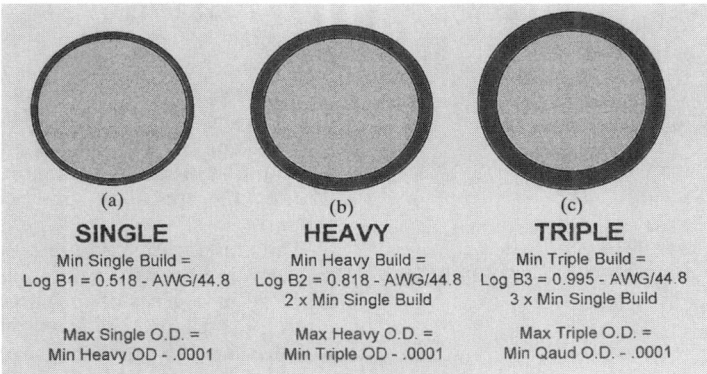

FIGURE 2.156 Wire and insulation concentricity. (*a*) Single insulation: minimum single build = log B_1 = 0.518 − AWG/44.8; maximum single OD = minimum heavy OD − 0.0001 in. (*b*) Heavy insulation: minimum heavy build = log B_2 = 0.818 − AWG/44.8 (double minimum single build); maximum heavy OD = minimum triply OD − 0.0001 in. (*c*) Triple insulation: minimum triple build = log B_3 = 0.995 − AWG/44.8 (triple minimum single build); maximum triple OD = minimum quad OD − 0.0001 in.

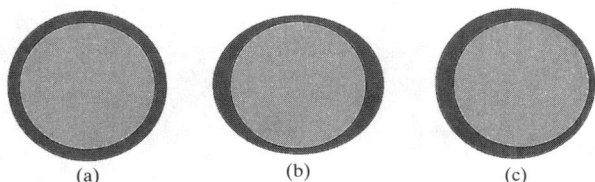

FIGURE 2.157 Wire and insulation dimensional terminology: (*a*) concentric, (*b*) out of round, and (*c*) eccentric.

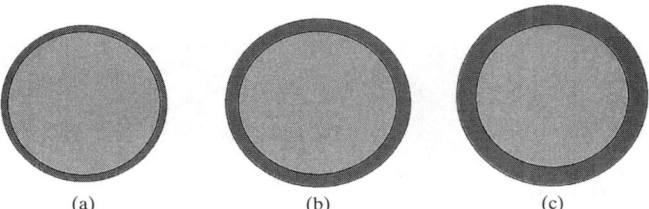

FIGURE 2.158 Wire insulation build designations. (*a*) Single insulation: minimum single build = log B_1 = 0.518 − AWG/44.8; maximum single OD = minimum heavy OD − 0.0001 in. (*b*) Heavy insulation: minimum heavy build = log B_2 = 0.818 − AWG/44.8 (double minimum single build); maximum heavy OD = minimum triply OD − 0.0001 in. (*c*) Triple insulation: minimum triple build = log B_3 = 0.995 − AWG/44.8 (triple minimum single build); maximum triple OD = minimum quad OD − 0.0001 in.

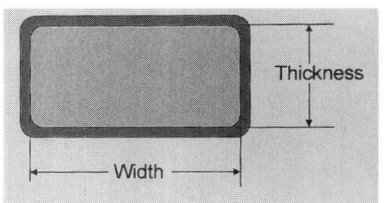

FIGURE 2.159 Square and rectangular wire and insulation dimensions. Sizes are expressed as thickness times width of the bare wire (i.e., 100 × 200).

2.10.2 Wire Testing

Elongation. Elongation is a measure of conductor softness. An elongation device is shown in Fig. 2.161. Here a 10-in wire sample is stretched at a rate of 1 ft/sec until it breaks. It is an indication of whether the annealing conditions were proper.

The amount of springiness or elastic memory in the wire is measured and reported in degrees of springback. (See Fig. 2.162.) Elongation is an indication of whether the annealing conditions were proper. The conductor softness is critical so that it can be easily formed into the final shape of the application.

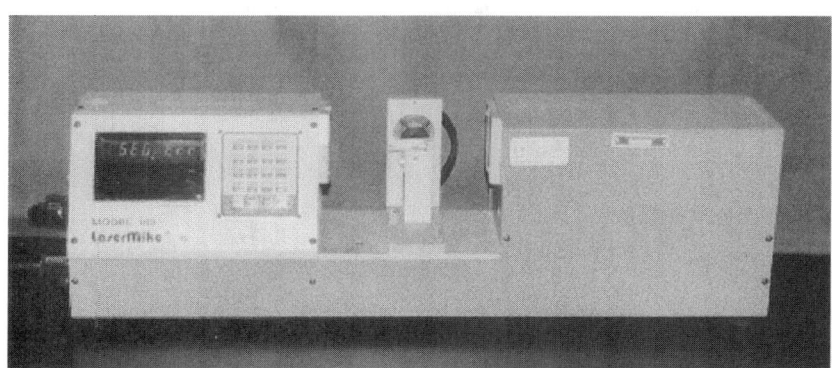

FIGURE 2.160 LaserMike machine for measuring wire and insulation.

Mandrel Flexibility. The magnet wire sample is elongated and wrapped around the mandrel. (See Fig. 2.163.) It is examined under magnification for cracks to bare. The insulation must be flexible to survive the bending and shaping while maintaining insulation integrity.

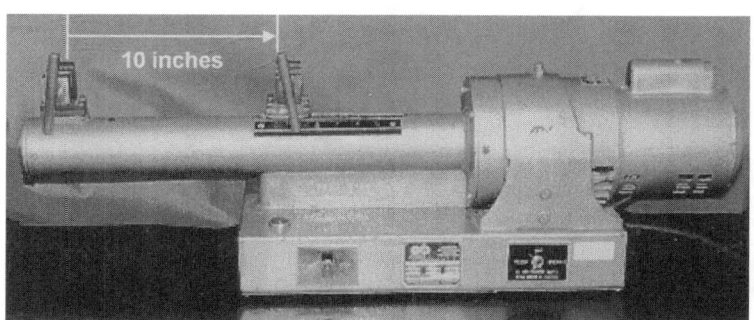

FIGURE 2.161 Round wire elongation device.

TABLE 2.76 Single-Film-Coated Standards

AWG size	AWG nominal bare wire diameter	Minimum film addition	Outside diameter			Lb./ 1000 FT	Ft/lb	Ohms/ 1000 ft	Ohms/ lb	WIRES/in^2
			Min. OD	Nom. OD	Max. OD					
14	0.0641	0.0016	0.0651	0.0658	0.0666	12.50	80.00	2.524	0.2019	230
15	0.0571	0.0015	0.0580	0.0587	0.0594	9.95	100.50	3.181	0.3197	288
16	0.0508	0.0014	0.0517	0.0524	0.0531	7.89	126.7	4.018	0.5093	363
17	0.0453	0.0014	0.0462	0.0468	0.0475	6.256	159.7	5.054	0.8073	455
18	0.0403	0.0013	0.0412	0.0418	0.0424	4.97	201.2	6.386	1.2846	572
19	0.0359	0.0012	0.0367	0.0373	0.0379	3.95	253.2	8.046	2.0370	715
20	0.0320	0.0012	0.0329	0.0334	0.0339	3.13	319.5	10.13	3.2364	896
21	0.0285	0.0011	0.0293	0.0298	0.0303	2.483	402.7	12.77	5.143	1119
22	0.0253	0.0011	0.0261	0.0266	0.0270	1.970	507.6	16.20	8.223	1403
23	0.0226	0.0010	0.0234	0.0238	0.0243	1.565	639.0	20.30	12.971	1751
24	0.0201	0.0010	0.0209	0.0213	0.0217	1.240	806.5	25.67	20.702	2204
25	0.0179	0.0009	0.0186	0.0190	0.0194	0.988	1012.1	32.37	32.763	2741
26	0.0159	0.0009	0.0166	0.0170	0.0173	0.784	1276.0	41.02	52.32	3460
27	0.0142	0.0008	0.0149	0.0152	0.0156	0.623	1605.0	51.44	82.57	4271
28	0.0126	0.0008	0.0133	0.0136	0.0140	0.495	2020.0	65.31	131.94	5407
29	0.0113	0.0007	0.0119	0.0122	0.0126	0.394	2538.0	81.21	206.12	6610
30	0.0100	0.0007	0.0106	0.0109	0.0112	0.312	3205.0	103.7	322.37	8417

TABLE 2.76 Single-Film-Coated Standards (*Continued*)

| AWG size | AWG nominal bare wire diameter | Minimum film addition | Outside diameter | | | Lb/ 1000 FT | Ft/lb | Ohms/ 1000 ft | Ohms/ lb | WIRES/in² |
			Min. OD	Nom. OD	Max. OD					
31	0.0089	0.0006	0.0094	0.0097	0.0100	0.248	4032.0	130.9	527.8	10628
32	0.0080	0.0006	0.0085	0.0088	0.0091	0.1966	5086.0	162.0	824.0	12913
33	0.0071	0.0005	0.0075	0.0078	0.0081	0.1570	6369.0	205.7	1310.2	16437
34	0.0063	0.0005	0.0067	0.0070	0.0072	0.1244	8039.0	261.3	2100.5	20408
35	0.0056	0.0004	0.0059	0.0062	0.0064	0.0989	10111.0	330.7	3343.8	26015
36	0.0050	0.0004	0.0053	0.0056	0.0058	0.0788	12690.0	414.8	5264.0	31888
37	0.0045	0.0003	0.0047	0.0050	0.0052	0.0624	16026.0	512.1	8207.0	40000
38	0.0040	0.0003	0.0042	0.0045	0.0047	0.0494	20243.0	648.2	13121.0	49383
39	0.0035	0.0002	0.0036	0.0039	0.0041	0.0393	25445.0	846.6	21542.0	65746
40	0.0031	0.0002	0.0032	0.0035	0.0037	0.0313	31949.0	1079.0	34473.0	81633
41	0.0028	0.0002	0.0029	0.0031	0.0033	0.02470	40486.0	1323.0	53563.0	104058
42	0.0025	0.0002	0.0026	0.0028	0.0030	0.01946	51387.0	1659.0	85252.0	127551
43	0.0022	0.0002	0.0023	0.0025	0.0026	0.01548	64599.0	2143.0	138437.0	160000
44	0.0020	0.0001	0.0020	0.0022	0.0024	0.01233	81103.0	2593.0	210300.0	206611
45	0.00176	0.0001	0.0018	0.0019	0.0020	0.00960	104167.0	3348.0	348750.0	277008
46	0.00157	0.0001	0.0016	0.0017	0.0018	0.00764	130890.0	4207.0	550654.0	346020
47	0.00140	0.0001	0.0015	0.0016	0.0017	0.00619	161551.0	5291.0	854766.0	390625

Source: Courtesy of Phelps Dodge.

TABLE 2.77 Half-Size Single-Film-Coated Standards

| AWG size | AWG nominal bare wire diameter | Minimum film addition | Outside diameter | | | Lb/ 1000 FT | Ft/lb | Ohms/ 1000 ft | Ohms/ lb | Wires/in^2 |
			Min. OD	Nom. OD	Max. OD					
14½	0.0606	0.0016	0.0616	0.0624	0.0631	11.17	89.52	2.824	0.252820	257
15½	0.0540	0.0015	0.0550	0.0556	0.0562	8.91	112.23	3.557	0.399214	323
16½	0.0481	0.0014	0.0490	0.0497	0.0504	7.078	141.2	4.482	0.633230	405
17½	0.0428	0.0014	0.0438	0.0444	0.0449	5.593	178.7	5.661	1.0122	507
18½	0.0381	0.0013	0.0390	0.0396	0.0402	4.450	224.7	7.143	1.6052	638
19½	0.0340	0.0012	0.0348	0.0354	0.0359	3.499	285.7	8.971	2.5639	798
20½	0.0303	0.0012	0.0312	0.0317	0.0322	2.810	355.8	11.297	4.0203	989
21½	0.0269	0.0011	0.0278	0.0283	0.0287	2.214	451.6	14.33	6.4724	1249
22½	0.0240	0.0011	0.0249	0.0253	0.0256	1.777	562.7	18.01	10.135	1562
23½	0.0214	0.0010	0.0222	0.0227	0.0231	1.405	711.7	22.64	16.114	1941
24½	0.0190	0.0010	0.0198	0.0202	0.0206	1.110	900.9	28.73	25.883	2451
25½	0.0169	0.0009	0.0176	0.0181	0.0186	0.8824	1133.0	36.31	41.149	3052
26½	0.0150	0.0009	0.0158	0.0162	0.0166	0.7000	1429.0	46.09	65.843	3810
27½	0.0134	0.0008	0.0141	0.0145	0.0148	0.5567	1796.0	57.74	113.41	4724
28½	0.0120	0.0008	0.0127	0.0131	0.0135	0.4503	2220.0	72.02	159.94	5827
29½	0.0107	0.0007	0.0113	0.0117	0.0120	0.3539	2825.0	90.58	255.95	7305
30½	0.0095	0.0007	0.0101	0.0105	0.0108	0.2827	3537.0	114.84	406.26	9070

Source: Courtesy of Phelps Dodge.

TABLE 2.78 Heavy-Film-Coated Standards

| AWG size | AWG nominal bare wire diameter | Minimum film addition | Outside diameter | | | Lb/ 1000 FT | Ft/lb | Ohms/ 1000 ft | Ohms/ lb | Wires/in² |
			Min. OD	Nom. OD	Max. OD					
14	0.0641	0.0032	0.0667	0.0674	0.0682	12.57	79.55	2.524	0.2008	221
15	0.0571	0.0030	0.0595	0.0302	0.0609	10.01	99.90	3.181	0.3178	276
16	0.0508	0.0029	0.0532	0.0538	0.0545	7.95	125.79	4.018	0.5054	344
17	0.0453	0.0028	0.0476	0.0482	0.0488	6.32	158.23	5.054	0.7997	429
18	0.0403	0.0026	0.0425	0.0431	0.0437	5.02	199.2	6.386	1.2721	536
19	0.0359	0.0025	0.0380	0.0386	0.0391	3.9	250.6	8.046	2.0165	668
20	0.0320	0.0023	0.0340	0.0346	0.0351	3.16	316.5	10.13	3.2057	835
21	0.0285	0.0022	0.0304	0.0309	0.0314	2.51	398.4	12.7	5.088	1041
22	0.0253	0.0021	0.0271	0.0276	0.0281	1.99	502.5	16.20	8.141	1303
23	0.0226	0.0020	0.0244	0.0248	0.0253	1.59	628.9	20.30	12.767	1613
24	0.0201	0.0019	0.0218	0.0222	0.0227	1.260	793.7	25.67	20.373	1993
25	0.0179	0.0018	0.0195	0.0199	0.0203	1.005	995.0	32.37	32.209	2475
26	0.0159	0.0017	0.0174	0.0178	0.0182	0.799	1252.0	41.02	51.34	3086
27	0.0142	0.0016	0.0157	0.0160	0.0164	0.634	1577.0	51.44	81.14	3858
28	0.0126	0.0016	0.0141	0.0144	0.0147	0.504	1984.0	65.31	129.58	4823
29	0.0113	0.0015	0.0127	0.0130	0.0133	0.401	2494.0	81.21	202.52	5917
30	0.0100	0.0014	0.0113	0.0116	0.0119	0.318	3145.0	103.7	326.10	7432
31	0.0089	0.0013	0.0101	0.0105	0.0108	0.254	3937.0	130.9	515.4	9070
32	0.0080	0.0012	0.0091	0.0095	0.0098	0.2019	4953.0	162.0	802.4	11080
33	0.0071	0.0011	0.0081	0.0085	0.0088	0.1611	6207.0	205.7	1276.8	13841
34	0.0063	0.0010	0.0072	0.0075	0.0078	0.1269	7880.0	261.3	2059.1	17778
35	0.0056	0.0009	0.0006	0.0067	0.0070	0.1010	9901.0	330.7	3274.3	22277
36	0.0050	0.0008	0.0057	0.0060	0.0063	0.0803	12453.0	414.8	5166.0	27778
37	0.0045	0.0008	0.0052	0.0055	0.0057	0.0641	15601.0	512.1	7889.0	33058
38	0.0040	0.0007	0.0046	0.0049	0.0051	0.0509	19646.0	648.2	12735.0	41649
39	0.0035	0.0006	0.0040	0.0043	0.0045	0.0403	24814.0	846.6	21007.0	54083
40	0.0031	0.0006	0.0036	0.0038	0.0040	0.0319	31348.0	1079.0	33824.0	69252

Source: Courtesy of Phelps Dodge.

TABLE 2.79 Half-Size Heavy-Film-Coated Standards

AWG size	AWG nominal bare wire diameter	Minimum film addition	Outside diameter			Lb/ 1000 ft	Ft/lb	Ohms/ 1000 ft	Ohms/ lb	Wires/in²
			Min. OD	Nom. OD	Max. OD					
14½	0.0606	0.0032	0.0632	0.0640	0.0647	11.24	88.97	2.8224	0.251246	244
15½	0.0540	0.0030	0.0565	0.0572	0.0578	8.97	113.8	3.557	0.396544	306
16½	0.0481	0.0029	0.0505	0.0512	0.0518	7.141	140.0	4.82	0.627643	381
17½	0.0428	0.0028	0.0452	0.0458	0.0463	5.653	176.9	5.661	0.999647	477
18½	0.0381	0.0026	0.0403	0.0409	0.0415	4.500	222.2	7.143	1.5873	598
19½	0.0340	0.0025	0.0362	0.0367	0.0372	3.539	282.6	8.971	2.5349	742
20½	0.0303	0.0023	0.0323	0.0329	0.0334	2.830	353.4	11.297	3.9919	924
21½	0.0269	0.0022	0.0289	0.0294	0.0298	2.241	446.2	14.33	6.3945	1157
22½	0.0240	0.0021	0.0259	0.0264	0.0268	1.797	556.4	18.01	10.022	1435
23½	0.0214	0.0020	0.0232	0.0237	0.0241	1.420	704.2	22.64	15.944	1780
24½	0.0190	0.0019	0.0207	0.0212	0.0216	1.130	885.0	28.73	25.425	2225
25½	0.0169	0.0018	0.0185	0.0189	0.0193	0.8679	1152.0	36.31	41.837	2799
26½	0.0150	0.0017	0.0166	0.0170	0.0173	0.7150	1399.0	46.09	64.462	3460
27½	0.0134	0.0016	0.0149	0.0153	0.0156	0.5677	1761.0	57.74	101.71	4272
28½	0.0120	0.0016	0.0135	0.0138	0.0141	0.4593	2177.0	72.02	156.80	5251
29½	0.0107	0.0015	0.0121	0.0124	0.0127	0.3609	2771.0	90.58	250.98	6504
30½	0.0095	0.0014	0.0108	0.0111	0.0114	0.2887	3464.0	114.85	397.82	8116

Source: Courtesy of Phelps Dodge.

FIGURE 2.162 Springback testing device.

Snap and snap flex measure the adherence of the insulation to the conductor. (See Fig. 2.164.) Snap a wire sample to break point or specified percentage performing elongation at a fast rate. The mandrel is wrapped and viewed under magnification.

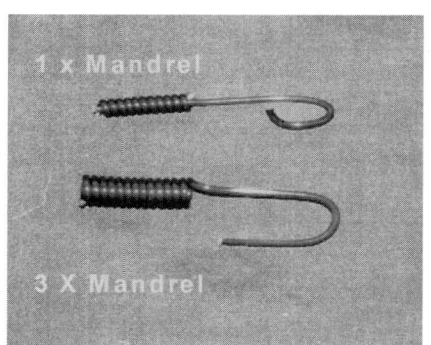

FIGURE 2.163 Mandrel testing of wire insulation flexibility.

Dynamic coefficient of friction measures slipperiness of the wire surface. It is dependent upon the external lubricant and is critical for processing. It is often referred to as *C of F* or *dynamic C of F*. (See Fig. 2.165.)

Electrical tests that are conducted include dielectric breakdown, high-voltage continuity, and conductor resistance.

Dielectric breakdown, which is a mathematical formula based on build and AWG number, measures the electrical insulating capability of the magnet wire. It is a critical application and design criterion. Nylon overcoat products cover 80 percent of standard dielectric minimums. A test for dielectric breakdown is the dielectric twist method (see Fig. 2.166). Here the number of twists and tension are dependent on wire size. Voltage is raised at a specified rate. It is usually 500 V/sec. The sample will fail when the voltage exceeds the insulation's dielectric breakdown. (See Fig. 2.167.)

High-voltage continuity is an electrical measurement of the uniformity of the insulation. A dc voltage potential is applied between the outside of the magnet wire and the conductor. A fault is detected when the voltage potential exceeds the insu-

MATERIALS

FIGURE 2.164 Snap testing device.

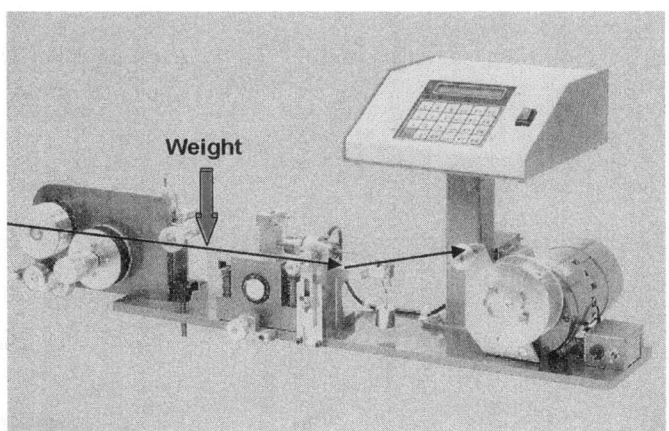

FIGURE 2.165 Coefficient of friction testing device.

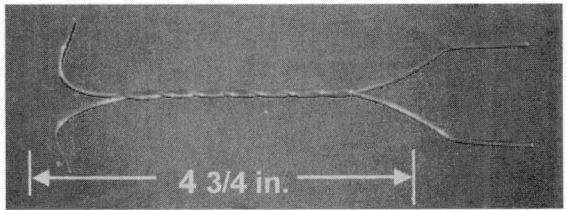

FIGURE 2.166 Dielectric twist method of testing for dielectric breakdown.

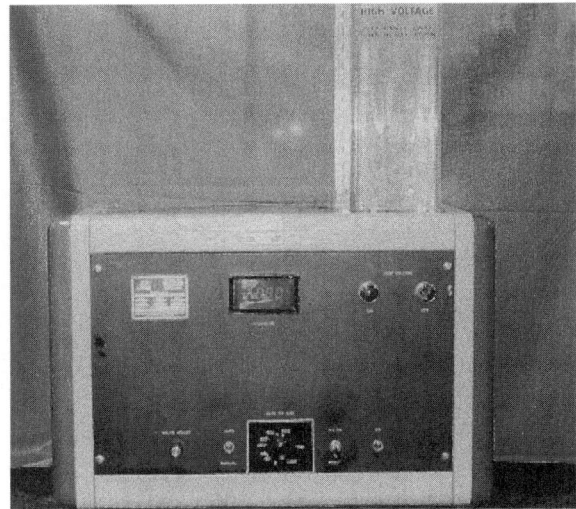

FIGURE 2.167 Dielectric breakdown testing device.

lation resistance. This test generally detects flaws in the bare conductor surface that the insulation did not adequately cover. (See Fig. 2.168.)

Conductor resistance is a measure of the resistance of the conductor, which is directly related to the cross-sectional area. This technique is used for dimensions on finer wire sizes where the resistance measurement is more accurate than physical contact measurements.

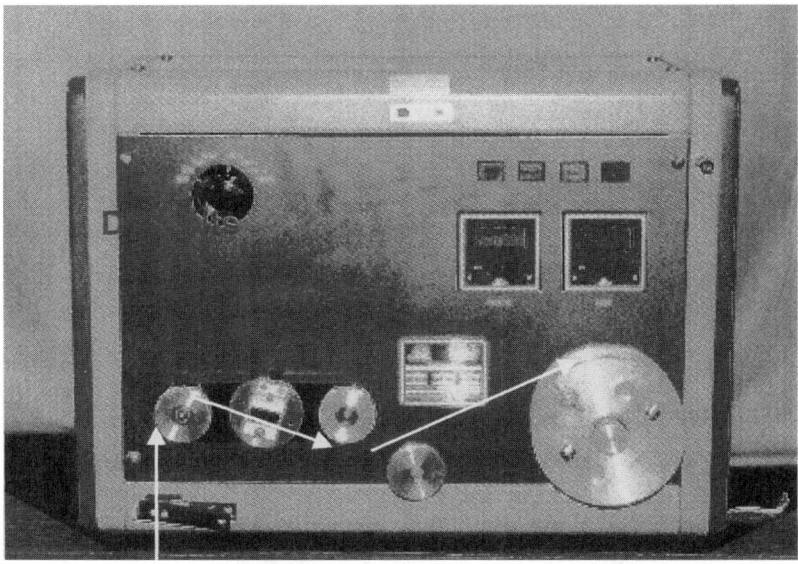

FIGURE 2.168 Bench dc high-voltage continuity testing device.

2.10.3 Cure Tests

Heat shock samples are prepared as in flexibility tests and subjected to oven heat exposure for a set time and temperature, and then allowed to cool. Samples are then examined under a microscope for cracks.

A toluene/alcohol boil is a 5-min boil test in 70/30 alcohol/toluene to examine for softening of the insulation. It is applicable to Formvar-type magnet wire.

The solderability test is the removal of insulation in molten solder at a set time and temperature. It is applicable to polyurethane and solderable esterimide–based products. It is designed to simulate the terminating processes.

Differential scanning calorimetry (DSC-T_gC) determines the glass transition temperature of the magnet wire insulation. (See Figs. 2.169 and 2.170.)

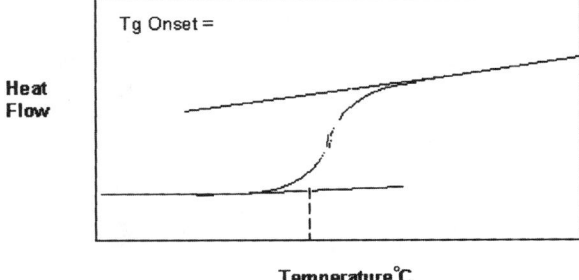

FIGURE 2.169 Differential scanning calorimetry (DSC) determination of glass transition temperature T_g, °C (T_gC).

FIGURE 2.170 Differential scanning calorimetry (DSC) cell.

TABLE 2.80 Lead Wire Application Chart

Insulation	Temp. rating, °C	Temp. range, °C	UL voltage rating, V	Dielectric strength, V/mil	Oil resistance	Ozone resistance	Abrasion	Varnish resistance	Flame resistance
Neoprene	90	−50–+90	300/600	300	Good	Good	Good	Excellent	Good
PVC/orlon Brd.	105	−35–+105	300	600	Excellent	Excellent	Excellent	Excellent	Excellent
PVC 1/32-in wall	105	−35–+105	600	60	Excellent	Excellent	Good	Excellent	Excellent
Hypalon	105	−45–+105	600	500	Good	Good	Good	Excellent	Good
Cross-linked polyethylene	105 125	−80–+125	300/600	800	Fair	Good	Good	Excellent	Good
EPDM	150	−67–+150	600	400	Fair	Good	Good	Excellent	Fair
Silicone rubber	150	−67–+150	300/600	300	Fair	Excellent	Poor	Poor	Good
Silicone rubber lass braid	150 200	−67–+200	600	300	Fair	Excellent	Excellent	Excellent	Excellent
Teflon-glass	250	−230–+280	300	1500		Excellent	Excellent		Excellent

Source: Courtesy of H. A. Holden Co.

2.11 LEAD WIRE AND TERMINATIONS*

Most motors are connected to a power source either by a stud, a screw, or a terminal inside a termination box or junction box on the motor or by routing lead wires through an appliance or by a cord set.

Studs, screws, and terminals must be of sufficient size to handle the motor current. Further, the attachment of the wire to the stud or screw must incorporate a cupped washer or binding head that retains the wire in place.

Many small motors utilize lead wires that are connected to the stator coils brought outside the motor and routed to the power source inside the appliance. The type of wire to be used in any situation depends on the operating temperature of the hottest spot, the rated voltage, and the application. The hottest spot is usually where it is attached to the stator windings. The wire size is selected based on the current requirements of the motor ampacities varied with the type of insulation being used. Charts are available from most wire manufacturers that list ampacities for the various types of insulation and wire sizes.

Table 2.80 shows some of the wire insulation types and lists common properties for each type that will aid in the selection process.

Some other things to consider are flexibility, agency approvals, and mechanical strength.

Flexibility is a property of the wire that allows it to bend. Stiff wire has one or a few strands of current-carrying metal, usually copper, to constitute the gauge size. Flexible wire has more strands of finer wire to constitute a gauge size. Generally speaking, the more flexible wire carries a higher price.

Many applications require that motor components be listed by same safety agency, such as UL, CSA, or VDE. These agencies specify tests for such things as abrasion, resistance, bending, dielectric strength, flammability, cut-through, and minimum wall thickness. These applications requirements should be determined before the lead wire type is selected.

There are also mechanical properties that need to be considered, including hardness, tensile strength, and elongation. While these numbers are good for comparing one material to another, the motor leads should not be subjected to these kinds of stresses in actual use.

---

*Section contributed by William H. Yeadon, Yeadon Engineering Services, PC.

CHAPTER 3
MECHANICS AND MANUFACTURING METHODS

Chapter Contributors

Larry C. Anderson	Bill Lawrence
Axis SPA	NMB Corporation
John S. Bank	Stanley D. Payne
Warren C. Brown	Derrick Peterman
Peter Caine	Karl H. Schultz
John Cocco	Chris Swenski
Phil Dolan	Harry J. Walters
Leon Jackson	Alan W. Yeadon
Roger O. LaValley	William H. Yeadon

3.1 MOTOR MANUFACTURING PROCESS FLOW*

The basic manufacturing processes for electric ac and dc motors are shown in this section. Within each process, there are significant variables, each depending upon the manufacturing equipment, size and variety of the parts, electrical efficiency requirements, and economics. Each one of these process variables is described in the following text.

3.1.1 AC Motor Manufacturing Process Flow

Figure 3.1 illustrates a basic ac motor manufacturing process flow. The first step is producing laminations. These laminations are separated into rotors and stators. The stator laminations, shown in Fig. 3.2, are then stacked into a core, and copper and/or aluminum wire is wound into the core, producing a wound stator core. An outer housing of some type is produced, and that is then wrapped around the wound stator core, making a wound stator assembly. The wound stator assembly is then sent to motor assembly.

---

* Sections 3.1 to 3.9 contributed by Karl H. Schultz, Schultz Associates, except as noted.

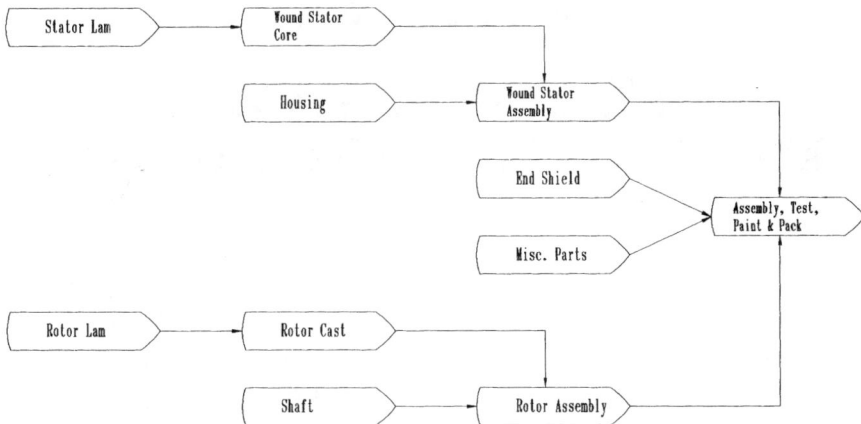

FIGURE 3.1 Ac motor manufacturing process flow.

The rotor laminations in Fig. 3.3 are also stacked and then aluminum die cast into a rotor casting, shown in Fig. 3.4. A shaft is then produced, and this is assembled into the rotor, making it a rotor assembly, shown in Fig. 3.5. The rotor assembly is sent to motor assembly.

Two end frames are produced and sent to motor assembly.

At the final operation, the wound stator assembly, rotor assembly, two end frames, and miscellaneous parts are assembled into a complete motor. The motor is then tested, painted, and packed for shipment.

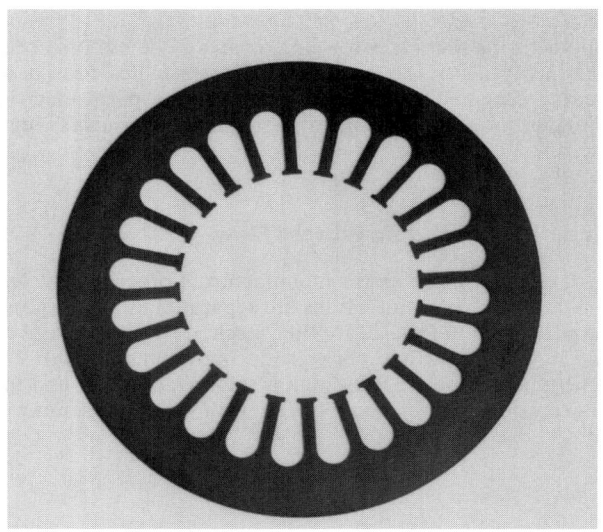

FIGURE 3.2 Stator laminations.

MECHANICS AND MANUFACTURING METHODS 3.3

FIGURE 3.3 Rotor laminations.

3.1.2 DC Motor Manufacturing Process Flow

The basic dc motor manufacturing process is illustrated in Fig. 3.6. Like ac motors, the first step is producing laminations for the pole piece and armature. The pole-piece lamination is stacked with several other components into a pole piece assembly. The pole piece on dc motors may be of solid steel, as shown in Fig. 3.7. A housing is produced, and when the pole pieces are inserted, it becomes a frame and field assembly, shown in Fig. 3.8. This frame and field assembly is then sent to motor assembly.

Brushes, with other components, are assembled into a brush assembly, as shown in Fig. 3.9, and this is then assembled on the frame and field assembly.

The armature lamination is stacked into a core, which is then assembled onto a shaft, and copper wire is inserted or wound onto the core. The coils may be connected to the commutator as they are wound, as in Fig. 3.10, or connected after the coils are inserted into the core and shaft assembly, as in Fig. 3.11. This is a completed armature assembly which then goes to final motor assembly.

The frame and field assembly, armature assembly, and miscellaneous parts are then assembled into a complete motor, as shown in Fig. 3.12. The motor is then tested, painted, and packed for shipment.

FIGURE 3.4 Rotor casting.

FIGURE 3.5 Rotor assembly.

3.2 END FRAME MANUFACTURING

The basic purpose of an end frame, sometimes denoted an *end bell, end shield,* or *bracket,* is to contain the shaft bearings and support the rotor assembly. It will also act as a heat transfer device. On open motors, the end frame will have slots for air to pass. On enclosed motors, the end frames will be solid, with no openings. A variety of end frames are shown in Figs. 3.13, 3.14, and 3.15.

Like housings, end frames come in cast-iron, steel, zinc, or aluminum castings.

Cast-iron castings are usually found on motors of 3 hp and larger. The service application is in the industrial market where severe conditions may exist. Materials are usually of about 30,000 lb/in^2 tensile strength and are free machining. The typical sequence of operations is a two-machine cell—a computer numerically controlled (CNC) machine prepares the bearing bore and end frame diameter, and a manual drill is used to prepare the holes for the housing attachment.

The steel material is usually SAE 1010 to 1020. This type of end frame may be found on all types and sizes of motors. A coil is processed through a stamping press, and each part is drawn into form as a stamping. This is usually a progressive die operation.

A self-aligning bearing is installed and lubricant is applied. Then the bearing is sized for the only machining process.

Zinc or aluminum end frames are found on most motor types and sizes and generally are castings. End frames are usually cast in a horizontal die caster. Because of its density, zinc is usually limited to end frames for motors 3 in in diameter or less. If the parts are small enough, more than one part is made at one time. This depends on the part and machine sizes. Also, on motors above ¼ hp, a steel bearing insert is usually die cast in the part. Following the die casting and part cooling, the part is trimmed. Many manufacturers have installed robots for this operation because of the heat and environmental conditions.

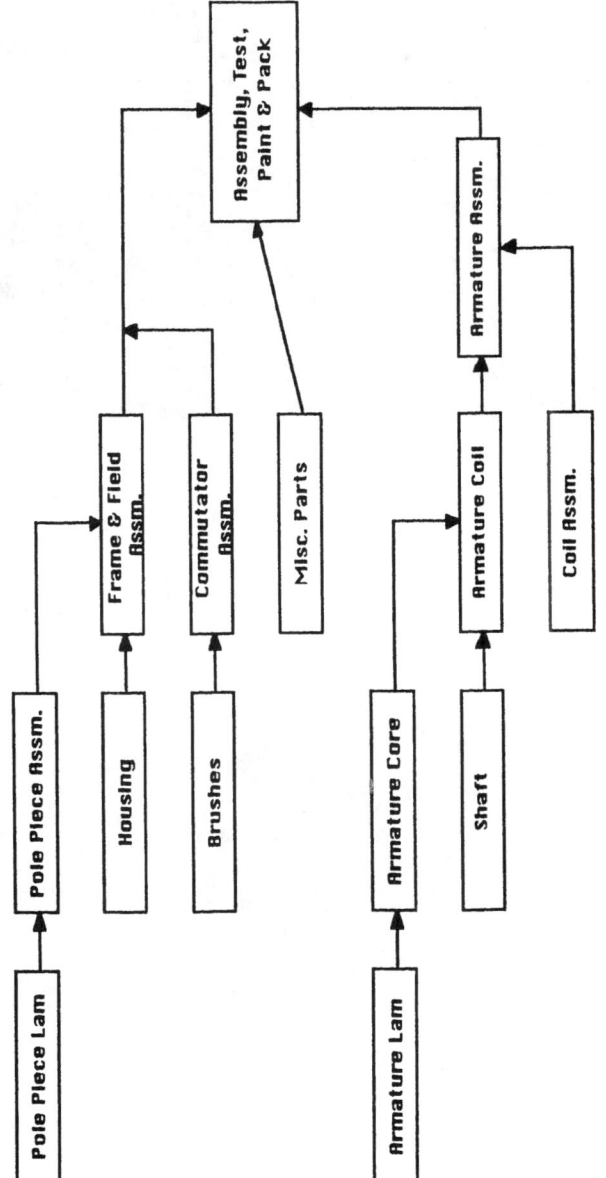

FIGURE 3.6 DC motor manufacturing process flow.

FIGURE 3.7 Solid steel pole piece.

FIGURE 3.8 Frame and field assembly.

FIGURE 3.9 Brush assembly.

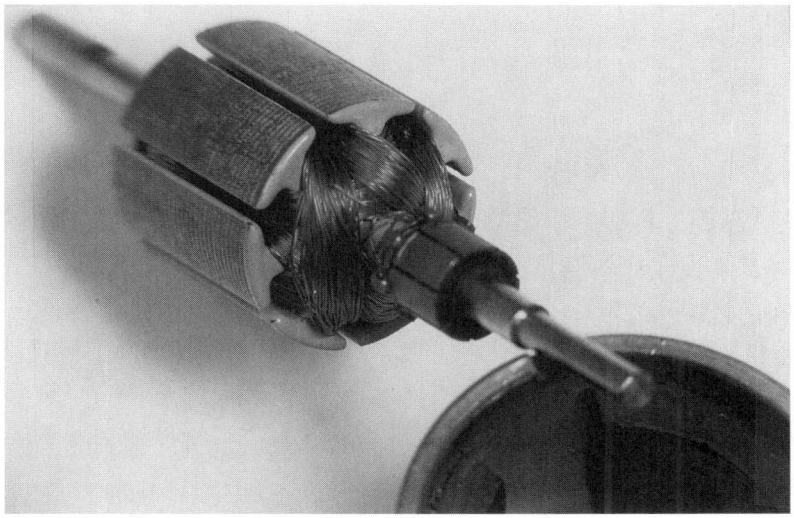

FIGURE 3.10 Winding a commutator.

FIGURE 3.11 Core and shaft assembly.

FIGURE 3.12 Complete motor.

FIGURE 3.13 End frame.

FIGURE 3.14 End frame.

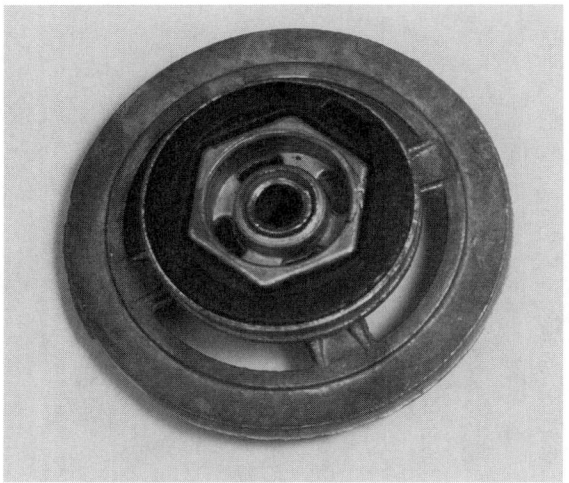

FIGURE 3.15 End frame.

The bearing bore and housing end frame diameter of the end frame are then machined. This is done on either a CNC lathe or a special automatic machine, depending on size and volume.

Some very small motors use an oil-soaked wick, as seen in Fig. 3.14, for lubrication. This is inserted after machining.

3.3 HOUSING MATERIALS AND MANUFACTURING PROCESSES

Housings, also known as frames, come in all types of materials and configurations. Basically, the housings are made in the same way for both ac and dc motors. The basic purpose of the housing is to cover the stator or pole-piece assembly, provide heat transfer and protection, provide a location for mounting the end frames, and serve as an attachment for other components, such as outlet boxes and lifting hooks.

3.3.1 Materials and Configurations

The housings come in cast iron; in rolled, wrapped, and tube steel; and in both cast and extruded tube aluminum.

Cast Iron. Castings are usually found on motors of 3 hp and larger. The service application is in the industrial market where severe conditions may exist. Materials are usually of about 30,000 lb/in^2 tensile strength and are free machining.

In most cases, the mounting feet are cast as part of the housing.

Steel. As mentioned, steel housings come in several configurations—rolled, wrapped, and tube. The material is usually SAE 1010 to 1020. This type of housing may be found on all types and sizes of motors.

Aluminum. This material is also found on most motor types and sizes. The cast housings may be produced for a size as large as NEMA 360 but are usually not found on motors rated below 3 hp.

The tubing may be found on the smallest motors up to about 25 hp. The material is usually SAE 6061.

3.3.2 Manufacturing Processes

Cast Iron. The typical sequence of a cast iron operation is as follows:

1. Machine and drill the mounting feet to be used as a locator for further machining operations.
2. Bore the inner diameter (ID).
3. Turn the end frame registers (optional—sometimes done as a wound stator assembly).
4. Drill and tap for the end frame attachment.
5. Mill for the outlet box attachment.

These machining operations can be completed on either manual machines or CNC machining centers. Usually machine-tool cells are incorporated.

Rolled Steel. A coil is processed through a stamping press and the shape is a flat form. This piece is then formed around a mandrel and welded. In some cases, the weld is a straight butt weld. In other cases, the rolled end attachment is interlocked mechanically with several weld beads.

The housing is then machine-faced to length. Next, a stamped mounting base is welded to the housing.

There are both highly automated and semimanual machines for this process.

Wrapped Steel. The manufacturing processes are the same as for a rolled housing, except that the stator core is used as the mandrel.

Tube Steel. A drawn-over-mandrel (DOM) tube or a hot-rolled seamless tube is processed in the following manner.

DOM. Cut to length, machine end frame diameter (optional—may be done as a wound stator assembly), and weld mounting feet.

Seamless tubing. Cut to length, machine end frame diameter (optional—may be done as a wound stator assembly), and weld mounting feet. Depending on the condition of the bore, it may have to be machined.

Aluminum Castings. Most aluminum castings are produced as a complete housing with mounting feet. These are machined like cast iron and with the same type of equipment. Some, however, are cast over a stator core. This process requires machining like cast iron, except that the bore is not machined.

Aluminum Tubing. The material is cut to length. Sometimes the end housing diameter is machined prior to stator core assembly. The mounting feet are then welded or screwed to the housing.

3.4 SHAFT MATERIALS AND MACHINING

3.4.1 Shaft Materials

Most motor manufacturers use SAE 1045 in either cold-rolled or hot-rolled steel (CRS or HRS). Other materials include sulfurized SAE 1117, SAE 1137, SAE 1144, hot-rolled SAE 1035, and cold-rolled SAE 1018. A ground stock of any material is used on special CNC Swiss turning machines.

Generally, the cold-rolled and sulfurized steels will cost about 15 percent more than HRS and will machine better. Machining trials need to be performed in order to justify the extra cost. Since all shaft-turning machines perform differently, there is no established material or machining practice.

Obviously, the hot-rolled plain carbon steel, on a cost-per-pound basis, is cheaper than cold-rolled sulfurized steel. But there are tradeoffs. The hot-rolled material has to be sized larger than cold-rolled because of the lack of outer diameter (OD) control in the rolling process. A manufacturer has to evaluate whether the larger-size and lower-material-cost hot-rolled bar stock is more or less costly than cold-rolled bar stock. Also, the hot-rolled material, by the very nature of its processing, has hard and soft spots, residual stresses, voids, and other material deficiencies, making machining more difficult. Again, machine trials need to be conducted to obtain the best cost option between CRS, HRS, nonsulfurized, and sulfurized materials.

Because of the difficulties with HRS, most motor manufacturers will use sulfurized CRS.

3.4.2 Machining Operations

Most manufacturers saw, shear, or turn the shaft length off the original bar stock.

Sawing is done with a band saw, machine back saw, or rotary saw, and the material is cut either as a separate piece or in bundles.

One process, to eliminate the saw-cut kerf material, is a shear cutoff process. It is very fast and noise has been eliminated. However, this meets with mixed results. In the shearing process, the end of the bar is deformed—the top of it is formed downward and the bottom has a burr, as illustrated in Fig. 3.16. This deformation has to be removed in the face-and-center operation, which is sometimes difficult and causes excess tool wear.

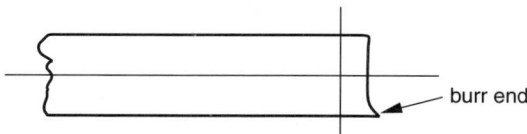

FIGURE 3.16 Shear cut-off process.

The third option is to cut off the shaft bar in a bar-turning machine. The bar-turning machine will complete the shaft diameter machining, and as a last operation a cutoff tool will remove the shaft from the bar.

Nearly all shafts for motors larger than ¼ hp have to be faced and centered for future machining operations. This operation is usually completed on one machine with a special face-and-center tool.

Both ends of the shaft are centered to provide a tool location in the lathe turning operation and in balancing as a rotor assembly. Facing is also done in order to provide a more precise length in turning and when face drivers are used in the turning operation.

Many motor manufacturers combine the bar cutoff and face-and-center operations.

Most motor manufacturers now use CNC turning machines because of their quick setup changeover capabilities, capability of completing a shaft in one operation, and ability to precisely turn a diameter to 0.0005-in tolerance and meet the surface finish requirements.

On motors greater than ½ hp, the bearing journal tolerances are generally 0.0005-in or higher. The ability to turn bearing journal diameters to a 0.0005-in tolerance has eliminated the subsequent grinding operation.

Some motor manufacturers that produce shafts larger than 2 in (3 hp and up) use a retractable jaw chuck in combination with a face driver, rather than a face driver alone, in order to maximize the machine horsepower yet provide the necessary precision. This type of chuck also works well on hot-rolled bar steel because it provides better clamping of the bar than do face drivers. The chuck jaws retract under the semifinish turning operation to allow turning under the jaws. Then the CNC machine completes the finish turning to size using the face drivers.

Most motor manufacturers combine keyway milling (on a manual machine) with the CNC lathe in a one-person cell.

Some motor manufacturers started incorporating CNC Swiss turning machines when they became available in the mid-1980s. These machines can machine a bar up to about 2 in in diameter and hold tolerances to 0.0003 in. They incorporate complete turning, including keyway milling, plus other special features such as threading and grooving. The process helps assist flexibility in short runs and in completing parts of extensive complexity. However, these machines require centerless ground stock, which is more expensive than CRS or HRS. Again, the economics will dictate the method of operation and equipment.

If the bearing journals require a size tolerance better than 0.0005 in, a separate grinding operation is usually required.

Other machining options are the use of manual multispindle machines for cutoff and turning and the use of grinders for grinding bearing journals and seal diameters. This option is usually used for shaft diameters 1 in and smaller. A high-volume option for 1-in and smaller shafts is a dedicated transfer line which uses ground bar stock.

Some motor manufacturers, particularly those that produce sizes of 5 hp and up, finish-machine the bearing journals and rotor diameter as a rotor assembly. This operation produces the best possible concentricity between the bearing journals and rotor diameter.

Few motor manufacturers have had success with postprocess gauging with feedback size compensation in the bearing journal finish-machining operations. However, this is expensive and is not always accurate because the part has to be clean.

Some people believe that once a shaft is removed from the turning operation, one can not use the centers for location in future operations. However, the method used is to set up a finished shaft (with or without rotor) in a lathe to indicate the drive end and both journals. If the output end is within 0.0005 in of true inner radius (TIR) and both journals with respect to each other are within 0.003 in TIR, turn the rotor OD as is. If not, adjust centers to get the acceptable TIR.

3.5 SHAFT HARDENING

In many instances it is desirable to harden shaft materials. Harder shafts take longer to wear out than softer shafts. Shafts that are too hard become brittle and subject to fracturing. The exact hardness required depends on the intended use of the motor and the life required. Generally, when the shaft is used with a sleeve bearing system, the shaft needs to be somewhere between 35 and 55 on the Rockwell C scale. The ability to harden a shaft depends on the material being used and the hardening process (see Table 3.1).

TABLE 3.1 Common Shaft Material Characteristics

Material	Tensile strength, (lb/in^2)	Hardenable	Characteristics
Carbon steels			
1018	62,000	Carburize	Will corrode
1050	105,000	Heat treat	Will corrode
1095	140,000	Heat treat	Will corrode
1117	71,000	Carburize	Good machinability
4140	148,000	Heat treat	Good machinability
Stainless steels			
303	90,000	No	Nonmagnetic—high wear when used with bronze bearings
416	120,000	Yes	Free machining
440	260,000	Yes	Corrosion resistant

Case Hardening.* Case hardening is a process of surface hardening involving a change in the composition of the outer layer of an iron-base alloy followed by appropriate thermal treatment. In order to harden low-carbon steel it is necessary to increase the carbon content of the surface of the steel so that a thin outer case can be hardened by heating the steel to the hardening temperature and then quenching it. The first operation is carburizing to impregnate the outer surface with sufficient carbon; the second operation is heat-treating the carburized parts so as to obtain a hard outer case and at the same time give the core the required physical properties. The term *case hardening* is ordinarily used to indicate the complete process of carburizing and hardening.

Some of the most common processes are described here.

Carbonitriding. A case-hardening process which causes simultaneous absorption of carbon and nitrogen by the surface.

Carburizing. A process in which carbon is introduced into a solid iron-base alloy while in contact with a carbonaceous material. Carburizing is frequently followed by quenching to produce a hardened case.

---

* This subsection from *Machinery's Handbook 23, Revised Edition,* Industrial Press, New York, 1988, p. 441.

Cyaniding. A process of case hardening an iron-base alloy by heating in a cyanide salt.

Nitriding. A process of case hardening in which an iron-base alloy of special composition is heated in an atmosphere of ammonia or while in contact with nitrogenous material.

3.6 ROTOR ASSEMBLY

The rotor assembly consists of a die-cast rotor and a shaft. Both components may be completely machined and assembled, partially machined and assembled, or a combination of both. The reasons for the various rotor assembly options are economics; size; unit volume; and desired electric motor efficiency, which relates to concentricies and the air gap between the rotor and stator.

3.6.1 Basic Assembly Process

The most efficient and economic assembly method is to assemble a nonmachined die-cast rotor with a completely machined shaft. Probably the most economical assembly method is to mechanically press-fit a shaft onto the rotor. This process can be hand operated or completely automated, with the process determined by unit volume and the variety of rotors and shafts. Less variety and smaller sizes lend themselves more to automation. Some companies have completely automated this assembly process.

Thus, in the basic process flow, illustrated in Fig. 3.17, a completely machined shaft and a nonmachined die-cast rotor are processed by being mechanically press-fit. This process is used for many very small motors. The resulting rotor assemblies probably will not have the best tolerance concentricities, thus affecting motor efficiencies and noise—but, as mentioned, they will be the least costly.

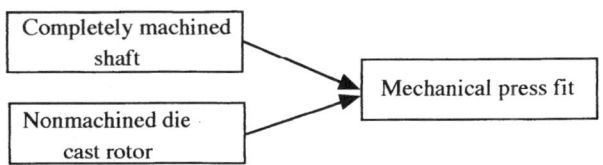

FIGURE 3.17 Rotor assembly process.

3.6.2 Rotor Machining

ID Machining. Cast-aluminum rotors tend to be banana-shaped due to the heat and sometimes due to lack of internal support in the casting process. Also, rotor laminations need to be rotated prior to die-casting in order to eliminate or reduce the lamination material camber, which will cause a banana shape; this is more prevalent in the heating and shrinking process than in the press-fit process. Part of the rotor core bore curve is imparted to the shaft. The rotor core is then turned to obtain the proper air gap. In service, the rotor heats up the aluminum bars, which expand more than the steel core, thus relieving the axial clamp and imparting the curve to the shaft. This can cause unbalance, increase slot-pitch noise, and generate structure-borne noise because of vibration. This effect is greater for long cores.

To solve this banana-shape problem, manufacturers ream, bore, or broach the core ID prior to shaft assembly. This surface then can be used for location when machining the rotor OD if required.

OD Machining—Rotor Only

Turning. Rotor OD machining is usually done on an expanding ID arbor. This allows turning the OD to the average bore diameter. If the rotor bore is machined prior to OD machining and used as a locator, there will be excellent concentricity between the bore and the OD. This possibly might eliminate machining the rotor OD when attached to the shaft, but laminations with the OD punched to size are needed.

Grinding. Rotors that have their ODs cut with a tool will have OD smearing. This causes lamination shorting at the air gap and will reduce efficiency and cause hot spots. Plunge grinding, with or without a shaft attached, will reduce smearing to a minimum. Some hermetic motor manufacturers use a centerless belt grinder to size and clean up the OD only. Sometimes, depending upon the application, the OD will be used as a locator to machine the ID.

OD Machining—Rotor on Shaft.

There are several schools of thought on how to finish-machine the rotor and shaft combination. One method is to allow stock on the bearing journals and then turn the rotor OD and journals in the same setup. In some CNC machines, the journals can be finished to size (not better than 0.0004 in) and finish [20 to 30 root mean square (RMS)] without grinding. This operation can also be completed with a plunge grinder, but the labor content makes it expensive. Completing the bearing journals and rotor OD in one setup is probably the best operation for obtaining consistent air gap.

3.6.3 Electrical Efficiency Improvement Processes

Most motor manufacturers need to have better electrical efficiencies than that provided by the basic assembly process (Sec. 3.6.1), and the machining processes for rotor assembly will affect the required efficiencies. This subsection examines many of the various processes that will improve electric motor efficiencies.

Rotor Machining. Most manufacturers machine the rotor outside diameter and shaft diameters after assembly, but there are other various ways to accomplish this process. The major interest is to achieve better electrical efficiencies. One must have the best concentricity between the shaft bearing diameters and the rotor outside diameter while leaving an equal amount of back-iron thickness. *Back-iron thickness* is defined as the distance between the rotor OD, which is the lamination, and the aluminum die-cast slot, as shown in Fig. 3.18.

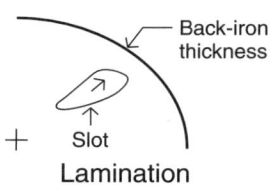

FIGURE 3.18 Rotor back-iron thickness.

Also, there is a concern that machining the rotor OD may "smear" the steel laminations and aluminum die-cast materials. This smearing will reduce electric motor efficiencies. In a turning operation, the cutting tools must be maintained in a sharp condition. This rotor OD turning operation can be achieved as a separate part or as a rotor assembly.

Rotor Grinding. Another process method to reduce smearing is to use a centerless abrasive belt grinder to grind the rotor outside diameter (not as a rotor assembly). The abrasive belt will not become loaded up with steel and aluminum as might occur with a hard-wheel grinder. This centerless grinding process also guarantees that the back-iron thickness will be uniform.

Lamination Punching. Some motor manufacturers punch the lamination to size, thus eliminating any rotor OD turning. However, the lamination dies must be maintained and the process monitored continuously.

Process Options. Several process options are shown in Figs. 3.19a through 3.19e. The optimally efficient process for electric motors is probably the one shown in Fig. 3.19e. Again, with the variety of machining options, a company must evaluate the requirements for electric motor efficiency and economics.

3.6.4 Options for Attaching Rotor to Shaft

There are basically four options in attaching the rotor to the shaft: (1) press-fitting, (2) heating and shrinking the rotor, (3) slip-fitting with adhesive, or (4) welding. The

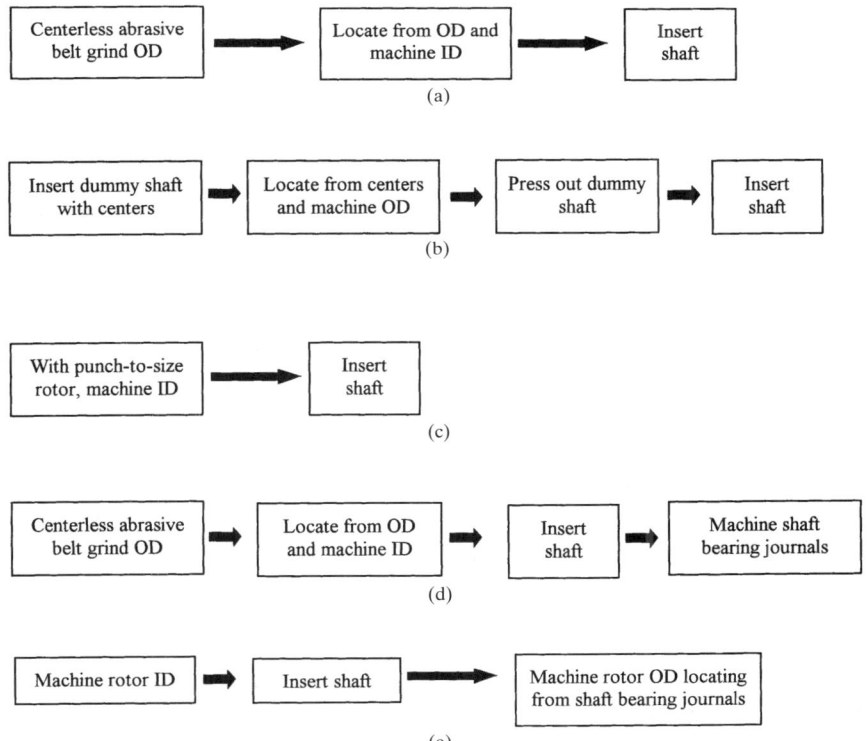

FIGURE 3.19 Options for optimum efficiency electric motor manufacturing process.

process selected is usually dictated by economics. These processes are discussed here (see Fig. 3.20).

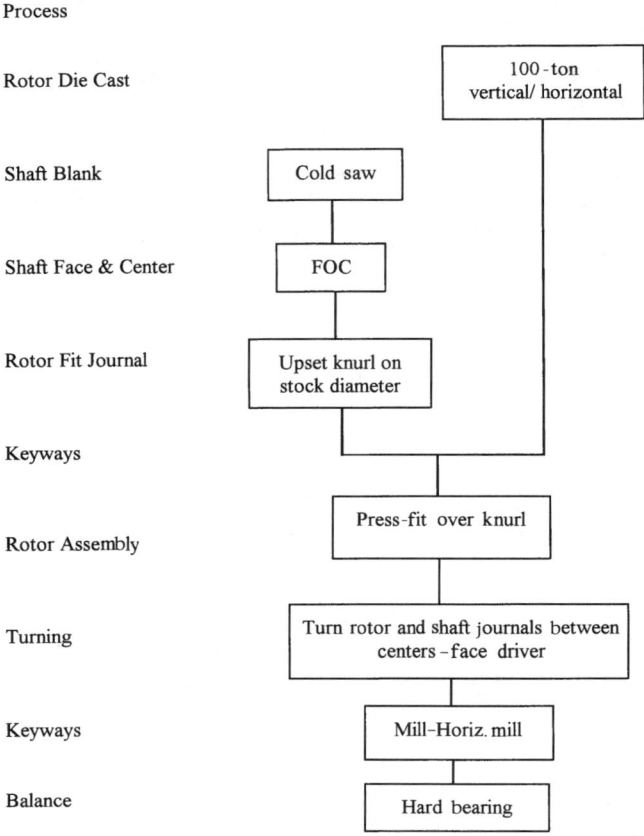

FIGURE 3.20 Rotor and shaft processing option.

Press-Fitting. The most basic and economical attachment process is pressing the rotor onto the shaft. This is usually done in a vertical hydraulic press. The rotor is placed in a holding fixture, and the shaft is placed into the rotor ID. Tolerance control of the rotor ID and the shaft OD must be maintained. Generally, the press fit should be in the range of 0.001 in per inch of shaft diameter minimum.

If the press fit is too tight, the shaft may bend. If the press fit is too loose, the shaft may turn on the rotor in application. Monitoring of the press hydraulic pressure during the press fit will provide a quality assurance check (preventing too tight or too loose a fit).

Usually the shaft rotor diameter will be upset in some manner—knurling, jab blocking, etc.—in order to ensure a press fit.

Heating and Shrinking the Rotor. Another attachment process is heating the rotor by induction or with some type of external heat source and dropping the rotor onto the shaft. The heating process requires energy, and one must be concerned with personnel handling hot parts. It also requires some in-process inventory in the rotor-heating process. This process provides a greater tightness between rotor and shaft than does press-fitting, plus it does not have the same potential for bent shafts as does the pressing operation.

For common shaft rotor tolerances, the rotor should be heated to between 400 and 450°F (204 and 232°C), but not above 700°F (371°C), as this temperature will start to affect the aluminum.

Generally, assembly of larger motors, over 5 hp, will use rotor heating because a very large hydraulic press is required for press-fitting.

Section 3.6.5 gives example calculations for determining shrink-fit dimensions.

Slip-Fitting with Adhesive. The rotor ID and shaft OD are sized to allow a slight slip fit of the rotor onto the shaft. It is usually on the order of 0.001 to 0.002 in of clearance. The exact clearance is a function of the adhesive and must be adjusted in accordance with the recommendations of the adhesive supplier. Parts must be clean and free of lubricants before assembly. A drop or two of adhesive is put on the shaft. It is then slipped into the rotor with a twisting motion. A fixture with a stop is necessary for proper shaft location. After assembly, the adhesive is given time to cure.

Welding. On 5 hp and higher motors, some manufacturers weld bead the final rotor-to-shaft attachment. Others use a key to ensure a locking condition.

Balancing. After the shaft and rotor are assembled, balancing is required. Most balancing operations are done by setting the rotor assembly with the bearing journals on support rollers and rotating the assembly to determine the out-of-balance condition in two planes.

There are two types of balancers, soft- and hard-bearing. Basically the difference is that a *soft-bearing* machine operates below the suspension's resonant frequency. *Hard-bearing* balancers are generally easier to use, safer, and provide a rigid work support.

Most balancing machines will determine the location and amount of weight that needs to be applied. Some motor manufacturers add an epoxy weight to the rotor core. However, a fast drying heat is required in order to speed up the hardening of the epoxy. Others design the rotor end casts with protrusions so that weights (washers) may be added. Very few drill or machine out weight because this can affect electrical efficiencies.

Balancing machines come in either manual- or automatic-load types, usually with computer controls.

3.6.5 Shrink-Fit Calculation Examples*

1. Determine the temperature differential ΔT, °F (± from room temperature).

$$\Delta T = \frac{\text{(differential expansion)/(basic shaft diameter, in)}}{\text{coefficient of thermal expansion}}$$

---

*Subsection contributed by Chris Swenski, Yeadon Energy Systems, Inc.

where the differential expansion is the total diameter change required. It includes the inference fit plus the slip clearance.

Some common coefficients of thermal expansion are listed in Table 3.2.

TABLE 3.2 Coefficients of Thermal Expansion for Common Materials

Material	Coefficient of thermal expansion, in/(in · $\Delta°F$)
Common steel	0.0000065
Nickel steel	0.0000070
Cast iron	0.0000062
Aluminum	0.0000124

2. Calculate the desired expansion and shrinkage to find temperature change required for 1020 CRS, where the shaft OD is Ø1.2500 and the rotor ID is Ø1.2480.

These diameters give a 0.002-in interference fit. The minimum desired slip fit clearance is 0.003 in, and the differential expansion is 0.005 in.

The temperature change ΔT required on these parts to give 0.005-in expansion is calculated as follows.

$$\Delta T = \frac{0.005/1.2500}{0.0000065} = 615.38°F \quad (3.1)$$

A 615.38°F change in temperature is required. Therefore, the total temperature would be 615.38°F plus ambient (72°F in this case). One could heat the rotor to 687.38°F (364.10°C). The shaft temperature could be reduced to shrink the shaft in order to reduce the heat needed for expansion of the rotor.

For instance, cool the shaft to −75°F (−59°C), and heat the rotor to 540.38°F (282.43°C) to get the required deferential expansion.

3. Another method is to use the maximum change in temperature to determine differential expansion. The total possible change in temperature using dry ice at −100°F (−73°C) to cool the shaft and an oven at 700°F (371°C) to heat the rotor is 800°F (444°C).

$$\Delta T = 800°F$$

$$\frac{\text{(Differential expansion)/(slot OD)}}{0.0000065} = 0.0065 \text{ in shrinkage growth} \quad (3.2)$$

Differential expansion = ΔT (coefficient of the thermal expansion) (shaft OD) = (800°F) [0.0000065 in/(in · Δ °F)] 1.250 in = 0.0065 in

0.0065 in − 0.002 in interference = 0.0045 in clearance at these temperatures.

For practical purposes, one may use a dry ice temperature of −75°F and an oven temperature of 650°F. This allows for the extremely fast warming of the shaft and cooling of the rotor while assembling.

A minimum of 0.003 in clearance was calculated for all fits. The usual finished interference between the parts ranges from 0.0005 to 0.003 in.

3.7 WOUND STATOR ASSEMBLY PROCESSING*

Wound stator assembly processing basically consists of attaching a wound stator core into a housing. However, there are many different assembly processes, depending upon the housing material, the size, and the electrical efficiency requirements.

3.7.1 Steel Pressing

The steel housings are pressed over the wound stator core. Sometimes the mounting base is welded or screwed to the housing before or after this operation. Usually a final attachment is made either by welding beads or by pinning, which requires drilling a hole into the wound stator core.

3.7.2 Cast-Iron Pressing

The cast-iron process, for motors up to about 25 hp, is the same as that used for steel housings, for motors above 25 hp. The housing is heated. For motors above 25 hp, the hydraulic press needed becomes very large, and the process is sometimes not economical.

3.7.3 Heating and Shrinking

Almost all aluminum housings are heated and shrunk onto the wound stator core. Usually, the housings are pinned to the wound stator core. The base is sometimes welded or screwed to the housing before or after this operation.

3.7.4 Electrical Efficiency Requirements

The mounting of the end frames to the housing is crucial in maintaining the best air gap (concentricies) possible. The end frame bearing housing is usually machined at the same time as the housing attachment diameter (see Fig. 3.21).

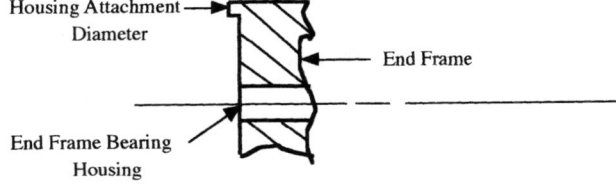

FIGURE 3.21 Mounting of end frame to housing.

---

* Sections 3.7 to 3.9 contributed by Karl H. Schultz, Schultz Associates.

In order to maintain the best possible concentricies for best air gap control, most manufacturers machine the housing end frame diameter as a wound stator assembly, locating off the bore (see Fig. 3.22).

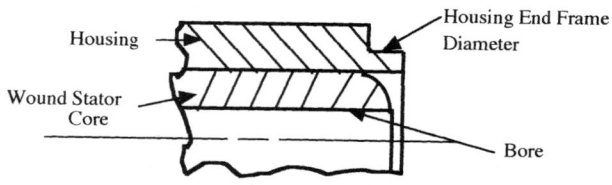

FIGURE 3.22 Machine housing.

3.8 ARMATURE MANUFACTURING AND ASSEMBLY

Armature manufacturing and assembly require significant hand labor, although the size and unit volumes will dictate the degree of automation. Following is the process and manufacturing flow.

3.8.1 Armature Core Assembly

1. Stack laminations to proper length. Sometimes this is done by weighing the stack. The outer end laminations are turned so that the burrs are on the inside. Two stacks are made.
2. Place the two stacks in a press. Locate a machined shaft on top of the stack and press into location.

3.8.2 Armature Coil Assembly

1. Insert insulation paper into the armature core lamination slots. This is done either manually or automatically.
2. Insert armature coils (usually rectangular-shaped copper wire) into the armature core lamination slots. This can be done manually or automatically.
3. Twist the ends of the armature coils. This requires a special machine.
4. Press the commutator onto the armature coil.
5. Press or stake the armature coil ends into the commutator.
6. Band the armature coil ends into the commutator.
7. Varnish.
8. Turn the commutator to achieve a very smooth finish. Sometimes a diamond tool is used.
9. Braze the commutator ends.

3.9 ASSEMBLY, TESTING, PAINTING, AND PACKING

The final motor assembly, testing, painting, and packing process is as varied as the other processes, depending upon the unit volume, size, and variety (see Fig. 3.23).

3.9.1 Assembly

Most very small motors up to about ¼ hp can be assembled automatically. Some are assembled in as little as 5 s by highly automated equipment costing hundreds of thousands of dollars. The success of high-volume automation is the quality of the parts. Part quality that can not be controlled will jam the machine and cause poor utilization.

Changeover from one motor size or configuration to another is not easily done on a high-volume automated machine, although there have been strides in recent years to provide for quick setup. High-volume automated assembly machines are best run without changeover because they need to be kept running as much as possible in order to justify their cost.

In a line loader, the operator gathers various parts needed to assemble a particular motor (end frames, rotor assembly, wound stator assembly, and miscellaneous parts) and places them on a tray, which moves down a conveyor to the assembly line.

The conveyor to the assembly line is controlled by an assembly operator, so the motor parts may be called for as they are needed.

There are several assembly stations, and each operator takes parts from the tray and completes their portion of the assembly. The operator then places that component on the tray, and it moves to the next station. Sometimes this process is done on a moving conveyor rather than trays.

Some low-volume or significantly sized motors are assembled in one-person cells. The components are set on pallets and/or in racks for the assembly operator's access. Usually the operator will assemble the complete motor.

3.9.2 Testing

A variety of electrical and mechanical tests are usually completed on a motor before shipping, some dictated by customer requirements. These tests can be any or all of the following: input voltage regulation, full load, no load, inertial load, equivalent circuit, locked rotor, low-voltage start, torque/speed curve, rotation direction, ac or dc high potential, insulation resistance, surge/impulse, vibration, acoustic noise, and temperature. Most test systems have preprogrammed menus so that the operator does not need to input the data.

After the initial setup, the operator loads and secures the motor into the fixture, makes the proper connections, and starts the test. All testing is automatically sequenced. After the test is completed, the measured results are compared to preprogrammed limits. The data is stored, printed, or transmitted as the user requires. The operator can observe whether the motor has passed or failed and can take appropriate action. Manual to completely automatic test equipment is available.

Sometimes mechanical tests are performed, such as for rotor assembly end play and tight bearings. Noise tests are also conducted, and most motor manufacturers enclose the entire test area in a sound booth so that any noise can be measured or heard.

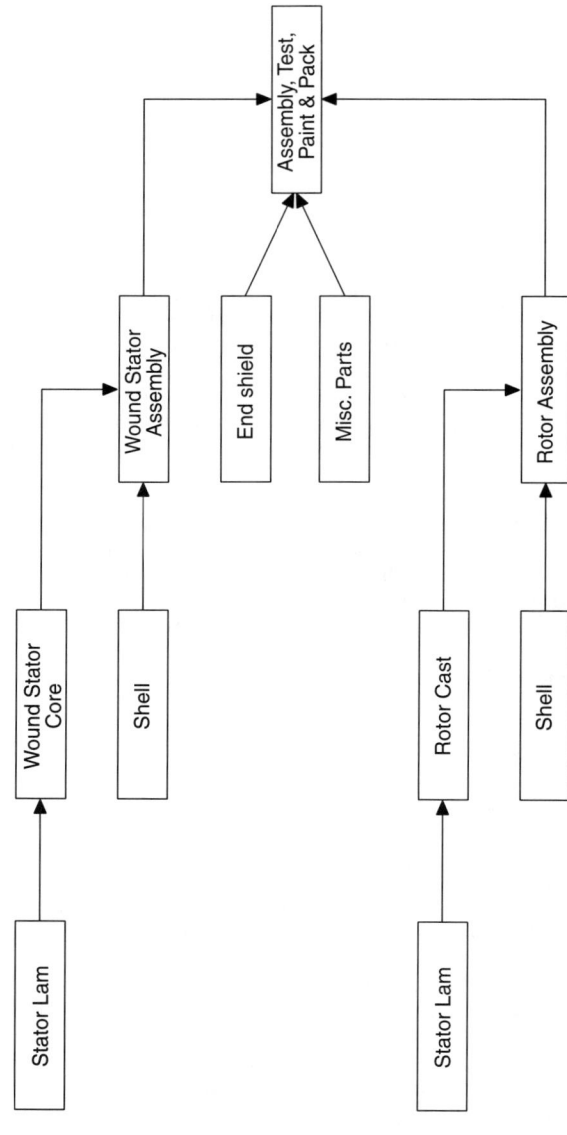

FIGURE 3.23 Assembly and testing.

3.9.3 Painting

Environmental regulations have largely restricted the type of paint that can be used. Most manufacturers have changed to a water-base or powder paint and have automated the process. Meeting customer color requirements requires motor manufacturers to install equipment that can be quickly changed over.

Some manufacturers paint components before assembly to eliminate masking. However, the components have to be handled properly in order to minimize marking at assembly.

3.9.4 Packing

Motors under ¼ hp are usually shrink-wrapped separately. Some have cardboard bases for support. Some are put on a pallet and shrink-wrapped as a container. Motors over ¼ hp are usually packed in cardboard boxes or on pallets. Most of the packaging process can be automated as much as possible.

3.10 MAGNETIC CORES*

Stator core assemblies, shown in Figs. 3.24 and 3.25, are insulated stacks of laminations lined up to close tolerances. The laminations themselves can be held to tolerances within 0.001 in (±0.0005 in) on the OD and ID. However, they must be assembled into a core in some fashion. Permanent-magnet motors have air gaps on the order of 0.015 to 0.040 in per side. Induction motors have air gaps on the order of 0.010 to 0.015 in. These gaps must be held very consistent to avoid performance and noise degradation. There are many methods for assembling stacks of laminations. This section discusses some of the most common stacking methods with their positive and negative attributes.

Loose armature, induction rotor, or outer rotor brushless dc motor laminations may be pressed onto a sleeve or shaft. Other methods include heat shrinking, ring staking, or adhesive bonding.

3.10.1 Welded Cores

The lamination stacks are generally fixtured off the stator bore and welded along the OD of the stack. The tolerance is now a result of the welding fixture, which allows for some shift in lamination placement. The eddy current core losses are increased because the welding short-circuits to the laminations at the weld joints. The core assembly may have to be machined to bring the OD back to an acceptable dimension. Weld depth should be kept to a minimum, and welds should be positioned behind the poles or teeth. When possible, a laser weld is best in the cases in which it provides adequate strength.

---

* Sections 3.10 to 3.10.3 contributed by William H. Yeadon and Alan W. Yeadon, Yeadon Engineering Services, PC.

 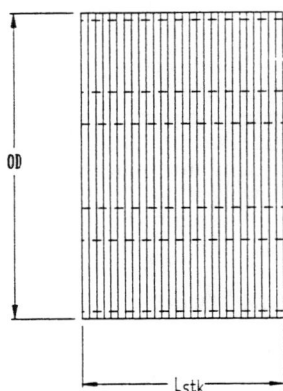

FIGURE 3.24 Outer rotor brushless dc stator core assembly (FEMD).

3.10.2 Bonded Cores

These cores are built by coating the laminations with an adhesive, aligning them on a fixture, and heating the cores to set up the adhesive. Here the tolerances are a function of the fixture tooling, and the OD may have to be machined to get it to an acceptable dimension. If anything, the adhesive assists in reducing eddy current losses; however, it takes up space, and the effective core length may be somewhat less than expected.

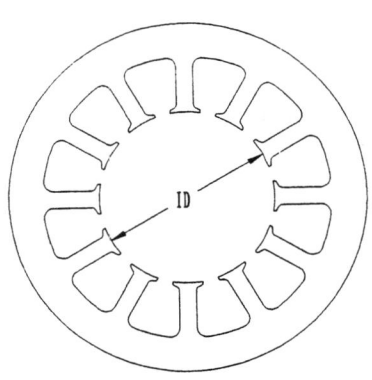 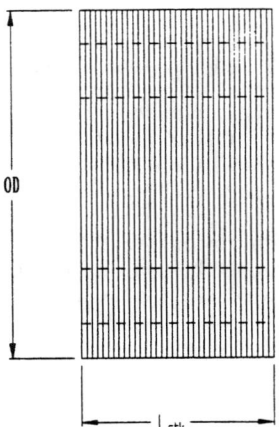

FIGURE 3.25 Inner rotor brushless dc stator core assembly.

MECHANICS AND MANUFACTURING METHODS

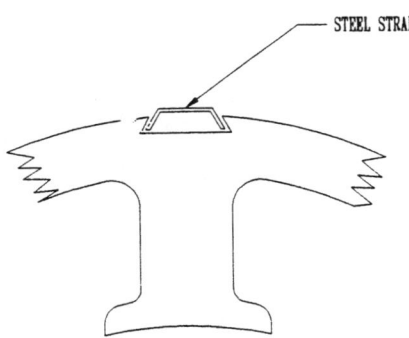

FIGURE 3.26 Cleated core.

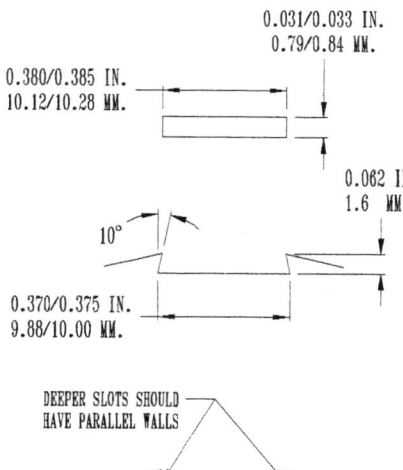

FIGURE 3.27 Cleat specs. Basic dimensions may vary, but tolerances should remain the same. (*Courtesy of Windamatic, Inc.*)

3.10.3 Cleated Cores

These cores are assembled by forcing steel straps into notches in the periphery of the laminations, as shown in Fig. 3.26. This method provides for good dimensional stability with very little increase in core loss. Cleats are typically folded over the ends of the cores and may have an adverse affect on electrical clearance. Cleat and notch tolerances are determined as shown in Fig. 3.27.

3.10.4 Lamination Design Consideration*

Perhaps the most significant single factor to be considered in automatic or semiautomatic winding systems is the lamination design. The second most important consideration is the winding specifications (wire size, turns, and winding configuration or pole configuration), which are covered in a later section in more detail. It is mentioned at this point only to emphasize the fact that winding specifications can be readily changed without a major penalty, whereas the lamination itself cannot be altered without paying a very large penalty in modifications to existing tooling or new and expensive toolings.

Lamination design is basically an electrical consideration; however, there are five major areas that should be considered when designing a lamination that directly affects the manufacturing process to achieve a completed stator assembly.

The OD shape or periphery of the lamination is always important from the standpoint of material savings, particularly when punching laminations from strip material. The shape of the cleat notch is an important consideration in that notch standardization should be maintained for all laminations, regardless of stator size. The number of cleating notches and their location then become the only variables. Cleating notches vary in number, ranging from 2 to 16 per stator depending on the size of the motor. Location of the cleat notches should be in line with a lamination

---

* Subsection contributed by Stanley D. Payne, Windamatic Systems, Inc.

tooth or slot opening, with lamination tooth alignment being preferable. On large stators having many cleats, it is essential that they be equally spaced about the periphery of the stator and always be two notches, 180°, apart.

It is always desirable to have the largest possible slot opening in a lamination; however, this can sometimes be in direct opposition to the desires of the designer from an efficiency standpoint and sometimes to considerations of noise in the final product. Hence, final selection of slot openings is always a compromise between the product parameters and the restrictions imposed on the manufacturing process. In the past, when hand winding was the only manufacturing process, the major consideration was the stiffness of the wire. Usually 17 AWG was considered the practical maximum wire size from an operator's standpoint.

The lamination designer then selected slot openings accordingly—usually in the 0.070- to 0.080-in category. With the advent of machine insertion, the operator restriction on wire size was removed. A reasonable maximum wire size was then 13 or 14 AWG; however, a portion of the placer tooling also occupies a part of the slot opening, as shown in Fig. 3.28. These two items resulted in larger slot openings, to the point where a majority of laminations are now in the 0.095- to 0.125-in category.

Even though the slot opening has been made larger, there are still compromises on wire sizes relative to the tooling opening selected. As a general rule, the larger the slot opening, the greater versatility when considering wire sizes.

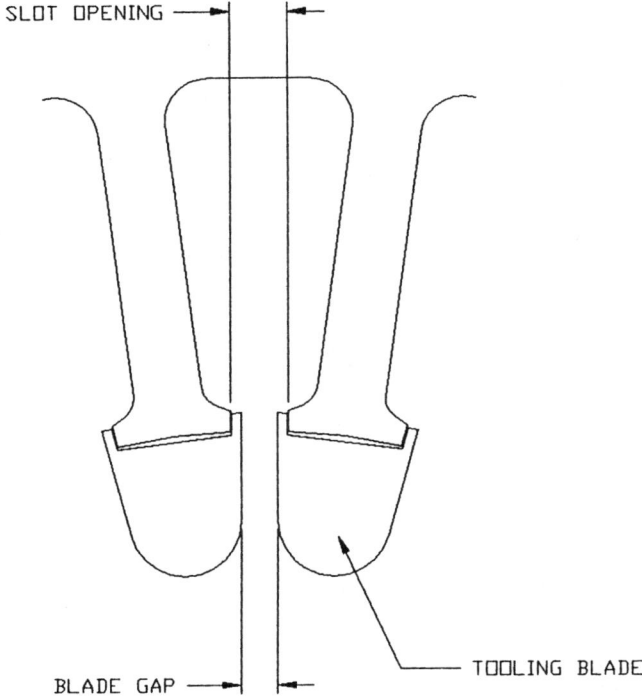

FIGURE 3.28 Slot configuration.

For automatic insertion of wedges, there is very definitely a desirable configuration. Figure 3.29 outlines this configuration for the bottom of the slot. Note that at this point we are not too much concerned about the overall shape of the slot, only the shape of the slot nearest the bore. Figure 3.29 gives a few basic dimensions defining the desirable shape.

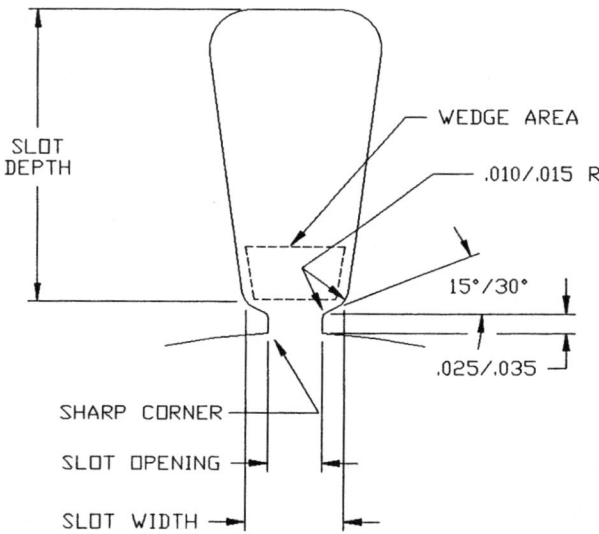

FIGURE 3.29 Blade gap illustration.

Referring to Fig. 3.29, the depth of the slot is defined as the length from the bottom of the slot to the back of the slot, and the width is defined as the width at the bottom of the slot prior to converging toward the slot opening. For ideal conditions for inserting wire, we would like to see this ratio approach infinity. That is, we would like to see a slot that is extremely narrow in width and very long in depth. Ideally, a slot would have a width equal to the wire diameter and a depth equal to the number of turns to be put into that particular slot. One can readily see that this is extremely impractical from the standpoint of material usage and lamination design, and would result in a motor of extremely large diameter. From a practical standpoint, based on experience gained from over 800 designs of placer or inserter tooling, it has been found that 4 is a practical value for this ratio. As with all rules of thumb, the number 4 is not sacred, but it is a practical rule in which any lamination design having a ratio in the area of 4 or greater can be expected to result in a relatively problem-free insertion. A lamination having a ratio of 3 or less, particularly when the ratio approaches 1, begins to present problems in process which are nonexistent in the higher-ratio designs. Most of these problems are involved with the wedge, such as wire behind the wedge, wedge dislocation, the wedge falling out of the slot, and many other undesirable characteristics.

The shape of the back of the slot is not crucial for most practical purposes. It is generally found in two basic configurations, a square bottom and a round bottom. The round bottom is the preferred shape due to the simplification of tooling.

The present basic practice in lamination design is to consider what is the best lamination design in order to obtain the desired electrical characteristics or performance of the motor. From an electrical and mechanical standpoint, there are some areas that must be different within a family of laminations which are important to the performance of the final product, such as having 24 slots for a 2-pole motor and 36 slots for a 6-pole motor. However, there are also certain other areas that can be standardized or grouped within a family of laminations, such as having two 36-slot 6-pole laminations, one for copper and one for aluminum windings, which differ only in the slot detail. Standardization of the slot opening and wedge area or base of the slot can result in identical placer toolings even though the laminations are not identical in all other respects.

The grouping of the family of laminations should begin with the bore, then further subdividing common bores into groups having the same number of slots. Examination of the laminations within the same subgroup will then show that there are usually very small differences in the significant areas, as previously mentioned—for example, the slot opening and the wedge entry area. Usually these small variations can be eliminated with little effect on the electrical performance of the final products. Standardization of laminations therefore can significantly reduce product costs through standardization of toolings, increased machine utilization, and reduced labor by eliminating tool changeovers.

Winding Specifications. Winding specifications—for example, wire size, number of turns, and pole configurations—are the second most crucial criteria, as mentioned previously for automatic or machine insertion of stators. Compatibility of the selected tooling opening, or blade gap, which is dependent on the slot opening and the wire size to be used, is essential. However, unlike lamination changes, wire sizes and numbers of turns can be varied to a certain degree without incurring the penalties of major costs or degraded product performance. As mentioned in the previous subsection, the larger the slot opening, the greater the versatility in wire-size selection.

The slot opening versus wire size situation can be best explained by the chart shown in Fig. 3.30, where the advantages and disadvantages become apparent.

There is an area on the chart that represents the *locking-wire* condition in which two wires attempting to pass one another lock. This condition can generate locking forces which can damage wire insulation, actually stall the insertion process, and in some cases cause tooling damage. The total locking force generated in the locking area depends on the number of turns (conductors) being inserted. Obviously, 1 turn cannot lock, 2 turns generates a small force, and a large number of turns, such as 50, generates a large force.

Empirical conclusions based on a large number of tooling designs and conditions indicate that, in general, a maximum turn count of 20 can be inserted per slot in the locking area without damage to wire or tooling.

The area between the locking-wire area and the maximum-wire-size curve is referred to as the *precision wind* area, so called as the wire must be in a single-layer or *precision* condition. This area also presents some restrictions on the number of turns due to *column height* and a condition similar to the previously mentioned locking situation. Usually, 35 turns for standard placer tooling and 50 to 55 turns with special tooling options are considered safe selections. Higher turn counts can result in wire damage and can involve extensive tooling development programs.

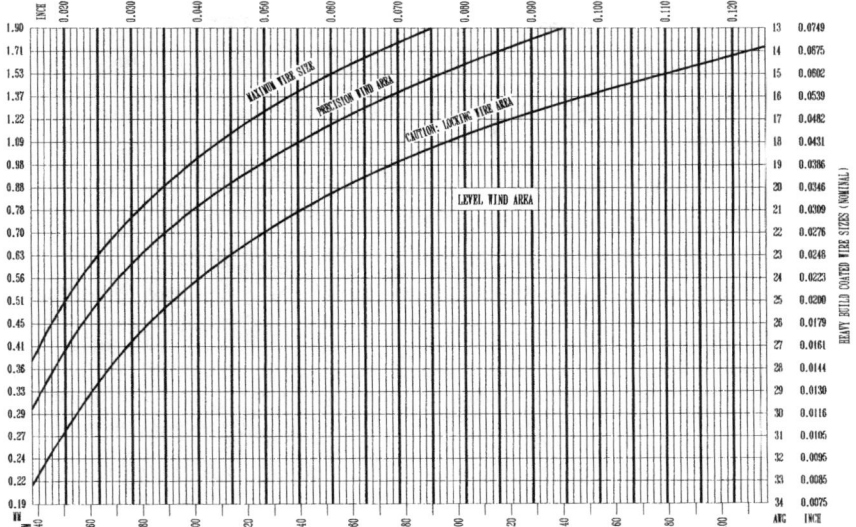

FIGURE 3.30 Blade gap chart. Note that allowance must be made for stack (core) skew or stagger. Maximum blade gap = iron gap − 0.030 in (0.8 mm).

The area below the locking area represents the *level wind* area and is considered the ideal situation for coil insertion. Slot fullness is usually the limiting condition for this area, except for special slot configurations. Anything below 70 percent is considered good, with 76 percent a practical maximum for standard toolings. Further discussion concerning slot fullness is covered in a later section.

Many developments in tooling design and special features have improved these restrictions and the quality obtained, but have not entirely eliminated them.

Once the wire has been inserted into the stator, the major material costs have been incurred and the basic quality of the stator assembly has been determined except for lead connecting. Compatibility of the slot opening and the wire size therefore becomes a major contributing factor to cost, through high or low scrap rates, which directly reflect the difficulty or ease of the manufacturing process and the final quality of the stator.

Pole Configuration. Pole configuration usually refers to the number of poles and the physical shape and location within a stator. There exist two major categories of pole configuration, lap winding and concentric.

The coil-insertion process has almost eliminated lap-type windings. Although it is possible to insert some lap windings, it is almost impossible to achieve the slot fills and production rates of equivalent concentric windings. Almost all of the lap-type motors, probably 90 to 95 percent, are hand wound.

This section therefore deals only with the single-phase and three-phase concentric-type windings. The vast majority of the single-phase motors produced today follow a standard two-, four-, or six-pole configuration, varying only in the number of concentric coils per pole, which is limited, and slot fullness. The tendency today is for higher slot fills, which in a single-phase motor are much more difficult to achieve than in a three-phase motor due to the nature of the main winding.

Three-phase motors present a much greater opportunity for variation in pole configuration. The industry standard has been the three-layer uniform design, having no shared or some shared slots, which is accomplished by varying the number of concentric coils per pole. In the past, some manufacturers have used a two-coil-sides-per-slot design in which all slots are shared, but this has not yet proven to be a popular approach.

The two-pole three-phase motors are fairly straightforward in a three-layer design, with some or all slots shared. This is fairly well limited due to the physical configuration inherent in a two-pole design.

The four-pole three-phase motors present a high degree of variation; however, the majority of the stators produced today are of the standard three-layer four-pole-per-layer uniform design. One fairly popular design is the European or consequent pole design. The main reason is that the 2-layer three-pole-per-layer configuration, usually in a nonsharing slot condition, requires only two insertions to complete the motor. Winding time is less to generate 6 total coils for a complete motor, rather than the normal 12 poles. Phase insulation for this type of motor is much simpler due to having only two layers, instead of three, and in some cases is eliminated.

However, there are some disadvantages. It is usually considered to be a less efficient motor, it requires more interpole or lead connections, and, finally, it has been known to require more copper than the three-layer design.

The six-pole three-phase motor also presents the same type of variations as the four-pole, but perhaps to a lesser degree. A six-pole European or consequent design is also used, but is less popular. The same basic criteria exists as in the four-pole, two insertions instead of three, and in this case 9 poles rather than the normal 18. Some of the same disadvantages also exist as for the four-pole.

The standard three-layer designs for two-, four-, and six-poles can be produced in a *gradient design*. The gradient design is achieved by decreasing the circumference of the pole for each succeeding phase or layer. The first layer or phase inserted would be the longest, the second slightly smaller, and the third slightly smaller yet. The net result is a savings in wire and better nesting of the end turns. This type of design is generally considered to be an *imbalanced-phase* winding and could have a negative effect on the performance. However, if properly designed, the advantages can outweigh the performance disadvantages.

Although gradient design is not generally used across the industry, the more progressive manufacturers are using this approach or looking at it very seriously. It is beginning to appear as an approach that could become standard practice in the future.

Slot Fill. Slot fill is usually given as a percentage figure that expresses the amount of wire in a slot in relation to the total slot area. Unfortunately, there are several methods presently in use for calculating slot fill.

Widely varying slot-fill percentages can be obtained for the same situation depending on the method used. Two general methods are circular mils and square wire. Electrical designers would normally use the circular-mil method because they are more interested in the actual cross section of the conductor. Process engineers or equipment suppliers would be more apt to use the square-wire method, which more nearly reflects the actual conditions with which they must work.

Variations within each of these basic methods will also occur depending on the method used to determine the slot area. The insulation in the slot—for example, the slot cell and wedge—will occupy a portion of the slot area.

Some calculations will take into consideration the slot insulation, while others will ignore it. Taking into consideration the slot insulation results in an *available slot area*.

For purposes of this section, *slot fill* (SF), is calculated as follows (see Fig. 3.31):

SF, % = (wire diameter)^2 × number of turns in available slot area

Wire diameter = bare wire plus insulation

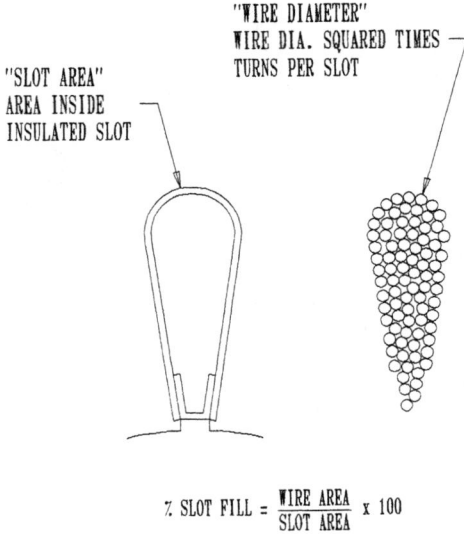

FIGURE 3.31 Percentage slot fill.

Available slot area is the slot area calculated from the punching, subtracting the insulation-occupied area and the area between the base of the wedge and the bore of the slot. This results in a true available slot area.

Today's trends and desires are to maximize slot fill. For purpose of brevity, we list some of the benefits here, and do not go into a lengthy explanation of reasons why these results are obtained.

- Greater performance efficiency
- Minimal material usage
- Smaller package for same performance

A few years ago, a slot fill of 50 to 65 percent was acceptable and efforts were concentrated on reducing labor. As material costs and volumes rose along with a greater consciousness of power efficiencies, the requirements for higher slot fills increased. Today, as a general rule, slot fills of 70 to 75 percent are fairly common and are produced in volume with relatively few production problems.

In some cases, slot fills of 80 to 81 percent have been achieved, but to do this requires special development and a concentrated effort in tailoring toolings for a particular application.

The desire for high slot fills is very great at the present time, and there are several theories or approaches under consideration which have not been developed, as yet, into practical production methods.

One area of effort is to develop a slot shape which approaches the ideal slot. As mentioned previously, the more nearly ideal slot from a production viewpoint involves a compromise of a larger-OD stator and therefore is less attractive than other approaches.

Perhaps the most attractive and promising approach is the *compaction process*. This process is exactly as its name implies—compacting the wires in the slot. Some manufacturers have taken tentative steps in this direction.

The theory is very simple and consists of inserting a first layer of wire into a slot, compacting or deforming the wires into the back of the slot to fill all of the void spaces between wires, then inserting a second layer of wire. As a simplified example, if a 92 percent slot fill is desired, a first insertion of 46 percent or half the total wires is inserted. By the process of compaction, these wires are then forced to occupy only 40 percent of the slot. The remainder of the slot, 60 percent, is now available for the next insertion of 46 percent of the wire, which results in a 76 percent slot-fill insertion attempt on the second pass. The second insertion falls within the range of feasibility based on present practice.

The use of the General Electric Electro-press process could also contribute to the feasibility of achieving the desired results or increasing the total percentage slightly. This can be accomplished by taking advantage of the fact that the Electro-press process has a tendency to straighten wires, eliminating the crossing in the slot that occupies additional space. Straightening of the wires prior to compaction would therefore assist in achieving higher slot fills.

The compaction and Electro-press slot-development approach should make slot fills in the 90 percent range accessible.

Phase Insulation. From a labor and materials standpoint, phase insulation is perhaps one of the most difficult, time-consuming, and expensive areas of stator manufacture. Phase insulation can be accomplished through several different methods. The main object of this particular section is to recommend the type of phase insulation and the point in the assembly line where phase insulation is to be installed. This assumes that phase insulation is required. A major savings can be accomplished if phase insulation is eliminated. Some motor manufacturers have eliminated this type of insulation, but not without some problems. For the most part, they have been successful. Unfortunately, at this time, the manufacturers who have eliminated phase insulation are in the minority. It has always been felt that phase insulation is required in order to guarantee a quality product and is an insurance policy of sorts. However, with today's improvements in wire insulation, slot-cell insulation, and manufacturing methods, the whole area of phase insulation should be very seriously considered for potential cost reduction. We would recommend that a program be instituted to investigate the possibility of the elimination of phase insulation.

At the present time, various methods of phase insulation are being used. One of the most popular is the H-type paper insulation, which must be installed between the layers or phases of windings during the process of winding the stator. This involves winding one layer of the stator, hand-inserting the H-type paper insulation, and then winding the next layer of wire into the stator. In three-phase stators, one more layer is yet to be inserted; therefore, a second hand insertion of phase paper must be undertaken, and then the third and final layer of winding is inserted into the stator. This is at best a slow operation and also very labor intensive, not to mention that the H paper is generally made of a polyester-type material with very high scrap rates.

A second method is to insert the first layer and then proceed by hand to add an adhesive tape to the end turns which acts as a phase insulation. The second layer is then inserted and has to be taped by hand prior to the insertion of the third layer. This is also a labor-intensive operation and requires special materials which sometimes are not compatible with the products being manufactured (for example, the adhesive tape).

The third method, the method which we prefer, is an operation in which all three layers (three-phase) or two layers (one-phase) of the motor are inserted in a continuous flow operation without interruption for phase insulation. A variable-wedge-length device is included on the equipment for notching wedges. In the end, the wedges constitute a part of the phase insulation. After the stator winding is completed, phase insulation is added into and between the end turns of the wire as required. This material can be of the same polyester as is used in the H paper; however, it is merely a strip of material, and scrap is virtually eliminated except for rounding some corners. There are several advantages to using this method:

- The winding operation is not interrupted.
- The equipment is not idle during phase insulation, and no unloading and reloading of the stator into the insertion flow system is required.
- The elimination of polyester scrap is an obvious cost savings.

The time required to insert the phase insulation is substantially less than that required for either of the first two methods.

There are some disadvantages to this method, one of which is that it must be a hand operation. Second, it requires some manipulation of the end turns with a hand tool in order to separate the windings where the insulation is to be inserted, which could possibly result in a deterioration of the quality. However, this type of process has been and is being used successfully in some major motor plants around the country. In conclusion, the best approach to phase insulation is to eliminate phase insulation. If this cannot be accomplished, then the second best is to automatically machine-insert phase insulation. However, at the present time this type of equipment is not yet very popular. In the interim, the process that is best utilized for minimizing cost is the hand insertion of polyester segments into the end turn at a station away from the winding and insertion of the stator.

Salient Pole Motors. Salient pole machines like universal motor fields, shaded-pole motor stators, and stepper motor stators are usually *needle-* or *gun-wound.* They may be wound with shroud tooling, as shown in Fig. 3.32, or they may be wound directly on molded plastic insulation which also serves as tooling, as shown in Fig. 3.33. In these cases, the minimum slot opening must take the needle size and path into account. The typical needle path is shown in Fig. 3.34. The slot opening must allow for this movement plus some clearance. The slot opening is determined as follows. Select the maximum coated wire diameter that will be used in the intended application. Next, determine the minimum allowable needle bore, outside dimensions, and clearances per Fig. 3.35.

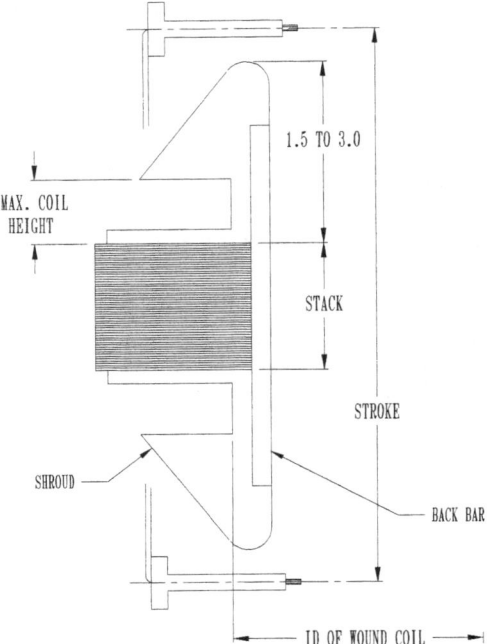

FIGURE 3.32 Typical two-pole motor with shrouds.

FIGURE 3.33 Plastic insulation and winding shroud combination.

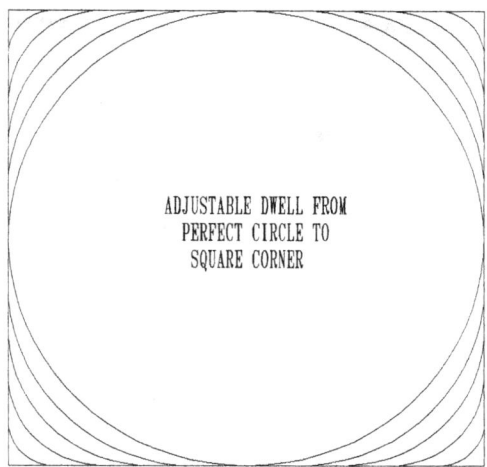

FIGURE 3.34 Typical needle winding path.

Stack-in-Die Cores. This method utilizes a punch in the edge to pierce a portion of each lamination about half of the way through the material, as shown in Fig. 3.36. Succeeding laminations are pressed into the previously punched lamination by inserting the protrusion of one into the recess of the other. Because of taper in the raw lamination steel, it is necessary to periodically rotate some of the laminations 180° to hold the stack square. This rotation usually is not necessary until the stack length L_{stk} reaches 2.00 in or more. Using this method allows the stack OD to be held within ±0.001 in up to about a 4.0-in stack length. As mentioned earlier, the laminations are annealed to promote grain growth, which reduces hysteresis losses. During

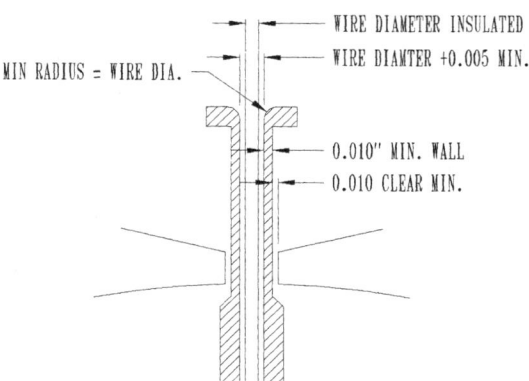

FIGURE 3.35 Needle, wire, and iron gap minimum dimensions and clearance relationships.

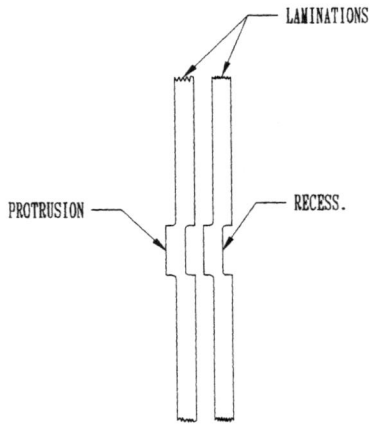

FIGURE 3.36 Stack-in-die laminations.

this process an oxide is also generally put on the surface of the lamination, which increases interlaminar resistances. This is extremely important in reducing eddy current losses, which are the predominant part of the core loss at higher frequencies. The laminations are stacked tightly together while in the punching die. Later, during the annealing process, some of the laminations in the center of the stack may not receive the same amount of coating as the laminations on the ends. This may result in higher than expected core loss. New processing methods tend to minimize this problem. A key to low core loss is the placement of the protrusions.

3.10.5 Manufacturing Rotor and Stator Stacks in the Stamping Die*

Laminations of all types (see Fig. 3.37) may be staked together into stacks of predetermined heights as they are stamped in progressive dies during the manufacturing process. This staking-in-the-die technique represents an opportunity for both improved quality and significant cost reduction to manufacturers of products that require rotor, stator, transformer, ballast, magnet assembly, backpole counterbalance, and other types of laminations that subsequently are laminated into stacks.

In addition, symmetrical laminations, such as for rotors for electric motors, may be rotated in the die prior to staking in order to compensate for material thickness variations or to produce a skew angle, as found in many motor designs.

Producing finished stacks of rotor and stator laminations in the die has many advantages to the manufacturers of electric motors, particularly the elimination of downstream manual or mechanical assembly operations such as riveting or welding. Staking in the die also can produce stacks of more uniform height.

A die capable of staking laminations together requires a specially designed die cavity and special staking punches. The staking punches create protrusions that cause the individual laminations to stick together as they become a stack in the die cavity. Whether as many as four or more staking punches should be used depends on the size and design of the rotor or stator part.

Absolute Count Stacking. The easiest way to control stack height is to program a microprocessor controller for a predetermined number of laminations in each stack. For example, a 1-in rotor using 0.025-in-thick material would require 40 laminations. By programming the die's staking punches, a 1-in stack of rotors may be produced with every 40 strokes of the press. The staking process produces tight stacks which then are allowed to drop from the die cavity onto a low-profile conveyor.

The integrity of the 1-in stack height is completely dependent on the thickness consistency of the 0.025-in coil stock. If the material begins to run thick during the stamping process, 40 laminations could produce an unacceptably long rotor stack. To

---

* Subsection contributed by Harry J. Walters and Phil Dolan, Oberg Industries, Inc.

FIGURE 3.37 Lamination stacks.

guard against this, the press operator must monitor the stack heights. If thick or thin material causes the stack to reach a height outside the acceptable tolerance range, the operator must adjust the lamination count accordingly.

A more accurate and reliable method of controlling stack height is to monitor the thickness of the coil material and let the controller dynamically calculate the number of laminations required for the proper stack height. The patented system uses a material thickness sensor that measures the inbound material and every lamination that will comprise the stack. By setting the control unit for a 1-in stack height (not on an absolute count of 40 laminations), the 1-in finished stack may contain 39, 40, or 41 laminations, depending on whether the 0.025-in material is running thick or thin.

There are several advantages to using a system of this sophistication. It produces rotor and stator stacks of consistently uniform height and frees the press operator from having to continuously measure stack heights to determine if they are within tolerance. The system also permits some motor manufacturers to use less uniform and less expensive steel stock because the controller determines the proper number of laminations needed to achieve the desired stack height.

Another consideration is that stack height can be changed on the control unit in a matter of seconds. This permits the manufacturer to shift production to a different motor length without turning off the stamping press.

Rotating Laminations in the Die. A material thickness monitoring system like that shown in Fig. 3.38 can compensate for stock variations throughout the length of the coil and automatically adjust the lamination count to maintain uniform stack height, but it cannot compensate for thickness variations across the stock width. For example, if the left side of the coil strip were consistently thicker than the right side, the stator and rotor stacks produced would lose their perpendicularity and would lean to the right. These stacks would be of inferior quality and would complicate subsequent manufacturing operations.

FIGURE 3.38 Material thickness monitoring system.

To solve this problem, a method has been patented using dies that have been designed to rotate the lamination in the staking die cavity. As each new lamination enters the cavity, the existing stack is rotated a fixed number of degrees as a function of the number of rotor slots. A rotor with two staking points, for example, may only be rotated 180° with respect to the following lamination. A rotor with four staking points could be rotated 90°.

Oberg Industries uses a belt system driven by a high-speed motor to rotate the die cavity. The motor is controlled by the same controller that monitors the staking and stack-height functions.

Rotation is not necessarily confined to rotors. If a stator is perfectly symmetrical, it also may be rotated in a staking die cavity. In addition, when motor manufacturers require loose rotor laminations, the laminations may be rotated without being staked together and loaded into stacking chutes. When these loose laminations finally are assembled, they will also exhibit improved perpendicularity and balance.

Many motor designs incorporate a skew angle in the rotor assembly to improve motor performance. In Oberg-produced dies that contain the rotating skewing cavity, the skew angle to the rotor stack is quickly set by entering the desired skew angle into the control system. Skew angles may be set in addition to the rotation or by themselves without any other rotation in the lamination.

When skewing, the consistency of material thickness throughout the length of the coil stock is a concern. Rotor stacks 1-in high of 0.025- ± 0.002-in laminations will have varied numbers of laminations to achieve proper height. To compensate for variations in coil-stock material thickness, the control unit adjusts the rotation on each lamination. The end result is a consistent skew offset, even though the stack may contain 39, 40, or 41 laminations.

Although staking, rotating, and skewing of the laminations is performed in the stamping die, the critical component of the proprietary system is the microprocessor control unit. The controller must have the capacity and speed to control the die's staking punches, the high-speed motor that drives the rotation cavity, and the material monitoring sensor, as well as permit the stamping press to operate at top speed. Some lamination die controllers may have to eliminate features to avoid significantly slowing the press speed.

System Benefits. Whether the motor manufacturer stamps its own laminations or buys them from a lamination stamper, staking, rotating, and skewing in the die offers several benefits. One of the benefits of the technology is that it requires less material handling. Staked lamination stacks eliminate much of the handling and moving associated with loose laminations. Some motor manufacturers are able to send stacks on conveyors to the next process area, directly from the press.

For many types of motors, staked rotor and stator stacks may eliminate welding and riveting operations. The labor and costs associated with these operations is eliminated, and there also is no need to replace welding and riveting equipment when it wears out.

One motor manufacturer, faced with the replacement of an obsolete welding line, invested instead in a staking die and controller. The company calculated a two-month payback on the investment and, in addition, was able to access much-needed floor space when the welding line was removed. The manufacturer intends to eliminate all welding in the plant within two years.

The rotor and stator stacks produced with the staking-and-rotation technique are consistently of higher quality than those produced from loose laminations. Some manufacturers using staking dies have realized a reduction in the costs associated with balancing and other motor-finishing operations. Motor performance also has been improved.

For manufacturers that stamp their own laminations, the benefits related to production flexibility may be substantial. Stack height and skew angle are changed easily, and combinations of height, skew angle, and rotation can be varied and adjusted by the operator.

Die and Controller Requirements. Rotation, skewing, and staking of laminations, the process used to produce the sample stacks shown in Fig. 3.39, presents several challenges to the die or controller supplier. The basic accuracy requirements of the rotational motion are recognized when one considers that a 10° skew angle in a 40-lamination stack results from the rotation of each individual lamination by 0.25°. Variations in material thickness can adjust that by 0.00005° or less. Also, extremely high accuracy is required in the stamping die to permit rotation of laminations while maintaining concentricity. The location of the stakes must be perfectly symmetrical

FIGURE 3.39 Sample lamination stacks.

for them to attach properly to the preceding rotated lamination. Rotation of square, rectangular, or other nonround shapes requires extreme accuracy of the rotational motion, since the punch is now penetrating a moving die section. It also requires mechanical devices that will prevent damage if the rotating chamber does not align with the punch.

The design and construction of the controller must take into account the fact that the system will be operated in a pressroom environment, and must minimize both additions to the operator's workload and intrusion into the already crowded work space. A simplified operator interface and a rugged, vibration-resistant package are basic to the operational success of a staking and stamping die.

Technology Limitations. There are areas within this staking and rotating technology where motor manufacturers face some limitations and cautions. First and most important, it should be noted that the staking process requires a sufficient open area on the face of the rotor and stator laminations to allow a stake protrusion to be made without distorting a critical dimension of the lamination. When designing a lamination with staking in mind, the advice of a die designer is essential.

Multipart dies are common in the motor industry, and staking dies have been built that produce as many as five rotors at a time.

Although staking and rotating dies for smaller-sized rotors may not require reduced press speeds, the rotation of the die cavity for larger-diameter rotors could force a press to run more slowly. Also, for annealed laminations, the electrical properties of staked stacks must be compared critically to those of stacks made from loose laminations. Motor manufacturers that stake stacks in the die generally have found few differences of consequence, but it is a factor to be considered when designing a motor.

Although the tightness and integrity of staked stacks usually are not problems, some handling precautions are advisable to prevent stack delamination. One manufacturer permits its stacks to drop almost 4 ft from the press into a collecting bin, but this kind of handling may not be suitable for some types of lamination designs.

The technology of staking, stacking, rotating, and skewing in the stamping die, along with other emerging technologies designed to reduce costs and improve quality, should position the small-motor industry to compete successfully in the global economy.

3.10.6 Electric Motor Stator Lacing*

Stator lacing is the process of tightly securing the field coil ends of an electric motor stator with a stitched cord (see Fig. 3.40). Lacing is typically used on long-life-expectancy or high-efficiency motors where the cost of failure is high. Generally, low-cost "throwaway" motors are not laced. There are several reasons for lacing motor stators:

- The lacing holds the thermal protector, coil ends, and leads in the proper position.
- Lacing extends a motor's life by preventing the wires in the coils from vibrating and causing fatigue failure during operation.

---

* Subsection contributed by Warren C. Brown, Link Engineering Company.

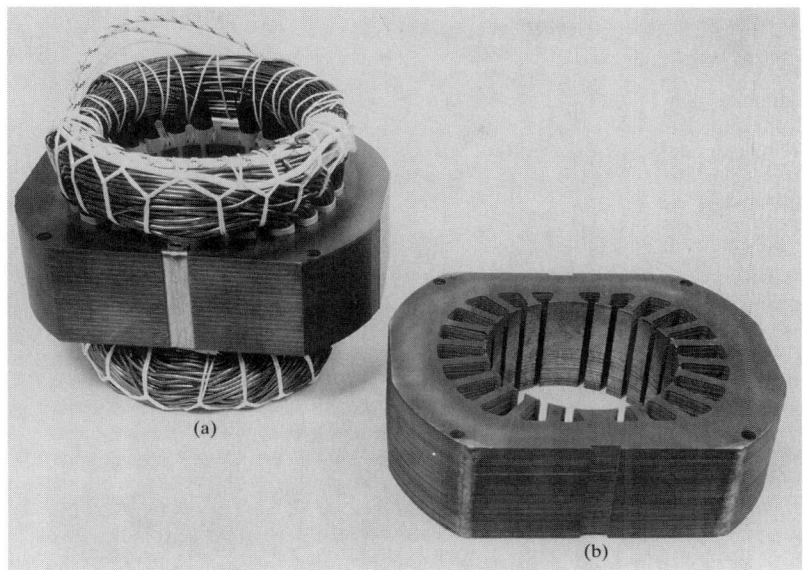

FIGURE 3.40 (*a*) Laced stator and (*b*) bare stator core.

- Lacing may be used to hold the coils in position and provide loops to hang the stator from a conveyor during the dip-and-bake varnishing operation.

The most obvious reason for lacing a stator automatically by machine is increased productivity. The lacing machine can lace more stators per hour than a person doing it manually, and usually at a substantially lower cost. The lacing machine also provides improved lacing quality and consistency over long time periods.

Another important advantage of machine lacing is the avoidance of carpal tunnel syndrome, a debilitating hand and wrist injury caused by repetitive strenuous handwork.

Several different styles of lacing machines are generally available to the motor manufacturer. The simplest machines lace one end of the stator at a time, and are referred to as *single-end lacers* (Fig. 3.41). Others have two needles and lace both ends of the stator at the same time, and are known as *double-end lacers* (Fig. 3.42). Link Engineering Company of Plymouth, Michigan, pioneered the development of the automatic stator-lacing machine with interchangeable tooling in the mid-1960s.

Lacing machines may be constructed so the stator is vertical and the lacing needle is horizontal, or with the stator horizontal and the lacing needle vertical.

Most lacing machines index the stator about its axis during the lacing cycle, but a few machines have been designed to clamp the stator and rotate a lacing head with the needle around the stator.

Some machines use a closed needle with an eye, like a sewing machine, but the vast majority use an open needle, similar to a crochet hook, to form the stitch. In general, the open-hook needle produces a "diamond" stitch, with diagonal coverage of the coils, whereas the closed-eye needle produces more of a radial stitch.

There is also a wide variety of lacing cords available. These cords may be made from fibers such as cotton, polyester, nylon, or other synthetics, and may be formed

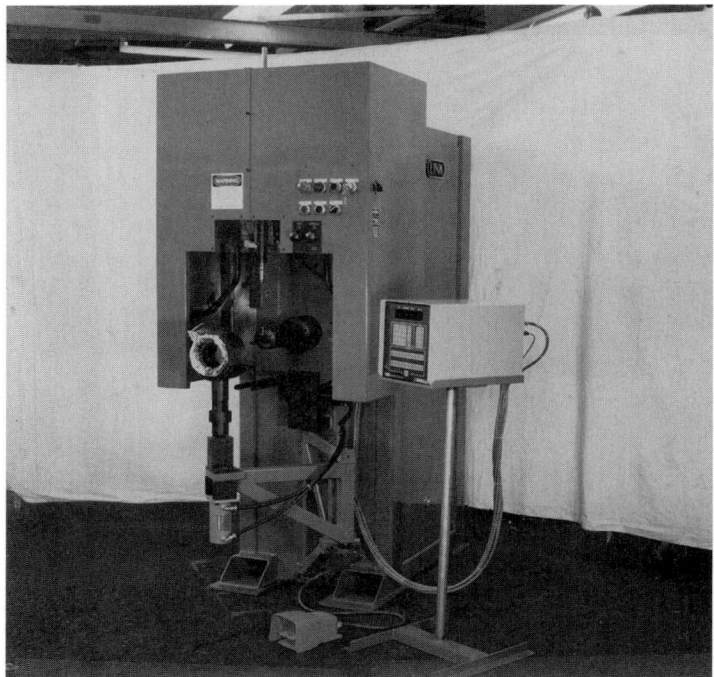

FIGURE 3.41 Single-end stator lacer.

by twisting or braiding into a round or flat-tape shape. Twisted round lacing cord is the most popular and least expensive type. Tensile strength of the cord is determined by its diameter, the material it is made from, and the weaving technique used to make it.

Most lacing cords are made from polyester, which shrinks when it is heated. The percentage of shrinkage may be specified from almost none up to about 15 percent. The higher-shrinkage cords tend to produce a tighter lacing when shrunk.

The size and type of the lacing cord, as well as the cord tension during the lacing and knot-tying cycle, play a critical role in achieving consistent high-quality lacing.

Functional Characteristics. There are several important characteristics of a stator-lacing machine that determine its performance in a demanding plant environment. One of the most important is *speed*. The faster a machine laces, the higher the throughput and the lower the cost per stator. To properly evaluate speed, however, the total lacing cycle must be considered. This includes loading a stator, positioning the leads if necessary, lacing, knot tying, removal of cord tails, and unloading the stator.

Typical time to manually unload a laced stator and load an unlaced stator is about 10 s. Automatic loading/unloading devices can be used to speed up the handling to and from the lacing machine.

Lead positioning may be done by an operator or by a lead clamp or lead wiper incorporated into the lacer. As the number and length of the leads increase, so does the difficulty encountered in lacing the stator.

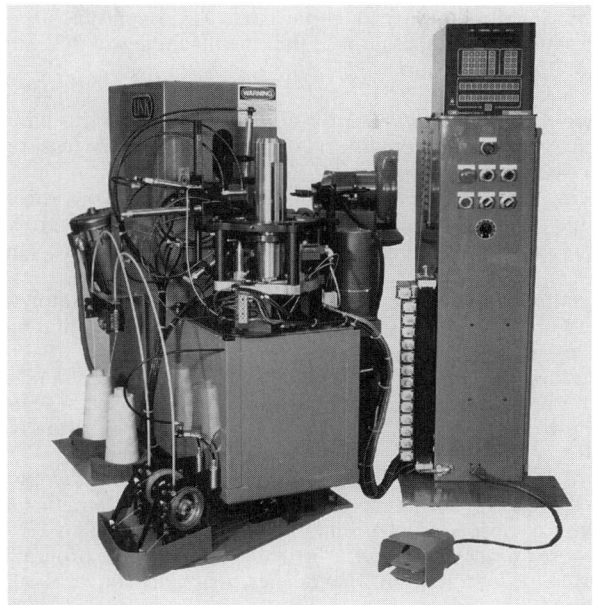

FIGURE 3.42 Link Model 940 double-end stator lacer with servo index and knot-tying system.

The actual stitching speed is usually a function of the coil end-turn size of the stator being laced. Large coils require greater movement of the relatively heavy needle mechanism and therefore require more time. Typical lacing speed for small stators is 2 stitches per second and for large stators about 1 stitch per second.

Manually tying a knot and burning off the tail typically takes about 5 s for each end of the stator. Automatic knot tying and tail burning typically takes about 4 s, and both ends are done simultaneously.

Setup time may be another important consideration in evaluating a lacing machine. If a line is dedicated to a single stator, or even if production runs are very long, with infrequent changeovers, it is not too important to be able to change from one stator to another quickly. But if runs are short, with only a few of each type of stator laced at one time, setup time can be more important than lacing speed. Changes in ID, OD, stack height, coil end-turn height, number of slots, or stitch pattern may require from a few seconds to 10 or 15 min each to make programming or mechanical adjustments to the machine. New servomotor-driven lacing machines, with computer control systems, offer dramatically reduced setup time, as little as 7 s. The servomotor lacing machines are ideal for motor manufacturers with small lot sizes which require frequent changeover.

The quality of the lacing is also a critical characteristic of the lacing machine. If the machine drops stitches, breaks the cord, makes loose stitches, forms loose knots, leaves long tails, or damages either the wire or lamination, it will not meet strict quality standards. Machine demonstrations and discussions with existing users can verify lacing quality.

The durability of the lacing machine is also important in determining its productivity. If the machine fails often, is difficult to get parts for, or takes a long time to repair, it will not meet overall throughput goals. A proven machine design from a reputable company is the best assurance that the machine will deliver uninterrupted performance on the plant floor.

Typical Lacing-Machine Features and Options. Although a basic stator-lacing machine can be a big improvement over hand lacing, many features and options make the process faster or more flexible.

A time-saving option is *automatic knot tying*. This device forms a secure, tight knot, then burns off the tail and vacuums it into a waste container. A device is also available to fully automate the cord cutting and clamping at the end of the knot-tying cycle.

Automatic stator lifting and lowering raises and lowers the stator so the operator can easily grasp it. This feature is particularly useful for stators that have short stack heights or are particularly heavy.

Automatic stator loading and unloading can take the form of a manually assisted arm and gripper or a fully automatic robotic handling device. These loading/unloading systems can be integrated into a fully automated line to virtually eliminate the requirement for an operator.

Broken-cord and *end-of-cord sensors* enable detection of the end of a spool of cord or a break in a cord. This is especially useful in fully automatic lines that do not have an operator to observe such cord faults.

Computer-based touch-screen control systems offer simple programming, graphic displays, internal documentation, diagnostics, machine statistics, and large data-storage capacity.

Nonradial slot lacing allows manufacturers to lace stators with odd slots that are not in line with the center of the stator core.

A *hanger loop* option forms two long loops in the lacing cord at opposite sides of the stator, so the stator can be hung from a conveyor for processing through a varnish bath.

Roller casters may be placed under the legs of the lacer and the electrical box to enable easy movement of the machine from one location in the plant to another.

Summary. The automatic stator-lacing machine has a proven track record of being a productive, reliable, cost-effective tool for motor manufacturers seeking to produce high-quality motors at a competitive price. Many evolutionary changes have led to a wide variety of models and options incorporating significant improvements in flexibility, reliability, and speed.

*3.11 BEARING SYSTEMS FOR SMALL ELECTRIC MOTORS**

Electric motors are devices that convert electrical energy into magnetic energy and finally into mechanical energy. The mechanical energy is generally transmitted from the rotor through a shaft that must be free to rotate in some type of bearing system (see Fig. 3.43). The choice of the bearing system is key to the motor's performance and life.

---

* Section contributed by William H. Yeadon and Alan W. Yeadon, Yeadon Engineering Services, PC, except as noted. Basic technical information courtesy of Nye Lubricants and NMB Corporation.

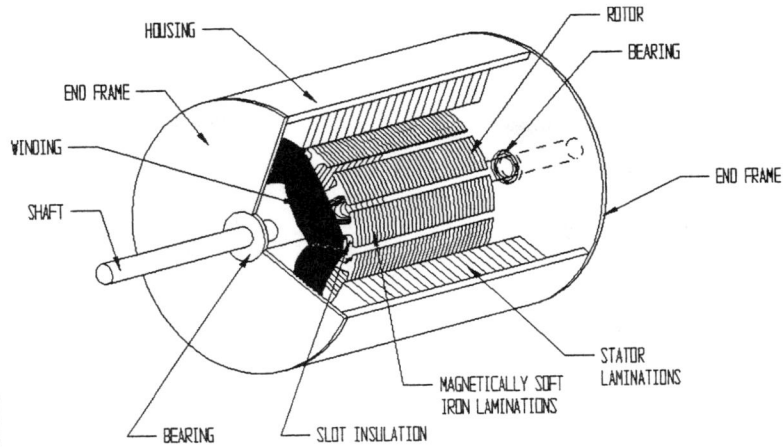

FIGURE 3.43 Electric motor components.

The system generally consists of a shaft, a bearing, and a lubricant arranged in a fashion that maintains a film of lubricant between the shaft and the bearing surface. The components and system are typically chosen to meet the requirement of the specific application.

This section first discusses the components, then systems, and finally the application.

3.11.1 Bearings

Bearing systems are used to support the rotor and shaft assembly so that it remains in a certain constant position relative to the stator and so as to reduce the friction between the shaft and the end frames. The most common bearings used in motors are *ball bearings* and *sleeve bearings*. Ball bearings are typically constructed as shown in Fig. 3.44. They consist of an inner race and outer race, balls, and a ball carrier. The races and balls are typically highly polished hardened steel. The ball carrier may be steel or plastic.

The *inner race* supports the shaft and rotates with it. The *outer race* is held stationary in the end frame. The *balls* provide a low-friction method of allowing the inner race to roll with respect to the outer race as the shaft turns. The *carrier* maintains proper spacing of the balls to evenly distribute the load.

Ball bearings are lubricated by injecting grease around the balls between the races. The grease may be contained by means of a shield or seal that fits between the races. *Sealed bearings* have a higher coefficient of friction, require more torque from the motor, and are more costly than *shielded bearings*. Therefore, they are used in applications such as pumps where it is necessary to keep moisture or corrosive agents out of the bearings.

Ball bearings need to be preloaded to keep the balls from moving freely in the axial direction. The amount of preload is listed in the bearing manufacturer's data for each type of bearing. Preloading is generally accomplished by means of a coil spring or wavy washer.

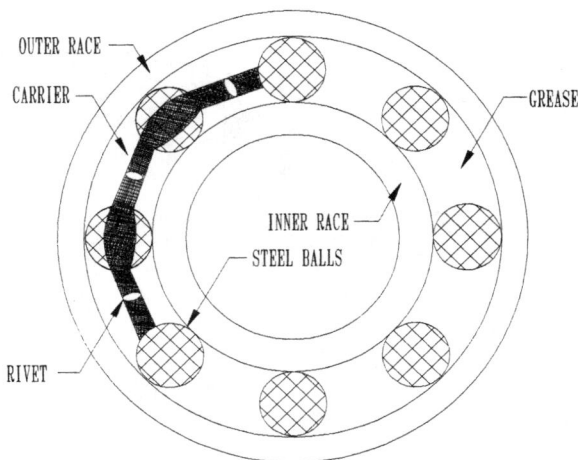

FIGURE 3.44 Ball bearing construction.

Ball bearings are generally purchased by grade number. The higher grade numbers have tighter part tolerances and lower radial play, and are more costly. High-grade bearings are used in applications where radial rotor or shaft movement must be minimized. The ABEC grades and tolerances are shown in Table 3.3.

TABLE 3.3 Ball-Bearing Grades

ABEC grade	Maximum radial runout		Mean diameter tolerance		
	Inner ring, in	Outer ring, in	Bore, in	OD, in	OD size, mm
1	0.0003	0.0006	+0.0000 −0.0003	+0.0000 −0.0003	0–18
				+0.0000 −0.00035	>18–30
3	0.0003	0.0004	+0.0000 −0.0002	+0.0000 −0.0003	0–30
5	0.00015	0.0002	+0.0000 −0.0002	+0.0000 −0.0002	0–30
7	0.0001	0.00015	+0.0000 −0.0002	+0.0000 −0.0002	0–30
9	0.000050 0.000050	0.000050 0.0001	+0.0000 −0.0001	+0.0000 −0.0001	0–18
			+0.0000 −0.0001	+0.0000 −0.00015	>18–30

Source: Courtesy of NMB Corporation.

3.11.2 Bearing Selection*

There are several important considerations which must be evaluated simultaneously when choosing the proper bearing for a particular device. The following subsections briefly discuss some of the more important ones.

Miniature and instrument ball bearings are normally made of either stainless steel or chrome alloy steel. The load ratings given are for chrome steel unless otherwise noted. Load ratings are affected by bearing material. Life calculations are affected by bearing material as well as lubrication selection.

Type of Cage. Two types of pressed-steel ball cages are available for most bearings, H (crown type) and R (two-piece ribbon type). These two cage types are interchangeable in most common applications. Cages made of molded and machined plastics are also available for some sizes (see Fig. 3.45).

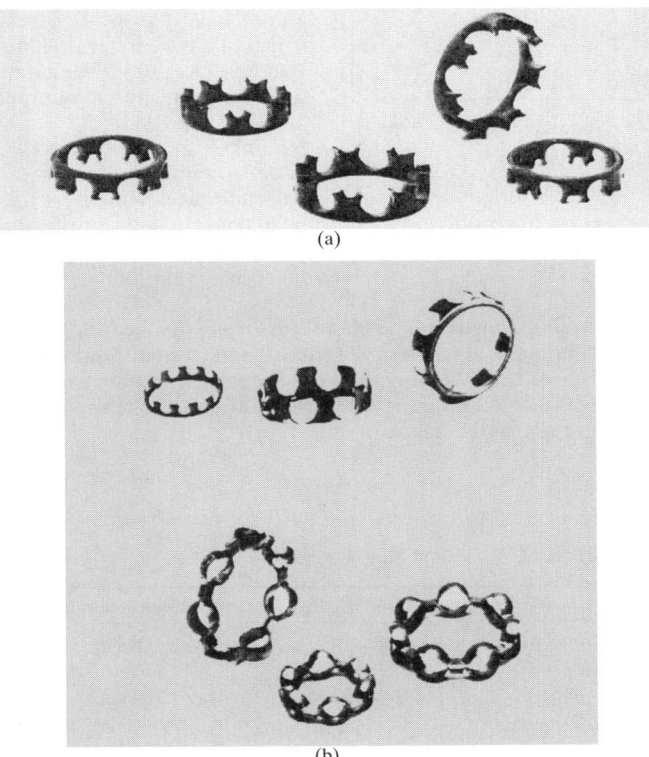

FIGURE 3.45 Molded and machined plastic ball bearing cages.

---

* Subsections 3.11.2 to 3.11.10 courtesy of NMB Corporation.

Shields and Seals. Shields are available for most sizes. These closures help to reduce the entrance of particulate contaminants into the bearing and reduce the amount of lubricant leakage. Radial clearance between the shield bore and the inner ring OD is approximately 0.002 to 0.005 in. The effect of shields on bearing torque or noise is insignificant.

Contacting seals made of synthetic rubber (type DD), as shown in Fig. 3.46, are available for most sizes. These seals provide the best protection from the entrance of contaminants or exit of lubricant, but they significantly increase operating torque. DD seals will withstand a slight amount of positive pressure differential.

FIGURE 3.46 Contacting seals (type DD, synthetic rubber).

FIGURE 3.47 Noncontacting seals (type LL, reinforced Teflon).

Noncontacting seals made of synthetic rubber (type SS) or reinforced polytetrafluoroethylene [PTFE (Teflon; type LL)], as shown in Fig. 3.47, are also available for most chassis sizes. This type of seal offers better sealing than a metal shield, while keeping operating torque at the lowest possible levels. LL seals will contact the inner ring in some cases, but the nature of the seal material serves to keep torque at a minimum.

Radial Play. Radial play is the free internal radial looseness between the balls and races. Radial play within a ball bearing is necessary to accommodate thermal expansion and the effects of interference fits and to control axial play. Table 3.4 suggests radial play ranges for some typical uses.

Starting and Running Torque. The operating torque of a bearing can be described as starting and running torque. *Starting torque* is the torque required to begin rotation from a bearing at rest. *Running torque* is the torque required to rotate one ring at a known speed while keeping the other ring stationary. The main contributors to bearing torque are seal and lubrication type.

TABLE 3.4 Radial Play Ranges

Typical application	Suggested radial play, in
Small high-speed precision electric motors	0.0005–0.0008
Tape guides and belt guides, low speed	0.0002–0.0005
Tape guides and belt guides, high speed	0.0005–0.0008
Precision gear trains, low-speed electric motors, synchros, and servos	0.0002–0.0005

Static C_{or} and Dynamic C_r Loads. In evaluating three static load conditions, any forces exerted during assembly and test must be considered along with vibration and impact loads sustained during handling, testing, shipment, and assembly. Dynamic loading includes built-in preload, weight of supported members, and the effect of any accelerations due to vibration or motion changes. The static and dynamic radial load ratings are shown for each chassis size in the tables that follow.

Speed of Operation. Although a very large bearing might be the best choice for long life due to its load-carrying capacity, it might very well fail early because of damage due to high centripetal forces or rubbing speeds generated by the rotational velocity. To determine whether a particular bearing will operate satisfactorily at the speed N_{max} required in a particular device, multiply the value given for that bearing by Eq. (3.3) by the proper factor taken from Table 3.5. This table takes into account lubricant, retainer type, and ring rotation.

$$\text{Manufacturer's speed rating} \leq \frac{N_{max}}{f_n} \quad (3.3)$$

Note that the N_{max} value should be used for reference only. Higher speeds than specified can be achieved through the accuracy of the device's components.

3.11.3 Optimum Lubricant

Selection of the lubricant is extremely important. Many lubricants are available for varying conditions and requirements.

Unless torque is a problem, the selection of a grease is much preferred in prelubricating bearings since it is less susceptible to migration and leakage. Grease can multiply the inherent bearing torque by a factor of 1.2 to 5.0, depending on the type and quantity of grease in the bearing. Table 3.5 gives a partial listing of the most common greases.

3.11.4 Ball-Bearing Components

To assist in selecting the bearing with the proper components (Fig. 3.48) for a particular design or use, an exploded view of a standard ball bearing with component

TABLE 3.5 f_n Versus Cage Type, Lubricant Type, and Ring Rotation

			Ring rotation			
	Metal cage, 2-piece or crown type		Acetal cage			
			Crown type		Full-section type	
Lubricant	Inner	Outer	Inner	Outer	Inner	Outer
Petroleum oil	1.0	0.8	2.0	1.2	4.0	2.4
Synthetic oil	1.0	0.8	2.0	1.2	4.0	2.4
Silicone oil	0.8	0.7	0.8	0.7	0.8	0.7
Nonchanneling grease	1.0	0.6	1.6	1.0	1.6	1.0
Channeling grease	1.0	0.8	2.0	1.2	2.4	1.3
Silicone grease	0.8	0.7	0.8	0.7	0.8	0.7

callouts is shown in Fig. 3.49. The part numbering system is shown in Table 3.6. To further illustrate the relative positioning of these components in the ball-bearing assembly, a cross section is shown in Fig. 3.50.

Basic Dimension Data. The dimensions and their associated symbols are shown in Fig. 3.51 and defined here. These dimensions establish bearing size and other

TABLE 3.6 Part Numbering System

Group	Factor	Designation	Description
1	Material	DD	Stainless-steel material which falls within the 400 series martensitic stainless-steel grouping. No code = chrome alloy steel (52100 or equivalent).
2	Type	RIF RI	RI, R, L = radial RIF, RF, LF = flanged radial F = flanged, tapered OD

3	Basic size	418
		5532
			Inch series first—one or two digits indicate OD in 16ths of an inch. The following two or three digits indicate the bore size in fractions of an inch, the first digit being the numerator and the second or the second and third digits being the denominator.
			Metric series first—two digits indicate OD in mm. Second two digits indicate ID in mm.
			Special size series
			  ZB = integral shaft
			  AS – __ = pulley-type assemblies, shaft assemblies, mechanical parts, tape guides, special pivot type, special bearings
			  X __ = following basic size indicates special ball complement assigned in numerical sequence, i.e., X1, X2, etc.

4	Features	ZZEE	Enclosures
			  Z = single metallic shield, removable
			  ZZ = double metallic shield, removable
			  D = single rubber seal, contact
			  DD = double rubber seal, contact
			  L = single glass-reinforced PTFE seal, noncontact
			  LL = double glass-reinforced PTFE seal, noncontact
			  LZ = glass-reinforced PTFE seal and shield with seal on flange side
			  ZL = shield and glass-reinforced PTFE seal with shield on flange side
			  DZ = rubber seal and shield
			  SSD21 = labyrinth seal, noncontact
			  H = single metallic shield, nonremovable
			  HH = double metallic shield, nonremovable
			  S = single rubber seal, noncontact
			  SS = double rubber seal, noncontact
			Extended inner ring
			  EE = Both sides

TABLE 3.6 Part Numbering System (*Continued*)

Group	Factor	Designation	Description

5	Anderon meter test and special designs		Anderon meter test
			  MT = motor quality
			  GT = extremely quiet, HDD spindle motor only
			  No code = noncritical application
			Special design
			  SD = special design bearing

6	Cage	H	H = crown R = ribbon J = acetal crown type MN = glass-fiber-reinforced molded nylon M7 = molded nylon
7	ABEC tolerance	A7 A7	A1 = ABEC 1 A3 = ABEC 3 A5 = ABEC 5 A7 = ABEC 7 *Note:* A1 miniature and instrument bearings of both the metric and inch configurations meet the tolerances of ABMA Standard 20 for ABEC 1 metric-series bearings.
8	Radial play	P25 P25	P followed by two, three, or four numbers indicates the radial play limits in ten-thousandths of an inch. *Example:* P25 indicates radial play of 0.0002 to 0.0005 in.

9	Lubricant	LY75
		LO1
			Lubricant letter codes are followed by a number to indicate specific type.
			  LO = oil
			  LG = greases
			  LY = other oils and greases
			  LD = dry, no lubrication (DD material only)

10	Lube quantity	L	X = 5–10% L = 10–15% T = 15–20% No code = 25–35% H = 40–50% J = 50–60% F = 100% A = void volume

bearing parameters so that designers may choose the ball bearing most suited to their requirements.

The symbols shown in Fig. 3.51 and used throughout this section are defined as follows:

d = inside diameter or bore
D = outside diameter (OD)
B = inner ring width
C = outer ring width

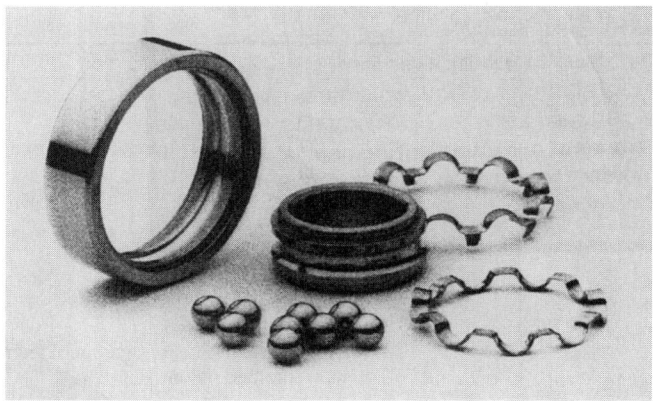

FIGURE 3.48 Ball bearing components.

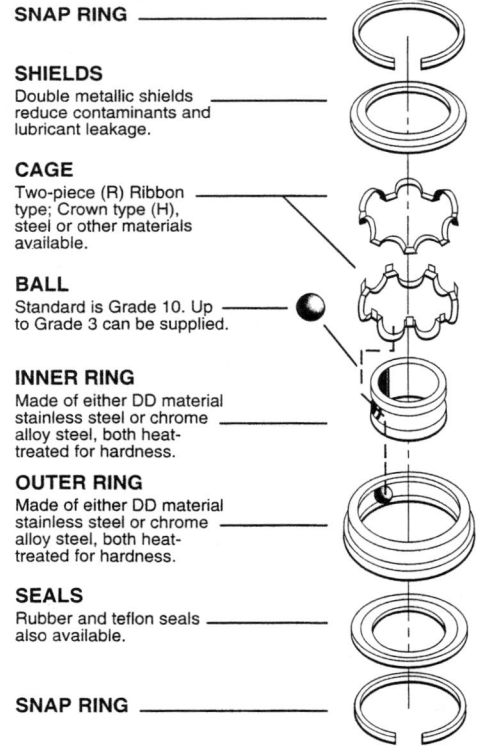

SNAP RING

SHIELDS
Double metallic shields reduce contaminants and lubricant leakage.

CAGE
Two-piece (R) Ribbon type; Crown type (H), steel or other materials available.

BALL
Standard is Grade 10. Up to Grade 3 can be supplied.

INNER RING
Made of either DD material stainless steel or chrome alloy steel, both heat-treated for hardness.

OUTER RING
Made of either DD material stainless steel or chrome alloy steel, both heat-treated for hardness.

SEALS
Rubber and teflon seals also available.

SNAP RING

FIGURE 3.49 Cross section of ball bearing.

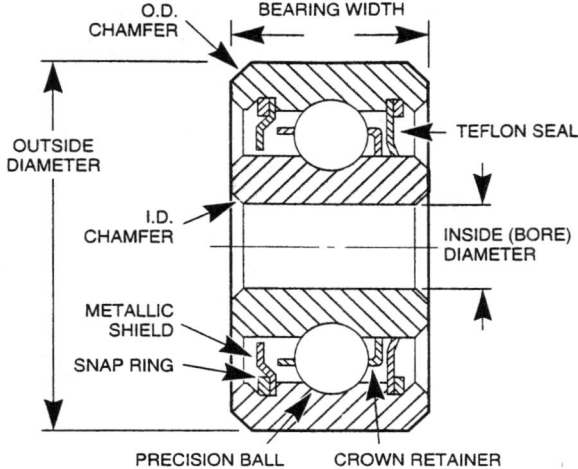

FIGURE 3.50 Bearing components.

D_f = flange outside diameter
B_f = flange width or thickness
L_i = inner ring reference diameter
L_o = outer ring reference diameter
r = maximum shaft of housing fillet radius that bearing corners will clear
Z = number of balls
D_W = nominal diameter of balls
N_{max} = maximum speed, rpm
f_n = cage and lubricant factor (See Table 3.5.)

3.11.5 Internal Bearing Geometry

When designing ball bearings for optimum performance, internal bearing geometry is a critical factor. For any given bearing load, internal stresses can be either high or low, depending on the geometric relationship between the balls and raceways inside the ball-bearing structure.

When a ball bearing is running under a load, force is transmitted from one bearing ring to the other through the ball set. Since the contact area between each ball and the rings is relatively small, even moderate loads can produce stresses of tens or even hundreds of thousands of pounds per square inch. Because internal stress levels have such an important effect on bearing life and performance, internal geometry must be carefully chosen for each application so bearing loads can be distributed properly.

Raceway, Track Diameter, and Track Radius. The *raceway* in a ball bearing is the circular groove formed in the outside surface of the inner ring and in the inside surface of the outer ring. When the rings are aligned, these grooves form a circular track that contains the ball set.

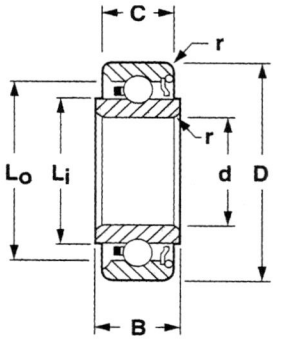

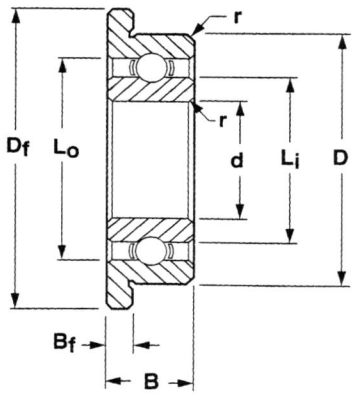

FIGURE 3.51 Bearing dimensions and symbols.

The track diameter and track radius are two dimensions that define the configuration of each raceway. *Track diameter* is the measurement of the diameter of the imaginary circle running around the deepest portion of the raceway, whether it be an inner or outer ring. This measurement is made along a line perpendicular to, and intersecting, the axis of rotation. *Track radius* describes the cross section of the arc formed by the raceway groove. It is measured when viewed in a direction perpendicular to the axis of the ring. In the context of ball-bearing terminology, track radius has no mathematical relationship to track diameter. The distinction between the two is shown in Fig. 3.52.

Radial and Axial Play. Most ball bearings are assembled in such a way that a slight amount of looseness exists between balls and raceways. This looseness is referred to as radial play and axial play. Specifically, *radial play* is the maximum distance that one bearing ring can be displaced with respect to the other, in a direction perpendicular to the bearing axis, when the bearing is in an unmounted state. *Axial play,* or *end play,* is the maximum relative displacement between the two rings of an unmounted ball bearing in the direction parallel to the bearing axis. Figure 3.53 illustrates these concepts.

Since radial play and axial play are both consequences of the same degree of looseness between the components in a ball bearing, they bear a mutual dependence. While this is true, both values are usually quite different in magnitude.

In most ball-bearing applications, radial play is functionally more critical than axial play. If axial play is determined to be an essential requirement, control can be obtained through manipulation of the radial-play specification.

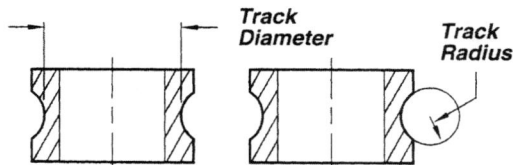

FIGURE 3.52 Distinction between track radius and track diameter.

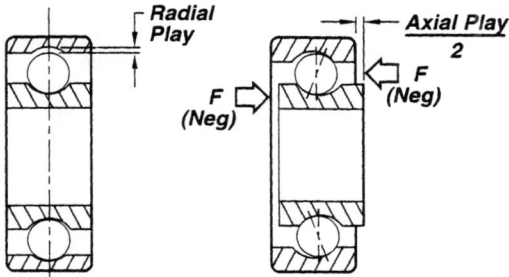

FIGURE 3.53 Distinction between radial and axial play.

Some general statements about radial play follow.

TABLE 3.7 Ball-Bearing Contact Angles

Ball size D_W	Radial-play code	
	P25	P58
0.025 in	18°	24½°
½ in (0.8 mm)	16½°	22°
1 mm	14½°	20°
¾₄ in	14°	18°
⅟₁₆ in	12°	16°
³⁄₃₂ in	9½°	13°
⅛ in	12½°	17°
⁵⁄₆₄ in	12°	16°
⁵⁄₃₂ in	11°	15°
³⁄₁₆ in	10°	14°

- The initial contact angle of the bearing is directly related to radial play—the higher the radial play, the higher the contact angle. Table 3.7 shows nominal contact angles, and Table 3.8 shows typical radial-play ranges. The contact angles in Table 3.7 are given for the mean radial play of the ranges shown—i.e., for P25 (0.0002 to 0.0005 in), the contact angle is given for 0.00035 in. Contact angle is affected by raceway curvature.

- For support of pure radial loads, a low level of radial play is desirable. Where thrust loading is predominant, higher radial-play levels are recommended.

- Radial play is affected by any interference fit between the shaft and bearing ID or between the housing and bearing OD.

Raceway Curvature. *Raceway curvature* is an expression that defines the relationship between the arc of the raceway's track radius and the arc formed by the slightly smaller ball that runs in the raceway. It is simply the track radius of the bearing raceway expressed as a percentage of the ball diameter. This number is a convenient index of fit between the raceway and ball. Figure 3.54 illustrates this relationship.

Track curvature values typically range from approximately 52 to 58 percent. The lower-percentage, tight-fitting curvatures are useful in applications where heavy loads are encountered. The higher-percentage, loose curvatures are more suitable

TABLE 3.8 Typical Radial-Play Ranges

Description	Radial-play range	NMB code
Tight	0.0001–0.0003 in	P13
Normal	0.0002–0.0005 in	P25
Loose	0.0005–0.0008 in	P58

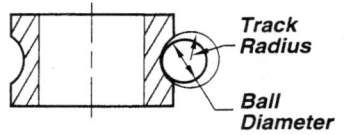

FIGURE 3.54 Relationship of track radius to ball diameter.

for torque-sensitive applications. Curvatures less than 52 percent are generally avoided because of excessive rolling friction that is caused by the tight conformity between the ball and raceway. Values above 58 percent are also avoided because of the high stress levels that can result from the small ball-to-raceway conformity at the contact area.

Contact Angle. The *contact angle* is the angle between a plane perpendicular to the ball-bearing axis and a line joining the two points where the ball makes contact with the inner and outer raceways. The contact angle of a ball bearing is determined by its free radial play-value, as well as its inner and outer track curvatures.

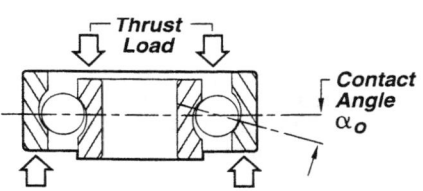

FIGURE 3.55 Contract angle for bearing loaded in pure thrust.

The contact angle of thrust-loaded bearings provides an indication of ball position inside the raceways. When a thrust load is applied to a ball bearing, the balls will move away from the median planes of the raceways and assume positions somewhere between the deepest portions of the raceways and their edges. Figure 3.55 illustrates the concept of contact angle by showing a cross-sectional view of a ball bearing that is loaded in pure thrust.

Free Angle and Angle of Misalignment. As a result of the previously described looseness, or play, which is purposely permitted to exist between the components of most ball bearings, the inner ring can be cocked or tilted a small amount with respect to the outer ring. This displacement is called the *free angle* of the bearing, and corresponds to the case of an unmounted bearing. The size of the free angle in a given ball bearing is determined by its radial play and track curvature values. Figure 3.56 illustrates this concept.

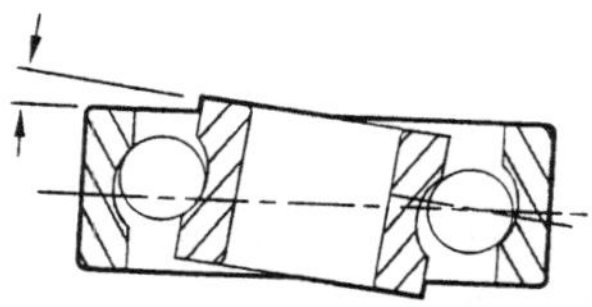

FIGURE 3.56 Free angle of bearing.

For the bearing mounted in an application, any misalignment present between the inner and outer rings (housing and shaft) is called the *angle of misalignment*. The misalignment capability of a bearing can have positive practical significance because it enables a ball bearing to accommodate small dimensional variations which may exist in associated shafts and housings. A maximum angle of misalignment of $\frac{1}{4}°$ is recommended before bearing life is reduced. Slightly larger angles can be accommodated, but bearing life will not be optimized.

3.11.6 Materials

Bearing Materials

Chrome steel. A bearing steel used for standard ball-bearing applications in uses and in environments where corrosion resistance is not a critical factor. The most commonly used ball-bearing steel in such applications is AISI 52100 or its equivalent. Due to its structure, this is the material chosen for extremely noise-sensitive applications.

DD400 0.7% C; 13% Cr. A 400-series martensitic stainless steel combined with a heat-treating process was exclusively developed for use in miniature and instrument bearings. Bearings manufactured from DD meet the performance specifications of such bearings using AISI 440C martensitic stainless steel, and it is equal or superior in hardness, superior in low-noise characteristics, and at least equivalent in corrosion resistance. These material characteristic advantages make for lower torque, smoother running, and longer-life bearings.

Cages and Retainer Materials. The retainer, also referred to as the *cage* or *separator,* is the component part of a ball bearing that separates and positions the balls at approximately equal intervals around the bearing's raceway. There are two basic types: the crown or open-end design and the closed ball-pocket design. The most common retainer is the two-piece closed retainer, commonly called a *ribbon retainer.*

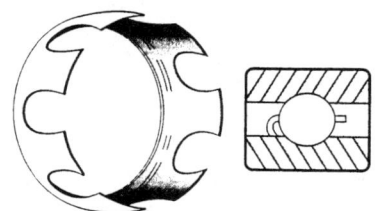

FIGURE 3.57 Standard one-piece crown retainer.

The open-end design, or crown retainer, as shown in Fig. 3.57, is of metal material. Crown retainers manufactured from molded plastics are available for some sizes. The metal retainer, constructed of hardened stainless steel, is very lightweight and has coined ball pockets which present a hard, smooth, low-friction contact surface.

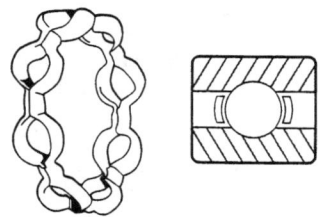

FIGURE 3.58 Metal two-piece closed-pocket ribbon retainer.

The closed-pocket design (two-piece construction) with clinching tabs, as outlined in Fig. 3.58, is a standard design for most miniature and instrument-sized ball bearings. The use of loosely clinched tabs is favorable for starting torque, and the closed-pocket design provides good durability required for various applications.

Shields and seals are necessary to provide optimum ball-bearing life by retaining lubricants and preventing contaminants from reaching central work surfaces. Different types of closures can be supplied on the same bearing, and nearly all are removable and replace-

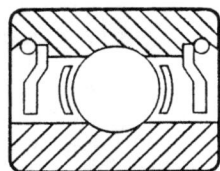

FIGURE 3.59 Two Z-type shields (removable).

able. They are manufactured with the same care and precision that goes into the ball bearings. The following are descriptions of the most common types of shields and seals available. Z- and H-type shields designate noncontacting metal shields. Z-type shields (Fig. 3.59) are the simplest form of closure and, for most bearings, are removable. H-type shields (Fig. 3.60) are similar to Z-types but are not removable.

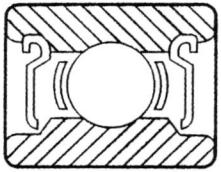

FIGURE 3.60 Two H-type shields (nonremovable).

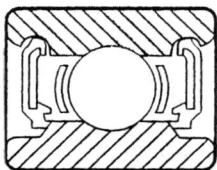

FIGURE 3.61 Two D-type seals (contacting rubber).

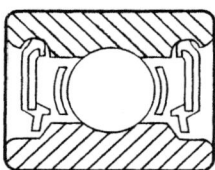

FIGURE 3.62 Two S-type seals (noncontacting rubber).

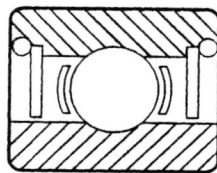

FIGURE 3.63 Two L-type seals (nonflexed Teflon).

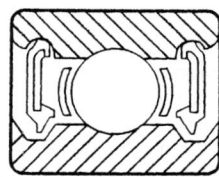

FIGURE 3.64 Two SSD21-type seals (labyrinth-design seal).

It is advantageous to use shields rather than seals in some applications because there are no interacting surfaces to create drag. This results in no appreciable increase in torque or speed limitations, and operation can be compared to that of open ball bearings. D-type contacting seals (Fig. 3.61) consist of a molded Buna-N rubber lip seal with an integral steel insert. While this closure type provides excellent sealing characteristics, several factors must be considered for its application. The material normally used on this seal has a maximum continuous operating temperature limit of 250°F (121°C). Although it is impervious to many oils and greases, consideration must be given to lubrication selection. It is also capable of providing a better seal than most other types by increasing the seal lip pressure against the inner ring OD. This can result in a higher bearing torque than with other types of seals and may cause undesirable seal lip heat buildup in high-speed applications. S-type noncontacting seals are constructed in the same fashion as the D-type seals. This closure type has the same temperature limitation of 250°F (121°C). It also is impervious to many oils and greases, but the same considerations should be noted on lubrication selection. The S-type seal (Fig. 3.62) is uniquely designed to avoid contact on the inner land, significantly reducing torque over the D-type configuration.

L-type seals (Fig. 3.63) are fabricated from glass-reinforced Teflon. When assembled, a very small gap exists between the seal lip and the inner ring OD. It is common for some contact to occur between these components, resulting in an operating torque increase. The nature of the seal material serves to keep this torque increase to a minimum. In addition, the use of this material allows high operating temperatures with this configuration.

The SSD21-type seals (Fig. 3.64) have the same operating characteristics as the D- and S-type seals, resulting in the same considerations of temperature limitation and lubricant selection. The SSD21-type seal is comprised of a noncontacting rubber seal combined with a labyrinth-design inner ring. The labyrinth-design configuration creates an extended path to the raceway, minimizing the tendency for contaminants to creep into the ball bearing.

3.11.7 Lubrication

Lubricant Types. *Oil* is the basic lubricant for ball bearings. Previously, most lubricating oil was refined from petroleum. Today, however, synthetic oils such as diesters, silicone polymers, and fluorinated compounds have found acceptance because of improvements in properties. Compared to petroleum-based oils, diesters in general have better low-temperature properties, lower volatility, and better temperature/viscosity characteristics. Silicones and fluorinated compounds possess even lower volatility and wider temperature/viscosity properties.

Virtually all petroleum and diester oils contain additives that limit chemical changes, protect the metal from corrosion, and improve physical properties.

Grease is an oil to which a thickener has been added to prevent oil migration from the lubrication site. It is used in situations where frequent replenishment of the lubricant is undesirable or impossible. All of the oil types mentioned in the next subsection can be used as grease bases to which are added metallic soaps, synthetic fillers, and thickeners. The operative properties of grease depend almost wholly on the base oil. Other factors being equal, the use of grease rather than oil results in higher starting and running torque and can limit the bearing to lower speeds.

Oils and Base Fluids. *Petroleum lubricants* have excellent load-carrying abilities, but are usable only at moderate temperature ranges [−25 to 250°F (−32 to 121°C)]. Greases that use petroleum oils for bases have a high dN capability. Greases of this type are recommended for use at moderate temperatures, light to heavy loads, and moderate to high speeds.

While *superrefined petroleum lubricants* are usable at higher temperatures than petroleum oils [−65 to 350°F (−54 to 177°C)], they still exhibit the same excellent load-carrying capacity. This further refinement eliminates unwanted properties, leaving only the desired chemical chains. Additives are introduced to increase the oxidation resistance, etc.

The *diesters* are probably the most common synthetic lubricants. They do not have the film-strength capacity of petroleum products, but do have a wide temperature range [−65 to 350°F (−54 to 177°C)] and are oxidation resistant.

Synthetic hydrocarbons are finding a greater use in the miniature and instrument ball-bearing industry because they have proved to be a superior general-purpose lubricant.

Silicone lubricants are useful over a wide temperature range [−100 to 400°F (−73 to 204°C)], but do not have the film strength of petroleum types and other synthetics. It has become customary in the instrument and miniature bearing industry, in recent years, to derate the dynamic load rating C_r of a bearing to one-third of its normal value if a silicone product is used.

Perfluorinated polyether oils and greases have found wide use where high temperature stability and/or chemical inertness are required. This specialty lubricant does not have the film strength of petroleum or diester products. However, it does have better film strength than silicone lubricants.

Lubrication Methods. Grease packing to approximately one-quarter to one-third of a ball bearing's free volume is one of the most common methods of lubrication. Volumes can be controlled to a fraction of a percent for precision applications by special lubricators. In some instances, people have used bearings that were to be lubricated 100 percent full of grease. Excessive grease is as detrimental to a bearing as insufficient grease. It causes shearing, heat buildup, and deterioration through constant churning which can ultimately result in bearing failure.

Centrifuging an oil-lubricated bearing removes excess oil and leaves only a very thin film on all surfaces. This method is used on very low torque bearings and can be specified for critical applications.

Operating Speed. When petroleum or synthetic ester oils are used, the maximum speed N_{max} is dictated by the ball cage material and design or the centripetal ball loads rather than by the lubricant.

For speed-limit values N_{max}, the N_{max}/f_n values shown in product listings must be multiplied by the f_n values shown in Table 3.5.

The following method may be used to select a lubricant.

Step 1. Define the temperature range of the application, including the environmental temperature plus any heat rise from motors, etc. Refer to Table 3.9 and select the proper lubricant base for the maximum and minimum operating temperature.

TABLE 3.9 Relationship Between Lubricants, dN Values, and Temperature Ranges

Type	dN	Temperature range
Silicone	200,000	−100 to 400°F
		(−73 to 204°C)
Diester	400,000	−65 to 350°F
		(−54 to 177°C)
Petroleum	600,000	−25 to 250°F
		(−32 to 121°C)

When selecting a base fluid type, the fluid with the greatest film support is the preferred choice. Refer to the description of lubricant types for individual capabilities.

Step 2. Determine the speed of the bearing and calculate the dN value (see next subsection, Speed Factor). Select the lubricant type that will operate within the dN speed factor, referring to Table 3.9.

Step 3. Knowing the dN value, determine the proper viscosity of the lubricating oil or the base oil of the grease (Fig. 3.65). Since grease is approximately 80 percent oil, it is necessary to determine the viscosity of the oil for any high-speed application. Improper selection can result in rapid deterioration of the base oil and failure of the unit.

Step 4. Once you have determined these factors, the lubricant selection has been narrowed to the type of base oil, the operating temperature, and the oil viscosity range for a particular dN value (see next subsection, Speed Factor). Next, determine whether a grease or oil is needed for the application. Then, individual lubricants should be examined to determine their suitability for the application.

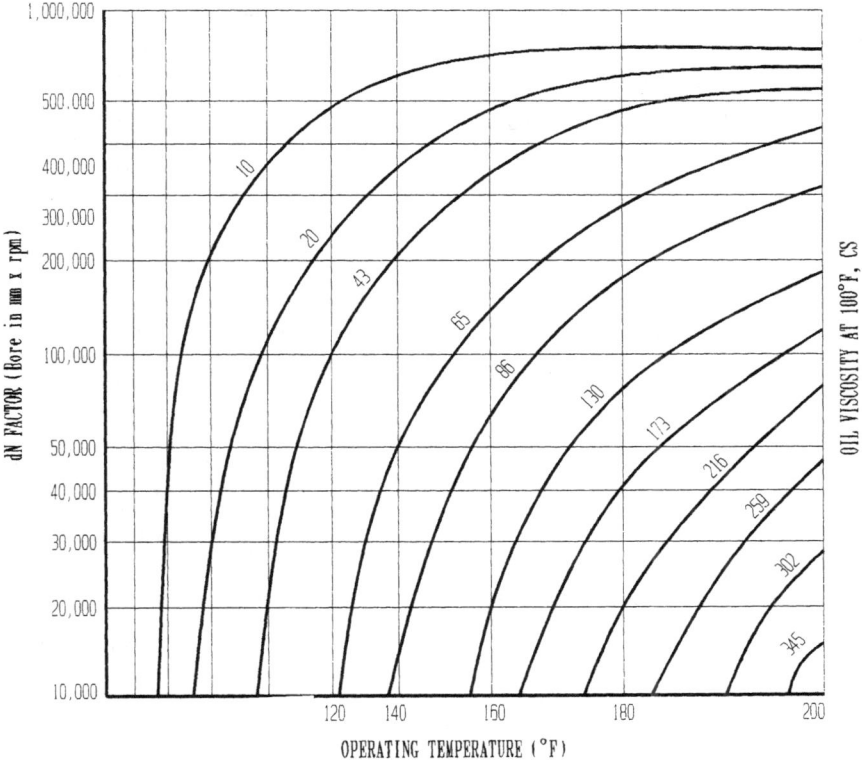

FIGURE 3.65 Speed factor. (*Courtesy of Exxon.*)

Speed Factor. The maximum usable operating speed of a grease lubricant is dependent on the type of oil. The speed factor is a function of the bore of the bearing d, mm, and the speed of the bearing N, rpm:

$$\text{speed factor} = d \times N = dN \tag{3.4}$$

There are many lubricants available for ball bearings. Refer to Table 3.10.

Dynamic Load Ratings and Fatigue Life

Dynamic Radial Load Rating. The dynamic radial load rating C_r for a radial ball bearing is a calculated, constant radial load which a group of identical bearings can theoretically endure for a rating life of 1 million revolutions. The dynamic radial load rating is a reference value only. The base rating-life value of 1 million revolutions has been chosen for ease of calculation. Since applied loading equal to the basic load rating tends to cause permanent deformation of the rolling surfaces, such excessive loading is not normally applied. Typically, a radial load that corresponds to 15 percent or more of the dynamic radial load rating is considered heavy loading for a ball bearing. In cases where loading of this degree is required, consult a bearing manufacturer's application engineer for information regarding bearing life and lubricant recommendations.

TABLE 3.10 Commonly Used Lubricant Types

Code	Brand name	Basic oil type	Operating temperature	Uses
LO1	Windsor L245X (MIL-L-6085A)	Ester oil	−60 to +250°F (−51 to 121°C)	Low-torque, low-speed instrument oil; rust preventative
LG20	Exxon Beacon 325	Channeling grease: mineral oil and sodium soap thickener	−60 to +250°F (−51 to 121°C)	General-purpose grease.
LG39	Exxon Andok C	Channeling grease: mineral oil and sodium soap thickener	−20 to +250°F (−29 to 121°C)	Low migration; general office equipment applications
LY48	Mobil 28 (MIL-G-81322)	Synthetic oil and clay thickener	−65 to +350°F (−54 to 177°C)	Good heat resistance with low torque; throttle body applications
LY75	Chevron SRI-2	Mineral oil and urea soap thickener	−20 to +350°F (−29 to 177°C)	Good heat resistance, high-speed grease; power tool and vacuum cleaner motor applications
LY83	Shell Alvania X2	Mineral oil and lithium soap thickener	−30 to 250°F (−29 to 121°C)	Long-life, general-use grease; power tool applications
LY121	Kyodo Multemp SRL	Ester oil and lithium soap thickener	40 to 300°F (3 to 149°C)	Low-noise, smooth-running grease; general motor applications

Rating Life. The rating life L_{10} of a group of apparently identical ball bearings is the life in millions of revolutions, or number of hours, that 90 percent of the group will complete or exceed. For a single bearing, L_{10} also refers to the life associated with 90 percent reliability. The median life L_{50}, the life which 50 percent of the group of ball bearings will complete or exceed, is usually not greater than 5 times the rating life.

Rating life is calculated as follows:

$$L_{10} = \left(\frac{C_r}{P_r} \right)^3$$

where L_{10} = rating life
C_r = dynamic radial load rating, kgf
P_r = dynamic equivalent radial load, kgf

The dynamic radial load rating C_r can be found from product listings. The dynamic equivalent load must be calculated according to the following procedure:

$$P_r = XF_r + YF_a \tag{3.5}$$

where X and Y are obtained from Table 3.11
F_r = radial load on the bearing during operation, kgf
F_a = axial load on the bearing during operation, kgf

TABLE 3.11 Axial Load Variables

Relative axial load, $F_a/(Z \times D_w^2)$	$F_a/F_r \leq e$		$F_a/F_r \geq e$		
	X	Y	X	Y	e
0.0175	1	0	0.56	2.30	0.19
0.0352	1	0	0.56	1.99	0.22
0.0703	1	0	0.56	1.71	0.26
0.105	1	0	0.56	1.55	0.28
0.143	1	0	0.56	1.45	0.30
0.211	1	0	0.56	1.31	0.34
0.352	1	0	0.56	1.15	0.38
0.527	1	0	0.56	1.04	0.42
0.703	1	0	0.56	1.00	0.44

Note: Z = number of balls
D_W = ball size, mm

The L_{10} life can be converted from millions of revolutions to hours using the rotation speed. This can be done as follows:

$$L_{10}, \text{ millions of revolutions} \times \frac{1{,}000{,}000}{\text{rpm} \times 60} = L_{10}, \text{ hours} \tag{3.6}$$

To convert pounds to kilograms force, divide by 0.45359:

$$\text{kgf} = \frac{\text{lb}}{0.4359 \text{ kgf/lb}}$$

Life Modifiers. For most cases, the L_{10} life obtained from the equation discussed previously will be satisfactory as a bearing performance criterion. However, for particular applications, it might be desirable to consider life calculations for different reliabilities and/or special bearing properties and operating conditions. Reliability adjustment factors, bearing material adjustment, and special operating conditions are discussed in the following subsections.

Bearing Material. Manufacturers recommend that radial load ratings published for chrome steel be reduced by 20 percent for stainless steel. This is a conservative approach to ensure that bearing capacity is not exceeded under the most adverse conditions. This is incorporated in the a_2 modifier, as shown in Table 3.12.

Reliability Modifier. Where a more conservative approach than conventional rating life L_{10} is desired, the American Bearing Manufacturers Association (ABMA) offers a means for such estimates. Table 3.12 provides selected modifiers a_2 for calculating failure rates down to 1 percent (L_1).

TABLE 3.12 Reliability Versus Material Life Modifier a_2

Required reliability, %	L_n	Value of a_2	
		Chrome	DD
90	L_{10}	1.00	0.50
95	L_5	0.62	0.31
96	L_4	0.53	0.27
97	L_3	0.44	0.22
98	L_2	0.33	0.17
99	L_1	0.21	0.11

Other Life Adjustments. The conventional rating life often has to be modified as a consequence of application abnormalities, whether they be intentional or unknown. Seldom are loads ideally applied. The following conditions all have the practical effect of modifying the ideal, theoretical rating life L_{10}.

- Vibration and/or shock-impact loads
- Angular misalignment
- High-speed effects
- Operation at elevated temperatures
- Fits
- Internal design

Oscillatory Service Life. Frequently, ball bearings do not operate with one ring rotating unidirectionally. Instead, they execute a partial revolution, reverse motion, and then repeat this cycle, most often in a uniform manner. Efforts to forecast a reliable fatigue life by simply relating oscillation rate to an "equivalent" rotational speed are invalid. The actual fatigue life of bearings operating in the oscillatory mode is governed by four factors: applied load, angle of oscillation, rate of oscillation, and lubricant.

Lubricant Life. In many instances a bearing's effective life is governed by the lubricant's life. This is usually the case where applications involve very light loads and/or very slow speeds.

With light loads and/or slow speeds, the conventional fatigue-life forecast will be unrealistically high. The lubricant's ability to provide sufficient film strength is sustained only for a limited time. This is governed by the following factors:

- Quality and Quantity of the lubricant in the bearing
- Environmental conditions (i.e., ambient temperature, area cleanliness)
- The load-speed cycle

3.11.8 Static Capacity

Static Radial Load Rating. The static radial load rating C_{or} is the radial load which a nonrotating ball bearing will support without damage, continuing to provide satisfactory performance and life.

The static radial load rating is dependent on the maximum contact stress between the balls and either of the two raceways. The load ratings shown were calculated in accordance with the ABMA standard. The ABMA has established the maximum acceptable stress level resulting from a pure radial load in a static condition to be 4.2 GPa (609,000 lb/in^2).

Static Axial Load Capacity. The static axial load capacity is the axial load which a nonrotating ball bearing will support without damage. The axial static load capacity varies with bearing size, bearing material, and radial play.

Radial static load ratings and thrust static load ratings in excess of published C_{or} values have practical applications where smoothness of operation and/or low noise are not of concern. Properly manufactured ball bearings, when used under controlled shaft and housing fitting practices, can sustain significantly greater permanent deformation, such as brinells, than deformations associated with normal static load ratings.

3.11.9 Preloading

Ball-bearing systems are preloaded for the following reasons:

- To eliminate radial and axial looseness
- To reduce operating noise by stabilizing the rotating mass
- To control the axial and radial location of the rotating mass and to control movement of this mass due to external force influences
- To reduce the repetitive and nonrepetitive runout of the rotational axis
- To reduce the possibility of damage due to vibratory loading
- To increase stiffness

Spindle motors and tape guides are examples of applications where preloaded bearings are used to accurately control shaft position when external loads are applied. As the name implies, a preloaded assembly is one in which a bearing load (normally a thrust load) is applied to the system so the bearings are already carrying a load before any external load is applied. There are essentially two ways to preload a ball-bearing system, by using a spring or by using a solid stack of parts.

Spring Preloading. For many applications, one of the simplest and most effective methods of applying a preload is by means of a spring. This can consist of a coil spring or a wavy washer which applies a force against the inner or outer ring of one of the bearings in an assembly.

When a spring is used, it is normally located on the nonrotating component; i.e., with shaft rotation, the spring should be located in the housing against the outer rings. Springs can be very effective when differential thermal expansion is a problem. In the spindle assembly shown in Fig. 3.66, when the shaft becomes very hot and

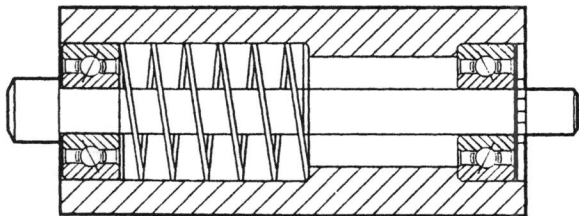

FIGURE 3.66 Spindle assembly using compression coil spring, with shaft rotation.

expands in length, the spring will move the outer ring of the left bearing and thus maintain system preload. Care must be taken to allow for enough spring movement to accommodate the potential shaft expansion.

Since, in a spring, the load is fairly consistent over a wide range of compressed length, the use of a spring for preloading negates the necessity for holding tight location tolerances on machined parts. For example, retaining rings can be used in the spindle assembly, thus saving the cost of locating shoulders, shims, or threaded members.

Normally, a spring preload would not be used where the assembly is required to withstand reversing thrust loads.

Solid-Stack Preloading. When precise location control is required, as in a precision motor (Fig. 3.67), or a flanged tape guide (Fig. 3.68), a solid preloading system is indicated. A solid-stack "hard" or "rigid" preload can be achieved in a variety of ways. Theoretically, it is possible to preload an assembly by tightening a screw, as shown in Fig. 3.68, or inserting shims, as shown in Fig. 3.69, to obtain the desired rigidity. It should be noted that care must be taken when using a solid-stack preloading system with miniature and instrument bearings. Overload of the bearings must be avoided so that the bearings are not damaged during this process.

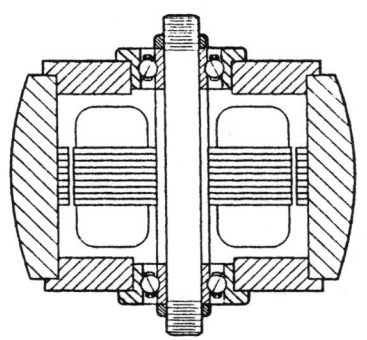

FIGURE 3.67 Rotor outer-ring spacer, with stator mount as inner ring.

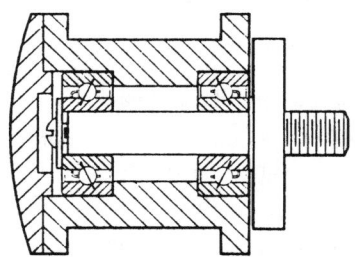

FIGURE 3.68 Typical tape-guide design using screw and washer for solid preloading by clamping inner rings, with outer-ring rotation.

Preload Levels. Preloading is an effective means of positioning and controlling stiffness because of the nature of the ball–raceway contact. Under light loads, the ball–raceway contact area is very small, and so the amount of yield or *definition* is substantial with respect to the amount of load. As the load is increased, the ball–raceway contact area increases in size (the contact is in the shape of an ellipse) and so provides increased stiffness or reduced yield per unit of applied load.

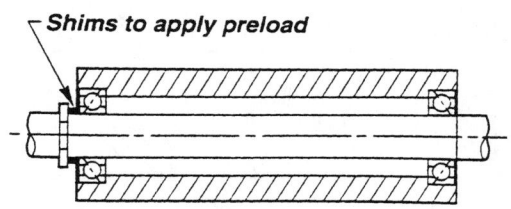

FIGURE 3.69 Shims to apply preload.

This is illustrated in the single-bearing deflection curve shown in Fig. 3.70. When two bearings are preloaded together and subjected to an external thrust load, the axial-yield rate for the pair is drastically reduced because of the preload and the interaction of the forces exerted by the external load and the reactions of the two bearings. As can be seen by the lower curve in Fig. 3.70, the yield rate for the preloaded pair is essentially linear.

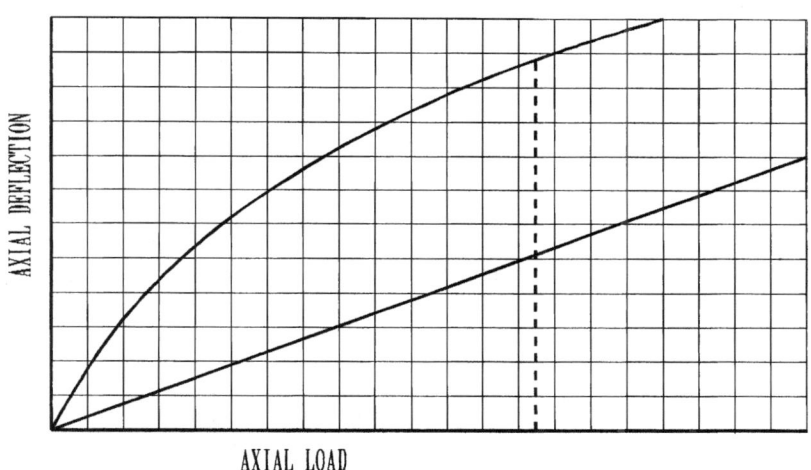

FIGURE 3.70 Single-bearing deflection curve.

Miniature and instrument bearings are typically built to accept light preloads normally ranging from 0.25 lb to not more than 10 lb.

3.11.10 Assembly and Fitting Procedure

The operating characteristics of a system can be drastically affected by the way in which the ball bearings are handled and mounted. A bearing which has been damaged due to excessive force or shock loading during assembly, or which is fitted too tight or too loose, may cause the device to perform in a substandard manner.

By following a few general guidelines during the design of mating parts and by observing some basic cautions in the assembly process, the possibility of producing malfunctioning devices can be considerably reduced.

Table 3.13 lists recommended fits for most normal situations. There are four cautions which must be observed.

1. When establishing shaft or housing sizes, the effect of differential thermal expansion must be accounted for. Table 3.13 assumes stable operating conditions, so if thermal gradients are known to be present or dissimilar materials are being used, the room temperature fits must be adjusted so that the proper fit is attained at operating temperature.

TABLE 3.13 Recommended Fits

Typical application	Shaft fit*	Shaft diameter	Housing fit*	Housing diameter
General application—inner ring rotation (inner ring press fit, outer ring loose fit)	0.0000–0.0004T	$d + 0.0000$ $d + 0.0002$	0.0000–0.0004L	$D + 0.0002$ $D + 0.0000$
General application—outer ring rotation (inner ring loose fit, outer ring press fit)	0.0000–0.0004L	$d - 0.0002$ $d - 0.0004$	0.0000–0.0004T	$D - 0.0002$ $D - 0.0004$
Tape-guide roller	0.0000–0.0004L	$d - 0.0002$ $d - 0.0004$	0.0001L–0.0003T	$D - 0.0001$ $D - 0.0003$
Drive motor (spring preload)	0.0001T–0.0003L	$d - 0.0001$ $d - 0.0003$	0.0000–0.0004L	$D + 0.0002$ $D - 0.0000$
Precision synchro or servo	0.0000–0.0002L	$d - 0.0001$ $d - 0.0003$	0.0000–0.0002L	$D + 0.0001$ $D - 0.0001$
Potentiometer	0.0001T–0.0003L	$d - 0.0001$ $d - 0.0003$	0.0000–0.0004L	$D + 0.0002$ $D - 0.0000$
Encoder spindle	0.0000–0.0002L	$d - 0.0001$ $d - 0.0003$	0.0000–0.0002T	$D - 0.0001$ $D - 0.0003$
Gear reducer	0.0000–0.0004L	$d - 0.0002$ $d - 0.0004$	0.0000–0.0004L	$D + 0.0002$ $D - 0.0000$
Light-duty mechanism	0.0000–0.0004L	$d - 0.0002$ $d - 0.0004$	0.0000–0.0004L	$D + 0.0002$ $D - 0.0000$
Clutches, brakes (inner race floats)	0.0000–0.0004L	$d - 0.0002$ $d - 0.0004$	0.0001T–0.0003L	$D + 0.0001$ $D - 0.0001$
Pulleys, rollers, cam followers (outer race rotates)	0.0000–0.0004L	$d - 0.0002$ $d - 0.0004$	0.0000–0.0004T	$D - 0.0002$ $D - 0.0004$

* T = tight fit; L = loose fit.

2. When miniature and instrument ball bearings are interference-fitted (either intentionally or as a result of thermal gradients), the bearing radial play can be estimated to be reduced by an amount equal to 80 percent of the actual diametrical interference fit. This 80 percent figure is conservative, but is of good use for design purposes. Depending on the materials involved, this factor will typically range from 50 to 80 percent. The following is an example of calculating loss of radial play:

Radial play of bearing:	0.0002 in
Total interference fit:	0.0003 in (tight)
80 percent of interference fit (0.0003 in × 80%):	0.00024 in
Theoretical resultant radial play of bearing	0.00004 in (tight)

Theoretically, this bearing could be operating with negative radial play. A bearing operated in an excessive negative radial-play condition will perform with reduced life. However, the preceding calculation is for design only, and does not take into account housing material, shaft material, or surface finish of the housing or shaft surfaces. As an example, if the finish of the shaft surface ring and shaft will be absorbed by the deformation of the shaft surface, this will serve to reduce the overall interference fit. Thus, the radial play of the bearing will not be reduced as much as is shown in the preceding calculation.

Table 3.13 is based on the use of bearings of ABEC 5 or better tolerance level.

3. If the outer or inner ring face is to be clamped or abutted against a shoulder, care must be taken to make sure that this shoulder configuration provides a good mounting surface:
   - The shoulder face must be perpendicular to the bearing mounting seat. The maximum recommended permissible angle of misalignment is ¼°.
   - The corner between the mounting diameter and the face must have an undercut or a fillet radius r no larger than that shown in Fig. 3.51.
   - The shoulder diameter must meet the requirements shown in Table 3.14.

4. Assembly technique is extremely critical. After the design is finalized and assembly procedures are being formulated, the bearing static capacity C_{or} becomes extremely important. It is easy, for instance, to exceed the 3-lb capacity of a DDRI-2 during assembly. After assembly to the shaft, damage can be done either

TABLE 3.14 Recommended Shoulder Diameter

Basic size	Minimum shaft shoulder diameter, in	Maximum housing shoulder diameter, in
DDRI-2	0.060	0.105
DDRI-2½	0.071	0.132
DDRI-3	0.079	0.164
DDRI-4	0.102	0.226
DDRI-3332	0.114	0.168
DDRI-5	0.122	0.284
DDRI-418	0.148	0.226
DDRI-518	0.153	0.284
DDRI-618	0.153	0.347
DDR-2	0.179	0.325
DDRI-5532	0.180	0.288
DDR-1640	0.210	0.580
DDRI-5632	0.210	0.288
DDRI-6632	0.216	0.347
DDR-3	0.244	0.446
DDR-1650	0.250	0.580
DDR-1950	0.250	0.700
DDR-1960	0.290	0.700
DDRI-614	0.272	0.352
DDRI-814	0.284	0.466
DDR-4	0.310	0.565
DDRI-1214	0.322	0.678
DDR-2270	0.325	0.810
DDR-2280	0.370	0.810
DDRI-8516	0.347	0.466
DDRI-1038	0.435	0.565
DDRI-1438	0.451	0.799
DDRI-1212	0.560	0.690
DDRI-1458	0.665	0.835
DDRI-1634	0.790	0.960

by direct pressure or by moment load while the bearing and shaft subassembly is being forced into a tight housing. A few simple calculations will underscore this point.

Adequate fixturing should always be provided for handling and assembling precision bearings. This fixturing must be designed so that when assembling the bearing to the shaft, force is applied only to the inner ring, and when assembling into the housing, force is applied only to the outer ring. Further, the fixturing must preclude the application of any moment or shock loads which would be transmitted through the bearing. Careful attention to this assembly phase of the total design effort can prevent many problems and provide savings when production starts.

3.12 SLEEVE BEARINGS*

There are two common types of sleeve bearings. The rigid steel-backed babbit bearing and the self-aligning sintered bearing are shown in Figs. 3.71a and 3.71b, respectively. *Babbit* is a soft alloy of tin and lead that has a low coefficient of friction. It is backed by a thin steel material to give the bearing rigidity. A wick notch and oil groove are normally formed in the bearing when it is manufactured.

Rigid bearings are pressed into the end frame and burnished or machined to size. An oil reservoir surrounds the bearing. If the bearing is a sintered bronze or sintered iron type, the oil wicks through the pores in the bearing by capillary action and provides lubricant to the shaft. In the case of steel-backed babbitt bearings, a wick of some type must be placed in a hole in the bearing so that oil from the reservoir can reach the shaft.

Self-aligning sintered bearings are held against the end frame by a retaining spring. Oil wicks from the oil reservoir through the pores in the material as in the rigid sintered bearings. Ball bearings are used where it is necessary to limit radial shaft play or where high side loads are expected in the application. These are the most costly bearings. Rigid bearings are used where moderate side loads are expected. Self-aligning bearings are used where light side loads are expected and starting friction has to be minimized.

The self-aligning bearing is the lowest-cost device, but it also carries the lowest load. The rigid sleeve bearing costs somewhat more than the self-aligning bearing, but it can carry a heavier load. Ball bearings can carry a much heavier load, but they also cost 5 to 20 times more than a sleeve bearing.

The most common shaft and bearing system consists of a steel shaft in a sintered bronze bearing. As a general rule, the harder the shaft, the longer the life of the system. Generally, Rockwell C scale 35 to 55 is a good range. Stainless steels are acceptable, except that the 300 series should be avoided with bronze bearings because excessive wear will result. When using bronze bearings, the load speed and shaft-to-bearing clearance must be considered. The pressure velocity factor PV must be calculated to ensure that it is within the rating of the material selected. Bronze is rated at $PV = 50,000$.

$$PV = \frac{w}{LD} \frac{\pi DN}{12} \qquad (3.7)$$

---

* Section contributed by William H. Yeadon, Yeadon Engineering Services, PC

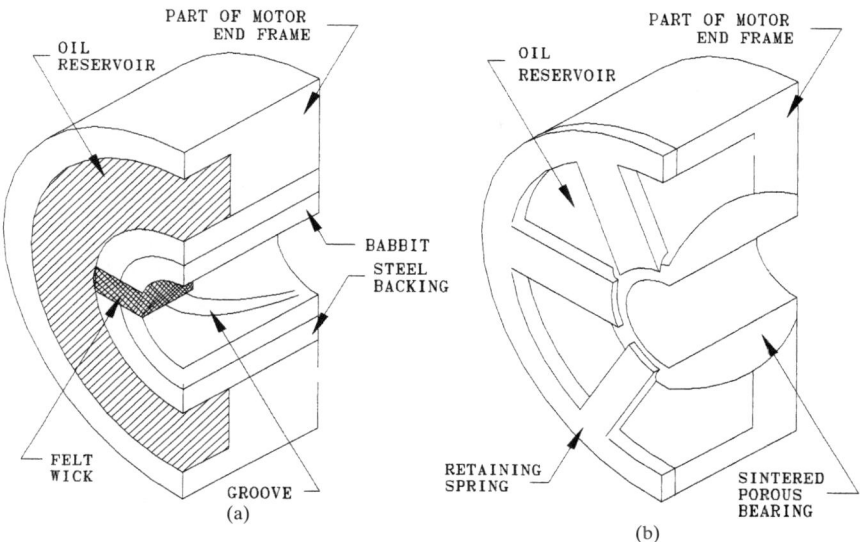

FIGURE 3.71 Sleeve bearings: (*a*) rigid and (*b*) self-aligning.

where P = load, lb/in^2
V = surface velocity of the shaft, ft/min
w = total bearing load, lb
L = bearing length, in
N = shaft speed, rpm
D = bearing ID, in

Additional lubrication is advised in order to reduce the coefficient of friction. These bearings may close down when installed and may need to be resized to maintain a shaft-to-bearing clearance between 0.0005 and 0.0015 in. The actual tolerance on bearing clearance will depend on the shaft diameter and the type of lubrication used. Sintered bearings should be resized with a hard, polished sizing pin or burnished roller to avoid closing down the pores that provide the lubrication to the bearing–shaft interface. There is a tendency to want to size the bearings with a reamer. This practice should be avoided, because the reamer cuts away the material and smears it across the pores. This results in closed pores and starves the shaft of lubrication. Shaft finish should be 16 μin or better to reduce wear and avoid pumping oil out of the bearing.

3.12.1 Lubrication

Lubricants are used in bearing systems to reduce the friction between surfaces sliding relative to each other. In the case of the electric motor, the shaft moves relative to the bearing, riding on a film of lubricant (oil), as shown in Fig. 3.72.

The velocity of the oil varies across the thickness of the film, as shown in Fig. 3.73. Near the stationary part of the bearing, point *B*, the velocity is zero. At the surface of the shaft, point *A*, it is equal to the shaft speed. The lubricant is thought to move in layers across its thickness.

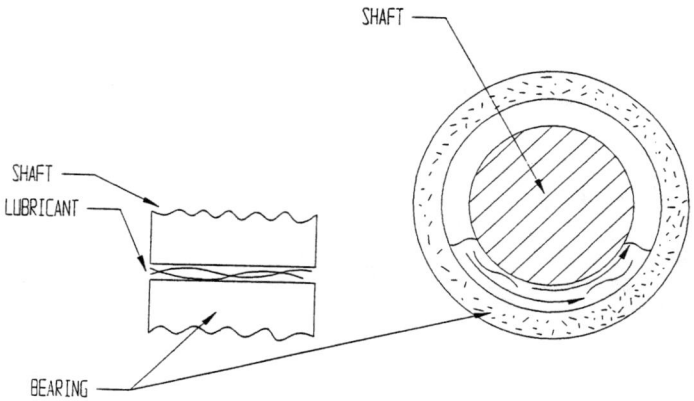

FIGURE 3.72 Hydrodynamic lubrication.

As the shaft rotates, a stress τ is developed in the layers of oil. This stress is known as the *shearing stress* and is measured in pounds per square foot or similar units. It is equal to the change in velocity $\Delta\upsilon$ of the liquid over its thickness $\Delta\gamma$ times the dynamic viscosity μ of the oil.

$$\tau = \mu \frac{\Delta\upsilon}{\Delta\gamma} \therefore \mu = \tau \frac{\Delta\gamma}{\Delta\upsilon} \tag{3.8}$$

Viscosity is defined as the property of a fluid which offers resistance to the relative motion of fluid molecules. The kinematic viscosity v is the ratio of dynamic viscosity to fluid density. It is measured in square feet per second or square centimeters per second (strokes) or saybolt universal seconds.

The *viscosity index* (VI) is a measure of how greatly the viscosity changes with temperature. Generally speaking, a high index means a small change in viscosity with temperature and a low index means a large change in viscosity with temperature.

It is evident from this information that the lubricant's viscosity has an effect on the amount of frictional loss in the electric motor.

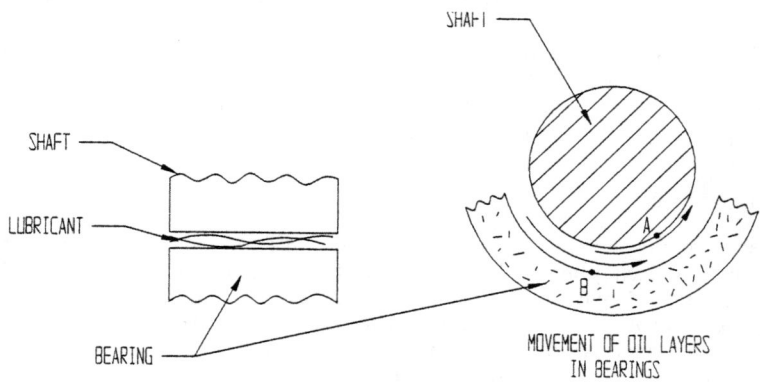

FIGURE 3.73 Velocity variation of oil layers.

In the case of the *rigid nonporous sleeve bearings* (Fig. 3.74), the rotating shaft picks up oil from the felt wick and forces it up the groove in the bearing. As the oil gets to the thin portion of the groove, it is forced out onto the surface of the bearing and provides a film between the bearing and the shaft.

Finally, we consider *porous sleeve bearings* (Fig. 3.75). In this system, the oil wicks through the small holes in the bearing and covers the shaft. As the shaft rotates, it

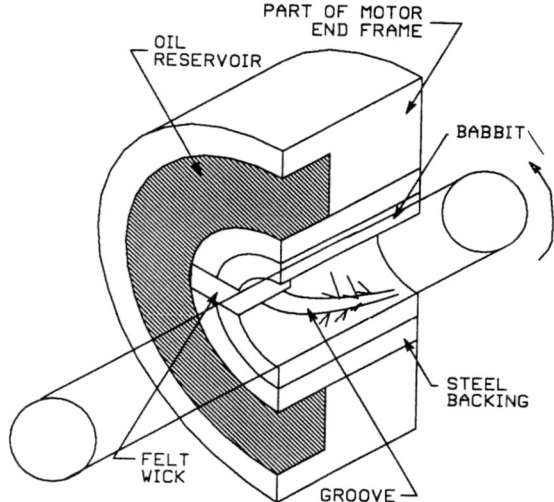

FIGURE 3.74 Oil flow in rigid nonporous sleeve bearing.

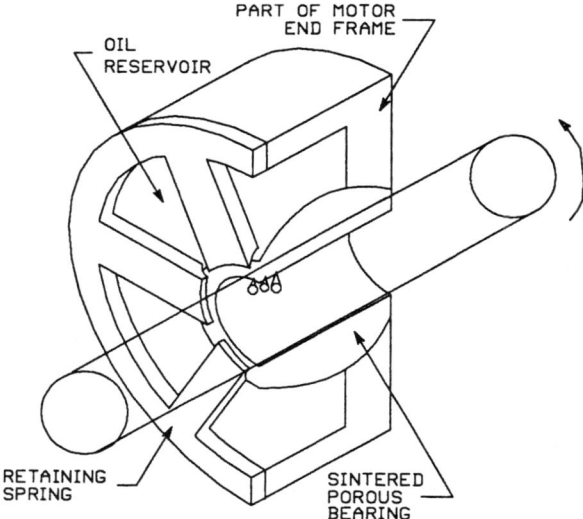

FIGURE 3.75 Oil flow in porous sintered sleeve bearing.

causes the oil to be forced onto the nonporous surfaces of the bearing, building up a film similar to that of the rigid nonporous sleeve bearing.

3.12.2 Application

The selection of the shaft and bearing were discussed earlier. The selection process is based on loads and speeds. Selection of the lubricant is usually based on two factors, system friction and temperature range.

Friction consists of viscous friction, static friction, and coulomb friction. *Viscous friction* is a function of the viscosity of the oil and changes with speed. *Static friction* is a retarding force that tends to prevent motion from starting. Once motion has started, static friction falls to zero. *Coulomb friction* is a torque force which has a constant amplitude and is not a function of velocity.

In a motor, these frictions translate into torques as follows:

Viscous friction

$$T_v(t) = B \frac{d\theta(t)}{dt} \tag{3.9}$$

Static friction

$$T_s(t) = \pm(F_s)\theta = 0 \tag{3.10}$$

Coulomb friction

$$T_c(t) = F_c \left[\frac{d\theta/dt}{d\theta/dt} \right] \tag{3.11}$$

where B = viscous friction coefficient
$(F_s)\theta = 0$ = static friction
F_c = coulomb friction coefficient
θ = angular displacement

The combination of these frictions are present in motor bearing systems, as graphically represented in Fig. 3.76.

The static friction $T_s(t)$ causes a high torque loss at starting (zero speed). There is a ripple effect as the oil film is formed. Thereafter, the coulomb friction $T_c(t)$, and viscous friction $T_v(t)$ cause the torque to increase in this nonlinear fashion.

Temperature considerations are important when selecting a motor lubricant. Extreme cold or heat can adversely affect motor performance if the proper lubricant is not selected. It is common to select a light (low-viscosity) oil for cold-temperature operation and a heavy (high-viscosity) oil for high-temperature operation. If a broad temperature range is necessary, it is common to select a synthetic oil that has operating characteristics well above and below the desired range.

Life at high temperatures has been limited because of oil degradation. Many potential applications have been abandoned or limited because of this problem. The availability of perfluoropolyether (PFPE) oils has made many of these applications more practical. These oils have been shown to be stable at these higher temperatures.

A few problems have occurred when applying these oils. Most of these problems occur because of interpretation and comparison of the specification sheets for lubricants.

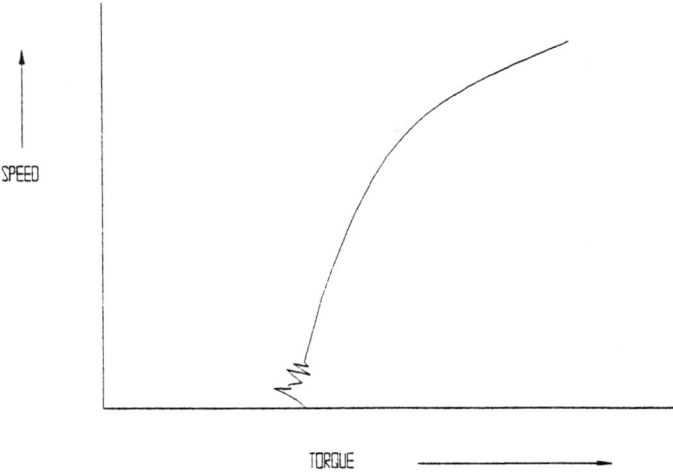

FIGURE 3.76 Frictional damping speed torque curve.

Specification sheets list temperature range and viscosity. The viscosity that is listed is generally the *kinematic viscosity* in centistokes (cSt). The comparison would lead a person to select a PFPE oil that is likely to be much heavier than necessary. This will result in higher friction and possibly higher bearing temperatures than with a normal oil. The trick is to predict the *dynamic viscosity* value of each oil and compare them. This is obtained by multiplying the kinematic viscosity by the density of the oil. When making this calculation, be certain that units are consistent.

Once the oil has been selected, an increase in friction may still be observed. The PFPE oils tend to stick to the shaft and fill the entire void between the shaft and bearing. Bearing systems are designed to provide a hydrodynamic wedge or wave of oil on which the shaft rides, as shown in Fig. 3.77.

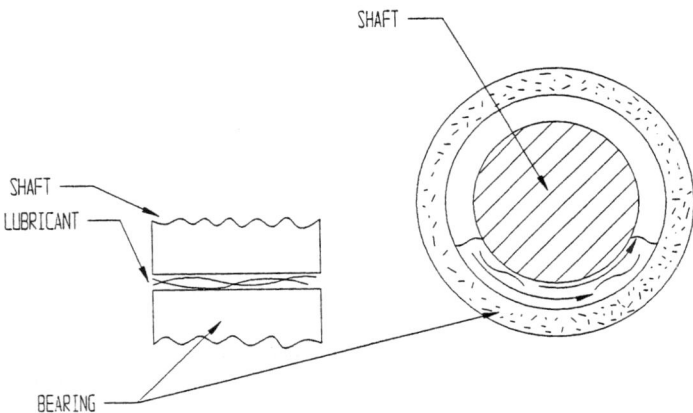

FIGURE 3.77 Shaft riding on wedge of oil.

This leaves an empty space over part of the bearing–shaft interface. When the PFPE oils fill this space, additional viscous friction results, which produces additional torque and heat losses. If this situation occurs, a few ten-thousandths of an inch additional clearance may be necessary between the shaft and bearing, or a bearing with a different porosity may be required. The exact value is best determined by experimentation.

Sleeve-bearing system life may be further extended by adding shaft slingers that throw the oil back into the oil reservoir.

Tables 3.15 and 3.16 list some oils commonly used in sintered bearings.

TABLE 3.15 Oils Used in Sintered Bearings

Oil	Temperature range	Viscosity at 40°C (104°F)	Application focus
Nye synthetic oil 310B	−20 to 125°C (−4 to 257°F)	500–560 cSt	High-viscosity polyalpha olefine (PAO) with additives to reduce friction and wear in sintered iron bearings.
Nye synthetic oil 132B	−60 to 120°C (−148 to 248°F)	20 cSt	PAO, plastic-compatible, light-viscosity oil for improved low-torque start-up
Nye synthetic oil 181B	−40 to 125°C (−40 to 257°F)	60 cSt	PAO, plastic-compatible, medium-high-viscosity oil; most commonly used viscosity for sintered bearings
Nye synthetic oil 188B	−40 to 150°C (−40 to 302°F)	35 cSt	Light-viscosity polyolester-based oil with copper pacifier and anti-wear additives for low-torque applications
Nye synthetic oil 634B*	−40 to 150°C (−40 to 302°F)	35 cSt	Light-viscosity polyolester-based oil with copper pacifier and anti-wear additives for low-torque applications
Nye synthetic oil 605*	−40 to 150°C (−40 to 302°F)	60 cSt	Medium-viscosity polyolester oil; excellent lubricity; contains copper pacifiers and load-bearing additives
Fluoroether oil 490	−75 to 225°C (−103 to 437°F)	90 cSt	Medium viscosity, exceptional lubricity and chemical inertness, low volatility, wide temperature capability
Fluoroether oil 491	−65 to 250°C (−85 to 482°F)	150 cSt	High viscosity, good film strength, exceptional lubricity and chemical inertness, very wide temperature capability.

* Ester-based oils may adversely affect some plastics, such as acrylonitrile-butadiene-styrene (ABS), polycarbonates, and polyphenylene oxides. If compatibility questions arise, contact the vendor prior to lubricant selection.

Source: Courtesy of Nye Lubricants.

TABLE 3.16 Krytox PFPE Lubricants for Sintered Bearings

GPL oil grade	103	104	105
XP oil grade (antirust)	1A3	1A4	1A5
ISO oil grade	32	68	150
Estimated useful range			
°C	−60 to 154	−51 to 179	−36 to 204
°F	−76 to 310	−60 to 355	−33 to 400
Oil viscosity, cSt			
20°C (68°F)	80	180	550
40°C (104°)	30	60	160
100°C (212°)	5	9	18
204°C (400°)	—	—	3
Oil viscosity index	121	124	134
Oil pour point			
°C	−60	−51	−36
°F	−76	−60	−33
Maximum oil volatility, % in 22 h			
@ 66°C (150°F)	1	1	1
@ 121°C (250°F)	7	3	2
@ 204°C (400°F)	—	—	10

Note: The GPL series oils are unadditized and are recommended for use in nonferrous applications. If rusting could occur, the XP oils should be used. The XP oils have a soluble antirust additive.

Source: Courtesy DuPont Corporation.

3.13 PROCESS CONTROL IN COMMUTATOR FUSING*

Fusing has been found to be the most critical process in electric motor manufacturing. This is particularly true for the new generation of asbestos-free, high-speed motors.

This section describes how to evaluate fusing results, the fusing process, the effects of commutator design and construction on the fusing process, and improvements to fusing-machine technology for improved process control.

Historically, the fusing process has been the single greatest source of rejects in armature manufacture. Thus, it is one essential area on which efforts to improve technology must be focused.

The process of fusing tang commutators is the focus for analysis here, although many of the observations, methods, and conclusions are also valid for fusing slot commutators. The electrical, mechanical, and thermal characteristics of fusing the two types of commutators remain the same, but the observations on the geometry and the thermal and electrical phenomena inside the tang are different for the slot type of commutator.

---

* Section contributed by Axis SPA.

To begin an analysis of the tang-fusing process, the objective of commutator fusing can be defined as establishing an electrical and mechanical connection between the wire and the commutator bar, with characteristics that guarantee the correct working of the motor in operating conditions for the expected life of the motor.

The process by which commutator fusing is done and solutions for optimizing the fusing process are presented. With this result in mind, we now examine the typical battery of tests that are performed either on the manufacturing floor or in the laboratory to qualify the fused commutator.

Generally speaking, fusing is accomplished by contacting the bar with a ground electrode, then compressing the tang and heating the tang with the fusing electrode, as in Fig. 3.78. This is discussed in detail later.

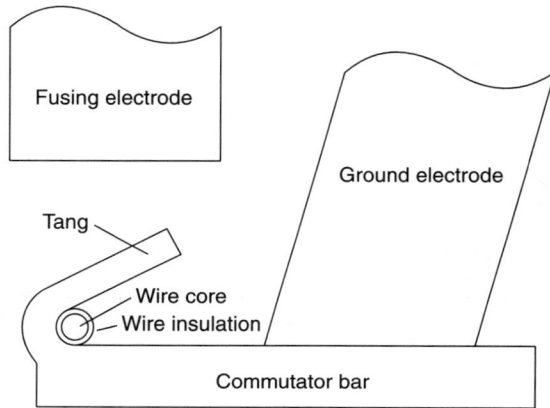

FIGURE 3.78 Evaluation of commutator fuse testing.

3.13.1 Evaluation of Commutator Fuse Testing

The first step is to define a list that represents the tests for qualifying fusing results. The following subsections present a brief description of each test and the relevance it has for analyzing the fusing process. The tests must be divided into two categories, qualifying checks and measurements.

Qualifying Checks or Observations. All of these tests are done randomly by visual inspection through a magnifying glass or a microscope. These tests require a low investment and can be performed on the manufacturing floor by the technician. They can be done quickly, but they do not offer any quantitative results as performed. Generally, it is very useful to test all the commutator tangs in order to verify the fusing uniformity.

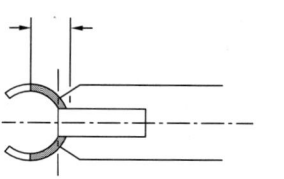

FIGURE 3.79 Well-closed tang with visible insulation peeling.

1. *Insulation peeling or "burn-back"* is observed, as in Fig. 3.79, where the wire extends from the sides of the fused tang. This test is carried out on the entire commutator to check the consistency of power applied during fusing.

2. *Bar oxidation* gives an indication of the heat stress caused to the commutator during fusing. This test may also be used to gauge the consistency of energy application during fusing, although it is heavily affected by external conditions such as the presence of cooling airflow.

3. *Tang roughness* gives an indication of the wear status of the electrode. The technician observes the top of the tang for pitting or scars. This occurs because the tungsten fusing-electrode surface breaks down due to the heat and pressure applied.

4. Observation of the *tang closing* checks for air gaps between the tang and the bar. A perfect closure, as in Fig. 3.83, prevents accumulation of extraneous material under the tang that would otherwise insulate the wire from the bar.

5. Subjective test of *resistance to mechanical stress*—stress the tang transversely or vertically, as shown in Fig. 3.80, to check the mechanical connection to the bar. When the joining force is reached the two parts separate, emitting an easily audible snap. When the tang is not connected, no sound is heard.

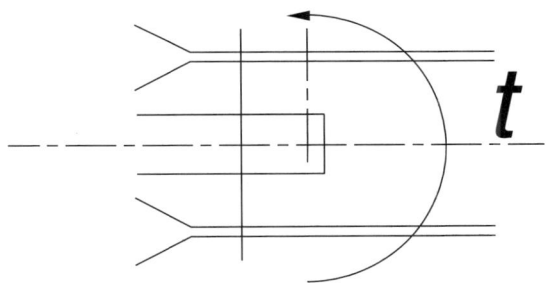

FIGURE 3.80   Resistance to mechanical stress.

6. Visual test of the *tang impression* on the bar—the degree of mechanical cohesion can be observed by lifting the tang in order to observe its impression on the bar. A well-marked impression does not always correspond to a good mechanical cohesion, while the presence of oxidation under the tang is a clear indication of noncohesion.

7. Visual test of the *wire impression* on the tang and bar—normally, with a good fuse, you can clearly see the impression of the wire under the tang and on the bar. The copper of the tang and bar surrounding the wire is a good indication that the wire has made good surface contact with the tang and the bar.

Measurements.   These tests are generally carried out either in the laboratory or directly on the production line. The benefit of testing on the line is that it gives the manufacturer the ability to create a database for quality control and for tracking serial parts, and it guarantees that each part has been measured against a set of tolerances.

1. *Electrical continuity* testing checks for electrical continuity between the commutator bars. It requires a simple and low-cost equipment setup; however, it does not distinguish poor from good electrical connections.

2. *Fuse resistance* testing is carried out on a commercial machine in the production line to measure each tang's fuse resistance. The test permits accurate evaluation of the electrical connection across the wire and bar.

3. Measurement of *final tang thickness* indicates whether the force and energy used during fusing have been maintained within required tolerances.

4. Measured *resistance to mechanical stress* testing, a more accurate measurement of tang-to-bar cohesion, can be obtained by using a dynamometer to measure the breaking load. In high-rpm applications, the armature is rotated to very high speed in order to accomplish a further check for bar lift.

5. Wire-squeezing testing evaluates the result of the motion of the tang during fusing. This test is destructive, because it requires that the tang be opened to measure the thickness of the wire. Normally, the optimal decrease in the wire diameter (minor diameter of ellipse-shaped wire) is 15 to 40 percent. Reduction of the wire diameter within this range ensures good wire stability while also ensuring that the wire is not excessively weakened.

6. Measurement of the *commutator temperature,* carried out immediately after fusing, indicates the thermal stress on the commutator. Generally, the commutator manufacturer indicates a maximum allowable temperature. This test has become very important with asbestos-free commutators.

3.13.2 Fusing Process

Basically, the fusing process requires compressing and heating by applying current flow through the parts that need to be fused.

The following is a description of the most important stages that occur during fusing (see Fig. 3.81):

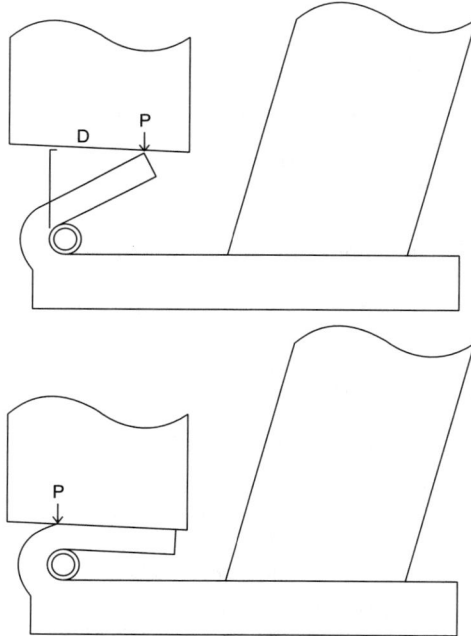

FIGURE 3.81 Stages 1 to 3 of fusing.

Stage 1. A properly dimensioned ground electrode is brought into contact with the commutator bar without damaging the commutator, by means of a pneumatic cylinder.

Stage 2. The fusing electrode is brought into contact with the commutator tang (Fig. 3.81) without excessively bending the tang. In physical terms, this requires that a well-proportioned ratio between the kinetic energy of the fusing head and the mechanical strength of the tang and the commutator be achieved. Since the mass of the fusing head can be reduced only within certain limits, the speed of the electrode becomes an important control variable in efforts to minimize the kinetic energy. In order to accomplish a low-speed approach in first contacting the tang, low electrode force must initially be applied. Then the force should be gradually increased to reach the preset values as required during the current-flow phase.

Stage 3. The electrode applies a resultant force F on the tang (Fig. 3.81), which, due to the increasing surface contact of the electrode, moves until it becomes balanced by the opposing deformation of the tang. When this balance is reached, the electrode stops moving and current should be applied. In particular, the geometry of the tang, the size of the wire, the wire position beneath the tang, and the sharpening angle of the electrode determine this balance position.

Stage 4. At the beginning of the power application, the current flows through the tang in a lengthwise direction (Fig. 3.82). The tang begins to heat at the elbow, the tang portion enveloping the wire, where the current density is higher. In this first stage, the wire insulating material is eliminated. The insulation evaporates easily because the tang is still open. When the temperature rises, the mechanical

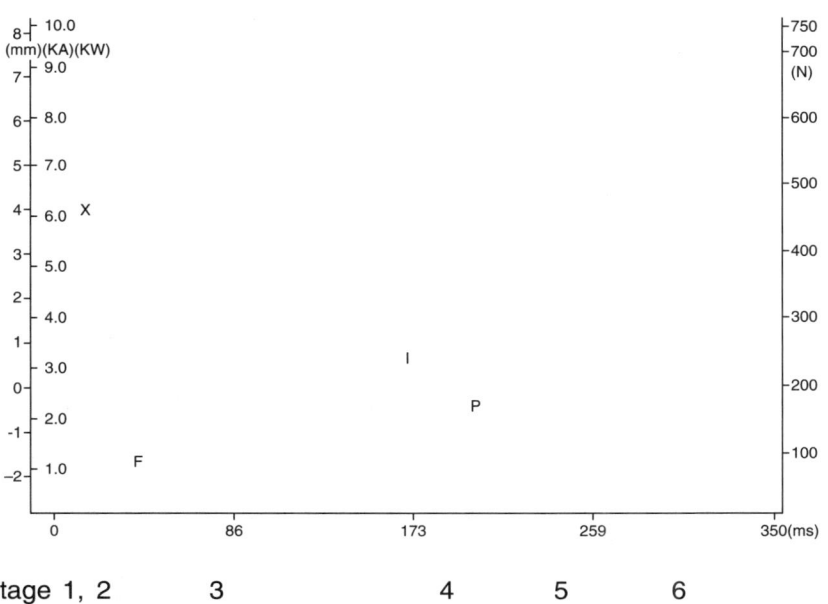

FIGURE 3.82 Stage 4 of fusing.

characteristics of the copper change; therefore, the tang becomes more plastic in its elbow (this is the point with the highest temperature), and the tang continues to close until it contacts the bar (Fig. 3.83).

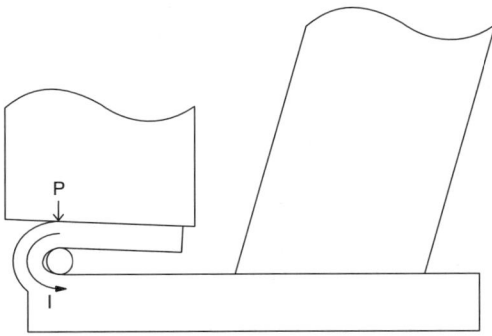

FIGURE 3.83 Stage 4 of fusing.

Stage 5. Once the tip of the tang contacts the bar, the current starts passing through the tang in a transverse direction. As the contact area with the bar increases, the current through this area also increases, thereby reducing the current density in the tang elbow. At this point the heat is mostly applied to the tip of the tang (Fig. 3.84).

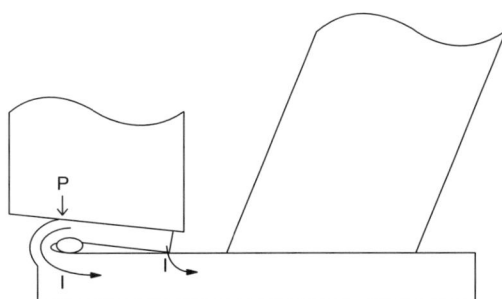

FIGURE 3.84 Stage 5 of fusing.

Stage 6. Once the current supply is terminated, the electrode continues to apply force to the tang for a preset time. During this time the temperature decreases and mechanical cohesion is completed (Fig. 3.85). Variables which influence cohesion are typically the electrode force, the time for applying the cohesion force, and the presence of forced cooling of the tang (e.g., by using low-temperature airflow).

Importance of the Wire and the Commutator in the Fusing Process. The geometric characteristics and the composition of the commutator and the winding wire are

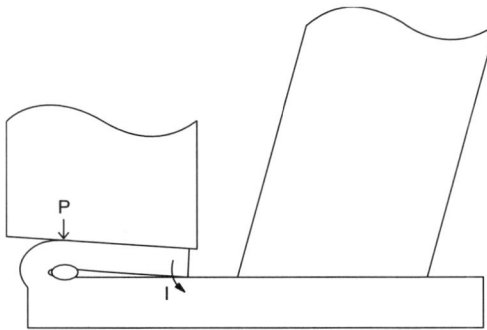

FIGURE 3.85 Stage 6 of fusing.

very important in achieving good results in the fusing process. The most important characteristics are the following:

1. *Tang cross section and length.* Together, these determine the minimum amount of force applied for "cold" fusing (see Stage 3 of the fusing process). The cross-sectional area and copper alloy determine the electrical resistance and thus the upper limit of the current to be applied during Stage 4 of the fusing process. If the current exceeds this limit, the tang will melt at the elbow.

2. *Top surface of the tang.* This surface determines the current value to be used in Stage 5 of the fusing process. An excessive current density causes surface melting of the tang and can result in a blown tang, while a low current density produces insufficient cohesion.

3. *Type of copper alloy.* The yield point and conductivity of the copper commutator varies with the silver content. This means the tang's mechanical resistance and temperature will vary as the copper alloy varies. Therefore, it is necessary to adjust the electrical and mechanical parameters accordingly.

4. *Wire diameter and position.* The wire hooked around the tang behaves as a support for the tang; thus, the wire affects the tang closing and therefore the forces necessary for the tang elbow. This effect often occurs as a result of wire hooking with insufficient wire tension or of double wire hooking and is more evident the shorter the tang.

5. *Wire insulation.* Wire insulation with a high thermal resistance requires a considerable increase in the current and fusing time to vaporize the insulation. In addition, using a lower force slows tang closure, thereby increasing the heating of the tang elbow.

6. *Commutator.* The inherent eccentricities in the commutator as supplied by the manufacturer can be eliminated as a negative influencing factor in the fusing process by measuring the electrode displacement relative to the bar rather than to the center of the armature shaft.

Commutators without asbestos are less resistant to thermal stress. To avoid damage, it is necessary to reduce the applied energy. The best results have been obtained by applying high power for a short time, in order to concentrate heating as much as possible to the tang area. Reducing the fusing time avoids excessive heat diffusion in the commutator bar and thus reduces defects caused by thermal stress. Less heat stress also results in lower bar-to-bar drop development after turning.

7. *Bar and tang geometry.* The relationship between tang mass m_t and bar mass m_b are correlated to the temperature T of the commutator bar at the end of fusing, as follows:

$$T = K \frac{m_t}{m_t + m_b} \qquad (3.12)$$

where K = temperature dependent coefficient

If the tang is small compared to the bar, fusing can be done with lower thermal stress on the commutator.

3.13.3 Improvements to Fusing-Machine Technology

1. During the fusing process, current distribution in various portions of the tang varies considerably, as has been described previously. This variation results in significant variations in the associated electrical resistance r_t of the tang. The heat Q applied to the tang by means of current i can be expressed as follows:

$$Q = \int r_t(t) * i^2(t)dt \qquad (3.13)$$

If heat application occurs by regulating only the RMS value of the current i, the variations in resistance r_t will cause unwanted variations in heat Q. To avoid this, for obtaining more precise heat application, it is necessary to regulate the power P applied during fusing.

$$Q = \int P(t)dt = \int v(t) * i(t)dt \qquad (3.14)$$

For this power regulation, and to obtain fusing in a sufficiently short time, medium-frequency dc inverters are the most adequate electrical power supply equipment that can be used. Normally, fusing cycle times are completed in 70 to 150 ms. DC medium-frequency inverters (1 to 1.2 kHz) regulate using closed-loop feedback every 415 to 500 μs. Line-frequency equipment regulates every 8,000 μs. Considering, for example, a fusing done in 72 ms, which corresponds to a little more than 4 cycles in American line frequency, line-frequency regulation is likely to be inadequate most of the time.

2. During the fusing process, preset electrode force profiles need to be applied to precisely control tang closure and to distribute the heat correctly in the various portions of the tang. To achieve force application in this way with necessary control accuracy, the fusing system should use closed-loop force control.

3. Quality fusing requires that the machine be designed and built to guarantee high repeatability and reliability. In particular, the fusing machine needs to have adequate cooling of the electrode to maintain consistent working temperature, design construction that guarantees accurate alignment of the electrode with the tang, and provisions for quick electrode replacement by the operator without affecting the precision and repeatability of the machine's performance.

3.13.4 Conclusions

Today, fusing-machine technology allows reliable and consistent fusing of commutators using low energy. In particular, this is required to process modern high-speed motors having asbestos-free commutators. Since the fusing process is also significantly affected by the commutator and wire-insulation characteristics, it should be carefully considered in new motor design.

3.14 ARMATURE AND ROTOR BALANCING*

Unbalance is defined as the uneven distribution of mass about the axis of rotation of a rotating body. Unbalance is the direct result of fixed or variable sources. *Fixed-source unbalance* is caused by nonsymmetrical design or manufacture, while *variable-source unbalance* is due to operational factors such as distortion or shifting of components by centripetal force. The result of the unbalance is a net centripetal force at the bearings, and this, in turn, causes vibrations in the system. *Balancing* is the procedure used to reduce the unbalance level to a level at which the resulting vibration is low enough that operation of the equipment is not impaired.

To define an unbalanced condition, four items must be identified. Unbalance magnitude U is the result of an unbalanced mass m located at a specific radius r. The location is defined first in terms of the angular location in degrees from a zero reference location, and finally by the axial location of the unbalance relative to the support bearings. In order to correct for unbalance in a rotor, you must identify these four parameters, represented graphically in Fig. 3.86. For this type of rotor, a single unbalance correction is adequate. For wider rotors, the resulting effects of multiple unbalances have to be resolved with two corrections at separate axial locations. This section explains these variations and provides guidelines to enable selection of appropriate production balancing methods.

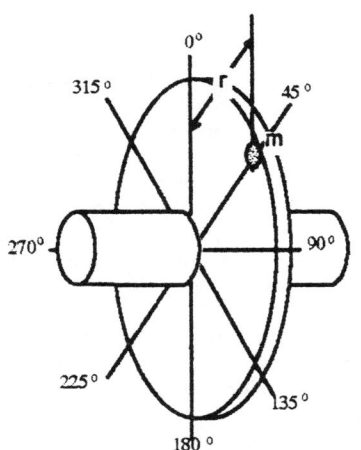

FIGURE 3.86 Location of unbalanced mass. Unbalance magnitude U is the result of unbalanced mass m located at angle 45° on radius r.

3.14.1 Constituents of a Rotating Object

Shaft Axis. The shaft axis (centerline) of the rotor, shown in Fig. 3.87, is defined by the centerlines of the support bearings. The shaft axis may or may not contain the center of mass. Parts are designed to rotate about the shaft axis of a rotor, which is also the rotational axis.

Principal Axis. The principal axis always includes the center of gravity (or mass center) of the part. By definition, the mass of the rotor is evenly distributed about

---

* Section contributed by Larry C. Anderson, American Hoffman Corporation.

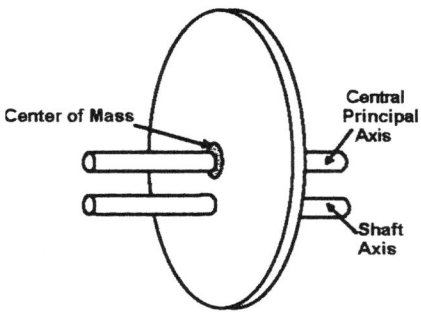

FIGURE 3.87 Misalignment of center of mass.

the principal axis. Unbalanced rotors attempt to rotate about the principal axis, but are restrained to rotate about the shaft axis by their support bearings. The vibrations and bearing loads due to the unbalance are directly proportional to this misalignment. Figure 3.87 presents an exaggerated illustration of how the mass of a rotor can be offset on the shaft axis. Note that the rotor is rotating around the shaft axis, but ideally, the center or principal axis is the true center of the rotor.

Center of Mass. The center of mass is generally considered to be at the geometric center of an object. In fact, the center of mass, or inertia, is the location of symmetric mass distribution and is not the same as the location of the geometric center.

In an unbalanced rotor, the principal axis does not coincide with the shaft axis. When a rotor is balanced, the principal axis coincides with the shaft axis, as shown in Fig. 3.88.

A rotor will always attempt to rotate about its principal axis but is constrained by the support bearings, which force it to rotate about the shaft axis. Displacement of these two axes results in bearing loads and vibration which is directly proportional to the misalignment.

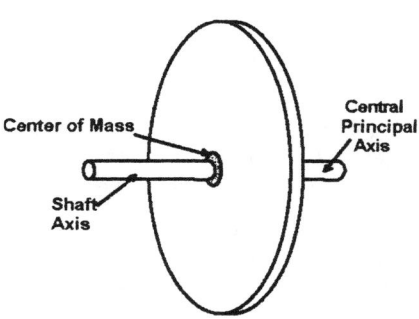

FIGURE 3.88 Balanced rotor with even mass distribution.

A rotating object is composed of components that have mass and flexibility and which absorb and dissipate energy when subjected to internal disturbances; the result is a unique pattern of motions called *rotor response*. The rotor response at operating speed directly affects the object's sensitivity to unbalance. Speeds close to resonant frequencies require much closer balance tolerances due to the increased rotor-response characteristics. The effects of speed and resonance are explained in more detail later.

3.14.2 Characteristics of Static and Dynamic Unbalance

Static Unbalance. Static unbalance is classified as single-plane unbalance, meaning that it is represented by a single vector quantity and corrected by a single correction mass applied opposite to the unbalance location and in the axial plane of the mass center of the rotor. When there is no way to add (or remove) mass in the same plane as the unbalance, the correction can be split into two equal corrections at the ends of the rotor to achieve the same results. Static unbalance can be detected by nonrotational equipment and is characterized by the fact that the principal axis is displaced parallel to the shaft axis, as shown in Fig. 3.89.

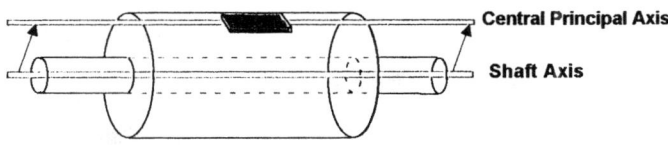

Static Unbalance

FIGURE 3.89 Static unbalance: rotor with centered single correction.

Couple Unbalance. Couple unbalance is a condition in which the principle axis intersects with the shaft axis at the center of mass (which means there is no static unbalance) but is inclined at an angle to the bearing axis, as shown in Fig. 3.90. Couple unbalance (also called *moment unbalance*) has equal unbalances spaced 180° from each other, at opposite ends. Generally, couple unbalance is the result of each bearing plane not being precisely perpendicular to the rotational axis. When the rotor spins, it vibrates with a twisting motion. Couple unbalance cannot be corrected in one plane; two corrections are required.

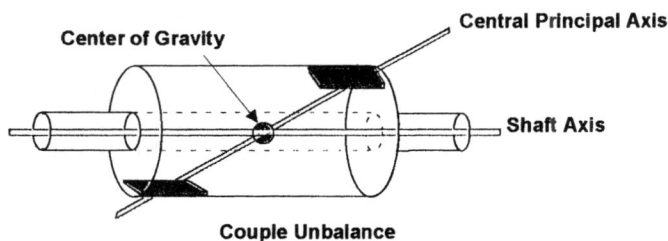

Couple Unbalance

FIGURE 3.90 Couple unbalance: equal forces at opposite ends, 180° apart.

Dynamic Unbalance. Dynamic unbalance follows the same condition as couple unbalance and can be measured only when the part is rotating. Dynamic unbalance is present when the central principal axis of inertia neither is parallel to nor intersects the shaft axis at the center of mass. This can only be corrected in two or more planes. Dynamic unbalance is the general case of a combination of static and couple unbalance.

Quasi-Static Unbalance. In quasi-static unbalance, there is a specific combination of static and couple unbalance such that the angular position of one couple component coincides with the angular position of the static unbalance. If the axial location of the unbalance can be used for unbalance correction, then a single correction is possible. Otherwise, it must be treated the same as dynamic unbalance. In some production situations, correction can be made at a single plane close enough to the source of the unbalance to enable efficient balancing correction to within the required tolerance. This rare case of couple and static unbalance is shown in Fig. 3.91.

Multiplane balancing is sometimes required when there is insufficient material or space available to make all the required unbalance corrections. In this case, a prebalanced operation in one or two auxiliary planes precedes final balancing. In some applications where the rotors are flexible, multiplane balancing is used to minimize the rotor's internal bending stresses.

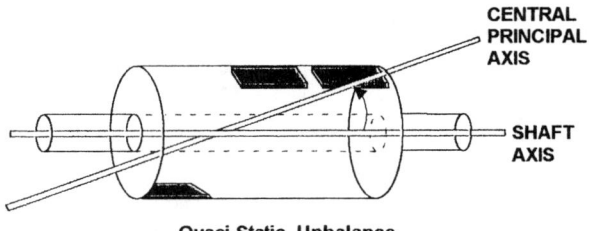

FIGURE 3.91 Quasi-static unbalance: combination of static and couple unbalance.

From a theoretical standpoint, the aim is always to achieve a zero-unbalance condition. In practice, the requirement is to reduce the unbalance to a point at which the unbalance forces have a negligible or nonharmful effect on part operation. Production situations require careful attention to the tolerances and correction procedures to achieve minimum cycle times with required service life and performance.

3.14.3 Effects of Increased Speed

The unbalance of a rotor does not change with speed but is solely due to its mass distribution about the bearing axis. The centripetal force generated by the unbalance increases with the square of the rotational speed change; therefore, the balance tolerance has to be determined for the maximum operating speed of the rotor.

The operating speed and bearing forces determine the bearing life, so the unbalance must be limited to achieve the required operating lifetime. The intensity of vibration (acceleration) must be limited to avoid noise and vibration apparent to the operator, which is proportional to the velocity of the vibration, which requires that the balancing tolerance be tightened as speed increases.

It is normally not necessary to balance a rotor at operating speed. In fact, the aim should always be to balance at the lowest practical speed. A lower balancing speed is preferable for a number of reasons. Faster speeds require a longer time to ramp up and ramp down and also require stronger, stiffer tooling arrangements and tighter clamping of the rotor assembly, which in turn increases the risk of distortion or damage. Lower speeds assure safety for the operator and require less elaborate machine guards. However, the performance of the balancing machine transducers and instrumentation improves with increasing speed due to the greater signal levels from the increase in centripetal force, illustrated in Fig. 3.92.

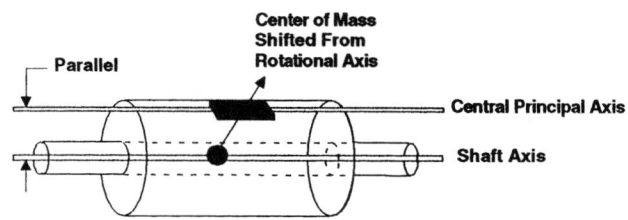

FIGURE 3.92 Effects of centripetal force on mass.

The requirements of tooling design, balance tolerance, cycle time, and machine guarding all interact when determining the optimum balancing speed.

3.14.4 Centripetal Force

What doesn't move does not need balancing.

According to Newton's laws, a body in motion tends to move in a straight line unless acted on by an outside force that causes it to deviate from its normal course. In the case of rotating bodies, a particle on the outside travels on a course tangent to the rotation of the body. Centripetal force acts on the particle to accelerate it while the particle is attached to the rotor toward the center of rotation. If the rotor is balanced, then there is an apparent even distribution of mass. However, if there is an excess of mass concentrated at one location on the rotor (i.e., an unbalance), the rotating mass attempts to force the unbalanced portion in a tangential direction, thus causing vibration and fatigue of the components. Rotational motion is different from straight-line motion. An automobile's acceleration stops when it reaches a constant speed. Turning a corner causes an instant acceleration even if the speed does not change; thus, acceleration is generated through the vehicle's tires. An unbalanced rotor experiences constant acceleration reacted through the bearings because the direction of the force is constantly changing. The velocity of the unbalanced mass is proportional to the speed, rpm, and the acceleration is proportional to the rate of change of the velocity, $(\text{rpm})^2$.

Components of centripetal force are the following:

- Mass (volume)
- Radius (distance)
- Velocity (speed)

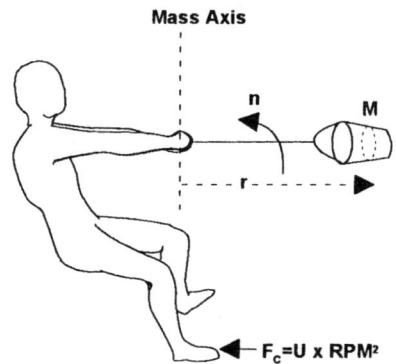

FIGURE 3.93 Components of centripetal force.

As an example, Fig. 3.93 shows a weight (mass) attached to a 3-ft string (radius) and twirled in a circle at 60 rpm (velocity). A greater mass produces a greater force, a longer string produces a greater force, and a faster speed produces a greater force proportional to the square of the speed.

Unbalance is independent of rotational speed and will not increase or decrease when standing still. At zero speed, the unbalance has no effect on the rotor. However, if the rotor is rotated, the unbalance will exert centripetal forces, causing vibration that becomes more violent as the speed increases. Centripetal force increases proportionately to the square of the speed increase. At double the speed, centripetal force quadruples; triple the speed and the force is nine times as much.

3.14.5 Resonance

Every mechanical object has properties of mass, stiffness, and damping which determine its natural frequency of oscillation. *Mass* is the volume of the material times its density. *Stiffness* depends on the elasticity of the material. *Damping* is a measure of the ability of the system to dissipate vibratory energy.

The natural frequency is directly proportional to the stiffness and inversely proportional to the mass. This is the frequency at which the object will tend to self-vibrate when rung by an impact.

Materials such as soft rubber have a high level of damping and a low stiffness and tend to absorb and dissipate vibration. Most hard materials have a higher stiffness and a lower level of damping. The damping factor determines the rate of energy loss to the surroundings. The damping factor is a nonlinear parameter and changes with speed. For a given structure, there is a frequency at which the damping factor approaches zero and therefore very little vibration energy is absorbed.

Resonance and critical speeds are frequencies that are governed by natural frequencies, damping, and vibratory forces. A *resonance* is a condition in a structure in which the frequency of a vibratory force, such as mass unbalance, is equal to a natural frequency of the system. If the vibratory force is caused by a rotating part, the resonance is called a *critical speed*.

A structure or object can be excited by one or more vibratory forces. Vibratory forces can be caused by various factors, including design, installation, manufacture, and wear, or the force can have a single constant frequency, as occurs with mass unbalance.

A rotational assembly with any finite unbalance acts as a vibration exciter and will produce a force as it is rotated. This is called the *excitation frequency*. When the natural frequency and the excitation frequency coincide, a state of resonance is said to exist. As rotational speed approaches the resonant frequency, the effects of the force increases. At resonant frequency, vibration amplitudes can become very large. If the rate of speed is close to the resonant frequency, a very low level of unbalance can still generate unacceptable vibration amplitudes.

As rotational speed reaches the resonant frequency, the support structure will vibrate directly with the exciting force (phase shift = $0°$). As the speed increases nearer to resonance, the phase begins to shift until at resonance there is a $90°$ phase shift. As the rotational speed continues to increase, the phase continues to change until it reaches opposition (phase shift = $180°$).

Balancing requires an exact knowledge of both the magnitude and the location of the unbalance, and so balancing speeds close to resonance are to be avoided. A small speed change will cause a large change in both the amount and the angle of the measured signal, and the results will be incorrect.

Sometimes equipment is designed to emphasize the resonant frequency. A tuning fork or piano string produces strong vibrations at the resonant frequency, which is beneficial; however, this is not the case with a stiff rotor, where the exact opposite condition is needed.

Vibrations that have large amplitudes can cause early fatigue failure. The energy expended by such vibrations causes significant power loss and speed reduction. In addition, noise levels from the vibration may be irritating to the operator as well as detrimental to the components surrounding the bearings.

It follows from this that as speeds and densities increase, keeping resonance away from the operating speed is a crucial part of the assembly designer's job. Ensuring that balancing speeds and tool design avoid resonance is a crucial part of the balancing-machine and tooling manufacturers' jobs.

3.14.6 Correcting for Unbalance

Correction can be accomplished by adding, removing, or moving material. The physical properties of the rotating device must be analyzed to determine the best method of correction. Methods of correcting unbalance in a production environment are as follows:

- Material removal by milling
- Material removal by drilling
- Material removal by cutting or clipping molded dimples
- Material addition by weighted epoxies
- Material addition by UV-cured epoxies
- Adjustments to molds and dies
- Relocation of the shaft axis (mass centering)

Whenever the unbalanced mass prohibits correction at a specified angle (as in an area where a correction cannot be made or the amount of added weight will not all fit on a designated angle), correction can be made by vectoring adjacent angles or components to implement a correction. The total unbalance is split into two vectors for adjacent components. For high levels of unbalance, it may be necessary to vector more than two components and use more than two weights.

When designing a rotor, it is important to estimate the maximum initial unbalance and to make allowance for adding sufficient correction mass at an appropriate location. The balance tolerance must be used to determine the amount of weight that can be added and the increments of weight in order to have enough resolution to bring the rotor into tolerance in one or two attempts.

A *correction plane* is determined by evaluating where mass can be added without affecting the mechanical operation of the rotor. Once a correction plane is selected, a correction method can be chosen. If weight is to be added, the weight mass and number of locations must be determined. Larger weight sizes permit more correction but limit the minimum value that can be achieved. If adhesive weights are to be used, a ridge, indentation, or roughened area is needed to prevent creep under loading.

Generally, weight removal is the preferred means of balancing small armatures and assemblies. Correction by drilling was the original preferred method; however this inefficient means of correction has been replaced almost exclusively by milling. Two of the most popular operations available on balancing machines are contour milling and component slot milling. *Contour milling,* shown in Fig. 3.94, involves the use of special milling cutters which are designed for the profile or curve of the stack. Material is removed at the exact polar location of the unbalance. Contour milling allows for larger amounts of unbalance in one cut, which reduces the cycle time of the balancing process. This is ideally suited for manufacturers of armatures with the same stack diameter but different numbers of poles, or for balancing rotors with a skewed stack.

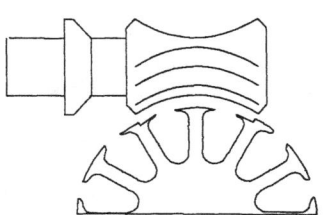

FIGURE 3.94 Contour mill cut.

Component slot milling is used when accuracy is required. Grooves machined into the centers of adjacent pole tips are milled either by a V-mill or slot cutter. Figure 3.95 shows this method of correction. The balancing machine calculates the unbalance vector using two or more components and determines how deep a cut should be made and whether the mill should transverse to adequately remove the

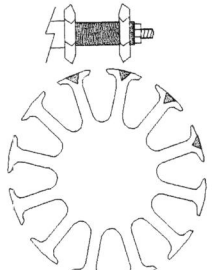

FIGURE 3.95 V-mill or slot mill cutter.

unbalance mass. This type of milling should be used in cases in which the laminations are too thin to allow adequate material removal through profile contour milling. This condition is exemplified by the edges of the pole tips becoming "feathered" because of a lack of material.

Weight removal offers several advantages over weight addition, especially for high-production applications: it is faster, it can be more accurately controlled, and it is easier to automate. However, hand milling has its advantages also. Whenever the armature has areas not easily accessed (such as fans), rotating components, or raised components, the armature or rotating assembly has to be corrected by hand. This is also true in low-production situations where automation is not justified.

To minimize cycle times it is important to achieve tolerance in the smallest number of steps. The accuracy of the balancing machine must be adequate to give a precise readout of the required correction needed. Production can benefit greatly when the balancer instrumentation can convert the basic unbalance data into specific correction masses at specific locations so the operator has step-by-step guidance.

3.14.7 Types of Balancing Machines

Balancing machines are used to determine the unbalance of a rotor in terms of both the magnitude of the mass unbalance and its angular position, in one or more planes. The system must support the rotor, spin it at a predetermined rpm, measure the forces that occur, and display the results in values useful for unbalance correction. Automated machines may even index the rotor to the required angle for a correction and add correction weights or remove unbalance mass.

For production balancing, the machine is typically set up to provide some indication of whether the rotor is within tolerance. If the rotor is out of tolerance, precise readout of the angle and amount are required. Hard-bearing balancing machines typically respond directly to the rotor unbalance and are not affected by speed. They are also permanently calibrated.

Since the force that a given amount of unbalance exerts at a given speed is always the same, the output from the sensing elements attached to the balancing machine is directly proportional to the unbalance in the rotor. The variation of the sensing element with balancing speed is known, and compensation for speed variations is automatic. The output is not influenced by bearing mass, rotor mass, or inertia, so that a permanent relationship between unbalance and sensing element output can be established, and the display of unbalance amounts can be directly in terms of the required correction mass. The output is affected by looseness or by a lack of stiffness in the tooling, and it is important that the tooling locates the housing (or bearing carrier) rigidly and without distortion.

The most popular method of obtaining a reference signal by which to determine the phase angle of the unbalance signal is through the use of an optical sensor that detects a reference mark representing zero degrees somewhere on the circumference of the rotor. A light source is reflected off the surface and picked up by the sensor. Through timing circuitry, an angle is established using the zero reference mark.

The first balancing machines were developed around 1900; they were trial-and-error devices consisting of a flexibly mounted set of bearings with some means of a drive. Some years later, mechanical indicators were developed. First the magnitude of vibration was indicated, then the high spot on the shaft was used as a crude reference for the angular position. Around 1940, electromechanical unbalance detection was developed, which was the predecessor to modern-day electronic balancing machines. Today, with the microprocessor, balancing machines have evolved into two sophisticated state-of-the-art, do-almost-anything machines categorized as soft-bearing and hard-bearing balancing machines

Soft-bearing balancing machines operate above the first critical speed, or resonance, of the machine/rotor system, as shown in Fig. 3.96. They determine unbalance in a rotor indirectly by measuring the displacement of the bearing bridge assembly that supports the rotor caused by the rotating part. Hence, the name *soft,* which actually refers to the relative stiffness of the bearing-support and spring system of the machine.

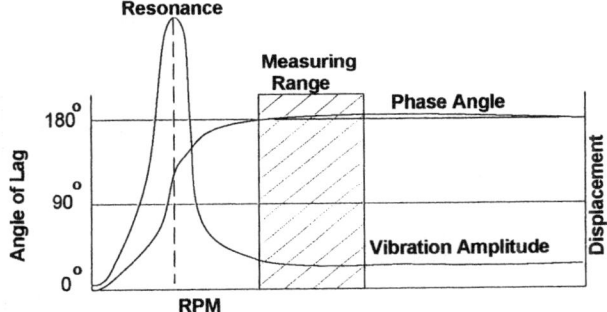

FIGURE 3.96 Soft-bearing balancing machines operate above resonance.

Every different part to be balanced on a soft-bearing machine causes a different displacement of the suspension system for a given amount of unbalance due to its weight, polar moments of inertia, and the like. This means that the machine has to be calibrated for each different workpiece type to be balanced. Soft-bearing machines were originally classified as trial-and-error balancing machines, because the first workpiece of each type had to be balanced to zero by trial and error before adding the known calibration weights. Modern electronics have simplified this procedure, but calibration of each rotor type is still required.

By contrast, *hard-bearing machines* operate below the critical speed of the machine/rotor system, as shown in Fig. 3.97. Hard-bearing machines directly measure the centripetal force created by the unbalance. Since a given amount of unbalance at any given speed causes the same force, regardless of rotor weight or shape, hard-bearing machines are classified as *permanently calibrated.* All that is required for setup is to input the geometric relationship between the bearing, or measuring, planes and the correction planes. This means that the unbalance in a rotor can be determined in the very first spin-up. No trial runs or calibration runs are required. This obviously is a major advantage over soft machines, resulting in reduced balancing time, greater productivity, increased accuracy due to the elimination of the possibility of operator error, and increased simplicity, which eliminates the need for highly skilled or experienced operators.

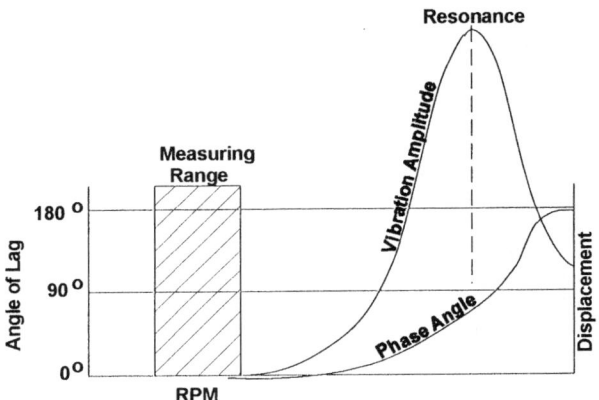

FIGURE 3.97 Hard-bearing balancing machines operate below resonance.

The mechanical design and features of each type of machine also affect its physical properties. This can have an impact or limitation on the balancing capability of the system. As previously stated, the soft-bearing machine operates above the first critical speed. Therefore, when balancing a part, the part must be accelerated through the critical speed to bring it to the machine's balancing speed. Accelerating a rotor through the critical speed of the machine can cause several problems. First of all, it is a safety hazard, because the rotor is much more susceptible to flying out of the machine due to the increased amplitude of the displacement. Second, it can damage the machine, as well as the workpiece, by hitting the travel limits of the suspension. Third, the measuring system needs more time to stabilize and take reliable readings. The threat of these problems can be reduced by locking up the suspension system until the rotor reaches balancing speed. However, this creates its own set of problems by increasing cycle time due to an additional step for the operator and by introducing its own safety problem—the operator must approach the machine to unlock the suspension while a rotor is spinning.

The displacement measuring principal causes other problems as well. For example, external shocks to the machine caused by other equipment will cause excursions of the vibratory system. These excursions add to the displacement caused by the unbalance, but are totally indistinguishable from them. Similarly, certain rotor types, such as fans and blowers or even armatures with fans on them, will cause these excursions. Being incapable of separating the unbalance from the external influences, windage in this case, the operator will overcorrect the unbalance. This either will produce a bad part or, if the mistake is caught during an audit cycle, will require additional steps to bring the part into tolerance, rendering the operation inefficient.

Another workpiece-related problem that affects only soft-bearing machines is difficulties, or even damage, resulting from rotors with high levels of initial unbalance. When this condition occurs, the unbalance present in the part will cause the vibratory system to hit the end stops, possibly causing damage, but always preventing the machine from measuring the unbalance. The part must then be prebalanced by some crude trial-and-error method so that the resulting displacement is within the measurable range of the machine.

The displacement measuring principle also affects the weight range of soft-bearing machines. The bearings and related support components vibrate in unison with the rotor, adding to its mass. This parasitic mass reduces machine sensitivity by damping or reducing the displacement of the vibratory system. As the weight of the rotor approaches the weight of the parasitic mass of the machine, displacement of the bearing and, what is more important, of the transducers is reduced. This limits the low end of the weight capacity of the machine. This can be overcome by having additional sets of bearing supports with less parasitic mass, allowing smaller, lighter parts to be balanced. This raises the cost of the soft machine and adds yet another cumbersome operation, limiting flexibility.

Hard-bearing machines use the entire bearing support as part of the spring system, eliminating parasitic mass altogether. Theoretically, hard-bearing machines have no lower weight limit. In actuality, however, they are limited by the physical size of the workpiece.

There is one further physical limitation of soft-bearing machines that is related to the type of workpiece to be balanced. This rotor type is classified as *cantilevered* or *overhung*. Balancing these parts requires special care on both types of machines. A cantilevered workpiece on a horizontal type of balance machine causes a negative load on the support furthest from the center of gravity (CG) of the part. On a hard machine, a counter–roller bearing must be used to hold the part securely in the machine. Measuring is not affected since the thrust bearing roller assembly is part of the vibratory system. On some types of soft machines, a downward load is required to ensure proper displacement of the vibratory system. This downward force can be created with an overhung part by special fixturing which holds the part so that the CG is between the bearing supports, creating a downward load on both supports. As in the case of using special roller carriages for small parts to extend the design limitations of the soft suspension, this arrangement allows balancing overhung parts, but at additional cost and time.

Balancing on a soft-bearing machine also requires repeatability of the balancing speed from run to run. The displacement of the suspension system on a soft machine is a function of the rotational speed. Therefore, it is essential that on subsequent runs the balancing speed be identical to the speed at which the machine was calibrated. It is clear from the preceding discussion that hard-bearing machines are the preferred choice for applications where frequent changeover and maximum flexibility are required. However, hard-bearing balancing machines have also become the norm in industries where the need for reliability and high-level accuracy outweigh all other considerations. These would include aircraft engines, turbomachinery, and computer disk drives, to name only a few. In general, hard-bearing balancing machines have become the industry standard for all types of balancing applications.

3.14.8 Measuring System Sensitivity

The capability of a measuring system is determined by its sensitivity, repeatability, and accuracy. It is a combination of the limitations of both the balancing machine itself and the intrinsic parameters of the rotor being measured.

The *sensitivity* of a balancing system is not the smallest display indication but the smallest amount that can be accurately measured. Sensitivity is typically measured in microinches.

The *repeatability* or *precision* of a system is the reproducibility of multiple measurements of the same part. For statistical evaluations, the repeatability test is carried out on a number of parts.

The *accuracy* of a system is the relation between the measured amount and the actual unbalance of the part. This is related to calibration of the system for a specific unbalance and linearity of the measurements with differing amounts of unbalance. The smaller the difference, the more accurate the system. Consistent readings are an indication of repeatability, not accuracy.

The question is "What is the least amount of unbalance that can be detected?"

Rotor operating characteristics play an important factor in determining the sensitivity of the machine. The initial unbalance, resonance, shaft concentricity and bearings of the rotor may limit the system. Bearing quality directly affects the noise level and therefore the signal-to-noise ratio of the unbalance signal. If the machine has to deal with a high initial unbalance, the measuring range has to be high to avoid overranging. The balancing machine's measuring range directly affects the amount of sensitivity. You can't measure ounces accurately if the range of the machine is in tons.

To find the limits for the system, a rotor and several removable calibration weights are needed. The rotor must have the capacity for mounting each of the calibration weights at several angular locations. The rotor should be balanced as closely as practical before measuring each of the calibration weights in each position. Plotting each set of data should produce a sine wave, with the amplitude representing the remaining unbalance and the average representing the mass of the calibration weights. The average values indicate the linearity of the system, and the variations of the amplitude from the measured residual unbalance indicate the lower limits of machine measurement. Detailed analysis of accuracy is a subject in itself and is beyond the scope of this handbook.

3.14.9 Tooling

Balance tooling is the cradle or work-holding device used to securely hold a part during the unbalance measurement and correction processes. The tooling must support and locate the rotor to ensure that the unbalance forces (signal) are transferred to the unbalance detection transducers. If the balance tooling does not repeat, locate, and clamp the rotor assembly, the result will be errors in amount, angle, and axial distribution (for two-plane balance) of the unbalance measurement.

Balance tooling may include other functions, such as autoindexing, remote angle, material addition/removal, and total indicated run-out (TIR), or contain the drive source for rotating the part. The tooling should be securely fastened to the measuring table or work supports on which the balancing operation is to be performed.

Typically, the armature will want to creep to one side or another while spinning. A stop or end thrust positioned so the armature shaft rides against it will maintain the rotating armature in a set position. When adjusted correctly, the end thrust will not affect the unbalance measurement of the rotor. Misalignment may cause restriction in the free spinning of the armature and/or bouncing in the cradle.

To ensure precision in any balancing operation, the rotor must be properly positioned with respect to the work supports and correction planes. This is called *referencing*. To ensure the desired accuracy, the operator must make sure the part is precisely located and rigidly supported. To make the rotor easier to load and unload, adjust the end thrust to allow the rotor to ride against it but also to compliment the pole sensors. A properly adjusted balancing cradle will ensure correct plane location, short load and unload times, and simple operation. Achieving these often contradictory requirements takes careful attention to detail.

The most advanced and accurate balancing machine is worthless if the tooling is not adequate. A good reference for accurate unbalance measurements starts with perfected tooling.

3.15 BRUSH HOLDERS FOR SMALL MOTORS*

Brush holders are crucial components of the overall commutation system in brush-type direct-current motors. Brush holders play a key role in determining the overall performance of the motor. When manufactured and installed properly, brush holders serve the following functions:

- Hold the carbon brush in place.
- Insulate the carbon brush and electrical system from the other conductive parts of the motor.
- Serve a crucial function in aligning the brushes against the commutator at the correct angle.
- Provide an unobstructed channel so that the brushes can move freely up and down to compensate for any commutator eccentricity.
- Maximize brush life, minimize brush noise or chatter, and maintain the integrity of the electrical connections.

For these reasons, careful thought and consideration must be given to the design, use and application of the brush holder. This section gives an overview of the different styles of brush holders as well as application and design considerations. The section focuses on brush holders designed for fractional-horsepower motors.

3.15.1 Different Styles of Brush Holders for Small Motors

Brush holders can be broadly categorized as either radial design or reaction design. *Reaction-design brush holders* establish the brush axis at an angle to the commutator. *Radial-design brush holders* establish the brush axis at a right angle perpendicular to the commutator. Within these categories, the particular design and size of a brush holder will depend on the size of the motor and the application for which the motor is intended. For small motors, the two principle physical designs are the cartridge-style brush holder, shown in Fig. 3.98, and the internal brush ring assembly.

FIGURE 3.98 Photo-cartridge-style brush holder.

---

* Section contributed by John S. Bank, Phoenix Electric Manufacturing.

Cartridge-Style Brush Holders. Cartridge-style brush holders are made to fit into the motor housing (like a cartridge) and have a threaded cap which screws onto the holder. The brush and coil spring are one assembled unit and are pushed into the brush holder. The cap is then screwed on, creating pressure on the coil spring, which in turn keeps the brush pressed against the rotating surface. The cartridge-style holder is made of plastic (usually a thermoset material, which offers the best overall electrical and insulating properties) into which a piece of extruded brass tubing is molded. The brass tubing is cut to size prior to molding and after molding is broached and deburred to ensure a smooth, close fit for the brush. The electrical connection is made by inserting a terminal clip into a specially designed molded space between the plastic body and the extruded brass tubing. A lead wire is crimped into the radial end of the terminal clip. This specially designed terminal clip is then inserted into the brush holder and mechanically locks into place. The current then flows through the terminal clip down the brass tubing into a shunt cap on the top of the brush assembly and through a copper pigtail into the brush itself. Some of the advantages of using a cartridge-style brush holder are the following:

- *Ease of replacing brushes.* Because the cartridge-style brush holder is accessible from the outside of the motor, all that is required when the brushes need to be replaced (or examined for wear) is to unscrew the brush cap from the brush holder and examine the brush. The motor does not have to be opened in a costly, time-consuming, and potentially damaging manner.
- *Use of larger brushes.* Cartridge-style brush holders facilitate the use of longer brushes since they extend to the outside diameter of the motor housing. Longer brushes mean longer brush life and less frequent replacement. Also, longer brushes lead to more stable brush contact with the commutator surface.
- *Less brush noise.* Generally speaking, cartridge-style brush holders can potentially limit the amount of brush noise that would otherwise occur, because the extruded brass tubing can be held to close tolerances. This limits lateral brush movement, adds stability to the brush, and can reduce brush chatter.
- *Better brush wear.* Because the extruded brass tubing is seamless and can be held to close tolerances, the brush is better able to move across the tubing surface. Typical tolerances for the interior brass tubing of a cartridge-style holder range from ±0.001 to 0.0015 in.
- *Lower tooling cost.* Typically, tooling to manufacture a custom-designed cartridge-style bush holder is much less expensive than that for a comparable internal brush ring assembly, thus allowing more variation and flexibility in design considerations.
- *Quality and durability.* Cartridge-style brush holders tend to be more durable and hence tend to have a longer life and be better suited for more rugged applications and more adverse environments.

Because of these advantages, cartridge-style brush holders are associated with higher motor performance and are generally used in higher-end applications where reliability and performance are paramount.

Internal Brush Ring Assemblies. Internal brush ring assemblies (IBRAs), shown in Fig. 3.99, come in various forms and designs. As the name implies, they are assembled units (generally assembled by the supplier) and fit inside the motor housing. The motor housing is completely enclosed. IBRAs usually are either two-pole or four-pole designs. IBRAs are typically made from brass stampings which are either

FIGURE 3.99 Internal brush ring assembly.

crimped or mounted (by screws or rivets) on a phenolic or plastic ring. A torsion spring is then mounted on a separate stud located next to the stamping on the ring. This torsion spring applies pressure to the brush and keeps it riding against the commutator. The electrical connection is made through a copper pigtail attached to the brush, which is in turn attached to another separate metal stud (either mounted in the ring or as a part of the brass stamping), on which a lead is fastened or soldered.

IBRAs are generally less expensive than functionally equivalent cartridge-style brush holders; however, they often present some design problems which the engineer must take into account when considering the end use of the motor. Once a particular design is tooled, design flexibility is limited since the stamping has a seam, which can interfere with a smooth brush ride if not properly fitted and deburred. Brush dust can accumulate in the seam, which can result in arcing. Also, a stamping cannot hold as tight a tolerance as an extrusion. A typical tolerance for the interior of a stamping is in the range of ±0.002 to 0.003 in. Also, brush replacement is difficult, time-consuming, and potentially damaging since the motor must be disassembled to access the brushes. For these reasons, IBRAs are usually used in throwaway-type applications where the motor is replaced rather than repaired.

Another type of IBRA is the one which forms an integral part of the motor housing itself. This is usually a single molded plastic part which provides several functions, such as holding a brush as well as forming the end bell. Pressure is applied to the brush by either a torsion spring or a leaf assembly. This type of design is better suited for high-volume production, because the tooling to design and manufacture this type of IBRA is expensive. In addition, once a particular design is tooled, modification of the tooling is expensive. Also, in certain high-temperature applications the plastic can melt or deform, thus causing the brush to hang up or stop in the brush channel. Tolerances are also an issue since it is difficult and expensive to hold tight tolerances in a plastic brush cannel. The advantages of using an integrally molded IBRA are the following:

- The unit cost is relatively inexpensive.
- They can be used in small motors where space is a premium.
- Since they serve one or more functions, parts can be consolidated.

3.16.2 Application and Design Considerations

As previously noted, the particular size, style, and design of a brush holder will depend on the size of the motor and the application for which the motor is intended. Required torque, voltage, and current will dictate the choice of grade of carbon for the brush and the required spring pressure for the brush. Brush-size and motor-size limitations will in turn determine the size of the brush holder. Other considerations include available space and life requirements.

Clearance Between the Brush and the Brush Holder. How the brush fits inside the brush holder is crucial to the correct functioning of the commutation system and thus the proper performance of the motor. Improper fit can lead to excessive brush wear, decreased brush life, and poor electrical contract. The amount of clearance between the brush and the interior of the brush holder depends on two factors. First, the interior of the brush holder (the brush channel) must be free of burrs, irregularities, and other impediments to ensure a good fit. Broaching and then deburring the brush channel is the best overall method of achieving this. Second, the amount of clearance will depend on the dimensions of the brush. While brushes are generally manufactured undersized and brush holders are generally manufactured oversized, there are currently no recognized standards in the United States for brush–brush holder clearance for small motors.

Table 3.17 is being considered by a National Electrical Manufacturers Association (NEMA) task force as the standard for brush–brush holder clearance for small motors. (It follows the format contained in International Electrotechnical Commission publication IEC 136.)

TABLE 3.17 Brush–Brush Holder Clearance

Phoenix electric brush size	Brush nominal value, in	Brush holder dimension, in	Tolerance, in Max.	Tolerance, in Min.
A	⅛—0.125	0.126/0.128	+0.003	+0.001
B	³⁄₁₆—0.1875	0.188/0.190	+0.0025	+0.0005
C	¼—0.250	0.251/0.253	+0.003	+0.001
D	⁵⁄₁₆—0.3125	0.313/0.315	+0.0025	+0.0005
E	⅜—0.375	0.376/0.378	+0.003	+0.001
F	⁷⁄₁₆—0.4375	0.438/0.440	+0.0025	+0.0005
G	½—0.500	0.501/0.503	+0.003	+0.001
H	⁹⁄₁₆—0.5625	0.563/0.565	+0.0025	+0.0005
I	⅝—0.625	0.626/0.628	+0.003	+0.001
K	¾—0.750	0.751/0.753	+0.003	+0.001

The coefficient of thermal expansion of the brush and brush holder should also be taken into consideration when determining the proper brush thickness, as high temperatures can cause dimensional distortion. In addition, it is prudent to check the brushes on a periodic basis to measure the thickness, as brushes can become worn over time.

Height of the Brush Holder. The clearance between the brush holder and the commutator is important in ensuring that the brush is held stable as the commutator rotates and that the brush remains in proper contact with the commutator. Instability or improper contact will increase brush wear, decrease brush life, subvert the integrity of the electrical system, and adversely affect motor performance. Obviously, the goal is to place the brush holder as close to the commutator as possible. This will minimize the angle that the brush will tilt as it wears down due to use. Again, there are no nationally recognized standards for brush holder–commutator clearance. The proper clearance depends on the size of the motor, but distances range from 0.020 to 0.125 in.

Location of the Brush Holder. The location of the brush holder relative to the commutator is important, as it will determine the location of the brush on the commutator. Brush holders should be spaced equally and should be located in the center of the commutator. Typically, in a large two-pole application the brush holders will be located on the side of the motor, so as to neutralize the effects of gravity. In large four- (or more) pole applications, spring pressure on the brushes located on the underside of the motor must be adjusted to compensate for gravity. The brushes should also be aligned with the commutator bars.

Angle of the Brush Holder. As previously mentioned, brush holders can be mounted either radially to the commutator or in a reactionary relationship to the commutator. A radially mounted brush holder is perpendicular (at a right angle) to the commutator. The principal advantages of this type of design are that it is easier to remove and install the brush and that there is no loss of downward brush-spring pressure. However, the brush tends to be less stable, and there is less brush contact with the commutator.

In a reaction-type design, the holder is mounted at an angle to the commutator. This type of design can be either leading (with the brush angled in a direction opposite to the direction the commutator rotates) or trailing (with the brush angled in the same direction as the direction that the commutator rotates). The principal advantage of this type of design is that it is generally more stable and increases the amount of brush that is in contact with the commutator. However, care must be taken to ensure that the brush is located at the correct angle to benefit from these advantages. The main disadvantage of this type of design is that the brush-spring pressure is dissipated as the angle of the brush and brush holder increases. The brush-spring pressure must be increased to compensate for this loss. In the vast majority of small motors, the brush holders are mounted radially to the commutator.

3.15.3 Summary

While often overlooked, brush holders form an integral part of the commutation system of a mechanically commutated motor. Careful thought must be given to the choice of style of the brush holders as well as to the dimensional aspects of the brush holders, especially as they relate to the brush and the commutator.

*3.16 VARNISH IMPREGNATION**

New varnish resins are having an increasingly larger impact on the final designs of new varnish impregnation machines and systems. Not only are process times and temperatures changing, but likewise the size of the systems is also changing compared with many older conventional systems now using heat-cured varnishes. This does not mean that a new trickle-varnishing system is necessarily better than a conventional conveyorized system. The following basic capital equipment goals still need to be met and justified when evaluating a new system:

- Can the new system meet the production level required?
- Can the new system be integrated into an existing line?
- Can the cost of the new system be justified? Does the return on investment meet the company's requirements?

---

* Section contributed by Peter Caine and Bill Lawrence, Oven Systems, Inc.

- Are there any environmental issues that may determine the type of equipment that is purchased?
- Will the machine produce a good product?

Following are descriptions of a number of different varnish impregnation methods and machines, as well as a description of the key areas in specifying, operating, and maintaining these systems.

Applying varnish to motors, transformers, and armatures is not a new process; consequently, most of the various methods of applying varnish have been around for a long time. The most common methods include dip-and-bake, dip-and-spin, and trickle systems. Some of the parameters that determine what type of system should be used are as follows:

- Type of varnish to be used—solvent based, water based, or low volatile organic compound (VOC)
- Part size and weight
- Quantity
- Conveying and handling preferences
- Available floor space

Smaller parts which are more likely to be processed in larger quantities may tend to be processed automatically on a continuously moving system—for example, armatures for hand power tools or vacuum cleaners.

Larger parts which are more difficult to handle and are typically manufactured in small lots or by the piece lend themselves to a more manual system—for example, large-frame industrial-duty motors.

Following are descriptions and examples of the various systems.

3.16.1 Dip-and-Bake Systems

The dip-and-bake system is probably one of the most common conventional methods of applying varnish. In this process, the complete stator and coil assembly is submerged in a tank of varnish, either under normal atmosphere or under vacuum or pressure.

Various parts may be either preheated or at ambient temperature depending on the type of varnish being used. The size of the part and configuration of the windings will determine the amount of time it takes for the varnish to fill the voids in the slots and windings. Submerging the part for various time intervals, weighing the part, and electrically testing the part after baking will determine the proper dip time. One disadvantage of dipping parts in varnish is that removal of the varnish in unwanted areas is required after baking, thus adding to the cost of the finished assembly.

Conventional dip-and-bake systems using convection heat typically have dip times in the range of 5 to 30 min for larger parts and bake times of 2 to 4 h at 350°F (205°C). Times range from 30 s to 5 min for smaller parts. Parts coated with water-based varnishes will operate at lower temperatures.

Production lines processing over 200 parts per hour most often can justify and utilize a dip-and-bake system.

Dipping systems primarily use three different processing methods.

1. *Indexing rack system.* In the first method, a part is set on a rack or lowered by a hoist into the varnish. It is then removed from the varnish, allowed to drain, and set into a chain-type conveyor system. Once the rack is loaded onto the chains, it

is conveyed through the oven and allowed to cure for the recommended time. Transformers are often manufactured by this method.

2. *Batch oven system.* The second method of curing uses the same dipping system, but the coated rack of parts is placed into a tray oven or batch oven. Midsized stators are often processed by this method. Curing is normally completed all at one level. Stacking the racks with separators is required if multiple levels of parts are to be cured in the oven.

3. *Continuous system.* In the third method, the part is hung on an overhead conveyor which either is indexed on timed intervals or is continuously moving. The conveyor travels through a preheat zone, goes on through the dip tank into the bake oven, and returns back to the loading/unloading area. Quite often, cooling by forced ambient air or chilled air is incorporated into the system as the final step in the process. Parts are hung on the conveyor either by single wire hooks or on a multipart fixture. Multipart fixtures and drag-through dip systems are very common and can process the highest volume of parts per hour.

3.16.2 Dip-and-Spin Systems

The dip-and-spin method is typically used to apply varnish to motor stators. The stator is hung from a hook or is set on a pallet with the bore up, and is then submerged in varnish. After dipping or submerging in varnish, the part is allowed to drain. After a short drain time, the stator is spun about its bore axis to remove excess varnish. The type of varnish used is normally mixed at a low viscosity, allowing only a thin film of varnish to be left on the part after spinning. Corrosion resistance and bonding of the lamination are reasons to use this type of varnishing system.

Dip-and-spin systems are often utilized in the processing of hermetic motors or motors that see a wet environment, such as pool pumps. In such cases a varnish coating over the entire core and coil is crucial.

A palletized conveying system can also be used. The parts are loaded on a pallet that is pushed or conveyed through a continuous oven. The pallets serve a dual purpose of supporting the parts and containing spilled varnish.

The heating and conveying process for the dip-and-spin method may be the same as for the conventional dip system.

3.16.3 Trickle Systems

Trickle systems take their name from the method by which the varnish is applied. Varnish is dispensed in controlled volumes and locations over an armature or stator. The benefits of trickling varnish are as follows:

- The ability to apply varnish where desired
- Reduction of postcuring cleanup operations
- The ability to fill the slots and coil with varnish
- Reduction of the volume of varnish used
- Reduction of the loss of varnish
- Minimization of processing times

These goals are met by choosing the right type of heating method, handling method, and varnish for the part being processed.

Trickle systems typically include a load station for placing parts on a moving conveyor; a preheating zone, which ensures better absorption of varnish; a trickle station or stations; a holding setup zone; a postheating zone; and, in some cases, a cooling zone after the varnish has catalyzed. Finally, the parts are taken off the line and sent to the next process.

Each individual function of a trickle system can make use of different methods of completing each of these tasks.

Heating Methods
- Convection—moving high-velocity heated air over the part
- Infrared—radiant heating by controlling the intensity of light supplied to the part
- Resistance—attaching a resistance heater to the leads of the part and energizing the coil
- Ultraviolet—curing a particular type of coating by UV energy

Conveying Methods
- Continuous belt or slats—used for large motors
- Continuous chain with fixture—used for armatures and stators
- Chain on edge with spindle—used for armatures
- Turret type—used for armatures and small parts
- Single-station or multistation spindle

Fixturing and Rotation. An important issue that has not been described earlier is the requirement that parts be held correctly and rotated.

Stators are most often held and conveyed by securing a fixture in the inner diameter or bore of the part. It is important that minimal contact be made so that bonding of the fixture to the part does not occur. An expanding mandrel of some sort is usually developed.

Armatures are normally held up to the commutator by the shaft, assuming that the shaft is long enough. Common methods of securing the armature are by a drill chuck, a collet, or a clamping device.

Most conventional trickle systems incorporate part rotation in the preheating, varnish application, and postheating stages of the process. Rotation is considered important for uniform part heating and uniform distribution of varnish over the circumference of the part.

Some manufacturers have been successful in applying varnish to stators without rotating the part. In this approach, the part is placed in a fixture with the bore facing up. Varnish is applied from the top only and is controlled so that minimal or no dripping occurs out of the bottom.

Varnish Application Methods

Meter-Mix Pumping Systems. These are pumping systems that have two cylinders pumping a fixed volume of varnish with each stroke. The two separate volumes are forced and mixed into a single tube and out a single dispensing head. This dispensing head then directs the varnish to the correct location.

Provisions need to be made to prevent the contents of the two separate cylinders from mixing inside the pumping equipment. Curing of the varnish in the tubing, mix head, and occasionally the cylinder has most likely happened at least once to anyone who has used this type of pumping system. Most pumps are designed so that they can be broken down and cleaned if this happens.

Meter-mix pumping systems can be used with most varnish formulations that need controlled and accurate mixing. These systems are by far the most reliable method of dispensing accurate volumes of varnish. Consequently, the cost of such systems is higher than those used with other methods. A manufacturer is most likely to use this type of pumping system when mixing has to occur just prior to dispensing the varnish, such as with a quick-setting catalyzed varnish or when the process requires strict control.

Peristaltic Pumps. Many older systems that used solvent-based varnishes on armatures utilized this type of pumping system. Often there are multiple feeder lines that continuously pump the varnish as the armatures travel past the tubing. Occasionally, a system indexes or cycles a number of parts past a fixed number of trickle heads. By doing this, the size of the equipment and system is often minimized.

Since the solvent-based varnishes cure by heat, and heat is generally kept away from the trickle area, the resin that drips off is often recycled back to the main holding tank, remixed, and reused.

With the new resins which catalyze after dispensing and heating, intermittent dispensing of a controlled volume of varnish on an individual part is becoming more common. Dispensing times and volumes based on the part size are normally determined by testing. The actual varnish dispensing time is based on the volume of varnish needed to properly coat the part and the rate at which the part can accept the varnish with minimal dripping or runoff. Part temperature is critical with these new resins. If a part is too hot, the varnish often thins too much when it touches the part. On longer core stacks, the varnish sometimes sets up in the slots and prevents a complete slot fill. If the temperature is too low, proper filling of the slots does not occur and the final varnish film will appear uneven.

Maintenance of peristaltic pumping systems requires periodic replacement of tubing. Since varnish is not exposed directly to the pumping components, the pumps remain relatively maintenance-free. Newer systems that dispense varnish to a single part at a number of precise locations require fewer pumps and pumping lines.

3.16.4 Process Times and Temperatures

Newer varnish formulations cure in minutes rather than hours. In the past, conventional varnishes required preheat temperatures of about 300°F (149°C) and bake temperatures of 350°F (177°C), whereas newer catalyzed varnishes have preheat temperatures in the range of 150 to 200°F (65 to 93°C) and bake temperatures just slightly higher. As a result of these lower cure temperatures, utility costs are also considerably reduced.

A generalized comparison of the energy cost for processing parts using a dip-and-bake resin and two available chemically cured trickle resins is shown in Tables 3.18 and 3.19. Identical 48-frame stator-coil assemblies were used in the

TABLE 3.18 Energy Costs of Using Three Varnish Resins

City	Gas cost, $/therm*	Dip and bake		Trickle, 100:3		Trickle, 100:1	
		Operating, $/h	Heat-up, $	Operating, $/h	Heat-up, $	Operating, $/h	Heat-up, $
Milwaukee	0.49	1.45	2.52	0.60	0.46	0.34	0.19
St. Louis	0.49	1.45	2.52	0.60	0.46	0.34	0.19
Dallas	0.40	1.18	2.05	0.49	0.38	0.28	0.16
Charlotte	0.55	1.63	2.82	0.68	0.52	0.39	0.22
Nashville	0.33	0.98	1.70	0.41	0.31	0.23	0.13

* Therm = 100,000 BTU. All prices as of May 1995 with taxes included.

TABLE 3.19 Annual Energy Operating Costs with Three Varnish Resins

City	Dip and bake	Trickle, 100:3	Trickle, 100:1
Milwaukee	$6490	$2665	$1441
St. Louis	6490	2565	1441
Dallas	5337	2096	1183
Charlotte	7370	2907	1647
Nashville	4432	1752	972
Average cost	$6024	$2377	$1377
Average cost per stator	$0.006	$0.002	$0.001

comparison and some data were estimated. The data are not exact and are for comparison purposes only.

3.16.5 Control Systems

With the shorter process times seen with new low-VOC varnishes, process control information has become a more dominant issue. Most systems can make use of programmable logic controller (PLC) systems. PLCs can simplify the design by reducing the number of electrical components in the control cabinet. They are also now more often utilized to collect data and system operating information.

PLCs can be used to monitor the following components on a system.

- Conveyor travel—speed and index times
- Part detection through photo eyes or proximity switches
- Auto/manual run modes
- Varnish dispensing—start and stop
- System operating conditions such as heating-system status or motor operation
- Safety alarm circuits for operator intervention

Recipes can also be developed to incorporate many of the listed items for individual parts, so an operator can simply pick a part number to run a certain recipe. Trickle heads can be integrated within a recipe to deposit varnish in correct locations. Linear actuators can be added for continuous varnish application from the trickle heads.

Strategically locating the control panel may enable all of these controls to be located in the same panel so that an operator can perform any necessary loading and unloading operations and still monitor the system.

3.16.6 Operational Maintenance Items

The operational maintenance of the system pertains to those items which need to be inspected by an operator or checked electrically to ensure that the system is operating within the control parameters.

These checks are not common to all of the previously mentioned systems.

Varnish Holding Tanks
- Periodic cleaning for buildup of varnish
- Heating and cooling systems
- Circulation systems
- Holding and fill tanks
- Automatic or manual liquid levels

Trickle Equipment
- Tubing line replacement
- Pump-speed settings
- Trickle head location and setup
- Trickle head volume and flow control (closeups)
- Photo eye settings for automatic detection of parts

3.16.7 Preventive Maintenance

- Care of conveyor systems
- Oven systems maintenance
  Fan bearings
  Motors
  Sheaves, belts, and drive units
  Burner flame safety and spark ignitors
- Maintenance of pumps varnish-dispensing

3.16.8 Conclusion

It is evident that it takes years to learn all the tricks of the trade in operating and maintaining the various varnish impregnation systems.

Fortunately, the systems are evolving into systems that are smaller and less complicated. A thorough evaluation of the size of a part, the quantity per hour being produced, and the desired final quality needs to be conducted to determine what type of system will best suit one's needs.

3.17 ADHESIVES*

Designers and manufacturers will discover many benefits and opportunities associated with adhesives and related materials in the assembly and fabrication of small electric motors. Adhesives can provide design advantages, improve overall product performance, speed assembly time, and reduce costs of small motor manufacturing as indicated in Fig. 3.100.

---

* Section contributed by John Cocco, Loctite Corporation.

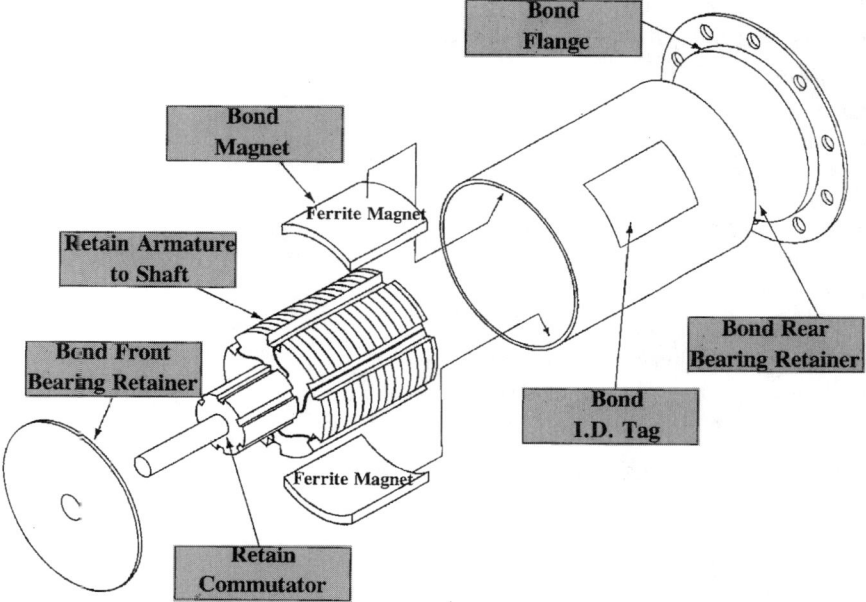

FIGURE 3.100 Small motor application guide. (*Courtesy of Loctite Corporation.*)

This section addresses a wide range of topics. Discussion includes an introduction to adhesives and sealants and an overview of existing chemistries. It addresses a variety of adhesive applications: threadlocking, gasketing, sealing porosity, retaining cylindrical assemblies, bonding, and specialty electronic and electrical applications. The section concludes with a discussion of processing equipment and safety issues.

3.17.1 Introduction to Adhesives

Use of adhesives and related sealing and coating technologies in small motor design and manufacturing has increased dramatically in recent years, replacing many mechanical techniques such as clipping motor magnets and press-fitting components onto shafts. Motor manufacturers' efforts to reduce costs and improve product performance and warranty life have contributed to the growing number of adhesive applications. Adhesive suppliers have also invested significant research in developing safe and reliable products that allow small motor manufacturers to move away from hostile chemical processes.

Adhesives offer significant benefits over mechanical fastening methods. Rather than concentrating stress at a single point, adhesives distribute stress load over a broader area, resulting in a more even stress distribution. A joint bonded with adhesive resists flex and vibration stresses better than a mechanical joint. Adhesives form a seal as well as a bond, eliminating corrosion. They easily join irregularly shaped surfaces, negligibly increase the weight of an assembly, and create virtually no change in part dimensions or geometry.

It is important to note that there is no single adhesive that will perform under all circumstances. Limitations of adhesives include setting and curing time (the amount of time it takes for the adhesive to fixture and strengthen fully), surface preparation requirements, and the potential need for joint disassembly. Proper adhesive selection and component design will often overcome these limitations.

A good adhesive is safe for the manufacturer and the end user and is environmentally friendly. It will adhere to the substrate, but will not attack or degrade it. A good adhesive is less costly than any mechanical fastener available. In its liquid state, the adhesive must be capable of flowing into the bond joint, and it must allow easy processing. Solidification or curing of the adhesive should occur in a predictable time frame and minimize work in process. The bonded joint must be strong enough to last the life of the assembly, and every assembly must offer consistent quality and performance.

Joint design is a critical factor in bond strength—a well-designed joint allows for the maximum possible bond area and combines mechanical locking methods with adhesive bonding. The adhesive application and curing processes must remain constant to ensure consistent performance joint after joint, assembly after assembly.

There are four types of failure that can occur in an adhesive joint. *Adhesion failure* occurs when the adhesive completely separates from the face of one substrate. When the adhesive splits from itself and remnants can be found on both substrates, the adhesive joint has experienced *cohesion failure*.

With adhesion failure, the weak point of the bond is the boundary layer between the bonded part and the adhesive. Either the material is not suited for bonding, or there is dirt on the bonding face. In both cases, bond strength will increase with suitable pretreatment of the surface.

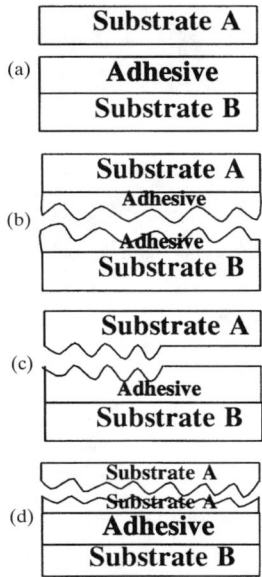

FIGURE 3.101 Mechanism of bond failure: (*a*) adhesive failure, (*b*) cohesive failure, (*c*) combination adhesive and cohesive failure, and (*d*) substrate failure. (*Courtesy of Loctite Corporation.*)

In a cohesion failure, the adhesive is overstressed by external factors—for example, temperature, aging, or stress spikes. Either the adhesive is not suitable for the application, or design changes should be made to the bonding geometry.

The third failure mode is a *combination of adhesive and cohesive failure*, while the fourth mode is *failure of the substrate* itself.

The best small motor designs that incorporate adhesives and sealants result when designers consider four important questions before specifying the final design and the mass production process. These questions are also important when troubleshooting an existing application.

1. *Does the adhesive bond to the substrate surfaces?* Will the material maintain adhesion over time? Figure 3.101 illustrates the various types of failure that may occur in an adhesive joint. In evaluating an adhesive design, it is important to identify the normal failure mode. Periodic monitoring of the normal failure mode will help to ensure consistency. For example, if the joint is evaluated and *cohesive* failure is determined to be the normal failure mode, then a sudden change to *adhesive* failure might indicate that surface contamination has been introduced to the manufacturing process.

In addition to matching the adhesive to the specific bonding surfaces, it is important to be aware of factors that influence wetting. *Wetting* is the degree to which an adhesive penetrates the surface of a substrate, spreading out and filling surface irregularities, as shown in Fig. 3.102. Good wetting is essential for developing reliable bonds. If wetting does not occur, the adhesive does not spread but forms a round droplet on the surface, the way water beads up on a newly waxed car, as shown in Fig. 3.103. Factors that affect surface wettability include substrate porosity, roughness, polarity, cleanliness, surface tension, and adhesive viscosity.

FIGURE 3.102 Proper adhesive wetting.

FIGURE 3.103 Improper adhesive wetting.

Cleanliness is crucial to wetting. All dirt, residues, and oils should be removed from the substrates using detergents, solvents, chemical cleaning systems, or mechanical abrasion.

2. *Can the adhesive be consistently dispensed into the bond joint and remain there until solidification occurs?* It is often difficult to ensure that adhesive is present in the bond joint. In small motor manufacturing, adhesive migration can cause a range of problems, from loss of electrical contact to failure of the painting processes. Figure 3.104 illustrates a common problem experienced when bonding cylindrical parts into blind holes; it also illustrates the solution, which takes advantage of the flow of both air and adhesive.

3. *Is the adhesive solidifying or curing?* Has the adhesive solidified enough to ensure its performance in the application? It is important that the individuals who control the manufacturing process have a basic understanding of the factors that affect adhesive curing. Frequently, in small motor manufacturing, chemically reacting adhesives are used. These products solidify or cure through a chemical reaction to form highly durable polymeric materials. For example, anaerobic adhesives, used to bond cylindrical commutators to shafts, will solidify faster at elevated temperatures or if trace amounts of copper are present on the shaft surface to catalyze the curing reaction.

4. *Will the adhesive perform throughout the assembly's life?* Will the adhesive meet the performance required for the manufacturing process and the end-use environment? Small motor assembly utilizes many commercially proven adhesive applications—for example, motor magnet bonding, threadlocking, cylindrical retaining, and laminate sealing and unitizing. Adhesive suppliers have created commercial formulations specifically designed to fulfill the requirements of these applications. Design engineers who specify proven adhesive formulations in their designs can

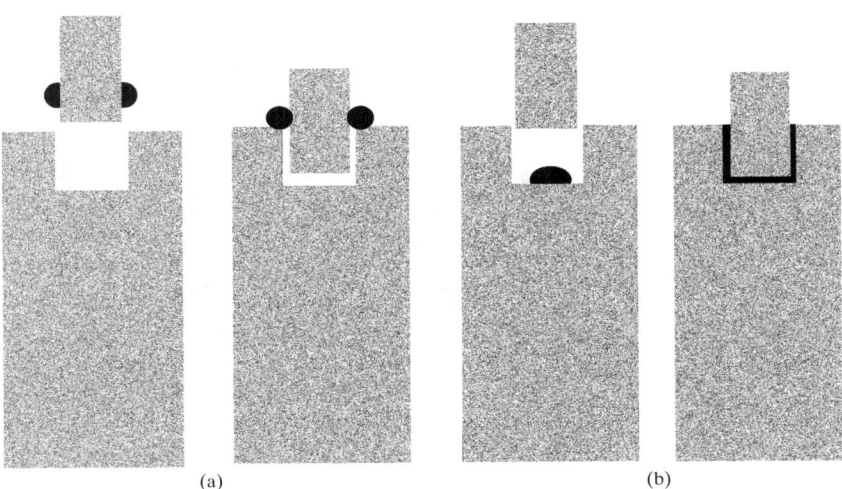

FIGURE 3.104 Blind hole example of adhesive placement. (*a*) Wrong—adhesive forced out of bond joint by air during assembly. (*b*) Right—adhesive forced into bond joint during assembly. (*Courtesy of Loctite Corporation.*)

minimize the risks associated with adhesive specification. However, since all designs possess unique requirements, both short- and long-term performance testing is suggested to ensure a design's success.

To ensure a good, long-lasting bond, the properties of the adhesive must match the requirements of the manufacturing process and those of the end-user application. For example, if the adhesive used is not formulated for solvent resistance, and the manufacturing or end-use environments subject the assembly to solvents, the bond will fail.

The manufacturer should consistently test devices for a series of performance factors once the adhesive is fully cured and the assembled device is ready for end use. A designer concerned about thermal performance, for example, must consider the amount of time that the exposed part will be subjected to high temperatures. Figure 3.105 graphically illustrates heat aging and hot strength. *Hot strength tests* are used to predict bond strength at a selected temperature. *Heat aging tests* are used to determine long-term durability after exposure to elevated temperatures.

Most reputable adhesive suppliers will assist in developing testing programs, and in many cases, will provide testing services at no charge.

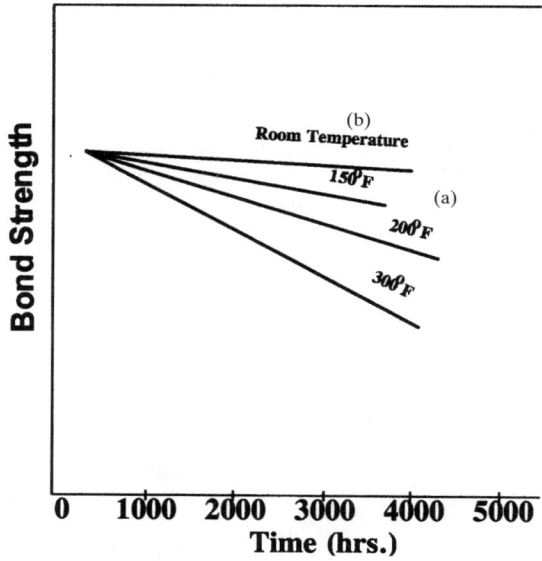

FIGURE 3.105 Hot strength and heat aging are both important measures. (*a*) Hot strength—bond tested at temperature. (*b*) Heat aging—bond tested at room temperature. (*Courtesy of Loctite Corporation.*)

3.17.2 Joint Design Guidelines

Understanding the basic types of mechanical stresses will help in joint design. Typical joint stresses are illustrated in Fig. 3.106.

- *Tensile stress.* Occurs when the force on a joint is pulling the substrates away from the bondline. Tensile stress tends to elongate the object.

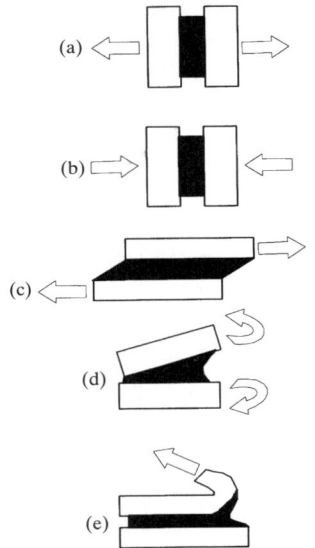

FIGURE 3.106 Types of joint stress. (*a*) Tensile stress tends to pull an object apart and tends to elongate an object. (*b*) Compressive stress tends to squeeze an object together. (*c*) Shear stress results in two objects sliding over one another. (*d*) Cleavage stress occurs when a joint is being opened at one end. (*e*) Peel stress occurs when a flexible substrate is being lifted or peeled from the other substrate. (*Courtesy of Loctite Corporation.*)

- *Compressive stress.* Occurs when the force on the joint pushes in toward the bond line, squeezing the assembly together. A bonded joint under compression is extremely strong as long as the adhesive used is resilient and flexible.
- *Shear stress.* Occurs when the two substrate surfaces slide over one another in opposite directions, with the majority of the stress concentrated at the edges of the joint.
- *Cleavage stress.* Occurs when the joint is being pried open at one end. Brittle adhesives have poor resistance to cleavage load.
- *Peel stress.* Occurs when a flexible substrate is being lifted or peeled away from the other substrate, with the load concentrated along a thin line at one edge of the bond.

Engineers must have a good understanding of how stress is distributed in order to design the strongest possible joint. Two universal guidelines will help during the design process:

- *Maximize shear/minimize peel and cleavage stress.* Shear stresses are distributed evenly across the bond, with the ends resisting more stress than the middle. When a cleavage or peel stress is applied to a joint, most of the stress is concentrated at one end, weakening the bond line and the life of the joint.
- *Maximize compression/minimize tensile stress.* When a bond experiences a tensile or compressive stress, the joint stress is evenly distributed across the entire bond. In most adhesive bonds, compressive strength is greater than tensile strength; therefore, an adhesive joint experiencing compressive force is less likely to fail than a joint undergoing tension.

Figure 3.107 shows several common bond joints that are successfully used in designing and assembling small electric motors.

3.17.3 Common Adhesive Chemistries

The following adhesive, sealant, and coating chemistries react to form solid bond joints. Chemically reacting adhesives have become extremely popular for small motor assembly because of their long-term durability and versatility.

Anaerobic adhesives are commonly used to bond cylindrical metal components such as fans, laminates, and commutators to armature shafts. They are also used as threadlockers and thread sealants designed to seal and minimize loosening of metal

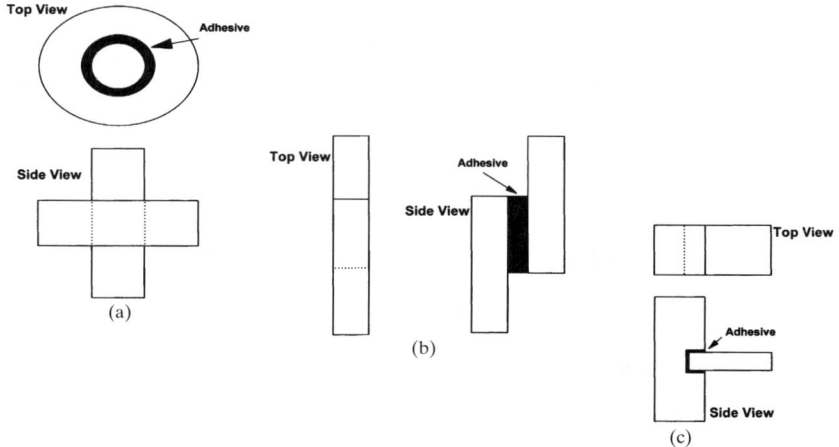

FIGURE 3.107 Common bond joint configurations: (*a*) cylindrical socket, (*b*) single lap, and (*c*) dado angle joint. (*Courtesy of Loctite Corporation.*)

fastener assemblies. They are single-component thermoset materials which cure at room temperature when deprived of oxygen in the presence of metal ions. Anaerobics offer high shear strength, good solvent and temperature resistance, rapid curing at room temperature, and easy dispensing. These products are typically used when gaps do not exceed 0.020 in and temperatures do not exceed 450°F (232°C).

Two-part, mix or no-mix *acrylics* are thermoset adhesives that bond well to lightly contaminated surfaces and are flexible and chemically resistant. They offer fast fixture, good adhesion to many substrates, high peel and impact strength, and good environmental resistance. The latest acrylic adhesives are rubber toughened for increased pliability and resilience. Acrylic adhesives are the most common motor magnet-bonding formulations.

Epoxies are common thermoset structural adhesives available in single-component heat-curing or two-part mix systems. They offer a number of benefits, including high bond strength on a wide variety of substrates, good gap-filling capabilities, excellent electrical properties, and good temperature and solvent resistance. Proper mixing and/or controlled handling and storage conditions are required to ensure successful use of these adhesives.

Cyanoacrylates, also known as *instant adhesives* or *superglues,* are one-part, solvent-free, room-temperature-curing thermoset adhesives. When pressed into a thin film between two surfaces, cyanoacrylates cure rapidly with excellent adhesion to most substrates. These adhesives are available in a wide variety of specialty formulations and viscosities, are easy to dispense via automated systems, and offer excellent bond strength, even after thousands of hours of exposure to temperatures as high as 250°F (121°C).

Extremely compatible with a wide variety of substrates, *urethanes* are available in one- or two-component systems, and cure to a flexible solid when exposed to ambient moisture. They offer excellent toughness and flexibility, even at low temperatures. Urethanes can cure at room temperature or in an oven, and offer good impact resistance. These products are typically used where gaps do not exceed 0.100 in and temperatures do not exceed 250°F (121°C).

Silicones are rubberlike polymers which cure at room temperature. They exhibit excellent resistance to high temperatures and moisture and are well suited for outdoor weathering applications as sealants and caulking compounds. Silicones have superior chemical and electrical resistance, and good gap-filling capabilities, and they are excellent for bonding glass to most other substrates. However, their low tensile strength limits their usefulness in carrying structural loads. It is important to maintain controlled curing conditions (e.g., ambient temperature and humidity) to ensure optimal results with silicones.

Many of these technologies can be adapted to employ UV or visible light to initiate curing. UV- and visible-light-curing adhesives contain photoinitiators which absorb light energy to begin the cure. This technology is excellent for manufacturing applications, as it allows the user to take as much time as necessary to position parts. The capability to cure on demand speeds processing and reduces total production costs. Upon exposure to a light source, the adhesive can be fully cured in less than a minute.

UV- and visible-light-curing formulations are becoming increasingly popular in small motor manufacturing for a number of reasons:

- They contain no solvents and do not require heat curing.
- They eliminate the need for holding racks and ovens.
- They allow flexible work-cell manufacturing configurations.
- They cure quickly and with precision control.

Light-curing products are currently replacing heat-curing and two-part epoxies, varnishes, trickle coatings, and many other assembly systems commonly found in small motor manufacture.

3.17.4 Adhesives in Small Electric Motor Assembly

Threadlocking. Mechanical fasteners and adhesives can work together to form a stronger bond than either method alone can provide. Design engineers who want to improve the safety and quality of an assembly will use a mechanical fastener in tandem with a threadlocking adhesive. The anaerobic threadlocker guarantees that the assembly will not fail or loosen and that corrosion will not reduce the life of the fastener.

Since threadlocking agents prevent loosening and movement and act as sealants, they can be used on small motor assemblies for tamperproofing, mounting bolts, locking assembly screws, bolting flanges, and setting adjustment screws. Threadlockers also offer excellent performance when used on severe-duty fasteners since they seal the bond joint and prevent corrosion. They offer excellent performance in high-temperature and high-vibration areas.

Threaded assemblies fail for two reasons. One cause is *tension relaxation*. Temperature changes, for example, cause bolts and substrate materials to expand and contract, reducing bolt tension and lowering the clamping force. The second cause of threaded assembly failure is *self-loosening* caused by sliding or vibration between contact surfaces.

Threadlocking adhesives prevent bolts from loosening by completely filling microscopic gaps between interfacing threads. In fact, threadlocking agents maintain clamp load better than lock washers in high-vibration environments. Liquid anaerobic threadlockers cure to a tough solid when they contact metal ions in the absence of air. When selecting a threadlocking adhesive, designers should consider end-use operating temperature, thread size, chemical and environmental factors, bolt reusability, and the need for a primer to speed curing.

When applying threadlockers, the total length of the thread must be wetted. Proper wetting depends on the size of the thread, the adhesive's viscosity, and the size of the parts. Nothing should be present on the assembly that will restrict curing. Assemblies must be free of all oils or cleaning systems that might impede curing.

Threadlocking adhesives offer a number of benefits. One distinct advantage is bolt reusability. Each adhesive has break-loose torque values ranging from low, which may be disassembled using normal tools, to high, which is difficult to disassemble. Threadlockers with low to medium strength can easily be loosened without damaging bolts. A bolt treated with threadlocker may be reused by removing old adhesive before applying new.

In addition to preventing movement, threadlocking adhesives can also be used to seal the joint, blocking out moisture, gases, fluids, and corrosives which can reduce the life of the assembly, as shown in Fig. 3.108.

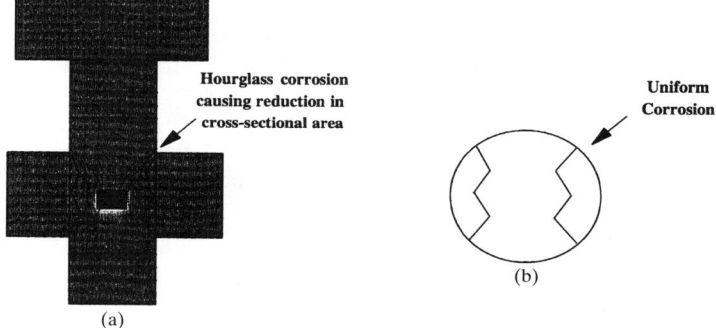

FIGURE 3.108 Benefits of adhesive to seal inner spaces in threadlocking. (*a*) Accelerated corrosion (crevice corrosion) causes nut lock or failure. (*b*) Uniform corrosion begins at exposed surfaces. (*Courtesy of Loctite Corporation.*)

Another benefit of threadlocking adhesives is the reduction of overall assembly costs by using standard bolts in place of special locking bolts. Liquid adhesive is effective regardless of bolt size and diameter, eliminating the need for elaborate inventories of specialized mechanical fasteners.

Anaerobic threadlocking adhesives are now available as coatings preapplied to threaded fasteners. When the fastener is assembled, microcapsules containing adhesive are crushed, releasing the liquid threadlocker. In addition to cost and logistical benefits, preapplied coatings improve the quality of the assembly by ensuring that a consistent amount of adhesive is dispensed every time.

When selecting an anaerobic threadlocker, small motor designers must take several factors into consideration. The continuous operating temperature and chemical and environmental conditions of the end-use environment will determine the performance criteria required of the adhesive. Thread size will dictate the adhesive's viscosity. The designer must know whether the part will be disassembled in the future in order to determine the appropriate adhesive strength. And the substrate materials used will determine whether a primer will be required to accelerate curing.

Sealing Pores and Laminations. Advanced product designs and new manufacturing techniques require modern methods for filling and sealing the voids, or *microporosity*, in substrate surfaces such as metal castings, powdered-metal parts, electronic components, plastic composites, weldments, and other porous materials. Pore sealants, also called *impregnation resins,* are used in a variety of applications, including sealing pores in metal housings and other structural components, unitizing commutators and other porous subcomponent parts, sealing electrical connectors, sealing magnets and other nonstructural components, and unitizing windings and laminate stacks.

Today, designers are creating new lightweight, thin-walled die castings for use in applications containing high-pressure fluids. These castings would not be possible without the sophisticated new vacuum-impregnation systems and sealants. Designers seeking to decrease the weight or cost of assemblies rely upon impregnation, which enables them to design thinner-walled die castings or to switch to lower-cost methods, such as powdered metal. Modern resin impregnation systems are so well accepted, effective, and economical that traditional leak testing of heavily machined castings has been phased out in favor of impregnation of parts.

There are two types of porosity. *Macroporosity* requires remelting because the gaps affect structural integrity. *Microporosity* does not affect structural strength and is the natural result of two physical phenomena which occur as molten metal solidifies—crystal formation and shrinkage, and gas absorption. Visible surface defects are usually not improved by pore sealing as the sealant washes out of the surface defects during processing, leaving them unchanged. Porosity that occurs on a microscopic scale can be readily sealed through impregnation. Even if a pore is large scale within the casting, it can be sealed if the surface opening is not large.

The most common reason for impregnation of castings is to enable parts to retain fluids under pressure. Properly impregnated, components such as hydraulic pump and motor parts are permanently sealed and can hold pressures up to the burst strength of the casting. Castings are also impregnated in preparation for metal-finishing operations such as painting or plating. In some instances, it may be desirable to seal pores in a casting so that corrosive fluids cannot enter. This is done to prevent corrosion that may originate within the porosity.

Castings should be fully machined, cleaned, and dried before impregnation, and should be at room temperature. Only after all machining is completed can the impregnation sealant reach into all areas that need to be sealed. The pores of clean, dry parts will not be blocked with foreign materials that might prevent thorough penetration of the impregnation sealant.

Powdered-metal parts should be impregnated after sintering but before any secondary operations. The pores are normally completely open at that point and can be filled by the impregnation sealant.

Plating, painting, or other finishing operations for either castings or powdered-metal parts should be done after the impregnation is complete and the sealant has been allowed to fully cure. The cured sealant in the impregnated parts will not be affected by the various cleaning and etching steps, even when strongly acidic solutions are used.

The unique self-curing capability of anaerobic sealants, along with the ability to regulate the rate of cure, has made anerobics the most reliable family of materials for sealing the porosity of metallic and nonmetallic parts. The ability to cure without heating the parts, combined with the use of an activator rinse to speed curing, gives the anaerobic sealants much greater capability in sealing a wide range of pore sizes, especially larger pores. Bleedout will not happen with anaerobic sealants, so part fouling does not occur and sealing performance is consistently high.

Certain formulations of cured anaerobic sealants are suitable for continuous service up to 392°F (200°C). Brief exposure to higher temperatures, as might occur in paint ovens, will not normally cause problems. These sealants can be used successfully in such high-temperature environments as automotive engine blocks, cylinder heads, and coolant pumps. The cured sealant is a thermoset plastic which will not melt, liquefy, or run out of parts at elevated temperatures. If operated beyond acceptable temperatures, the cured sealant will slowly lose weight and turn to ash.

Impregnation sealants are tested by the manufacturer for chemical resistance to determine their suitability to environments exposed to fuels, lubricants, coolants, cleaners, and other chemicals that may be encountered in automotive, aerospace, and general industrial environments.

Gasketing (Flange Sealing). Gaskets (flange sealants) prevent leakage of liquids or gases by forming impervious barriers between mating flanges. In small motor applications, flange sealants are used for gasketing fluid- and air-tight components, sealing crimped components, sealing flange couplings for improved power transmission, and ensuring the performance of hermetically sealed components.

There are three types of flange gaskets: conventional *compression gaskets* of cork, paper, rubber, metal, and other asbestos-free materials; *cured-in-place liquid compression gaskets* cured in seconds with UV light prior to assembly; and *formed-in-place liquid gaskets* cured after the parts are assembled. All three types of gaskets must create and maintain seals, remain impervious to fluid flow, and be compatible with the machinery.

Liquid formed-in-place (FIP) gasketing materials are used to dress conventional rubber, paper, or cork gasketing materials, and can often completely replace cut gaskets. These liquid anaerobic gasket dressings fill surface imperfections in the mating flanges and extend the life of the gasket. When parts are assembled, the flange sealant flows into voids, gaps, and scratch marks, forming a durable seal after curing. Formed-in-place gaskets offer a number of advantages over precut compression gaskets, including improved reliability, reduced costs, and easier application and service. Two common types of FIP materials are room-temperature-vulcanizing (RTV) silicones and anaerobic compounds.

Anaerobic FIP materials are generally used on rigid joints made of cast aluminum or iron, such as pumps, engines, and transmissions. Anaerobic gaskets offer several benefits over traditional sealing systems. They will not relax, so retorquing is never necessary. These materials allow metal-to-metal contact, and therefore do not require shimming. Anaerobics will not begin to cure before flanges are mated, and any excess material which squeezes out of the flange faces will not cure and can be easily wiped away. These high-shear-strength materials demonstrate excellent solvent resistance and chemical compatibility, and eliminate the need for large inventories of precut gaskets.

RTV silicones cure by reacting with atmospheric moisture and are typically used on high-movement joints made of stamped steel or molded plastic. RTV silicones are applied as beads to the flange area of a component, and flow over the flange to form a gasket that completely fills voids, scratches, and other surface imperfections. They provide a small amount of squeeze-out on both the inside and outside of the joint, forming a fillet around the flange which acts as a secondary seal. FIP silicones are capable of filling large gaps, flex easily in response to the movement of the flange, adhere well to many surfaces, and withstand a wide range of temperatures. As with anaerobic materials, silicones eliminate the need for large and costly gasket inventories.

Silicone cured-in-place (CIP) gasketing involves positioning a compression gasket as a permanent part of one flange surface. The gasket is created by a tracing machine which applies precise beads of silicone to the flange surfaces. The beads are cured and bonded to the component flange in 30 s with UV light, or in 7 to 14 days with slow moisture curing. Sealing is achieved through compression of the cured gasket during assembly of the flange joint. CIP compression gaskets offer many advantages over die-cut rubber, die-cut foam rubber, or molded gaskets. Equipment manufactured with CIP-sealed access ports is easy to produce and service. CIP gaskets reduce labor costs, improve product quality, reduce gasket inventories, and allow flexibility in the manufacturing process.

Retaining Cylindrical Assemblies. The term *retaining compounds* describes adhesives used in cylindrical assemblies joined by inserting one part into another. A typical example is a bearing mounted in an electric motor housing. Retaining compounds enable buyers of new bearings to salvage worn housings.

Incorporating retaining compounds into the assembly of small electric motors provides a number of performance benefits. Retaining compounds can augment, improve, or replace normal press fits and can increase the strength of heavy press fits. They also eliminate distortion when installing drill bushings. Retaining creates an essential seal which eliminates fretting corrosion and seizure. Finally, retaining compounds help to reduce stress in parts and enhance the combined performance of substrate materials.

Retaining compounds are used in a variety of small motor applications: mounting bearings in housings or on shafts; attaching pulleys, cams, fans, and gears; replacing or augmenting interference fits, keyways, and splines; retaining commutators and mounting flanges; and assembling armatures. Retaining compounds work extremely well on cylindrical components made of dissimilar materials, as well as on parts that must be protected from corrosion and backlash.

Retaining compounds are primarily anaerobic materials, used to bond metal to metal or metal to a nonmetallic substrate. However, for nonmetallic substrates such as plastics, cyanoacrylate adhesives can be used.

Two key formulation variables exist for anaerobic retaining compounds. The viscosity of the adhesive determines its ability to fill gaps in the assembly—the higher the viscosity of the material, the greater its ability to fill gaps. Like threadlockers, anaerobic retaining compounds are formulated in different strengths—from low, which may be easily disassembled, to high, which is difficult to disassemble. An assembly treated with the appropriate-strength anaerobic retaining compound may be disassembled and reused.

As with all adhesives, it is extremely important that substrates be thoroughly cleaned before retaining compounds are applied. Contamination will greatly inhibit curing, especially on metallic substrates.

Retaining compounds are crucial in two types of cylindrical assemblies. In bonded slip-fit assemblies, the shaft is inserted into the hub with no force required. Cured adhesive located between the shaft and the hub transmits the load or torque. In bonded interference-fit assemblies, friction and the adhesive combine to transmit the load or torque. There are two types of interference fit assemblies—bonded shrink fits, where the hub is heated to allow insertion of the shaft, and bonded press fits, where axial force allows insertion of the shaft

In selecting an appropriate retaining compound, designers should consider the type of assembly being bonded, the thermal expansion of the individual substrates, the radial clearance of the bonded slip-fit assemblies, the required shear strength and ease of disassembly, and the continuous operating temperature of the end-use environment.

Bonding. A number of adhesive technologies are commonly used for bonding a variety of substrate surfaces to one another. Common surface-bonding opportunities for small motor assembly include flat-face bonding, plastic and sheet-metal bonding, motor magnet bonding, affixing nameplates or identification tags, attaching the housing to the base, bonding armature segments, and bonding the cooling fins to the heat sink.

When bonding assemblies with large gaps, high-viscosity adhesives can act as fillers, occupying space between substrate surfaces. However, there is an inverse relationship between the size of the gap and the overall performance of the adhesive. Large gaps reduce the adhesive's bond strength and durability and increase the fixture time required, diminishing the adhesive's effectiveness.

Magnet Bonding. Acrylic adhesives are now used extensively to bond magnet segments in motor manufacturing. Millions of motors are operating successfully in demanding environments using only adhesive systems to reliably secure their magnets.

Magnet bonding has evolved significantly over the past decade. Ten to fifteen years ago, it was common to use spring steel clips to fix magnet segments into position. In addition to the relatively high cost of materials, magnet clipping was difficult to automate and required relatively complex parts-handling systems and insertion. In addition, clips could loosen or shift, allowed corrosion to occur between the magnet and the motor assembly, and did nothing to prevent noise and vibration.

As cost and performance became increasingly important, many motor manufacturers who were skeptical of adhesives' bonding performance began to use epoxy adhesives to prevent corrosion and reduce noise, but still augmented the bond using clips. Adhesive suppliers began to develop formulations specifically for magnet-bonding applications. Specialty acrylic adhesives that were tough and environmentally resistant were introduced to bond magnets without clips.

Adhesives proved successful in motor magnet bonding because they were designed to process quickly and bond reliably in less-than-ideal industrial environments. In addition, adhesive bonds last through the severest environments that a motor may encounter.

Factors relating to the magnet and the housing play a critical role in successful magnet bonding. Magnets must be dust-free and be formed or machined to ensure that gaps between the magnet and the housing remain small. Most magnet-bonding adhesives solidify faster and are stronger when gaps are less than .010 in. For curved magnet segments, some motor manufacturers have moved from single-radius segments to the tri-arc configuration seen in Fig. 3.109. This design helps to reduce movement of the magnet in the fixture, and can help reduce the gap due to tolerance differences between the housing and the magnet. Examination of the tri-arc bond joints shows that the adhesive fixtures or solidifies quickly at the two points of contact on the magnet surface.

The surface and shape of the housing can be crucial to the success of the bonding design. The surfaces of housings are often fabricated so that they resist corrosion and can be painted. Conversion coating processes such as chromating, phosphating, galvanizing, and anodizing are used to fabricate housing surfaces. In most cases, prepaint coatings such as zinc phosphate or chromic acid anodizing are the best for adhesive bonds. Coating processes that leave weak surface layers can be problematic. Such processes include galvanizing or yellow zinc dichromating. Some adhesive manufacturers have developed primers that repair weak coating surfaces.

The method used to fabricate the housing will often affect its ultimate dimensional tolerance. Tighter control of housing dimensions can be used to ensure small bondline gaps. Drawn or extruded housings are typically the most stable from a dimensional standpoint. Magnet bonded housings are commonly roll-formed housings, which are more difficult to control dimensionally. When rolled housings are

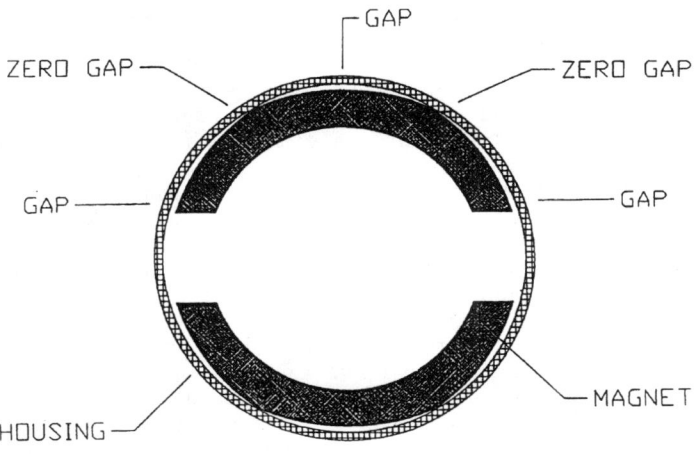

FIGURE 3.109 Triarc magnet assembly.

used, it is important that the seam be smooth and properly fitted. Misalignment or tabs that cause large gaps are unacceptable. The surface finish range that optimizes adhesive strength is a 62- to 125-μin RA finish.

3.17.5 Other Applications

Adhesive chemistry and technology have been used in a variety of less traditional applications. Specialty adhesives have been formulated to provide a number of benefits in electrical and electronic device assembly, including thermal conductivity for heat sinks, coil termination and strain relief, potting and encapsulation, conformal coatings, wire tacking, and armature balancing.

3.17.6 Processing Equipment

The capital expenditure for setting up production-line bonding systems will pay for itself in materials and labor cost savings, particularly on high-volume production lines. A broad range of dispensing equipment is available, from manual systems for small-batch production or repairs to fully automated systems. The adhesive used determines the type of dispensing equipment required.

Viscosity is one decisive factor in choosing a dispensing system. The adhesive's flow properties determine the control technique and the valve choice for adhesive application. In order to properly wet the bond faces, the properties of both the adhesive and the surfaces must be controlled. With sufficient experience, an adhesive manufacturer can integrate individual components of each dispensing unit into a complete system that can cover most dispensing tasks.

Increasingly, automatic dispensing systems are being used to ensure a well-bonded joint by consistently applying the correct quantity of adhesive in the proper place. Automatic systems can easily be monitored and checked for correct positioning and adhesive quantity.

3.17.7 Safety

Employee safety and environmental friendliness are two issues of concern throughout the industrialized world. Manufacturers are striving to develop production processes which ensure compliance with the stricter safety and environmental standards in place today. The trend toward safety has created both challenges and opportunities for the adhesives, sealants, and coatings industry.

Although most adhesives have evolved to meet stringent safety requirements, improper handling and process design will yield unnecessary problems and expenses.

It is possible to achieve a safe and trouble-free adhesive process if safe handling and use issues are attended to during the process design stage. Implementing an appropriate safety and ventilation strategy can make the difference between success and failure.

Extensive information concerning the safe handling of chemicals in the workplace exists, but it is scattered and difficult to gather, filter, and apply to specific operations. Your adhesive supplier should be able to provide guidance on material safety; safe shipping and storage of unopened containers; minimizing and monitoring exposure to potentially dangerous materials; and waste collection, storage, and disposal.

3.18 MAGNETIZERS, MAGNETIZING FIXTURES, AND TEST EQUIPMENT*

This section is intended to provide practical guidelines for analyzing and selecting equipment for magnetizing, demagnetizing, and measurement systems. An overview of the information required to analyze a typical application and some of the reasoning for each item is provided. A checklist of required information and descriptions of various magnetizing and calibration systems follows. In many cases, several different magnetizing systems could be recommended for any given application. The most important aspect of the analysis is developing a complete picture of the requirements. This can happen only with an ongoing dialog between the product design and manufacturing engineers and the equipment supplier.

In the early product development stage, magnet suppliers can assist project engineers in the selection of the appropriate magnet material for any given project. With the extreme range of magnet materials available today (see Table 3.20), a close working relationship with magnet producers and equipment suppliers is increasingly important.

Magnetizer equipment manufacturers can assist you in selecting the right equipment after the magnet material has been selected.

3.18.1 Checklist

The following items should be discussed by the project engineers and the product suppliers:

- Magnet material
- Grade
- Characteristics

---

* Section contributed by Roger O. LaValley, Magnetic Instrumentation.

TABLE 3.20 Magnetizing Force for Some Common Permanent-Magnet Materials

Material	Coercive force H_c, Oe	Required magnetizing force, Oe
Alnico I	400	2,000
Alnico II	460	2,500
Alnico II	550	2,500
Alnico V	620	3,000
Alnico VDG	650	3,500
Alnico IV	700	3,500
Alnico V-7	700	3,500
Alnico VI	750	4,000
Alnico XII	950	4,000
Alnico VII	1,100	5,000
Alnico VIII	1,500	6,500
Ceramic 1	1,850	10,000
Ceramic 5	2,400	10,000
Ceramic 8	2,950	12,000
Samarium cobalt (RE_2Co_5 and RE_2Co_{17})	4,500–12,000	20,000–100,000
Neodymium	3,500–13,000	20,000–45,000

- Operating point
- Orientation
- Physical dimensions
- Configuration or shape
- Polar configuration
- Magnetizing and calibration requirements
- Percentage of demagnetization and required accuracy
- Parts handling per cycle rate and system control
- Operating environment
- Measuring and testing Options

3.18.2 Magnet Material

What does one need to know about the magnet material? The magnet grade and energy product, MGOe; the required magnetizing force for saturation H_c and H_{ci}; the B_r; whether the material is nonoriented or oriented; and the operating point of the magnet or assembly. This information is used to determine the field strength required for magnetizing and demagnetizing, as well as the preferred direction of the field and calibration requirements (see Fig. 3.110).

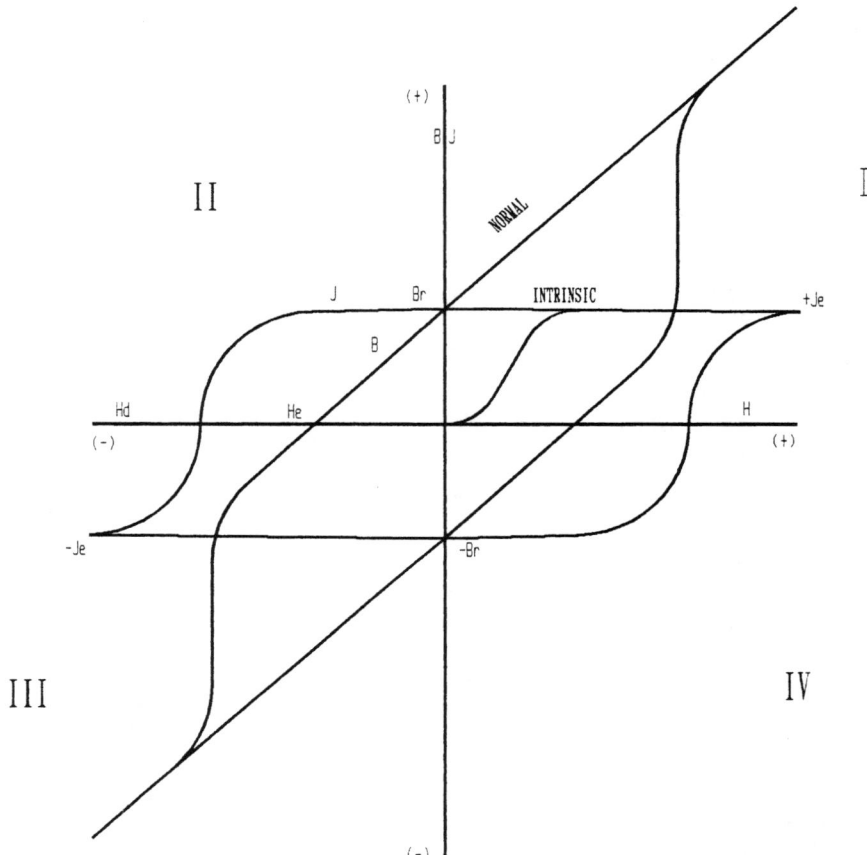

FIGURE 3.110 Magnet hysteresis loop.

3.18.3 Physical Dimensions and Shape

What physical information is required? In short, as much as is available: shape, diameter, axial length, and a drawing of the magnet with tolerance information (see Fig. 3.111). This information helps to narrow down the size of the magnetizing fixture and the amount of energy required.

3.18.4 Polar Configuration

The polar configuration information relates to how the magnetizing field will be presented to the magnet or magnet assembly (see Fig. 3.112). Some of the items to be discussed are the number of poles, spacing between poles (if crucial), physical location or alignment of the poles on the magnet, requirement for skewed-angle poles, and requirements for alignment with respect to other items in magnet assembly. This

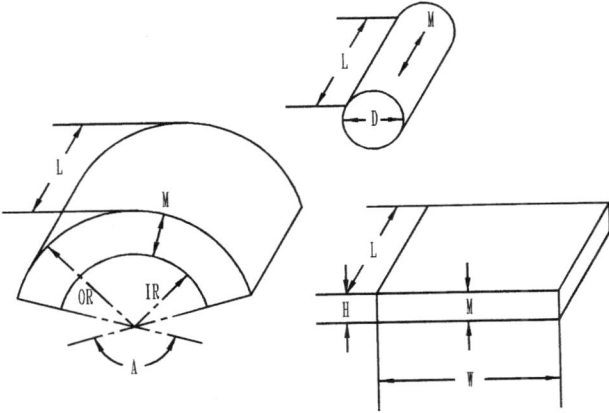

FIGURE 3.111 Magnet shape, dimensions and magnetizing direction.

information helps to determine if the equipment required is a simple solenoid, solid iron and copper, or a laminated-steel multipole magnetizing fixture.

3.18.5 Magnetizing State

The magnetizing state is the step in the manufacturing cycle when the magnet will be magnetized. Typical information should include the individual magnet or magnet structure, the materials in the structure, how many magnets are in the structure, structure dimensions, and a drawing of the structure. This information also helps to determine the overall size of the magnetizing fixture, and whether there is any requirement for a long pulse or additional field above the material's recommended field (see Fig. 3.113).

3.18.6 Percentage of Demagnetization and Required Accuracy

This item relates to the operating point of the magnet or magnetic device and the tolerance or accuracy of the set-point requirement. Information to include is as follows:

- The percentage of demagnetization (i.e., 5 or 20 percent from saturation). The percentage of demagnetization is directly related to the amount of energy required.
- The accuracy requirement and acceptance band for the final set point.

The accuracy and tolerance window gives an idea of the resolution required by successive demagnetizing pulses and the amount of time it may take to reach this level. This will impact the processing cycle time.

3.18.7 Parts Handling, Cycle Rate, and System Control

The decision here is how the magnet or assembly will be handled in the manufacturing (magnetizing) process.

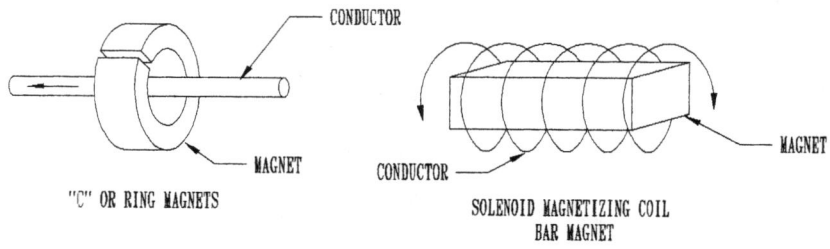
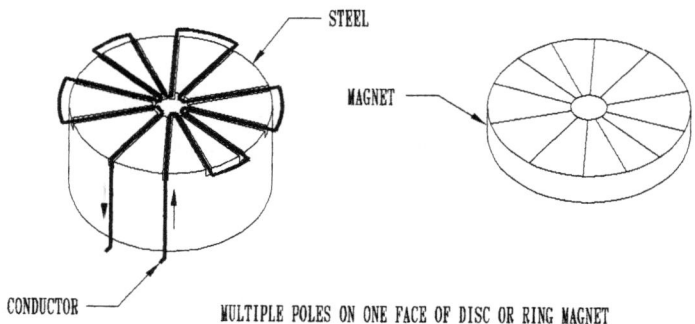
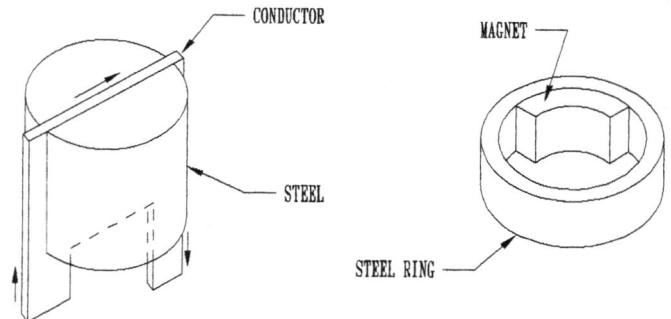

FIGURE 3.112 Polar configuration.

- One-at-a-time or bulk magnetization
- Manual loading by an operator
- Automated system
- Cycle rate
- PLC or system computer control

The number of units magnetized in one cycle has a direct impact on the size and complexity of the magnetizing fixture and the energy-level requirement of the magnetizer.

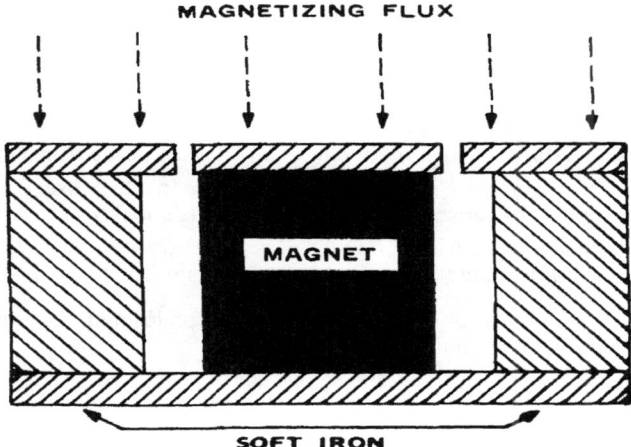

FIGURE 3.113 Magnet magnetized in an assembly.

The cycle rate can have an impact on the decision to offer a dc electromagnet, capacitive discharge, or half-cycle magnetizer. It may also impact the system cooling requirements and fixture design. The cycle rate is one of the key factors in deciding between a half-cycle or capacitive-discharge system.

If PLC or computer control is desired, one needs to understand what external controls may be required to interface with automation equipment or system controllers and software.

3.18.8 Operating Environment

The operating environment considerations are as follows:

- Whether the equipment will be used in a laboratory, clean room, or production environment
- Any unusual temperature or humidity considerations
- Available input power

A laboratory magnetizer normally requires much slower cycle rates, potentially less energy, and less automation. In many cases, a laboratory magnetizer is not easily transferred into production.

Clean room requirements can be very restrictive and need to be identified early in the design cycle.

The selection of system enclosures to accommodate temperature and humidity extremes may be a consideration.

3.18.9 Measuring and Testing Options

There are many different ways to assure the quality of the magnetizing process, magnet material, or magnet assembly. The following options provide various levels of control over the manufacturing and magnetizing process.

- A current monitor with a comparator circuit to measure and compare the magnetizing current at acceptable levels
- Gaussmeters with comparator circuits to measure either the magnetizing pulse or the residual flux density in the magnet
- Fluxmeters with comparator circuits in conjunction with search coils, embedded in magnetizing fixtures, to measure the total flux density of the magnet or assembly
- Temperature monitors to measure temperature rise in magnetizing fixtures and to halt the magnetizing process if the temperature rise goes above a safe level
- A PLC system controller or computer in conjunction with the preceding equipment to control the operation and store statistical information

Individual manufacturing and quality assurance philosophies typically drive the decision to add these various testing options.

3.18.10 Magnetizer Types

Magnetizers may be generalized into four categories: the permanent-magnet magnetizer, the dc magnetizer or electromagnet, the half-cycle impulse magnetizer, and the stored-energy or capacitive-discharge magnetizer.

Permanent-Magnet Magnetizer. The older-style permanent-magnet magnetizers consisted of a large U-shaped permanent magnet (usually alnico V) having adjustable pole pieces. The maximum air gap length is usually about 1.25 in. With the gap adjusted for a length of 1 in, a field of about 3000 Oe is obtainable.

The advent of high-energy permanent-magnet material has enhanced the development of permanent-magnet magnetizers.

DC Magnetizer (Electromagnet). The dc magnetizer utilizes one or two coils of wire wound on an iron frame having a C-shaped configuration. Adjustable pole pieces allow air gaps of various lengths to be used. Direct current is applied to the coils, and the resultant electromagnetic force produces a magnetic field in the air gap.

DC magnetizers are limited to charging straight or slightly curved magnets. Many C-shaped and U-shaped configurations cannot be fully saturated on this type of magnetizer; since the main magnetizing path is straight, the leakage field around the pole pieces is curved.

The duty cycle of the dc magnetizer is usually rather short, due to the large amount of heat generated by the current passing through the coils.

The dc magnetizer is useful primarily where fields having a straight-line characteristic are required. The relatively long time base of the magnetizing field (2 s or longer) is important where magnet assemblies having high eddy current losses are to be charged. An example of this would be large magnets contained in a cast-aluminum housing.

Half-Cycle Magnetizer. The advantage of a half-cycle magnetizer, illustrated in Fig. 3.114, is its ability to generate pulses at a very fast rate. The half-cycle magnetizer operates by picking off one-half cycle from the power line. It will pass a part of a half-cycle with 1, 2, or 3 cycles from a 110- to 600-Vac 50- or 60-Hz line through a magnetizing fixture or pulse transformer.

The half-cycle magnetizer is generally a fixed installation because of the necessity of connecting to a heavy power line.

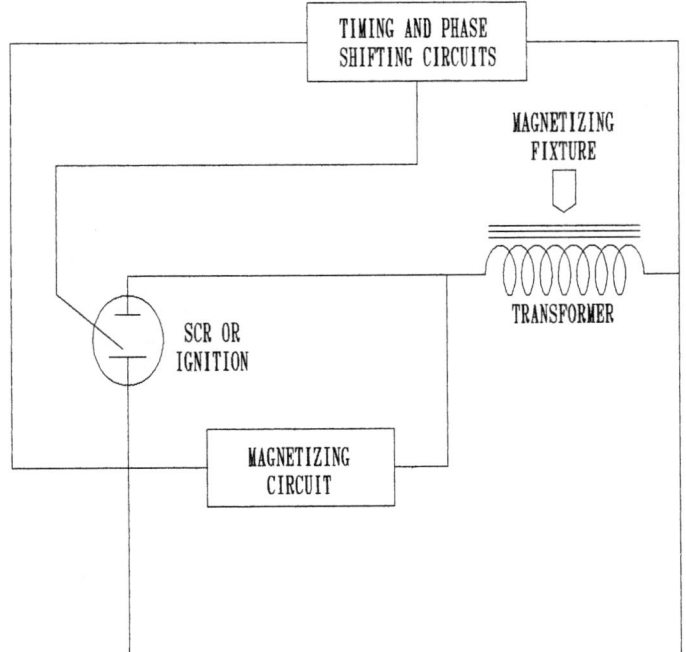

FIGURE 3.114 Basic elements of a half-cycle magnetizer.

The half-cycle magnetizer utilizes wound fixtures or current transformers with single-turn fixtures in conjunction with timing- and discharge-control circuitry. Duration of the applied pulse is dependent on the line frequency and adjustment of the control circuits. Pulse duration up to 8 ms may be obtained from a 60-Hz line. The magnetizing force which may be obtained is, of course, dependent on the amount of current that can be pulled off the power line. Rapid operation can be obtained with a half-cycle magnet magnetizer, since there is no capacitor bank requiring several seconds to recharge; however, water cooling of the ignitrons or silicon controlled rectifiers (SCRs) and magnetizing fixtures is generally required where rapid operation is desired.

Capacitive-Discharge Impulse Magnetizer. This magnetizer operates on a stored-energy principle. Voltage is stored in a capacitor bank, usually requiring several seconds, and the stored energy is then discharged through a unidirectional switch (ignitron or SCR) into a charging fixture or charging transformer. The duration and waveform of the pulse is dependent on the capacity of the bank, the inductance of the fixture or transformer, and, to some extent, the resistance of the fixture. Pulse lengths of 100 µs to several tens of milliseconds may be obtained.

The basic elements of a capacitive-discharge magnetizer are shown in Fig. 3.115.

Depending on the pulse duration, which can be varied with capacitive-discharge magnetizers, charging fields in excess of 100,000 Oe can be developed.

One of the outstanding features of the impulse magnetizer is its ability to saturate curved and multipole magnets, as well as the straight bar and disk types. The magne-

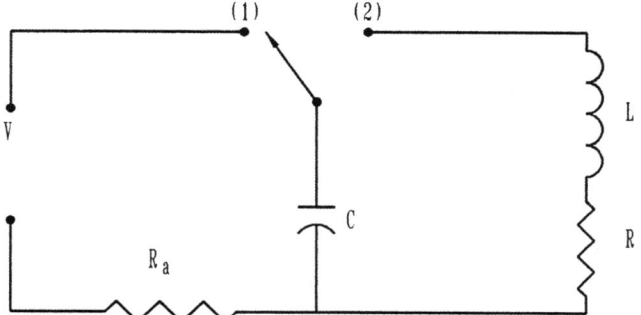

FIGURE 3.115 Basic elements of capacitive-discharge magnetizers.

tizing field around the secondary charging conductor is circular in shape and therefore closely follows the configuration of C-shaped and U-shaped magnets (see Fig. 3.116).

Selection of Magnetizer. Analyzing the features of the four major types of magnetizing systems leads to the following conclusions. Where economy is a consideration, long magnetizing pulses are acceptable, and only two pole structures (generally) are to be magnetized, and where the required magnetizing forces do not exceed 10,000 Oe, the electromagnet will prove satisfactory. Self-heating of the electromagnet can become a limiting factor when the time that power is applied to the coils is prolonged. Water cooling of the coils can help overcome this drawback. Certain types of multipole magnets can be magnetized using special pole pieces; however, the configuration of the magnetizing field generally precludes placing more than two poles on a structure.

The half-cycle magnetizer is perhaps the most rapid means of magnetizing large production volumes of parts of similar magnetic structure. The half-cycle magnetizer can produce many pulses per minute, providing increased throughput over impulse

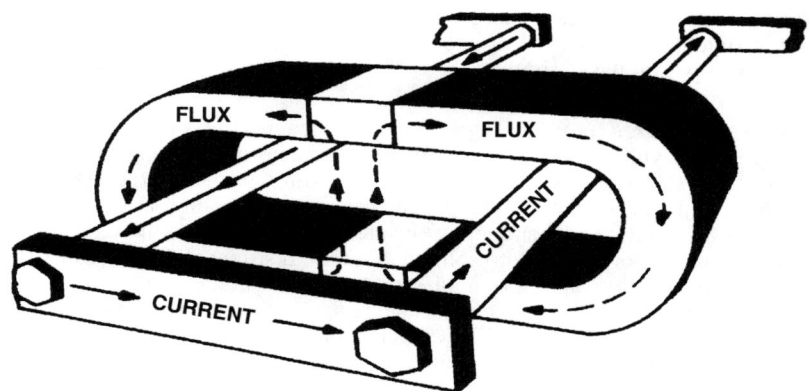

FIGURE 3.116 Magnetizing two U-shaped magnets simultaneously.

magnetizers. However, self-heating of the fixture can necessitate water cooling to prevent fixture damage. The relatively long pulse length of the half-cycle magnetizer is useful where large volumes of highly permeable materials are to be magnetized.

The impulse magnetizer has most of the attributes of the electromagnet and the half-cycle system with some minor limitations. Pulse length and peak current can be varied by means of changing storage bank capacity and changing storage voltage levels. Speed of operation is not as rapid as the half-cycle method, since 1 to 20 s may be required to charge the capacitor bank. Cost can vary depending on the energy-storage level.

Extremely high currents, and consequently high fields, may be obtained with capacitive-discharge magnetizers. The power-line requirement for this type of unit is relatively low, ranging from 1 A for the low-power models to 20 to 30 A for the largest magnetizers required. Line demand is minimal in most cases, with little or no line effects. And almost any type of magnet and pole configuration can be accommodated.

Figure 3.117 shows a typical pulse from a half-cycle or capacitive-discharge impulse magnetizer. One of the most important parameters, as far as waveform is concerned, is the rise time to the peak magnetizing current. Current rise time determines the eddy current losses and resultant heat developed in the magnetizing fixture or in the structure being magnetized.

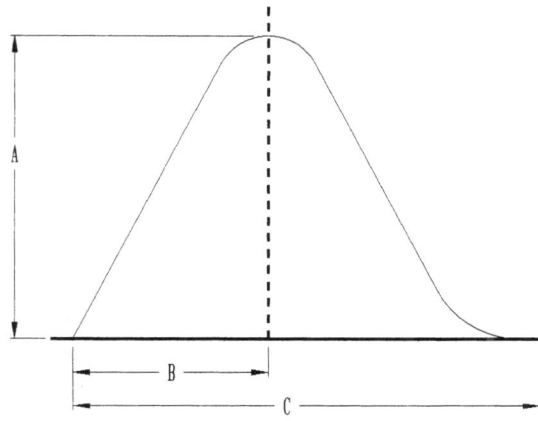

FIGURE 3.117 Magnetizing pulse characteristics: A = peak magnetizing current, B = rise time to peak, and C = pulse duration.

3.18.11 Fixture Design for Impulse and Half-Cycle Magnetizers

The simplest type of magnetizing fixture used is the wire-wound solenoid. The number of turns, wire size, and length of the solenoid are all determined by the size of the magnet which is to be magnetized and the voltage and capacitance level of the capacitor bank. The waveform of the current pulse applied through the coil should closely approximate half of a sine wave for maximum efficiency Figure 3.118 shows a solenoid-type magnetizing fixture.

Where high energy levels are applied to a coil, especially where short pulse lengths at high peak currents are involved, proper physical bonding of the coil turns

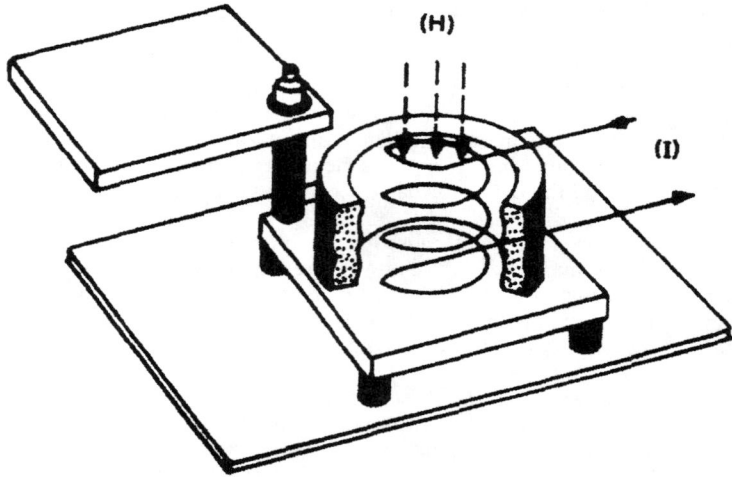

FIGURE 3.118 Solenoid-type wire-wound fixture showing relationship of flux *H* to current *I*.

is extremely important. The stress forces applied to the coil can be large in some cases, and insufficient bonding can cause the coil to be physically damaged.

A solenoid may be termed an *air-wound* solenoid or it may be surrounded with a ferrous return path to enable higher magnetizing forces to be achieved by the low-reluctance return path. The ferrous return path may provide additional physical strength.

The impulse or half-cycle magnetizer's arc is widely used in the magnetization of straight bar magnets. The large amount of energy which can be produced by this type of magnetizer enables the saturation of unusually large volumes of magnetic materials. Magnetization of straight magnets is usually accomplished in a cavity-type fixture employing a coil wound like a solenoid and a soft-iron return path. This structure has high efficiency, since virtually all of the magnetizing force produced passes through the cavity along the axis of the coil.

Magnets can be magnetized as individual units, in groups, or as completed magnetic assemblies. Saturation of a completed structure is generally preferred, since the danger of contamination by ferrous particles during handling is reduced, and demagnetization by open circuiting is prevented. *Open-circuiting* a magnet refers to the removal of the keeper or working-gap assembly from the pole ends. This, in effect, applies a demagnetizing force to the magnet and in some cases (determined by the magnet configuration) can cause a loss of as much as 6 percent of residual induction.

The impulse and half-cycle magnetizers also lend themselves to the utilization of multipole wound magnetizing fixtures, diagrammed in Fig. 3.119. The multipole fixture is used for the magnetization of various sensors, generator rotors, alternator rotors, synchronous motor rotors, multipole stator assemblies, and a wide variety of multipole configurations of a cylindrical, flat, or curved nature.

Where a magnet surface is continuous, not segmented, or has salient poles, the pole location and the polar embrace are determined by the pole-tip configuration of the charging fixture.

The polar embrace of a structure which incorporates pole pieces is determined by the pole pieces themselves. It is usually advantageous, although not always neces-

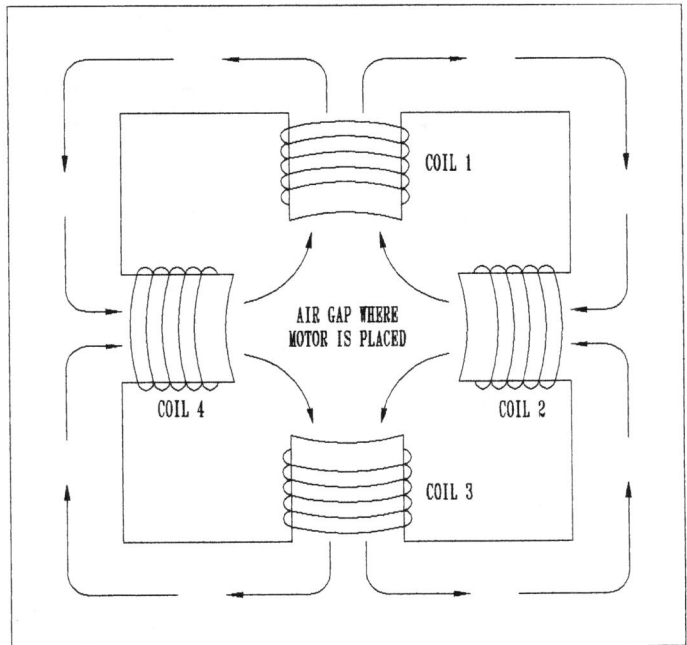

FIGURE 3.119 Four-pole laminated-steel wire-wound fixture.

sary, to make the multipole charging fixture from layers of ferrous material rather than carve it from a solid block of iron. Stock used may be transformer iron or cold-rolled steel sheets, ranging in thickness form 0.014 to 0.060 in. This minimizes eddy current losses and results in more efficient use of the magnetizing energy. The laminated sheets should be insulated from each other by suitable varnish or other material. Some shorting between laminations may result when the structure is slotted and the cavity is turned, but this will cause only minor losses.

Multipole magnetizing fixtures can also be made as the so-called single turn, copper and iron fixtures. Instead of many turns of wire on each salient pole (for example, 20 turns of number 16 wire per pole), a heavy conductor (say, having a cross section of 0.5 in) is convoluted around each pole. This generally entails the necessity of milling and hard-soldering the copper conductor sections. A further accessory is also required, the low-impedance single-turn fixture. Magnetizing several poles simultaneously can also be achieved using this type of fixture in conjunction with pulse transformers. A typical four-pole fixture is shown in Fig. 3.120.

A multipole cylindrical magnet can be charged by either method; however, the method used will have an impact on the flux distribution characteristics of the magnet or assembly. It is important to know how this may impact the end product performance. During the early product development stage, testing of different magnetizing techniques will allow the design engineer to determine whether better performance can be achieved with one technique over another. As an example, a two-pole motor can be magnetized from the outside in using a wire-wound

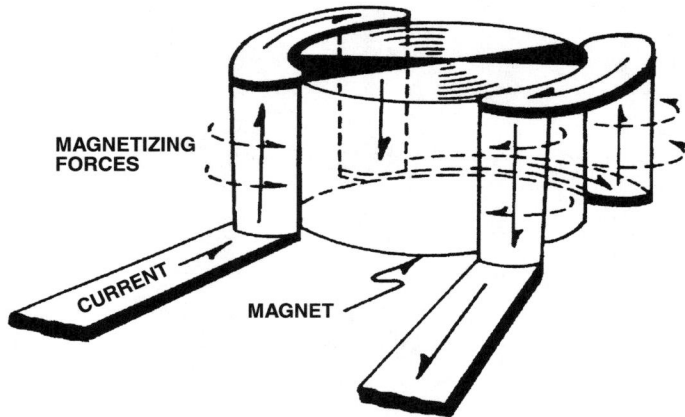

FIGURE 3.120 Copper and iron four-pole fixture.

laminated-steel fixture, as shown in Fig. 3.121. In this case, a completed motor could be inserted between the two poles and magnetized from the outside in—that is, all magnetizing windings are external to the motor housing.

Another alternative is to use a hairpin copper and iron single-turn fixture, as shown in Fig. 3.122. In this case, the motor shell, consisting of the steel housing and magnet segments, is inserted over the copper conductors and the internal steel pole

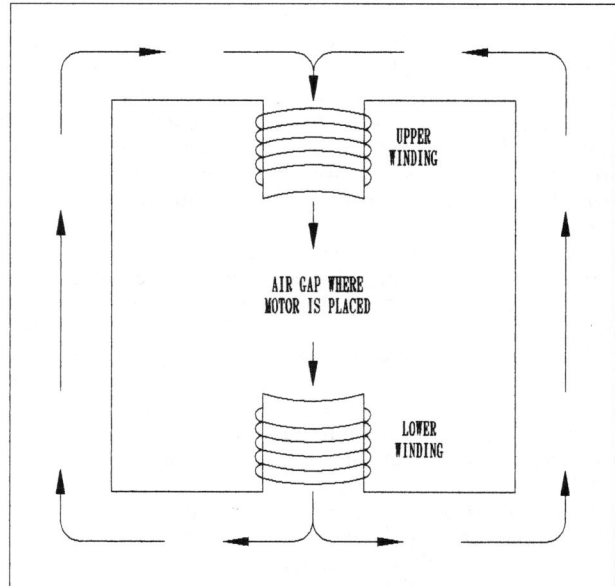

FIGURE 3.121 Two-pole wire-wound magnetizing fixture.

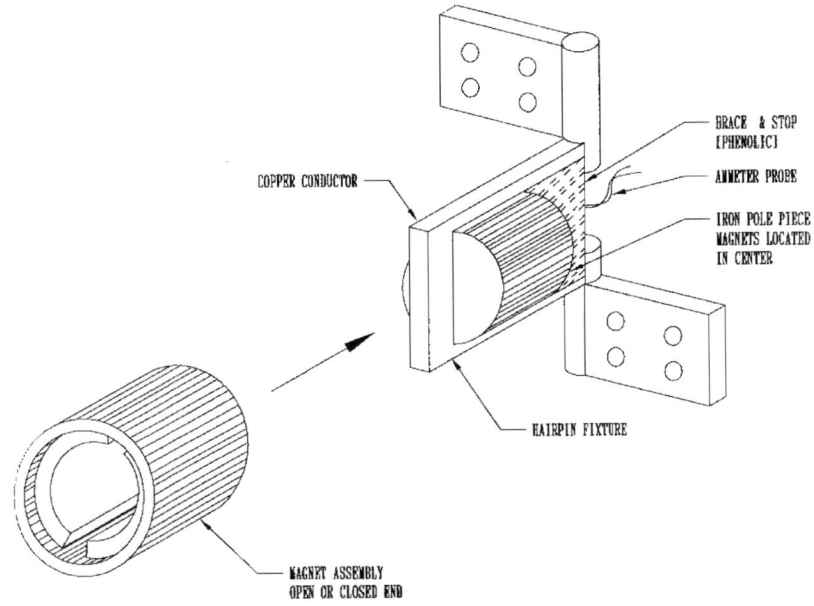

FIGURE 3.122 Two-pole hairpin magnetizing fixture.

piece. The current-carrying conductors are actually placed between the magnet segments—hence, the term "magnetizing from the inside out."

The hairpin fixture is designed to fit exactly inside the motor assembly, with the outside dimensions of the fixture matched to the inside dimensions of the motor. When the energy stored in the magnetizer is released, flux is developed in the hairpin fixture that will magnetize the motor assembly.

Magnetizing from the inside can also be performed with wire-wound hairpin fixtures in which multiple turns of smaller wire are used in place of the single-turn copper construction.

The hairpin fixture can be designed with a shim in its center to accept a gaussmeter probe. This permits flux density measurements to be made immediately after magnetizing, without removing the motor assembly from the fixture.

3.18.12 Conclusion

As one can see, there are many different ways to magnetize any given assembly. With the number of choices available, it is imperative that a clear picture of the overall system requirements be developed and communicated to the magnet and equipment suppliers. An early involvement can save time and money in moving the project into production and in ensuring that the system meets the production goals.

3.19 CAPACITIVE-DISCHARGE MAGNETIZING*

3.19.1 Getting the Most Out of Your Magnets

Magnets are useless until they are magnetized. No clever or innovative electric motor design can overcome the fact that if the magnets are not magnetized, the motor will not work. While enormous resources are often committed to electric motor design, optimization, and production efficiencies, it is easy to overlook the magnetizing process. Improperly magnetized magnets can offset gains made in design and production engineering. Advancements in higher-energy-product magnets also create new challenges to motor producers who chose to take advantage of these materials. This section will assist the electric motor producer in finding the optimal magnetizing process for motor production, improving the performance and reliability of the final product.

A magnet is magnetized when it is exposed to a large magnetic field. Large magnetic fields require large electric currents, which require large amounts of energy. A magnetizing system converts energy from a power supply into magnetic fields used to magnetize the magnet. Magnetizing systems consist of two basic components: the magnetizer, which stores and releases energy in the form of an electric current, and the magnetizing coil, which converts the large electrical current into magnetic fields to magnetize the magnets. The magnetizer can be thought of as an energy storage and release system, while the magnetizing coil directs the energy from the magnetizer into useful work that magnetizes the magnets.

Let us first consider the magnetizer, which stores energy to produce currents. There are several ways to store electrical energy:

- Batteries
- Utility power lines
- Inductors
- Rotating devices (generators)
- Flywheels
- Electrical capacitors

Given the large fields usually required for the magnetizing process, pulse currents rather than continuous currents are used to create magnetizing fields. Because of the brief but intense currents used in the magnetizing process, electrical capacitors have been the usual component used to store electrical energy in magnetizing systems. Figure 3.123 shows a block diagram of a basic magnetizing system. The instantaneous energy in a pulsed magnetic field can be extremely high. Five million watts is not uncommon.

3.19.2 Storing Electrical Energy in Capacitors

Since magnetizing systems usually use electrical capacitors to store energy, they are often referred to as *capacitive-discharge magnetizing* (CDM) systems. The energy stored in a capacitor or bank of capacitors is mathematically expressed by the formula

---

* Section contributed by Derrick Peterman and Leon Jackson, LDJ Electronics.

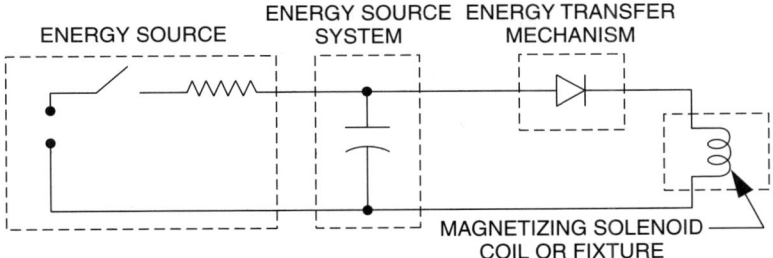

FIGURE 3.123 Block diagram of magnetizing system.

$$E = \frac{1}{2} CV^2 \qquad (3.15)$$

where C = capacitance, F
  V = voltage across capacitor, V
  E = energy stored in capacitor, J or W.

Commercially available capacitors usually do not have sufficient energy storage capacity. Therefore, individual capacitors are assembled into capacitor banks using series and parallel combinations of capacitors. Figure 3.124 shows a CDM system with Y capacitors connected in series and X capacitors connected in parallel. In a typical CDM system, a capacitor bank will produce from a few hundred to tens of thousands of joules. Table 3.21 relates the energy of some typical CDM systems, their storage energy, their materials, and the approximate sizes of magnets they can magnetize.

The specifications for capacitor banks used as energy storage for CDM systems should include the following:

- Energy storage capacity
- Electrical voltage
- Maximum instantaneous current
- Estimated life (number of charges and discharges)
- Internal resistance

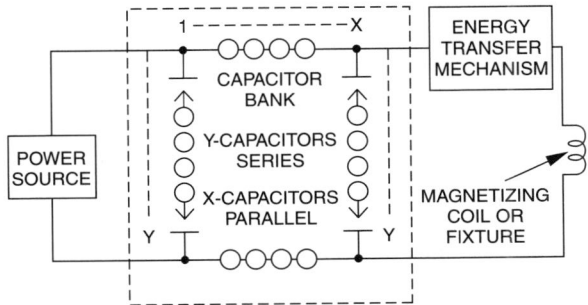

FIGURE 3.124 Connection of capacitors in magnetizing system.

TABLE 3.21 Typical Capacitor Specifications

Specification	Capacitor type	
	Electrolytic	Oil-filled
Typical voltage, V	300–450	1000–45,000
Typical capacity, µF	2000–4000	100–1000
Typical bank voltage, V	300–1200	1000–450
Energy storage, J/kg	163	23
Energy storage, J/cm^3	0.19	0.034
Typical charge/discharge cycles	4 million	6 million
Capacitors required per 5000-J bank	32	10
Weight of 5000-J bank, kg	31	220
Size of 5000-J bank, m^3	0.04	0.15
	3500 µF	160 µF
Typical capacitor	300 V dc	2,500 V dc
	811 cm^3	14,730 cm^3
	0.96 kg	22 kg

Special precautions must be taken to ensure that the capacitors can withstand the repeated charging and discharging required in CDM systems. This is especially important in high-speed production applications. Also, the instantaneous current may exceed 10,000 A, which can cause internal heating in capacitors. Internal heat buildup can cause capacitor failure.

Two types of capacitors are used for energy storage in CDM systems, electrolytic and oil-filled capacitors.

Electrolytic Capacitors. Due to the materials used in the manufacture of electrolytic capacitors, the maximum voltage rating is about 500 V dc. Higher voltages can be reached by using them in series.

A major limitation of electrolytic capacitors is their inability to be charged in the reverse direction. This limits them to an exponential waveform that is discussed later in more detail.

The size of an electrolytic capacitor with a 3500 µF capacity at 300 V is approximately 6.7 cm diameter by 14 cm long. It weighs 1 kg and stores 157.5 J of energy. Electrolytic capacitors are usually much cheaper than oil-filled capacitors.

Oil-Filled Capacitors. Because of the materials and manufacturing methods used, oil-filled capacitors are usually much larger in size than electrolytic capacitors and can be charged in either direction, allowing them to produce either an exponential waveform or a half-sine oscillating waveform. Oil-filled capacitors contain less capacitance for the same size of package. Typical values might be 2500 V at a capacitance value of 1600 µF, weighing 22 kg and storing 500 J of energy. The size of this capacitor is 13 × 34 × 33 cm. Table 3.21 lists a number of factors that pertain to electrolytic and oil-filled capacitors used in CDM systems.

Most voltages used by CDM systems require extreme care, both in operator safeguards and fixture design. The higher operating voltages of oil-filled capacitors increase the precautions that must be taken. Also, the cost of components increases as the voltage increases.

Electrolytic capacitor systems weigh less and occupy less space than oil-filled capacitor systems. Energy storage per unit of weight or volume is usually not a critical factor in choosing an energy storage system. However, the lower energy density of oil-filled capacitors can contribute to a longer capacitor life and increased charge/discharge cycles for oil-filled capacitor energy storage systems. This advantage of lower energy density becomes more crucial when the application requires large peak currents and rapid cycle rates, which tend to shorten capacitor life.

In addition to longer capacitor life, oil-filled capacitors have the advantage of producing either an exponential, half-sine, or oscillating waveform. Since electrolytic capacitors can produce only the exponential waveform, most of the energy from the capacitor bank is dissipated into the magnetizing coil each cycle, while the half-sine waveform dissipates only a small portion of the capacitor-bank energy into the magnetizing coil. The half-sine energy that is not dissipated by the magnetizing coil recharges the capacitor bank in the opposite direction. With proper electronic circuitry, it can then be discharged through the magnetizing coil in the next cycle. The advantages are savings in input energy and reduced heating in the magnetizing coil.

3.19.3 Connecting the Magnetizer to the Magnetizing Coil

The switch to transfer energy from the energy storage system to the magnetizing coil can range from very elementary to very sophisticated. In one application, the switch was a short piece of magnet wire about 2 mm in diameter. The wire was clamped between two blocks. The energy storage system was a bank of lead-acid batteries, with a 2-mm length of copper coil in series with a magnetizing coil. In fractions of a second, as the current built up in the circuit, the magnet wire became hot to the point of violent disintegration. When the magnet wire disintegrated, it worked as a switch to shut off the electric current so that neither the magnetizing coil nor the energy source would be damaged. By choosing the size of magnet wire, the peak current could be controlled. With this technique, safety glasses are an absolute must because the disintegrating wire causes copper to be splattered about during each magnetizing cycle.

A far less spectacular but preferable switching method is to use a solid-state electronic switch, such as an SCR. SCRs are easy to control electronically and are very reliable. In some CDM systems, an ignitron vacuum tube is used. The ignitron tube has some disadvantages:

- It has a limited life.
- It uses mercury, which is a hazardous material.
- It may require water cooling.

However, the ignitron will take more punishment in terms of exceeding its electrical ratings than an SCR. If its specified electrical ratings are exceeded, an SCR may be destroyed. If the ratings of an ignitron are exceeded, it may sustain only minor damage and continue to operate.

CDM systems can be designed to produce magnetic field pulses specific to the application. The factors to be considered are the following:

- The magnitude of the magnetic field required
- The duration of the magnetic field required
- The size of the magnets or magnetic assembly
- The cycle rate of the magnetic field pulse
- Whether the magnet will be magnetized by itself or as part of an assembly

3.19.4 Magnitude of the Magnetic Field

The magnitude of the field required is dependent on the coercive field of the material. Virtually all magnet producers and suppliers provide this data. The *coercive field* is a measure of how difficult it is to both magnetize and demagnetize the magnet. The higher the coercive field of the magnet, the more difficult it is to magnetize. Table 3.22 indicates the range of coercive field values for different types of magnets. A rule of thumb is that the applied field should be 1½ to 4 times the coercive force of the magnetic material. The coercive force of high-energy materials may reach 15,000 Oe or more; therefore, magnetic fields of more than 50,000 Oe may be required to reach magnetic saturation. The magnetic fields in a fixture can be measured using either a gaussmeter or a fluxmeter.

TABLE 3.22 Magnetizing Force Required for Saturation of Common Commercial Magnets

Material	Decade introduced	Coercivity, Oe	Magnetizing field, Oe	Resistivity, $\Omega \cdot cm$
Alnico	1940s	500–1,500	1,500–5,000	50
Ferrite	1950s	2,300–4,000	5,000–10,000	1×10^6
SmCo	1970s	6,000–20,000	6,000–50,000	53
NdFeB	1980s	11,000–25,000	15,000–50,000	150

3.19.5 Duration of the Pulsed Magnetic Field

The time that the magnetic field must be applied to the magnet material is more complicated, and unlike the coercive field of a magnet, cannot simply be looked up in a data table or specification sheet.

Many magnetic materials are metallic, and metals conduct electric current. As the magnetic field changes, eddy currents are produced in the material, which produces an opposing magnetic field. The magnetic field in the metallic material (i.e., the magnet) is the sum of the magnetic field generated by the magnetizing coil minus the field generated by the electric current flowing in the magnet to oppose this field. This effect is commonly known as *Lenz's law*.

We can understand this phenomenon mathematically. Lenz's law can be expressed as follows:

$$e = -k \frac{d\phi}{dt} \qquad (3.16)$$

where e is the voltage generated by Lenz's law to produce the eddy currents, $d\phi$ is the change of magnetic flux caused by the CDM current pulse, dt is the time interval of the magnetic flux change, and k is a simple proportionality constant. Note that the negative sign in front of k is a result of the eddy currents opposing the change in magnetic flux.

It can be shown that eddy currents are proportional to the rate of change of magnetic field with respect to time by

$$I_{\text{eddy}} = K \frac{dH_{\text{CDM}}}{dt} \quad (3.17)$$

where I_{eddy} is the magnitude of the eddy current and dH_{CDM} is the change in the magnetic field produced by the magnetizer over a time interval dt. From Eq. (3.17), we see that if

$$\frac{dH_{\text{CDM}}}{dt} \quad (3.18)$$

the rate of change of the magnetic field produced by the magnetizer is high, the size of the eddy currents will also be high. Since the eddy currents act to oppose the applied field, large eddy currents can cancel out a substantial fraction of the field produced by the magnetizing fixture. This problem can be overcome by simply reducing the rate at which the magnetic field increases. This requires a longer magnetizing pulse width. However, lengthening the duration of the magnetic field pulse causes heating in the magnetizing coil.

While eddy currents complicate the magnetization process, they can be dealt with effectively. Since eddy currents depend on the resistivity of the magnet, ferrite magnets, which are electrical insulators, are highly immune to the effects of eddy currents during the magnetization process. Unfortunately, alnico, samarium cobalt (SmCo) and neodymium iron boron (NdFeB) magnets are all decent conductors, as indicated in Table 3.22, and so eddy currents are produced in these materials during pulsed magnetization. Since alnico magnets have relatively low coercive fields, the effect of eddy currents can be overcome by simply using large enough applied fields to overcome the opposing fields that result from the eddy currents. Eddy currents are more difficult to overcome with SmCo and NdFeB magnets.

The difficulty with magnetizing SmCo and NdFeB magnets is that since these magnets have such large coercivities, it is highly impractical to simply "overpower" the effect of the eddy currents. This creates more problems than it solves. To begin with, extremely large amounts of energy must be stored and released to overcome eddy currents in this manner. As we shall soon see, this requires a very large magnetizer, which is both expensive and takes up potentially valuable space on the factory floor. Huge currents flowing through the magnetizing fixture dissipate tremendous heat—heat that can damage the fixture, even possibly causing catastrophic failure. If the magnets are in a steel assembly, large forces due to the rapidly charging magnetic fields can cause flexing of the steel assembly, which can break the magnets inside. At the very least, additional time between magnetizing firings is required for the fixture to cool. This reduces production throughput. An effective method of magnetizing rare earth magnets balances the large currents required to produce a large field against longer pulse times with reduced eddy current loss effects.

A seldom-mentioned advantage of bonded rare earth magnets is that they are nonconductive; therefore, eddy currents are usually small in these magnets. The rubber or plastic binders are insulating and electrically isolate the conducting grains from one another. Bonded magnets are easier to magnetize than similar-sized bulk magnets for this reason.

When balancing the high peak currents required to magnetize rare earth magnets against long pulse durations to minimize eddy currents, the following formulas are helpful.

$$i_p = V\sqrt{\frac{C}{L}}\, e^{-a} \qquad (3.19)$$

$$a = -\frac{R}{4\omega L}\tan^{-1}\left(\frac{4\omega L}{R}\right) \qquad (3.20)$$

$$\omega = \sqrt{\frac{1}{LC} - \left(\frac{R}{2L}\right)^2} \qquad (3.21)$$

These equations describe the peak current i_p in the magnetizing coil and the angular frequency ω of the current pulse. The duration t_p of the current pulse is then

$$t_p = \frac{2\pi}{\omega} \qquad (3.22)$$

The symbols R, L, and C refer to the electrical *resistance* of the magnetizing circuit, the *inductance* of the magnetizing coil, and the *capacitance* of the magnetizer. The resistance is usually small compared to the inductance, and can therefore be ignored for most analysis. In this case, the peak current i_p is simply expressed as

$$i_p = V\sqrt{\frac{C}{L}} \quad \text{when } R \to 0 \qquad (3.23)$$

The inductance is proportional to the number of turns squared on the magnetizing coil. The capacitance is the capacitance of the capacitor bank in the CDM system. In typical CDM systems, the time to the peak magnetizing field will vary from less than 1 ms to more than 20 ms. The capacitance, inductance, and resistance of the CDM system determine the magnetic field pulse duration.

3.19.6 Size of the Magnets or Magnetic Assembly

Because it takes energy to create magnetic fields, a magnetic field over a large volume will require more energy than the same field over a smaller volume. Thus, the size of the magnet or magnetic assembly to be magnetized is important. One can describe the magnetizing process as a transfer of energy from the magnetizer to the magnets. The energy difference between an unmagnetized magnet and a magnetized magnet is proportional to the square of the magnetic induction times the volume of the magnet. The energy requirements for a given magnet material scale linearly with the volume of the magnet. Table 3.23 provides the energy requirements of typical magnetizing applications.

TABLE 3.23 Energy Requirements of Typical Magnetizing Applications

Energy storage, J	Size and material of magnets or assemblies
2,500	Large alnico; medium ferrites; small SmCo and NdFeB
15,000	Virtually all alnico assemblies; large ferrite; medium SmCo and NdFeB
30,000	Almost all ferrites; large SmCo and NdFeB

3.19.7 Cycle Rate of the Magnetic Field Pulse

The cycle rate of the magnetic field pulse is an important consideration for the magnetizing system. For high-volume electric motor production, the magnetizing system must be designed to handle a high cycle rate. For prototype and laboratory use where the cycle rate is of little importance, a more versatile magnetizing system than that used for production purposes is usually preferred. The rate of magnetizer charging is an important consideration in the choice of the capacitors used.

Any energy that is not dissipated in the magnetizing coil as heat or energy to magnetize the magnets must be restored in the capacitor bank. This fact has advantages and disadvantages. Since only a small part (as little as 10 percent) of the energy is dissipated in the magnetizing coil or used to magnetize the magnets, about 90 percent of the energy will be returned to the capacitor bank. This greatly reduces the source energy needed to recharge the capacitor bank. However, the capacitor bank is now charged in the opposite direction, and the next magnetic pulse will be in the opposite direction unless the proper electronic circuitry is employed to reverse the current through the magnetizing coil.

Since electrolytic capacitors cannot be charged in the reverse direction, a CDM system that produces a half-sine waveform must use oil-filled capacitors. The additional cost of the electronics and oil-filled capacitors can double the capital cost of a CDM system. The operating energy savings are usually not sufficient to warrant the additional capital cost of using the half-sine waveform, unless technical reasons warrant its use. If magnetizing-coil heating is a serious problem, the use of oil-filled capacitors and the half-sine waveform may well be the best choice. Where magnetizing-coil heating is not a problem, even when refrigerator cooling is required, electrolytic capacitors and the exponential waveform are usually the most cost-effective solution.

The cycle rate mostly affects the heat generated in the magnetizing coil. The cycle rate for production CDM systems varies from a few parts per hour to 1 or 2 s per part. For most applications, with cycle rates above 200 parts per hour, the magnetizing fixture will require cooling. However, the space available for the magnetizing coil and the magnitude of the magnetic field required can greatly alter this general rule. The most convenient cooling method is by pumping a refrigerated liquid through the magnetizing fixture.

3.19.8 Magnetizing Magnets Individually or as an Assembly

A magnet which is built into the final assembly, such as a dc motor, is usually harder to magnetize than the magnet by itself. This can result in a larger CDM system, lower cycle rates, or more complex magnetizing coils and fixtures.

Metallic magnet assembly housing will dissipate energy during the magnetizing process in the form of eddy currents. This is especially true in the case of magnetic metals, such as steel alloys (including magnetic stainless steels) and laminations. Nonmagnetic metals such as aluminum are also a source of eddy current loss, but to a lesser extent than magnetic steels.

However, the advantages gained by assembling unmagnetized magnets are usually more than worth the payment required in magnetizing problems. It is difficult to assemble magnets once they are magnetized. In fact, this is a potentially dangerous operation to do by hand. This is especially true with large magnets and magnets with high energy products, which are very difficult to handle when magnetized.

3.19.9 Magnetizing Coils

The magnetizing coil is usually the weak link in a CDM system. It is on the receiving end of an electrical energy jolt that can reach an instantaneous value of 5 million W. The magnetic forces caused by the high currents in the magnetizing coil windings can cause physical movement of the windings, resulting in shorted turns or causing the windings to open.

The sheer number of electrons flowing in the windings can cause the coil winding to blow apart if the windings have sharp bends. Heat buildup can soften encapsulating epoxy, allowing physical movement of the conductors. Heat buildup can also weaken the winding insulation, creating the possibility of shorted turns. The high current pulses (a few hundred to more than 10,000 A) will cause instantaneous heating of weak spots in the magnetizing coil winding. These hot spots can be at connection points or at flaws in the copper conductor.

The magnetizing coil must be designed to do the following:

- Withstand a few hundred to several thousand volts.
- Accept a few hundred to several thousand amperes.
- Withstand the mechanical forces caused by the electrical current flow.
- Produce the magnitude of magnetic field required for the material being magnetized.
- Generate a magnetic field pulse of sufficient duration to allow the magnetic field to fully saturate all of the magnetic material.
- Have a sufficient cooling system so that the required production cycle rates can be realized.
- Be safe when operators come in contact with it. The safety issues must deal with both the electrical voltages and physical forces.

The high voltages involved require that special attention be given to insulation. The proper wire should be used, with insulation specifications that meet or exceed the maximum voltage of the CDM system. Special care must be taken when making connections. All metal parts of the fixture should be securely grounded with wire equal or superior to the magnetizing coil conductors in size and insulation specifications. The operator should be precluded from coming in contact with the magnetizing conductors. This can be done by encapsulating the magnetizing coil in epoxy, placing the magnetizing coil out of reach of the operator during the magnetizing field pulse, or both. It is usually necessary to encapsulate the magnetizing coil in high-strength epoxy. This encapsulation serves two purposes:

- It keeps the conductors from moving during the magnetic field pulses.
- It provides safety for the operator.

The magnitude of the magnetic field is determined by the amount of energy applied to the magnetizing coil. The duration of the pulse is greatly affected by the number of turns on the magnetizing coil. The main variables of the magnetizing coil are the number of turns and its size. The general rule for a magnetizing coil of fixed size is that when the turns increase, the conductor size decreases, the peak field decreases, and the pulse duration increases.

The design of magnetizing coils and fixtures requires careful consideration of conductive materials that may be in the magnetic circuit. Conductive materials, the size of the part to be magnetized, and its shape should also be taken into account when designing magnetizing fixtures and coils. All coils and fixtures should be impregnated with epoxy to ensure that conductors cannot move during the magnetic field pulse.

In some magnetizing applications, it is advantageous to use steel poles to increase the magnetic field produced by the magnetizing coil. Care must be taken when using steel pole pieces. The steel pole piece will increase the inductance of the magnetizing coils; introducing steel pole pieces into a magnetizing process may, therefore, require some change in the magnetizer capacitance. Since steel pole pieces are a source of eddy current loss, stacked steel laminations are recommended instead of solid pole pieces. Stacked laminations reduce the eddy current losses and allow larger fields to be produced.

Multiple poles can be magnetized with the proper magnetizing fixture design. Magnetizing coils can be wound around two pole pieces. The magnet to be magnetized completes the magnetic circuit. When the current pulse passes through the magnetizing coils, north and south poles are created in the magnet. Multiple poles can be created on a single magnet by repeating this design along the length or circumference of the magnet. Figure 3.125 illustrates a simple example of pole creation at the ends of a magnet.

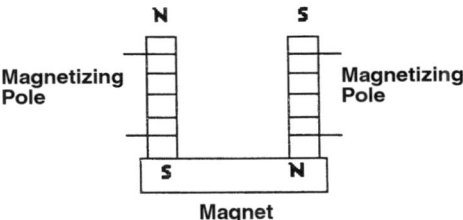

FIGURE 3.125 A simple example of magnetizing poles along the length of a magnet. The horizontal magnet completes a magnetic circuit between the two vertical magnetizing pole pieces.

In general, the fields generated by the magnetizing coils should be oriented in the optimal direction to magnetize the magnet. This optimal direction is determined by the magnet design. If the magnetizing fields do not magnetize the magnet in the orientation specified in the design, the motor may very likely not perform as designed. A poorly conceived magnetizing coil design can rob a motor of performance.

CHAPTER 4
DIRECT-CURRENT MOTORS

Chapter Contributors
Andrew E. Miller
Earl F. Richards
Alan W. Yeadon
William H. Yeadon

This chapter covers methods of calculating performance for direct-current (dc) mechanically commutated motors. Section 4.1 discusses the electromagnetic circuit for dc motors and series ac motors. Sections 4.2 and 4.3 establish some common geometry and symbols and discuss commutation for dc motors. Section 4.4 presents permanent-magnet direct-current (PMDC) calculation methods. Section 4.5 presents series dc and universal ac/dc performance calculations. Sections 4.6 and 4.7 discuss methods for calculating the performance of shunt- and compound-connected dc motors. Finally, Secs. 4.8 and 4.9 discuss dc motor windings and automatic armature winding.

4.1 THEORY OF DC MOTORS*

4.1.1 DC Series Motors

A series motor operating on direct current has characteristics similar to those when it is operated on ac current at power-system frequencies. However, it is best to describe dc and ac operation separately so that comparisons can be made.

The general equivalent electrical circuit of the series dc motor and its physical construction is shown in Figs. 4.1 and 4.2. The motor consists of a stator having a concentrated field winding (Fig. 4.3) connected in series by way of a commutator to a

---
* Section contributed by Earl F. Richards.

wound armature (Fig. 4.4). One of the first things to be considered in the operation of the motor are the motor and generator action, which exist simultaneously in the armature circuit of the motor. These two principles are (1) the instantaneous electromotive force (emf), which is induced in the armature conductors when moving with a velocity v within a magnetic field, and (2) the force produced on the conductors as the result of carrying an electric current in this same magnetic field.

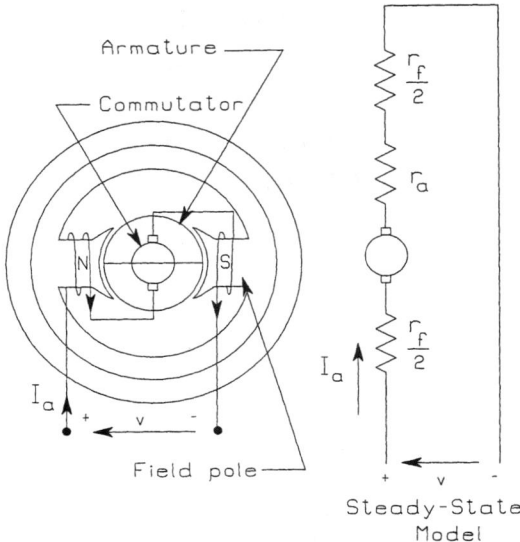

FIGURE 4.1 Series motor diagram: r_a = armature resistance measured at brushes, r_f = main field resistance, and v = applied voltage.

FIGURE 4.2 Series motor.

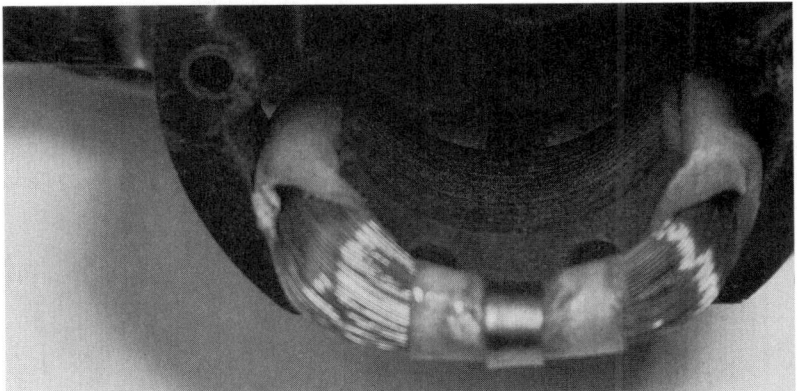

FIGURE 4.3 One pole of a series motor field winding.

FIGURE 4.4 Wound armature.

It is known that the instantaneous force on a conductor of length ℓ carrying a current i in a magnetic field B is:

$$f = Bi\ell \sin\theta \quad \text{N} \tag{4.1}$$

Or, in vector notation:

$$f = \ell\,(i \times B) \tag{4.2}$$

where θ is the angle between the direction of the magnetic field and the direction of current flow in the conductor, B is in webers per square meter (teslas), i is in amperes, and ℓ is in meters. Motors, by design, have the armature conductors and magnetic flux at quadrature to one another. Therefore, the force becomes:

$$f = Bi\ell \tag{4.3}$$

Assume a situation in Fig. 4.5 where a conductor of length ℓ is located in a magnetic field and is free to move in the x direction perpendicular to the field. From the preceding discussion, a force is produced on the conductor, causing it to move in the x direction. Then:

$$V - ir = e \quad \text{or} \quad Vi - i^2 r = ei \tag{4.4}$$

So, differential electrical energy into the conductor less differential $i^2 r$ loss equals differential mechanical output energy:

$$dW_{\text{elect}} = dW_{\text{mech}} \tag{4.5}$$

$$ei\, dt = f\, dx \quad \text{where } f = Bi\ell \tag{4.6}$$

$$ei\, dt = Bi\ell\, dx \tag{4.7}$$

$$e = B\ell \frac{dx}{dt} = B\ell v \tag{4.8}$$

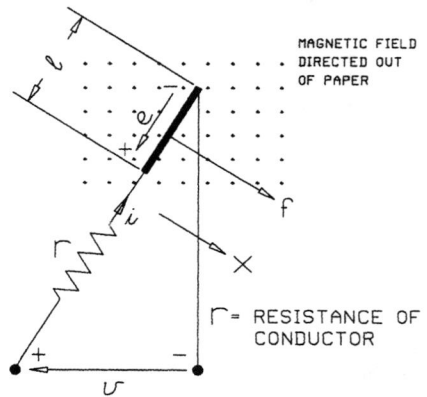

FIGURE 4.5 Conductor moving in a magnetic field.

This force causes movement in the conductor, which in turn causes a voltage to be induced into that conductor which is opposite to the direction of the original current. This is an important concept in the operation of motors in general and one which is used to discuss the operation of the series motor. This induced voltage is usually called a *counter-emf (cemf) voltage* because of its opposition to the applied voltage.

This example also indicates that a reversible energy or power exchange is possible (i.e., between the mechanical and electrical systems). Therefore, the same machine may operate as a motor or a generator, depending on the flow of energy in the armature.

In the motor mode, the field and armature of a dc series motor are supplied with the same current by an applied voltage, and a magnetic field (flux) is produced in the magnetic circuit. Since the armature conductors (coils) are located in this field, each of the conductors in the field experiences a force (torque) tending to make it move (rotate), and as we have just indicated, a countervoltage (cemf) is produced opposing the applied voltage. Other than the cemf which the armature produces, we must recognize that the armature circuit (coils) also produces a magnetic field of its own. The armature, because of its commutator and brush construction, has a unidirectional current and therefore produces a fixed-direction magnetomotive force (mmf), measured in ampere-turns. This mmf is the product of the effective coil turns on the armature and the current through those turns. It must be understood here that the armature winding must be considered in developing those ampere-turns (i.e., there may be parallel paths through the armature, with the possibility of the armature coils being wound in series or parallel arrangements; hence, there will also be a division of the total armature current in each of the windings). This topic is addressed later. The following is a discussion of the action of the mmf produced by the armature.

Armature Reaction. The conductors in the armature carry a current proportional to the load. The magnetic field produced by this current reacts with the main field produced by the same current flowing in the field coils. Figure 4.6 indicates two so-called belts of armature conductors (coil sides) under each pole face. Each of the conductors comprising these belts carries current in the same direction and hence produces additive mmf. In addition, there are conductors which also carry unidirectional currents but are not under the pole arcs. The important consideration here is the effect of the presence of magnetic material in the pole pieces, armature core, and armature teeth. The flux paths through the armature are influenced by the reluctances of the paths. It is obvious that the reluctance of the flux paths under the pole pieces is less than that of the paths adjacent to the brush area, which constitute a material of much greater reluctance, namely air.

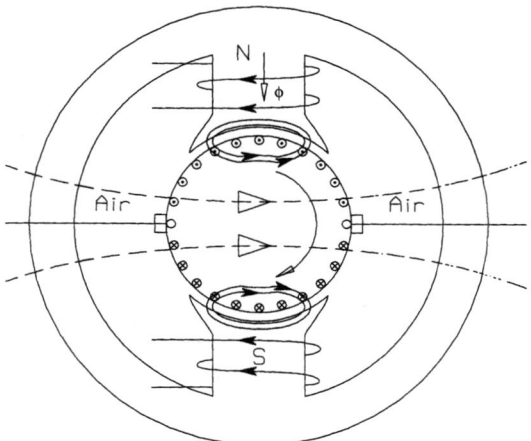

FIGURE 4.6 Armature magnetic field.

Figure 4.6 indicates that the brushes are on the mechanical neutral axis (i.e., halfway between the poles). The general direction of the armature mmf is along the brush axis at quadrature to the main field. The armature conductors adjacent to the poles produce flux densities in the air gap which are equal and opposite at the pole tips. Keep in mind that the flux density produced by the armature is directly related to the armature current. (Also, a uniformly distributed flux density in the air gap is attributed to the main field and is directly related to this same armature current.) Now the net mmf (or flux in the air gap) is the result of both the main field mmf and the armature field mmf. The resultant air gap flux is now increased at one pole tip and reduced at the other pole tip because of the armature reaction. The flux distortion in the air gap is illustrated in Fig. 4.7. In this figure, two poles and the armature conductors beneath the poles have been unfolded to illustrate the distortion more clearly. Ampere's circuital law is useful here in determining the armature mmf. The resulting air gap mmf is the result of the superposition of the field and armature mmfs. MMF drops in the poles and armature iron are considered negligible compared to the air gap mmf. For reference, positive direction for the mmfs is assumed to be a flux out of a north pole. The armature mmf is shown as a linear relationship, but actually, because the armature slots are discrete, this relationship is actually made up of small discrete stair-step transitions. However, it is shown as a smooth curve here for easier analysis.

It is obvious that the air gap flux varies along the pole face. Another observation from the distortion mmf pattern is that harmonics are very present in the air gap mmf. Also, because of symmetry, only odd harmonics can exist in an analysis of the air gap mmf. It must also be remembered that in the case of a series ac motor, the variation of flux distortion would be approximately the same, only pulsating.

The results of a Fourier analysis of the air gap flux harmonics for the case when $(N_f I_f)/(N_a I_a) = 7/5$ are given in Figs. 4.8 to 4.10. The figures show the fundamental, the fundamental plus third and fifth harmonics, and finally the fundamental and odd harmonics, up to and including the fifteenth harmonic. The magnitudes of the odd harmonics beginning with the fundamental are 8.94, 0.467, 1.41, 0.924, −0.028, −0.02, 0.523, and 0.0094. That is, the series can be represented as follows:

$$\phi_{\text{air gap}} = 8.94 \cos \theta + 0.467 \cos 3\theta + 1.41 \cos 5\theta + 0.924 \cos 7\theta$$

$$- 0.028 \cos 9\theta - 0.02 \cos 11\theta + 0.523 \cos 13\theta + 0.0094 \cos 15\theta + \cdots \quad (4.9)$$

Of course, changing the pole arc or magnitudes of the armature and field mmfs will change the harmonic content. The change in flux density across the air gap produces two effects: (1) a reduction in the total flux emanating from each pole, and (2)

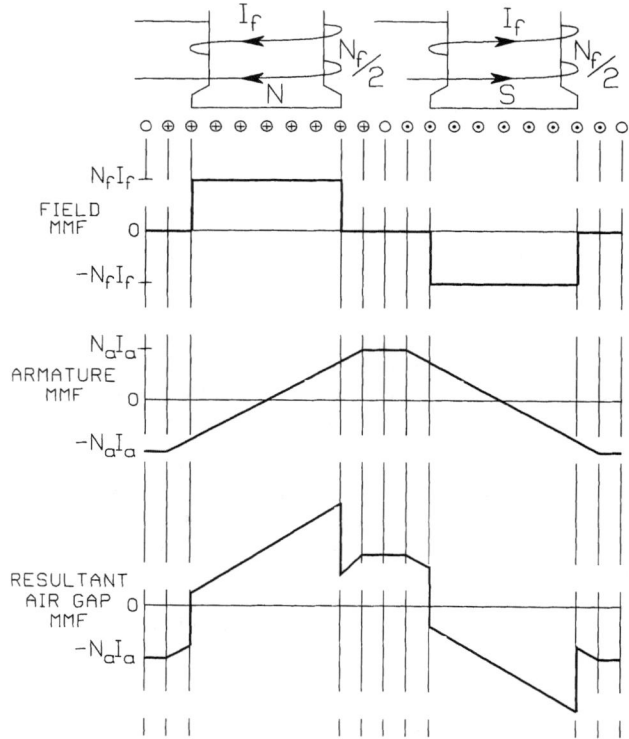

FIGURE 4.7 Flux distortion in the air gap.

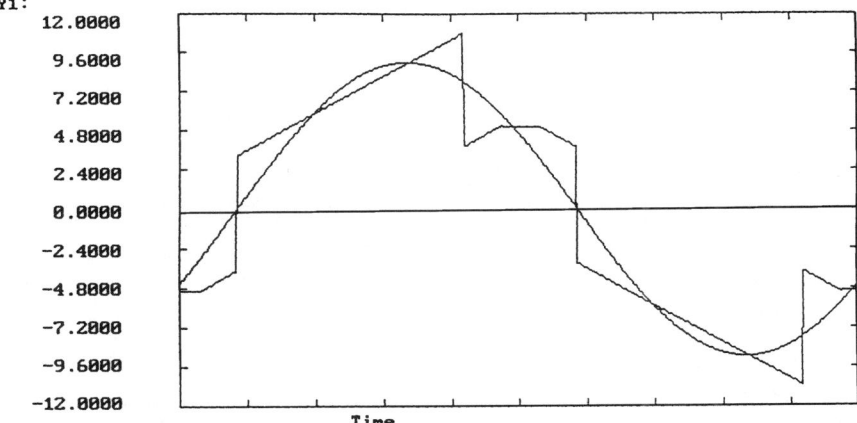

FIGURE 4.8 Fundamental frequency of air gap flux.

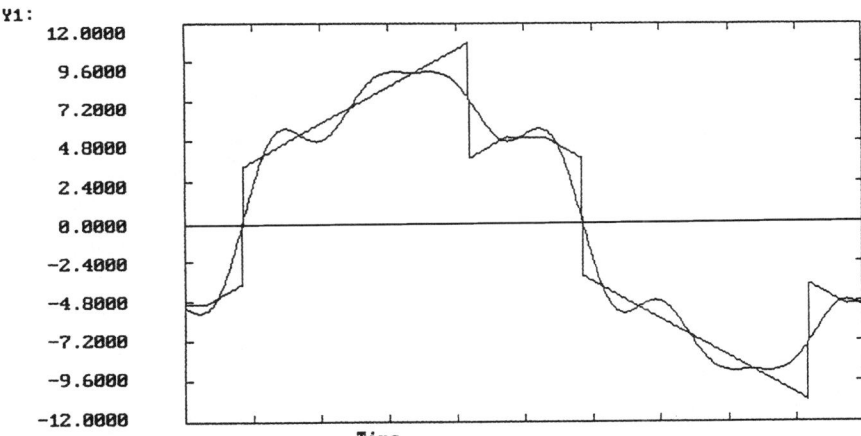

FIGURE 4.9 Air gap flux due to fundamental plus third and fifth harmonics.

a shift in the electrical neutral axis, lending to commutation problems due to the flux distortion.

The flux distortion which the armature produces has been called a *cross-magnetizing armature reaction,* and rightly so. The net effect is illustrated in Fig. 4.11, showing the resultant field where the armature cross field is at right angles to the main field. The result is a distortion of the net flux in the motor; the second effect, which was indicated previously, is a reduction in the total main field flux. This reduction of flux is not too obvious; in fact, it would almost appear from Fig. 4.8 that the vector addition of these two mmfs would lead to an increase in flux which would occur with the brushes on the mechanical neutral axis.

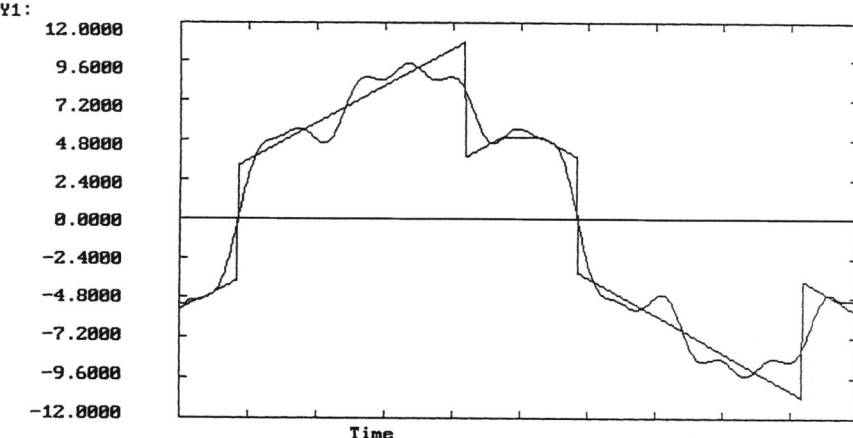

FIGURE 4.10 Air gap flux due to odd harmonics through fifteenth.

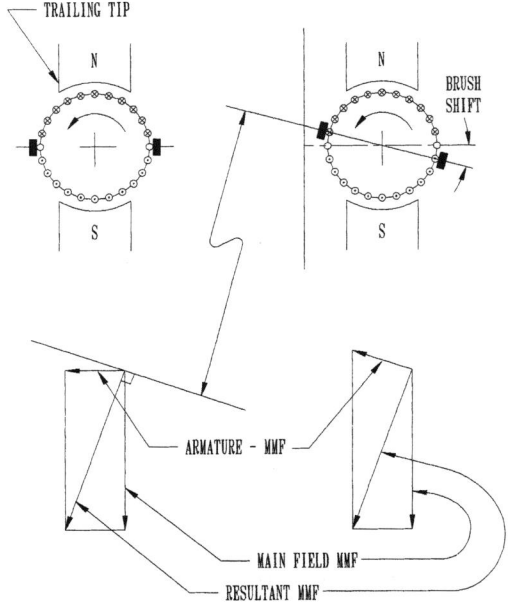

FIGURE 4.11 Field of armature with and without brush shift.

This resulting decrease in flux can be attributed to magnetic saturation. Figure 4.12 illustrates how this comes about when one operates the motor at the knee of the saturation curve. The area ABC is proportional to the reduction in flux at the pole tip with a decrease in flux density, while the area CDE is proportional to the increase in flux at the other pole tip. Note that the reduction in flux exceeds the increase in flux because of the saturation effect. This is sometimes called the *demagnetizing effect of armature reaction.* This demagnetization effect can be on the order of about 4 percent.

Both armature reaction distortion and flux can be reduced or eliminated by compensation windings in the pole face or by increasing the reluctance at the pole tips; the latter by either lamination design or a nonuniform air gap.

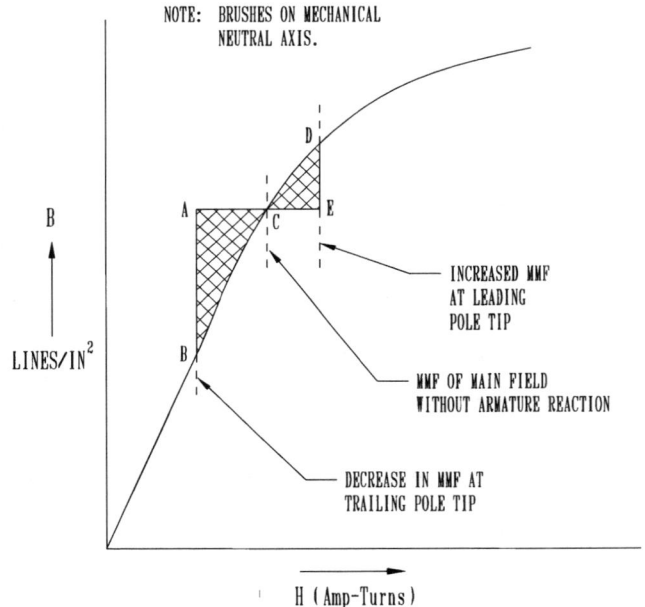

FIGURE 4.12 Armature reaction resulting in main pole flux reduction.

Because the purpose of the commutator and brushes is to change the current in a short-circuited coil from, say, a current of $+I$ to $-I$ in the length of time the coil is short-circuited by the brushes, some arcing at the brushes is expected. One desires to keep any voltage induced into the coil to a minimum in order to keep this arcing as small as possible.

When the brushes are on the mechanical neutral position and the main field becomes distorted, the coils being commutated will have an induced voltage from the distorted air gap flux. A shifting of the brushes to a new neutral position is suggested to keep the induced voltage to a minimum and keep the brush arcing low. It should be recognized that the amount of distortion of the field is a function of the ampere-turns of the armature and hence is dependent on the motor load. One then realizes that the electrical neutral position is a function of the load and that shifting brushes to an electrically neutral position is therefore not a matter of shifting them to a unique point in space.

Brush shifting causes another effect—that is, a demagnetization effect on the air gap flux. This effect is in addition to the demagnetizing effect which occurs because of saturation. When the brushes are shifted, the pole axis of the armature is shifted.

The result is that the angle between the main field and the armature field is greater than 90°. This process is also illustrated in Fig. 4.11.

In summary, there are two processes which cause reduction of the air gap flux: (1) reduction due to cross-magnetization when the brushes are on the mechanical neutral position, which changes the flux distribution across the pole face and the net flux because of saturation effects, and (2) demagnetization resulting from a brush shift and change of the armature pole orientation with respect to the main field, resulting in the armature field mmf having a component in direct opposition to the main field mmf.

Reactance Voltage and Commutation. It was indicated previously that during commutation the current in the shorted coil(s) under a brush must reverse and change direction. The self-inductance of the shorted winding by Lenz's law induces an emf in the shorted winding to oppose the change in coil current. This voltage is sometimes called the *reactance voltage*. (Actually, it may be only a portion of the reactance voltage, as is shown later.) This voltage slows down the reversal of current and tends to produce sparks or arcing as the trailing commutator bar leaves a brush. The reactance voltage hinders good commutation. The magnitude of this voltage depends directly on the square of the number of coil turns, the current flowing, and the armature velocity; it is inversely proportional to the reluctance of the magnetic path.

When shifting brushes to seek an electrical neutral position on a motor, the shifting is done in the opposite direction of armature rotation. It must be remembered that the reactance voltage is an $e = L(di/dt)$ voltage and is simply due to the current change during commutation. It has nothing to do with induced voltage from air gap flux. Shifting the brushes really does not help the reactance voltage. Theoretically, there is no induced voltage on the neutral axis from the motor flux (in fact, $d\lambda/dt$ should be zero here since the coil sides are moving parallel to the flux). In order to counteract the reactance voltage and induce a voltage opposite to the reactance voltage, the brushes must be shifted further backward than the magnetic neutral axis. At this point a $d\lambda/dt$ voltage is induced into the commutated coil by the field flux, which counteracts the reactance voltage. The result is that two voltages in opposite polarities are induced into the shorted coil, thereby reducing brush arcing.

An important consideration is that there can be another component of the reactance voltage which can occur when two coil sides are in the same slot and both are undergoing commutation. There is then the following self-inductance term:

$$\text{Reactance voltage} = L_{11}\frac{di_1}{dt} + M_{12}\frac{di_2}{dt} \qquad (4.10)$$

where the coil being considered is the coil carrying current i_1 and the coil in the same slot undergoing commutation is carrying current i_2. If there is more coupling between coils, more mutual terms can exist.

Brushes are a very important consideration, and contrary to normal electrical principles, one would assume that keeping the brush resistance low would assist in reducing arcing at the brush commutator bar interface. This is far from the truth. In fact, resistance commutation is now an accepted technology. Brushes normally have a graphite or carbon formulation and hence introduce, by their characteristics, resistance into the interface. If constant current density could be achieved at a brush for all loads and speeds, an ideal condition would exist for commutation. Figure 4.13 illustrates what would occur if ideal conditions are assumed. The assumptions are as follows:

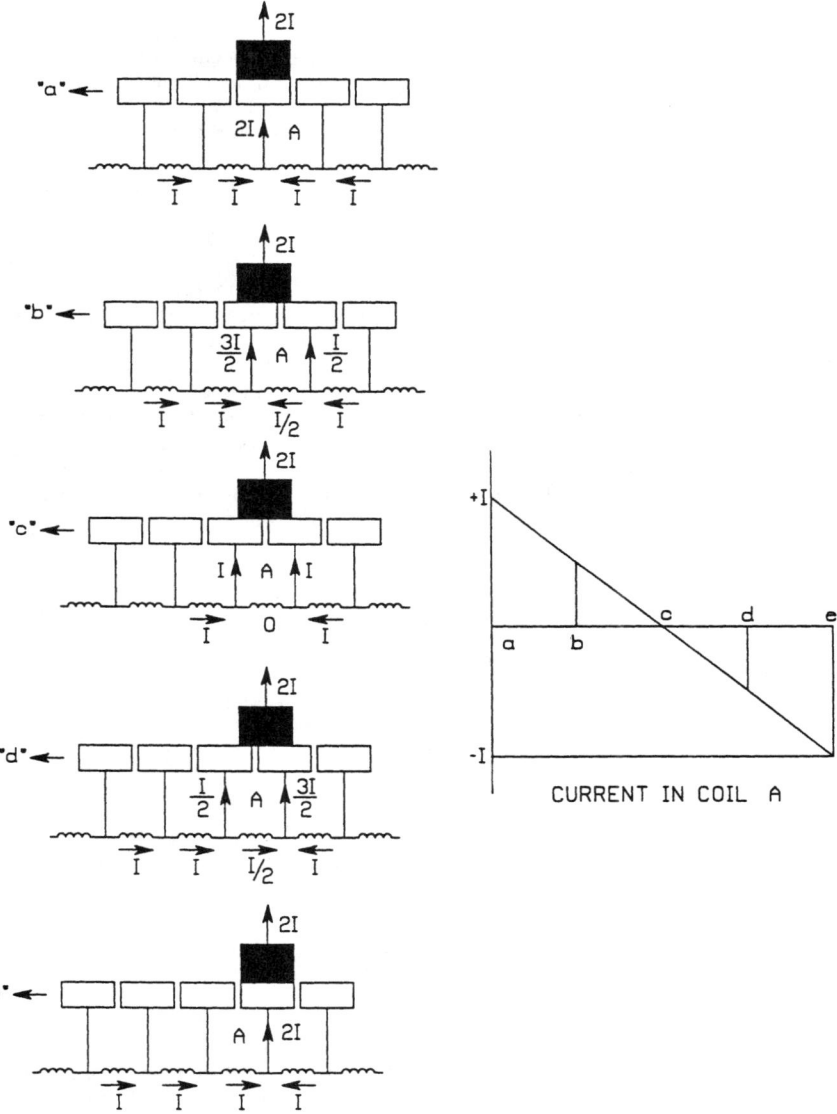

FIGURE 4.13 Linear commutation.

- The brush width is equal to that of a commutator segment.
- The current density under a brush is constant.
- The reactance voltage and resistance are zero.
- The commutator bar insulation is small compared to the width of a commutator segment.

Note that as the commutator passes beneath the brush, for the assumptions given, the contact resistance is inversely proportional to the contact area. The sequence in Fig. 4.13 illustrates how the current change occurs in an armature coil under these ideal conditions. This change in surface contact changes the brush contact resistance; hence, with constant current density, the change in current flow changes linearly according to contact resistance. Remember, the resistance is inversely proportional to the contact area. This ideal commutation has been termed *linear commutation* because of the linear transfer of current in the commutated coil, as indicated in Fig. 4.13.

If the brush width is greater than that of one commutator segment, linear commutation still exists because the resistance still changes linearly; hence, the current density remains constant. The commutation period is longer; hence, the rate of change of current di/dt is reduced, and hence the reactance voltage is reduced also.

All of the assumptions made in the previous analysis cannot occur, and *nonlinear commutation* results. The armature coils have resistance, have self- and mutual inductance, and hence linear commutation cannot be attained.

If these nonlinearities do occur, then either *undercommutation* or *overcommutation* occurs. These are illustrated in Fig. 4.14. Since brush heating is a product of brush resistance and the square of the current, linear commutation is preferred, as it minimizes brush heating. That is, the squared value of the three forms of current commutation in Fig. 4.13 is smallest for linear commutation.

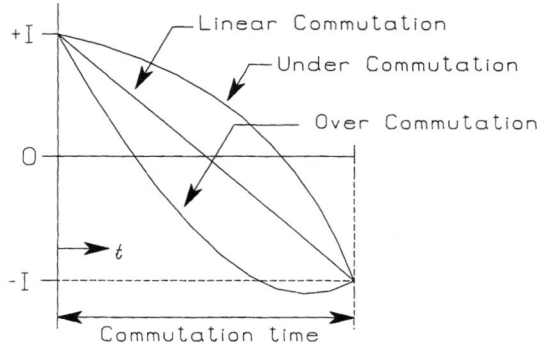

FIGURE 4.14 Overcommutation, undercommutation, and linear commutation.

It is possible to develop a circuit model for the commutation process, as shown in Fig. 4.15. The factors have previously been described, and they are restated here. Consider the subject coil of turns N_1, and self-inductance L_{11}, and the current i_1 which is being commutated.

1. The voltage of self inductance (reactance voltage) of the commutated coil is

$$e_{L1} = L_{11} \frac{di_1}{dt} \tag{4.11}$$

where i_1 is the current being commutated and L_{11} is the self-inductance of the coil being commutated.

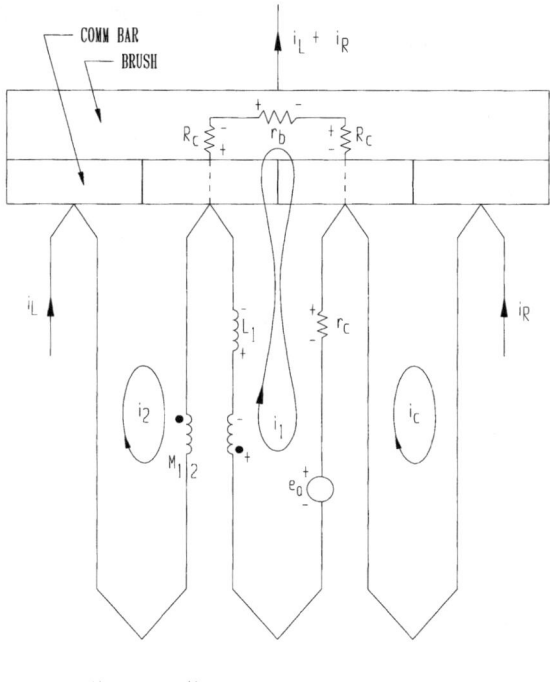

$$L_1 \frac{di_1}{dt} + M_{12} \frac{di_2}{dt} + (r_c + r_b + 2R_c)i_1 + e_0 = 0$$

WHERE: R_c = BRUSH CONTACT RESISTANCE
r_c = COIL RESISTANCE
r_b = BRUSH RESISTANCE

FIGURE 4.15 Commutation voltages.

2. The voltage of mutual inductance of an adjacent coil in the same slot carrying current i_2 and undergoing commutation and linked to the subject coil is

$$e_{M1} = M_{12} \frac{di_2}{dt} \qquad (4.12)$$

where M_{12} is the mutual inductance between adjacent coils.

3. The voltage induced into the subject coil of turns N_1 when cutting flux ϕ_a due to the armature mmf is

$$e_a = N_1 \frac{d\phi_a}{dt} \qquad (4.13)$$

Sometimes this is called *commutating voltage;* it opposes the reactance voltage. This can be controlled by brush overshifting to make e_a cancel the voltages e_{M1} and e_{L1}.

4. There are voltage drops due to coil resistance r_c; R_c and R_c, the left and right contact resistances at the brush; and the brush resistance between commutator segments r_b.

The commutation process is not simple, and brush composition and interface film, measurements, and conditions all can be made to assist in the commutation process. Maintaining a suitable interface film between the brush and the copper commutator is extremely important. This interface is formed of copper oxide and free particles of graphite film; it provides a general resistance commutation and supplies lubricant to reduce surface friction and heat between the commutator and the brush.

Torque-Speed Characteristics of DC Series Motors. The electrical equivalent steady-state circuit is the most appropriate method for analyzing the motor. This equivalent circuit is repeated in Fig. 4.16 for convenience.

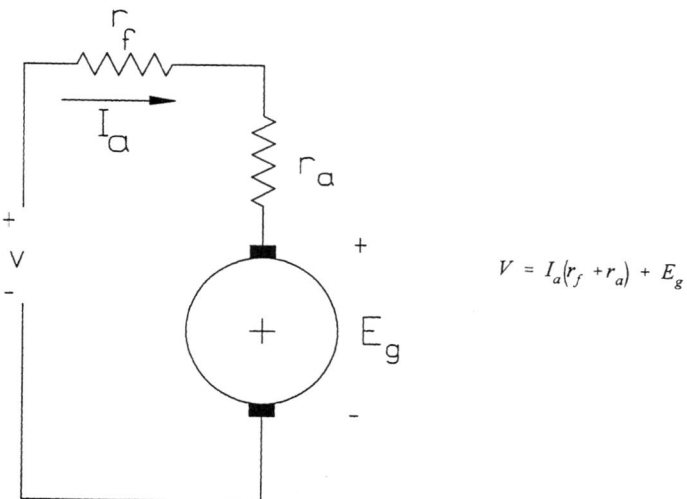

FIGURE 4.16 Electric circuit of series dc motor.

Writing Kirchhoff's voltage law around the loop gives:

$$V = I_a(r_f + r_a) + E_{cemf} \quad (4.14)$$

$$= I_a(r_f + r_a) + E_g \quad (4.15)$$

The counter-emf E_g was addressed earlier when discussing the voltage induced into a conductor of length ℓ and moving with a velocity v perpendicular to a magnetic field of flux density B. The counter-emf voltage E_g was expressed as e and given by:

$$e = B\ell v \quad (4.8)$$

It is necessary to modify this and express e in terms of the motor parameters. Beginning with Faraday's law:

$$E_g = E_{cemf} = N \frac{\Delta \phi}{\Delta t} \quad V \tag{4.16}$$

where N is the number of conductors per armature path when Z is the total number of conductors (coil sides) on the armature.
Then:

Z = number of slots × coils/slot × turns/coil × 2 (conductors/turn)
a = number of parallel paths through the armature
P = number of poles
n = armature speed, rpm
ϕ = flux per pole, maxwells

Then:

$$E_g = E_{cemf} = \frac{Z \phi n P}{60a} \times 10^{-8} = K_e \phi n \quad V \tag{4.17}$$

where ϕ is in lines (maxwells) and

$$K_e = \frac{ZP}{60a} \times 10^{-8} \tag{4.18}$$

From the equivalent circuit, the developed power of the motor is $E_g I_a$; that is:

$$P_{dev} = K_e \phi n I_a \quad W \tag{4.19}$$

$$P_{dev} = \frac{K_e \phi n I_a}{746} \quad hp \tag{4.20}$$

$$T_{dev} = \left(\frac{33{,}000}{2\pi n} \right) \left(\frac{K_e \phi n I_a}{746} \right) \quad lb \cdot ft \tag{4.21}$$

$$T_{dev} = 7.04 \left(\frac{ZP}{60a} \right) \phi I_a \times 10^{-8} \quad lb \cdot ft \tag{4.22}$$

$$T_{dev} = 22.5 \left(\frac{ZP}{a} \right) \phi I_a \times 10^{-8} \quad oz \cdot in \tag{4.23}$$

The equivalent circuit voltage equation then becomes

$$V = I_a(r_f + r_a) + K_e \phi n \tag{4.24}$$

And the motor speed equation becomes

$$n = \frac{V - I_a(r_f + r_a)}{K_e \phi} \tag{4.25}$$

For no load speed:

$$n_L \approx \frac{V}{K_e \phi} \tag{4.26}$$

For stalled torque:

$$\lambda = K_e \phi I_{a,\text{stalled}} \tag{4.27}$$

$$= K_e \phi \frac{V}{r_a} \tag{4.28}$$

Efficiency of DC Series Motors. Earlier discussion has indicated that a number of losses occur in the series motor. These losses can be summarized as follows:

1. Copper loss in the armature winding $I_a^2 r_a$
2. Copper loss in the series field winding $I_a^2 r_f$
3. Brush contact loss
4. Friction (brush and bearing friction) and windage
5. Core loss (hysteresis and eddy current)
6. Stray load loss (losses in addition to those above)

A comment should be made on the brush contact loss, listed as number 3. Experimentation and Institute of Electrical and Electronic Engineers (IEEE) specifications have suggested as an approximation that

$$\text{Brush contact loss} = \text{brush voltage drop} \times \text{armature brush current} \tag{4.29}$$

This expression was arrived at for carbon or graphite brushes by observing the brush voltage drop as a function of current density. At high- and low-current steady-state densities, the voltage falls off and approaches zero. However, in between these limits the voltage drop is approximately constant at about 1 V per brush. Therefore, for a pair of brushes this becomes 2 V. On small machines this voltage drop tends to increase, since the tendency is for the voltage drop to increase for lower brush and commutator temperatures.

Carbon and graphite materials have a resistivity many times that of copper and have a negative temperature coefficient. This negative temperature coefficient is attributed to the rise in brush voltage drop in small machines.

Stray load loss, as the term suggests, is a function of motor load and changes in load. Changes in load produce changes in armature current and hence affect (1) magnetic saturation in the magnetic circuit, (2) armature reaction changes, and (3) eddy current loss changes.

Figure 4.17 presents a diagram displaying these losses. By definition:

$$\text{Overall efficiency} = \frac{\text{output}}{\text{input}} = \frac{\text{output}}{\text{output} + \text{losses}} \tag{4.30}$$

$$= \frac{\text{useful mechanical output}}{\text{total electrical input}} \tag{4.31}$$

Sometimes both electrical and mechanical efficiencies are of interest in order to determine where improvements can be made.

$$\text{Mechanical efficiency} = \frac{\text{useful mechanical output}}{\text{mechanical output} + \text{rotational losses}} \tag{4.32}$$

$$\text{Electrical efficiency} = \frac{\text{electrical power output}}{\text{electrical power output} + \text{electrical losses}} \tag{4.33}$$

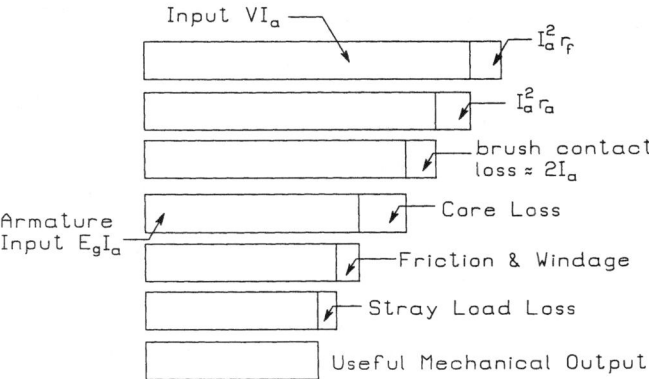

FIGURE 4.17 Losses in a series dc motor.

The conditions for maximum efficiency can be related to those losses which are considered to be constant and those that vary with the motor load current. If the losses are segregated as follows:

K_1 = constant losses

K_2 = losses which vary linearly with I_a

K_3 = losses which vary as the square of I_a

Then the efficiency is

$$\text{eff} = \frac{\text{input} - \text{losses}}{\text{input}} = \frac{VI_a - (K_1 + K_2 I_a + K_3 I_a^2)}{VI_a} \quad (4.34)$$

$$\frac{d\text{eff}}{dI_a} = \frac{VI_a(V - K_2 - 2K_3 I_a) - [VI_a - (K_1 + K_2 I_a + K_3 I_a^2)]V}{(VI_a)^2} \quad (4.35)$$

Equating this expansion to zero gives the condition for maximum efficiency.

$$VI_a[V - K_2 - 2K_3 I_a] - [VI_a - (K_1 + K_2 I_a + K_3 I_a^2)]V = 0 \quad (4.36)$$

$$-2K_3 I_a^2 + K_1 + K_3 I_a^2 = 0 \quad (4.37)$$

$$K_3 I_a^2 = K_1 \quad (4.38)$$

Thus, for maximum efficiency the constant losses must be equal to those that vary as the square of the armature current. This is typical for all different pieces of rotational electrical equipment. The constant losses usually are considered to be the core losses, friction, and windage. Usually the brush loss is small.

4.1.2 AC Series Motors

One of the first considerations when considering ac operation of a series-wound motor is: Does the motor develop a unidirectional torque when operated on ac supply voltages? We know the answer is yes. Since the armature and field are connected in series, a current reversal in the field also produces a current reversal in the armature. The torque is therefore in the same direction. As indicated previously, the construction must be an ac type with a completely laminated magnetic path. If it were not laminated it would react as a solid-core inductor, with extremely high eddy current losses.

The basic phasor and electrical circuit diagram of a series ac motor is shown in Fig. 4.18. The common current in both armature and field produce a motor flux which is nearly in phase with the current, the small difference being the hysteresis and eddy current effects, which are not accounted for in the phasor diagram. Seven voltages are required to overcome the applied voltage. These are the voltages produced by the leakage reactances and resistances of both armature and field, the transformer voltage E_{Tf}, the commutator brush drop, and the counter-emf voltage E_g. When the armature is rotating, the armature conductors, which are moving through an alternating field flux, generate an alternating voltage E_g which is in phase with the flux, as indicated on the diagram.

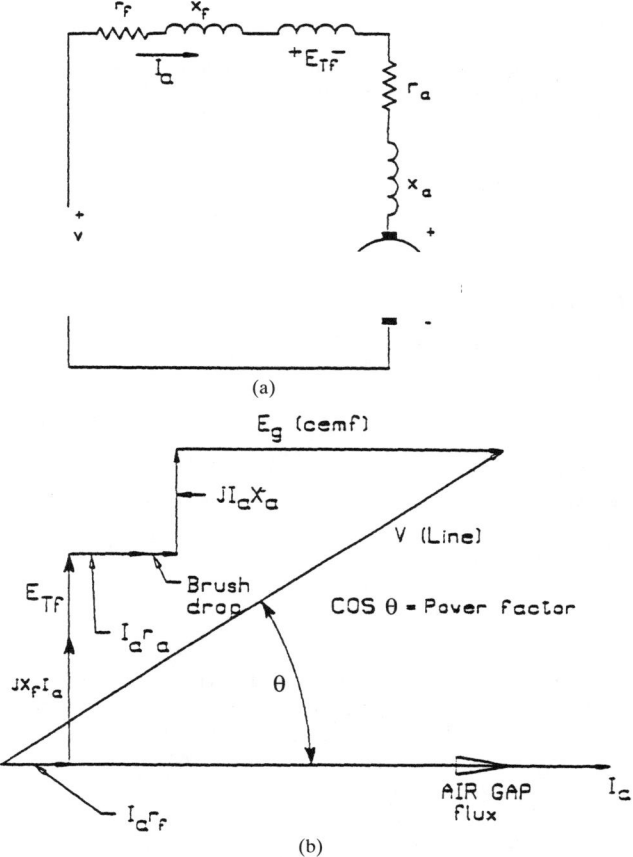

FIGURE 4.18 Series ac motor: (*a*) equivalent circuit and (*b*) phasor diagram.

This voltage is proportional to the magnitude of the flux and the rate at which it is cutting the flux (Faraday's law). Since the flux is a sinusoidally varying quantity, the voltage will also vary sinusoidally at the same frequency as the flux wave. The transformer voltage E_{Tf} has been included and reflects the Faraday voltage induced into the field by the core flux. A similar voltage would be present in the armature, but it is small in comparison to E_{Tf} and has not been included in the phasor diagram. The phasor sum of these voltages then represents the applied voltage V and the basic phasor diagram of the ac series motor. Here all voltages are root mean square (RMS) quantities. As previously mentioned, the effects of hysteresis and eddy currents have not been taken into account, nor have the effects of commutation and brush shifting. The power factor angle θ is also indicated on the diagram and represents the power factor at the input motor terminals.

Several things are apparent from the phasor diagram:

1. At starting, the generated voltage E_g is zero. Hence, the starting current is limited only by the impedances of the armature and field and the transformer voltage E_{Tf}.
2. The motor draws a lagging power factor $\cos \theta$.
3. As the motor speed increases at a constant line voltage, the motor current decreases and the power factor improves, which is a good reason to run a motor at high armature velocities.
4. A reduction in frequency also improves the power factor by reducing the reactance drops in both armature and field.

At this point it should be obvious that dc and ac performance of series motors must differ due to reactance and transformer voltages and also due to ac losses (additional hysteresis and eddy current losses—that is, core losses—and possible changes in ac resistance as compared to dc resistance of the armature and field). Note that in the development of the phasor diagram leakage reactance voltages for both the field and armature, as well as a transformer voltage in the main field, have been included. It is appropriate to develop some mathematical formulations and basis for these voltages. In reality, the leakage reactance and transformer voltage result from an application of Faraday's law and are broken down into two parts to obtain each voltage. This Faraday voltage is expressed in instantaneous form and consists of two parts, as follows.

$$e_f = e_{Tf} + e_\ell \tag{4.39}$$

where e_f = Faraday voltage
 e_{Tf} = transformer voltage
 e_ℓ = leakage reactance voltage

Then:

$$e_f = \frac{d}{dt}(\lambda_M + \lambda_\ell) = N_f \frac{d}{dt}(\phi_M + \phi_\ell) \tag{4.40}$$

where ϕ_M is the flux within the complete magnetic motor circuit and ϕ_ℓ is the leakage flux.

Then:

$$\lambda_f = N_f \left(\frac{\mathscr{F}}{\mathscr{R}_M} + \frac{\mathscr{F}}{\mathscr{R}_\ell} \right) \tag{4.41}$$

where $\mathcal{F}$ is the ampere-turns of the main field and $\mathcal{R}_M$ and $\mathcal{R}_\ell$ are the reluctances of the motor magnetic circuit and the leakage path, respectively.

Then:

$$\lambda_f = N_f \left(\frac{N_f}{\mathcal{R}_M} i + \frac{N_f}{\mathcal{R}_\ell} i \right) \quad (4.42)$$

$$= \frac{N_f^2}{\mathcal{R}_M} i + \frac{N_f^2}{\mathcal{R}_\ell} i \quad (4.43)$$

Then:

$$e_f = \frac{N_f^2}{\mathcal{R}_M} \cdot \frac{di}{dt} + \frac{N_f^2}{\mathcal{R}_\ell} \cdot \frac{di}{dt} \quad (4.44)$$

$$= L_f \frac{di}{dt} + L_\ell \frac{di}{dt} \quad (4.45)$$

where L_f and L_ℓ are the inductances associated with the two fluxes present in the main field winding.

Now, in the steady-state response E_f can be written as follows:

$$E_f = E_{Tf} + E_\ell = j\omega L_f I + j\omega L_\ell I \quad (4.46)$$

$$= jx_f I + jx_\ell I \quad (4.47)$$

where jx_ℓ is the leakage reactance, and the voltage and current are expressed in RMS values.

If we assume that the applied voltage is sinusoidal, the Faraday voltage also will be sinusoidal, and it is possible to express the transformer voltage E_{Tf} in terms of other variables.

By Faraday's law,

$$e_{Tf} = N_f \frac{d\phi_M}{dt} \quad (4.48)$$

$$\phi_M = \frac{1}{N_f} \int e_{Tf} \, dt \quad (4.49)$$

if we represent

$$e_{Tf} = \sqrt{2} E_{Tf} \sin(\omega t + \alpha) \quad (4.50)$$

$$\phi_M = \frac{\sqrt{2} E_{Tf}}{N_f \omega} \cos(\omega t + \alpha) + \phi_c \quad (4.51)$$

where E_{Tf} is the rms value.

The constant of integration ϕ_c is a transient quantity and disappears in the steady state analysis. Hence:

$$\phi_M = \frac{\sqrt{2} E_{Tf}}{N_f \omega} \cos(\omega t + \alpha) \quad (4.52)$$

$$\phi_{M,\max} = \frac{\sqrt{2} E_{Tf}}{N_f \omega} \quad (4.53)$$

Or

$$E_{Tf} = \frac{2\pi f N_f \phi_{M,\max}}{\sqrt{2}} = 4.44 f N_f \phi_{M,\max} \tag{4.54}$$

This is the general transformer voltage relationship common in transformer theory.

Development of the Equivalent Circuit for a Universal Motor Considering Brush Shift.
Begin by considering a two-energy-source system (i.e., a main field and an armature) with no brushes (Fig. 4.19), where:

- v_a = voltage applied to armature
- v_f = voltage applied to field
- r_a = armature resistance
- r_f = field resistance
- i_a = armature current
- i_f = field current

With no brushes, we have the following:

$$v_a = r_a i_a + p\lambda_a \tag{4.55}$$

$$V_f = r_f i_f + p\lambda_f \tag{4.56}$$

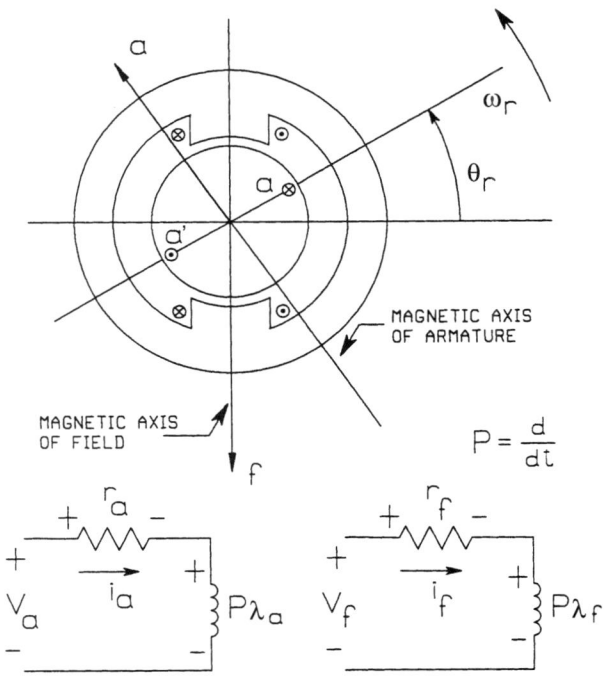

FIGURE 4.19 Consideration of universal motor brush shift.

where

$$p = d/dt \quad \text{(differential operator)}$$

So, where

$$\lambda_a = \mathcal{L}_{aa}i_a + \mathcal{L}_{af}i_f \quad (4.57)$$

$$\lambda_f = \mathcal{L}_{fa}i_a + \mathcal{L}_{ff}i_f \quad (4.58)$$

all $\mathcal{L}$s could be functions of θ_r.
Then:

$$v_a = r_a i_a + i_a p\mathcal{L}_{aa} + \mathcal{L}_{aa}pi_a + i_f p\mathcal{L}_{af} + \mathcal{L}_{af}pi_f \quad (4.59)$$

$$v_f = r_f i_f + i_a p\mathcal{L}_{fa} + \mathcal{L}_{fa}pi_a + i_f p\mathcal{L}_{ff} + \mathcal{L}_{ff}pi_f \quad (4.60)$$

Now:

$$\lambda_{aa} = N_a\phi_a = \frac{N_a^2}{\mathcal{R}_{aa}} i_a = \mathcal{L}_{aa}i_a \quad \text{and} \quad \lambda_{af} = \frac{N_a N_f}{\mathcal{R}_{af}} i_f = \mathcal{L}_{af}i_f \quad (4.61)$$

where

$$\mathcal{L}_{aa} = \frac{N_a^2}{\mathcal{R}_{aa}} \quad \mathcal{L}_{af} = \frac{N_a N_f}{\mathcal{R}_{af}} \quad (4.62)$$

With $\mathcal{L}_{aa}$ and the plot of $\mathcal{R}_{aa}$ versus θ_r (Fig. 4.20):

$$\mathcal{L}_{aa} = \frac{L_{\max} + L_{\min}}{2} + \frac{L_{\max} - L_{\min}}{2} \cos(2\theta_r) \quad (4.63)$$

See Figs. 4.21 and 4.22.

$$\mathcal{L}_{af} = \mathcal{L}_{fa} = -L\cos(\theta_r) \quad \text{where } L = \frac{N_a N_f}{R_{af}} \quad (4.64)$$

The self-inductance of the field is a constant with a round armature.

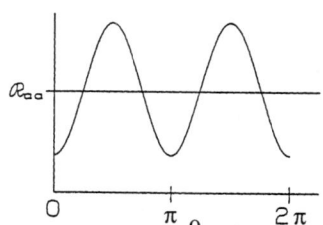

FIGURE 4.20 Plot of $\mathcal{R}_{aa}$.

$$\mathcal{L}_{ff} = \text{constant} = L_{ff} \quad \mathcal{L}_{ff} = \frac{N_f^2}{\mathcal{R}_f} \quad (4.65)$$

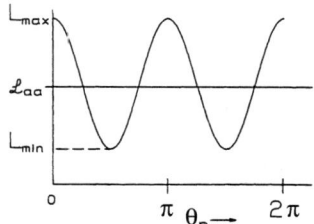

FIGURE 4.21 Mutual terms between armature and field.

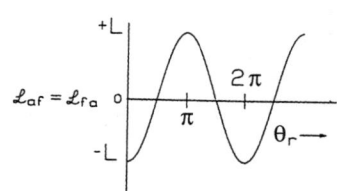

FIGURE 4.22 Plot of $\mathcal{L}_{af}$.

Using the expressions for $\mathcal{L}_{aa}$, $\mathcal{L}_{af}$, $\mathcal{L}_{fa}$, and $\mathcal{L}_{ff}$ gives the following:

$$v_a = r_a i_a + \omega_r (L_{min} - L_{max}) \sin(2\theta_r) + \left[\frac{L_{max} + L_{min}}{2} + \frac{L_{max} - L_{min}}{2} \cos(2\theta_r) \right] p i_a$$

$$+ \omega_r L \sin(\theta_r) i_f - L \cos(\theta_r) p i_f \quad (4.66)$$

And

$$v_f = r_f i_f + \omega_r L \sin(\theta_r) i_a - L \cos(\theta_r) p i_a + L_{ff} p i_f \quad (4.67)$$

Now, add brushes at an angle θ_{ra}. If the brushes were added with no shift, the resulting magnetic field of the armature would be at quadrature to the f axis. This would fix θ_r at 90°. Since the brushes are shifted to θ_{ra}, θ_r is fixed at $90° + \theta_{ra}$. Also, shifting the brushes causes a voltage to be induced in the field winding from the armature current, so $\mathcal{L}_{fa} \neq 0$. In fact, $\mathcal{L}_{fa}$ = constant = L_{fa} with $\theta_r = 90° + \theta_{ra}$. Then:

$$v_a = r_a i_a - \omega_r (L_{min} - L_{max}) \sin(2\theta_{ra}) i_a$$

$$+ \left[\frac{L_{max} + L_{min}}{2} - \frac{L_{max} - L_{min}}{2} \right] \cos(2\theta_{ra}) p i_a$$

$$+ \omega_r L \cos(\theta_{ra}) i_f + L \sin(\theta_{ra}) p i_f \quad (4.68)$$

And

$$v_f = r_f i_f + L_{fa} p i_a + L_{ff} p i_f \quad (4.69)$$

Now, defining the constants:

$$L_{af1} = L \cos(\theta_{ra}) \qquad L_{aa1} = (L_{min} - L_{max}) \sin(2\theta_{ra}) \quad (4.70)$$

$$L_{af2} = L \sin(\theta_{ra}) \qquad L_{aa2} = \left[\frac{L_{max} + L_{min}}{2} - \frac{L_{max} - L_{min}}{2} \cos(2\theta_{ra}) \right] \quad (4.71)$$

The equations in matrix form now are as follows:

$$\begin{bmatrix} v_f \\ v_a \end{bmatrix} = \begin{bmatrix} r_f + pL_{ff} & -pL_{fa} \\ \omega_r L_{af1} + pL_{af2} & r_a + \omega_r L_{aa1} + pL_{aa2} \end{bmatrix} \begin{bmatrix} i_f \\ i_a \end{bmatrix} \quad (4.72)$$

Or

$$v_f = r_f i_f + L_{ff} p i_f - L_{fa} p i_a \quad (4.73)$$

$$v_a = \omega_r L_{af1} i_f + L_{af2} p i_f + r_a i_a + \omega_r L_{aa1} i_a + L_{aa2} p i_a \quad (4.74)$$

From the preceding, the equivalent circuit can be drawn (Fig. 4.23). From this equivalent circuit, the transient response of the series motor can be calculated.

4.1.3 Permanent-Magnet DC Motors (Shunt PM Field Motors)

The general equivalent electrical circuit of a PMDC motor and its physical construction are shown in Figs. 4.24 and 4.25. The motor consists of a stator (Fig. 4.26) having permanent magnets attached to a soft steel housing and a commutator connected through brushes to a wound armature (Fig. 4.27). One of the first things to be

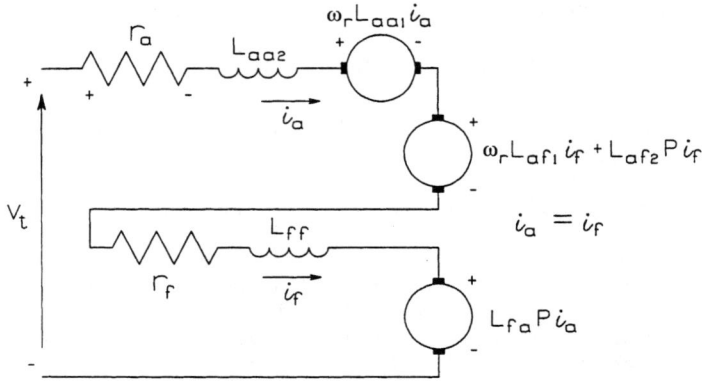

FIGURE 4.23 Equivalent circuit of the series motor.

considered in the operation of the motor are the motor and generator action, which exist simultaneously in the armature circuit of the motor. These two principles are (1) the instantaneous emf, which is induced in the armature conductors when moving with velocity v within a magnetic field, and (2) the force produced on the conductors as the result of carrying an electric current in this same magnetic field.

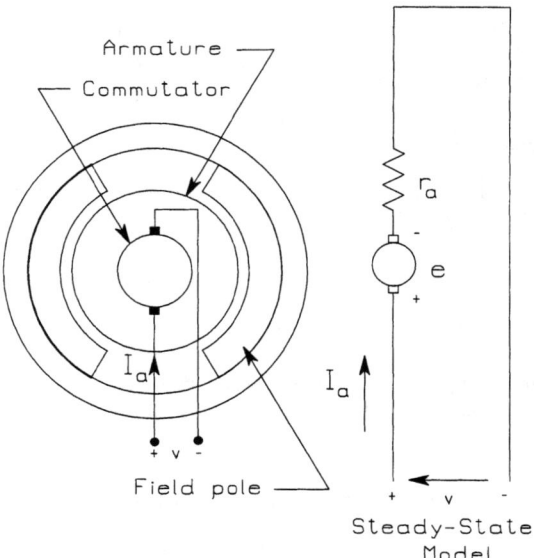

FIGURE 4.24 Permanent-magnet dc motor. r_a = armature resistance measured at brushes, e = motor counter-emf, and v = applied voltage.

FIGURE 4.25 Permanent-magnet dc motor.

FIGURE 4.26 Stator.

It is known that the instantaneous force on a conductor of length ℓ carrying a current i in a magnetic field B is:

$$f = Bi\ell \sin \theta \quad \text{N} \tag{4.1}$$

Or, in vector notation:

$$f = \ell(i \times B) \tag{4.2}$$

FIGURE 4.27 Commutator connected to wound armature.

where θ is the angle between the direction of the magnetic field and the direction of current flow in the conductor, B is in webers per square meter (teslas), i is in amperes, and ℓ is in meters. Motors, by design, have the armature conductors and magnetic flux at quadrature to one another. Therefore the force becomes:

$$f = Bi\ell \tag{4.3}$$

Assume a situation in Fig. 4.28 where a conductor of length ℓ is located in a magnetic field and is free to move in the x direction perpendicular to the field. From the preceding discussion, a force is produced on the conductor, causing it to move in the x direction. Then:

$$V - ir = e \quad \text{or} \quad Vi - i^2 r = ei \quad \text{(power)} \tag{4.4}$$

So, differential electrical energy into the conductor less differential $i^2 r$ loss equals differential mechanical output energy:

$$dW_{elect} = dW_{mech} \tag{4.5}$$

$$ei\,dt = f\,dx \quad \text{where } f = Bi\ell \tag{4.6}$$

$$ei\,dt = Bi\ell\,dx \tag{4.7}$$

$$e = B\ell \frac{dx}{dt} = B\ell v \tag{4.8}$$

Thus, the force causes movement of the conductor, which in turn causes a voltage to be induced into that conductor which is opposite to the direction of the original current. This is an important concept in the operation of motors in general and one which is used to discuss the operation of the dc motor. This induced voltage is usually called a *counter-emf* (*cemf*) *voltage* because of its opposition to the applied voltage.

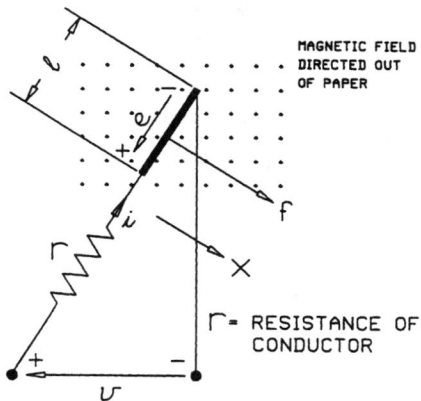

FIGURE 4.28 Conductor moving in a magnetic field.

This example also indicates that a reversible energy or power exchange is possible (i.e., between the mechanical and electrical systems). Therefore the same machine may operate as a motor or a generator, depending on the flow of energy in the armature.

In the motor mode, the armature of the PMDC motor is supplied with a current by the applied voltage, and a magnetic field (flux) is produced by the permanent magnets. Since the armature conductors (coils) are located in this field, each of the conductors in the field experiences a force (torque) tending to make it move (rotate), and as we have just indicated, a countervoltage (cemf) is produced opposing the applied voltage. Other than the cemf which the armature produce, we must recognize that the armature circuit (coils) also produces a magnetic field of its own. The armature, because of its commutator and brush construction, has a unidirectional current and therefore produces a fixed-directed mmf, measured in ampere-turns. This mmf is the product of the effective coil turns on the armature and the current through those turns. It must be understood here that the armature winding must be considered in developing those ampere-turns (i.e., there may be parallel paths through the armature, with the possibility of the armature coils being wound in series or parallel arrangements; hence, there will also be a division of the total armature current in each of the windings). This topic is addressed later. The following is a discussion of the action of the mmf produced by the armature.

Armature Reaction. The conductors in the armature carry a current proportional to the load. The magnetic field produced by this current reacts with the PM field. Figure 4.29 indicates two so-called belts of armature conductors (coil sides) under each pole face. Each of the conductors comprising these belts carries current in the same direction and hence produces additive mmf. In addition, there are conductors which also carry unidirectional currents but are not under the pole arcs. The important consideration here is the effect of the presence of the PM material in the pole pieces, armature core, and armature teeth. The flux paths through the armature are influenced by the reluctances of the paths. It is obvious that the reluctance of the flux paths under the pole pieces is less than that of the paths adjacent to the brush area, which constitute a material of much greater reluctance, namely a large air gap.

Figure 4.29 indicates that the brushes are on the mechanical neutral axis (i.e., halfway between the poles). The general direction of the armature mmf is along the brush axis at quadrature to the main field. The armature conductors adjacent to the poles produce flux densities in the air gap which are equal and opposite at the pole tips. Keep in mind that the flux density produced by the armature is directly related to the armature current together with a uniformly distributed flux density in the air gap attributed to the PM field. Now the net mmf (or flux in the air gap) is the result of both the PM mmf and the armature field mmf. The resultant air gap flux is now increased at one pole tip and reduced at the other pole tip because of the armature reaction. The flux distortion in the air gap is illustrated in Fig. 4.30. In this figure, two poles and the armature conductors beneath the poles have been unfolded to illustrate the distortion more clearly. Ampere's circuital law is useful here in determining the armature mmf. The resulting air gap mmf is the result of the superposition of the PM and armature mmfs. MMF drops in the poles and armature iron are considered negligible compared to the air gap mmf. For reference, positive direction for the mmf is assumed to be a flux out of a north pole. The armature mmf is shown as a linear relationship, but actually, because the armature slots are discrete, this relationship is actually made up of small discrete stair-step transitions. However, it is shown as a smooth curve for easier analysis.

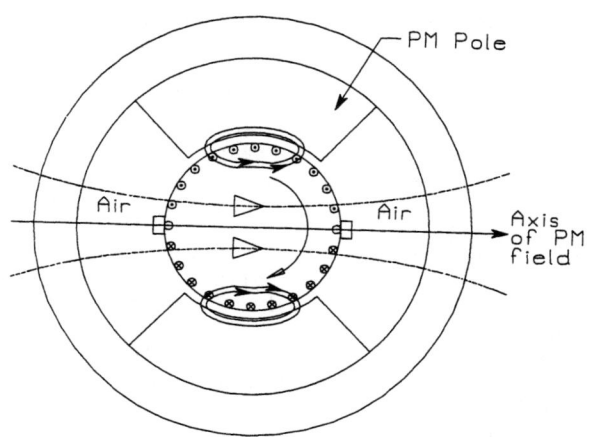

FIGURE 4.29 Armature magnetic field.

It is obvious that the air gap flux varies along the pole face. Another observation from the distortion mmf pattern is that harmonics are very present in the air gap mmf. Also, because of mmf symmetry, only odd harmonics can exist in an analysis of the air gap mmf.

$$\frac{\text{mmf PM}}{\text{mmf armature}} = \frac{7}{5} \qquad (4.75)$$

The results of a Fourier analysis of the air gap flux harmonics for this case are given in Figs. 4.31, 4.32, and 4.33. The figures show the fundamental, the fundamental plus third and fifth harmonics, and finally the fundamental and odd harmonics, up to and including the fifteenth harmonic. The magnitudes of the odd harmonics beginning with the fundamental are 8.94, 0.467, 1.41, 0.924, −0.028, −0.02, 0.523, and 0.0094. That is, the series can be represented as follows:

$$\phi_{\text{air gap}} = 8.94 \cos \theta + 0.467 \cos 3\theta + 1.41 \cos 5\theta + 0.924 \cos 7\theta$$
$$- 0.028 \cos 9\theta - 0.02 \cos 11\theta + 0.523 \cos 13\theta + 0.0094 \cos 15\theta + \cdots \qquad (4.9)$$

Of course, changing the PM pole arc or magnitudes of the armature and PM mmfs will change the harmonic content. The change in flux density across the air gap produces two effects: (1) a reduction in the total flux emanating from each pole, and (2) a shift in the electrical neutral axis, lending to commutation problems due to the flux distortion.

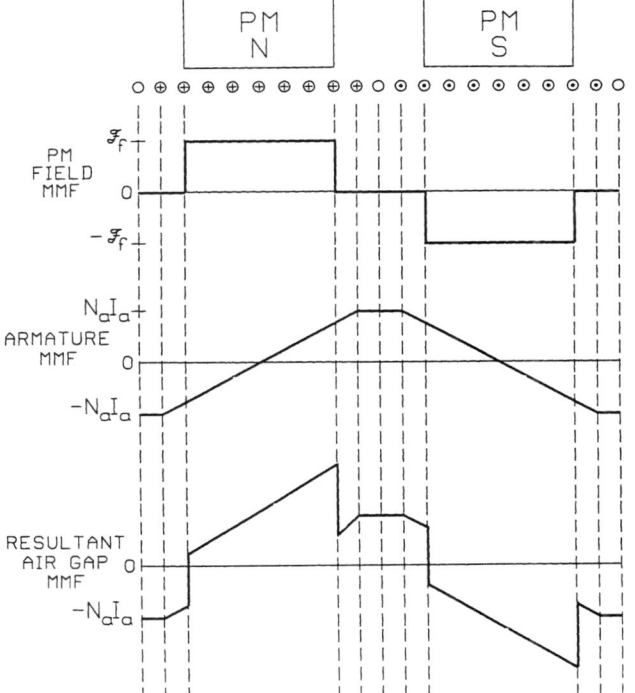

FIGURE 4.30 Flux distortion in the air gap mmf.

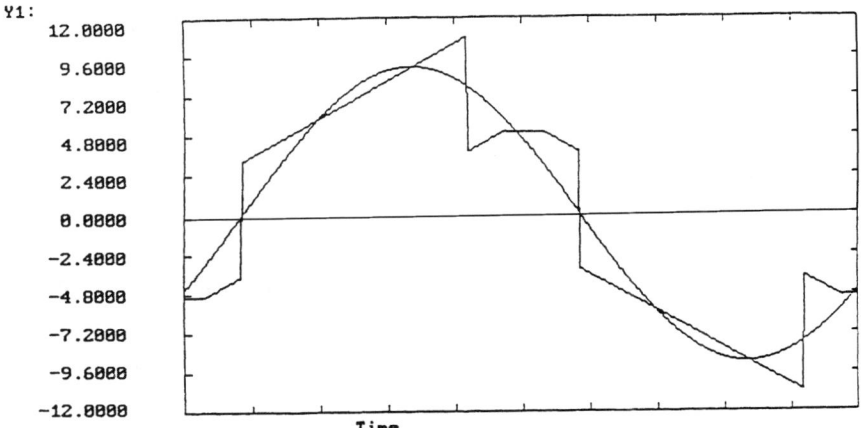

FIGURE 4.31 Fundamental frequency of air gap flux.

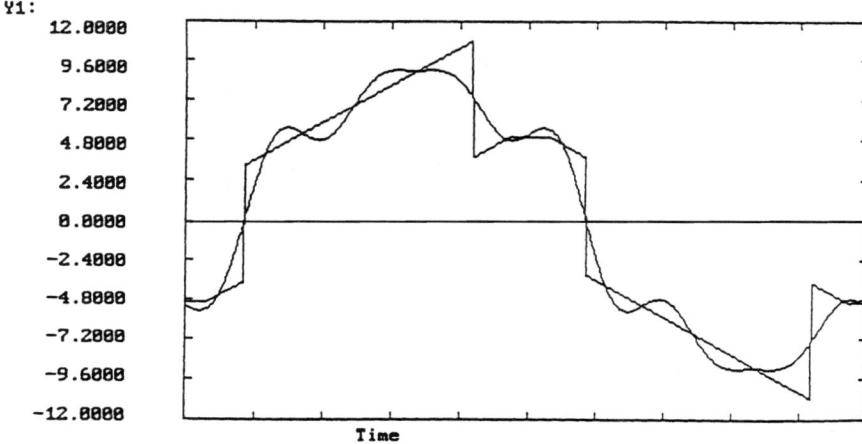

FIGURE 4.32 Air gap flux due to fundamental plus third and fifth harmonics.

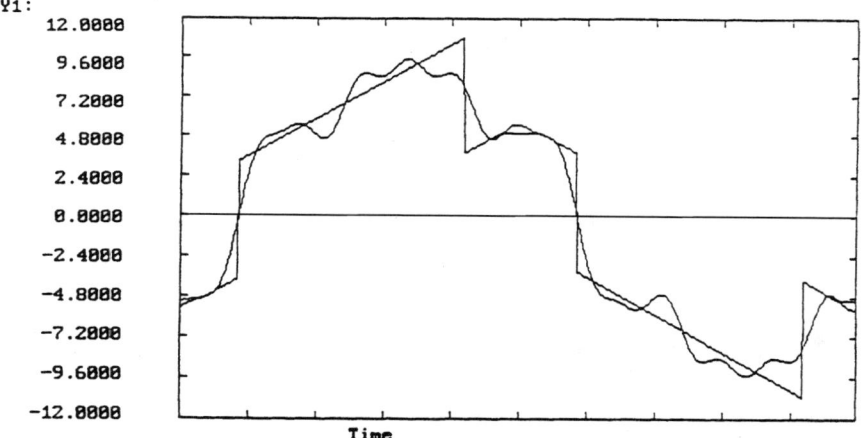

FIGURE 4.33 Air gap flux due to odd harmonics through fifteenth.

The flux distortion which the armature produces has been called a *crossmagnetizing armature reaction,* and rightly so. The net effect is illustrated in Fig. 4.34, showing the resultant field where the armature cross field is at right angles to the main field. The result is a distortion of the net flux in the motor; the second effect, which was indicated previously, is a reduction in the total main field flux. This reduction of flux is not too obvious; in fact, it would almost appear from Fig. 4.34 that the vector addition of these two mmfs would lead to an increase in flux which would occur with the brushes on the mechanical neutral axis.

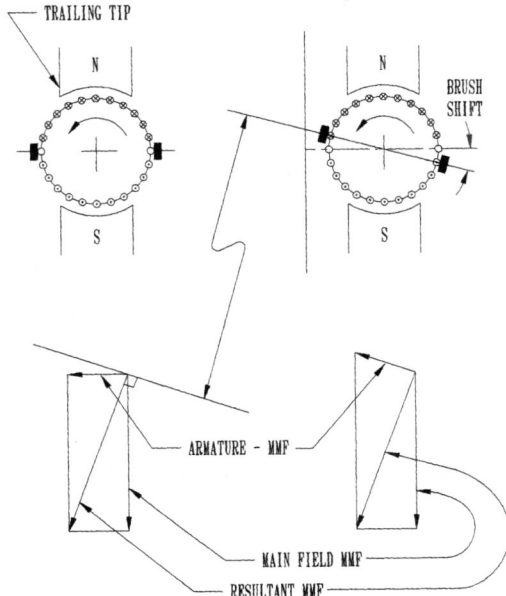

FIGURE 4.34 Field of armature with and without brush shift.

This resulting decrease in flux can be attributed to magnetic saturation. Figure 4.35 illustrates how this comes about when one operates the motor at the knee of the saturation curve. The area ABC is proportional to the reduction in flux at the pole tip with a decrease in flux density, while the area CDE is proportional to the increase in flux at the other pole tip. It can be observed that the reduction in flux exceeds the increase in flux because of the saturation effect. This is sometimes called the *demagnetizing effect of armature reaction;* it is a serious consideration when using PM as a field structure. This problem is discussed later.

Because the purpose of the commutator and brushes is to change the current in a short-circuited coil from, say, a current of $+I$ to $-I$ in the length of time the coil is short-circuited by the brushes, some arcing at the brushes is expected. One desires to keep any voltage induced into the coil to a minimum in order to keep this arcing as small as possible.

When the brushes are on the mechanical neutral position and the main field becomes distorted, the coils being commutated will have an induced voltage from the distorted air gap flux. A shifting of the brushes to a new neutral position is suggested to keep the induced voltage to a minimum and keep the brush arcing low. It should be recognized that the amount of distortion of the field is a function of the ampere-turns of the armature and hence is dependent on the motor load. One then realizes that the electrical neutral position is a function of the load and that shifting brushes to an electrically neutral position is therefore not a matter of shifting them to a unique point in space.

Brush shifting causes another effect—that is, a demagnetization effect on the air gap flux. This effect is in addition to the demagnetizing effect which occurs because

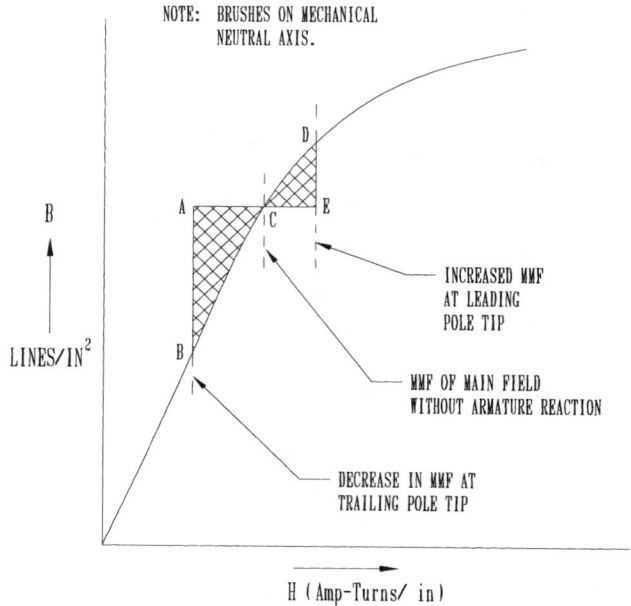

FIGURE 4.35 Armature reaction resulting in main pole flux reduction.

of saturation. When the brushes are shifted, the pole axis of the armature is shifted. The result is that the angle between the main field and the armature field is of greater than 90 degrees. This process is also illustrated in Fig. 4.34.

In summary, there are two processes which cause reduction of the air gap flux: (1) reduction due to cross-magnetization when the brushes are on the mechanical neutral position, which changes the flux distribution across the pole face and the net flux because of saturation effects, and (2) demagnetization resulting from a brush shift and change of the armature pole orientation with respect to the PM field, resulting in the armature field mmf having a component in direct opposition to the PM field mmf.

Reactance Voltage and Commutation. It was indicated previously that during commutation the current in the shorted coil(s) under a brush must reverse and change direction. The self-inductance of the shorted winding by Lenz's law induces an emf in the shorted winding to oppose the change in coil current. This voltage is sometimes called the *reactance voltage*. (Actually, it may be only a portion of the reactance voltage, as is shown later.) This voltage slows down the reversal of current and tends to produce sparks or arcing as the trailing commutator bar leaves a brush. The reactance voltage hinders good commutation. The magnitude of this voltage depends directly on the square of the number of coil turns, the current flowing, and the armature velocity; it is inversely proportional to the reluctance of the magnetic path.

When shifting brushes to seek an electrical neutral position on a motor, the shifting is done in the opposite direction of armature rotation. It must be remembered that the reactance voltage is an $e = L(di/dt)$ voltage and is simply due to the current change during commutation. It has nothing to do with induced voltage from air gap flux. Shifting the brushes really does not help the reactance voltage. Theoretically, there is no induced voltage on the neutral axis from the motor flux (in fact, $d\lambda/dt$ should be zero here since the coil sides are moving parallel to the flux). In order to counteract the reactance voltage and induce a voltage opposite to the reactance voltage the brushes must be shifted further backward than the magnetic neutral axis. At this point a $d\lambda/dt$ voltage is induced into the commutated coil by the PM field flux, which counteracts the reactance voltage. The result is that two voltages in opposite polarities are induced into the shorted coil thereby reducing brush arcing.

An important consideration is that there can be another component of the reactance voltage which can occur when two coil sides are in the same slot and both are undergoing commutation. There is then the following mutual inductance term which must be added to the self-inductance term:

$$\text{Reactance voltage} = L_{11}\frac{di_1}{dt} + M_{12}\frac{di_2}{dt} \quad (4.10)$$

where the coil being considered is the coil carrying current i_1 and the coil in the same slot undergoing commutation is carrying current i_2. If there is more coupling between coils, more mutual terms can exist.

Brushes are a very important consideration, and contrary to normal electrical principles, one would assume that keeping the brush resistance low would assist in reducing arcing at the brush commutator bar interface. This is far from the truth. In fact, resistance commutation is now an accepted technology. Brushes normally have a graphite or carbon formulation and hence introduce, by their characteristics, resistance into the interface. If constant current density could be achieved at a brush for all loads and speeds, an ideal condition would exist for commutation. Figure 4.36 illustrates what would occur if ideal conditions are assumed. The assumptions are as follows:

- The brush width is equal to that of a commutator segment.
- The current density under a brush is constant.
- The reactance voltage and resistance are zero.
- The commutator bar insulation is small compared to the width of a commutator segment.

Note that as the commutator passes beneath the brush, for the assumptions given, the contact resistance is inversely proportional to the contact area. The sequence in Fig. 4.36 illustrates how the current change occurs in an armature coil under these ideal conditions. This change in surface contact changes the brush contact resistance; hence, with constant current density, the change in current flow changes linearly according to contact resistance. Remember, the resistance is inversely proportional to the contact area. This ideal commutation has been termed *linear commutation* because of the linear transfer of current in the commutated coil, as indicated in Fig. 4.36.

If the brush width is greater than that of one commutator segment, linear commutation still exists because the resistance still changes linearly; hence, the current density remains constant. This has an advantage in that the commutation period is longer; hence, the rate of change of current di/dt is reduced, and hence the reactance voltage is reduced also.

All of the assumptions made in the previous analysis cannot occur, and *nonlinear commutation* results. Armature coils have resistance, have self- and mutual inductance and hence linear commutation cannot be attained.

If these nonlinearities do occur, then either *undercommutation* or *overcommutation* occurs. These are illustrated in Fig. 4.37. Since brush heating is a product of brush resistance and the square of the current, linear commutation is preferred, as it

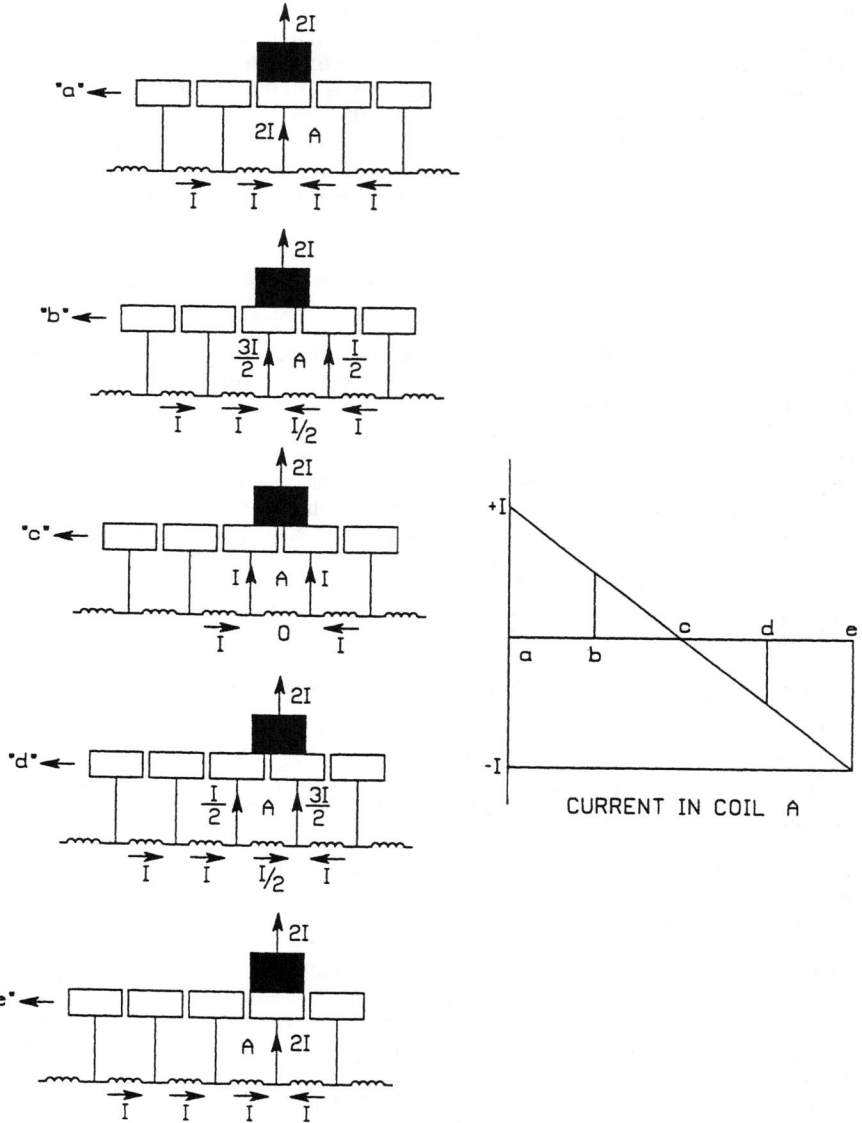

FIGURE 4.36 Linear commutation.

minimizes brush heating. That is, the squared value of the three forms of current commutation in Fig. 4.37 is smallest for linear commutation.

It is possible to develop a circuit model for the commutation process, as shown in Fig. 4.38. The factors have previously been described, and they are restated here. Consider the subject coil of turns N_1 and self-inductance L_1, and the current i_i which is being commutated.

DIRECT-CURRENT MOTORS

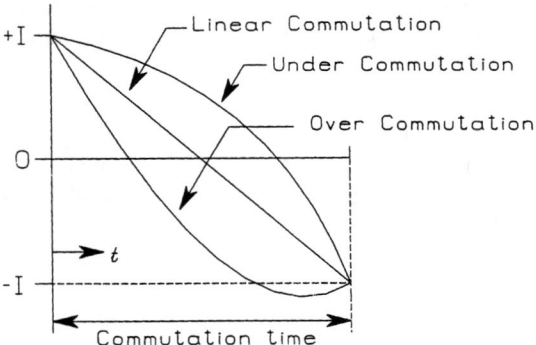

FIGURE 4.37 Overcommutation, undercommutation, and linear commutation.

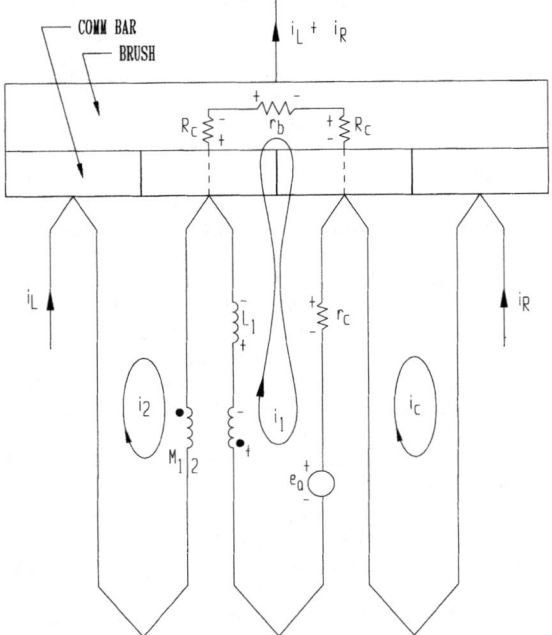

FIGURE 4.38 Coils under commutation.

1. The voltage of self inductance (reactance voltage) of the commutated coil is

$$e_{La} = L_1 \frac{di_a}{dt} \tag{4.76}$$

where i_1 is the current being commutated and L_1 is the self-inductance of the coil being commutated.

2. The voltage of mutual inductance of an adjacent coil in the same slot carrying current i_2 and undergoing commutation and linked to the subject coil is

$$e_{ma} = M_{12} \frac{di_b}{dt} \qquad (4.77)$$

where M_{12} is the mutual inductance between adjacent coils.
3. The voltage induced into the subject coil of turns N when cutting flux ϕ_a due to the armature mmf is

$$e_a = N_1 \frac{d\phi_a}{dt} \qquad (4.13)$$

Sometimes this is called *commutating voltage;* it opposes the reactance voltage. This can be controlled by brush overshifting to make e_a cancel the voltages e_{ma} and e_{La}.
4. There are voltage drops due to coil resistance r_c; R_c and R_c, the left and right contact resistances at the brush; and the brush resistance between commutator segments r_b.

The commutation process is not simple, and brush composition and interface film, measurements, and conditions all can be made to assist in the commutation process. Maintaining a suitable interface film between the brush and the copper commutator is extremely important. This interface is formed of copper oxide and free particles of graphite; it provides a general resistance commutation and supplies lubricant to reduce surface friction and heat between the commutator and the brush.

We have previously discussed demagnetization which occurs at the leading tip of a motor due to armature reaction.

The subsection on armature reaction shows that the mmf of this reaction could be closely represented as a triangular wave (Fig. 4.30). We are concerned in PM motor design with the demagnetizing effect of this armature reaction. In addition to the normal design demagnetization curve shown in Fig. 4.39, a set of intrinsic curves is also indicated. These curves are extremely important because they represent limits on the amount of demagnetization which can be tolerated. The value of $-H$ required to remove the magnetization is given the value H_{ci}. The PM of the motor normally does not span 180° electrical; therefore we are not faced with the total ampere-turns of demagnetization of the armature reaction, only that component which exists at the pole tip. If we let β represent the angle subtended from the center of magnet to the magnet tip, a trigonometric relationship can be used to calculate the demagnetization armature reaction at the magnet tip. If $N_a I_a$ represents the maximum value of a triangular wave, the demagnetization ampere-turns at the magnet tip equals $(\beta/90°)(N_a I_a)$. If this value is less than the design value of the radial magnet length (i.e., $l_m \cdot H_{ci}$), loss of magnetization at the pole tip can occur.

One important consideration when using permanent magnets in machinery is the fear of demagnetization of the magnet. This must be given important consideration. The shape of the demagnetization curve of the permanent magnet is an important property in this respect. For example, in Fig. 4.39, for the alnico magnet with an operating point on a load line (see the next subsection, Permanent Magnets for DC Motors), only a small amount of demagnetizing ampere-turns per inch will completely demagnetize the magnet, whereas the ceramic or rare earth magnets' characteristics have much more latitude. However, there is a considerable difference in the available flux densities among the alnico, ceramic, and rare earth magnets;

hence, for required large flux densities the alnico magnets are superior. Demagnetization results primarily from armature reaction, which can be large because of large in-rush currents upon motor starting or possibly because of control applications involving dynamic braking. These currents cause a large demagnetizing effect across the pole face, as previously described in the armature reaction discussion. Another important consideration when using magnets is the loss of magnetic properties as the temperature of the magnet increases. Thus, the temperature range of motor operation must be considered when employing magnets.

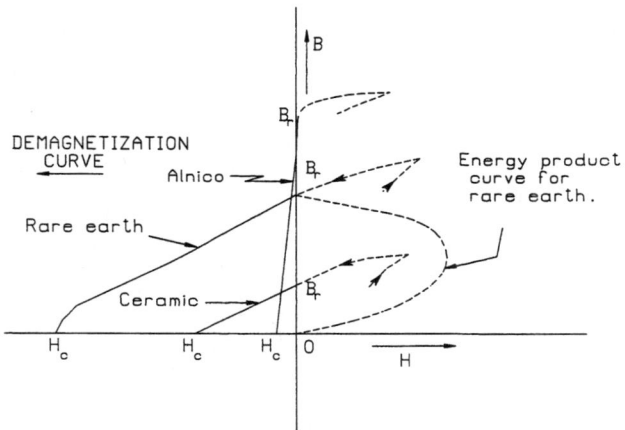

FIGURE 4.39 Typical demagnetization curves for three magnets.

Magnets for PMDC motors are selected by their characteristic demagnetization curves—that is, their residual flux density B_r, their energy product BH, and their coercive force H_c.

Ceramic magnets have a characteristic and a permeability very nearly equal to those of air. That is, μ_0 approaches 3.2 in English units. This is an important consideration because of the large effective reluctance of the magnetic path, and removal of the armature does not change the operating point of the magnet appreciably or affect the performance of the magnet.

The air gap flux is not adjustable with permanent-magnet materials, of course. This means that changes in motor speed and torque must be made by changes in armature current.

The performance of PMDC motors is very similar to that of wound-field DC motors, and similar equations exist. This development is the next topic to be addressed.

$$L_1 \frac{di_a}{dt} + M_{12} \frac{di_b}{dt} + (r_c + r_b + 2R_c)i_a + e_a = 0 \qquad (4.78)$$

where R_c = brush contact resistance
r_c = coil resistance
r_b = brush resistance

Permanent Magnets for DC Motors. A permanent magnet (PM), when magnetized by an external source for use in a PMDC motor, has a remaining residual flux density and responds to a normal hysteresis characteristic. The useful portion of a magnet in PM operation is in the second quadrant of the hysteresis loop and is usually called the *demagnetization curve* of the magnet. It represents the relationship between B and H of the magnet once it has been magnetized. Typical demagnetization curves for three PM materials are shown in Fig. 4.39. In this figure, B_r is the residual flux density and H_c is the coercive force. For proper utilization of a permanent magnet a user makes use of the energy product curve $B \cdot H$, shown in Fig. 4.39 for the rare earth magnet only.

Figure 4.40 presents a series magnetic circuit in which a PM, an air gap, and magnetic material have been introduced into a closed magnetic circuit. Ampere's law around the circuit (Ohm's law for magnetic circuits) is as follows:

$$H_m \ell_m + H_g \ell_g + H_s \ell_s = 0 \qquad (4.79)$$

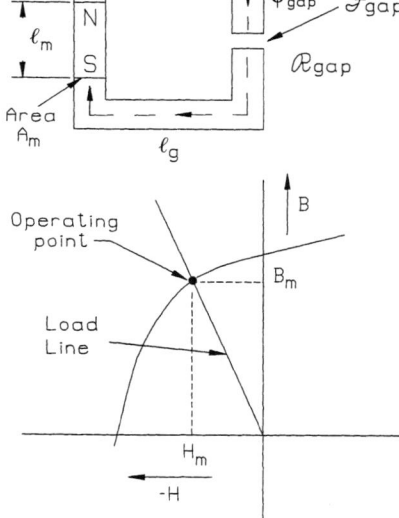

FIGURE 4.40 Permanent magnet with air gap introduced.

where H_m, H_g, and H_s are the magnetic field intensities of the PM, air gap, and steel, respectively. The lengths of the magnetic paths in the PM, air gap, and steel are respectively ℓ_m, ℓ_g, and ℓ_s. By neglecting any leakage flux, the flux is the same throughout the series circuit. If we assume the cross-sectional areas of the PM, air gap, and steel are respectively A_m, A_g, and A_s, then:

$$B_m A_m = B_q A_q = B_s A_s = \text{core flux} \qquad (4.80)$$

where B_m, B_g, and B_s are the respective flux densities. Since the relationship between B and H in a material is the permeability μ, we can state:

$$H_m \ell_m + \frac{B_g}{\mu_g} \ell_g + \frac{B_s}{\mu_s} \ell_s = 0 \qquad (4.81)$$

Or

$$H_m \ell_m + \frac{B_m A_m}{\mu_g A_g} \ell_g + \frac{B_m A_m}{\mu_s A_s} \ell_s = 0 \qquad (4.82)$$

Then:

$$H_m = -\left(\frac{\ell_g A_m}{\mu_g \ell_m A_g} + \frac{\ell_s A_m}{\mu_s \ell_m A_s}\right) B_m = -\frac{A_m}{\ell_m}\left[\frac{\ell_g}{\mu_g A_g} + \frac{\ell_s}{\mu_s A_s}\right] B_m \qquad (4.83)$$

This is the equation of a straight line on the BH demagnetization curve. If $\mu_s \gg \mu_g$, the expression reduces to:

$$\frac{B_m}{H_m} \approx -\frac{\mu_g \ell_m A_g}{\ell_g A_m} \qquad (4.84)$$

where

$$-\frac{\mu_g \ell_m A_g}{\ell_g A_m} \tag{4.85}$$

is the slope of the line.

Thus, the values of B_m and H_m can be found from the demagnetization curves and the cross-sectional area of the magnet, the air gap, and the respective lengths of the magnetic paths. If flux fringing is neglected, usually $A_g = A_m$, and the slope becomes:

$$\frac{B_m}{H_m} = -\frac{\mu_g \ell_m}{\ell_g} \tag{4.86}$$

This becomes an approximate expression for the so-called load line of the PM, and the intersection with the demagnetization curve becomes the operating point.

As previously indicated, the energy product $B \cdot H$, which has units of watt-seconds per cubic meter, is important in PM utilization. In a linear medium it can be shown rather easily that $B \cdot H$ is an expression for energy density in a given volume.

$$W = \frac{\text{energy in a volume}}{\text{volume}} \tag{4.87}$$

$$= \frac{\mathscr{F}_m \phi_m}{A_m \ell_m} = \frac{1}{2}\left(\frac{\phi_m}{A_m}\right)\left(\frac{\mathscr{F}_m}{\ell_m}\right) = \frac{1}{2} B_m H_m \tag{4.88}$$

It may also be of interest to look at the energy density which is present when operating at the maximum energy product as it relates to the air gap and PM energy density.

Looking at the expression (magnitudes only)

$$\frac{B_m}{H_m} \approx \frac{\mu_g \ell_m A_g}{\ell_g A_m} \tag{4.89}$$

If $B_m = B_g$ flux densities in the PM and the air gap are equal. Then:

$$B_g = \left(\frac{\mu_g \ell_m A_g}{\ell_g A_m} H_m\right) = \left(\frac{\mu_g V_m}{V_g} H_m\right) \quad \text{if} \quad A_g = A_m \tag{4.90}$$

where V_m and V_g are the volumes of the PM and the air gap.

$$B_g = \frac{B_g}{H_g}\frac{V_m}{V_g} H_m = \frac{B_m}{H_g}\frac{V_m}{V_g} \cdot H_m \tag{4.91}$$

$$B_g H_g V_g = B_m H_m V_m \tag{4.92}$$

This equation then indicates that the point of operation for maximum available energy occurs at the maximum energy product $B \cdot H$.

If designing for the maximum energy product BH, consider the magnet characteristic shown in Fig. 4.41, where B_d and H_d are corresponding values giving the maximum energy product. Knowing the slope of the load line, the ratio B_d/H_d is known. If, as previously suggested, the area of the magnet A_m and the area of the air gap A_g are equal, then:

$$\frac{B_d}{H_d} = -\mu \frac{\ell_m}{\mu_g \ell_g} = \text{known} \quad \text{(slope)} \tag{4.93}$$

From this ℓ_m can be obtained.

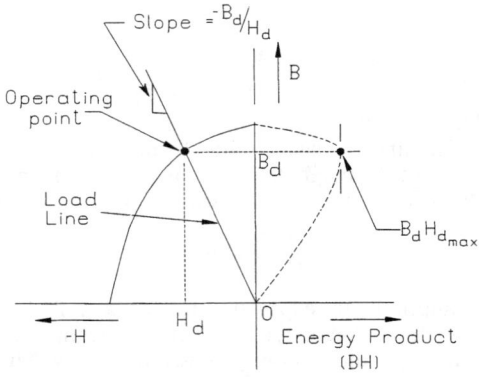

FIGURE 4.41 Designing for maximum energy.

It is appropriate at this point to introduce some examples with magnetic circuits utilizing PMs.

Example 1. Suppose the following parameters are given for Fig. 4.42. Design the magnet for the maximum energy product.

Use the demagnetization curve given in Fig. 4.43.

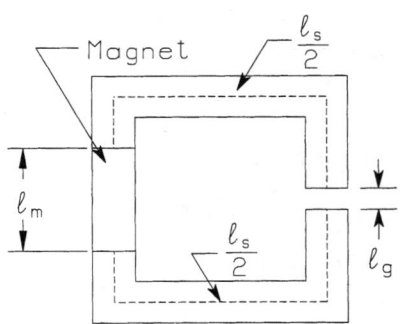

FIGURE 4.42 Figure for example PM circuit calculations.

Assume: $A_m = A_s = A_g = 5 \text{ cm}^2$ (4.94)

$\ell_g = 0.5 \text{ cm} \qquad \ell_s = 8 \text{ cm}$ (4.95)

$\mu_s = \mu_0 \mu_r \quad \text{where } \mu_r = 3500$ (4.96)

$= (4\pi \times 10^{-7})(3500)$ (4.97)

Operating temperature = 20°C (68°F)

From previous development:

$$H_d = -\left(\frac{\ell_g}{\mu_g} + \frac{\ell_s}{\mu_s}\right)\frac{B_d}{\ell_m}$$

if $A_m = A_s = A_g$ (4.98)

$$H_d = -\left(\frac{0.5}{4\pi \times 10^{-7}} + \frac{8}{(4\pi \times 10^{-7})(3500)}\right)\frac{B_d}{\ell_m} \qquad (4.99)$$

$$= -\left(0.5 + \frac{8}{3500}\right)\frac{B_d}{\ell_m 4\pi \times 10^{-7}} \qquad (4.100)$$

$$= -(0.5 + 2.27 \times 10^{-3})\frac{B_d}{\ell_m\, 4\pi \times 10^{-7}} \qquad (4.101)$$

Note the magnitude of the second term in comparison to the first term in the parentheses. Therefore, the approximation as previously indicated is appropriate.

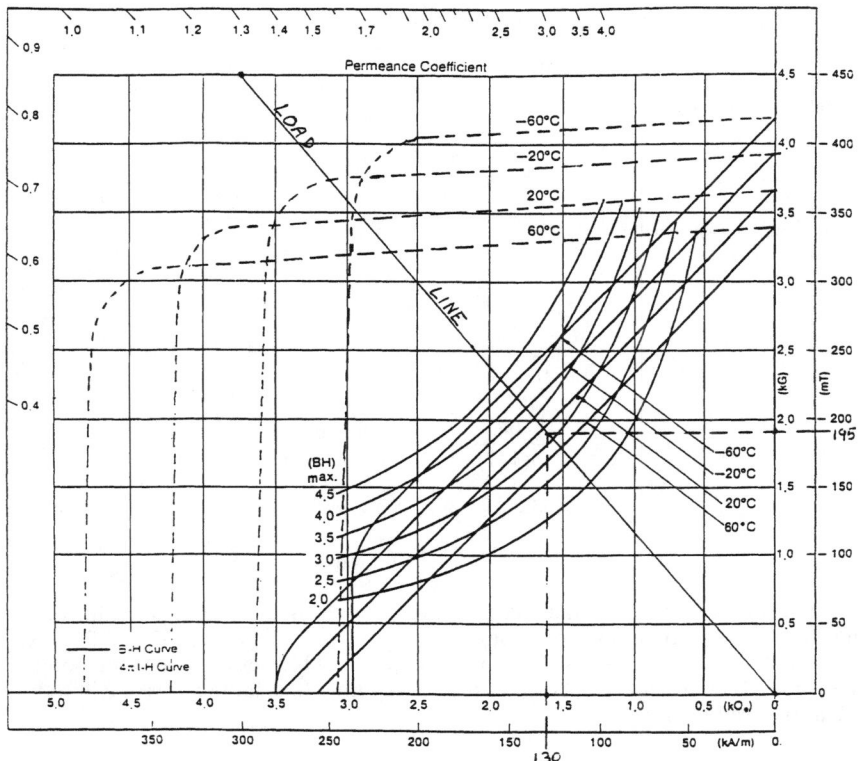

FIGURE 4.43 Magnet performance curve.

$$H_d \approx \frac{0.5 B_d}{\ell_m \, 4\pi \times 10^{-7}} = \frac{3.98 \times 10^5}{\ell_m} B_d \qquad (4.102)$$

From Fig. 4.43, at 20°C the maximum energy product is approximately 3.25, with $H_d \approx 130$ kA/m and $B_d \approx 0.195$ T.
Therefore:

$$\ell_m \approx \frac{3.98 \times 10^5}{130 \times 10^3} \times 0.195 = 0.6 \text{ cm} \qquad (4.103)$$

Example 2. Suppose we have a 48-V PMDC motor running at 1400 rpm with an armature resistance of 1.5 Ω. The motor constant is 75 (V · s)/Wb. If the power input to the motor is 200 W and the rotational efficiency is 80 percent, calculate the magnetic flux per pole.
Using the SI system

$$\omega = 1400 \times \frac{2\pi}{60} = 147 \text{ rad/s} \qquad (4.104)$$

The power developed therefore must be

$$200 + 200(1.0 - 0.8) = 240 \text{ W}$$

The developed torque is then

$$T_{\text{dev}} = \frac{P_D}{\omega} = 1.63 \text{ N} \cdot \text{m} \quad (4.105)$$

Using the equation developed in Example 3

$$\omega = \frac{V}{K\phi_p} - \frac{V_a}{(K\phi_p)^2} T \quad (4.106)$$

$$\phi_p^2 - \frac{V}{K\omega}\phi_p + \frac{r_a T}{K^2 \omega} = 0 \quad (4.107)$$

$$\phi_p^2 - \frac{48}{75(147)}\phi_p + \frac{(1.5)(1.63)}{(75)^2(147)} = 0 \quad (4.108)$$

$$\phi_p^2 - 0.004\phi_p + 2.9 \times 10^{-6} \quad (4.109)$$

$$\phi_p = 3.04 \times 10^{-3} \text{ Wb} \quad (4.110)$$

Note: This value of flux when divided by the magnetic area represents a value of flux density and an operating point on the demagnetization curve of the PM. Any demagnetization mmf due to armature reaction would lower the operating point on the curve. Keeping the reluctance of the air gap high (i.e., large air gap) will reduce the armature reaction.

Torque-Speed Characteristics of PMDC Motors. The electrical equivalent steady-state circuit is the most appropriate method for analyzing the motor. This equivalent circuit is repeated for convenience in Fig. 4.44, and the armature brush drop is included.

Writing Kirchhoff's voltage law around the loop gives:

$$V = I_a r_a + E_g + E_b \quad \text{V} \quad (4.111)$$

$$VI_a = I_a^2 r_a + E_g I_a + I_a E_b \quad \text{W} \quad (4.112)$$

The counter-emf E_g was addressed earlier when discussing the voltage induced into a conductor of length ℓ and moving with a velocity v perpendicular to a magnetic field of flux density B. The counter-emf voltage E_g was expressed as e and given by:

$$e = B\ell v \quad (4.8)$$

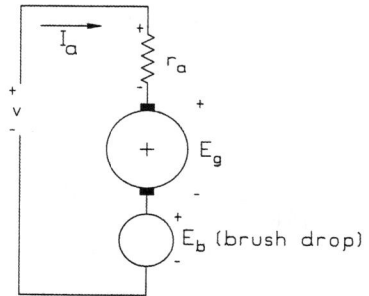

FIGURE 4.44 Armature Circuit for the PMDC motor.

It is necessary to modify this and express e in terms of the motor parameters. Beginning with Faraday's law:

$$E_g = E_{cemf} = N \frac{\Delta \phi}{\Delta t} \quad \text{V} \tag{4.16}$$

where N is the number of conductors per armature path when Z is the total number of conductors (coil sides) on the armature.

Then:

Z = number of slots × coils/slot × turns/coils × 2 (conductors/turn)
a = number of parallel paths through the armature
P = number of poles
n = armature speed, rpm
ϕ = flux, Wb

Then:

$$E_g = E_{cemf} = \frac{Z \phi n P}{60a} \times 10^{-8} = K_e \phi n \quad \text{V} \tag{4.17}$$

where ϕ is the flux per pole in lines (maxwells) and

$$K_e = \frac{ZP}{60a} \times 10^{-8} \tag{4.18}$$

From the equivalent circuit, the developed power of the motor is $E_g I_a$, that is:

$$P_{dev} = K_e \phi n I_a \quad \text{W} \tag{4.19}$$

$$P_{dev} = \frac{K_e \phi n I_a}{746} \quad \text{hp} \tag{4.20}$$

$$T_{dev} = \left(\frac{33,000}{2\pi n} \right) \left(\frac{K_e \phi n I_a}{746} \right) \quad \text{lb} \cdot \text{ft} \tag{4.21}$$

$$T_{dev} = 7.04 \left(\frac{ZP}{60a} \right) \phi I_a \times 10^{-8} \quad \text{lb} \cdot \text{ft} \tag{4.22}$$

$$T_{dev} = 22.5 \left(\frac{ZP}{a} \right) \phi I_a \times 10^{-8} \quad \text{oz} \cdot \text{in} \tag{4.23}$$

The equivalent circuit voltage equation (neglecting brush drop) then becomes

$$V = I_a r_a + K_e \phi n \tag{4.113}$$

And the motor speed equation becomes

$$n = \frac{V - I_a r_a}{K_e \phi} \tag{4.114}$$

For no load speed:

$$n_{nl} \approx \frac{V}{K_e \phi} \tag{4.115}$$

For stalled torque:

$$T_{stall} = K_e \phi I_{a,stall} \tag{4.116}$$

$$= K_e \phi \frac{V}{r_a} \tag{4.117}$$

Example 3. Given a permanent magnet motor whose armature resistance is 0.8 Ω and that operates on a 24-v supply, assume the air gap flux is 0.002 Wb and the motor torque and voltage constant is 100. The motor is operating under a load torque of 1 N · m (neglect brush drop).

$$V = I_a r_a + E_g = I_a r_a + K\phi_p \omega \tag{4.118}$$

$$\omega = \frac{V}{K\phi_p} - \frac{r_a}{K\phi_p} I_a \quad \text{but} \quad T = K\phi_p I_a \tag{4.119}$$

$$I_a = \frac{T}{K\phi_p} \tag{4.120}$$

Then:

$$\omega = \frac{V}{K\phi_p} - \frac{r_a}{(K\phi_p)^2} T \tag{4.121}$$

Where T = developed torque, N · m
 ω = motor speed, rad/sec
 ϕ_p = flux per pole, Wb
 K = motor torque and voltage constant

Then:

$$\omega = \frac{24}{(100)(0.002)} - \frac{0.8}{(100 \times 0.002)^2}$$

$$= 120 - 20 = 100 \text{ rad/s} \tag{4.122}$$

Or

$$n = 100 \, \frac{60}{2\pi} = 955 \text{ rpm} \tag{4.123}$$

The stall torque can be found by setting $\omega = 0$.
Then:

$$T_{stall} = \frac{V(K\phi_p)}{r} = \frac{24(100 \times 0.002)}{0.8}$$

$$= 6 \text{ N} \cdot \text{m} \tag{4.124}$$

Efficiency of DC Motors Earlier discussion has indicated that a number of losses occur in PMDC motors. These losses can be summarized as follows:

1. Copper loss in the armature winding $I_a^2 r_a$
2. Brush contact loss

3. Friction (brush and bearing friction) and windage
4. Core loss (hysteresis and eddy current)
5. Stray load loss (losses in addition to those above)

A comment should be made on the brush contact loss listed as number 2. Experimentation and IEEE specifications have suggested as an approximation that:

$$\text{Brush contact loss} = \text{brush voltage drop} \times \text{armature brush current} \quad (4.29)$$

This expression was arrived at for carbon or graphite brushes by observing the brush voltage drop as a function of current density. At high- and low-current steady-state densities, the voltage falls off and approaches zero. However, in between these limits the voltage drop is approximately constant at about 1 V per brush. Therefore, for a pair of brushes this becomes 2 V. On small machines this voltage drop tends to increase, since the tendency is for the voltage drop to increase for lower brush and commutator temperatures.

Carbon and graphite materials have a resistivity many times that of copper and have a negative temperature coefficient. This negative temperature coefficient is attributed to the rise in brush voltage drop in small machines.

Stray load loss, as the term suggests, is a function of motor load and changes in load. Changes in load produce changes in armature current and hence affect (1) magnetic saturation in the magnetic circuit, (2) armature reaction changes, and (3) eddy current loss changes.

Figure 4.45 presents a display of these losses. By definition:

$$\text{Overall efficiency} = \frac{\text{output}}{\text{input}} = \frac{\text{output}}{\text{output} + \text{losses}} \quad (4.30)$$

$$= \frac{\text{useful mechanical output}}{\text{total electrical input}} \quad (4.31)$$

Sometimes both electrical and mechanical efficiencies are of interest in order to determine where improvements can be made.

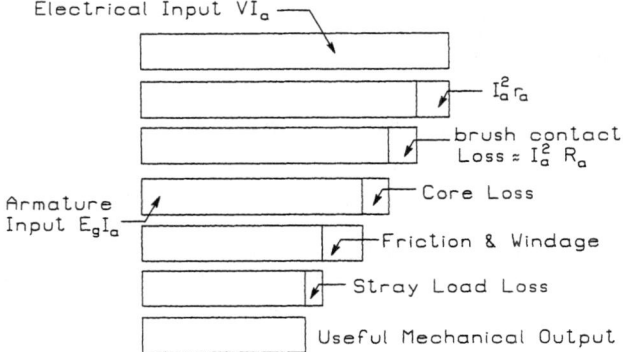

FIGURE 4.45 Efficiency of a permanent-magnet dc motor.

$$\text{Mechanical efficiency} = \frac{\text{useful mechanical output}}{\text{mechanical output} + \text{rotational losses}} \quad (4.32)$$

$$\text{Electrical efficiency} = \frac{\text{electrical power output}}{\text{electrical power output} + \text{electrical losses}} \quad (4.33)$$

The conditions for maximum efficiency can be related to those losses which are considered to be constant and those that vary with the motor load current. If the losses are segregated as follows:

K_1 = constant losses

K_2 = those losses which vary linearly with I_a

K_3 = those losses which vary as the square of I_a

then the efficiency is

$$\text{eff} = \frac{\text{input} - \text{losses}}{\text{input}} = \frac{VI_a - (K_1 + K_2 I_a + K_3 I_a^2)}{VI_a} \quad (4.34)$$

$$\frac{d\text{eff}}{dI_a} = \frac{VI_a(V - K_2 - 2K_3 I_a) - [VI_a - (K_1 + K_2 I_a - K_3 I_a^2)]V}{(VI_a)^2} \quad (4.35)$$

Equating this expansion to zero gives the condition for maximum efficiency.

$$VI_a[V - K_2 - 2K_3 I_a] - [VI_a - (K_1 + K_2 I_a + K_3 I_a^2)]V = 0 \quad (4.36)$$

$$-2K_3 I_a^2 + K_1 + K_3 I_a^2 = 0 \quad (4.37)$$

$$K_3 I_a^2 = K_1 \quad (4.38)$$

Thus, for maximum efficiency the constant losses must be equal to those that vary as the square of the armature current. This is typical for all different pieces of rotational electrical equipment. The constant losses usually are considered to be the core losses, friction, and windage. Usually the brush loss is small.

4.2 LAMINATION, FIELD, AND HOUSING GEOMETRY*

4.2.1 Universal (Series-Wound) Motor Construction

Figures 4.46 and 4.47 show the general construction of a two-pole-wound field universal motor. Figure 4.48 shows a larger four-pole series motor designed to run on direct current.

The stator (field) and armature laminations are stacked, insulated, and wound with magnet wire. The armature (Fig. 4.47) is wound in consecutive continuous loops or coils. Each coil end is connected to a commutator bar. The field is wound with

---

* Sections 4.2 to 4.4 contributed by Alan W. Yeadon, Yeadon Engineering Services, PC.

DIRECT-CURRENT MOTORS

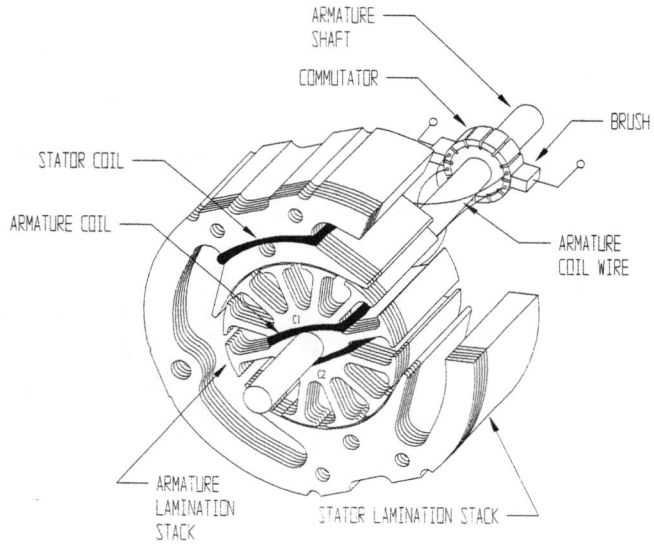

FIGURE 4.46 Universal motor construction.

FIGURE 4.47 Universal motor.

FIGURE 4.48 Four-pole dc series motor.

magnet wire and the coils are connected in series such that they will provide opposite magnetic polarity. The field (Fig. 4.47) is connected in series with the armature through the brush-commutator connection.

Armature Geometry. We start by analyzing the physical dimensions of the armature lamination, which gives us the following:

Gross slot area. This tells us how much cross-sectional area is available to accommodate the armature windings (before adding insulation and slot pegs).

Net slot area. This is the net available area to accommodate the copper wire after the slot insulators and pegs are subtracted from the gross slot area.

Magnetic path length. For the armature, this is the distance that the flux will travel and over which the mmf will be dropped.

Magnetic path area. This is the cross-sectional area of the magnetic path as seen from the flux's point of view. The smaller the path area, the higher the flux density, and the more mmf will be dropped along the path.

Weight of steel. This must be known to calculate the inertias of the armature assembly.

Inertia. This is used in determining start-up acceleration or load matching.

Slot constants. This is used in estimating reactance. This is largely a function of lamination geometry.

Using trigonometry, we break the armature lamination up into several sections in order to calculate the gross slot area. The radii used in the calculations are readily available from most lamination drawings supplied by the manufacturer.

Using Fig. 4.49, determine the following dimensions:

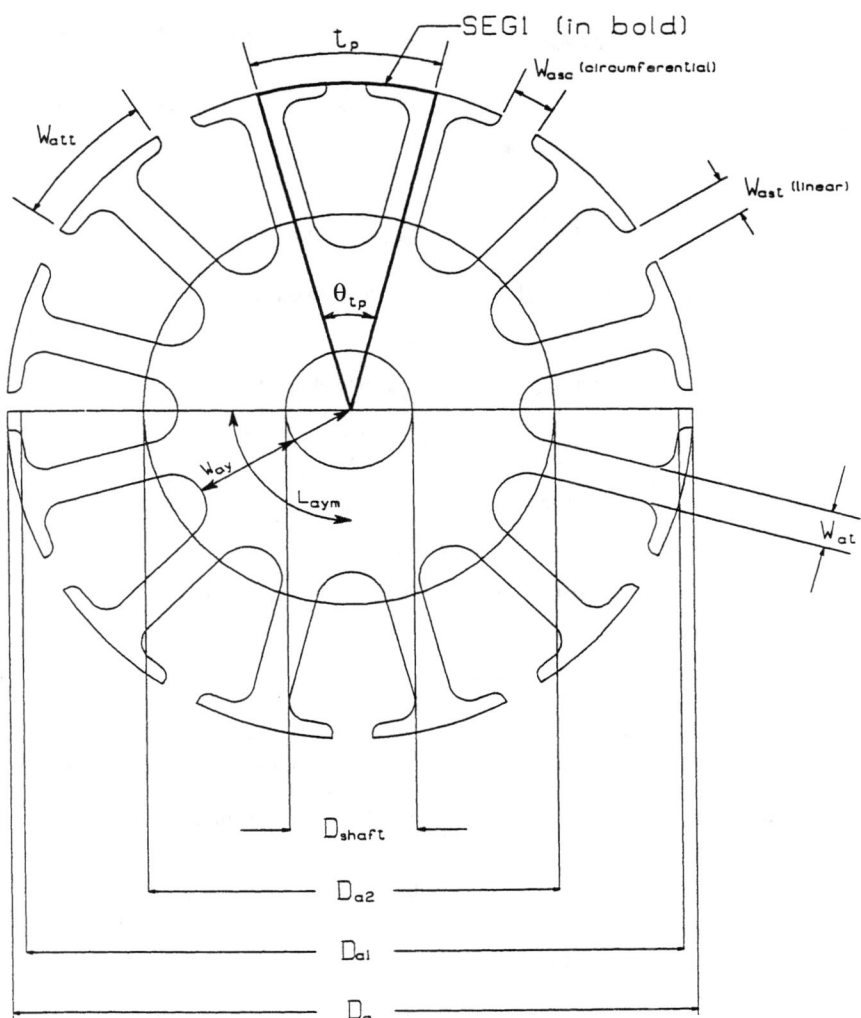

FIGURE 4.49 Armature lamination.

R_a = outside radius of the armature lamination
R_{a1} = inside radius of the armature teeth
R_s = radius of the shaft
W_{ast} = width of the armature slot top in straight-line distance
R_{a2} = radius of the circle on which the arc segments of the bottoms of the slots are centered
N_{at} = number of armature teeth
L_{stk} = length of the armature stack
θ_{pole} = angle of pole arc

4.50 CHAPTER FOUR

To find the tooth pitch angle θ_{tp}, between two consecutive tooth centers, in degrees:

$$\theta_{tp} = \frac{360°}{N_{at}} \tag{4.125}$$

To convert to radians:

$$\theta_{tpr} = \theta_{tp}\frac{\pi}{180} = \theta_{tp}(0.017453) \tag{4.126}$$

Tooth pitch t_p, the circumferential distance between two tooth centers, in inches, is

$$t_p = R_a\theta_{tpr} \tag{4.127}$$

Width of the armature slot along the circumference W_{asc}, in, is

$$W_{asc} = 2R_a \sin_r^{-1}\left(\frac{W_{ast}}{2R_a}\right) \tag{4.128}$$

Width of the tooth tip along the circumference W_{att}, in, is

$$W_{att} = t_p - W_{asc} \tag{4.129}$$

Area of the lamination consisting of the pie-shaped segment between tooth centers A_{SEG1}, in^2, is

$$A_{SEG1} = \frac{R_a^2\theta_{tpr}}{2} \tag{4.130}$$

Now that we know the area of the SEG1 section, we can use right triangles to determine the areas of smaller sections and subtract them from the SEG1 section (Fig. 4.50). Due to symmetry, only half of the SEG1 section needs to be calculated.

For the triangle TR1, at the bottom of the slot, the angle is half of the tooth pitch angle:

$$\theta_{TR1} = \frac{\theta_{tp}}{2} \tag{4.131}$$

The length of the base of TR1 is

$$L_{TR1} = \frac{R_{as}}{\tan \theta_{TR1}} \tag{4.132}$$

The angle within TR1 that the arc of radius R_{as} sweeps is

$$\theta_{as} = 90° - \theta_{TR1} \tag{4.133}$$

The area of triangle TR1 is

$$A_{TR1} = \frac{R_{as}^2}{2 \tan \theta_{TR1}} \tag{4.134}$$

The area of the slot SEG2 which is in TR1 is

$$A_{SEG2} = \frac{R_{as}^2\theta_{as}(0.017453)}{2} \tag{4.135}$$

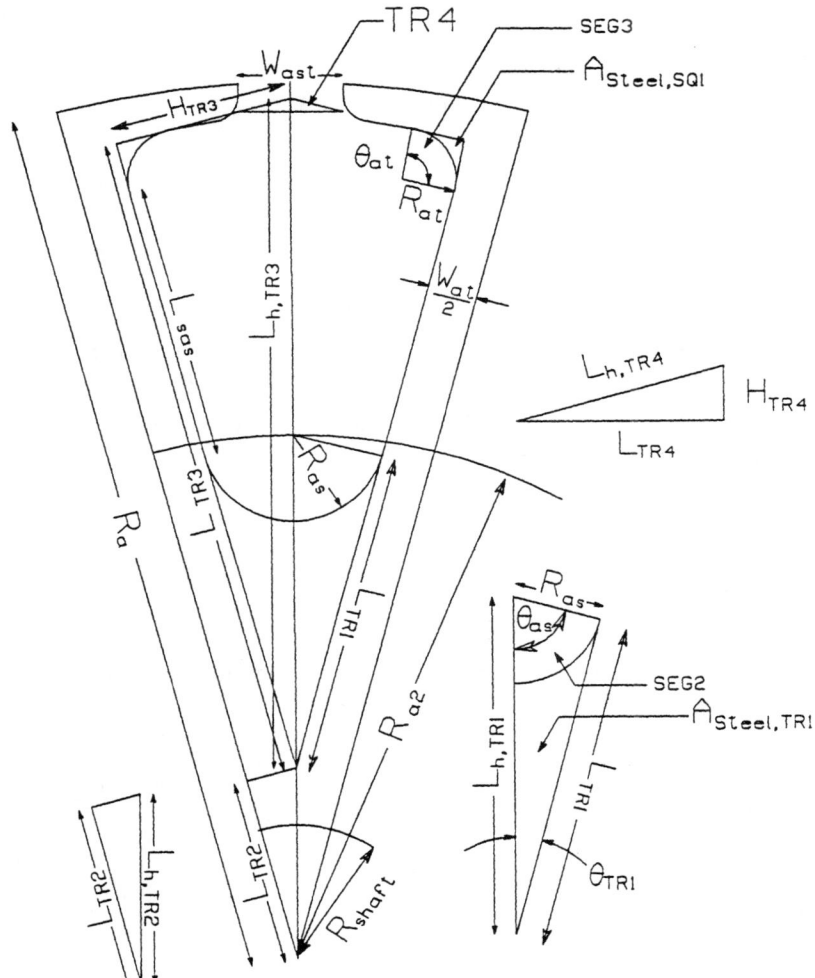

FIGURE 4.50 Armature lamination Slot.

The area of the steel inside TR1 will be needed later:

$$A_{\text{steel,TR1}} = A_{\text{TR1}} - A_{\text{SEG2}} \tag{4.136}$$

To determine the net slot area, we need to know the length of the arc of SEG2, in inches

$$L_{\text{arc,SEG2}} = R_{as}\theta_{as}(0.017453) \tag{4.137}$$

The area of the square SQ1 at the joint of the tooth and the tooth tip is the square of the radius R_{at}, since this arc segment is approximately 90°.

4.52 CHAPTER FOUR

$$A_{SQ1} = R_{at}^2 \qquad (4.138)$$

Using the circle segment of R_{at} over θ_{at} (approximately 90°) as SEG3:

$$A_{SEG3} = \frac{R_{at}^2 \theta_{at}(0.017453)}{2} \qquad (4.139)$$

Area of the steel in SQ1:

$$A_{steel,SQ1} = A_{SQ1} - A_{SEG3} \qquad (4.140)$$

Length of the arc of SEG3, in: is

$$L_{arc,SEG3} = R_{at}\theta_{at}(0.017453) \qquad (4.141)$$

The following calculations are to determine the area of the steel near the center of the armature lamination. Length of the hypotenuse of TR1, in:

$$L_{h,TR1} = \frac{L_{TR1}}{\cos \theta_{TR1}} \qquad (4.142)$$

Length of the hypotenuse of TR2, in:

$$L_{h,TR2} = R_{a2} - L_{h,TR1} \qquad (4.143)$$

Length of the base of TR2, in:

$$L_{TR2} = L_{h,TR2} \cos \theta_{TR1} \qquad (4.144)$$

The length of the side of the armature slot with no arcs in it is between SQ1 and TR1. This length, in inches, is

$$L_{sas} = R_{a1} - R_{at} - L_{TR1} - L_{TR2} \qquad (4.145)$$

TR3 is the large triangle that is made from the intersections of the slot center line, the extended side of the slot, and the inside of the tooth tip. Length of TR3, in:

$$L_{TR3} = L_{TR1} + L_{sas} + R_{at} \qquad (4.146)$$

Height of TR3, in:

$$H_{TR3} = L_{TR3} \tan \theta_{TR1} \qquad (4.147)$$

Area of the large triangle TR3, in²:

$$A_{TR3} = \frac{H_{TR3} L_{TR3}}{2} \qquad (4.148)$$

Using the small triangle TR4 inside TR3, the length of the base of TR4, in, is half the linear width of the top of the armature slot.

$$L_{TR4} = \frac{W_{ast}}{2} \qquad (4.149)$$

Length of the hypotenuse of TR4, in:

$$L_{h,TR4} = \frac{L_{TR4}}{\cos \theta_{as}} \qquad (4.150)$$

Height of TR4, in:

$$H_{TR4} = \frac{L_{TR4}}{\tan \theta_{as}} \qquad (4.151)$$

Area of TR4, in^2:

$$A_{TR4} = \frac{L_{TR4}H_{TR4}}{2} \qquad (4.152)$$

The gross slot area (GSA) is found by subtracting the area of TR4 and the steel from the large triangle TR3 and then doubling it. GSA, in^2:

$$A_{GS,a} = 2[A_{TR3} - (A_{steel,SQ1} + A_{steel,TR1} + A_{TR4})] \qquad (4.153)$$

The net slot area (NSA) is the area left after the armature slot insulation is put in. In order to determine the area that the insulation takes up, we need to know the length around the slot.

Length of the inside of the armature tooth that is not included in any other dimensions as of yet L_{iat}, in:

$$L_{iat} = H_{TR3} - R_{at} - L_{h,TR4} \qquad (4.154)$$

Length of the armature slot insulation L_{asi}, in:

$$L_{asi} = 2(L_{ait} + L_{arc,SEG2} + L_{arc,SEG3} + L_{sas}) \qquad (4.155)$$

Net slot area that the armature winding can occupy $A_{NS,a}$, in^2:

$$A_{NS,a} = A_{GS,a} - L_{asi}T_{asi} \qquad (4.156)$$

where T_{asi} is the thickness of the armature slot insulation.

Weight of Armature Steel. The stacking factor percentage is typically 92 to 95 percent, or:

$$F_{stk} = \frac{\text{lamination thickness} \times \text{number of laminations}}{\text{actual stack length}} \qquad (4.157)$$

Effective magnetic armature stack length, L_{stke}, in:

$$L_{stke} = L_{stk} \times F_{stk} \qquad (4.158)$$

Length of the hypotenuse of the large TR3 $L_{h,TR3}$, in:

$$L_{h,TR3} = \frac{L_{TR3}}{\cos \theta_{TR1}} \qquad (4.159)$$

Distance from the center of the lamination to the top of the usable armature slot R_{ast}, in:

$$R_{ast} = L_{h,TR3} + L_{h,TR2} - H_{TR4} \qquad (4.160)$$

Height of the armature slot top (where there is no wire, near the outside edge) H_{ast}, in:

$$H_{ast} = R_a - R_{ast} \qquad (4.161)$$

Area of the armature slot top A_{ast}, in^2:

$$A_{ast} = H_{ast}W_{ast} \tag{4.162}$$

Cross-sectional area of the armature shaft A_{shaft}, in^2:

$$A_{shaft} = \pi R_{shaft}^2 \tag{4.163}$$

Total area of the steel in the armature lamination $A_{steel,arm\ lam}$, in^2:

$$A_{steel,arm\ lam} = N_{at}\left[A_{SEG1} - \left(A_{GSA} + A_{ast} + \frac{A_{shaft}}{N_{at}}\right)\right] \tag{4.164}$$

which takes into account the area of the shaft that is inside each large SEG1. Volume of the steel in the armature stack V_{as}, in^3:

$$V_{as} = L_{stke}A_{steel,arm\ lam} \tag{4.165}$$

Weight of the steel in the armature stack W_{tas}, oz:

$$W_{tas} = 4.536\ V_{as} \tag{4.166}$$

where 4.536 is the density of steel in ounces per cubic inch.
Volume of the armature shaft V_s, in^3:

$$V_s = A_{shaft}L_{shaft} \tag{4.167}$$

Weight of the steel armature shaft W_{ts}, oz:

$$W_{ts} = 4.536V_s \tag{4.168}$$

4.2.2 Magnetic Circuits

Magnetic Paths in the Armature. The *armature yoke* is the area between the bottom of the slot and the shaft. The effective width of the armature yoke is larger for a magnetic shaft and smaller for a nonmagnetic one. For the initial calculations, use $L_{stke} = 1.0$.

Width of the armature yoke (Figs. 4.51 and 4.52) for a magnetic shaft, in inches:

$$W_{ay} = R_{a2} - R_{as} \tag{4.169}$$

Width of the armature yoke for a nonmagnetic shaft, in inches:

$$W_{ay} = R_{a2} - R_{as} - R_{shaft} \tag{4.170}$$

Only one of the preceding formulas will be used for each design.
Height of the armature tooth tip, which is the same as its magnetic path length, in inches:

$$L_{attm} = H_{att} \tag{4.171}$$

The magnetic path length of the armature tooth is from the tooth tip to the yoke, in inches:

$$L_{atm} = R_a - R_{a2} + R_{as} - H_{att} \tag{4.172}$$

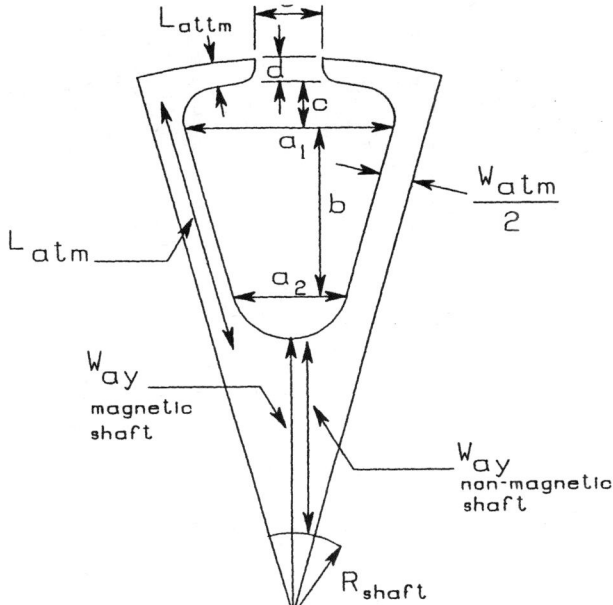

FIGURE 4.51 Armature magnetic paths and reactance dimensions.

Area that the flux "sees" through the armature tooth path, the magnetic area of the armature tooth, in square inches:

$$A_{atm} = W_{at} L_{stke} \quad (4.173)$$

The total magnetic area of armature teeth under a pole is the magnetic area of the armature tooth times the pole arc angle divided by the tooth pitch angle. This turns out to be the effective number of teeth under a pole times the magnetic area of one tooth, in square inches:

$$A_{T,atm} = \frac{\theta_{pole}}{\theta_{tp}} A_{atm} \quad (4.174)$$

where the ratio $\theta_{pole}/\theta_{tp}$ must be rounded up to the next-higher integer. This rounding up assumes that if any part of the tooth tip is over the pole, the flux will use that tooth as part of its magnetic path. Figure 4.53 shows a finite element analysis (FEA) of a universal motor with only the field coils turned on. Note that the flux lines utilize the teeth where there is any part of the tooth tip over the pole.

Magnetic area of the armature tooth tip, in square inches:

$$A_{attm} = W_{att} L_{stke} \quad (4.175)$$

The total magnetic area of armature tooth tips under a pole is the effective number of teeth under a pole times the effective number of tooth tips under a pole, in square inches:

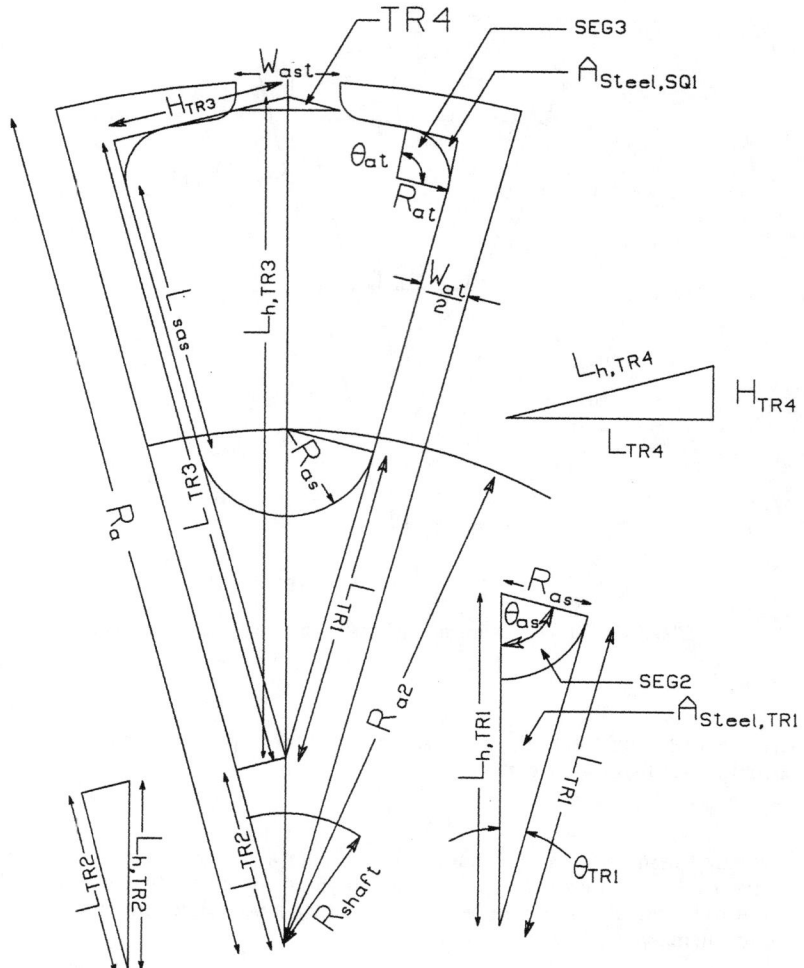

FIGURE 4.52 Armature lamination slot.

$$A_{T,\text{attm}} = \frac{\theta_{\text{pole}}}{\theta_{\text{tp}}} A_{\text{attm}} \tag{4.176}$$

Magnetic area of the armature yoke, in square inches:

$$A_{\text{aym}} = W_{\text{ay}} L_{\text{stke}} \tag{4.177}$$

The magnetic path length of the armature yoke (Fig. 4.54) depends on the shaft material. For a nonmagnetic shaft, the path length in inches is

$$L_{\text{aym}} = \left(\frac{\pi}{2}\right)\left(R_{\text{shaft}} + \frac{W_{\text{ay}}}{2}\right) \tag{4.178}$$

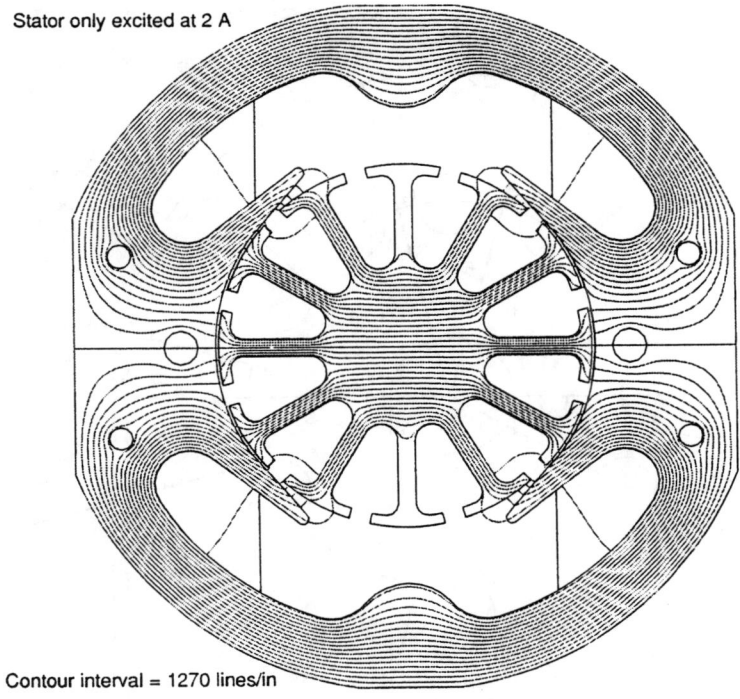

FIGURE 4.53 Finite element analysis showing flux paths through stator teeth.

And for a magnetic shaft

$$L_{\text{aym}} = \left(\frac{\pi}{2}\right)\left(\frac{R_{\text{shaft}} + W_{\text{ay}}}{2}\right) \qquad (4.179)$$

The magnetic path area of the shaft is accounted for in the armature yoke magnetic path calculations.

4.2.3 Armature Conductors

In order to determine the resistance of the armature as seen by the brushes, the following parameters need to be set:

S_{ac} = number of teeth over which one coil is wound
T_{pca} = number of turns of wire per armature coil
CPS = number of coil sides in each armature slot

The length of one turn in the winding must be determined. There are three different methods of determining the length of a turn.

Armature End-Turn Length—Method 1. This method uses the geometry of the lamination to estimate the average end-turn length. The accuracy of this method is

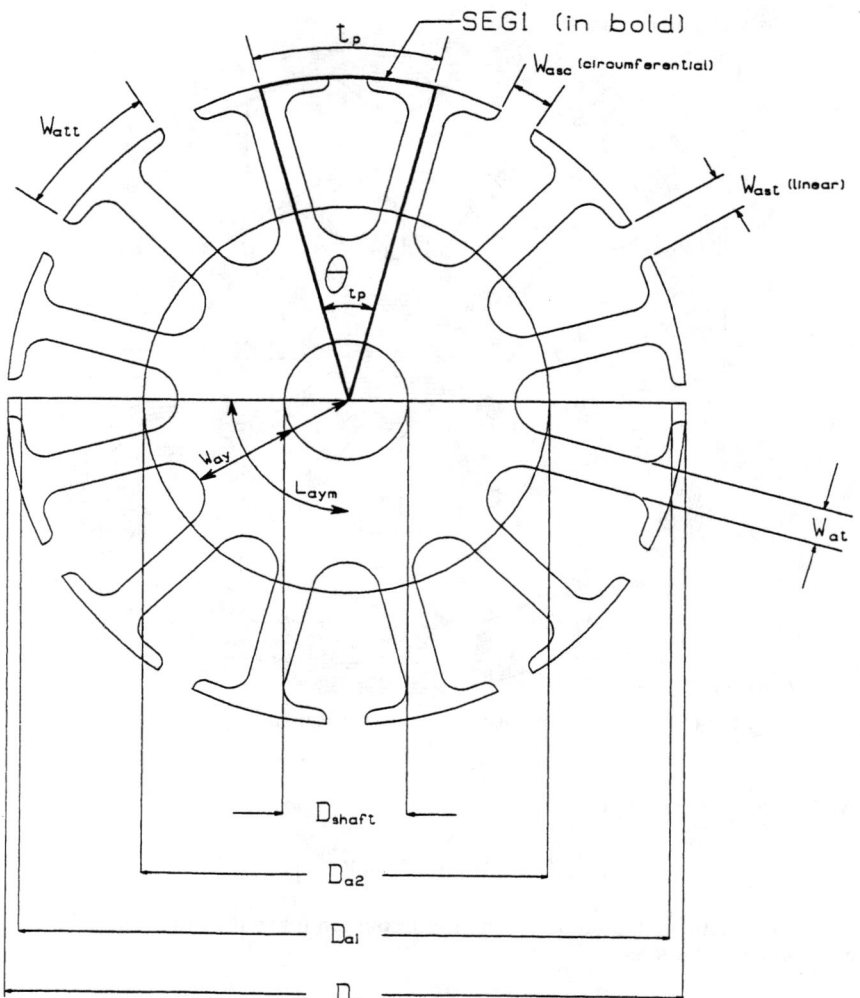

FIGURE 4.54 Armature lamination slot.

dependent on the capabilities of the winding machinery. Using geometry and the width of the armature yoke for a magnetic shaft, the mean end-turn length is found by averaging the distances around the lamination at the radii of the top of the slot and the bottom of the slot. The mean end-turn length of one side of one turn, in inches, is

$$L_{\text{meta}} = \frac{S_{\text{ac}}(R_{\text{ast}})\theta_{\text{tp}}(0.017453) + 2W_{\text{ay}} \sin\left[(S_{\text{ac}}/2)\,\theta_{\text{tp}}\right]}{2} \tag{4.180}$$

In order to understand the other methods of determining end turn length, the formulas for armature resistance are discussed first.

DIRECT-CURRENT MOTORS

Armature Resistance. The mean turn length is the sum of the mean end-turn length for each of two ends plus twice the stack length, in feet (because wire resistances are given in ohms per foot):

$$L_{mta} = \frac{2L_{stk} + 2L_{meta}}{12} \tag{4.181}$$

Length of copper per armature coil, in feet:

$$L_{ac} = L_{mta} \times T_{pca} \tag{4.182}$$

Resistance of one armature coil, in ohms:

$$R_{ac} = L_{ac}\rho_a F_{wsa} \tag{4.183}$$

where ρ_a is the armature wire resistance per foot and F_{wsa} is the armature winding stretch factor, usually 1.02 to 1.10.

The number of coils to be considered for armature resistance depends on the number of commutating coils, so the number of active coils for the armature is

$$N_{ca} = N_{at} - 1 \tag{4.184}$$

where N_{at} is odd, and

$$N_{ca} = N_{at} - 2 \tag{4.185}$$

where N_{at} is even.

The *net armature resistance* (Fig. 4.55) is the combination of half the number of active coils in parallel with half the number of active coils, in ohms:

$$R_{at} = \frac{(N_{ca}/2)R_{ac}}{2} \tag{4.186}$$

Armature End-Turn Length—Method 2. This method assumes that you already have a motor and know R_{at}, T_{pca}, F_{wsa}, and ρ_a. From this, the L_{meta} can be calculated and then used for calculations for other motors with the same lamination. The following four formulas determine the mean end-turn length for one end of one turn.

$$R_{ac} = \frac{2R_{at}}{N_{ca}/2} \tag{4.187}$$

$$L_{ac} = \frac{R_{ac}}{\rho_a F_{wsa}} \tag{4.188}$$

$$L_{mta} = \frac{L_{ac}}{T_{pca}} \tag{4.189}$$

$$L_{meta} = \frac{12L_{mta} - 2L_{stk}}{2} \tag{4.190}$$

Armature End-Turn Length—Method 3. Measure what looks like the average end turn on an existing motor.

Armature Slot Fill. The maximum armature slot fill depends on the winding equipment. A full slot could be anywhere from 40 to 55 percent. The slot fill is calculated by assuming that the wires are lying side by side as if they were square. Using

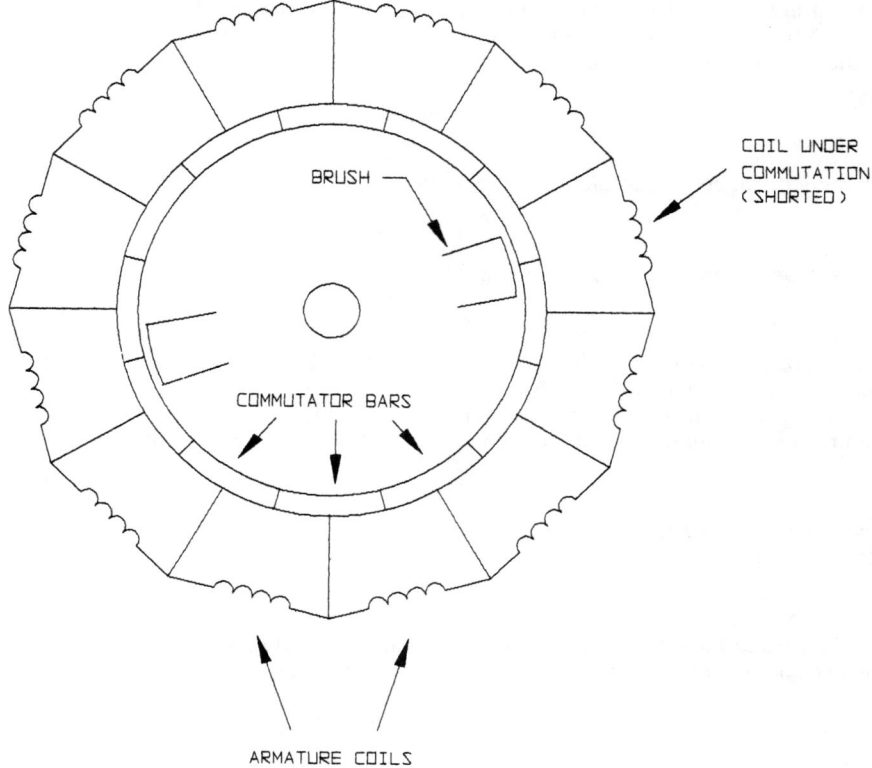

FIGURE 4.55 Schematic of armature current paths.

the outside diameter of the armature wire as D_{aw} (this depends on insulation thickness), the percentage of armature slot fill is

$$F_{as} = \frac{(D_{aw})^2 T_{pca} \text{CPS}}{A_{NSA}} \times 100 \tag{4.191}$$

Armature Copper Weight. Using pound-feet from the wire tables, the weight of the armature copper W_{tac}, is:

$$W_{tac} = L_{ac} N_{at} \left(\frac{\text{lb}}{\text{ft}}\right)\left(\frac{\text{CPS}}{2}\right) \tag{4.192}$$

Armature Inertia. The inertia of the shaft J_s, oz · in · s², is

$$J_s = (0.0184) R_s^4 \, L_{\text{shaft}} \tag{4.193}$$

Assuming the steel is distributed over a solid cylinder, the inertia of the armature lamination steel, J_{as}, oz · in · s², is

$$J_{as} = \frac{(W_{tas}) R_a^2}{772} - J_s \frac{L_{stke}}{L_s} \tag{4.194}$$

where the factor 772 is 2g, in s².

Assuming that the copper in the end turns effectively distributes the total copper weight over a solid cylinder, the inertia of the copper in the armature J_{ac}, $oz \cdot in \cdot s^2$, is

$$J_{ac} = \frac{(W_{tac} \times 16)R_{ast}^2}{772} - (0.02095)R_s^4 L_{stk} \qquad (4.195)$$

The inertia of the commutator is split into the inertia of the copper bars and the insulation on the inside (see Fig. 4.69 in Sec. 4.3.2). The commutator inertia J_c, $oz \cdot in \cdot s^2$, is

$$J_{ccu} = (0.02095)L_{comm,copper}\left[\left(\frac{D_c}{2}\right)^4 - \left(\frac{d_{ci}}{2}\right)^4\right] \qquad (4.196)$$

$$J_{ci} = (0.0005086)L_{comm,ins}\left(\frac{d_{ci}}{2}\right)^4 \qquad (4.197)$$

$$J_c = J_{ccu} + J_{ci} \qquad (4.198)$$

where $L_{comm,copper}$ = length of commutator copper measured parallel to shaft
$L_{comm,ins}$ = length of commutator insulation material
D_c = outside diameter of commutator
d_{ci} = inside diameter of commutator copper

The total armature inertia J_a, $oz \cdot in \cdot s^2$, is

$$J_a = J_s + J_{as} + J_{ac} + J_c \qquad (4.199)$$

Armature Balance. ISO Standard 1941 gives guidelines for acceptable balance and vibration limits for electric machines based on their usage. In general, the lower the unbalance, the longer the motor life. Unbalance affects bearing wear by causing the shaft to pound against the bearing, breaking down the surface and pumping out the lubricant. It affects brush life adversely by causing the brushes to instantaneously lose contact with the commutator surface, resulting in electrical arcing, wear, and higher levels of EMI. If this occurs near a resonant speed, the condition becomes severe and the motor instantaneously ceases operation because the brush lifts completely away from the commutator, causing an open circuit.

Armatures should be aligned in the magnetic field such that there is a net magnetic force pulling the armature toward the commutator end of the motor. This reduces end bounce and improves brush life.

4.2.4 Field (Stator)

The geometry of the stator lamination requires breaking it down into several small areas in the same manner as used for the armature.

Using Figs. 4.56, 4.57, and 4.58 and the XY coordinate system as shown, this subsection uses the given dimensions to determine the field slot area.

The following dimensions are needed from the lamination drawing. Some manufacturers distribute drawings with only a few dimensions on them, but will give these detailed drawings if requested.

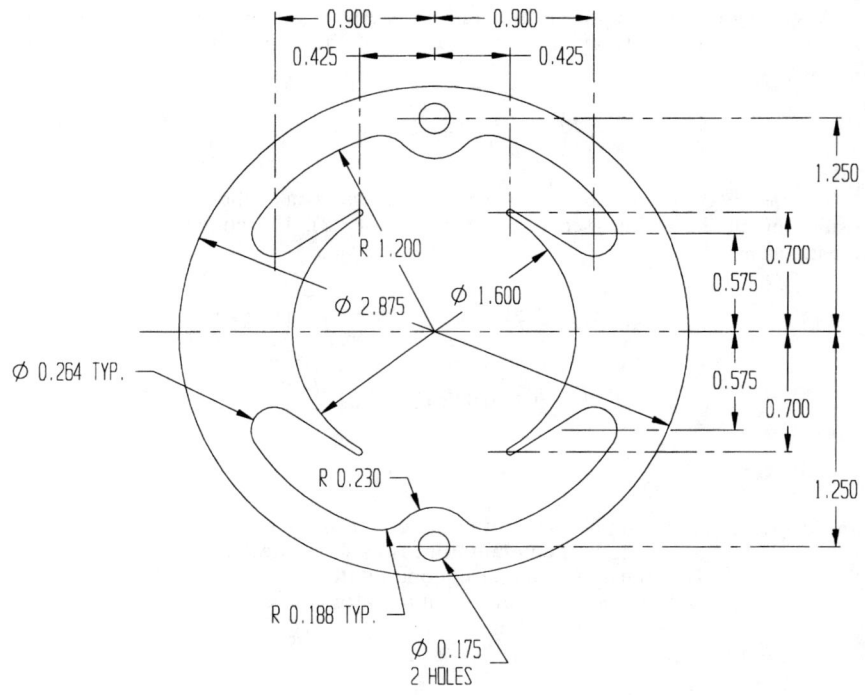
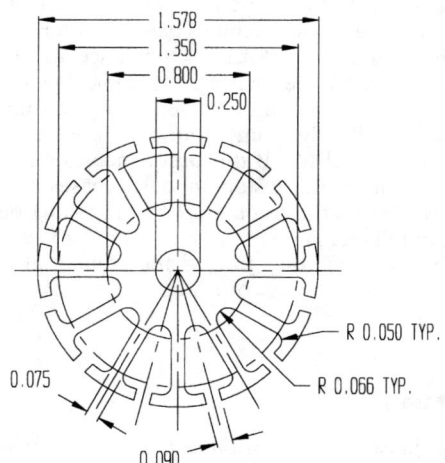

FIGURE 4.56 Example of a lamination drawing.

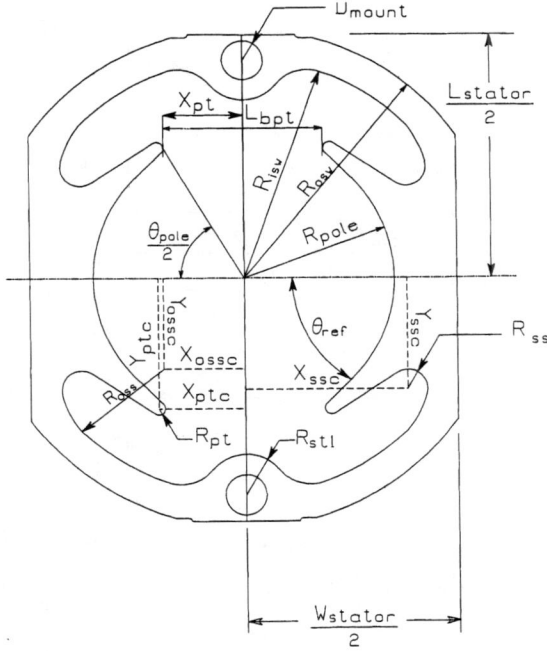

FIGURE 4.57 Stator (field) lamination.

L_{bpt} = distance between opposite pole tips
X_{ptc}, Y_{ptc} = coordinates of pole tip center
R_{pt} = radius of pole tip
X_{ossc}, Y_{ossc} = coordinates of center of arc of outside of stator slot
R_{oss} = radius of outside of stator slot
X_{ssc}, Y_{ssc} = coordinates of center of radius of stator slot center
R_{ss} = radius of stator slot
R_{osw} = radius of outside of stator wall
R_{isw} = radius of inside of stator wall, which blends with arc of outside of stator slot

Depending on which stator dimension is given, the X coordinate for the center of the pole tip, in inches, is

$$X_{ptc} = \frac{L_{bpt}}{2} + R_{pt} \qquad (4.200)$$

Or, the distance between opposite pole tips, in inches, is

$$L_{bpt} = 2(X_{ptc} - R_{pt}) \qquad (4.201)$$

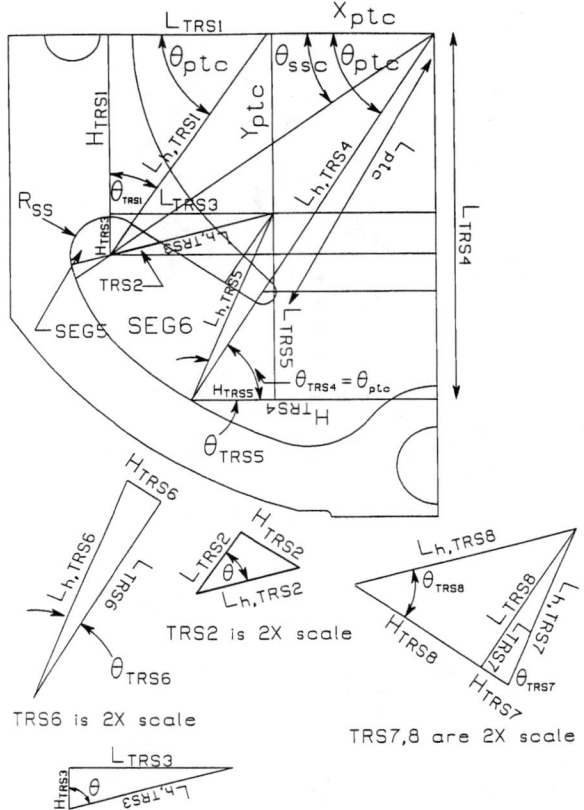

FIGURE 4.58 Stator (field) slot area.

The pole arc angle is twice the angle from the center of the lamination to the tip of the pole closest to the other pole. The pole arc angle, in degrees, is

$$\theta_{\text{pole}} = 2\tan^{-1}\left(\frac{Y_{\text{ptc}}}{L_{\text{bpt}}/2}\right) \tag{4.202}$$

Angle from the center of the stator to the center of the pole tip, in degrees:

$$\theta_{\text{ptc}} = \tan^{-1}\left(\frac{Y_{\text{ptc}}}{X_{\text{ptc}}}\right) \tag{4.203}$$

Distance from the center of the stator to the pole tip center, in inches:

$$L_{\text{ptc}} = \sqrt{X_{\text{ptc}}^2 + Y_{\text{ptc}}^2} \tag{4.204}$$

Using stator triangle TRS1, which has the hypotenuse parallel to L_{ptc}, the height is

$$H_{\text{TRS1}} = Y_{\text{ssc}} \tag{4.205}$$

DIRECT-CURRENT MOTORS

Angle between the height and the hypotenuse, in degrees:

$$\theta_{TRS1} = 90° - \theta_{ptc} \tag{4.206}$$

For later use, the angle from the center of the stator to the center of the stator slot, in degrees, is

$$\theta_{ssc} = \tan^{-1}\left(\frac{Y_{ssc}}{X_{ssc}}\right) \tag{4.207}$$

The triangle TRS3 has the height of the difference of Y coordinates of the stator slot center and the outside of the stator slot center, in inches:

$$L_{TRM4} = R_{ss} \tag{4.208}$$

And the length, in inches:

$$L_{TRS3} = X_{ssc} - X_{ossc} \tag{4.209}$$

Angle between the hypotenuse and the height of TRS3, in degrees:

$$\theta_{TRS3} = \tan^{-1}\left(\frac{L_{TRS3}}{H_{TRS3}}\right) \tag{4.210}$$

Length of the hypotenuse of TRS3, in inches:

$$L_{h,TRS3} = \frac{H_{TRS3}}{\cos \theta_{TRS3}} \tag{4.211}$$

The small TRS2 has the leg from the center of the stator slot with its length, in inches:

$$L_{TRS2} = R_{ss} \tag{4.212}$$

Angle between the length of TRS2 and the hypotenuse, in inches:

$$\theta_{TRS2} = \theta_{TRS3} - \theta_{TRS1} \tag{4.213}$$

Height of the triangle TRS2, in inches:

$$H_{TRS2} = L_{TRS2} \tan \theta_{TRS2} \tag{4.214}$$

Length of the hypotenuse of TRS2, in inches:

$$L_{h,TRS2} = \frac{L_{TRS2}}{\cos \theta_{TRS2}} \tag{4.215}$$

Area of TRS2, in square inches:

$$A_{TRS2} = \frac{L_{TRS2} H_{TRS2}}{2} \tag{4.216}$$

The arc segment SEG5 at the end of the stator slot traverses the angle, in degrees:

$$\theta_{SEG5} = 180° - \theta_{TRS2} \tag{4.217}$$

Area of SEG5, in square inches:

$$A_{SEG5} = \theta_{SEG5}(0.017453)\frac{R_{ss}^2}{2} \qquad (4.218)$$

Length of the arc of SEG5, in inches:

$$L_{arc,SEG5} = \theta_{SEG5}(0.017453)R_{ss} \qquad (4.219)$$

Using $\theta_{TRS4} = \theta_{ptc}$, the height of the very large TRS4, in inches:

$$H_{TRS4} = R_{isw}\cos\theta_{ptc} \qquad (4.220)$$

Length of TRS4, in inches:

$$L_{TRS4} = R_{isw}\sin\theta_{ptc} \qquad (4.221)$$

Height of TRS5, in inches:

$$H_{TRS5} = H_{TRS4} - X_{ossc} \qquad (4.222)$$

Length of TRS5, in inches:

$$L_{TRS5} = L_{TRS4} - Y_{ossc} \qquad (4.223)$$

Angle between the height of TRS5 and the hypotenuse, in inches:

$$\theta_{TRS5} = \tan^{-1}\left(\frac{L_{TRS5}}{H_{TRS5}}\right) \qquad (4.224)$$

Angle of the stator triangle TRS6, in degrees, is

$$\theta_{TRS6} = \theta_{TRS5} - \theta_{ptc} \qquad (4.225)$$

since the angle of TRS4 is the same as the angle of the pole tip center.
Length of TRS6, in inches:

$$L_{TRS6} = R_{isw} - R_{pt} - L_{ptc} \qquad (4.226)$$

Height of TRS6, in inches:

$$H_{TRS6} = L_{TRS6}\tan\theta_{TRS6} \qquad (4.227)$$

Length of the hypotenuse of TRS6, in inches:

$$L_{h,TRS6} = \frac{L_{TRS6}}{\cos\theta_{TRS6}} \qquad (4.228)$$

Area of TRS6, in square inches:

$$A_{TRS6} = \frac{L_{TRS6}H_{TRS6}}{2} \qquad (4.229)$$

Using triangle TRS7, the angle between the height and the hypotenuse, in degrees:

$$\theta_{TRS7} = 90° - \theta_{TRS6} \qquad (4.230)$$

Length of the hypotenuse of TR7, in inches:

$$L_{h,TRS7} = R_{oss} - L_{h,TRS6} \qquad (4.231)$$

Height of TRS7, in inches:
$$H_{TRS7} = L_{h,TRS7} \cos \theta_{TRS7} \qquad (4.232)$$

Length of TRS7, in inches:
$$L_{TRS7} = L_{h,TRS7} \sin \theta_{TRS7} \qquad (4.233)$$

Area of TRS7, in square inches:
$$A_{TRS7} = \frac{L_{TRS7} H_{TRS7}}{2} \qquad (4.234)$$

TRS8 shares a leg with TRS7, so the length of TRS8, in inches, is
$$L_{TRS8} = L_{TRS7} \qquad (4.235)$$

Length of the hypotenuse of TRS8, in inches:
$$L_{h,TRS8} = L_{h,TRS3} - L_{h,TRS2} \qquad (4.236)$$

The angle between the height of TRS8 and the hypotenuse is the same as the third angle of TRS2, in degrees:
$$\theta_{TRS8} = 90° - \theta_{TRS2} \qquad (4.237)$$

Height of TRS8, in inches:
$$H_{TRS8} = L_{h,TRS8} \cos \theta_{TRS8} \qquad (4.238)$$

Area of TRS8, in square inches:
$$A_{TRS8} = \frac{L_{TRS8} H_{TRS8}}{2} \qquad (4.239)$$

The large SEG6 traverses the same angle of the large triangle which would be comprised of TRS7 and TRS8. Angle for SEG6, in degrees:
$$\theta_{SEG6} = 180° - \theta_{TRS7} - \theta_{TRS8} \qquad (4.240)$$

Length of the arc of SEG6, in inches:
$$L_{arc,SEG6} = \theta_{SEG6}(0.017453) R_{oss} \qquad (4.241)$$

Area of SEG6, in square inches:
$$A_{SEG6} = \theta_{SEG6}(0.017453) \frac{R_{oss}^2}{2} \qquad (4.242)$$

Gross slot area of the stator slot, in square inches:
$$A_{GS,f} = A_{TRS2} + A_{SEG5} + A_{TRS6} + (A_{SEG6} - A_{TRS7} - A_{TRS8}) \qquad (4.243)$$

Length of the stator slot insulation, in inches,
$$L_{ssi} = H_{TRS6} + H_{TRS7} + H_{TRS8} + H_{TRS2} + L_{arc,SEG5} + L_{arc,SEG6} \qquad (4.244)$$

The net slot area available for the field winding on one pole is the GSA minus the area of the insulation, in square inches:

$$A_{NS,f} = A_{GS,f} - L_{ssi} T_{ssi} \qquad (4.245)$$

where T_{ssi} is the thickness of the stator slot insulation in inches.

Constants for Stator Reactance Calculations. As with the armature, the geometry of the stator slot must be known in order to calculate the reactance. Note that the field slot can be turned to resemble the armature slot. Use Figs. 4.58 and 4.59 to determine the following values which will be needed later.

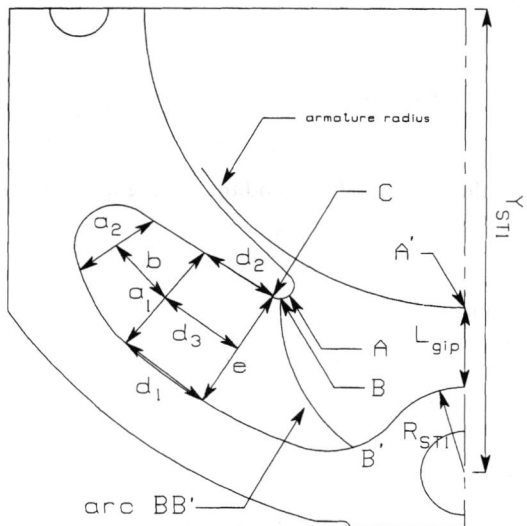

FIGURE 4.59 Dimensions for stator reactance.

The values of a and b, as similar to the armature, in inches:

$$a_1 \approx R_{oss} - L_{TRS7} \qquad (4.246)$$

$$a_2 \approx 2R_{ss} \qquad (4.247)$$

$$b = \frac{H_{TRS2} + H_{TRS8} + H_{TRS6}}{2} \qquad (4.248)$$

Since the stator slot has a different shape than the armature slot, the value d_3 for the stator is the average of d_1 and d_2, in:

$$d_2 = b \qquad (4.249)$$

$$d_1 \approx \frac{L_{arc,SEG6}}{2} \qquad (4.250)$$

DIRECT-CURRENT MOTORS

$$d_3 = \frac{d_1 + d_2}{2} \tag{4.251}$$

Length of the line across the wide opening of the slot e, in:

$$e = L_{TRS6} \tag{4.252}$$

There is no c dimension as in the armature slot dimensions.

Length of the interpolar air gap, which is the distance between the armature and the stator yoke at their closest position, in inches:

$$L_{gip} = Y_{ST1} - R_{ST1} - R_a \tag{4.253}$$

where Y_{ST1} is the Y coordinate of the center of radius R_{ST1}.

The points on the pole tip A, B, and C are used as the starting points of three arcs which are drawn with the edge of the armature as the center. These arcs were derived from flux mapping using curvilinear squares, which will not be discussed here. C is the point where line e intersects the pole. A is the point where the radius of the pole tip would be parallel to the armature tooth tip. B is the point on the slot side of the pole tip which is halfway between A and C. The point B' is the point that the average leakage flux would travel to on the stator yoke.

Arc length BB' is the length of the arc of radius $R_{BB'}$ (which is the distance from A' to B) shortened by the radius R_{ST1}. Arc length AC is the arc around the pole tip of about 90°.

$$R_{BB'} = \sqrt{X_{ptc}^2 + (Y_{ptc} + R_{pt} - R_a)^2} \tag{4.254}$$

$$BB' \approx \frac{\pi}{2} R_{BB'} - R_{ST1} \tag{4.255}$$

$$AC \approx \frac{\pi}{2} R_{pt} \tag{4.256}$$

4.2.5 Magnetic Paths in the Field (Stator)

Using the quarter-stator drawing in Fig. 4.60, the portion of the stator is divided up into seven magnetic paths. In order to determine the widths and lengths of these paths, the dimensions of four triangles need to be determined. All magnetic path widths are multiplied by 2 because this analysis is done on a per-pole basis and the dimensions being used are from a drawing of half of one pole. For the initial calculations, use $L_{stke} = 1.0$.

Width of the stator yoke, in inches:

$$W_{SY} = R_{OSW} - R_{ISW} \tag{4.257}$$

magnetic length of the stator yoke, in inches:

$$L_{SY} = \left(R_{ISW} + \frac{W_{SY}}{2}\right)(90° - \theta_{ssc})(0.017453) \tag{4.258}$$

Magnetic area of the stator yoke, in square inches:

$$A_{SY} = 2W_{SY}L_{stke} \tag{4.259}$$

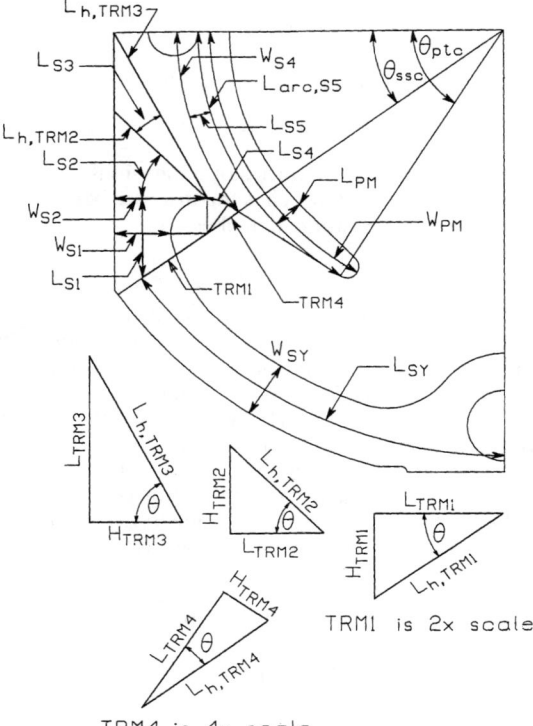

FIGURE 4.60 Stator magnetic path lengths and widths.

Angle of triangle TRM1 between the length and the hypotenuse, in degrees:

$$\theta_{TRM1} = \theta_{ssc} \qquad (4.260)$$

Width of stator path 1, in inches:

$$W_{S1} = \frac{W_{stator}}{2} - X_{ssc} - R_{ss} \qquad (4.261)$$

Length of triangle TRM1, in inches:

$$L_{TRM1} = W_{S1} + R_{ss} \qquad (4.262)$$

Height of TRM1, in inches:

$$H_{TRM1} = \frac{L_{TRM1}}{\tan \theta_{TRM1}} \qquad (4.263)$$

Magnetic length of stator path 1, in inches:

$$L_{S1} = H_{TRM1} + R_{ss} \qquad (4.264)$$

Magnetic area of stator path 1, in square inches:

$$A_{S1} = 2W_{S1}L_{stke} \quad (4.265)$$

The length of triangle TRM2 is the same as the width of the stator magnetic path 2, in inches:

$$L_{TRM2} = W_{S2} = W_{S1} + R_{ss} \quad (4.266)$$

The height of TRM2 is related to the length of TRM3. Length of TRM3, in inches:

$$L_{TRM3} = H_{TRS1} - R_{ss} \quad (4.267)$$

Height of TRM2, in inches:

$$H_{TRM2} = \frac{L_{TRM3}}{2} \quad (4.268)$$

Length of the hypotenuse of TRM2, in inches:

$$L_{h,TRM2} = \sqrt{L_{TRM2}^2 + H_{TRM2}^2} \quad (4.269)$$

Angle between the length and the hypotenuse of TRM2, in degrees:

$$\theta_{TRM2} = \tan^{-1}\left(\frac{H_{TRM2}}{L_{TRM2}}\right) \quad (4.270)$$

The magnetic path length of stator section 2 is the radial length of the arc which has the radius of half the length of the hypotenuse, in inches:

$$L_{S2} = \frac{L_{h,TRM2}}{2} \theta_{TRM2}(0.017453) \quad (4.271)$$

Magnetic area of stator path 2, in square inches:

$$A_{S2} = 2W_{S2}L_{stke} \quad (4.272)$$

Height of TRM3, in inches:

$$H_{TRM3} = L_{TRM2} \quad (4.273)$$

Length of the hypotenuse of TRM3, in inches:

$$L_{h,TRM3} = \sqrt{L_{TRM3}^2 + H_{TRM3}^2} \quad (4.274)$$

Angle between the hypotenuse and the height of TRM3, in degrees:

$$\theta_{TRM3} = \tan^{-1}\left(\frac{L_{TRM3}}{H_{TRM3}}\right) \quad (4.275)$$

Width of stator section 3, in inches:

$$W_{S3} = L_{h,TRM2} \quad (4.276)$$

The magnetic path length of stator section 3 is the radial length of the arc which has the radius of half the length of the hypotenuse of TRM3, in inches:

$$L_{S3} = \frac{L_{h,TRM3}}{2} [\theta_{TRM3} - \theta_{TRM2}](0.017453) \quad (4.277)$$

Magnetic area of stator section 3, in square inches:

$$A_{S3} = 2W_{S3}L_{stke} \qquad (4.278)$$

Using the small TRM4 which is inside the stator slot, the length of TRM4, in inches:

$$L_{TRM4} = R_{ss} \qquad (4.279)$$

Angle between the length of TRM4 and the hypotenuse, in degrees,

$$\theta_{TRM4} = 90° - \theta_{ssc} - \theta_{TRS1} \qquad (4.280)$$

Length of the hypotenuse of TRM4, in inches:

$$L_{h,TRM4} = \frac{L_{TRM4}}{\cos \theta_{TRM4}} \qquad (4.281)$$

Radius at which the width of stator path 4 is centered, in inches:

$$R_{S4} = \sqrt{X_{ssc}^2 + Y_{ssc}^2} - L_{h,TRM4} \qquad (4.282)$$

The width of magnetic stator path 4 is the length of the arc that is swung from the hypotenuse of TRM4 to the horizontal axis, in inches:

$$W_{S4} = R_{S4}\theta_{ssc}(0.017453) \qquad (4.283)$$

The length of magnetic stator path 4 is the length of the arc with radius R_{ss} that is swung from the vertical to the leg of TRM4, in inches:

$$L_{S4} = R_{ss}\theta_{TRS1}(0.017453) \qquad (4.284)$$

Magnetic area of stator path 4, in square inches:

$$A_{S4} = 2W_{S4}L_{stke} \qquad (4.285)$$

In order to find the width of stator path 5, the length of the arc from the far side of the pole tip to the horizontal must be determined, in inches:

$$L_{arc,S5} = [L_{ptc} + R_{pt}]\theta_{ptc}(0.017453) \qquad (4.286)$$

The width of stator path 5 is the average of the arc length in Eq. (4.286) and the width of stator path 4, in inches:

$$W_{S5} = \frac{W_{S4} + L_{arc,S5}}{2} \qquad (4.287)$$

Magnetic length of stator path 5, in inches:

$$L_{S5} = R_{S4} - (L_{ptc} + R_{pt}) \qquad (4.288)$$

Magnetic area of stator path 5, in square inches:

$$A_{S5} = 2W_{S5}L_{stke} \qquad (4.289)$$

Width of the magnetic path consisting of the actual pole, in inches:

$$W_{PM} = L_{ptc}\left[\frac{\theta_{pole}}{2}\right](0.017453) \qquad (4.290)$$

Magnetic length of the pole itself, in inches:

$$L_{PM} = (L_{ptc} + R_{pt}) - R_{pole} \qquad (4.291)$$

Magnetic area of the pole, in square inches:

$$A_{PM} = 2W_{PM}L_{stke} \qquad (4.292)$$

4.2.6 Field Conductors

In order to determine the resistance and slot fill of the field, the following parameters need to be determined by the designer:

T_{pcf} = turns of wire per field coil
D_{fw} = diameter of field wire

The length of one turn of the field winding must be determined. As with the armature winding, there are three different methods of determining end-turn length for the field.

Field End-Turn Length—Method 1. The mean end-turn length for the field is the average of the length of the shortest turn which is wound tightly across the pole and the length of the turn on the outer edges of each slot. The shortest turn is the straight distance across the pole, and the longest turn is the length of the arc swung across the outer edges of the slots. The field winding mean end-turn length, in inches, is

$$L_{metf} = \frac{2(H_{TRS1} - R_{ss}) + \pi L_{TRS4}}{2} \qquad (4.293)$$

where π is 180° converted to radians. The end-turn length will vary slightly with slot fill, winding equipment, and the overhang of the field slot insulation relative to the stack.

In order to use method 2 of determining end-turn length, which is similar to the armature method, you first need to know how the field resistance is determined.

Resistance of Field Windings. The *field resistance* is the total resistance of two field coils in series. The mean turn length of one field coil, in feet (because wire resistances are given in ohms per foot), is

$$L_{mtf} = \frac{2L_{stk} + 2L_{metf}}{12} \qquad (4.294)$$

Length of copper per field coil, in feet:

$$L_{fc} = L_{mtf} \times T_{pcf} \qquad (4.295)$$

Resistance of one field coil, in ohms:

$$R_{fc} = L_{fc}\rho_f F_{wsf} \qquad (4.296)$$

where ρ_f is the field wire resistance per foot, found from the wire tables for the known wire gauge, and F_{wsf} is the winding stretch factor, usually 1.02 to 1.10.

The net field resistance is the series combination of two field coils, in ohms:

$$R_f = 2R_{fc} \tag{4.297}$$

Field End-Turn Length—Method 2. As with the armature, this method assumes that you already have a motor and know R_f, T_{pcf}, F_{wsf}, and ρ_f (from wire tables). From this, L_{metf} can be calculated and then used for calculations for other motors with the same lamination.

The following four formulas give the mean end-turn length for one end of one field turn, in inches.

$$R_{fc} = \frac{R_f}{2} \tag{4.298}$$

$$L_{fc} = \frac{R_{fc}}{\rho_f F_{wsf}} \tag{4.299}$$

$$L_{mtf} = \frac{L_{fc}}{T_{pcf}} \tag{4.300}$$

$$L_{metf} = \frac{12L_{mtf} - 2L_{stk}}{2} \tag{4.301}$$

Field End-Turn Length—Method 3. Measure what looks like the average end turn on existing motor with the same lamination.

Field (Stator) Slot Fill. The maximum field slot fill depends on the winding equipment. A full slot could be anywhere from 80 to 100 percent. These percentages are greater than for the armature because there is only one coil in the field slot and it is possible to use the slot insulation to extend the area of the slot. The field slot fill is calculated by assuming that the wires are lying side by side as if they were square. Using the outside diameter of the field wire as D_{fw} (found in the wire tables taking into account insulation thickness), the percentage of winding slot fill is

$$F_{fs} = \frac{D_{fw}^2 T_{pcf}}{A_{NSf}} \times 100 \tag{4.302}$$

Field Copper Weight. Using pound-feet from the wire tables, the weight of the field copper W_{tfc}, lb, is

$$W_{tfc} = 2 \frac{L_{fc}(\text{lb/ft})}{F_{wsf}} \tag{4.303}$$

Turns Ratio. The field-to-armature turns ratio is the ratio of the total number of field turns to the total number of series-connected turns in one path of the armature.

$$\psi_{fa} = \frac{PT_{pcf}}{(N_{ca}/4) \, 2T_{pca}} \tag{4.304}$$

where P is the number of poles and N_{ca} is the number of active coils in the armature.

This ratio represents the mmf produced by the field in relation to the mmf produced by the armature. If the ratio is too high, the field produces strong mmf, which induces a higher voltage in the commutating coil. This high voltage causes arcing problems during commutation. If the ratio is too low, the armature mmf weakens the pole too much and a loss of torque results. A good range for this ratio is 1.00 to 1.30.

4.2.7 Magnet Circuit

This subsection derives a curve for the relationship of flux across the air gap versus the related mmf drops. This curve tells you how many ampere-turns are necessary to produce a specified air gap flux. Special consideration is taken to account for saturation and for the air gap.

The *air gap* is the area between the pole on the stator and the tooth tips on the armature. The air gap is not uniform because of the spaces between the tooth tips and because the pole is not perfectly round. Figure 4.61 shows that the air gap is larger at the pole tips than at the center.

First, take into account the longer air gap at the pole tips. On the manufacturer's drawing, the pole arc changes to a straight line at a specified angle θ_{ref}, which is usually about 45°. When measuring from the center of the armature, at $90° - \theta_{ref}$, the air gap starts to get longer and the length continues to increase out to the pole tip. (*Longer* refers to the magnetic length.) The average length of the air gap over this part of the arc is, in inches:

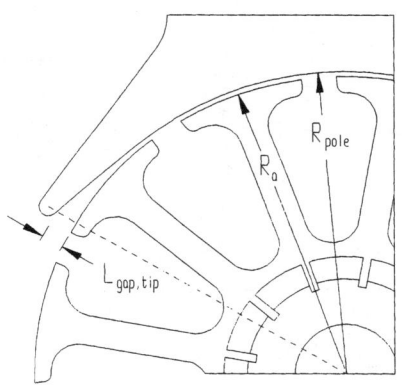

FIGURE 4.61 Nonuniform air gap.

$$L_{gap,tip} = \frac{R_{pole} + (L_{ptc} - R_{pt})}{2} - R_a \quad (4.305)$$

The next formula gives an effective mechanical air gap length by using a weighted average over half the pole arc. The effective mechanical air gap length, in inches, is

$$L_{gap,eff,mech} = \frac{[(\theta_{pole}/2) - (90° - \theta_{ref})]L_{gap,tip} + (90° - \theta_{ref})(R_{pole} - R_a)}{\theta_{pole}/2} \quad (4.306)$$

The magnetic length must account for the spaces between the armature teeth. Carter's coefficient K_C accounts for the flux that fringes from the pole to the sides of the armature tooth tips (Fig. 4.62). Carter's coefficient can be determined by first using an available graph of the width of armature slot along the circumference divided by the effective mechanical air gap length versus σ_C, where σ_C is an arbitrary variable name. Using σ_C:

$$K_C = \frac{W_{att} + W_{asc}}{W_{att} + W_{asc}(1 - \sigma_C)} \quad (4.307)$$

Another method for determining Carter's coefficient (with no graph available) is by the following two formulas:

$$\lambda = \frac{2}{\pi}\left[\tan_r^{-1}\left(\frac{W_{asc}}{2L_{gap,eff,mech}}\right) - \frac{L_{gap,eff,mech}}{W_{asc}}\ln\left(1 + \frac{W_{asc}^2}{4L_{gap,eff,mech}^2}\right)\right] \quad (4.308)$$

$$K_C = \frac{t_p}{t_p - \lambda W_{asc}} \quad (4.309)$$

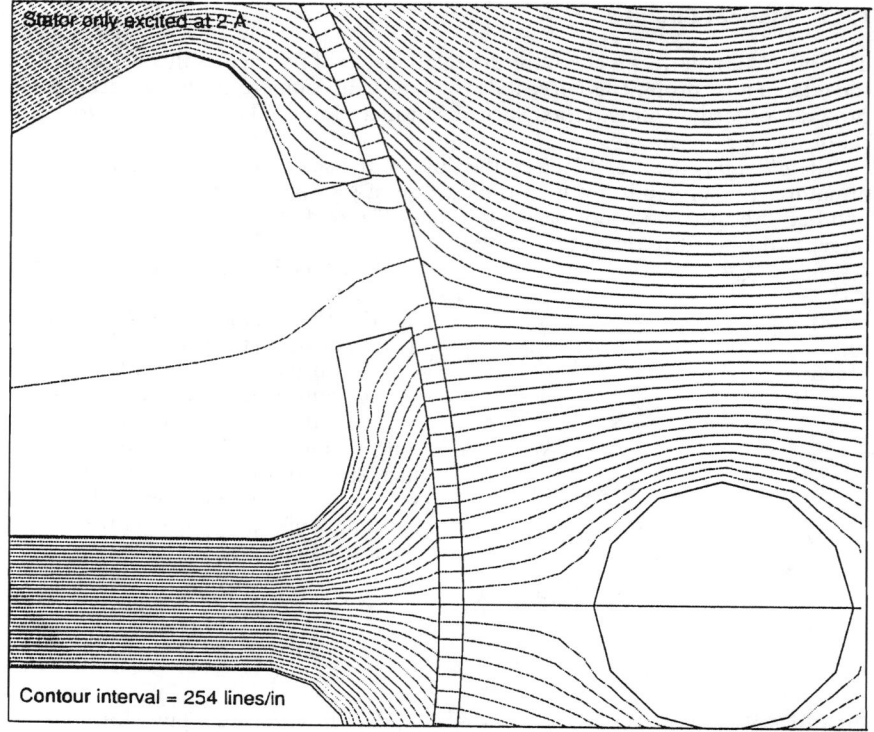

FIGURE 4.62 Fringing of flux at air gap.

Effective magnetic air gap length, in inches:

$$L_{gap} = L_{gap,eff,mech} K_C \qquad (4.310)$$

The magnetic width of the air gap is the length of the arc along the center of the magnetic air gap, in inches:

$$W_{gap} = \left[R_{pole} - \left(\frac{R_{pole} - R_a}{2} \right) \right] \theta_{pole}(0.017453) \qquad (4.311)$$

Magnetic area of the air gap, in square inches:

$$A_{gap} = W_{gap} L_{stke} \qquad (4.312)$$

Figure 4.63 shows the analogous electrical circuit for the magnetic circuit of a universal motor. Each resistor is associated with an mmf drop, similar to an emf (or voltage) drop in an electrical circuit. The resistor in the middle of the armature slot represents an mmf drop that occurs only at high saturation. The Trickey factor K_{tr} accounts for this mmf drop and is discussed later.

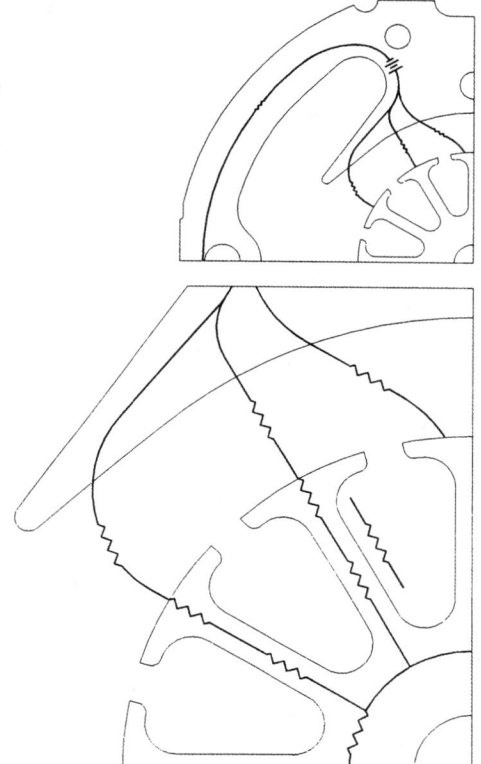

FIGURE 4.63 Schematic of magnetic circuits.

Plot of Flux per Pole versus Ampere-Turns Excitation. Flux is required for an electric motor to produce torque. In order to produce flux, mmf must be supplied. Each combination of armature and stator laminations will produce a different amount of flux for a given supplied mmf. A curve of air gap flux (in kilolines ϕ_{gap}) versus mmf in ampere-turns of excitation can be determined by determining the sum of all mmf drops across the magnetic circuit (lamination set) for each of several values of flux. In order to determine the mmf drop across a specific area, the flux density must be determined. From the *BH* curve for the material being used, the mmf drop in ampere-turns can be determined for the specific area.

For a given ϕ_{gap}, klines, the flux densities for the various paths are as follows: B_{gap}, kline/in², in the air gap:

$$B_{gap} = \frac{\phi_{gap}}{A_{gap}} \tag{4.313}$$

B_{att}, kline/in², in the armature tooth tips:

$$B_{att} = \frac{\phi_{gap}}{A_{T,attm}} \tag{4.314}$$

B_{at}, kline/in², in the armature teeth:

$$B_{at} = \frac{\phi_{gap}}{A_{T,atm}} \tag{4.315}$$

B_{ay}, kline/in², in the armature yoke:

$$B_{ay} = \frac{\phi_{gap}}{A_{aym}} \tag{4.316}$$

For the following calculations, assume that the flux is in the return path in the stator.

B_{SY}, kline/in², in the stator yoke:

$$B_{SY} = \frac{\phi_{gap}}{A_{SY}} \tag{4.317}$$

B_{S1}, kline/in², in stator section 1:

$$B_{S1} = \frac{\phi_{gap}}{A_{S1}} \tag{4.318}$$

B_{S2}, kline/in², in stator section 2:

$$B_{S2} = \frac{\phi_{gap}}{A_{S2}} \tag{4.319}$$

B_{S3}, kline/in², in stator section 3:

$$B_{S3} = \frac{\phi_{gap}}{A_{S3}} \tag{4.320}$$

B_{S4}, kline/in², in stator section 4:

$$B_{S4} = \frac{\phi_{gap}}{A_{S4}} \tag{4.321}$$

B_{S5}, kline/in², in stator section 5:

$$B_{S5} = \frac{\phi_{gap}}{A_{S5}} \tag{4.322}$$

B_{pole}, kline/in², in the pole:

$$B_{pole} = \frac{\phi_{gap}}{A_{PM}} \tag{4.323}$$

In order to calculate the total mmf drops in the magnetic circuit, you need to read the magnetic field intensity H, (A · turn)/in, from the BH curve for each section. The mmf drop for the section is the intensity times the length of the section.

Field intensity H_{gap}, (A · turn)/in, in the air gap:

$$H_{gap} = \frac{B_{gap}}{3.19} \tag{4.324}$$

MMF drop across the air gap, in ampere-turns:

$$\mathscr{F}_{gap} = H_{gap} L_{gap} \tag{4.325}$$

For the following equations, the value of H is read from the graph for each different flux density B.

$$\mathscr{F}_{att} = H_{att} L_{attm} \tag{4.326}$$

Special consideration must be taken to account for the fact that when the flux density reaches saturation in the armature teeth, the flux will set up on the sides of the teeth. At about 110 kline/in^2, the armature teeth saturate, and any additional flux will exist along the sides of the teeth as well as in the teeth themselves (Fig. 4.64). The mmf drop due to the flux on the sides of the tooth can be considered by adding it to the mmf drop in the tooth itself. Also, as B_{at} becomes greater than 110,000 line/in^2, the initial calculation of B_{at} becomes $B_{at,apparent}$.

$$B_{at} > 110{,}000 \text{ line/in}^2 = B_{at,apparent} = B_{ata} \tag{4.327}$$

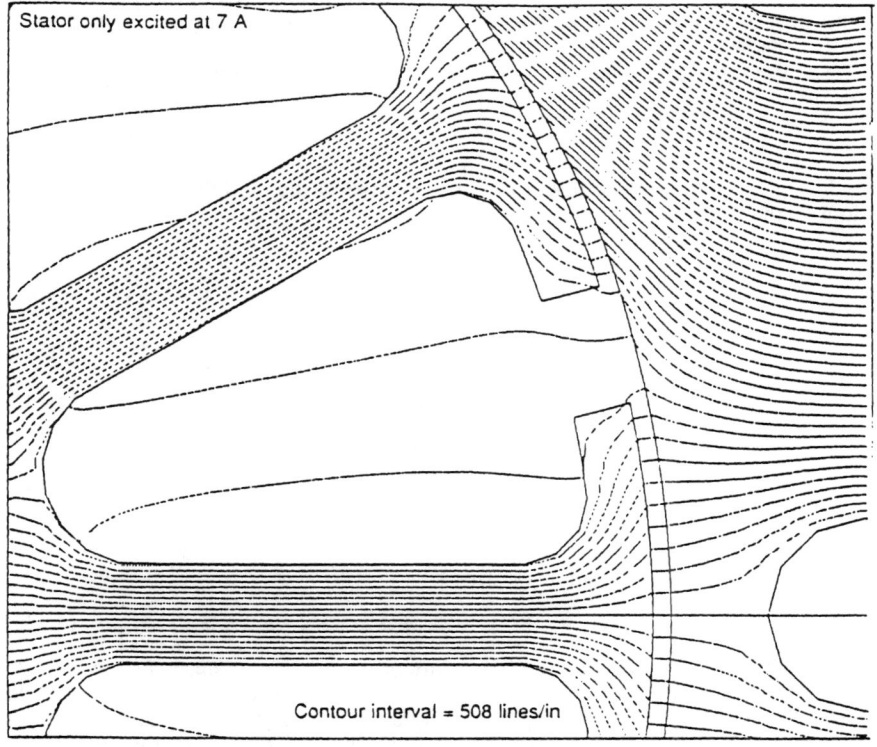

FIGURE 4.64 Flux in air beside teeth (Trickey factor).

The Trickey factor K_{tr} must be determined:

$$K_{tr} = \frac{t_p}{W_{at}F_{stk}} - 1 \qquad (4.328)$$

The Trickey factor is the number of tooth widths in the air beside the teeth that the flux occupies at saturation. Note that it accounts for the air between the laminations by dividing by the stacking factor.

A little extra effort is required to determine the field intensity H for a saturated tooth. Using the BH curve for the material and the $B_{at,apparent}$ which is over 110 kline/in^2, determine the apparent field intensity H_{temp} in the armature teeth. A new apparent flux density is B_{atal}.

$$B_{ata} = \mu H_{temp} + 3.19 K_{tr} H_{temp} \qquad (4.329)$$

where $\quad \mu = B_{ata}/H_{temp}$

Therefore $\quad B_{ata} = \mu H_{temp}$

H_{temp} is altered until $B_{atal} = B_{ata}$. B_{at} then equals the new μH_{temp}.

B_{at} can also be derived graphically. See P. H. Trickey, vol. I-3-109 (1971). Using this new flux density, the new value for H_{at} can be determined from the curve for the mmf drop in ampere-turns across the armature teeth for an apparent flux density >110 kline/in^2.

The last mmf drop to calculate for the armature is that of the armature yoke $\mathscr{F}_{ay}$, A · turn:

$$\mathscr{F}_{ay} = H_{ay}L_{aym} \qquad (4.330)$$

Using the same method as before for the values of the mmf drops across the sections of the stator:

$$\mathscr{F}_{sy} = H_{SY}L_{SY} \qquad (4.331)$$

$$\mathscr{F}_{S1} = H_{S1}L_{S1} \qquad (4.332)$$

$$\mathscr{F}_{S2} = H_{S2}L_{S2} \qquad (4.333)$$

$$\mathscr{F}_{S3} = H_{S3}L_{S3} \qquad (4.334)$$

$$\mathscr{F}_{S4} = H_{S4}L_{S4} \qquad (4.335)$$

$$\mathscr{F}_{S5} = H_{S5}L_{S5} \qquad (4.336)$$

$$\mathscr{F}_{pole} = H_{pole}L_{pole} \qquad (4.337)$$

Sum of the mmf drops, in ampere-turns:

$$\mathscr{F}_{total} = \mathscr{F}_{gap} + \mathscr{F}_{att} + \mathscr{F}_{at} + \mathscr{F}_{ay} + \mathscr{F}_{SY} + \mathscr{F}_{S1} + \mathscr{F}_{S2} + \mathscr{F}_{S3} + \mathscr{F}_{S4} + \mathscr{F}_{S5} + \mathscr{F}_{pole} \qquad (4.338)$$

The value $\mathscr{F}_{total}$ represents the field ampere turns per pole necessary to provide the specified air gap flux ϕ_{gap}. This means that the product of the number of field turns T_{pcf} and the field current I_{line} must equal $\mathscr{F}_{total}$. A curve of air gap flux versus ampere-turns of excitation can be plotted by varying ϕ_{gap}. Remember that each curve is specific to the lamination set and that it does not account for any loss in mmf due to armature reaction.

DIRECT-CURRENT MOTORS

Reactances. Reactances are largely a function of lamination geometry. In order to calculate reactances, the slot dimensions and lengths must be known. The following subsections examine methods for calculating field and armature reactances.

Field Leakage Reactance. The first formula is from Puchstein (1961) for field leakage reactance, in ohms.

$$X_{f,\text{total}} = 2\pi f \phi_c T_{\text{pcf}}^2 L_{\text{mtf}} \times 10^{-8} \tag{4.339}$$

where L_{mtf} is the mean turn length in inches of the field and ϕ_c is the number of lines per ampere-conductor per inch of mean turn length. A good value to use for ϕ_c is 6.

The next formula is from Trickey for field leakage reactance.

$$X_{f/\text{pole}} = (2\pi f(2T_{\text{pcf}})^2 \times 10^{-8}) \left[\frac{3.19 L_{\text{stke}}}{2P} (K_{\text{sf1}} + K_{\text{sf2}}) + \frac{S_{\text{fc}}}{P} \right] \tag{4.340}$$

where t_{pcf} = field turns per pole
K_{sf1} = slot constant for part containing coil
K_{sf2} = slot constant for pole tip flux
S_{fc} = span of field coil, in

You also need to know the following in order to calculate armature reactances:

$\psi_p = \theta_{\text{pole}}/\text{pole pitch} = \theta_{\text{pole}}/180°$
N_{as} = number of armature slots = N_{at}
Z_s = armature conductors in a slot = $T_{\text{pca}}(\text{CPS})$
Z_a = total armature conductors = $T_{\text{pca}}(\text{CPS})N_{\text{at}}$
f = line-current frequency, Hz

The span of the field coil is the arc length from the center of the field winding in the slot to the same position in the other slot, in inches.

$$S_{\text{fc}} = (\sqrt{Y_{\text{ssc}}^2 + X_{\text{ssc}}^2}) \frac{\pi}{180} 2 \left(\frac{\theta_{\text{ssc}} + \theta_{\text{ptc}}}{2} \right) \tag{4.341}$$

The slot constants K_{sf1} and K_{sf2} are shown here.

$$K_{\text{sf1}} = \frac{d_3}{e} + F \tag{4.342}$$

$$K_{\text{sf2}} = 1.5 \left(\frac{AC}{BB'} \right) \tag{4.343}$$

The values for $a_1, a_2, b, d_1, d_2, d_3, e, AC$, and BB that are used to calculate F are shown in Fig. 4.65.

$$F = 0.28 - 0.14 \left(\frac{a_1}{a_2} \right) + \left[\frac{b/a_2}{2.87(a_1/a_2) + 0.008} \right] + 0.08 \tag{4.344}$$

Trickey's total field reactance must be multiplied by the number of poles.

$$X_{f,\text{total}} = 2X_{f/\text{pole}} \tag{4.345}$$

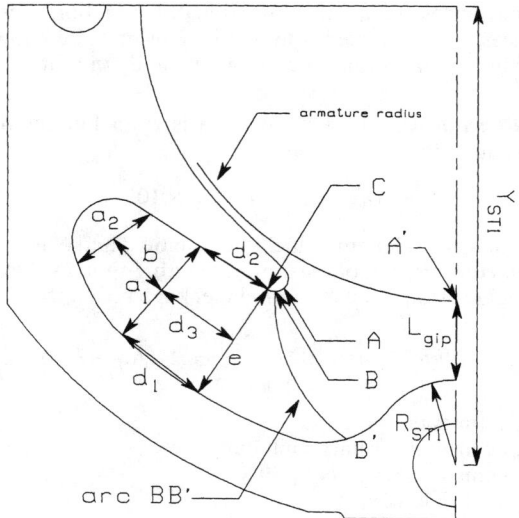

FIGURE 4.65 Dimensions for stator reactances.

Using either of the preceding two methods for field leakage reactance, find the total reactance.

$$X_{\text{total}} = X_{\text{ar},T} + X_{f\text{total}} \tag{4.346}$$

Note: Trickey's method does not consider end-turn leakage.

Armature Reactances. In order to calculate reactances, the slot constants of the armature slots must be determined. Figure 4.66 shows the dimensions which are needed to calculate slot constants. You also must use the actual value for L_{stke}. These dimensions were calculated previously.

Armature Leakage Reactances Under a Pole, Ω

$$X_{\text{apf/path}} = (10.5) \frac{2f}{PL_{\text{gap}}} \left(\frac{Z_a}{2P}\right)^2 D_a L_{\text{stke}} \psi_p^3 \times 10^{-8} \tag{4.347}$$

For inductors in parallel:

$$X_{\text{apf},T} = \frac{X_{\text{apf/path}}}{2} \tag{4.348}$$

Armature Slot Leakage Reactance, Ω

$$X_{\text{as}} = \frac{20 Z_s^2 L_{\text{stke}} f N_{\text{as}} \psi_p}{P^2 \times 10^8} K_{\text{sa}} \tag{4.349}$$

where K_{sa} = slot constant for round-bottom armature slots. Use the following formulas.

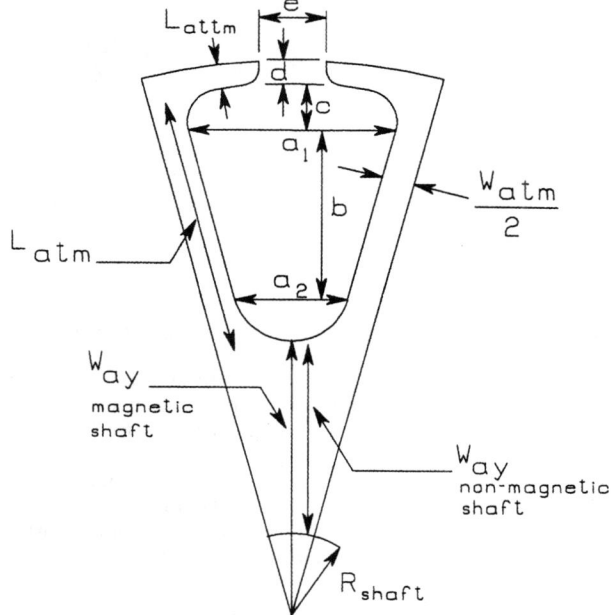

FIGURE 4.66 Armature magnetic paths and reactance dimensions.

$$F = 0.28 - 0.14 \left(\frac{a_1}{a_2}\right) + \left[\frac{b/a_2}{2.87(a_1/a_2) + 0.08}\right] + 0.08 \quad (4.344)$$

$$K_{sa} = F + \frac{d}{e} + \frac{2c}{e + a_1} \quad (4.350)$$

where $a_1, a_2, b, c, d,$ and e are dimensions of the armature slot found previously and shown in Fig. 4.66.

Armature End Leakage Reactance, Ω

$$X_{ae} = \frac{3.19 f (2 L_{\text{meta}}) Z_a^2 N_{as}}{P^2 10^8} \quad (4.351)$$

Armature Interpolar Leakage Reactance, Ω

$$X_{\text{gip/path}} = 10.5 \frac{f L_{\text{stke}}}{L_{\text{gip}}} \left(\frac{Z_a}{2P}\right)^2 (D_a + L_{\text{gip}})(1 - \psi_p^3) \times 10^{-8} \quad (4.352)$$

$$X_{\text{gip},T} = \frac{X_{\text{gip/path}}}{2} \quad (4.353)$$

where L_{gip} = length of gap between armature OD and ID of stator yoke.

Armature Tooth Tip Leakage Reactance, Ω

$$X_{att} = \frac{14.7(Z_s)^2 L_{stke} f N_{as}(1-\psi_p)}{P^2 10^8} \log_{10}\left[1 + \frac{\pi(t_p - W_{ast})}{2W_{ast}}\right] \quad (4.354)$$

Total Armature Reactance, Ω

$$X_{ar,T} = X_{apf,T} + X_{as} + X_{ae} + X_{gip,T} + X_{att} \quad (4.355)$$

4.3 COMMUTATION*

4.3.1 A Practical Approach to Commutation and Brush Selection

There are many factors that affect commutation. Among these are inductance, brush friction, surface speed of the commutator, brush material, commutator finish, brush pressure, contact film, induced emf, current density, armature balance, and concentricity of the commutator to the shaft.

Perfect *linear commutation,* as shown in Fig. 4.67, would occur when the current in the coil under commutation goes from a peak $+I$ through 0 to a peak $-I$ as the brush position changes from full contact on one bar to complete shorting of one coil to full contact on the next bar. This condition would provide a relatively constant current density in the brush.

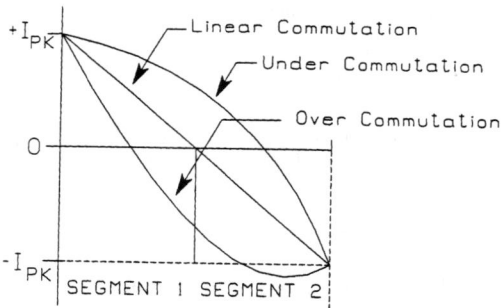

FIGURE 4.67 Overcommutation, undercommutation, and linear commutation.

Undercommutation (Fig. 4.67) occurs when the current stays at $+I_{PK}$ for most of brush travel and is abruptly changed to $-I_{PK}$ as the brush leaves the segment.

This effect is a result of the winding inductance trying to maintain the current in the coil during commutation. The current density in the brush varies from low to very high as the brush moves from segment 1 to segment 2.

Overcommutation (Fig. 4.67) occurs when the emf induced in the coil during commutation causes the current reversal to take place before the brush contacts the adjacent segment. In this case the current density in the brush is very high initially and drops to very low.

---

*Section contributed by Alan W. Yeadon and William H. Yeadon, Yeadon Engineering Services, PC.

In the cases of both undercommutation and overcommutation, the high brush current densities cause excessive heating and wear.

Considering the list of things that affect commutation, it is obvious that perfect commutation is impossible. However, there are some things to keep in mind when designing a motor which will tend to minimize commutation problems.

Brush pressure is important. Figure 4.68 shows relative brush and commutator wear verses brush pressure. At low pressures, the wear is mostly electrical in nature because the poor contact results in high-resistance spots with localized heating and arcing. At high pressures, the wear is mostly mechanical and is due to the parts in effect grinding themselves away because of friction losses. For fractional-horsepower motors using commonly available brush and commutator materials, the "ideal" range is from 3 to 9 lb/in^2. A good approach to this is to select 6 lb/in^2 as a nominal value and design the brush rigging and spring system to maintain as close to 6 lb/in^2 over the usable brush length as possible. One can use a constant force spring or, if this is not practical, try to select a spring that starts out with a pressure of 7 to 8 lb/in^2 and has a pressure of 4 to 5 lb/in^2 at the end of the usable brush length. In applications where the brushes and commutators are submerged in a liquid, the pressure needs to be increased to overcome hydroplaning. A good range is 10 to 16 lb/in^2, with 12 to 14 lb/in^2 being preferred over the usable brush length.

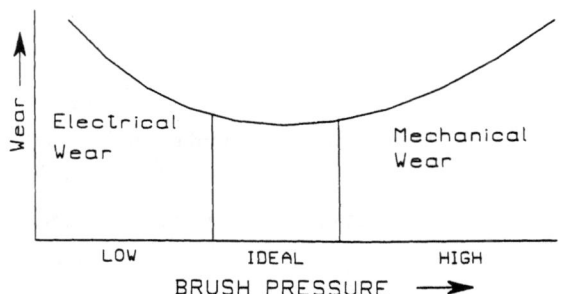

FIGURE 4.68 Brush pressure versus wear.

The next thing to do is to ensure that the machine is commutating in the neutral zone. The neutral zone W_n is defined as the mechanical arc distance between the pole tips.

$$W_n = \left(1 - \frac{\theta_{pole} P}{360}\right) \frac{\pi D_a}{P} \tag{4.356}$$

where θ_{pole} = pole angle, degrees
 D_a = outside diameter of amature
 P = number of poles

The magnetic neutral zone is located at some angle shifted from the mechanical neutral position. The angle of shift is a function of the armature mmf. This angle can be estimated from experimental data on similar motors. Try 15° to 20° to start. A better way is to make a finite element analysis and estimate the angle from the flux plot with the motor at full load current. Next, consult a manufacturer's catalog and select a standard commutator that is about half the diameter at the armature lamination being utilized. Select a brush material that the manufacturer recommends for the type of application in which the commutation system will be used. Design the brush thickness so that it will cover about one bar pitch of the commutator. Set the brush width to cover as much of the commutator bar as possible, taking into consideration armature end play and the required mechanical and electrical clearances.

Calculate the commutation zone W_c. It is defined as the circumferential distance the coil under commutation travels before current reversal is complete.

$$W_c = \left[\frac{(t_{br} - t_{cin})C_b}{\pi D_c} + \frac{C_b}{N_{at}} + \left(\frac{C_b}{P} - \frac{C_b}{N_{at}} S_{ac} \right) - \frac{a}{P} \right] \frac{\pi D_a}{C_b} \qquad (4.357)$$

where t_{br} = brush thickness
t_{cin} = insulation thickness between bars
D_a = outside diameter of armature
P = number of poles
a = number of parallel armature paths
C_b = number of commutator bars
S_{ac} = number of teeth spanned by an armature coil
N_{at} = number of armature teeth
D_c = finished outside diameter of commutator

Then calculate the commutation–to–neutral zone ratio ψ_{cn}.

$$\psi_{cn} = \frac{W_c}{W_n} \qquad (4.358)$$

The result should be less than 1 for good commutation. According to Puchstein (1961), the optimum value is 0.667, but this seems practical only in larger machines. The precise calculation for this ratio is shown in Sec. 4.3.2.

Calculate the full load current of the motor. Then calculate the current density in the brush.

$$\gamma_{br} = \frac{I_{line}}{t_{br} W_{br}} \qquad (4.359)$$

This value should be less than the value recommended by the brush supplier for the type of brush selected.

If the current density is not acceptable, select a brush material with a higher rating, or increase the brush thickness until it is acceptable. If the brush thickness is increased, the commutation–to–neutral zone ratio must be recalculated. If it is now greater than 1, increase the commutator diameter until it is less than 1. Note that changing the brush area will change the brush pressure. Because of this, it will be necessary to recheck the spring to ensure that the pressure remains in the preferred range.

Other tips on improving commutation include reducing the motor speed to as small a value as possible and keeping the coil inductances low. When designing the brush rigging, keep 0.002 to 0.003 in total side-to-side clearance and top-to-bottom clearance between the brush and its guide. Less clearance may cause binding problems and may not allow the brush to remain in contact with the commutator. Try to extend the brush guide to within 0.020 in of the commutator surface if practical. This will minimize brush movement, which will improve contact and reduce arcing. Balance the armature to its ISO 1941 requirement. Keep the shaft straight and the bearings lined up and concentric to the field core.

Brush Dust and Slot Packing. As the commutator and brushes move relative to each other, they wear, and fine particles of dust are released into the motor. Depending on the brush materials used, this dust may contain copper and silver, but it is mostly carbon. In any event, the material is conductive.

As the particles are dislodged from the brushes and the commutator, they are moved to the edge of the commutator slots and forced down into them. If the surface speed of the commutator is sufficient, the tangential forces will throw the particles from the slots out into the surrounding area. This is the desired effect.

If the surface speed is too slow, or a contaminant, such as oil, is present, the particles will be forced into the slots until the slots are packed with the material. Since the material is conductive, the effect of this packing is to short the bars together, creating one or more shorted coils. This adversely affects motor performance and life. To avoid this situation, keep oils away from the commutator by using slingers and antiwetting agents. Run the motor at sufficient speed to throw the particles from the slots. The actual armature minimum speed required depends on the mass of the particles and the commutator diameter, but an old rule of thumb is to keep the speed above 1500 rpm.

Excessive dust buildup can cause other problems as well, such as shorting the brush rigging to the end frames. This can be minimized by designing the brush rigging to trap the dust or by isolating the surrounding area with insulating materials.

4.3.2 Commutation System Design

A lot of effort goes into designing a motor that has good commutation. Good commutation saves on brush wear and reduces sparking. In order to calculate commutation parameters, we need to know the following (refer to Figs. 4.69 and 4.70):

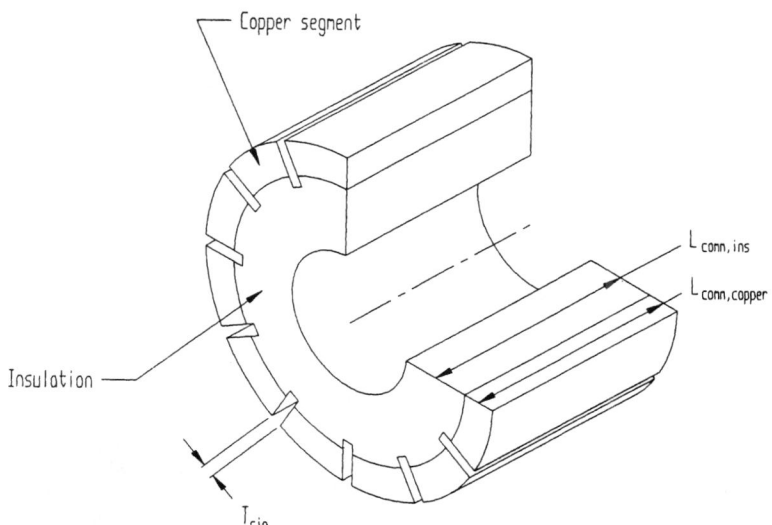

FIGURE 4.69 Commutator construction.

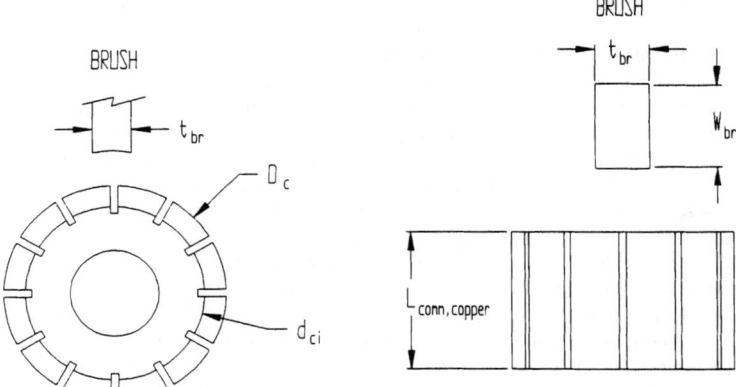

FIGURE 4.70 Round commutator dimensions.

C_b = number of bars on the commutator
D_c = outside diameter of commutator
t_{br} = brush thickness (thickness which is tangential to the commutator)
T_{cin} = insulation thickness between commutator bars
a = number of current paths in armature
P = number of poles
N_{at} = number of armature teeth
S_{ac} = number of teeth spanned by a coil
D_c = commutator diameter

Width of the commutation zone W_c and the neutral zone W_n:

$$W_c = \left[\frac{(t_{br} - T_{cin})C_b}{\pi D_c} + \frac{C_b}{N_{at}} + \left(\frac{C_b}{P} - \frac{C_b}{N_{at}}\right)S_{ac} - \frac{a}{P}\right]\frac{\pi D_a}{C_b} \quad (4.360)$$

$$W_n = \left(1 - \frac{\theta_{pole}P}{360}\right)\frac{\pi D_a}{P} \quad (4.361)$$

The figure of merit for the commutation–to–neutral zone ratio should be less than 1.0 for good commutation.

$$\psi_{cn} = \frac{W_c}{W_n} \quad (4.362)$$

Flat commutators are commonly used where long brushes are required for long motor life. Flat commutators allow the length of the brush to extend the motor length instead of the motor diameter. Figure 4.71 shows a flat commutator and the dimensions used to calculate commutation parameters. The following dimensions (which can be substituted into the commutation parameters just given) need to be determined for the commutator and corresponding brush:

DIRECT-CURRENT MOTORS

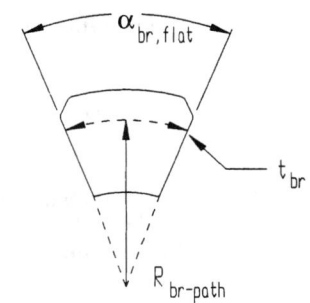

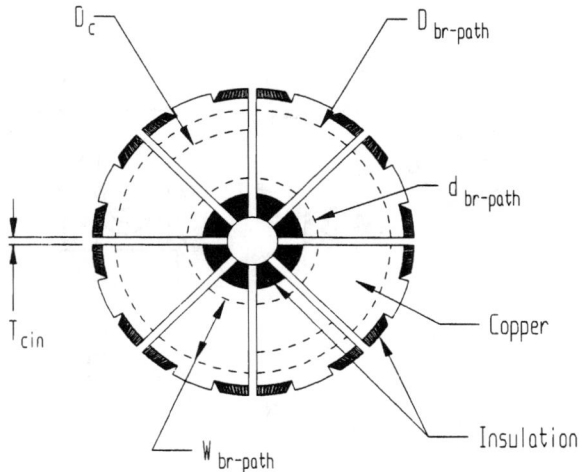

FIGURE 4.71 Flat commutator dimensions.

$\alpha_{br,flat}$ = angle which the brush encompasses
$D_{br,path}$ = outside diameter of the path the brush travels on the commutator
$d_{br,path}$ = inside diameter of the path the brush travels on the commutator
t_{cin} = thickness of the insulation between commutator bars

The following formulas calculate the values to be used in the commutation zone formula.

Geometric mean diameter of the brush path, to replace the diameter of the commutator, in inches:

$$D_C = 0.707 \sqrt{\left(\frac{D_{br,path}}{2}\right)^2 + \left(\frac{d_{br,path}}{2}\right)^2} \qquad (4.363)$$

Radius of the brush path, in inches:

$$R_{br,path} = \frac{D_C}{2} \qquad (4.364)$$

The thickness of the brush for a flat commutator is the arc distance across the brush at the radius of the brush path, in inches:

$$t_{br} = \left(\frac{\pi}{180}\right)(\alpha_{br,flat})(R_{br,path}) \tag{4.365}$$

Commutation is the process of switching the direction of current in a coil. If the direction of the coil were not switched, the motor would not turn. Figure 4.72 shows a current-carrying coil rotating in a magnetic field. The reaction of the flux from the magnets (or from the field poles in the universal motor) and the flux from the armature coil sides causes the coil to rotate until a steady-state position is reached. The direction of current in the coil needs to be reversed in order to keep the coil turning.

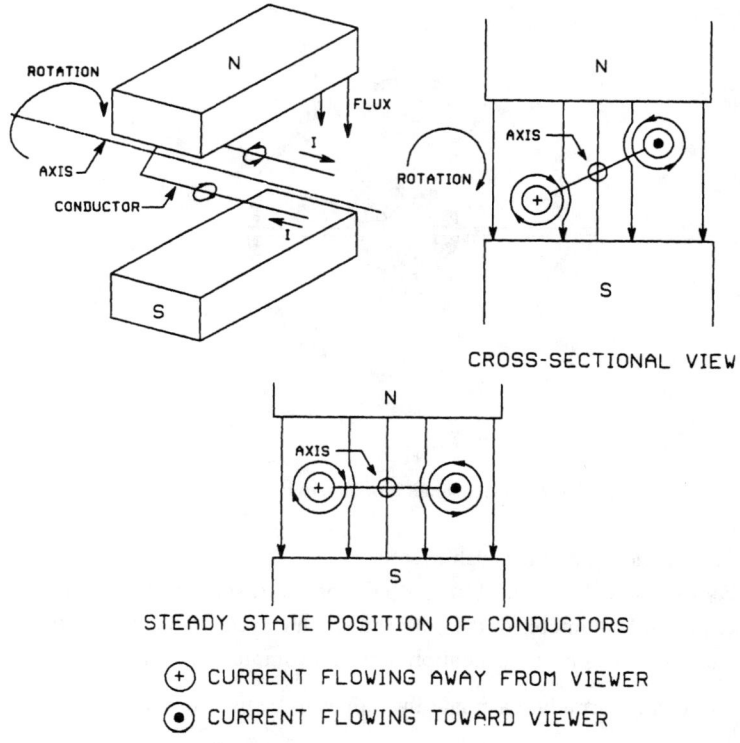

FIGURE 4.72 Coil rotating in magnetic field.

Figure 4.73 shows the magnetic field resulting from the stator winding being energized.

Figure 4.74 shows a rotating armature with two coils. Coil 1 has end points on commutator bars C1 and C1 which are about to be commutated by Brush 1. The current in the coils is in a steady state at this point. Figure 4.75 shows the brushes immediately before they will short out their respective coils. Figure 4.76 shows the coils being short-circuited by the brushes. The current direction in the coils is in the pro-

DIRECT-CURRENT MOTORS

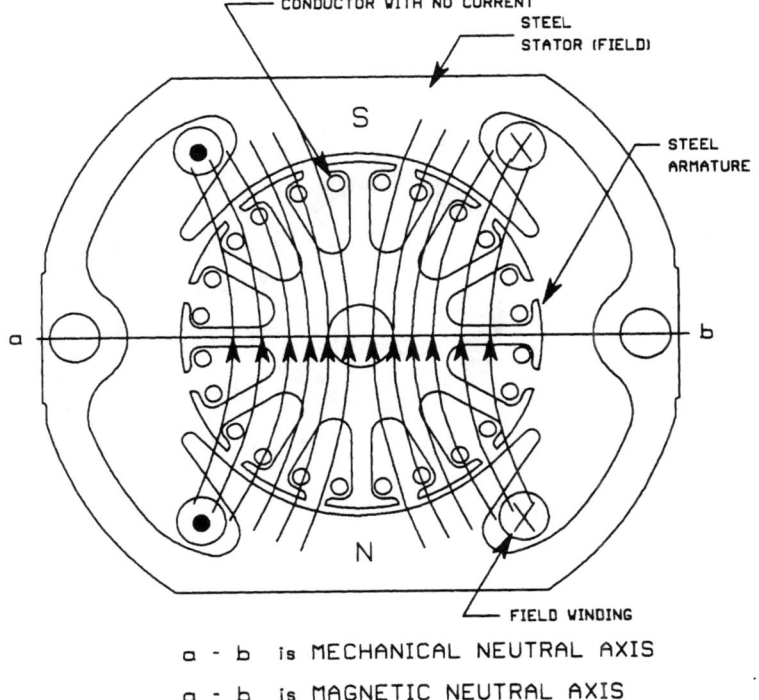

a - b is MECHANICAL NEUTRAL AXIS
a - b is MAGNETIC NEUTRAL AXIS

⊗ - DENOTES CURRENT FLOWING AWAY FROM VIEWER
⊙ - DENOTES CURRENT FLOWING TOWARD VIEWER

FIGURE 4.73 Magnetic field with only stator energized.

cess of being switched. Note that the coil sides (which are into the paper) are now moving parallel to the magnetic flux lines shown in Fig. 4.73. This is beneficial because a short-circuited coil moving across a magnetic field will cause voltages to be induced across the ends of the coil (which are the commutator bars). Figure 4.77 shows the position of the brushes after the coil has been commutated. The current in this coil is now in the opposite direction in a steady state.

Example. A universal motor with a C-shaped lamination is produced to run at 240 V ac. The motor currently has the following attributes:

$\theta_{pole} = 100°$
$D_a = 1.238$
$t_{br} = 0.250$
$S_{ac} = 5$ teeth
$t_{cin} = 0.015$
$D_c = 0.89$
$C_b = 24$

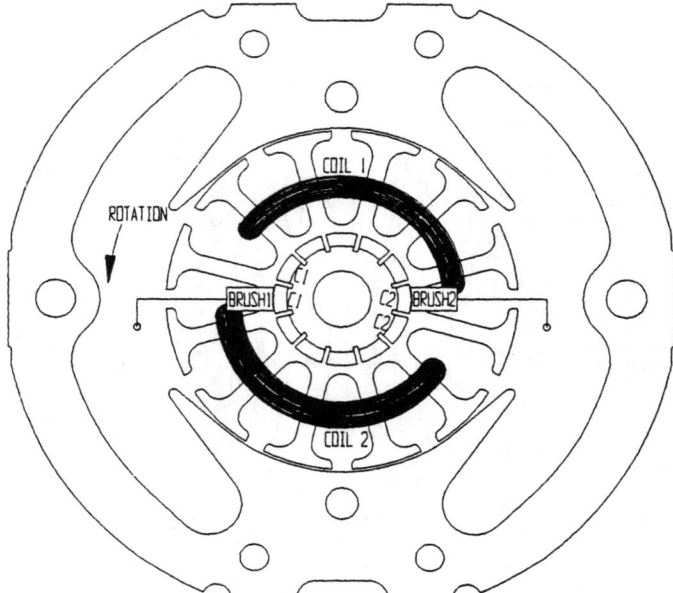

FIGURE 4.74 Armature coils with steady-state current.

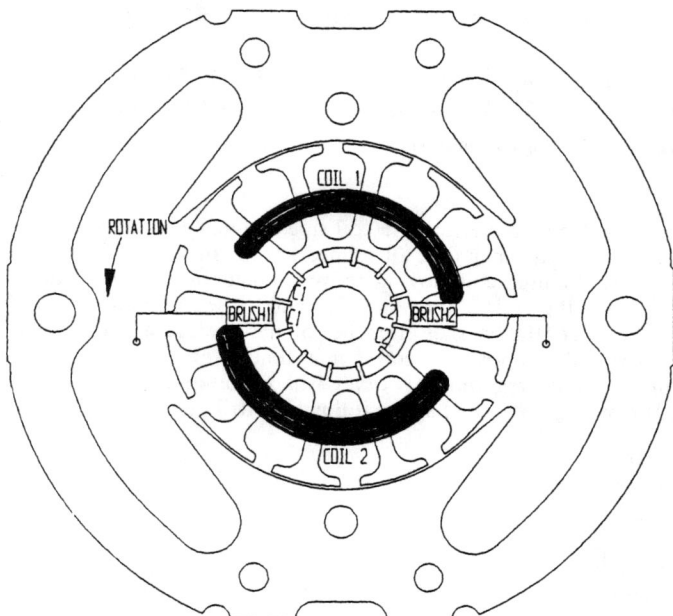

FIGURE 4.75 Coils just about to be commutated.

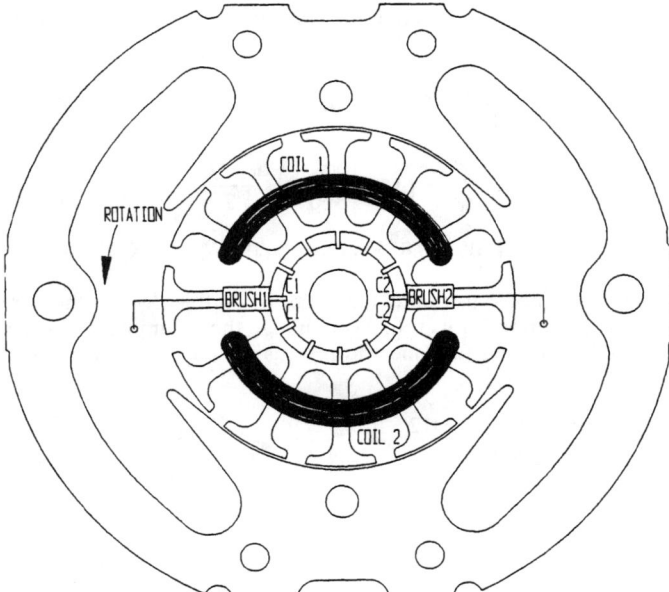

FIGURE 4.76 Coils being commutated (shorted by brush).

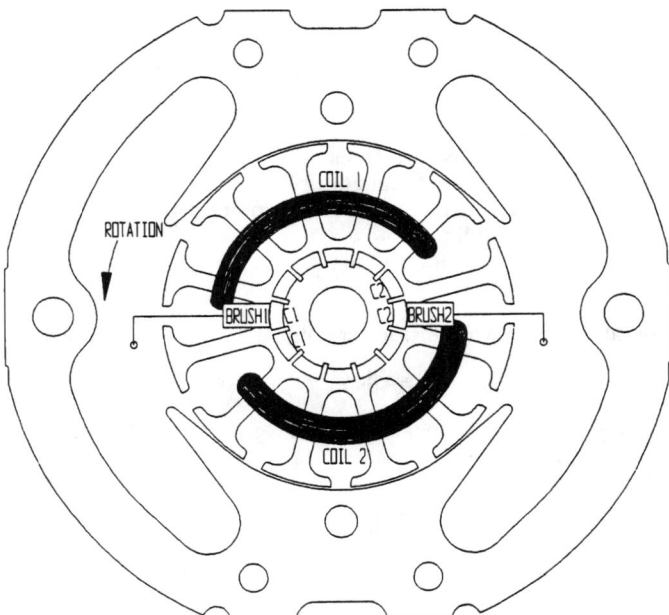

FIGURE 4.77 Coils with new current direction in steady state.

For this motor, the widths of the commutation zone and the neutral zone are as follows:

$$W_c = \left[\frac{(0.250 - 0.015)24}{\pi 0.890} + \frac{24}{12} + \left(\frac{24}{2} - \frac{24}{12}(5)\right) - \frac{2}{2}\right]\frac{\pi 1.238}{24} \quad (4.366)$$

$$W_n = \left(1 - \frac{100(2)}{360}\right)\frac{\pi 1.238}{2} \quad (4.367)$$

Solving shows $W_n = 0.864$, $W_c = 0.813$, and $\psi_{cn} = W_c/W_n = 0.94$. These are acceptable values.

Now, change the motor to run at 115 V ac. To do this, double the wire sizes and halve the turns to keep the slots full. Also, change the commutator to 12 bars with a diameter of 0.875. The neutral zone width remains the same, but the commutation zone becomes

$$W_c = \left[\frac{(0.250 - 0.015)12}{\pi 0.875} + \frac{12}{12} + \left(\frac{12}{2} - \frac{12}{12}(5)\right) - \frac{2}{2}\right]\frac{\pi 1.238}{12} \quad (4.368)$$

$$W_c = 0.656$$

$$\psi_{cn} = W_c$$

$$W_n = 0.759$$

Even better commutation!

Reactance voltage of commutation, in volts:

$$V_{rc} = \frac{1.33 C_b S \frac{I_L}{P_a} T_{pca}^2 (K_{st} L_{stk} + 0.2 L_{meta}) * 10^{-8}}{t_{br}/t_{com}} \quad (4.369)$$

Where S = motor speed, rpm
P_a = number of armature paths
t_{com} = thickness of one commutator bar
T_{pca} = turns per armature coil being commutated
$K_{st} = 1.0$ if S_{ac} is full pitch
$K_{st} = 0.5$ if S_{ac} is less than a full pitch
I_L = line current

In ac operation there is a transformer voltage in the armature, in volts per coil:

$$E_{ta} = 4.44 \phi_{max} f T_{pca} \times 10^{-8} \quad (4.370)$$

where ϕ_{max} is the air gap flux in peak lines at the given current. The transformer voltage does not exist in dc operation.

Short-circuit voltage:

$$E_S = 2Bv\, L_{stk} T_{pca} \times 10^{-8} \quad (4.371)$$

where

$$B \approx \frac{3.19 Z_a I_{ap}}{(L_{gip}/2)\, 2P} \quad (4.372)$$

$$I_{ap} = \frac{2I}{P_a} \tag{4.373}$$

and

$$v = S\left(\frac{\pi D_a}{60}\right) \tag{4.374}$$

where v = velocity of armature surface, in/s
Permissible commutation voltage:

$$E_{pc} = \sqrt{E_{ta}^2 + (E_S + 2.5V_{rc})^2} \tag{4.375}$$

E_{pc} should be less than or equal to 5 to 7 V. This is difficult to achieve in small high-speed motors.

4.3.3 Flashover and Ring Fire

Under steady-state conditions, the field and armature fluxes and currents are relatively constant. Commutation voltages can be calculated using the previously described formulas, and a suitable commutation system can be devised. If, however, the motor is expected to see rapid load changes or voltage reversal, the steady-state commutation system may no longer be suitable. Sudden load changes that cause the motor to rapidly accelerate or decelerate will cause transformer voltages to be induced in the windings.

There are two components to these voltages. There is a voltage et_{ae} caused by changing field distortion armature flux. It is largest near the pole tips and falls to zero under the pole. Since the commutated coil is nearest the pole tip, this voltage shows up at the brushes. If it is high enough, the voltage exceeds the insulation dielectric value between bars nearest the brush. This will result in arcing or flashover between bars. This voltage has the value

$$et_{ae} = \frac{T_{pc}}{10^8} \frac{P}{P_a} \frac{d\phi_a}{dt} \tag{4.376}$$

where T_{pc} = turns per coil
P = poles
P_a = armature paths
$d\phi_a/dt$ = time rate of change of armature flux at the pole tips

The second is a voltage et_{ap} induced in the armature coils under a main pole. This results from a change in armature mmf caused by rapidly changing armature current as the load changes. Its value is

$$et_{ap} = \frac{T_{pc}}{10^8} \frac{P}{P_a} \frac{d\phi_c}{dt} \tag{4.377}$$

where $d\phi_c/dt$ is the rate of change in flux in a coil under a pole with respect to time.

These voltages are superimposed and seen at all bars. If these voltages are high enough, the dielectric between all bars may break down, resulting in arcing or flashover between all bars. This phenomenon is sometimes referred to as *ring fire*. It should be pointed out here that the magnitude and polarity of these voltages is dependent on whether the load is increasing or decreasing and at what rate.

Flashover and ring fire may be observed on starting some motors yet have very little effect on their life because it is intermittent and not seen under normal operation.

It is obvious from the formulas that the best way to control flashover on motors with variable loads is to limit T_{pc}. This usually means more armature slots and larger commutators with more bars. For a more detailed explanation of flashover, see Puchstein (1961) and Gray (1926).

4.4 PMDC MOTOR PERFORMANCE

This section is intended to give the PMDC motor designer a method to calculate PMDC motor performance given the material magnetic and electrical properties and physical dimensions.

The basic construction of a PMDC motor is as shown in Fig. 4.78. To calculate the performance for this motor, one must predict the air gap flux and calculate the no-load speed and current and the stall torque and current. A straight line drawn between the no-load speed and the stall torque represents the *speed-torque curve* of the motor. A straight line drawn between the no-load current and the stall current represents the *current-torque performance curve*. Examples of such curves are shown in Figs. 4.79 and 4.80.

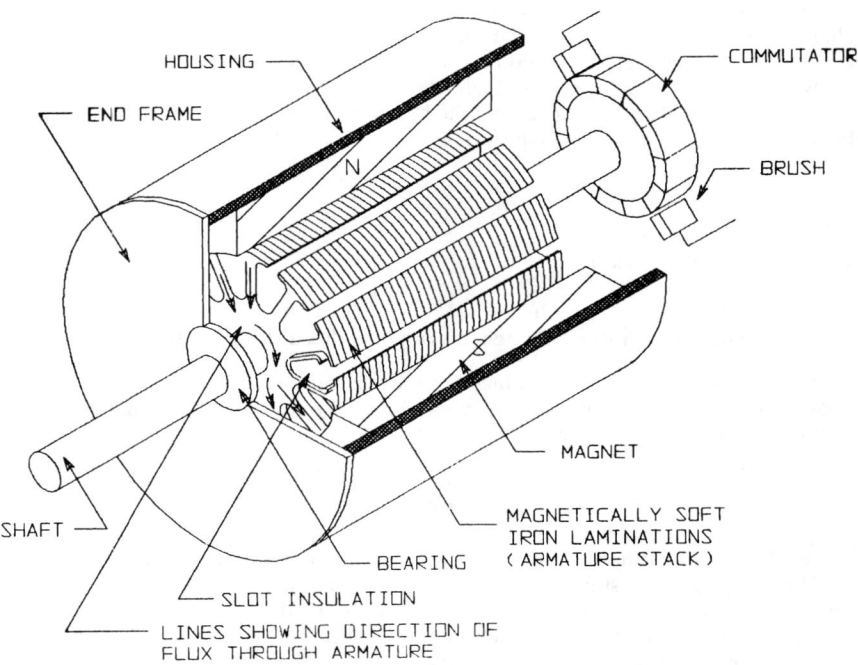

FIGURE 4.78 PMDC motor construction.

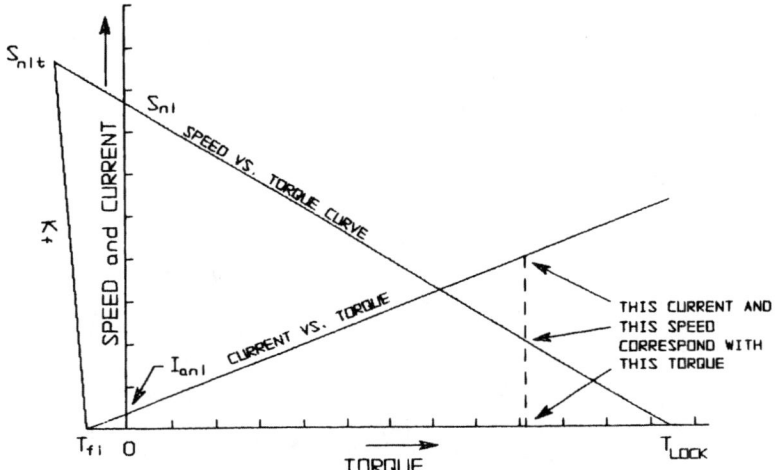

FIGURE 4.79 PMDC motor performance curves.

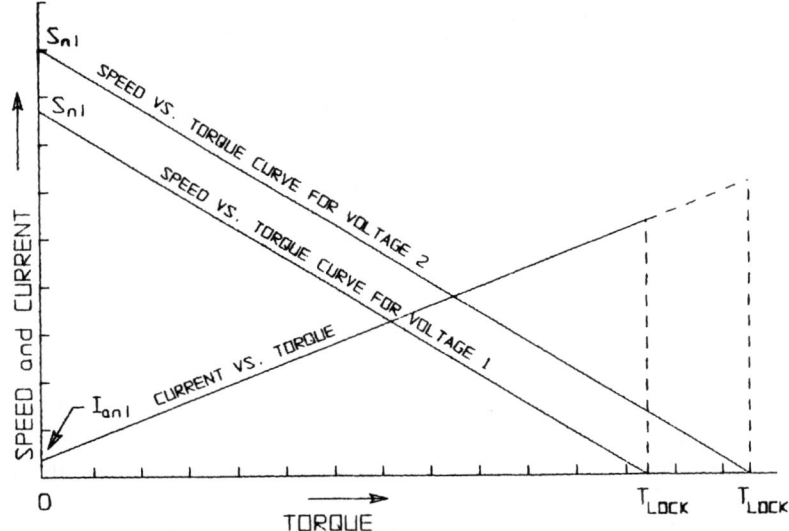

FIGURE 4.80 PMDC motor performance varying with voltage.

4.4.1 Predicting Air Gap

A typical approach to predicting air gap flux is described here. The magnets are generally attached to a magnetically soft steel housing. When they are charged, they set up a nearly constant flux in the air gap between the magnets and the armature. In order to determine the motor performance, you need to know the amount of air gap flux linking the armature conductors, the number of conductors, the number of poles, and the current in the armature. Figure 4.81 shows the direction of the flux due to the magnets. You can determine the flux in the air gap by the following procedure:

- Finding the permeance coefficient of the magnetic circuit
- Determining the flux density in the magnet
- Finding the total flux
- Factoring out the leakage flux

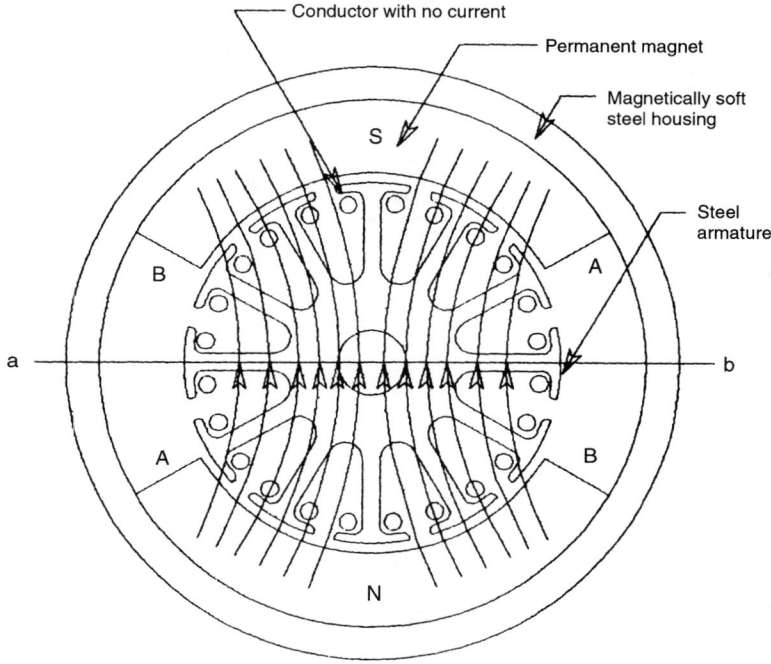

FIGURE 4.81 Flux due to magnets.

The remaining flux interacts with the armature conductors and produces the motor torque.

The permeance coefficient is determined by the geometry of the cross section of the motor:

$$P_c = \frac{\sigma}{R_f} \left(\frac{L_{mr} A_g}{A_m L_g} \right) \qquad (4.378)$$

where σ = flux leakage factor (typically 1.05 to 1.15)
R_f = reluctance factor (typically 1.1 to 1.3)
L_{mr} = radial length (thickness) of the magnet
L_g = length of the air gap in the radial direction
A_m = area of the magnet
A_g = area of the air gap

The area of the magnet is not necessarily the same as the area of the air gap, because the magnets typically overhang the armature. A_g is usually smaller than A_m.

The permeance coefficient determines the load line of the magnet on its normal demagnetization curve. These curves are commonly supplied by the magnet manufacturer. A typical curve is shown in Fig. 4.82; it is the upper left quadrant of the hysteresis loop shown in other chapters of this handbook. The H and B axes must be appropriately scaled for this technique to be accurate.

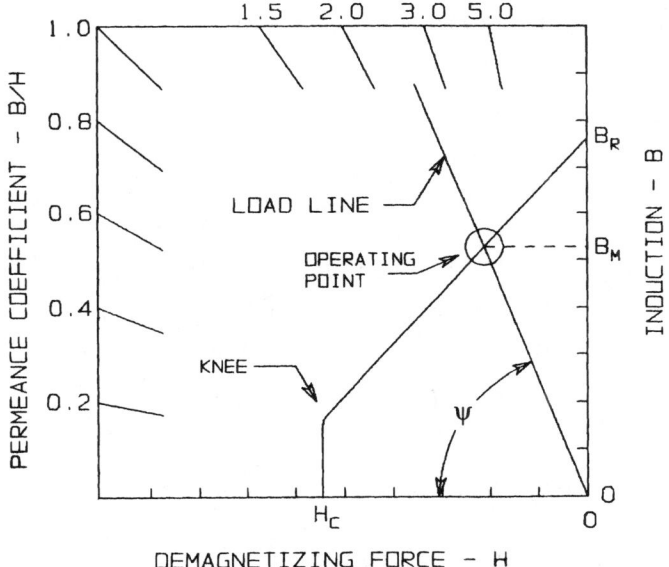

FIGURE 4.82 Finding magnet load point.

To plot the load line, take the arctangent of the permeance coefficient, calculate the angle $\psi = \tan^{-1} P_c$, and plot the line as shown in Fig. 4.82. The flux density in the magnet B_m can be found by finding the intersection of the load line and the normal curve and reading the induction from the right vertical axis. The flux in the magnet is found by multiplying the flux density by the area of the magnet:

$$\phi_m = B_m A_m \tag{4.379}$$

The air gap flux can then be found by dividing the magnet flux by the leakage factor:

$$\phi_g = \frac{\phi_m}{\sigma} \tag{4.380}$$

This is a ballpark approach used by Ireland (1968) and Puchstein (1961). The effects of magnet overhang should be included as they add some additional flux.

The method used here also predicts the permeance coefficient, but the effect of slots in the armature on the magnetic air gap length is accounted for using Carter's coefficient.

The effects of magnet overhang are predicted by calculating the permeance coefficient of each section of the overhang using methods developed by Roters (1941).

The magnetic air gap length L_{gl} is determined by Carter's method, but first you must calculate the circumferential width of the armature slot, as follows.

Given the following dimensions (Figs. 4.83 and 4.84):

R_a = outside radius of the armature lamination
W_{ast} = width of the armature slot top in straight-line distance
N_{at} = number of armature teeth
R_{pole} = radius of the permanent-magnet pole face

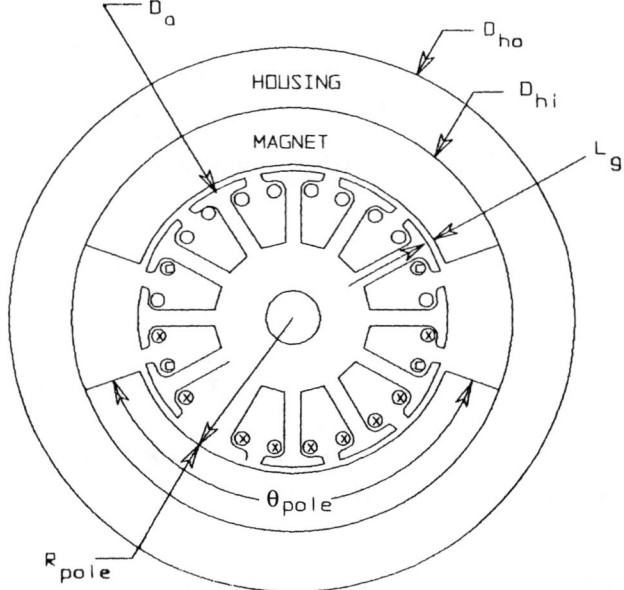

FIGURE 4.83 PMDC motor cross-section.

To find the tooth pitch angle, in degrees, between two consecutive tooth centers:

$$\theta_{tp} = \frac{360°}{N_{at}} \quad (4.125)$$

To convert to radians:

$$\theta_{tpr} = \theta_{tp} \frac{\pi}{180} = \theta_{tp} (0.017453) \quad (4.126)$$

Tooth pitch, the circumferential distance between two tooth centers, in inches:

$$t_p = R_a \theta_{tpr} \quad (4.127)$$

DIRECT-CURRENT MOTORS

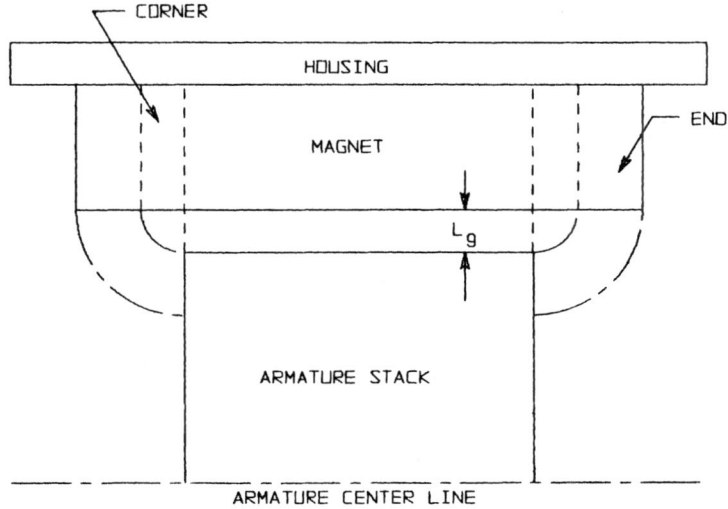

FIGURE 4.84 Permeances for stack, corner and end.

Width of the armature slot along the circumference, in inches:

$$W_{asc} = 2R_a \sin_r^{-1}\left(\frac{W_{ast}}{2R_a}\right) \tag{4.128}$$

Mechanical length of the air gap L_g, in:

$$L_g = R_{pole} - R_a \tag{4.381}$$

The magnetic length of the air gap is longer than the mechanical length because the lines of flux fringe through the armature slot toward the armature teeth. The following is a mathematical method for determining Carter's coefficient K_C, which accounts for this fringing.

$$\lambda = \frac{2}{\pi}\left[\tan_r^{-1}\left(\frac{W_{asc}}{2L_g}\right) - \frac{L_g}{W_{asc}}\ln\left(1 + \frac{W_{asc}^2}{4L_g^2}\right)\right] \tag{4.382}$$

$$K_C = \frac{t_p}{t_p - \lambda W_{asc}} \tag{4.383}$$

Effective magnetic air gap length, in inches:

$$L_{gl} = L_g K_C \tag{4.384}$$

Next, calculate the leakage factor σ. This is determined by finding the total permeance factors, including the leakage permeance factors at the ends of the magnet and along the edges of the magnet.

Now that you know the magnetic air gap length, determine the magnetic areas of the air gap between the armature and the magnet.

Geometric mean radius of the magnet, in inches:

$$R_{mm} = 0.707\sqrt{R_{hi}^2 + R_{pole}^2} \qquad (4.385)$$

where R_{hi} is the radius of the inside of the housing.

The effective magnet area is the arc distance of the pole at the geometric mean radius times the mechanical stack length, in square inches:

$$A_{ms} = \frac{\pi}{180}\, \theta_{pole} R_{mm} L_{stk} \qquad (4.386)$$

where θ_{pole} = magnet pole arc, degrees
L_{stk} = length of armature lamination stack

The average air gap radius is the distance from the center of the armature to the center of the air gap, in inches:

$$R_{gm} = \frac{R_{pole} + R_a}{2} \qquad (4.387)$$

Air gap area over the stack, in square inches:

$$A_{gs} = \frac{\pi}{180}\, \theta_{pole} R_{gm} L_{stk} \qquad (4.388)$$

The permeance factor (or permeance path) for the area between the magnet and the armature stack is the ratio of the area of the air gap over the stack to the magnetic air gap length, in units of inches (these units may not make sense at this time, but they cancel out when calculating the leakage factor σ):

$$P_{gs} = \frac{A_{gs}}{L_{gl}} \qquad (4.389)$$

To account for the flux at the corner, as shown in Figs. 4.84 and 4.85, the effective air gap length at the corner, in inches, has been empirically determined to be

$$L_{gc} = 1.021 L_{gl} \qquad (4.390)$$

The effective radius of the air gap at the corner is the distance from the center of the armature to the pole face minus 65 percent of the magnetic air gap length, in inches:

$$R_{mc} = R_{pole} - 0.65 L_{gl} \qquad (4.391)$$

The circumferential length of the gap at the corner, as shown in Fig. 4.86, is the arc length of the radius of the air gap at the corner swung along the arc of the magnet, in inches:

$$L_c = \frac{\pi}{180}\, \theta_{pole} R_{mc} \qquad (4.392)$$

Area of the gap at the corner, in square inches,

$$A_{gc} = 0.76 L_c L_{gl} \qquad (4.393)$$

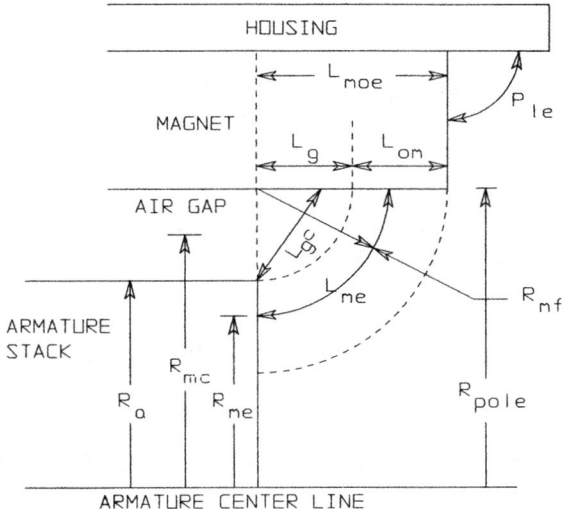

FIGURE 4.85 Permeances for corner and end.

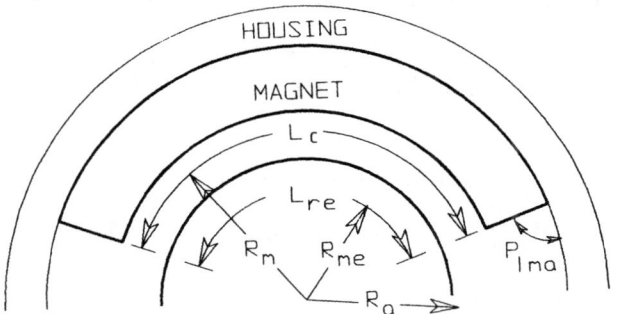

FIGURE 4.86 Permeances of corner and mean flux path.

Permeance factor of the gap at the corner:

$$P_{gc} = \frac{A_{gc}}{L_{gc}} \quad (4.394)$$

Area of the magnet at the corner, in square inches:

$$A_{mc} = \frac{\pi}{180}\, \theta_{pole} R_{mm} L_g \quad (4.395)$$

The flux at the end of the magnet segment will now be accounted for. The overhang length per end L_{moe} contributes flux to the armature stack depending on the lengths of the armature radius and the air gap.

$$L_{moe} = \frac{L_{ma} - L_{stk}}{2} \tag{4.396}$$

If the overhang per end is equal to or greater than the armature radius R_a plus the air gap length L_g, the mean flux path radius R_{mf}, in (Fig. 4.85) is

$$R_{mf} = \sqrt{L_g(L_g + R_a)} \tag{4.397}$$

This formula for R_{mf} assumes that the magnet overhang in excess of $R_a + L_g$ contributes a negligible amount of useful flux to the armature stack. If the overhang is less than $R_a + L_g$, then use Eq. (4.398):

$$R_{mf} = \sqrt{L_g L_{moe}} \tag{4.398}$$

The radius from the center of the armature to the mean flux path of the end R_{me}, in, is

$$R_{me} = R_a + L_g - R_{mf} \tag{4.399}$$

The circumferential length of the mean flux path L_{re}, in (Fig. 4.85), is

$$L_{re} = \frac{\pi}{180} \theta_{pole} R_{me} \tag{4.400}$$

To determine the area of the gap at the end, you must find the length of magnet overhang which produces the flux you are accounting for. Figure 4.85 shows L_{om}, in:

$$L_{om} = L_{moe} - L_g \tag{4.401}$$

Area of the gap at the end, in square inches:

$$A_{ge} = L_{om} L_{re} \tag{4.402}$$

Length of the mean flux path for the end, in inches:

$$L_{me} = \frac{\pi}{180} (90°) R_{mf} \tag{4.403}$$

Area of the magnet at the end, in square inches:

$$A_{me} = \frac{\pi}{180} \theta_{pole} R_{mm} L_{om} \tag{4.404}$$

Permeance factor for the gap at the end:

$$P_{ge} = \frac{A_{ge}}{L_{me}} \tag{4.405}$$

Now, consider the flux leakage at the ends of the magnet and the sides of the magnet. The permeance factor for the flux leakage at the axial end of the magnet (leakage off the end of the overhang) is

$$P_{le} = (0.0181511) K_{ml} \theta_p R_{mm} \tag{4.406}$$

where K_{ml} is the magnetic leakage constant based on the material type and orientation. This constant accounts for energy product BH_{max} of the magnet and its ability to hold orientation.

The permeance factor for the leakage along the axial length of the magnet accounts for the flux which leaks off the sides of the straight edges of the magnet and into the housing:

$$P_{lma} = (1.04)K_{ml}L_{ma} \qquad (4.407)$$

The leakage factor is the ratio of the sum of all permeance factors to the sum of all of the useful flux permeance factors:

$$\sigma = \frac{(P_{gs} + P_{gc} + P_{ge} + P_{le} + P_{lma})}{(P_{gs} + P_{gc} + P_{ge})} \qquad (4.408)$$

Permeance coefficient over the stack:

$$P_{cs} = \frac{L_{mr}A_{gs}\sigma}{A_{ms}L_{gl}R_f} \qquad (4.409)$$

where R_f is the reluctance factor (initially set equal to 1.5 and calculated later).
Flux density over the stack, in gauss:

$$B_{ps} = \frac{B_r}{1 + (M_x/P_{cs})} \qquad (4.410)$$

where M_x is the slope of the demagnetization curve.
The flux over the stack is the flux density in gauss converted to lines per square inch times the area of the magnet over the stack.
Flux over the stack, in lines:

$$\phi_{ms} = (6.4516)B_{ps}A_{ms} \qquad (4.411)$$

In a similar manner to the method for determining flux over the stack, you determine the useful flux supplied by the corner segment of the magnet to the armature. The permeance coefficient for the corner is

$$P_{cc} = \frac{L_{mr}A_{gc}\sigma}{L_{gc}A_{mc}R_f} \qquad (4.412)$$

Flux density of the magnet segment at the corner, in gauss:

$$B_{pc} = \frac{B_r}{1 + (M_x/P_{cc})} \qquad (4.413)$$

where M_x is the slope of the demagnetization curve.
The useful flux supplied by the magnet at the corner is the flux density in gauss converted to lines per square inch times the area of the magnet over the corner. Flux over the corner, in lines:

$$\phi_{mc} = (6.4516)B_{pc}A_{mc} \qquad (4.414)$$

Determine the flux supplied by the magnet at the end as follows:

$$P_{ce} = \frac{L_{mr}A_{ge}\sigma}{L_{me}A_{me}R_f} \qquad (4.415)$$

$$B_{pe} = \frac{B_r}{1 + (M_x/P_{ce})} \qquad (4.416)$$

$$\phi_{me} = (6.4516)B_{pe}A_{me} \tag{4.417}$$

The total flux supplied by the magnet is the sum of the individual sections. Considering two overhanging ends per magnet, the total flux supplied to the armature, in lines, is

$$\phi_T = \phi_{ms} + 2(\phi_{mc} + \phi_{me}) \tag{4.418}$$

The air gap flux, in lines, is the total flux divided by the leakage factor:

$$\phi_g = \frac{\phi_T}{\sigma} \tag{4.419}$$

4.4.2 Armature Calculation

Now that you know the air gap flux, you must solve the magnetic circuit by determining the flux densities in the sections of the armature and shell and then the mmf drops. First, analyze the geometry of the armature laminations. This analysis gives you the following:

Gross slot area. This tells you how much cross-sectional area is available to accommodate the armature windings (before adding insulation and slot pegs).

Net slot area. This is the net available area to accommodate the copper wire after the slot insulators and pegs are subtracted from the gross slot area.

Magnetic path length. For the armature, this is the distance that the flux will travel and over which the mmf will be dropped.

Magnetic path area. This is the cross-sectional area of the magnetic path as seen from the flux's point of view. The smaller the path area, the higher the flux density, and the more mmf will be dropped along the path.

Weights of steel. This must be known to calculate the inertias of the armature assembly.

Inertia. This is used in determining start-up acceleration or load matching.

Using trigonometry, you break the armature lamination up into several sections in order to calculate gross slot area. The radii used in the calculations are readily available from most lamination drawings supplied by the manufacturer.

4.4.3 Armature Slot Calculations

To start the calculations, first determine the following from a drawing of a lamination similar to that shown in Fig. 4.87 or 4.88:

R_a = outside radius of the armature lamination
R_{al} = inside radius of the armature teeth
R_s = radius of the shaft
W_{ast} = width of the armature slot top in straight-line distance
R_{a2} = radius of the circle on which the arc segments of the bottoms of the slots are centered

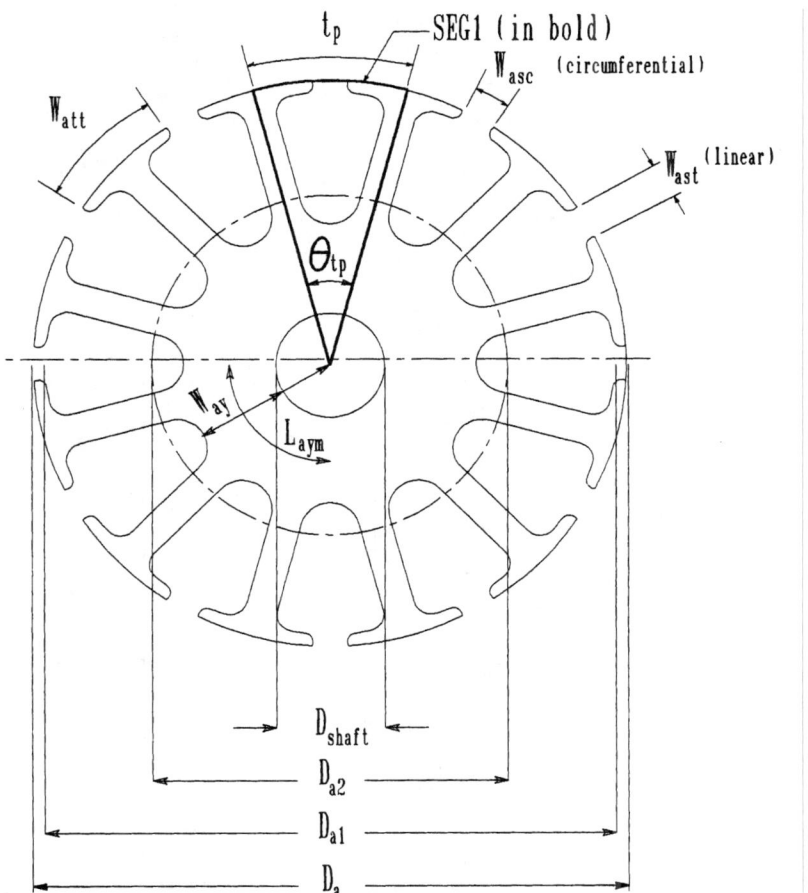

FIGURE 4.87 Armature lamination with round-bottom slots.

N_{at} = number of armature teeth
L_{stk} = length of the armature stack
θ_{pole} = angle of pole arc

To find the tooth pitch angle, in degrees, between two consecutive tooth centers:

$$\theta_{tp} = \frac{360°}{N_{at}} \tag{4.125}$$

To convert to radians:

$$\theta_{tpr} = \theta_{tp}\frac{\pi}{180} = \theta_{tp}(0.017453) \tag{4.126}$$

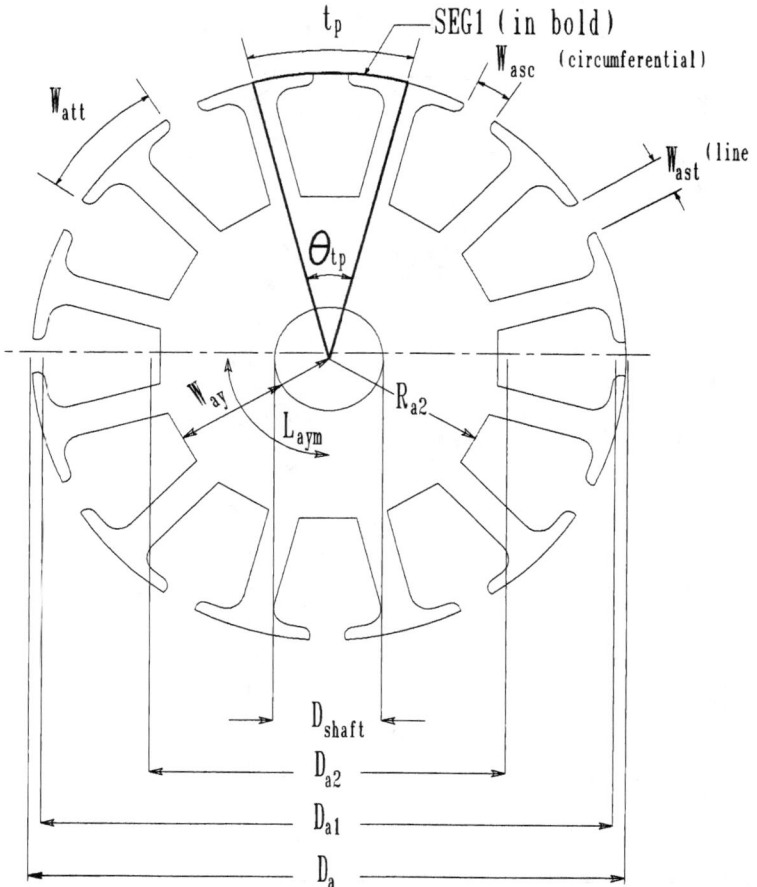

FIGURE 4.88 Armature lamination with flat-bottom slots.

Tooth pitch, the circumferential distance between two tooth centers, in inches:

$$t_p = R_a \theta_{tpr} \tag{4.127}$$

Width of the armature slot along the circumference, in inches:

$$W_{asc} = 2R_a \sin_r^{-1}\left(\frac{W_{ast}}{2R_a}\right) \tag{41.28}$$

Width of the tooth tip along the circumference, in inches:

$$W_{att} = t_p - W_{asc} \tag{4.129}$$

The method used to calculate areas and paths of the armature is the same as that used for the series motor. This information can readily be calculated by any modern computer-aided design (CAD) program.

4.4.4 Magnetic Circuit

Special consideration is taken to account for saturation and for the air gap.

The air gap is the area between the pole and the tooth tips on the armature. The mechanical L_g and magnetic L_{gl} air gap lengths were calculated earlier.

The magnetic width of the air gap is the length of the arc along the center of the magnetic air gap, in inches:

$$W_{gap} = \left[\frac{R_{pole} + R_a}{2} \right] \theta_{pole}(0.017453) \tag{4.420}$$

Magnetic area of the air gap, in square inches:

$$A_{gap} = W_{gap} L_{stke} \tag{4.312}$$

Magnetic width of the housing, in inches (Fig. 4.89):

$$W_{hm} = D_{ho} - D_{hi} \tag{4.421}$$

where D_{ho} = outside diameter of housing
D_{hi} = inside diameter of housing

The magnetic area of the housing is the magnetic width of the housing times the axial length of the magnet L_{ma}. Magnetic area of the housing, in inches:

$$A_{hm} = W_{hm} L_{ma} \tag{4.422}$$

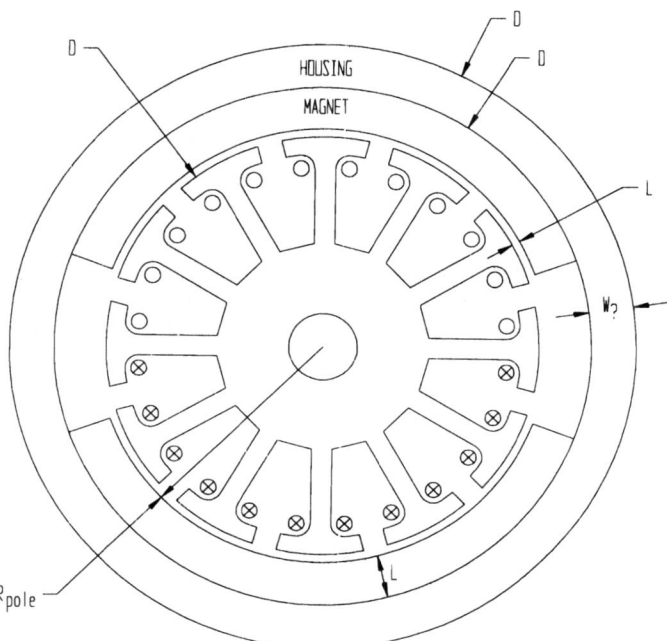

FIGURE 4.89 Housing magnetic paths.

Flux is required for an electric motor to produce torque. In order to produce flux, mmf must be supplied. In the case of a PMDC motor, the mmf is supplied by the magnet. In order to determine the mmf drop across a specific area, the flux density must be determined. From the BH curve for the material being used, the magnetic field intensity in ampere-turns per inch can be determined. The mmf drop in ampere-turns for each specific area can be determined by multiplying the field intensity by the magnetic length for each specific area.

For the previously determined flux in the air gap ϕ_g, kline, the flux densities for the various paths are as follows:

B_{gap}, kline/in², in the air gap:

$$B_{\text{gap}} = \frac{\phi_{\text{gap}}}{A_{\text{gap}}} \quad (4.313)$$

If B_{at} >100 kline/in², it needs to be corrected using the Tricky method presented in the Plot of Flux per Pole Versus Ampere-Turns Excitation subsection of Sec. 4.2.7.
B_{att}, kline/in², in the armature tooth tips:

$$B_{\text{att}} = \frac{\phi_{\text{gap}}}{A_{T,\text{attm}}} \quad (4.314)$$

B_{at}, kline/in², in the armature teeth:

$$B_{at} = \frac{\phi_{\text{gap}}}{A_{T,\text{atm}}} \quad (4.315)$$

B_{ay}, kline/in², in the armature yoke:

$$B_{ay} = \frac{\phi_{\text{gap}}}{A_{\text{aym}}} \quad (4.316)$$

For the following calculation, assume that the flux is in the return path in the housing. Flux density in the housing, in kilolines per square inch:

$$B_h = \frac{\phi_{\text{gap}}}{A_{\text{hm}}} \quad (4.323)$$

In order to calculate the total mmf drops in the magnetic circuit, you need to read the magnetic field intensity H (A · turn)/in, from the BH curve for each section. The mmf drop for the section is the intensity times the length of the section.

Field intensity H_{gap}, (A · turn)/in, in the air gap:

$$H_{\text{gap}} = \frac{B_{\text{gap}}}{3.19} \quad (4.324)$$

MMF drop across the air gap, in ampere-turns:

$$\mathscr{F}_{\text{gap}} = H_{\text{gap}} L_{gl} \quad (4.424)$$

For the following equations, the value of H is read from the graph for each different flux density B.

$$\mathscr{F}_{\text{att}} = H_{\text{att}} L_{\text{attm}} \quad (4.326)$$

The last mmf drop to caculate for the armature is that of the armature yoke $\mathcal{F}_{ay}$, A · turn:

$$\mathcal{F}_{ay} = H_{ay}L_{aym} \tag{4.330}$$

The mmf drop in the housing is calculated by reading the field intensity for the housing H_h from the appropriate BH curve. Note that it is common to have the housing of different material than the armature. Make sure that the proper curve is being used. The mmf drop in the housing is taken over a distance of the arc of 30° using the radius to the center of the housing material.

Magnetic length of the housing, in inches:

$$L_{hm} = \frac{\pi}{3P}\left[D_{hi} + \left(\frac{D_{ho} + D_{hi}}{2}\right)\right] \tag{4.425}$$

MMF drop in the housing, in ampere-turns:

$$\mathcal{F}_h = H_h L_{hm} \tag{4.426}$$

The armature reaction and brush shift α cause demagnetization of the field.

Field distortion due to armature reaction, in ampere-turns:

$$\mathcal{F}_{dl} = K_{dl}\left(\frac{\theta_{pole}}{360}\right)\left(\frac{Z_{ae}I_{line}}{(P)(a)}\right) \tag{4.427}$$

where P = number of poles
$\quad\quad\quad\;\, a$ = number of parallel paths in armature
$\quad\quad\quad\;\, I_{line}$ = line current, A
$\quad\quad\quad\;\, Z_{ae}$ = effective number of armature conductors (calculated later)
$\quad\quad\quad\;\, K_{dl}$ = empirically determined constant whose value depends on materials; use 0.15 if no data is available

Field distortion due to brush shift, in ampere-turns:

$$\mathcal{F}_{bl} = \frac{\alpha Z_{ae}I_{line}}{(180)(a)(P)} \tag{4.428}$$

where α = brush shift angle, degrees.

These values are mmf drops which are added to the total mmf drops in the steel, which were calculated previously.

To continue the analysis of the mmf drops, the sum of the mmf drops, in ampere-turns, is

$$\mathcal{F}_{total} = \mathcal{F}_{gap} + \mathcal{F}_{att} + \mathcal{F}_{at} + \mathcal{F}_{ay} + \mathcal{F}_h + \mathcal{F}_{dl} + \mathcal{F}_{bl} \tag{4.429}$$

Now you know the total mmf drops in the magnetic circuit. Calculate the mmf supplied by the magnet.

Total magnet area, in square inches:

$$A_{mt} = \frac{\pi}{180}\,\theta_{pole}R_{mm}L_{ma} \tag{4.430}$$

Average flux density in the magnet, in gauss:

$$B_{pm} = \frac{\phi_t}{(6.4516)A_{mt}} \tag{4.431}$$

Magnetizing intensity due to the magnet, using M_x as the slope of the magnet curve, in ampere-turns per inch:

$$H_m = \frac{(B_r - B_{pm})}{M_x} \; (2.021) \tag{4.432}$$

The mmf due to the magnet is the field intensity times the radial length. MMF in ampere-turns,

$$\mathcal{F}_m = H_m L_{mr} \tag{4.433}$$

If the total mmf drops in the magnetic circuit $\mathcal{F}_{total}$ do not equal within reason the mmf supplied by the magnet, then choose another total magnet flux ϕ_t and reiterate the procedure. Remember during the iteration process to use the same value for line current (preferably locked rotor current) when you recalculate the mmf drops due to armature reaction and brush shift.

Once the total mmf drops equal the supplied mmf, calculate the average permeance coefficient.

$$\mathcal{P}_{c,av} = \frac{B_{pm} M_x}{B_x - B_{pm}} \; (2.021) \tag{4.434}$$

Average reluctance factor (previously estimated at 1.5):

$$\mathcal{R}_{f,av} = \frac{\mathcal{F}_{total}}{\mathcal{F}_{gap}} \tag{4.435}$$

4.4.5 Output

The number of armature conductors can be found by using all of the conductors in the slots of the armature.

$$Z_a = (T_{pca})(N_{as})(\text{CPS}) \tag{4.436}$$

Or, to account for the fact that the conductors are not spanning the entire flux path, two distribution factors are introduced. The *winding distribution factor* K_{w1} accounts for the spread of the winding as compared to a full 180°.

$$K_{w1} = \sin\left(\frac{90 S_{ac} P}{N_{at}}\right) \tag{4.437}$$

The *pole arc distribution factor* K_{w2} accounts for the spread of the pole arc as compared to a full 180°.

$$K_{w2} = \sin\left(\frac{90 \theta_{pole} P}{360}\right) \tag{4.438}$$

The effective number of armature conductors Z_{ac} can be calculated by multiplying the actual number of armature conductors times the winding distribution factor and the pole arc distribution factor.

$$Z_{ac} = (Z_a)(K_{w1})(K_{w2}) \tag{4.439}$$

Effective turns are also a function of the number of coils shorted by the brushes. These formulas for K_w take that effect into account in many small motors.

Z_{ae} may be substituted in the following equations for Z_a.

The locked rotor current is found as follows:

$$I_{lock} = \frac{E_t}{R_{at} + R_b} \tag{4.440}$$

where E_t = terminal voltage
R_{at} = terminal resistance of the armature
R_b = brush resistance

Torque developed by the motor at locked rotor, in oz · in:

$$T_{dev} = (2.256 \times 10^{-7})\phi_g Z_{ae} I_{lock} \frac{P}{a} \tag{4.441}$$

Since ϕ_g is relatively constant for a PMDC motor, the only variable in this torque equation is the current I. The developed torque can be described by defining a torque constant K_t (oz · in)/A:

$$K_t = (2.256 \times 10^{-7})\phi_g Z_{ae} \frac{P}{a} \tag{4.442}$$

The developed torque equation now becomes

$$T_{dev} = K_t I_{load} \tag{4.443}$$

where I_{load} is the armature current at the particular load.

The actual output locked-rotor torque is less the internal motor friction (determined in Fig. 4.90). Internal motor friction T_{fi} is determined from test data on a similar motor.

Locked rotor torque, in oz · in:

$$T_{lock} = T_{dev} - T_{fi} \tag{4.444}$$

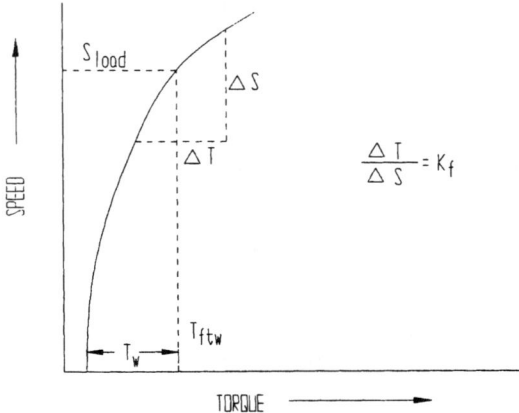

FIGURE 4.90 K_f curve.

The true no-load speed can be calculated using the preceding values for terminal voltage, number of conductors, and air gap flux.

True no-load speed, in rpm:

$$S_{nlt} = \frac{(6 \times 10^9) E_t a}{Z_{ae} \phi_g P} \quad (4.445)$$

The true no-load speed is the speed the motor would run at if it had no internal friction or windage. It is the speed at which the generated voltage of the armature equals the terminal voltage.

The no-load current is determined by dividing the friction and windage torque by the torque constant.

No-load current, in amps:

$$I_{nl} = \frac{T_{fi} + S_{nlt} K_f}{K_t} \quad (4.446)$$

where K_f, (oz · in)/rpm, is the open circuit damping coefficient which was previously determined from a similar motor, as shown in Fig. 4.90. It is determined by driving the motor and measuring the reaction torque.

The actual no-load speed S_{nl} can be calculated as follows:

$$S_{nl} = [E_t - I_{nl} R_{at}] \frac{(6 \times 10^9) a}{P Z_{ae} \phi_g} \quad (4.447)$$

Now that the end points of the speed-torque and current-torque curves are known, the motor performance can be plotted as in Fig. 4.79.

For specific load points, the speed, torque, and current can be taken from the curves and more performance parameters can be calculated. The output power is a function of the load speed S_{load} and the load torque T_{load}.

Output power:

$$P_{out} = \frac{S_{load} T_{load}}{1,008,000} \quad \text{hp} \quad (4.448)$$

$$P_{out} = \frac{(746) S_{load} T_{load}}{1,008,000} \quad \text{W} \quad (4.449)$$

Input power, in watts:

$$P_{in} = E_t I_{load} \quad (4.450)$$

Efficiency of the motor at the load conditions, in percent:

$$\eta = \frac{P_{out}}{P_{in}} \quad (4.451)$$

Losses. The power lost in copper windings as heat is the square of the line current times the armature resistance.

Power lost in the windings as heat, in watts:

$$P_{L,copper} = I_{load}^2 R_{at} \quad (4.452)$$

Power lost in the brushes, in watts:

$$P_{L,\text{brush}} = I_{\text{load}}^2 R_b \qquad (4.453)$$

Friction and windage losses, in watts:

$$P_{f+w} = \frac{(T_{\text{fi}} + S_{\text{load}} K_f) S_{\text{load}}}{1{,}008{,}000} \times 746 \qquad (4.454)$$

The loss which has not been accounted for, the *stray loss,* is the input power − (output power + losses).

Current Densities in the Brushes and Conductors. Current densities are in units of amperes per square inch. Using the diameter of the bare wire $D_{\text{aw,bare}}$ the current density in the armature wire is found as follows:

$$\gamma_a = \frac{I_{\text{line}}/2}{\pi(D_{\text{aw,bare}}/2F_{\text{wsa}})^2} \qquad (4.455)$$

Current density in the brushes, in amperes per square inch:

$$\gamma_{\text{br}} = \frac{I_{\text{line}}}{t_{\text{br}} W_{\text{br}}} \qquad (4.456)$$

where W_{br} is the width of the brush, measured in the direction parallel to the armature shaft.

Brushes should be limited to about 75 to 100 A/in^2, depending on vendor specification.

Motor Constants. Following are some of the motor constants and figures of merit for a PMDC motor.

Loaded acceleration, in radians per second squared:

$$\alpha_{\text{load}} = \frac{T_{\text{load}} - T_{\text{fe}}}{J_a + J_{\text{load}}} \qquad (4.457)$$

where T_{load} = motor output torque, oz · in
T_{fe} = external friction torque, oz · in
J_a = armature inertia, oz · in · s^2
J_{load} = load inertia, oz · in · s^2

The torque constant K_t has already been determined.
Back-emf constant K_{be}, Volts/krpm (English units):

$$K_{\text{be}} = (0.74) K_t \qquad (4.458)$$

Back-emf constant K_{bm}, Volts/(rad · s) (metric units):

$$K_{\text{bm}} = (7.06 \times 10^{-3}) K_t \qquad (4.459)$$

Mechanical time constant, in seconds:

$$T_m = \frac{R_t J_a}{K_{\text{bm}} K_t} \qquad (4.460)$$

Zero-impedance (leads short-circuited) damping coefficient, in ounce-inches per radian per second:

$$K_d = \frac{K_t K_{bm}}{(R_{bnl}/2) + R_{at}} \quad (4.461)$$

where R_{bnl} is the no-load brush resistance (empirically determined).
Motor constant K_m:

$$K_m = \frac{K_t}{\sqrt{R_t}} \quad (\text{oz} \cdot \text{in})/(\text{A}\sqrt{\Omega}) \text{ or } (\text{oz} \cdot \text{in})/\sqrt{W} \quad (4.462)$$

Magnetic-to-electric loading ratio for the load point:

$$M = \frac{Pa\phi_g}{Z_a I_{load}} \quad (4.463)$$

where ϕ_g is in lines. M should be greater than 50 for good commutation.

4.5 SERIES DC AND UNIVERSAL AC PERFORMANCE*

The process is the same for both for series dc and universal ac performance, except that in direct current there is no core loss in the field. The reactance values go to 0. Transformer voltages go to 0, and the power factor goes to 1. The following describes the ac method.

Procedure

1. From the dimensions of a known lamination set, calculate the magnetic circuit areas, lengths, and volumes.
2. Plot a flux f versus magnetomotive force (mmf) curve for the lamination set on a per-unit-length basis, including leakage flux.
3. Assume a field winding and an armature winding.
4. Select a line voltage.
5. Select a fundamental current I_{fund} for the motor and build a matrix of I_{fund} versus the following:

    I_{load} = load current
    F_{dl} = field distortion mmf
    F_{bl} = field distortion due to brush shift
    F_{net} = net mmf drop
    f_{temp} = air gap flux per unit length
    f_{act} = actual air gap flux
    C_{real} = real (or resistive) components of the phasor diagram less back-emf
    E_{tf} = field transformer voltage

---

* Section contributed by Alan W. Yeadon and Andrew E. Miller, Yeadon Engineering Services, PC.

V_{im} = imaginary (or reactive) components of the phasor diagram
q_{pf} = power factor angle
PF = power factor
V_{real} = real components of phasor diagram
E_g = back-emf or generated voltage
S_{load} = load speed
T_{dev} = developed torque
T_{fw} = friction and windage torque
T_{CL} = torque loss due to core loss
T_{load} = output torque at calculated load speed
P_{out} = output power, Watts
P_{in} = input power, Watts
Eff = motor efficiency

4.5.1 Performance Calculations

Using the lamination set of Fig. 4.91, and knowing the stack length L_{stk}, we calculate the magnetic circuit lengths, areas, and volumes.

4.5.2 Air Gap Flux Versus MMF Drop

We will create an f versus mmf curve per inch of stack, so for these calculations we will assume a 1.0-in stack height. Therefore, the magnetic area of the air gap in square inches is

$$A_{gap} = W_{gap} \cdot 1.0 \tag{4.464}$$

We must first make an assumption for the maximum value of air gap flux. We will use the following formula:

$$f_{max} = B_{max} \cdot A_{tatm} \cdot 1.1 \tag{4.465}$$

where: f_{max} = maximum air gap flux, kline
B_{max} = maximum flux density, line/in^2
$A_{T,atm}$ = total armature tooth area per pole, in^2

Starting with $f_g = 0$ and iterating toward f_{max}, we can now create our graph. See Figs. 4.92 and 4.93 for an illustration of the flux paths. For any value of f_g, the following formulas apply.

B_{gap}, kline/in^2, in the air gap:

$$B_{gap} = \frac{\phi_{gap}}{A_{gap}} \tag{4.313}$$

B_{att}, kline/in^2, in the armature tooth tips:

$$B_{att} = \frac{\phi_{gap}}{A_{T,attm}} \tag{4.314}$$

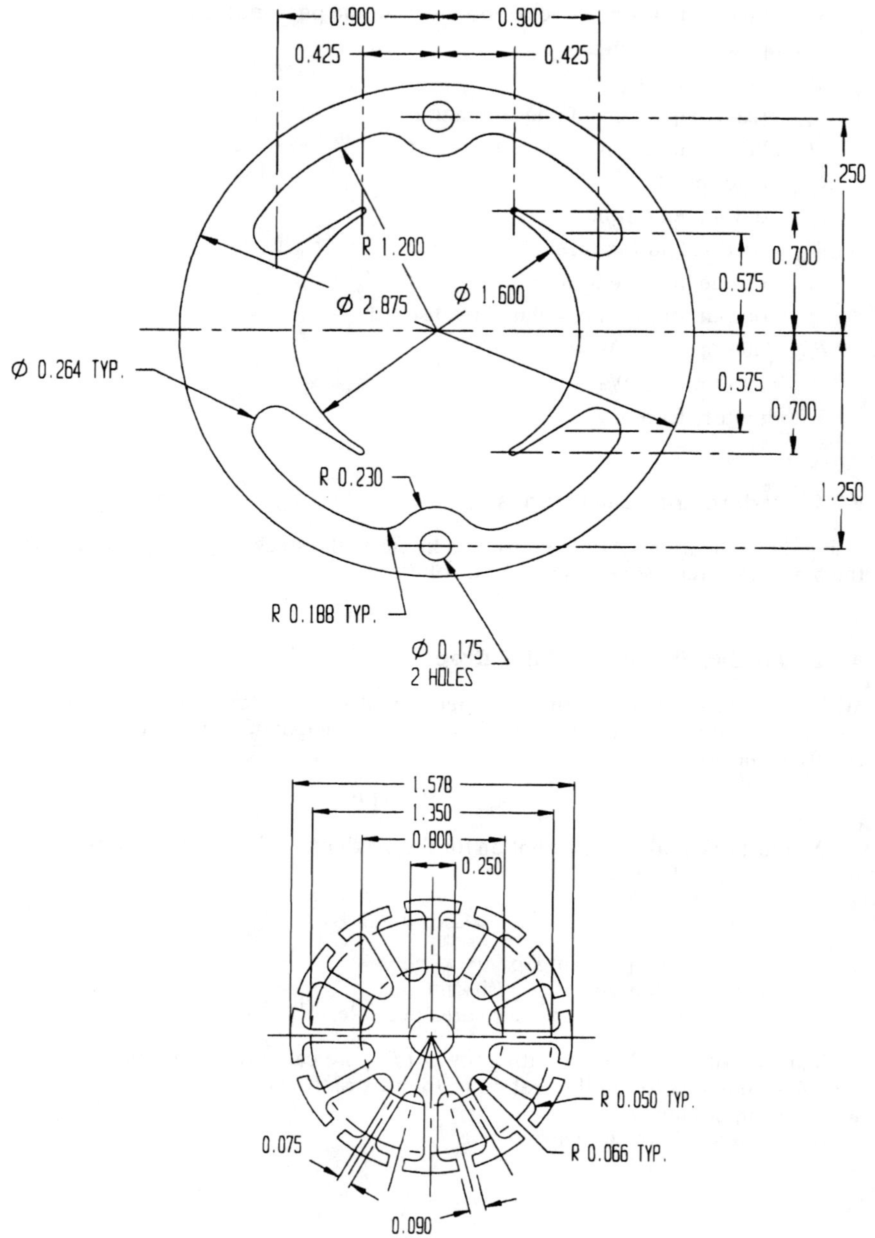

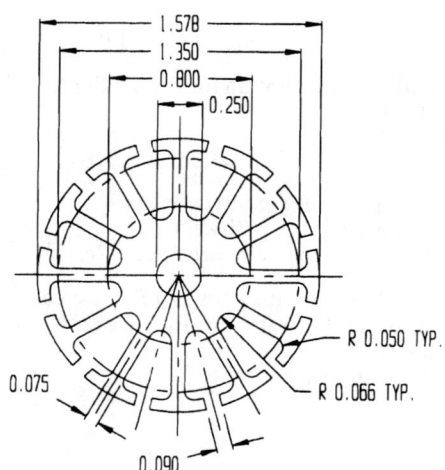

FIGURE 4.91 Laminations of sample motor.

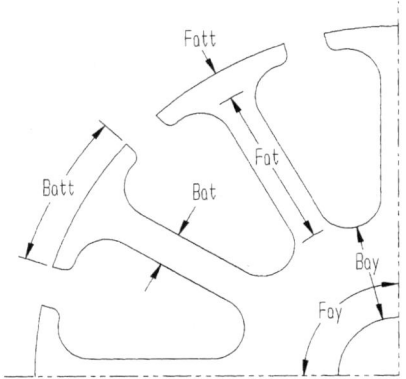
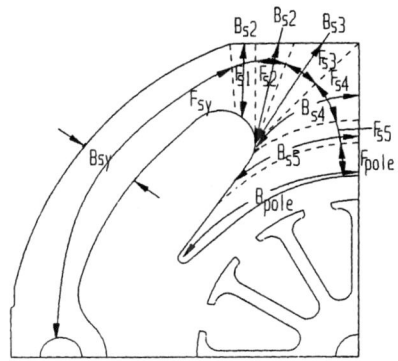

FIGURE 4.92 Flux paths in the armature. **FIGURE 4.93** Flux paths in the stator.

B_{at}, kline/in², in the armature teeth:

$$B_{at} = \frac{\phi_{gap}}{A_{T,attm}} \tag{4.315}$$

B_{ay}, kline/in², in the armature yoke:

$$B_{ay} = \frac{\phi_{gap}}{A_{aym}} \tag{4.316}$$

To determine the magnetic circuit variables in the stator, we must first calculate the leakage factor.

4.5.3 Calculation of Leakage Factor σ

The leakage factor is calculated based on several permeances. See Fig. 4.94 for a graphical representation of the key variables involved. First, calculate R_{ms}, in, the mean stator radius:

$$R_{ms} = 0.707 \cdot \sqrt{\left(\frac{W_{stator}}{2}\right)^2 + R_{pole}^2} \tag{4.466}$$

The average air gap radius is the distance from the center of the armature to the center of the air gap, in inches:

$$R_{gm} = \frac{R_{pole} + R_a}{2} \tag{4.467}$$

Air gap area over the stack, in square inches:

$$A_{gs} = R_{gm} \cdot \theta_p \cdot \frac{\pi}{180} \cdot L_{stk} \tag{4.468}$$

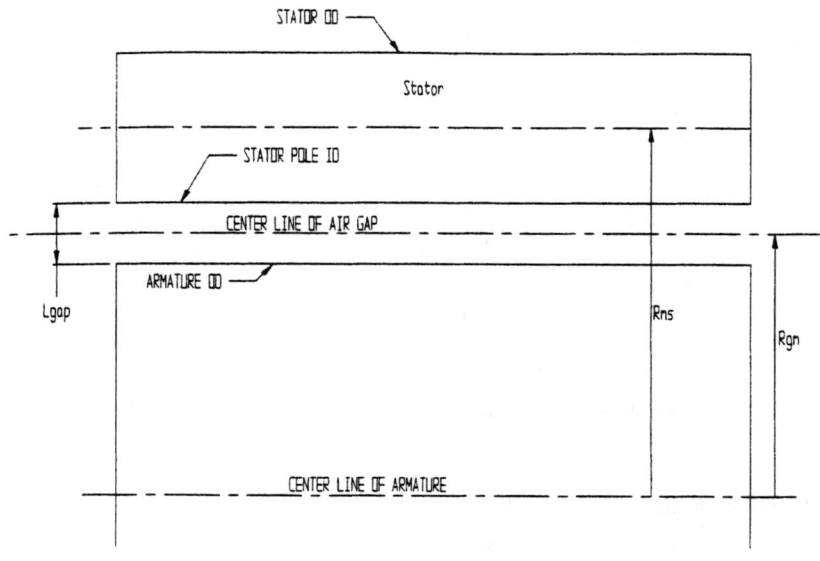

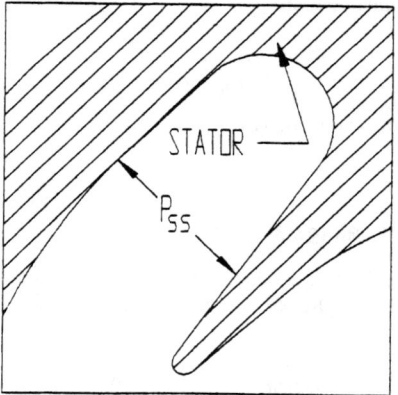

FIGURE 4.94 Permeance paths in the universal motor.

The permeance for the area between the magnetizing field coil and the armature stack is the ratio of the area of the air gap over the stack to the magnetic air gap length, in inches:

$$\mathcal{P}_{gs} = \frac{A_{gs}}{Lg_{mag}} \cdot \mu_0 \tag{4.469}$$

where μ_0 is a constant defined as the permeability of free space, the value of which is constant for all of the permeance calculations and will cancel out when we calculate the leakage factor.

Flux leakage at the ends and sides of the stator:

$$\mathcal{P}_{le} = 1.04 \cdot \frac{\pi}{180} \cdot \theta_p \cdot R_{mm} \cdot \mu_0 \qquad (4.470)$$

The permeance for the leakage along the axial length of the stator accounts for the flux which leaks off the sides of the straight edges and in the housing:

$$\mathcal{P}_{lsa} = 1.04 \cdot L_{stk} \cdot \mu_0 \qquad (4.471)$$

Permeance for the leakage from the stator pole back onto the stator yoke:

$$\mathcal{P}_{ss} = \frac{(\theta_p/4) \cdot (\pi/180) \cdot (R_{pole} \cdot 2R_{pt}) \cdot L_{stke}}{R_{isw} - (R_{pole} \cdot 2R_{pt})} \cdot \mu_0 \qquad (4.472)$$

The leakage factor is the ratio of the sum of all permeances to the sum of the useful flux permeance:

$$\sigma_1 = \frac{\mathcal{P}_{gs} + \mathcal{P}_{le} + \mathcal{P}_{lsa} + \mathcal{P}_{ss}}{\mathcal{P}_{gs}} \qquad (4.473)$$

Note that this value for the leakage factor works well at about half the maximum flux f_{max}. There will be no leakage when there is no flux (or $f_g = 0$) and more leakage when the flux increases. The actual leakage factor will increase as flux increases. The method used by the author is to use linear interpolation from $s = 0$ to s_1 as f_g goes from 0 to $f_{max}/2$. A maximum leakage factor s_{max} must then be assumed. s_{max} can be determined from finite element analysis or from approximation. Normally, one is safe to assume a maximum leakage factor of 1.2 to 1.3. Linear interpolation can be used so that $s = s_1$ to $s = s_{max}$ as f_g increases from $f_{max}/2$ to f_{max}.

To calculate the magnetic circuit variables in the stator, we take each section of the stator as follows:

B_{sy}, kline/in^2, in the stator yoke:

$$B_{sy} = \frac{\phi_{gap} \cdot \sigma}{A_{sy}} \qquad (4.474)$$

where s is the leakage factor at that flux.

B_{s1}, kline/in^2, in stator section 1:

$$B_{s1} = \frac{\phi_{gap} \cdot \sigma}{A_{s1}} \qquad (4.475)$$

B_{s2}, kline/in^2, in stator section 2:

$$B_{s2} = \frac{\phi_{gap} \cdot \sigma}{A_{s2}} \qquad (4.476)$$

B_{s3}, kline/in^2, in stator section 3:

$$B_{s3} = \frac{\phi_{gap} \cdot \sigma}{A_{s3}} \qquad (4.477)$$

B_{s4}, kline/in^2, in stator section 4:

$$B_{s4} = \frac{\phi_{gap} \cdot \sigma}{A_{s4}} \qquad (4.478)$$

B_{s5}, kline/in², in stator section 5:

$$B_{s5} = \frac{\phi_{gap} \cdot \sigma}{A_{s5}} \quad (4.479)$$

B_{pole}, kline/in², in the stator pole:

$$B_{pole} = \frac{\phi_{gap} \cdot \sigma}{A_{pm}} \quad (4.480)$$

In order to calculate the total mmf drops in the magnetic circuit, we need to read the magnetic field intensity H, (A · turn)/in, from the BH curve for each section. The mmf drop for the section is the intensity times the length of the section.

Field intensity H_{gap}, (A · turn)/in, in the air gap:

$$H_{gap} = \frac{B_{gap}}{3.19} \quad (4.324)$$

MMF drop across the air gap, in ampere-turns:

$$\mathcal{F}_{gap} = H_{gap} \cdot L_{gap} \quad (4.481)$$

For the following equations, the value of H is read from the BH curve for each different flux density B.

MMF drop across the armature teeth:

$$\mathcal{F}_{at} = H_{at} \cdot L_{atm} \quad (4.482)$$

MMF drop across the armature tooth tips:

$$\mathcal{F}_{att} = H_{att} \cdot L_{attm} \quad (4.483)$$

MMF drop across the armature yoke, in ampere-turns:

$$\mathcal{F}_{ay} = H_{ay} \cdot L_{aym} \quad (4.484)$$

MMF drop across each section of the stator, in ampere-turns:

$$\mathcal{F}_{sy} = H_{sy} \cdot L_{sy} \quad (4.485)$$

$$\mathcal{F}_{pole} = H_{pole} \cdot L_{pole} \quad (4.486)$$

$$\mathcal{F}_{s2} = H_{s2} \cdot L_{s2} \quad (4.487)$$

$$\mathcal{F}_{s1} = H_{s1} \cdot L_{s1} \quad (4.488)$$

$$\mathcal{F}_{s3} = H_{s3} \cdot L_{s3} \quad (4.489)$$

$$\mathcal{F}_{s4} = H_{s4} \cdot L_{s4} \quad (4.490)$$

$$\mathcal{F}_{s5} = H_{s5} \cdot L_{s5} \quad (4.491)$$

Sum of the mmf drops, in ampere-turns:

$$\mathcal{F}_{total} = \mathcal{F}_{gap} + \mathcal{F}_{att} + \mathcal{F}_{at} + \mathcal{F}_{ay} + \mathcal{F}_{sy} + \mathcal{F}_{s1} + \mathcal{F}_{s2} + \mathcal{F}_{s3} + \mathcal{F}_{s4} + \mathcal{F}_{s5} + \mathcal{F}_{pole} \quad (4.492)$$

Calculated results for the sample motor are shown in Table 4.1, and the curve is shown in Fig. 4.95.

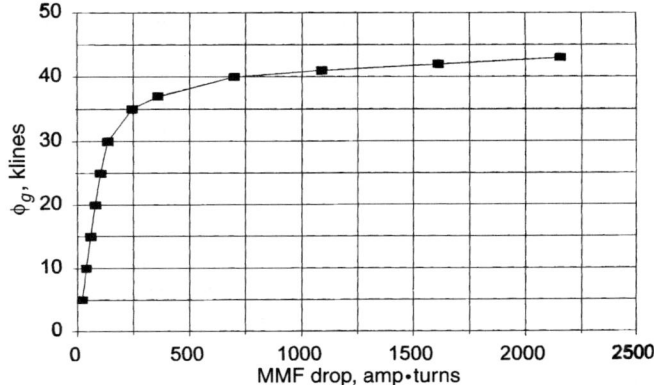

FIGURE 4.95 Air gap flux versus mmf drop for 1-in stack.

We can now calculate the armature and field resistances based on the actual stack height of the motor.

Resistance per armature coil, in ohms:

$$R_{ac} = L_{cuc} \cdot \rho_a \cdot F_{wsa} \tag{4.493}$$

where R_{ac} = resistance per armature coil
L_{cuc} = total armature coil length, ft
ρ_a = resistivity of the armature wire, Ω/ft
F_{wsa} = armature wire stretch factor

The total armature resistance is the product of the number of active coils times the resistance per armature coil, or:

$$R_{arm} = \frac{N_{ac} \cdot R_{ac}}{P_a \cdot 2} \tag{4.494}$$

The field resistance R_f is calculated by the following formula:

$$R_f = 2 \cdot L_{fc} \cdot \rho_f \cdot F_{wsf} \tag{4.495}$$

where R_f = field resistance, Ω
L_{fc} = total field coil length, ft
ρ_f = resistivity of the field wire, Ω
F_{wsf} = field wire stretch factor

The total winding resistance is the sum of the armature plus the field resistance:

$$R_{tot} = R_{arm} + R_f \tag{4.496}$$

The effective number of conductors can be calculated by multiplying the total number of conductors by the following two distribution factors:

$$K_{w2} = \sin\left(\frac{90 \cdot \theta_p \cdot P}{360}\right) \tag{4.497}$$

TABLE 4.1 Air Gap Flux Versus MMF Drop Calculations

Component or factor	Air gap flux, kline											
	5	10	15	20	25	30	35	37	40	41	42	43
Air gap												
Flux density, lines	3092	6184	9275	12367	15459	18551	21642	22879	24734	25353	25971	26589
Field intensity	969	1938	2908	3877	4846	5815	6784	7172	7754	7948	8141	8335
MMF drop	18.12	36.25	54.37	72.50	90.62	108.75	126.87	134.12	144.99	148.62	152.24	155.87
Armature tooth tips												
Flux density	3.99	7.98	11.97	15.96	19.95	23.93	27.92	29.52	31.91	32.71	33.51	34.31
Field intensity	0.90	1.54	1.69	1.81	1.91	2.19	2.48	2.62	2.81	2.88	2.93	2.97
MMF drop	0.07	0.12	0.13	0.14	0.15	0.17	0.19	0.20	0.22	0.22	0.23	0.23
Armature teeth												
Apparent flux density	13.70	27.40	41.10	54.79	68.49	82.19	95.89	101.37	109.59	112.33	115.07	117.81
Apparent field $_p$ intensity, H_{tem}	1.75	2.44	3.29	4.58	7.08	12.10	32.66	57.13	134.62	170.50	233.24	295.98
Actual flux density (using Trickey factor)	13.70	27.40	41.10	54.79	68.49	82.19	95.89	101.37	109.59	111.21	113.92	116.63
Actual field intensity	1.75	2.44	3.29	4.58	7.08	21.10	32.66	57.13	134.62	155.01	206.89	269.00
MMF drop	0.60	0.84	1.13	1.57	2.43	4.15	11.20	19.59	46.16	53.15	70.94	92.24
Armature yoke												
Flux density	15.06	30.12	45.18	60.24	75.30	90.36	105.42	111.45	120.48	123.49	126.51	129.52
Field intensity	1.79	2.66	3.44	5.63	9.26	19.72	83.16	158.04	384.42	502.12	646.76	837.81
MMF drop	0.68	1.02	1.32	2.16	3.55	7.56	31.98	60.61	147.42	192.56	248.03	321.30
Leakage (based on 1.0525 at $\phi_{max}/2$)	1.01	1.02	1.03	1.04	1.05	1.05	1.07	1.08	1.09	1.09	1.09	1.10
Stator yoke												
Flux density	15.21	30.72	46.52	62.63	79.03	95.19	112.78	119.95	130.86	134.53	138.23	141.94
Field intensity	1.79	2.71	3.51	5.90	10.45	30.57	180.75	363.61	922.71	1788.8	2947.1	4111.5
MMF drop	0.69	1.04	1.35	2.26	4.01	11.72	69.32	139.44	353.86	686.03	1130.2	1576.7
Stator section 1												
Flux density	7.94	16.03	24.29	32.69	41.26	49.69	58.87	62.62	68.31	70.23	72.16	74.09
Field intensity	1.54	1.81	2.21	2.87	3.30	3.84	5.47	5.90	7.02	7.64	8.26	8.88
MMF drop	1.31	1.54	1.88	2.44	2.80	3.26	4.61	5.01	5.96	6.49	7.01	7.54

TABLE 4.1 Air Gap Flux Versus MMF Drop Calculations (*Continued*)

Component or factor	Air gap flux, kline													
	5	10	15	20	25	30	35	37	40	41	42	43		
Stator section 2														
Flux density	5.51	11.13	16.86	22.70	28.64	34.50	40.88	43.48	47.43	48.76	50.10	51.45		
Field intensity	1.24	1.66	1.83	2.10	2.54	2.98	3.29	3.38	3.61	3.74	3.88	4.02		
MMF drop	0.14	0.18	0.20	0.23	0.28	0.33	0.36	0.37	0.40	0.41	0.43	0.45		
Stator section 3														
Flux density	4.99	10.08	15.27	20.56	25.94	31.25	37.02	39.37	42.95	44.16	45.37	46.59		
Field intensity	1.13	1.62	1.79	1.95	2.33	2.76	3.11	3.23	3.36	3.41	3.45	3.52		
MMF drop	0.11	0.16	0.18	0.19	0.23	0.27	0.31	0.32	0.33	0.34	0.34	0.35		
Stator section 4														
Flux density	4.94	9.97	15.10	20.32	25.65	30.89	36.60	38.92	42.46	43.66	44.86	46.06		
Field intensity	1.11	1.62	1.79	1.94	2.31	2.73	3.09	3.21	3.34	3.39	3.43	3.47		
MMF drop	0.08	0.12	0.13	0.14	0.17	0.20	0.23	0.24	0.25	0.25	0.25	0.26		
Stator section 5														
Flux density	3.60	7.28	11.02	14.84	18.72	22.55	26.72	28.42	31.00	31.87	32.75	33.63		
Field intensity	0.81	1.52	1.66	1.78	1.87	2.09	2.39	2.53	2.74	2.81	2.88	2.93		
MMF drop	0.03	0.05	0.06	0.06	0.06	0.07	0.08	0.08	0.09	0.09	0.10	0.10		
Stator pole section (with 10% leakage)														
Flux density	2.85	5.75	8.72	11.73	14.80	17.83	21.13	22.47	24.51	25.20	25.89	26.59		
Field intensity	0.64	1.30	1.57	1.68	1.78	1.85	1.99	2.09	2.23	2.28	2.32	2.38		
MMF drop	0.06	0.12	0.14	0.15	0.16	0.17	0.18	0.19	0.20	0.20	0.21	0.21		
Total mmf drops (using *BH* curves at 60 Hz)	21.89	41.44	60.89	81.85	104.46	136.65	245.24	360.18	699.89	1088.3	1610.0	2155.3		

Note: Air gap flux versus mmf excitation, 1-inch stack, *BH* curve at 60 Hz, where:

$L_{gap} = 0.0187$ $L_{s1} = 0.8495$ $L_{xc5} = 0.0335$ $W_{pm} = 0.8861$ $W_{s4} = 0.5115$ $A_{asym} = 0.3320$ $A_{s3} = 1.0114$ $L_{atm} = 0.3429$ $A_{Tatm} = 1.2534$

$L_{sy} = 1.4305$ $L_{s2} = 0.1108$ $W_{gap} = 1.6172$ $W_{s1} = 0.3180$ $W_{s5} = 0.7007$ $A_{pm} = 1.7723$ $A_{s4} = 1.0231$ $T_p = 0.3943$ $A_{Tatm} = 0.3650$

$L_{aym} = 0.3835$ $L_{s3} = 0.0987$ $W_{sy} = 0.5006$ $W_{s2} = 0.4580$ $A_{gap} = 1.6172$ $A_{s1} = 0.6360$ $A_{s5} = 1.4014$ $F_{stk} = 0.95$ $W_{at} = 0.0730$

$L_{pm} = 0.0893$ $L_{s4} = 0.0742$ $W'_{ay} = 0.3320$ $W_{s3} = 0.5057$ $A_{sy} = 0.2503$ $A_{s2} = 0.9160$ $L_{attm} = 0.0781$ $\phi_{max} = 53.0$ $K_{tr} = 4.6257$

$$K_{w1} = \sin\left(\frac{90 \cdot \text{span} \cdot P}{N_{at}}\right) \tag{4.498}$$

where span = number of armature teeth spanned by one coil
θ_p = pole arc, degrees

So:

$$Z_{\text{eff}} = Z_{\text{tot}} \cdot K_{w1} \cdot K_{w2} \tag{4.499}$$

where Z_{eff} = effective number of conductors
Z_{tot} = total number of conductors

We will start by assuming a fundamental current I_{fund} at which all of our load points will be calculated. I_{fund} will then be incremented to give a complete speed-torque-current curve.

$$I_{\text{load}} = \frac{I_{\text{fund}}}{0.958} \tag{4.500}$$

$$F_{\text{bl}} = \frac{\theta_B \cdot Z_{\text{tot}} \cdot I_{\text{fund}}}{180 \cdot P \cdot P_a} \tag{4.501}$$

$$F_{\text{dl}} = 0.15 \cdot \frac{\theta_p}{360} \cdot \frac{Z_{\text{tot}} \cdot I_{\text{fund}}}{P \cdot P_a} \tag{4.502}$$

where F_{dl} = mmf drop due to field distortion
F_{bl} = mmf drop due to brush contact
θ_B = brush shift angle, degrees
P = number of poles
P_a = number of paths

$$F_{\text{net}} = I_{\text{fund}} \cdot \text{TPC}_f - F_{\text{dl}} - F_{\text{bl}} \tag{4.503}$$

where F_{net} is the net MMF drop. The air gap flux f_g can then be obtained from the plot of f versus F (see Fig. 4.96), using F_{net}, and must be scaled according to stack height.

$$f_{\text{gact}} = L_{\text{stke}} \cdot f_g \tag{4.504}$$

where L_{stke} is the magnetic stack height, obtained from:

$$L_{\text{stke}} = L_{\text{stk}} \cdot F_{\text{stk}} \tag{4.505}$$

where L_{stk} = mechanical stack height, in
F_{stk} = stacking factor

A phasor diagram (see Fig. 4.96) will be used to calculate the power factor and the sums of the reactive and resistive voltages. The real components of the phasor diagram consist of resistive voltages, the back-emf voltage E_g, and the brush voltage drop V_{br}. We'll assume a V_{br} for now. Then, let C_{real} be the sum of the real components of the vector that we have so far:

$$C_{\text{real}} = I_{\text{fund}} \cdot R_{\text{tot}} + V_{\text{br}} \tag{4.506}$$

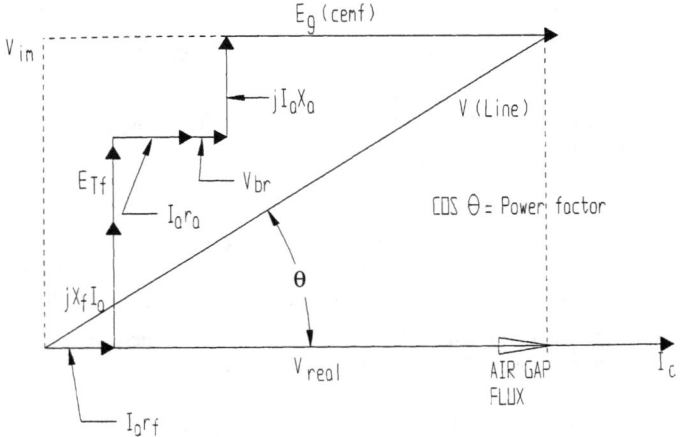

FIGURE 4.96 Universal motor phasor diagram.

Stator field transformer voltage E_{tf}:

$$E_{tf} = \frac{\phi_{gact} \cdot 2 \cdot P \cdot \text{TPC}_f \cdot \text{frequency}}{45 \cdot 10^6} \quad (4.507)$$

where E_{tf} = field transformer voltage
frequency = line-current frequency
TPC_f = turns per coil of the field

The imaginary components of the phasor diagram consist of transformer voltage and reactance voltages. Now, let C_{im} be the sum of the imaginary components of the phasor diagram.

$$C_{im} = E_{tf} + I_{fund} \cdot X_{tot} \quad (4.508)$$

where X_{tot} is the total of the reactances of the motor.
We can now calculate the power factor angle and power factor.

$$\theta_{PF} = \sin^{-1}\left(\frac{C_{im}}{V_{line}}\right) \quad (4.509)$$

$$\text{PF} = \cos \theta_{PF} \quad (4.510)$$

$$V_{real} = V_{line} \cdot \text{PF} \quad (4.511)$$

where V_{real} is the total voltage drop in the real direction.
We can now calculate the back-emf voltage E_g:

$$E_g = V_{real} - C_{real} \quad (4.512)$$

Load speed, in rpm, can now be calculated:

$$S_{load} = \frac{E_g \cdot 60 \cdot 2 \cdot 10^8}{Z_{tot} \cdot 2 \cdot \phi_{gact}} \quad (4.513)$$

Total developed torque, in ounce-inches:

$$T_{dev} = \frac{1351.68 \cdot [(Z_{tot} \cdot P)/(60 \cdot Pa)] \cdot \phi_{gact} \cdot I_{fund}}{10^8} \quad (4.514)$$

The output torque is the difference between the total developed torque and the friction and windage torque T_{fw} where:

$$T_{fw} = S_{load} \cdot K_{fe} + T_{fi} \quad (4.515)$$

where T_{fw} = torque due to friction and windage, oz · in
K_{fe} = slope of the friction and windage torque versus speed line, (oz · in)/rpm
T_{fi} = intercept of the friction and windage torque versus speed line, oz · in

Torque due to core loss must be calculated for each section of the stator and armature. In general, the formula for core loss is as follows:

$$CL = Vol_{steel/section} \cdot \rho_{steel} \cdot \frac{W}{lb} \quad (4.516)$$

where: CL = core loss, W
$Vol_{steel/section}$ = volume of steel of each section
ρ_{steel} = density of the lamination steel
W/lb = core loss per pound from the steel manufacturers' core loss curve

Since core loss is dependent on frequency, that core loss in the armature will depend on the load speed and the following formula:

$$\text{Frequency} = \frac{S_{load} \cdot P}{120} \quad (4.517)$$

For each section of the stator, core loss must be calculated at the line-current frequency. Torque loss, oz · in, due to core loss can then be calculated from CL_{tot}, W, the sum of all core losses.

$$T_{CL} = \frac{CL_{tot} \cdot 1{,}008{,}000}{746 \cdot S_{load}} \quad (4.518)$$

4.5.4 Friction and Windage

Next, we must calculate the torque loss due to friction and windage. Friction and windage is more complicated to determine for a universal motor than it is for a PMDC or brushless dc (BLDC) motor. In order to calculate the friction and windage constants for a universal motor, the motor must be driven with another motor, usually a PMDC or BLDC motor. To calculate the friction and windage constants of the driver motor, the motor must be disconnected from any load sources, such as a dynamometer. Then, by recording speed versus current, the friction and windage constants can be determined as follows:

1. Multiply current, A, by the torque constant K_t, (oz · in)/A, to get friction and windage torque, oz · in.
2. Plot speed, rpm, versus friction and windage torque, oz · in.

3. The slope of the best line fit going through this curve will be K_{fe}, (oz · in)/rpm. If desired, K_f, oz · in · s, can be calculated by multiplying by 60/2π.
4. The intercept of the line is T_{fi}, oz · in.

To determine the friction and windage constants for the universal motor, connect the driver motor's shaft to the test motor's shaft with a coupling. Repeat the test to obtain a curve like that shown in Fig. 4.97, record the speed versus current, and calculate the constants as follows:

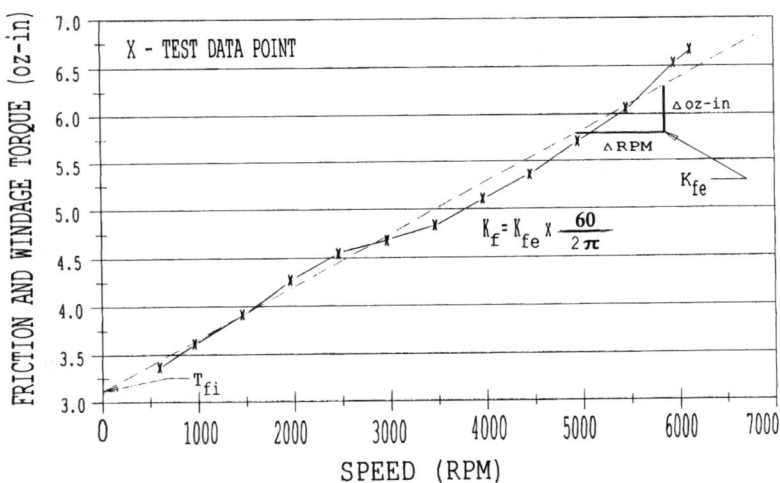

FIGURE 4.97 Friction and windage torque.

1. Multiply current, A, by the *driver motor's* torque constant to get the total friction and windage torque, oz · in.
2. Calculate the driver motor's friction and windage torque from the following equation:

$$T_{f+w} = \text{speed} \cdot K_{fe} + T_{fi} \quad (4.519)$$

3. Subtract the driver motor's friction and windage torque from the total friction and windage torque to get the test motor's friction and windage torque.
4. The slope of the best line fit going through this curve will be K_{fe}, (oz · in)/rpm, for the test motor. If desired, K_f, oz · in · s, can be calculated by multiplying by 60/2π.
5. The intercept of the line is T_{fi}, oz · in, for the test motor.
6. Alternately, a polynomial may be used to determine the curve.

The load torque is then calculated by subtracting the friction and windage torque and the torque loss due to core loss from the total developed torque:

$$T_{\text{load}} = T_{\text{dev}} - T_{\text{fw}} - T_{\text{CL}} \quad (4.520)$$

Power, in watts, can be calculated from the following two formulas:

$$P_{out} = \frac{S_{load} \cdot T_{load} \cdot 746}{1,008,000} \quad (4.521)$$

$$P_{in} = V_{line} \cdot PF \cdot I_{fund} \quad (4.522)$$

The efficiency can be calculated from the following formula:

$$Eff = \frac{P_{out}}{P_{in}} \cdot 100 \quad (4.523)$$

Table 4.2 shows the calculated results for the sample motor. Figure 4.98 shows the speed-torque curve of the calculated results.

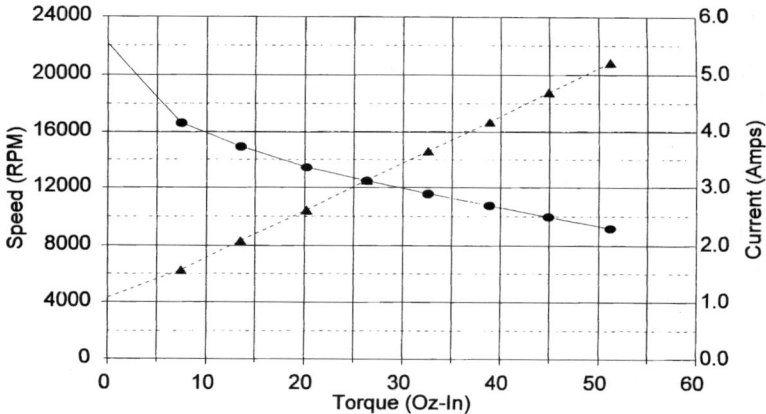

FIGURE 4.98 Calculation of speed-torque curve of sample motor (● = speed, ▲ = current).

4.6 SHUNT-CONNECTED DC MOTOR PERFORMANCE*

Shunt-connected motors are constructed similarly to series-connected motors, as shown in Fig. 4.99. The shunt motor has performance characteristics much like those of the PMDC motor. It has a straight-line speed-torque curve over a limited range of performance. In this motor the permanent magnets are replaced by wound field coils, much as they are in the series dc motor. A series motor generally has a relatively few turns of large wire in the field. The shunt motor has a large number of turns of fine wire in comparison. Further, the field winding is connected across the armature—that is to say, in parallel with it, as shown in Fig. 4.100. The armature then needs to have a larger number of turns, as it sees full line voltage.

---

* Sections 4.6 to 4.8 contributed by William H. Yeadon, Yeadon Engineering Services, PC.

TABLE 4.2 Performance Calculations for Sample Universal Motor

Parameter	Value									
I_{fund}, A	0.00	0.50	1.00	1.50	2.00	2.50	3.00	3.50	4.00	4.50	5.00
I_{load}, A	0.00	0.52	1.04	1.57	2.09	2.61	3.13	3.65	4.18	4.70	5.22
F_{di}, A · turn	0.0	7.2	14.3	21.5	28.7	35.8	43.0	50.1	57.3	64.5	71.6
F_{bi}, A · turn	0.0	15.0	30.0	45.0	60.0	75.0	90.0	105.0	120.0	135.0	150.0
F_{net}, A · turn	0.0	47.8	95.7	143.5	191.3	239.2	287.0	334.9	382.7	430.5	478.4
ϕ_{temp}, kline	0.0	11.6	23.1	30.3	32.5	34.7	35.7	36.6	37.2	37.6	38.0
ϕ_{gact}, kline	0.0	13.8	27.4	38.6	41.2	42.4	43.4	44.2	44.7	45.2	
C_{real}, V	2.00	5.05	8.10	11.15	14.20	17.25	20.30	23.35	26.40	29.45	32.50
E_{tf}	0.00	0.01	0.02	0.03	0.03	0.03	0.03	0.03	0.03	0.03	0.03
V_{im}, V	0.00	5.41	10.81	16.21	21.61	27.01	32.40	37.80	43.19	48.59	53.98
θ_{PF}, degrees	0.0	2.6	5.2	7.8	10.4	13.0	15.7	18.4	21.1	23.9	26.7
PF	1.000	0.999	0.996	0.991	0.984	0.974	0.963	0.949	0.933	0.914	0.893
V_{real}, V	120.00	119.88	119.51	118.90	118.04	116.92	115.45	113.89	111.96	109.72	107.17
E_g, V	118.00	114.83	111.41	107.75	103.84	99.67	95.24	90.54	85.56	80.27	74.67
S_{load}, rpm	∞	46131.5	22606.8	16628.1	14939.0	13429.9	12471.8	11586.3	10760.2	9982.3	9182.6
T_{dev}, oz · in	0.00	1.68	6.66	13.14	18.79	25.08	30.97	36.97	42.99	48.91	54.96
T_{fq}, oz · in	∞	13.41	7.27	5.71	5.27	4.88	4.63	4.39	4.18	3.98	3.77
T_{load}, oz · in	N/A	−11.73	−0.61	7.43	13.52	20.20	26.34	32.57	38.81	44.94	51.19
P_{out}, W	N/A	N/A	N/A	91.40	149.48	200.80	243.13	279.32	309.07	331.98	347.89
P_{in}, W	N/A	N/A	N/A	178.35	236.08	292.30	346.63	398.62	447.83	493.75	535.86
Eff	N/A	N/A	N/A	51.25%	63.32%	68.70%	70.14%	70.07%	69.01%	67.24%	64.92%

where

V_{line} = 120 V
span = 5 teeth
N_{at} = 12 teeth
P_a = 2 poles
P_a = 2 paths
θ_P = 122°

N_{CB} = 12 bars
C_s = 2 sides
TPC_a = 45 turns
Z_{tot} = 1080 conductors
K_{w1} = 0.9659
K_{w2} = 0.8746

Z_{eff} = 912.4 conductors
θ_B = 20°
TPC_f = 140 turns
L_{stk} = 1.25 in
F_{stk} = 0.95
L_{stke} = 1.1875 in

frequency = 60 Hz
R_{am} = 2.634 Ω
R_f = 3.466 Ω
R_{tot} = 6.1 Ω
V_{br} = 2.0 V
X_{tot} = 10.79 V

K_f = 0.002493 oz · in · s
T_{fi} = 1.37 oz · in

FIGURE 4.99 Shunt-wound dc motor field.

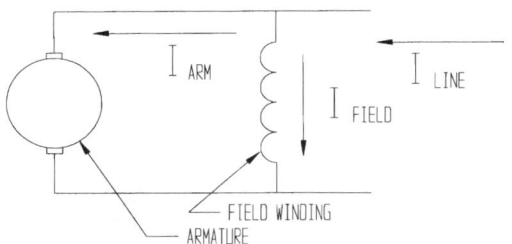

FIGURE 4.100 Shunt-connected dc motor schematic diagram.

To calculate the performance of this motor, one first needs to calculate the field flux. This is done by calculating the shunt winding resistance and using it to calculate the shunt current. Then, from the known ampere-turns one can calculate the flux and determine the mmf in the magnetic circuit. This procedure is the same as for the PMDC motor except that the magnet is replaced by a coil which produces the field mmf.

The motor geometry and magnetic paths for the field and armature are calculated in the same manner as for the series-wound dc motor.

Windings are selected such that the ampere-turns in the armature are about 75 percent of the ampere-turns in the field at full load.

The required no-load flux ϕ_{nl} is determined by the desired no-load speed S_{nl} and is calculated as follows:

$$\phi_{nl} = \frac{E_g P_a (60 \times 10^8)}{P Z_a S_{nl}} \tag{4.524}$$

where E_g = armature generated voltage or cemf
       P = number of poles
       P_a = number of parallel armature paths
       Z_a = active armature conductors

$$E_g = V_L - I_{anl}R_{at} \tag{4.525}$$

where V_L = line voltage
       I_{anl} = estimated armature no-load current
       R_{at} = armature terminal resistance

It is obvious that this type of motor limits the maximum no-load speed S_{nl}. The constant flux causes the speed to stop increasing when $E_g = V_L$.

When designing the field winding one must add in enough full-load ampere-turns to account for the ampere-turns lost as a result of field distortion caused by armature reaction and brush shift.

According to Puchstein (1961), field distortion F_{dl}, A · turn, is calculated as follows:

$$F_{dl} = 0.3(1 - \psi_p)\frac{Z_a I_a}{2P} \tag{4.526}$$

Where: ψ_p = pole arc ÷ pole pitch
       Z_a = active armature conductors
       I_a = current per armature path
       P = number of poles

Field distortion F_{bl}, A · turn, caused by brush shift is calculated as follows:

$$F_{bl} = \frac{0.0055 \theta_b Z_a}{P} \tag{4.527}$$

where

θ_b = brush shift angle, degrees

Shunt windings produce a nearly constant flux over a limited range of loads. This allows for a fairly straight-line speed-torque curve (Fig. 4.101).

As the load increases, the armature reaction causes F_{dl} to increase to a point where the soft-iron pole tips are demagnetized. This reduces the main pole flux, which results in a reduction of torque. This field distortion can be seen in the finite element shown in Fig. 4.102.

To avoid commutation problems, shunt motors should be operated well above the point on the speed-torque curve where field distortion becomes severe.

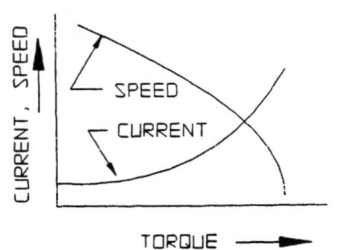

FIGURE 4.101 Shunt-connected dc motor field performance characteristics.

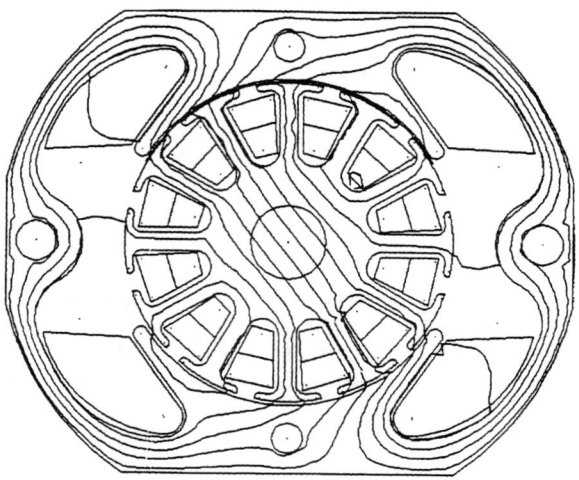

FIGURE 4.102 Magnetic field of shunt-connected dc motor with field and armature energized.

4.7 COMPOUND-WOUND DC MOTOR CALCULATIONS

Compound-wound dc motor construction is similar to series-wound dc motor construction, as shown in Fig. 4.103. The major difference is that this motor has both a series and a shunt field. The shunt field limits the maximum motor speed, in a manner similar to that in the PMDC and shunt-wound motors. The series winding increases starting torque and limits the starting current of the motor. The inductance of the series winding also helps to reduce flashover under rapid load changes.

There are two common types of compound motor connections. The *long-shunt connection* has the shunt winding across the power source, as shown in Fig. 4.104. Here the shunt field flux is a function of the line voltage and the shunt field resistance.

The *short-shunt connection* is shown in Fig. 4.105. In this configuration, the shunt winding is across the armature but in series with the series winding. Here the shunt field current is a function of the voltage across the armature and the voltage drop in the series winding. At stall there is no generated voltage in the armature, so the shunt winding sees a voltage V_{sh}.

$$V_{line} = V_{se} + V_{sh+Ra} \tag{4.528}$$

$$V_{line} = I_{se}R_{se} + I_{se}\left[\frac{R_{sh}R_a}{R_{sh}+R_a}\right] \tag{4.529}$$

This is a function of the series winding resistance R_{se} and parallel combination of the shunt winding resistance R_{sh} and the armature resistance R_a.

As the speed increases, the generated voltage component is added, which effectively increases R_a, causing the equivalent parallel resistance to increase. This results in an increase in voltage across the shunt winding. This causes an increase in the field

DIRECT-CURRENT MOTORS

FIGURE 4.103 Compound-wound dc motor field.

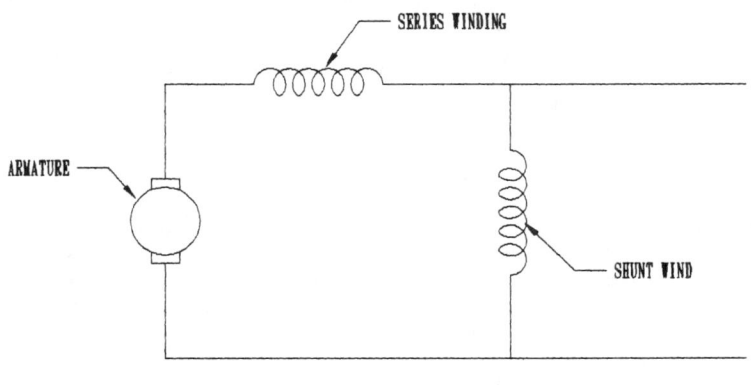

FIGURE 4.104 Compound motor long-shunt connection diagram.

flux. At the same time, the current through the series winding is decreasing, which results in a decrease in field flux.

The shunt winding turn counts for the short and long connections would have to be different to get the same resultant and field flux because of the different voltages across them.

There are two different types of compound motors in common use. They are the cumulative compound motor and the differential compound motor. In the *cumulative compound motor,* the field produced by the series winding aids the field pro-

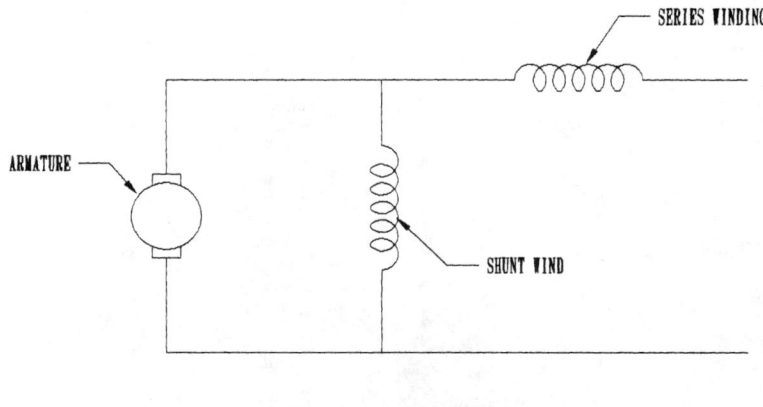

FIGURE 4.105 Compound motor short-shunt connection diagram.

duced by the shunt winding. The speed of this motor falls more rapidly with increasing current than does that of the shunt motor because the field flux increases.

In the *differential compound motor,* the flux from the series winding opposes the flux from the shunt winding. The field flux, therefore, deceases with increasing load current. Because the flux decreases, the speed may increase with increasing load. Depending on the ratio of the series-to-shunt field ampere-turns, the motor speed may increase very rapidly.

4.7.1 Performance Calculations

Calculation of motor performance for this motor is done in a fashion similar to that for the series motor discussed earlier in this chapter. All of the slot areas, magnetic paths, and other geometric properties are virtually identical to those of the series motor. The flux and mmf drops for this motor must be calculated at each desired speed or load point. This is necessary because the series winding causes the total field flux to change as the load current changes. The changes for the cumulative connection are not as dramatic as those of the series motor because the shunt field adds a constant component of flux. The flux change for the differential connection can, however, be very dramatic.

4.7.2 Cumulative Connected Motor

It is typical practice to initially proportion the available field winding space to allow 80 percent for the shunt winding and 20 percent for the series winding. The ratio of series field ampere-turns to shunt field ampere-turns is chosen in the range of 15 to 30 percent. The full-load current is estimated. From this the wire sizes are chosen to keep the I^2R less and current densities within reasonable limits.

Select the no-load speed and calculate the necessary flux.

No-load flux ϕ_{gnl}:

DIRECT-CURRENT MOTORS

$$\phi_{gnl} = \frac{E_g P_a (60 \times 10^8)}{PZ_{act} S_{nl}} \quad (4.530)$$

Where E_g = back emf at no load, estimated to be $V_{line} - I_a R_a - I_a R_{se}$
P = number of poles
P_a = number of parallel armature paths
Z_{act} = effective number of armature conductors
S_{nl} = desired no-load speed, rpm

Next, estimate the full-load current I_{fl} required for the desired output power.

$$W_o = \text{hp} \times 745.7 \quad (4.531)$$

$$W_{in} = \frac{W_o}{\eta} \quad (4.532)$$

Where W_o = watts output
W_{in} = watts input
η = assumed efficiency

This typically ranges between 25 percent on very small motors to 90 percent on very large motors. You need to have a reasonable estimate for the type of product upon which you are working.

$$I_{fl} = \frac{W_{in}}{V_{line}} \quad (4.533)$$

Use this value to solve the magnetic circuit for the air gap flux ϕ_{gfl}.
The back-emf E_g is estimated from

$$E_g = V_{line} - I_{se} R_a - I_{se} R_{se} \quad (4.534)$$

where $I_{se} = I_{fl} - I_{shunt}$

Once E_g is known, the full-load speed N_{fl} becomes

$$N_{fl} = \frac{E_g P_a (60 \times 10_8)}{PZ_{act} \phi_{gfl}} \quad (4.535)$$

The full-load torque T_{fl} can be found as follows:

$$T_{fl} = \frac{W_a 5250}{(745.7) N_{fl}} \quad \text{lb} \cdot \text{ft} \quad (4.536)$$

4.7.3 Differentially Connected Motor Performance

The procedure for calculating the performance of differentially connected motors is the same as for cumulative motors except for the full load air gap flux ϕ_{gfl}. In this case, the magnetic circuit is solved by subtracting flux due to the ampere-turns of the series winding from the shunt field flux. From the preceding formulas, one can see that a smaller full load air gap flux ϕ_{gfl} will cause the full-load speed to increase over the no-load speed.

One can see that if the efficiency is estimated incorrectly, the calculated performance results will be incorrect. This is, in effect, an iterative process which can be done rapidly and accurately with the aid of modern computers.

4.8 DC MOTOR WINDINGS

There are two common armature winding schemes in PMDC motors; lap winding and wave winding. *Lap windings* are wound as shown in Fig. 4.106. This configuration is for a 12-slot armature with a 2-pole field. In this case, the winding pitch is slots 1 to 6, or a span of 5 teeth. The coil pitch should be greater than the arc of the field coil (or permanent magnet) for good commutation. The first coil goes on slots 1 to 6. The next coil is wound on slots 5 to 12. Then keep shifting 1 slot at a time until all 12 coils are in place. Coil ends are likewise connected to commutator bars in sequence as each coil is wound. As these coils are wound in sequence, the outer coils are necessarily larger than the inner coils because the end turns overlap each other. This results in mechanical imbalance.

FIGURE 4.106 Completed lap winding pattern.

Mechanical imbalance can be overcome to some extent by using a double flier and winding two coils at once, as shown in Fig. 4.107.

Coils 1 to 6 and 12 to 7 are wound simultaneously. Next, coils 2 to 7 and 1 to 8 are wound simultaneously. This process continues until all 12 coils are in place. Coil ends are connected to the commutator bars in sequence as each coil is completed. This is the most common type of winding for PMDC motors.

The lap winding is also used in motors having more than two poles, but one pair of brushes is required for each pair of poles. The number of pairs of brushes can be reduced by using a winding method called *wave winding*.

In the case of wave winding, the coil ends are not tied to an adjacent commutator bar after each coil is completed. Instead, coils are wound so that a coil under one pole is connected to a coil 180 electrical degrees away which is under a like pole. This allows one pair of brushes to commutate two pairs of poles.

If a two-pole motor is being designed, the winding type is necessarily a lap configuration. A four-pole motor may use lap winding or wave winding. With lap windings, wave windings, or even slot armatures, a pair of brushes is required to commutate each pair of poles. In the case of an odd-slot four- or six-pole motor, the coils can be commutated by a single pair of brushes. The one-pair brush rigging in a

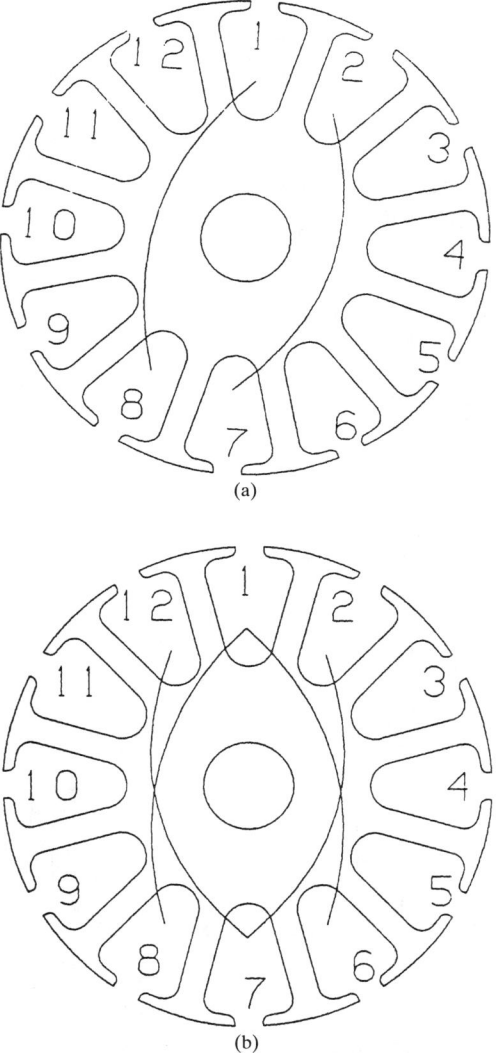

FIGURE 4.107 Double-flier lap winding pattern: (*a*) first set of coils wound, and (*b*), second set of coils wound.

wave-wound odd-slot motor will be less complicated than the one brush pair per pair of poles rigging would be. However, the current density in these brushes would be higher because now they would carry all of the current. The area of each brush would have to be increased to get the current densities to reasonable levels.

In lap winding, the finish wire is connected to the next adjacent bar. The wave winding has its start and finish wires connected approximately 360 electrical degrees apart instead of adjacent. Diagrams for the lap and wave winding patterns are shown in Figs. 4.108 and 4.109, respectively.

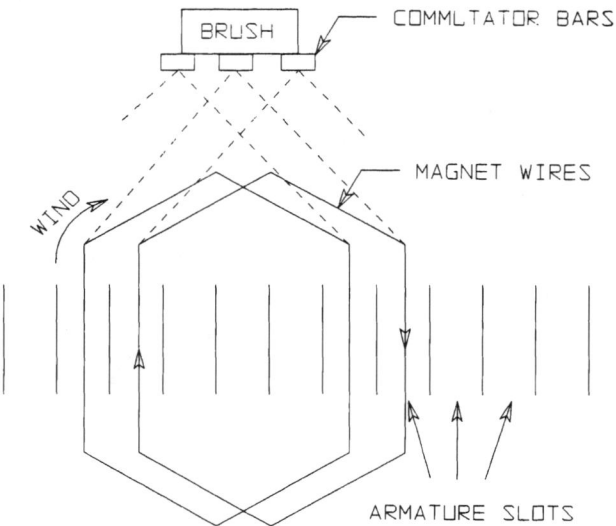

FIGURE 4.108 Diagram for lap winding pattern.

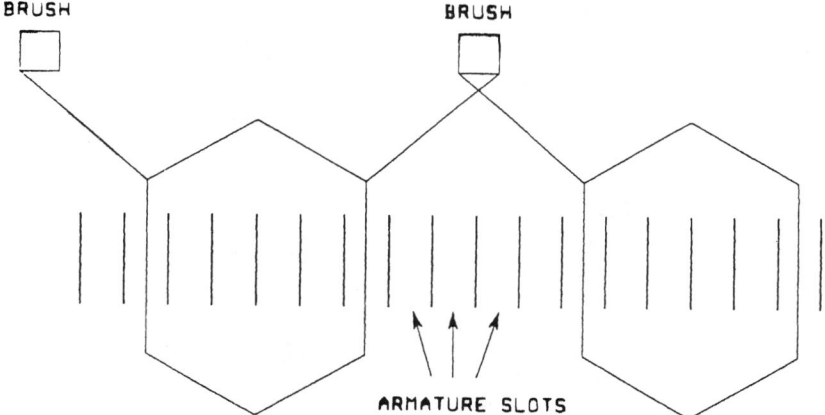

FIGURE 4.109 Diagram for wave winding pattern.

4.9 AUTOMATIC ARMATURE WINDING PIONEERING THEORY AND PRACTICE*

4.9.1 Background

In the 1930s, the electrical manufacturing industry began to realize and attempt to meet the demand for small electrical motors.

---

* Section courtesy of Globe Products, Dayton, Ohio.

In the low-voltage range, these included 6- or 12-V dc motors for automotive fans, windshield wipers, window operators, etc.; and 15- to 30-V ac motors for toy trains, erector sets, and other toys. In the household voltage range, these included motors for sweepers, mixers, and hand tools.

These motors were of a two-pole, series-connected arrangement, having a slotted armature, with coils wound into nearly diametrically spaced slots, with start and finish wires connected to adjacent bars on a commutator. Large production rooms were filled with workers using simple tools and equipment to insulate the slots, place the coils therein, wedge them in place, connect them to the commutator, and so on.

The advent of World War II nearly halted the production of these motors for peacetime uses, requiring production of motors to the more exacting requirements of the armed services for navigation and munitions control devices. More handwork was required, and mechanized apparatus was more restricted in meeting the extremely rigid specifications set by the services.

Following the end of the war, peacetime requirements surged. At the same time, labor tended to hold back production rates. It became necessary for the machines to pace the operators, whereas the operators could previously be depended on to pace the production from simple machines and fixtures.

4.9.2 Semiautomatic Winding Methods

For years, the armature had been held in a fixture fitted with guide wings, which was rotated (end over end) for the desired number of turns within a coil, stopped, and positioned by the operator. A lead loop was pulled and laid on a hook or under a clip. The armature was indexed in the fixture to the new coil location, and the process was repeated.

Technique and kinds of lead loops as well as sequence of indexing varied between different applications and according to the ingenuity of the manufacturer's designer. Certain patterns became geographically popular.

Then the method of holding the armature stationary and spinning the wire in with a flier became popular. The operator could now index more simply, and could hold the lead loop with a finger while starting the next coil.

All these methods involved hand operations which could not efficiently be mechanized to produce the full variety of armature patterns and lead shapes to which each manufacturer had become accustomed.

4.9.3 Automating the Wind Cycle

It became necessary to standardize the armature production industry with mechanisms in mind, rather than involving the human element. Only then could automatic machines gain acceptance.

As customers became interested in automatic equipment, their specific armature patterns were analyzed, and new patterns substuted, permitting mechanization. The writer recalls spending long tedious hours in conference with engineering groups in each customer instance, illustrating the equivalency of our recommendations to their favorite customary method. Eventually, one by one, they became sufficiently convinced to ask for test samples for actual proof.

During the past quarter century the appropriateness of the automatic equivalent winding patterns, along with the economic pressure for automation, has brought an end to the era of hand-wound armatures. Today's designers following industry-standard armature practices do not even realize that such practices were developed by Globe in order that armature production could be automated.

4.9.4 Armature Analysis Method

The following theory is recorded in order that this tried and proven method of armature analysis can be relied upon, with confidence, in the application of further improvement methods.

In general, the small series-connected motor, frequently called the *universal motor,* consists of a two-pole field surrounding the armature, as shown in Fig. 4.110.

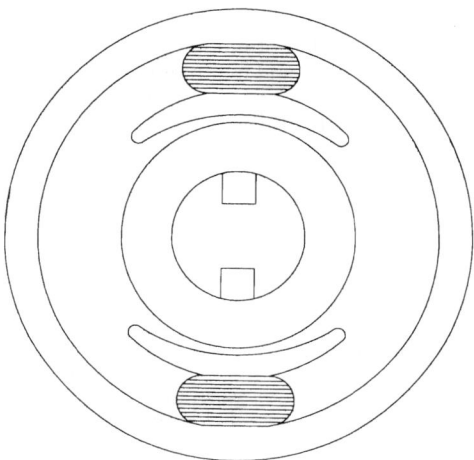

FIGURE 4.110 Universal motor field winding.

The field yoke may be circular, as illustrated, with coils on each pole; or it may extend from pole to pole only on one side of the armature, as in Fig. 4.111, in which case there is usually only one field coil on the leg of the yoke. In many cases, permanent-magnet fields are used, wherein no field coils are required.

The armature may have 12 slots and a 12-bar commutator, as illustrated in Fig. 4.112, wherein there will be 1 coil per slot (2 coil sides per slot).

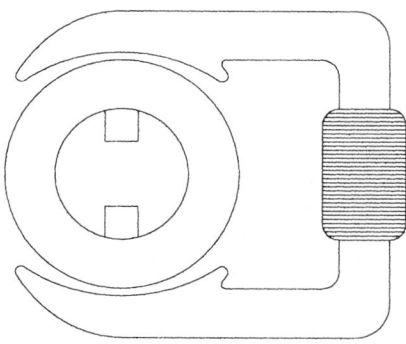

FIGURE 4.111 Single-coil-type universal motor field winding.

There may be from 8 to 12 slots; and in rare instances numbers beyond this range.

Also, there may be two coils per slot (four coil sides in each slot) in which case there are twice as many commutator bars as slots. In rare cases, there may be three coils per slot.

The brushes may be located in line with the center of the poles, as illustrated in Fig. 4.112, or shifted in angularity therefrom at the will of the designer, usually for convenience of space, access, or other reasons. In such cases the coil leads are similarly extended; see Fig. 4.113.

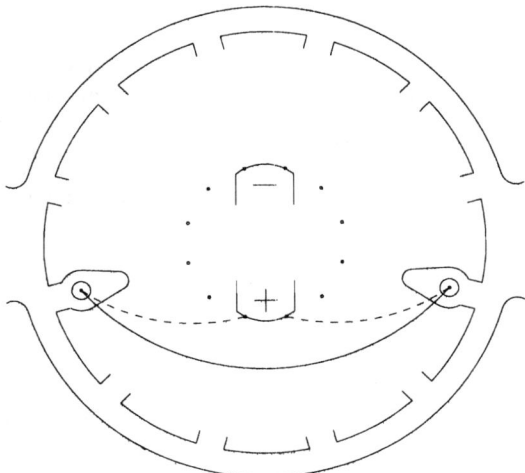

FIGURE 4.112 Armature winding showing brushes located in center pole.

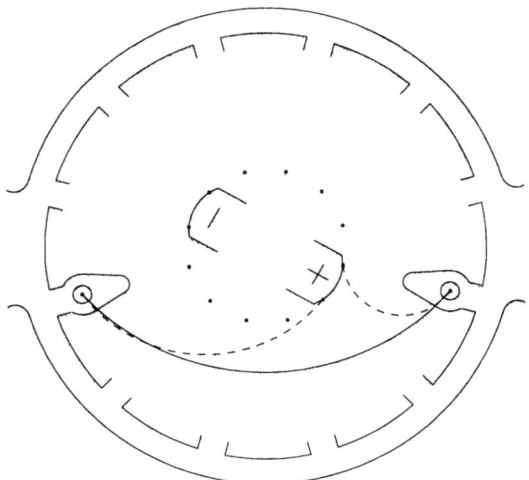

FIGURE 4.113 Armature winding showing brushes located offset from center pole.

The single-coil 12-slot armature with in-line brushes has been selected for simplicity in introducing this analysis. Also, the coils are shown to be wound as from a single wire continuing from coil to coil in sequence, as from a single-flier winder. The result is a single-lap appearance to the finished armature, with small start coils, medium single-overlap coils, and large double-overlap final coils, producing inherent electrical and mechanical unbalance in the armature. This is in contrast to double-flier-wound armatures which show a double-lap finish with similar equal and opposite coils, resulting in good balance.

In the end views, such as in Fig. 4.114, looking at the commutator end, the outer two rings of dots represent eventual coil sides, while the inner ring of dots represents the commutator bars. Brushes are indicated inside the commutator ring for clarity, whereas they normally extend to the outside.

The side views, as in Fig. 4.114, illustrate the armature as though it were unrolled and laid out flat. This is usually the only view used in analyzing the winding pattern.

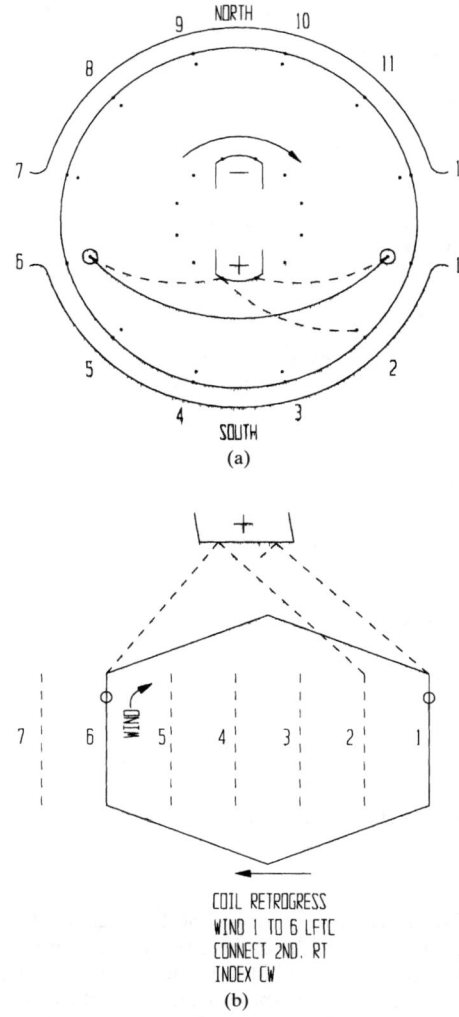

FIGURE 4.114 Armature lap winding, first coil retrogresses: (*a*) end view, and (*b*) side view.

As the armature rotates, the brushes, in contacting the commutator bars, each will touch two adjacent bars. This shorts out the coil attached there, and that coil is said to be *under commutation*. In other words, prior to becoming shorted the current flowed in one direction, but after clearing from the short, the current will flow in the opposite direction through the coil. This shorting must occur while the coil sides are not under the poles. Otherwise the induced voltage in the shorted coil would overheat the coil, also causing severe sparking and burning of the brushes and bars.

Note in Fig. 4.114, that the commutating coils are indicated by a small circle on the coil sides. Also, the direction of current flow in the active coils is indicated by the arrowhead (from + toward −), Fig. 4.116.

The direction the motor would run is indicated by the circular arrow in the end view when full conditions are disclosed.

4.9.5 Winding Specifications

Normally, a 12-slot armature is wound to a 1-to-6 pitch (one slot short of diametric). In the armature selected for analysis, shown in Figs. 4.114 to 4.125, the direction of wind is away from the commutator in slot 1, towards slot 6. In a double-flier winder this is *left flier top coming* (LFTC).

The *retrogressive wind* (Fig. 4.114) connects the finish wire to the second bar to the right (connect second right), continuing to slot 2 to wind the next coil, 2 to 7 (Fig. 4.116). Please note that the coils advance in the opposite direction from the path in forming the lead. Thus, *retrogressive*.

The *progressive wind* (Fig. 4.115) requires the finish wire connection third bar right, and continues to slot 12 for the 12-to-5 coil (Fig. 4.117).

The index of the armature in Fig. 4.114 is clockwise (CW), looking at the shaft end (the end opposite the commutator). Indexing the armature CW causes the counterclockwise (CCW) advance of coils.

4.9.6 Armature Polarity

The polarity of the armature becomes obvious with the winding of the second coil, as shown in Figs. 4.116 and 4.117.

Tracing the path of current flow away from the positive brush in Fig. 4.116, through the active coil 2 to 7, the direction through the coil sides within the coil commutated by the positive brush is away from the commutator.

Figure 4.117 shows the opposite current direction. Therefore, the armature of Fig. 4.117 will run in the opposite direction from that of Fig. 4.116.

Note that this discussion is in terms of a steady dc condition. However, the relationship still exists for any instantaneous condition of ac motors. Thus, the same method of analysis is used in both ac and dc universal motors.

Figures 4.118 and 4.119 show each armature wind continued to the point of the first overlap. Note in Fig. 4.118 that two coil sides appear in slot 6, while in Fig. 4.119, two coil sides appear in slot 1.

Figures 4.120 and 4.121 show the addition of coil 7 to 12 (12 to 7 progressive), which coil is opposite the commutated coil 1 to 6. The coil 7 to 12 is commutated by the negative brush while the coil 1 to 6 is commutated by the positive brush.

Figure 4.122 shows the continuance of wind to the first double-overlap coil 8 to 1. Also, Fig. 4.123 shows the first double-overlap coil 11 to 6 in place.

Figures 4.124 and 4.125 show the wind completed.

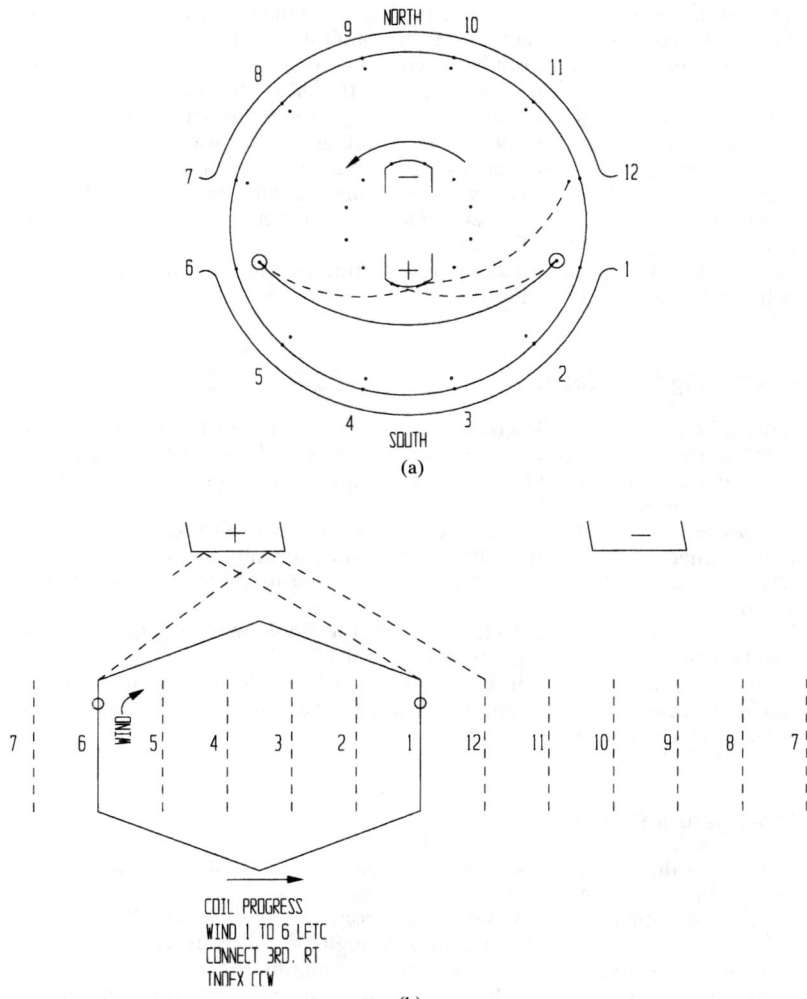

FIGURE 4.115 Armature lap winding, first coil progresses: (*a*) end view, and (*b*) side view.

DIRECT-CURRENT MOTORS 4.147

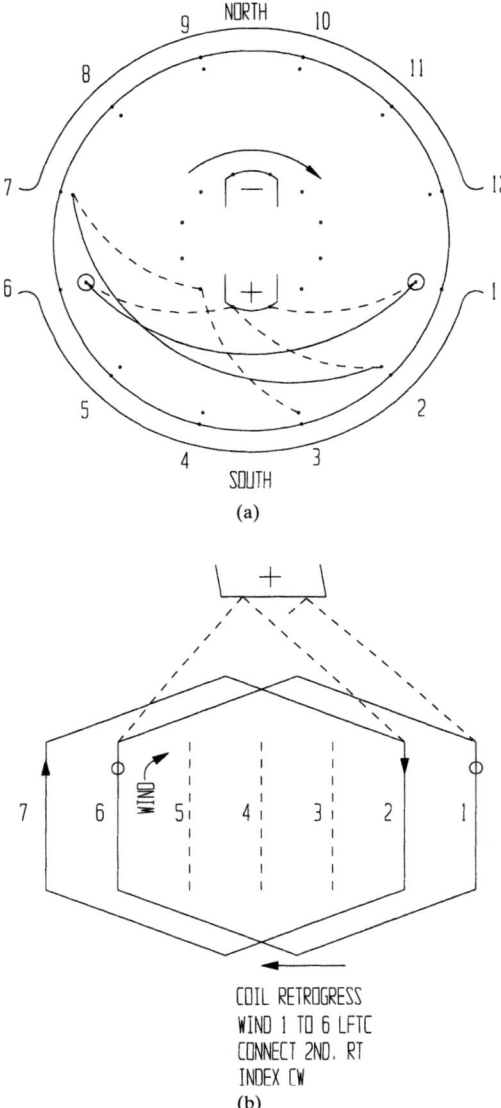

FIGURE 4.116 Armature lap winding, second coil retrogresses: (*a*) end view, and (*b*) side view. Polarity and connection pattern are now indicated.

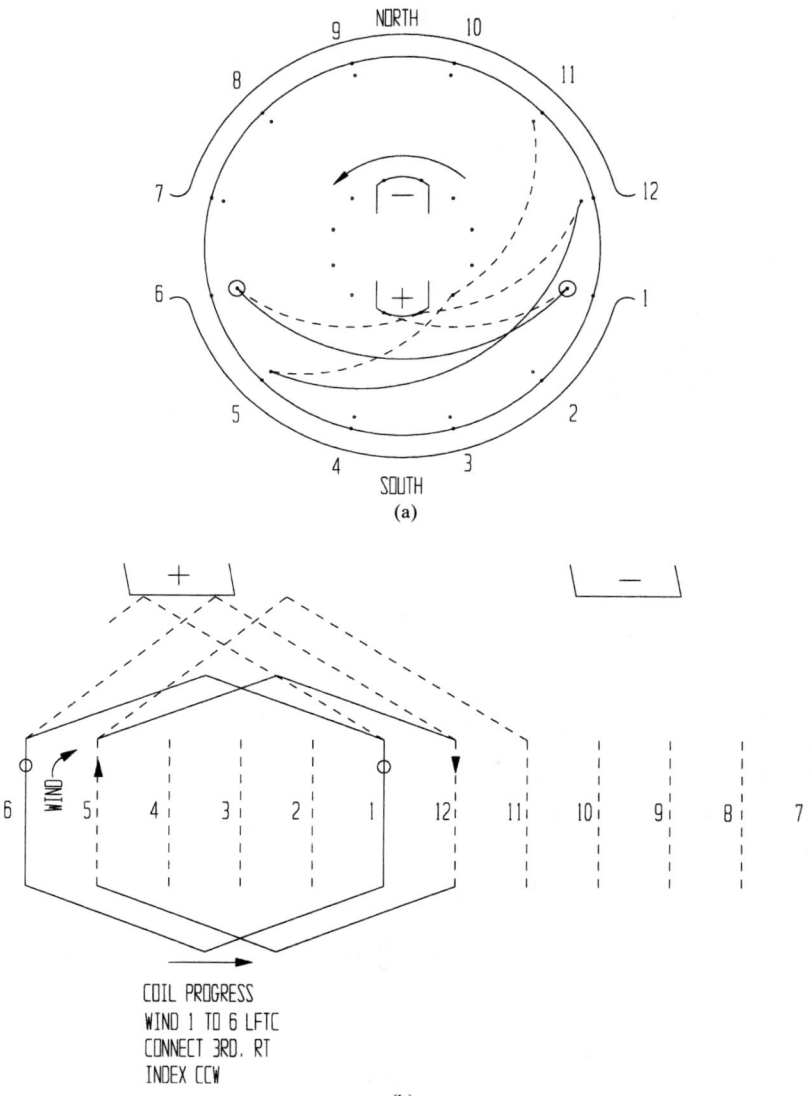

FIGURE 4.117 Armature lap winding, second coil progresses: (*a*) end view, and (*b*) side view. Polarity and connection pattern are now indicated.

DIRECT-CURRENT MOTORS

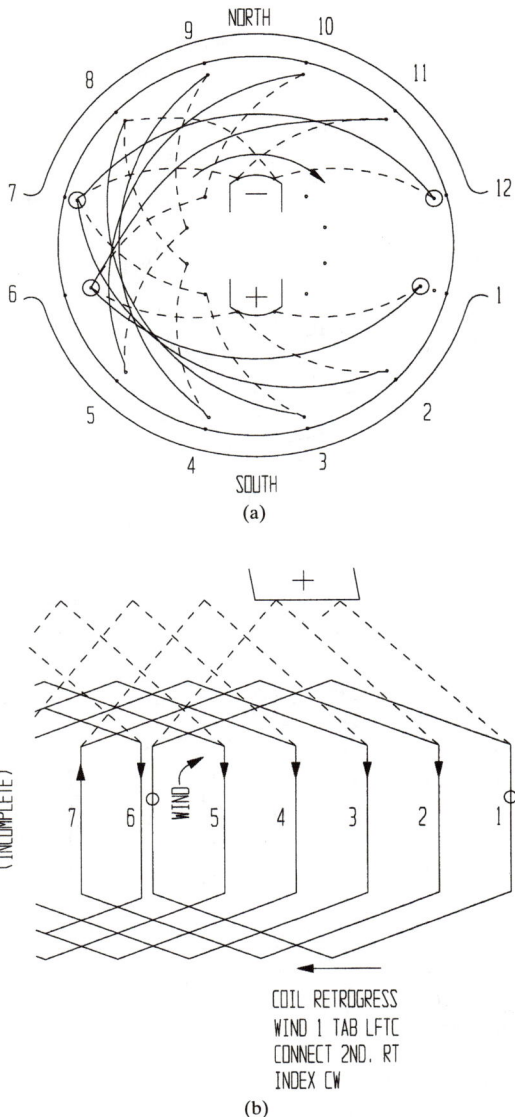

FIGURE 4.118 Armature lap winding, first overlap, coil retrogresses: (*a*) end view, and (*b*) side view.

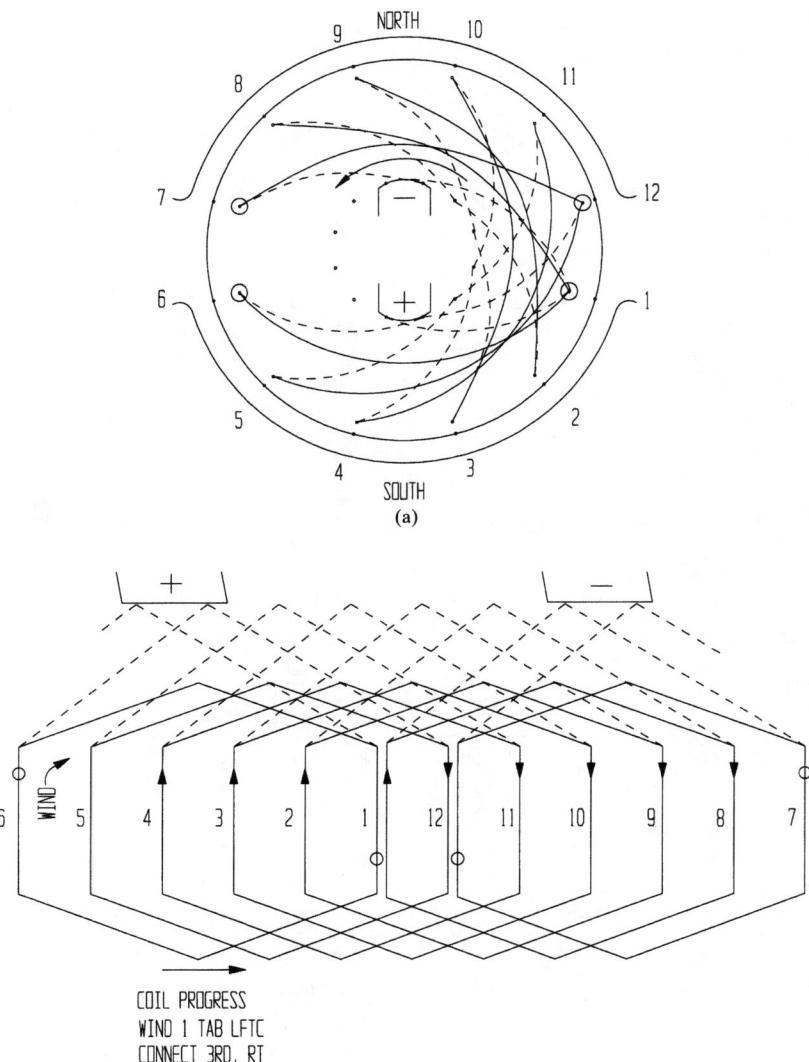

FIGURE 4.119 Armature lap winding, first overlap, coil progresses: (*a*) end view, and (*b*) side view.

DIRECT-CURRENT MOTORS

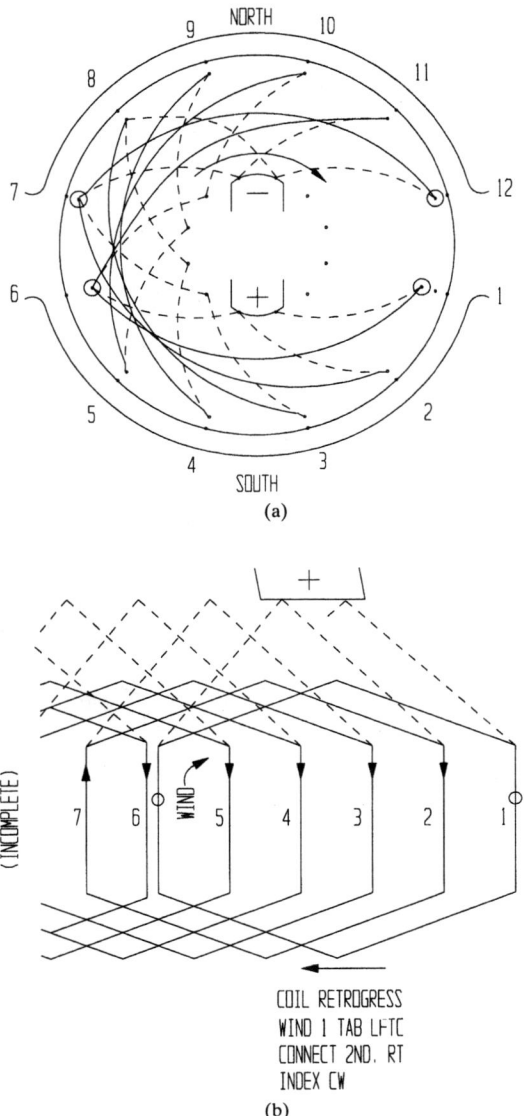

FIGURE 4.120 Armature lap winding, second commutating coil retrogresses: (*a*) end view, and (*b*) side view.

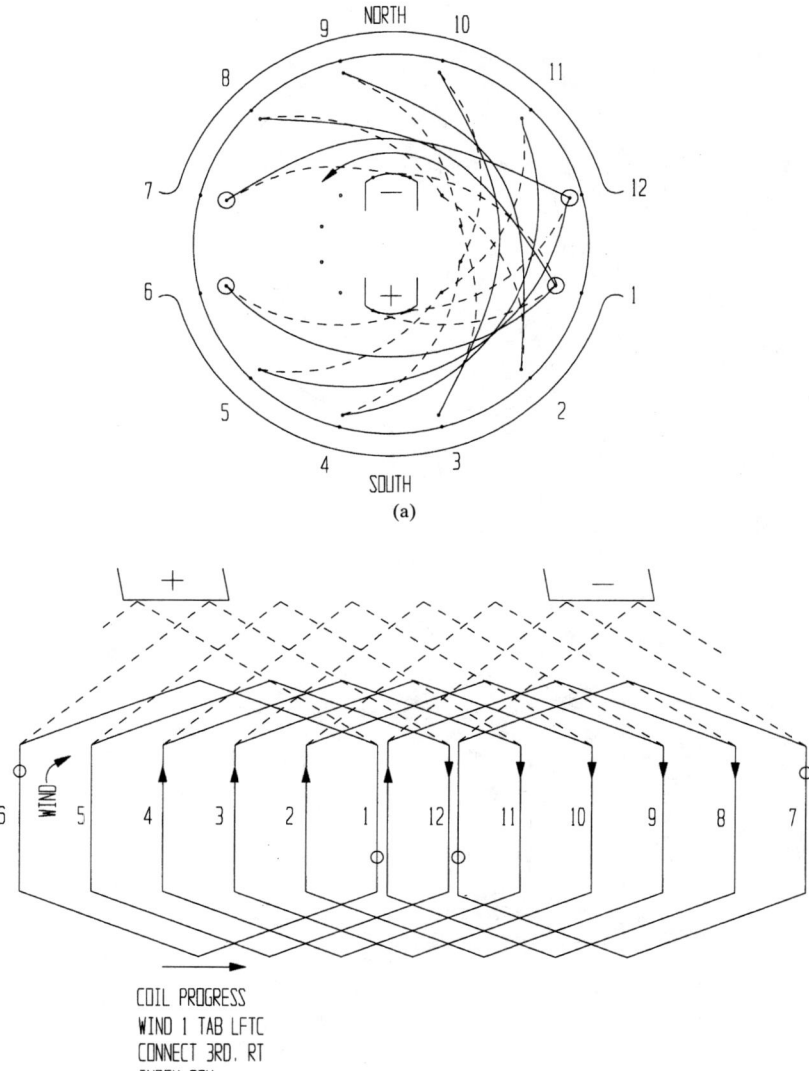

FIGURE 4.121 Armature lap winding, second commutating coil progresses: (*a*) end view, and (*b*) side view.

DIRECT-CURRENT MOTORS

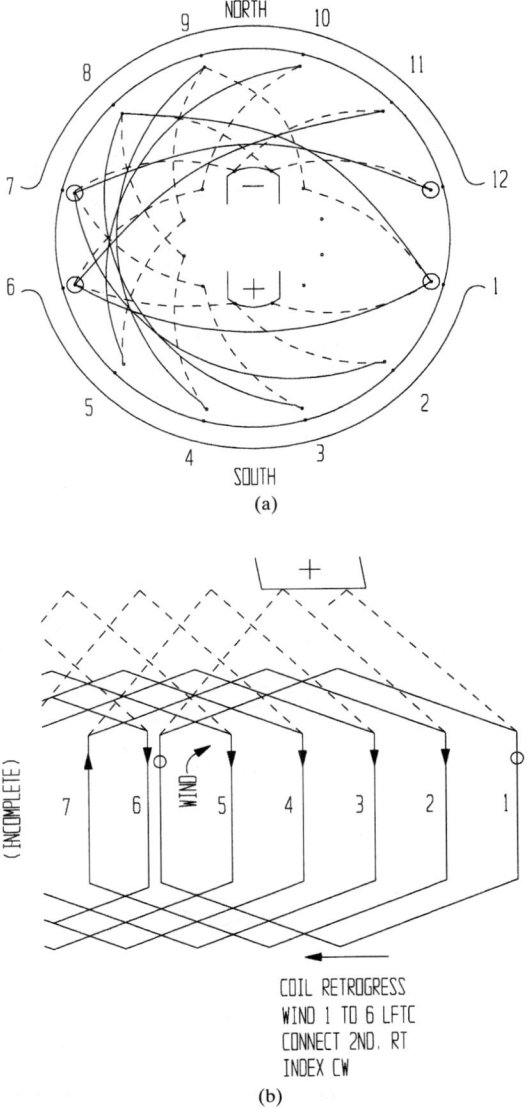

FIGURE 4.122 Armature lap winding, first double overlap, coil retrogresses: (*a*) end view, and (*b*) side view.

4.154 CHAPTER FOUR

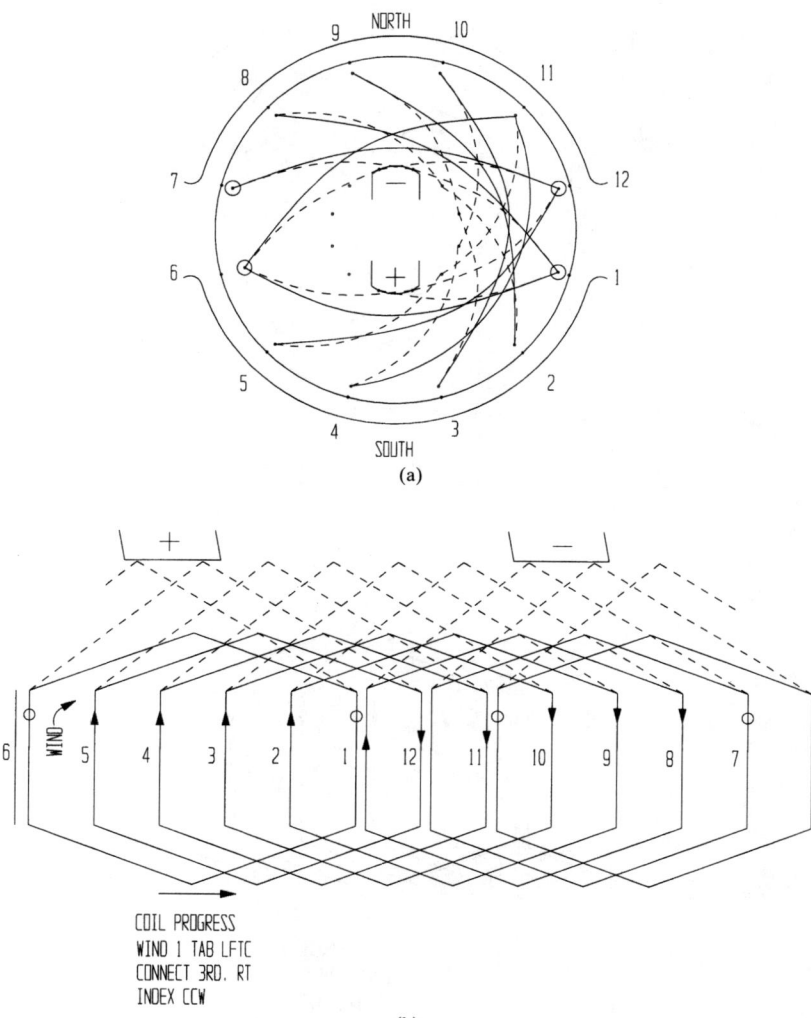

FIGURE 4.123 Armature lap winding, first double overlap, coil progresses: (*a*) end view, and (*b*) side view.

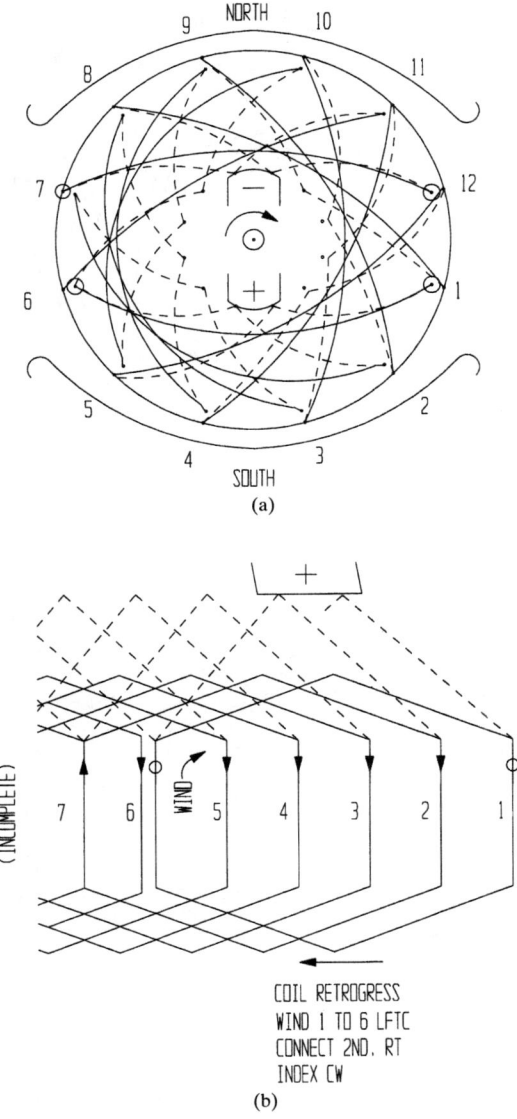

FIGURE 4.124 Armature lap winding, completed progressive wind, coil retrogresses: (*a*) end view, and (*b*) side view.

4.156 CHAPTER FOUR

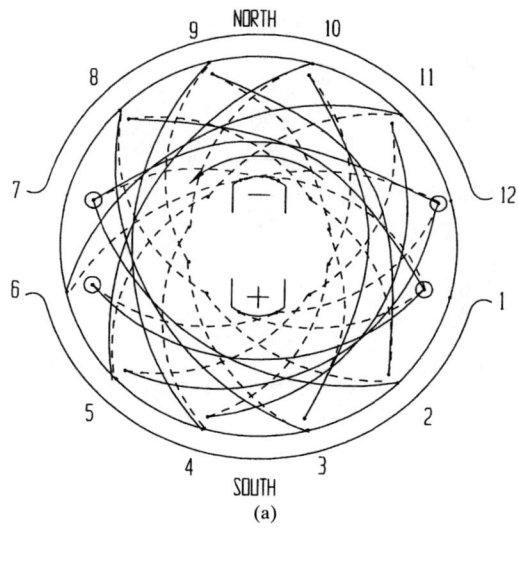

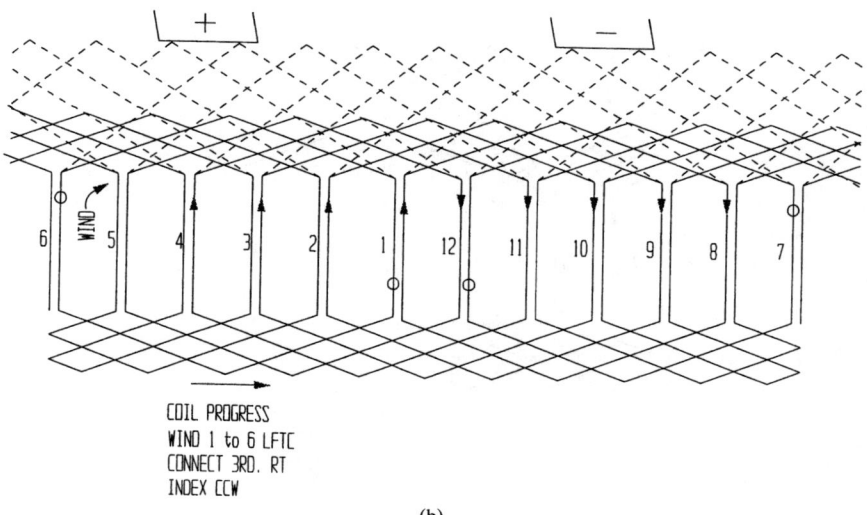

FIGURE 4.125 Armature lap winding, completed progressive wind, coil progresses: (*a*) end view, and (*b*) side view.

4.9.7 Anchored Lead Loops

Prior to the late 1950s, most armatures were wound with lead loops which were hand-sorted and laid into slots in the commutator bars. It was necessary for automatic machines to mechanically form such loops in a manner that they would be secure and identifiable.

Several arrangements were used, but the most popular was the *anchored loop*. The anchored loop could be formed in several ways, but the usual way was as shown in Figs. 4.126 through 4.131. The leads might be shifted to the right as shown or to the left, depending upon the requirements.

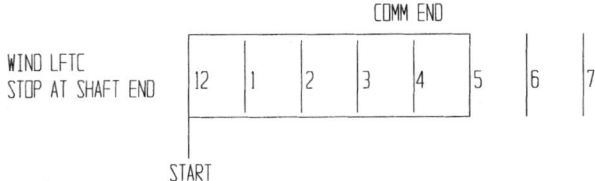

FIGURE 4.126 Forward pattern anchored lead sequence.

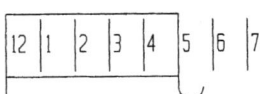

FIGURE 4.127 Index CCW.

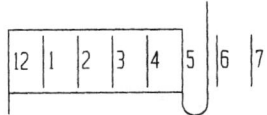

FIGURE 4.128 Flier reverse (to commutator end).

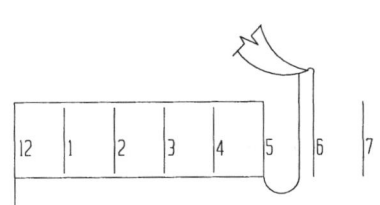

FIGURE 4.129 Extending lead hook flier return (to shaft end).

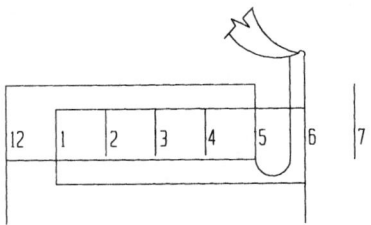

FIGURE 4.130 Winding second coil.

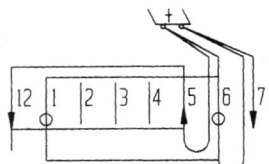

FIGURE 4.131 Repeating second loop (compare to Fig. 4.117).

This pattern became known as the *forward pattern* (Figs. 4.134b), and its mirror image (Fig. 4.134a) as the *reverse pattern*. The choice usually depended upon angular brush location and required polarity. Please note that two extra lead wires (canceling each other) appear in each slot.

4.9.8 Selection for Polarity

The key items controlling polarity are (1) brushes, (2) direction of wind, and (3) direction of index. Changing any one of these will reverse the polarity, as in Fig. 4.132. Reversing any two (like a double negative) will keep the polarity the same, as in Fig. 4.133. Changing all three will reverse the polarity, as in Fig. 4.134. Please note that in this latter instance the brush locations are out of line from the poles, as is the usual case.

Variations of the anchored loop were often used for convenience.

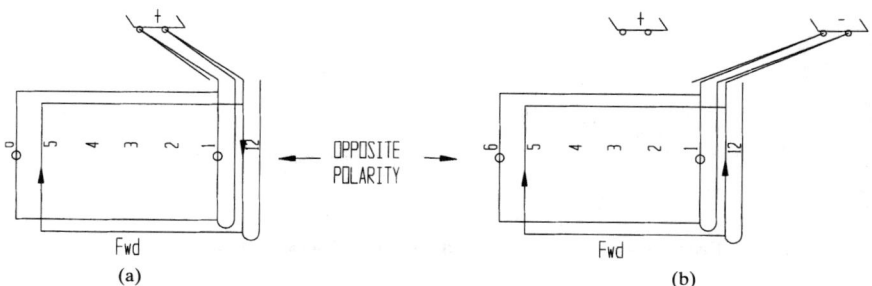

FIGURE 4.132 Polarity of winding connections, opposite polarity: (*a*) forward pattern, and (*b*) forward pattern.

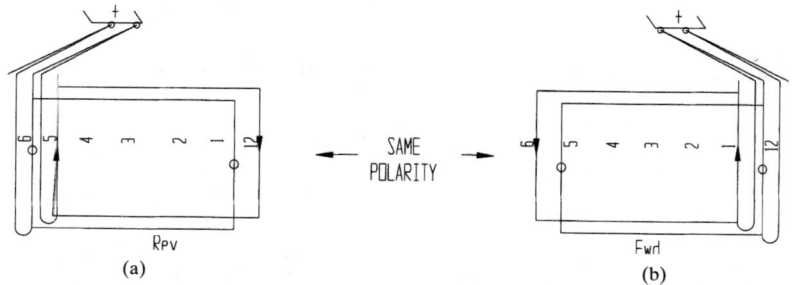

FIGURE 4.133 Polarity of winding connections, same polarity: (*a*) reverse pattern, and (*b*) forward pattern.

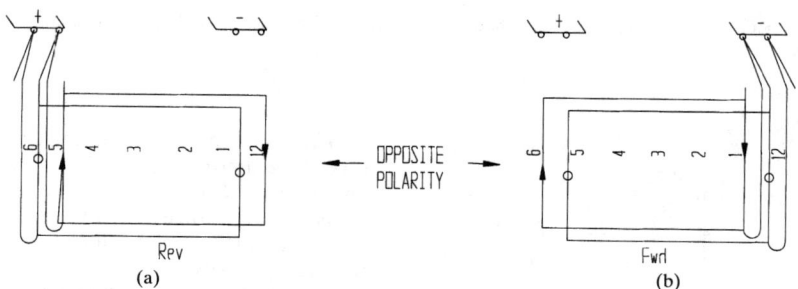

FIGURE 4.134 Polarity of winding connections, brush location 90° to left: (*a*) reverse pattern, and (*b*) forward pattern.

4.9.9 Starting Wire Connection

While the wire starting at the shaft end has not been shown as connected in Fig. 4.126, in practice it was finally laid toward the commutator in slot 5, along with a finish wire, and connected to the bar preceding the first loop.

4.9.10 Tang-Type Commutators

In the late 1950s, the tang hook on the commutator bar became popular. Simultaneously, the *hot-staking* method became practical. In this method, the finish wire (or lead loop) was laid over the selected tang, by the flier, and the subsequent coil was then wound. There was no chance to remove the insulation from the wire to permit soldering. Therefore, the hot-staking method of pressure and heat application disintegrated the insulation and pressed the tang tightly around the lead. Good electrical contact was thus possible.

Another difficulty was the unsupported wire extending from the coil to the tang. Winding subsequent coils tended to stretch and even break that wire. The first application involved heavy wire which could withstand this pressure. For fine wire, a pattern was selected wherein the lead was connected at a tang sufficiently removed so that the wire laid tangent to the shaft.

A third problem was shorts at the crossovers of the leads which approached and left the tangs (see Fig. 4.125). The anchoring pattern (Fig. 4.131) solved both these latter problems in most instances.

4.9.11 Typical Polarity Matching

Two basic winds, with brushes shifted from pole centerline, are shown in Fig. 4.135.

The standard forward and reverse patterns are shown as adapted to match Fig. 4.136.

Special anchored loop patterns, in which loops are drawn before indexing, are shown in Fig. 4.137.

Figure 4.138 shows front leads drawn to distant tangs where they will be supported by tangent contact with the shaft. A special *armature diverter* is used to spin the armature, presenting the proper tang, and then return to the original position for normal indexing when engaging the lead.

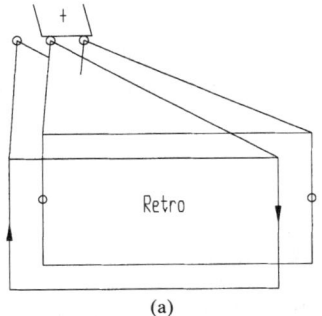
(a)

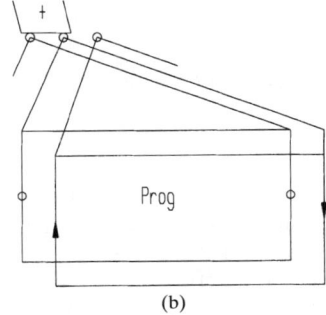
(b)

FIGURE 4.135 Automotive winding patterns: (*a*) retrogressive, and (*b*) progressive.

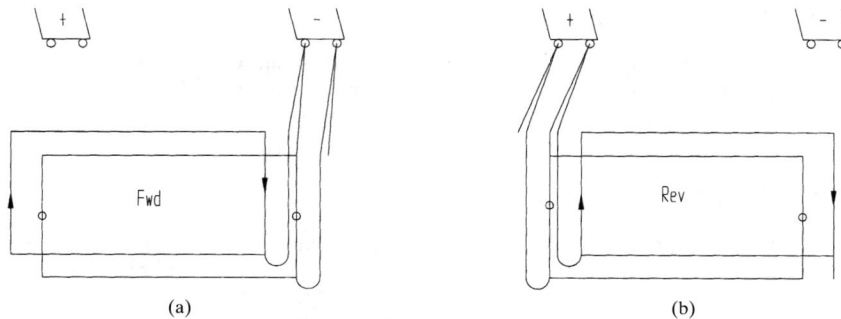

FIGURE 4.136 Standard anchored-lead winding patterns: (*a*) forward pattern, and (*b*) reverse pattern.

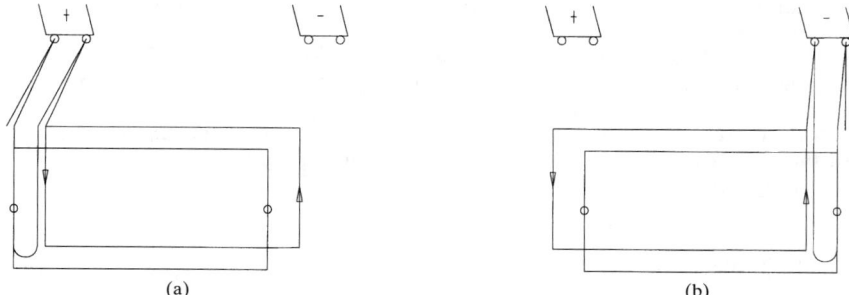

FIGURE 4.137 Special anchored-lead winding patterns: (*a*) reverse pattern, and (*b*) forward pattern.

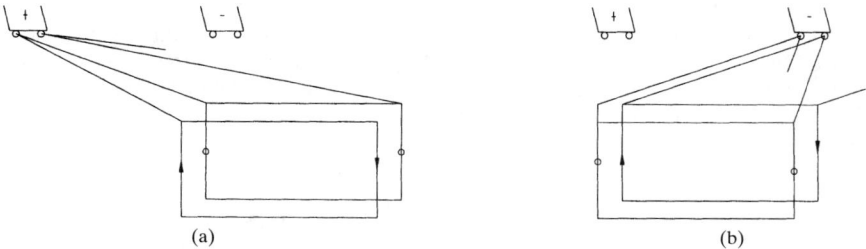

FIGURE 4.138 Tangential front-lead winding patterns: (*a*) reverse pattern, and (*b*) forward pattern.

A recently developed arrangement involves carrying the lead wire almost halfway around the shaft, engaging the tang with a reverse wrap (alpha connection), and then continuing around the shaft to a slot advanced (fwd) or retracted (rev) from the previous coil (Fig. 4.139). All conditions can be met with LFTC wind and CCW armature rotation, requiring only the proper tang selection and progressed (fwd) or retrogressed (rev) slot selection. The leads are anchored around the shaft,

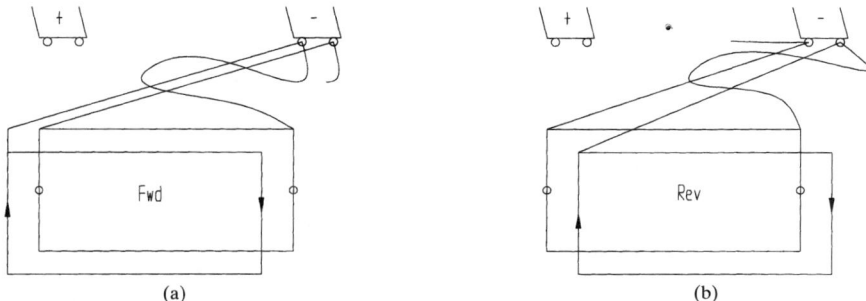

FIGURE 4.139 Alpha-connected shaft-anchored-lead winding patterns: (*a*) forward pattern, and (*b*) reverse pattern.

not taking extra wire and slot space for the anchoring turn, and with the alpha connection giving safe crossover locations in the lead system.

It should be noted that winding over tangs, as in Figs. 4.136*a* and 4.137*b*, becomes difficult because, when reversing the flier, the wire does not follow the form into the slot, requiring an extra guide provision. However, Figs. 4.136*b* and 4.137*a*, represent patterns in which the wire is guided into the slot by the form.

4.9.12 Double-Flier Winding

The foregoing discussion presents the principles used by the single-flier winding apparatus. Double-flier winding uses all these principles in duplicate. Figure 4.140 shows the winding of coils 1 to 6 and 7 to 12 simultaneously. Figure 4.141 shows the winding of coils 2 to 7 and 8 to 1 after CW indexing.

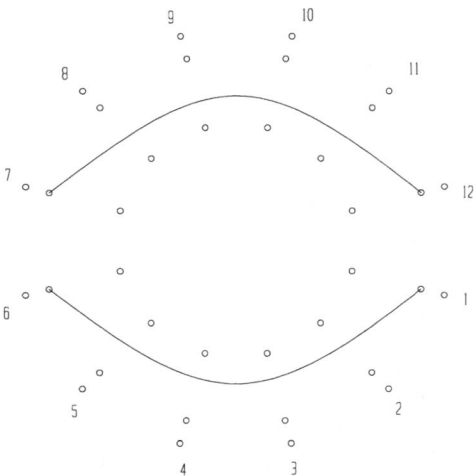

FIGURE 4.140 Double-flier balanced wind: first wind.

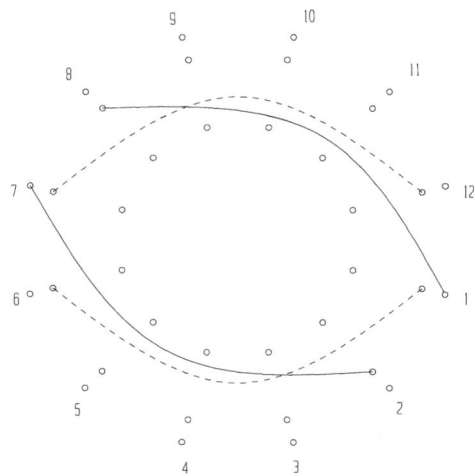

FIGURE 4.141 Double-flier balanced wind: second wind.

4.9.13 Odd Slot in Double-Flier Winding

It is necessary to reduce an odd-slot to an even-slot wind by winding one coil alone, leaving an even number of coils to be wound using both fliers. In Globe winders, the right flier is disengaged while the left flier winds out the odd coil. Then the right flier reengages to complete the winding.

Figure 4.142 shows that the coil 1 to 6 was wound first; then coils 2 to 7 and 7 to 1 were simultaneously wound. In this instance the coil span is ½ slot short of diametric. However, in rare instances, the pitch is 1½ slots short of diametric, as shown in Fig. 4.143. Less copper is used, but poorer commutating is experienced. However, in this case the winding forms are offset and, for CW indexing, are closer together at the top than at the bottom; while for CCW indexing, they are closer together at the bottom.

4.9.14 Two Coils Per Slot

In appliance motors (110 or 220 V), a larger number of bars and coils is desired to keep the voltage between bars down. Without increasing the number of slots, the bars can be doubled by winding a first coil, pulling a lead, then winding a second coil in the same slot (Fig. 4.144). Matching this involves the two-coil forward pattern (Fig. 4.145).

Note that in forming the lead, after the first coil 2 to 7, a reverse (CW) index is used so that the loop emerging from slot 1 returns from the hook into slot 2, thus to wind the second 2-to-7 coil. This is referred to as a *straddle loop* and usually is formed shorter than the next (straight) loop, whose sides both emerge from slot 2. Thus, when pulled as a lead loop, it can be readily identified.

Figure 4.146 illustrates the alpha-connected shaft-anchored pattern equivalent (called the *alpha pattern* for short).

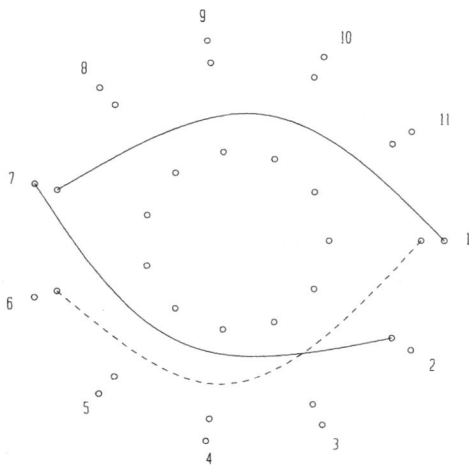

FIGURE 4.142 Odd slot: coil 1 to 6 wound first, then coils 2 to 7 and 7 to 1 wound simultaneously.

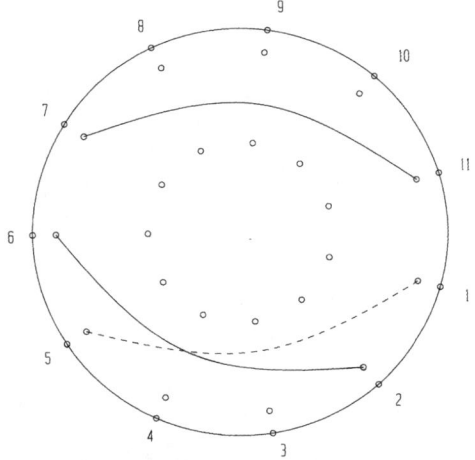

FIGURE 4.143 Odd slot: pitch is 1½ short of diametric.

4.9.15 Opposite Brush Selection

When the proposed Globe winding involves connecting to the opposite brush from the customer's sample or design, this is done by connecting to a commutation segment located 180° from that to which the sample is connected.

However, odd-slot armatures having only one coil per slot will not have a commutator segment at the 180° position. For instance, a 15-slot armature would have

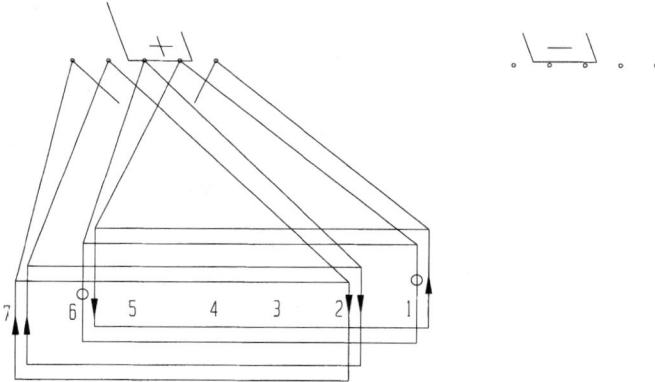

FIGURE 4.144 Two coils per slot LFTC index CW.

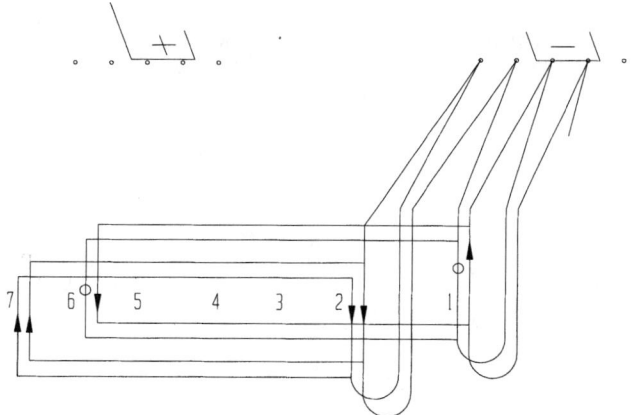

FIGURE 4.145 Two-coil forward pattern.

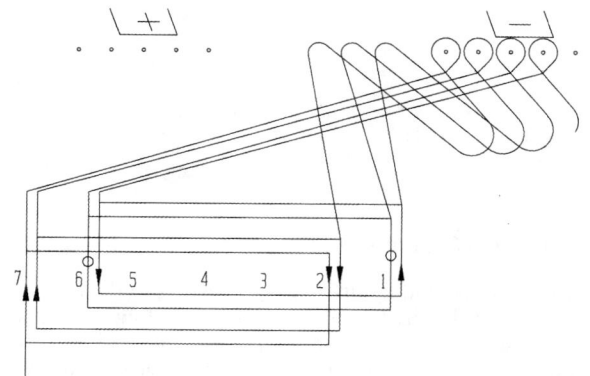

FIGURE 4.146 Two-coil progressive (fwd) alpha-connected shaft-anchored pattern.

segments at 168° or 192° from the sample connected segment. The coil would be commutated either 12° early or 12° late if connected to one of these "opposite" segments. Brush sparking and loss of power would result from this displacement.

The solution is for the customer to compensate by placing the commutator at ½ slot change in location so a segment will be at the 180° location for the selected pattern.

When two coils per slot are involved, or when even-slot armatures are involved, there will be a segment at this 180° location, and no special commutator location need be requested.

4.9.16 Four-Pole Armatures

The foregoing discussion concerns *two-pole armatures* with the start and finish of each coil connected to adjacent segments. The *two-pole* designation means that the coil sides are in nearly opposite slots; the result being that the windings naturally draw in toward the bottom of the slots. The *adjacent segment* connection means that the armature is *lap* wound.

The vast majority of small-appliance and tool motor armatures fall into this classification. However, many slower-speed or heavier duty motors have four or more poles. Also, a different connection pattern (known as *wave* connection) may be used.

It is not the intent here to weigh the relative merits of two-pole versus four-pole armatures, nor the merits of wave versus lap connections. Rather, you must be able to understand how to analyze a design you encounter, so you can determine how the equipment can wind it or its equivalent. In recommending an equivalent, you have the opportunity to accommodate desirable features and attachments.

Winding Pitch. Each armature coil is wound with its sides in slots spaced approximately one pole span, and this is just the same kind of span as in the two-pole armatures. This is fundamental because a coil must be commutated while its coil sides are passing between poles, with one side coming out from under one polarity pole while the other side comes out from under an opposite polarity pole.

Thus, in a 17-slot 4-pole armature, the coil pitch would be 1 to 5 (4 slot spans) or just less than 90° mechanical; see Fig. 4.147. Please note that this is just less than 180° electrical, since there are 720° electrical in a four-pole motor.

This is obvious from the end (upper) view, sometimes called the "rose" pattern. The periphery (360° mechanical) includes two pairs of poles, each of which pair provides 360° electrical.

Obviously, a quarter-span coil will draw against the sides of the slots instead of toward the bottom. Special tooling design and wire handling is necessary to fill the slots of this coil span. The problem is greater if six or more poles are involved.

Wave Versus Lap Winding. Lap winding is illustrated in Fig. 4.115 (and all the foregoing figures); the finish wire is connected to the first bar forward of its starting wire. In the 12-slot, 1-coil-per-slot armature, this would be 30° electrical forward of its start. This is progressive lap winding. However, in Fig. 4.114, the finish lead connects to the first bar back, or 30° electrical back of its start, which is retrogressive lap winding.

Referring to Fig. 4.147, the four-pole progressive lap wind finish again connects to the first bar forward of its start. The wave connection has the start and finish leads connected approximately 360° electrical apart instead of adjacent. When the finish is connected 360° electrical forward to the first bar forward, as shown in Fig. 4.148, it is a progressive wave-wound connection.

4.166 CHAPTER FOUR

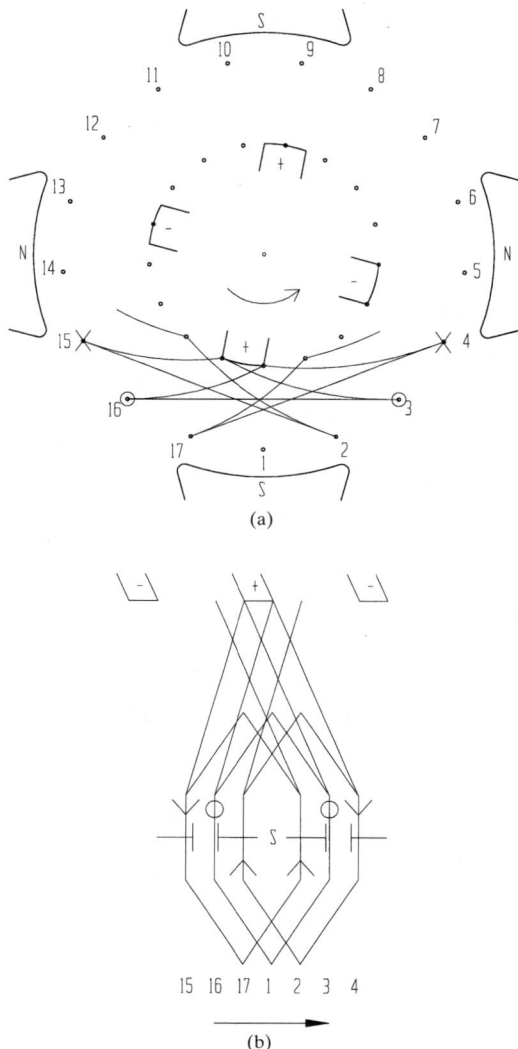

FIGURE 4.147 Lap progressive conventional pattern: (*a*) end view, and (*b*) side view.

Figure 4.149 is different in that the coils as shown in the (lower) spread pattern are wound clockwise instead of counterclockwise as in Fig. 4.148. Note that the start wire enters slot 7 of coil 3 to 7, emerging from slot 3; the finish connects to the first bar back of 360° electrical and is a retrogressive wave winding. From there it continues, entering slot 15, emerging from slot 11, and connecting one bar back of the start to 7.

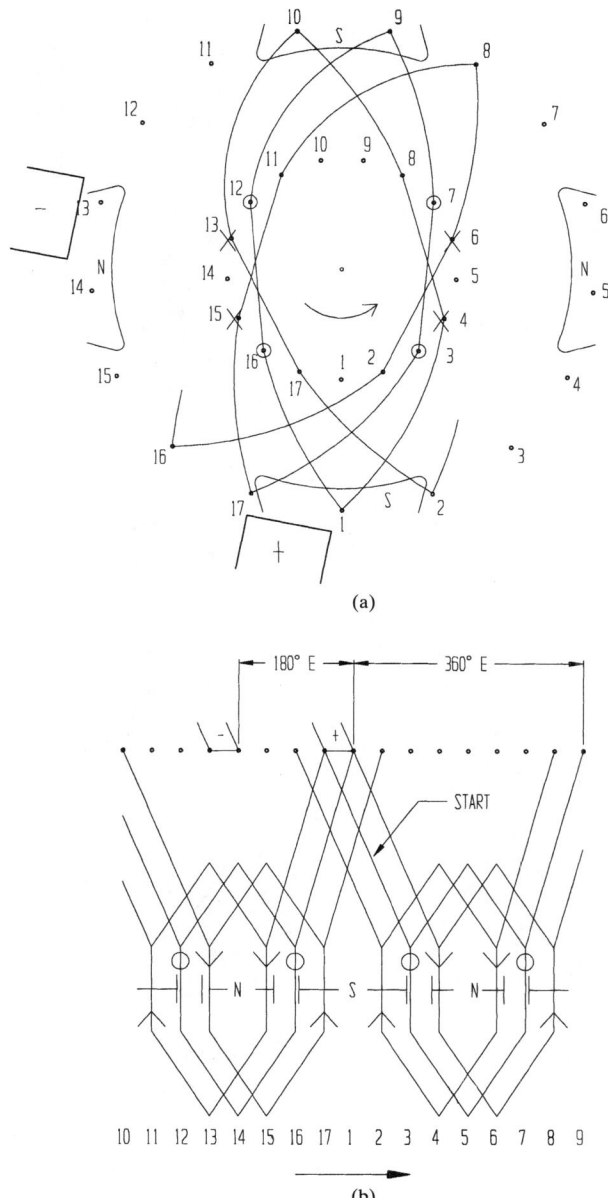

FIGURE 4.148 Wave progressive conventional pattern: (*a*) end view, and (*b*) side view.

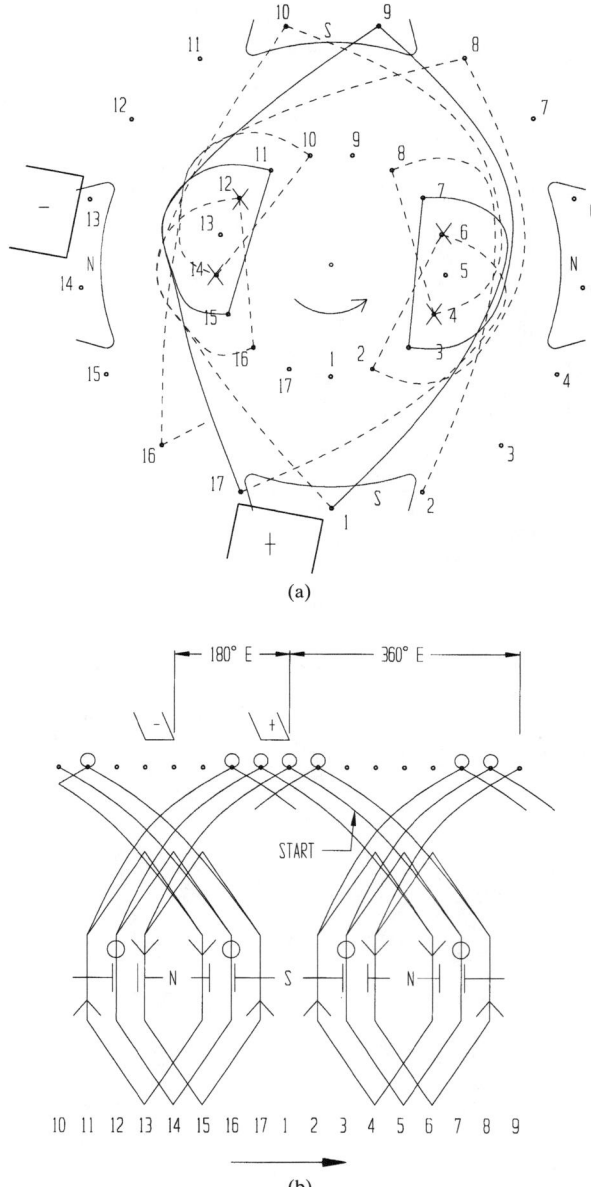

FIGURE 4.149 Wave retrogressive globe shaft-anchored pattern: (*a*) end view, and (*b*) side view.

Polarity. Note that Fig. 4.148, which shows a conventional winding, is matched in polarity by Fig. 4.149, in which the lead will lie against the shaft for support and can be wound with the Alpha connection and the shaft-anchored pattern. The alpha-rotator mechanism, when provided with a set of four-pole wave cams, provides the index for this family of four-pole wave windings according to the shaft-anchored pattern of Fig. 4.149, whether progressive or retrogressive.

Keep in mind that in the rose pattern (end view), only the ends of the coil sides are seen. Current flowing from the positive brush (1), Fig. 4.148, enters slot 4 going away and is represented by the X as the feather-end of an arrow. Returning in the other coil side, in slot 8, it is represented by a dot as the front tip of an approaching arrow. Note that the current continues to bar 10, and then to coil 13 to 17, and so on.

The spread pattern is shown fully developed in Figs. 4.150, 4.151, and 4.152, corresponding respectively to Figs. 4.147, 4.148, and 4.149. The commutator is shifted ½ slot in Fig. 4.149 (also in Fig. 4.152) to maintain commutation performance for this retrogressive wind. Note that coils 3 to 7 and 11 to 15 are commutating instead of coils 3 to 7 and 12 to 16, as in the progressive wind of Fig. 4.148. Such a shift in the commutator is often required when using an alternate winding pattern in an odd-slot armature.

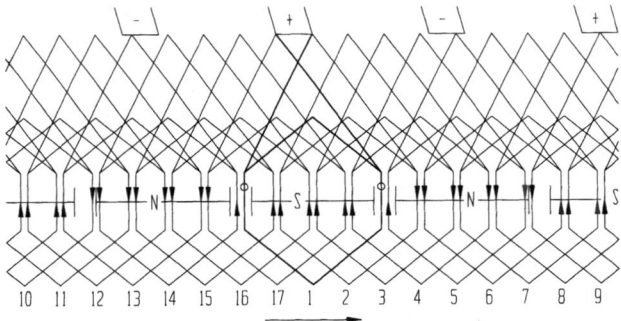

FIGURE 4.150 Four-pole lap wind progressive conventional pattern.

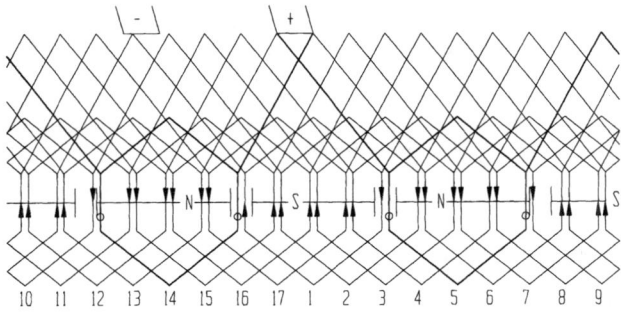

FIGURE 4.151 Wave wind progressive conventional pattern.

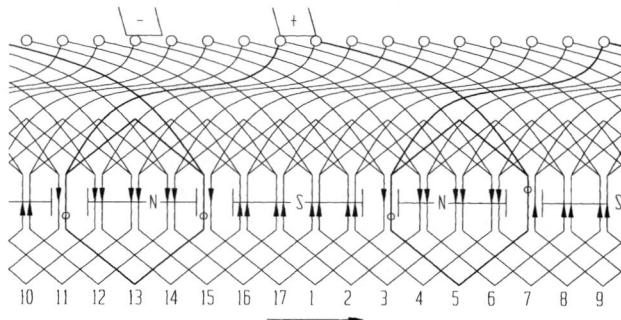

FIGURE 4.152 Wave wind retrogressive globe shaft-anchored pattern.

Brushes.   In lap windings a brush is needed for each pole in order to commutate all of the coils as they move from pole to pole. Thus, a pair of brushes is needed for a two-pole lap, two pairs for a four-pole lap, etc.

In wave windings, however, with an odd number of slots, all coils can be commutated as they move from pole to pole by a single pair of brushes regardless of whether there are four or six poles, or more. The proof of this is that when tracing through two coils in series on a four-pole (three coils in series on a six-pole, etc.) armature. The finish lead connects to the next bar to the start lead. It will be to the next beyond for progressive wind or next back for retrogressive wind.

Note that if you continue to trace the circuit through all the coils in series, the finish from the last coil will connect to the start of the first coil. This is as though the entire group of coils was wound by one single uncut wire whose finish connects to its start (see Figs. 4.151 and 4.152).

This is not so, however, if there is an even number of slots. When tracing through two coils of an even-slot four-pole wave (three coils of a six-pole, etc.), the finish will connect incorrectly to the same bar as its start, or correctly to the second bar away, depending on the lead span (look forward to see the special cases in Figs. 4.163 through 4.168).

Four brushes will be required to commutate the necessary coils in a four-pole, even-slot, wave-wound armature.

Automatic Winding.   For automated winding and connecting, the consecutive coils of a wave pattern are not in adjacent pairs of slots. Instead, the winder must lay the finish wire of a coil on the designated commutator tang, then index to the coil slot into which the wire leaving that same tang must go.

For instance after starting on bar 16 (Fig. 4.148), coil 2 to 6 would be wound. The finish must be laid on tang 8, then indexed to wind coil 11 to 15, then connected to tang 17, and so on.

Single-Flier Winding.   In a single-flier winder, the four-pole odd-slot armatures could be completely wound in a regular pattern as previously discussed and as indicated in Figs. 4.153 through 4.158 (neglecting the dotted line designation). In other words, after proceeding in the described pattern through 17 coils (Figs. 4.157 and 4.158), the finish wire will attach in the regular pattern to the same tang to which the original start was connected. Figures 4.159 and 4.160 show 19-slot 1–5 pitch 4-pole retrogressive and progressive windings, respectively.

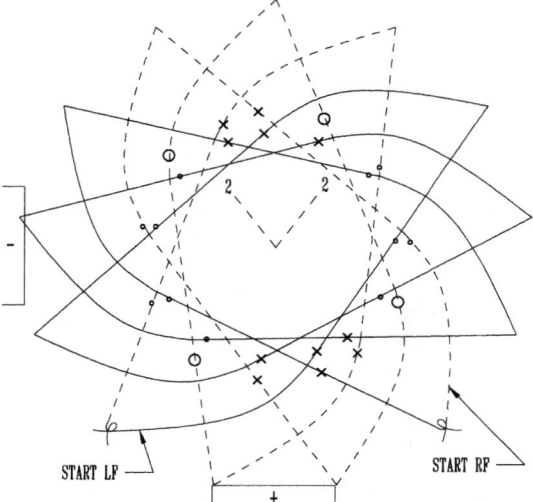

FIGURE 4.153 Four-pole odd-slot retrogressive winding: 13 slot, 1–4 pitch.

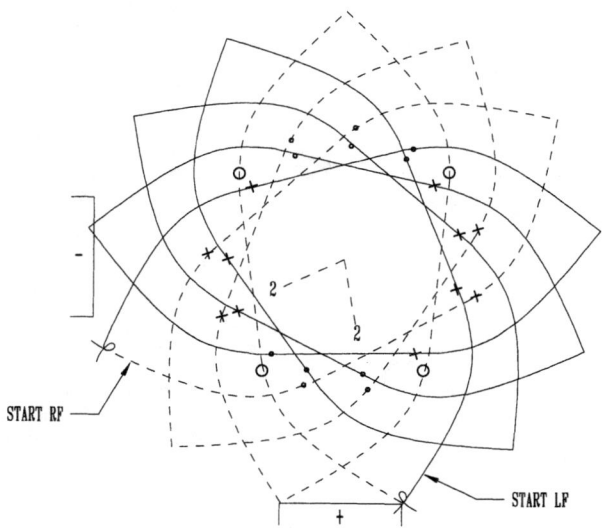

FIGURE 4.154 Four-pole odd-slot progressive winding: 13 slot, 1–4 pitch.

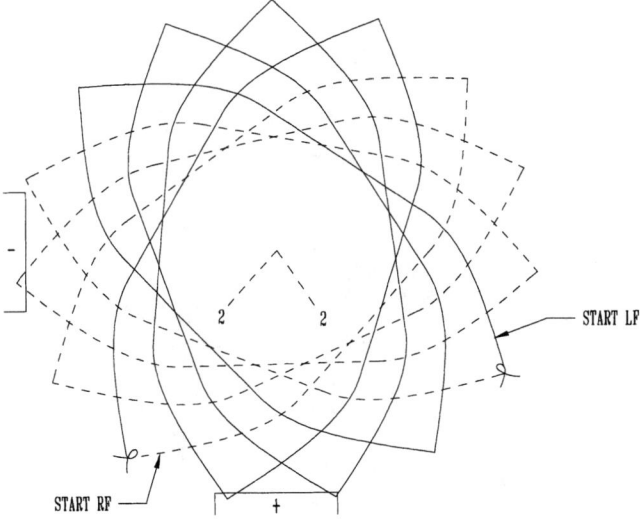

FIGURE 4.155 Four-pole odd-slot retrogressive winding: 15 slot, 1–4 pitch.

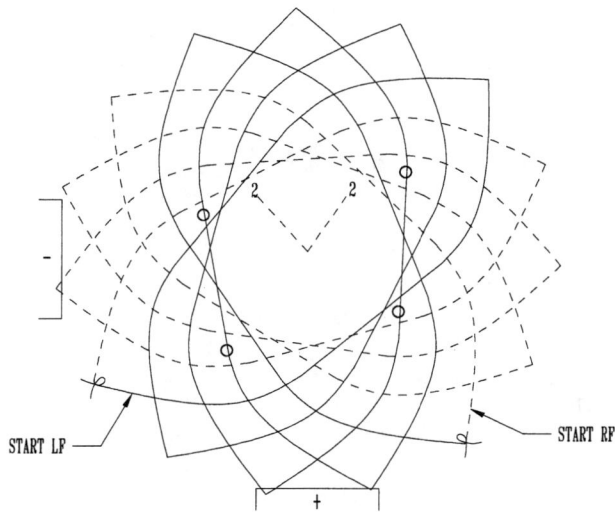

FIGURE 4.156 Four-pole odd-slot progressive winding: 15 slot, 1–4 pitch.

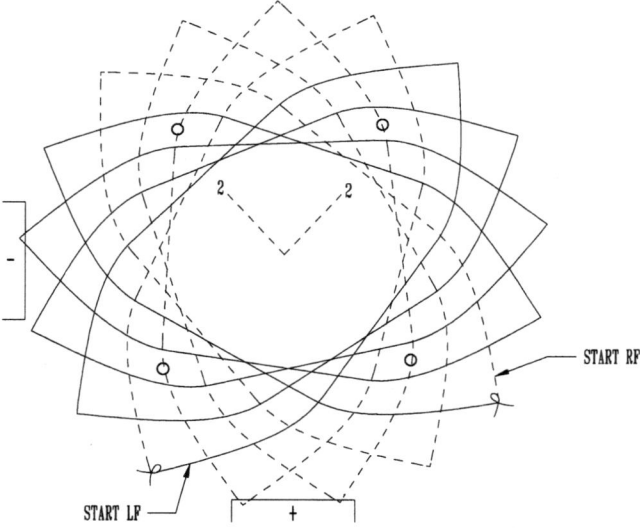

FIGURE 4.157 Four-pole odd-slot retrogressive winding: 17 slot, 1–5 pitch.

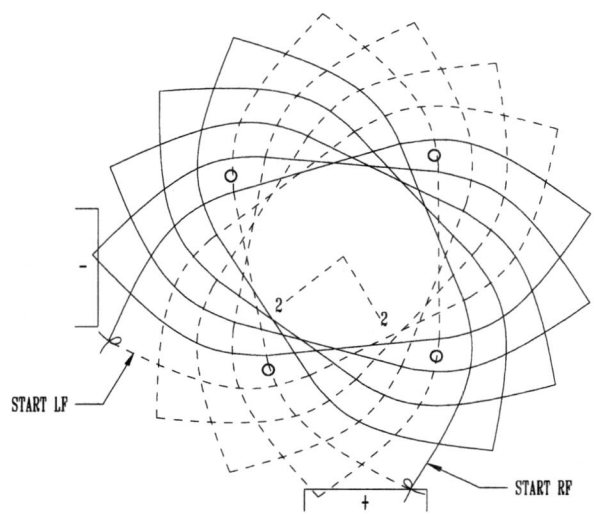

FIGURE 4.158 Four-pole odd-slot progressive winding: 17 slot, 1–5 pitch.

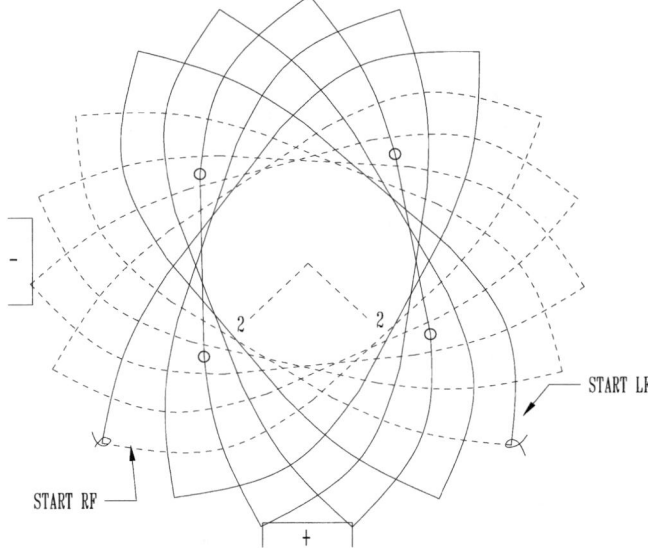

FIGURE 4.159 Four-pole odd-slot retrogressive winding: 19 slot, 1–5 pitch.

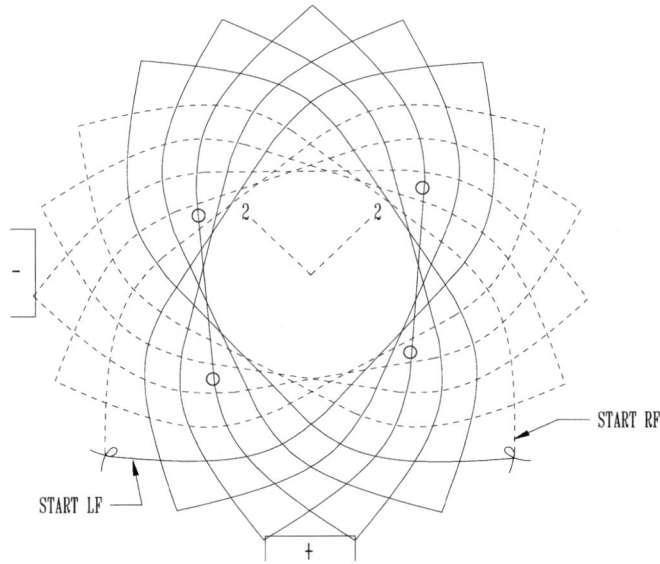

FIGURE 4.160 Four-pole odd-slot progressive winding: 19 slot, 1–5 pitch.

This is also true of four-pole even-slot armatures, if the slots are in multiples of 4 (see Figs. 4.161 and 4.162; also Figs. 4.165 and 4.166). However, after winding 7 coils of a 14-slot 4-pole armature (Fig. 4.163), the finish of that seventh coil would fall on the same tang as the start of its first coil. Then an irregularity must be programmed to tie off that lead, move to the next adjacent tang with a new start, wind in the intermediate coils in the same regular pattern, and thus complete the armature.

Double-Flier Winding. Very readily it can be seen that the even-slot four-pole armatures lend themselves to double-flier winding.

In Figs. 4.163 through 4.168, the solid line shows the windings from the left flier, while the dotted line shows those for the right flier. In the center of each rose, the short center lines indicate the axis of each of the two first coils. In other words, these are the centerlines of the two winding forms at the start.

The pattern shows one flier to progress or retrogress, always skipping a coil space. The other flier places coils into those skipped spaces to simultaneously complete the winding.

This is not true in odd-slot four-pole windings. At about halfway through the program, one flier will start laying a coil over one previously wound by the other flier if the skipping plan is followed.

Figures 4.153 through 4.158 show normal progression, wherein the left flier winds the first coil then progresses to the coil marked 2 at the centerline angle shown. The right flier then can start at the centerline shown for it by the dotted line. With the forma at this quadrature position, the windings can be simultaneously completed in the usual fashion.

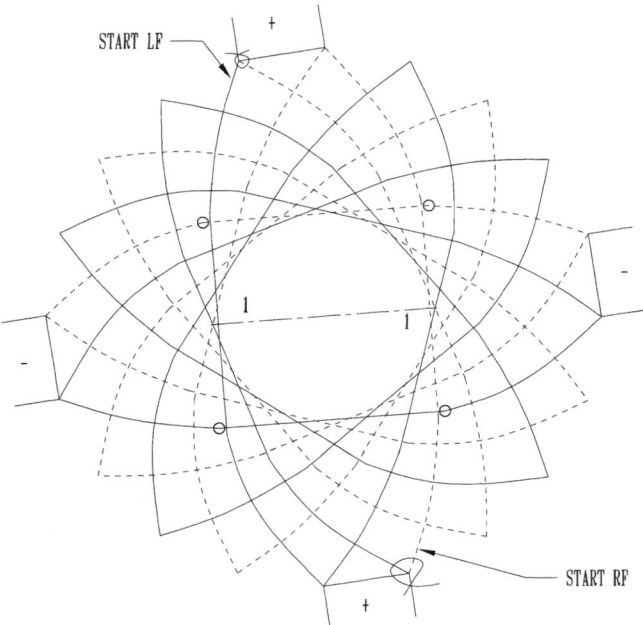

FIGURE 4.161 Four-pole even-slot retrogressive winding: 20 slot, 1–6 pitch.

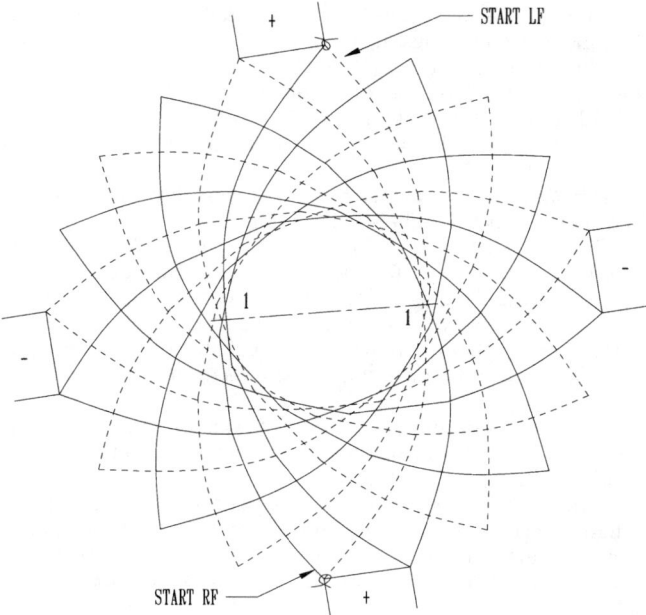

FIGURE 4.162 Four-pole even-slot progressive winding: 20 slot, 1–6 pitch.

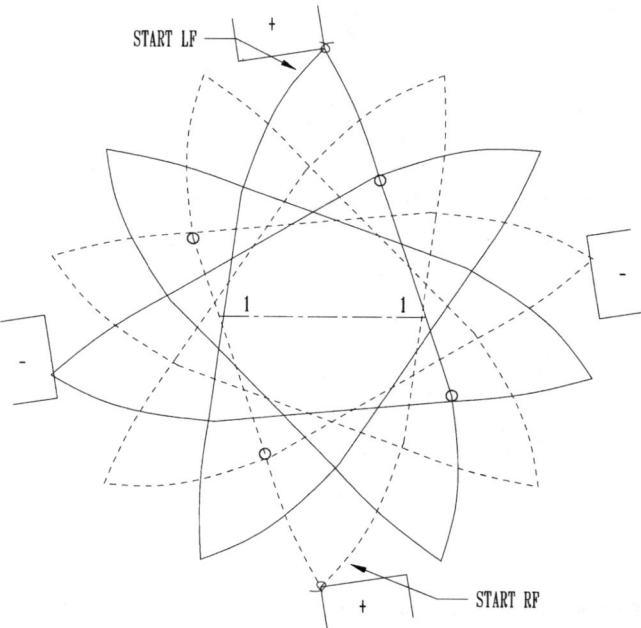

FIGURE 4.163 Four-pole even-slot retrogressive winding: 14 slot, 1–4 pitch.

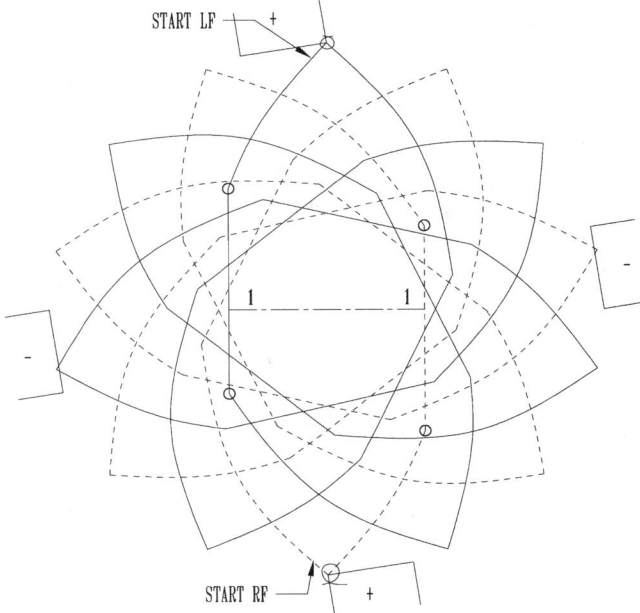

FIGURE 4.164 Four-pole even-slot progressive winding: 14 slot, 1–4 pitch.

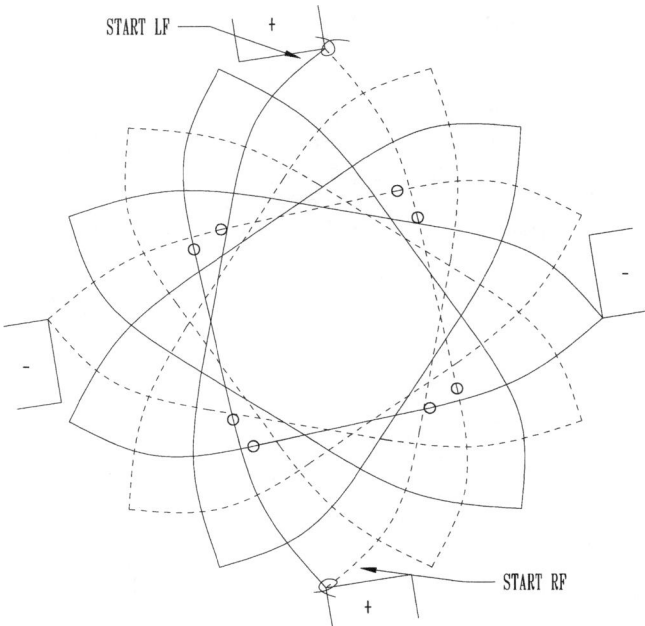

FIGURE 4.165 Four-pole even-slot retrogressive winding: 16 slot, 1–5 pitch.

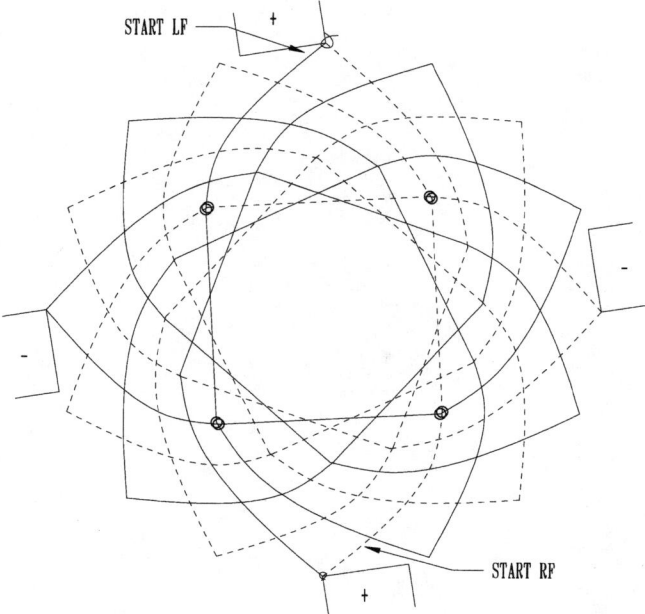

FIGURE 4.166 Four-pole even-slot progressive winding: 16 slot, 1–5 pitch.

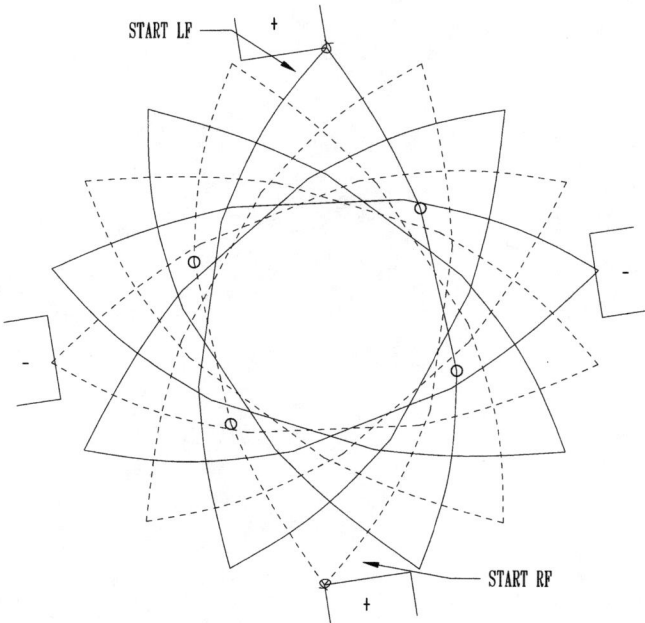

FIGURE 4.167 Four-pole even-slot retrogressive winding: 18-slot, 1–5 pitch.

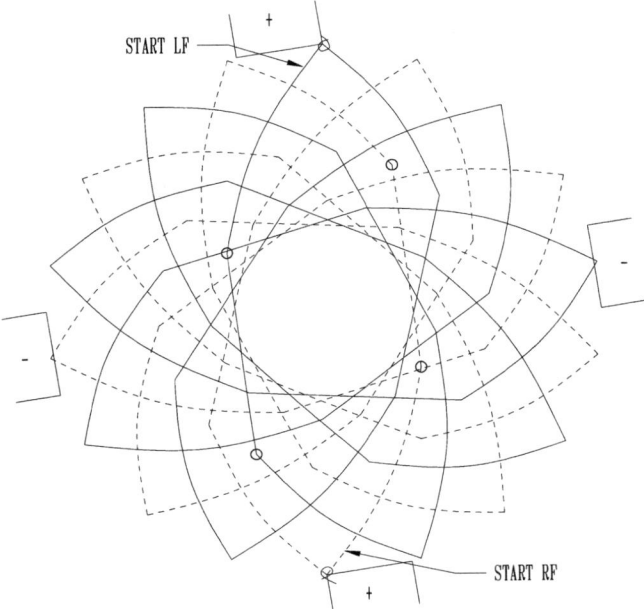

FIGURE 4.168 Four-pole even-slot progressive winding: 18 slot, 1–5 pitch.

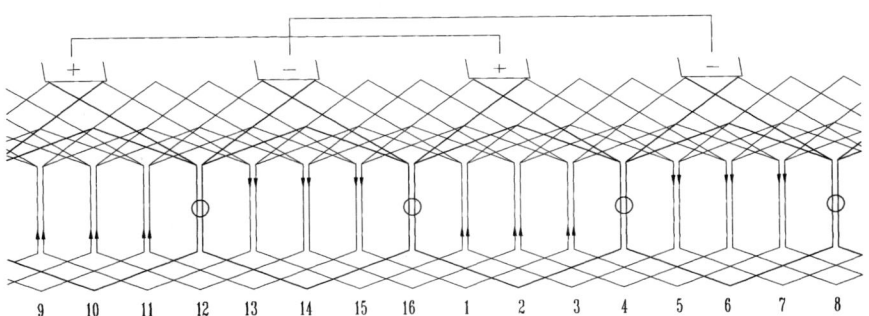

FIGURE 4.169 Four-pole even-slot retrogressive lap winding: 16 slot, 1–5 pitch.

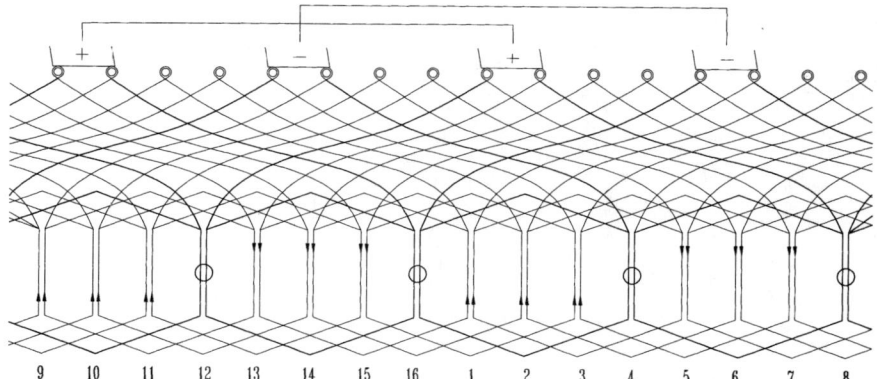

FIGURE 4.170 Equivalent globe progressive wave alpha-connected shaft-anchored winding.

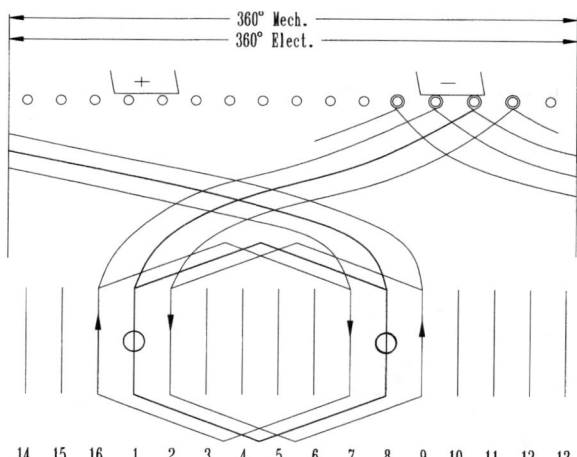

FIGURE 4.171 Two-pole progressive alpha-connected shaft-anchored winding.

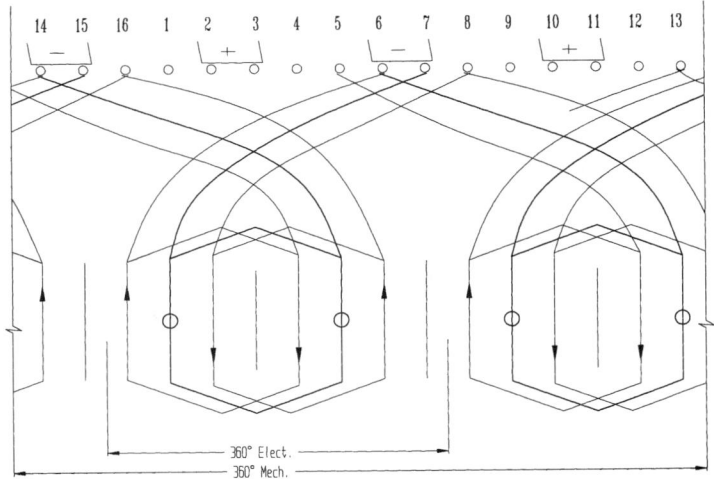

FIGURE 4.172 Four-pole progressive alpha-connected shaft-anchored winding.

Special off-center tooling must be developed to match this quadrature spacing to the opposed spindles of a double-flier winder. One solution is to wind two separate armatures in single-flier fashion in the double-flier machine.

Translation from Lap to Wave Winding. In some instances, the lap design can be converted to an equivalent wave pattern with the advantage of supporting the leads against the shaft. Also, an improvement is possible in making the alpha connection to the commutator tang.

Figure 4.169 shows a 16-slot lap winding pitched 1 to 5, with connections directly out from the coil in the usual pattern.

Figure 4.170 shows the exact wave equivalent using the alpha connection and shaft-anchored lead, which is possible using the alpha rotator with special four-pole wave cams.

Similarity in Two-Pole and Four-Pole Patterns. Figure 4.171 shows the schematic spread pattern of the two-pole alpha-connected shaft-anchored patented arrangement.

Figure 4.172 shows the similar four-pole pattern. Note that duplicate brushes are used in tandem for commutating and that the coils become tandem in effect to conform to the four poles in the stator.

However, comparing the two-pole (Fig. 4.171) to only 360° electrical (half) of the four-pole (Fig. 4.172), and recognizing that two negative brushes, even though separate, are one and the same in polarity, the two patterns become fully equivalent.

CHAPTER 5
ELECTRONICALLY COMMUTATED MOTORS

Chapter Contributors

Duane C. Hanselman
Dan Jones
Douglas W. Jones
William H. Yeadon

5.1 BRUSHLESS DIRECT-CURRENT (BLDC) MOTORS*

This chapter covers performance calculations and characteristics of the most common electronically commutated motors. The first motor to be covered is the brushless direct-current (BLDC) motor in several configurations.

Next, several configurations of the stepper motor are discussed. This is followed by the switched-reluctance motor, which is really a specialized configuration of the variable-reluctance stepper motor.

Brushless direct-current motors (BLDCs) are so named because they have a straight-line speed-torque curve like their mechanically commutated counterparts, permanent-magnet direct-current (PMDC) motors. In PMDC motors, the magnets are stationary and the current-carrying coils rotate. Current direction is changed through the mechanical commutation process.

In BLDC motors (Fig. 5.1), the magnets rotate and the current-carrying coils are stationary. Current direction is switched by transistors. The timing of the switching sequence is established by some type of rotor-position sensor. A typical rotor assembly for an inner-rotor configuration with sensors and commutation magnet is shown in Fig. 5.2. The three white devices mounted on the printed-circuit board are Hall-effect switches. They are positioned next to the larger-diameter magnet wheel, which causes them to switch high and low as the wheel changes from north to south as it

---

* Section contributed by William H. Yeadon, Yeadon Engineering Services, PC, except as noted.

rotates. The angular position of the Hall devices are adjusted to provide the optimum firing angle for the application. Figure 5.3 shows via magnetic field–viewing film the relative position of the motor magnet transition zone with respect to the Hall device. The stator assembly for this is shown in Fig. 5.4. This motor has a low-energy-product solid-plastic ring magnet with magnetic poles superimposed on it. Another type of motor has a rotor with solid magnet arcs, as shown in Fig. 5.5.

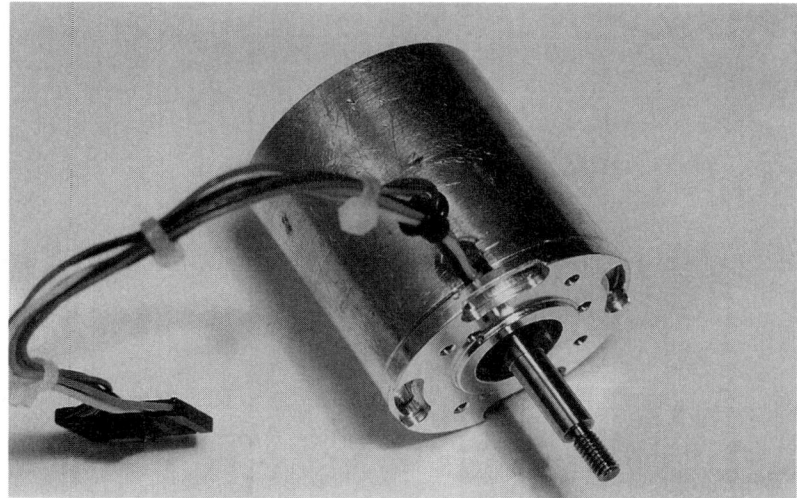

FIGURE 5.1 BLDC motor.

FIGURE 5.2 BLDC rotor assembly.

FIGURE 5.3 BLDC rotor showing magnetic pole transitions.

FIGURE 5.4 BLDC stator assembly.

FIGURE 5.5 BLDC rotor assembly with core segment magnets.

5.1.1 Basic Configuration of BLDC Motors

Outer-Rotor Motors. These motors are generally used where relatively high rotor inertia is beneficial to system performance. Common applications are computer disk drives and cooling fans. Construction is shown in Figs. 5.6 and 5.7.

The rotor assembly in Fig. 5.6 consists of a flexible magnet with four poles magnetized on it. It is enclosed by a magnetically soft steel cup or housing. A shaft which

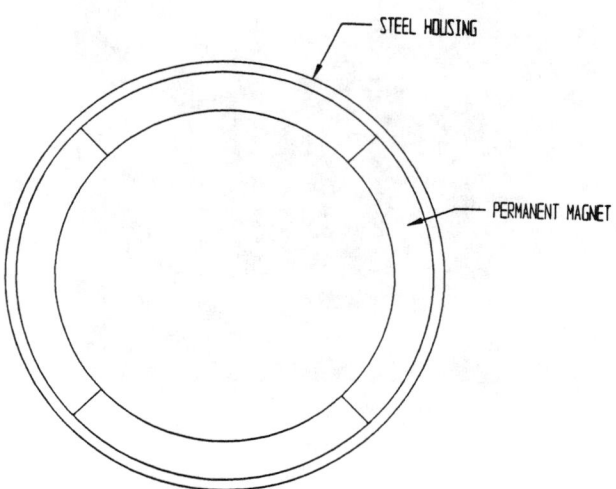

FIGURE 5.6 BLDC outer rotor assembly.

ELECTRONICALLY COMMUTATED MOTORS

FIGURE 5.7 Outer-rotor BLDC motor rotor and stator assembly.

allows the rotor to turn with respect to the stator is attached to the center of the steel housing.

The stator assembly (Figs. 5.8 and 5.9) consists of a lamination stack with coils of wire wrapped around the pole pieces. This is supported by a mounting base which also contains a bearing support and a control circuit. This motor's poles are alternately magnetized N-S-N-S.

As the rotor turns, the currents are turned off in one winding set and turned on in the other set by the Hall switch. This results in an S-N-S-N magnetization in which the S poles are induced at the interpoles and the poles where no current exists in the windings. This keeps the rotor turning. A close observation of this structure shows

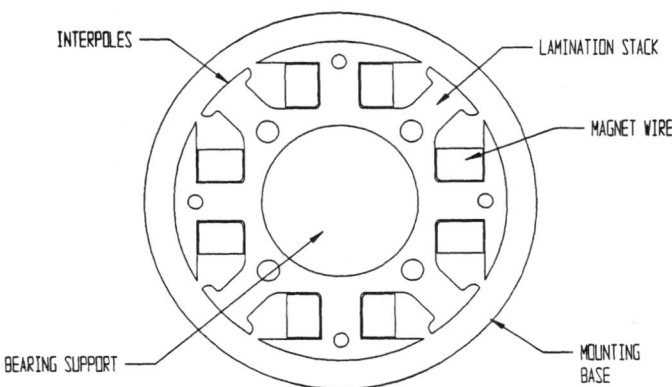

FIGURE 5.8 Outer-rotor BLDC stator diagram.

FIGURE 5.9 Outer-rotor BLDC wound stator assembly.

that it is in fact a single-phase motor and, as such, will have rotor positions where there is zero torque. This could result in failure to start. This is overcome by placing interpoles between the wound poles, as shown in Fig. 5.10. The magnet provides some position bias, which always results in rotation when the pole is energized.

Some other methods used to start these single-phase motors are shown in Figs. 5.11 and 5.12. In the case of Fig. 5.11, the reluctance torque produced by the magnet

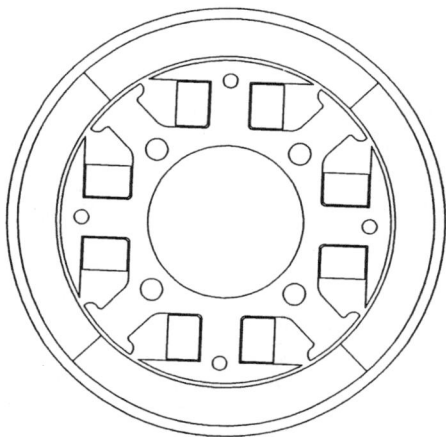

FIGURE 5.10 Outer-rotor motor diagram.

causes the rotor to line up over the pole. When the coils are energized, the field resulting from the energized coil appears to move across the face of the pole in the direction of the wider section of the pole that is closest to the gap. This starts rotation, which may then be maintained by alternately reversing the winding currents. The structure shown in Fig. 5.12 is another method of providing an unbalanced magnetic circuit which causes the rotor to have a preferred resting position in the unenergized state that is different from that of the energized state.

All of the motors just discussed are single-phase motors. Although this is probably the case for most outer-rotor motors, it is quite possible to build a multiphase motor in this configuration, as shown in Fig. 5.13.

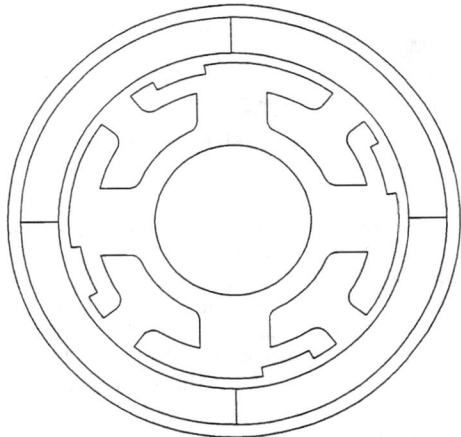

FIGURE 5.11 Stator with reluctance notches.

FIGURE 5.12 Stator with reluctance holes.

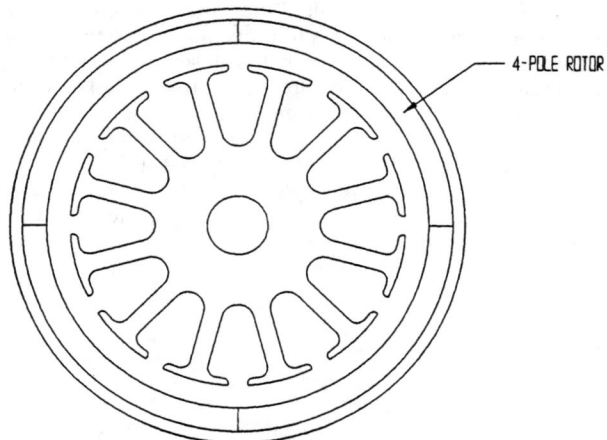

FIGURE 5.13 Multiphase outer-rotor motor.

Inner-Rotor Motors.   If one simply inverts the outer-rotor motor diagrammed in Fig. 5.8, the device shown in Fig. 5.14 is generated. This is also a single-phase device, and it would operate in the same fashion as the outer-rotor motor.

Some observations need to be made about these motors. The outer-rotor motor has much more magnetic material than the inner-rotor device, which means it is capable of more flux when the identical materials are used. It would be necessary to use a higher-energy-product magnet to get the same performance from an inner-rotor motor.

The inertia of the inner-rotor motor is lower because of its smaller rotor diameter. Therefore, it accelerates more rapidly than the outer-rotor motor.

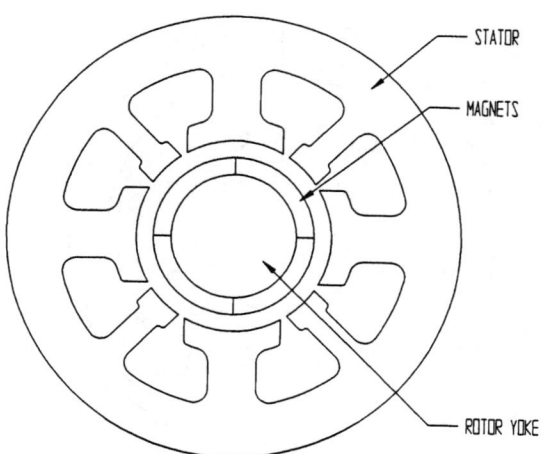

FIGURE 5.14 Single-phase inner-rotor motor.

Most inner-rotor motors have multiple phases in an effort to reduce the starting problems associated with single-phase motors. The stators may have salient poles or distributed windings. Figure 5.15 illustrates a three-phase four-pole salient-pole machine. Here the rotor is a magnetically soft steel core with magnet arc segments bonded to it. As the phases are switched in sequence A-B-C, the rotor moves to line up with the subsequent phase. When the rotor has moved 180° electrical, it is necessary to reverse the current in the windings, starting with phase A, in order to properly polarize the stator poles with respect to the magnet poles.

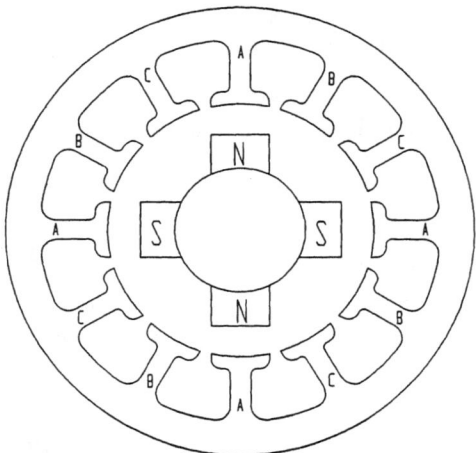

FIGURE 5.15 Multiphase inner-rotor motor.

In order to properly switch the coil currents at the right time and in the right order, the control electronics must know the rotor position. This is usually accomplished by means of Hall-effect devices, encoders, or counter-emf sensing.

Salient-pole machines have inherently high torque perturbation characteristics because of the abrupt permeance changes as the rotor moves from pole to pole. This can be reduced by distributing the windings over several stator teeth per pole. This is illustrated in Fig. 5.16.

Special Configurations—Slotless Brushless Inner-Rotor Motors. One variation of the distributed-winding multiphase brushless motor is the slotless brushless motor. As shown in Fig. 5.17, the winding distribution is similar to that of the stator shown in Fig. 5.16. However, the stator section has no teeth. The advantage of this type of motor is that there are no variations in permeance as the rotor moves. Thus, there are no torque perturbations or cogging. There is, however, a price to be paid. The lack of teeth increases the effective air gap, thereby lowering the available flux.

Another variation is the axial airgap motor. This motor is illustrated in Fig. 5.18. It has a stator with triangular coils (Fig. 5.19) and a disc rotor which is alternately magnetized NS-NS-NS-NS through its thickness. The stator coils (Fig. 5.20) are mounted to a nonmagnetic substrate, while the rotor disc is mounted to a magnetically soft steel. The advantage of this motor is that, like the other slotless motor, it is low cogging, but it also has relatively high inertia.

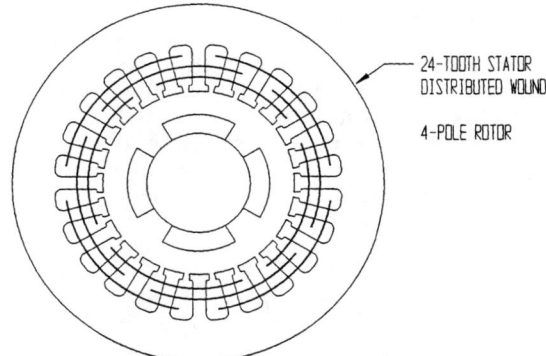

FIGURE 5.16 Inner rotor distributed-wound stator.

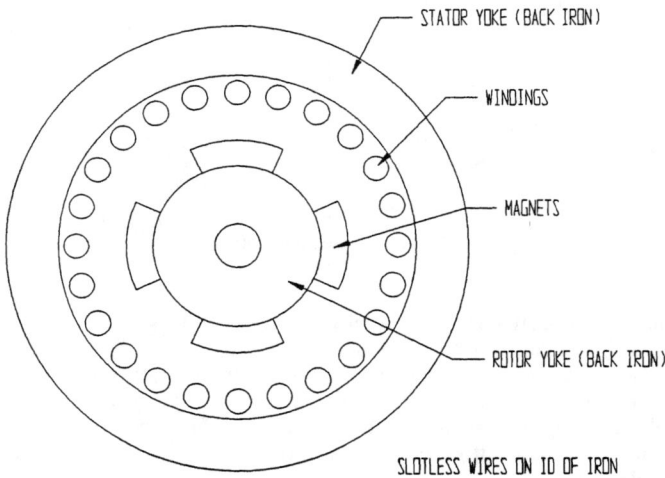

FIGURE 5.17 Inner rotor slotless stator.

Summary. Brushless DC motors can be put into three basic categories according to their structure:

- Outer-rotor motors
- Inner-rotor motors
- Special-configuration motors

Table 5.1 presents a relative comparison of the inner- and outer-rotor types. This assumes the same physical size motor operating at the same speed and torque.

FIGURE 5.18 Axial air gap brushless dc motor.

5.1.2 Sizing and Shaping the Motor

Mechanical Parameters. All designs must start with some parameters and assumptions. Generally, there is a specification of some type which includes output performance, available power input, mechanical mounting method, and means of connecting the load. First, look at the intended application and determine whether an inner- or outer-rotor motor would be the first choice. Some common indicators are charted in Table 5.2.

Establishing Motor Cross-Section Dimensions. It should be noted here that magnetic-circuit design is an interactive process and that the following rules are just guidelines to establish a starting point. At this point, the outside diameter of the motor and the shaft diameter have been selected. The next task is to roughly proportion the magnetic-circuit components to establish a starting point for the design. First look at the outer-rotor motor (Fig. 5.21). Then look at the inner-rotor motor (Fig. 5.22).

The shaft and bearing space has already been established, so the space for active material that remains is the area between the shaft hole and the outside diameter. Next, decide on the air gap. Gaps on the order of 0.020 to 0.040 in per side are very common. The smaller the air gap, the more flux that is available, but the higher the value of cogging torque. Manufacturing tolerances on stack diameter and magnet thickness generally dictate a gap somewhere in the range stated. Start with the smallest gap your factory can comfortably hold. If in doubt, try 0.030 in to start.

In order to set the rotor and stator magnetic circuit dimensions, you must make some assumptions with respect to the level at which you will work the materials. This is accomplished by setting the flux densities in the magnetic circuit as outlined in Table 5.3.

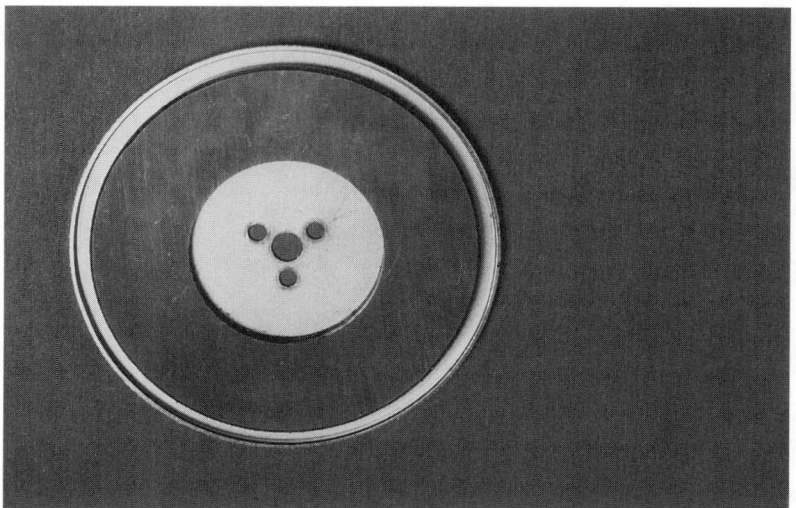

FIGURE 5.19 Axial air gap rotor with magnets.

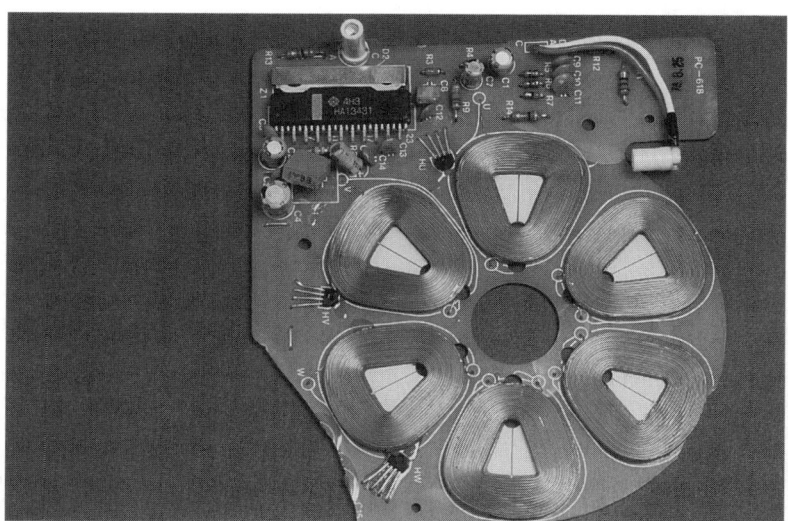

FIGURE 5.20 Axial air gap stator with windings.

TABLE 5.1 Comparison of Outer-Rotor and Inner-Rotor Motors

Outer rotor	Inner rotor
Single phase or multiphase.	Single phase or multiphase.
Can be wound with dc armature-winding equipment.	Requires ac stator-winding equipment (more expensive).
Bifilar windings can be employed on single-phase motors; 50% copper utilization.	Salient-pole machines utilize more copper.
Available copper space on distributed-winding motors is less than on inner-rotor types.	Distributed-winding motors provide smoother performance and better copper utilization.
Shorter end turns yield lower inductance and less copper loss.	Longer end turns yield higher inductance and more copper loss.
Greater rotor inertia.	Lower rotor inertia.
Less torque perturbation.	More torque perturbation.
Slower acceleration.	Fast acceleration.
Lower-energy magnets can be used.	Higher-energy magnets required.

Rotor Back Iron. In the outer-rotor motor, this is usually a steel cup of some kind. To determine its thickness, keep in mind that the flux has a long steel path that requires mmf to overcome. Set the flux density equal to 75 kline/in^2. Then:

$$B = \frac{\phi}{A} \quad A = \frac{\phi}{B}$$

But you don't yet know how much flux will be needed. You do know, however, that it will be constant on a per-pole basis. This allows you to proportion the steel housing and lamination sections based on the relative flux densities given in Table 5.3. Taking the nominal values and setting the stator teeth as 1.0, you can define ratios for the remaining parts. There are still some problems to overcome before you can proceed. You have not yet decided on the number of poles or teeth.

Poles and Teeth. Some general rules on selecting the number of poles, teeth, and phases are presented in Table 5.4.

TABLE 5.2 Inner-Rotor Versus Outer-Rotor Motor Applications

Requirement	Inner rotor	Outer rotor
Rapid acceleration	Very good	Poor
Heat dissipation	Very good	Poor
Low cogging	Okay	Good
Pump application	Okay	Good
Disk-drive application	Poor	Very good
Fan application	Poor	Very good
High side load	Good	Poor
Use with speed reducers	Good	Poor to okay
Reversible	Very good	Poor

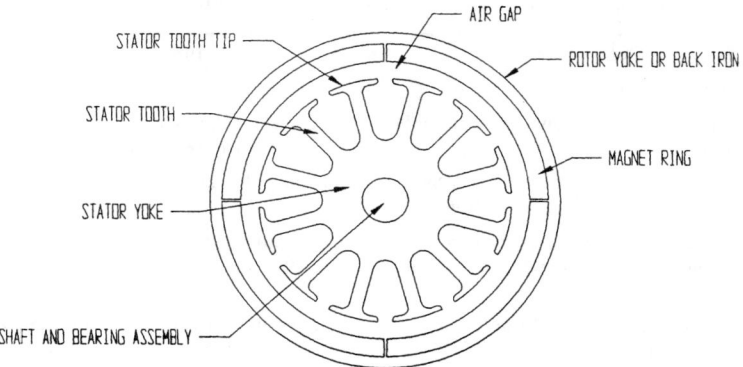

FIGURE 5.21 Outer-rotor motor

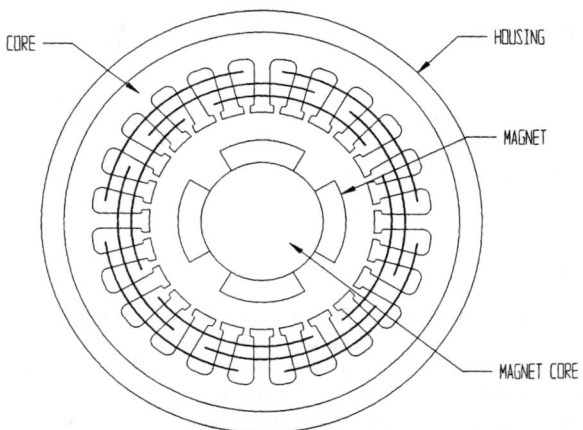

FIGURE 5.22 Inner-rotor motor

Number of Phases. Single-phase motors have poor conductor utilization, high torque ripple, and null zones that may create starting problems, but they are easy to wind and low cost and require only one or two power switches.

Two-phase motors also have poor conductor utilization, but the null zones are eliminated, the torque ripple is greatly reduced, and the cost is higher because a minimum of four power switches are required.

Three-phase motors have better conductor utilization, no starting problems, and greatly reduced torque ripple; they can get by with as few as three power switches, but they generally cost more to wind.

Increasing the number of phases to four or greater realizes small gains in copper utilization and torque ripple, but the costs of winding and power switches usually outweigh the gains.

TABLE 5.3 Preferred Flux Densities

Magnetic path section	Flux density	
	kline/in^2	T
Rotor back iron	70–80	1.09–1.24
Stator tooth tips	50–70	0.78–1.09
Stator teeth	110–125	1.71–1.94
Stator yoke section	80–100	1.24–1.55

As a starting point then, assuming that performance is as important as cost, the three-phase motor is a good choice.

Number of Poles. Some general rules of thumb to consider when selecting the number of poles are the following:

1. As the total number of poles increases, the requirement for rotor and stator back iron decreases because the total flux is spread over more poles reducing the density.
2. Since less back iron is needed, more space is available for windings, allowing for a reduction in copper losses.
3. More poles have more parts and cost more money.

Two-pole motors require substantial back iron and long winding spans, which means more expensive winding equipment. Four-pole motors require about half the back iron and have shorter coil spans. Although the equipment required may be the same as for two-pole motors, four-pole motors have shorter end turns—therefore less mutual inductance, less copper loss, and an easier time in winding.

TABLE 5.4 Effects of Changing Number of Poles, Teeth, and Phases

	Effect on design factors				
Change	Cogging	Speed	Torque	Active material utilization	Cost
Number of poles					
Increased	Decreases	Decreases	Increases	Increases	Increases
Decreased	Increases	Increases	Decreases	Decreases	Decreases
Number of teeth					
Increased	Decreases	No change	No change	Increases	Increases
Decreased	Increases	No change	No change	Decreases	Decreases
Number of phases					
Increased	Decreases	No change	No change	Increases	Increases
Decreased	Increases	No change	No change	Decreases	Decreases

Six-pole and higher motors allow for some additional gains in copper utilization and reduced back iron, but manufacturing costs and control-circuit costs start increasing disproportionately. So, barring a requirement for high torque, low speed or low cogging, a four-pole motor is also a good place to start.

Number of Teeth. So far, you have picked three phases and four poles. The next major item is the number of teeth. There are many combinations of teeth and poles that will work for motors having two, three, four and more phases. Common numbers of teeth for two-phase motors are 8, 12, 16, 24, and 48. Common three-phase numbers are 6, 9, 12, 15, 24, 36, and 48. There are obviously other numbers of teeth that will work for these motors. These are just some of the common ones. As the number of teeth selected increases, the number of slots available for winding also increases, resulting in more coils per pole. In general, one should choose the lowest number of teeth possible that will provide a reasonable winding pattern.

Some things to keep in mind as combinations of teeth, poles, and phases are chosen are the following:

- *Cogging.* A torque perturbation occurs every time a magnet pole tip passes a stator tooth. Even numbers of teeth and poles cause a greater perturbation than uneven numbers because at any given time there are more pole tips passing teeth. This can be addressed by using an uneven number of teeth versus poles, which gives fewer poles passing teeth at a given time, or by skewing the stator or the magnets.
- *Manufacturability.* Increasing the number of teeth to improve performance has its limits. As the number of teeth increases, the laminations may become difficult to punch, or the teeth may bend easily as the slots are wound.

For this example, choose 12 teeth. Therefore, you have chosen 3 phases, 4 poles and 12 teeth.

To estimate the size of the respective components, look at the motor section in Fig. 5.23 and note the flux paths.

The flux from any one pole of this four-pole motor travels through the teeth to the stator yoke, then splits through the yoke, goes back through the teeth under the

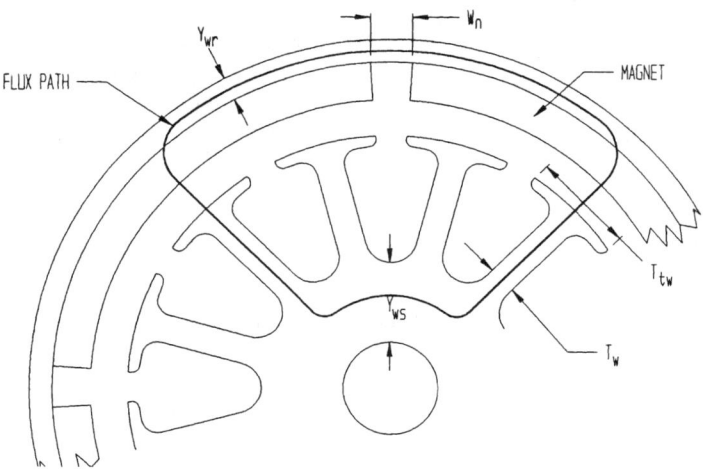

FIGURE 5.23 Outer-rotor motor flux path.

adjacent magnet pole across the air gap, goes through the magnet, splits through the rotor yoke, and goes back to the originating magnet. The magnet sets up the flux; all the flux goes across the gap (except for leakage), into the tooth tips, and through the teeth. When proportioning the sizes, refer to Table 5.5.

TABLE 5.5 Nominal Flux Densities

Magnetic-circuit component	Nominal flux density, kline/in²	
	Outer rotor	Inner rotor
Stator teeth	115	115
Stator tooth tips	60	60
Stator yoke	90	75
Rotor yoke	75	90

Note that the stator and rotor yoke densities are inverted for the inner- and outer-rotor motors. This is because the short path can afford a higher density without resulting in too much mmf drop.

Since the flux density is inversely proportional to the area ($B = \phi/A$), and the area is proportioned to the width of the magnetic-circuit component times the length of the stack L_{stk}, the widths may be ratioed directly according to the flux density in Table 5.5.

Ratio all components to the teeth as follows:

$$A_{sy} = Y_{ws} \cdot L_{stk}$$

where L_{stk} = stack length

$$\therefore Y_{ws} \propto \frac{115}{(90)(2)} \approx 0.64 \times \text{total tooth width}$$

The density ratio is divided by 2 because of the flux split.

$$A_{tt} = T_{tw} \cdot L_{stk}$$

$$\therefore T_{tw} \propto \frac{115}{60} \approx 2$$

$$A_{ry} = Y_{wr} \cdot L_s$$

$$\therefore Y_{wr} \propto \frac{115}{75(2)} \approx 0.75 \times \text{total tooth width}$$

In the case of the outer-rotor motor, the length of the shell L_s is usually longer then the stator stack, so Y_{wr} is more in the range of 30 to 40 percent of total tooth width.

To estimate the width of the teeth, assume that the area available after subtracting magnet area, air gap area, and stator yoke is to be divided up so that the slot area is approximately 2.5 times the tooth area. In this case, tooth area is the tooth width times its length in the radial direction. Assume straight teeth with parallel sides. Then take the total tooth width and divide it by the number of teeth per pole you have chosen. In this case, the number is 3. These dimensions may need to be modified to balance mmf drops once the magnetic circuit is solved.

Tooth-Tip Shapes. At this point, there are some things that need to be considered about the shape of the tooth tips. The outer-rotor motor (Fig. 5.21) has a stator that is much like a brush-type dc-motor armature, and they can be wound with similar equipment. The tooth-tip thickness T_{tt} (Fig. 5.24) should be large enough to keep the region from saturating as the magnet pole tips pass the edge of the tooth tip. This will reduce the torque perturbations and audible noise. A radius should be added at position 1 to allow for easy entry of the magnet wire and at position 2 to reduce magnetic saturation of this region. The width of the slot opening W_{so} should be wide enough to allow the largest-diameter wire you intend to use to be easily wound in place. It

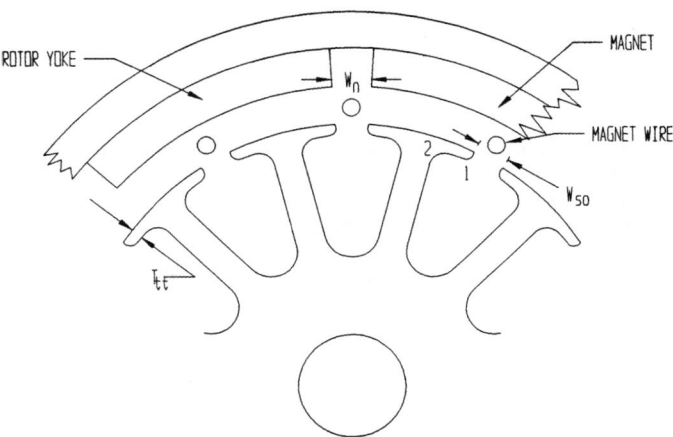

FIGURE 5.24 Outer-rotor tooth shapes.

should be kept small enough so that the magnet wire will be easily retained after the winding operation is complete.

The inner-rotor motor (Fig. 5.25) follows the same rules with respect to tip thickness as the outer rotor design. Some of the significant differences occur with the width of the slot opening and the shape of the slot. If the stator is going to be wound with a needle or gun winder, W_{so} must be sufficient to allow for the needle width and lateral motion. If the stator coils are going to be wound on forms and then inserted into the stator with ac-motor-type inserters, the following rules apply (Fig. 5.26):

1. The slot depth D to the slot width W_s, D/W_s ratio should be as great as possible, with 4 being a good value and 3 being the minimum value.
2. The slot bottom shape is not critical. Round or square will work, but round is easier to fill.
3. The angle α should be between 15 and 30 degrees.

An important rule to remember when establishing part dimensions is that the width of any part should be 1.5 times the thickness of the material. Failure to follow this rule results in part distortion and die wear. The material should be selected based on cost, induction, and core loss requirements.

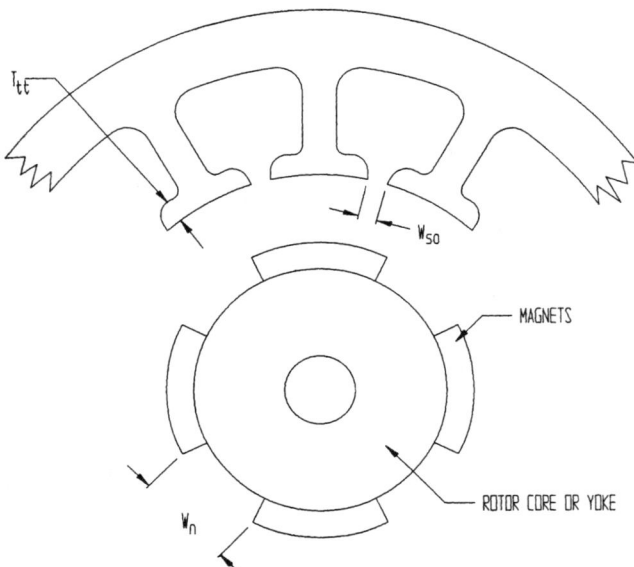

FIGURE 5.25 Inner-rotor tooth shapes.

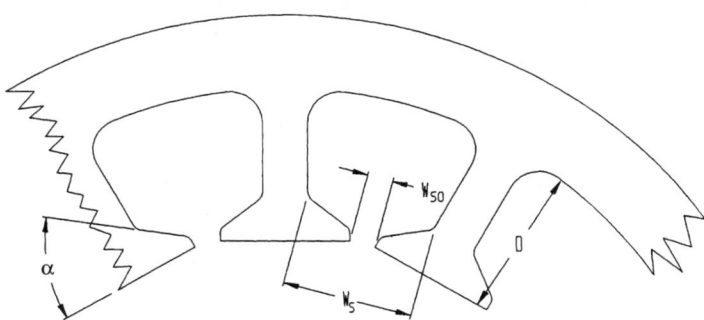

FIGURE 5.26 Inner-rotor tooth proportions.

Structural Magnetic Materials. There are some parts of the motor magnetic circuit that generally wind up being used for mechanical structural purposes as well. The outer-rotor motor in Fig. 5.23 has a rotor yoke which is used to contain the magnets as well as serve as a flux path. The inner-rotor motor in Fig. 5.24 has a rotor core to which the magnets are bonded. This core serves as the magnetic rotor yoke as well as the means of attaching the output shaft to the magnets. These materials are usually a magnetically soft steel of the type ANSI CRS 1008 to 1026. It is not a requirement that theses parts be laminated because the direction of the flux in them does not change; therefore, there are no core losses. A typical magnetization curve of this material is shown in Fig. 5.27.

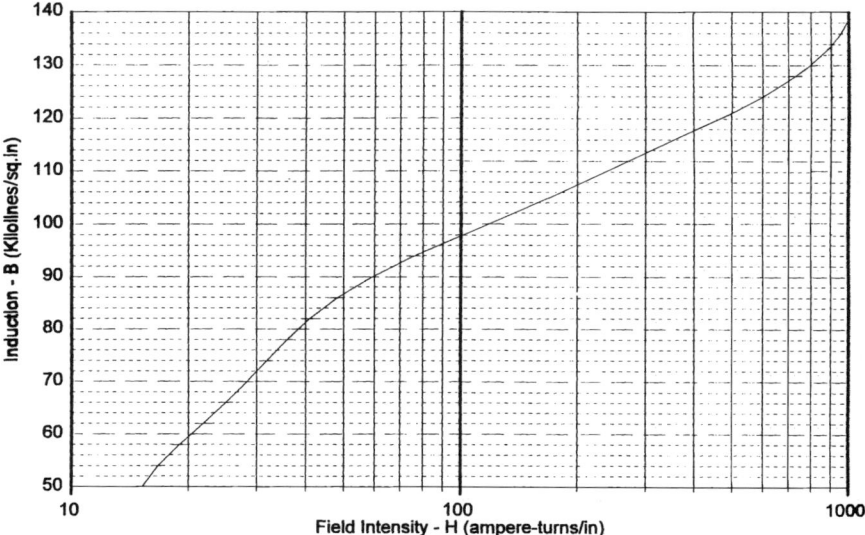

FIGURE 5.27 Unannealed 1020 CRS housing material.

Rotor Inertia. The following method for calculating rotor inertia is based on the assumption that all parts of the rotor rotate around the center of the shaft.

For the outer-rotor motor in Fig. 5.28, calculate the inertia of each part separately, then add the parts together for the total rotor inertia. First, calculate the inertia of the shaft:

$$J_{\text{shaft}} = 0.0184 \left(\frac{D_s}{2}\right)^4 L_{\text{shell}}$$

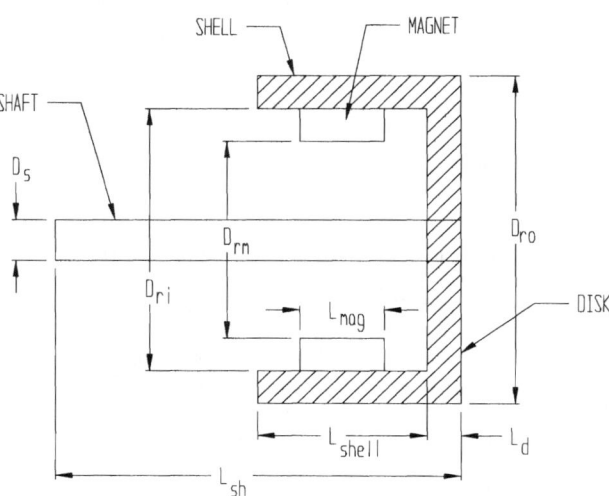

FIGURE 5.28 Outer-rotor inertia.

Then calculate the inertia of the disk:

$$J_D = 0.0184\left(\frac{D_{\text{ro}}}{2}\right)^4 L_d - 0.0184\left(\frac{D_s}{2}\right)^4 L_d$$

Next, calculate the inertia of the shell:

$$J_{\text{shell}} = 0.0184\left(\frac{D_{\text{ro}}}{2}\right)^4 L_{\text{shell}} - 0.0184\left(\frac{D_{\text{ri}}}{2}\right)^4 L_{\text{shell}}$$

Then calculate the inertia of the magnets (see Table 5.6 for magnetic density):

$$J_{\text{mag}} = \left(\frac{D_{\text{ri}}}{2}\right)^4 \frac{L_{\text{mag}}(\text{mag density})}{772} - \left(\frac{D_{\text{rm}}}{2}\right)^4 \frac{L_{\text{mag}}(\text{mag density})}{772}$$

TABLE 5.6 Densities of Common Magnetic Materials

	Density		
Material	lb/in³	oz/in³	g/cm³
Bonded ferrite	0.134	2.14	3.71
Sintered ferrite	0.177	2.83	4.90
Alnico	0.264	4.22	7.31
Samarium cobalt	0.302	4.83	8.35
Neodymium-iron-boron	0.271	4.34	7.50

Finally:

$$J_{\text{rotor}} = J_s + J_D + j_{\text{shell}} + J_{\text{mag}}$$

Now, consider the rotor of the inner-rotor motor (Fig. 5.29).
First:

$$J_{\text{shaft}} = 0.0184\left(\frac{D_s}{2}\right)^4 L_{\text{shaft}}$$

Then:

$$J_{\text{mag}} = \left(\frac{D_{\text{ro}}}{2}\right)^4 \frac{L_{\text{mag}}\,(\text{mag density})}{772} - \left(\frac{D_{\text{rm}}}{2}\right)^4 \frac{L_{\text{mag}}\,(\text{mag density})}{772}$$

Assuming the hub is steel, it follows that:

$$J_{\text{hub}} = 0.0184\left(\frac{D_{\text{rm}}}{2}\right)^4 L_{\text{hub}} - 0.0184\left(\frac{D_s}{2}\right)^4 L_{\text{hub}}$$

$$J_{\text{rotor}} = J_{\text{shaft}} + J_{\text{mag}} + J_{\text{hub}}$$

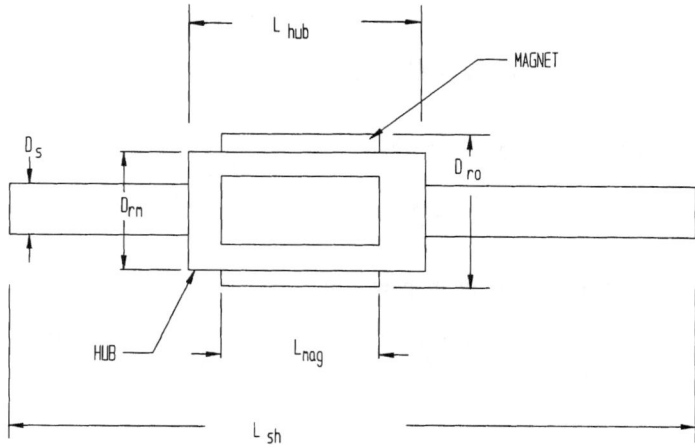

FIGURE 5.29 Inner-rotor inertia

5.1.3 Stator Winding Design Considerations*

Brushless dc motors initially were designed in large numbers for spindle drives in Winchester disk drives. The early designs were three-phase, later moving to two-phase and then one-phase, due to the very slow start-up requirements, very small friction loads, and the need to reduce unit cost at all levels. The industrial and machine tool markets started with and continue to use three-phase BLDC motors in their variable-speed, variable-load, high-start-up applications. The overwhelming popularity of three-phase BLDC motors focuses this subsection toward three-phase windings. Many of the initial design activities for various winding patterns can be traced back to the 1920s and earlier based on work done on three-phase ac windings.

This subsection reviews the various winding line connections, the key winding patterns and hookups, various winding constants, and winding selection and design techniques.

Basic Winding Configurations. There are other basic decisions that must be made by the design engineer before a BLDC motor design can commence. Previously defined is the number of phases, which is three here. Next in importance is the number of poles. The use of two poles is waning, and the use of six or eight poles is increasing. Four-pole BLDC motors are among the most popular used today. Two- and four-pole BLDC motor designs are used here, but the rules for two and four poles can be extended to higher pole counts. The number of stator slots (and teeth) and the winding pattern are key design decisions. This section is dedicated to reviewing the important parameters of these two design decisions.

In a three-phase motor there are three windings or phases positioned 120° electrical apart. Figure 5.30 shows the location of 6 coils in a representative 12-slot stator. A two-pole rotor (not shown) will rotate as the three windings are energized in sequence A-B-C, as A-A', B-B', and then C-C' are energized sequentially. The three-phase winding always develops positive starting torque, no matter where the rotor starts its motion.

---

* Subsection contributed by Dan Jones, Incremotion Associates.

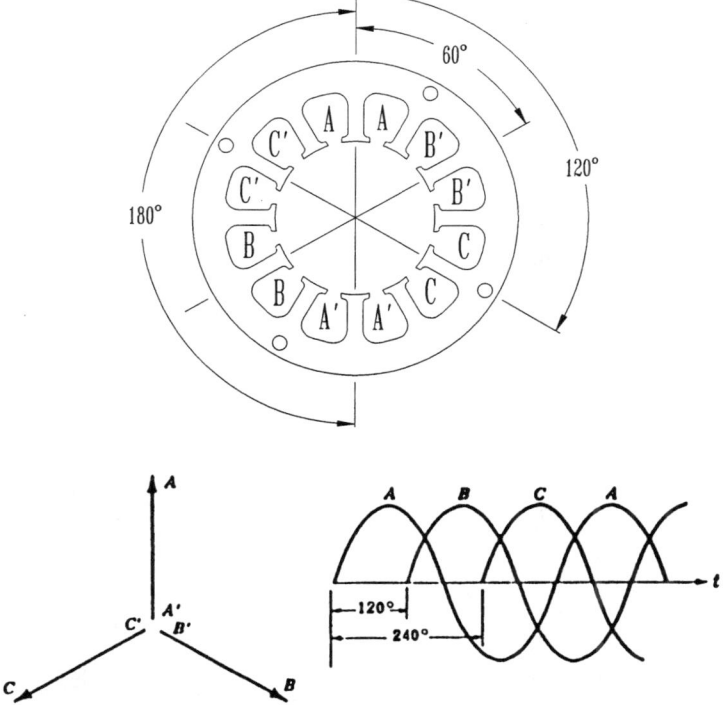

FIGURE 5.30 Basic 3-phase winding layout.

There are many winding line connections that can be used in three-phase drive systems. Figure 5.31 illustrates the various configurations. The *half-wave wye* is the simplest three-phase line configuration (Fig. 5.31a). It uses three power lines and one return line (four leads). The excitation is shown adjacent to the schematic in Fig. 5.31a. Only 33 percent (one lead) of the half-wave wye windings are energized at any time in operation. The second wye winding, the *full-wave wye* (Fig. 5.31b) has only three leads but 66 percent (two leads) of the windings are in operation simultaneously. The excitation scheme is shown to the right of the schematic.

The third major winding connection pattern is the *delta*, shown in Fig. 5.31c. It possesses the same excitation scheme as the full-wave wye. The delta winding configuration has been used more extensively than the wye in fractional-horsepower (≤746 W) motor applications. The wye is more popular with the larger-sized integral-horsepower BLDC motor users. The final winding to be reviewed is the *independent* winding line connection (Fig. 5.31d). In this scheme, each winding is independent of its neighbor. The excitation scheme is more complicated, but each winding can be operated in parallel, thereby distributing the total current. The windings are still situated 120° electrical away from each other. This winding configuration has seen limited use to date.

The most popular winding line configurations are the full-wave wye and the delta. In a balanced wye configuration, the line and coil (phase) currents are equal,

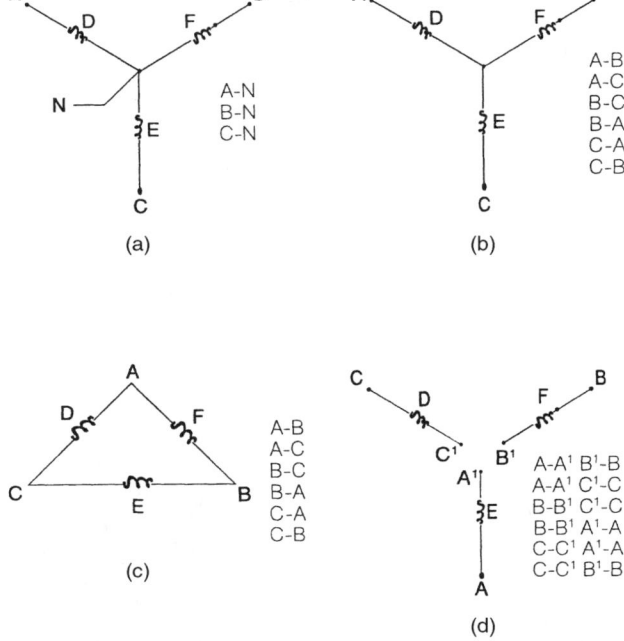

FIGURE 5.31 Popular 3-phase BLDC motor winding line connections: (a) wye (half wave), (b) wye (full wave), (c) delta, and (d) independent.

the neutral current is zero, and the line-to-line voltage is $\sqrt{3}$ times the phase voltage. In a balanced delta connection, the line-to-line and coil (phase) voltages are equal, but coil currents are $1/\sqrt{3}$ times the line-to-line currents.

Key Winding Patterns. There are many types of winding patterns that can be utilized. Four major winding patterns are listed here:

1. Constant integral pitch—lap winding (full)
2. Variable pitch—concentric winding
3. Constant fractional pitch—lap winding for even or odd stator slots
4. Half-pitch

Each of these winding patterns has two coils per stator slot. There is one winding type designated, a consequent pole winding where there is only a single coil per slot. Consequent pole windings are very popular in single-phase ac motors of fractional-horsepower size.

Table 5.7 is revised from Veinott and Martin (1987). It displays the various stator slot and rotor pole combinations along with the maximum number of parallel circuit combinations with a specific slot and pole combination. For purposes of simplicity, either 12- or 24-slot stators are used here to illustrate the various winding patterns. In one case, a 15-slot stator is used to illustrate an odd-slot fractional-pitch lap winding.

TABLE 5.7 Slots Versus Poles Versus Parallel Hookups Versus Coils per Pole per Phase

Slots or coils	2 Poles		4 Poles		6 Poles		8 Poles	
	Circuits	Coils per phase per pole	Circuits	Coils per phase per pole	Circuits	Coils per phase per pole	Circuits	Coils per phase per pole
6	2	1	—	—	—	—	—	—
9	1	1½	—	—	—	—	—	—
12	2	2	4	1	—	—	—	—
15	1	2½	1	1¼	—	—	—	—
18	2	3	2	1½	6	1	—	—
21	1	3½	1	1¾	NP*	X	—	—
24	2	4	4	2	NP	X	8	1
27	1	4½	1	2¼	3	1½	1	1⅛
30	2	5	2	2½	NP	X	2	1¼
33	1	5½	1	2¾	NP	X	1	1⅜
36	2	6	4	3	6	2	4	1½
39	1	6½	1	3¼	NP	X	1	1⅝
42	2	7	2	3½	NP	X	2	1¾
45	1	7½	1	3¾	3	2½	1	1⅞
48	2	8	4	4	NP	X	8	2

*NP = not practical.

If one uses a 12-slot stator winding, there are two full-pitch integral lap windings available, one for two poles and the other for four poles.

$$P \times \text{Ph} = nS \tag{5.1}$$

where P = number of poles
Ph = number of phases
S = number of stator slots
n = integer number 1, 2, 3, ... n

If $P = 2$, Ph = 3, and $S = 12$, then $n = 2$, an integer. Figure 5.30 shows the basic winding-slot pattern for a 12-slot 2-pole 3-phase 2-coils-per-pole-per-phase ($n = 2$) configuration. The coil pattern for this winding configuration is shown in Figs. 5.32 and 5.33 as a series wye line configuration and as a parallel wye line configuration, respectively. There are really 12 coils used in this design, but only 6 are shown. There are two 1- to 7-throw coils—one inserted on the right side (CW direction), the second inserted on the left side (CCW direction), and doubles on the other five coils also, inserted as described previously. Note the position of the teeth for the 12-slot stator. The slot pitch (adjacent slot to slot) is 360/12 or 30° mechanical or electrical. The angular location for the phase 2 winding (CW direction) is only 60°. It is supposed to be 120°. Symmetry solves the problem if 180° (polarity change) is added to the 60° to achieve 240° mechanical or electrical. So the 1, 2, 3 winding phase hookups displayed in Fig. 5.32 will yield a CCW rotation.

The series and parallel hookup options are very important from a practical aspect of magnet wire size selection. It is easier to pack smaller-sized magnet wires in a

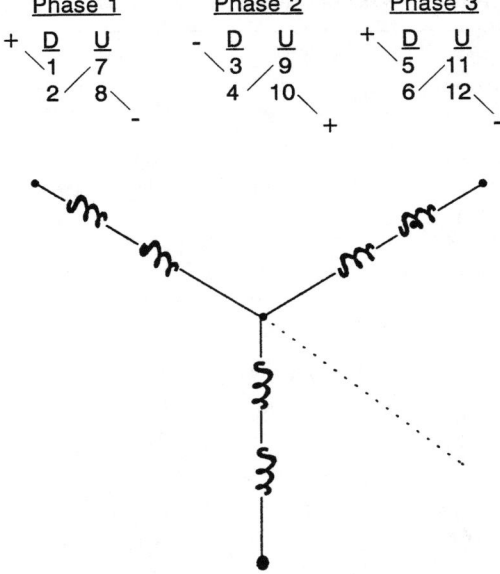

FIGURE 5.32 Series hookup for 2-pole wye-winding 12-slot brushless dc motor.

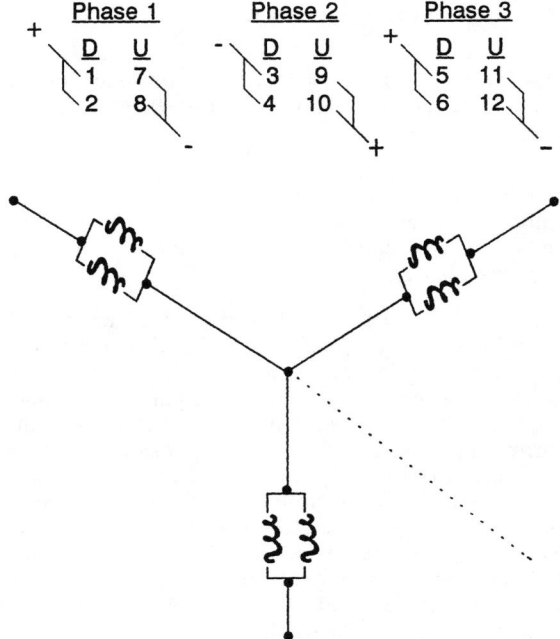

FIGURE 5.33 Parallel hookup for 2-pole 3-phase wye-winding 12-slot brushless dc motor, constant-pitch lap winding pattern.

BLDC stator slot than larger ones. Since total turns per phase are directly proportional to torque, putting all the needed turns (N per phase) in a single coil with smaller magnet wires and then paralleling the coils will yield extra space for more turns. This is a packing factor consideration.

The disadvantage of using parallel coils is that both sets of coil leads must be used to properly connect the coils (see Fig. 5.33). Figure 5.34 shows a representation of how two coils would be inserted into adjacent stator slots. Remember that there are 2 slots per coil per phase for this 2-pole 12-stator-slot winding pattern.

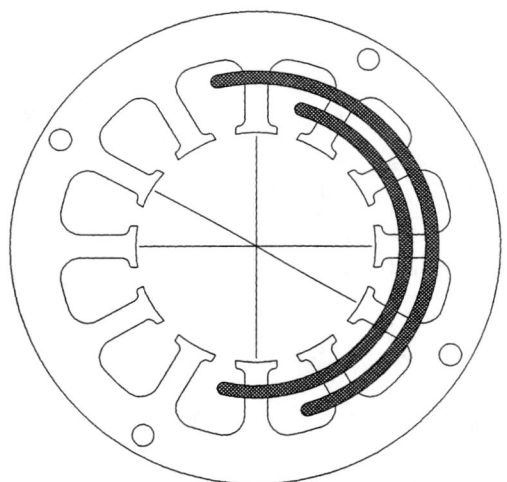

FIGURE 5.34 12-slot lamination, 2 poles, 3-phase constant-pitch lap pattern.

Figure 5.35 shows the winding pattern for a constant-pitch 4-pole 12-slot BLDC stator. There is now only one coil per pole per phase ($n = 1$) in this winding, as shown in the following equation:

$$4 \text{ poles} \times 3 \text{ phase} = 12 \text{ slots}$$

$$n = 1 \text{ coil per pole per phase} \tag{5.2}$$

The four coils per phase can be connected as four coils in a series hookup or four coils in a parallel hookup. It is also possible to connect two coils in series and two coils in parallel to achieve a series-parallel hookup. Figure 5.36 displays two adjacent coils inserted into the proper stator slots. Note the shorter length of these coils because the end turns are shorter while the segments of the turns (conductors) within the appropriate stator slots remain the same length. The shorter the end-turn length (which doesn't create any torque), the better the motor design.

The variable-pitch winding was developed to reduce the stator end-turn height caused by the numerous adjacent coil crossovers by nesting the coils inside each other, as shown in Fig. 5.37. This pattern can be used only when coils per phase per pole n is 2 or greater. Figure 5.37 displays an $n = 3$ condition. The actual winding pattern is shown in Fig. 5.38. Since $n = 2$, there will be two different winding lengths or

5.28 CHAPTER FIVE

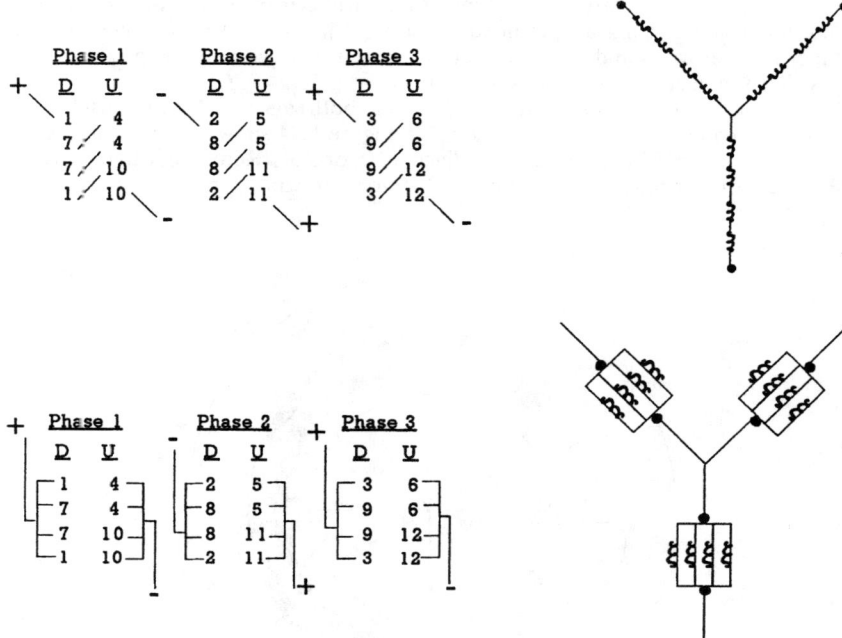

FIGURE 5.35 Winding pattern for a constant-pitch 4-pole 12-slot stator, 3-phase hookup, 30° mechanical, 60° electrical.

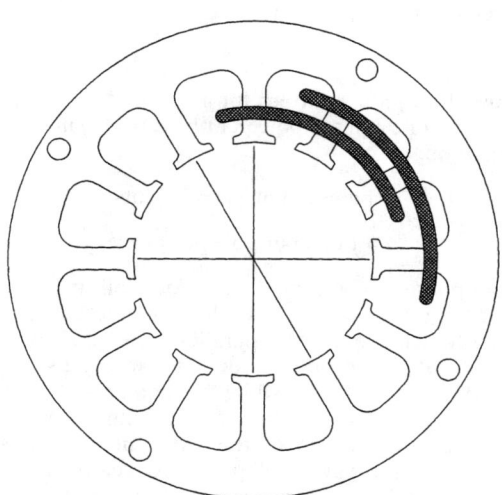

FIGURE 5.36 Constant-pitch 4-pole lap winding pattern with adjacent coil layout.

throws. The following equation describes the method used to determine the two variable winding pitches:

$$VSP_1 = \frac{\text{number of stator slots}}{\text{number of poles}} + 1$$

$$VSP_2 = \frac{\text{number of stator slots}}{\text{number of poles}} - 1 \tag{5.3}$$

$VSP_1 = 7$ and $VSP_2 = 5$. The average of these two variable-pitch coils should equal the integral winding pitch for a 3-phase 2-pole 12-slot design, which is 6. Figure 5.39 displays the location of two adjacent coils placed in the proper stator slots. This winding pattern will reduce end-turn height and coil lengths by 10 to 15 percent. The concentric winding with variable pitch has been used extensively in larger integral-horsepower units, particularly by the various electric winding repair houses.

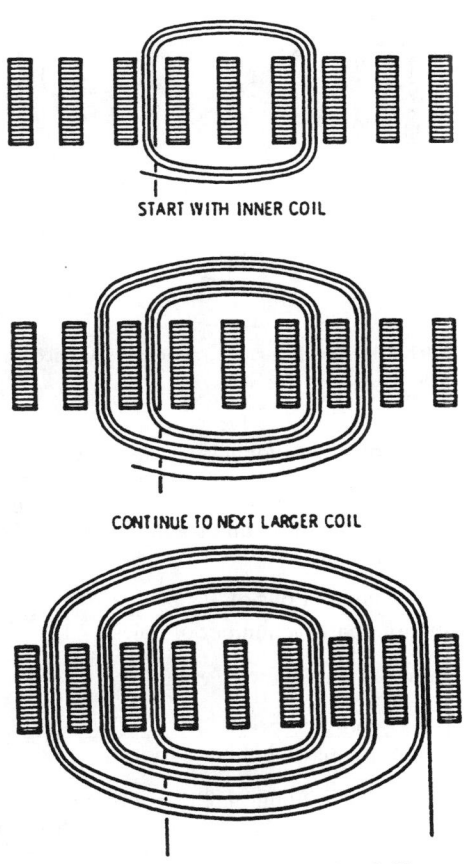

FIGURE 5.37 Side view of variable-pitch concentric winding pattern.

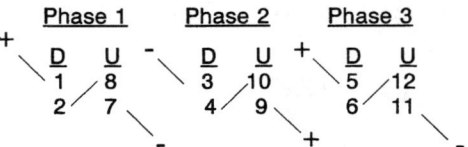

FIGURE 5.38 Variable-pitch concentric winding pattern, 2 poles, 3-phase series hookup.

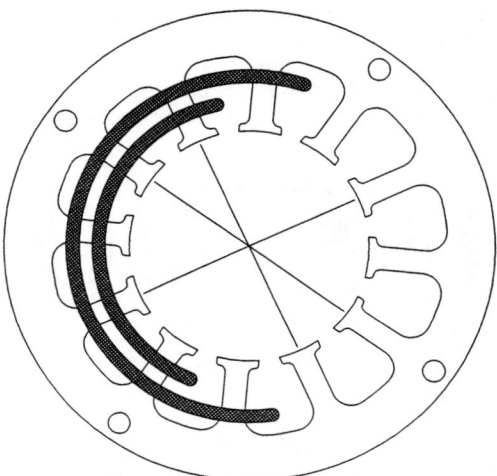

FIGURE 5.39 2-pole 3-phase variable-pitch concentric pattern.

The third winding pattern is fractional-pitch winding, which is used in many applications, particularly odd stator slots where cogging torque must be reduced. Table 5.7 identifies the fractional pitch as 1¼ for a 15-slot 4-pole configuration. This pattern is popular for resolver windings and some low-cog BLDC motors. The following equation yields the pitch:

$$\frac{\text{number of slots}}{\text{number of phases} \times \text{number of poles}} = \frac{15}{12} = 1\tfrac{1}{4} \quad (5.4)$$

The winding throw is shown in the following equation:

$$\frac{\text{number of slots}}{\text{number of phases} \times \text{number of poles}} = \frac{15}{4} = 3\tfrac{3}{4} \quad (5.5)$$

Equation (5.5) identifies the winding pitch or throw as either 3 or 4. Three is selected for the pattern shown in Fig. 5.40. The starting points for the coils must meet the pattern of coil placements shown. This pattern places two adjacent coils strategically and equidistantly around the stator with each of the three phases having one set of two adjacent coils and three sets of single coils (pattern from Veinott and Martin, 1987).

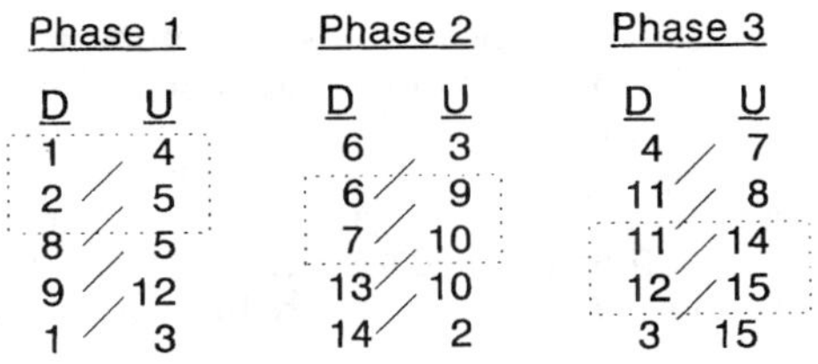

```
     Phase 1        Phase 2        Phase 3
     D    U         D    U         D    U
     1    4         6    3         4    7
     2    5         6    9         11   8
     8    5         7   10         11  14
     9   12        13   10         12  15
     1    3        14    2          3  15
```

FIGURE 5.40 Constant fractional-pitch odd-slot 4-pole, 3-phase series hookup. Recommended coil groups: 2-1-1-1-2-1-1-1-2-1-1-1 = 15.

As with any winding, there are advantages counterbalanced by disadvantages. The following equation defines the tooth or slot pitch of the 15-slot stator.

$$SP = \frac{360°}{\text{number of slots}}$$

$$SP = \frac{360}{15} = 24° \text{ mechanical} \tag{5.6}$$

Now, since this design is a four-pole BLDC design, there are two full electrical cycles for one full mechanical cycle. The next equation defines the relationship between electrical and mechanical degrees for any four-pole design.

Degrees electrical = degrees mechanical × number of pole pairs

$$\therefore \text{ Degrees electrical} = 24° \times 2 = 48° \text{ electrical} \tag{5.7}$$

A 48° electrical pitch does not equal the desired 60° pitch, so there must be torque loss.

Figure 5.41 illustrates the winding pattern for a 12-slot 2-pole fractional-pitch winding. A full winding pitch would possess a value of 6 with a throw of 1 to 7. This winding pattern has a winding pitch of ⅚ (fractional) or a throw of 1 to 6. This group of fractional-pitch windings has a pitch less than 1 when an even stator slot count is employed. This winding pattern is used in larger three-phase ac motors to decrease the harmonic content of both the voltage and mmf waveforms. This technique is very similar to that of short-pitch lap windings used in brush dc motors.

The final winding type is the half-pitch winding, which has the simplest winding pattern. The coil is wound directly around the stator tooth with a winding pitch of 1

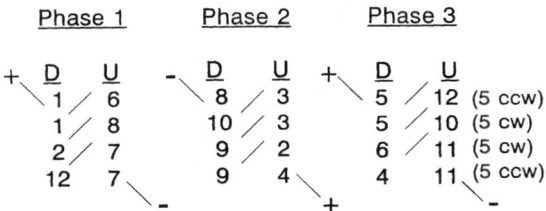

FIGURE 5.41 Constant fractional-pitch even-slot 2-pole 3-phase series hookup.

and a throw of 1 to 2. It is used extensively in step motors and switched-reluctance motors. It is also used in PM BLDC motors used in disk drives. A nine-slot eight-pole winding is very popular, as well as a nine-slot six-pole. This winding by definition is also a fractional-pitch winding. It is the most cost-effective and simplest winding with the shortest mean length of turn (MLT) and therefore the lowest resistance per coil. It does suffer from reduced torque, as all fractional pitch windings do. The various winding factors that determine the reduced torque values are reviewed in the next subsection.

Winding Factors for Different Winding Patterns. There are a number of winding factors that adjust for the peak magnetomotive force (mmf) and the winding-generated flux ϕ which directly leads to adjustments to the winding back emf (K_e) and peak developed torque of the BLDC motor. These factors can be identified as follows:

- Chord factor (pitch factor)
- Distribution factor (breadth factor)
- End-turn factor (coil-length factor)

The pitch factor K_p and the distribution factor K_d are the factors discussed in this subsection. The end-turn factor K_{ET} is discussed in a later subsection. The pitch factor is defined by the following equation:

$$K_p = \sin \frac{\text{winding pitch} \times 90}{\text{slots per pole}}$$

$$K_p = \sin\left(\frac{6 \times 90}{12/2} = \frac{540}{6}\right)° = \sin 90° = 1 \tag{5.8}$$

which is the pitch factor for a 2-pole 12-slot integral-pitch winding.
The distribution factor K_d is delineated by Eq. (5.9).

$$K_d = \frac{\sin(n\alpha/2)}{n \sin(\alpha/2)} \tag{5.9}$$

where n = slots per pole per phase

$$\alpha = \frac{\text{number of poles} \times 180°}{\text{number of slots}}$$

$$K_d = \frac{\sin 2(15°)}{2 \sin 15°}$$

$$K_d = \frac{\sin 30°}{2 \sin 15°} = \frac{0.500}{0.5176} = 0.966$$

for a 12-slot 2-pole 3-phase integral-pitch winding with 2 slots per pole per phase. Veinott and Martin (1987) developed a table with the distribution factor for each major winding pattern, summarized in Table 5.8.

Table 5.9 contains the tabulated results for the K_p and K_d values for the six winding patterns presented plus a 24-slot 4-pole integral-pitch winding pattern illustrated in Fig. 5.42.

TABLE 5.8 Summary of Distribution Factors

Slots per pole per phase, or coils per group	2 Phase	3 Phase Conventional	3 Phase Consequent pole
1	1.000	1.000	1.000
2	0.924	0.966	0.866
3	0.911	0.960	0.844
4	0.906	0.958	0.837
5	0.904	0.957	0.833
6	0.903	0.956	0.831
∞	0.900	0.955	8.270
1½	0.911	0.960	0.844
All other fractional slot windings	0.900	0.955	0.827

TABLE 5.9 Summary of K_p and K_d Factors

Stator slots	Poles	Winding type	Winding pitch	Winding throw	Slot pitch, ° mechanical	Slots per pole per phase n	K_d	K_p	$K_d K_p$
12	2	Integral	1	1-7	30	2	0.966	1.000	0.966
12	4	Integral	1	1-4	30	1	1.000	1.000	1.000
12	2	Variable	1 (average)	1-8/1-6	30	2	0.966	1.000	0.966
15	4	Fractional	1¼	1-4	24	1	1.000	0.924	0.924
12	2	Fractional	⅝	1-6	30	1	1.000	0.966	0.966
9	8	Half-pitch	1⅛	1-2	40	1	1.000	0.985	0.985
24	4	Integral	1	1-7	15	2	0.966	1.000	0.966

Filling the Stator Slots. The first item in filling the stator slot is to compute the area of the slot. There are many types of stator slot shapes but the trapezoidal (constant-tooth-width) slot shown in Fig. 5.30 and the round (variable-tooth-width) slot are the most popular slot shapes. One can use basic trigonometry to determine the slot area or obtain the actual slot area from the lamination vendor.

There are three methods used to compute slot area and the total volume of copper magnet wire used. They use the following units:

- Square inches (in^2)
- Square mils (mil^2)
- Circular mils (cmil)

The circular mils method uses the nominal diameter of the insulated wire in mils or thousandths of an inch and takes the square of this diameter, which is the circular mil value.

$$\text{Area, cmil} = (\text{nominal diameter, mil})^2 \qquad (5.10)$$

For 18 AWG (American Wire Gauge), the nominal insulated single-build wire diameter from the Phelps Dodge magnet wire chart (Table 2.76) is 41.8 mil. The

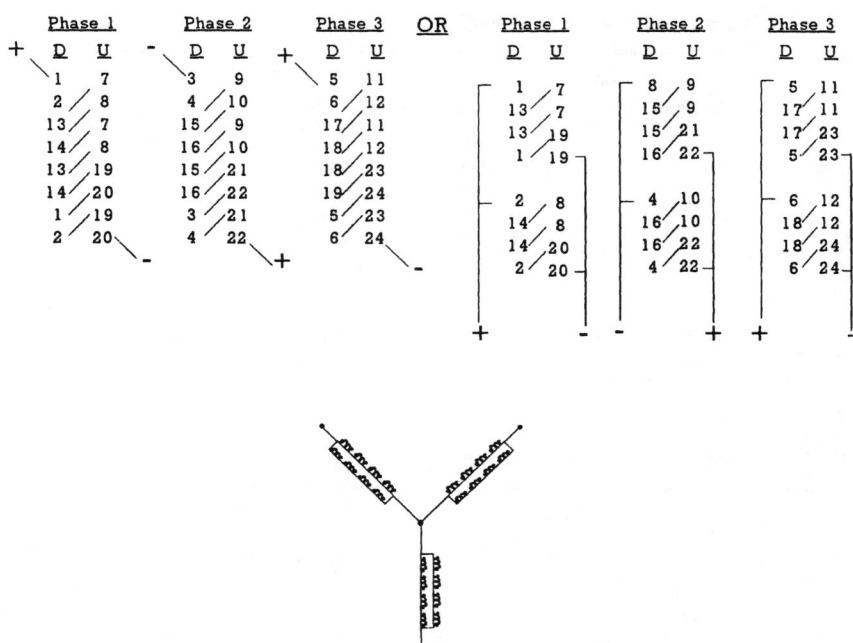

FIGURE 5.42 24 stator slots, 4 poles, 3-phase hookups 30° mechanical, 60° electrical.

wire area is 1747 cmil per Eq. (5.10). The square mils value is smaller and can be computed by modifying Eq. (5.10) to Eq. (5.11). In many cases, the wire charts compute the magnet wire's nominal diameter without insulation coating (bare wire diameter). It is strongly recommended that the insulated wire dimensions be used.

$$\text{Area, mil}^2 = \frac{(\text{nominal diameter, mils})^2}{1.27324} \tag{5.11}$$

The Phelps Dodge magnet wire chart (Table 2.76) also shows wires per square inch. The more important parameter is turns (conductors) per square inch. It yields the value of total number of turns—or, more appropriately, conductors—that can be placed in a slot, assuming 100 percent fill. Now, the most copper fill in terms of a percentage of actual turns per square inch versus 100 percent fill turns per square inch that this author has actually done by hand-insertion methods is 73 percent with 37 AWG and 63 percent with 21 AWG.

The practical limit is somewhere between 40 and 50 percent of this theoretical value depending on the type of winding machine, tooling used on the winding machine, length of stator stack, size of stator slot, etc. If one wanted to use 22 AWG, the turns (conductors) per linear inch would be 37.5 and the turns (conductors) per square inch would be 1410.

Total conductors (theoretical) = (SA) (turn/in^2)

Total conductors (theoretical) = 0.2192 × 1410 = 309   (5.12)

where SA = slot area

The total conductors must be an even number because 2 conductors equal 1 turn. Here the maximum value is 308 conductors with 22 AWG magnet wire. That would be 308/2 or 154 turns, since two coils are represented in every stator slot. Based on a practical slot fill of 45 percent, the maximum number of turns would probably be 69.3, or 69 turns per coil. The actual number of turns to achieve the desired performance has not yet been determined but for the AWG size selected, 69 turns or 138 conductors is the maximum practical limit.

The next important parameter to establish is the coil resistance. First, one must establish the MLT in the stator coil.

$$\text{MLT} = \frac{(2\text{SAL}) + 2[(\text{PI/NP})(\text{DR}/2)] + (2\text{TW})}{K_{\text{ET}}} \quad (5.13)$$

where   SAL = stack axial length
           PI = π
           NP = number of poles
           DR = slot OD + slot ID
           TW = tooth width
           K_{ET} = end-turn coil factor, ≤1.00

The value of K_{ET} is based on experience but a guide for its use is that the K_{ET} value approaches unity as the design increases in pole count and decreases in number of turns. For example, an 8-pole design using 10 turn/coil of 22 AWG would have a K_{ET} value of 0.97. A 2-pole design using 60 turn/coil of 22 AWG would have a K_{ET} value of 0.88.

The MLT for the example is 5.894 in/turn and the resistance per coil is 0.549 Ω, which can be computed using the formula in the following equation.

$$R_{\text{coil}} = \text{MLT} \times \text{total turns} \times \frac{\Omega}{\text{ft}} \quad (5.14)$$

The MLT value is converted into turns per feet to resolve the units and the value of ohms per foot for 22 AWG is 0.162 Ω.

Stator Winding Computations.   Two major motor performance parameters are required to be computed in order to establish overall motor performance. They are the torque constant K_t and the phase resistance R_{phase}. From these two parameters, all motor performance can be essentially derived.

The K_t value is derived for the case of RMS torque based on a trapezoidal torque versus position profile and near-perfect commutation. There is an adjustment of 10 percent to achieve a reasonable value for RMS torque. This value can be adjusted based on actual torque waveforms if required.

The following equation computes the K_t value per phase as described.

$$K_{t,\text{PH}} = \frac{T}{I} = \frac{K_1 K_2 \text{NPZ}\phi_g}{\text{Ph}K_3} \quad (5.15)$$

where K_1 = constant for units, cgs system
$K_2 = K_p K_d$ (winding factors)
K_3 = factor for series or parallel winding hookup
NP = number of poles
Z = total conductors
ϕ_g = gap flux, kmaxwell
Ph = number of phases = 3

For trapezoidal torque waveforms:

$$K_{t,\,wye} = 1.73\, K_{t,\,PH} \tag{5.16}$$

Phase resistance for a parallel wye winding configuration is computed as follows:

$$R_{PH} = \frac{\text{coil resistance}}{\text{coils/phase}} \tag{5.17}$$

$$R_{LL} = 2\, R_{PH} \quad \text{for a wye winding} \tag{5.18}$$

Torque constant and resistance computations are as follows:

1. $K_t = 21.046$ (oz · in)/A
   $K_t = 1.315$ (lb · in)/A
   $K_t = 0.1486$ (N · m)/A
2. $R_{PH} = 0.1375\ \Omega$
   $R_{LL} = 0.275\ \Omega$

5.1.4 Design Equations*

Design Approach. In the design equations that follow, the approach is to start with basic motor geometrical constraints and a magnetic circuit describing magnet flux flow. From this circuit, the magnet operating point is found, as are the important motor dimensions and current required to generate a specific motor output power at some rated speed. Given the desired counter-emf at rated speed, the number of turns per phase are found. From the winding information, phase inductances and resistances are computed.

Radial Flux Motor Design. The radial flux topology considered here is shown in Fig. 5.43.

Fixed Parameters. Many unknown parameters are involved in the design of a brushless PM motor. As a result, it is necessary to fix some of them and then determine the remaining ones as part of the design. Which parameters to fix is up to the designer. Usually, one has some idea about the overall motor volume allowed, the desired output power at some rated speed, and the voltage and current available to drive the motor. Based on these assumptions, Table 5.10 shows the fixed parameters assumed here.

The parameters given in Table 5.10 are grouped according to function. The required power or torque at rated speed, the peak counter-emf, and the maximum

---
* Subsection contributed by Duane C. Hanselman, University of Maine at Orono.

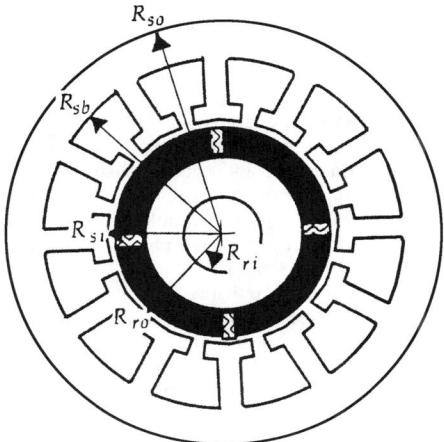

FIGURE 5.43 Radial-flux motor topology showing geometrical definitions.

conductor current density are measures of the motor's input and output. Topological constraints include the number of phases, magnet poles, and slots per phase. The air gap length, magnet length, outside stator radius, outside rotor radius, motor axial length, core loss, lamination stacking factor, back-iron mass density, conductor resistivity and associated temperature coefficient, conductor-packing factor, and magnet fraction are physical parameters. Magnet remanence, magnet recoil permeability, and maximum steel flux density are magnetic parameters. Shoe parameters include the slot-opening width and shoe depth fraction. Finally, the winding approach must be specified.

Of the parameters in the table, it is interesting to note that the stator outside radius, motor axial length, and rotor outside radius are considered fixed. The stator outside radius and axial length are fixed because they specify the overall motor size. The rotor outside radius is fixed because one often wishes to either specify the rotor inertia, which increases as R_{ro}^4, or to maximize R_{ro}, since torque increases as R_{ro}^2. Clearly, as R_{ro} increases for a fixed R_{so}, the area available for conductors decreases, forcing one to accept a higher conductor current density to achieve the desired torque. Secondarily, by specifying the rotor outside radius, the design equations follow in a straightforward fashion, and no iteration is required to find an overall solution.

Geometric Parameters. From the parameters given in Table 5.10 and the dimensional description shown in Figs. 5.43 and 5.44, it is possible to identify important geometric parameters. The various radii are associated by

$$R_{sb} = R_{so} - w_{bi}$$
$$R_{si} = R_{sb} - d_s = R_{ro} + g \qquad (5.19)$$
$$R_{ri} = R_{ro} - l_m - w_{bi}$$

TABLE 5.10 Fixed Parameters for the Radial Flux Topology

Parameter	Description
P_{hp} or T	Power, hp, or rated torque, N · m
S_r	Rated speed, rpm
E_{max}	Maximum counter-emf, V
J_{max}	Maximum slot current density, A/m²
N_{ph}	Number of phases
N_m	Number of magnet poles
N_{sp}	Number of slots per phase; $N_{sp} \geq N_m$
g	Air gap length, m
l_m	Magnet length, m
R_{so}	Outside stator radius, m
R_{ro}	Outside rotor radius, m
L	Motor axial length, m
$\Gamma(B, f)$	Steel core loss density versus flux density and frequency
k_{st}, p_{bi}	Lamination-stacking factor and steel mass density
ρ, β	Conductor resistivity and temperature coefficients
k_{cp}	Conductor-packing factor
a_m	Magnet fraction t_m/t_p
B_r	Magnet remanence, T
μ_R	Magnet recoil permeability
B_{max}	Maximum steel flux density, T
w_s	Slot opening, m
α_{sd}	Shoe depth fraction $(d_1 + d_2)/w_{tb}$
Winding approach	Lap or wave, single or double layer, or other

The pole pitch at the inside surface of the stator is related to the angular pole pitch by

$$\tau_p = R_{si}\theta_p \tag{5.20}$$

where

$$\theta_p = \frac{2\pi}{N_m} \tag{5.21}$$

is the angular pole pitch in mechanical radians, and the coil pitch at the rotor inside radius is

$$\tau_c = \alpha_{cp}\tau_p \tag{5.22}$$

where α_{cp} is given by Eq. (5.23):

$$\alpha_{cp} = \frac{\tau_c}{\tau_p} = \frac{\text{int }(N_{spp})}{N_{spp}} \tag{5.23}$$

Likewise, the slot pitch at the rotor inside radius is

$$\tau_s = R_{si}\theta_s \tag{5.24}$$

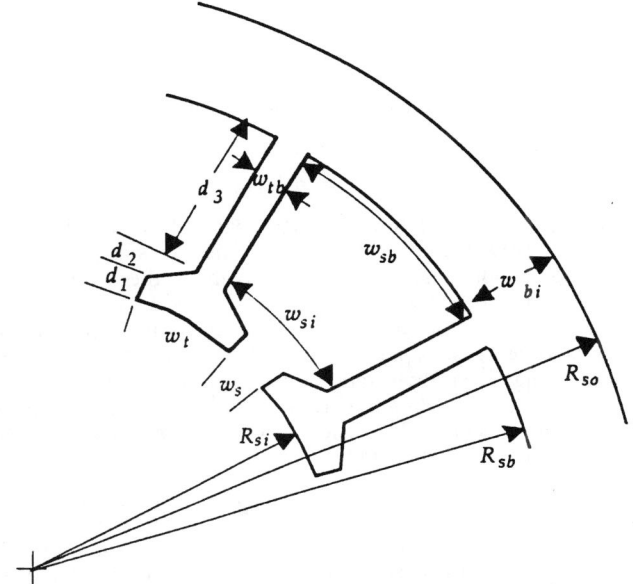

FIGURE 5.44 Slot geometry for the radial-flux motor topology.

where

$$\theta_s = \frac{2\pi}{N_s} \qquad (5.25)$$

is the angular slot pitch in mechanical radians. Knowledge of the slot opening gives the tooth width at the stator surface of

$$w_t = \tau_s - w_s \qquad (5.26)$$

The width of the slot bottom is given by

$$w_{sb} = R_{sb}\theta_s - w_{tb} \qquad (5.27)$$

Given that $d_s = d_1 + d_2 + d_3$ and

$$d_1 + d_2 = \alpha_{sd} w_{tb} \qquad (5.28)$$

The conductor slot depth is

$$d_3 = d_s - \alpha_{sd} w_{tb} \qquad (5.29)$$

And the slot cross-sectional area available for conductors is

$$A_s = d_3 \left[\theta_s \left(R_{sb} - \frac{d_3}{2} \right) - w_{tb} \right] \qquad (5.30)$$

In addition, the slot width just beyond the shoes is

$$w_{si} = (R_{si} + \alpha_{sd}w_{tb})\theta_s - w_{tb} \tag{5.31}$$

From this expression it is possible to define the slot fraction as

$$\alpha_s = \frac{w_{si}}{w_{si} + w_{tb}} \tag{5.32}$$

As shown in Fig. 5.44, the stator teeth have parallel sides and the slots do not. However, the situation where the slots have parallel sides and the teeth do not is equally valid. A trapezoidal-shaped slot area maximizes the winding area available and is commonly implemented when the windings are wound randomly, when they are wound turn by turn without any predetermined orientation in a slot (Hendershot, 1990). On the other hand, a parallel-sided slot with no shoes is more commonly used when the windings are fully formed prior to insertion into a slot.

The unknowns in the preceding equations are the back-iron widths of the rotor and stator w_{bi}. Given these two dimensions, all other dimensions can be found. In particular, the total slot depth is given by

$$d_s = R_{sb} - R_{ro} - g \tag{5.33}$$

which must be greater than zero. In addition, the inner rotor radius R_{ri} must be greater than zero. If either of these constraints is violated, then R_{ro} or R_{so} must be changed.

Magnetic Parameters. The unknown geometric parameters w_{tb} are determined by the solution of the magnetic circuit. The air gap flux and flux density are given by Eqs. (5.34) and (5.35), respectively, and can be evaluated using the fixed and known geometric parameters given previously.

$$\phi_g = \frac{1}{1 + \mu_R k_c k_{ml}/P_c} \phi_r \tag{5.34}$$

In terms of magnet and air gap flux densities, this expression becomes

$$B_g = \frac{C_\phi}{1 + \mu_R k_c k_{ml}/P_c} B_r \tag{5.35}$$

The flux from each magnet splits equally in both the stator and rotor back irons and is coupled to the adjacent magnets. Thus, the back iron must support one-half of the air gap flux; that is, the back-iron flux is

$$\phi_{bi} = \frac{\phi_g}{2}$$

If the flux density allowed in the back iron is B_{max} from the table of fixed values, then the preceding equation dictates that the back-iron width must be

$$w_{bi} = \frac{\phi_g}{2B_{max}k_{st}L} \tag{5.36}$$

where k_{st} is the lamination-stacking factor.

Since there are $N_{sm} = N_{spp}N_{ph}$ slots and teeth per magnet pole, the air gap flux from each magnet travels through N_{sm} teeth. Therefore, each tooth must carry $1/N_{sm}$

of the air gap flux. If the flux density allowed in the teeth is also B_{max}, the required tooth width is

$$w_{tb} = \frac{\phi_g}{N_{sm}B_{max}k_{st}L} = \frac{2}{N_{sm}} w_{bi} \tag{5.37}$$

Using Eqs. (5.36) and (5.37), all geometric parameters can be found.

Electrical Parameters. The electrical parameters of the motor include resistance, inductance, counter-emf, and current. All of these parameters are a function of how the motor is wound. It is assumed that no matter what winding approach is used, all coils making up a phase winding are connected in series. This assumption maximizes the counter-emf and minimizes the current required per phase to produce the required rated torque.

Before any parameters can be found, it is necessary to convert the rated motor speed to radians per second. Then, if the motor output is specified in terms of horsepower, it must be converted into an equivalent torque. Since there are 746 W/hp, the equivalent torque is

$$T = \frac{746 P_{hp}}{\omega_m} \tag{5.38}$$

where ω_m, rad/s, is the rated mechanical speed.

Torque. To find the electrical parameters, it is necessary to specify the relationship between torque and the other motor parameters. The torque developed by a single phase when $N_{spp} = 1$ is

$$T = (N_m B_g L n_s i) R_{ro}$$

where the product in parentheses is the force produced by the interaction of N_m magnet poles providing air gap flux density of B_g, with each pole interacting with n_s conductors each carrying a current i exposed to B_g over a length L. In this situation, where there may be more than one slot per pole per phase, n_s must be replaced by the number of turns per pole per phase $n_{tpp} = N_{spp} n_s$, which gives a torque expression of

$$T = N_m B_g L R_{ro} N_{spp} n_s i$$

If $N_{spp} > 1$, the air gap flux density must be modified by the distribution factor and pitch factor. Moreover, if the magnets are skewed, the skew factor given in Eq. (5.39) must be included.

$$k_s = 1 - \frac{\Theta_{se}}{2\pi} \tag{5.39}$$

Inclusion of these terms gives a final torque expression of

$$T = N_m k_d k_p k_s B_g L R_{ro} N_{spp} n_s i \tag{5.40}$$

Counter-EMF. Now, using Eq. (5.40) and the input-output power relationship $T\omega = e_b i$, the peak counter-emf at rated speed ω_m is

$$e_{max} = \frac{T \omega_m}{i} = N_m k_d k_p k_s B_g L R_{ro} N_{spp} n_s \omega_m \tag{5.41}$$

From which the number of turns per slot require to produce E_{max} is

$$n_s = \text{int}\left(\frac{E_{max}}{N_m k_d k_p k_s B_g L R_{ro} N_{spp} \omega_m}\right) \quad (5.42)$$

where once again int $(\cdot)$ returns the integer part of its argument because the number of turns must be an integer. Due to the truncation involved in Eq. (5.42), the peak counter-emf may be slightly less than E_{max}. The actual peak counter-emf achieved can be found by substituting the value computed in Eq. (5.42) back into Eq. (5.41).

Current. Given the desired torque, the required current can be specified in a number of ways. Conductor current, slot current, phase current, or their associated current densities can be found. In addition, these can be specified when any number of phases are conducting simultaneously. Moreover, the peak or RMS value can be specified. And finally, the shape of the current is a function of the counter-emf waveform as well as of the implemented motor drive scheme. As a result, the peak slot current and peak slot-current density under the assumption that only one phase is producing the desired torque will be computed. These values represent a worst-case condition, since more than one phase is usually contributing to the motor torque at one time. In addition, the phase current is computed under the assumption that all phases are contributing equally and simultaneously to the motor torque.

Solving the torque expression Eq. (5.40) for the total slot current $I_s = n_s i$ gives

$$I_s = \frac{T}{N_m k_d k_p k_s B_g L R_{ro} N_{spp}} \quad (5.43)$$

If all N_{ph} phases are conducting current simultaneously and the counter-emf is a square wave, the phase current is also a square wave having a peak and RMS value of

$$I_{ph} = \frac{I_s}{N_{ph} n_s} \quad (5.44)$$

This current value is useful for estimating the ohmic or $I^2 R$ losses of the motor when producing the rated output. In an actual motor, the RMS phase current is greater than Eq. (5.44), since the counter-emf is never an exact square wave. Therefore, computations using Eq. (5.44) are optimistic.

The slot current given by Eq. (5.43) is distributed among n_s conductors occupying the slot cross-sectional area given by Eq. (5.30). Part of this area is occupied by conductor insulation, inevitable gaps between slot conductors, and additional insulation pieces placed around the slot periphery, called slot liners. As a result, only some fraction of the total cross-sectional area is occupied by slot conductors themselves. This fraction is taken into account by specifying a conductor-packing factor:

$$k_{cp} = \frac{\text{area occupied by conductors}}{\text{total area}}$$

Typically, k_{cp} is less than 50 percent, but it can be higher under special circumstances. The exact value of this parameter is known only through experience.

Using Eqs. (5.30) and (5.43) and the conductor-packing factor given in Table 5.10, the slot and conductor current density is

$$J_c = \frac{I_s}{k_{cp} A_s} \quad (5.45)$$

This current density must be compared with the maximum allowable current density J_{max} given in Table 5.10. If J_c exceeds J_{max}, some compromise must be made. The easiest way to decrease the current density is to increase the available slot area by increasing the difference $R_{so} - R_{ro}$. Since a higher current density implies higher I^2R losses, the value of J_{max} is limited only by the ability to cool the motor and the maximum allowable motor temperature. The choice of J_{max} is usually based on experience. For comparison purposes, typical copper residential wiring has a rated peak current density between 4 and 10 MA/m^2. This range of current densities is also typical for motor windings, with the lower end being acceptable for totally enclosed motors and the upper end acceptable for forced-air-cooled motors.

Based on the slot cross-sectional area, the number of turns required, and the conductor current density, it is straightforward to choose a wire gauge suitable for the motor windings. Because of the variety of wire types, insulation types and thicknesses, and slot liners available, this additional analysis is beyond the scope of this text.

Resistance. The phase resistance and inductance of the motor windings are functions of the winding approach chosen, the end-turn layout, and N_{spp}. The phase resistance determines the ohmic or I^2R losses of the motor, and the phase inductance determines the maximum rate of change in phase current, since $di/dt = v_p/L$, where v_p is the phase voltage.

Resistance is determined as shown in the previous section on winding types.

Inductance. The phase inductance has three components due to the slots and end turns. Writing the air gap inductance on a per slot basis gives

$$L_g = \frac{n_s^2 \mu_R \mu_o L \tau_c k_d}{4(l_m + \mu_R k_c g)} \qquad (5.46)$$

where k_d has been included to compensate the air gap inductance roughly for distributed windings. The slot leakage inductance for the trapezoidal slots shown in Fig. 5.44 is

$$L_s = n_s^2 \left[\frac{\mu_o d_3 L}{3 w_{sb}} + \frac{\mu_o d_2 L}{(w_s + w_{sb})/2} + \frac{\mu_o d_1 L}{w_s} \right] \qquad (5.47)$$

The first term in Eq. (5.47) is the distributed inductance of the winding area. Because the width of the slot varies with radius, w_{sb} in the denominator must be replaced with an average or effective radius. Since the slot depth of this area is d_3, the effective slot width is A_s/d_3. The second term in Eq. (5.47) is the inductance of the sloping portion of the shoe. Here w_{sb} must be interpreted as w_{si}. The final term in Eq. (5.47) is the inductance of the shoe tip and requires no correction. Applying these corrections to Eq. (5.47) gives a slot leakage inductance per slot of

$$L_s = n_s^2 \left[\frac{\mu_o d_3^2 L}{3 A_s} + \frac{\mu_o d_2 L}{(w_s + w_{si})/2} + \frac{\mu_o d_1 L}{w_s} \right] \qquad (5.48)$$

The approximate end-turn inductance is given as

$$L_e = \frac{n_s^2 \mu_o \tau_c}{8} \ln\left(\frac{\tau_c^2 \pi}{4 A_s} \right) \qquad (5.49)$$

As earlier with the phase resistance, given N_{sp} slots per phase and one end turn per slot, the total phase inductance is

$$L_{ph} = N_{sp}(L_g + L_s + L_e) \qquad (5.50)$$

Performance. To compute the efficiency it is necessary to compute the ohmic winding loss and the core loss. Of these, the core loss is the most difficult to compute accurately. The magnets and rotor back iron experience little variation in flux and therefore do not generate significant core loss. On the other hand, the stator teeth and stator back iron experience flux reversal on the order of B_{max} at the fundamental electrical frequency. With knowledge of B_{max} and f_e, the core loss of the stator can be roughly approximated. In reality, various areas of the stator experience different flux density magnitudes as well as different flux density waveforms, making it difficult to use traditional core-loss curves based on a sinusoidal flux density waveform.

The ohmic motor loss P_r is equal to the sum of that from each phase.

$$P_r = N_{ph}I_{ph}^2 R_{ph} \qquad (5.51)$$

This ohmic power loss is optimistic since it assumes an ideal square-wave counter-emf and simultaneous square-wave conduction of all phases.

Thus, one can expect the ohmic loss to be significantly greater than that given in Eq. (5.51). More accurate estimation of the ohmic power loss requires knowledge of the motor drive scheme and more accurate prediction of the counter-emf.

Before considering core loss, it is interesting to consider the area over which this heat is generated. Without developing a thermal model for the motor or conducting a thermal analysis, it is at least beneficial to identify the density at which heat leaves the slot conductors and passes into the stator teeth and back iron. Using $L(2d_3 + w_{sb})$ as the slot area in contact with the conductors, the heat density in watts per square meter leaving the slot conductor is

$$q_s = \frac{P_r}{L(2d_3 + w_{sb})N_s} \qquad (5.52)$$

Clearly, the greater q_s is, the higher the operating temperature of the motor will be.

Using core-loss data for the stator material, the core loss is given approximately by

$$P_{cl} = \rho_{bi} V_{st} \Gamma(B_{max}, f_e) \qquad (5.53)$$

Where ρ_{bi}, kg/m^3, is the mass density of the back iron; V_{st} is the stator volume; and $\Gamma(B_{max}, f_e)$, W/kg, is the core loss density of the stator material at the flux density B_{max} and frequency f_e. In Eq. (5.53), the stator volume is given with sufficient accuracy by

$$V_{st} = [\pi(R_{so}^2 - R_{si}^2) - N_s A_s]Lk_{st} \qquad (5.54)$$

where A_s is the slot cross-sectional area given in Eq. (5.30).

The efficiency of the motor producing rated torque at rated speed is

$$\eta = \frac{T\omega_m}{T\omega_m + P_r + P_{cl} + P_s} \cdot 100\% \qquad (5.55)$$

where P_s is the stray loss, composed of windage, friction, and other less-dominant loss components. Depending on motor speed and construction, P_s typically decreases the efficiency on the order of several percent. If desired, the loss incurred in driving the motor can be included in Eq. (5.55), giving a more realistic total system efficiency.

Finally, summing the ohmic and core losses and dividing by the stator peripheral area gives an estimate of the maximum heat density to be removed from the motor:

$$q_{st} = \frac{P_r + P_{cl}}{2\pi R_{so} L} \quad (5.56)$$

5.1.5 Clean-Sheet Design Approach*

Methodology. Assuming that no similar design exists, the following method may be used to design a motor. First, obtain information about the available power source, the physical size constraints, speed-torque requirements, starting torque, and cost constraints.

Next, assume an efficiency. If you don't have a number you like based on your experience, choose one from Table 5.11 as a place to start.

TABLE 5.11 Assumed Efficiency

Motor diameter, in	Assumed efficiency
<2.0	65%
2.0–4.0	75%
>4.0	85%

Note: Some small outer-rotor motors are in the range of 40 percent efficient.

Since the speed and torque are known, the output power can be calculated. Having chosen an efficiency, the input watts can be determined. Find the losses by subtracting the output from the input. Proportion the losses as follows:

I^2R	65 percent
Friction and windage	20 percent
Iron	10 percent
Stray load	5 percent

From this find the current and estimate the resistance. Next, find the counter-emf and torque constant.

Assume a current density of about 5000 A/in² and calculate the wire size. Assume 65 percent slot fill or less and estimate the turns. From diameter and length constraints, proportion dimensions as shown in Sec. 5.1.2.

Knowing from Sec. 5.1.3 that the torque constant is

$$K_t = \left(\frac{KNPz\phi_g}{\text{phases}} \right) 1.732$$

where K is a constant for units and winding distribution.

---

* Subsection contributed by William H. Yeadon, Yeadon Engineering Services, PC.

Solve the magnetic circuit, calculate the performance, and adjust the design parameters as required to meet the specified performance.

If there is a known motor available, simply ratio its dimension as a first pass according to the following power ratio:

$$\frac{P_0 \text{ new motor}}{P_0 \text{ old motor}} = \frac{D^2L \text{ new motor}}{D^2L \text{ old motor}}$$

Summary. Clean-sheet methodology:

1. Determine physical size
2. Determine available input.
3. Select poles, phases, teeth, and winding type.
4. Assume efficiency:
   2 in   65 percent
   4 in   75 percent
   4 in   85 percent
5. Calculate input watts.
6. Determine losses.
7. Proportion losses:
   I^2R 65 percent
   F_w 20 percent
   F_e 10 percent
   Stray   5 percent
8. Find current and estimate resistance from speed-torque point; calculate K_b and K_t.
9. Proportion magnetic-circuit components.
10. Choose maximum allowable stack length L_{stk}.
11. Assume $\gamma = 5000$ A/in².
12. Calculate wire size.
13. Assume 65 percent slot fill.
14. Calculate turns.
15. Since K_t is known and

$$K_t = \left[\frac{(2.24 \times 10^{-7})NPz\phi_g}{\text{phases}}\right] 1.732 \quad \text{oz·in·A}$$

calculate ϕ_g.
16. Assume P_c.
17. Select magnet type to produce ϕ_g.
18. Solve magnetic circuit.
19. Adjust component sizes if necessary.
20. Calculate performance and evaluate design.

Alternative method:

$$\frac{P_0 \text{ new motor}}{P_0 \text{ old motor}} = \frac{D^2L \text{ new motor}}{D^2L \text{ old motor}}$$

5.2 STEP MOTORS*

5.2.1 Variable-Reluctance Step Motors

The construction of variable-reluctance (VR) motors is generally as shown in Fig. 5.45. There is a stator assembly consisting of an insulated lamination stack with copper coils wound around the teeth. The stator assembly is positioned within a housing or main frame such that its location is secured. The rotor assembly consists of a steel magnetic core, a steel output shaft, and bearings. The rotor assembly is centrally located inside the stator assembly by end frames or bearing supports.

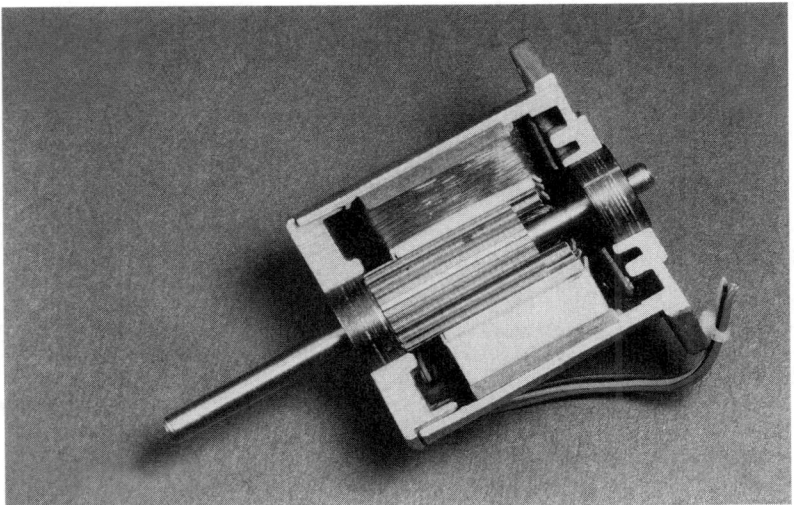

FIGURE 5.45 Cutaway view of a variable-reluctance step motor.

Motor Operation. When a stator phase is energized, the soft-iron rotor is electromagnetically attracted to the stator poles (Fig. 5.46). A rotor tooth attempts to line up with the nearest energized stator pole. A magnetic flux path exists from stator pole to rotor tooth, through the rotor to a different rotor tooth and stator pole, and around the stator back iron. When the rotor teeth are directly lined up with the stator poles, the rotor is in a position of minimum reluctance to the magnetic flux. If an external torque is applied to the rotor, a counteracting torque is developed which tends to restore the rotor to a position of minimum reluctance. A rotor "step" takes place when one stator phase is deenergized and the next phase in sequence is energized, thus creating a new position of minimum reluctance for the rotor.

Variable-reluctance step motors are capable of high stepping rates because of their high torque–to–rotor inertia ratio. The many possible combinations of rotor and stator teeth and phases lead to the availability of a wide range of step angles.

---

* Section contributed by William H. Yeadon, Yeadon Engineering Services, PC, except as noted.

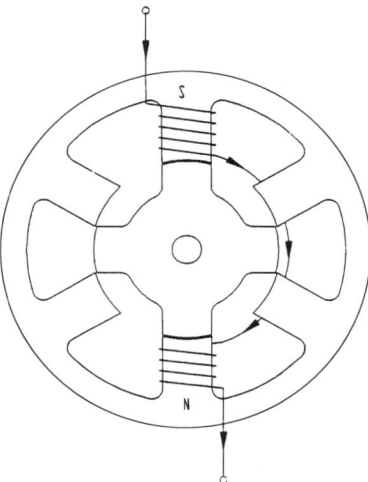

FIGURE 5.46 Wiring diagram for one phase of a variable-reluctance step motor.

5.2.2 Permanent-Magnet-Rotor Step Motors

The PM step motor is illustrated in Fig. 5.47. It consists of two sets of stamped steel cups with diagonal teeth facing the rotor. Each set of cups circumscribes a coil of wire. The two sets are positioned with respect to each other such that they circumscribe the rotor but they are offset from each other by one-half of a tooth pitch.

FIGURE 5.47 Permanent-magnet step motor.

The permanent-magnet-rotor step motor is commonly referred to as the *stamped-construction* or *sheet-metal step motor*. It is sometimes called simply a *PM step motor* but should not be confused with the hybrid permanent-magnet step motor. The rotor in a stamped-construction motor is a smooth cylindrical permanent magnet radially magnetized with alternating N and S poles (Fig. 5.48). The stator has two cup-shaped halves with formed stator teeth (Fig. 5.49). Each half contains a circular, bobbin-wound coil. Because of this simple design, the price is low, but step accuracy and speed may not equal the performance of other step-motor types.

Stamped-construction motors are produced with two types of coil windings which require different types of drive circuits:

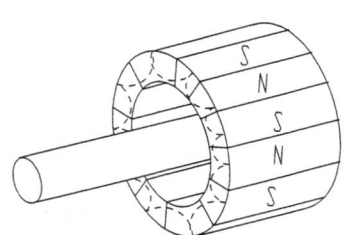

FIGURE 5.48 Magnetization pattern of permanent-magnet step-motor rotor.

- *Bipolar* (monofilar)—a single winding per bobbin. Motor operation requires a drive circuit which can reverse the direction of current in the winding. Either a bridge inverter or a circuit with positive and negative supplies may be used.

- *Unipolar* (bifilar)—two windings per bobbin. Motor operation requires a drive circuit which energizes only one of the two windings at a time. In addition, the direction of current flow in each winding is such that fluxes of opposite polarity are produced. Frequently, the two windings are connected inside the motor to form the equivalent of a single center-tapped winding. Only one power supply and two switches per bobbin are sufficient to operate a motor with this type of winding.

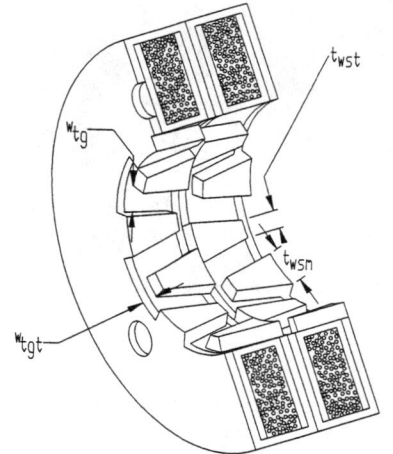

FIGURE 5.49 Cutaway view of permanent-magnet step-motor stator.

Motor Operation. The individual stator halves have uniformly spaced teeth. The two halves are offset by one-half of a tooth pitch. The rotor contains the same number of teeth as each stator half. When one stator half is energized, the rotor aligns its poles with the stator teeth of opposite polarity. A flux path exists from stator tooth to opposite rotor pole, between adjacent rotor poles, back to the opposite (adjacent) stator tooth, and through the stator back iron. When the energization is switched from one stator half to the other, the rotor moves to a new position of alignment one-fourth of a pole pitch from the previous position. The next step occurs when the original stator half is energized with opposite polarity. A four-step switching sequence produces a movement of one pole pitch. The same motion may be achieved by energizing both stator halves at once and allowing the rotor to align itself with the resultant magnetic field. Typical motors of this type have either 12 or 24 poles per stator half, and thus produce step angles of 15° or 7.5°, respectively.

The presence of the permanent-magnet rotor results in a lower operating power requirement than that of the VR motor. It also contributes to good damping (settling) characteristics because of the unenergized detent torque. This type of construction is not suitable for small step angles.

5.2.3 Hybrid Permanent-Magnet Step Motors

The hybrid step motor is generally constructed as shown in Fig. 5.50. It has a stator assembly similar to that of the VR motor, but the rotor consists of three sections. Two pieces are similar to the VR step-motor rotor, but a magnet is placed between them, and they are offset circumferentially from each other by one-half tooth pitch.

FIGURE 5.50 Hybrid step motor.

This motor is termed a *hybrid* because it uses elements of both variable-reluctance and permanent-magnet-rotor step motors. The commonly known version is the 1.8° step-angle motor. It was originally designed as an ac two-phase synchronous inductor motor for low-speed applications.

Its stator construction is similar to that of a variable-reluctance step motor with salient poles (multiple teeth per pole). The phase windings may be either monofilar or bifilar coils, as discussed for the stamped-construction motor. The rotor contains a cylindrical permanent magnet axially magnetized and enclosed on each end by a soft-iron cup with uniformly spaced teeth. As for the variable-reluctance motor, the number of stator phases and differing number of stator and rotor teeth determine the step angle.

Motor Operation. When a stator phase consisting of two or more poles is energized, the teeth in each rotor half align themselves with the nearest unlike stator teeth on the energized poles. A flux path exists from the stator pole rotor half, axially through the rotor, back to a different stator pole, and through the back iron. A new position of rotor alignment is produced each time one phase is deenergized and the next is energized in the same type of sequence as described for the stamped-construction motor. Similar operation is obtained when multiple motor phases are energized.

The permanent-magnet rotor provides the hybrid step motor with good damping characteristics. The hybrid motor has a higher torque per input watt ratio than the stamped construction motor, and it exhibits better step-angle accuracy. Typical step-angle values lie in the range of 0.9° to 3.6°.

5.2.4 Step-Motor Performance

If a step motor is turned on and the rotor is mechanically rotated, a holding-torque curve such as the one shown in Fig. 5.51 will be produced. The torque will go positive and negative as it moves through its detent position. Figure 5.51 shows a possible step error caused by motor friction and magnetic structure inaccuracy.

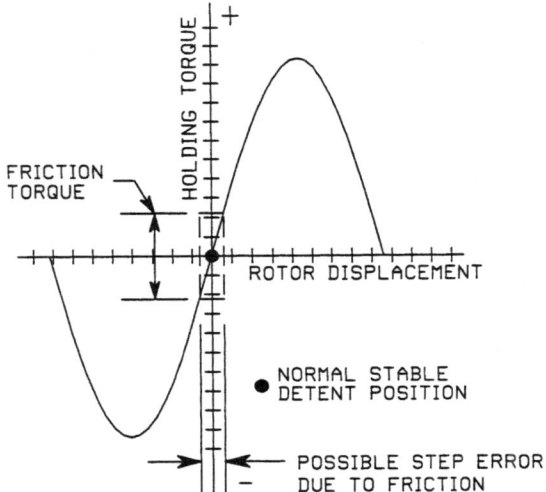

FIGURE 5.51 Holding torque curve.

If phase A is turned on and plotted for several cycles, and then B and C are alternately plotted from the same starting position, the curve in Fig. 5.52 is produced. The maximum available running torque is at the points where the curves produced by the alternate phases intersect.

5.2.5 Sizing the Motor

Basic information is required before any step motor can be designed. Listed here is information for an example specification sheet.

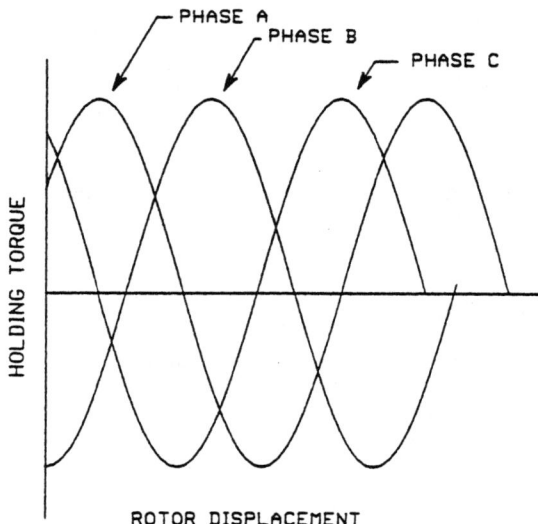

FIGURE 5.52 Holding torque, curves showing relative position of each phase.

Example Specification Sheet

Step angle, degrees

Step accuracy, percent

Rotor inertia, oz·in·s^2

Maximum holding torque, oz·in

Maximum operating speed, steps/s

Number of phases

Voltage

Drive configuration
- Bipolar
- Unipolar
- Chopper

Shaft size, in

Maximum motor diameter, in

From the specification sheet the type of step motor can be selected. Some general characteristics of each type are shown in Table 5.12 to assist with the selection.

5.2.6 Designing VR Step Motors

In step motors, the key to efficient performance is to match the load inertia to the step-motor rotor inertia.

ELECTRONICALLY COMMUTATED MOTORS

TABLE 5.12 Typical Step-Motor Characteristics

	Motor type		
Specification	VR	Hybrid	PM
Step angle	0.6618–30°	0.45–5.0°	3.75–45°
Phases	3, 4, 5	2, 5	2, 4
Drive type	Unipolar	Bipolar	Unipolar/bipolar
Rotor inertia	Low	Medium	High

First, size the motor from the torque requirement as follows (see Fig. 5.53):

$$T = KD_r^2 L_{stk}$$

where T = torque
   K = output coefficient proportional to product of electric and magnetic loading ≈ 0.65 for step motors <4.0 in OD (rough initial guess)
   D_r = rotor OD
   L_{stk} = rotor stack length

Next, size the rotor from the inertia requirement.
The rotor assembly consists of a shaft and a rotor body with salient poles on its periphery (Fig. 5.54). Since the shaft and rotor body all rotate about the same axis, the inertia J for each part may be calculated separately; then they may be added together to get the total inertia.

$$J_{rs} = J_{rotor} + J_{shaft}$$

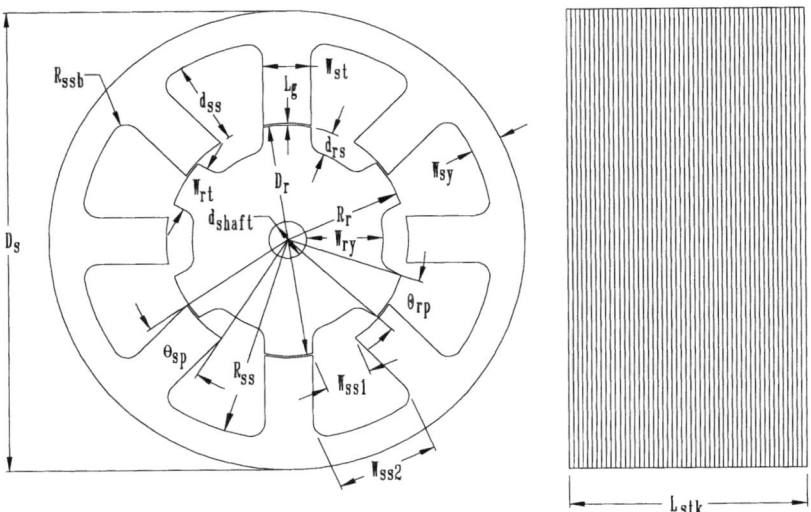

FIGURE 5.53 Variable-reluctance step-motor lamination stack.

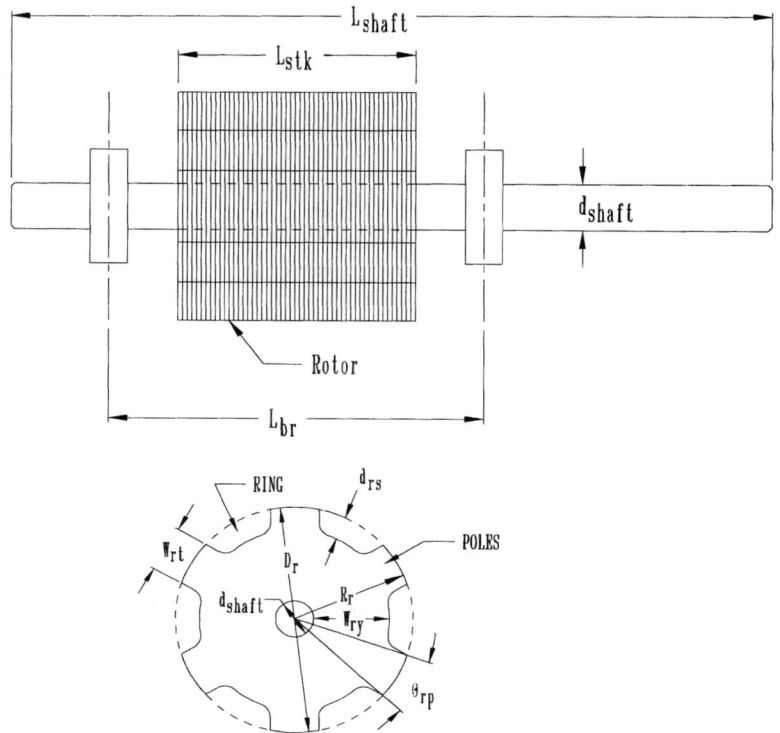

FIGURE 5.54 Rotor and stator assembly.

where J_{rs} is the inertia of the rotor and shaft assembly.

$$J_{shaft} = 0.0184 \left(\frac{d_{shaft}}{2} \right)^4 L_{shaft}$$

where 0.0184 is a coefficient for steel and units of inertia are ounce-inches-seconds squared. Then:

$$J_{rotor} = J_{poles} + J_{cylinder}$$

$$J_{cylinder} = 0.0184 \left(\frac{D_r - 2d_{rs}}{2} \right)^4 L_{stk} - 0.0184 \left(\frac{d_{shaft}}{2} \right)^4 L_{stk}$$

$$J_{poles} = J_{ring} \left(\frac{Wt_{poles}}{Wt_{ring}} \right)$$

$$J_{ring} = 0.0184 \left(\frac{D_r}{2} \right)^4 L_{stk} - 0.0184 \left(\frac{D_r - 2d_{rs}}{2} \right)^4 L_{stk}$$

$$Wt_{ring} = [\pi(R_r)^2 - (R_r - d_{rs})^2 \pi] L_{stk} \rho_{steel}$$

$$Wt_{poles} = [d_{rs} W_{rt} L_{stk}] N_r \rho_{steel}$$

where the density of steel $\rho_{steel} = 4.528$ oz/in^3 and all inertias are in ounce-inches-seconds squared.

Since the shaft is known, calculate its inertia first and subtract it from the total allowable inertia. Then select the largest rotor diameter that is practical for the allowable motor diameter and length. The larger the diameter, the more torque available per unit length. Try to match the load inertia as reflected to the motor. At this point in the design you have not selected the d_{rs} dimension, so estimate it as 30 percent of R_r. Now that d_{rs} is known, save the rest of the allowable motor diameter for the stator. Note here that the stator OD should normally be greater than $2D_r$. Now all major dimensions required to produce the torque are known except for the air gap length.

Sizing the Air Gap. The air gap length in VR step motors should be kept between 0.0015 and 0.003 inch per side. This requires very tight machining tolerances.

Selecting the Number of Phases and Poles

Three phase

- Always has positive torque
- High torque ripple
- Poor peak torque–to–average torque ratio
- Low number of power transistors (3 minimum)

Four phase

- Low torque ripple
- Good peak torque–to–average torque ratio

Five phase

- Lower torque ripple
- More expensive controller

Next, select the phases. This is usually a requirement of the drive system to be used. However, some people prefer a three-phase motor because it utilizes the largest percentage of iron and copper at a given time, while others prefer four or more phases because of the reduced torque ripple at slewing operation.

It is now necessary to select the number of stator and rotor teeth required to achieve the desired step angle. The step angle is defined as the difference in stator and rotor tooth pitch expressed in degrees.

Example. **12 stator teeth and 8 rotor teeth**

$$\text{Step angle} = \frac{360°}{8} - \frac{360°}{12} = 45° - 30° = 15°$$

The number of rotor teeth must differ from the number of stator teeth by the number of poles. It is generally desirable to have fewer rotor teeth than stator teeth because fewer rotor teeth yield lower inertia and flux per pole. Consequently, the torque is a function of the tooth pitch of the member having the largest number of teeth.

The number of poles should be made as large as possible because the torque varies directly with the number of poles. Other combinations are possible and may prove to be a better choice for a given application. Some considerations when selecting pole combinations are the following:

- More stator poles mean more windings.
- More stator poles mean a smaller step angle.

- More stator poles mean a more expensive controller.
- More stator poles mean higher core loss for a given shaft speed.

Stepping Rate. This is given as

$$SR = 0.539 \sqrt{\frac{T_s(386)}{J_t}}$$

where T_s = stall torque, oz·in
J_t = motor plus load inertia, oz·in·s²

For convenience, Tables 5.13 to 5.20 contain lists of step angles for various combinations of poles, teeth, and phases for VR step motors.

$$S_r = S_s \pm P$$

$$S_s = mPn$$

$$\theta_s = \left(\frac{360°}{S_s} - \frac{360°}{S_r}\right)n$$

where m = number of phases
P = number of poles per phase
S_s = number of stator slots at the air gap
S_r = number of rotor slots
θ_s = step angle, degrees
n = number of small stator teeth in one tooth pitch of each major stator tooth S_{sm}, $n \geq 1$, where $S_{sm} = mP$

See Tables 5.13 through 5.20 for step angles.

TABLE 5.13 Two-Pole Three-Phase Step Angles

	$P = 2, m = 3, S_{sm} = 6$		
n	S_s	S_r	θ_s
1	6	4	30.0000
1	6	8	15.0000
2	12	10	12.000
2	12	14	8.5714
3	18	16	7.5000
3	18	20	6.0000
4	24	22	5.4545
4	24	26	4.6154
5	30	28	4.2857
5	30	32	3.7500
6	36	34	3.5294
6	36	38	3.1579
7	42	40	3.0000
7	42	44	2.7273
8	48	46	2.6087
8	48	50	2.4000
9	54	52	2.3077
9	54	56	2.1429
10	60	58	2.0690
10	60	62	1.9355

Source: From Trickey (1965).

TABLE 5.14 Two-Pole Four-Phase Step Angles

$P = 2, m = 4, S_{sm} = 8$			
n	S_s	S_r	θ_s
1	8	6	15.0000
1	8	10	9.0000
2	16	14	6.4286
2	16	18	5.0000
3	24	22	4.0909
3	24	26	3.4615
4	32	30	3.0000
4	32	34	2.6471
5	40	38	2.3684
5	40	42	2.1429
6	48	46	1.9565
6	48	50	1.8000
7	56	54	1.6667
7	56	58	1.5517
8	64	62	1.4516
8	64	66	1.3636
9	72	70	1.2857
9	72	74	1.2162
10	80	78	1.1589
10	80	82	1.0976

Source: From Trickey (1965).

TABLE 5.15 Four-Pole Three-Phase Step Angles

$P = 4, m = 3, S_{sm} = 12$			
n	S_s	S_r	θ_s
1	12	8	15.0000
1	12	16	7.5000
2	24	20	6.0000
2	24	28	4.2857
3	36	32	3.7500
3	36	40	3.0000
4	48	44	2.7273
4	48	52	2.3077
5	60	56	2.1429
5	60	64	1.8750
6	72	68	1.7647
6	72	76	1.5789
7	84	80	1.5000
7	84	88	1.3636
8	96	92	1.3043
8	96	100	1.2000

Source: From Trickey (1965).

TABLE 5.16 Four-Pole Four-Phase Step Angles

$P = 4, m = 4, S_{sm} = 16$			
n	S_s	S_r	θ_s
1	16	12	7.5000
1	16	20	4.5000
2	32	28	3.2143
2	32	36	2.5000
3	48	44	2.0455
3	48	52	1.7308
4	64	60	1.5000
4	64	68	1.3235
5	80	76	1.1842
5	80	84	1.0714
6	96	92	0.9783
6	96	100	0.9000
7	112	108	0.8333
7	112	116	0.7759
8	128	124	0.7258
8	128	132	0.6818

Source: From Trickey (1965).

TABLE 5.17 Six-Pole Step Angles

$P = 6, m = 3, S_{sm} = 18$			
n	S_s	S_r	θ_s
1	18	12	10.0000
1	18	24	5.0000
2	36	30	4.0000
2	36	42	2.8571
3	54	48	2.5000
3	54	60	2.0000
4	72	66	1.8182
4	72	78	1.5385
5	90	84	1.4286
5	90	96	1.2500
6	108	102	1.1765
6	108	114	1.0526

Source: From Trickey (1965).

Lamination Design. After selecting the phases, poles, and teeth required for the motor, it is necessary to complete the stator and rotor lamination designs. For stators with one tooth per pole, use the following method for determining sizes, except be certain to leave a gap between poles at the stator ID large enough to insert the windings. If a needle winder is being used, this may be determined as follows:

W_{so} = max wire OD + 0.010 in clearance for max wire + 0.025 in needle wall

+ 0.010 in clearance for needle movement + (2 × slot insulation thickness)

TABLE 5.18 Six-Pole Four-Phase Step Angles

	$P = 6, m = 4, S_{sm} = 24$		
n	S_s	S_r	θ_s
1	24	18	5.0000
1	24	30	3.0000
2	48	42	2.1429
2	48	54	1.6667
3	72	66	1.3636
3	72	78	1.1538
4	96	90	1.0000
4	96	102	0.8824
5	120	114	0.7895
5	120	126	0.7143
6	144	138	0.6522
6	144	150	0.6000

Source: From Trickey (1965).

TABLE 5.19 Eight-Pole Three-Phase Step Angles

	$P = 8, m = 3, S_{sm} = 24$		
n	S_s	S_r	θ_s
1	24	16	7.5000
1	24	32	3.7500
2	48	40	3.0000
2	48	56	2.1429
3	72	64	1.8750
3	72	80	1.5000
4	96	88	1.3636
4	96	104	1.1538

Source: From Trickey (1965).

TABLE 5.20 Eight-Pole Four-Phase Step Angles

	$P = 8, m = 4, S_{sm} = 32$		
n	S_s	S_r	θ_s
1	32	24	3.7500
1	32	40	2.2500
2	64	56	1.6071
2	64	72	1.2500
3	96	88	1.0227
3	96	104	0.8654
4	128	120	0.7500
4	128	136	0.6618

Source: From Trickey (1965).

Sizing the Poles. Generally, the steel widths throughout the magnetic current should be about equal to yield the same flux density. The stator yoke width W_{sy} (Fig. 5.53) may be a little larger to reduce the mmf drop in that region since it is such a long flux path. For the stator section, the pole arc θ_{sp} of Fig. 5.53 and the gap between the poles should be about equal; therefore:

$$\theta_{sp} = \frac{360°}{2N_s}$$

The pole sides should parallel each other from the stator bore to the intersection with the stator yoke. This leaves a large trapezoidal area for winding space between the poles. The stator yoke width W_{sy} should be equal to or greater than one-half the pole width.

$$W_{sy} \geq \frac{W_{st}}{2}$$

In VR step motors, it is common to make $W_{sy} = W_{st}$.
The rotor pole arc θ_{rp} should be equal to the stator pole arc.

$$\theta_{rp} = \theta_{sp}$$

Then the rotor pole width W_{rt} should be made equal to the stator tooth width.

$$W_{rt} = W_{st}$$

The depth of the rotor tooth should be at least 20 times the gap length.

$$d_{rs} \geq 20 L_g$$

But the rotor yoke width W_{ry} should not be allowed to be less than one-half of the rotor pole width.

$$W_{ry} \geq \frac{W_{rt}}{2}$$

In cases where a magnetic shaft material is used, one-half of the shaft diameter may be counted as part of the rotor yoke width.

Once the rotor and stator laminations are laid out, look at them for uniformity and proportions. With one pole pair excited, the flux density in the poles should be 20 to 30 percent higher than in the yoke. The effective stator yoke width may be increased either by reducing d_{ss} or by adding a housing of magnetic material. Reducing d_{ss} has the unpleasant side effect of reducing the available copper area, so adding a steel housing may be the preferred option.

For stator laminations with multiple teeth per pole (Fig. 5.55), the tooth width should be narrowed and becomes W_{st2}.

$$W_{st2} \approx 0.75\, W_{st}$$

The tooth width of the rotor W_{rt} and stator W_{stt} should be identical and should occupy between 40 and 60 percent of the tooth pitch of the member having the larger number of teeth.

$$W_{rt} = W_{stt}$$

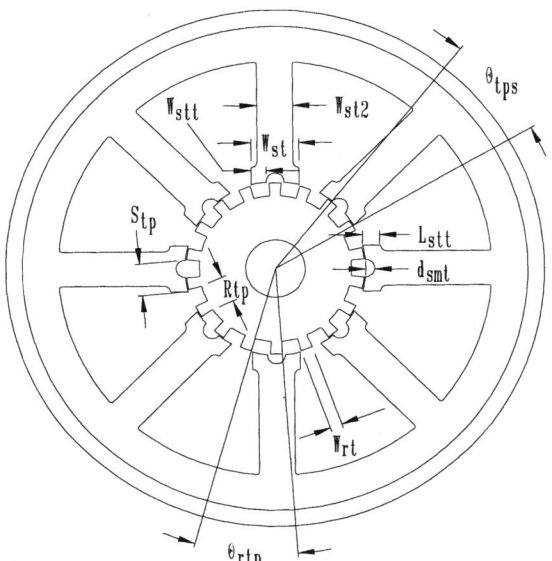

FIGURE 5.55 Cross section of 5° variable-reluctance step motor.

The rotor tooth pitch angle θ_{rtp} is found as follows:

$$\theta_{rtp} = \frac{360°}{S_r} \quad \text{or} \quad \frac{2\pi}{S_r} \text{ rad}$$

The tooth pitch is then

$$R_{tp} = R_r \theta_{rtp} \quad \text{rad}$$

Then:

$$W_{rt} = 0.4 R_{tp} \text{ to } 0.6 R_{tp}$$

The stator tooth W_{stt} may limit W_{rt} because it is necessary to leave enough space between the pole sets to allow for the winding needle as in the previous case.

The tooth pitch angle θ_{stp} on the stator is nominally

$$\theta_{stp} = \frac{360°}{S_s} \quad \text{or} \quad \frac{2\pi}{S_s} \text{ rad}$$

The tooth pitch is then:

$$S_{tp} = (R_r + L_g)(\theta_{stp}) \quad \text{rad}$$

Then:

$$W_{stt} \approx 0.4 S_{tp} \text{ to } 0.6 S_{tp}$$

In the case of small motors, these dimensions are somewhat limited by mechanical considerations. The minimum width of any stamped part should be 1.5 times the material thickness.

The stator minor tooth depth d_{smt} should be a minimum of 20 times the air gap length.

$$d_{smt} \geq 20 L_g$$

The stator tooth tip length L_{stt} should be about 2 times the length of the stator minor tooth depth.

$$L_{stt} \approx 2 d_{smt}$$

It should be noted here that there are stator lamination designs that require a large number of teeth, but the size of the ID would make it impossible to have enough space between the poles to insert the winding needle. In these cases, the designer should eliminate some of the teeth but keep the original spacing. For example, the 1.8° step motor requires 48 stator teeth on 8 major stator teeth and 50 rotor teeth. Most motors of this step angle drop 1 minor tooth from each major tooth so that the stator actually has 40 stator teeth on a 48-tooth spacing.

Estimating Slot Area and Winding Space for Straight-Sided Teeth. Before a winding can be calculated, it is necessary to determine the available slot area. First, calculate the gross slot area (GSA), then subtract the insulation area, winding needle area, and slot wedge area (Fig. 5.56).

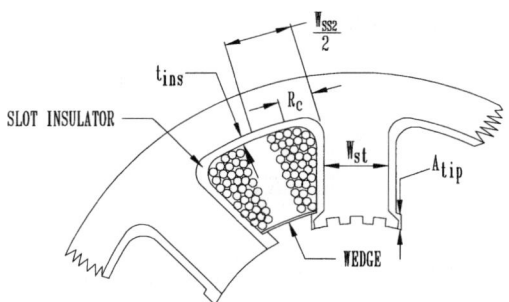

FIGURE 5.56 Stator slot.

To calculate GSA, first find the stator slot depth d_{ss} (refer to Fig. 5.53).

$$d_{ss} = \frac{D_s - D_r - 2(L_g + W_{sy})}{2}$$

Next, calculate the GSA by calculating the total area of the motor inside the stator yoke A_{s1} and subtracting the area occupied by the rotor and air gap A_{s2}. Then divide by the number of slots N_s; finally, subtract the area of the pole A_{T1}.

$$\text{GSA} = \left[\frac{A_{s1} - A_{s2}}{N_s} \right] - A_{T1}$$

$$A_{s1} = \pi R_{ss}^2$$

$$r_{syi} = \frac{D_s - 2W_{sy}}{2}$$

$$A_{s2} = \pi(R_r + L_g)^2$$

$$A_{T1} = [R_{ss} - (R_r + L_g)]W_{st}$$

To find the net slot area (NSA), subtract the area occupied by the insulation A_{ins}, the area needed for the winding needle A_{ned} and the area needed for the slot wedge A_{weg} (refer to Fig. 5.56).

$$A_{ned} = [d_{ss} - 2t_{ins}]W_{ned}$$

where W_{ned} = width of winding needle including clearance
t_{ins} = thickness of slot insulator

The area the insulator takes up in the slot A_{ins} is found as follows:

$$A_{ins} = [2d_{ss} + W_{ss2}]t_{ins}$$

where W_{ss2} is the arc length of one slot inside the stator yoke.

$$W_{ss2} = R_{ss}\theta_s$$

$$\theta_s = \frac{2\pi}{N_s} - \theta_{sp}$$

where θ_s and θ_{sp} are in radians.

$$A_{weg} = W_{weg}t_{weg}$$

where W_{weg} = width of slot wedge
t_{weg} = thickness of slot wedge

$$\text{NSA} = \text{GSA} - [A_{ned} + A_{ins} + A_{weg}]$$

To find the slot area available for windings per pole, divide NSA by 2.

$$A_{wp} = \frac{N_{sa}}{2}$$

Calculating Resistance. To determine resistance the mean turn length (MTL), the number of turns, and the gauge of wire must be known.

$$\frac{\text{MTL}}{\text{pole}} = 2L_{stk} + \text{METL}$$

where METL is the mean end-turn length per pole (includes both ends). Refer to Fig. 5.57.

$$\text{METL} = 2[W_{st} + 2S_{rc}]$$

where S_{rc} is the mean length of turn at the coil corner.

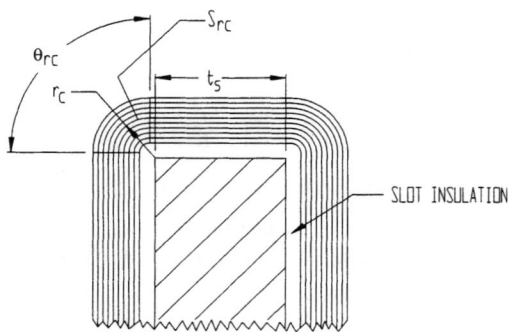

FIGURE 5.57 Side view of end turn.

$$S_{rc} = R_c \, \theta_{rc}$$

$$\theta_{rc} = \frac{\pi}{2} = 1.57 \text{ rad}$$

Referring to Fig. 5.56:

$$R_c = \frac{(W_{ss2}/2) - (W_{ned}/2)}{2}$$

The resistance per pole R_p is found by finding the total wire length in feet and, from Table 2.76, multiplying by the value for ohms per 1000 ft for the wire gauge being used.

$$R_p = \frac{N_p \, \text{MTL}}{12} \, \frac{\Omega}{1000 \text{ ft}}$$

Then the resistance per phase R_{ph} is

$$R_{ph} = R_p \, \frac{\text{poles}}{\text{phase}}$$

when assuming a series connection.

Slot fill S_{fill} is then the available winding area A_{wp} divided by the area of a wire times the number of turns per pole N_p.

$$S_{fill} = \frac{A_{wp}}{N_p (\text{OD}_w \pi/2)^2}$$

For poles with multiple teeth, it is necessary to take the area of the tooth tip A_{tip} into account.

Estimating Rated Torque. A formula for predicting peak torque comes from Trickey. Here:

$$T_{pk} = (0.1129 \times 10^{-5}) \left(\frac{P}{2}\right) \left[\frac{(D_r/2) + (L_g/2)}{L_g}\right] L_{stk} \left(\frac{N_p}{P} \, 2\right)^2 I^2 \quad \text{oz·in}$$

where P = poles per phase
       I = current per phase

The number of turns per pole N_p can be estimated by determining the ampere-turns required to nearly saturate the steel chosen for the motor laminations, then selecting the wire size that will keep current densities reasonable. Stack length may have to be adjusted. Actual winding selection is heavily dependent on the drive type.

Once the number of turns has been calculated, determine the wire size that can be used.

Look up the wire sizes in Table 2.76. Then calculate the resistance per phase R_{ph}. Divide this value into V_s to determine the maximum current possible.

$$A_{wp} = N_p(OD_w)^2$$

$$\therefore OD_w = \sqrt{\frac{A_{wp}}{N_p}}$$

$$I_{max} = \frac{V_s}{R_{ph}}$$

If this value times N_p does not equal or exceed the required $N_p i$, some adjustments will have to be made.

Verifying the Design. Note that up to this point, many assumptions have been made about flux densities, dimensions, and ampere-turns per pole. It is now time to select the materials and verify the reasonableness of the design.

Select the lamination steel by determining the importance of core loss in this application. If it is a high-speed, high-switching-frequency, or high-efficiency motor, start by selecting a medium-grade material such as M-19, 29 gauge. If it is a very cost-sensitive application, try cold-rolled steel, 24 gauge. In military or space applications, 0.006-in vanadium permendur might be required.

The loose assumption made previously can obviously lead to a less-than-perfect design. The next step is to model the motor in a boundary element analysis (BEA) or finite element analysis (FEA) software package and more accurately calculate the flux and torque. These packages give only static performance, but the results can be utilized by the designer to more accurately predict dynamic performance.

Figures 5.58, 5.59 and 5.60 show the effects of various rotor and stator materials on motor torque. Figure 5.58 shows an M-19 stator core with a cast-iron rotor, an M-19 rotor, and a cold-rolled steel (CRS) rotor. Note that the M-19 and CRS rotors have nearly identical torques, while the cast-iron rotor has much lower torques. In Fig. 5.59, the stator is made of cast iron. It has less torque than the M-19 stator with all three rotors, but the M-19 and CRS rotors provide more torque than the cast-iron rotor. Figure 5.60 shows the same plots except the stator is now CRS. The M-19 stator provides the best overall torque, notwithstanding the rotor material being M-19 or CRS. These curves do not take core loss into account. The M-19 material provides lower core loss.

Figures 5.58, 5.59, and 5.60 were derived from a BEA of a VR step motor with 2.5-in OD stator with $N_s = 8$ and $N_r = 6$. The torque was calculated for a 1.0-m stack length with 1 phase turned on. The net current for 1 coil side is 350 A for this analysis. Note the "×E1" on the torque scale to correctly read the torque values. The values of torque and rotation are negative because the software considers CCW rotation to be in the positive direction. To convert these torque values to the torque of an actual motor, multiply these values by $0.0254 \times L_{stk}$, where L_{stk} is in inches. Figure 5.61 shows a plot of flux linkage versus current of the motor with the rotor in the aligned and unaligned positions.

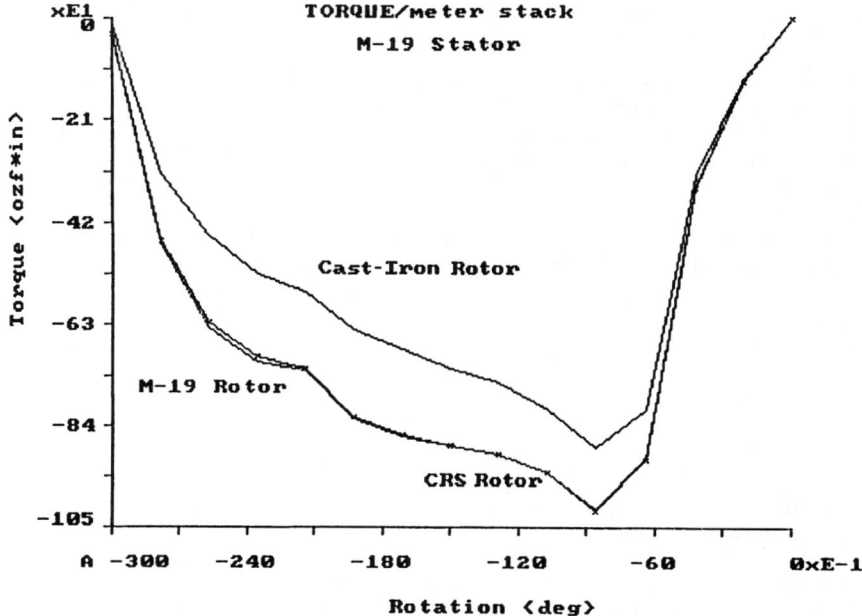

FIGURE 5.58 Torque versus rotation for M-19 stator with alternative rotor materials.

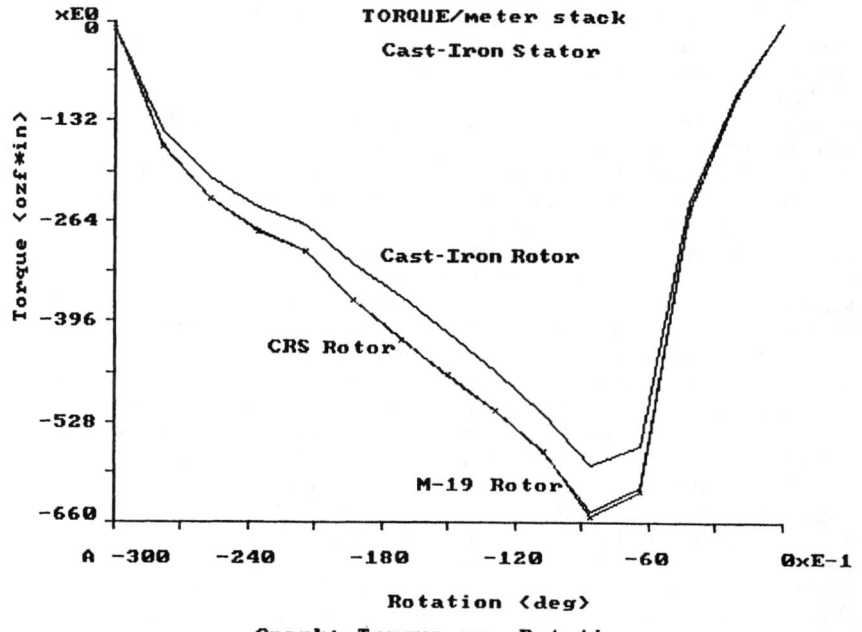

FIGURE 5.59 Torque versus rotation for cast-iron stator with various rotor materials.

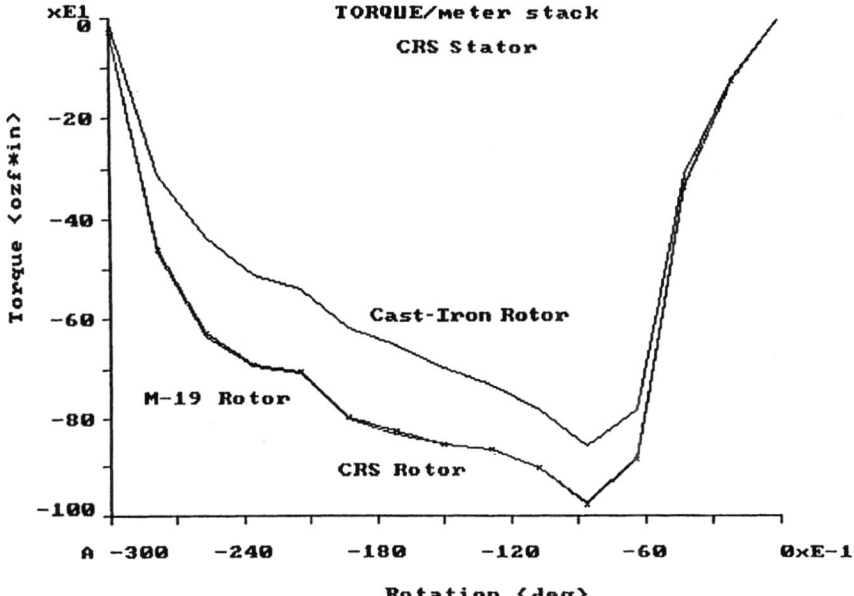

FIGURE 5.60 Torque versus rotation for CRS stator with various rotor materials.

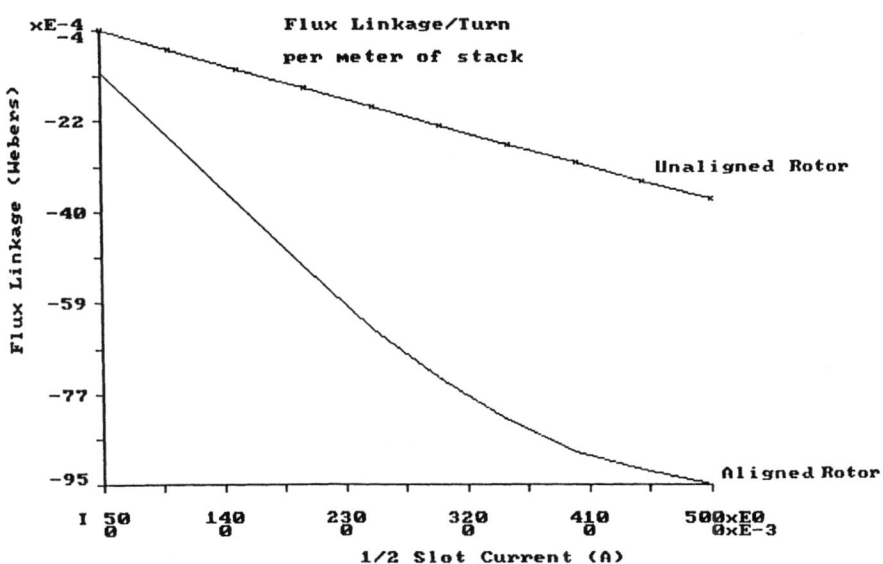

FIGURE 5.61 Flux linkage versus current.

5.2.7 Designing PM Step Motors

A typical PM step motor is shown in Fig. 5.62 and was described earlier. Since this is a step motor, rotor inertia is an important item in the design. If a rotor or load inertia has been specified, this component should be designed first.

The rotor assembly (Fig. 5.63) consists of a steel shaft, an aluminum hub, and a cylindrical ring magnet. Since all components rotate about the same center, the individual inertias can be calculated and then added together to obtain the total inertia.

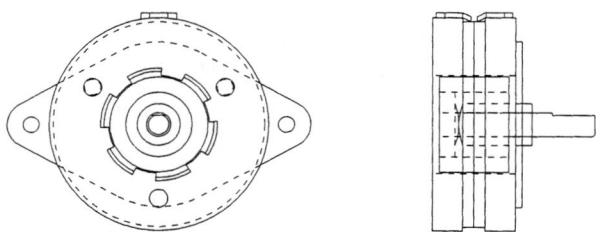

FIGURE 5.62 Permanent-magnet step motor.

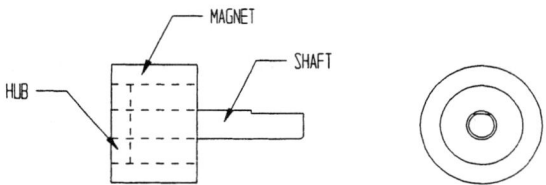

FIGURE 5.63 Permanent-magnet step-motor rotor assembly.

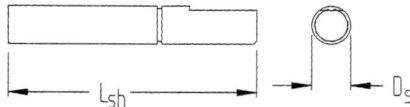

FIGURE 5.64 Permanent-magnet step-motor shaft.

The shaft inertia J_{shaft}, oz·in·s^2 (Fig. 5.64), is

$$J_{\text{shaft}} = 0.0184 \left(\frac{D_s}{2} \right)^4 L_{\text{sh}}$$

for a steel shaft.

The inertia of the hub J_{hub}, oz·in·s^2 (Fig. 5.65), is

$$J_{\text{hub}} = 0.00636 \left(\frac{D_{\text{rm}}}{2} \right)^2 L_{\text{hub}} - 0.00636 \left(\frac{D_s}{2} \right)^4 L_{\text{hub}}$$

for an aluminum hub.

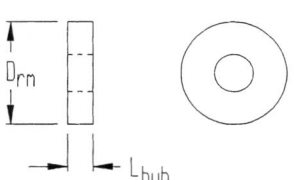
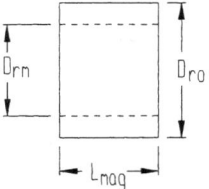
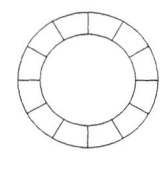

FIGURE 5.65 Permanent-magnet step-motor hub dimensions.

FIGURE 5.66 Permanent-magnet step-motor magnet dimensions.

The inertia of the magnet J_{mag}, oz·in·s² (Fig. 5.66), is

$$J_{mag} = \left(\frac{D_{ro}}{2}\right)^4 \frac{L_{mag}(\text{mag density})}{772} - \left(\frac{D_{rm}}{2}\right)^4 \frac{L_{mag}(\text{mag density})}{772}$$

The densities of some common magnetic materials are listed in Table 5.21. The total rotor inertia J_{rotor} is then:

$$J_{rotor} = J_{shaft} + J_{hub} + J_{mag}$$

TABLE 5.21 Densities of Common Magnetic Materials

Material	Density		
	lb/in³	oz/in³	g/cm³
Bonded ferrite	0.134	2.14	3.71
Sintered ferrite	0.177	2.83	4.90
Alnico	0.264	4.22	7.31
Samarium cobalt	0.302	4.83	8.35
Neodymium-iron-boron	0.271	4.34	7.50

When choosing the rotor assembly component dimensions, some practical items must be taken into consideration. The shaft diameter and length are usually specified. The rotor magnet ID must be large enough to accommodate a bearing plus clearances. The rotor magnet is generally made of low-energy isotropic material. The material near the ID is magnetized at right angles to the poles at the surface to carry the pole flux (Fig. 5.67).

If the rotor magnet is of ceramic material, the practical limit of rotor thickness is ⅛ in, which means that the rotor OD D_{ro} is

$$D_{ro} = D_{rm} + 0.25 \text{ in}$$

Air Gap. The air gap length L_g on these motors is much larger than on VR step motors. An L_g of 0.010 to 0.015 in per side is common.

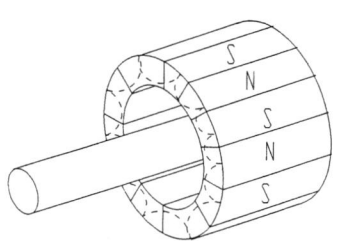

FIGURE 5.67 Permanent-magnet rotor magnetization pattern.

Stator Pole Pieces. The ID of the stator D_{si} is found as follows:

$$D_{si} = D_{ro} + 2L_g$$

Next, select the material for the pole pieces. Generally, a soft low-carbon steel on the order of 1010 or 1020 is used. This material is easy to stamp and form and is very low cost. From the required resolution, select the number of stator teeth. The step angle θ_s is equal to 360° divided by the total number of stator teeth N_s.

$$\theta_s = \frac{360°}{N_s}$$

$$\therefore N_s = \frac{360°}{\theta_s}$$

For example: 360°/15° = 24 teeth total
Then the number of teeth on one phase set N_{sp} becomes

$$N_{sp} = \frac{N_s}{2}$$

For example: 24/2 = 12
Each phase set consists of two pole pieces P_{pc} so each pole piece has one-half of the teeth N_{pc}

$$N_{pc} = \frac{N_{sp}}{2}$$

For example: 12/2 = 6
The size and shape of the teeth need to be determined. The greater the tooth area, the greater the air gap flux and torque. So make the tooth area in the bore as large as possible. It is necessary to limit the tooth-to-tooth leakage flux, so a gap must exist between them. Synchronous motors exist where the teeth have parallel sides. The teeth in step motors are tapered to improve step accuracy. Design the teeth as follows:
Find the circumference of the stator ID C_{si}.

$$C_{si} = \pi D_{si}$$

For example: π 1.04 = 3.267
Determine the space available for teeth by dividing C_{si} by the total teeth N_{sp}.

$$L_{space} = \frac{C_{si}}{N_{sp}}$$

For example: 3.267/12 = 0.272
Set the major tooth width t_{wsm} (Fig. 5.68) to something less than the available space. If you assume that an 0.050-in gap R_{pg} is created between poles on the rotor by the magnetizing fixture, subtract this amount from the available space to get t_{wsm}.

$$t_{wsm} = L_{space} - R_{pg}$$

For example: 0.272 − 0.050 = 0.222
The teeth should then be tapered so that there is a uniform gap between the teeth W_{tg}. This should be a minimum of 10 times the air gap length L_g as a rule of thumb.

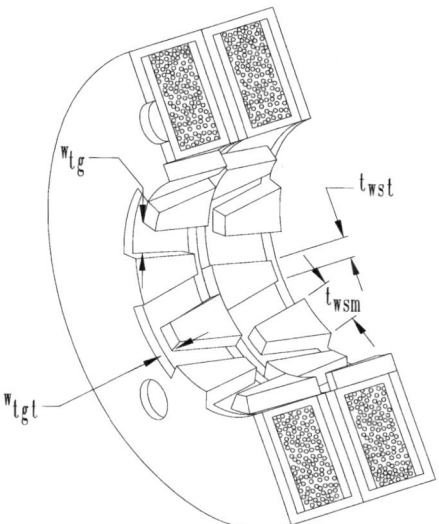

FIGURE 5.68 Stator of permanent-magnet step motor

$$W_{tg} \geq 10\, L_g$$

The tooth tip cannot overlap the housing edge because it would create a magnetic short circuit. The reluctance of the path to the housing would be much lower than the path to the rotor. Sizing the dimension between the housing and the tooth tip W_{tgt} should follow a rule similar to the one for finding the distance between the teeth. Since the tooth is much narrower at the tip than at the base, the effect of leakage would be much less; so a rule of 5 times the air gap length is generally satisfactory.

$$W_{tgt} \geq 5\, L_g$$

The actual size and shape of the teeth depends on the motor length as well as the rotor diameter. The motor length is often limited by the application specification. The stator outside diameter D_{so} should be about twice the inside diameter of the stator.

$$D_{so} \approx 2\, D_{si}$$

The thickness of the material should be as thin as possible for ease of forming but not so thin as to be easily deformed in normal handling. Since these parts are in the main flux path, they should not be made so thin as to add significant reluctance to the circuit. A rule of thumb is to select a material in the range of 0.025 to 0.075 in as a place to start. Material in the range of 0.040- to 0.050-in CRS is a good choice.

When the pole pieces are assembled, the inside pieces P_{c1} and P_{c2} (Fig. 5.69) that are joined together require some kind of locating device. It is important that these pieces be offset from each other by one-half tooth pitch. The more accurate the alignment, the better the step accuracy will be. Pole pieces (P_{c3} and P_{c4}) need to be accurately located one tooth pitch from their mating halves. Assembly of this unit is the key to its performance. Keep the parts symmetrical and uniform for best results.

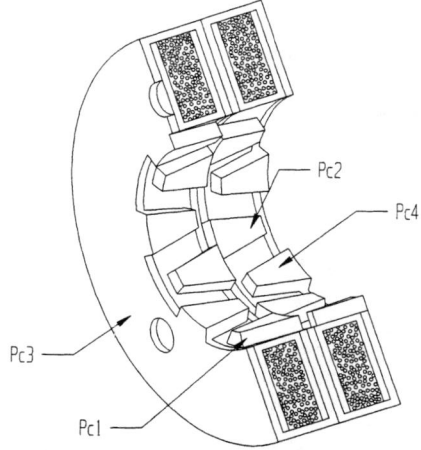

FIGURE 5.69 Permanent-magnet step-motor stator piece.

Windings. This motor supplies the drive mmf from coils wound on bobbins and inserted between pole pieces 1 and 3 and 2 and 4, respectively. The bobbin (Fig. 5.70) is usually made of a nylon material. The OD D_{bso} is slightly smaller than the stator OD.

$$D_{bso} = D_{so} - (2 \times \text{material thickness}) - \text{clearance}$$

The inside diameter D_{bsi} is the stator ID plus clearance for the material.

$$D_{bsc} = D_{si} + (2 \times \text{material thickness}) + \text{clearance}$$

The bobbin should have a wall thickness in the range of 0.025 in. The bobbin width is a function of the motor length. The available winding area A_{wnd} is

$$A_{wnd} = \left[\frac{D_{so} - (D_{bsi} + 2T_{mtl})}{2} \right] (w_b - 2T_{mtl})$$

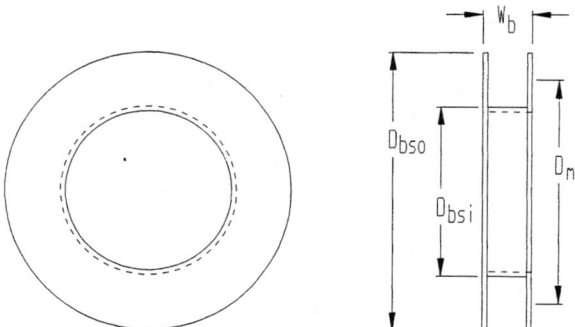

FIGURE 5.70 Permanent-magnet step-motor bobbin.

Winding Resistance. The resistance R_{ph} is simply the mean turn length (MTL) times the number of turns N_{ph} times the resistivity ρ of the wire gauge being used.

$$R_{ph} = \frac{(\text{MTL, in})(N_{ph}\, \rho_\Omega)}{(12 \text{ in/ft})(1000 \text{ ft})}$$

$$\text{MTL} = \pi\, D_{mt}$$

$$D_{mt} = \frac{D_{bso} - D_{bsi}}{2}$$

ELECTRONICALLY COMMUTATED MOTORS

Estimating the Torque. The torque is a function of the air gap flux ϕ_g, the number of poles, and the stator ampere-turns. Kuo (1979) did an exhaustive treatise in which he determined that the average torque T_{av}, N·m, is

$$T_{av} = \frac{2}{\pi} P^2 \frac{F_1 \phi_m}{\sigma}$$

where P = number of pole pairs
 F_1 = stator mmf, A·turn
 ϕ_m = Flux under one rotor pole, Wb
 σ = leakage factor

Or:

$$T_{av} = 90 \, P^2 \, F_1 \, \phi_2 \quad \text{oz·in}$$

F_1 is just the coil current times the number of turns in the coil.

Calculating σ, F_1, and ϕ is difficult and time consuming. The permeances of the main flux paths and leakage flux paths must be determined. The load line on the magnet curve must be established. The mmf drops in the circuit must be calculated. Typically, the mmf drops do not equal the mmf supplied, so it is necessary to iterate to a solution. The following method may be used to estimate the air gap flux and torque. However very accurate predictions generally require the use of a three-dimensional boundary or finite element analysis.

- Find the permeance coefficient (P_c) of the magnet
- Determine the flux density in the magnet
- Find the total flux

$$P_c = \frac{\sigma}{R_f} \left(\frac{L_{mr} A_g}{A_m L_g} \right)$$

Where: σ = leakage factor (typically 1.05 to 1.15)
 R_f = reluctance factor (typically 1.1 to 1.3)
 L_{mr} = radial length of magnet
 L_g = radial length of air gap
 A_m = area of magnet
 A_g = area of air gap

The area of the magnet is not necessarily the same as the area of the air gap, because motors typically have magnets somewhat shorter than the stator. A_g is usually larger than A_m.

The permeance coefficient determines the load line of the magnet on its normal demagnetization curve. These curves are commonly supplied by the magnet manufacturer. A typical curve is shown in Fig. 5.71 and is the upper-left quadrant of the hysteresis loop shown in other sections of this handbook.

To plot the load line, take the arctangent of the permeance coefficient, calculate the angle $\psi = \tan^{-1} P_c$, and plot the line as shown in Fig. 5.71. The flux density in the magnet B_m can be found by finding the intersection of the load line and the normal curve and reading the induction from the right vertical axis. The flux in the magnet is found by multiplying the flux density by the area of the magnet A_m, cm²:

$$\phi_m = B_m A_m$$

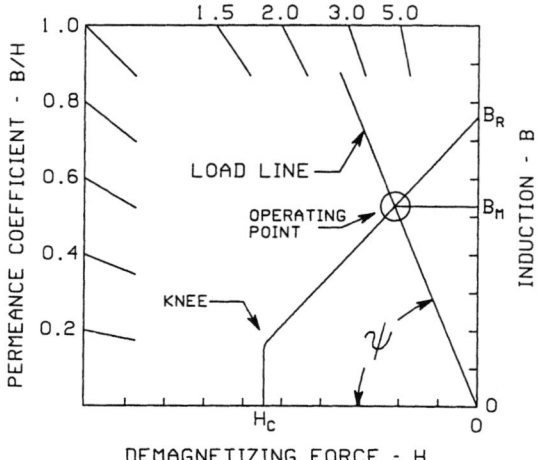

FIGURE 5.71 Magnetic load line and operating point.

Then:

$$\phi_g = \frac{\phi_m}{\sigma}$$

These curves usually give H in kilooersteds and B in kilogauss. This gives ϕ_m in kilolines, which must be converted to webers by multiplying by 10^{-5} before substituting into the torque equation.

5.2.8 Designing Hybrid Step Motors

A typical hybrid step motor is illustrated in Fig. 5.72.

The method for designing this motor is the same as that for the VR step motor. The major difference is that the rotor (Fig. 5.73) has two pieces that are hobbed, extruded, or punched as laminations. These pieces are partially hollowed out (Fig. 5.74) and partially enclose a cylindrical magnet (Fig. 5.75). When these parts are built into a rotor and shaft assembly (Fig. 5.76), the two end pieces are offset from each other by one-half tooth pitch circumferentially.

The rotor inertia calculation now changes somewhat. Follow the method outlined in the VR step-motor section, except subtract out the hollow portion of the end pieces and add in the magnet inertia. The inertia of the hollowed portion is

$$J_{\text{hollow}} = 0.0184 \left(\frac{D_{\text{mc}}}{2}\right)^4 d_{\text{mag}} - 0.0184 \left(\frac{d_{\text{shaft}}}{2}\right)^4 d_{\text{mag}} \quad \text{oz·in·s}^2$$

The inertia of the magnet is

$$J_{\text{mag}} = \left(\frac{D_{\text{mo}}}{2}\right)^4 \frac{L_{\text{mag}}(\text{mag density})}{772} - \left(\frac{D_s}{2}\right)^4 \frac{L_{\text{mag}}(\text{mag density})}{772} \quad \text{oz·in·s}^2$$

ELECTRONICALLY COMMUTATED MOTORS **5.75**

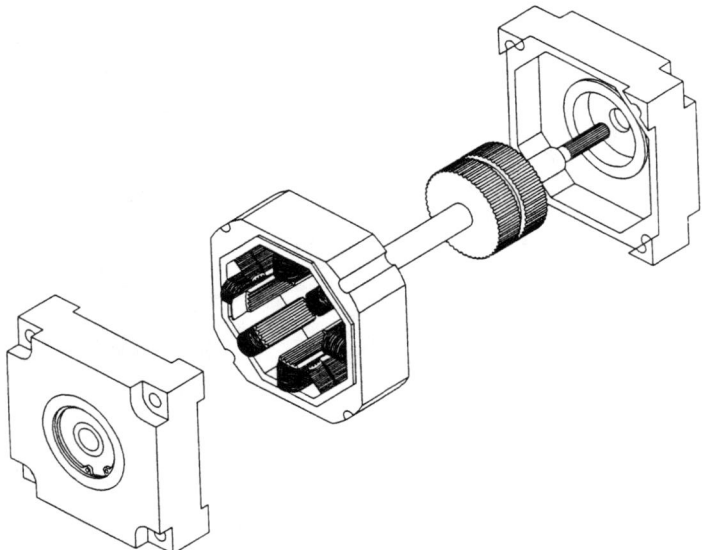

FIGURE 5.72 Hybrid step-motor construction.

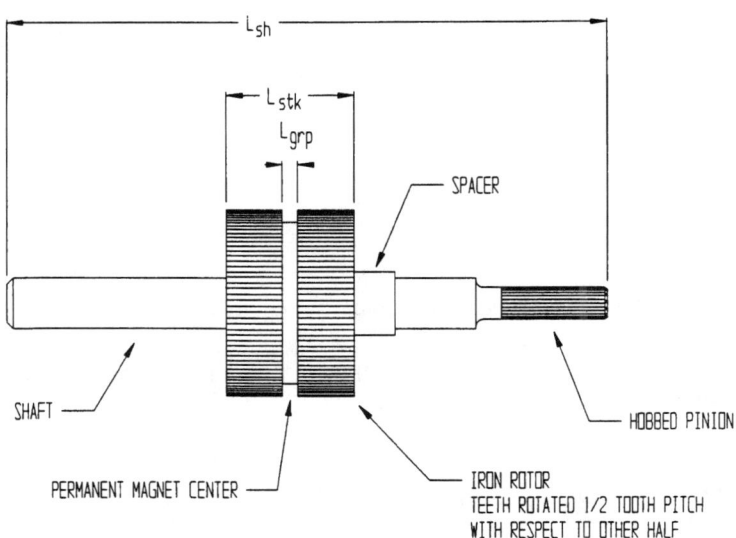

FIGURE 5.73 Hybrid rotor construction.

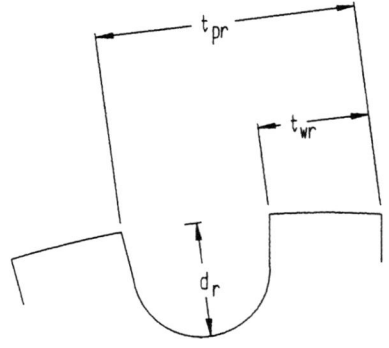

FIGURE 5.74 Iron rotor pole piece.

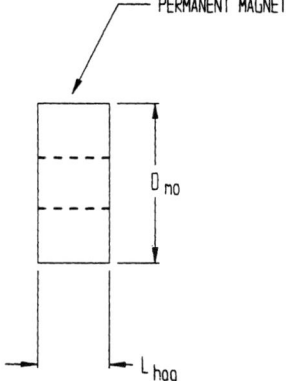

FIGURE 5.75 Permanent magnet.

Refer to Table 5.21 for magnetic material densities. The total inertia is now:

$$J_{tot} = J_{shaft} + J_{rotor\ pieces} - J_{hollow} + J_{mag}$$

Some rules to follow when designing the rotor are listed here. Refer to Figs. 5.74 and 5.75.

- Select a magnet material that will provide the needed flux and has the necessary demagnetization characteristics. Alnico is a common magnet material for these rotors; however, rare earth materials are used often because of their high resistance to demagnetization.

- Set d_{mag} to a depth that will avoid driving the end pieces into saturation. Start with

$$d_{mag} = \frac{L_{ep}}{2}$$

- Make L_{ep} as long as possible for maximum air gap flux.
- Keep $L_{grp} \geq 10\ L_g$.
- D_{mo} should be kept to a size that will allow sufficient rotor yoke section to avoid saturation at the teeth of the overhang region. Start with

$$D_{mo} = D_{ro} - 4\ d_{rs}$$

Hybrid Motor Torque. According to Kuo (1979) the torque consists of three components which include the mutual torque T_m, the variable reluctance torque T_{vr}, and the detent torque T_D.

$$T = T_m + T_{vr} + T_D$$

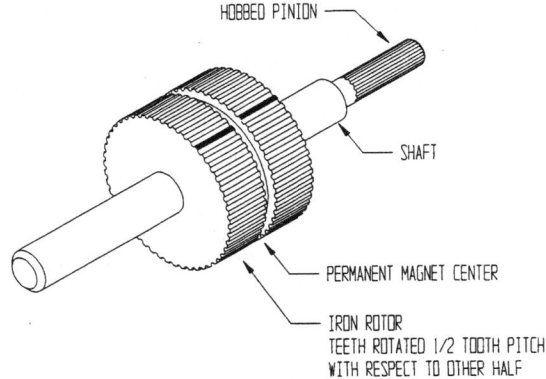

FIGURE 5.76 Hybrid motor with half tooth pitch offset.

where
$$T_m = -\frac{N_r}{2}(NI)B_0 A_m \frac{P_1}{P_0}\sin\theta_e$$

$$T_{VR} = \frac{N_r}{2}(NI)^2 \frac{P_1}{P_0}\sin 2\theta_e$$

$$T_D = -\frac{N_r}{2}\left(\frac{P_m}{P_0}\right)2_{F_m}\sin 4\theta_e$$

N_r = number of rotor teeth
NI = ampere-turns
B_0 = flux density of rotor magnet
A_m = cross-sectional area of magnet
P_1 = fundamental component of air gap permeance
P_0 = average value of air gap permeance
P_m = permeance magnitude of permanent magnet
P_4 = magnitude of fourth harmonic of air gap permeance
F_m = mmf of permanent magnet
θ_e = rotor position in electrical angle = $N_r\theta$
θ = rotor position in mechanical angle

According to Chai (1995), the torque is

$$T = -K_t I\left(\sin\theta_e - \frac{NIP_L}{f_m P_m}\sin 2\theta_e\right) + \Delta T$$

These methods take into account the harmonic torques and require complex models for accurate results. As an initial try, the method used for the PM step motor will get you into the ballpark. For accurate prediction of torque, the VR motor should be modeled on a two-dimensional BEA or FEA software package. The PM and hybrid motors should be modeled on a three-dimensional BEA or FEA package.

Windings. PM step motors are simply wound on bobbins with one (monofilar) or two (bifilar) strands of wire per bobbin. They are then connected to operate as follows. The monofilar windings are operated on a bipolar driver. The bifilar windings are operated on a unipolar driver. The bipolar driver allows for greater copper utilization than the unipolar driver because only half the copper is being used at any one time in the unipolar system.

Hybrid and VR step motors may utilize a variety of winding schemes to obtain varying results. Some of the common connections are listed here.

- Figure 5.77 shows a VR three-phase step motor with the windings connected to produce an S-N pole arrangement for one-phase-on operation. Here the flux divides and goes around the complete stator yoke.

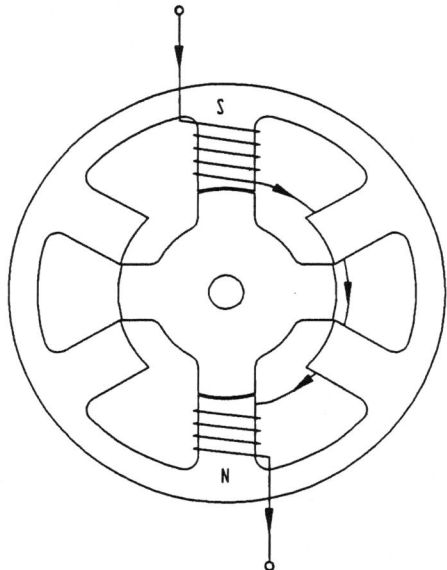

FIGURE 5.77 6/4 3-phase variable-reluctance step motor.

- Figure 5.78 shows a connection for two-phase-on operation of a four-phase motor. This motor has long flux paths, similar to the one-phase-on motor.
- Figure 5.79 shows another two-phase-on motor, except that the windings are connected to provide short flux paths. Here the flux exits the north pole and goes through the rotor to an adjacent south pole. This reduces the stator yoke mmf drop and core losses, making for a more efficient motor.
- Figure 5.80 shows a four-phase motor wound with a bifilar winding to allow for unipolar operation.

These winding configurations all have peculiar characteristics along with advantages and disadvantages and should be selected based on the system requirements.

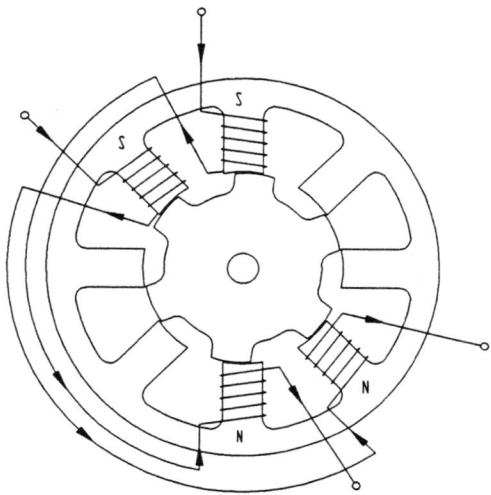

FIGURE 5.78 2-phase-on 4-phase step motor with long flux paths.

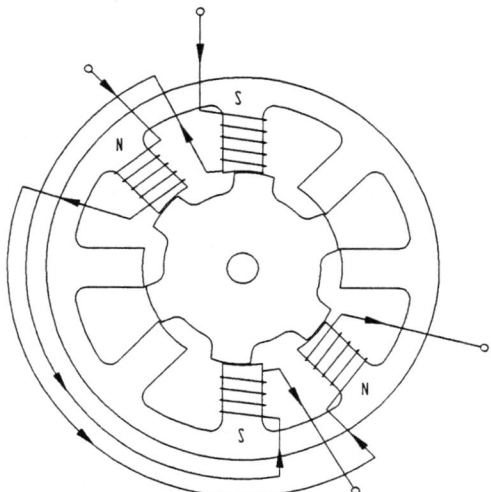

FIGURE 5.79 2-phase-on 4-phase step motor with short flux paths.

5.2.9 Measurement of Step Motor Characteristics

This subsection discusses the most common step-motor characteristics and methods of determining them.

Resistance. Resistance is measured using a Wheatstone bridge or equivalent.

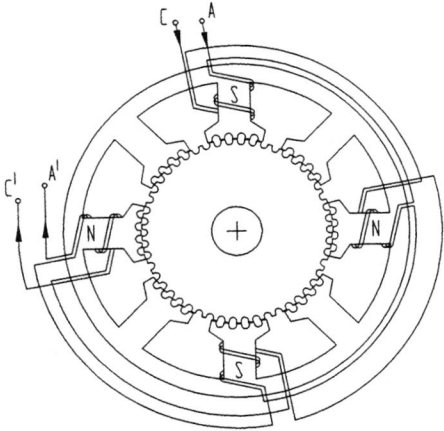

FIGURE 5.80 4-phase variable-reluctance step motor winding connections. Polarity as shown with current in A; current reverses with polarity in C.

Phase Inductance. The inductance of each phase winding is measured using an impedance bridge capable of generating a 1000-Hz ac voltage signal. The rotor position must be in the null or rotor-stator tooth-aligned position. The VR step motor may require a dc voltage signal as well to ensure proper alignment. Other methods may be used, but the values may differ and the method should be identified.

Holding or Static Torque. Apply rated dc current to one or two phases of the motor using a suitable torque measuring device, such as a dynamometer or torque watch. Determine the amount of torque required to cause slow rotation of the shaft. A typical curve is shown in Fig. 5.81.

As subsequent phases are plotted relative to each other, the complete torque profile as shown in Fig. 5.82 is produced. The intersection of the torque curves is the maximum continuous torque that can be expected.

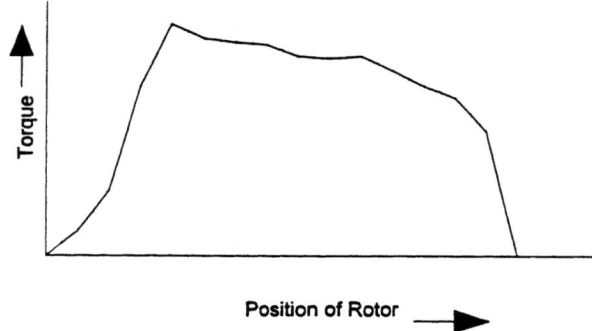

FIGURE 5.81 Torque versus rotor position: one phase on.

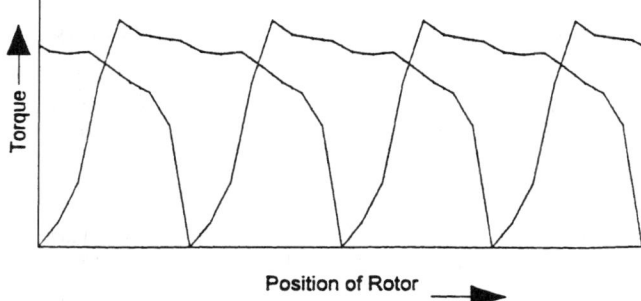

FIGURE 5.82 Torque versus rotor position: four phases on.

Pull-In Torque. A particular step rate is selected but is not yet gated to the stepmotor drive circuit, and the brake is then energized. The step pulses are then gated to the step motor, and the motor attempts to accelerate itself and the torque transducer and brake into a synchronous speed.

As it succeeds, a larger brake torque is required. If the step motor fails to accelerate, the step pulses are turned off. The brake torque is reduced slightly, and the pulses are again gated to the stepping motor. This process is repeated until the stepping motor is just able to gain synchronism at the set brake torque. The torque is measured after the motor and test setup have reached synchronous speed.

Pull-Out Torque. Set the brake torque to a minimum load and ramp the step motor up to some specified step rate. The brake torque is slowly increased until the stepping motor loses synchronism. The torque is measured just before the motor loses synchronism. Record the minimum and maximum torques. Use the average value to generate the pull-out-torque curve. This test is repeated at different step rates until sufficient data has been recorded to plot the pull-out curve.

In the running mode with a constant voltage supply of increasing frequency, the motor accelerates with decreasing torque (Fig. 5.83). The speed-torque curve resembles that of a universal motor. In this case, the torque decreases because the inductance limits the winding current as the switching frequency increases. Core losses are also increased as frequency increases. Employing a control system that overcomes the inductance and increases coil current at high speeds will significantly improve the output torque.

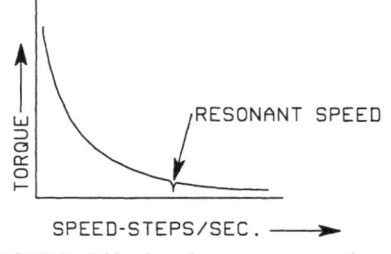

FIGURE 5.83 Speed-torque curve of step motor.

Inertia. Rotor inertia may be determined by removing the rotor and suspending it from a thin steel rod having a known torsion constant c. Displace the rotor slightly and time the period of oscillation f. The moment of inertia J is calculated from $J = (C/4)^2 f^2$.

Note: Most manufacturers publish this data. There are also test devices available that give a direct readout of inertia.

Residual or Detent Torque. With no power applied and leads separated (not touching), use a suitable measuring device to determine the torque required to cause slow rotation.

Step Accuracy. Accuracy from step to step measured in degrees of arc can be determined using an encoder having at least 10 times the resolution of the error to be measured. The encoder output is compared with the theoretical individual step angle to determine noncumulative error plus or minus, step to step.

Step accuracy is affected by part and assembly tolerances, friction, and magnetic balance. The tighter the tolerance, the better the step accuracy. The lower the friction, the better the step accuracy. The better the magnetic-circuit balance, the better the step accuracy.

Counter-EMF (PM Step Motors). The step-motor counter-emf can be determined by driving the rotor at a constant known speed with the stator winding deenergized and open circuited. A periodic voltage will be generated in the stator windings whose magnitude is proportional to speed and is generally sinusoidal. The average value of the voltage E_g divided by rotor speed ω equals the counter-emf constant.

$$K_e, \text{V}/(\text{rad·s}) = \frac{E_g, \text{V}}{\omega, \text{rad/s}}$$

Friction. There are two components of friction, static friction T_s and coulomb friction T_f. Friction may be measured using a rotary torque transducer and a dc servo/tachometer. T_s is the torque required to start rotation. T_f is the torque required to drive the step motor as speed varies after rotation starts.

Damping. The friction and windage torque T_f and T_w can be determined by plotting a K_f curve. This is done by driving the motor shaft from no-load speed to stall and measuring the stator reaction torque. Figure 5.84 shows a K_f curve for a step motor. The value of torque at zero speed is the internal motor friction T_f. K_f is the change in torque ΔT to the change in speed ΔS. Note that the slope of K_f changes at high speeds. The friction and windage torque T_{f+w} is the negative torque of the rotor at speed S_{load}. The K_f value is also called the *damping coefficient*.

Torque Constant. To obtain the torque versus current plot necessary to calculate the torque constant K_t, position the rotor at the maximum torque angle point. Increase the winding current and measure torque and current until the stator is driven into magnetic saturation. Plot the data and calculate K_t = peak torque, (oz·in)/(winding current, A).

Natural Frequency ω_0. This is the frequency at which the rotor will oscillate freely. It is a function of the torque, the pole pairs, and the inertia of the motor.

$$\omega_0 = \sqrt{\frac{TP}{J_r}} \quad \text{rad/s}$$

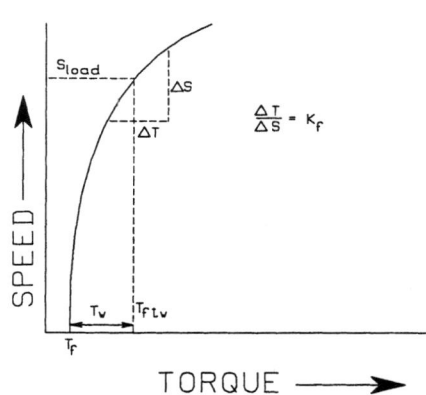

FIGURE 5.84 K_f curve for step motor.

Where: T = torque, oz·in
 P = number of pole pairs
 J_r = rotor inertia, oz·in·s^2

The motor will oscillate and lose synchronism at this frequency or a harmonic of this frequency.

Resonance. A phenomenon known as *midfrequency resonance* occurs in step motors operated in the open-loop mode. When this frequency range is reached, the motor develops velocity perturbations about the nominal drive velocity which reach catastrophic proportions, resulting in the loss of synchronism. The motor stalls as a result. If proper damping is used, it is possible to get through this region and operate above the resonant point.

Damping. This is the process of adding retarding torque to the motor system such that velocity perturbations are restrained. This can be done by adding friction or, electronically, by applying current to a previous phase.

5.2.10 Stepping-Motor Operation*

As discussed, stepping motors come in two varieties, permanent magnet and variable reluctance (hybrid motors are indistinguishable from PM motors from the controller's point of view). Lacking a label on the motor, you can generally tell the two apart by feel when no power is applied. PM motors tend to cog as you twist the rotor with your fingers, while VR motors almost spin freely (although they may cog slightly because of residual magnetization in the rotor). You can also distinguish between the two varieties with an ohmmeter. VR motors usually have three (sometimes four) windings, with a common return, while PM motors usually have two independent windings, with or without center taps. Center-tapped windings are used in unipolar PM motors.

Stepping motors come in a wide range of angular resolutions. The coarsest motors typically turn 90° per step, while high-resolution PM motors are commonly able to handle 1.8° or even 0.72° per step. With an appropriate controller, most PM and hybrid motors can be run in half-steps, and some controllers can handle smaller fractional steps or microsteps.

For both PM and VR stepping motors, if just one winding of the motor is energized, the rotor (under no load) will snap to a fixed angle and then hold that angle until the torque exceeds the holding torque of the motor, at which point the rotor will turn, trying to hold at each successive equilibrium point.

VR Motors. If your motor has three windings, typically connected as shown in the schematic diagram in Fig. 5.85, with one terminal common to all windings, it is most likely a VR stepping motor. In use, the common wire typically goes to the positive supply, and the windings are energized in sequence.

Figure 5.85 shows a cross section of a 30°-per-step VR motor. The rotor in this motor has four teeth and the stator has six poles, with each winding wrapped around two opposite poles. With winding 1 energized, the rotor teeth marked X are attracted to this winding's poles. If the current through winding 1 is turned off and winding 2 is turned on, the rotor will rotate 30° degrees clockwise so that the poles marked Y line up with the poles marked 2.

---
* Subsection contributed by Douglas W. Jones, University of Iowa.

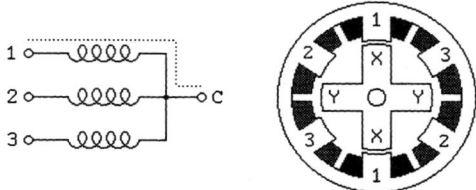

FIGURE 5.85 Variable-reluctance connection diagram.

To rotate this motor continuously, we just apply power to the three windings in sequence. Assuming positive logic, where 1 means turning on the current through a motor winding, the following control sequence will spin the motor illustrated in Fig. 5.85 clockwise 24 steps or 2 revolutions:

Winding 1	100100100100100100100100
Winding 2	010010010010010010010010
Winding 3	001001001001001001001001
Time →	

The section on midlevel control in Chap. 10 provides details on methods for generating such sequences of control signals, while the section on control circuits (Sec. 10.8) discusses the power switching circuitry needed to drive the motor windings from such control sequences. There are also VR stepping motors with four and five windings, requiring five or six wires. The principle for driving these motors is the same as that for the three-winding variety, but it becomes important to work out the correct order to energize the windings to make the motor step nicely.

The motor geometry illustrated in Fig. 5.85, giving 30° per step, uses the fewest number of rotor teeth and stator poles that perform satisfactorily. Using more motor poles and more rotor teeth allows construction of motors with smaller step angles. Toothed faces on each pole and a correspondingly finely toothed rotor allow for step angles as small as a few degrees.

Unipolar Motors. Unipolar stepping motors, both PM and hybrid stepping motors with five or six wires, are usually wired as shown in the schematic in Fig. 5.86, with a center tap on each of two windings. In use, the center taps of the windings are typically wired to the positive supply, and the two ends of each winding are alternately grounded to reverse the direction of the field provided by that winding.

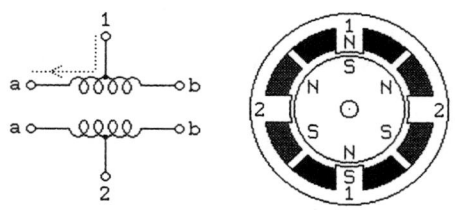

FIGURE 5.86 Unipolar motor.

Figure 5.86 shows a cross section of a 30°-per-step PM or hybrid motor—the difference between these two motor types is not relevant at this level of abstraction. Motor winding 1 is distributed between the top and bottom stator pole, while motor winding 2 is distributed between the left and right motor poles. The rotor is a permanent magnet with six poles, three south and three north, arranged around its circumference.

For higher angular resolutions, the rotor must have proportionally more poles. The 30°-per-step motor in Fig. 5.86 is one of the most common PM motor designs, although 15°- and 7.5°-per-step motors are widely available. PM motors with resolutions as good as 1.8° per step are made, and hybrid motors are routinely built with 3.6° and 1.8° per step, with resolutions as fine as 0.72° per step available.

As shown in Fig. 5.86, the current flowing from the center tap of winding 1 to terminal a causes the top stator pole to be a north pole while the bottom stator pole is a south pole. This attracts the rotor into the position shown. If the power to winding 1 is removed and winding 2 is energized, the rotor will turn 30°, or one step.

To rotate the motor continuously, we just apply power to the two windings in sequence. Assuming positive logic, where 1 means turning on the current through a motor winding, the following two control sequences will spin the motor illustrated in Fig. 5.86 clockwise 24 steps or 4 revolutions:

Winding 1a	100010001000100010001
Winding 1b	001000100010001000100
Winding 2a	010001000100010001000
Winding 2b	000100010001000100010
	Time →
Winding 1a	110011001100110011001
Winding 1b	001100110011001100110
Winding 2a	011001100110011001100
Winding 2b	100110011001100110011
	Time →

Note that the two halves of each winding are never energized at the same time. Both sequences shown will rotate a permanent magnet one step at a time. The top sequence powers only one winding at a time, as illustrated in Fig. 5.86; thus, it uses less power. The bottom sequence involves powering two windings at a time and generally produces a torque about 1.4 times greater than the top sequence while using twice as much power.

The section on midlevel control in Chap. 10 provides details on methods for generating such sequences of control signals, while the section on control circuits (Sec. 10.8) discusses the power switching circuitry needed to drive the motor windings from such control sequences.

The step positions produced by the two preceding sequences are not the same; as a result, combining the two sequences allows half-stepping, with the motor stopping alternately at the positions indicated by one or the other sequence. The combined sequence is as follows:

Winding 1a	110000011100000111000000111
Winding 1b	000111000001110000011100000
Winding 2a	011100000111000001110000001
Winding 2b	000001110000011100000011100
	Time →

Bipolar Motors. Bipolar PM and hybrid motors are constructed with exactly the same mechanism as is used on unipolar motors, but the two windings are wired more simply, with no center taps. Thus, the motor itself is simpler, but the drive circuitry needed to reverse the polarity of each pair of motor poles is more complex. The schematic in Fig. 5.87 shows how such a motor is wired, while the motor cross section shown here is exactly the same as the cross section shown in Fig. 5.87.

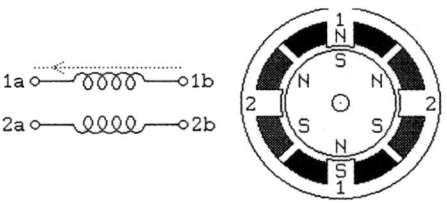

FIGURE 5.87 Bipolar motor.

The drive circuitry for such a motor requires an *H-bridge* control circuit for each winding; these are discussed in more detail in Sec. 10.8. Briefly, an H bridge allows the polarity of the power applied to each end of each winding to be controlled independently. The control sequences for single stepping such a motor are shown here, using + and − symbols to indicate the polarity of the power applied to each motor terminal:

```
Terminal 1a    +---+---+---+---  ++--++--++--++--
Terminal 1b    --+---+---+---+-  --++--++--++--++
Terminal 2a    -+---+---+---+--  -++--++--++--++-
Terminal 2b    ---+---+---+---+  +--++--++--++--+
               Time →
```

Note that these sequences are identical to those for a unipolar permanent magnet motor, at an abstract level, and that above the level of the H-bridge power-switching electronics, the control systems for the two types of motor can be identical.

Note that many full H-bridge driver chips have one control input to enable the output and another to control the direction. Given two such bridge chips, one for winding, the following control sequences will spin the motor identically to the control sequences given previously:

```
Enable 1       1010101010101010 1111111111111111
Direction 1    1x0x1x0x1x0x1x0x 1100110011001100
Enable 2       0101010101010101 1111111111111111
Direction 2    x1x0x1x0x1x0x1x0 0110011001100110
               Time →
```

To distinguish a bipolar PM motor from other four-wire motors, measure the resistances between the different terminals. It is worth noting that some PM stepping motors have four independent windings, organized as two sets of two. Within each set, if the two windings are wired in series, the result can be used as a high-voltage bipolar motor. If they are wired in parallel, the result can be used as a low-voltage bipolar motor. If they are wired in series with a center tap, the result can be used as a low-voltage unipolar motor.

Multiphase Motors. A less common class of PM stepping motor is wired with all windings of the motor in a cyclic series, with one tap between each pair of windings in the cycle. The most common designs in this category use three-phase and five-phase wiring. The control requires half of an H bridge for each motor terminal, but these motors can provide more torque from a given package size because all or all but one of the motor windings are energized at every point in the drive cycle. Some 5-phase motors have high resolutions on the order of 0.72° per step (500 steps per revolution).

With a 5-phase motor, there are 10 steps per repeat in the stepping cycle, as shown here:

Terminal 1	+++------+++++------++
Terminal 2	--+++++------+++++---
Terminal 3	+------+++++------++++
Terminal 4	+++++------+++++------
Terminal 5	----+++++------+++++-
	Time →

Here, as in the bipolar case, each terminal is shown as being connected to either the positive or negative bus of the motor power system. Note that at each step, only one terminal changes polarity. This change removes the power from one winding attached to that terminal (because both terminals of the winding in question are of the same polarity) and applies power to one winding that was previously idle. Given the motor geometry suggested by Fig. 5.88, this control sequence will drive the motor through two revolutions.

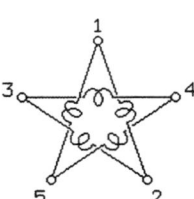

FIGURE 5.88 Multiphase motor

To distinguish a five-phase motor from other motors with five leads, note that if the resistance between two consecutive terminals of the five-phase motor is R, the resistance between nonconsecutive terminals will be $1.5R$.

Note that some 5-phase motors have 5 separate motor windings, with a total of 10 leads. These can be connected in the star configuration shown in Fig. 5.88, using five half-bridge driver circuits, or each winding can be driven by its own full bridge. While the theoretical component count of half-bridge drivers is lower, the availability of integrated full-bridge chips may make the latter approach preferable.

Stepping-Motor Physics. In any presentation covering the quantitative physics of a class of systems, it is important to be aware of the units of measurement used. This presentation of stepping-motor physics assumes standard physical units, as outlined in Table 5.22.

A force of 1 lb will accelerate a mass of 1 slug at 1 ft/s². The same relationship holds between the force, mass, time, and distance units of the other measurement systems. Most people prefer to measure angles in degrees, and the common engineering practice of specifying mass in pounds or force in kilograms will not yield correct results in the formulas given here. Care must be taken to convert such irregular

TABLE 5.22 Standard Physical Units

	System		
Measurement	English	CGS	MKS
Mass	Slug	Gram (g)	Kilogram (kg)
Force	Pound (lb)	Dyne (dyn)	Newton (N)
Distance	Foot (ft)	Centimeter (cm)	Meter (m)
Time	Second (s)	Second (s)	Second (s)
Angle	Radian (rad)	Radian (rad)	Radian (rad)

units to one of the standard systems outlined in Table 5.22 before applying the formulas given here.

Statics. For a motor that turns S rad per step, the plot of torque versus angular position for the rotor relative to some initial equilibrium position will generally approximate a sinusoid. The actual shape of the curve depends on the pole geometry of both rotor and stator, and neither this curve nor the geometry information is given in the motor data sheets this author has seen. For PM and hybrid motors, the actual curve usually looks sinusoidal, but looks can be misleading. For VR motors, the curve rarely even looks sinusoidal; trapezoidal and even asymmetrical sawtooth curves are not uncommon.

For a three-winding VR or PM motor with S rad per step, the period of the torque versus position curve will be $3S$; for a five-phase PM motor, the period will be $5S$. For a two-winding PM or hybrid motor, the most common type, the period will be $4S$, as illustrated in Fig. 5.89.

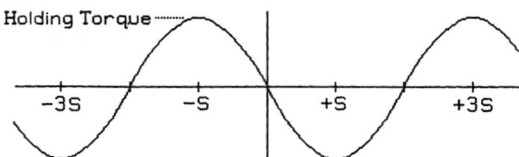

FIGURE 5.89 Holding torque versus rotor position.

Again, for an ideal two-winding PM motor, this can be mathematically expressed as follows:

$$T = -h \sin\left(\frac{\pi/2}{S} \theta\right)$$

where T = torque
  h = holding torque
  S = step angle, rad
  θ = shaft angle, rad

But remember, subtle departures from the ideal sinusoid described here are very common. The single-winding holding torque of a stepping motor is the peak value of the torque versus position curve when the maximum allowed current is flowing through one motor winding. If you attempt to apply a torque greater than this to the motor rotor while maintaining power to one winding, it will rotate freely.

It is sometimes useful to distinguish between the electrical shaft angle and the mechanical shaft angle. In the mechanical frame of reference, 2 rad is defined as one full revolution. In the electrical frame of reference, a revolution is defined as one period of the torque versus shaft-angle curve. Throughout this subsection, θ refers to the mechanical shaft angle, and $(\pi/2)/S$ gives the electrical angle for a motor with 4 steps per cycle of the torque curve.

Assuming that the torque versus angular-position curve is a good approximation of a sinusoid, as long as the torque remains below the holding torque of the motor, the rotor will remain within ¼ period of the equilibrium position. For a two-winding PM or hybrid motor, this means the rotor will remain within one step of the equilibrium position. With no power to any of the motor windings, the torque does not always fall to zero. In VR stepping motors, residual magnetization in the magnetic circuits of the motor may lead to a small residual torque, and in PM and hybrid stepping motors, the combination of pole geometry and the permanently magnetized rotor may lead to significant torque with no applied power.

The residual torque in a PM or hybrid stepping motor is frequently referred to as the *cogging torque* or *detent torque* of the motor because a naive observer will frequently guess that there is a detent mechanism of some kind inside the motor. The most common motor designs yield a detent torque that varies sinusoidally with rotor angle, with an equilibrium position at every step and an amplitude of roughly 10 percent of the rated holding torque of the motor, but a quick survey of motors from one manufacturer (Phytron) shows values as high as 23 percent for one very small motor to a low of 2.6 percent for one midsized motor.

Half-Stepping and Microstepping. So long as no part of the magnetic circuit saturates, powering two motor windings simultaneously will produce a torque versus position curve that is the sum of the torque versus position curves for the two motor windings taken in isolation. For a two-winding PM or hybrid motor, the two curves will be S rad out of phase, and if the currents in the two windings are equal, the peaks and valleys of the sum will be displaced $S/2$ rad from the peaks of the original curves, as shown in Fig. 5.90.

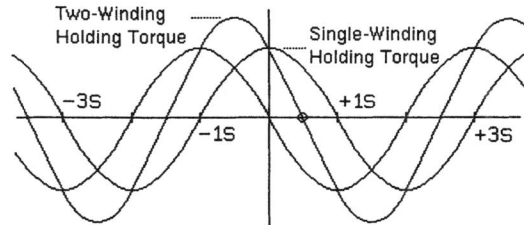

FIGURE 5.90 Holding torque versus rotor position, one and two phases on.

This is the basis of half-stepping. The two-winding holding torque is the peak of the composite torque curve when two windings are carrying their maximum rated current. For common two-winding PM or hybrid stepping motors, the two-winding holding torque is

$$h_2 = 2^{0.5} h_1$$

where h_1 = single-winding holding torque
 h_2 = two-winding holding torque

This assumes that no part of the magnetic circuit is saturated and that the torque versus position curve for each winding is an ideal sinusoid.

Most PM and VR stepping-motor data sheets quote the two-winding holding torque and not the single-winding figure; in part, this is because it is larger, and in part, it is because the most common full-step controllers always apply power to two windings at once.

If any part of the motor's magnetic circuits is saturated, the two torque curves will not add linearly. As a result, the composite torque will be less than the sum of the component torques, and the equilibrium position of the composite may not be exactly $S/2$ rad from the equilibria of the original.

Microstepping allows even smaller steps by using different currents through the two motor windings, as shown in Fig. 5.91.

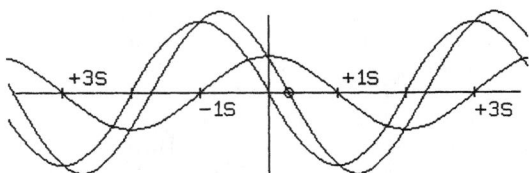

FIGURE 5.91 Relative holding-torque position under microstep control.

For a two-winding VR or PM motor, assuming nonsaturating magnetic circuits, and assuming perfectly sinusoidal torque versus position curves for each motor winding, the following formula gives the key characteristics of the composite torque curve:

$$h = (a^2 + b^2)^{0.5}$$

$$x = \frac{S}{\pi/2} \arctan \frac{b}{a}$$

where a = torque applied by winding with equilibrium at 0 rad
b = torque applied by winding with equilibrium at S rad
h = holding torque of composite
x = equilibrium position, rad
S = step angle, rad

In the absence of saturation, the torques a and b are directly proportional to the currents through the corresponding windings. It is quite common to work with normalized currents and torques, so that the single-winding holding torque or the maximum current allowed in one motor winding is 1.0.

Friction and the Dead Zone. The torque versus position curve shown in Fig. 5.89 does not take into account the torque the motor must exert to overcome friction. Note that frictional forces may be divided into two large categories, *static* or *sliding friction,* which requires a constant torque to overcome, regardless of velocity, and *dynamic friction* or *viscous drag,* which offers a resistance that varies with velocity. Here, we are concerned with the impact of static friction. Suppose the torque needed to overcome the static friction on the driven system is half the peak torque of the motor, as illustrated in Fig. 5.92.

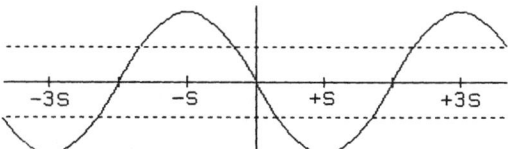

FIGURE 5.92 Holding torque versus rotor position showing relative static friction.

The dotted lines in Fig. 5.92 show the torque needed to overcome friction; only that part of the torque curve outside the dotted lines is available to move the rotor. The curve showing the available torque as a function of the shaft angle is the difference between these curves, as shown in Fig. 5.93.

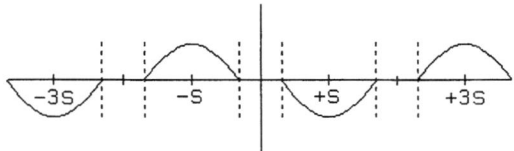

FIGURE 5.93 Curve of available torque once friction torque is removed.

Note that the consequences of static friction are twofold. First, the total torque available to move the load is reduced; second, there is a dead zone about each of the equilibria of the ideal motor. If the motor rotor is positioned anywhere within the dead zone for the current equilibrium position, the frictional torque will exceed the torque applied by the motor windings, and the rotor will not move. Assuming an ideal sinusoidal torque versus position curve in the absence of friction, the angular width of these dead zones is

$$d = 2\left(\frac{S}{\pi/2}\right) \arcsin \frac{f}{h} = \frac{S}{\pi/4} \arcsin \frac{f}{h}$$

where d = width of dead zone, rad
S = step angle, rad
f = torque needed to overcome static friction
h = holding torque

The important thing to note about the dead zone is that it limits the ultimate positioning accuracy. For the example, where the static friction is half the peak torque, a 90°-per-step motor will have dead zones 60° wide. This means that successive steps may be as large as 150° and as small as 30°, depending on where in the dead zone the rotor stops after each step.

The presence of a dead zone has a significant impact on the utility of microstepping. If the dead zone is 70° wide, then microstepping with a step size smaller than 70° may not move the rotor at all. Thus, for systems intended to use high-resolution microstepping, it is very important to minimize static friction.

Dynamics. Each time you step the motor, you electronically move the equilibrium position S rad. This moves the entire curve illustrated in Fig. 5.89 a distance of S rad, as shown in Fig. 5.94.

The first thing to note about the process of taking one step is that the maximum available torque is at a minimum when the rotor is halfway from one step to the next. This minimum determines the *running torque,* the maximum torque the motor can drive as it steps slowly forward. For common two-winding PM motors with ideal sinusoidal torque versus position curves and holding torque h, this will be $h/(2^{0.5})$. If the motor is stepped by powering two windings at a time, the running torque of an ideal two-winding PM motor will be the same as the single-winding holding torque.

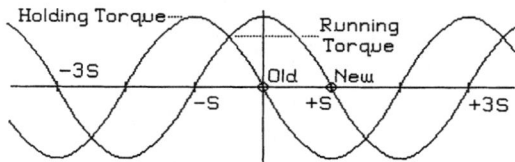

FIGURE 5.94 Holding-torque curves showing maximum torque when moved S rad.

It should be noted that at higher stepping speeds, the running torque is sometimes defined as the *pull-out torque.* That is, it is the maximum frictional torque the motor can overcome on a rotating load before the load is pulled out of step by the friction. Some motor data sheets define a second torque figure, the *pull-in torque.* This is the maximum frictional torque that the motor can overcome to accelerate a stopped load to synchronous speed. The pull-in torques documented on stepping-motor data sheets are of questionable value, because the pull-in torque depends on the moment of inertia of the load used when it is measured, and few motor data sheets document this.

In practice, there is always some friction, so after the equilibrium position moves one step, the rotor is likely to oscillate briefly about the new equilibrium position. The resulting trajectory may resemble the one shown in Fig. 5.95. Here, the trajectory of the equilibrium position is shown as a dotted line, while the solid curve shows the trajectory of the motor rotor.

Resonance. The resonant frequency of the motor rotor depends on the amplitude of the oscillation; but as the amplitude decreases, the resonant frequency rises to a

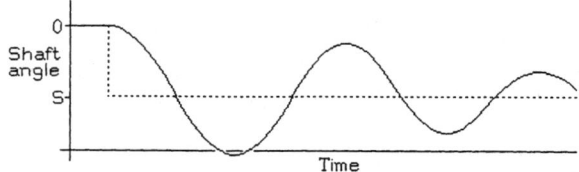

FIGURE 5.95 Trajectory of rotor oscillation.

well-defined small-amplitude frequency. This frequency depends on the step angle and on the ratio of the holding torque to the moment of inertia of the rotor. Either a higher torque or a lower moment will increase the frequency.

Formally, the small-amplitude resonance can be computed as follows. First, recall Newton's law for angular acceleration:

$$T = \mu A$$

where T = torque applied to rotor
μ = moment of inertia of rotor and load
A = angular acceleration, rad/s^2

Assume that, for small amplitudes, the torque on the rotor can be approximated as a linear function of the displacement from the equilibrium position. Therefore, Hooke's law applies:

$$T = -k\theta$$

where k = spring constant of the system, torque units/rad
θ = angular position of rotor, rad

You can equate the two formulas for the torque to get:

$$\mu A = -k$$

Note that acceleration is the second derivative of position with respect to time:

$$A = \frac{d^2\theta}{dt^2}$$

So you can rewrite this in differential equation form:

$$\frac{d^2}{dt^2} = -\left(\frac{k}{\mu}\right)\theta$$

To solve this, recall that, for

$$f(t) = a \sin bt$$

the derivatives are

$$\frac{df(t)}{dt} = ab \cos bt$$

$$\frac{d^2 f(t)}{dt^2} = -ab^2 \sin bt = -b^2 f(t)$$

Note that, throughout this discussion, it is assumed that the rotor is resonating. Therefore, it has an equation of motion something like:

$$\theta = a \sin (2\pi ft)$$

where a = angular amplitude of resonance
f = resonant frequency

This is an admissible solution to the preceding differential equation if you agree that

$$b = 2\pi f$$

$$b^2 = \frac{k}{\mu}$$

Solving for the resonant frequency f as a function of k and μ, you get:

$$f = \frac{(k/\mu)^{0.5}}{2\pi}$$

It is crucial to note that it is the moment of inertia of the rotor plus any coupled load that matters. The moment of the rotor, in isolation, is irrelevant. Some motor data sheets include information on resonance, but if any load is coupled to the rotor, the resonant frequency will change.

In practice, this oscillation can cause significant problems when the stepping rate is anywhere near a resonant frequency of the system; the result frequently appears as random and uncontrollable motion.

Resonance and the Ideal Motor. Up to this point, the discussion has dealt only with the small-angle spring constant k for the system. This can be measured experimentally, but if the motor's torque versus position curve is sinusoidal, it is also a simple function of the motor's holding torque. Recall that:

$$T = -h \sin\left(\frac{\pi/2}{S}\theta\right)$$

The small-angle spring constant k is the negative derivative of T at the origin.

$$k = -\frac{dT}{d\theta} = -\left(-h\,\frac{\pi/2}{S}\cos 0\right) = \frac{\pi}{2}\frac{h}{S}$$

Substituting this into the formula for frequency:

$$f = \frac{[(\pi/2)(h/S)/\mu]^{0.5}}{2} = \left(\frac{h}{8\pi\mu S}\right)^{0.5}$$

Given that the holding torque and resonant frequency of the system are easily measured, the easiest way to determine the moment of inertia of the moving parts in a system driven by a stepping motor is indirectly from the preceding relationship.

$$\mu = \frac{h}{8f^2 S}$$

For practical purposes, it is usually not the torque or the moment of inertia that matters, but rather the maximum sustainable acceleration. Conveniently, this is a simple function of the resonant frequency. Starting with the Newton's law for angular acceleration:

$$A = \frac{T}{\mu}$$

You can substitute this formula for the moment of inertia as a function of resonant frequency, and then substitute the maximum sustainable running torque as a function of the holding torque to get:

$$A = \frac{h/(2^{0.5})}{h/(8\pi f^2 S)} = \frac{8\pi S f^2}{2^{0.5}}$$

Measuring acceleration in steps per second squared instead of in radians per second squared, this simplifies to

$$A_{\text{steps}} = \frac{A}{S} = \frac{8\pi f^2}{2^{0.5}}$$

Thus, for an ideal motor with a sinusoidal torque versus rotor position function, the maximum acceleration in steps per second squared is a trivial function of the resonant frequency of the motor and rigidly coupled load.

For a two-winding PM or VR motor, with an ideal sinusoidal torque versus position characteristic, the two-winding holding torque is a simple function of the single-winding holding torque:

$$h_2 = 2^{0.5} h_1$$

where h_1 = single-winding holding torque
h_2 = two-winding holding torque

Substituting this into the formula for resonant frequency, you can find the ratios of the resonant frequencies in these two operating modes:

$$f_1 = (h_1/\cdots)^{0.5}$$

$$f_2 = (h_2/\cdots)^{0.5} = (2^{0.5} h_1/\cdots)^{0.5} = 2^{0.25} (h_1/\cdots)^{0.5} = 2^{0.25} f_1 = 1.189 \cdots f_1$$

This relationship holds only if the torque provided by the motor does not vary appreciably as the stepping rate varies between these two frequencies.

In general, as discussed later, the available torque will tend to remain relatively constant up until some cutoff stepping rate, and then it will fall. Therefore, this relationship holds only if the resonant frequencies are below this cutoff stepping rate. At stepping rates above the cutoff rate, the two frequencies will be closer to each other.

Living with Resonance. If a rigidly mounted stepping motor is rigidly coupled to a frictionless load and then stepped at a frequency near the resonant frequency, energy will be pumped into the resonant system, and the result of this is that the motor will literally lose control. There are three basic ways to deal with this problem.

Controlling Resonance in the Mechanism. Use of elastomeric motor mounts or elastomeric couplings between motor and load can drain energy out of the resonant system, preventing energy from accumulating to the extent that it allows the motor rotor to escape from control.

Or, viscous damping can be used. Here, the damping will not only draw energy out of the resonant modes of the system, it will also subtract from the total torque available at higher speeds. Magnetic eddy current damping is equivalent to viscous damping for our purposes.

Figure 5.96 illustrates the use of elastomeric couplings and viscous damping in two typical stepping-motor applications, one using a lead screw to drive a load, and the other using a tendon drive.

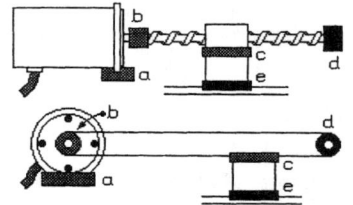

FIGURE 5.96 Damping in step-motor systems.

Elastomeric motor mounts are shown at *a* and elastomeric couplings between the motor and load are shown at *b* and *c*. The end bearing for the lead screw or tendon, at *d*, offers an opportunity for viscous damping, as do the ways on which the load slides, at *e*. Even the friction found in sealed ball bearings or Teflon on steel ways can provide enough damping to prevent resonance problems.

Controlling Resonance in the Low-Level Drive Circuitry. A resonating motor rotor will induce an alternating current voltage in the motor windings. If some motor winding is not currently being driven, shorting this winding will impose a drag on the motor rotor that is exactly equivalent to using a magnetic eddy current damper.

If some motor winding is currently being driven, the ac voltage induced by the resonance will tend to modulate the current through the winding. Clamping the motor current with an external inductor will counteract the resonance. Schemes based on this idea are incorporated into some of the drive circuits illustrated in later sections of this handbook.

Controlling Resonance in the High-Level Control System. The high-level control system can avoid driving the motor at known resonant frequencies, accelerating and decelerating through these frequencies and never attempting sustained rotation at these speeds.

Recall that the resonant frequency of a motor in half-stepped mode will vary by up to 20 percent from one half-step to the next. As a result, half-stepping pumps energy into the resonant system less efficiently than full stepping. Furthermore, when operating near these resonant frequencies, the motor control system may preferentially use only the two-winding half steps when operating near the single-winding resonant frequency, and only the single-winding half steps when operating near the two-winding resonant frequency. Figure 5.97 illustrates this.

The darkened curve in Fig. 5.97 shows the operating torque achieved by a simple control scheme that delivers useful torque over a wide range of speeds despite the fact that the available torque drops to zero at each resonance in the system. This solution is particularly effective if the resonant frequencies are sharply defined and well separated. This is the case in minimally damped systems operating well below the cutoff speed defined in the next subsection.

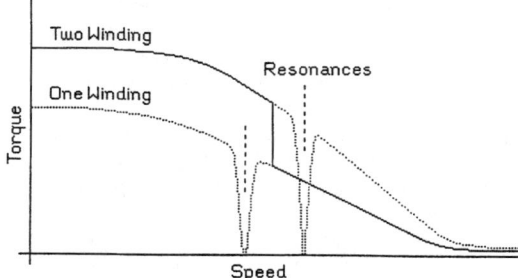

FIGURE 5.97 Variation of resonant speed based on winding excitation.

Torque Versus Speed. An important consideration in designing high-speed stepping-motor controllers is the effect of the inductance of the motor windings. As with the torque versus angular-position information, this is frequently poorly documented in motor data sheets; indeed, for VR stepping motors, it is not a constant. The inductance of the motor winding determines the rise and fall time of the current through the windings. While one might hope for a square-wave plot of current versus time, the inductance forces an exponential, as illustrated in Fig. 5.98.

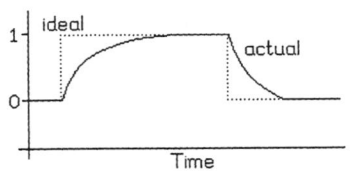

FIGURE 5.98 Effect of inductance on torque.

The details of the current versus time function through each winding depend as much on the drive circuitry as they do on the motor itself. It is quite common for the time constants of these exponentials to differ. The rise time is determined by the drive voltage and drive circuitry, while the fall time depends on the circuitry used to dissipate the stored energy in the motor winding.

At low stepping rates, the rise and fall times of the current through the motor windings has little effect on the motor's performance, but at higher speeds, the effect of the inductance of the motor windings is to reduce the available torque, as shown in Fig. 5.99.

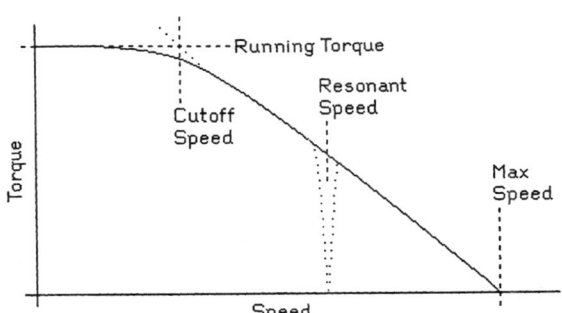

FIGURE 5.99 Reduction of torque at higher speeds caused by inductance.

The motor's maximum speed is defined as the speed at which the available torque falls to zero. Measuring maximum speed can be difficult when there are resonance problems, because these cause the torque to drop to zero prematurely. The cutoff speed is the speed above which the torque begins to fall. When the motor is operating below its cutoff speed, the rise and fall times of the current through the motor windings occupy an insignificant fraction of each step, while at the cutoff speed, the step duration is comparable to the sum of the rise and fall times. Note that a sharp cutoff is rare; therefore, statements of a motor's cutoff speed are, of necessity, approximate.

The details of the torque versus speed relationship depend on the details of the rise and fall times in the motor windings, and these depend on the motor control system as well as the motor. Therefore, the cutoff speed and maximum speed for any particular motor depend, in part, on the control system. The torque versus speed curves published in motor data sheets occasionally come with documentation on the motor controller used to obtain those curves, but this is far from universal practice.

Similarly, the resonant speed depends on the moment of inertia of the entire rotating system, not just the motor rotor, and the extent to which the torque drops at resonance depends on the presence of mechanical damping and on the nature of the control system. Some published torque versus speed curves show very clear resonances without documenting the moment of inertia of the hardware that may have been attached to the motor shaft in order to make torque measurements.

The torque versus speed curve shown in Fig. 5.99 is typical of the simplest of control systems. More complex control systems sometimes introduce electronic resonances that act to increase the available torque above the motor's low-speed torque. A common result of this is a peak in the available torque near the cutoff speed.

Electromagnetic Issues. In a PM or hybrid stepping motor, the magnetic field of the motor rotor changes with changes in shaft angle. The result of this is that turning the motor rotor induces an ac voltage in each motor winding. This is referred to as the *counter-emf* because the voltage induced in each motor winding is always in phase with and counter to the ideal waveform required to turn the motor in the same direction. Both the frequency and amplitude of the counter-emf increase with rotor speed; therefore, counter-emf contributes to the decline in torque with increased stepping rate.

VR stepping motors also induce counter-emf. This is because the reluctance of the magnetic circuit declines as the stator winding pulls a tooth of the rotor toward its equilibrium position. This decline increases the inductance of the stator winding, and this change in inductance demands a decrease in the current through the winding in order to conserve energy. This decrease is evidenced as a counter-emf.

The reactance (inductance and resistance) of the motor windings limits the current flowing through them. Thus, by Ohm's law, increasing the voltage increases the current, and therefore increases the available torque. The increased voltage also serves to overcome the counter-emf induced in the motor windings, but the voltage cannot be increased arbitrarily. Thermal, magnetic, and electronic considerations all serve to limit the useful torque that a motor can produce.

The heat given off by the motor windings is due to simple resistive losses, eddy current losses, and hysteresis losses. If this heat is not conducted away from the motor adequately, the motor windings will overheat. The simplest failure this can cause is insulation breakdown, but it can also heat a permanent magnet rotor above its Curie temperature, resulting in loss of magnetization. This is a particular risk with many modern high-strength magnetic alloys.

Even if the motor is attached to an adequate heat sink, increased drive voltage will not necessarily lead to increased torque. Most motors are designed so that the magnetic circuits of the motor are near saturation at the rated current flow through the windings. Increased current will not lead to an appreciably increased magnetic field in such a motor.

Given a drive system that limits the current through each motor winding to the rated maximum for that winding, but uses high voltages to achieve a higher cutoff torque and higher torques above cutoff, there are other limits that come into play. At high speeds, the motor windings must, of necessity, carry high-frequency ac signals. This leads to eddy current losses in the magnetic circuits of the motor, and it leads to skin-effect losses in the motor windings.

Motors designed for very high speed running should, therefore, have magnetic structures using very thin laminations or even nonconductive ferrite materials, and they should have small-gauge wire in their windings to minimize skin-effect losses.

Common high-torque motors have large-gauge motor windings and coarse core laminations. At high speeds, such motors can easily overheat, and they should therefore be derated accordingly for high-speed running.

It is also worth noting that the one way to demagnetize something is to expose it to a high-frequency high-amplitude magnetic field. Running the control system to spin the rotor at high speed when the rotor is actually stalled, or spinning the rotor at high speed against a control system trying to hold the rotor in a fixed position, will expose the rotor to a high-amplitude high-frequency field. If such operating conditions are common—particularly if the motor is run near the Curie temperature of the permanent magnets—demagnetization is a serious risk, and the field strengths (and expected torques) should be reduced accordingly.

5.3 SWITCHED-RELUCTANCE MOTORS*

Switched-reluctance motors (SRMs) are designed in a manner very similar to that for VR step motors. The major difference is that SRMs are designed for continuous running, whereas VR step motors are designed for a high torque-to-inertia ratio and high step accuracy.

As a starting point, the SRM has a rotor diameter D_r (Fig. 5.100) which is approximately one-half of the stator diameter.

5.3.1 Selection of Poles and Phases

The stroke angle is a function of the number of poles, phases, and rotor teeth. It may be calculated in the same fashion as for the step motor.

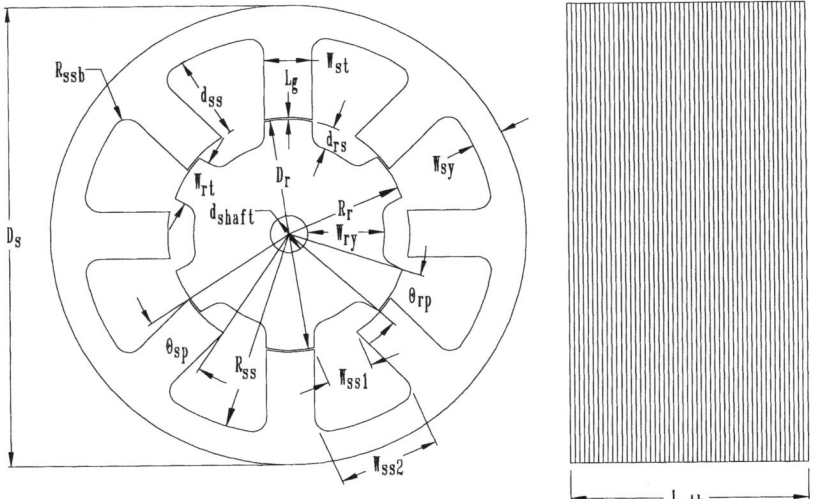

FIGURE 5.100 Variable-reluctance step-motor lamination stack

---

* Section contributed by William H. Yeadon, Yeadon Engineering Services, PC.

Selection of the number of poles and phases should be based on the following considerations:

- ·One- and two-phase motors have dead zones and need help starting.
- Three-phase motors always generate positive torque but have high torque ripple.
- Four- and five-phase motors have low torque ripple but require more expensive controllers.
- Higher numbers of phases and poles require higher switching frequencies, which increases core losses.

5.3.2 Sizing the Air Gap

The air gap L_g (Fig. 5.100) should be chosen to give the greatest torque per stroke. This approach will, however, tend to make the air gap length small. T. J. E. Miller (1993) recommends that it be about 0.05 percent of the rotor diameter. This tends to produce air gaps in the 0.007- to 0.010-in range for small motors. This can lead to manufacturing problems and increased costs in commercial-type motors because tight tolerances will be required on many parts. Holding a 0.010-in air gap is very feasible in most manufacturing situations and is a good initial selection. Small air gaps increase torque but also increase audible noise.

5.3.3 Lamination Properties

Generally speaking, the lamination should be sized in the same manner as for the VR step motor, with the following exceptions:

- The width of the stator yoke W_{sy} may be thinner than for the step motor but should be equal to or greater than the half-tooth width W_{st}.

$$W_{st} \geq 0.5\ W_{st}$$

- The rotor pole arc θ_{rp} and stator pole arc θ_{sp} should be the same, except that the rotor tooth can be larger than the stator tooth by some small percentage in an effort to reduce tooth tip saturation.

Slot and winding space as well as resistance are calculated in the same manner as for the VR step motor.
The design should be verified by using a finite element analysis program.

5.3.4 Performance Characteristics

Run in the open-loop mode, these motors have performance characteristics similar to those of VR stepper motors. Static torque curves can be plotted by turning on the phase current of one phase Ph_1, rotating the shaft through one stroke angle, and measuring the stator reaction torque (Fig. 5.101).

As subsequent phases are plotted relative to each other, the complete torque profile, as shown in Fig. 5.102, is produced. The intersection of the torque curves is the maximum continuous torque that can be expected.

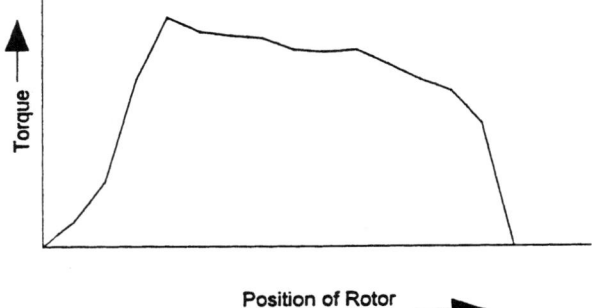

FIGURE 5.101 Torque versus rotor position: one phase on.

In the running mode, with a constant voltage supply of increasing frequency, the motor accelerates with decreasing torque (Fig. 5.103). The speed-torque curve resembles that of a universal motor. In this case, the torque decreases because the inductance limits the winding current as the switching frequency increases. Core losses are also increased as frequency increases. Employing a control system that overcomes the inductance and increases coil current at high speeds will significantly improve the output torque.

Input power P_{in} is the motor terminal voltage times the current for a dc motor.

$$P_{in} = VI$$

Output power, in watts:

$$P_{out} = \frac{S \times T}{5250} \times 746$$

where T = torque, lb·ft
S = speed, rpm

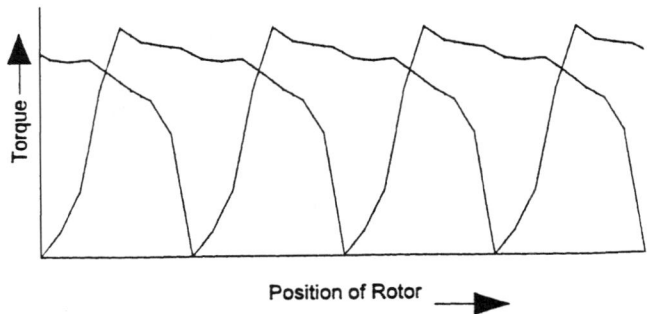

FIGURE 5.102 Torque versus rotor position: four phases on.

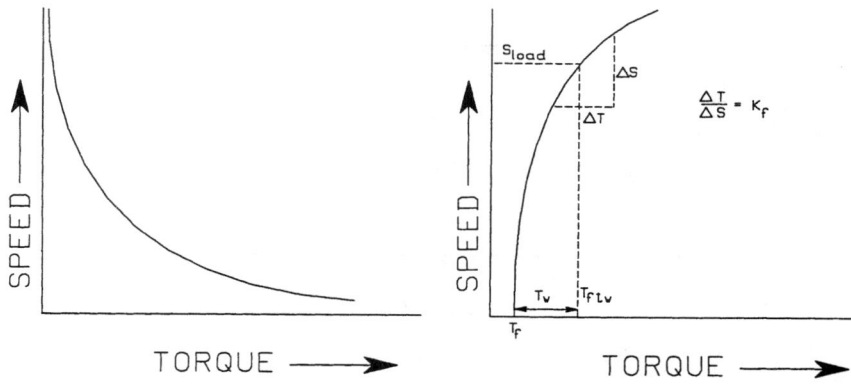

FIGURE 5.103 Speed-torque curve of switched-reluctance motor.

FIGURE 5.104 K_f curve for switched-reluctance motor.

Efficiency η is then:

$$\eta = \frac{P_{out}}{P_{in}}$$

Losses are therefore:

$$P_{loss} = P_{in} - P_{out}$$

Losses consist of:

$$P_{loss} = W_{f+w} + W_{copper} + W_{core}$$

where W_{f+w} = friction and windage loss
$W_{copper} = I^2R$ (heat) losses in windings
W_{core} = hysteresis and eddy current losses

The friction and windage torque T_f and T_w can be determined by plotting a K_f curve. This is done by driving the motor shaft from no-load speed to stall and measuring the stator reaction torque. Figure 5.104 shows a K_f curve for a SRM. The value of torque at zero speed is the internal motor friction T_f. K_f is the change in torque ΔT to the change in speed ΔS. Note that the slope of K_f changes at high speeds. The friction and windage torque T_{f+w} is the negative torque of the rotor at speed S_{load}.

The loss in watts due to friction and windage W_{f+w}, can be determined from the P_{out} formula:

$$W_{f+w} = \frac{S \times T_{f+w}}{5250} \times 746$$

where T_f = bearing friction, lb·ft
T_w = windage resistance caused by rotor moving air, lb·ft

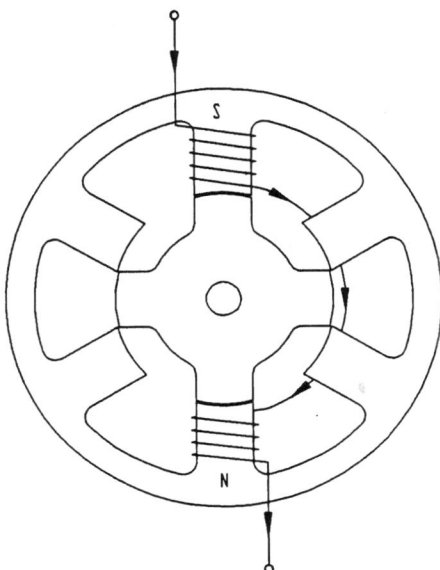

FIGURE 5.105 6/4 3-phase switched-reluctance motor.

The copper and core loss can be determined simply as follows:

$$W_{copper} = I^2 R$$

$$W_{core} = P_{loss} - W_{f+w} - W_{copper}$$

where W_{copper} is for all phases which are turned on at any given time.

5.3.5 Design Procedures

Applications. The switched-reluctance motor has some advantages that make it particularly suited for certain applications. The stator is a relatively low-cost component because it is easy to wind. Figure 5.105 shows a 6/4 3-phase motor with 2 poles excited. This is wound on a gun winder and is a simple series connection.

Figure 5.106 shows an 8/6 motor with 2 phases wound. Here the adjacent phases are the same polarity, and the flux goes through the rotor to the opposite poles at 180°.

Figure 5.107 shows the same stator and rotor configuration as shown in Fig. 5.106, except that the adjacent poles are connected in the opposite polarity. Here the flux goes through only the top portion of the rotor and uses only ⅙ of the stator yoke. This configuration reduces core losses, increases flux, and increases output torque. The stator yoke W_{sy} needs to be wider for this connection because the yoke carries all of the pole flux.

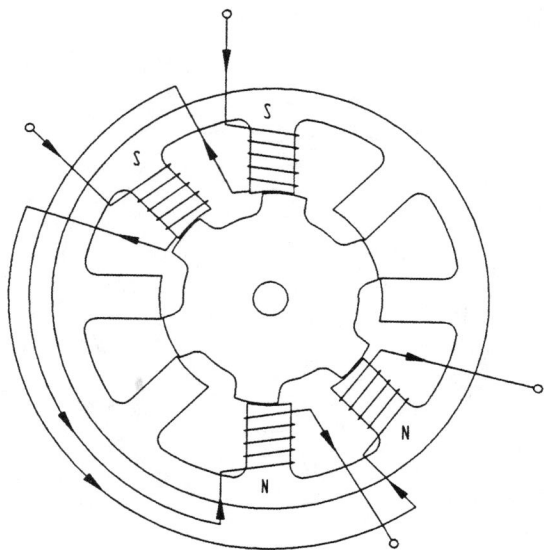

FIGURE 5.106 8/6 4-phase switched-reluctance motor.

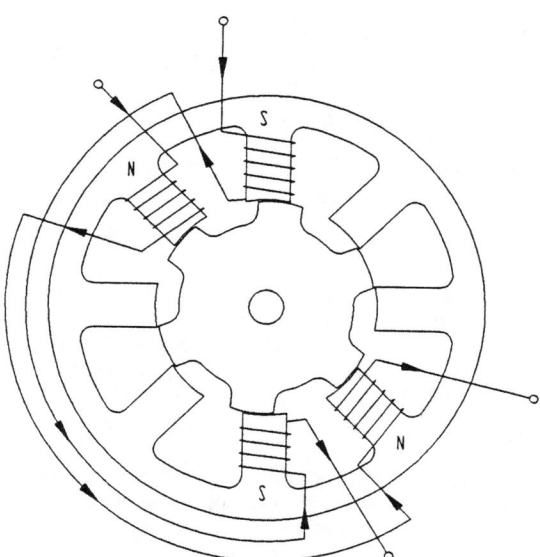

FIGURE 5.107 8/6 switched-reluctance motor with short flux paths.

These windings can be simple series connections with the controller reversing the current in the phases as necessary. An alternative is to wind two strands of wire on each pole and connect them in opposite polarities. This is called a *bifilar winding*. It has the advantage of using a simpler controller but has only half the space available for copper per active phase.

SRMs can be operated at relatively high speeds, which makes them candidates for hand tools, power tools, lawn care equipment, kitchen appliances, and any application where a universal motor is used. There are some applications that use battery-operated PMDC or BLDC motors that have high speed requirements that could be replaced by switched-reluctance motors.

The switched-reluctance motor has no brush commutator set; thus, it has no sparking. This opens the door for use in explosion-proof pumps.

Noise. There are, however, a few drawbacks. The SRM's small air gap and saliency lead to opportunities for noise (refer to Fig. 5.108).

As the rotor pole tips are drawn toward an energized stator pole, a force exists that attempts to bend the stator pole in the direction of the rotor pole. The rotor moves past the aligned position, starts to slow down, and is snapped to the next stator pole.

This pole also attempts to bend in the direction of the rotor. The result is vibrations set up in both the rotor and the stator teeth. Further, the air gap flux in the aligned position attempts to pull the rotor and stator poles together. This causes the

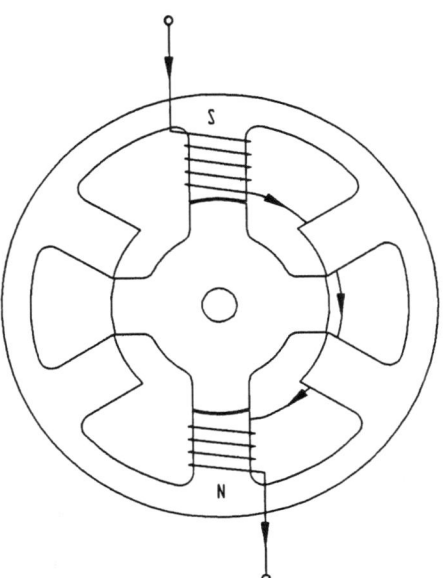

FIGURE 5.108 Cross section of switched-reluctance motor.

OD of the stator to collapse minutely, resulting in the round stack becoming slightly oval. The radial deflection Δ can be estimated from the pole force as follows:

$$\Delta \approx \frac{1.8F}{E\,(t/R)^3}$$

where F = radial force per unit axial length
  E = modulus of elasticity
  t = radial thickness of the stator yoke
  R = mean radius of the stator yoke

Generically, the radial force on a pole face area can be estimated using the flux density in the pole face area:

$$F_a \approx \frac{B^2}{\mu_0}$$

Or, to calculate F_a in pounds:

$$F_a = (1.4 \times 10^8) B^2 A$$

where B = flux density in the pole face, Wb/in^2
  A = area of the pole face, in^2

When the air gap is very small and the flux density is very high, it is possible to magnetically cause rotor-stator interference, as well as noise.

Another cause of noise is the rotor being offset in the stator bore. This results in

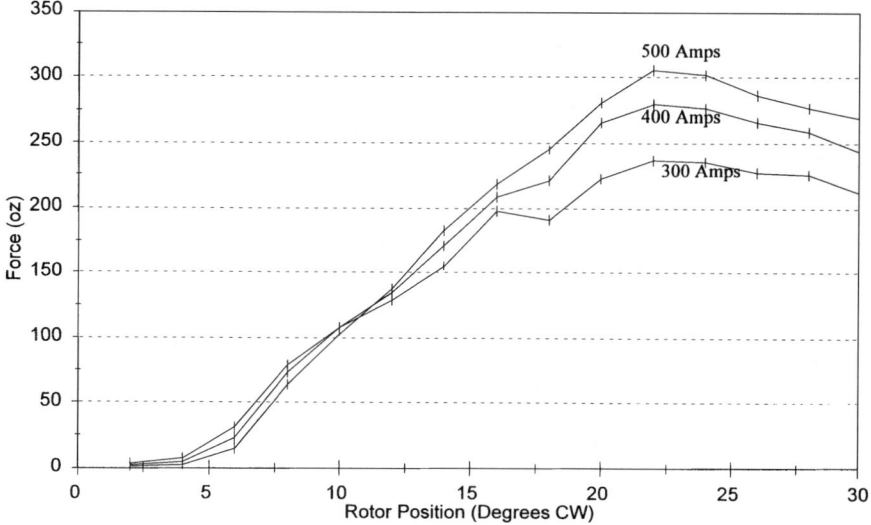

FIGURE 5.109 Force on offset rotor versus position for 2.5 OD 8/6 1.25 stack switched-reluctance motor. Rotor is offset by $\tfrac{1}{2}\,L_g$ toward one active pole. Forces with rotor centered are very small. *(Data compiled using boundary element analysis software from Integrated Engineering Software.)*

a shaft deflection that tends to bend the shaft in the shape of a bow. Figure 5.109 shows the increase in force or side pull at various stator currents as the rotor is offset by half the air gap length.

Some things that can be done to reduce noise are the following:

- Lower the flux density (i.e., use a longer stack).
- Increase motor mass.
- Stiffen the poles by adding large radii in the corners.
- Increase rotor inertia.
- Increase the shaft diameter to reduce deflection.
- Increase the air gap.
- Employ control system damping schemes.
- Keep the rotor centered in the stator.

CHAPTER 6
ALTERNATING-CURRENT INDUCTION MOTORS

Chapter Contributors

Brad Frustaglio
Andrew E. Miller
Earl F. Richards
Chris A. Swenski
William H. Yeadon

6.1 INTRODUCTION*

This chapter covers the most common types of ac induction motors, including polyphase and single-phase motors. Section 6.2 develops the theory of operation first for polyphase and single-phase operation, including combined winding operations for split-phase capacitor-start and capacitor-run motors.

Next, the section develops the two-phase motor and develops single-phase motor theory using the revolving-field and symmetrical-components approach. Finally, Sec. 6.3 develops the three-phase polyphase motor theory.

Then, Sec. 6.4 develops geometry, flux path, and performance calculations for the most common configurations of single-phase motors. Next, the section discusses the shaded-pole motor using the cross-field theory as developed by P. H. Trickey and S. S. L. Chang. Modified calculation procedures are included. Finally, the section finishes the calculations by developing geometry, flux path, and performance calculations for the polyphase motor.

6.1.1 Stator and Coil Assemblies

A typical stator and coil assembly is shown in Figs. 6.1, 6.2, and 6.3. The stator cores discussed in Chap. 3 are wound with magnet wire in order to produce the magnetomotive force (mmf) that produces the torque.

---

* Section contributed by William H. Yeadon, Yeadon Engineering Services, PC.

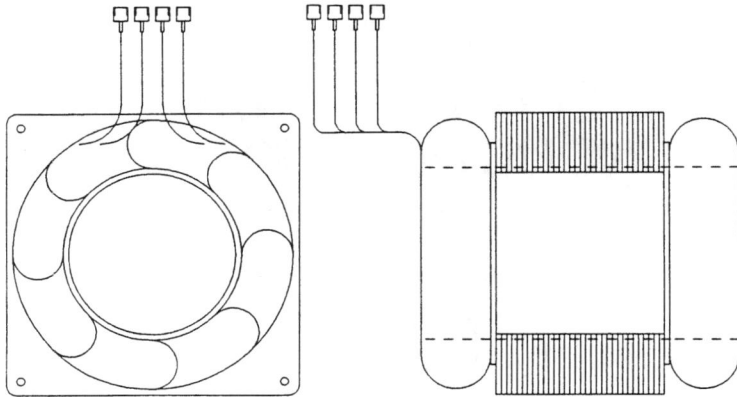

FIGURE 6.1 Stator and coil assembly diagram.

6.1.2 Windings

Stator and coil assemblies for induction motors usually have distributed wound cores. The objective here is to produce a sinusoidal air gap mmf. A coil of wire, as shown in Fig. 6.4, produces a square-wave mmf. A distributed winding improves this condition.

Next is an example of how sinusoidal distribution is determined. This is followed by tables showing near-sinusoidal winding distributions for some common pole-tooth combinations.

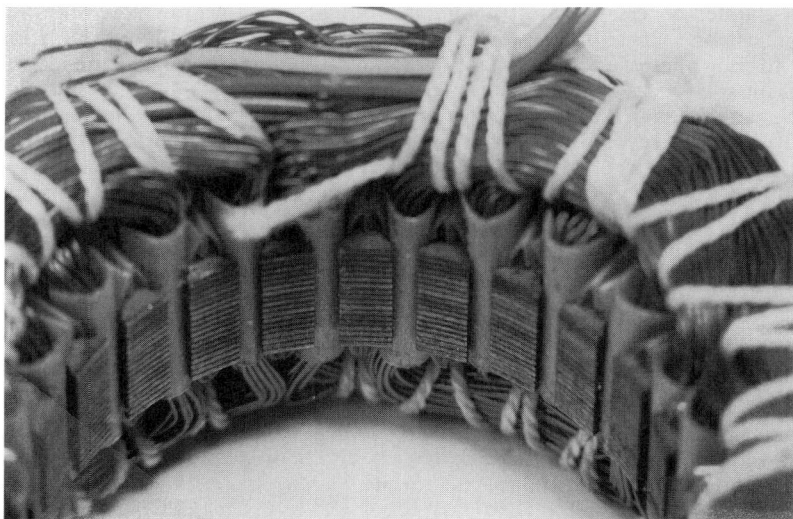

FIGURE 6.2 Stator and coil assembly lead end.

FIGURE 6.3 Stator and coil assembly opposite lead end.

Example. For any slot combination, the effective turn and sinusoidal distribution can be determined as follows

6 Poles, 1 Phase, in 36-Slot Stator

1. Full pitch (90°) = slots ÷ poles = 36 ÷ 6 = 6 teeth
2. Winding span = 5 teeth and 3 teeth

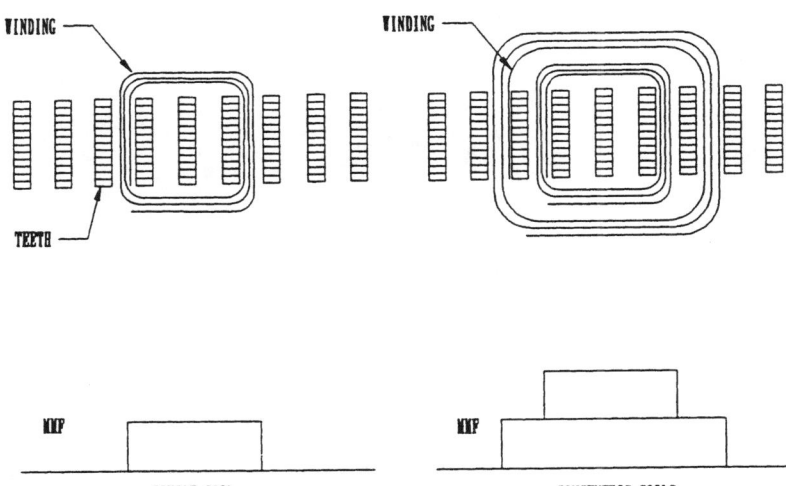

FIGURE 6.4 Air gap mmf waves versus winding distribution.

3. Effective turn factor:

$$5 \text{ teeth} = \frac{5}{6} \, 90° = 75°; \quad \sin 75° = 0.9659$$

$$3 \text{ teeth} = \frac{3}{6} \, 90° = 45°; \quad \sin 45° = 0.7071$$

$$\text{Sum} = 1.6730$$

4. Percentage of total pole turns per coil = effective turn factor per coil ÷ sum × 100

$$5 \text{ teeth} = \frac{0.9659}{1.6730} = 0.5773 \times 100 = 57.73\%$$

$$3 \text{ teeth} = \frac{0.7071}{1.6730} = 0.4226 \times 100 = 42.26\%$$

5. Winding factor K_w (sometimes called *winding pitch factor* K_p) = $\sum$ percentage of total turns per coil × effective turn factor per coil

$$5 \text{ teeth} = 0.5576 = 0.5773 \times 0.9659$$

$$3 \text{ teeth} = 0.2988 = 0.4226 \times 0.7071$$

$$\text{Sum} = 0.8564$$

$$K_w/\text{pole} = 0.8564$$

Tables 6.1 and 6.2 show sinusoidal winding distributions for even and odd slot combinations, respectively.

Polyphase motors can be wound in a lap winding configuration (Fig. 6.5) or a concentrically distributed winding configuration (Fig. 6.6). Single-phase motors are typically wound and inserted in the manner shown in Fig. 6.7. Figure 6.8 shows a typical

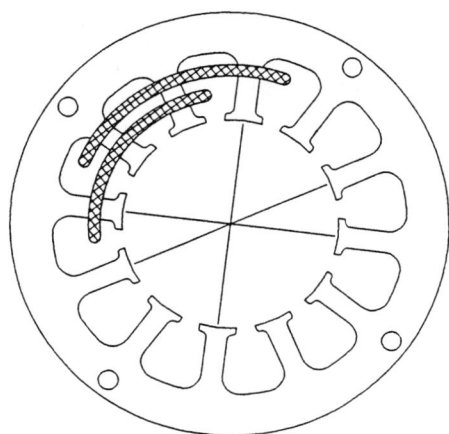

FIGURE 6.5 Lap-wound polyphase configuration.

TABLE 6.1 Sinusoidal Winding Distribution—Even Tooth Spans

Slots per pole	\multicolumn{9}{c}{Tooth span}	K_w
	2	4	6	8	10	12	14	16	18	
18		6.3	9.0	11.6	13.8	15.7	17.0	17.8	9.0	0.808
			9.6	12.4	14.7	16.7	18.1	18.9	9.6	0.835
				13.7	12.4	16.4	18.4	20.0	20.9	10.6	0.873
					18.9	21.3	23.2	24.3	12.3	0.909
						26.3	28.6	29.9	15.2	0.944
16		7.9	11.3	14.4	17.2	18.9	20.0	10.3		0.812
			12.4	15.7	18.5	20.5	21.8	11.1		0.848
				17.9	21.1	23.4	24.9	12.7		0.889
					25.7	28.5	30.3	15.5		0.928
						38.4	40.8	20.8		0.963
12	6.8	13.2	18.6	22.8	25.4	13.2				0.789
		14.1	20.0	24.5	27.3	14.1				0.829
			23.3	28.5	31.8	16.4				0.883
				37.2	41.4	21.4				0.936
					65.9	34.1				0.977
9	12.1	22.7	30.6	34.6						0.795
		25.7	34.8	39.5						0.855
			47.8	52.2						0.929
8	15.3	28.0	36.8	19.9						0.795
		33.1	43.4	23.5						0.870
			64.8	35.2						0.950
6	26.8	46.4	26.8							0.804
		63.4	36.6							0.914
4	60.8	39.2								0.822
										1.000

wound coil ready for insertion. Figure 6.9 shows the stator core assembly ready for coil insertion. Figure 6.10 shows the partially inserted stator and coil assembly.

6.2 THEORY OF SINGLE-PHASE INDUCTION-MOTOR OPERATION*

One of the most important concepts in the analysis and design of all electrical rotating machinery, whether it be a generator or motor, is to obtain a circuit model for performance calculations. Single-phase induction motors are no exception.

An equally important concept, which is necessary for induction or synchronous motor rotation, is the existence of a revolving magnetic field in the air gap of the motor. Without this there can be no torque development for producing rotation. Ideally, a constant-magnitude rotating mmf is preferred. This latter concept of a rotat-

---
* Sections 6.2 and 6.3 contributed by Earl F. Richards.

TABLE 6.2 Sinusoidal Winding Distribution—Odd Tooth Spans

Slots per pole	Tooth span									K_w
	3	5	6	9	11	13	15	17	19	
18	4.6	7.5	10.2	12.5	14.5	16.0	17.1	17.6		0.794
		7.8	10.6	13.2	15.2	16.8	17.9	18.5		0.820
			11.5	14.2	16.5	18.2	19.5	20.1		0.854
				16.1	18.6	20.6	22.0	22.7		0.892
					22.2	24.6	26.2	27.0		0.927
16	5.8	9.4	12.7	15.4	17.6	19.2	19.9			0.797
		10.0	13.4	16.4	18.7	20.4	21.1			0.829
			14.9	18.2	20.8	22.6	23.5			0.868
				21.4	24.5	26.5	27.6			0.910
					31.1	33.8	35.1			0.946
12	10.3	16.5	21.4	25.0	26.8					0.809
		18.3	24.0	27.8	29.9					0.854
			29.3	34.1	36.6					0.910
				48.2	51.8					0.969
9	18.5	28.3	34.7	18.5						0.821
		34.7	42.6	22.7						0.893
			65.3	34.7						0.961
8	27.6	33.2	39.2							0.815
		45.8	54.2							0.813
6	42.3	57.7								0.856
		100.0								0.965
4	100.0									0.923

ing magnetic field is important in the understanding not only of single-phase induction motors but of all rotating ac electromechanical devices.

In all multiphase symmetrically balanced motor windings, the existence of a rotating magnetic field can be mathematically demonstrated and verified rather easily. The case of the single-phase motor is a bit more difficult to show, but it can be derived mathematically.

Let us explore this rotating revolving field concept by beginning with a two-phase symmetrically balanced induction motor operating on a two-phase power system. This is a motor having at least two sinusoidally (or as near as possible) distributed stator windings, each of the windings being displaced 90° electrical apart in the stator slots. This means a four-pole motor would have four stator windings located 90° electrical apart, which is equivalent to 45° mechanical or spatial.

6.2.1 Two-Phase Operation

One might wonder, "why start with a two-phase motor to explain the single-phase motor?" However, a single-phase motor is designed to be "fooled" into acting like an unbalanced two-phase motor on starting, even though it is operating on a single-phase source when running and starting windings exist on the stator.

ALTERNATING-CURRENT INDUCTION MOTORS

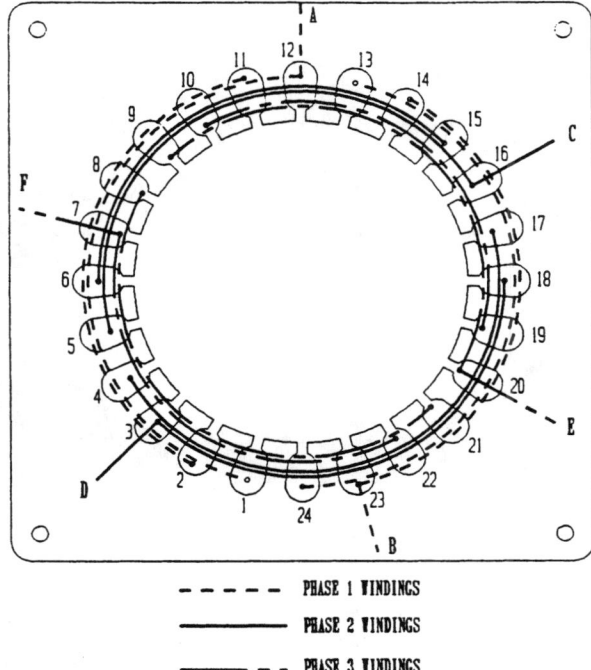

– – – – – PHASE 1 WINDINGS
———— PHASE 2 WINDINGS
——— – – PHASE 3 WINDINGS

FIGURE 6.6 Concentric-wound polyphase configuration.

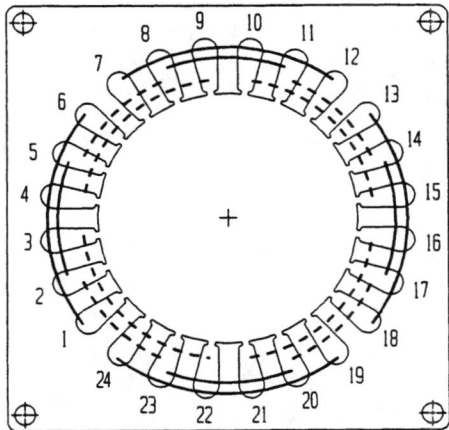

FIGURE 6.7 Concentric winding for a single-phase four-pole motor with two coils per pole. Solid lines indicate main winding coil sets; dashed lines indicate auxiliary winding coil sets.

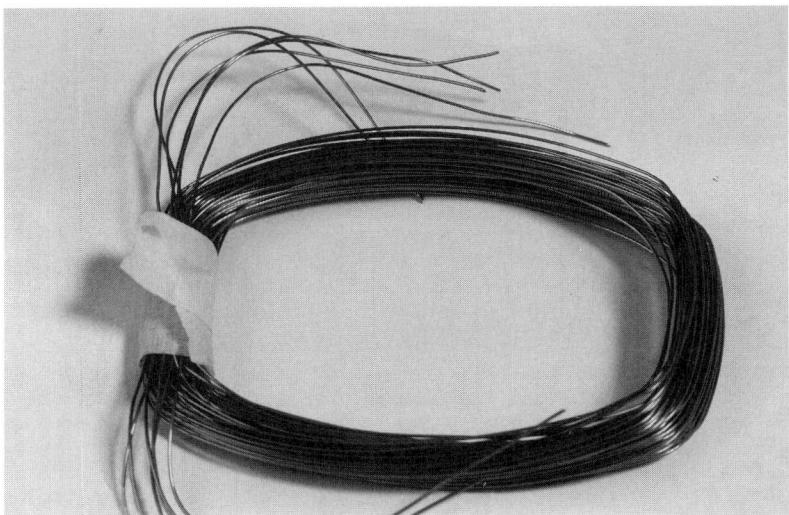

FIGURE 6.8 Typical wound coil ready for insertion.

Consider Figure 6.11, showing a conceptual diagram of a balanced two-pole two-phase motor having sinusoidally distributed windings *a* and *b* located 90° electrical apart on the stator. The dashed lines indicate a sinusoidal winding distribution.

Directions of the magnetic axes of both the *a* and *b* windings for the arbitrarily assumed positive direction of current i_a and i_b are shown. We recognize that the individual mmf waves will always lie on the two stationary axes but certainly will change

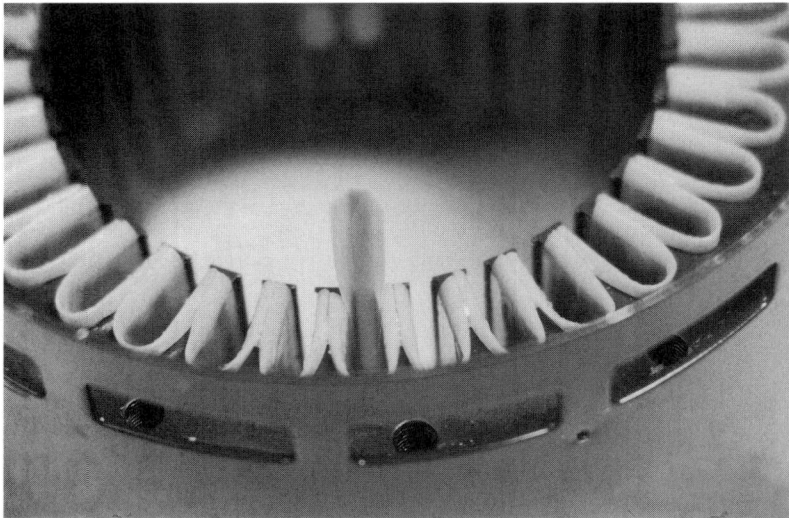

FIGURE 6.9 Stator core assembly ready for coil insertion.

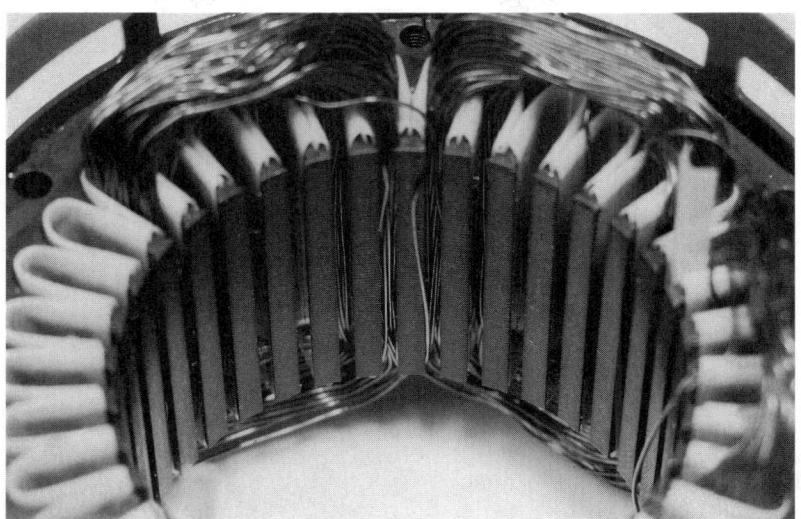

FIGURE 6.10 Partially inserted stator/coil assembly.

in magnitude and direction when sinusoidal currents flow in the windings. Let us call the special position around the inside of the stator ϕ_s, measured from the assumed positive direction of the a winding axis in a counterclockwise (CCW) direction. Assume each winding has an effective number of sinusoidally distributed turns N_s. Then the mmfs in the air gaps (two per winding axis) of each winding can be expressed with the assumptions of an infinite-permeance magnetic circuit and the mmf distributed evenly across the two air gaps in a sinusoidal fashion, as follows:

$$\text{mmf}_a = \frac{N_S}{2} i_a \cos \phi_S \tag{6.1}$$

and

$$\text{mmf}_b = \frac{N_S}{2} i_b \sin \phi_S \tag{6.2}$$

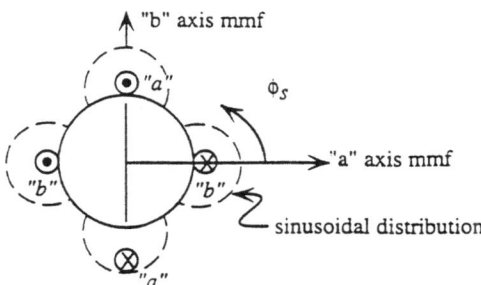

FIGURE 6.11 Conceptual diagram of a balanced two-pole two-phase motor.

where i_a and i_b are two currents supplied from a balanced two-phase source. Hence, i_a and i_b can be expressed as

$$i_a = \sqrt{2}I \cos \omega_e t \tag{6.3}$$

$$i_b = \sqrt{2}I \sin \omega_e t \tag{6.4}$$

where I is the RMS value of the currents and $\omega_e = 2\pi f_e$ is the frequency associated with the source voltage.

By using the trigonometric identity

$$\cos x \cos y = \frac{1}{2} \cos(x+y) + \frac{1}{2} \cos(x-y)$$

$$\text{mmf}_a = \frac{N_S}{2} \sqrt{2}I \left[\frac{1}{2} \cos(\omega_e t + \phi_S) + \frac{1}{2} \cos(\omega_e t - \phi_S) \right] \tag{6.5}$$

Here $\omega_e t$ and ϕ_S are respectively functions of time and displacement about the air gap on the stator. Let us jump the gun here and determine what must prevail to make mmf_a be a constant magnitude.

Certainly, if $(\omega_e t + \phi_S)$ and $(\omega_e t - \phi_S)$ are constants, we will have fulfilled our objective. So then let us proceed:

$$(\omega_e t + \phi_S) = A \quad \text{and} \quad (\omega_e t - \phi_S) = B$$

where both A and B are constants.

If we take the derivatives with respect to time of both these equations, we have

$$\omega_e + \frac{d\phi_S}{dt} = 0 \tag{6.6}$$

and

$$\omega_e - \frac{d\phi_S}{dt} = 0 \tag{6.7}$$

What can this mean? Equation (6.7) indicates that if we traverse around the stator CCW at a constant velocity $\omega_e = \omega_s$ and Eq. (6.6) indicates that if we traverse around the stator CW at velocity $\omega_e = -\omega_s$, the arguments in Eq. (6.5) will be constant.

Really, what we have shown is that winding a alone has produced two revolving mmfs of half amplitude, with one revolving CCW and the other CW. Think for a moment in terms of a single running winding of a single-phase motor—are these not identical? We will pursue this later when we discuss single-phase operation.

Now let us turn our attention to the b winding. By a similar procedure with useful trigonometric identities, we can expand as follows:

$$\text{mmf}_b = \frac{N_S}{2} \sqrt{2}I \left[\frac{1}{2} \cos(\omega_e t - \phi_S) - \frac{1}{2} \cos(\omega_e t + \phi_S) \right] \tag{6.8}$$

By the same analysis followed for mmf_a, we see that again we have two mmf waves rotating CCW and CW of half amplitude.

What happens when we find the total air gap mmf mmf_t with both windings energized simultaneously?

$$\text{mmf}_t = \text{mmf}_a + \text{mmf}_b$$

$$= \frac{N_S}{2}\sqrt{2}I\left[\frac{1}{2}\cos(\omega_e t + \phi_S) + \frac{1}{2}\cos(\omega_e t - \phi_S)\right.$$

$$\left. + \frac{1}{2}\cos(\omega_e t - \phi_S) - \frac{1}{2}\cos(\omega_e t + \phi_S)\right] \tag{6.9}$$

$$= \frac{N_S}{2}\sqrt{2}I\left[\cos(\omega_e t - \phi_S)\right] \tag{6.10}$$

which indicates that if we traverse around the stator CCW at a velocity $\omega_e = \omega_S$, mmf_t is a constant. This result is a single rotating mmf_t of constant amplitude produced from mmf_a and mmf_b. Remember, this is the result for the assumed positive direction assumed at the start. If, for example, we had assumed

$$i_b = -\sqrt{2}I\sin\omega_e t$$

it is clear from Eq. (6.10) that the result would have been

$$\text{mmf}_t = \frac{N_S}{2}\sqrt{2}I\left[\cos(\omega_e t + \phi_S)\right] \tag{6.11}$$

which is a rotating mmf_t having the same constant magnitude but rotating in a CW direction. Think again for a moment how you reverse direction in a single-phase motor having a running and starting winding. Do you not reverse the current in one of the windings (i.e., either the start or the run winding but not both)?

Thus, the results we have obtained so far are as follows:

- With a single balanced symmetrical stator winding we obtained two rotating, half-amplitude mmf waves, one going clockwise and one going counterclockwise. (Later we will discuss the double-revolving field analysis of the single-phase motor and will use this concept.)
- With two symmetrically balanced windings, 90° electrical apart and operated from a two-phase balance source, a constant-amplitude rotating magnetic field is obtained. The direction of rotation is a function of the current direction in the two windings.

Let us now relate what happens to the rotating mmf in considering two cases for the two-phase motor with the single-phase motor in mind.

1. Let us assume both balanced symmetrical windings are excited from the same voltage source. This is equivalent, of course, to single-phase excitation. Return to Eqs. (6.1), (6.2), and (6.9), with the excitation source equal to

$$i_a = i_b = \sqrt{2}I\cos\omega_e t \tag{6.12}$$

Then mmf_a and mmf_b become

$$\text{mmf}_a = \frac{N_S}{2}i_a\cos\phi_S$$

and
$$\text{mmf}_b = \frac{N_S}{2} i_a \sin \phi_S$$

$$\text{mmf}_a = \frac{N_S}{2} \sqrt{2} I \cos \omega_e t \cos \phi_S$$

and
$$\text{mmf}_b = \frac{N_S}{2} \sqrt{2} I \cos \omega_e t \sin \phi_S$$

$$\text{mmf}_a = \frac{N_S}{2} \sqrt{2} I \left[\frac{1}{2} \cos(\omega_e t + \phi_S) + \frac{1}{2} \cos(\omega_e t - \phi_S) \right] \quad (6.13)$$

$$\text{mmf}_b = \frac{N_S}{2} \sqrt{2} I \left[\frac{1}{2} \sin(\omega_e t + \phi_S) - \frac{1}{2} \sin(\omega_e t - \phi_S) \right] \quad (6.14)$$

As before, the total air gap is $\text{mmf}_t = \text{mmf}_a + \text{mmf}_b$.

$$\text{mmf}_t = \frac{N_S}{2} \sqrt{2} I \left[\frac{1}{2} \cos(\omega_e t + \phi_S) + \frac{1}{2} \cos(\omega_e t - \phi_S) \right.$$

$$\left. + \frac{1}{2} \sin(\omega_e t - \phi_S) \frac{1}{2} \sin(\omega_e t + \phi_S) \right]$$

$$\text{mmf}_t = \frac{N_S}{2} \sqrt{2} I \left[\frac{1}{2} \cos(\omega_e t + \phi_S) + \frac{1}{2} \sin(\omega_e t + \phi_S) \right.$$

$$\left. + \frac{1}{2} \cos(\omega_e t - \phi_S) - \frac{1}{2} \sin(\omega_e t - \phi_S) \right]$$

$$\text{mmf}_t = \frac{N_S}{2} \sqrt{2} I \left[\frac{1}{\sqrt{2}} \cos(\omega_e t + \phi_S - 45°) + \frac{1}{\sqrt{2}} \cos(\omega_e t - \phi_S + 45°) \right] \quad (6.15)$$

Hence, we again have two equal oppositely rotating mmfs. The first term of mmf_t rotates CCW. In terms of a single-phase motor, what does this mean? There can be no starting torque with equal excitation current, which implies that the two windings (start and run windings) are identical in electrical characteristics.

2. What if the two windings have different electrical characteristics and a different number of turns? Consider the following:

Let N_a, i_a, and $N_b i_b$ be the turns and currents in the two windings with $N_a \neq N_b$ and $i_a \neq i_b$ where

$$i_a = \sqrt{2} I_a \cos \omega_e t \quad \text{and} \quad i_b = \sqrt{2} I_b \cos \omega_e t$$

The total mmf_t starting from Eq. (6.9) is then:

$$\text{mmf}_t = \frac{N_a}{2} \sqrt{2} \left\{ \frac{I_a}{2} \left[\cos(\omega_e t + \phi_S) + \cos(\omega_e t - \phi_S) \right] \right\}$$

$$+ \frac{N_b}{2} \sqrt{2} \left\{ \frac{I_b}{2} \left[\cos(\omega_e t - \phi_S) - \cos(\omega_e t + \phi_S) \right] \right\}$$

By combining terms:

$$\text{mmf}_t = \frac{\sqrt{2}}{2}\left[\frac{N_a I_a - N_b I_b}{2}\cos(\omega_e t + \phi_S) + \frac{N_a I_a + N_b I_b}{2}\cos(\omega_e t + \phi_S)\right] \quad (6.16)$$

What does this mean? We have two unequal mmf magnitudes, the first being smaller and rotating CW and the second being larger and rotating CCW. The net mmf hence rotates CCW at ω_e. Does this not suggest a method by which we can start the single-phase motor? All we need is two windings having different electrical characteristics to get the motor moving, since one mmf is larger than the other. In the single-phase motor, one winding (start) is opened. The other winding (run) produces the double revolving field by Eq. (6.5), and we will show that the torque produced by the CW and CCW mmfs will result in a net torque which will accelerate the motor in the direction that the start winding gives in initial rotation.

We have now shown that a single-phase motor can produce a revolving field under starting and running conditions. Let us now proceed to develop the dynamic equations for the solution of the transient response. From these equations we will determine the steady-state circuit which can be used for performance calculations.

6.2.2 Dynamic Equations and Steady-State Equivalent Circuit

Consider an orthogonal two-winding unbalanced winding stator. The windings, in terms of a single-phase motor, would be the starting and running windings. These windings are unsymmetrical but have sinusoidally distributed stator coils. Figure 6.12 is representative of this motor. The two rotor windings are the same (symmetrical).

Applying Kirchhoff's voltage law to the four windings in Fig. 6.12 yields the following:

$$v_{as} = r_{as}i_{as} + \frac{d\lambda_{as}}{dt} \quad (6.17)$$

$$v_{bs} = r_{bs}i_{bs} + \frac{d\lambda_{bs}}{dt} \quad (6.18)$$

$$v_{ar} = r_{ar}i_{ar} + \frac{d\lambda_{ar}}{dt} \quad (6.19)$$

$$v_{br} = r_{br}i_{br} + \frac{d\lambda_{br}}{dt} \quad (6.20)$$

where $r_{ar} = r_{br}$
λ = total flux linkage between stator and rotor windings

Now, if we are considering a typical two-winding single-phase motor, the rotor will be a round squirrel cage, giving a uniform air gap. If we assume that the magnetic circuit of the motor is magnetically linear, which approaches linearity if the magnetic-circuit permeance is large, then all the distribution of ampere-turns will appear across the air gaps (two air gaps per winding).

The flux linkage equations required to solve the previous voltage equations are the following:

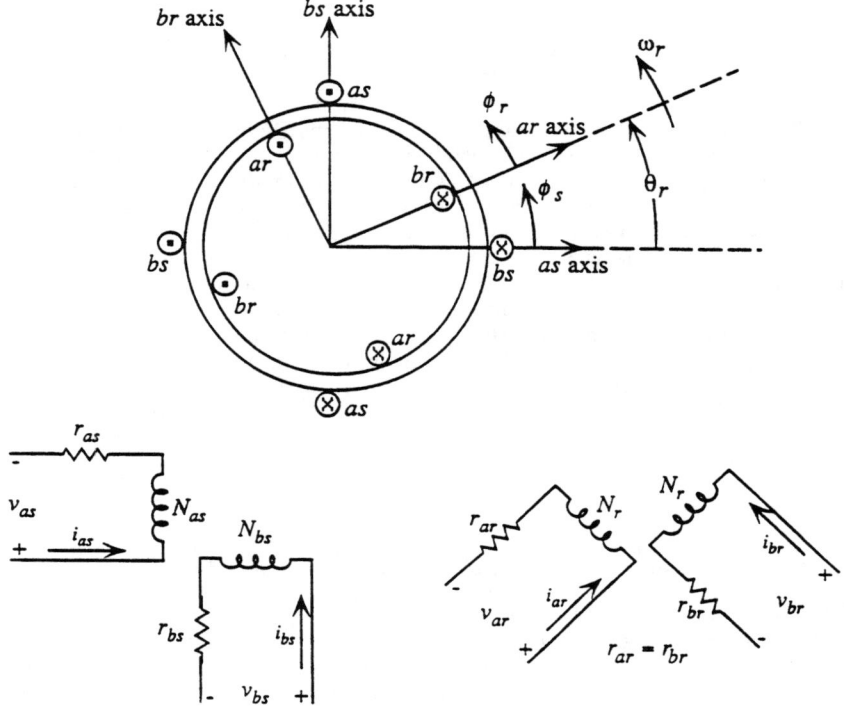

FIGURE 6.12 Orthogonal unbalanced two-winding stator.

$$\lambda_{as} = L_{asas}i_{as} + M_{asbs}i_{bs} + M_{asar}i_{ar} + M_{asbr}i_{br} \tag{6.21}$$

$$\lambda_{bs} = M_{bsas}i_{as} + L_{bsbs}i_{bs} + M_{bsar}i_{ar} + M_{bsbr}i_{br} \tag{6.22}$$

$$\lambda_{ar} = M_{aras}i_{as} + M_{arbs}i_{bs} + L_{arar}i_{ar} + M_{arbr}i_{br} \tag{6.23}$$

$$\lambda_{br} = M_{bras}i_{as} + M_{brbs}i_{bs} + M_{brar}i_{ar} + L_{brbr}i_{br} \tag{6.24}$$

The notation used here means, for example, M_{bsar} is the mutual inductance relative to the b winding of the stator when the a winding of the rotor is energized. In these equations, L is the self-inductance and M is the mutual inductance between the respective windings. Because of the orthogonality of the windings and also the squirrel-cage rotor, there are mutual inductance terms in the flux linkage equations in both stator and rotor which are zero.

$$M_{asbs} = M_{bsas} = 0 \tag{6.25}$$

and

$$M_{arbr} = M_{brar} = 0 \tag{6.26}$$

Therefore, there are only three terms on the right side of Eqs. (6.21) to (6.24) for each of the four equations. Now it should be apparent that the mutual inductances between the rotor and stator windings are functions of rotor position θ_r. With sinusoidally distributed windings, we are able to express these inductances for the assumed stator and rotor winding currents as follows:

$$M_{asar} = M_{aras} = \frac{N_{as}N_r}{R_m} \cos\theta_r = M_{asr}\cos\theta_r \tag{6.27}$$

$$M_{asbr} = M_{bras} = -\frac{N_{as}N_r}{R_m} \sin\theta_r = -M_{asr}\sin\theta_r \tag{6.28}$$

$$M_{bsar} = M_{arbs} = \frac{N_{bs}N_r}{R_m} \sin\theta_r = M_{bsr}\sin\theta_r \tag{6.29}$$

$$M_{bsbr} = M_{brbs} = \frac{N_{bs}N_r}{R_m} \cos\theta_r = M_{bsr}\cos\theta_r \tag{6.30}$$

R_m is the reluctance of the air gaps for a sinusoidal distribution of flux in the air gap. N_{as} and N_{bs} are the effective turns of the stator windings, and N_r is the effective turns of the two rotor windings. The self-inductances of the stator and rotor can be expressed with leakage inductances included as follows:

$$\frac{N_{as}^2}{R_m} \quad \frac{N_{bs}^2}{R_m} \quad \frac{N_r^2}{R_{lr}}$$

These are respectively the magnetizing inductances L_{mas}, L_{mbs}, and L_{mr}. Let L_{las}, L_{lbs}, and L_{lr} be the leakage inductances expressed in terms of their respective leakage reluctances:

$$L_{las} = \frac{N_{as}^2}{R_{las}} \tag{6.31}$$

$$L_{lbs} = \frac{N_{bs}^2}{R_{lbs}} \tag{6.32}$$

$$L_{lr} = \frac{N_r^2}{R_{lr}} \tag{6.33}$$

Then:

$$L_{asas} = L_{las} + L_{mas} \tag{6.34}$$

$$L_{bsbs} = L_{lbs} + L_{mbs} \tag{6.35}$$

$$L_r = L_{lr} + L_{mr} \tag{6.36}$$

In condensed matrix form, these become:

$$\begin{bmatrix} \overline{\lambda}_{abs} \\ \overline{\lambda}_{abr} \end{bmatrix} = \begin{bmatrix} \overline{L}_s & \overline{M}_{sr} \\ \overline{M}_{sr}^t & \overline{L}_r \end{bmatrix} \begin{bmatrix} \overline{i}_{abs} \\ \overline{i}_{abr} \end{bmatrix} \tag{6.37}$$

$$\overline{L}_s = \begin{bmatrix} L_{las} + L_{mas} & 0 \\ 0 & L_{lbs} + L_{mbs} \end{bmatrix} \tag{6.38}$$

$$\overline{L}_r = \begin{bmatrix} L_{lr} + L_{mr} & 0 \\ 0 & L_{lr} + L_{mr} \end{bmatrix} \tag{6.39}$$

where

$$\overline{M}_{sr} = \begin{bmatrix} M_{asr}\cos\theta_r & -M_{asr}\sin\theta_r \\ M_{bsr}\sin\theta_r & M_{bsr}\cos\theta_r \end{bmatrix} \quad (6.40)$$

$\overline{M}_{sr}^t$ is the transpose of $\overline{M}_{sr} = \begin{bmatrix} M_{asr}\cos\theta_r & M_{bsr}\sin\theta_r \\ -M_{asr}\sin\theta_r & M_{bsr}\cos\theta_r \end{bmatrix} \quad (6.41)$

Keep in mind that our goal is to develop the equivalent steady-state circuit for the single-phase motor; however, in this process we will also have available the differential equations to obtain the transient response. Our first step will be to remove the dependency of the mutual inductances [Eq. (6.10)] on rotor position. We will make an orthogonal q–d transformation to perform this step. The new set of variables will transform both stator and rotor machine parameters to a stationary axis q–d with the q axis aligned with the mmf$_a$ axis of the stator. This is the simplest of other available transformations which could also be used. Consider Fig. 6.13 for the stator transformation. The stator transformation is given for the stator voltages in terms of the new set of axis q–d. This transformation is indicated for voltage, but it could also represent currents and flux linkages in the two stator windings. This transformation for the voltages is

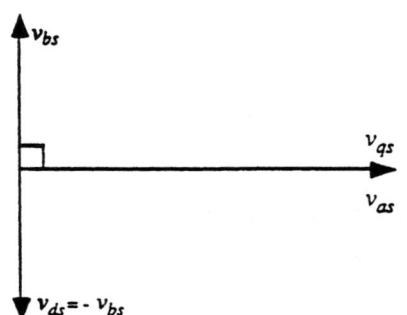

FIGURE 6.13 Stator transformation to q–d axis. The subscripts q and d represent the new voltage variables.

$$v_{qs} = v_{as} \quad (6.42)$$

$$v_{ds} = -v_{bs} \quad (6.43)$$

$$\begin{bmatrix} v_{qs} \\ v_{ds} \end{bmatrix} = \begin{bmatrix} 1 & 0 \\ 0 & -1 \end{bmatrix} \begin{bmatrix} v_{as} \\ v_{bs} \end{bmatrix} \quad (6.44)$$

The subscript s here represents stator.

To simplify the matrix notation, this can be written as follows:

$$\overline{v}_{qds} = \overline{K}_s \overline{v}_{abs}$$

where

$$\overline{K}_s = \begin{bmatrix} 1 & 0 \\ 0 & -1 \end{bmatrix} \quad (6.45)$$

Note that in Fig. 6.13 the stationary axis q is in the same direction as the assumed positive direction of mmf$_a$, whereas the assumed positive direction of the d axis is the negative of the positive direction of the mmf$_b$ axis. Note that the q–d axis is orthogonal and also note that

$$\overline{K}_s = (\overline{K}_s)^{-1} = (\overline{K}_s)^t \quad (6.46)$$

where t indicates the transpose of $\overline{K}_s$, and $(\overline{K}_s)^{-1}$ indicates the inverse of $\overline{K}_s$.

Consider now the rotor transformation as indicated in Fig. 6.14. It may be helpful to also refer to Fig. 6.11. By taking components of the rotor voltages on the q–d axis in terms of θ_r, we have:

$$v_{qr} = v_{ar} \cos \theta_r - v_{br} \sin \theta_r$$

$$v_{dr} = -v_{ar} \sin \theta_r - v_{br} \cos \theta_r$$

Or, in concise matrix form as for the stator:

$$\overline{v}_{qdr} = \begin{bmatrix} \cos \theta_r & -\sin \theta_r \\ -\sin \theta_r & -\cos \theta_r \end{bmatrix} \overline{v}_{abr} = \overline{K}_r \overline{v}_{abr} \qquad (6.47)$$

The subscript r here indicates a rotor variable.
Again:

$$\overline{K}_r = (\overline{K}_r)^{-1} = (\overline{K}_r)^t$$

And we have represented voltages, but this could represent currents or flux linkages.

6.2.3 Stator Transformation

Let us now transform the stator voltage equations, Eqs. (6.17) and (6.18), and the flux linkage equations, Eqs. (6.21) and (6.22), using the stator transform, Eq. (6.45). Again writing in concise matrix form, these become:

$$\overline{v}_{abs} = \overline{r}_s \overline{i}_{abs} + p\overline{\lambda}_{abs} \qquad (6.48)$$

where p = derivative operation d/dt
and

$$\overline{r}_s = \begin{bmatrix} r_{as} & 0 \\ 0 & r_{bs} \end{bmatrix}$$

and

$$\overline{\lambda}_{abs} = \overline{L}_s \overline{i}_{abs} + \overline{M}_{sr} \overline{i}_{abr} \qquad (6.49)$$

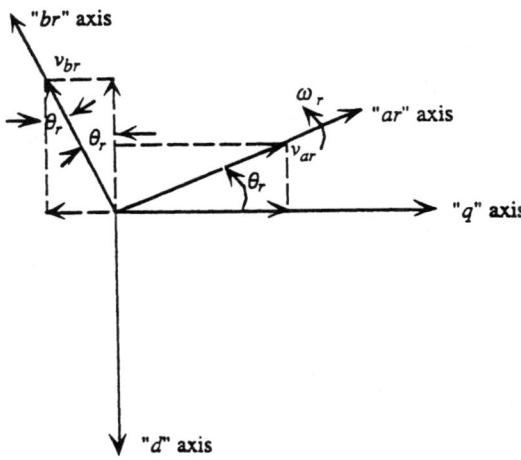

FIGURE 6.14 Rotor transformation to q–d axis.

where $\overline{L}_s$ and $\overline{M}_{sr}$ are given in Eqs. (6.38) and (6.40). Substituting Eq. (6.45) into Eq. (6.48), we have the following:

$$(\overline{K}_s^{-1})\,\overline{v}_{qds} = \overline{r}_s(\overline{K}_s)^{-1}\,\overline{i}_{qds} + p(\overline{K}_s)^{-1}\,\overline{\lambda}_{qds}$$

$$\overline{v}_{qds} = (\overline{K}_s)\,\overline{r}_s(\overline{K}_s)^{-1}\,\overline{i}_{qds} + (\overline{K}_s)\,p(\overline{K}_s)^{-1}\,\overline{\lambda}_{qds} \qquad (6.50)$$

$$\overline{v}_{qds} = \overline{r}_s\overline{i}_{qds} + P\overline{\lambda}_{qds}$$

Advantage has been taken here because $\overline{r}_s$ is a diagonal matrix and $\overline{K}_s$ is a constant matrix. We now have the stator voltage equations [Eqs. (6.17) and (6.18)] in terms of the q–d variables. The flux linkage equations [Eqs. (6.21) and (6.22)] for the stator become:

$$(\overline{K}_s)^{-1}\overline{\lambda}_{qds} = \overline{L}_s(\overline{K}_s)^{-1}\overline{i}_{qds} + \overline{M}_{sr}(\overline{K}_s)^{-1}\overline{i}_{qdr}$$

$$\overline{\lambda}_{qds} = \overline{L}_s\overline{i}_{qds} + \overline{M}_{absr}\overline{i}_{qdr} \qquad (6.51)$$

where

$$\overline{M}_{absr} = \begin{bmatrix} M_{asr} & 0 \\ 0 & M_{bsr} \end{bmatrix} = \overline{M}^t_{absr} \qquad (6.52)$$

Note that Eq. (6.49) has now become Eq. (6.54) in the q–d form and independent of θ_r. M_{asr} and M_{bsr} are defined in Eqs. (6.27) to (6.30) and are constants.

6.2.4 Rotor Transformation

Again writing voltage equations [Eqs. (6.19) and (6.20)] and flux linkage equations [Eqs. (6.23) and (6.24)] for the rotor terms, we have the following:

$$\overline{v}_{abr} = \overline{r}_r\overline{i}_{abr} + p\overline{\lambda}_{abr} \qquad (6.53)$$

$$\overline{\lambda}_{abr} = (\overline{M}_{sr})^t\overline{i}_{abr} + \overline{L}_r\overline{i}_{abr} \qquad (6.54)$$

where

$$\overline{r}_r = \begin{bmatrix} r_{ar} & 0 \\ 0 & r_{br} \end{bmatrix}$$

$\overline{L}_r$ and $\overline{M}^t_{sr}$ are given in Eqs. (6.39) and (6.41). Substituting the transformation equation Eq. (6.47) for the rotor into Eq. (6.53) we have the following:

$$(\overline{K}_r)^{-1}\overline{v}_{qdr} = \overline{r}_r(\overline{K}_r)^{-1}\overline{i}_{qdr} + p(\overline{K}_r)^{-1}\overline{\lambda}_{qdr}$$

$$\overline{v}_{qdr} = \overline{r}_r\overline{i}_{qdr} + (\overline{K}_r)p(\overline{K}_r)^{-1}\overline{\lambda}_{qdr}$$

$$\overline{v}_{qdr} = \overline{r}_r\overline{i}_{qdr} + (\overline{K}_r)[p(\overline{K}_r)^{-1}]\overline{\lambda}_{qdr} + (\overline{K}_r)[(\overline{K}_r)^{-1}p\overline{\lambda}_{qdr}]$$

$$\overline{v}_{qdr} = \overline{r}_r\overline{i}_{qdr} + \omega_r\begin{bmatrix} 0 & -1 \\ 1 & 0 \end{bmatrix}\overline{\lambda}_{qdr} + p\overline{\lambda}_{qdr} \qquad (6.55)$$

Also the flux linkage equation Eq. (6.54) becomes:

$$(\overline{K}_r)^{-1}\overline{\lambda}_{qdr} = (\overline{M}_{sr})^t(\overline{K}_s)^{-1}\overline{i}_{qdr} + \overline{L}_r(\overline{K}_r)^{-1}\overline{i}_{qdr}$$

$$\overline{\lambda}_{qdr} = (\overline{K}_r)(\overline{M}_{sr})^t(\overline{K}_s)^{-1}\overline{i}_{qdr} + (\overline{K}_r)\overline{L}_r(\overline{K}_r)^{-1}\overline{i}_{qdr}$$

$$\overline{\lambda}_{qdr} = \overline{M}^t_{absr}\overline{i}_{qds} + \overline{L}_r\overline{i}_{qdr} \qquad (6.56)$$

where $\overline{M}_{absr}^t$ is defined in Eq. (6.52). Note again that we have removed all dependency on rotor position θ_r. We have now converted the machine motor parameters in terms of the q–d reference parameters. They are related through Eqs. (6.45) and (6.47).

Equations (6.50), (6.51), (6.55), and (6.56) represent eight simultaneous differential equations which we could solve for the transient response in the q–d reference frame. The transient response equations in terms of machine parameter was mentioned previously, before the q–d transformation. Of course, both give the same result when the transformation equations Eqs. (6.45) and (6.47) are used.

We now proceed to the equivalent circuit commonly used for the single-phase motor, where the rotor quantities are usually referred to the stator by the turns ratios of the stator and rotor windings. Let us represent the primed values in terms of the equivalent $(q$–$d)_r$ rotor variables referred to the $(q$–$d)_s$ frame. Consider the normal voltage, current, and flux linkages in terms of turns ratio of both stator and rotor windings.

$$i'_{qr} = \frac{N_r}{N_{as}} i_{qr} \tag{6.57}$$

$$i'_{dr} = \frac{N_r}{N_{bs}} i_{dr} \tag{6.58}$$

$$v'_{qr} = \frac{N_{as}}{N_r} v_{qr} \tag{6.59}$$

$$v'_{dr} = \frac{N_{bs}}{N_r} v_{dr} \tag{6.60}$$

$$\lambda'_{qr} = \frac{N_{as}}{N_r} \lambda_{qr} \tag{6.61}$$

$$\lambda'_{dr} = \frac{N_{bs}}{N_r} \lambda_{dr} \tag{6.62}$$

We now will introduce Eqs. (6.57) to (6.62) into the voltage and flux linkage equations Eqs. (6.50), (6.51), (6.55), and (6.56). It is probably easier to write out the eight equations.

The voltage equations:

$$v_{qs} = r_{as}i_{qs} + p\lambda_{qs}$$

$$v_{ds} = r_{bs}i_{ds} + p\lambda_{ds}$$

$$\frac{N_r}{N_{as}} v'_{qr} = r_{ar} \frac{N_{as}}{N_r} i'_{qr} - \omega_r \left(\frac{N_r}{N_{bs}}\right) \lambda'_{dr} + p\left(\frac{N_r}{N_{as}}\right) \lambda'_{qr}$$

$$v'_{qr} = \left(\frac{N_{as}}{N_r}\right)^2 r_{ar}i'_{qr} - \omega_r \left(\frac{N_{as}}{N_{bs}}\right) \lambda'_{dr} + p\lambda'_{qr}$$

$$\left(\frac{N_r}{N_{bs}}\right) v'_{dr} = r_{br} \left(\frac{N_{bs}}{N_r}\right) i'_{dr} + \omega_r \left(\frac{N_r}{N_{as}}\right) \lambda'_{qr} + p\left(\frac{N_r}{N_{bs}}\right) \lambda'_{dr}$$

$$v'_{dr} = \left(\frac{N_{bs}}{N_r}\right)^2 r_{br}i'_{dr} + \omega_r \left(\frac{N_{bs}}{N_{as}}\right) \lambda'_{qr} + p\lambda'_{dr}$$

The flux linkage equations:

$$\lambda'_{qs} = (L_{las} + L_{mas})i_{qs} + M_{asr}\left(\frac{N_{as}}{N_r}\right)i'_{qr}$$

$$\lambda_{qs} = L_{las}i_{qs} + L_{mas}(i_{qs} + i'_{qr})$$

where

$$L_{mas} = M_{asr}\left(\frac{N_{as}}{N_r}\right) \quad \text{[see Eqs. (6.31) to (6.36)]}$$

$$\lambda_{ds} = (L_{lbs} + L_{mbs})i_{ds} + M_{bsr}\frac{N_{bs}}{N_r}i'_{dr}$$

$$= L_{lbs}i_{ds} + L_{mbs}(i_{ds} + i'_{dr})$$

where

$$M_{bsr}\frac{N_{bs}}{N_r} = L_{mbs} \quad \text{[see Eqs. (6.31) to (6.36)]}$$

$$\frac{N_r}{N_{as}}\lambda'_{qr} = M_{asr}i_{qs} + (L_{lr} + L_{mr})\left(\frac{N_{as}}{N_r}\right)i'_{qr}$$

$$\lambda'_{qr} = \frac{N_{as}}{N_r}M_{asr}i_{qs} + (L_{lr} + L_{mr})\left(\frac{N_{as}}{N_r}\right)^2 i'_{qr}$$

$$\lambda'_{qr} = L_{mas}i_{qs} + (L_{lr} + L_{mr})\left(\frac{N_{as}}{N_r}\right)^2 i'_{qr}$$

$$\lambda' = L_{lr}\left(\frac{N_{as}}{N_r}\right)^2 i'_{qr} + L_{mas}(i_{qs} + i'_{qr})$$

where

$$(L_{mr})\left(\frac{N_{as}}{N_r}\right)^2 = L_{mas} \quad \text{[see Eqs. (6.31) to (6.36)]}$$

Finally:

$$\frac{N_r}{N_{bs}}\lambda'_{dr} = M_{bsr}i_{ds} + (L_{lr} + L_{mr})\left(\frac{N_{bs}}{N_r}\right)i'_{dr}$$

$$\lambda'_{dr} = M_{bsr}\left(\frac{N_{bs}}{N_r}\right)i_{ds} + (L_{lr} + L_{mr})\left(\frac{N_{bs}}{N_r}\right)^2 i'_{dr}$$

$$= L_{lr}\left(\frac{N_{bs}}{N_r}\right)^2 i'_{dr} + L_{mbs}(i_{ds} + i'_{dr})$$

where

$$M_{bsr}\left(\frac{N_{bs}}{N_r}\right) = L_{mbs} \qquad L_{mr}\left(\frac{N_{bs}}{N_r}\right)^2 = L_{mbs}$$

Rewriting the eight equations:

$$v_{qs} = r_{as}i_{qs} = p\lambda_{qs} \tag{6.63}$$

$$v_{ds} = r_{bs}i_{ds} + p\lambda_{ds} \tag{6.64}$$

$$v'_{qr} = \left(\frac{N_{as}}{N_r}\right)^2 r_{ar}i'_{qr} - \omega_r \left(\frac{N_{as}}{N_{bs}}\right)\lambda'_{dr} + p\lambda'_{qr} \tag{6.65}$$

$$v'_{dr} = \left(\frac{N_{bs}}{N_r}\right)^2 r_{br}i'_{dr} + \omega_r \left(\frac{N_{bs}}{N_{as}}\right)\lambda'_{qr} + p\lambda'_{dr} \tag{6.66}$$

$$\lambda_{qs} = L_{las}i_{qs} + L_{mas}(i_{qs} + i'_{qr}) \tag{6.67}$$

$$\lambda_{ds} = L_{lbs}i_{ds} + L_{mbs}(i_{ds} + i'_{dr}) \tag{6.68}$$

$$\lambda'_{qr} = L_{lr}\left(\frac{N_{as}}{N_r}\right)^2 i'_{qr} + L_{mas}(i_{qs} + i'_{qr}) \tag{6.69}$$

$$\lambda'_{dr} = L_{lr}\left(\frac{N_{bs}}{N_r}\right)^2 i'_{dr} + L_{mbs}(i_{ds} + i'_{dr}) \tag{6.70}$$

Note: All of the rotor variables have been referred to the stator (i.e., all rotor voltages, currents, and flux linkages have primed values). Also note that the coefficients of all variables are constants in the steady state.

Two decoupled equivalent circuits could be made here for the eight equations. However, since we are pursuing a single-phase steady-state circuit, we will not do this. To obtain the sinusoidal steady-state equivalent we only need to set

$$p = \frac{d}{dt} = j\omega_e$$

where ω_e is the radian frequency of the applied voltage. Under steady state we can write the voltages and currents in steady-state notation with a superscript tilde. Also, if we are using a squirrel-cage rotor, the voltages $\tilde{V}'_{qr}$ and $\tilde{V}'_{dr}$ are zero. Equations (6.63) to (6.66) now become

$$\tilde{V}_{qs} = r_{as}\tilde{I}_{qs} + j\omega_e[L_{las}\tilde{I}_{qs} + L_{mas}(\tilde{I}_{qs} + \tilde{I}'_{qr})] \tag{6.71}$$

$$\tilde{V}_{ds} = r_{bs}\tilde{I}_{ds} + j\omega_e[L_{lbs}\tilde{I}_{ds} + L_{mbs}(\tilde{I}_{ds} + \tilde{I}'_{dr})] \tag{6.72}$$

$$0 = r_{ar}\left(\frac{N_{as}}{N_r}\right)^2 \tilde{I}'_{qr} - \omega_r\left(\frac{N_{as}}{N_{bs}}\right)\left[L_{lr}\left(\frac{N_{bs}}{N_r}\right)^2 \tilde{I}'_{dr} + L_{mbs}(\tilde{I}_{ds} + \tilde{I}'_{dr})\right]$$

$$+ j\omega_e\left[L_{lr}\left(\frac{N_{as}}{N_r}\right)^2 \tilde{I}'_{qr} + L_{mas}(\tilde{I}_{qs} + \tilde{I}'_{qr})\right] \tag{6.73}$$

$$0 = r_{br}\left(\frac{N_{bs}}{N_r}\right)^2 \tilde{I}'_{dr} + \omega_r\left(\frac{N_{bs}}{N_{as}}\right)\left[L_{lr}\left(\frac{N_{as}}{N_r}\right)^2 \tilde{I}'_{qr} + L_{mas}(\tilde{I}_{qs} + \tilde{I}'_{qr})\right]$$

$$+ j\omega_e\left[L_{lr}\left(\frac{N_{bs}}{N_r}\right)^2 \tilde{I}'_{dr} + L_{mbs}(\tilde{I}_{ds} + \tilde{I}'_{dr})\right] \tag{6.74}$$

Solutions of Eqs. (6.73) and (6.74) could be considered for the starting performance of a single-phase motor. For example, we could let the a winding and the b winding be the starting winding. In normal single-phase operation, the b winding would be open at approximately 70 percent of synchronous speed. From the solution of these equations, one could obtain the transient torque, speed, current, etc. on starting. If we choose to look at the steady-state performance, only the a winding will be energized. Let us pursue this route under these conditions:

1. $\tilde{V}_{qr} = \tilde{V}_{dr} = 0$ (squirrel-cage rotor not energized)
2. $\tilde{I}_{ds} = 0$ (winding b open)

Equations (6.71) to (6.74) can be put in matrix form and simplified by the following substitutions.
Let:

$$j\omega_e(L_{las} + L_{mas}) = j(X_{las} + X_{mas}) = jX_{as} = j\omega_e L_{as} \tag{6.75}$$

$$j\omega_e(L_{lbs} + L_{mbs}) = j(X_{lbs} + X_{mbs}) = jX_{bs} = j\omega_e L_{bs} \tag{6.76}$$

$$r_{ar}\left(\frac{N_{as}}{N_r}\right)^2 = r'_{ar} \tag{6.77}$$

$$r_{br}\left(\frac{N_{bs}}{N_r}\right)^2 = r'_{br} \tag{6.78}$$

$$L_{lr}\left(\frac{N_{as}}{N_r}\right)^2 = L'_{lar} \tag{6.79}$$

$$L_{lr}\left(\frac{N_{bs}}{N_r}\right)^2 = L'_{lbr} \tag{6.80}$$

$$j\omega_e(L'_{lar} + L_{mas}) = j(X'_{lar} + X_{mas}) = jX'_{ar} = j\omega_e L'_{ar} \tag{6.81}$$

$$j\omega_e(L'_{lbr} + L_{mbs}) = j(X'_{lbr} + X_{mbs}) = jX'_{br} = j\omega_e L'_{br} \tag{6.82}$$

$$L'_{lar} + L_{mas} = L'_{ar} \tag{6.83}$$

$$L'_{lbr} + L_{mbs} = L'_{br} \tag{6.84}$$

$$\frac{N_{as}}{N_{bs}} = a \tag{6.85}$$

$$\begin{bmatrix} \tilde{V}_{qs} \\ \tilde{V}_{ds} \\ 0 \\ 0 \end{bmatrix} = \begin{bmatrix} r_{as} + j\omega_e L_{as} & 0 & j\omega_e L_{mas} & 0 \\ 0 & r_{bs} + j\omega_e L_{bs} & 0 & j\omega_e L_{mbs} \\ j\omega_e L_{mas} & -\omega_r a L_{mbs} & r'_{ar} + j\omega_e L'_{ar} & -\omega_r a L'_{br} \\ \omega_r/a\, L_{mas} & j\omega_e L_{mbs} & \omega_r/a\, L'_{ar} & r'_{br} + j\omega_e L'_{br} \end{bmatrix} \begin{bmatrix} \tilde{I}_{qs} \\ 0 \\ \tilde{I}_{qr} \\ \tilde{I}_{dr} \end{bmatrix} \tag{6.86}$$

Equation (6.86) can now be solved for $\tilde{V}_{qs}$ in terms of $\tilde{I}_{qs}$. Note that here we have three equations in three unknown currents which can be solved for $\tilde{V}_{qs}$ in terms of $\tilde{I}_{qs}$. Also keep in mind that here $\tilde{V}_{qs} = \tilde{V}_{as}$ and $\tilde{I}_{qs} = \tilde{I}_{as}$, where these voltages are the winding a voltage and current. The solution for $\tilde{V}_{qs} = \tilde{V}_{as}$ is as follows:

$$\tilde{V}_{as} = \left(r_{as} + j\omega_e L_{as} - j\omega_e L_{mas}\left\{\frac{L_{mas}[(\omega_r^2 - \omega_e^2)L'_{ar} + j\omega_e r'_{ar}]}{Z_r^2 + \omega_r^2 L'^2_{ar}}\right\}\right) \tilde{I}_{as} \tag{6.87}$$

where
$$Z'_r = r'_{ar} + j\omega_e(L'_{lar} + L_{mas})$$
$$r'_{ar}/r'_{br} = (N_{as}/N_{bs})^2$$
$$L'_{ar}/L'_{br} = (N_{as}/N_{bs})^2$$
$$L_{mas}/L_{mbs} = (N_{as}/N_{bs})^2$$

We now have to either show that Eq. (6.87) leads to the well-known equivalent circuit shown in Fig. 6.15 or assume the circuit is correct and work from the equivalent circuit and arrive at Eq. (6.87). It can be done either way and is long either way, but this author has found it more direct to use the latter approach.

Therefore, the input impedance of Fig. 6.15 can be written as follows:

$$Z = r_{as} + j\omega_e L_{las} + \frac{Z_1 Z_m}{Z_1 + Z_m} + \frac{Z_2 Z_m}{Z_2 + Z_m}$$

where

$$Z_1 = \frac{r'_{ar}}{} + j\frac{\omega_e L'_{lar}}{}$$

$$Z_2 = \frac{r'_{ar}}{} + j\frac{\omega_e L'_{lar}}{}$$

$$Z_m = j\frac{\omega_e L_{mas}}{2}$$

$$s = \frac{\omega_e - \omega_r}{\omega_e}$$

$$2 - s = \frac{\omega_e + \omega_r}{\omega_e}$$

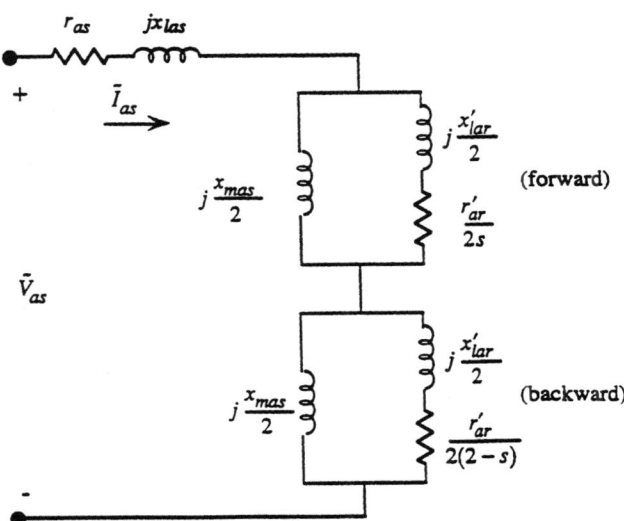

FIGURE 6.15 Single-phase motor equivalent circuit.

Then:
$$Z = r_{as} + j\omega_e L_{as} + \frac{Z_m(Z_1 Z_m + Z_2 Z_m + 2Z_1 Z_2)}{Z_1 Z_2 + Z_1 Z_m + Z_2 Z_m + Z_m^2}$$

So:
$$j\omega_e L_{as} = j\omega_e(L_{las} + L_{mas})$$
$$Z = r_{as} + j\omega_e L_{las} + \frac{Z_m(Z_1 Z_m + Z_2 Z_m + 2Z_1 Z_2)}{Z_1 Z_2 + Z_1 Z_m + Z_2 Z_m + Z_m^2} - j\omega_e L_{mas}$$

First, look at the denominator by substituting for Z_1, Z_2, and Z_m and simplifying:

$$Z_1 Z_2 + Z_1 Z_m + Z_2 Z_m + Z_m^2 = \frac{1}{4} \frac{r_{ar}'^2}{s(2-s)} + j\frac{1}{2}\omega_e L_{ar}' \frac{r_{ar}'}{} - \frac{1}{4}\omega_e^2 L_{ar}'^2$$

$$= (r_{ar}'^2 + j2\omega_e L_{ar}' r_{ar}' - \omega_e^2 L_{ar}'^2) + \omega_r^2 L_{ar}'^2$$

$$= Z_r^2 + \omega_r^2 L_{ar}'^2 \qquad (6.88)$$

In this process we have multiplied by $4(2-s)(s)$ to simplify.
Note here that

$$4(2-s)(s) = 4\left(1 - \frac{\omega_r^2}{\omega_e^2}\right) = 4\left(\frac{\omega_e^2 - \omega_r^2}{\omega_e^2}\right)$$

Comparing Eq. (6.88) with Eq. (6.87), we see that the denominators are the same. Now, look at the numerator and remember that in simplifying we multiplied the denominator by $4(2-s)(s)$. We now have:

$$Z = r_{as} + j\omega_e L_{as} + \frac{4(2-s)(s)[Z_m(Z_1 Z_m + Z_2 Z_m + 2Z_1 Z_2)]}{Z_r^2 + \omega_r^2 L_{ar}'^2} - j\omega_e L_{mas}$$

$$Z = r_{as} + j\omega_e L_{as} + \frac{4(2-s)(s)[Z_m(Z_1 Z_m + Z_2 Z_m + 2Z_1 Z_2)] - j\omega_e L_{mas}(Z_r^2 + \omega_r^2 L_{ar}'^2)}{Z_r^2 + \omega_r^2 L_{ar}'^2}$$

$$Z = r_{as} + j\omega_e L_{as} + 4(2-s)(s)\left\{\frac{j\omega_e L_{mas}}{2}[2(Z_1 Z_2 + Z_1 Z_m + Z_2 Z_m + Z_m^2) - Z_m(Z_1 + Z_2) - 2Z_m^2]\right\}$$

$$+ \frac{-j\omega_e L_{mas}(Z_r^2 + \omega_r^2 L_{ar}'^2)}{Z_r^2 + \omega_r^2 L_{ar}'^2}$$

$$Z = r_{as} + j\omega_e L_{as}$$
$$+ \frac{j\omega_e L_{mas}(Z_r^2 + \omega_r^2 L_{ar}'^2) - 4(2-s)(s)[(j\omega_e L_{mas}/2)(Z_m)(Z_1 + Z_2 + 2Z_m)] - j\omega_e L_{mas}(Z_r^2 + \omega_r^2 L_{ar}'^2)}{Z_r^2 + \omega_r^2 L_{ar}'^2}$$

$$Z = r_{as} + j\omega_e L_{as} + \frac{-4(2-2)(s)[(j\omega_e L_{mas}/2)(Z_m)(Z_1 + Z_2 + 2Z_m)]}{Z_r^2 + \omega_r^2 L_{ar}'^2}$$

$$Z = r_{as} + j\omega_e L_{as} + \frac{-4(2-s)(s)[(j\omega_e L_{mas}/2)\,(j\tfrac{1}{2}\omega_e L_{mas}\{[r'_{ar}/(2-s)(s)] + j\omega_e L'_{ar}\})]}{}$$

$$= r_{as} + j\omega_e L_{as} - j\omega_e L_{mas}\left\{\frac{L_{mas}[(\omega_r^2 - \omega_e^2)L'_{ar} + j\omega_e r'_{ar}]}{Z_r'^2 + \omega_r^2 L_{ar}'^2}\right\} \qquad (6.89)$$

This is the impedance function of Eq. (6.87). Therefore, we have shown that Fig. 6.15 is the equivalent steady-state circuit representing the single-phase motor operating on the running winding.

Useful Trigonometric Identities

$$\cos x \cos y = \frac{1}{2}\cos(x+y) + \frac{1}{2}\cos(x-y)$$

$$\sin x \sin y = \frac{1}{2}\cos(x-y) - \frac{1}{2}\cos(x+y)$$

$$\sin x \cos y = \frac{1}{2}\sin(x+y) + \frac{1}{2}\sin(x-y)$$

$$\cos(x \pm y) = \cos x \cos y \mp \sin x \sin y$$

$$\sin(x \pm y) = \sin x \cos y \pm \cos x \sin y$$

It may be good to again recap what we have done so far. We have completed the following points:

1. Addressed the necessary conditions for the existence of a single constant-rotation mmf produced by balanced two-phase orthogonal stator windings.
2. Shown the production of a double revolving field produced by a single stator winding.
3. Implied that if we have two unbalanced orthogonal stator windings, each will on its own produce a double revolving field.
4. Shown through a q–d transformation of the voltage and flux equations that the criteria for determining both transient and steady-state response of the unbalanced two-winding motor can be obtained. Also, that a transient and steady-state circuit model could be developed (i.e., this would correspond to a single-phase motor operating on both main and auxiliary windings).
5. Developed from scratch the steady-state equivalent circuit of the single-phase motor operating on the running winding only, using the q–d transform approach. The purpose in using the complex approach is to show that the popular steady-state model can be rigorously developed mathematically.

To continue, we now would like to make use of the steady-state model developed in point 5—that is, given the motor parameters, calculate the motor performance. After performing this task, we would like to expand the model in point 5 to include the performance of both main and auxiliary windings (i.e., combined winding performance).

6.2.5 Single-Winding Motor Calculations from the Equivalent Circuit

We have previously arrived at the steady-state equivalent circuit of the single-phase motor using the double-revolving-field approach. The circuit is repeated in

Fig. 6.16, with a change to more common notation—the subscript m indicates the main winding.

$$s = \frac{n_s - n_r}{n_s} = \frac{\omega_s - \omega_r}{\omega_s}$$

where ω_s = synchronous angular velocity
ω_r = angular velocity of the rotor

$$2 - s = \frac{n_s + n_r}{n_s} = \frac{\omega_s + \omega_r}{\omega_s}$$

$$r_2' = \left(\frac{N_m}{N_r}\right)^2 r_2 \qquad x_2' = \left(\frac{N_m}{N_r}\right)^2 x_2$$

N_m and N_r are the effective turns of the stator and rotor, respectively.

The analysis is sometimes simplified by redefining the apparent resistances and reactances of the forward and backward fields to simplify the calculations. The redrawn circuit then appears as shown in Fig. 6.17, where

$$Z_f = R_f + jx_f = \frac{1}{2} \frac{jx_m[(r_2'/s) + jx_2']}{(r_2'/s) + j(x_2' + x_m)}$$

and

$$Z_b = R_b + jx_b = \frac{1}{2} \frac{jx_m\{[r_2'/(2-s)] + jx_2'\}}{[r_2'/(2-s)] + j(x_2' + x_m)}$$

Then the input impedance is

$$Z_{in} = Z_m + Z_f + Z_b$$

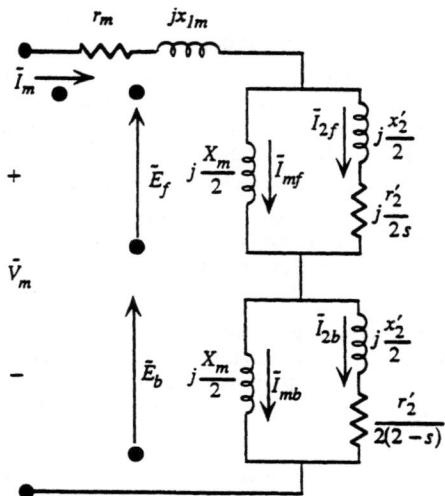

FIGURE 6.16 Single-phase motor equivalent circuit with common notation.

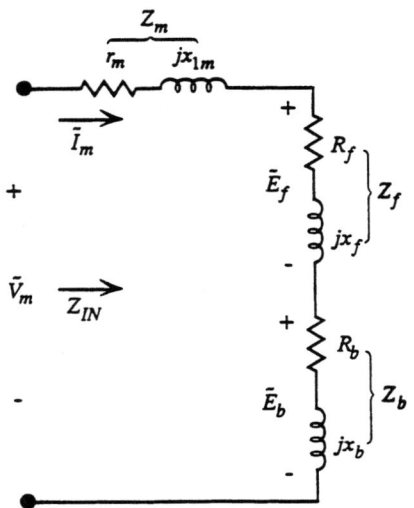

FIGURE 6.17 Simplified single-phase motor equivalent circuit.

Input current is then:

$$\tilde{I}_m = \frac{\tilde{V}_m}{Z_m + Z_f + Z_b}$$

The power input to the motor is

$$P_{in} = \mathcal{R}_e(\tilde{V}\tilde{I}_m^*) = |\tilde{V}\tilde{I}_m|\cos\theta$$

where θ = power factor angle
$\mathcal{R}_e$ = real component of complex power

The stator copper loss is

$$P_{s,Cu} = |\tilde{I}_m|^2 r_1$$

The air gap power input to both forward and backward fields is

$$P_{ag} = P_{in} - P_{s,Cu}$$

The air gap power inputs for both fields are

$$P_{agf} = I_{2f}^2\left(\frac{r_2'}{2s}\right) = \left[\frac{jx_m}{(r_2'/s) + j(x_2' + x_m)}\tilde{I}_m\right]^2\left(\frac{r_2'}{2s}\right) = |\tilde{I}_m|^2 R_f \quad (6.90)$$

$$P_{agb} = I_{2b}^2\left(\frac{r_2'}{2(2-s)}\right) = \left\{\frac{jx_m}{[r_2'/2(2-s)] + j(x_2' + x_m)}\tilde{I}_m\right\}^2\left(\frac{r_2'}{2(2-s)}\right) = |\tilde{I}_m|^2 R_b$$

(6.91)

The torques developed by both forward and backward fields can now be calculated:

$$T_f = \frac{P_{agf}}{\omega_s} \quad \text{N·m}$$

$$T_b = \frac{P_{agb}}{\omega_s} \quad \text{N·m}$$

Net forward torque = $T_f - T_b$.

The rotor copper loss caused by both the forward and backward rotating fields is equal to the slip times the power transferred from stator to the rotor air gap power. Therefore:

Forward field rotor copper loss = sP_{agf}

Backward field rotor copper loss = $(2-s)P_{agb}$

Total rotor copper loss = $sP_{agf} + (2-s)P_{agb}$

So, developed power output P_d is as follows:

$$P_d = (P_{agf} + P_{agb}) - [sP_{agf} + (2-s)P_{agb}]$$
$$= (P_{agf})(1-s) - P_{agb}(1-s) \qquad (6.92)$$
$$= (P_{agf} - P_{agb})(1-s)$$

The developed torque of the motor is then:

$$T_{dev} = \frac{(P_{agf} - P_{agb})(1-s)}{\omega_s(1-s)} = \frac{P_{agf} - P_{agb}}{\omega_s}$$

The shaft power output P_{out} is P_d minus rotational loss P_r and core loss P_c:

$$P_{out} = P_d - (P_r + P_c)$$

The output torque T_{out} then becomes

$$T_{out} = \frac{P_{out}}{(1-s)(\omega_s)}$$

And the percent efficiency η is

$$\eta \frac{P_{out}}{P_{in}} \times 100$$

6.2.6 Steady-State Combined-Winding Performance

When designing a single-phase motor, thought must be given to the auxiliary winding in starting, and in the case of capacitor or two-value-capacitor motors, two-winding performance must be considered. The following points come to mind:

1. Does the motor have sufficient starting torque? Does it have sufficient torque when switching out the starting winding to bring the motor up to speed?
2. If the motor includes a capacitor, what is the voltage across the capacitor?
3. What is the performance when both auxiliary and running windings are energized at all times, as with a permanent-split-capacitor (PSC) or two-value-capacitor motor?
4. What is the starting current?

The following assumptions are made here:

- The auxiliary and main windings are electrically orthogonal to each other.
- Each winding produces a double revolving field. Therefore, we have two forward and two backward fields produced by the auxiliary and running windings.
- Following the previous discussion of operations of a single winding, then both the running and auxiliary winding have a forward and backward equivalent circuit, with one additional change. Each of the four revolving fields induces a speed voltage in the other winding. That is, the forward-revolving fields and the backward-revolving fields induce voltages in the forward and backward circuits of the other, and vice versa. This amounts to four dependent voltage sources. These speed-voltage terms can be seen in the q–d transformation matrix equation, Eq. (6.86), where ω_r appears.

- The two windings are arranged such that the forward field of the auxiliary winding induces a speed voltage in the main winding that lags the forward field by 90°. This implies that the main winding is displaced 90° ahead of the auxiliary winding in space, causing a forward-revolving field.

Before proceeding with the double-revolving-field theory, it is worth mentioning that there are other approaches to this problem—namely, the cross-field theory; the use of symmetrical components; and the q–d transformation approach, which is mentioned but never pursued in the preceding sections.

The voltage induced by forward and backward fields can then be represented as follows. The forward and backward voltages induced in the main field winding due to the forward and backward fields of the main winding are

$$\tilde{E}_{fm} = (R_f + jx_f)\tilde{I}_m = Z_f \tilde{I}_m \tag{6.93}$$

$$\tilde{E}_{bm} = (R_b + jx_b)\tilde{I}_m = Z_b \tilde{I}_m \tag{6.94}$$

where $\tilde{I}_m$ is the steady-state main-winding current.

The forward and backward voltages induced in the auxiliary field winding due to forward and backward fields of the auxiliary winding are

$$\tilde{E}_{fa} = a^2 Z_f \tilde{I}_a$$

$$\tilde{E}_{ba} = a^2 Z_b \tilde{I}_a$$

where a is the turns ratio of auxiliary winding turns to main winding turns

$$a = \frac{N_a}{N_m}$$

Note that these four voltages are current dependent and appear in the forward and backward field circuits and are indicated in Fig. 6.18.

We next concern ourselves with the speed voltages induced into the main and auxilary windings by the auxilary and main double revolving fields, respectively. It has previously been indicated that the voltage induced in the main winding by the forward revolving field of the axiliary field lags 90° behind the voltage $\tilde{E}_{fa}$. On the other hand, the voltage induced into the main winding by the backward revolving field of the auxiliary winding leads the voltage $\tilde{E}_{fb}$ by 90°. Remember also that the magnitudes of voltages $|\tilde{E}_{fm}|$ and $|\tilde{E}_{bm}|$ are related to $|\tilde{E}_{fa}|$ and $|\tilde{E}_{ba}|$ by $1/a$. Then:

$$\tilde{E}_1 = \frac{-j\tilde{E}_{fa}}{a} = -jaZ_f \tilde{I}_a \tag{6.95}$$

$$\tilde{E}_2 = \frac{\tilde{E}_{ba}}{a} = jaZ_b \tilde{I}_a \tag{6.96}$$

$$\tilde{E}_3 = ja\tilde{E}_{fm} = jaZ_f \tilde{I}_m \tag{6.97}$$

$$\tilde{E}_4 = -ja\tilde{E}_{bm} = -jaZ_b \tilde{I}_m \tag{6.98}$$

Figure 6.18 is redrawn as Fig. 6.19 to simplify the circuit.

Included in the circuit is one impedance Z_e which represents an external component which may be added to the auxiliary winding to enhance performance. For example, it could be (1) resistance split-phase operation, (2) capacitance which is

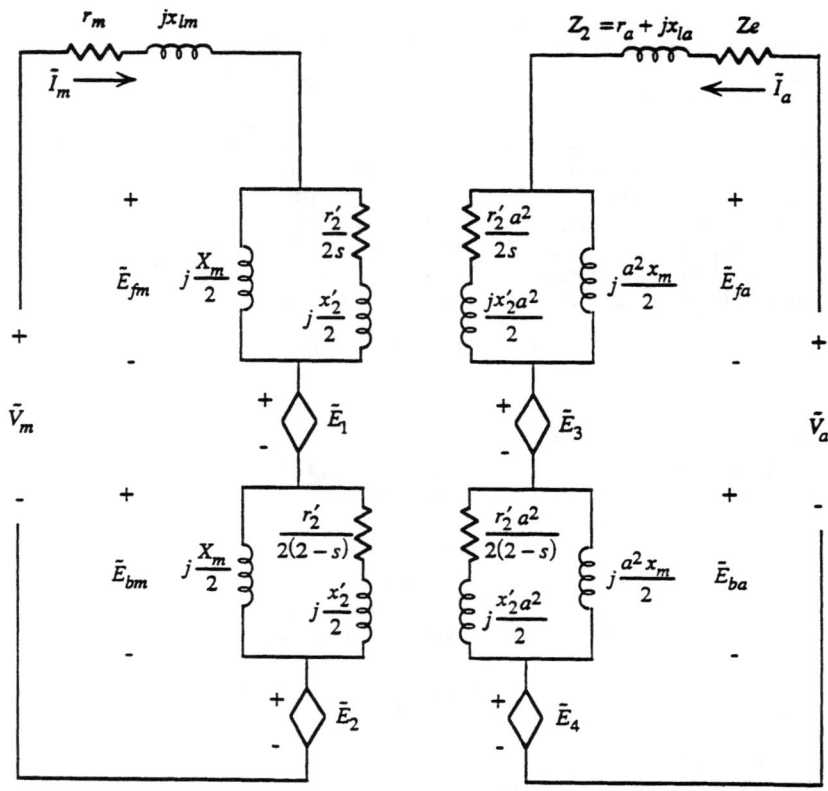

FIGURE 6.18 Single-phase motor equivalent circuit for combined winding performance.

removed after starting (capacitor start motor), (3) a change in capacitance (two-value-capacitor motor), or (4) a fixed capacitance (PSC motor).

r'_2 and x'_2 are now the resistance and reactance, respectively, of the rotor referred to the main winding; that is

$$r'_2 = \left(\frac{N_m}{N_r}\right)^2 r_2 \qquad x'_2 = \left(\frac{N_m}{N_r}\right)^2 x_2$$

and

$$a = \frac{N_a}{N_m}$$

N_a = effective turns of auxiliary winding
N_m = effective turns of main winding
N_r = effective turns of rotor winding

Writing the voltage equations around each winding gives the following:

ALTERNATING-CURRENT INDUCTION MOTORS

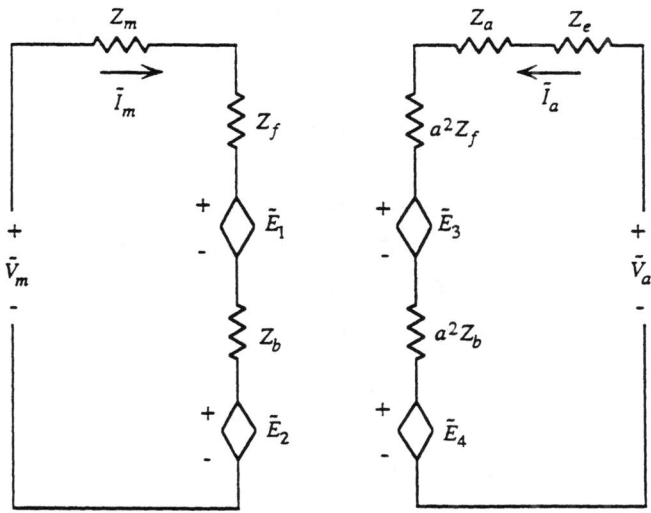

FIGURE 6.19 Simplified single-phase motor equivalent circuit.

$$\tilde{V}_m = Z_m\tilde{I}_m + Z_f\tilde{I}_m + Z_b\tilde{I}_m + \tilde{E}_1 + \tilde{E}_2$$

$$\tilde{V}_a = Z_a\tilde{I}_a + a^2Z_f\tilde{I}_a + a^2Z_b\tilde{I}_a + Z_a\tilde{I}_a + Z_e\tilde{I}_a + \tilde{E}_3 + \tilde{E}_4$$

$$\tilde{V}_m = (Z_m + Z_f + Z_b)\tilde{I}_m - jaZ_f\tilde{I}_a + jaZ_b\tilde{I}_a$$

$$\tilde{V}_a = (Z_a + a^2Z_f + a^2Z_b + Z_e)\tilde{I}_a + jaZ_f\tilde{I}_m - jaZ_f\tilde{I}_m$$

$$\tilde{V}_m = (Z_T)\tilde{I}_m - ja(Z_f - Z_b)\tilde{I}_a$$

$$\tilde{V}_a = ja(Z_f - Z_b)\tilde{I}_m + Z'_T\tilde{I}_a$$

where $Z_T = Z_m + Z_f + Z_b$
$Z'_T = (Z_a + Z_e) + a^2(Z_f + Z_b)$

In matrix form:

$$\begin{bmatrix}\tilde{V}_m \\ \tilde{V}_a\end{bmatrix} = \begin{bmatrix} Z_T & -ja(Z_f - Z_b) \\ ja(Z_f - Z_b) & Z'_T \end{bmatrix}\begin{bmatrix}\tilde{I}_m \\ \tilde{I}_a\end{bmatrix} \qquad (6.99)$$

$$\begin{bmatrix}\tilde{I}_m \\ \tilde{I}_a\end{bmatrix} = \frac{\begin{bmatrix} Z'_T & ja(Z_f - Z_b) \\ -ja(Z_f - Z_b) & Z_T \end{bmatrix}}{Z'_T Z_T + a^2(Z_f - Z_b)^2}\begin{bmatrix}\tilde{V}_m \\ \tilde{V}_a\end{bmatrix} \qquad (6.100)$$

$$\tilde{I}_m = \frac{Z'_T\tilde{V}_m + ja(Z_f - Z_b)\tilde{V}_a}{} \qquad (6.101)$$

$$\tilde{I}_a = \frac{-ja(Z_f - Z_b)\tilde{V}_m + Z_T\tilde{V}_a}{Z'_T Z_T + a^2(Z_f + Z_b)^2} \qquad (6.102)$$

In most cases here $\tilde{V}_m = \tilde{V}_a$ when supplied from the same voltage source, in which case the total line current would be $\tilde{I}_m + \tilde{I}_a = \tilde{I}$.

$$P_{in} = \mathcal{R}_e[\tilde{V}_m \tilde{I}^*] = |\tilde{V}_m \tilde{I}| \cos \theta$$

where θ is the power factor angle between $\tilde{V}$ and $\tilde{I}$.

The stator copper loss is

$$P_{Cu} = I_m^2 r_m + I_a^2 r_2$$

The total power to the revolving fields of the main and auxiliary windings is

$$\begin{aligned}P_{forward} &= \mathcal{R}_e[(\tilde{E}_{fm} + \tilde{E}_1)\tilde{I}_m^*] + \mathcal{R}_e[(\tilde{E}_{fa} + \tilde{E}_3)\tilde{I}_a^*] \\ &= \mathcal{R}_e[(Z_f \tilde{I}_m - jaZ_f \tilde{I}_a)\tilde{I}_m^*] + \mathcal{R}_e[(a^2 Z_f \tilde{I}_a + jaZ_f)\tilde{I}_a^*] \\ &= \mathcal{R}_e[(Z_f \tilde{I}_m^2 - jaZ_f \tilde{I}_m^*)\tilde{I}_m^*] + \mathcal{R}_e[(a^2 Z_f \tilde{I}_a^2 + jaZ_f \tilde{I}_m)\tilde{I}_a^*] \\ &= (I_m^2 + a^2 I_a^2)R_f + 2aI_m I_a R_f \sin\theta\end{aligned} \quad (6.103)$$

where θ is the angle between $\tilde{I}_a$ and $\tilde{I}_m$

$$\theta = \theta_a - \theta_m$$

By the same procedure, the backward power can be found.

$$P_{backward} = (I_m^2 + a^2 I_a^2)R_b - 2aI_m I_a R_b \sin\theta$$

Therefore, the net air gap power is

$$P_{net} = P_{forward} - P_{backward} = (I_m^2 + a^2 I_a^2)(R_f - R_b) + 2a(R_f + R_b)I_m I_a \sin\theta \quad (6.104)$$

The power developed P_d is

$$P_d = (1-s)P_{net}$$

The power output P_{out} is

$$P_{out} = P_d - \text{core loss} - \text{rotational losses}$$

The shaft torque is

$$T_s = \frac{P_{out}}{\omega_r}$$

where $\omega_r = (1-s)\omega_s$

6.2.7 Symmetrical Component Approach to Single-Phase Motors

For applications where unbalanced conditions occur, C. L. Fortesque devised a process whereby unbalanced systems can be reformed into a system of balanced networks. The number of balanced networks required is the same as the number of degrees of freedom which occur in the unbalanced system. For example, an unbalanced three-phase power system, which has three degrees of freedom, requires three balanced networks.

In the case of an unbalanced two-winding single-phase motor having a main and auxiliary winding, two networks are required. Fortesque named these two balanced networks the *positive* and *negative sequence networks* simply because they rotate in opposite directions. In terms of the motor in question, this is the same as the double-revolving-field theory, where we have two fields for each winding, one going CCW and the other CW. The CCW rotation is usually referred to as the *positive sequence*, whereas the CW rotation is referred to as the *negative sequence*.

The technique is powerful in that through the transformation we end up with balanced networks which we are able to handle. In the case of the two-degree network, the two components of each of the CCW (forward) and the CW (backward) are orthogonal to each other.

If we define the CCW rotation of the fields of both the main and auxiliary windings as progressing from the auxiliary to the main winding, we can then draw these components—for example, in terms of the currents in these windings, I_m and I_a. On the other hand, the CW rotation is a transition from the main winding to the auxiliary winding. Consider the motor with unbalanced stator windings shown in Fig. 6.20.

$$V_m = Z_{im}I_m + tE_m \tag{6.105}$$

$$V_a = Z_{ia}I_a + E_a \tag{6.106}$$

The sequence relationships are represented in Fig. 6.21.

Now, Fortesque defined the currents as follows:

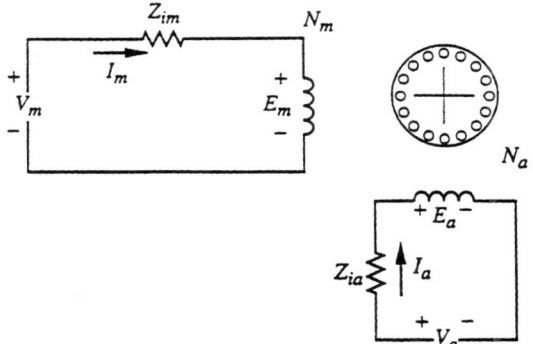

FIGURE 6.20 Motor circuit with unbalanced windings.

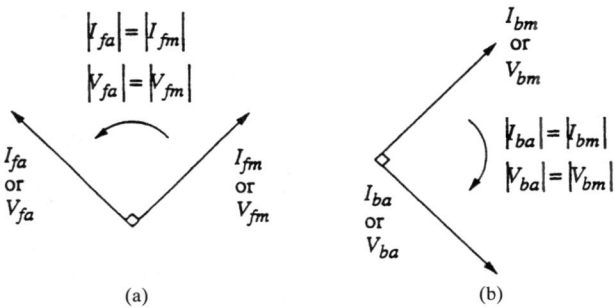

FIGURE 6.21 Sequence relationship of motor with unbalanced windings: (*a*) forward (+), and (*b*) backward (−).

$$I_m = I_{fm} + I_{bm} \qquad (6.107)$$

$$I_a = I_{fa} + I_{ba} \qquad (6.108)$$

By similar reasoning, the voltage equations are

$$V_m = V_{fm} + V_{bm} \qquad (6.109)$$

$$V_a = V_{fa} + V_{ba} \qquad (6.110)$$

At this point, the normal procedure is to pick either the main or auxiliary winding as the reference winding—i.e., to convert either the sequence components of the main in terms of the auxiliary or vice versa. The normal procedure is to use the latter approach, which can be seen in Fig. 6.21.

$$I_{fa} = \frac{jI_{fm}}{a} \qquad (6.111)$$

$$V_{fa} = jaV_{fm} \qquad (6.112)$$

$$I_{ba} = \frac{-jI_{bm}}{a} \qquad (6.113)$$

$$V_{ba} = -jaV_{bm} \qquad (6.114)$$

We have now resolved the auxiliary winding currents and voltages in terms of the orthogonal of forward and backward components. That is:

$$\begin{bmatrix} I_m \\ I_a \end{bmatrix} = \begin{bmatrix} 1 & 1 \\ j/a & -j/a \end{bmatrix} \begin{bmatrix} I_{fm} \\ I_{bm} \end{bmatrix} \qquad (6.115)$$

or

$$\begin{bmatrix} I_{fm} \\ I_{bm} \end{bmatrix} = \frac{1}{2} \begin{bmatrix} 1 & -ja \\ 1 & ja \end{bmatrix} \begin{bmatrix} I_m \\ I_a \end{bmatrix} \qquad (6.116)$$

Similarly, the voltage equations are

$$\begin{bmatrix} V_m \\ V_a \end{bmatrix} = \begin{bmatrix} 1 & 1 \\ ja & -ja \end{bmatrix} \begin{bmatrix} V_{fm} \\ V_{bm} \end{bmatrix} \qquad (6.117)$$

or

$$\begin{bmatrix} V_{fm} \\ V_{bm} \end{bmatrix} = \frac{1}{2} \begin{bmatrix} 1 & -j/a \\ 1 & j/a \end{bmatrix} \begin{bmatrix} V_m \\ V_a \end{bmatrix} \qquad (6.118)$$

The second matrix allows us to find the forward (+) and the backward (−) sequence components and voltages. The reference forward and backward loop circuit model then becomes as shown in Fig. 6.22.

Figure 6.22a, the forward sequence, is one of two forward motor models referred to the main winding. Figure 6.22b, the backward sequence, is one of two backward motor models referred to the main winding. This is because both auxiliary and main windings have a forward and backward field, for a total of four models. Note that Z_f and Z_b as defined here are twice the values defined in Fig. 6.17.

Now, the total counter-emf generated in the forward and backward fields in Fig. 6.22 is equal to the sum $E_{fm} + E_{bm}$, indicated as E_m and E_a in Fig. 6.20. That is:

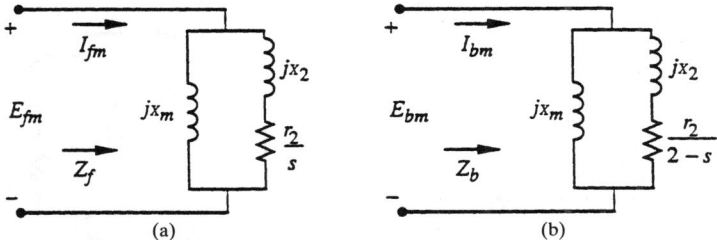

FIGURE 6.22 Field loop circuits: (*a*) forward, and (*b*) backward.

$$E_m = E_{fm} + E_{bm} = Z_f I_{fm} + Z_b I_{bm} \tag{6.119}$$

and

$$E_a = jaE_{fm} - jaE_{bm} = jaZ_f I_{fm} - jaZ_b I_{bm} \tag{6.120}$$

Substitution of Eqs. (6.107), (6.108), (6.119), and (6.120) into Eqs. (6.105) and (6.106) yields:

$$V_m = Z_{im}(I_{fm} = I_{bm}) + Z_f I_{fm} + Z_b I_{bm} \tag{6.121}$$

$$V_a = Z_{ia}\left(\frac{jI_{fm}}{a} - \frac{jI_{bm}}{a}\right) + jaZ_f I_{fm} - jaZ_b I_{bm} \tag{6.122}$$

After collecting terms and multiplying V_a by $-j/a$:

$$V_m = (Z_{im} + Z_f)I_{fm} + (Z_{im} + Z_b)I_{bm}$$

$$-\frac{j}{a}V_a = \left(\frac{Z_{ia}}{a^2} + Z_f\right)I_{fm} - \left(\frac{Z_{ia}}{a^2} + Z_b\right)I_{bm}$$

Now, adding and also subtracting Eqs. (6.121) and (6.122) and dividing by two yields:

$$\frac{1}{2}\left(V_m - \frac{jV_a}{a}\right) = \left[\frac{(Z_{ia}/a^2) + Z_{im}}{2} + Z_f\right]I_{fm} - \left(\frac{Z_{ia}}{a^2} - \frac{Z_{im}}{2}\right)I_{bm} \tag{6.123}$$

$$\frac{1}{2}\left(V_m - \frac{jV_a}{a}\right) = \left[\frac{(Z_{ia}/a^2) - Z_{im}}{2}\right]I_{fm} - \left[\frac{(Z_{ia}/a^2) + Z_{im}}{2} + Z_b\right]I_{bm} \tag{6.124}$$

We recognize by comparison that the left sides of Eqs. (6.123) and (6.124) are the symmetrical components V_{fm} and V_{bm} of the applied voltages referred to the main winding [see Eqs. (6.117) and (6.118)].

At this point an equivalent circuit (Fig. 6.23) of Eqs. (6.125) and (6.126) can be drawn by first simplifying. Let

$$Z_A = \frac{1}{2}\left(\frac{Z_{ia}}{a^2} + Z_{im}\right) \tag{6.125}$$

$$Z_B = \frac{1}{2}\left(\frac{Z_{ia}}{a^2} - Z_{im}\right) \tag{6.126}$$

$$Z_A - Z_B = Z_{im}$$

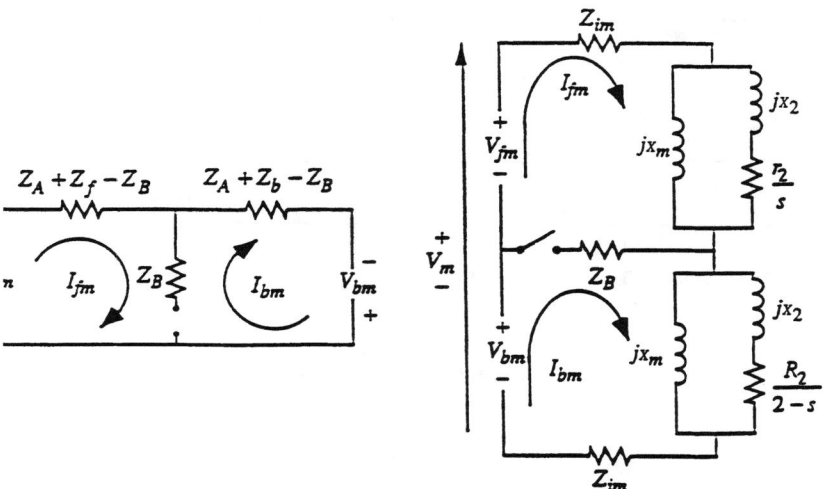

FIGURE 6.23 Combined equivalent circuit including forward and backward field components.

Now

$$V_{fm} = (Z_A + Z_f)I_{fm} - Z_B I_{bm} \tag{6.127}$$

$$V_{bm} = Z_B I_{fm} + (Z_A + Z_b)I_{bm} \tag{6.128}$$

Note that a switch has been added in Fig. 6.23, indicative of the starting switch in the auxiliary winding of the motor. Several special cases can be applied to Fig. 6.23; for example:

1. With the switch open, $I_{fm} = I_{bm}$; by Eqs. (6.115) and (6.116), $I_a = 0$; and from Eqs. (6.107) and (6.108), $I_m = 2I_{fm} = 2I_{bm}$. The circuit then reduces to $2Z_{im}$ in series with Z_f and Z_b. Figure 6.23 then reduces to the equivalent circuit of Fig. 6.16 by division of the circuit impedances by the 2 factor.
2. The impedance Z_B can include a capacitance, giving rise to a capacitor motor.
3. If $Z_B = 0$, the motor has balanced windings, since $Z_B = Z_{ia}/a^2 - Z_{im} = 0$, and the circuit reduces to balanced operation, since the forward and backward circuits are decoupled from each other.

The currents in the two networks shown in Fig. 6.23 can be found if the motor constants are known, along with the applied voltages. In addition, if slip is assumed, the losses and torques (starting and running) can be found. The currents are

$$I_{fm} = \frac{(Z_A + Z_B)V_{fm} + Z_B V_{bm}}{(Z_A + Z_f)(Z_A + Z_b) - Z_B^2} \tag{6.129}$$

$$I_{bm} = \frac{(Z_A + Z_f)V_{bm} + Z_B V_{fm}}{(Z_A + Z_f)(Z_A + Z_b) - Z_B^2} \tag{6.130}$$

The power determined by the forward field is

$$P_f = 2I_{fm}^2 R_f \qquad (6.131)$$

The power determined by the backward field is

$$P_b = 2I_{bm}^2 R_b \qquad (6.132)$$

The net power developed $= P_f - P_b$.

Note: In Eqs. (6.131) and (6.132), a 2 factor is required, because we have one forward and one backward field in each of the windings.

The starting torque can be found from I_{bm} and I_{fm} with $s = 1$ and $R_f = R_b$ as follows:

$$T_{\text{start}} = 2(I_{fm}^2 - I_{bm}^2)R \qquad (6.133)$$

where $R = R_f$ or R_b with $s = 1$.

6.2.8 Single-Phase Induction Motor Design Procedure

C. Veinott (1959) presents a method for predicting desired motor performance characteristics, given:

Motor voltage and frequency
Number of poles
Stator punching dimensions
Rotor punching dimensions
Rotor finished dimensions
Stack length
Stacking factor
BH curves for laminations
Main and auxiliary winding data
Rotor squirrel-cage data winding skew

The problem is to predict the performance of a motor built to a proposed design in order to make a decision as to whether the design must be modified to meet the performance specifications.

The approach is to calculate the parameters of the equivalent circuit model plus the losses and compute the performance based on the revolving field theory.

The procedure to accomplish this is as follows:

1. Find the flux per pole from voltage, frequency, and winding data.
2. Find the effective cross sections and lengths of the magnetic circuit (i.e., stator yoke, rotor yoke, stator teeth, rotor teeth, and air gap).
3. From the cross sections, find flux densities in each component.
4. From *BH* data, find the field intensities in each component; from magnetic pole lengths, compute the mmf required to drive the flux calculated in step 1 through each component; and hence calculate the total mmf required. Calculate the saturation factor as a ratio of total ampere-turns to the air gap ampere-turns.
5. Calculate the resistance of both main and auxiliary windings. The average length of one conductor is obtained by adding the stack length to the average coil throw (ACT) multiplied by an empirical correction factor.

6. Calculate r_2', the rotor resistance referred to the main winding. This calculation takes into account the additional length of the rotor bars due to skew and the nonuniform current distribution in the end rings.
7. Calculate the core losses using the flux densities calculated earlier and published loss characteristics for the core material.
8. Obtain friction and windage from a table corresponding to the synchronous speed and frame size.
9. Calculate the leakage reactance x_1 and x_2'. This reactance includes the sum of the slot, end leakage zigzag, belt, and skew reactance; after the total is found, x_1 is assumed to be equal to x_2'.

- Compute the slot reactance from the *slot constants* for both rotor and stator. These constants are a measure of the magnetic permeance between the slot sides per unit length, taking into account the increased mmf applied across the slot as more and more ampere conductors are included as one moves up the slot. Both stator and rotor reactances are included in one calculation.
- Calculate zigzag leakage for the tooth-face dimensions, using a zigzag constant.
- End leakage is a function of the average conductor throw (ACT). Compute this empirically.
- Belt leakage depends on the average number of slots per pole in both rotor and stator and the size of the air gap. Calculate this empirically.
- Calculate skew leakage from the skew factor of the rotor.
- Calculate the magnetizing reactance x_m from the equation derived from the ratio of the emf induced by the air gap flux to the current required to produce that field.

The constants for the equivalent circuit model have now been calculated.

6.3 THREE-PHASE INDUCTION MOTOR DYNAMIC EQUATIONS AND STEADY-STATE EQUIVALENT CIRCUIT

The three-phase symmetrical induction motor has three identical sinusoidally distributed windings displaced electrically 120° apart per set of poles. To simplify our approach as much as possible, consider a two-pole machine as depicted in Fig. 6.24 and compare this to Fig. 6.12 in the two-phase development. (Note that here we do not consider balanced windings on the stator, which were deliberately introduced for single-phase operation.)

Because of the addition of two additional axes plus the 120° phase displacement, the analysis becomes more complex due to the geometrical relationships; however, the approach parallels that for the two-phase development. Because of the additional winding, when we perform a transformation, as we did for the two-phase machine, it will require one more dimension in our transformation. However, because we are considering only balanced three-phase operation, it will not enter or complicate our analysis.

Let us now write the three-phase machine equations equivalent to Eqs. (6.17) to (6.24) for the two-phase machine.

$$v_{as} = r_s i_{as} + \frac{d\lambda_{as}}{dt} \qquad (6.134)$$

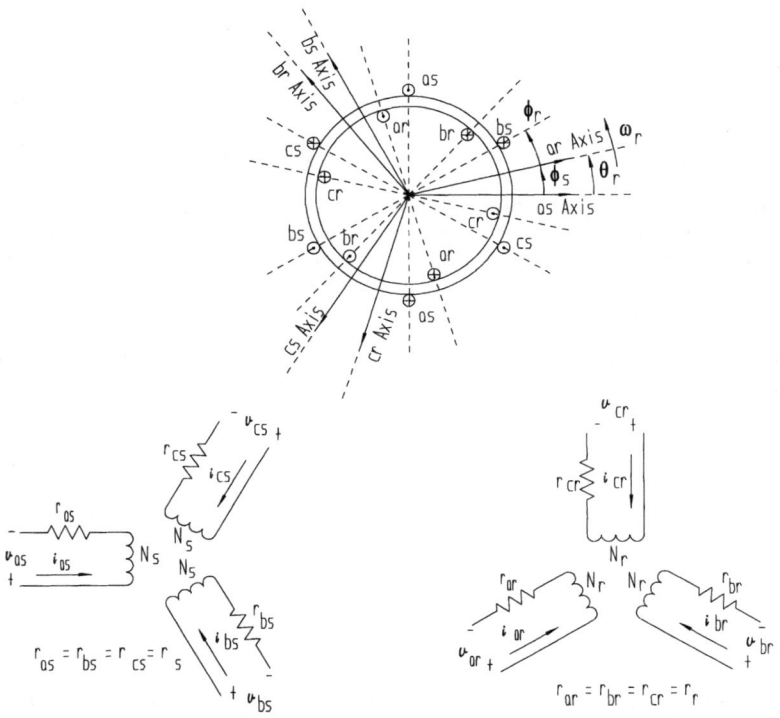

FIGURE 6.24 Balanced three-phase voltages and currents. N_s = stator turns per phase; N_r = rotor turns per phase.

$$v_{ar} = r_r i_{ar} + \frac{d\lambda_{ar}}{dt} \qquad (6.135)$$

$$v_{bs} = r_s i_{bs} + \frac{d\lambda_{bs}}{dt} \qquad (6.136)$$

$$v_{br} = r_r i_{br} + \frac{d\lambda_{br}}{dt} \qquad (6.137)$$

$$v_{cs} = r_s i_{cs} + \frac{d\lambda_{cs}}{dt} \qquad (6.138)$$

$$v_{cr} = r_r i_{cr} + \frac{d\lambda_{cr}}{dt} \qquad (6.139)$$

In the following equations, the L terms include both the leakage and magnetizing inductances, and the M terms are mutual inductance terms.

$$\lambda_{as} = L_{asas}i_{as} + M_{asbs}i_{bs} + M_{ascs}i_{cs} + M_{asar}i_{ar} + M_{asbr}i_{br} + M_{ascr}i_{cr} \qquad (6.140)$$

$$\lambda_{bs} = M_{bsas}i_{as} + L_{bsbs}i_{bs} + M_{bscs}i_{cs} + M_{bsar}i_{ar} + M_{bsbr}i_{br} + M_{bscr}i_{cr} \qquad (6.141)$$

$$\lambda_{cs} = M_{csas}i_{as} + M_{csbs}i_{bs} + L_{cscs}i_{cs} + M_{csar}i_{ar} + M_{csbr}i_{br} + M_{crcr}i_{cr} \qquad (6.142)$$

$$\lambda_{ar} = M_{aras}i_{as} + M_{arbs}i_{bs} + M_{arcs}i_{cs} + L_{arar}i_{ar} + M_{arbr}i_{br} + M_{arcr}i_{cr} \qquad (6.143)$$

$$\lambda_{br} = M_{bras}i_{as} + M_{brbs}i_{bs} + M_{brcs}i_{cs} + M_{brar}i_{ar} + L_{brbr}i_{br} + M_{brcs}i_{cr} \qquad (6.144)$$

$$\lambda_{cr} = M_{cras}i_{as} + M_{crbs}i_{bs} + M_{crcs}i_{cs} + M_{crar}i_{ar} + M_{crbr}i_{br} + L_{crcr}i_{cr} \qquad (6.145)$$

Unlike in the two-phase motor, where the stator windings and the rotor winding terms are orthogonal and hence zero, in the three-phase motor these terms are not orthogonal and therefore exist. However, because of the fixed relations between stator windings and also between rotor windings, these mutual inductances are constants. The mutual inductances between stator and rotor windings are functions of θ_r.

Since both stator windings are fixed at 120° electrical apart, and assuming the material coupling varies as the cosine of the angle between them, the mutual inductances are as follows:

$$M_{asbs} = M_{ascs} = M_{bsas} = M_{bscs} = M_{csas} = M_{csbs} = \frac{N_s^2}{R_m}\cos\phi_s = \frac{N_s^2}{R_m}\cos 120° = -\frac{1}{2}M_s$$
$$(6.146)$$

where

$$M_s = \frac{N_s^2}{R_m}$$

Here M_s is the maximum value of the mutual coupling—namely, where their axes are the same. Now the rotor windings again have the same relationship to each other as the stator windings; hence, the mutual inductors between rotor windings can be written as follows:

$$M_{arbr} = M_{arcr} = M_{brar} = M_{brcr} = M_{crar} = M_{crbr} = -\frac{1}{2}M_r \qquad (6.147)$$

where

$$M_r = \frac{N_r^2}{R_m}$$

Also, because of the symmetrical balanced windings, the self-inductances and the leakage inductances are as follows:

$$L_{asas} = L_{bsbs} = L_{cscs} = L_{ls} + L_m = L_{ls} + \frac{N_s^2}{R_m} = L_{ls} + M_s \qquad (6.148)$$

$$L_{arar} = L_{brbr} = L_{crcr} = L_{er} + L_r = L_{er} + \frac{N_r^2}{R_m} = L_{er} + M_r \qquad (6.149)$$

We now turn our attention to the mutual coupling between stator and rotor windings. By referring to Fig. 6.24 and being aware of axis orientation, we have the following:

$$M_{asar} = M_{aras} = \frac{N_s N_r}{R_m}\cos\theta_r = M_{sr}\cos\theta_r \qquad (6.150)$$

$$M_{asbr} = M_{bras} = \frac{N_s N_r}{R_m} \cos(\theta_r + 120°) = M_{sr} \cos(\theta_r + 120°) \tag{6.151}$$

$$M_{ascr} = M_{aras} = \frac{N_s N_r}{R_m} \cos(\theta_r - 120°) = M_{sr} \cos(\theta_r - 120°) \tag{6.152}$$

$$M_{bsar} = M_{arbs} = \frac{N_s N_r}{R_m} \cos(\theta_r - 120°) = M_{sr} \cos(\theta_r - 120°) \tag{6.153}$$

$$M_{bsbr} = M_{brbs} = \frac{N_s N_r}{R_m} \cos \theta_r = M_{sr} \cos \theta_r \tag{6.154}$$

$$M_{bscr} = M_{crbs} = \frac{N_s N_r}{R_m} \cos(\theta_r + 120°) = M_{sr} \cos(\theta_r + 120°) \tag{6.155}$$

$$M_{csar} = M_{arcs} = \frac{N_s N_r}{R_m} \cos(\theta_r + 120°) = M_{sr} \cos(\theta_r + 120°) \tag{6.156}$$

$$M_{csbr} = M_{brcs} = \frac{N_s N_r}{R_m} = \cos(\theta_r - 120°) = M_{sr} \cos(\theta - 120°) \tag{6.157}$$

$$M_{cscr} = M_{crcr} = \frac{N_s N_r}{R_m} \cos \theta_r = M_{sr} \cos \theta_r \tag{6.158}$$

In the development of the two-phase and single-phase motors we made a transformation to remove the time-varying mutual inductances, then refined the rotor quantities in terms of stator terms. This time we choose to reverse the order, which makes no difference; however, this author believes it to be simpler. Here we refer the rotor quantities in stator terms as we did for the two-phase motor.

$$i'_{ar} = \frac{N_r}{N_s} i_{ar} \tag{6.159}$$

$$v'_{ar} = \frac{N_s}{N_r} v_{ar} \tag{6.160}$$

$$\lambda'_{ar} = \frac{N_s}{N_r} \lambda_{ar} \tag{6.161}$$

$$i'_{br} = \frac{N_r}{N_s} i_{br} \tag{6.162}$$

$$v'_{br} = \frac{N_s}{N_r} v_{br} \tag{6.163}$$

$$\lambda'_{br} = \frac{N_s}{N_r} \lambda_{br} \tag{6.164}$$

$$i'_{cr} = \frac{N_r}{N_s} i_{cr} \tag{6.165}$$

$$v'_{cr} = \frac{N_s}{N_r} v_{cr} \tag{6.166}$$

$$\lambda'_{cr} = \frac{N_s}{N_r} \lambda_{cr} \tag{6.167}$$

Making these substitutions in the rotor voltage equations and multiplying through by N_s/N_r yields

$$v'_{ar} = r_r \left(\frac{N_s}{N_r}\right)^2 i'_{ar} + \frac{d}{dt} \lambda'_{ar}$$

$$v'_{br} = r_r \left(\frac{N_s}{N_r}\right)^2 i'_{br} + \frac{d}{dt} \lambda'_{br}$$

$$v'_{cr} = r_r \left(\frac{N_s}{N_r}\right)^2 i'_{cr} + \frac{d}{dt} \lambda'_{cr}$$

Similarly, by substituting Eqs. (6.159) to (6.167) into the flux linkage equations [Eqs. (6.140) to (6.145)] and multiplying through by N_s/N_r, λ_{as}, λ_{bs}, λ_{cs}, λ'_{ar}, λ'_{br}, and λ'_{cr} can be obtained. It is convenient to write the voltage and flux linkage equations in matrix form at this point.

$$\overline{v}_{abcs} = r_s \overline{I} \overline{i}_{abcs} + \frac{d}{dt} \overline{\lambda}_{abcs} \tag{6.168}$$

where $\overline{I}$ is the identity matrix

$$\overline{v}'_{abcr} = r_r \left(\frac{N_s}{N_r}\right)^2 \overline{I} \overline{i}'_{abcr} + \frac{d}{dt} \overline{\lambda}_{abcs} = r'_r \overline{I} \overline{i}'_{abcr} + \frac{d}{dt} \lambda'_{abcr} \tag{6.169}$$

$$\overline{\lambda}_{abcs} = \begin{bmatrix} L_{es} + M_s & -1/2\, M_s & -1/2\, M_s \\ -1/2\, M_s & L_s + M_s & -1/2\, M_s \\ -1/2\, M_s & -1/2\, M_s & L_s + M_s \end{bmatrix} \overline{i}_{abcs}$$

$$+ M_s \begin{bmatrix} \cos \theta_r & \cos(\theta_r + 120°) & \cos \theta_r - 120° \\ \cos(\theta_r - 120°) & \cos \theta_r & \cos(\theta_r + 120°) \\ \cos(\theta_r + 120°) & \cos(\theta_r - 120°) & \cos \theta \end{bmatrix} \overline{i}_{abcs} \tag{6.170}$$

$$\overline{\lambda}'_{abcr} = M_s \begin{bmatrix} \cos \theta_r & \cos(\theta_r + 120°) & \cos(\theta + 120°) \\ \cos(\theta_r + 120°) & \cos \theta_r & \cos(\theta_r - 120°) \\ \cos(\theta_r - 120°) & \cos(\theta_r + 120°) & \cos \theta \end{bmatrix} \overline{i}_{abcs}$$

$$+ \begin{bmatrix} L'_{er} + M_s & -1/2\, M_s & -1/2\, M_s \\ -1/2\, M_s & L'_{er} + M_s & -1/2\, M_s \\ -1/2\, M_s & -1/2\, M_s & L'_{er} + M_s \end{bmatrix} \overline{i}'_{abcr} \tag{6.171}$$

where

$$r'_r = r_r \left(\frac{N_s}{N_r}\right)^2 \tag{6.172}$$

$$L'_{er} = \left(\frac{N_s}{N_r}\right)^2 \tag{6.173}$$

$$\left(\frac{N_s}{N_r}\right) = \frac{M_s}{M_r} \tag{6.174}$$

ALTERNATING-CURRENT INDUCTION MOTORS

Note that these six equations are dependent on θ_r; hence, a change in variable is appropriate, as was done in the two-phase approach. However, since we have three stator and three rotor variables, the transformation must be of third order. This was mentioned earlier. This transformation removes the dependency on θ_r.

One suitable power-invariant set of transforms for the stator and rotor variables, respectively, follows.

Stator:

$$\overline{v}_{qdos} = \overline{K}_s \overline{v}_{abcs} \quad (6.175)$$

$$\overline{i}_{qdos} = \overline{K}_s \overline{i}_{abcs} \quad (6.176)$$

$$\overline{\lambda}_{qdos} = \overline{K}_s \overline{\lambda}_{abcs} \quad (6.177)$$

$$\overline{K}_s = \frac{2}{3} \begin{bmatrix} 1 & -1/2 & -1/2 \\ 0 & -\sqrt{3}/2 & \sqrt{3}/2 \\ V_2 & 1/2 & 1/2 \end{bmatrix} \quad (6.178)$$

$$\overline{K}_s^{-1} = \begin{bmatrix} 1 & 0 & 1 \\ -1/2 & -\sqrt{3}/2 & 1 \\ -1/2 & \sqrt{3}/2 & 1 \end{bmatrix} \quad (6.179)$$

Rotor:

$$\overline{v}'_{qdos} = \overline{K}_r \overline{v}'_{abcr} \quad (6.180)$$

$$\overline{i}'_{qdos} = \overline{K}_r \overline{i}'_{abcr} \quad (6.181)$$

$$\overline{\lambda}'_{qdos} = \overline{K}_r \overline{\lambda}'_{abcr} \quad (6.182)$$

$$\overline{K}_r = \frac{2}{3} \begin{bmatrix} \cos\theta_r & \cos(\theta_r + 120°) & \cos(\theta_r - 120°) \\ -\sin\theta_r & -\sin(\theta_r + 120°) & -\sin(\theta_r - 120°) \\ 1/2 & 1/2 & 1/2 \end{bmatrix} \quad (6.183)$$

$$\overline{K}_r^{-1} = \begin{bmatrix} \cos\theta_r & \sin\theta_r & 1 \\ \cos(\theta_r + 120°) & -\sin(\theta_r + 120°) & 1 \\ \cos(\theta_r - 120°) & -\sin(\theta_r - 120°) & 1 \end{bmatrix} \quad [(6.184)]$$

If these transformations [Eqs. (6.179), (6.183) and (6.184)] are introduced into Eq. (6.171), and the zero sequence is designated since we are considering a balanced three-phase machine, we have the following for all six voltage equations.

$$\begin{bmatrix} v_{qs} \\ v_{ds} \\ v_{os} \\ v'_{qr} \\ v'_{or} \\ v'_{o4} \end{bmatrix} = \begin{bmatrix} r_s + p(L_{es} + M_s) & 0 & 0 & PM_s & 0 & 0 \\ 0 & r_s + p(L_{es} + M_s) & 0 & 0 & PM_s & 0 \\ 0 & 0 & v_s + pL_{es} & 0 & 0 & 0 \\ PM_s & -\omega_r M_s & 0 & r'_r + p(L'_{er} + M_s) & -\omega_r(L'_{er} + M_s) & 0 \\ -\omega_r M_s & PM_s & 0 & \omega_r(L'_{er} + M_s) & r'_r + p(L'_{er} + M_s) & 0 \\ 0 & 0 & 0 & 0 & 0 & r'_r + pL'_{e4} \end{bmatrix} \begin{bmatrix} i_{qs} \\ i_{ds} \\ i_{os} \\ i'_{qr} \\ i'_{dr} \\ i'_{or} \end{bmatrix}$$

(6.185)

where $p = d/dt$, the derivative operator.

Now, to get the steady-state equivalent circuit in terms of the machine variables, we let $P = j\omega$; by the transformation between abc and qdo we find that:

$$\overline{v}'_{qr} = \overline{v}_{as}$$

$$\bar{I}_{qs} = \bar{I}_{as}, \bar{I}'_{qr}, \bar{I}'_{ar}$$

$$\bar{V}'_{qr} = \bar{V}'_{ar}$$

$$\bar{I}_{ds} = j\bar{I}_{as}$$

$$\bar{I}'_{dr} = j\bar{I}'_{ar}$$

Since in the equivalent circuit we are interested in only one phase for balanced operation, all we need in the preceding matrix are the voltage equations $\bar{V}_{qs}$ and V'_{qr}. Writing these equations and making the substitutions, we have the following (capitals indicate steady-state quantities):

$$\bar{V}_{as} = [r_s + j\omega_e(L_{es} + M_s)]\bar{I}_{as} + j\omega_e M_s \bar{I}'_{ar} \tag{6.186}$$

$$\bar{V}'_{ar} = j\omega_e M_s \bar{I}_{as} - \omega_r M_s(j\bar{I}_{as}) + [r'_r + j\omega_e(L'_{er} + M_s)]\bar{I}'_{ar} - \omega_r(L'_{er} + M_s)(j\bar{I}'_{ar}) \tag{6.187}$$

$$\bar{V}_{as} = (r_s + jx_{er})\bar{I}_{as} + jx_s(\bar{I}_{as} + \bar{I}'_{ar}) \tag{6.188}$$

$$\bar{V}'_{ar} = jx_s\bar{I}_{as} - j\frac{\omega_r}{\omega_e}x_s\bar{I}_{as} + [r'_r + j(x'_{er} + x_s)]\bar{I}'_{ar} - j\frac{\omega_r}{\omega_e}(x'_{er} + x_s)\bar{I}'_{ar} \tag{6.189}$$

$$\bar{V}'_{ar} = jx_s\left(1 - \frac{\omega_r}{\omega_e}\right)\bar{I}_{as} + \left[r'_r + j(x'_{er} + x_s)\left(1 - \frac{\omega_r}{\omega_e}\right)\right]\bar{I}'_{ar} \tag{6.190}$$

$$\left(1 - \frac{\omega_r}{\omega_e}\right) = \frac{\omega_e - \omega_r}{\omega_e} = \text{slip } (s) \tag{6.191}$$

$$\bar{V}'_{ar} = jx_s(s)\bar{I}_{as} + [r'_r + j(x'_{er} + x_s)(s)]\bar{I}'_{ar} \tag{6.192}$$

$$\frac{\bar{V}'_{ar}}{s} = (jx_s)(\bar{I}_{as} + \bar{I}'_{ar}) + \left(\frac{r'_r}{-s} + jx'_{er}\right)\bar{I}'_{ar} \tag{6.193}$$

If we are not considering a double-fed motor that is a squirrel-cage induction motor, $\bar{V}'_{ar} = 0$.

Using Eqs. (6.186) through (6.193), we can draw the equivalent steady-state circuit shown in Fig. 6.25 for one phase of the three-phase squirrel-cage motor, not considering core losses.

It is interesting to note that the magnetizing reactance here is 3/2 as large as the magnetizing reactance for the two-phase or single-phase motor.

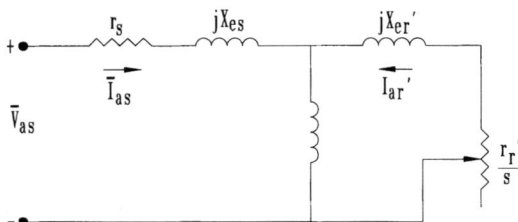

FIGURE 6.25 Simplified equivalent circuit of one phase of a polyphase motor.

6.4 SINGLE-PHASE AND POLYPHASE INDUCTION MOTOR PERFORMANCE CALCULATIONS*

The following procedures provide methods of calculating the performance of single-phase induction motors. Three types of single-phase motors are discussed. They are the split-phase, capacitor-start, and permanent-split-capacitor (PSC) motors. The general procedure for calculating these types of motors is identical. Differences lie in the starting conditions on the split-phase and capacitor-start motors. The PSC motor has additional performance calculations that have to be taken into account.

A single-phase motor by itself has a field that pulsates north-south, south-north through the rotor but does not revolve. The rotor bars are not cut by any flux lines. No voltage or current is induced, and the motor will not start. Auxiliary means are necessary to start the motor. This can be accomplished by adding an auxiliary winding in space quadrature and by varying reactance by varying turns and resistance or by adding capacitance or inductance. The split-phase motor contains an auxiliary winding that is displaced in space by 90° electrical and connected in parallel with the main winding. Typically, the auxiliary winding has 60 to 70 percent of the number of turns of the main winding and is approximately three gauges smaller in wire size. Once the motor reaches 75 to 80 percent of synchronous speed, the auxiliary winding is switched out and the motor runs on the main winding only. The capacitor-start motor is wired in the same manner as the split-phase motor and also contains a switch that cuts out the auxiliary winding after starting is achieved; however, a capacitor is placed in series with the auxiliary winding. The capacitor creates a displacement of the currents in the two windings by 90° in time. This displacement produces the effect of a rotating field and thus a starting torque. The capacitor-start motor produces a greater locked rotor torque than its split-phase counterpart. The locked rotor torque is proportional to the product of three main factors: (1) the sine of the phase displacement angle between the currents in the two windings, (2) the product of the main-winding current multiplied by the auxiliary-winding current, and (3) the number of turns in the auxiliary winding. All three of these factors are more favorable in the capacitor-start motor.

In a PSC motor, the auxiliary winding and capacitor are used continuously. No starting switch or relay is needed. These motors are generally used only in special-duty applications. A PSC motor is well suited for fan applications because of its ability to vary its speed. The continuously running capacitor improves performance by creating magnetic field conditions similar to those of a polyphase motor. These balanced field conditions can be obtained at only one load point. Generally, the designer will balance the motor at full load.

The following calculation procedure for the single-phase motor is based on the cross-field theory by P. H. Trickey. Figure 6.26 shows a single phase motor with a cross-field flux. At standstill, the pulsating stator field induces a rotor current that induces an in-phase component of rotor flux. When the rotor is moving, according to Flemming's law, a current is induced which produces quadrature rotor flux ϕ. The in-phase rotor current interacting with ϕ produces useful motor torque. At a standstill there is no cross-field flux, and therefore no torque. Both the revolving-field (as discussed in Sec. 6.2) and the cross-field theories demonstrate that single-phase motors

---

* Section contributed by Brad Frustaglio, Chris A. Swenski, and William H. Yeadon, Yeadon Energy Systems, Inc.

will not start by themselves. Auxiliary methods must be employed. This section covers several of them. Calculations can begin by solving for the rotor and stator geometrical constants, plus the winding factors for the motor. Next, the slot, zigzag, end, and belt leakage reactance can be calculated and summed. The open-circuit reactance can then be calculated, along with the flux, mmf drops, and saturation factors. Primary and secondary resistance can be calculated using properties of the wire for the primary resistance and using dimensions and properties of the rotor bar material for the secondary resistance. Calculation of the iron loss and an educated estimation of the friction and windage loss will round out the preliminary calculations. Once all of the preceding values of the motor have been calculated, one can begin the performance calculation procedure. Magnetizing, secondary, and secondary cross-field current can be calculated. The overall performance of the motor, such as its torque, efficiency, and power factor, can be evaluated. As previously discussed, the single-phase motor inherently has zero starting torque. Calculation procedures are given for evaluating the starting conditions for the split-phase and the capacitor-start motor at the end of the performance calculations.

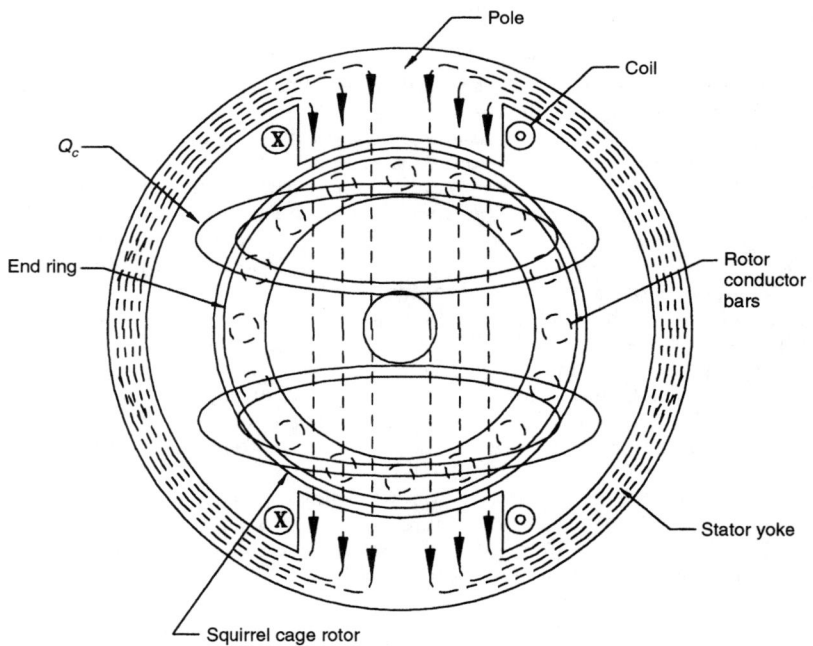

ⓧ Indicates current flowing away from viewer

ⓞ Indicates current flowing toward viewer

FIGURE 6.26 Single-phase motor with cross-field flux (main pole).

ALTERNATING-CURRENT INDUCTION MOTORS

Polyphase induction motor calculations follow in a manner similar to that of the single-phase routine. As with the single-phase motor, once all of the motor constants and reactance values are calculated, the performance of the motor can be evaluated.

The calculation procedure for PSC motors will give performance results at any load. The procedure outlined for calculating the performance of a PSC motor is similar in nature to that for a two-phase motor. First, calculate the reactances and resistances of the motor, just as for the polyphase motor. Next, calculate a series of motor constants based on the reactances and resistances. The currents in the motor can be solved from these constants. Finally, the performance of the motor can be evaluated. The section later discusses how to proportion the windings and the value of capacitance to obtain near-two-phase motor performance at one desired load point.

6.4.1 General Procedure for Calculating the Performance of Single-Phase Induction Motors

Variables may be identified from the following list below or from the figures.

A_B = area of rotor bar
A_{c1} = area of primary core
A_{c2} = area of secondary core
A_g = area of air gap
A_r = area of end ring
AT_{c1} = ampere-turns per pole of primary core
AT_{c2} = ampere-turns per pole of secondary core
AT_{gap} = ampere-turns per pole of air gap
AT_{total} = total ampere-turns per pole
AT_{t1} = ampere turns per pole of primary tooth
AT_{t2} = ampere turns per pole of secondary tooth
A_{t1} = area of primary tooth
A_{t2} = area of secondary tooth
B = coil span
B_{t1} = density of primary tooth
B_{t2} = density of secondary tooth
B_{c1} = density of primary core
B_{c2} = density of secondary core
C = total number of series conductors
C_{nd} = conductivity of rotor cage (Cu = 1.0; Al = 0.5 ~ 0.58)

C_r = mean rotor core circumference
C_s = mean stator core circumference
D_1 = inside diameter of stator
D_2 = outside diameter of rotor
D_a = diameter of mean air gap
D_i = inside diameter of resistance ring
D_r = mean diameter of end ring
D_{ring} = diameter of resistance ring at center of rotor bars
E = applied voltage
F = slot correction factor
F_p = full pitch
f = frequency
G = real power developed
G_1 = minimum air gap
H_{c1} = magnetizing force of primary core
H_{c2} = magnetizing force of secondary core
H_g = magnetizing force of air gap
H_{t1} = magnetizing force of primary tooth
H_{t2} = magnetizing force of secondary tooth

I_1 = primary current
I_2 = secondary current
I_3 = cross-field secondary current
I_m = magnetizing current
K_{ring} = end-ring constant
k_1 = primary slot constant
k_2 = secondary slot constant
k_m = air gap area factor
k_p = pitch factor
k_r = flux factor
k_w = winding distribution factor
L_{c1} = primary core length
L_{c2} = secondary core length
L_g = physical length of air gap
L_{met} = length of mean end turn
L_{t1} = length of primary teeth
L_{y1} = length of primary yoke
L_{y2} = length of secondary yoke
l_g = effective length of air gap
MTL = mean turn length
m = number of phases
N_{sp} = average number of slots per pole
n = number of coils per group
P_L = permeance of leakage path
P_t = effective phase turns divided by effective main turns
P_{tc} = percentage of total pole turns per coil
P_{xbelt} = belt leakage permeance
P_{xend} = end leakage permeance
P_{xslot} = slot leakage permeance
P_{xzz} = zigzag leakage permeance
p = number of poles
R = effective resistance
R_s = effective resistance of auxiliary phase
r_1 = primary resistance
r_2 = secondary resistance
S = speed divided by synchronous speed
S_1 = primary slots
S_2 = secondary slots
S_{fm} = saturation factor (main phase)
S_k = skew of rotor bars measured circumferentially, in
S_{kf} = skew factor
S_{tr} = number of strands of wire in parallel
s = full-load slip
T = run torque
T_{BD} = breakdown torque
T_{f1} = width of primary tooth face
T_{f2} = width of secondary tooth face
T_{fctr} = effective turn factor
T_{p1} = primary tooth pitch
T_{p2} = secondary tooth pitch
T_s = starting torque (capacitor start)
T_{sp} = starting torque (split phase)
t_{fctr} = effective turn factor per coil
t_{w1} = primary tooth width
t_{w2} = secondary tooth width
V_{ph} = volts per phase
W_{in} = total input, W
W_{out} = output, W
w = total series conductors
w_2 = core width, in
W_{fec} = cross-field iron loss
W_{fem} = main field iron loss
W_{f+w} = friction and windage loss
X = total leakage reactance
X_c = capacitor reactance
X_s = start reactance
X_r = run reactance
X_{aux} = auxiliary phase reactance
X_o = open circuit reactance
Z = impedance of main
θ_p = total pole pitch
ϕ = total flux per pole
η = efficiency

See Fig. 6.27 for a representation of some of these variables.

Variables to Calculate
Primary tooth pitch T_{p1}:

$$T_{p1} = \frac{D_1 \pi}{s_2}$$

Secondary tooth pitch T_{p2}:

$$T_{p2} = \frac{D_2 \pi}{s_2}$$

Total pole pitch θ_p:

$$\theta_p = \frac{\pi(D_1 + D_2)}{2p}$$

Calculation of Constants
Slot correction factor F (see Figs. 6.29 to 6.37):

$$F = 0.28 - \left[0.14\left(\frac{y_1}{y_2}\right)\right] + \frac{d_3/y_2}{2.87(y_1/y_2) + 0.18}$$

Gap factor K_g (see Fig. 6.28):

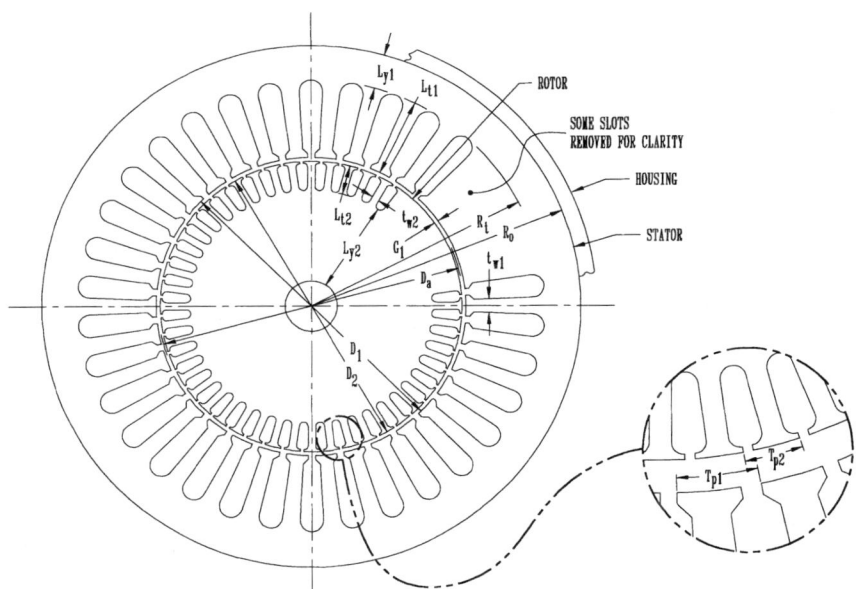

FIGURE 6.27 Lamination nomenclature.

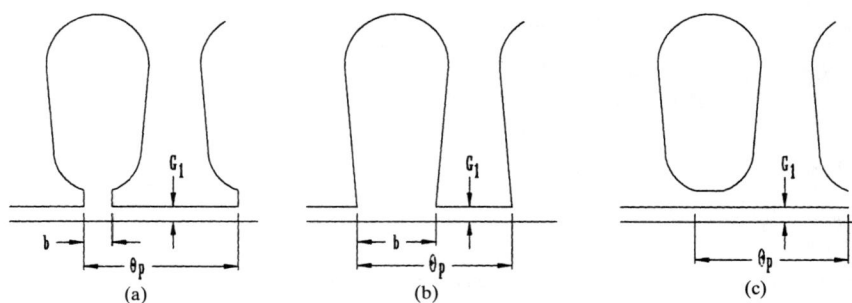

FIGURE 6.28 Gap factor slot shapes: (*a*) semiclosed, (*b*) open, and (*c*) tunnel.

Semiclosed slot

$$K_g = \frac{\theta_p(4.4G_1 + 0.75b)}{\theta_p(4.4G_1 + 0.75b) - b^2}$$

Open slot

$$K_g = \frac{\theta_p(5G_1 + b)}{\theta_p(5G_1 + b) - b^2}$$

Tunnel slot

$$K_g = 1.03$$

Effective gap length l_g:

$$l_g = L_g \times k_g$$

Air gap area factor k_m:

$$k_m = \frac{w_2 \theta_p}{l_g \, \text{SF} \, p}$$

Sinusoidal Winding Distribution Factor k_w
Full pitch 90° F_p:

$$F_p = \frac{S_1}{p}$$

Effective turn factor per coil t_{fctr}:

$$t_{\text{fctr}} = \sin\left[\left(\frac{B}{F_p}\right)90°\right]$$

Effective turn factor T_{fctr}:

$$T_{\text{fctr}} = \sum \left[t_{\text{fctr}}(\text{coil}_1) + t_{\text{fctr}}(\text{coil}_2) + \cdots + t_{\text{fctr}}(\text{coil}_n)\right]$$

Percentage of total pole turns per coil P_{tc}:

$$P_{tc} = \frac{T_{fctr}}{\sum} \times 100$$

Winding factor per pole k_w/pole:

$$K_w/\text{pole} = \text{sum}\,[P_{tc}t_{fctr}(\text{coil}_1) + P_{tc}t_{fctr}(\text{coil}_2) + \cdots + P_{tc}t_{fctr}(\text{coil}_n)]$$

Primary and Secondary Slot Constant Calculations k_1 and k_2

Note: See figures for variable descriptions. k_1 represents the primary (stator) slot constant, while k_2 represents the secondary (rotor) slot constant. They are found using the same set of equations, being careful to use the equation which most closely resembles that of the slot in question.

Round-bottom slot constant k_1 or k_2 (note that F is different for the two constants):

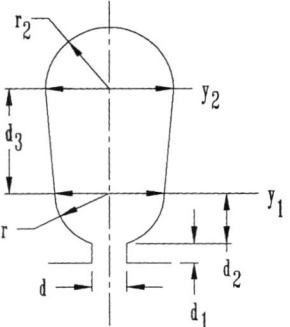

FIGURE 6.29 Round-bottom slot, shape A.

Slot shape A (see Fig. 6.29)

$$k_1 \text{ or } k_2 = F + \frac{d_2}{d} + \frac{4d_2}{3d + y_1}$$

Slot shape B (see Fig. 6.30)

$$k_1 \text{ or } k_2 = F + \frac{d_1}{d} + \frac{2d_2}{d + y_2}$$

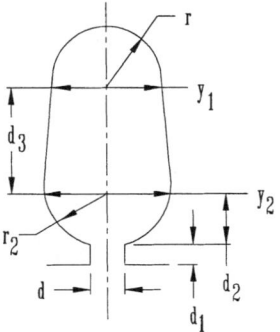

FIGURE 6.30 Round-bottom slot, shape B.

Round-bottom closed slots (see Fig. 6.31)

$$k_1 \text{ or } k_2 = \frac{0.284 - 0.143(y_1/y_2) + (d_3/y_1)}{2.87(y_1/y_2) + 0.96}$$

Square-bottom slot constant k_1 or k_2:
Open Slot (see Fig. 6.32)

$$k_1 \text{ or } k_2 = \frac{d_1}{d} + \frac{d_3}{y_1}$$

Bridged slot (see Fig. 6.33)

$$k_1 \text{ or } k_2 = \frac{d_1}{0.02} + \frac{d_3}{y_1}$$

Flat-bottom slot constant k_1 or k_2:
Slot shape A (see Fig. 6.34)

$$k_1 \text{ or } k_2 = \frac{d_1}{d} + \frac{d_3}{y_1}$$

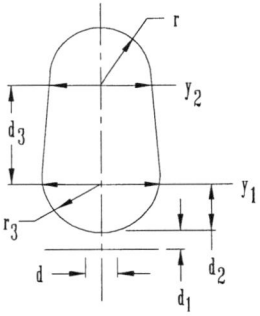

FIGURE 6.31 Round-bottom closed slot.

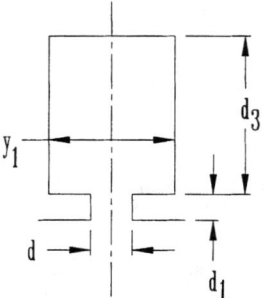

FIGURE 6.32 Square-bottom open slot.

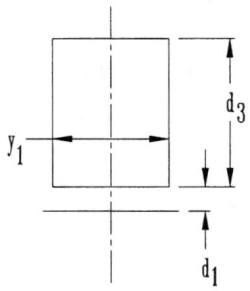

FIGURE 6.33 Square-bottom bridged slot.

Slot shape B (see Fig. 6.35)

$$k_1 \text{ or } k_2 = \frac{d_1}{d} + 1.5\left(\frac{d_2}{d+r}\right) + \frac{d_3}{y_1}$$

Slot shape C (see Fig. 6.36)

$$k_1 \text{ or } k_2 = \frac{d_1}{d} + \frac{4d_2}{3d+y_1} + \frac{d_3}{y_1}$$

Flat-bottom closed slot (see Fig. 6.37)

$$k_1 \text{ or } k_2 = \frac{y_1}{y_2} + \frac{d_3}{y_2} + 1$$

Leakage Reactance Calculations
Slot leakage permeance P_{xslot}:

$$P_{xslot} = \left(\frac{6.38\,w_2}{S_1}\right)\left[k_1 + \left(\frac{S_1}{S_2}\right)k_2\right]$$

Zigzag leakage permeance P_{xzz}:

$$P_{xzz} = \left(\frac{2.126\,w_2}{S_1\,G_1}\right)\left[\frac{(T_{f1}+T_{f2})^2}{4(T_{p1}+T_{p2})}\right]$$

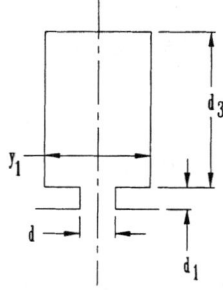

FIGURE 6.34 Flat-bottom slot, shape A.

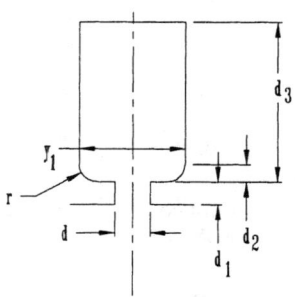

FIGURE 6.35 Flat-bottom slot, shape B.

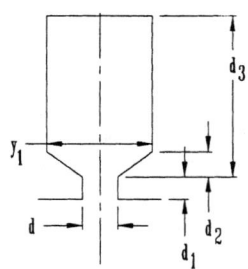

FIGURE 6.36 Flat-bottom slot, shape C.

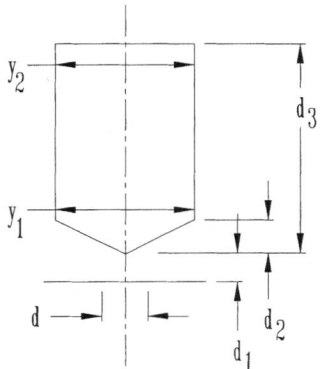

FIGURE 6.37 Flat-bottom closed slot.

End leakage permeance P_{xend}:

$$P_{xend} = \left(\frac{D_1 B}{S_1 p}\right)(1.1 + 0.1p)(2)$$

Belt leakage permeance P_{xbelt}:

$$P_{xbelt} = 0.00239 \, k_m \left(\frac{20p + 0.6}{S_1 + S_2}\right)$$

Skew leakage permeance P_{xskew}:

$$P_{xskew} = 0.645 k_m k_p \left[1 - \frac{\sin(R_s/\theta_p)\,90°}{\pi R_s/2\theta_p}\right]^2$$

Permeance of leakage path P_L:

$$P_L = P_{xslot} + P_{xzz} + P_{xend} + P_{xbelt} + P_{xskew}$$

Permeance of leakage path P_L:

$$P_L = P_{xslot} + P_{xzz} + P_{xend} + P_{xbelt} + P_{xskew}$$

Permeance of Main Path P_m:

$$P_m = 0.645 k_m \left[\frac{\sin(S_k/\theta_p)\,90°}{\pi \, S_k/2\theta_p}\right]$$

Permeance of all paths P_o:

$$P_o = P_m + \frac{P_L}{2}$$

Total leakage reactance X:

$$X = 2\pi f(Ck_w)^2 10^{-8} \, P_L \left[\frac{1 + (P_m/P_o)}{2}\right]$$

Open-circuit leakage reactance X_o:

$$X_o = 2\pi f(Ck_w)^2 10^{-8} \, P_o$$

Start reactance X_s:

$$X_s = 0.80X$$

Run reactance X_r:

$$X_r = 1.20X$$

Motor Constant Calculations for Single-Phase Motors

Primary resistance r_1:

$$r_1 = \frac{C(\text{MTL})(\text{resistance}/1000\text{ ft})}{12{,}000 \, S_{tr}}$$

where MTL = $2w_2 + 2L_{met}$

Secondary resistance r_2:

$$r_2 = (Ck_w)^2(10^{-6})(0.693)(2)\left[\frac{\sqrt{(w_2)^2 + S_k^2}\,(100)}{S_2 A_B C_{nd}} + \frac{63.7 D_r K_{ring}}{A_r p^2 C_{nd}}\right]$$

where (see Fig. 6.38)

$$K_{ring} = \frac{p}{2}\left(1 - \frac{D_i}{D_{ring}}\right)\left[\frac{1 + (D_i/D_{ring})^p}{1 - (D_i/D_{ring})^p}\right]$$

Calculation of Magnetizing Current and Reactance

Total flux per pole ϕ:

$$\phi = \frac{E(10^8)}{2.22 f C k_w}$$

Areas of Various Flux Paths

Primary tooth area A_{t1}:

$$A_{t1} = \left(\frac{S_1}{p}\right) t_{w1} w_2$$

Primary core area A_{c1}:

$$A_{c1} = (R_o - R_t) w_2$$

Secondary tooth area A_{t2}:

$$A_{t2} = \left(\frac{S_2}{p}\right) t_{w2} w_2$$

Secondary core area A_{c2}:

$$A_{c2} = 2L_{y2} w_2$$

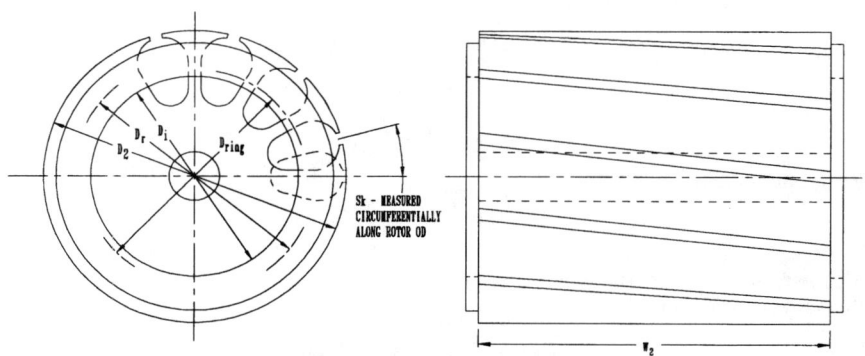

FIGURE 6.38 Resistance ring dimensions.

Area of air gap A_g:

$$A_g = \frac{l_g D_a \pi}{p}$$

Flux Densities in Various Paths
Primary tooth density B_{t1}:

$$B_{t1} = \left(\frac{\phi}{A_{t1}}\right)\left(\frac{\pi}{2}\right)$$

Primary core density B_{c1}:

$$B_{c1} = \frac{\phi}{A_{c1}}$$

Secondary tooth density B_{t2}:

$$B_{t2} = \left(\frac{\phi}{A_{t2}}\right)\left(\frac{\pi}{2}\right)$$

Secondary core density B_{c2}:

$$B_{c2} = \frac{\phi}{A_{c2}}$$

Air gap density B_g:

$$B_g = \left(\frac{\phi}{A_g}\right)\left(\frac{\pi}{2}\right)$$

Magnetizing Forces. The magnetizing force of each path other than the air gap is found by referring the preceding densities to the proper saturation curve and reading across to find the respective magnetizing force.

Primary tooth H_{t1}
Primary core H_{c1}
Secondary tooth H_{t2}
Secondary core H_{c2}
Air gap magnetizing force H_g

$$H_g = 0.313 B_g$$

Lengths of Various Flux Paths
Primary tooth length L_{t1}: see Fig. 6.27
Primary core length L_{c1}:

$$L_{c1} = \frac{C_s}{2p}$$

Secondary tooth length L_{t2}: see Fig. 6.27
Secondary core length L_{c2}:

$$L_{c2} = \frac{C_r}{2p}$$

Ampere-Turns per Pole
Primary tooth AT_{t1}:

$$AT_{t1} = H_{t1}L_{t1}$$

Primary core AT_{c1}:

$$AT_{c1} = H_{c1}L_{c1}$$

Secondary tooth AT_{t2}:

$$AT_{t2} = H_{t2}L_{t2}$$

Secondary core AT_{c2}:

$$AT_{c2} = H_{c2}L_{c2}$$

Air gap AT_{gap}:

$$AT_{gap} = H_g l_g$$

Total ampere-turns per pole AT_{total}:

$$AT_{total} = AT_{t1} + AT_{c1} + AT_{t2} + AT_{c2} + AT_{gap}$$

Main phase saturation factor S_{fm}:

$$S_{fm} = \frac{AT_{total}}{AT_{gap}}$$

Calculation of Iron Losses (Fe Loss)
Volume of primary teeth V_{t1}:

$$V_{t1} = A_{t1}L_{t1}S_1$$

Volume of primary core V_{c1}:

$$V_{c1} = A_{c1}L_{c1}$$

Find rotor and stator iron weights by multiplying V_{c1} and V_{t1} by density, then obtain the watts lost per pound in the teeth and core from Fig. 6.39. The sum of these two is the total iron loss W_{fe}.

Main field iron loss W_{fem}:

$$W_{fem} = 0.55 W_{fe}$$

Cross-field iron loss W_{fec}:

$$W_{fec} = 0.45 W_{fe}$$

Friction and Windage Losses. Friction and windage losses W_{f+w} are obtained by testing similar machines or by calculating bearing friction and windage from the bearing supplier's information.

Performance Calculations for Single-Phase Induction Motors
Full-load slip s:

$$s = \frac{3I_2^2 r_2}{3I_2^2 r_2 + W_{out} + W_{f+w}}$$

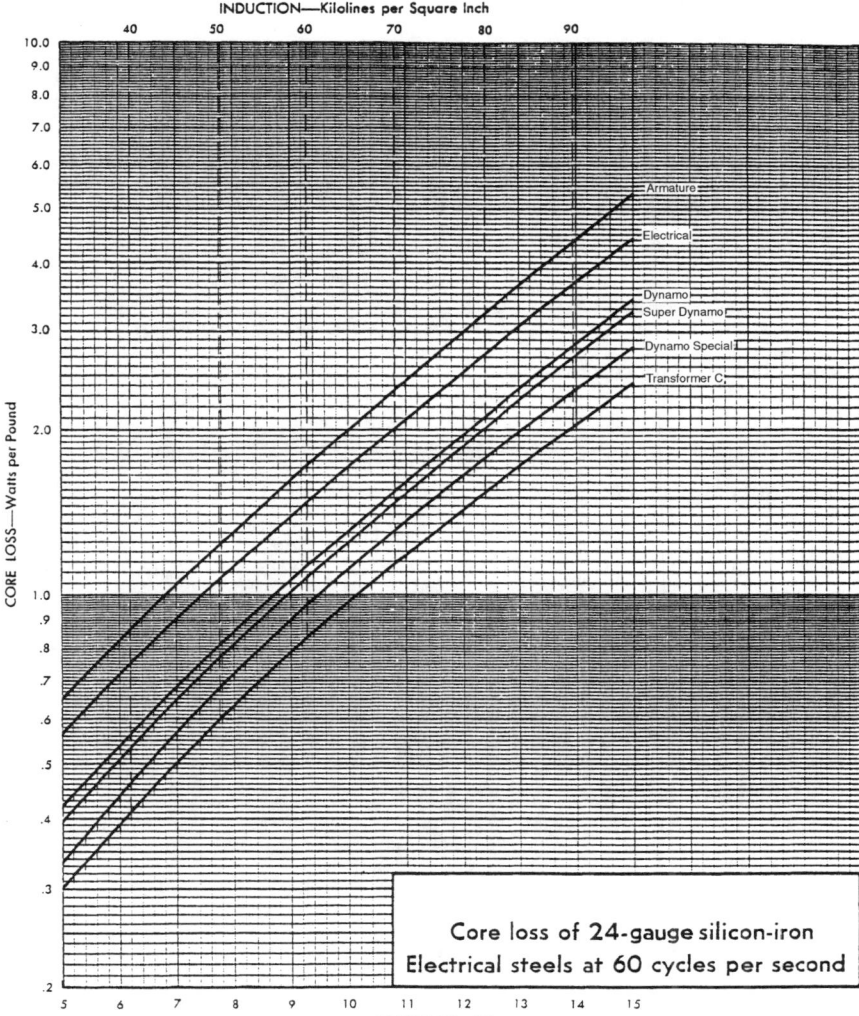

FIGURE 6.39 Core loss curve.

Magnetizing current I_m:

$$I_m = \frac{E}{X_o}$$

Primary current I_1:

$$I_1 = \frac{\sqrt{\{(1-S^2)E - (E\,r_2^2/X_0^2) - [(1-S^2)r_1 + (2-k_w^2)r_2](W_{\text{fem}}/E)\}^2 + [2(I_m r_2)]^2}}{\sqrt{[(1-S^2)r_1 + (2-k_w^2)r_2]^2 + [(1-S^2)X - (2r_1+r_2)(r_2/X_0)]^2}}$$

Secondary current I_2:

$$I_2 = \frac{\sqrt{[(1-S^2)Ek_w]^2 + [(I_m r_2)k_w]^2}}{\sqrt{[(1-S^2)r_1 + (2-k_w^2)r_2]^2 + [(1-S^2)X - (2r_1+r_2)(r_2/X_0)]^2}}$$

Cross-field secondary current I_3:

$$I_3 = \frac{S(I_m r_2)k_w}{\sqrt{[(1-S^2)r_1 + (2-k_w^2)r_2]^2 + [(1-S^2)X - (2r_1+r_2)(r_2/X_0)]^2}}$$

Real power developed G:

$$G = \frac{[(1-S^2)(Ek_w)^2 r_2 + (I_m r_2 k_w)^2 r_2]S^2}{\{[(1-S^2)r_1 + (2-k_w^2)r_2](W_{\text{fem}}/E)\}^2 + [(1-S^2)X - (2r_1+r_2)(r_2/X_0)]^2}$$

Input W_{in}:

$$W_{\text{in}} = I_1^2 r_1 + I_2^2 r_2 + I_3^2 r_2 + W_{\text{fem}} + G$$

Output W_{out}:

$$W_{\text{out}} = G - [W_{\text{fec}} + W_{f+w}]$$

Flux factor k_r:

$$k_r = \frac{X_o - X}{X_o}$$

Torque T:

$$T = \frac{112.6\, W_o}{\text{rpm}}$$

Breakdown torque T_{BD}:

$$T_{\text{BD}} = \left(\frac{E}{115}\right)^2 \frac{p}{4} \frac{60}{f} \frac{345 - 9.2[(r_1 - r_2)/X]^2}{(r_1 + r_2) + X} k_r \frac{1}{16} \quad \text{lb} \cdot \text{ft}$$

Efficiency η:

$$\eta = \frac{W_{\text{out}}}{W_{\text{in}}}$$

Power factor PF:

$$\text{PF} = \frac{\text{input}}{EI_1}$$

Apparent Efficiency = $PF\eta$

Starting Torque Calculations for Split-Phase Motors. See the split-phase schematic in Fig. 6.40.
Effective resistance of main R:

$$R = r_1 + r_2 k_r$$

Note: Reactance of auxiliary phase X_{aux} is found in a manner similar to that for X, using the following reactance equations, but substituting appropriate values where they might change.

Slot leakage permeance P_{xslot}:

$$P_{xslot} = \left(\frac{6.38 w_2}{S_1}\right)\left[k_1 + \left(\frac{S_1}{S_2}\right) k_2\right]$$

Zigzag leakage permeance P_{xzz}:

$$P_{xzz} = \left(\frac{2.126 w_2}{S_1 G_1}\right)\left[\frac{(T_{f1} + T_{f2})^2}{4(T_{p1} + T_{p2})}\right]$$

End leakage permeance P_{xend}:

$$P_{xend} = \left(\frac{D_1 B}{S_1 p}\right)(1.1 + 0.1 p)(2)$$

Belt leakage permeance P_{xbelt}:

$$P_{xbelt} = 0.00239 \, k_m \left(\frac{20p + 0.6}{S_1 + S_2}\right)$$

Skew leakage permeance P_{xskew}:

$$P_{xskew} = 0.645 \, k_m k_p \left[1 - \frac{\sin(R_s/\theta_p) \, 90°}{\pi R_s/2\theta_p}\right]^2$$

Permeance of leakage path P_L:

$$P_L = P_{xslot} + P_{xzz} + P_{xend} + P_{xbelt} + P_{xskew}$$

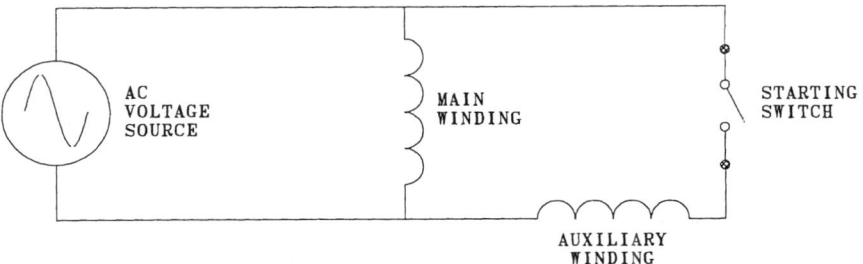

FIGURE 6.40 Single-phase split phase diagram.

Permeance of main path P_m:

$$P_m = 0.645 k_m \left[\frac{\sin(S_k/\theta_p)\,90°}{\pi S_k/2\theta_p} \right]$$

Permeance of all paths P_o:

$$P_o = P_m + \frac{P_L}{2}$$

Total auxiliary leakage reactance X_{aux}:

$$X_{\text{aux}} = 2\pi f (Ck_w)^2 10^{-8} P_L \left[\frac{1 + (P_m/P_o)}{2} \right]$$

Effective resistance of auxiliary phase R_s:

$$R_s = \left[r_2 + \left(\frac{X_{\text{aux}}}{X} \right) r_2 \right] k_r + \frac{E\,\text{PF}}{I_1}$$

Impedance of main Z:

$$Z = \sqrt{\left(\frac{X_s}{p}\right)^2 + (r_1 + r_2)^2}$$

Split-phase starting torque T_{sp}:

$$T_{\text{sp}} = \left(\frac{1.88\,p\,E^2 P_r r_2 S_{kf}}{f} \right) \left[\frac{R_s X - R X_s}{(R^2 + X^2)(R_s^2 + X_s^2)} \right]$$

Starting Torque Calculations for Capacitor-Start Motors. See the capacitor-start schematic in Fig. 6.41.

Reactance of capacitor X_c:

$$X_c = \frac{10^6}{2\pi f (\mu F)}$$

Capacitor-start starting torque T_s:

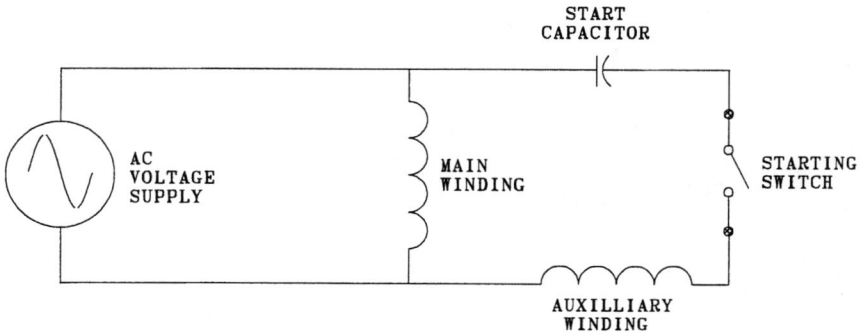

FIGURE 6.41 Single-phase capacitor-start diagram.

ALTERNATING-CURRENT INDUCTION MOTORS

$$T_s = \left(\frac{1.88pE^2P_rr_2S_{kf}}{f}\right)\left\{\frac{R_sX - RX_s - X_c}{(R^2 + x^2)[R_s^2 + (X_s - X_c)^2]}\right\}$$

6.4.2 General Procedure for Calculating the Performance of Permanent-Split-Capacitor (PSC) Motors

Variables

A_B = area of rotor bar
A_r = area of end ring
a = winding ratio
B = coil span
C = total number of series conductors
C_{nd} = conductivity of rotor cage (Cu = 1.0; Al = 0.5 ~ 0.58)
D_i = inside diameter of end ring
D_r = mean diameter of end ring
D_{ring} = diameter of end ring at center of rotor bars
D_1 = inside diameter of stator
D_2 = outside diameter of rotor
E_a = auxiliary-phase voltage
E_c = capacitor voltage
F = slot correction factor
f = frequency
G = width of air gap
G_1–G_{25} = motor constants based on resistance, reactance, and turns ratio
I_1 = primary current
I_{2a} = main-phase secondary current
I_{2m} = auxiliary-phase secondary current
I_a = auxiliary-phase locked-rotor current
I_L = line current
I_m = magnetizing current
j = conjugate symbol ($\sqrt{-1}$)
K_1 = flux factor
K_{p1} = pitch factor
k_d = distribution factor

K_g = gap factor
k_m = air gap area factor
k_w = winding distribution factor
k_p = winding pitch factor
k_1 = primary slot constant
k_2 = secondary slot constant
L_g = physical length of air gap
L_{met} = length of mean end turn
L_{meta} = length of mean end turn of auxiliary phase
l_g = effective length of air gap
MTL = mean turn length
MTL$_a$ = mean turn length of auxiliary winding
m = number of phases
N_{1h}–N_{4h} = intermediate calculation values
N_{1v}–N_{5v} = intermediate calculation values
N_r = run speed
N_s = synchronous speed
n = number of coils per group
P_{xskew} = skew leakage permeance
P_{xbelt} = belt leakage permeance
P_{xslot} = slot leakage permeance
P_{xzz} = zigzag leakage permeance
P_{xend} = end leakage permeance
PF = power factor
P_L = permeance of leakage path
P_m = permeance of main path
P_o = permeance of all paths
p = number of poles
R_c = capacitor resistance
R_s = effective resistance of auxiliary path

r_1 = primary resistance
r_{1a} = auxiliary winding resistance (hot)
r_{1ac} = resistance of auxiliary winding and capacitor
r_{1u} = intermediate resistance value
r_2 = secondary resistance
r_{2u} = intermediate resistance value
r_{3u} = intermediate resistance value
S = run speed divided by synchronous speed
SF = saturation factor (find as for polyphase motors)
S_k = skew of rotor bars measured circumferentially, in
S_{tr} = number of strands of wire in parallel
S_{tra} = number of strands of wire in parallel in auxiliary winding
S_1 = number of primary slots
S_2 = number of secondary slots
T = run torque
T_a = average torque

T_{DF} = double-frequency pulsating torque
T_{f1} = width of primary tooth face
T_{f2} = width of secondary tooth face
T_{p1} = primary tooth pitch
T_{p2} = secondary tooth pitch
U = intermediate calculation values
W_{fe} = iron loss
W_{f+w} = friction and windage loss
W_{in} = input, W
W_{loss} = loss, W
W_{out} = output, W
w_2 = core width
X = total leakage reactance
X_{3u} = intermediate reactance value
X_{4u} = intermediate reactance value
X_c = capacitor reactance
X_o = open-circuit leakage reactance
Z_c = capacitor impedance
η = efficiency
θ_p = pole pitch

Variables to Calculate
Pole pitch θ_p:

$$\theta_p = \frac{\pi(D_1 + D_2)}{2p}$$

Gap factor K_g (see Fig. 6.28):
Semiclosed slot

$$K_g = \frac{\theta_p(4.4G + 0.75b)}{\theta_p(4.4G + 0.75b) - b^2}$$

Open slot

$$K_g = \frac{\theta_p(5G + b)}{\theta_p(5G + b) - b^2}$$

Tunnel slot

$$K_g = 1.03$$

Effective gap length l_g:

$$l_g = L_g k_g$$

Primary and secondary tooth pitch T_{p1} and T_{p2}:

$$T_{p1} = \frac{D_1 \pi}{S_1}$$

$$T_{p2} = \frac{D_2 \pi}{S_2}$$

Calculation of Winding and Slot Constants
Pitch factor k_p:

$$k_p = \sin(\text{pitch} \cdot 90°)$$

where pitch is expressed as a fraction of the full pitch, such as ⅝, etc.
Distribution factor k_d:

$$k_d = \frac{\sin(B/2)}{n \sin(B/2n)}$$

for distributed even-grouped windings.
Total winding distribution factor k_w:

$$k_w = k_p k_d$$

Slot correction coefficient F:

$$F = 0.28 - \left[0.14\left(\frac{y_1}{y_2}\right)\right] + \frac{d_3/y_2}{2.87(y_1/y_2) + 0.18}$$

Air gap area factor k_m:

$$k_m = \frac{w_2 \theta_p}{l_g \text{SF } p}$$

Primary and Secondary Slot Constant Calculations k_1 *and* k_2:

Note: See figures for variable descriptions. k_1 represents the primary (stator) slot constant, while k_2 represents the secondary (rotor) slot constant. They are found using the same set of equations, being careful to use the equation which most closely resembles that of the slot in question.

Round-bottom slot constant k_1 or k_2 (note that F is different for the two constants):
   Slot shape A (see Fig. 6.29)

$$k_1 \text{ or } k_2 = F + \frac{d_2}{d} + \frac{4d_2}{3d + y_1}$$

   Slot shape B (see Fig. 6.30)

$$k_1 \text{ or } k_2 = F + \frac{d_1}{d} + \frac{2d_2}{d + y_2}$$

Round-bottom closed slot (see Fig. 6.31)

$$k_1 \text{ or } k_2 = \frac{0.284 - 0.143(y_1/y_2) + (d_3/y_1)}{2.87(y_1/y_2) + 0.96}$$

Square-bottom slot constant k_1 or k_2:
Open Slot (see Fig. 6.32)

$$k_1 \text{ or } k_2 = \frac{d_1}{d} + \frac{d_3}{y_1}$$

Bridged slot (see Fig. 6.33)

$$k_1 \text{ or } k_2 = \frac{d_1}{0.02} + \frac{d_3}{y_1}$$

Flat-bottom slot constant k_1 or k_2:
Slot shape A (see Fig. 6.34)

$$k_1 \text{ or } k_2 = \frac{d_1}{d} + \frac{d_3}{y_1}$$

Slot shape B (see Fig. 6.35)

$$k_1 \text{ or } k_2 = \frac{d_1}{d} + 1.5\left(\frac{d_2}{d+r}\right) + \frac{d_3}{y_1}$$

Slot shape C (see Fig. 6.36)

$$k_1 \text{ or } k_2 = \frac{d_1}{d} + \frac{4d_2}{3d + y_1} + \frac{d_3}{y_1}$$

Flat-bottom closed slot (see Fig. 6.37)

$$k_1 \text{ or } k_2 = \frac{y_1}{y_2} + \frac{d_3}{y_2} + 1$$

Leakage Reactances
Slot leakage permeance P_{xslot}:

$$P_{xslot} = \left(\frac{3.19 w_2 m}{S_1}\right)\left[k_1 + \left(\frac{S_1}{S_2}\right)k_2\right]$$

Zigzag leakage permeance P_{xzz}:

$$P_{xzz} = \left(\frac{1.063 w_2 m}{S_1 G}\right)\left[\frac{(T_{f1} + T_{f2})^2}{4(T_{p1} + T_{p2})}\right]$$

End leakage permeance P_{xend}:

$$P_{xend} = \left(\frac{D_1 B m}{S_1 p}\right)(1.1 + 0.1p)$$

Belt leakage permeance P_{xbelt}:

$$P_{xbelt} = 0.00118 \, m k_m \left(\frac{20p + 0.6}{S_1 + S_2}\right)$$

Skew leakage permeance P_{xskew}:

$$P_{\text{xskew}} = 0.324 m k_m k_p \left[1 - \frac{\sin(R_s/\theta_p)\, 90°}{\pi R_s/2\theta_p}\right]^2$$

Permeance of leakage path P_L:

$$P_L = P_{\text{xslot}} + P_{\text{xzz}} + P_{\text{xend}} + P_{\text{xbelt}} + P_{\text{xskew}}$$

Permeance of main path P_m:

$$P_m = 0.324 m k_m \left[\frac{\sin(S_k/\theta_p)\, 90°}{\pi S_k/2\theta_p}\right]$$

Permeance of all paths P_o:

$$P_o = P_m + \frac{P_L}{2}$$

Total leakage reactance X:

$$X = 2\pi f (C k_w)^2 10^{-8} P_L \left[\frac{1 + (P_m/P_o)}{2}\right]$$

Open-circuit leakage reactance X_o:

$$X_o = 2\pi f (C k_w)^2 10^{-8} P_o$$

Primary resistance r_1:

$$r_1 = \frac{(C)(\text{MTL})(\text{resistance}/1000\text{ ft})}{12{,}000 S_{\text{tr}}}$$

where $\text{MTL} = 2w_2 + 2L_{\text{met}}$

Secondary resistance r_2:

$$r_2 = (C k_w)^2 (10^{-6})(0.693)(2) \left[\frac{\sqrt{(w_2)^2 + S_k^2}\,(100)}{S_2 A_B C_{\text{nd}}}\right] + \left(\frac{63.7 D_r}{A_r p^2}\right)\left(\frac{K_{\text{ring}}}{C_{\text{nd}}}\right)$$

where (see Fig. 6.38)

$$K_{\text{ring}} = \frac{p}{2}\left(1 - \frac{D_i}{D_{\text{ring}}}\right)\left[\frac{1 + (D_i/D_{\text{ring}})^p}{1 - (D_i/D_{\text{ring}})^p}\right]$$

Auxiliary winding resistance r_{1a}:

$$r_{1a} = \frac{(C)(\text{MTL}_a)(\text{resistance}/1000\text{ ft})}{12{,}000 S_{\text{tra}}}$$

where $\text{MTL}_a = 2w_2 + 2L_{\text{meta}}$

Resistance of auxiliary winding and capacitor r_{1ac}:

$$r_{1ac} = r_{1a} + R_c$$

Reactance of capacitor X_c:

$$X_c = \frac{10^6}{2\pi f(\mu F)}$$

Turns ratio a:

$$a = \frac{\text{effective conductors in auxiliary winding}}{\text{effective conductors in main winding}}$$

Magnetizing current I_m:

$$I_m = \frac{E_a}{X_o}$$

Intermediate resistance to reactance ratios r_{1u}, r_{2u}, and r_{3u}:

$$r_{1u} = \frac{r_1}{X_o}$$

$$r_{2u} = \frac{r_2}{X_o}$$

$$r_{3u} = \frac{r_{1ac}}{a^2 X_o}$$

Flux factor K_r:

$$K_r = \frac{X_o - X}{X_o}$$

Pitch factor K_{p1}:

$$K_{p1} = \sqrt{K_r}$$

Intermediate reactance ratios X_{3u} and X_{4u}:

$$X_{4u} = \frac{X + (X_c/a^2)}{X_o}$$

$$X_{3u} = X_{4u} + K_r$$

Motor constants based on resistance, reactance, and turns ratio G_1–G_{25}:

$$G_1 = K_p X_{3u}$$

$$G_2 = K_p r_{3u}$$

$$G_3 = \frac{K_p}{a}$$

$$G_4 = \frac{K_p^2}{a}$$

$$G_5 = \frac{K_p}{a^2}$$

$$G_6 = \left(\frac{K_p}{a}\right) r_{1u}$$

$$G_7 = \left(\frac{K_p}{a}\right) r_{3u}$$

$$G_8 = \left(\frac{K_p}{a}\right) X_{3u}$$

$$G_9 = \left(\frac{X_{4u}}{r_{2u}}\right)$$

$$G_{10} = \frac{(X_{4u} K_p)}{r_{2u}}$$

$$G_{11} = \frac{r_{3u}}{r_{2u}}$$

$$G_{12} = \frac{r_{3u} K_p}{r_{2u}}$$

$$G_{13} = \left(\frac{r_{1u}}{a^2}\right) K_p$$

$$G_{14} = \left(\frac{r_{1u}}{a^2}\right)\left(\frac{1}{r_{2u}}\right)$$

$$G_{15} = \frac{K_p r_{1u}}{a^2 r_{2u}}$$

$$G_{16} = 2 r_{3u} + r_{2u} X_{3u}$$

$$G_{17} = X_{3u} + X_{4u} - r_{2u} r_{3u}$$

$$G_{18} = \frac{(1 - K_r) K_p}{r_{2u} a^2}$$

$$G_{19} = \frac{1 - K_r}{r_{2u} a^2}$$

$$G_{20} = \frac{2 r_{1u} + r_{2u}}{a^2}$$

$$G_{21} = \frac{2 - K_r - r_{1u} r_{2u}}{a^2}$$

$$G_{22} = \frac{(r_{1u} X_{4u}/r_{2u}) + [(1 - K_r)/r_{2u}] r_{3u}}{I_m}$$

$$G_{23} = \frac{[(1 - K_r)/r_{2u}] (-X_{4u}) + (r_{1u}/r_{2u}) r_{3u}}{I_m}$$

$$G_{24} = \frac{2 - K_r - r_{1u} r_{2u} - X_{3u} + (2 r_{1u} + r_{2u}) r_{3u} + K_r}{I_m}$$

$$G_{25} = \frac{(2 r_{1u} + r_{2u})(X_{3u}) + [(2 - K_r) - r_{1u} r_{2u}] r_{3u} - r_{1u} K_r}{I_m}$$

Intermediate Calculation Values

$$N_{1h} = G_{16} - G_4 S - G_9 (1 - S^2)$$

$$N_{2h} = G_2 + G_3 S - G_{10}[1 - S^2]$$

$$N_{3h} = G_{20} - G_4 S - G_{19}(1 - S^2)$$

$$N_{4h} = G_{13} - G_8 S - G_{18}(1 - S^2)$$

$$U = G_{24} G_{22}(1 - S^2)$$

$$N_{1v} = G_{17} + G_{11}(1 - S^2)$$

$$N_{2v} = G_1 + G_6 S + G_{12}(1 - S^2)$$

$$N_{3v} = G_{21} + G_{14}(1 - S^2)$$

$$N_{4v} = G_5 - G_7 S + G_{15}(1 - S^2)$$

$$N_{5v} = G_{25} - G_{23}(1 - S^2)$$

Current Calculations
Primary current I_1:
$$I_1 = \frac{\sqrt{[G_{16} - (1-S^2)G_9 - G_4S]^2 + [G_{17} + (1-S^2)G_{11}]^2}}{\sqrt{[G_{24} - (1-S^2)G_{22}]^2 + [G_{25} + (1-S^2)G_{23}]^2}}$$

Main-phase secondary current I_{2m}:
$$I_{2m} = \frac{\sqrt{[G_2 - (1-S^2)G_{10} - SG_3]^2 + \{[G_1/(1-S^2)G_{12}] + SG_6\}^2}}{\sqrt{[G_{24} - (1-S^2)G_{22}]^2 + [G_{25} + (1-S^2)G_{23}]^2}}$$

Auxiliary-phase locked-rotor current I_a:
$$I_a = \frac{\sqrt{[G_{20} - (1-S^2)G_{19} + SG_4]^2 + [G_{21} + (1+S^2)G_{14}]^2}}{\sqrt{[G_{24} - (1-S^2)G_{22}]^2 + [G_{25} + (1-S^2)G_{23}]^2}}$$

Auxiliary-phase secondary current I_{2a}:
$$I_{2a} = \frac{\sqrt{[G_{13} - (1-S^2)G_{18} + SG_8]^2 + [G_5 + (1-S^2)G_{15} - SG_7]^2}}{\sqrt{[G_{24} - (1-S^2)G_{22}]^2 + [G_{25} + (1-S^2)G_{23}]^2}}$$

Capacitor impedance Z_c:
$$Z_c = \sqrt{R_c^2 + X_c^2}$$

Capacitor voltage E_c:
$$E_c = I_a Z_c$$

Line current I_L:

$$I_L = \frac{\sqrt{[G_{16} - (1-S^2)G_9 + G_4S + G_{20} - (1-S^2)G_{19} + SG_4]^2 + [G_{17} + (1-S^2)G_{11} + G_{21} + (1-S^2)G_{14}]^2}}{\sqrt{[G_{24} - (1-S^2)G_{22}]^2 + [G_{23} + (1-S^2)G_{23}]^2}}$$

Losses, W_{loss}:
$$W_{\text{loss}} = I_1^2 r_1 + I_{2m}^2 r_2 + I_a^2 r_{1a} + I_a^2 R_c + I_{2a}^2 r_2 a^2 + W_{\text{fe}} + W_{f+w}$$

Output W_{out}:
$$W_{\text{out}} = SX_o K_p a \frac{N_{2h} N_{3h} + \{[G_1/(1-S^2)G_{12}] + SG_6\}N_{3v} - N_{1h}N_{4h} + N_{1v}N_{4v}}{U^2 + N_{5v}^2 W_{\text{loss}}}$$

Input W_{in}:
$$W_{\text{in}} = W_{\text{loss}} + W_{\text{out}}$$

Run torque T:
$$T = \frac{7.04 W_{\text{out}}}{N_r} \quad \text{lb} \cdot \text{ft}$$

Power factor PF:

ALTERNATING-CURRENT INDUCTION MOTORS

$$\text{PF} = \frac{W_{in}}{E_a I_L}$$

Efficiency η:

$$\eta = \frac{W_{out}}{W_{in}}$$

Average torque T_a:

$$T_a = \left(\frac{0.94_p}{16f}\right)\left(\frac{X_o K_p a}{U^2 + N_{5v}^2}\right)\{[(N_{1h} - N_{2h})(-N_{4h}) - (N_{1v} - N_{2v})(N_{4v})]$$
$$- [(N_{3h} - N_{4h})(-N_{2h}) - (N_{3v} - N_{4v})(N_{2v})]\} \quad \text{lb} \cdot \text{ft}$$

Double-frequency pulsating torque T_{DF}:

$$T_{DF} = T_a + \left(\frac{0.94_p}{16f}\right)\left(\frac{X_o K_p a}{U^2 + N_{5v}^2}\right) j\{[(N_{1h} - N_{2h})(N_{4v}) - (N_{1v} - N_{2v})(N_{4h})]$$
$$+ [(N_{3h} - N_{4h})(N_{2v}) - (N_{3v} - N_{4v})(N_{2h})]\} \quad \text{lb} \cdot \text{ft}$$

Method for Balancing a PSC Motor. See the single-phase permanent-split-capacitor diagram in Fig. 6.42.

Variables used in the following PSC balancing equations:

A = power component of main winding primary current
C_{VA} = capacitor voltamperes
E = line voltage
E_c = capacitor voltage
I_c = line current of current capacitor motor
I_{1a} = auxiliary phase current
I_{mp} = magnitude of polyphase current (calculated as for two-phase motor)

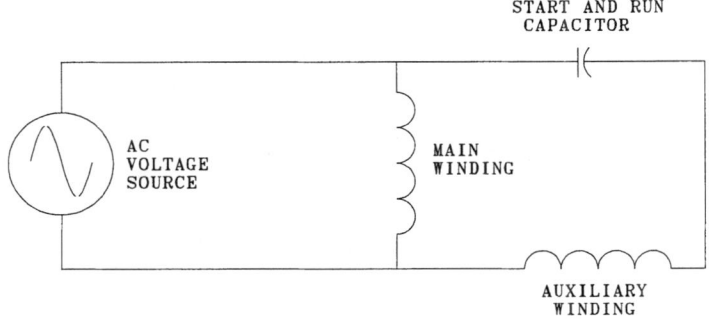

FIGURE 6.42 Single-phase permanent-split-capacitor diagram.

K = ratio of effective conductors in capacitor winding to effective conductors in main winding

K_a = ratio of area of conductor in main winding to area of conductor in capacitor winding

PF = power factor

R_B = reactive component of main winding primary current

R_c = effective resistance of capacitor

r_1 = resistance of main winding

X_c = effective reactance of capacitor

Calculation Procedure

1. Design the main winding to achieve the necessary maximum torque.
2. Calculate the performance as a two-phase motor.
3. Solve for K.
   - K_a must be a function of the cubed root of 2 since wire sizes vary in this ratio.
   - Assume K_a will be one of these values: 1.26, 1.59, or 2.00.
   - Set the value of K on the right side of the equation to K_a.
   - Solve for K, substitute this value for the assumed value, and perform a second iteration.
4. Solve for X_c.
5. Design the capacitor from X_c and correct preceding solutions if R_c is too far in error.
6. Calculate the capacitor voltage E_c and the capacitor voltamperes.
7. Calculate performance equations based on the primary loss, capacitor loss, and power factor calculated.

Use the calculation procedures described in the polyphase section to compute locked-rotor torque. If it is not satisfactory, it may be necessary to reduce K, increase the microfarads, or increase the rotor resistance.

The procedure described will design for the proper value of capacitance to achieve the balance point. However, it is not possible to balance a motor at any load desired. The turns ratio and the capacitance will have to be varied to achieve balanced operation at a desired load point. However, there will be a value of capacitance at any load point that will give a minimum component of backward field.

PSC Motor Balancing Calculation Equations

Power component of main winding primary current A:

$$A = I_{mp} \text{PF}$$

Reactive component of main winding primary current R_B:

$$R_B = \sqrt{I_{mp}^2 - A^2}$$

Ratio of effective conductors in capacitor winding to the effective conductors in main winding K:

$$K = \frac{ER_B - K_a r_1 I_{mp}^2 - (R_c/K)I_{mp}^2}{EI_{mp}\text{PF} - r_m I_{mp}^2}$$

Effective reactance of capacitor X_c:

$$X_c = \frac{KE - R_c R_B - K K_a R_B (r_1 + K^2) R_B r_1}{I_{mp} \text{PF}}$$

Line current of capacitor motor I_c:

$$I_c = \frac{I_{mp}\sqrt{K^2 + 1}}{K}$$

Power factor PF:

$$\text{PF} = \frac{A + (R_B/K)}{I_c}$$

Primary copper loss $L_{p,\text{Cu}}$:

$$L_{p,\text{Cu}} = \frac{K - K_a}{2K}$$

Line current of current capacitor motor L_c:

$$L_c = \frac{I_{mp} R_c}{K^2}$$

Capacitor voltage E_c:

$$E_c = I_{1a} X_c$$

Capacitor voltamperes C_{VA}:

$$C_{\text{VA}} = E_c I_{1a}$$

A connection diagram of a single-speed PSC motor is shown in Fig. 6.43. For some applications, it is adequate to bring only three leads out of the motor by using an internal connection. The capacitor is often called a *run capacitor*, even though it remains connected to the motor during both starting and running operation.

PSC motors are commonly used for multispeed applications. Three common connections are shown in Figs. 6.44 and 6.45. Figure 6.44 represents a T-connected motor. Figure 6.45 represents an L-connected motor. Speed is selected by connecting the power source between the common lead and one of the speed leads. The lead colors shown are commonly used, but others may be substituted.

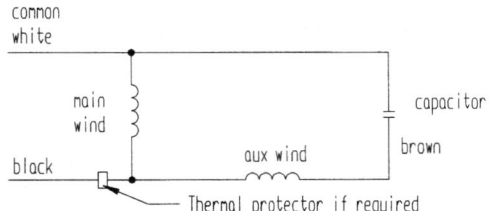

FIGURE 6.43 PSC wiring diagram.

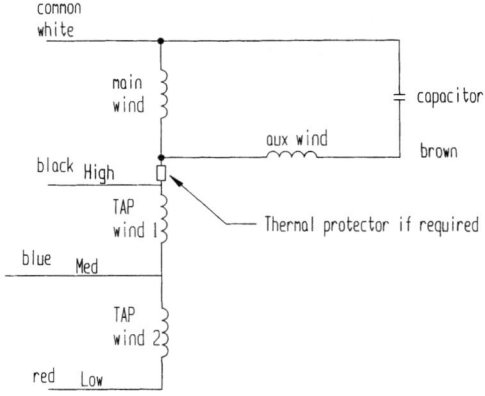

FIGURE 6.44 T-connected multispeed PSC motor.

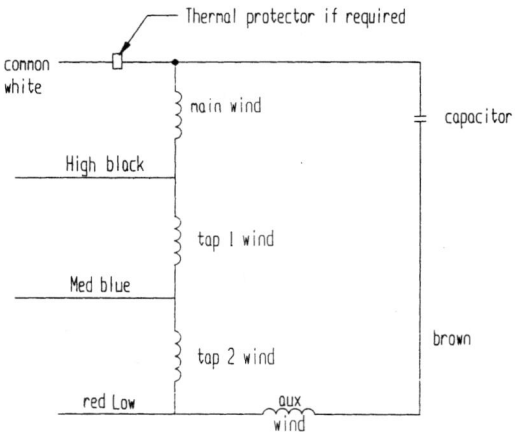

FIGURE 6.45 L-connected multispeed PSC motor.

6.4.3 GENERAL PROCEDURES FOR CALCULATING PERFORMANCE OF SHADED-POLE MOTORS*

Shaded-pole motors generally consist of a squirrel-cage rotor inside a stator which contains a shading band. The shading band acts as an auxiliary winding in this type of induction motor. In the 1890s it was discovered that a squirrel-cage motor with a salient-pole single-phase field will run if a portion of the pole is short-circuited with a winding or coil (Trickey, 1936). The theory behind the shaded-pole motor is that the shading coil causes the flux in that portion of the pole to lag a small fraction of

---

* Subsections 6.4.3 and 6.4.4 contributed by Andrew E. Miller, Yeadon Engineering Services, PC.

the time cycle behind the flux in the main part of the pole, causing the points of maximum flux to progress around the motor, resulting in a rather poor rotating field which draws the rotor around with it, as in any induction motor. If there were no shading coil on the salient pole, the flux in that pole and in the air gap under it would simply alternate from north to south, and no rotation would be produced. The rotating field draws the rotor from the main pole toward the shading pole. There are two typical configurations of shaded pole motors, round-frame and C-frame. See Fig. 6.46 for a *round-frame* shaded-pole motor and Fig. 6.47 for a *C-frame* or *skeleton-frame* shaded-pole motor.

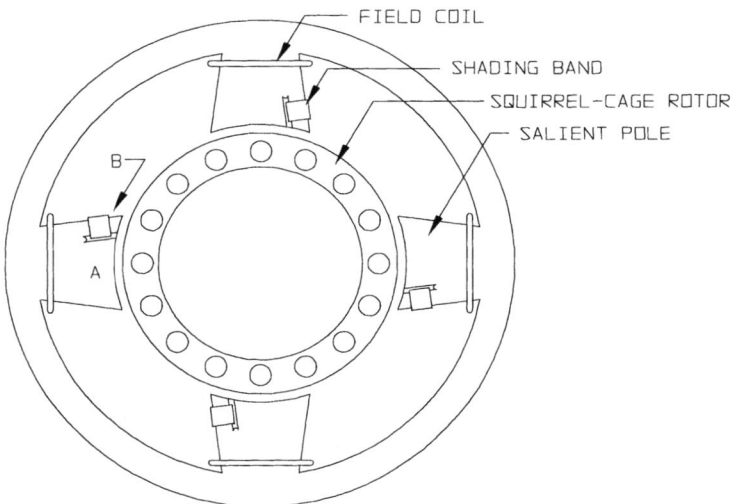

FIGURE 6.46 Round-frame shaded-pole motor.

Construction. The shaded-pole stator has a magnet wire winding and a shading band. The shading band is made of a horseshoe-shaped copper strip that is inserted after the stator lamination stack is formed. The ends of the strip can then be welded together. A drawing of a round-frame stator lamination is shown in Fig. 6.48. According to Chang (Chang and Karr, 1949; 1950), the pitch of the shading coil pitch θ_s should be a minimum of 46.5° electrical, with a proper range between 46.5 and 60°. Trickey suggests using between one-quarter and one-half of the pole pitch for the shaded pole, or 45 to 90° electrical.

The rotor is usually a squirrel-cage type, consisting of steel laminations with copper or aluminum bars. After the rotor lamination stack is formed, the conducting material is cast into the slots and molded at the ends to form end rings, thus connecting all of the bars at each end. Copper conductors have lower resistance, but aluminum is easier to cast and is therefore more commonly used today.

Slot Combinations. Choosing a good combination of stator poles and rotor slots can be a challenge. Too few conductors can result in high zigzag leakage reactance, which reduces the motor output. According to Chang, the optimum number of rotor slots is about 11 to 13 per pole pair.

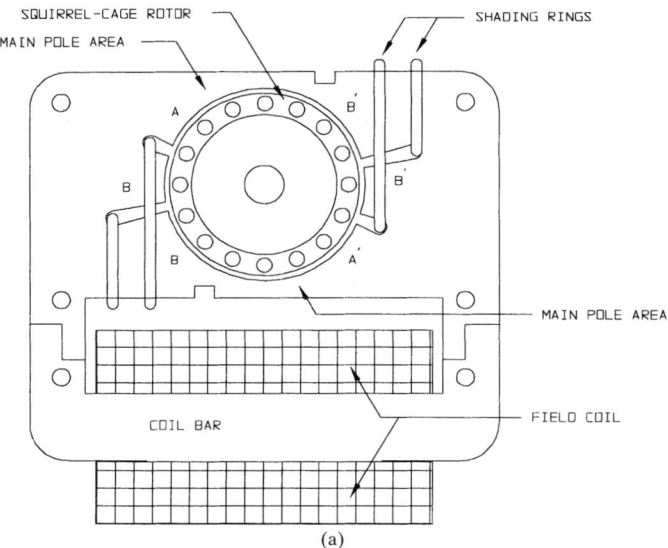

FIGURE 6.47 C-frame shaded-pole motor.

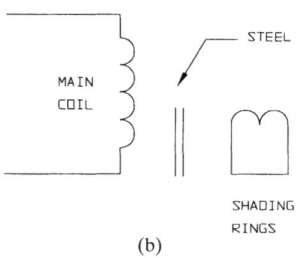

FIGURE 6.47 (*Continued*) C-frame shaded-pole motor.

Performance Calculations. Shaded-pole motors typically have low starting torque, low power factor, and low efficiency. A typical speed-torque curve is shown in Fig. 6.49.

The maximum rotor speed (or *synchronous speed*) is given by the following equation:

$$\text{Syn} = \frac{120f}{P}$$

where f = frequency, Hz
P = number of poles

Figure 6.50 shows an equivalent circuit of a shaded-pole motor, as constructed by P. H. Trickey. Trickey uses the cross-field theory of induction motors for his equivalent circuit, as opposed to the revolving-field theory. All of the equations in this subsection are derived from the cross-field theory. From this equivalent circuit, using Kirchhoff's voltage laws, the following four equations can be derived:

$$I_{2a}(r_{2a} + jX_{2a}) - j(I_1 - I_a - I_{2a})X_{ma} + S(I_1 - I_{2m})X_{mm} - SI_{2m}X_{2m} = 0$$

$$I_a(r_a + jX_a) - j(I_1 - I_a)X_{a1} - j(I_1 - I_a - I_{2a})X_{ma} = 0$$

$$I_1(r_1 + jX_1) + j(I_1 - I_{2m})X_{mm} + j(I_1 - I_a)X_{a1} + j(I_1 - I_a - I_{2a})X_{ma} = E$$

$$I_{2m}(r_{2m} + jX_{2m}) - j(I_1 - I_{2m})X_{mm} - S(I_1 - I_a - I_{2a})X_{ma} + SI_{2a}X_{2a} = 0$$

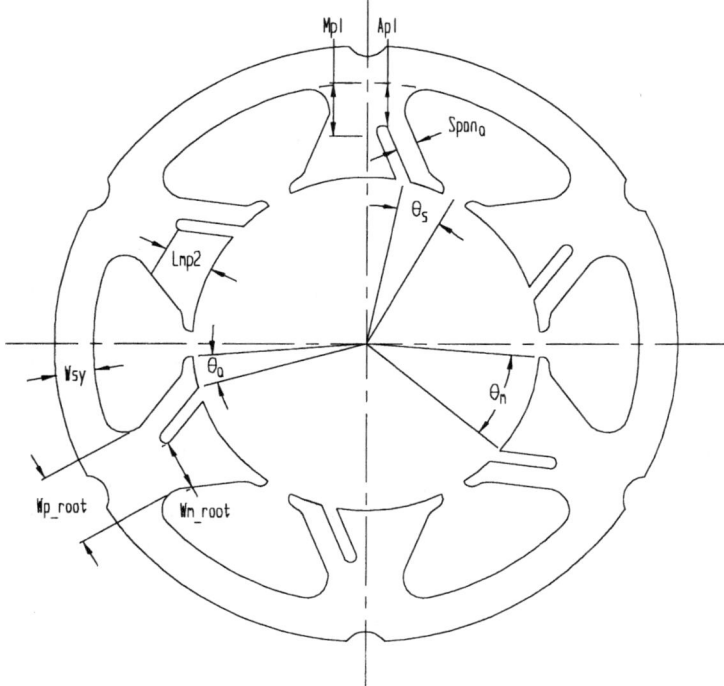

FIGURE 6.48 Shaded-pole laminations.

where r_1 = primary winding resistance
r_a = shading band resistance reflected onto main winding
X_1 = reactance due to main field
X_{ma} = reactance of fundamental mutual shaded-field air gap flux
X_{mm} = reactance of fundamental mutual main-field air gap flux
X_a = reactance of the shading coil leakage flux
X_{a1} = sum of reactance caused by skew through shaded field flux, reactance of harmonics of shaded-field air gap flux, and reactance of magnetic-bridge or pole-tip flux not linking the rotor conductors
S = ratio of rotor speed to synchronous speed at fundamental frequency

To solve these equations in terms of the various currents, the reactances and resistances of the motor must first be calculated. This step is much more complicated than it may sound. We begin by calculating steel areas and volumes. Unless otherwise mentioned, all length dimensions are in inches.

$$L_{g,\text{axial}} = W_s + K_{\text{FPPP}} L_{g,\text{mech}}$$

where W_s = short stack length (either rotor stack length or stator stack length)
$L_{g,\text{mech}}$ = mechanical air gap length
$L_{g,\text{axial}}$ = axial air gap length
K_{FPPP} = air gap constant defined as:

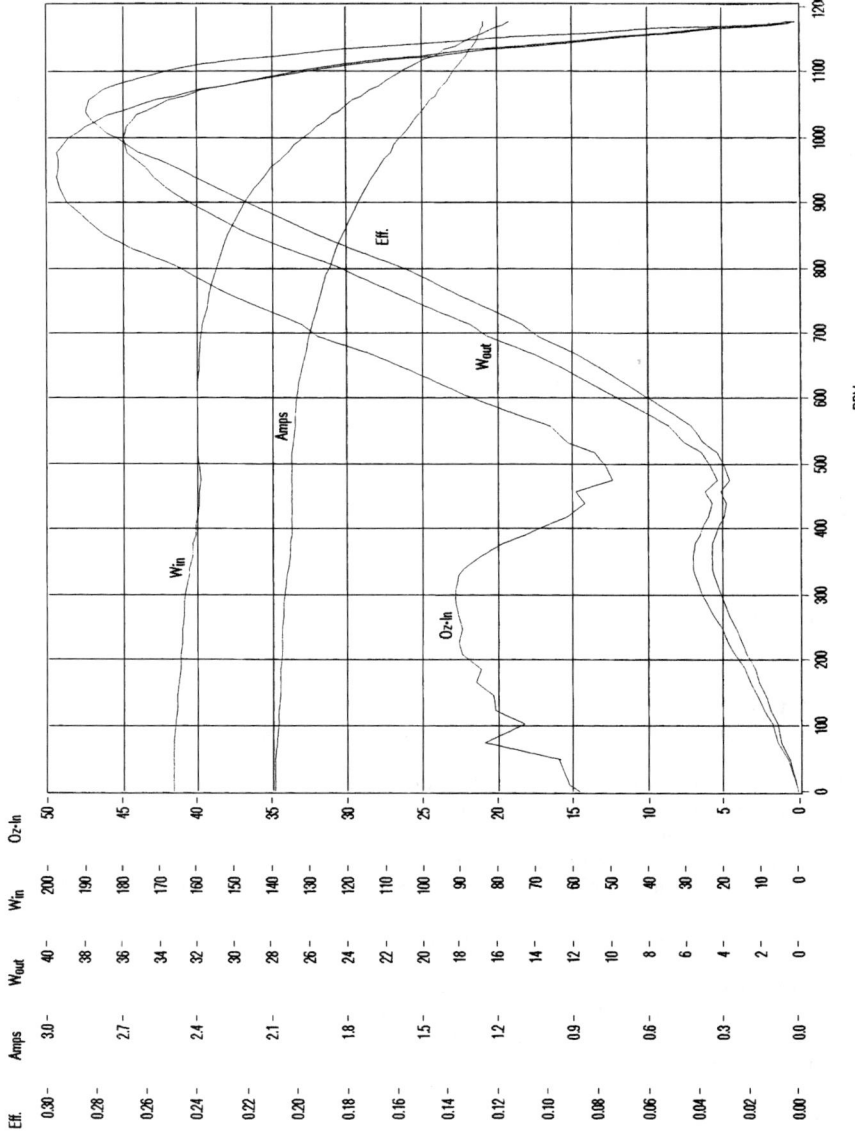

FIGURE 6.49 Typical shaded-pole motor performance curve.

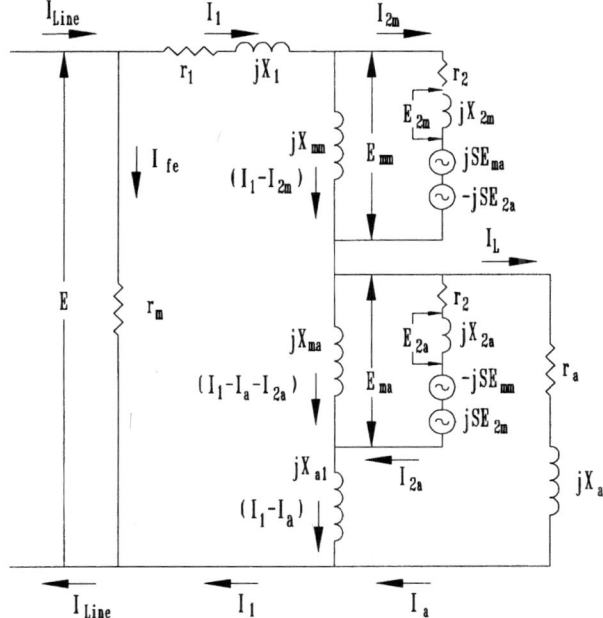

FIGURE 6.50 Shaded-pole motor equivalent circuit.

$$K_{\text{FPPP}} = 1 + 0.138\left(\frac{L_{oe}}{L_{g,\text{mech}}}\right) - \frac{0.0039(L_{oe}/L_{g,\text{mech}})^2}{1 + 0.0159(L_{oe}/L_{g,\text{mech}})}$$

where L_{oe} is the length of the overhang per end. Then:

$$A_{\text{pole},m} = L_{g,\text{axial}} L_{\text{pole},m}$$

where $L_{\text{pole},m}$ = arc length of the main pole (see Fig. 6.51)
$A_{\text{pole},m}$ = magnetic area of the air gap

Next, the width of the rotor yoke W_{ry} must be calculated. If the shaft material is magnetic, then:

$$W_{ry} = \frac{\text{OD}_{\text{rot}}}{2} - D_{\text{rotor slot}}$$

where OD_{rot} = outside diameter of the rotor
$D_{\text{rotor slot}}$ = depth of one rotor slot

If the shaft material is nonmagnetic, then:

$$W_{ry} = \frac{\text{OD}_{\text{rot}} - \text{ID}_{\text{rot}}}{2} - D_{\text{rotor slot}}$$

where ID_{rot} is the inside diameter of the rotor (or bore diameter).

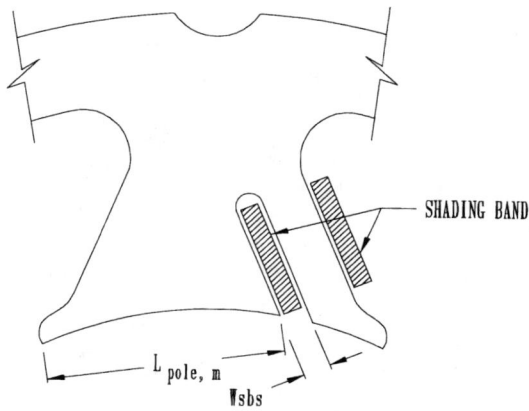

FIGURE 6.51 Main pole length.

To calculate the magnetic length of the air gap, we must first calculate Carter's coefficient K_c, as in the following formula:

$$K_c = \frac{\text{tp}_{\text{rot}}(5L_{g,\text{mech}} + W_{20})}{\text{tp}_{\text{rot}}(5L_{g,\text{mech}} + W_{20}) - W_{20}^2}$$

where W_{20} = rotor slot opening
tp_{rot} = rotor tooth pitch (rotor circumference divided by number of rotor slots)

Then, the effective air gap length $L_{g,\text{mag}}$ is found by multiplying the mechanical air gap length by Carter's coefficient, or:

$$L_{g,\text{mag}} = L_{g,\text{mech}} K_c$$

where $L_{g,\text{mag}}$ is the effective length of the air gap to account for slot openings and iron saturation.

Next, we calculate the magnetic areas and flux path lengths. Naturally, the magnetic circuit will be slightly different for a C-frame shaded-pole motor (Fig. 6.52) than for a round frame. The area that will be affected most is the area of the stator yoke A_{sy}. The other areas of interest are the area of the main and primary portion of the stator pole $A_{m,\text{root}}$ and $A_{p,\text{root}}$, respectively; the area of the rotor yoke A_{ry}; and the area of the rotor tooth A_{rt}. For the C-frame motor, the stator yoke area is divided into two sections and approximated using the following set of formulas:

$$W_{sy1} = \frac{W_{\text{side}}}{2}$$

$$W_{sy2} = \frac{W_{\text{sidl}}}{2}$$

$$A_{sy1} = 2L_{\text{stk},S} F_{\text{stk},S} W_{sy1}$$

$$A_{sy2} = 2L_{\text{stk},S} F_{\text{stk},S} W_{sy2}$$

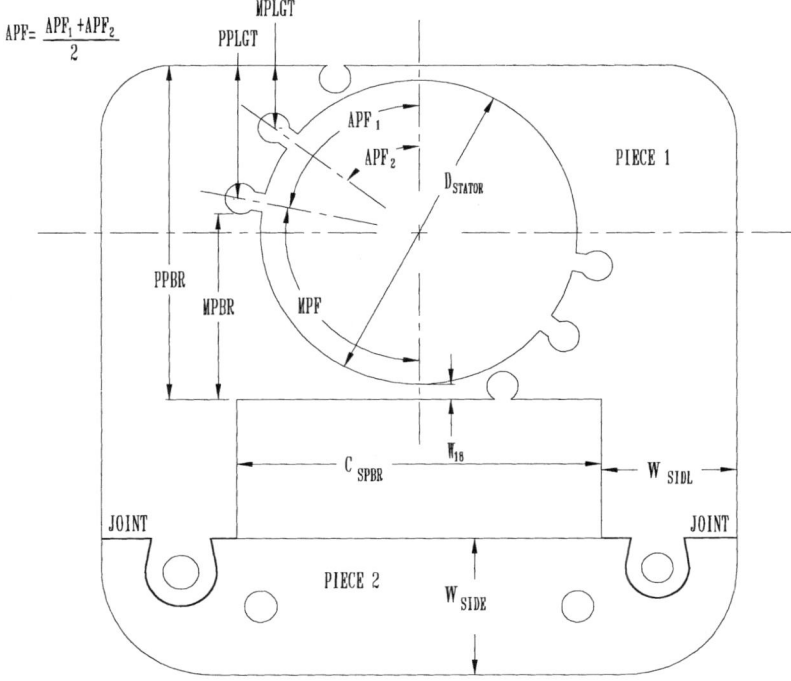

FIGURE 6.52 C-frame dimensions.

For the round-frame motor (Fig. 6.48), the formula is

$$A_{sy} = 2L_{stk,S}F_{stk,S}W_{sy}$$

where $L_{stk,S}$ = length of stator stack
$F_{stk,S}$ = stator lamination stacking factor to account for air between laminations (generally around 0.95)
W_{sy} = width of stator yoke

$$A_{m,root} = W_{m,root}L_{stk,S}F_{stk,S}$$

$$A_{p,root} = W_{p,root}L_{stk,S}F_{stk,S}$$

where $W_{m,root}$ and $W_{p,root}$ are the widths of the main and primary portion of the stator pole at the root.

$$A_{ry} = 2L_{stk,R}F_{stk,R}W_{ry}$$

$$A_{rt} = \frac{W_{rt}N_{rt}F_{stk,R}L_{stk,R}}{P}$$

where $L_{stk,R}$ = rotor stack length
$F_{stk,R}$ = rotor lamination stacking factor

N_{rt} = number of rotor teeth
P = number of stator poles

For round-frame motors, we also include the trapezoidal tooth portion of the pole, as in the following two equations:

$$W_{mp2} = \frac{2W_{m,root} + L_{pole,m}}{3}$$

$$A_{mp2} = W_{mp2} L_{stk,S} F_{stk,S}$$

To calculate their respective magnetic volumes, the flux path lengths must now be calculated. Once again, there is a difference in the stator yoke calculations for the two stator shapes. For the C-frame motor, the following equations are used:

$$L_{sy1} = \frac{C_{spbr} + W_{sid,L}}{2}$$

$$L_{sy2} = \frac{U_{lgth} + W_{m,root} - 2L_{sy1}}{2}$$

where C_{spbr} = width of the bobbin-carrying section
$W_{sid,L}$ = width of the side "legs" of the stator

For the round frame, these equations are used:

$$L_{sy,mech} = \frac{\pi(OD_{stat} - W_{sy})}{2P}$$

$$L_{sy,mag} = L_{sy,mech} - \frac{W_{p,root}}{4}$$

where $L_{sy,mech}$ and $L_{sy,mag}$ are the mechanical and magnetic lengths of the stator yoke.

We now calculate the flux densities in the motor. The motor is again broken up into different sections in order to calculate the flux density in each section. We start by estimating the total flux from the following formula:

$$\text{Flux} = \frac{45(V/f)}{C/1000} (0.97)$$

where V = motor voltage
f = frequency
flux = motor flux, kline

$$B_{sy} = \frac{\text{flux}}{A_{sy}}$$

$$B_{pp} = \frac{\text{flux}}{A_{p,root}}$$

$$B_{mp} = \frac{\text{flux}}{A_{m,root}}$$

$$B_{mp2} = \frac{\text{flux}}{A_{mp2}}$$

$$B_{ry} = \frac{\text{flux} \times \text{Ass}_{Kp}}{A_{ry}}$$

$$B_{rt} = \frac{\frac{\pi}{2}\text{Ass}_{Kp} \times \text{flux}}{A_{rt}}$$

$$B_{gap} = \frac{\text{flux} \times \text{Ass}_{Kp}}{A_{pole,m}}$$

where B_{sy} = flux density in the stator yoke (B_{sy1} and B_{sy2} for C-frame motors)
 B_{pp} = flux density in root section of primary portion of stator poles
 B_{mp} = flux density in root section of main portion of stator poles
 B_{mp2} = flux density in tooth section of main portion of stator poles
 A_{mp2} = area or tooth section of main portion of stator poles
 B_{ry} = flux density in rotor yoke
 B_{rt} = flux density in rotor teeth
 B_{gap} = flux density in air gap
 Ass_{Kp} = factor to allow for voltage drop due to main winding leakage, defined as:

$$\text{Ass}_{Kp} = \frac{\text{voltage in rotor}}{\text{terminal voltage}} \approx \left(1 - \frac{P}{100}\right)(0.85)$$

Now the *BH* curve(s) for the steel laminations are needed. A value of *H* (field intensity) must be recorded for each *B* value. The mmf drops can then be calculated by multiplying the field intensity (in amp-turns per inch) by the path length of the section. In other words:

$$NI_x = H_x L_x$$

where x = each section from the previous set of equations
 L_x = length of section x
 H_x = field intensity of section x
 NI_x = mmf drop in section x, A · turns

The two exceptions to the previous formula are the mmf drop in the air gap and the mmf drop in the rotor teeth. The following two equations are used for those sections:

$$NI_{gap} = 313 B_{gap} L_{g,mag}$$

$$NI_{rt} = H_{rt} L_{rt} \frac{\pi}{2}$$

The total mmf drop is the sum of the mmf drops for each section, or:

$$NI_{tot} = NI_{gap} + NI_{sy} + NI_{pp} + NI_{mp} + NI_{mp2} + NI_{ry} + NI_{rt}$$

Note: For motors with more than one stator piece (like the C-frame motor in Fig. 6.47), there will also be an mmf drop across the joint. In this case, an estimate will have to be made for the length and area, and then the mmf drop is calculated as with any other section.

The main saturation factor S_{FM} is found by dividing the total mmf drop over the mmf drop in the air gap, or:

$$S_{FM} = \frac{NI_{tot}}{NI_{gap}}$$

This brings us to the reactance equations. Before we begin, we must define a couple of multiplication factors. These are as follows:

$$\theta_{skew} = \frac{\pi \times skew}{pole\ pitch}$$

$$C_{skew} = \frac{\sin(\theta_{skew}/2)}{\theta_{skew}/2}$$

where skew = circumferential skew along the outside of the rotor, in (Fig. 6.53)
C_{skew} = Trickey's skew factor

$$L_{pole,m} = \frac{\theta_m}{360} \pi ID_s$$

$$\alpha_M = \frac{L_{pole,m}}{pole\ pitch}$$

$$\beta_M = \alpha_M \pi$$

$$C_{fm} = \sin\left(\frac{\beta_M}{2}\right)$$

where θ_m = arc angle of the main pole, ° mechanical
β_M = main pole pitch, rad electrical
C_{fm} = Trickey's main-pole multiplication factor

Similarly, for the auxiliary pole:

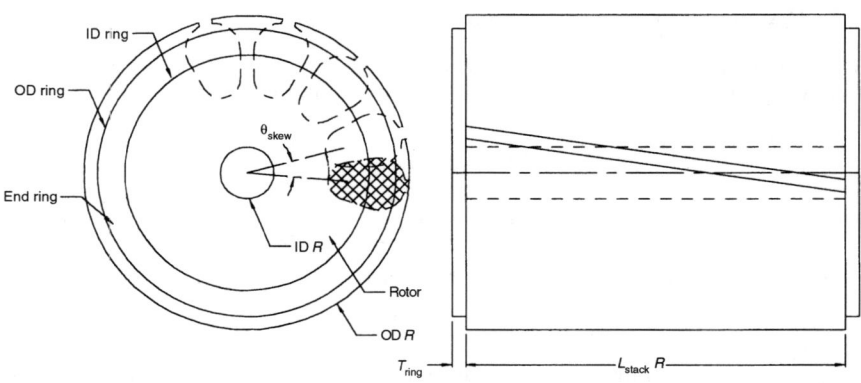

FIGURE 6.53 Rotor skew.

$$L_{\text{pole},a} = \frac{\theta_a}{360} \pi ID_s$$

$$\alpha_A = \frac{L_{\text{pole},a}}{\text{pole pitch}}$$

$$\beta_A = \alpha_A \pi$$

$$C_{\text{fa}} = \sin\left(\frac{\beta_A}{2}\right)$$

where θ_a = arc angle of the auxiliary pole, ° mechanical
C_{fa} = Trickey's auxiliary-pole correction factor

$$C_{\text{kw}} = CC_{\text{fm}}$$

where C = number of conductors [defined by $C = 2\,(TPC(P))$]
C_{kw} = effective number of conductors

$$X_{\text{mm}} = C^2 f 2\pi\,(10^{-8})(0.647)\left(\frac{L_{g,\text{axial}} \times \text{pole pitch}}{L_{g,\text{mag}} S_{\text{FM}} P}\right) C_{\text{skew}} \alpha_M$$

$$X_{\text{ma}} = f 2\pi\,(10^{-8})\left(\frac{C_{\text{fa}} C}{C_{\text{fm}}}\right)^2 0.647 \left(\frac{L_{g,\text{axial}} \times \text{pole pitch}}{L_{g,\text{mag}} S_{\text{FA}} P}\right) C_{\text{skew}} \alpha_A$$

where S_{FM} = main saturation factor (as calculated previously or by an alternate method)
S_{FA} = auxiliary saturation factor (assumed to be 1.157 or calculated)
X_{mm} = reactance of fundamental mutual main-field air gap flux
X_{ma} = reactance of fundamental mutual shaded-field air gap flux

For the reader who would like a more in-depth analysis, Chang uses harmonics much more intensively than Trickey does. He calculates reactances at each harmonic and then takes the total of all of the harmonics, with some adding and some subtracting. For example, Chang calculates the mutual reactance of the nth harmonic using the following formulas:

$$C_{\text{skew},n} = \frac{\sin(n\theta_{\text{skew}}/2)}{n\theta_{\text{skew}}/2}$$

$$X_{\text{mn}} = 2\pi f C_n^2\,(10^{-8})\left(\frac{0.647 L_{\text{stk},R} \times \text{pole pitch}}{L_{g,\text{mag}} P n^2}\right) C_{\text{skew},n}$$

for the nth harmonic equivalent circuit, where $n = 1, 3, 5$, etc. The total mutual reactance would then be the sum of each harmonic, or:

$$X_m = \sum X_{m1} + X_{m3} + X_{m5} \cdots$$

This section adheres to Trickey's formulas in an attempt to minimize the overall complexity of the subject.
Additional reactances are as follows:

$$X_{\text{skap}} = X_{\text{ma}}\left(\frac{1 - C_{\text{skew}}}{C_{\text{skew}}}\right)$$

$$X_{aap} = \frac{X_{ma}}{C_{skew}} \left(\frac{1.232\alpha_A}{C_{fa}^2} - 1 \right)$$

where X_{skap} = reactance in auxiliary portion of stator pole due to rotor skew
X_{aap} = additional reactance in auxiliary portion of stator pole

$$X_{slt2} = \frac{C^2 f 2\pi \, (6.38 L_{stk,R} K_{S2} C_{fm}^2)(10^{-8})}{N_{rot}}$$

where K_{S2} = rotor slot constant
X_{slt2} = rotor slot leakage

$$X_{zz2} = \left(\frac{C^2 f 2\pi \, (10^{-8})(6.38 L_{stk,R} W_{rtt})}{N_{rot} 4 L_{g,mech}} \right) (\alpha_M + \alpha_A) \, C_{fm}^2$$

where W_{rtt} = width of rotor tooth at tip (approximately $tp_{rot} - W_{20}$)
X_{zz2} = zigzag reactance

$$X_{skmp} = X_{mm} \left(\frac{1 - C_{skew}}{C_{skew}} \right)$$

$$X_{amp} = \frac{X_{mm}}{C_{skew}} \left(\frac{1.232\alpha_M}{C_{fm}^2} - 1 \right)$$

where X_{skmp} = reactance in main portion of stator pole due to rotor skew
X_{amp} = additional reactance in main portion of stator pole

Combining terms, we have:

$$X_{2m} = X_{slt2} + X_{zz2} + X_{skmp}$$

$$X_{2a} = X_{slt2} + X_{zz2} + X_{skap}$$

where X_{2m} = total reactance in main portion of pole
X_{2a} = total reactance in auxiliary portion of pole

$$X_a = f 2\pi \, (10^{-8}) \left(\frac{C_{fa} C}{C_{fm}} \right)^2 \left(\frac{2\text{span}_a}{P} + \frac{0.7975 L_{stk,S} \text{SlK}_{sa}}{P \text{Sap}} \right)$$

where SlK_{sa} = Trickey's shading slot constant, approximately $0.6667 + 0.01/W_{sbs}$,
where W_{sbs} = width of shading slot (Fig. 6.54)
Sap = number of shading slots per pole
X_a = shading coil leakage flux reactance

$$X_{brg} = \frac{C^2 f 2\pi \, (10^{-8})(0.7975 L_{stk,S})(T_{stt}/W_{sg})}{P}$$

where T_{stt} = thickness of stator tooth (or pole) tip
W_{sg} = width of stator gap (distance between pole tips)
T_{stt}/W_{sg} = Trickey's bridge constant
X_{brg} = reactance of stator bridge

$$X_{a1} = X_{aap} + X_{skap} + X_{brg}$$

where X_{a1} = total reactance in the auxiliary

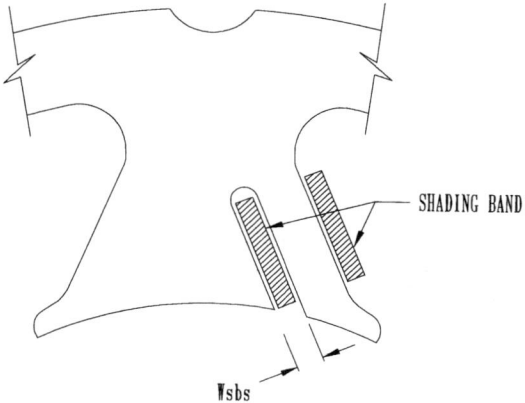

FIGURE 6.54 Shaded section of round-frame motor.

There is a difference here in equations for the C-frame motor and for the round-frame motor. The following equations are only for the round-frame motor.

$$X_{\text{slt1}} = \frac{1.545 C^2 f 2\pi L_{\text{stk},s} K_{s1} (10^{-8})}{P}$$

where X_{slt1} = reactance of the stator slot
K_{s1} = Trickey's stator slot constant, defined approximately as follows:

$$K_{s1} = \frac{1.5 D_{10}}{W_{10}} + \frac{0.667 D_{13}}{W_{12} + W_{13}}$$

where $D_{10}, D_{13}, W_{10}, W_{12},$ and W_{13} are defined in Fig. 6.55.

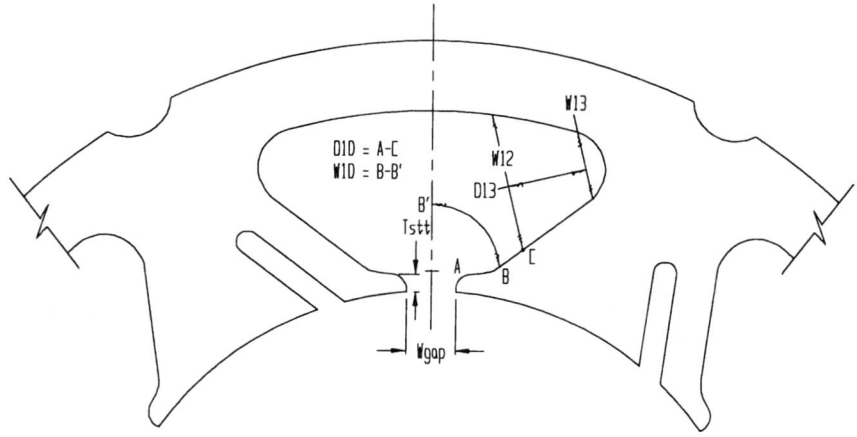

FIGURE 6.55 Dimensions for stator reactance.

$$\theta_p = \frac{2\pi}{P}$$

$$D_{coil} = \frac{L_{mc} - L_{stk,S}}{\theta_p}$$

$$X_{end1} = \frac{1.9C^2 f 2\pi\,(10^{-8})\,D_{coil}\sin(\theta_p)}{P}$$

where θ_p = angle of one pole, rad mechanical
D_{coil} = equivalent diameter of end turn
X_{end1} = reactance of coil at end

Similarly, the following equations are only for C-frame motors.

$$C_{sp2} = \frac{C_{spbr}}{2}$$

$$F_{PPPP} = \left(1 + 0.138\,\frac{W_{side}}{C_{sp2}}\right) - \left[\frac{0.0039\,(W_{side}/C_{sp2})^2}{1 + 0.0159\,(W_{side}/C_{sp2})}\right]$$

$$L_{g,axial2} = L_{stk,S} + 2F_{PPP}C_{sp2}$$

$$K_{sx} = 0.162\left(\frac{C_{spbr}}{2W_{18}} - 1\right)^{0.6}$$

$$X_{slot} = C^2 f 2\pi\,(10^{-8})\left(\frac{1.0633333\,L_{g,axial2}}{2C_{spbr}} + 3.19 L_{g,axial2}K_{sx}\right)$$

$$X_{slt1} = \frac{X_{slot}}{4}$$

$$X_{sc1} = 0.553 W_{side} + 0.276 L_{stk,S} + 0.327 C_{spbr}$$

$$X_{sc11} = 4.98 W_{side}\log\left(1 + \frac{2W_{sidel}}{C_{spbr}}\right) + 2.34 L_{stk,S}\log\left(1 + \frac{2W_{sidel}}{C_{spbr}}\right) + W_{sidel}$$

$$X_{end1} = C^2 f 2\pi\,(10^{-8})\left(\frac{X_{sc1} + X_{sc11}}{4}\right)$$

The rest of the equations apply to both types of motors. Adding terms again gives us the following equation:

$$X_1 = X_{slt1} + X_{end1} + X_{skmp} + X_{amp} + X_{brg}$$

where X_1 = reactance due to skew, harmonics, and leakage in the main

We now calculate the rotor resistance at the base temperature of 20°C, $R_{2,cld}$. Rotor resistance is made up of the bars and the ring. We calculate each of these and then take the sum. We begin with the bars, where the total area of the rotor bars is found by multiplying the area of one bar by the number of bars, or:

$$A_{bar,tot} = A_{bar}N_{rot}$$

$$\text{Bar} = \frac{L_{bar}}{A_{bar,tot}\,\text{Con}_{bar}}$$

where L_{bar} = length of rotor bar
Con_{bar} = unit conductivity of rotor bar material (Cu = 1.0)
bar = bar resistance on a per-unit basis

The following equations are used for the ring resistance:

$$K_{ring} = \frac{P}{2}\left(1 - \frac{ID_{ring}}{DE_2}\right)\frac{1 + (ID_{ring}/D_{E2})^P}{1 - (ID_{ring}/D_{E2})^P}$$

where D_{E2} = diameter of rotor halfway through rotor slots
K_{ring} = Trickey's ring factor

$$A_{ring} = T_{ring}\frac{OD_{ring} - ID_{ring}}{2}$$

where T_{ring} = axial thickness of rotor ring
OD_{ring} and ID_{ring} = outside and inside diameters of rotor ring
A_{ring} = area of rotor ring

$$\text{Ring} = \frac{0.637 D_{E2} K_{ring}}{A_{ring} P^2 Con_{ring}}$$

where Con_{ring} = unit conductivity of bar material (Cu = 1.0)
ring = ring resistance on a per-unit basis

The total rotor resistance is found by taking the sum of the bar and the ring, and converting from a per-unit basis to an actual basis by multiplying by the resistivity of copper. Graphically, the equation is as follows:

$$R_{rot} = 0.6693(\text{bar} + \text{ring})\, 2$$

where 0.6693 = resistivity of copper, $\Omega \cdot$ in
2 = number of phases (even though a shaded-pole motor is generally single-phase)
R_{rot} = actual rotor resistance at 20° C

Next, we convert the *actual* rotor and stator resistances into *effective* resistances, by reflecting them as follows:

$$R_{2,cld} = R_{rot}\left[\frac{C(C_{fm} + C_{fa}/2)}{1000}\right]^2$$

$$R_{1,cld} = \left(\frac{L_{mc}F_{ws}}{12}\right)\left(\frac{C}{1000}\right)\rho_{wire}$$

Now, we calculate the resistance of the shading coil. The following equations are used:

$$A_{shc} = W_{shc} T_{shc}$$

$$R_{a,cld} = 1.385\,\frac{L_{m,shc}}{A_{shc}}\, N_2 P \left(\frac{C}{C_a}\right)^2 (10^{-6})$$

where W_{shc} = width of shading coil
T_{shc} = thickness of shading coil
A_{shc} = cross-sectional area of shading coil
$L_{m,shc}$ = total length of shading coil (up and down the stack)

R_{as} = resistance of shading coil
$R_{2,cld}$ = effective rotor resistance at 20° C
$R_{a,cld}$ = effective auxiliary resistance at 20° C
C_a = total number of shading coil conductors
N_2 = shading-coil turns per pole
F_{ws} = wire stretch factor
ρ_{wire} = resistivity of wire, $\Omega/(1000\ \text{ft})$
$R_{1,cld}$ = winding resistance at 20° C

These resistances can be converted to the actual motor operating temperature by using the following correction factor:

$$T_{mult} = 1 + \frac{T_{max} - 20}{254.5}$$

where T_{max} = maximum motor operating temperature, °C

$$R_{1,hot} = R_{1,cld} T_{mult}$$

$$R_{2,hot} = R_{2,cld} T_{mult}$$

$$R_{a,hot} = R_{a,cld} T_{mult}$$

We now combine terms again to simplify the calculations.

$$X_{02} = X_{2m} + X_{mm}$$

$$X_{03} = X_{2a} + X_{ma}$$

The previously calculated values allow us to solve the current equivalencies much more quickly. As an alternative to complex linear algebra, Trickey has solved the equations and broken them down into a series of B constants. The B constants allow a relatively quick calculation of the various currents. The following equation defines the total current going through the windings:

$$i_m = \frac{V}{X_{02}}$$

where V = source voltage

Referring to the equivalent circuit diagram and to the original four current equations, we now divide through by X_{02}. Rearranging and multiplying the last three equations by j yields a set of equations that can be put into matrix form.

$$\left(\frac{R_{1,hot}}{X_{02}} + j\frac{X_1 + X_{mm} + X_{ma} + X_{a1}}{X_{02}}\right)I_1 + \left(-j\frac{X_{a1} + X_{ma}}{X_{02}}\right)I_a$$

$$+ \left(-j\frac{X_{mm}}{X_{02}}\right)I_{2m} + \left(-j\frac{X_{ma}}{X_{02}}\right)I_{2a} = \frac{V}{X_{02}} = i_m$$

$$\left(\frac{X_{a1} + X_{ma}}{X_{02}}\right)I_1 + \left(j\frac{R_{a,hot}}{X_{02}} - \frac{X_a + X_{a1} + X_{ma}}{X_{02}}\right)I_a + (0)I_{2m} + \left(-\frac{X_{ma}}{X_{02}}\right)I_{2a} = 0$$

$$\left(\frac{X_{mm}}{X_{02}} - jS\frac{X_{ma}}{X_{02}}\right)I_1 + \left(jS\frac{X_{ma}}{X_{02}}\right)I_a + \left(j\frac{R_{2,hot}}{X_{02}} - 1\right)I_{2m} + \left(jS\frac{X_{03}}{X_{02}}\right)I_{2a} = 0$$

$$\left(\frac{X_{ma}}{X_{02}} + jS\frac{X_{mm}}{X_{02}}\right)I_1 + \left(-\frac{X_{ma}}{X_{02}}\right)I_a + (-jS)I_2m + \left(j\frac{R_{2,\text{hot}}}{X_{02}} - \frac{X_{03}}{X_{02}}\right)I_{2a} = 0$$

After solving these equations in terms of known values, in order to simplify calculations, Trickey created a set of equations using the B constants. The equations for the 20 B constants are shown at the end of this subsection.

We then define eight more variables to simplify the final equations.

$$U = B_{19} + (1 - S^2)B_{17}$$

$$W = B_{20} + (1 - S^2)B_{18}$$

$$N_{1h} = (1 - S^2)B_{17} + B_{19}$$

$$N_{1v} = (1 - S^2)B_{18} + B_{20}$$

$$N_{2h} = B_1 + SB_{10} + (1 - S^2)B_{12}$$

$$N_{2v} = B_{15}$$

$$N_{3h} = B_9 + SB_3 + (1 - S^2)B_{14}$$

$$N_{3v} = B_{11} + (1 - S^2)B_6 - SB_8$$

$$N_{4h} = B_8 - (1 - S^2)B_2 - SB_{11}$$

$$N_{4v} = (1 - S^2)B_4 + SB_9 + B_3$$

where S = ratio of speed to synchronous speed ($S = 1 - $ slip SL)

We can now rewrite the current equations as follows:

$$I_1 = i_m \frac{N_{1h} + N_{1v}}{U + jW}$$

$$I_a = i_m \frac{N_{2h} + jN_{2v}}{U + jW}$$

$$I_{2m} = i_m \frac{N_{3h} + jN_{3v}}{U + jW}$$

$$I_{2a} = i_m \frac{N_{4h} + jN_{4v}}{U + jW}$$

The line current I_{line} is calculated by the following formulas:

$$I_{1h} = \left(\frac{UN_{1h}}{U^2 + W^2} + \frac{WN_{1v}}{U^2 + W^2}\right)i_m + I_{FE}$$

$$I_{1v} = \sqrt{I_1^2 - \left[\left(\frac{UN_{1h}}{U^2 + W^2} + \frac{WN_{1v}}{U^2 + W^2}\right)i_m\right]^2}$$

$$I_{\text{line}} = \sqrt{I_{1v}^2 + I_{1h}^2}$$

where I_{FE} = core loss, W, divided by terminal voltage = P_{FE}/V

Slip-Dependent Calculations

Obviously, solving the equivalent circuit is only part of the problem in calculating the performance of a motor. Speed, torque, and power are also very useful to know. The following equations all take place at a given slip SL. The easiest variable to calculate is, of course, speed:

$$S = 1 - SL$$

$$\text{Speed} = S \times \text{syn}$$

Experience has shown that torque is best calculated using two different equations. For speeds above breakdown, we first calculate power and then derive torque from the output power, as in the following set of equations:

$$P_1 = \frac{(X_{mm}/X_{02})[N_{4h}(N_{1h} - N_{3h})] + N_{4v}(N_{1v} - N_{3v})}{U^2 + W^2}$$

$$P_2 = \frac{(X_{ma}/X_{02})[(N_{1h} - N_{4h} - N_{2h})N_{3h} + N_{3v}(N_{1v} - N_{4v} - N_{2v})]}{U^2 + W^2}$$

$$P_3 = P_1 - P_2$$

$$P_{dev} = P_3 S i_m V$$

$$P_{out} = P_{dev} - S^2(P_{HiCy} + P_{f+w})$$

$$T_{load} = \frac{1,008,000 P_{out}}{\text{speed} \times 746}$$

where P_1, P_2, and P_3 = temporary power variables to shorten equations, W
P_{dev} = total developed power, W
P_{out} = actual output power, W
P_{HiCy} = high-cycle energy loss
P_{f+w} = friction and windage loss at synchronous speed
T_{load} = load torque, oz · in

Unfortunately, these equations tend to predict torque values too high after breakdown. Therefore, after breakdown we use the following equation for torque:

$$T_{load} = \frac{(I_{2a}^2 + I_{2m}^2) R_{2,hot} (112.7)(12)}{\text{syn (SL)}}$$

To calculate efficiency, we must first know input torque. Input torque is equal to developed torque plus losses. We must include the I^2R losses in each section, as in the following set of equations:

$$P_{cop,M} = I_{line}^2 R_{1,hot}$$

$$P_{cop,A} = I_a^2 R_{a,hot}$$

$$S_{cop,M} = I_{2m}^2 R_{2,hot}$$

$$S_{cop,A} = I_{2a}^2 R_{2,hot}$$

$$P_{in} = P_{cop,M} + P_{cop,A} + S_{cop,M} + S_{cop,A} + P_{dev}$$

where P_{in} = input power, W

From here, calculating efficiency, power factor, and horsepower is simple:

$$\text{Eff} = \frac{P_{out}}{P_{in}}$$

$$\text{PF} = \frac{P_{in}}{VI_{line}}$$

$$\text{HP} = \frac{P_{out}}{746}$$

Equations for Trickey's **B** *Constants*

$$B_1 = \left(\frac{R_{2,\text{hot}}}{X_{02}}\right)^2 \left(\frac{x_{a1} + x_{ma}}{X_{02}}\right)$$

$$B_2 = \left(\frac{X_{ma}}{X_{02}}\right)\left(\frac{X_a}{X_{02}}\right)$$

$$B_3 = B_2 \left(\frac{R_{2,\text{hot}}}{X_{02}}\right)$$

$$B_4 = \left(\frac{R_{a,\text{hot}}}{X_{02}}\right)\left(\frac{X_{ma}}{X_{02}}\right)$$

$$B_5 = \left(\frac{R_{a,\text{hot}}}{X_{02}}\right)\left(\frac{X_{03}}{X_{02}}\right)$$

$$B_6 = B_5 \left(\frac{X_{mm}}{X_{02}}\right)$$

$$B_7 = \left(\frac{R_{2,\text{hot}}}{X_{02}}\right)\left[\left(\frac{R_{2,\text{hot}}}{X_{02}}\right)\left(\frac{X_a + X_{a1} + X_{ma}}{X_{02}}\right) + \left(\frac{R_{a,\text{hot}}}{X_{02}}\right)\left(1 + \frac{X_{03}}{X_{02}}\right)\right]$$

$$B_8 = \left(\frac{R_{a,\text{hot}}}{X_{02}}\right)\left(\frac{X_{ma}}{X_{02}}\right)\left(\frac{R_{2,\text{hot}}}{X_{02}}\right)$$

$$B_9 = \left(\frac{R_{a,\text{hot}}}{X_{02}}\right)\left(\frac{X_{mm}}{X_{02}}\right)\left(\frac{R_{2,\text{hot}}}{X_{02}}\right)$$

$$B_{10} = \left(\frac{R_{2,\text{hot}}}{X_{02}}\right)\left(\frac{X_{mm}}{X_{02}}\right)\left(\frac{X_{ma}}{X_{02}}\right)$$

$$B_{11} = \left(\frac{R_{2,\text{hot}}}{X_{02}}\right)\left(\frac{X_{mm}}{X_{02}}\right)\left(\frac{X_a + X_{a1} + X_{ma}}{X_{02}}\right)$$

$$B_{12} = \left(\frac{X_{ma}}{X_{02}}\right)^2 - \left(\frac{X_{03}}{X_{02}}\right)\left(\frac{X_{a1} + X_{ma}}{X_{02}}\right)$$

$$B_{13} = \left(\frac{X_{ma}}{X_{02}}\right)^2 - \left(\frac{X_{03}}{X_{02}}\right)\left(\frac{X_a + X_{a1} + X_{ma}}{X_{02}}\right)$$

$$B_{14} = \left(\frac{X_{mm}}{X_{02}}\right)\left[\left(\frac{X_{ma}}{X_{02}}\right)^2 - \left(\frac{X_{03}}{X_{02}}\right)\left(\frac{X_a + X_{a1} + X_{ma}}{X_{02}}\right)\right]$$

$$B_{15} = \left(\frac{R_{2,hot}}{X_{02}}\right)\left[\left(1 + \frac{X_{03}}{X_{02}}\right)\left(\frac{X_{a1} + X_{ma}}{X_{02}}\right) - \left(\frac{X_{ma}}{X_{02}}\right)^2\right]$$

$$B_{16} = \left(\frac{R_{2,hot}}{X_{02}}\right)\left[\left(1 + \frac{X_{03}}{X_{02}}\right)\left(\frac{X_a + X_{a1} + X_{ma}}{X_{02}}\right) - \left(\frac{X_{ma}}{X_{02}}\right)^2 - \left(\frac{R_{a,hot}}{X_{02}}\right)\left(\frac{R_{2,hot}}{X_{02}}\right)\right]$$

$$B_{17} = \left(\frac{R_{a,hot}}{X_{02}}\right)\left[\left(\frac{X_{ma}}{X_{02}}\right)^2 + \left(\frac{X_{mm}}{X_{02}}\right)^2\left(\frac{X_{03}}{X_{02}}\right) - \left(\frac{X_{03}}{X_{02}}\right)\left(\frac{X_1 + X_{mm} + X_{ma} + X_{a1}}{X_{02}}\right)\right]$$
$$+ \left(\frac{R_{1,hot}}{X_{02}}\right)\left[\left(\frac{X_{ma}}{X_{02}}\right)^2 - \left(\frac{X_{03}}{X_{02}}\right)\left(\frac{X_a + X_{a1} + X_{ma}}{X_{02}}\right)\right]$$

$$B_{18} = \left(\frac{R_{1,hot}}{X_{02}}\right)\left(\frac{R_{a,hot}}{X_{02}}\right)\left(\frac{X_{03}}{X_{02}}\right) + \left(\frac{X_1 + X_{mm}}{X_{02}}\right)\left[\left(\frac{X_{ma}}{X_{02}}\right)^2\right.$$
$$\left. - \left(\frac{X_{03}}{X_{02}}\right)\left(\frac{X_a + X_{a1} + X_{ma}}{X_{02}}\right)\right] + \left(\frac{X_{ma}}{X_{02}}\right)^2\left(\frac{X_a}{X_{02}}\right) - \left(\frac{X_{mm}}{X_{02}}\right)^2\left[\left(\frac{X_{ma}}{X_{02}}\right)^2\right.$$
$$\left. - \left(\frac{X_{03}}{X_{02}}\right)\left(\frac{X_a + X_{a1} + X_{ma}}{X_{02}}\right)\right] - \left(\frac{X_a}{X_{02}}\right)\left(\frac{X_{03}}{X_{02}}\right)\left(\frac{X_{a1} + X_{ma}}{X_{02}}\right)$$

$$B_{19} = \left(\frac{R_{2,hot}}{X_{02}}\right)\left\{\left[\left(\frac{R_{2,hot}}{X_{02}}\right)\left(\frac{X_a + X_{a1} + X_{ma}}{X_{02}}\right) + \left(\frac{R_{a,hot}}{X_{02}}\right)\left(1 + \frac{X_{03}}{X_{02}}\right)\right]\right.$$
$$+ \left(\frac{X_a + X_{a1} + X_{ma}}{X_{02}}\right)\left(\frac{X_{mm}}{X_{02}}\right)^2\right\} + \left(\frac{R_{2,hot}}{X_{02}}\right)\left[\left(\frac{X_a}{X_{02}}\right)\left(\frac{X_{ma}}{X_{02}}\right)^2\right.$$
$$+ \left(\frac{X_{a1} + X_{ma}}{X_{02}}\right)\left(1 + \frac{X_{03}}{X_{02}}\right)\left(\frac{X_{a1} + X_{ma}}{X_{02}}\right)\right]$$
$$+ \left(\frac{R_{2,hot}}{X_{02}}\right)\left[\left(\frac{R_{a,hot}}{X_{02}}\right)\left(\frac{R_{2,hot}}{X_{02}}\right)\left(\frac{X_1 + X_{mm} + X_{ma} + X_{a1}}{X_{02}}\right)\right]$$
$$- \left(\frac{R_{2,hot}}{X_{02}}\right)\left\{\left(\frac{X_1 + X_{mm}}{X_{02}}\right)\left[\left(1 + \frac{X_{03}}{X_{02}}\right)\left(\frac{X_a + X_{a1} + X_{ma}}{X_{02}}\right) - \left(\frac{X_{ma}}{X_{02}}\right)^2\right]\right.$$
$$\left. - \left(\frac{X_{a1} + X_{ma}}{X_{02}}\right)\left(1 + \frac{X_{03}}{X_{02}}\right)\left(\frac{X_a + X_{a1} + X_{ma}}{X_{02}}\right)\right\}$$

$$B_{20} = \left(\frac{R_{2,hot}}{X_{02}}\right)\left\{\left(\frac{R_{1,hot}}{X_{02}}\right)\left[\left(1 + \frac{X_{03}}{X_{02}}\right)\left(\frac{X_a + X_{a1} + X_{ma}}{X_{02}}\right) - \left(\frac{X_{ma}}{X_{02}}\right)^2\right.\right.$$
$$\left.\left. - \left(\frac{R_{2,hot}}{X_{02}}\right)\left(\frac{R_{a,hot}}{X_{02}}\right)\right]\right\} + \left(\frac{R_{2,hot}}{X_{02}}\right)\left[\left(\frac{R_{2,hot}}{X_{02}}\right)\left(\frac{X_a}{X_{02}}\right)\left(\frac{X_{a1} + X_{ma}}{X_{02}}\right)\right.$$
$$\left. + \left(\frac{R_{a,hot}}{X_{02}}\right)\left(1 + \frac{X_{03}}{X_{02}}\right)\left(\frac{X_1 + X_{mm} + X_{ma} + X_{a1}}{X_{02}}\right)\right]$$

$$+ \left(\frac{R_{2,\text{hot}}}{X_{02}}\right)\left\{\left(\frac{R_{2,\text{hot}}}{X_{02}}\right)\left(\frac{X_1 + X_{\text{mm}}}{X_{02}}\right)\left(\frac{X_a + X_{a1} + X_{\text{ma}}}{X_{02}}\right)\right.$$

$$\left. - \left(\frac{R_{a,\text{hot}}}{X_{02}}\right)\left[\left(\frac{X_{\text{mm}}}{X_{02}}\right)^2 + \left(\frac{X_{\text{ma}}}{X_{02}}\right)^2\right]\right\}$$

6.4.4 General Procedure for Calculating the Performance of Polyphase Motors

Variables may be identified from the following list or from the figures.

A_B = area of rotor bar
A_{c1} = area of primary core
A_{c2} = area of secondary core
A_g = area of air gap
A_r = area of end ring
A_{t1} = area of primary tooth
A_{t2} = area of secondary tooth
AT_{c1} = ampere-turns per pole of primary core
AT_{c2} = ampere-turns per pole of secondary core
AT_{gap} = ampere-turns per pole of air gap
AT_{total} = total ampere-turns per pole
AT_{t1} = ampere turns per pole of primary tooth
AT_{t2} = ampere turns per pole of secondary tooth
B = slot span
B_{c1} = density of primary core
B_{c2} = density of secondary core
B_g = density of air gap
B_{t1} = density of primary tooth
B_{t2} = density of secondary tooth
b = width of slot opening
C = series conductors per phase
C_{nd} = conductivity of cage (Cu = 1.0; Al = 0.5 ~ 0.58)
C_r = mean rotor core circumference
C_s = mean stator core circumference

D_1 = outside diameter of stator
D_2 = inside diameter of rotor
D_a = diameter of mean air gap
D_i = inside diameter of end ring
D_r = mean diameter of end ring
D_{ring} = diameter of end ring at centers of rotor bars
E = applied voltage
F = slot correction coefficient
f = line frequency
G_1 = minimum air gap
H_{c1} = magnetizing force of primary core
H_{c2} = magnetizing force of secondary core
H_g = magnetizing force of air gap
H_{t1} = magnetizing force of primary tooth
H_{t2} = magnetizing force of secondary tooth
I_1 = primary current
I_2 = secondary current
I_w = in-phase (real) component of reactive current
I_x = imaginary component of reactive current
K_g = gap factor
k_1 = primary slot constant
k_2 = secondary slot constant
k_d = distribution factor
k_m = air gap area factor

k_p = pitch factor
k_s = slot leakage constant
k_w = winding distribution factor
L_{c1} = primary core length
L_{c2} = secondary core length
L_g = physical length of air gap
L_{met} = length of mean end turn
L_n = actual core length × stacking factor
L_{t1} = length of primary teeth
L_{t2} = length of secondary teeth
L_{y1} = length of primary yoke
L_{y2} = length of secondary yoke
l_g = effective gap length
m = number of phases
M_p = magnetizing permeance
MTL = mean turn length
N_B = speed at breakdown
N_r = run speed
N_s = synchronous speed
n = number of coils per group
PF = power factor
P_{xbelt} = belt leakage permeance
P_L = permeance of leakage path
P_o = permeance of all paths
P_m = permeance of main path
P_{xslot} = slot leakage permeance
P_{xzz} = zigzag leakage permeance
P_{xenc} = end leakage permeance
p = number of poles
R_o = outside radius of stator
R_t = radius to outside of stator slots

r_1 = primary resistance
r_2 = secondary resistance
S = run speed divided by synchronous speed
s = full-load slip
s_m = slip at breakdown
S_1 = number of stator slots
S_2 = number of rotor slots
SF = saturation factor
S_k = skew of rotor bars measured circumferentially, in
S_{tr} = number of strands of wire in parallel
T_B = breakdown torque
T_{p1} = primary tooth pitch
T_{p2} = secondary tooth pitch
T_s = starting torque
V_{ph} = volts per phase
V_{t1} = volume of primary teeth
W_{in} = total input, W
W_{fe} = iron loss
W_{f+w} = friction and windage loss
W_{out} = output, W
w = effective series conductors
w_2 = length of rotor stack
X = total leakage reactance
X_m = magnetizing reactance
X_r = run reactance
X_s = start reactance
Z = impedance
η = efficiency
θ_p = pole pitch
ϕ = total flux per pole

See Fig. 6.27 for a graphic representation of some variables.

Variables to Calculate:
Pole Pitch θ_p:

$$\theta_p = \frac{\pi(D_1 + D_2)}{2p}$$

Gap factor K_g (see Fig. 6.28):

Semiclosed slot

$$K_g = \frac{\theta_p(4.4G_1 + 0.75b)}{\theta_p(4.4G_1 + 0.75b) - b^2}$$

Open slot

$$K_g = \frac{\theta_p(5G_1 + b)}{\theta_p(5G_1 + b) - b^2}$$

Tunnel slot (K_g)

$$K_g = 1.03$$

Effective gap length l_g:

$$l_g = L_g k_g$$

Primary and secondary tooth pitch T_{p1} and T_{p2}:

$$T_{p1} = \frac{D_1 \pi}{S_1}$$

$$T_{p2} = \frac{D_2 \pi}{S_2}$$

Calculations of Winding and Slot Constants
Winding pitch factor k_p:

$$k_p = \sin(\text{pitch} \cdot 90°)$$

where pitch is expressed as a fraction of the full pitch, such as ⅚, etc.
Distribution factor k_d:

$$k_d = \frac{\sin(B/2)}{n \sin(B/2n)}$$

for distributed even-grouped windings
Total winding distribution factor k_w:

$$k_w = k_p k_d$$

Slot correction coefficient F:

$$F = 0.28 - \left[0.14\left(\frac{y_1}{y_2}\right)\right] + \frac{d_3/y_2}{2.87(y_1/y_2) + 0.18}$$

Air gap area factor k_m:

$$k_m = \frac{w_2 \theta_p}{l_g \, \text{SF} \, p}$$

Primary and Secondary Slot Constant Calculations k_1 and k_2

Note: See figures for variable descriptions. k_1 represents the primary (stator) slot constant, while k_2 represents the secondary (rotor) slot constant. They are found using the same set of equations, being careful to use the equation which most closely resembles that of the slot in question.

Round-Bottom Slot Constant k_1 or k_2 (note that F is different for the two constants):
  Slot shape A (see Fig. 6.29)

$$k_1 \text{ or } k_2 = F + \frac{d_2}{d} + \frac{4d_2}{3d + y_1}$$

Slot shape B (see Fig. 6.30)

$$k_1 \text{ or } k_2 = F + \frac{d_1}{d} + \frac{2d_2}{d + y_2}$$

Round-bottom closed slot (see Fig. 6.31)

$$k_1 \text{ or } k_2 = \frac{0.284 - 0.143(y_1/y_2) + (d_3/y_1)}{2.87(y_1/y_2) + 0.96}$$

Square-bottom slot constant k_1 or k_2:
  Open slot (see Fig. 6.32)

$$k_1 \text{ or } k_2 = \frac{d_1}{d} + \frac{d_3}{y_1}$$

Bridged slot (see Fig. 6.33)

$$k_1 \text{ or } k_2 = \frac{d_1}{0.02} + \frac{d_3}{y_1}$$

Flat-bottom slot constant k_1 or k_2:
  Slot shape A (see Fig. 6.34)

$$k_1 \text{ or } k_2 = \frac{d_1}{0.02} + \frac{d_3}{y_1}$$

Slot shape B (see Fig. 6.35)

$$k_1 \text{ or } k_2 = \frac{d_1}{d} + 1.5\left(\frac{d_2}{d + r}\right) + \frac{d_3}{y_1}$$

Slot shape C (see Fig. 6.36)

$$k_1 \text{ or } k_2 = \frac{d_1}{d} + \frac{4d_2}{3d + y_1} + \frac{d_3}{y_1}$$

Flat-bottom closed slot (see Fig. 6.37)

$$k_1 \text{ or } k_2 = \frac{y_1}{y_2} + \frac{d_3}{y_2} + 1$$

Slot leakage constant k_s:

$$k_s = F + \frac{d_1}{d} + \frac{2d_2}{d + y_1}$$

Leakage Reactances

Slot leakage permeance P_{xslot}:

$$P_{xslot} = \left(\frac{3.19\, w_2 m}{S_1}\right)\left[k_1 + \left(\frac{S_1}{S_2}\right)k_2\right]$$

Zigzag leakage permeance P_{xzz}:

$$P_{xzz} = \left(\frac{1.063\, w_2 m}{S_1 G_1}\right)\left[\frac{(T_{f1} + T_{f2})^2}{4(T_{p1} + T_{p2})}\right]$$

End leakage permeance P_{xend}:

$$P_{xend} = \left(\frac{D_1 B\, m}{S_1 p}\right)(1.1 + 0.1p)$$

Belt leakage permeance P_{xbelt}:

$$P_{xbelt} = 0.00118\, m\, k_m \left(\frac{20p + 0.6}{S_1 + S_2}\right)$$

Skew leakage permeance P_{xskew}:

$$P_{xskew} = 0.324\, m\, k_m k_p \left[1 - \frac{\sin(R_s/\theta_p)\, 90°}{\pi R_s/2\theta_p}\right]^2$$

Permeance of leakage path P_L:

$$P_L = P_{xslot} + P_{xzz} + P_{xend} + P_{xbelt} + P_{xskew}$$

Permeance of main path P_m:

$$P_m = 0.324\, m\, k_m \left[\frac{\sin(S_k/\theta_p)\, 90°}{\pi S_k/2\theta_p}\right]$$

Permeance of all paths P_o:

$$P_o = P_m + \frac{P_L}{2}$$

Total leakage reactance X:

$$X = 2\pi f (Ck_w)^2 10^{-8} P_L \left[\frac{1 + (P_m/P_o)}{2}\right]$$

Open-circuit leakage reactance X_o:

$$X_o = 2\pi f (Ck_w)^2 10^{-8} P_o$$

Start reactance X_s:

$$X_s = 0.80X$$

Run reactance X_r:

$$X_r = 1.20X$$

Calculation of Magnetizing Current and Reactance

Total flux per pole ϕ:

$$\phi = \frac{E(10^8)}{2.22 f C k_w}$$

Areas of Various Flux Paths

Primary tooth area A_{t1}:

$$A_{t1} = \left(\frac{s_1}{p}\right) t_{w1} w_2$$

Primary core area A_{c1}:

$$A_{c1} = (R_o - R_t) w_2$$

Secondary tooth area A_{t2}:

$$A_{t2} = \left(\frac{s_2}{p}\right) t_{w2} w_2$$

Secondary core area A_{c2}:

$$A_{c2} = 2 L_{y2} w_2$$

Air gap area A_g:

$$A_g = \frac{l_g D_a \pi}{p}$$

Flux Densities in Various Paths

Primary tooth density B_{t1}:

$$B_{t1} = \left(\frac{\phi}{A_{t1}}\right)\left(\frac{\pi}{2}\right)$$

Primary core density B_{c1}:

$$B_{c1} = \frac{\phi}{A_{c1}}$$

Secondary tooth density B_{t2}:

$$B_{t2} = \left(\frac{\phi}{A_{t2}}\right)\left(\frac{\pi}{2}\right)$$

Secondary core density B_{c2}:

$$B_{c2} = \frac{\phi}{A_{c2}}$$

Air gap density B_g:

$$B_g = \left(\frac{\phi}{A_g}\right)\left(\frac{\pi}{2}\right)$$

Magnetizing Forces. The magnetizing force of each path other than the air gap is found by referring the preceding densities to the proper saturation curve and reading across to find the respective magnetizing force.

Primary tooth H_{t1}
Primary core H_{c1}
Secondary tooth H_{t2}
Secondary core H_{c2}
Air gap magnetizing force H_g

$$H_g = 0.313$$

Lengths of the Various Flux Paths
Primary tooth length L_{t1}: see Fig. 6.27
Primary core length L_{c1}:

$$L_{c1} = \frac{C_s}{2p}$$

Secondary tooth length L_{t2}: see Fig. 6.27
Secondary core length L_{c2}:

$$L_{c2} = \frac{C_r}{2p}$$

Ampere-Turns per Pole
Primary tooth AT_{t1}:

$$AT_{t1} = H_{t1}L_{t1}$$

Primary core AT_{c1}:

$$AT_{c1} = H_{c1}L_{c1}$$

Secondary tooth AT_{t2}:

$$AT_{t2} = H_{t2}L_{t2}$$

Secondary core AT_{c2}:

$$AT_{c2} = H_{c2}L_{c2}$$

Air gap AT_{gap}:

$$AT_{gap} = H_g l_g$$

Total ampere-turns per pole AT_{total}:

$$AT_{total} = AT_{t1} + AT_{c1} + AT_{t2} + AT_{c2} + AT_{gap}$$

Saturation factor SF:

$$SF = \frac{AT_{total}}{AT_{gap}}$$

Magnetizing reactance X_m:

$$X_m = \frac{2.03 f w_2 \theta_p l_g}{SF\, k_g G_1 p}$$

Magnetizing permeance M_p:

$$M_p = \frac{2.03\theta_p l_g}{\text{SF } k_g p G_1}$$

Primary resistance r_1:

$$r_1 = \frac{(C)(\text{MTL})(\text{resistance}/1000 \text{ ft})}{12{,}000 \, S_{tr}}$$

where $\text{MTL} = 2w_2 + 2L_{met}$
Secondary resistance r_2:

$$r_2 = (Ck_w)^2 \, (10^{-6})(0.693)(2) \left[\frac{\sqrt{(w_2)^2 + S_k^2}\,(100)}{S_2 A_B C_{nd}} + \left(\frac{63.7(D_r)}{A_r p^2} \right) \left(\frac{K_{ring}}{C_{nd}} \right) \right]$$

where (see Fig. 6.38)

$$K_{ring} = \frac{p}{2} \left(1 - \frac{D_i}{D_{ring}} \right) \left[\frac{1 + (D_i/D_{ring})^p}{1 - (D_i/D_{ring})^p} \right]$$

Calculation of Iron Losses W_{fe}
Volume of primary teeth V_{t1}:

$$V_{t1} = A_{t1} L_{t1} p$$

Volume of primary core V_{c1}:

$$V_{c1} = A_{c1} C_s$$

Find rotor and stator iron weights by multiplying V_{c1} and V_{t1} by density, then obtain the watts lost per pound from Fig. 6.39 and multiply by the volumes to obtain losses in the teeth and core. The sum of these two is the total iron loss W_{fe}.

Friction and Windage Losses.
Friction and windage losses are obtained by testing similar machines or by calculations from the bearing manufacturers.

Performance Calculations
Real component of reactive current I_w:

$$I_w = \frac{\text{HP}(746)}{3 \, V_{ph} \eta}$$

Assume an efficiency η of, say, 80 percent.
Imaginary component of reactive current I_x:

$$I_x = \frac{\text{HP}(746)}{3 \, V_{ph} I_m X}$$

Primary current I_1:

$$I_1 = \sqrt{I_w^2 + (I_x + I_m)^2}$$

Secondary current I_2:

$$I_2 = \sqrt{I_w^2 + I_x^2}$$

ALTERNATING-CURRENT INDUCTION MOTORS

Input W_{in}:

$$W_{in} = I_1^2 r_1 m + I_2^2 r_2 m + W_{fe} + \frac{(1-S)(I_2^2 r_2 m)}{S}$$

Output W_{out}:

$$W_{out} = \frac{(1-S)(I_2^2 r_2 m)}{S} - W_{f+w}$$

Losses W_{loss}:

$$W_{loss} = W_{in} - W_{out}$$

Synchronous speed N_s:

$$N_s = \frac{120f}{p}$$

Run speed N_r:

$$N_r = (1-S)\left(\frac{120f}{p}\right)$$

Power factor PF:

$$PF = \frac{I_w}{\sqrt{I_w^2 + (I_x + I_m)^2}} = \frac{W_{in}}{EI_m}$$

Full-load slip s:

$$s = \frac{3 I_2^2 r_2}{3 I_2^2 r_2 + W_{out} + W_{f+w}}$$

Impedance Z:

$$Z = \sqrt{X_z^2 + (r_1 + r_2)^2}$$

where $X_z = X_s$ for start conditions
        $X_z = X_r$ for run conditions

Starting torque T_s:

$$T_s = \frac{m V_{ph}^2 r_2 (7.04)}{Z^2 N_s} \quad \text{lb} \cdot \text{ft}$$

Breakdown torque T_B:

$$T_B = \frac{m V_{ph}^2 p}{34.08 f (r_1 + \sqrt{r_1^2 + X^2})} \quad \text{lb} \cdot \text{ft}$$

Slip at breakdown s_m:

$$s_m = \frac{r_2}{\sqrt{r_1^2 + X^2}}$$

Speed at breakdown N_B:

$$N_B = S_m N_s$$

CHAPTER 7
SYNCHRONOUS MACHINES

Chapter Contributors

Chris A. Swenski
William H. Yeadon

This chapter covers some of the ac synchronous motors commonly encountered in the industry. While it could be said that the electronically commutated motors discussed in Chap. 5 are also synchronous motors, this chapter is confined to the typical ac versions. While larger polyphase machines are well covered by others, little information is available on these smaller motors.

7.1 INDUCTION SYNCHRONOUS MOTORS*

These motors are built in a manner very similar to that for induction motors. They may have polyphase windings or be designed as single-phase motors, such as capacitor-start, split-phase, or shaded-pole types.

The rotors have a dual construction that allows for induction motor starting characteristics and salient-pole synchronous running conditions.

These rotors may be made from induction motor stampings with some of the teeth removed, as shown in Fig. 7.1. They are then die-cast in the same manner as an induction motor rotor (Fig. 7.2).

Some motors use a permanent magnet in conjunction with an induction motor rotor. Figures 7.3 and 7.4 show such a motor. This is a four-pole shaded-pole motor. Here the field coils are connected such that they directly produce two like poles and induce two opposite poles at 90° in the unwound space between the coils. In these

---

* Sections 7.1 to 7.3 contributed by William H. Yeadon, Yeadon Engineering Services, PC.

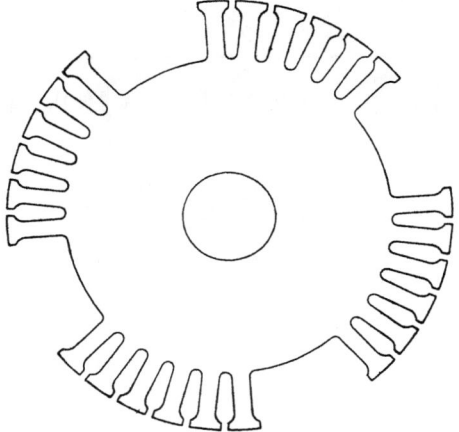

FIGURE 7.1 Induction synchronous motor lamination.

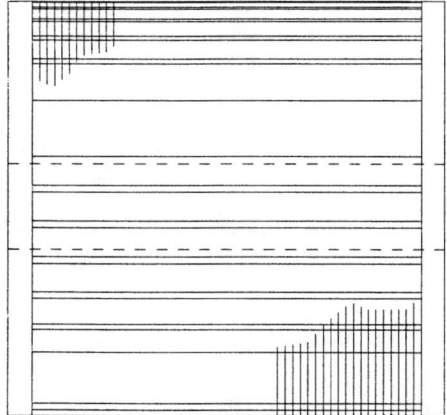

FIGURE 7.2 Four-pole induction synchronous rotor assembly.

figures, the rotor is shown to be enclosed by the stator. Figure 7.3 shows the permanent-magnet part of the rotor, while Fig. 7.4 shows the induction rotor end.

The rotor is shown alone in Figs. 7.5 and 7.6, with a piece of magnetic viewing film over the permanent-magnet portion. Figure 7.5 demonstrates the position of the magnetic poles, of which there are four on the rotor. Figure 7.6 shows that the induction rotor portion is a laminated structure with copper-wire bars swedged over copper-plate end rings.

FIGURE 7.3 Shaded-pole synchronous motor, permanent-magnet rotor end.

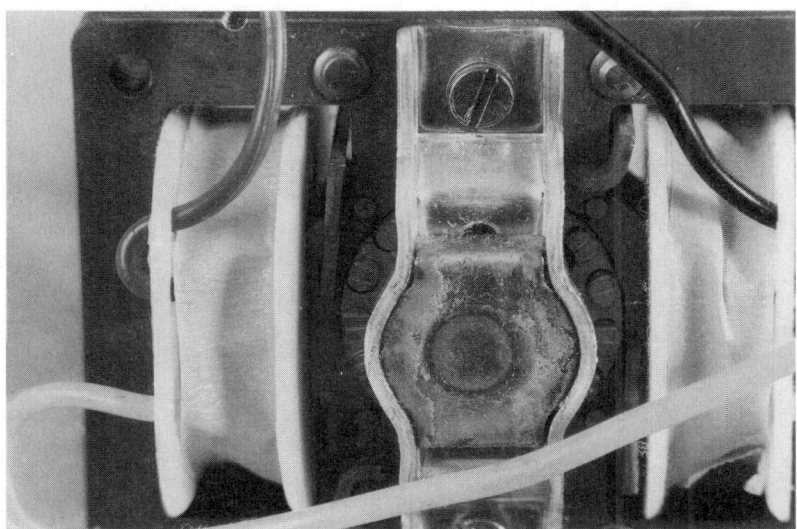

FIGURE 7.4 Shaded-pole synchronous motor, induction rotor end.

FIGURE 7.5 Shaded-pole synchronous motor showing magnetized poles.

FIGURE 7.6 Shaded-pole synchronous motor rotor showing induction motor bars and end rings.

7.2 HYSTERESIS SYNCHRONOUS MOTORS

These motors have a rotor made of a cobalt alloy or another material that can be magnetized semipermanently by the stator field. They have a rather weak second-quadrant demagnetization curve which can be easily demagnetized. The demagnetization curves of some of these materials are shown in Figs. 7.7, 7.8, and 7.9.

Figure 7.10 shows a motor utilizing a wound-field distributed stator (Fig. 7.11) and a cobalt hysteresis ring rotor (Fig. 7.12). This motor is connected and run like a permanent-split-capacitor (PSC) motor. It produces a speed-torque curve like the one shown in Fig. 7.13.

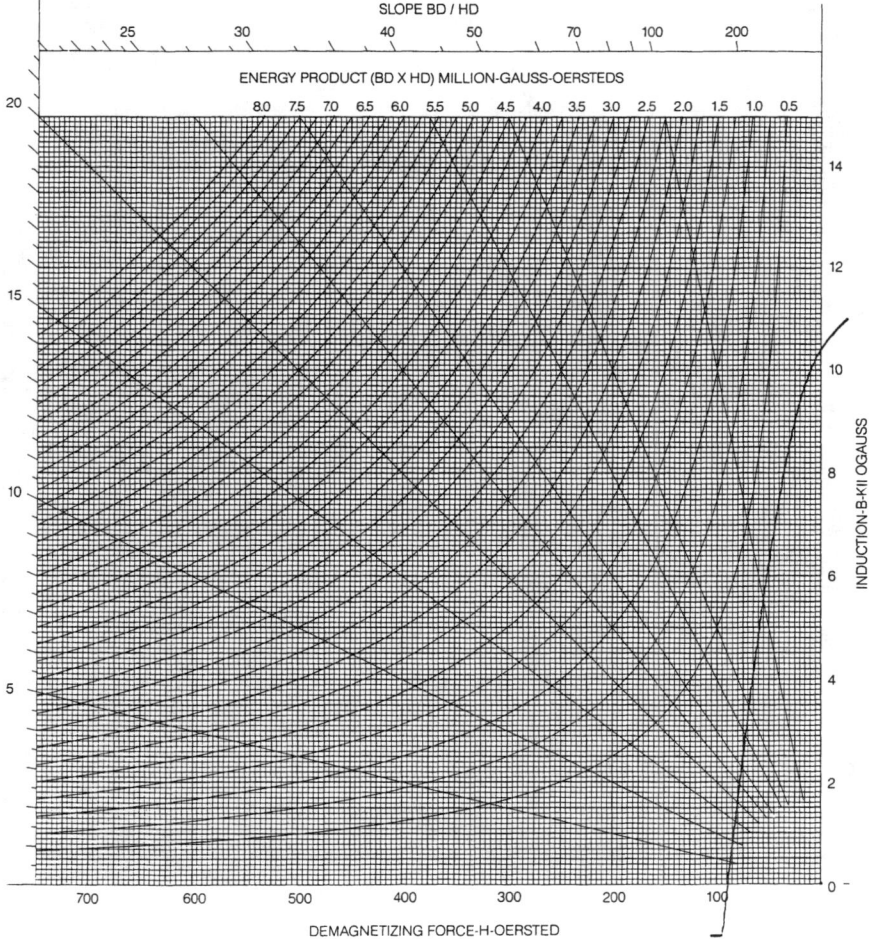

FIGURE 7.7 0.32-MGOe hysteresis material. $BR = 10.4$ kG, $BD = 7.0$ kG, $HC = 88$ Oe, $HD = 45$ Oe, $BH_{max} = 0.315$ MGOe, $BD/HD = 155.6$. *(Courtesy of Arnold Engineering Company.)*

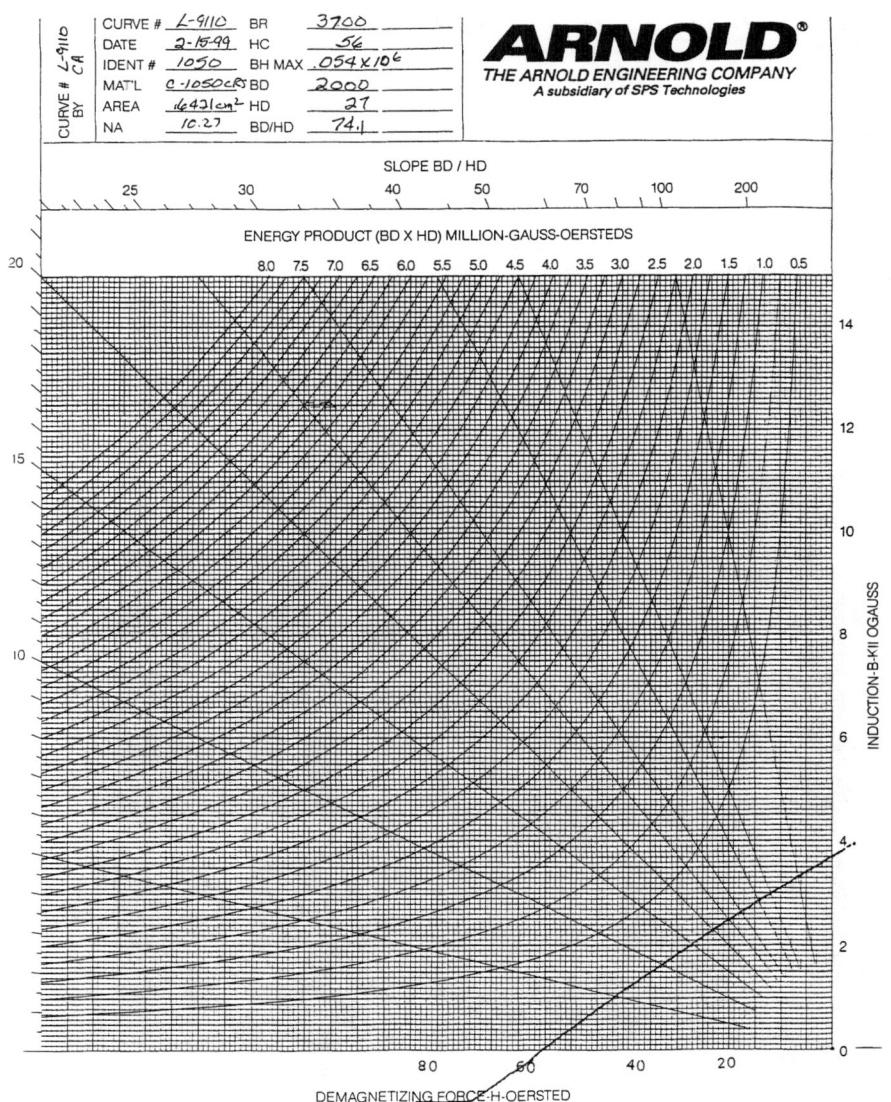

FIGURE 7.8 0.054-MGOe hysteresis material: hardened A151050. $BR = 37.00$ kG, $BD = 2.0$ kG, $HC = 56$ Oe, $HD = 27$ Oe, $BH_{max} = 0.054$ MGOe, $BD/HD = 24.1$. *(Courtesy of Arnold Engineering Company.)*

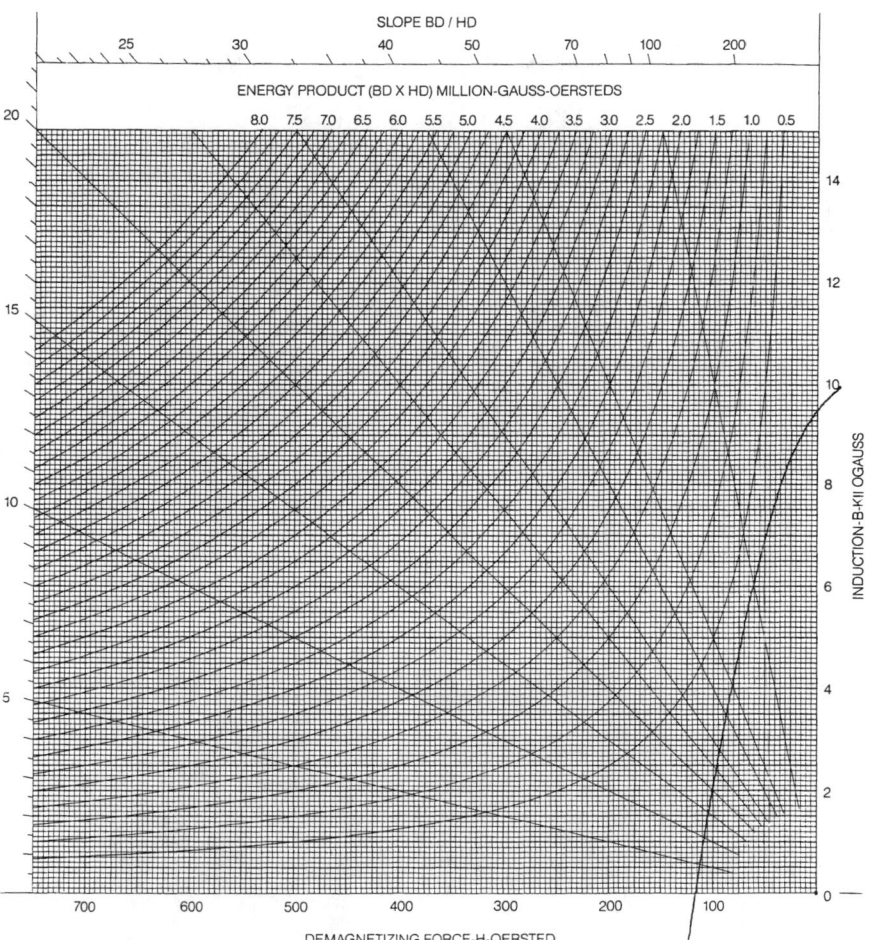

FIGURE 7.9 0.36-MGOe hysteresis material. $BR = 9.5$ kG, $BD = 6.0$ kG, $HC = 116$ Oe, $HD = 60$ Oe, $BH_{max} = 0.360$ MGOe, $BD/HD = 100.0$. *(Courtesy of Arnold Engineering Company.)*

FIGURE 7.10 Wound-field motor.

FIGURE 7.11 Distributed wound stator.

FIGURE 7.12 Cobalt hysteresis ring rotor.

Many timer motors and value actuators use a hysteresis ring but use the shaded-pole type of stator to provide starting torque. Figures 7.14, 7.15, and 7.16 show a timer motor which utilizes this principle. The shaded-pole stator provides the starting torque, but it also makes the motor unidirectional.

7.3 PERMANENT-MAGNET SYNCHRONOUS MOTORS

Many of these motors are used in clocks or timing devices. Figure 7.17 shows a typical clock motor. Note that the stator portion has an uneven distribution of magnetic poles (Fig. 7.18). The purpose of this is to give the rotor a preferred starting point while providing an apparent shift in field during starting due to the uneven reluctance of the stator. Some of these motors have a spring return mechanism to reverse the rotation just in case it starts turning the wrong way.

Other PM synchronous motors are essentially PM stepper motors run as PSC motors. The motor shown in Fig. 7.19 has a stator consisting of two sets of coils with the teeth offset from each other by one-half tooth pitch (Fig. 7.20). The rotor has magnetized poles along its length, as shown by magnetic viewing film (Fig. 7.21). One stator half serves as the main field winding. The other serves as the auxiliary phase. They are connected as in PSC motors, with a capacitor in series with the auxiliary winding.

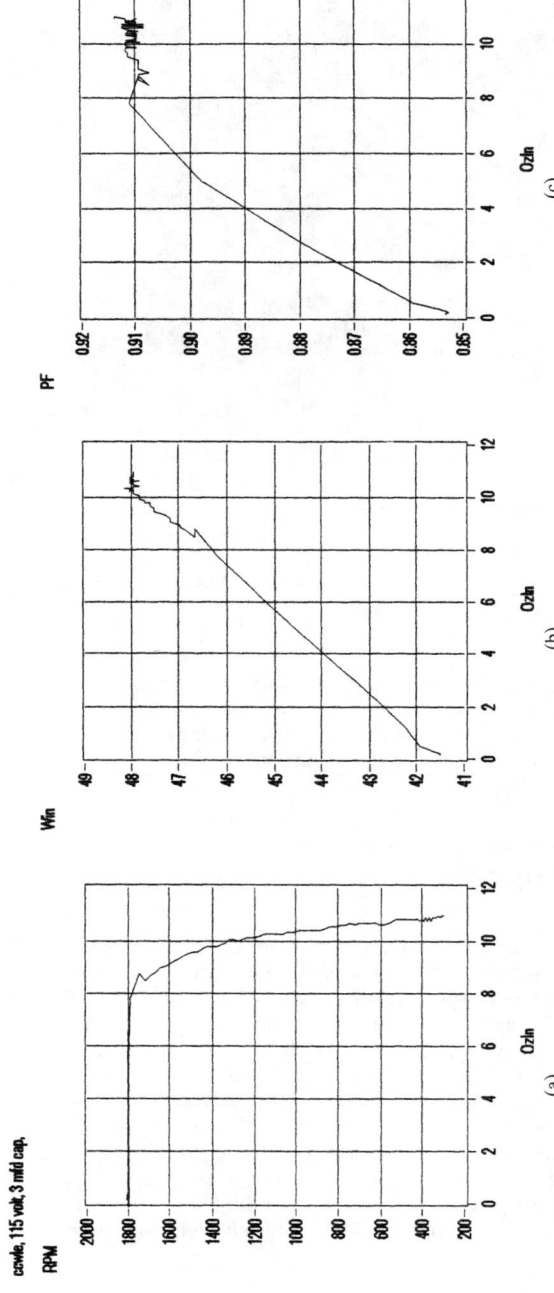

FIGURE 7.13 Synchronous motor speed-torque curves: Oz · in versus (a) rpm, (b) W_{in}, (c) PF, (d) amps, (e) horsepower, and (f) efficiency.

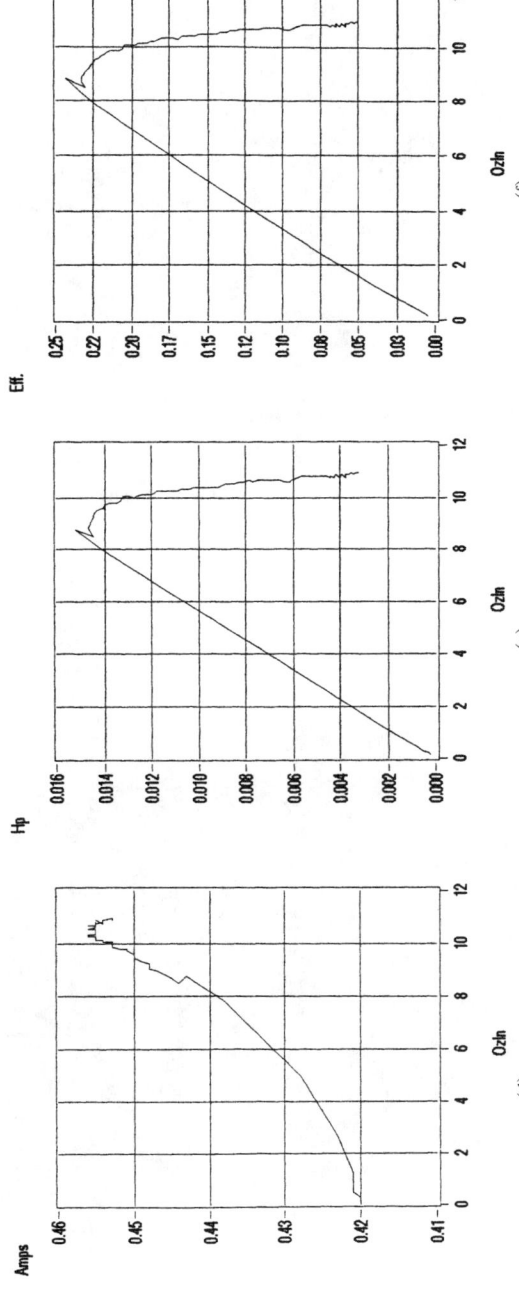

FIGURE 7.13 (*Continued*) Synchronous motor speed-torque curves: Oz · in versus (*a*) rpm, (*b*) W_{in}, (*c*) PF, (*d*) amps, (*e*) horsepower, and (*f*) efficiency.

FIGURE 7.14 Timer or actuator motor by Cramer Motor Company. *(Courtesy of MH Rhodes, Inc.)*

FIGURE 7.15 Timer stator. *(Courtesy of MH Rhodes, Inc.)*

FIGURE 7.16 Timer rotor.

FIGURE 7.17 Typical clock motor.

FIGURE 7.18 Uneven distribution of magnetic stator poles.

FIGURE 7.19 Permanent-magnet synchronous motor.

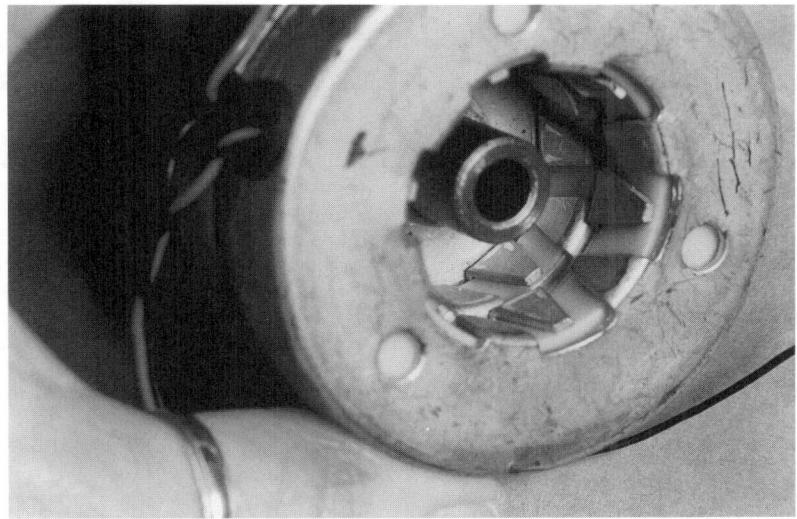

FIGURE 7.20 Half-tooth-pitch offset consisting of two sets of coils with offset teeth.

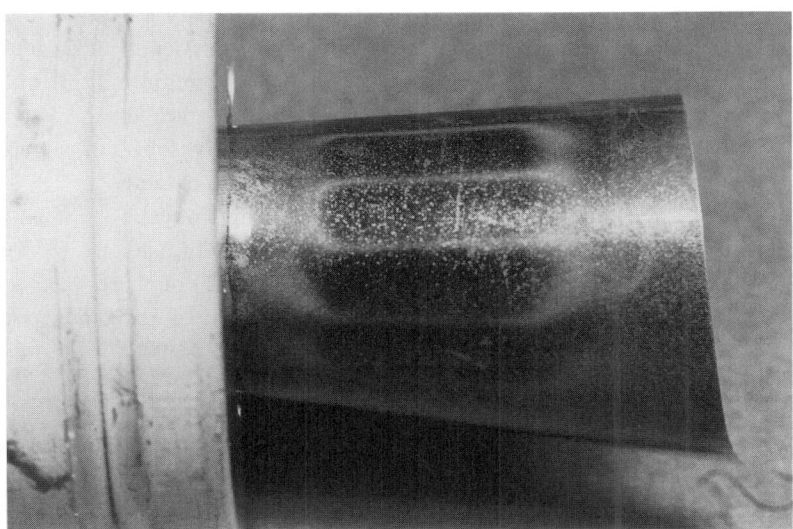

FIGURE 7.21 Rotor with magnetized poles.

7.4 PERFORMANCE CALCULATION AND ANALYSIS*

P. H. Trickey suggests a method for calculating the performance of these synchronous motors which is summarized here.

Generally, this method applies to motors using cobalt ring rotors. Motor parameters are calculated using the same methods as for ac induction motors, except for rotor losses. Winding resistance, iron losses, friction and windage losses, and stator leakage reactance use exactly the same methods as induction motors. Air gap leakage reactance is calculated as one-half the value of zigzag and one-third the value of belt leakage that would have been obtained using a rotor with the same number of slots as the stator.

The hysteresis power is equal to the stator input power minus the stator losses, friction and windage losses, and rotor parasite losses. Stator losses include I^2r_1 plus core losses. Parasitic losses include hysteresis and eddy current losses of minor loops resulting from flux variation at tooth slot openings, losses resulting from harmonics of a nonsinusoidal winding distribution, and double-frequency backward field hysteresis and eddy current losses.

Variables

B = slot span, ° electrical
b = slot opening
C = series conductors per phase
C_1 = flux form coefficient
C_x = correction factor
C_{t1} = maximum ampere-turns per inch (teeth)
C_{y1} = maximum ampere-turns per inch (yoke)
D_1 = stator ID
D_2 = rotor OD
F = slot correction coefficient
f = frequency, Hz
G_1 = minimum air gap
HP = output power, hp
h_1 = straight portion of end-turn extension
I = primary current (assumed)
I_d = direct axis current
I_q = quadrature axis current
k_d = distribution factor
k_g = gap factor

k_p = pitch factor
k_s = slot leakage constant
k_w = total winding distribution factor
k_{zz} = zigzag leakage coefficient
k_1 = stator (primary) slot constant
k_2 = secondary slot constant
L_g = physical length of air gap
L_{met} = mean end-turn length of coil
L_s = stator stack length
l_g = effective length of air gap
L_{t1} = length of teeth
L_{y1} = length of yoke
MTL = mean turn length
m = number of phases
m_e = length of remaining portion of coil extension
N_s = synchronous speed
n = coils per group
P_b = belt leakage permeance factor
PF = power factor
p = number of poles

---

* Section contributed by Chris A. Swenski, and William H. Yeadon, Yeadon Engineering Services, PC.

R_s = rotor skew, in
r_1 = resistance per phase, Ω
S_p = span (average coil throw)
SF = saturation factor
SF_d = direct axis saturation factor
SF_q = quadrature axis saturation factor
S_{tr} = number of strands of wire in parallel
s_1 = stator (primary) slots
s_2 = rotor (secondary) slots
T = torque, oz·in
T_{p1} = primary tooth pitch
T_{p2} = secondary tooth pitch
T_{f1} = width of primary tooth face
T_{f2} = width of secondary tooth face
V_{ph} = volts per phase
W_{in} = total input, W
W_L = Loss, W
W_{out} = output, W

W_{ph} = watts per phase
w = total series conductors
w_2 = axial rotor stack length
X_d = total direct axis reactance
X_{fctr} = reactance factor
X_{pc} = primary end reactance
X_{ps} = primary slot reactance
X_q = total quadrature axis reactance
X_l = total leakage reactance
X_{1d} = primary direct axis reactance
X_{1q} = primary quadrature axis reactance
Y = end-winding length coefficient
θ_p = total pole pitch
θ_{pd} = direct axis pole pitch
θ_{pq} = quadrature axis pole pitch
ξ = efficiency
ϕ_2 = flux factor

Note: See figures for those variables not listed here, but used in the following equations.

Calculation of Constants

Pitch factor k_p:

$$k_p = \sin(\text{pitch} \cdot 90°)$$

where pitch is expressed as a fraction of the full pitch, such as ⅚, etc.

Distribution factor k_d:

$$k_d = \frac{\sin(B/2)}{n \sin(B/2n)}$$

Total winding distribution factor k_w:

$$k_w = k_p k_d$$

Slot correction coefficient F:

$$F = 0.28 - \left[0.14\left(\frac{y_1}{y_2}\right)\right] + \frac{d_3/y_2}{2.87(y_1/y_2) + 0.08} + 0.08$$

Slot leakage constant k_s:

$$k_s = F + \frac{d_1}{d} + \frac{2d_2}{d + y_1}$$

Gap factor k_g (see Fig. 7.22):
  Semiclosed slot

$$k_g = \frac{\theta_p(4.4G_1 + 0.75b)}{\theta_p(4.4G_1 + 0.75b) - b^2}$$

  Open slot

$$k_g = \frac{\theta_p(5G_1 + b)}{\theta_p(5G_1 + b) - b^2}$$

  Tunnel slot

$$k_g = 1.03$$

Effective gap length l_g:

$$l_g = L_g k_g$$

Zigzag reactance correction factor C_x:

$$C_x = \frac{T_{p1}L_s}{T_{f1}l_g} - 1$$

Primary and Secondary Slot Constant Calculations k_1 *and* k_2

Note: See figures for variable descriptions. k_1 represents the primary (stator) slot constant, while k_2 represents the secondary (rotor) slot constant. They are found using the same set of equations, being careful to use the equation which most closely resembles that of the slot in question.

Round-bottom slot constant k_1 or k_2 (note that F is different for the two constants):
  Slot shape A (see Fig. 7.23)

$$k_1 \text{ or } k_2 = F + \frac{d_1}{d} + \frac{2d_2}{d + y_1}$$

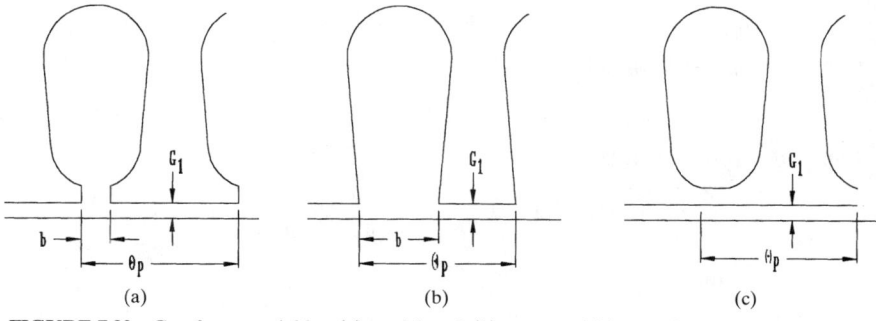

FIGURE 7.22 Gap factor variables: (*a*) semiclosed, (*b*) open, and (*c*) tunnel.

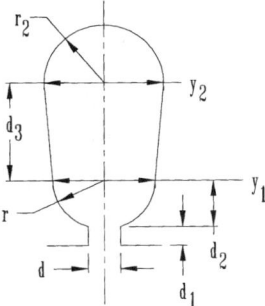

FIGURE 7.23 Round-bottom slot, shape A.

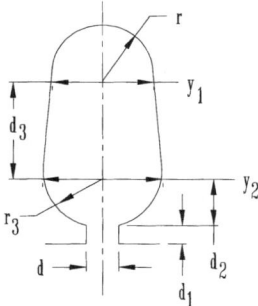

FIGURE 7.24 Round-bottom slot, shape B.

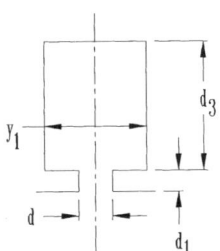

FIGURE 7.25 Square-bottom open slot.

Slot shape B (see Fig. 7.24)

$$k_1 \text{ or } k_2 = F + \frac{d_1}{d} + \frac{2d_2}{d + y_1}$$

Square-bottom slot constant k_1 or k_2:
  Open slot (see Fig. 7.25)

$$k_1 \text{ or } k_2 = \frac{d_1}{d} + \frac{d_3}{y_1}$$

Bridged slot (see Fig. 7.26)

$$k_1 \text{ or } k_2 = \frac{d_1}{0.02} + \frac{d_3}{y_1}$$

Flat-bottom slot constant k_1 or k_2:
  Slot shape A (see Fig. 7.27)

$$k_1 \text{ and } k_2 = \frac{d_1}{d} + \frac{d_3}{y_1}$$

Slot shape B (see Fig. 7.28)

$$k_1 \text{ and } k_2 = \frac{d_1}{d} + 1.5 \frac{d_2}{d+r} + \frac{d_3}{y_1}$$

Slot shape C (see Fig. 7.29)

$$k_1 \text{ and } k_2 = \frac{d_1}{d} + \frac{4d_2}{3d + y_1} + \frac{d_3}{y_1}$$

Variables to Calculate
Primary tooth pitch T_{p1}:

$$T_{p1} = \frac{D_1 \pi}{s_1}$$

Secondary tooth pitch T_{p2}:

$$T_{p2} = \frac{D_2 \pi}{s_2}$$

Total pole pitch θ_p:

$$\theta_p = \frac{\pi(D_1 + D_2)}{2p}$$

Zigzag leakage coefficient k_{zz}:
  Open slot

$$k_{zz} = \frac{(T_{f1} - T_{f2})^2}{4(T_{p1} + T_{p2})}$$

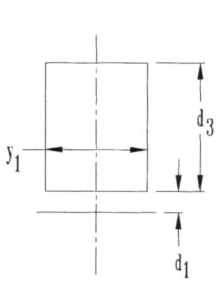

FIGURE 7.26 Square-bottom bridged slot.

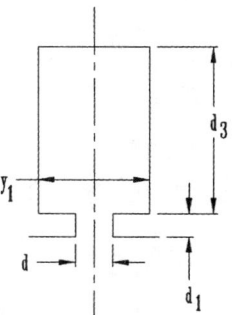

FIGURE 7.27 Flat-bottom slot, shape A.

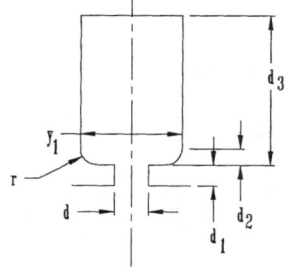

FIGURE 7.28 Flat-bottom slot, shape B.

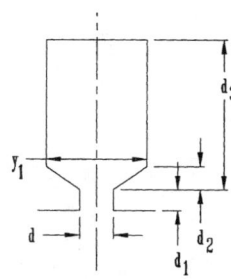

FIGURE 7.29 Flat-bottom slot, shape C.

Closed slot

$$k_{zz} = \frac{1.063 W_{ph}}{G_1 s_1}$$

Direct axis pole pitch θ_{pd} (see Fig. 7.30):

$$\theta_{pd} = \frac{20 p k_s}{s_1} + \left[\left(\frac{20 p C_x}{s_1}\right) k_{zz}\right] + \frac{4\pi D_1 S_p (1.1 + 0.1 p)}{2 s_1 L_s} + P_b$$

where belt leakage permeance factor P_b is

$$P_b = 0 \quad \text{for three-phase}$$

$$P_b = 0.95 \left(\frac{D_1}{p m k_1}\right) \quad \text{for two-phase}$$

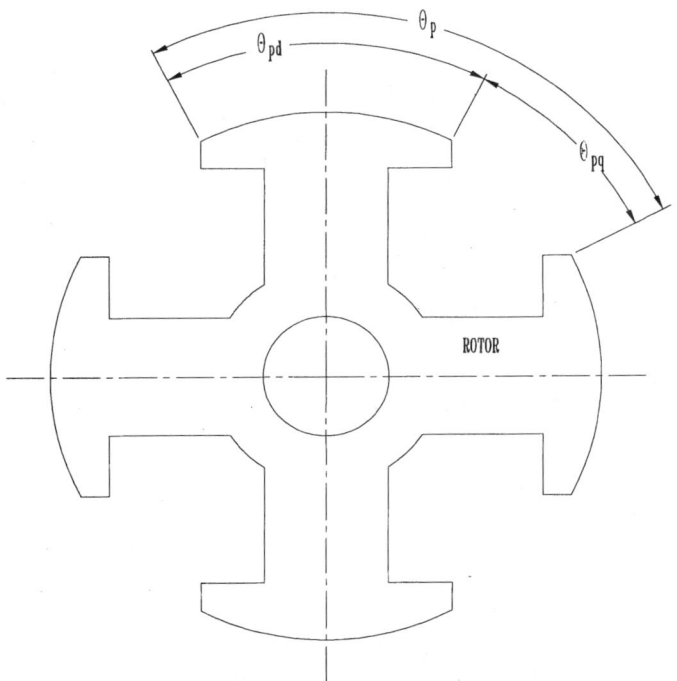

FIGURE 7.30 Rotor cross section showing θ_{pd} and θ_{pq}. Dimensions in inches, not degrees, measured along the circumference of the rotor.

Quadrature axis pole pitch θ_{pq} (see Fig. 7.30):

$$\theta_{pq} = \frac{20pk_s}{s_1} + 1.25\left[\left(\frac{20pC_x}{s_1}\right)k_{zz}\right] + \frac{4\pi D_1 S_p(1.1 + 0.1p)}{2s_1 L_s} + \frac{(\sin 1.5)[(S_p + p\pi)/s_1]}{k_p}$$

$$+ Q\left[\frac{2.03\theta_p(L_s + 2G_1)}{G_1 k_1 L_s}\right]$$

where

$$Q = 1 - \left[\frac{\sin(\theta_{Rs}/2)}{\pi\theta_{Rs}/360}\right]^2 \quad \text{and} \quad \theta_{Rs} = \left(\frac{R_s}{\theta_p}\right)180°$$

Saturation Calculations
Saturation factor SF:

$$SF = \frac{\text{total ampere} - \text{turns}}{\text{airgap ampere} - \text{turns}}$$

Flux factor Φ_2:

$$\Phi_2 = \frac{\sqrt{\{[\sqrt{I^2 - (W_{in}/V_{ph}m)^2} \cdot r_1] - (W_{in}/V_{ph})X_1\}^2 + \{(V_{ph} - (W_{in}/V_{ph}m)r_1) - \sqrt{[I^2 - (W_{in}/V_{ph}m)^2} \cdot X_1]\}^2}}{\left(\dfrac{45.0 \times 10^6}{Ck_w}\right)}$$

Area of the Air Gap:
Direct axis area A_{gd}

$$A_{gd} = \left(\frac{\theta_{pd}}{\theta_p}\right)\left[\frac{\pi(D_1 + D_2)}{2p}\right](L_s + 2G_1F_2)$$

Quadrature axis area A_{gq}

$$A_{gq} = \left(\frac{\theta_{pq}}{\theta_p}\right)\left[\frac{\pi(D_1 + D_2)}{2p}\right](L_s + 2G_1F_2)$$

Direct axis saturation factor SF_d:

$$SF_d = \frac{C_{y1}L_{y1} + C_{t1}L_{t1} + 491.41(F_2/A_{gd})l_g}{491.41(F_2/A_{gd})l_g}$$

Quadrature axis saturation factor SF_q:

$$SF_q = \frac{C_{y1}L_{y1} + C_{t1}L_{t1} + 491.41(F_2/A_{gq})l_g}{491.41(F_2/A_{gq})l_g}$$

Direct and Quadrature Axis Reactances

Reactance factor X_{fctr}:

$$X_{fctr} = \frac{f(Ck_w)^2 w_2 m}{p(10^8)}$$

Total direct axis reactance X_d:

$$X_d = X_{fctr}[(CmC_1/SF_d) + \theta_{pd}]$$

Total quadrature axis reactance X_q:

$$X_q = X_{fctr}[(Cf/SF_q) + \theta_{pq}]$$

Primary direct axis reactance X_{1d}:

$$X_{1d} = X_{fctr}\theta_{pd}$$

Primary quadrature axis reactance X_{1q}:

$$X_{1q} = X_{fctr}\theta_{pq}$$

Primary Resistance
Mean turn length MTL:

$$MTL = 2L_s + 2L_{met}$$

Primary resistance r_1:

$$r_1 = \frac{(C)(MTL)(\text{resistance}/1000 \text{ ft})}{12{,}000 S_{tr}}$$

Reactances
Primary slot X_{ps}

$$X_{ps} = \frac{w^2 f}{10^8} \cdot \frac{L_s k_1 k_s)}{s_1}$$

Primary end X_{pc} (see Fig. 7.31):

$$X_{pc} = \frac{w^2 f}{10^8} \cdot \frac{(h_1 + 0.5 m_e) k_p^2}{p}$$

Zigzag X_{zz}:

$$X_{zz} = \frac{w^2 f}{10^8} \cdot \frac{\pi D_1 L_s k_s}{s_1(s_1 + s_2) G_1 k_g}$$

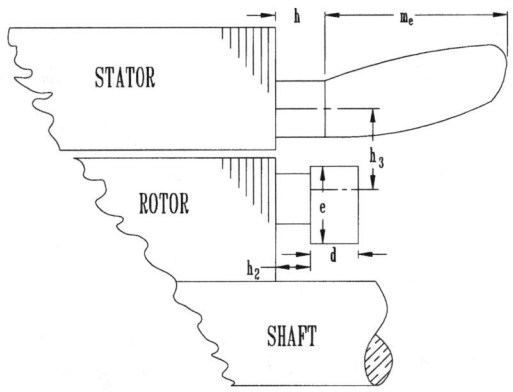

FIGURE 7.31 Dimensions for end-turn leakage reactance.

Primary Leakage Reactance. The primary leakage reactance X_1 is the sum of the preceding reactances:

$$X_1 = X_{ps} + X_{pc} + X_{zz}$$

Friction and Windage Losses. Friction and windage losses are obtained by testing similar machines (same frame size, bearing size, rpm, enclosure and cooling fan).

Output Calculations

Speed N_s:

$$N_s = \frac{120f}{p}$$

Losses W_L

$$W_L = (I^2 m r_1) + (\text{Fe loss}) + (F + W)$$

Output power HP:

$$\text{HP} = \frac{\text{inputs} - \text{losses}}{745.7}$$

Input W_{in}:

$$W_{in} = \text{HP}(745.7) + \text{losses}$$

Torque T:

$$T = \frac{\text{HP}(1{,}008{,}000)}{\text{rpm}} \quad \text{oz·in}$$

Output W_{out}:

$$W_{out} = \text{HP}(745.7)$$

Efficiency ξ:

$$\xi = \frac{\text{HP}(745.7)}{\text{HP}(745.7) + (I^2 m r_1) + (\text{Fe loss}) + (F + W)}$$

Power factor PF:

$$\text{PF} = \frac{\text{HP}(745.7) + \text{losses}}{m V_{ph} I}$$

Variables for direct and quadrature axis current calculations θ, δ, ψ:

$$\theta = \cos^{-1} \text{PF}$$

$$\delta = \tan^{-1} \left[\frac{\text{PF} + (r_1/x_q) \sin \theta}{\sin \theta + (V_{ph}/I x_q) - (r_1/x_q) \cos \theta} \right]$$

$$\psi = \theta + \delta$$

where δ = torque angle

Direct axis current I_d:

$$I_d = I(\sin \psi)$$

Quadrature axis current I_q:

$$I_q = I(\cos \psi)$$

CHAPTER 8
APPLICATION OF MOTORS

Chapter Contributors

Earl F. Richards
William H. Yeadon

There are many things that must be considered when deciding what type and size of motor is most suitable to one's application. They can be broken into three general categories—namely, mechanical, electrical, and thermal considerations. The main reference in common use is the standard by the National Electrical Manufacturers Association (NEMA).

NEMA Standard MG1 covers standard frame sizing, dimensions, mounting configurations, electrical ratings, nominal speeds, and performance requirements of fractional- and integral-horsepower ac and dc motors. Table 8.1 lists some of the main topics.

NEMA Standard MG7 covers similar information on servomotors and control systems. Some of the major topics are listed in Table 8.2.

8.1 MOTOR APPLICATION REQUIREMENTS*

Any application of a motor must meet some specific output requirement for a given input. Furthermore, the motor must fit in the allotted space and mount securely to the device.

NEMA standards provide recommended mounting configurations for almost all applications and define performance ranges for them. Since these configurations are currently available, it behooves one to start the selection process there. There are,

---

* Sections 8.1 to 8.3 contributed by William H. Yeadon, Yeadon Engineering Services, PC.

TABLE 8.1 Main Topics in NEMA Standard MG1

Classification according to size	Small motors for shaft-mounted fans and blowers
Classification according to application	
Classification according to electrical type	Small motors for belted fans and blowers built in frames 56 and smaller
Classification according to environmental protection and methods of cooling	Small motors for air conditioning condensers and evaporator fans
Classification according to variability of speed	Application data
Rating, performance, and testing	Small motors and sump pumps
Complete machines and parts	Small motors for gasoline-dispensing pumps
Classification of insulation systems	Small motors for oil burners
DC motors and generators	Small motors for home laundry equipment
AC motors and generators	Motors and jet pumps
AC generators and synchronous motors	Small motors for coolant pumps
Single-phase motors	Index
Motors for hermetic refrigeration compressors	

however, many applications where a standard motor just won't fit. The ever-shrinking computer devices and appliances are a few examples. In these cases, special shafts, mounting hardware, and cooling provisions may be necessary. Listed here are some characteristics that must be considered.

I. Mechanical requirements
    A. Shaft size and features
    B. Shaft materials
        1. Compatibility with application
        2. Materials
        3. Finish
        4. Hardness
    C. Bearings
        1. Ball or sleeve
        2. Side load
        3. Axial load
        4. Lubrication temperature range
        5. Service life
    D. Mounting system
        1. Flange
        2. Bolts
        3. Studs
        4. Assembly method

TABLE 8.2 Main Topics in NEMA Standard MG7

Referenced standard, definitions, and safety standards	Safety standards for construction and guide for selection, installation, and operation of motion-control systems
Definitions	
Control definitions	Motion-control systems
Feedback device definitions	Construction
Motor definitions	Motors

        5. Strength
        6. Vibration resistance and damping
    E. Motor finish
        1. Corrosion resistance
        2. Material compatibility
    F. Operating environment
        1. Dust
        2. Moisture or vapors
        3. Heat (required insulation class)
II. Electrical requirements
    A. Power supply type
        1. AC
            a. Voltage
            b. Frequency
        2. DC—voltage
    B. Electronic drive requirements
        1. Voltage
        2. Number of phases
        3. Control scheme
            a. Open loop
            b. Closed loop
            c. Vector control
            d. Pulse width modulation
            e. Phase locked loop
            f. Linear amplifier
            g. Pulse frequency modulation
            h. Silicon controlled rectifier (SCR) controls
    C. Electromagnetic interference (EMI) or electromagnetic compatibility (EMC) requirements
III. Output requirements
    A. Duty cycle
        1. Continuous
        2. Intermittent duty—time on, time off, number of cycles
    B. Rated load torque and speed
    C. Load acceleration characteristics
    D. Velocity profile

The preceding requirements must be determined before selecting a proper motor that will meet the application requirements. Next, a profile of the load characteristics needs to be determined. These characteristics very widely. A given load profile may have a wide requirement range as soon as such things as life, temperature, and humidity vary. For example, if one is to apply a motor to a fan load, the fan characteristics must be determined. Typically, a fan is required to move a certain volume of air. That is to say, it must move so many cubic feet per minute (cfm). Figure 8.1 shows a range of cfm values for a fan superimposed on a multispeed motor speed-torque curve.

The solid lines show the motor speed-torque curves at high, medium, and low speeds. The dotted lines show fan performance under changing conditions of system pressure. These curves can be obtained from the fan manufacturer. When applying a motor to such a system, the nominal curve is applied to the motor such that it produces the proper volume of airflow at each of the selected speeds. Next, the operating temperature of the motor must be estimated and the performance curves

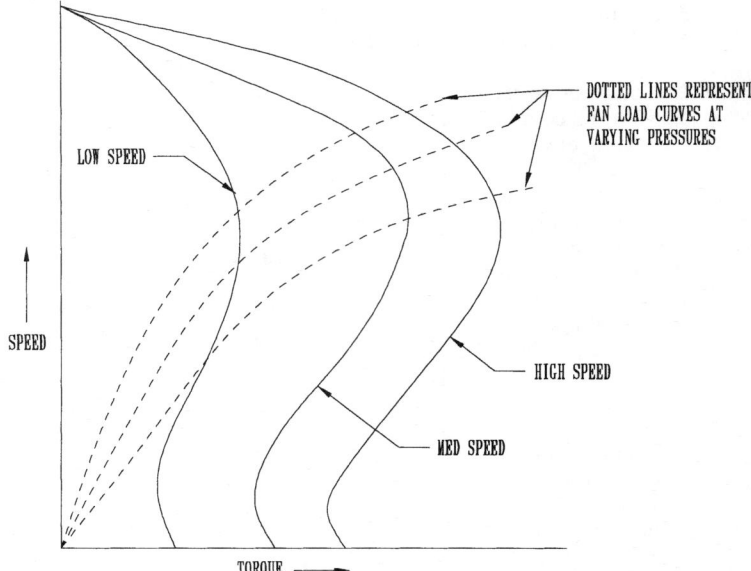

FIGURE 8.1 Three-speed ac motor with superimposed fan curves.

adjusted accordingly. As the temperature of an ac motor increases, motor torque at a given speed generally decreases. The fan curves are again superimposed on the hot speed-torque curves to ensure that the air delivery stays within reasonable limits. The same approach is used with dc motors.

If the load is a pump or some device with high starting-torque requirements, this superimposition of load curves must also include the region at and near the stalled-rotor condition.

Some ac motors have very weak low-speed curves or high third-harmonic cusp components, such as is shown in Fig. 8.2. In this case it is possible for the fan or pump load to reach the cusp area and remain at that speed because sufficient motor torque does not exist to accelerate the load through the cusp region.

8.2 VELOCITY PROFILES

Many devices require that the motor adapt to a changing load, or change speed within a certain time, or stop and reverse direction. During acceleration and deceleration, additional torque in excess of the normal running torque is required to overcome the motor and system inertia. The torques and currents necessary to accomplish these moves or changes may generally be determined as follows for dc motors having linear speed, torque, and current characteristics.

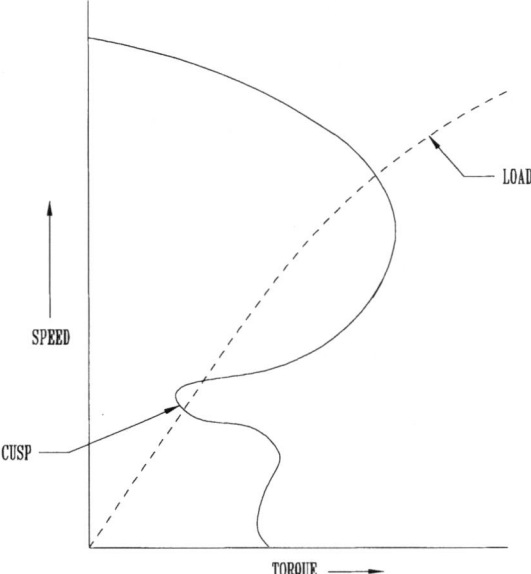

FIGURE 8.2 Motor speed-torque curve showing excessive third harmonic cusp.

8.2.1 Acceleration—Motor Subjected to a Step Voltage

To find the speed S at a known time t

$$S = S_{nl}\left(1 - \frac{T_L}{T_{st}}\right)\left[1 - \exp\left(-\frac{t}{T_m}\right)\right]$$

where S_{nl} = no-load speed
T_L = load torque, oz · in
T_{st} = motor stall torque, oz · in
T_m = motor mechanical time constant
t = time, s

Alternate method:

$$S = \frac{1}{0.1047 K_d}\left[(E_t - V_{br})\frac{K_t}{R_t} - T_L\right]\left[1 - \exp\left(\frac{K_d}{J_t}\right)\right]$$

where E_t = terminal voltage
V_{br} = brush voltage drop
K_d = zero source impedance damping coefficient (leads shorted), oz · in · s
J_t = rotor plus load inertia, oz · in · s
K_t = torque constant, (oz · in)/A

(V_{br} is subtracted only in the case of a brush-type dc motor.)

To find the rate of change in speed at a known time t:

$$\frac{dS}{dt} = \frac{S_{nl}}{T_m}\left(1 - \frac{T_L}{T_{st}}\right)\exp\left(-\frac{t}{T_m}\right)$$

To find the number of revolutions R_v at a known time t:

$$R_v = \frac{S_{nl}}{60}\left(1 - \frac{T_L}{T_{st}}\right)\left\{t - T_m\left[1 - \exp\left(-\frac{t}{T_m}\right)\right]\right\}$$

8.2.2 Acceleration—Motor Subjected to a Ramped Voltage

To find the speed at a known time t:

$$S = \frac{E_s}{0.001 K_b}\left\{t - T_m\left[1 - \exp\left(-\frac{t}{T_m}\right)\right]\right\}$$

where K_b = back-emf constant, V/krpm
E_s = slope of ramp voltage, V/s

To find the rate of change in speed at a known time t:

$$\frac{dS}{dt} = \frac{E_s}{0.001 K_b}\left[1 - \exp\left(-\frac{t}{T_m}\right)\right]$$

To find the number of revolutions R_v at a known time t:

$$R_v = \frac{E_s}{(0.001)(60)K_b}\left(\frac{t^2}{2} - T_m\left\{t - T_m\left[1 - \exp\left(-\frac{t}{T_m}\right)\right]\right\}\right)$$

8.2.3 Deceleration—Leads Open-Circuited or Short-Circuited

To find the speed at a known time t:

$$S = \frac{(T_{ft} + 0.1047\, K_x S_o)\exp\left(-\frac{K_x}{J_t}t\right) - T_{ft}}{0.1047\, K_x}$$

where T_{ft} = motor friction plus load friction, oz · in
$K_x = K_f$ (leads open-circuited)
$K_x = K_d + K_f$ (leads short-circuited)
K_f = infinite source impendence (leads open-circuited) damping coefficient, oz · in · s
S_o = speed, rpm, at time $t = o$

To find the rate of change in speed at a known time t:

$$\frac{dS}{dt} = -\left[\frac{T_{ft} + (0.0147\, K_x S_o)}{0.1047\, J_t}\right]\exp\left(-\frac{K_x}{J_r}t\right)$$

Time to stop t_s, s:

$$t_s = \frac{J_t}{K_x} \ln\left(1 + \frac{0.1047 K_x S_o}{T_{ft}}\right)$$

To find the number of revolutions R_v at a known time t to a maximum time to stop t_a:

$$R_v = \frac{J_t}{60 K_x}\left(\frac{T_{ft}}{0.1047 K_x} + S_o\right)\left[1 - \exp\left(-\frac{K_x}{J_t} ts\right)\right] - \frac{T_{ft} t_s}{(60)(0.1047) K_x}$$

8.2.4 Deceleration—Plugging Step Voltage (Reversal of Voltage Polarity)

$$S = -\frac{E_{tr} K_t + T_{ft} R_t}{0.001 K_t K_b}\left[1 - \exp\left(-\frac{t}{T_m}\right)\right] + S_o \exp\left(-\frac{t}{T_m}\right)$$

where E_{tr} = voltage of opposite polarity but not necessarily = E_f
R_t = terminal resistance, Ω

Alternate method:

$$S = -\frac{1}{0.1047 K_d}\left[\frac{E_{tr} K_t}{R_t} + T_{ft}\right]\left[1 - \exp\left(-\frac{t}{T_m}\right)\right] + S_o \exp\left(-\frac{K_d}{J_t} t\right)$$

Time to stop t_s:

$$t_s = T_m \ln\left(1 + \frac{0.001 K_b K_t S_o}{E_{tr} K_t + T_{ft} R_t}\right)$$

Alternate method:

$$t_s = T_m \ln\left\{1 + \left[\frac{0.1047 K_d S_o}{(E_{tr} K_t / R_t) + T_{ft}}\right]\right\}$$

A method of linear approximation to determine torque current and heating may be used in many cases. The example that is shown next may be modified to meet any particular problem. Note that it does not take thermal capacitance into account. In some situations thermal capacitance may be significant. See Sec. 8.4 for a method that does account for thermal capacitance.

Define a varying load and determine its effect on the heating of the motor. Assume the load and velocity parameters specified here. Refer to the velocity profile shown in Fig. 8.3.

Using motor parameters from available data:

K_t = 9.80 (oz · in)/A J_a = 2.32 × 10^{-3} oz · in · s^2
R_t = 2.51 T_{fi} = 0.85 oz · in
R_{th} = 5.50°C/W

where R_{th} is the motor thermal resistance.

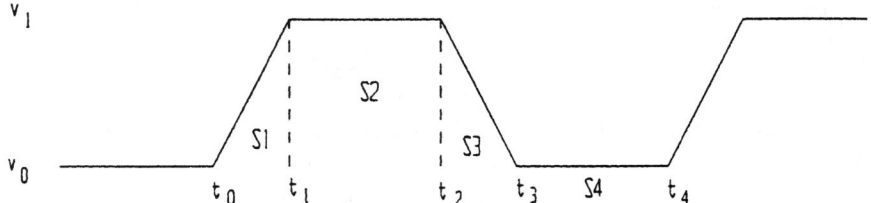

FIGURE 8.3 Velocity profile.

Calculate the required torque for each section of the velocity profile. For section 1, the torque to accelerate must be included.

$$T_{s1} = T_{fi} + T_L + J_a\alpha + Jl\alpha$$

$$\alpha = \frac{v_1 - v_0}{t_1 - t_0} = \frac{152 - 0}{0.05 - 0} = \frac{157 \text{ rad/s}}{0.05 \text{ s}} = 3142 \text{ rad/s}^2$$

$$T_{s1} = 0.85 + 10.0 + (2.32 \times 10^{-3})(3142) + (4.32 \times 10^{-3})(3142)$$

$$= 0.85 + 10.0 + 7.29 + 13.57$$

$$= 31.71 \text{ oz} \cdot \text{in}$$

$$I_{s1} = \frac{T_{s1}}{K_t} = \frac{31.71 \text{ oz} \cdot \text{in}}{9.80 \text{ (oz} \cdot \text{in)/A}} = 3.235 \text{ A}$$

$$W_{L,s1} = I_{s1}^2 R_t = (3.235)^2 (2.51) = 26.26 \text{ W}$$

Find the energy Q_{s1}.

$$Q_{s1} = W_{L,s1} t_{s1} = (26.26)(0.05) = 1.313 \text{ W} \cdot \text{s}$$

For section 2, only the friction torque is required.

$$T_{s2} = T_{fi} + T_L = 0.85 + 10.0 = 10.85 \text{ oz} \cdot \text{in}$$

$$I_{s2} = \frac{T_{s2}}{K_t} = \frac{10.85}{9.80} = 1.107 \text{ A}$$

$$W_{L,s2} = I_{s2}^2 R_t = (1.107)^2 (2.51) = 3.08 \text{ W}$$

$$Q_{s2} = W_{L,s2} (t_2 - t_1) = (3.08)(0.15 - 0.05) = 0.308 \text{ W} \cdot \text{s}$$

For section 3, the inertia must be included, but here the frictional torques assist in the deceleration.

$$T_{s3} = T_{fi} + T_L + J_a\alpha + J_1\alpha$$

$$\alpha = \frac{v_0 - v_1}{t_3 - t_2} = \frac{0 - 157}{0.2 - 0.15} = \frac{-157}{0.05} = -3142 \text{ rad/s}^2$$

$$T_{s3} = 0.85 + 10.0 + (-3142)(2.32 \; 10^{-3}) + (-3142)(4.32 \; 10^{-3})$$

$$= 10.85 - 7.29 - 13.57$$

$$T_{s3} = -10.01 \text{ oz} \cdot \text{in}$$

This just means that the power supply must produce 10.01 oz · in of torque in the other direction to get the motor to stop in the required time.

$$I_{s3} = \frac{T_{s3}}{K_t} = \frac{-10.01 \text{ oz} \cdot \text{in}}{9.80 \text{ (oz} \cdot \text{in)/A}} = -1.021 \text{ A}$$

$$W_{s3} = I^2 R_t = (-1.021)^2 \; 2.51 = 2.62 \text{ W}$$

$$Q_{s3} = W_{s3} t_{s3} = (2.62)(0.05) = 0.131 \text{ W} \cdot \text{s}$$

For section 4, it is obvious that the torque, losses, and energy are zero. The total energy is

$$Q_t = Q_{s1} + Q_{s2} + Q_{s3} + Q_{s4}$$

$$= 1.31 + 0.308 + 0.131 + 0.0$$

$$Q_t = 1.749 \text{ W} \cdot \text{s}$$

Average power dissipated (lost) $P_{L,av}$ or $W_{L,av}$ is:

$$W_{L,av} = \frac{Q_t}{t_4} = \frac{1.749}{0.3} = 5.83 \text{ W}$$

Temperature rise = $W_{L,av} R_{th}$ = (5.83 W) × 5.5°C/W = 32°C average rise

Is this a good choice of motor? Assume that the motor has Class B insulation. Then the maximum operating temperature of the copper wire is 130°C. Assume there is a hot spot inside the copper windings that is 10°C hotter than the outside of the windings, just to be safe. Now the maximum temperature becomes 130°C − 10°C = 120°C. If the ambient temperature of the device is 40°C, then you are allowed a maximum temperature rise of 120°C − 40°C = 80°C. You calculated a 32°C rise under your load profile. You could run the motor 48°C hotter. Therefore, you have chosen more motor than you need. Select a smaller motor.

8.3 CURRENT DENSITY

The ultimate goal of motor selection is to find the most cost-effective motor that will meet the performance and service life objectives of the application. This must be done while keeping the motor from exceeding its thermal limits. Motors for many applications are selected by the methods described previously, and while the average losses cause heating that does not exceed the insulation class of the motor, yet failures from apparent overheating occur. Typically, these motors exhibit charred winding insulation with shorted turns. However, the rest of the motor does not appear to have seen excessive heating.

In many cases, the failure occurs when the motor sees high currents for short durations, such as on acceleration, deceleration, plug reversal, or momentary overloads. When the current densities are calculated under these conditions, they exceed the range of the normal running conditions, which are shown in Table 8.3. Keep in mind that these current densities are guidelines and may not be appropriate in all cases and that motor temperature must be kept within the insulation class of the motor. These densities are commonly exceeded, but the wire insulation grade must be capable of handling the heating.

TABLE 8.3 Current Densities

Motor type	Density range, A/in^2
Enclosed	2000–4000
Ventilated	6000–8000

These high currents force the electrons to the outer surface of the conductor, causing localized excessive heating at the magnet wire insulation and copper junction. This results in insulation degradation and failure. The insulation in these spots is likely to flake off, resulting in shorted turns.

The solution to this problem is to reduce the current density by increasing the diameter of the magnet wire being used. There are times when the motor's slot fill is already too high to allow for a larger wire. In this case, the higher-temperature magnet-wire insulation may be adjusted to solve the problem.

8.4 THERMAL ANALYSIS FOR A PMDC MOTOR*

This section develops an approximate equivalent electrical circuit analog for the thermal characteristics of the permanent-magnet dc (PMDC) motor. Surely it can be understood why the exact equivalent cannot be found with the many variable parameters included; however, the circuit analog will assist in the thermal analysis of the motor and will indicate where problems may exist or improvements may be made.

The heat balance in any thermal system states that the heat added to a system must equal the heat transmitted from the system plus the heat stored in the system. In the steady-state mode, the heat transmitted from the system is equal to the heat added to the system, so that the heat stored is a constant; hence, the temperature of the system is a constant. The question then, is whether this is a safe operating temperature in terms of the system components. In the case of a motor, is this a safe operating condition—for example, is it within the safe temperature insulation class for the motor windings?

A lumped parameter electric circuit analog is used here; however, we must be aware that in order to accurately represent the motor, a distributed parameter model would have to be used. By using lumped parameters, the model parameters are characterized such that materials that offer resistance to heat flow have negligi-

---

* Sections 8.4 to 8.8 contributed by Earl F. Richards.

ble heat capacitance and materials that are represented by heat capacitance have negligible resistance to heat flow.

There are three methods by which heat can flow from one body to another. These are conduction, convection, and radiation. Fortunately, because of the relatively low temperatures involved, we do not have to consider *radiation*.

Conduction of heat is familiar—for example, the transfer of heat in solid objects, where heat flows from the high-temperature side to the low-temperature side. In the motor, one can envision this flow from the motor winding to the armature teeth by way of the motor winding slots. *Convection,* on the other hand, involves heat transfer by fluids over heated surfaces. In our case, the most common fluid medium is air. The transfer of heat from a rotating armature into the air gap involves the convection process.

It is probably obvious that convection heat transfer is more complex and involves more loosely defined parameters that are dependent on surface velocities, fluid densities, and specific heat. As velocities over a surface are increased, surface conditions change and heat transfer rates increase in accordance with Reynolds numbers. There is a maximum velocity, at which the heat transfer is also at a maximum.

To arrive at a circuit model, it is necessary to explore these heat transfer characteristics. We will have to use heat transfer surface coefficients which have, for the most part, been found experimentally in laboratories. Values are included by way of tables and later by curves. In some cases, experimenters have applied empirical expressions to calculate the parameters. These references are in the bibliography.

8.4.1 Theory

In developing electric circuit analogs, one has a choice in what variables in the system are to be represented by what electrical circuit parameters. In may cases, this choice can lead to a more understandable model. In thermal systems (i.e., motors) we have already mentioned that we have heat flow and heat storage. Hence, a two-parameter model is in order. Here again, we have a choice—are we interested in a *transient* or a *steady-state* model?

The beginning point is with the mathematical expressions for heat transfer.

For conduction:

$$q = K(T_1 - T_2) = K\Delta T \tag{8.1}$$

where q = heat flow rate, J/s or W
    K = proportionality constant, J/(s · °C) or W/°C
    ΔT = Temperature differential between surfaces, °C

$$T_1 > T_2$$

The right side of Eq. (8.1) can be written in a more convenient form as

$$q, \text{W} = \frac{\Delta T, (°C)}{1/K, °C/W} = \frac{\Delta T, °C}{R_{TH}, °C/W} \tag{8.2}$$

where R_{TH} is called the *thermal resistance* and is equal to

$$R_{TH} = \frac{1}{K} = \frac{L}{kA} \tag{8.3}$$

where L = length of thermal path, in
  A = cross-sectional area of path, in^2
  k = thermal conductivity, W/(in · °C)

Note the resemblance to the expression for electrical resistance

$$R = \frac{pl}{a}$$

where l = length of path (conductor)
  a = cross-sectional area (area of conductor)
  p = electrical resistivity constant

In most cases, motors will be composed of many different materials whose thermal resistances will be different. Also, in most cases there will be multiple thermal paths, and series and parallel arrangements will be required.

Thermal bonding between surfaces of different materials is extremely important. Unless close thermal contact is made, a dead-air film can lead to a high thermal resistance, which will affect the heat transfer to a very high degree.

Remember that in many cases the temperature differential will involve the ambient temperature, which will affect the thermal resistance. For example, a pump motor submerged in a well has a different thermal resistance than one in an air environment.

When convection heat transfer is considered, surface area is of concern; however, it can be treated in a fashion similar to that for conduction thermal resistance in order to assist in the model development. If in Eq. (8.3) the proportionality constant K is considered to be a function only of the surface conditions and the surface area for convection, then:

$$R_{TH} = \frac{1}{hA_s}\left(\frac{°C}{W}\right) \quad (8.4)$$

where h = surface heat transfer coefficient, W/(°C · in^2)
  A_s = surface area, in^2

The major problem is determining an acceptable value of h for the many conditions which could exist at the surface. It is very possible that an internal or external motor fan can be incorporated in the motor system design. It is also known, for example, that rotation of the rotor causes an air-pumping action on the surface which is a function of the peripheral velocity of the rotor. On the outer side of the motor shell, only natural convection may occur. All of these conditions cause heat transfer from the hotter surfaces to cooler fluids (air). Thermal heat transfer also occurs from the hotter internal motor air to the cooler stator components.

It is necessary to define all heat sources (losses) in the PMDC motor, because they will become the power sources in our electrical circuit model. It will be found that current sources become the best choice to correspond to these losses. Therefore, it is necessary to have a value or good estimate of what losses will be. The losses considered are as follows:

- Copper loss I^2R occurring in the armature
- Friction and windage losses
- Brush loss (heating and friction)
- Core loss in the armature

In the armature, the copper loss occurs partly in the air gap armature slots and partly in the end turns on both ends of the armature where the heat transfer characteristics are different. Some of the losses are speed dependent. So far, the thermal resistances for conduction and convection have been considered. If we are considering only steady-state heat transfer, this is all that is needed. To complete the transient heat transfer, we need to consider the heat-storage elements. These are elements in the motor which have appreciable mass. Moreover, the specific heat of each mass is important in determining the thermal heat capacity of each element. Physics defines the thermal capacity of a mass as follows:

$$\text{Thermal capacity} = \text{mass} \times \text{specific heat}$$

$$C_{TH} = M c_p \tag{8.5}$$

where C_{TH} = thermal capacitance, J/°C = (W · s)/°C
M = body mass, lb
c_p = specific heat, J/(lb · °C)

For each mass element of the motor, a transient thermal balance equation is required. That is, repeating what was previously stated:

Heat stored in the element = heat added to the element
− heat transmitted from the element

Stated mathematically:

$$C_{TH} = \frac{dT}{dt} = P_{added} - P_{trans} \tag{8.6}$$

where T = temperature, °C
t = time, s
P_{added} = heat, W
P_{trans} = heat, W

If we now refer to Eq. (8.2) as the relationship for heat conduction involving thermal resistance and to Eq. (8.5) for the thermal capacitance, and substitute into Eq. (8.6), we have:

$$C_{TH} = \frac{dT}{dt} = P_{added} - \frac{\Delta T}{R_{TH}} \quad \text{W}$$

or

$$P_{added} = C_{TH} \frac{dT}{dt} + \frac{\Delta T}{R_{TH}} \tag{8.7}$$

Let us first look at an electrical RC circuit (Fig. 8.4) in the form of a node voltage equation.

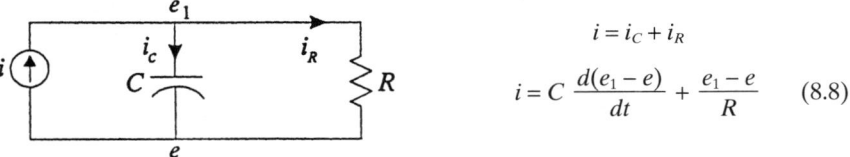

$$i = i_C + i_R$$

$$i = C \frac{d(e_1 - e)}{dt} + \frac{e_1 - e}{R} \tag{8.8}$$

FIGURE 8.4 RC circuit.

It can be seen how an analog between the thermal system and the electrical circuit can be made. If Eqs. (8.7) and (8.8) are compared, the analogies are as shown in Table 8.4.

In the interest of simplicity, it is usually convenient to assume the reference node e to be zero. In the final analysis, the temperature rise can then be found by subtracting the ambient temperature.

TABLE 8.4 Comparison of Thermal System and Electrical Circuit

Thermal system	Electrical circuit
P_{added} = heat loss, W	i = current source, A
T = temperature, °C	e = voltage, V
C_{TH} = thermal capacitance, (W · s)/°C	C = capacitance, F
R_{TH} = thermal resistance, °C/W	R = resistance, Ω

With the modeling indicated, the various currents in the electrical analog will become power flow (watts), and the nodes in the electrical circuit will become the temperatures at various points. In the PMDC motor two very important nodes of concern could be the armature and the permanent-magnet temperature.

If we are to develop the thermal analog, we must be able to calculate for each motor component the thermal resistance and thermal capacitance, or both. In order to do this, we need to be able to obtain or calculate the physical dimensions of the motor. In many cases, approximations will be required. It may be appropriate to trace the heat flow from the various losses which occur in the motor. The flow chart shown in Fig. 8.5 traces the heat flow.

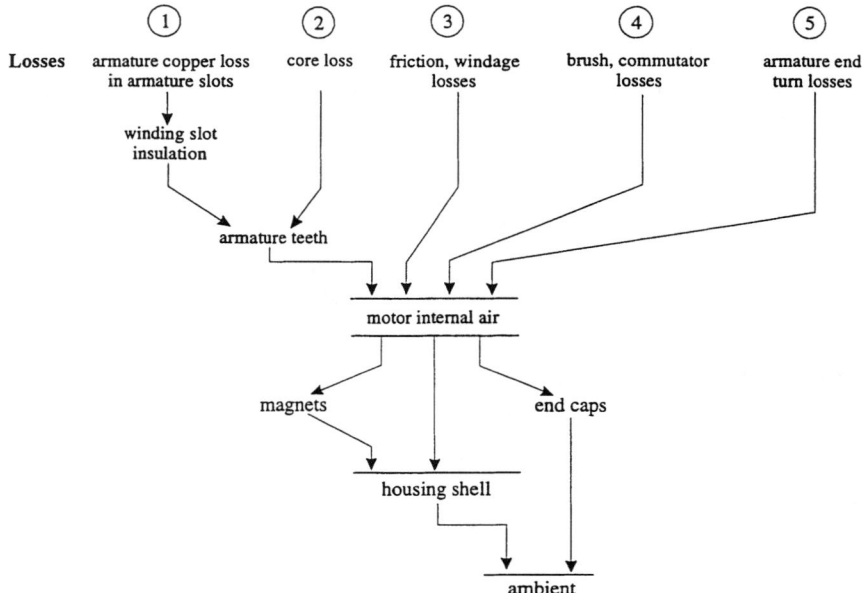

FIGURE 8.5 Heat flow in a motor.

Note from the heat flow in Fig. 8.5 that some elements in the conduction paths require both thermal resistances and thermal capacitances, while convection elements simply require thermal resistances. All three of these have been defined in Eqs. (8.3), (8.4), and (8.5). From the heat flow related in Fig. 8.5 it is obvious that we expect the electrical circuit analog to include series and parallel circuit elements made up of R and C values. All of the power losses are to be represented as current sources; conduction elements will have parallel R_{TH} and C_{TH} and convection elements only R_{TH}. This then allows us to draw the electrical analog from the heat flow diagram in Fig. 8.5, using the ambient temperature as reference 0. To make the electrical analog look more like an electrical circuit, the elements have been rearranged in Fig. 8.6. The subscripts TH of thermal resistances and capacitances have been replaced by number subscripts in the circuit.

Note that the heat transfer from the internal air to the motor housing has been broken down into two parts. One part includes the magnets plus the housing shell covered by the magnets, and the other part includes the section of the housing not covered by the magnets. In the circuit, single-component subscripts indicate conduction elements, while double-component subscripts indicate convection elements.

By using operational forms for the capacitances (i.e., $1/C_s$ where s is the Laplace operator d/dt), a formal solution could be reached. It is a complex solution; however, many software programs would make this a very simple solution. Any attempt at network reduction and simplification by neglecting rather insignificant parameters could be appropriate. However, in doing so one must not lose sight when eliminating significant nodes, which represent temperatures at various points, or one may lose the purpose of the evaluation.

It would be useful at this point to attempt a solution to a steady-state problem, which means the thermal capacitances can be eliminated.

A PMDC motor has been tested in the laboratory and the temperature rises have been recorded at the steady-state conditions. Having all of the physical dimensions and data on the motor, let us now develop a model and compare the results.

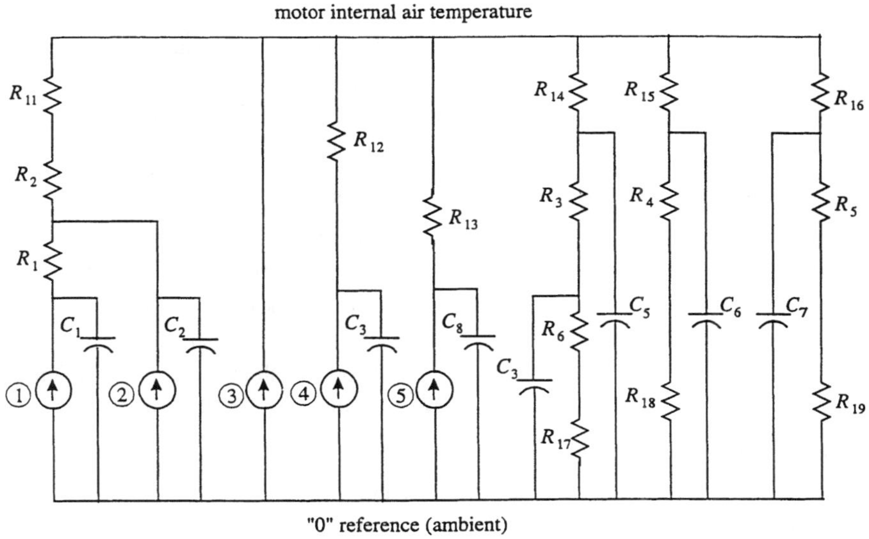

FIGURE 8.6 PMDC motor electrical analog of the thermal properties.

In order to calculate the thermal resistances and thermal capacitances, it is necessary to obtain the thermal conductivity and heat transfer surface coefficients.

In most texts and tables of thermal conductivity, values are given in units of Btu/(h · ft · °F). In U.S. motor technology units of W/(in · °C) are preferred. Many texts list constants for common materials used in motor construction at reasonable motor operating temperatures. These are listed in Table 8.5 (conversion factor is 0.044).

TABLE 8.5 Thermal Conductivity Values

Material	Value Btu/(h · ft · °F)	W/(in · °C)
Aluminum	123	5.41
Copper	215	9.46
Silicon steel	15	0.66
Air	0.014	0.0006
Insulators	0.079	0.0035
Ceramic magnets	0.34	0.015
Low-carbon steels	35	1.54

The values for convection surface conditions are much more dependant on air velocities than are conductivity constants on temperature.

Following in the figures in this section are several graphs from the bibliography showing values for heat transfer coefficients.

EXAMPLE. 2.1 hp PMDC, running at 1380 rpm, 65 V at 8 A.
Motor data:

Shell

Length	8.8 in
ID	3.375 in
Thickness	0.212 in
Material	1018 steel

Armature

Diameter	2.5 in.
Stack length	4 in.
Material	Electrical M-54
Commutator	0.7 × 1.5 in

Magnets

Number	2
Material	Ceramic
Arc	135°
Size	0.4 × 4.25 in

Slots

Number	16
Depth	0.5 in (round bottom)
Tooth thickness	0.120 in.
Opening	0.120 in.
Tooth width at surface	0.37 in
Insulation	0.0125 in

End caps

Material	Aluminum, 4 holes; effective area 60 percent
Thickness	0.25 in.
Diameter	3.875 in.

APPLICATION OF MOTORS

Let us calculate the thermal resistances in order (in all cases it is required to refer to Tables 8.6 to 8.9 and Figs. 8.7 to 8.19 for the best transfer constants).

$$u = \text{speed, m/s}$$

$$c = \text{cooling coefficient, } (m^2 \cdot °C)/W$$

TABLE 8.6 Heat Conductivities of Some Metals and Insulating Materials

Material	Value
Copper	9.75
Brass (70 Cu: 30 Zn)	2.80
Aluminum	5.35
Organic varnishes	0.0035–0.0045
Shellac	0.0063
Air	0.0006
Cotton tape, untreated	0.0019
Cambric, varnished	0.004
Mica, sheet	0.014
Micafolium	0.0044
Micanite (for commutator)	0.0104
Paper, board, varnished	0.0073
Paper, kraft tissue	0.001
Pressboard	0.0046
Silk, untreated	0.0012
Silk, varnished	0.0042
Micarta with paper and cotton fillers	0.0046–0.008
Micarta with fiberglass and asbestos filler	0.006–0.010
Micarta (0.125 in)	0.0066
Asbestos, paper (0.025 in)	0.0040
Quartz	0.0380
Asphalt	0.00175
Powdered graphite (through 20- to 40-mesh sieve)	0.034
Powdered graphite (through 40-mesh sieve)	0.011
Rope paper	0.0029
Rope paper, impregnated in oil	0.0037
Rope paper, impregnated in varnish	0.0043
Cement paper	0.0034
Fish paper	0.0046
Cotton, untreated	0.0012–0.00175
Cotton, impregnated in varnish	0.0061–0.0068

Note: Laminations in radial direction = 0.80–1.2; laminations in axial direction = 0.013–0.014.

To obtain h, W/(°C · in²), for Eq. (8.4):

1. Divide speed, ft/min, by 197 to get u, m/s.
2. Calculate c.
3. Multiply c by 1550 to get c, (°C · in²)/W.
4. Invert c to get h, W/(in² · °C).

TABLE 8.7 Heat Transfer Coefficients for Air as Cooling Medium

End winding and parts of winding in radial vents	Lamination in radial vents	Lamination in axial channels	Field winding
0.025–0.10 W/(in² · °C) 600–6000 ft/min	0.025–0.07 W/(in² · °C) 800–400 ft/min	Length of channel = 2 in 0.03–0.11 W/(in² · °C) 1200–5000 ft/min	0.008–0.025 W/(in² · °C) 800–8000 ft/min
		Length of channel = 20 in. 0.02–0.08 W/(in² · °C) 1200–5000 ft/min	

TABLE 8.8 Typical Values of Cooling Coefficient

Part	c	u	Remarks
Cylindrical surface of dc armature	$\dfrac{0.015 \text{ to } 0.035}{1+0.1u}$	Armature peripheral speed	Smaller values for large, open machines
Cylindrical surfaces of stator and rotor	$\dfrac{0.03 \text{ to } 0.05}{1+0.1u}$	Relative peripheral speed	Lower values for forced cooling
Back of stator core	0.025 to 0.04	Zero	
Stationary field coils	$\dfrac{0.14 \text{ to } 0.16}{1+0.1u}$	Armature peripheral speed	Based on total coil surface
	$\dfrac{0.06 \text{ to } 0.08}{1+0.1u}$		Based on exposed coil surface only
Rotating field coils	$\dfrac{0.08 \text{ to } 0.12}{1+0.1u}$	Armature peripheral speed	Based on total coil surface
	$\dfrac{0.06 \text{ to } 0.08}{1+0.1u}$		Based on exposed coil surface only
Commutator	$\dfrac{0.015 \text{ to } 0.025}{1+0.1u}$	Commutator peripheral speed	Surface includes a proportion for risers
Ventilating ducts in core	$\dfrac{0.08 \text{ to } 0.2}{u}$	Air velocity in ducts	Taken as about 10 percent of peripheral speed of core

APPLICATION OF MOTORS

TABLE 8.9 Weights and Thermal Properties of Materials

Material	P_e lb/in³	Thermal capacity, (W · s)/(lb · °C)	Conductivity, K_f W/(in³ · °C) (if insulation, taken across grain or fiber)
Aluminum, commercial	0.093	433	5.0
Copper, commercial	0.32	180	8.8
Steel, solid, electrical sheet	0.283	225	1.15
Steel, 0.014 laminations	0.28	225	1.10
Across laminations	—	—	0.045
Air at rest, 1 atm (constant pressure)	0.428	453	0.0007
Asbestos paper	0.08	373	0.006
Cotton, compressed	—	690	0.0008
Fullerboard, varnished	0.035	700	0.0035
Impregnating compound	0.035	1,000	0.006
Impregnated cotton, paper	0.035	800	0.005
Paraffin	0.029	1,125	0.007
Mica, molded	0.09	390	0.003
Paper	0.035	700	0.0033
Varnished cambric, cotton	0.035	700	0.0065

Conductance of slot insulation R_1:

$$R_1 = \frac{L}{kA} = \frac{0.0125}{(0.0035)A} \tag{8.9}$$

where

A = number of slots × stack length × 1.5 × slot depth

$\quad = 16 \times 4 \times 1.5 \times 0.5 = 48$ in²

$R_1 = 0.074$

Conductance of teeth R_2:

$$R_2 = \frac{L}{kA} \tag{8.10}$$

where

A = number of teeth × stack length × tooth width

$L = 0.7 \times$ tooth depth

$$R_2 = \frac{0.7\,(0.5)}{1.54(16 \times 4 \times 0.37)} = 0.010$$

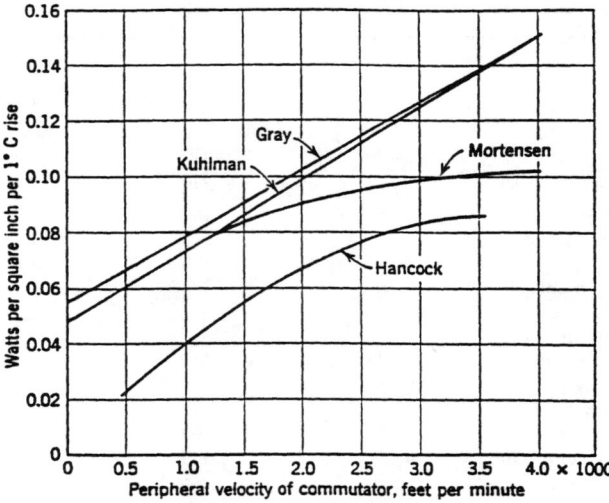

FIGURE 8.7 Temperature rise in commutators. *(From Roye, 1978.)*

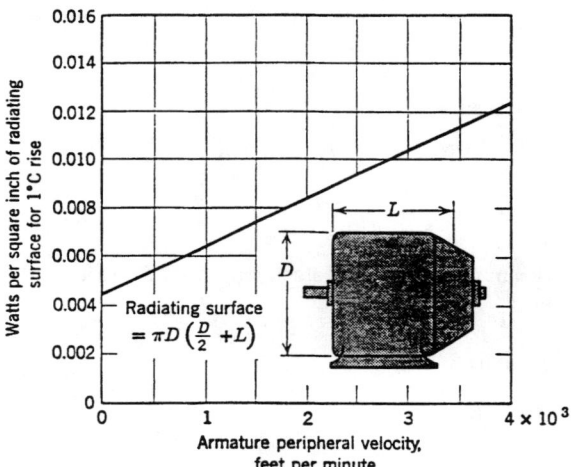

FIGURE 8.8 Heating in fully enclosed machines. *(From Roye, 1978.)*

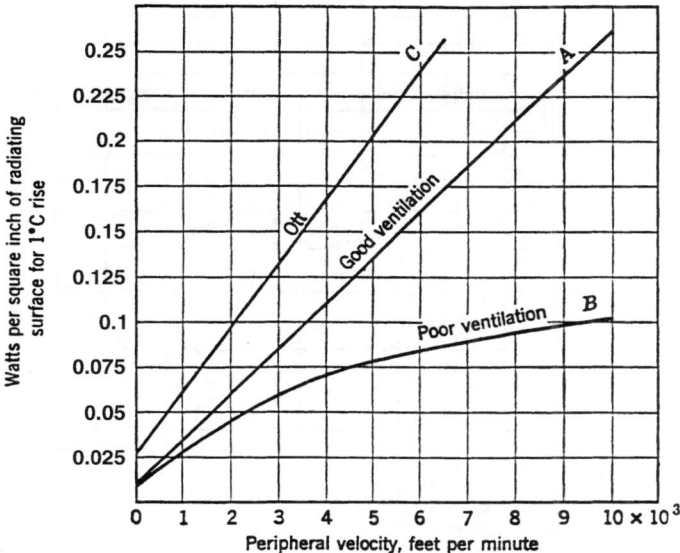

FIGURE 8.9 Watts radiated per square inch of cylinder surface in open motors. *(From Roye, 1978.)*

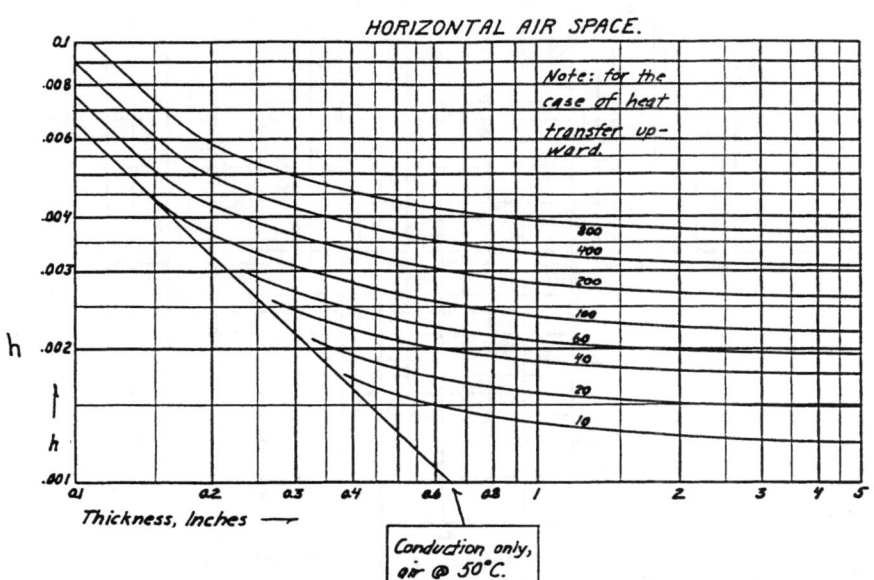

FIGURE 8.10 Heat transfer across a horizontal air space (heat transfer upward). $h = \text{W}/(\text{in}^2 \cdot {}^\circ\text{C})$. *(From Figel and Labahn, 1972.)*

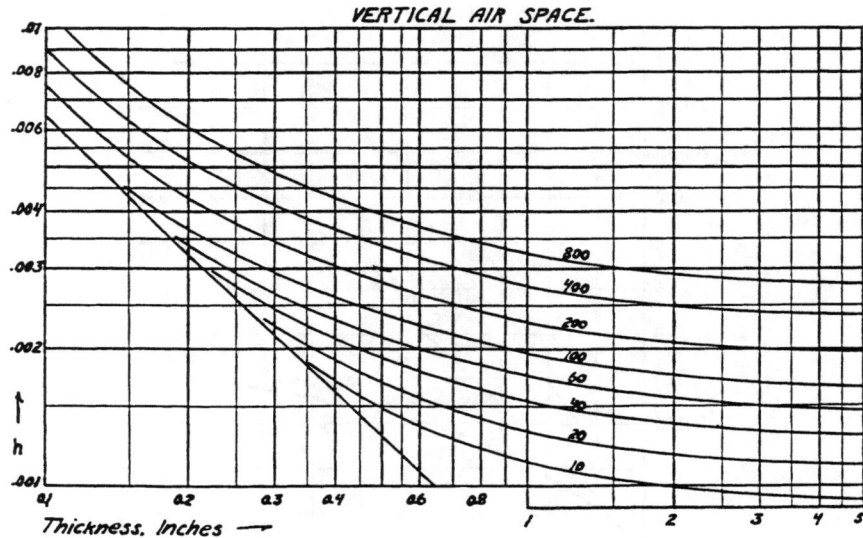

FIGURE 8.11 Heat transfer across a vertical air space (heat transfer sideward). Radiation not included. Curve numbers show temperature drop across space, °C; cold side at 25°C. *(From Figel and Labahn, 1972.)*

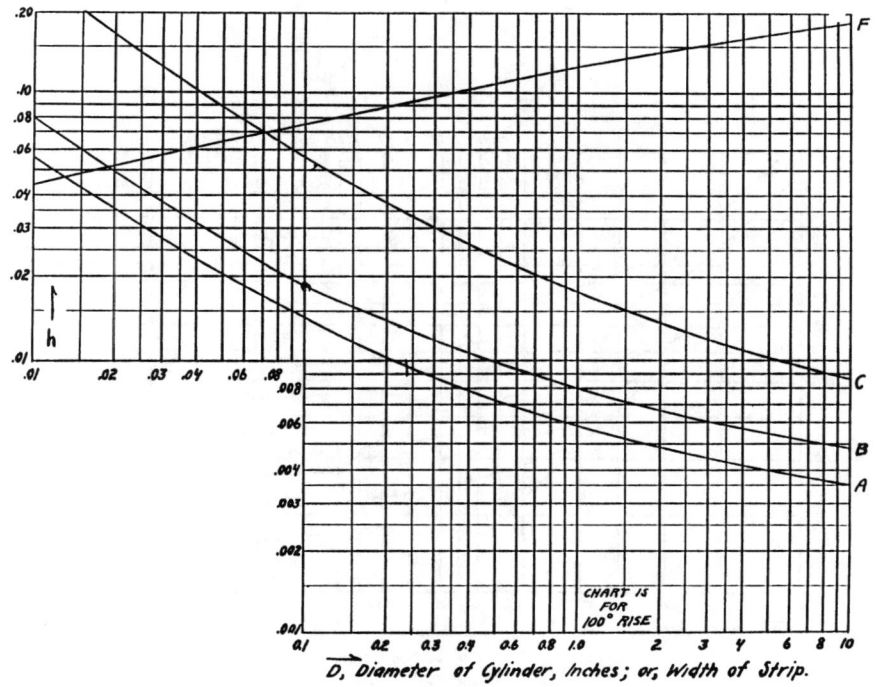

FIGURE 8.12 Natural convection coefficients in gases, where ambient temperature is 25°C, temperature rise is 100°C, hot body is at 125°C, and $h = W/in^2 \cdot °C$). A = horizontal cylinder in air, B = thin strips in air, C = horizontal cylinder in H_2, and F = computed fictive film thickness for case A, ordinates in inches. For curves other than 100°C rise, find new $h' = h(\Delta T/100)^{0.25}$. *(From Figel and Labahn, 1972.)*

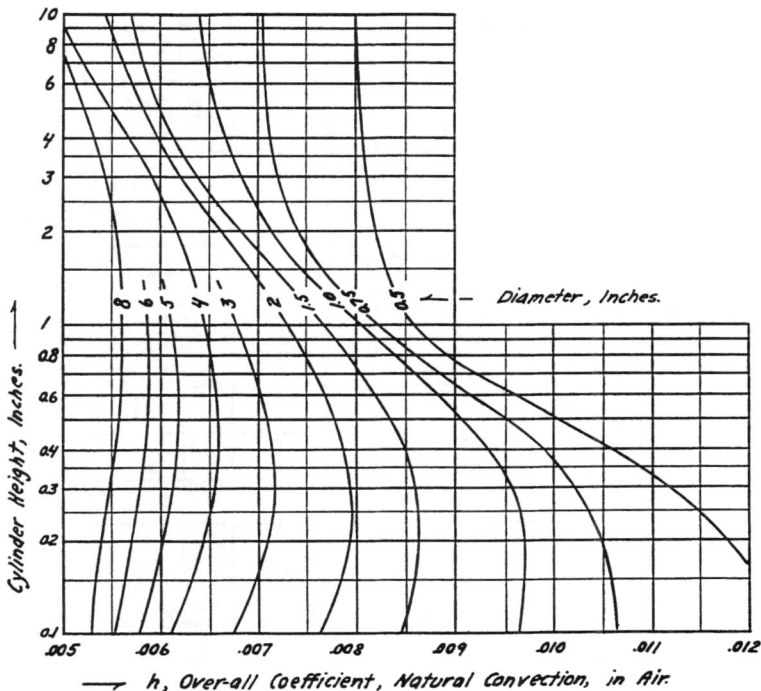

FIGURE 8.13 Vertical cylinder synthetic convection coefficients for 100°C temperature rise. $h = \text{W}/(\text{in}^2 \cdot °\text{C})$. *(From Figel and Labahn, 1972.)*

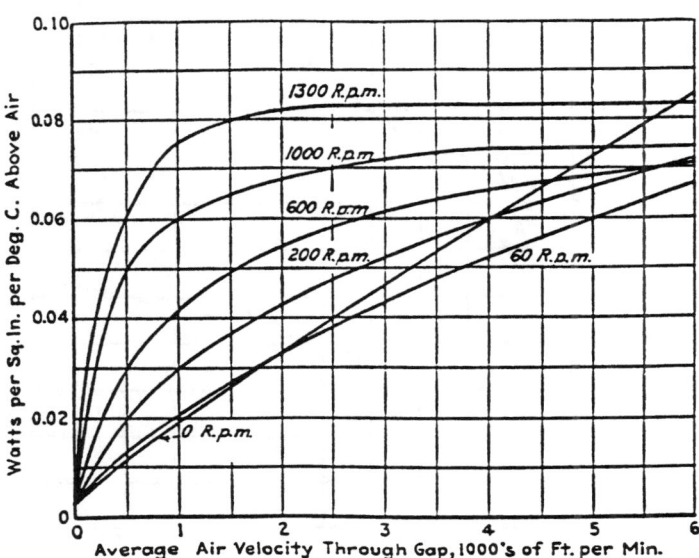

FIGURE 8.14 Dissipation of heat from the surface of a 25-in-diameter rotor with forced axial ventilation through a 0.5-in air gap. *(From Packer, 1951.)*

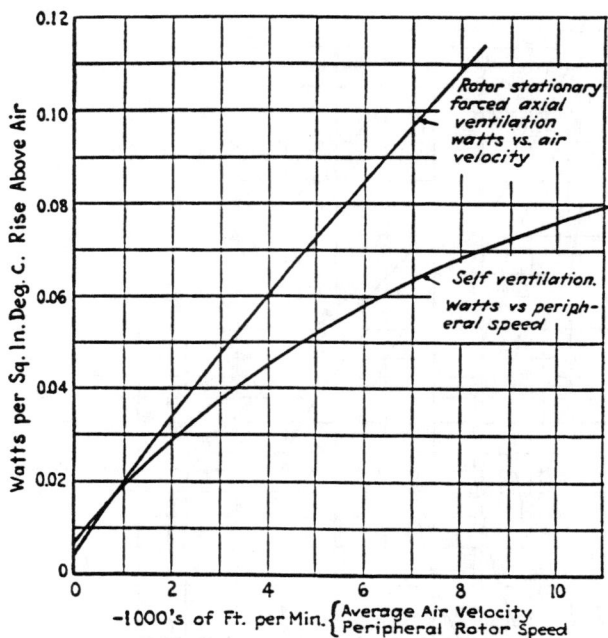

FIGURE 8.15 Dissipation of heat from rotor surface. *(From Packer, 1951.)*

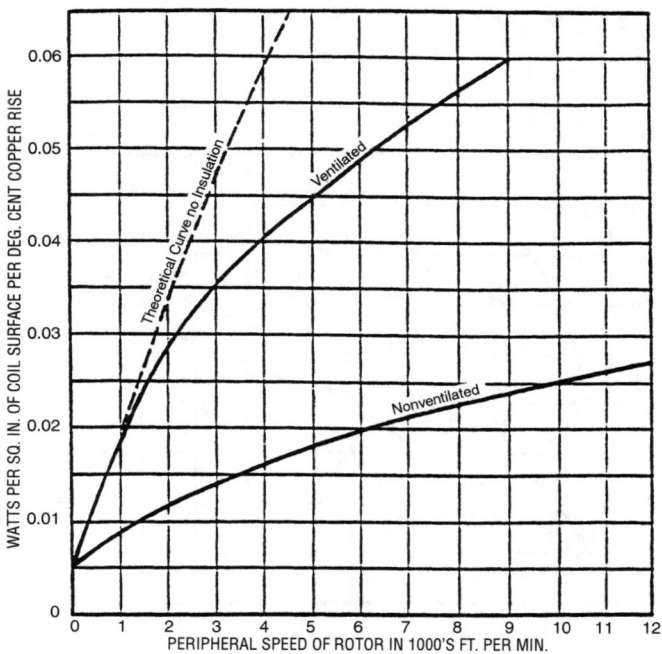

FIGURE 8.16 Dissipation of heat from the surface of a 25-in-diameter 36-in-long rotor. *(From Packer, 1951.)*

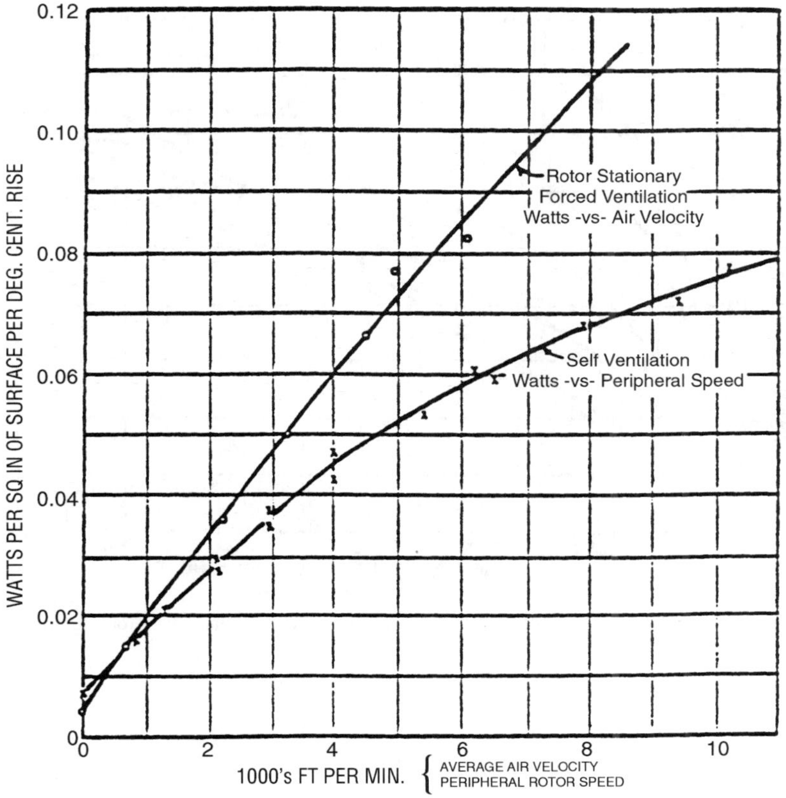

APPLICATION OF MOTORS

FIGURE 8.17 Dissipation of heat from the surface of a 25-in-diameter 36-in-long rotor. *(From Packer, 1951.)*

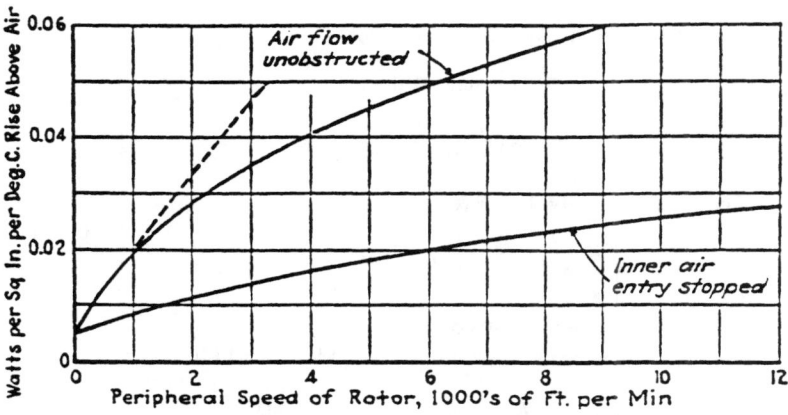

FIGURE 8.18 Dissipation of heat from the surfaces of rotating end windings (temperature rise of copper above air). *(From Packer, 1951.)*

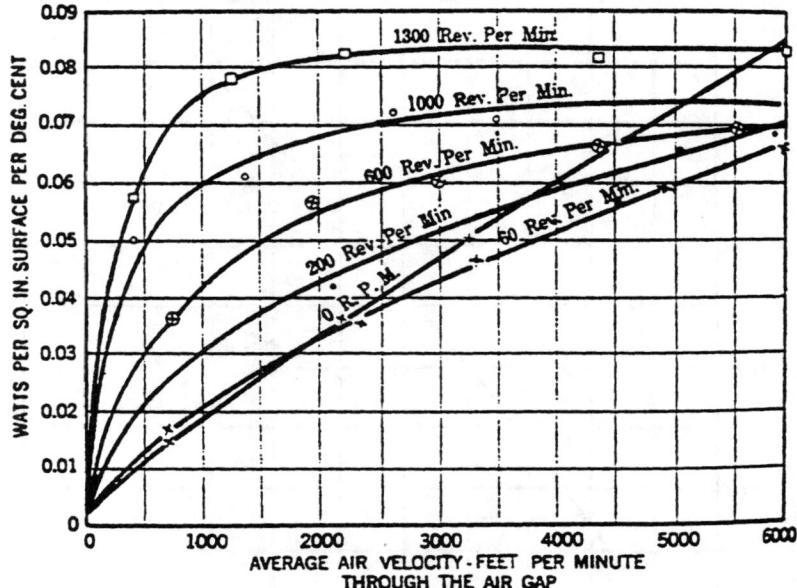

FIGURE 8.19 Dissipation of heat from the surface of a 25-in-diameter rotor with forced axial ventilation through a 0.5-in air gap. *(From Packer, 1951.)*

Thermal conductance of magnets R_3:

$$R_3 = \frac{L}{kA} \qquad (8.11)$$

where $A = \pi(3.375 - 0.4)(135/180)(4.25) = 29.7$ in^2

L = magnet thickness = 0.4

$$R_3 = \frac{0.4}{(0.015)(29.7)} = 0.89$$

Thermal conductance of end caps R_4:

$$R_4 = \frac{L}{kA} \qquad (8.12)$$

where L = thickness of end caps = 0.25
A = area of end caps = $2 \times \pi(3.875)^2/4 \times 0.6 = 14.14$ in^2

$$R_4 = \frac{0.25}{(5.41)(14.14)} = 0.003$$

Conductance of shell not covered by magnets R_5:

$$R_5 = \frac{L}{kA} \qquad (8.13)$$

where $L = 0.212$ in

$A = \pi$ (diameter of shell) (length) − (area covered by magnets)
$= \pi(3.8)(88) - 29.7 = 75.7$ in^2

$$R_5 = \frac{0.212}{1.54(75.7)} = 0.0018$$

Conductance of shell covered by magnets R_6:

$$R_6 = \frac{L}{kA}$$

where $A = 29.7$ in^2 (from R_3)

$$R_6 = \frac{0.212}{1.54(29.7)} = 0.0046$$

The convection thermal resistances are as follows.
Convection from armature teeth to air gap R_{11}:

$$R_{11} = \frac{1}{hA} \qquad (8.14)$$

where $A =$ total area or armature teeth $= 16 \times 4 \times 0.37 = 23.68$ in^2

The peripheral velocity of the armature at 1380 rpm is

$$v = \omega r = 1380\left(\frac{\pi}{30}\right)\left(\frac{2.5}{12}\right) = 30.1 \text{ ft/s} = 1806 \text{ft/min}$$

$$R_{11} = \frac{1}{h(23.68)} = \frac{1}{(0.075)(23.68)} = 0.56$$

Convection from brush, commutator to air gap, R_{12}:

$$R_{12} = \frac{1}{hA} \qquad (8.15)$$

where $A = \pi (1.5)(0.7) = 3.3$ in^2

$$R_{12} = \frac{1}{0.08(3.3)} = 3.78 \qquad \text{Peripheral velocity} = 1083 \text{ ft/s}$$

Convection from end turns to air gap R_{13}:

$$R_{13} = \frac{1}{hA} \qquad (8.16)$$

where $A = 2(\pi 2.5^2/4) - \pi(1.5)^2/4 = 8.0$ in^2

Assume peripheral velocity = 1806 × (0.75) = 1354 ft/min

$$R_{13} = \frac{1}{0.035(8)} = 3.57$$

Convection of air gap to magnets R_{14}:

$$R_{14} = \frac{1}{hA} \qquad (8.17)$$

where $A = 29.7$ in^2 (from R_3)

$$R_{14} = \frac{1}{0.035(29.7)} = 0.96$$

Convection of air gap to end cap R_{15}:

$$R_{15} = \frac{1}{hA} \qquad (8.18)$$

where $A = 14.4$ in^2 (from R_4)

$$R_{15} = \frac{1}{0.035(14.4)} = 1.98$$

Convection of air gap to shell not covered by magnet R_{16}:

$$R_{16} = \frac{1}{hA} \qquad (8.19)$$

where $A = 26.3$

$$R_{16} = \frac{1}{0.075(26.3)} = 0.51$$

convection of shell covered by magnets to ambient air R_{17}:

$$R_{17} = \frac{1}{hA} \qquad (8.20)$$

where $A = (3.8)(8.8)(0.75) = 78.8$ in^2

$$R_{17} = \frac{1}{0.008(7877)} = 1.58$$

Convection of end caps to ambient air R_{18}:

$$R_{18} = \frac{1}{hA} \qquad (8.21)$$

where $A = 14.4$ in^2 (from R_4)

$$R_{18} = \frac{1}{0.008(14.4)} = 0.070$$

convection of shell not covered by magnets R_{19}:

$$R_{19} = \frac{1}{hA} \tag{8.22}$$

where $A = 75.7$ in^2 (from R_5)

$$R_{19} = \frac{1}{0.008(75.7)} = 1.65$$

Figure 8.20 includes the previously calculated values for the thermal resistances. It is obvious that circuit reduction can be made to determine a solution for the temperatures at various nodes in the network. The solution is also simplified because all of the current sources (watt losses) are known. Since the current at the air gap node must satisfy the condition that the current summation be zero, the watts into the air gap should be equal to the watts out of the air gap.

As suggested earlier, it is necessary to know or be able to estimate the motor losses. Certainly, if one were designing a motor, performance calculations would be made and losses known. Also, laboratory testing could be done to obtain the losses.

From laboratory measurements and performance calculations, the following motor losses were found for the motor under the following operating conditions:

Motor speed	1380 rpm
Armature current	8 A
Motor voltage	65 V

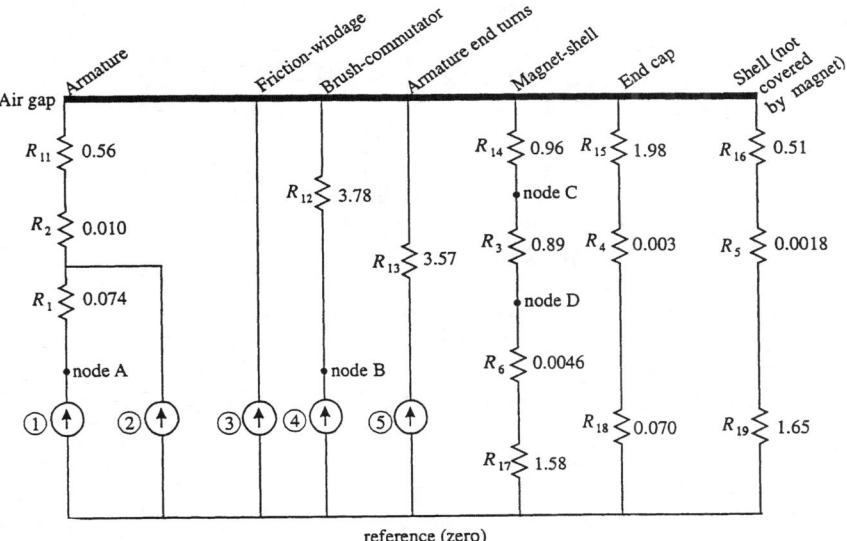

FIGURE 8.20 Electrical analog of thermal model with calculated thermal resistance.

8.30 CHAPTER EIGHT

The losses for these conditions are as follows:

Copper loss	95 W
Core loss	17 W
Friction and windage loss	25 W
Brush and commutator loss	12 W
Total losses	149 W

It is necessary to divide the total copper loss between the copper loss in the armature slot and the end-turn copper loss. The ratio of the copper in the slots to end turns is $8/4.66 = 1.71$. Therefore, of the 95 W copper loss, $8/12.66 \times 95 = 60$ W occurs in the slots, and 35 W occurs in the end turns. The current sources then become

$1 = 60$ W
$2 = 17$ W
$3 = 25$ W
$4 = 12$ W
$5 = 35$ W

From Fig. 8.20, the right three branches can be reduced to a single thermal resistance by series and parallel addition. These three branches become

$$R_{16} + R_5 + R_{19} = 2.16$$

$$R_{15} + R_4 + R_{18} = 2.05$$

$$R_{14} + R_3 + R_6 + R_{17} = 3.43$$

The parallel combination of these three is

$$\frac{1}{R_{eq}} = \frac{1}{2.16} + \frac{1}{2.05} + \frac{1}{3.43} = 1.24$$

$$R_{eq} = 0.806$$

The total current is equal to the sum of the losses, which is 149. The temperature difference across the R_{eq} branch is $T_{eq} = 149 (R_{eq}) = 149 (0.806) = 120°C$.

The temperature of the motor copper (node A) is

$$T_{eq} + 0.57 (77) + 0.074 (60) = 120 = 43.9 + 4.44 = 168.3°C$$

The temperature of the commutator (node B) is

$$T_{eq} + 3.78 (12) = 120 + 45.3 = 165.3°C$$

The temperature of the magnet (node C) is

$$T_{eq} - \frac{T_{eq}}{3.43}(0.96) = 120 - 33.6 = 86.4°C$$

The temperature of the shell (node D) is

$$T_{eq} - \frac{T_{eq}}{3.43}(1.85) = 120 - \frac{120}{3.43}(1.85) = 120 - 64.7 = 55.2°C$$

TABLE 8.10 Calculated and Measured Temperatures of Motor Components

Material	Temperature, °C	
	Calculated	Measured
Copper windings	168.3	166
Commutator	165.3	141
Magnets	86.4	73
Shell	55.2	66

The calculated values and measured values for comparison are given in Table 8.10.

8.4.2 Conclusion

If network reduction (i.e., application of network theorems) were performed on the network shown in Fig. 8.6, the result would be single RC network with a current source equal to the total motor losses (see Fig. 8.21).

It is not difficult to recognize this form if we just consider the heat balance which must exist. That is, the losses which are occurring in the motor must be stored in the thermal capacitance of the material, less that which is conducted to the ambient environment.

If we initially begin with the motor at ambient temperature, then energize the motor, and further assume that the motor losses remain constant at a given speed, load and current—the currents i_c and i_R in Fig. 8.21—will change as shown in Fig. 8.22.

R_{TH} and C_{TH} can be found from laboratory measurements on a motor by the following procedure:

1. Block the rotor; monitor the motor windings, shell, etc. with thermocouples; and monitor current, voltage, and power input until the motor reaches a steady-state

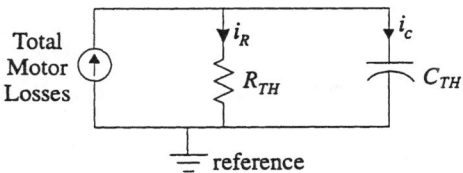

FIGURE 8.21 Resultant electrical analog.

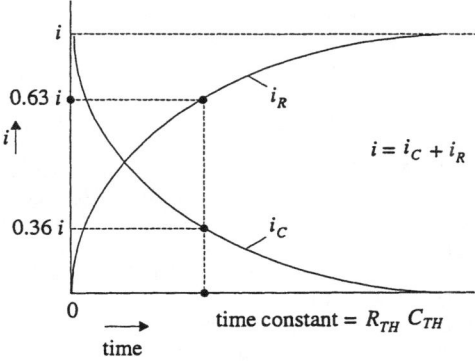

FIGURE 8.22 Change in thermal flow with time.

condition. That is, power into the motor becomes equal to the power conducted to the ambient environment.

2. Then:

$$R_{TH} = \frac{\text{winding temperature rise, C°}}{\text{power at steady state, W}}$$

3. With the power input at the magnitude just found, the motor at ambient temperature, and monitoring of the winding temperature and the time, take data to plot temperature rise versus time until a steady-state condition is reached.
4. Plot the data and determine the time constant in seconds. This will give a form of Fig. 8.22.
5. $R_{TH} \, C_{TH}$ = time constant, s

$$\therefore C_{TH} = \frac{\text{time constant}}{R_{TH}} \quad \text{J/C° or (W · s)/C°}$$

8.5 SUMMARY OF MOTOR CHARACTERISTICS AND TYPICAL APPLICATIONS

The most common motors, their characteristics, and their typical applications are listed in Tables 8.11 and 8.12. Keep in mind that there are thousands of variations of motors and applications, and only general categories are labeled here.

8.6 ELECTROMAGNETIC INTERFERENCE (EMI)

Many motors are used in sensitive applications that require that specific limits of radiated or conducted electromagnetic interference not be exceeded.

Electromagnetic interference (EMI), sometimes referred to as *radio-frequency interference* (RFI), is a phenomenon which can directly or indirectly contribute to a degradation in performance of electronic receivers or systems. EMI consists of undesirable voltages and currents that reach the affected device either by conduction through the power lines or by radiation through the air and cause the device to exhibit undesirable performance.

AC induction motors do not generally have any EMI problems unless they are powered by some type of electronic drive.

Mechanically commutated motors and electronically controlled motors may emit significant amounts of EMI. The emissions are regulated by various governmental agencies throughout the world. The International Electrotechnical Commission (IEC) has three technical committees (TCs) that work on EMI standards. They are the International Committee on Radio Interference (CISPR); TC77, which is concerned with electromagnetic compatibility (EMC) of electrical equipment; and TC65, which is concerned with immunity standards. The CISPR 10 and CISPR 23 standards are results of the committee's work.

In the United States, the Federal Communications Commission (FCC) regulates EMI. In Europe, the European Commission (EC), through the European Standards

TABLE 8.11 Summary of DC Motors

Motor Type	Eff, %	Characteristics	Typical applications
Iron-core PMDC	40–90	Straight-line S-T curve, good controllability, battery operation, good thermal properties	Servo systems, printers, actuators, pumps, automotive, medical equipment, vending, robotics
Moving-coil hollow-rotor PMDC	40–85	Straight-line S-T curve, good controllability, battery operation, poor thermal properties, high acceleration rate, low inertia	High-speed printers, high-speed servo systems, robotics
Brushless PMDC	30–90	Straight-line S-T curve, good controllability, requires a controller and Hall effect switches or back-emf sensors, battery operation, good thermal properties, long life	Servo systems, robotics, medical equipment, disk drives, pumps
Stepping motor	N/A	Moves in increments of 15 to 0.9°, open-loop controllability (no position sensors necessary)	Printers, actuators, disk drives, robotics, automotive
Switched reluctance	30–90	High acceleration rate, high speed, no back-emf generated, requires a feedback device and controller	Servo systems, pumps, vacuum cleaners, printers
Series-connected wound field	40–60	High starting torque, high no-load speed, steep S-T curve, poor speed regulation, can be made to operate on ac or dc	Power tools, fans, vacuum cleaners, appliances, sewing machines, saws
Shunt-connected wound field	50–70	Good speed regulation, flat S-T curve, controllability	Conveyors, saws, large power tools, pumps
Compound-connected wound field	60–80	Very high starting torque, flat S-T curve in operating range, good controllability	Large power tools, pumps, conveyors, traction drives

TABLE 8.12 Summary of AC Motors

Motor type	Eff, %	Characteristics	Typical applications
Split phase	50–70	Moderate starting torque, high starting current	Pumps, blowers, tools, fans, appliances, furnaces
Capacitor start	60–75	High starting torque, low starting current	Pumps, blowers, fans, tools, appliances, air conditioning
Permanent split capacitor	35–70	Low starting torque, low starting current, multispeed	Fans, furnaces, air conditioning, blowers, actuators, garage door openers
Shaded pole	20–35	Low torque, low starting current, multispeed, low efficiency	Fans, blowers, appliances, actuators, vending machines
Three phase	70–90	High efficiency, very high starting torque, low starting current, requires three-phase power supply	Machinery, industrial pumps and equipment, large compressors

Committee (CEN) and the European Committee for Electrotechnical Standards, has adopted harmonized EMC standards. They cover Class 1 (residential, commercial and light industry) and Class 2 (heavy industry) standards for testing. Some standards of concern here are listed in Table 8.13.

8.7 ELECTROMAGNETIC FIELDS AND RADIATION

In a general sense, EMI is caused by the radiation of unwanted electromagnetic fields. They are set up as follows.

A current I flowing in a wire sets up an electric field as well as a magnetic field. The electric field E (Fig. 8.23) is perpendicular to the magnetic field and is along the length of the wire. If a pair of conductors carrying time-varying current of the same

TABLE 8.13 European Harmonized Emission Standards

Standard	Description
EN 50081-1: 1991	EMC generic emission standard, Part 1: Residential, commercial, and light industry
EN 50065-1: 1990	Signaling on low-voltage electrical installations in the frequency range 3–148.5 kHZ, Part 1: General requirements, frequency bands, and electromagnetic disturbances
EN 55011: 1989	CISPR 11 (1990) ed. 2: Limits and methods of measurement of radio disturbance characteristics of industrial, scientific, and medical (ISM) radio-frequency equipment
EN 55013: 1988	CISPR 13 (1975) ed. 1 + Amdt. 1 (1983): Limits and methods of measurement of radio disturbance characteristics of broadcast receivers and associated equipment
EN 55014: 1986	CISPR 14 (1985) ed. 2: Limits and methods of measurement of radio interference characteristics of household electrical appliances, portable tools, and similar electrical apparatus
EN 55015: 1986	CISPR 15 (1985) ed. 3: Limits and measurement of radio interference characteristics of fluorescent lamps and luminaries
EN 55022: 1986	CISPR 22 (1985) ed. 1: Limits and measurements of radio interference characteristics of information technology equipment
EN 60555-2: 1986	IEC 555-2 (1982) ed. 1 + Amdt. 1 (1985): Disturbances in supply systems caused by household appliances and similar electrical equipment, Part 2: Harmonics
EN 60555-3: 1986	IEC 55-3 (1982) ed. 1: Disturbances in supply systems caused by household appliances and similar electrical equipment, Part 3: Voltage fluctuations

Source: James Klouda, Elite Electronics Engineering.

magnitude but opposite in direction are placed in close proximity to each other, the electric fields will be 180° out of phase and will effectively cancel each other. Complete cancellation will occur only when the conductors occupy the same space. Since this is a physical impossibility, some fields will always exist, but they may be reduced to acceptable levels.

At radio frequencies, waves travel down transmission lines designed to minimize loss. Properly designed lines minimize radiation because of the cancellation effect just discussed. If a line is properly terminated, the energy is all dissipated in the load (Fig. 8.24). If a line is left open (Fig. 8.25), some of the energy is radiated into space. Making the transmission line more imperfect causes a greater amount of energy to be radiated (Fig. 8.26).

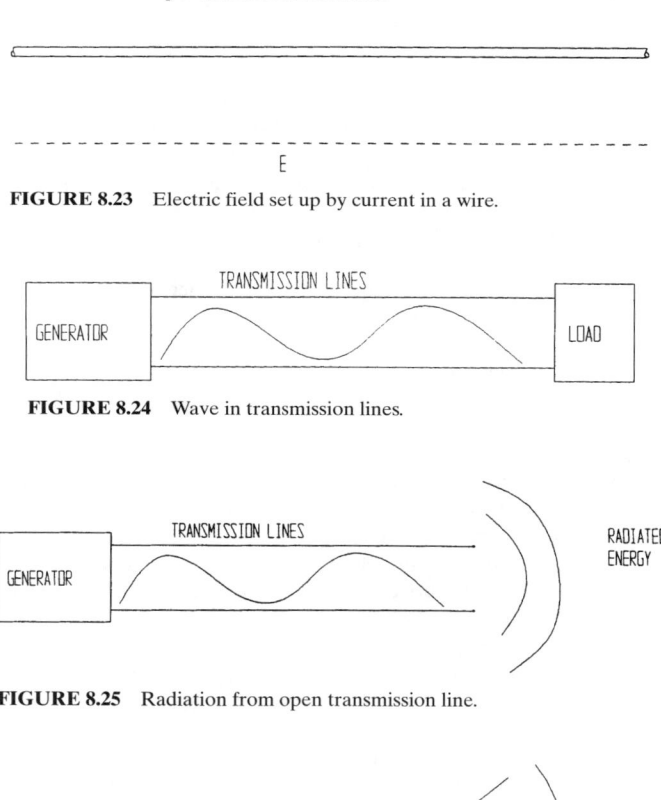

FIGURE 8.23 Electric field set up by current in a wire.

FIGURE 8.24 Wave in transmission lines.

FIGURE 8.25 Radiation from open transmission line.

FIGURE 8.26 Radiation from imperfect transmission line.

The wavelength λ is a function of the frequency f. They are related as follows:

$$\lambda = \frac{v_c}{f}$$

where V_c = speed of light
$= 3 \times 10^8$ m/s
$= 186{,}000$ mi/s
$= 982 \times 10^6$ ft/s
$= 1.178 \times 10^{10}$ in/s

For complete radiation a perfect transmission line needs to be perfectly terminated into a half-wave dipole antenna. While perfection is impossible, a practical antenna can be achieved by splitting the transmission line at right angles, such that the length from tip to tip is one-half wavelength λ, as shown in Fig. 8.27.

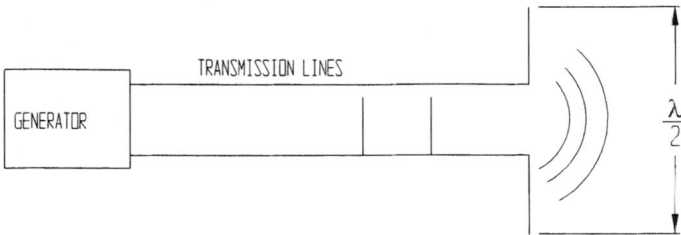

FIGURE 8.27 Radiation from half-wave dipole antenna.

The radiated EMI of concern covers frequencies from 0.01 to 1000 MHz. This yields wave lengths from

$$\lambda_1 = \frac{v_c}{f} = \frac{1.178 \times 10^{10} \text{ in/s}}{0.01 \times 10^6 \text{ Hz}} = 1{,}178{,}496 \text{ in}$$

to

$$\lambda_2 = \frac{1.178 \times 10^{10} \text{ in/s}}{1000 \times 10^6 \text{ Hz}} = 11.78 \text{ in}$$

Antennas are usually designed for good radiation patterns. Practically, they are cut at $5\lambda/8$, $\lambda/2$, or $\lambda/4$. Antennas shorter then $\lambda/4$ exhibit capacitance which must be offset by some kind of inductive loading for good radiation. In electric motors, the lead wires can act as antennas. To properly radiate EMI energy in the frequency range of concern, lead lengths ℓ would range from

$$\ell_1 = \frac{\lambda}{2} = \frac{1{,}178{,}496}{2} = 589{,}248 \text{ in}$$

to

$$\ell_2 = \frac{\lambda}{4} = \frac{11.78}{4} = 2.945 \text{ in}$$

It is unlikely that typical motor leads would radiate the lower frequency, but it is obvious that the higher frequencies could easily utilize motor leads as antennas. The length and position of the lead should be adjusted to minimize radiation of the frequencies of concern.

EMI may be radiated from an antenna or by a conducted device from the power lines.

8.8 CONTROLLING EMI

EMI may be controlled in a number of ways. In the case of mechanically commutated motors, proper design of the commutation system is a must. The commutation–to–neutral zone ratio should be kept to <1.0, and bar-to-bar voltages should be kept to a minimum. Lead lengths should not be a multiple or submultiple of any EMI frequency. This is to minimize the possibility of them becoming antennas. Printed-circuit boards should not have tracks at right angles that could form a dipole antenna. The leads may be wound into twisted pairs. Full-pitch armature windings, high-resistance brushes, proper brush pressure, and proper armature balance will also reduce EMI.

Should these design guidelines alone fail to achieve acceptable EMI levels, it may be necessary to add filters or enclose the motor in shielding materials.

Electronically commutated motors have a different problem in that the EMI is generated by the controller and driver. In this case, adding filters to the motor leads may cause undesirable motor and drive performance. Shielding the power leads may be necessary.

CHAPTER 9
TESTING

Barry Landers
Rockwell Automation

9.1 SPEED-TORQUE CURVE

A typical computerized test station consists of a power analyzer, dynamometer controller, digital indicator, and multimeter (see Figs. 9.1 and 9.2). This kind of test station will quickly generate speed-torque curves and summary sheets, as shown in Fig. 9.3, or full sets of curves for speed versus torque, efficiency, current, watts, and power factor, as shown in Fig. 9.4. Different inertial compensations can easily result in an entire family of curves and can greatly affect results. By using a dynamometer controller to take the motor from idle to a predetermined minimum speed and back to idle at a controlled rate, inertial effects will nearly disappear simply by interpolating for the same speed points and averaging the two sets of data. Points from the resulting curves will then compare very well with static measurements made at the same performance point. Finally, an accurate stall point should consist of an average of at least three sets of locked rotor data at different rotor positions.

In order to obtain accurate test results, the maximum motor torque should equal at least 30 percent of the dynamometer rating. Voltage drops during the test will drastically affect results and require heavier wiring than specified by the National Electrical Code. A good rule of thumb would increase wiring in conduit by six wire gauges and bench wiring by four wire gauges. Voltage control devices such as power stats must also minimize voltage drops, either by voltage regulation or by selecting a rating of at least four times the maximum motor current. The power supply frequency should not vary more than 0.5 percent from the rated value. In addition, polyphase systems should not exceed 0.5 percent voltage unbalance between phases.

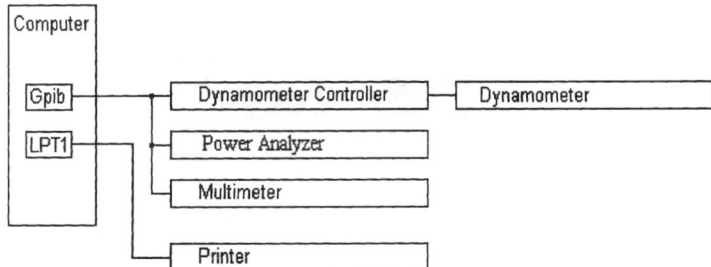

FIGURE 9.1 Typical motor test bench diagram.

FIGURE 9.2 Typical test station.

9.1.1 Acceleration Test

This test determines the acceleration characteristics of a motor for a given amount of inertia and load torque. The inertia test stand shown in Fig. 9.5 includes a set of inertia wheels that will provide a range of inertias from 50 to 4000 lb·in^2 in 50-lb·in^2 increments and an analog tachometer to monitor speed. In order to minimize oscillations, the setup uses low-backlash couplings between the motor, the inertia wheel shaft, and the dynamometer. An oscilloscope monitors the voltage from the analog tachometer and provides voltage and time data to a PC through the general-purpose interface bus (GPIB) connection. At this point, the PC calculates the speeds based on the tachometer voltage constant and plots the acceleration curve (see Fig. 9.6).

TESTING

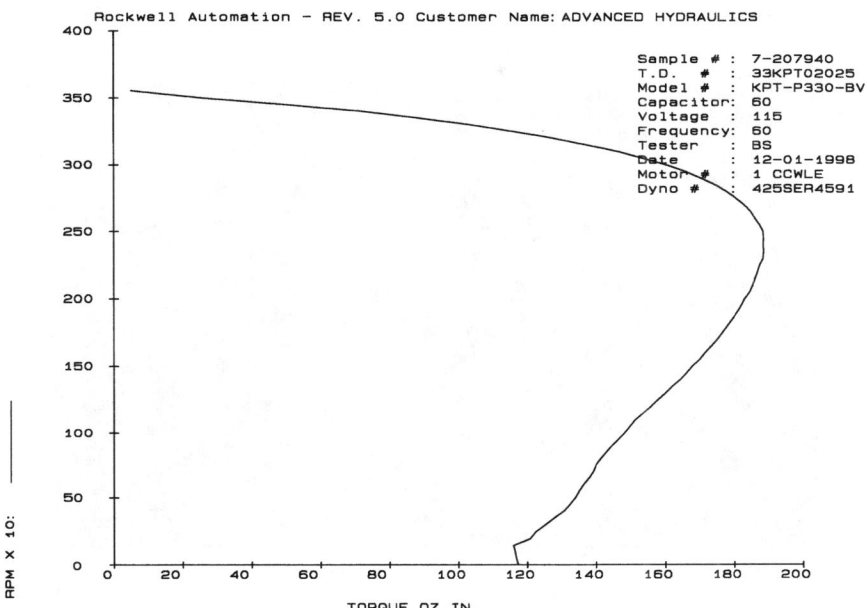

FIGURE 9.3 Speed-torque curve and summary sheet generated by a computerized test station.

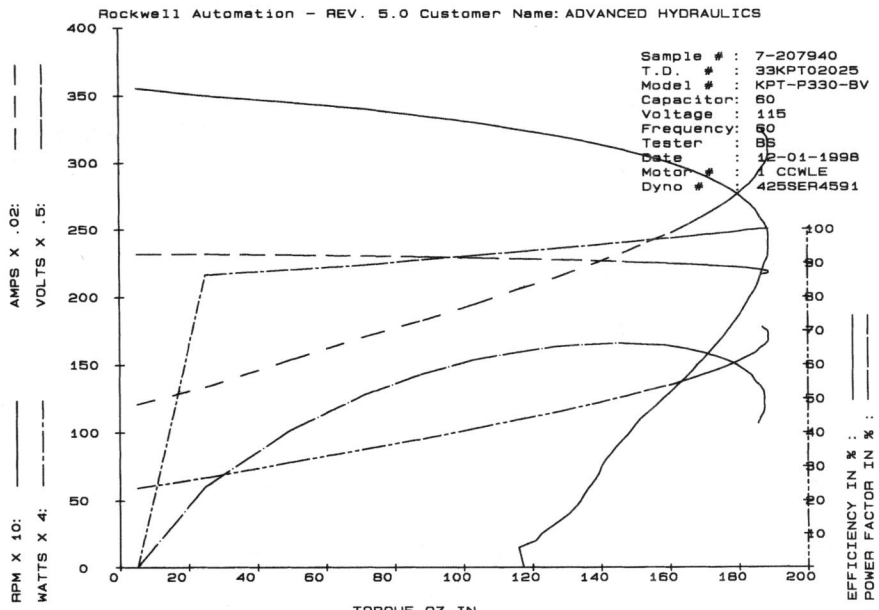

FIGURE 9.4 Speed-torque, efficiency, current, watts, and power factor curves generated by a computerized test station.

9.4 CHAPTER NINE

FIGURE 9.5 Inertia test stand.

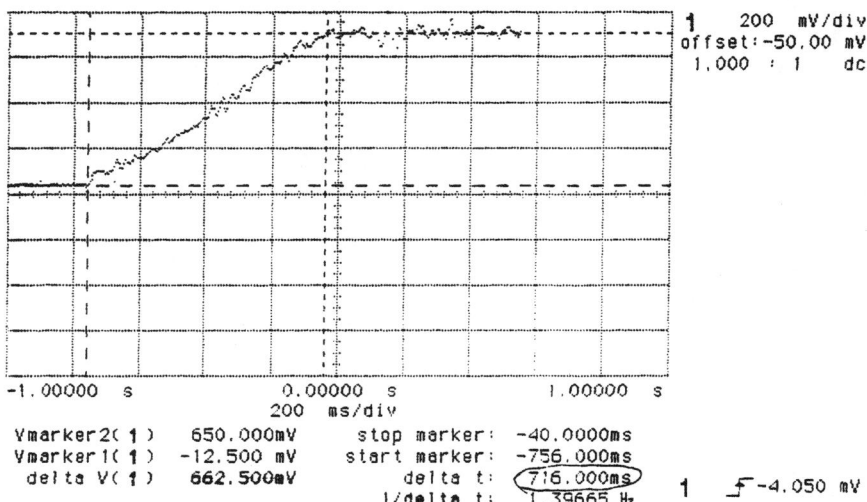

FIGURE 9.6 Acceleration plot.

AC motors and open-loop dc motors could use the bench shown in Fig. 9.1 to acquire and document additional data such as torque, efficiency, current, and watts. However, closed-loop motors require a different approach, as outlined in the dc motor test section on speed profiles.

9.1.2 Good Test Practices

1. Review test requirements for consistency with UL and other requirements to avoid wasted test time. Obtain clarification as needed.
2. Measure the resistance of the motor windings using a precision multimeter or bridge and record ambient temperature. Ensure that resistance meets winding specifications prior to any tests.
3. Perform high-potential tests per UL requirements prior to energizing the motor. In the absence of other information, a good rule of thumb would apply a voltage equal to (1000 plus twice the rated voltage) times 1.2 for at least 1 s.
4. Select stable capacitors within at least 1 percent of specified values and check for drift rather than relying on old measurements marked on the capacitor. Always discharge the capacitors prior to checking the value to avoid damage to the meter. Switch between all selections on capacitor decade boxes when bleeding charge.
5. Confirm that the dynamometer has appropriate cooling operating.
6. Use brass-tipped setscrews in the coupling and check tightness periodically to avoid bad data.
7. Select appropriate ranges for equipment that does not autorange.
8. Minimize coupling backlash and torsional deflection.
9. Monitor display to confirm at least the initial printout.
10. If the motor has a ground wire, connect it to the workstation ground.
11. Recheck motor connections against specifications prior to test.
12. Adjust speed-torque test time and minimum speed to minimize motor oscillation below breakdown.

9.2 AC MOTOR THERMAL TESTS

Thermal tests for ac motors determine the thermal protector trip and reset temperatures under locked and running condition, as well as the leveling temperatures for running conditions. See Fig. 9.7 for test setup. These tests usually run at nameplate-rated voltage, frequency, and current, but UL may require tests at different voltages, as listed in Table 9.1.

Running conditions include full load, running overload, full-voltage idle, and reduced-voltage idle heat runs. The running test definitions follow.

Full-load heat run: A dynamometer or other load device maintains the motor at the rated torque or current, as determined by customer and agency requirements. The test continues until the winding temperatures reach equilibrium for 60 min.

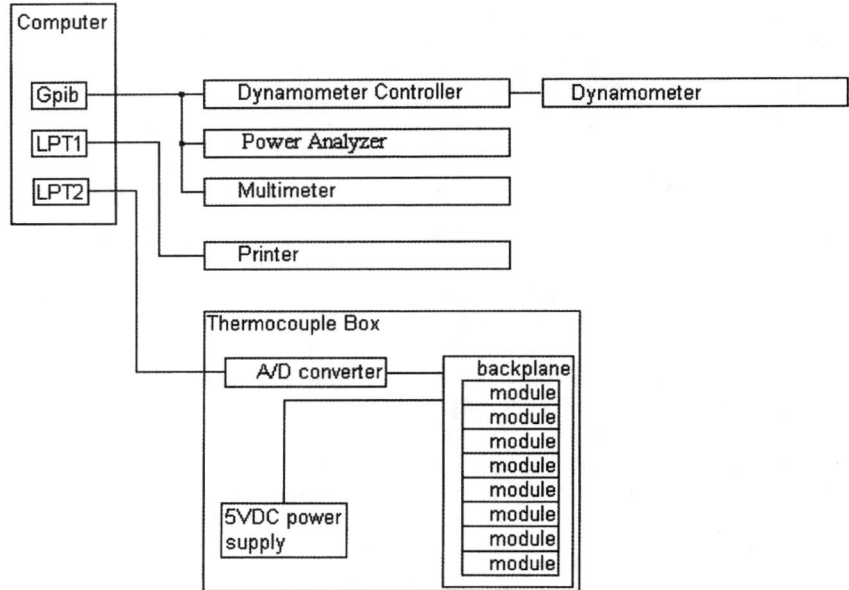

FIGURE 9.7 Typical motor test bench diagram.

Running overload: A dynamometer holds the motor current per the nameplate rating until the motor windings attain thermal equilibrium for 60 min or until the protector trips. The dynamometer controller then increases the torque to provide a current increment per Table 9.1, and maintains this current until the winding temperature reaches equilibrium. This sequence repeats until the thermal protector opens or the motor stalls. (*Note:* For running-overload tests, UL requires performing the first test at nameplate current. If the no-load current of the motor at rated voltage equals or exceeds the nameplate current, the first running-overload test occurs at idle. If the thermal protector opens at the nameplate current—above the no-load current—then the motor test must also occur under no-load condition at the nominal test voltage. If the protector opens at idle, then the idle test repeats for

TABLE 9.1 Nominal Voltages and Currents for Running-Overload Test

Nameplate current	Current increment	Nameplate voltage	Nominal voltage
0.1–10 A	0.2 A	110, 115, or 120	120
10–15 A	0.3 A	200 or 208	208
15–20 A	0.4 A	208–220 or 208–230	240
		220, 230, or 240	240
		265 or 277	277
		440, 460, or 480	480
		550, 575, or 600	600

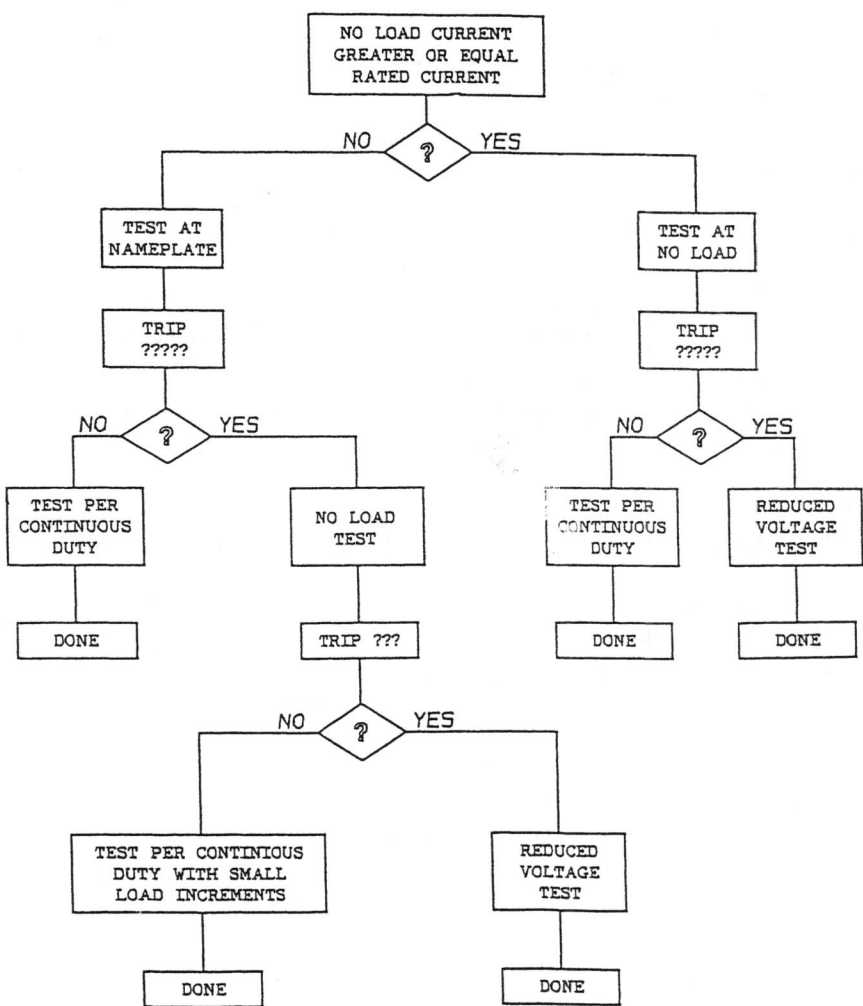

FIGURE 9.8 Logic diagram for idle thermal test.

voltages reduced by 10-V increments until the winding temperature stabilizes without opening the protector. See Fig. 9.8 and Table 9.1 for these test requirements.)

Idle heat run: The motor operates without load at nameplate voltage until the windings reach thermal equilibrium for 60 min.

Reduced-voltage idle heat run: The motor operates without load at voltages reduced by 10-V increments below rated voltage until the windings reach thermal equilibrium for 60 min without tripping the thermal protector.

9.2.1 Test Conditions

UL 2111 allows the following maximum temperatures for Class B insulation with the various thermal tests:

Locked rotor	225°C in first hour
	200°C in the second hour
Full-load heat run	165°C
Running overload	165°C if protector opens
	175°C if protector does not open

IEC 34-1 requires a maximum temperature rise of 85°C above room temperature as measured by thermocouple or 90°C as measured by resistance. For Class F insulation, the maximum temperatures increase to 110 and 115°C, respectively.

For full-load, running-overload, and idle heat runs, UL requires a temperature rise variation at temperature equilibrium of no more than ±1.0°C for 2 h.

9.2.2 Equipment Required

Dynamometer sufficient to handle the long-term dissipation
Dynamometer controller
Power analyzer
Speed-torque indicator
Correct size and type of coupling between motor and dynamometer
Motor test stand and means of securing stand to stationary mount
Type J thermocouples and accompanying chart equipment
Printer
Computer hardware and software
Locking bar for locked-rotor test (optional)
Powerstat
Fan for quick cooling of the test motor

9.2.3 Test Setup

Motor Setup Requirements

1. Mount the motor in a test stand with the protector at the six o'clock down position with the motor connected to the dynamometer.
2. Couple the motor to a dynamometer with a thermally insulating coupling. For locked-rotor tests, either apply enough torque with a dynamometer to overcome the motor starting torque or use a thermally insulated lock bar to stall the motor.
3. Align and secure test stand to dynamometer.
4. Connect thermocouples from the motor to chart equipment with electrical isolation.
5. Connect motor to capacitor or relay (if specified) and to power source per outline.

6. Confirm that the dynamometer cooling and dissipation rating meets the test needs before operating the dynamometer.

For locked-rotor tests or tests on intermittent-duty motors, let the test run with the motor off for a short time to clearly document a room temperature start. Other tests may start without delay. All locked-rotor tests, regardless of thermal rating, must start with winding temperatures within 5°C of room temperature. For other thermal tests, only intermittent-duty motors must begin test with windings within 5°C of room temperature.

Computer-controlled tests can produce many times as much information as the old manual tests, while providing greater accuracy, drastically reduced test times, and at least an order-of-magnitude productivity increase.

9.3 DC MOTOR TESTING

Figures 9.9 and 9.10 show a typical computerized dc motor test bench. Everything except thermal tests should begin with a motor temperature of 25 ± 5°C. DC motor thermal tests determine safe operating area and thermal resistance values. Both tests determine the operating conditions for an 85°C winding temperature rise above room temperature by thermocouple or 90°C by resistance for Class B insulation. For Class F insulation, the maximum temperatures increase to 110 and 115°C, respectively.

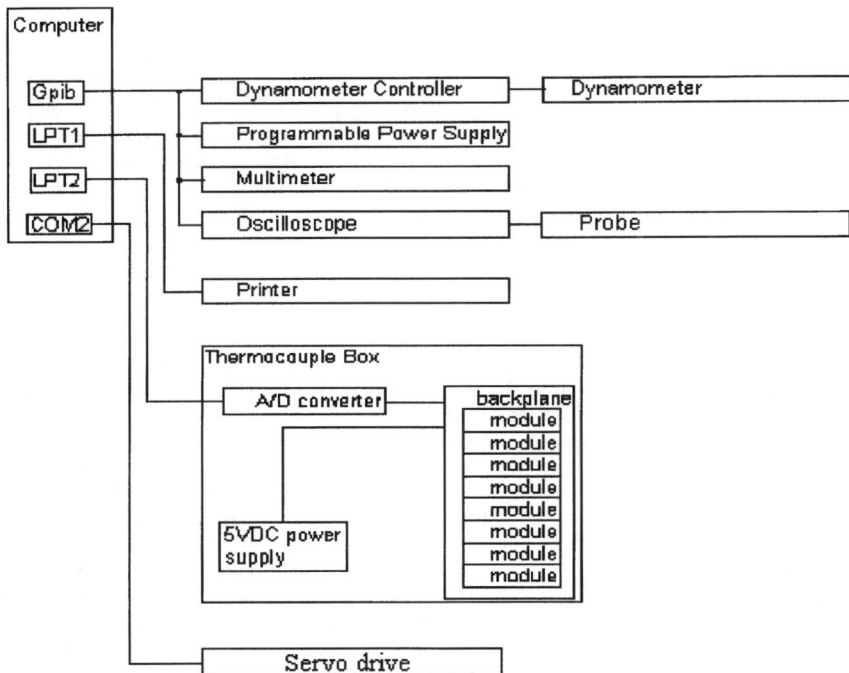

FIGURE 9.9 Typical computerized dc motor test bench diagram.

FIGURE 9.10 Typical computerized dc motor test bench.

9.3.1 Voltage Constant Test

The K_e test checks the voltage constant in volts per thousand revolutions per minute (V/krpm) for a backdriven dc test motor. Any motor capable of maintaining an exact speed under a varying load can serve as a backdrive motor. For brush dc motors, measure the dc voltage generated by the test motor with a multimeter (generally in both rotations). For brushless dc motors, acquire the peak-to-peak voltage with an oscilloscope and divide by twice the drive speed in krpm to obtain the K_e.

9.3.2 Terminal Resistance Test

While multimeters can accurately measure brushless dc motor resistance, brush dc motors experience variations in brush contact drop as well as resistance changes based on the relative position of the brush to the commutator bars. Brush dc motors therefore require averaging several locked-rotor measurements to provide a stable reading, or, preferably, a dynamic measurement, while backdriven at low revolutions per minute. For either test, attach the motor terminals to a dc power supply and set the current limit high enough to reduce contact drop fluctuations and low enough to minimize heating during the test. In the absence of a specification, use a current limit of 25 percent of the rated motor current. Acquire the voltage necessary to drive the current through the motor for calculation of resistance ($R = V/I$). For dynamic measurements, reverse either the polarity of the voltage or the rotation of the drive motor and average the two measurements to remove the counter-emf contribution. Backdriven speeds of 30 to 100 rpm work reasonably well and will generally provide better repeatability than a locked test.

9.3.3 Speed-Torque Test

The speed-torque test provides a curve of the speed and the test motor torque (see Fig. 9.11). Couple the shaft of the test motor to the dynamometer and attach the motor terminals to a programmable power supply. Set the supply voltage limit at the rated motor voltage and the current limit to the rated peak current. Acquire voltage and current from the programmable power supply and speed and torque from the dynamometer controller.

By using the dynamometer controller to take the motor from idle to a predetermined maximum torque and back to idle speed at a controlled rate, inertial effects will nearly disappear simply by interpolating for the same speed points and averaging the two sets of data. Points from the resulting curves will then compare very well with static measurements made at the same performance point. In most cases, a dc motor curve will not include the locked point, since a test at stall will often risk demagnetizing the motor and/or damaging the commutator.

9.3.4 Demagnetization Test

The demagnetization test determines the amount of current the test motor can draw before reducing the K_e by 5 percent. Couple the shaft of the test motor to a dynamometer and attach the motor terminals to a dc power supply. Use an oscilloscope to monitor a current probe or current shunt on the positive output of the dc supply. Set the supply voltage limit to the rated motor voltage and the current limit well beyond the calculated demagnetization point. Using the oscilloscope to determine the current, quickly apply torque until reaching the desired current. Remove all torque immediately and repeat two more times. Recheck the K_e of the test motor

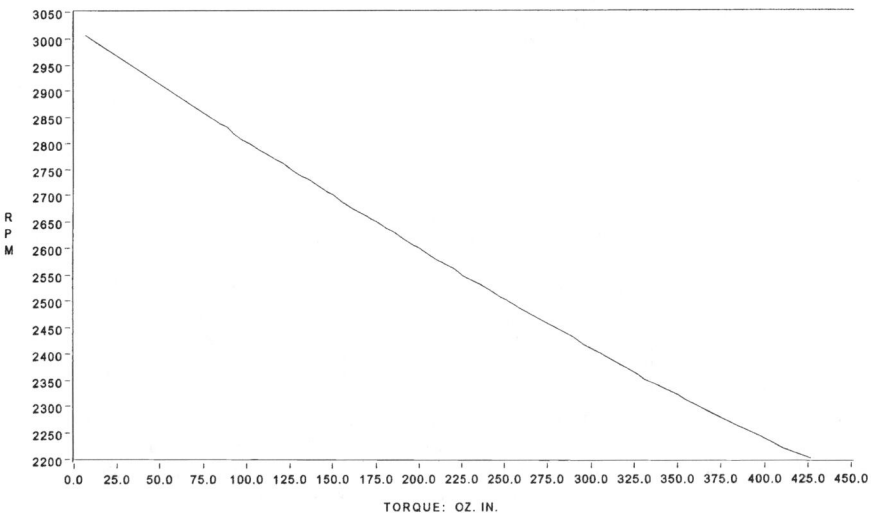

FIGURE 9.11 DC motor speed-torque curve.

after allowing the motor to fully return to room temperature, generally after about 30 to 60 min. A reduction in K_e greater than 5 percent means the test motor has demagnetized. Otherwise, repeat the test at a higher current (typically in 5 percent increments).

9.3.5 Thermal Resistance Test

The thermal resistance test determines the temperature (°C) rise per watt loss of the test motor. Place the test motor into a stand and lock the shaft. For brush dc motors, route wires from the commutator and under the bearings to the outside for the winding resistance measurement, to avoid errors introduced by the brushes and the contact drop. Measure the cold winding resistance with the multimeter and record the ambient temperature. Attach a thermocouple to the shell to monitor the temperature rise. Attach the motor terminals to a programmable power supply. Slowly increase the voltage until reaching the rated current. Hold the rated current for 1 h after the shell temperature levels. Quickly detach the motor terminals and take the hot winding resistance with the multimeter. Record the ambient temperature. Repeat the test for the rated running condition, preferably with a different motor for brush dc tests to avoid the possible effects of a burned commutator. The thermal resistance constant equals approximately:

$$R_{th} = \frac{(R_{hot}/R_{cold}) - 1 + (amb_{hot} - amb_{cold})}{\text{winding const} \times \text{voltage}_{hot} \times \text{current}_{hot}}$$

Use a winding constant for copper of 0.00393.

where
R_{th} = thermal resistance
R_{hot} = hot winding resistance
R_{cold} = cold winding resistance
amb_{cold} = cold ambient temperature
amb_{hot} = hot ambient temperature
$voltage_{hot}$ = hot winding voltage
$current_{hot}$ = hot winding current

9.3.6 Safe Operating Area Curve Test

The safe operating area curve (SOAC) determines the boundaries of safe operation. Use the running thermal resistance constant to estimate the winding temperature and maintain safe operating temperatures (usually below 85°C plus ambient). Place the test motor into a stand and couple the test motor to the dynamometer. Attach a thermocouple to the shell to monitor the temperature rise. Attach the motor terminals to a programmable power supply. Set the supply voltage limit at the rated motor voltage, and the current limit to the rated peak current. Start the current at the rated continuous motor current with the appropriate torque on the dynamometer. Adjust the voltage on the power supply to obtain the desired shell temperature when level. Acquire the speed and torque from the dynamometer controller, and the voltage and current from the power supply. Adjust the torque as needed to keep the winding temperature at 85°C plus ambient. Repeat the test at the next desired level. Acquire data at different speeds and torques to plot the SOAC. The winding temperature rise equals approximately:

$$\text{Temp}_{\text{winding}} = R_{\text{th}} \quad (\text{watts lost})$$

$$\text{Temp}_{\text{winding}} = R_{\text{th}} \times \left(\frac{\text{voltage} \times \text{current} - \text{speed} \times \text{torque}}{1351.7} \right)$$

where torque is in ounce-inches. Record the cold winding resistance at the beginning of the test, the hot resistance at the end of the test, and the ambient temperature for each measurement.

9.3.7 Holding Tests

The holding torque test determines the current, speed, and temperature at a specified torque. Place the test motor into a stand and couple the test motor to the dynamometer. Attach a thermocouple to the shell to monitor the temperature rise. Attach the motor terminals to a programmable power supply. Set the voltage on the power supply to the rated motor voltage, the current to the rated continuous current, and the dynamometer to the desired torque. Run at the desired torque until the shell temperature levels. Acquire the speed and torque from the dynamometer controller, and the voltage and current from the power supply. Record the cold winding resistance at the beginning of the test, the hot resistance at the end of the test, and the ambient temperature for each measurement.

The holding speed test determines the current, torque, and temperature at a specified speed. The holding current test determines the speed, torque, and temperature at a specified current. The holding temperature test determines the speed, torque, and current at a specified temperature. The PC monitors the specific parameter and controls the test to hold it constant, similar to the holding torque test. The application determines which test will provide the most relevant data, and the PC automatically controls the test and acquires the data.

The PC then prints data sheets for all tests, prints summary sheets for SOAC and holding tests, and graphs for SOAC, speed-torque, and holding tests. For brushless dc testing, the servo drive takes the place of the programmable power supply. The voltage divided by the current provides an estimate of resistance. During the SOAC test the PC uses the servo drive to control speed and the dynamometer controller to control torque to obtain the desired temperature.

9.3.8 Torque Ripple*

The output torque of a dc motor at low speeds appears constant, but closer examination reveals a cyclic component called *torque ripple*, as illustrated in Fig. 9.12. This torque ripple results from the switching action of the commutator, from the armature reluctance torque and sometimes from the bearings.

Torque ripple usually constitutes a very small percentage of the rated output torque and proves negligible for most uses. However, torque ripple may become critical in some applications, thereby requiring a means of measurement.

The apparatus illustrated in Fig. 9.13 can accurately measure torque ripple, as long as the moment of inertia of the measuring device remains much smaller than the motor moment of inertia (otherwise inertia filtering invalidates the ripple measurement).

---

* Adapted from the *Electro-Craft Handbook: DC Motors, Speed Controls, Servo Systems* (1980).

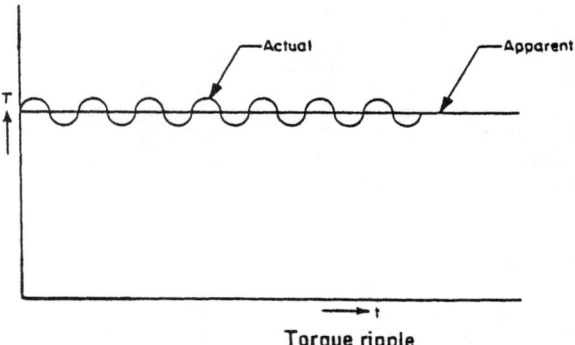

FIGURE 9.12 Torque ripple test.

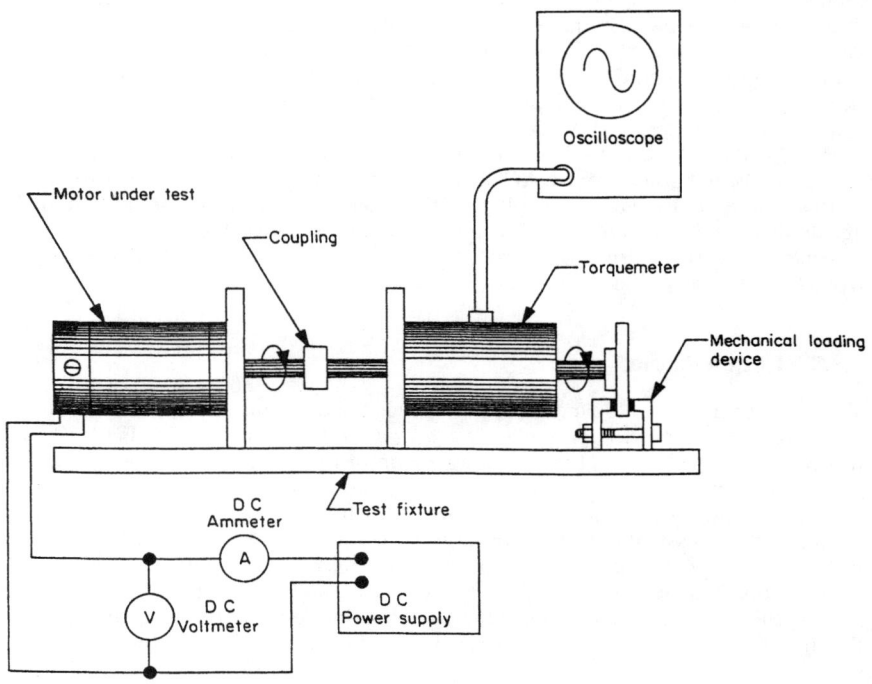

FIGURE 9.13 Torque ripple test setup.

The percent peak-to-peak ripple torque equals:

$$T_r = \frac{\text{peak-to-peak torque ripple (100)}}{\text{average output torque}}$$

9.4 MOTOR SPECTRAL ANALYSIS

This section provides examples of the value of sound, vibration, and current spectral analysis for electric motors. Since spectral analysis encompasses a tremendous range of diverse technical areas, requiring many books to adequately cover, this section provides primarily hands-on snapshots of a few uses for motors. The Vibration Institute in Willowbrook, Illinois, offers a wide range of literature and training for those beginning in this field.

Figure 9.14 illustrates a typical time signal, with time on the horizontal axis and amplitude (usually in volts) on the vertical axis. An oscilloscope commonly provides this type of display. After a fast Fourier transform (FFT), the horizontal axis changes to frequency, as shown in Fig. 9.15. However, the basic analysis techniques remain the same, regardless of whether the electrical signal derives from sound, vibration, current, flux, surface finish, roundness, or any other parameter containing data of a periodic nature. For variable-speed motors, the waterfall display in Fig. 9.16 simplifies the tracking of suspect frequencies with speed. Fixed frequency problems such as resonances and critical speeds stand out clearly as compared to speed-related problems.

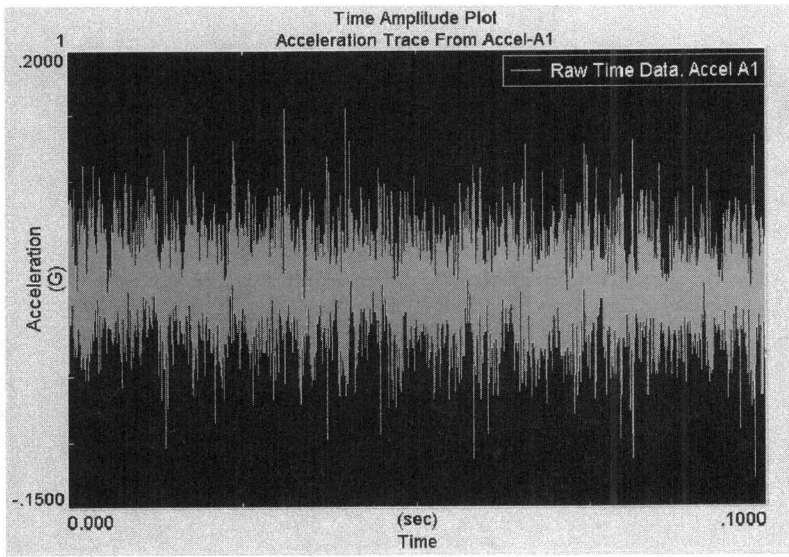

FIGURE 9.14 Time history amplitude graph. (*Courtesy of Pemtech, Inc.*)

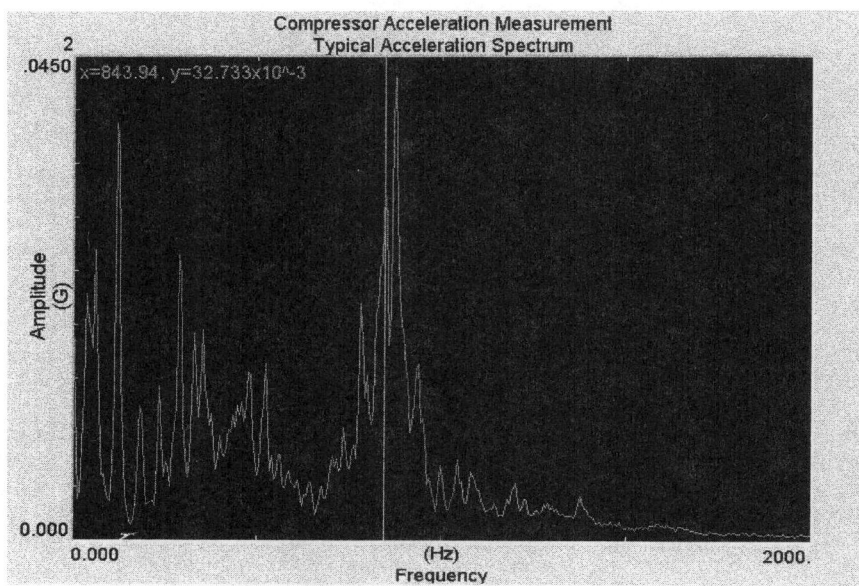

FIGURE 9.15 Spectral amplitude graph. (*Courtesy of Pemtech, Inc.*)

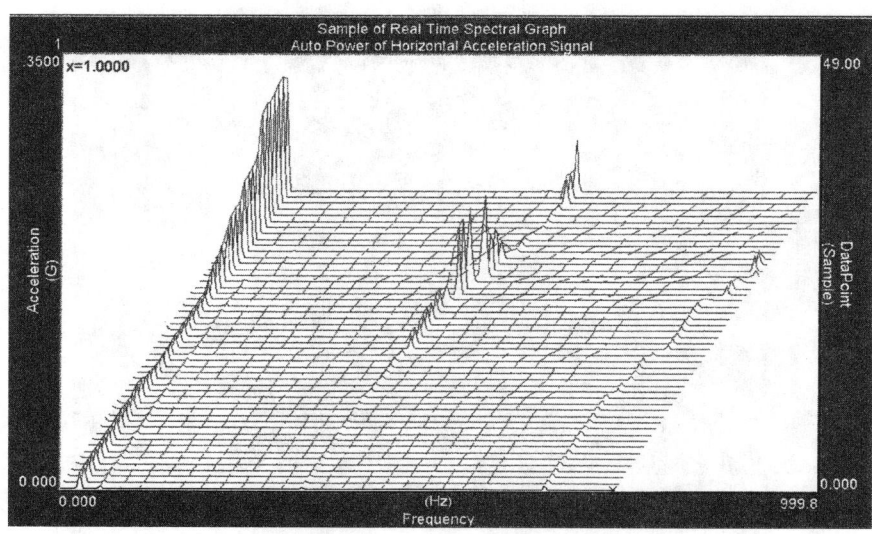

FIGURE 9.16 Real-time waterfall display. (*Courtesy of Pemtech, Inc.*)

9.4.1 Ball Bearing Analysis

The spectrum in Fig. 9.17 illustrates a motor with predominantly inner race and ball defect frequencies in the sound spectra. As shown in Fig. 9.18, subsequent microscopic examination revealed false brinells on the side of the inner race closest to the rotor. Since the preload spring forces the balls toward the rotor, the inboard orientation suggests that the damage occurred following assembly. However, this type of mark results from extended periods of vibration on a nonoperating bearing, such as might occur during shipment, but not during assembly. If at all possible, motors should ship with horizontal shafts to avoid exciting the axial rotor-spring resonance and to take advantage of the typically higher radial stiffness.

Qualification of motor packaging on shaker and oscillatory tables will prevent false brinelling from inadequate packaging; Figs. 9.19 and 9.20 are examples of this equipment. Shock testing will further qualify packaging to prevent housing distortion or true brinells in bearings. Swept sine wave tests will reveal resonances in products or packaging that shipping vibration could excite. Sine dwell tests at the resonance points will then ensure that normal transportation vibration would not damage the product at these frequencies. Spectral analysis of motors before and after testing will reveal damage by changes in bearing frequencies or frequencies associated with other motor components that produce periodic vibration.

Motors with true brinells tend to produce sound spectra with strong inner race and ball defect frequencies and lesser outer race defect frequencies. However, the

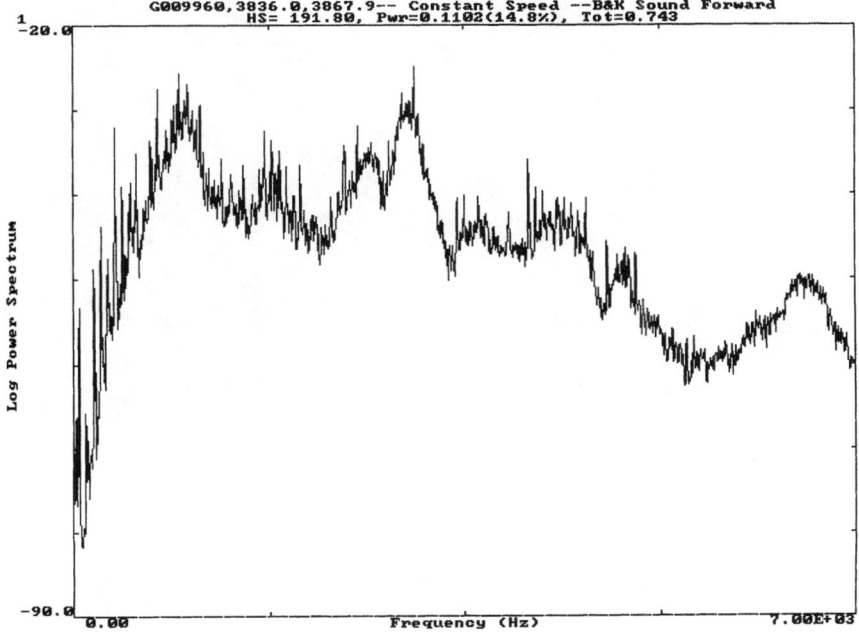

FIGURE 9.17 Log power spectrum of bearing showing inner race and ball defects.

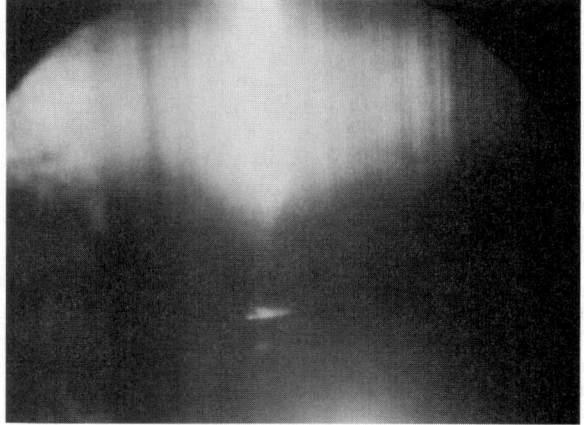

FIGURE 9.18 False brinell on inner race of bearing.

FIGURE 9.19 Shaker table.

nature of the defect will vary the contribution of each component and often will cause modulations that produce families of sidebands mingled with harmonics. These sidebands can greatly complicate bearing spectral analysis, particularly when combined with large numbers of harmonics and sidebands from gears or other components. The simple sound spectrum in Fig. 9.21 includes mostly inner race harmonics generated by a severe true brinell as shown in Fig. 9.22.

FIGURE 9.20 Oscillatory table.

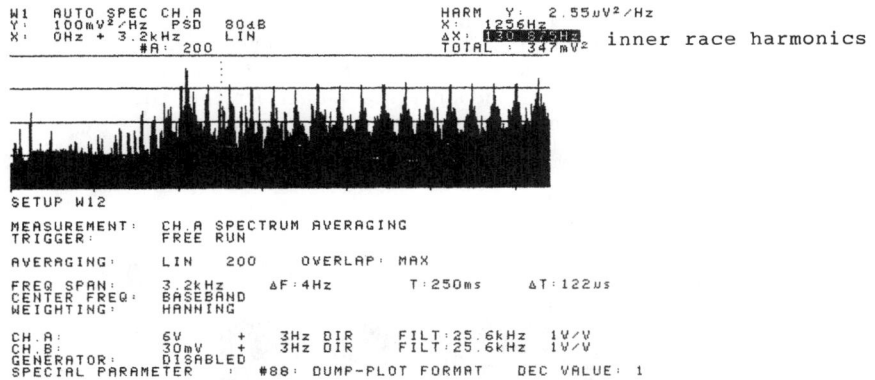

FIGURE 9.21 Sound spectrum with inner race harmonics, $\Delta x = 130.875$ Hz.

Bearing contamination generally creates considerable white noise, combined with mixed bearing frequencies. For example, the spectra in Fig. 9.23 consist mainly of white noise with some inner race and ball defect harmonics. This bearing noise resulted from contamination denting (see Fig. 9.24) caused by very fine paper dust contamination shown on the bearing housing in Fig. 9.25. Harder particles such as graphite brush dust will produce larger denting until the bearing mills the particle size down. At this point, ball indentations will produce large numbers of transfer marks on the raceways. Because of the smaller race curvature lending less support at

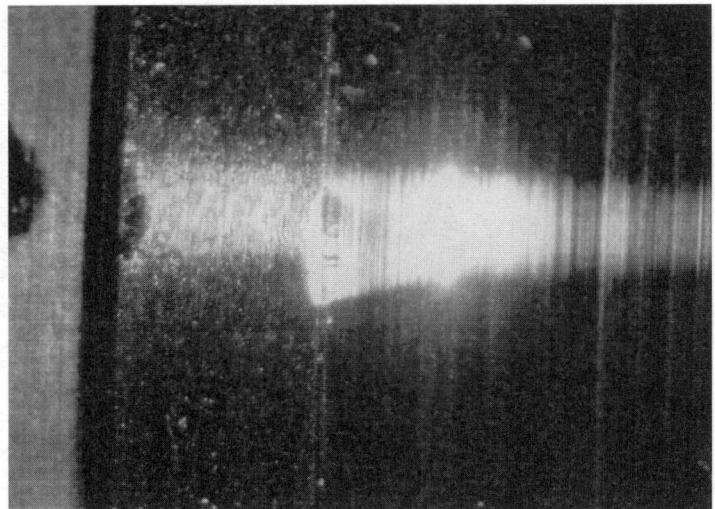

FIGURE 9.22 True brinell.

the point of contact, the inner race usually suffers more damage than the outer race. Consequently, the ball and inner race defect frequencies tend to dominate the spectra and will often produce ball sidebands next to the ball harmonics, but separated by the difference between the ball and inner race frequencies. This type of spectrum (see Fig. 9.26) almost always indicates hard particle contamination.

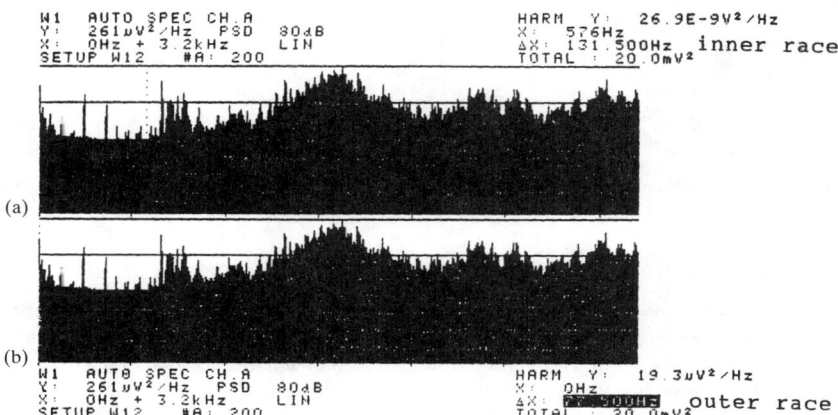

FIGURE 9.23 Spectra consisting mainly of white noise with some inner race and ball defect harmonics: (*a*) inner race, $\Delta x = 131.500$ Hz, and (*b*) outer race, $\Delta x = 77.500$ Hz.

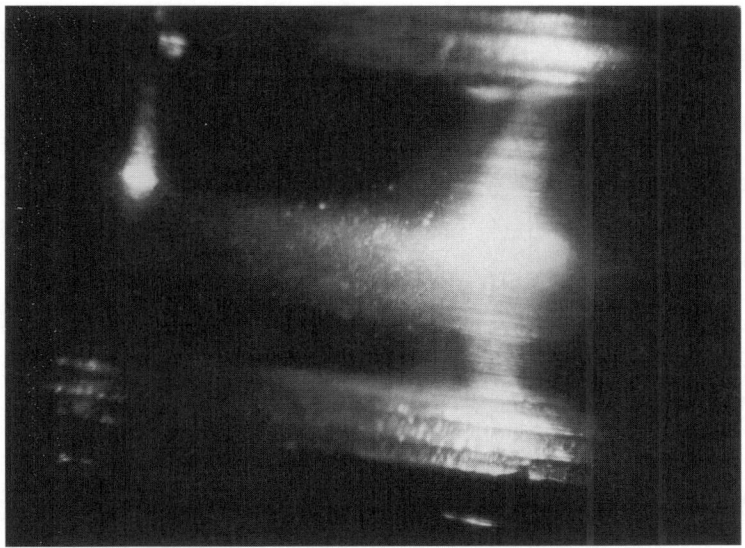

FIGURE 9.24 Denting caused by contamination.

FIGURE 9.25 Paper dust contamination shown on the bearing housing.

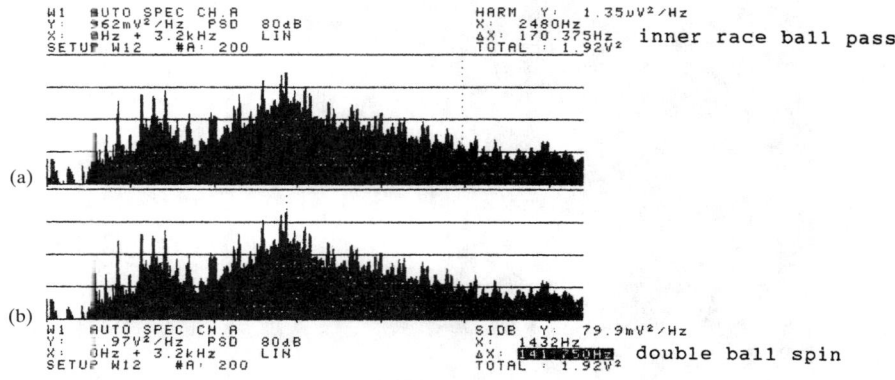
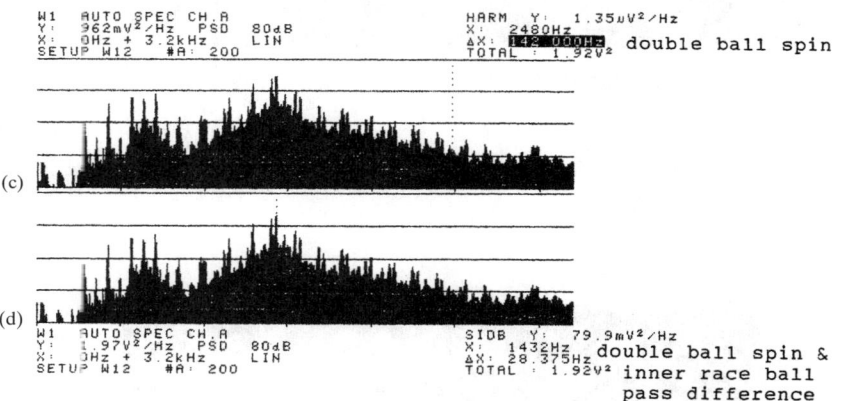

FIGURE 9.26 Spectra indicating hard particle contamination: (a) inner race ball pass, $\Delta x = 170.375$ Hz, (b) double ball spin, $\Delta x = 141.750$ Hz, (c) double ball spin, $\Delta x = 142.000$ Hz, and (d) double ball spin and inner race ball pass differential, $\Delta x = 28.375$ Hz.

Although the bearing defect frequency calculations in Fig. 9.27 appear straightforward, the presence of hundreds of intermingled harmonics, sidebands with unrelated carriers, and difference frequencies can greatly complicate analysis. In addition, field returns often generate bearing frequencies at variance with new bearings because of the effects of wear and slippage. The spectra in Fig. 9.28 further demonstrate unexpected harmonics and sidebands resulting from unusual defects. The ball bearings in an optical encoder had 15 out of 18 cage ball pockets rubbing on the shield, producing a mixture of cage harmonics and sidebands as well as a 15-times cage harmonics and sidebands. A change in internal shaft and housing shoulders solved the problem. Various unusual defects can take hours to diagnose, depending on the complexity of the product involved.

NHBB SSRI-1458 KC Ball Bearing

$D = 0.7543$ (pitch diameter)
$B = 12°$ (contact angle)
$d = 0.0625$ (ball diameter)
$n = 18$ (balls)
$\varnothing = 59.25$ Hz (spindle rotational speed)

$\text{ORBP} = (n/2) \varnothing (1- (d/D) \cos B)$ (outer race defect)

$\quad = (18/2) (59.25)(1- (0.0625/0.7543) \cos 12°)$

$\quad = 490$ Hz

$(\text{ORBP} = 8.271 \varnothing)$

$\text{IRBP} = (n/2) \varnothing (1+ (d/D) \cos B)$ (inner race defect)

$\quad = (18/2) (59.25)(1+ (0.0625/0.7543) \cos 12°)$

$\quad = 576.5$ Hz

$(\text{IRBP} = 9.729 \varnothing)$

$\text{DBS} = (D/d) \varnothing (1- (d/D)^2 \cos^2 B)$ (ball defect)

$\quad = (0.7543/0.0625) (59.25)(1-(0.0625/0.7543)^2 \cos^2 12°)$

$\quad = 710.4$ Hz

$(\text{DBS} = 11.990 \varnothing)$

$\text{FTF} = (\varnothing/2) (1- (d/D) \cos B)$ (cage defect)

$\quad = (59.25/2) (1- (0.0625/0.7543) \cos 12°)$

$\quad = 27.2$ Hz

$(\text{FTF} = 0.459 \varnothing)$

FIGURE 9.27 Bearing defect frequency calculations.

9.4.2 Magnetic Noise

Even though ball bearing defects cause the most concern in noise issues, magnetic noise remains the dominant contributor to overall motor sound. For ac motors, varying the voltage provides the easiest method of separating magnetic and mechanical noise. Since magnetic sound varies roughly as the square of the voltage, cutting the

9.24 CHAPTER NINE

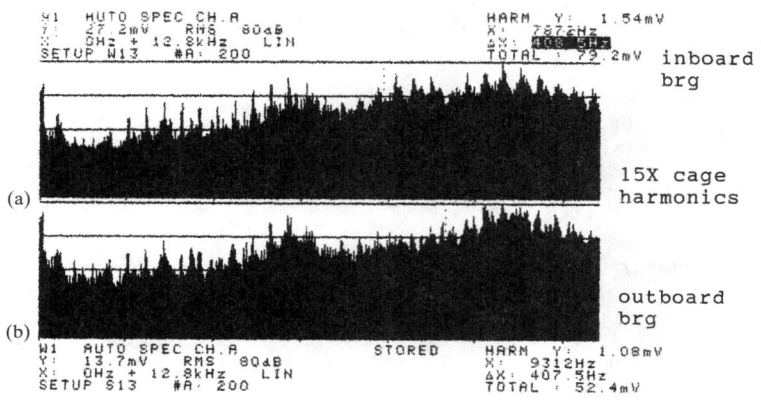

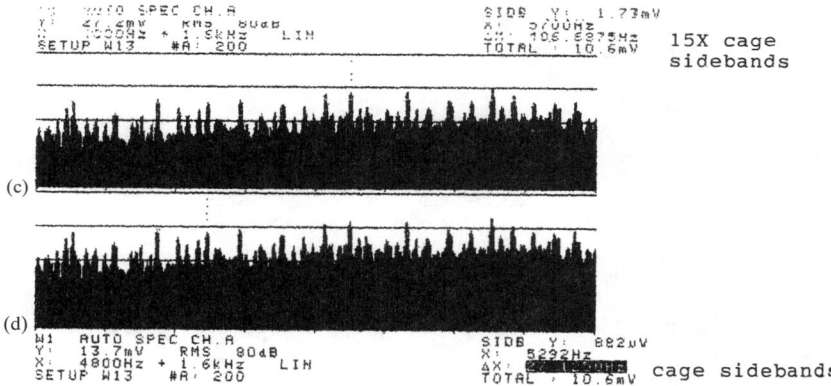

FIGURE 9.28 Unexpected harmonics and sidebands resulting from unusual defects: (*a*) inboard bearing, 15 times cage harmonics, $\Delta x = 408.5$ Hz, (*b*) outboard bearing, $\Delta x = 407.5$ Hz, (*c*) 15 times cage sidebands, $\Delta x = 106.6375$ Hz, and (*d*) cage sidebands, 27.1250 Hz.

voltage in half should reduce the total sound from 189 to 47 mV², versus the actual result of 40 (see Fig. 9.29). This indicates that magnetic noise constitutes well over 90 percent of the total noise from this motor. Further examination of stator and rotor harmonics reveals that rotor frequencies dominate the sound spectrum. The zoomed spectra in Fig. 9.30 reveal slip frequency sidebands that confirm the diagnosis of rotor magnetic frequencies as the dominant contributor. These slip frequencies equal the number of poles times the induced rotor current frequency, which equals the synchronous speed minus the rotational speed.

Efforts for noise reduction can focus on the exact contributors as opposed to trial-and-error methods. Furthermore, separation of mixed noises permits prioritiza-

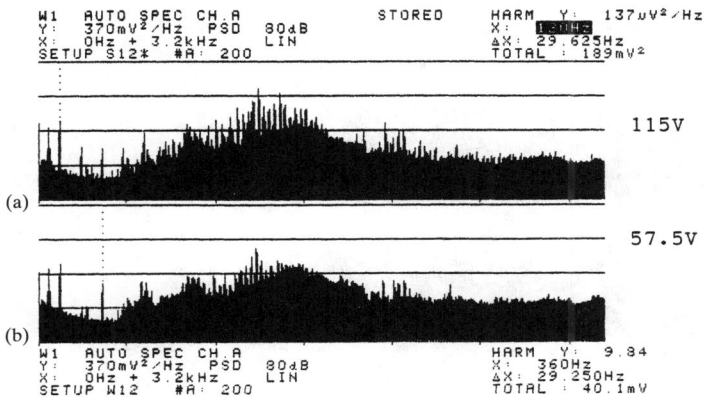

FIGURE 9.29 Magnetic and mechanical noise: (a) 115 V, $x = 120$ Hz, $\Delta x = 29.625$ Hz, and (b) 57.5 V, $x = 360$ Hz, $\Delta x = 29.250$ Hz.

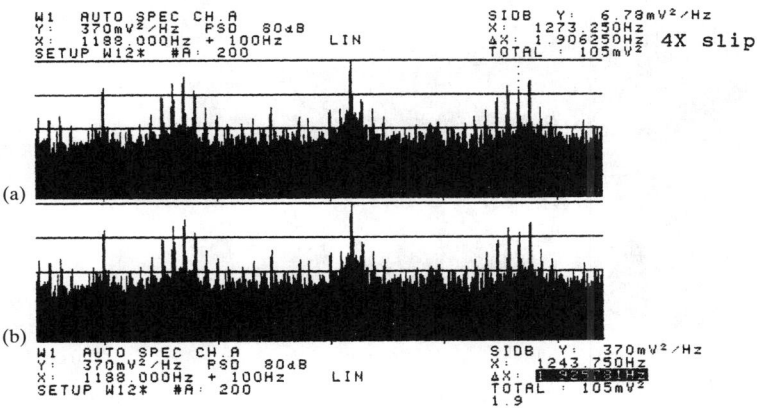

FIGURE 9.30 Zoomed spectra: (a) 4 times slip, $x = 1273.250$ Hz, $\Delta x = 1.906250$ Hz, and (b) $x = 1243.750$ Hz, $\Delta x = 1.925781$ Hz.

tion of methods to reduce noise and predictions of effectiveness. For example, the spectra in Fig. 9.31 consist of 80 percent gear pump pinion frequencies and less than 20 percent rotor magnetic frequencies. Even completely eliminating the motor noise would not achieve the required customer unit noise reductions. These data allowed an objective means of negotiating the best methods of meeting the customer's sound target, while avoiding subjective arguments. In a similar vein, the spectra in Fig. 9.32 contain mostly twice rotational speed harmonics resulting from misalignment introduced when a customer applied a third bearing to the motor shaft. The noise problem disappeared with the correction of the misalignment.

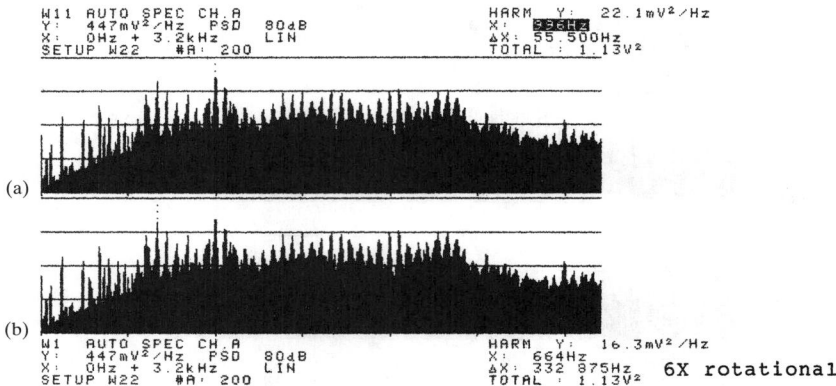

FIGURE 9.31 Gear pump frequencies: (a) $x = 996$ Hz, $\Delta x = 55.550$ Hz, and (b) 6 times rotational, $x = 664$ Hz, $\Delta x = 332.875$ Hz.

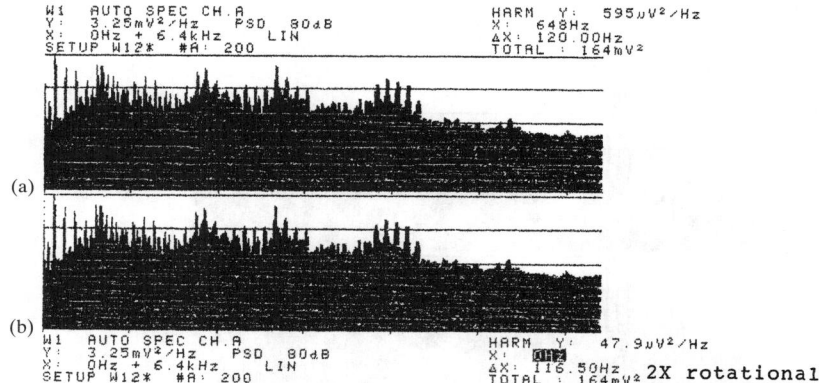

FIGURE 9.32 Noise resulting from additional bearing: (a) $x = 648$ Hz, $\Delta x = 120.00$ Hz, and (b) 2 times rotational, $x = 0$ Hz, $\Delta x = 116.50$ Hz.

Rotor magnetic vibration can cause fretting corrosion in oversize fits of the bearing to the casting. This process gradually increases the noise of motors in the field and can result in premature failure from loss of end play and excessive bearing loads caused by differential axial thermal expansion during operation. Figure 9.33 includes a sound spectrum from a motor with fretting corrosion in the aluminum bearing housing. Additions of damping solved this problem less expensively than reducing the magnetic vibration directly.

Stator magnetic noise becomes more of an issue with intermittent-duty motors than with continuous-duty motors because of the higher flux densities. Improved lamination bonding can provide some reductions. In some cases, improper bonding can cause excessive noise, as illustrated in Fig. 9.34. The 24 stator slots times rotational spike generated over half the noise from the motor, and other harmonics pushed this contribution to over 80 percent. Rebonding the stator solved the problem.

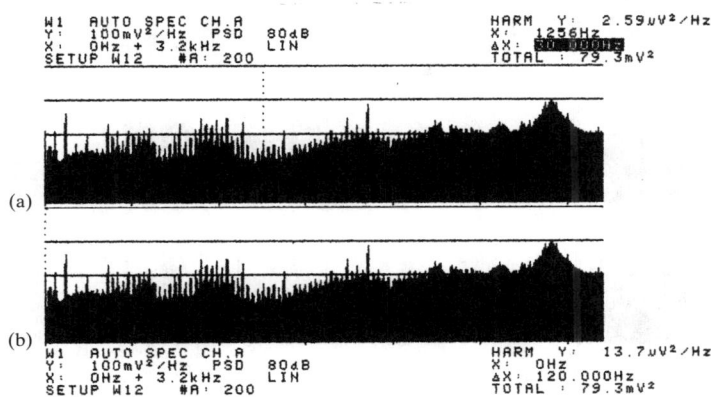

FIGURE 9.33 Noise resulting from fretting collision: (a) $x = 1256$ Hz, $\Delta x = 30.000$ Hz, and (b) $x = 0$ Hz, $\Delta x = 120.000$ Hz.

Brushless dc motors eliminate the brush noise and much of the commutation noise produced by brush dc motors. However, magnet spacing becomes even more critical to quiet operation. The brushless dc motor sound in Fig. 9.35 included a four-times rotational click caused by uneven magnet spacing. After tightening the production fixture and print, the new rotors nearly eliminated the clicking and solved the problem.

These case histories document a few of the ways in which spectral analysis can solve problems, improve customer relationships, and objectively deal with noise and vibration issues. Although spectral analysis can become a powerful tool within a company, attempts to improperly apply spectral analysis can very quickly turn it into a tremendous waste of time and resources.

9.5 RESONANCE CONTROL IN SMALL MOTORS

All structures have natural frequencies of vibration at which exciting forces become amplified. This amplification occurs when the excitation frequency coincides with the natural frequency in a condition called *resonance*. With variable-speed drives, many exciting frequencies change with motor speed and will result in resonance each time they cross a natural frequency. The strength of the resonance varies with the excitation amplitude and proximity to the natural frequency, as well as with the damping, mass, and stiffness associated with the natural frequency. Figure 9.36 illustrates the relationships of these variables.

Resonances often vary erratically and seem to randomly appear and disappear. Under certain conditions, the resonance will grow over a period of minutes and cause a sirenlike pure tone. Figure 9.37 includes a motor sound spectrum with as much as 99 percent of the total sound contained within a 24-Hz band. Increasing the end-cap thickness reduced the amplitude somewhat and increased the frequency, but did not eliminate the pure tone. Cutting variable stiffness pockets in the thicker end cap caused the end cap to ripple at different frequencies and spread the resonance noise over a band of 300 to 400 Hz (see Fig. 9.38). However, reduced pure

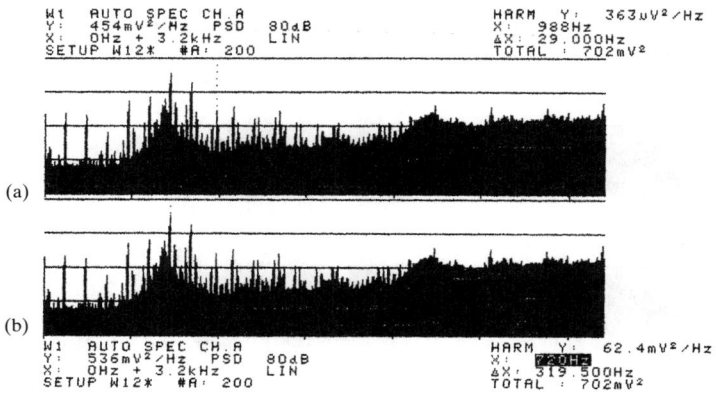

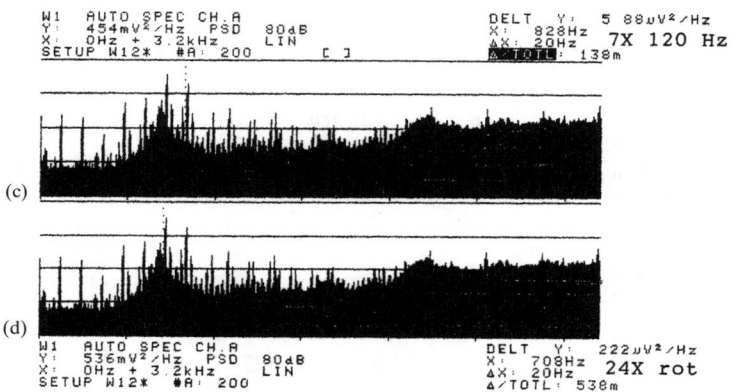

FIGURE 9.34 Noise caused by improper stator bonding: (*a*) $x = 998$ Hz, $\Delta x = 29.000$ Hz, (*b*), $x = 720$ Hz, $\Delta x = 319.500$ Hz, (*c*) 7 times rotational, 120 Hz, $x = 828$ Hz, $\Delta x = 20$ Hz, and (*d*) 24 times rotational, $x = 708$ Hz, $\Delta x = 20$ Hz.

tones still appeared sporadically and required further structural changes for suitable control. Since this type of noise problem occurs seemingly at random and varies radically from motor to motor, field noise complaints often seem unpredictable and, in some motors, defy verification.

Modal analysis offers a method of experimentally determining the type of movement, or mode of vibration, associated with a natural frequency. Figure 9.39 includes an animation of the mode previously described, where the end cap vibrates axially like the bottom of an oil can. This natural frequency can easily dominate the sound produced by a motor, because the radial stiffness typically exceeds axial stiffness by an order of magnitude. The proximity of the bearing to this natural frequency can

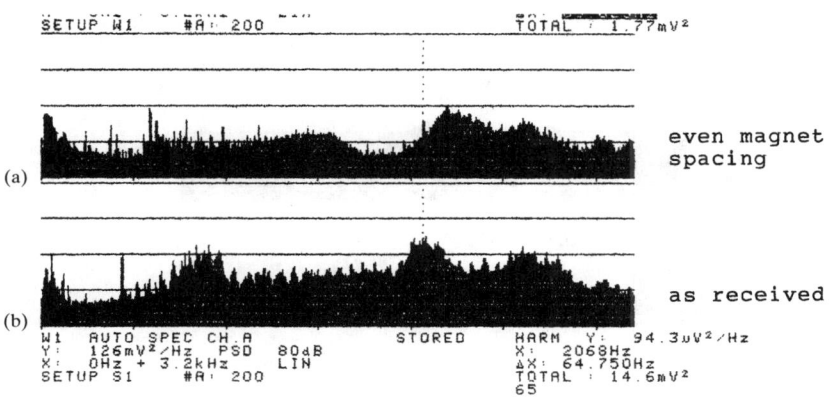

FIGURE 9.35 Noise spectra of a brushless dc motor: (*a*) even magnet spacing, $\Delta x = 65.625$ Hz, and (*b*) as received, $\Delta x = 64.750$ Hz.

greatly amplify the noise of an otherwise acceptable bearing. Efforts to control this noise by tightening bearing vibration requirements will usually result in much higher bearing costs. However, adjustments to the casting design prior to tooling could easily reduce the amplification of the natural frequency enough to avoid bearing changes, with little or no effect on casting cost.

Modal analysis generally begins by exciting the structure with a shaker or, more commonly, an impact hammer. A typical impact hammer, shown in Fig. 9.40, includes a force transducer to measure the applied force and a modally tuned handle to avoid contaminating data with handle natural frequencies. An accelerometer (often triax-

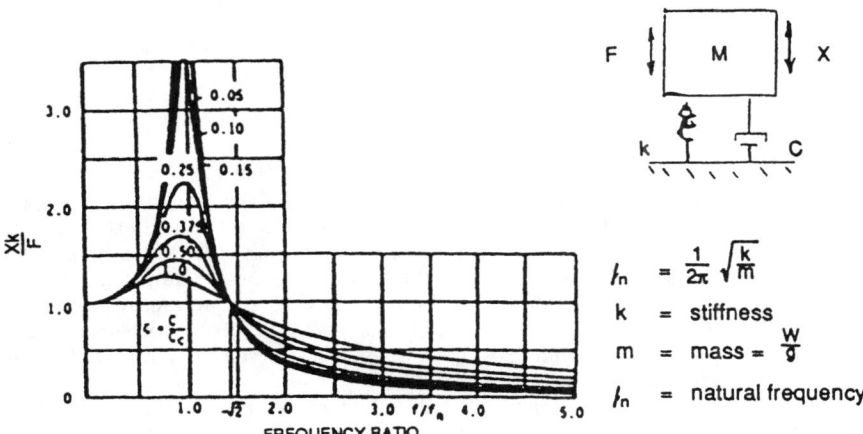

FIGURE 9.36 Amplification of vibration at resonance. (*From Ronald Eshelman,* Vibration Control, *courtesy of the Vibration Institute.*)

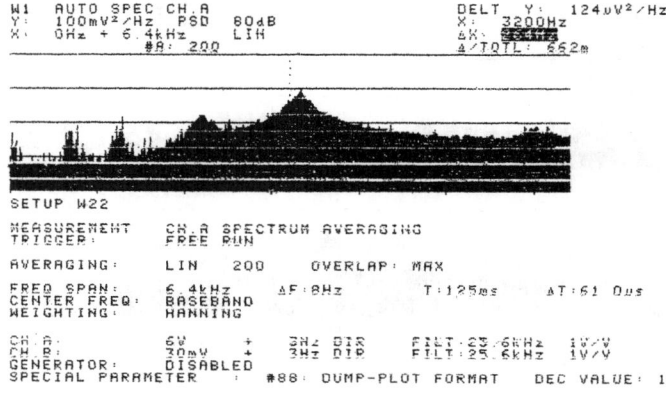

FIGURE 9.37 Spectrum of sirenlike tone, $\Delta x = 264$ Hz.

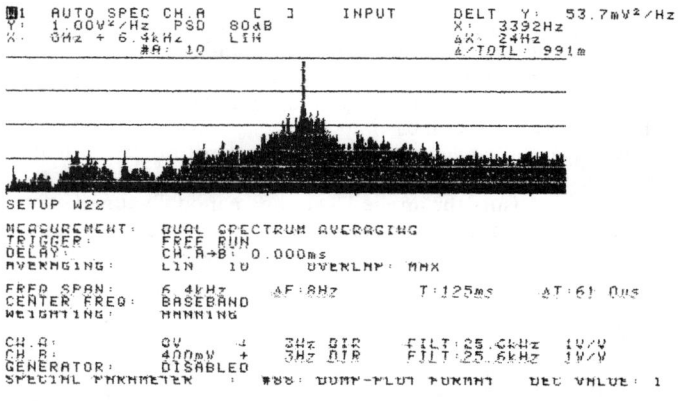

FIGURE 9.38 Spectrum of sporadic pure tone, $\Delta x = 24$ Hz.

ial) measures the response at predetermined locations on the structure, and a spectrum analyzer ratios the force and response to produce *frequency response functions* (FRFs), as shown in Fig. 9.41. At this point, modal analysis software uses the FRF and geometry information to determine the various modes of vibration, including their frequency, damping, amplitude, and phase. This information allows empirical confirmation of finite element analysis predictions of natural frequencies made during the design process and minimizes the number of prototype iterations required to optimize the structure.

Figure 9.42 illustrates the sound spectrum of a motor with 94 percent of the total noise resulting from a shaft-bending resonance. The 1334-Hz harmonics resulted from the clipped sine wave caused by the bearing clearance truncating the shaft movement that normally would result from this resonance. The resonance noise var-

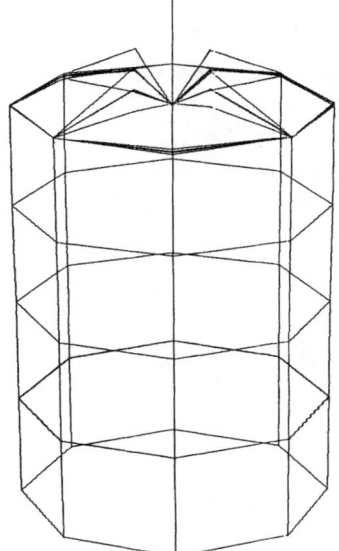

FIGURE 9.39 Model animation analysis of a motor (shaft vertical), showing axial oilcanning of the head (end cap) at the frequency of the high-frequency spike.

ied drastically in amplitude and often disappeared, seemingly at random, for long periods. The lack of higher-order shaft-bending resonances in the FRF in Fig. 9.43 reemphasizes that the 1334-Hz harmonics derived from signal distortion and not higher-frequency bending resonances. This type of resonance often does not show up after assembly into the customer unit because of the mass and stiffness additions. While increased shaft stiffness will usually resolve this problem, added damping often provides a more practical solution in small motors.

Motors with tachometer armatures assembled to the same shaft have a torsional natural frequency between the motor armature inertia and the tachometer armature inertia. The FRF in Fig. 9.44 illustrates a 1232-Hz torsional resonance on a motor with a high-gain controller. This closed-loop system remained stable under low gain but became unstable under high gains because of the torsional oscillation amplified by the resonance. An increased joint stiffness solved the problem by reducing the amplitude and increasing the frequency, as shown in Fig. 9.45. Further improvements, as revealed in Fig. 9.46, resulted from an increased shaft stiffness that provided more than enough margin for manufacturing variations. Figure 9.47 includes a Bode plot of the same type of motor with a load inertia added to the shaft extension. Similar techniques apply to the control of this additional potential resonance. Figure 9.48 shows a frequency response test setup.

Customers sometimes encounter noise and vibration issues that relate to isolators or other components in their unit that go into resonance. In many cases, customers will perceive these systems issues as motor problems. Although the motor contributes to the total system as a component, motor changes often will not significantly affect this type of problem or will cost far more than addressing the main contributor. In these instances, the motor manufacturer may choose to identify the system problem through system testing in order to preserve a long-term relationship. This testing usually consists of acoustic or vibration spectral analysis, followed by surveys of system natural frequencies with an impact hammer or shaker test.

Figure 9.49 includes a swept sine wave test performed by a shaker on a unit with typical isolators. After mounting the unit solidly to the shaker, a constant 0.5g sine wave excitation began at the lowest frequency of interest and increased uniformly to the highest frequency under investigation. Many isolators include two or three frequencies that result in significant vibration amplification. Furthermore, these high levels can produce not only noise problems, but also physical damage to the motor, the unit, or the product in the unit.

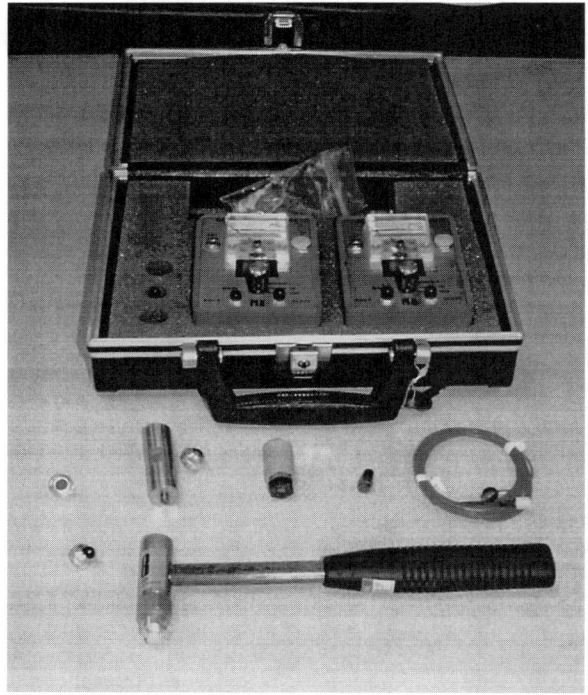

FIGURE 9.40 Impact hammer.

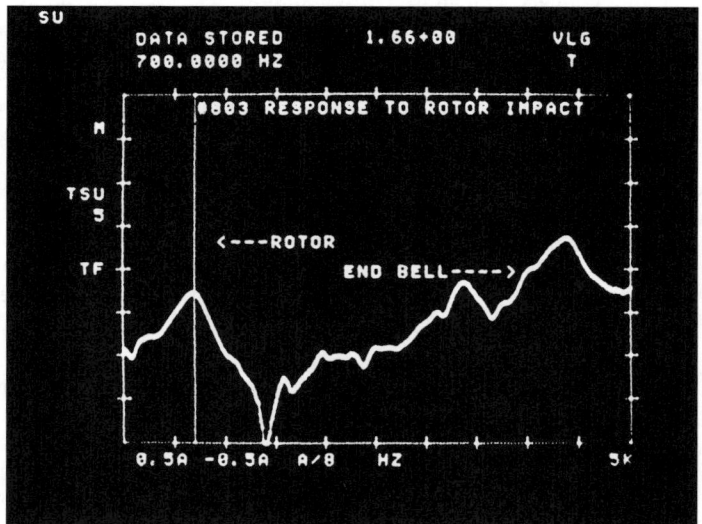

FIGURE 9.41 Frequency response function of impact hammer.

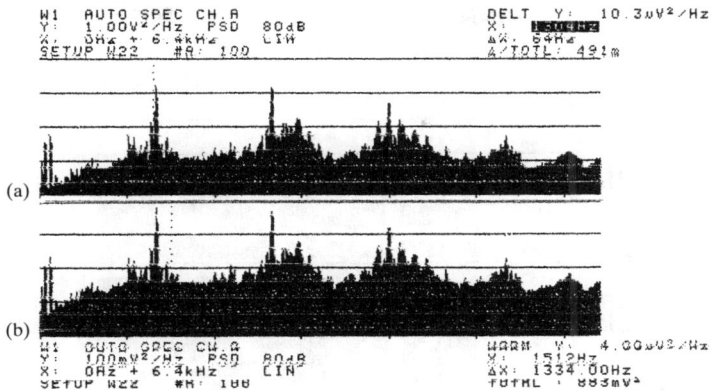

FIGURE 9.42 Sound spectra of shaft-bending resonance: (*a*) $x = 1304$ Hz, and (*b*) $x = 1334.00$ Hz.

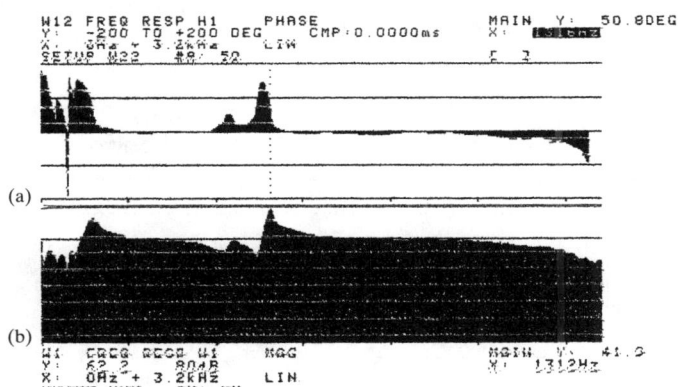

FIGURE 9.43 Frequency response functions of shaft-bending resonance: (*a*) $x = 1316$ Hz, and (*b*) $x = 1312$ Hz.

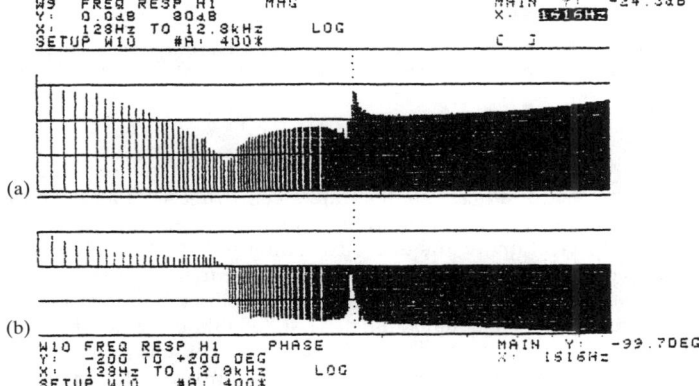

FIGURE 9.44 Frequency response functions of torsional resonance of armature and tachometer: (*a*) $x = 1616$ Hz, and (*b*) $x = 1616$ Hz.

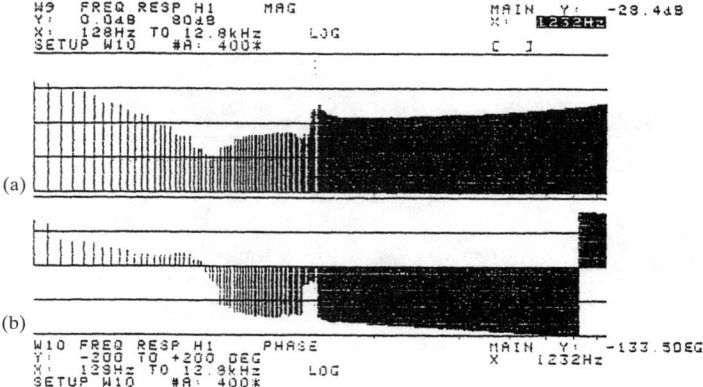

FIGURE 9.45 Frequency response functions of unit with increased joint stiffness: (a) $x = 1232$ Hz, and (b) $x = 1232$ Hz.

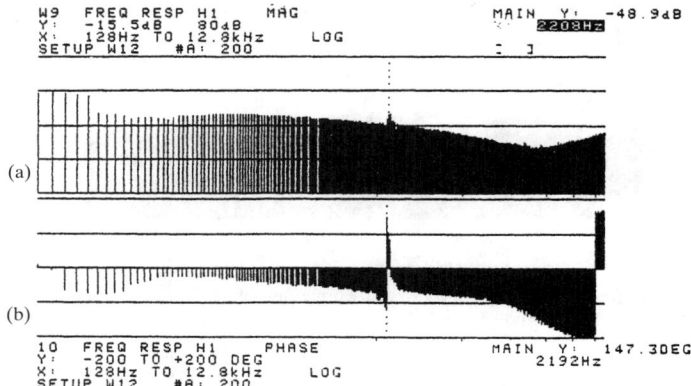

FIGURE 9.46 Frequency response functions of unit with enough additional stiffness to accommodate manufacturing tolerances: (a) $x = 2208$ Hz, and (b) $x = 2192$ Hz.

The vibration spectra in Fig. 9.50 illustrate a centrifuge with and without mount resonance modulation. The isolation mounts included three different natural frequencies with amplification factors often exceeding 10 to 1. As exciting frequencies from the variable-speed drive crossed these natural frequencies, the mount would go into resonance and create a subsynchronous wave within the liquid in the bowl. The rotational speed of this wave equaled the drive speed minus the natural frequency of the mount. At certain centrifuge speeds, this wave would tear apart the blood cells and destroy the product. The intended use of this centrifuge for operating rooms made high reliability and performance essential. The customer thought the problem resulted from differences between prototype and production motors. Spectral anal-

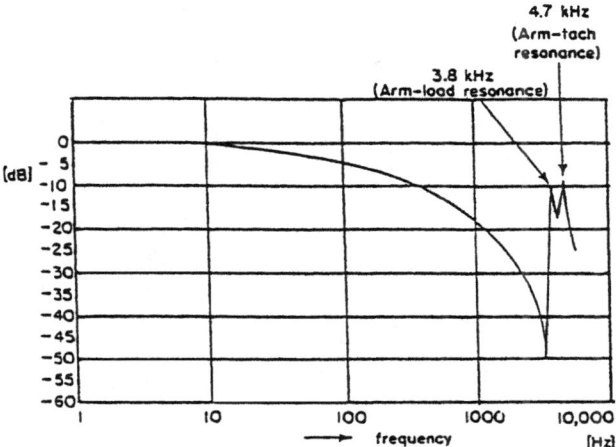

FIGURE 9.47 Bode plot of motor with load inertia added. (*From the* Electro-Craft Handbook, *courtesy of Electro-Craft Motion Control.*)

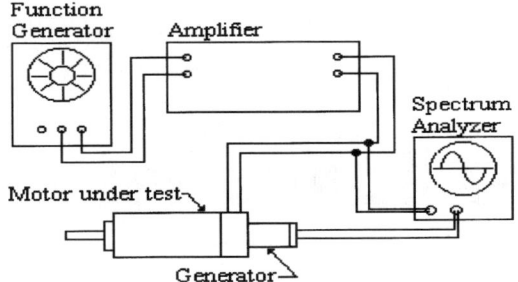

FIGURE 9.48 Frequency response test setup

ysis and shaker tests provided an objective way of identifying the problem and minimizing the controversy that often prevails with this kind of problem. The customers then fine-tuned their mount and changed their software to quickly take the centrifuge through dangerous speed ranges before the resonance could build to damaging levels. Shipping vibration can also excite these low-frequency isolator resonances, leading to fretting in clearance fits, false brinells in ball bearings, and destroyed optical encoders. In many cases, isolators require shipping restraints to avoid this kind of damage. Qualification shaker tests of packaging and products can prevent major field problems with new products or with changes to old products that affect these natural frequencies.

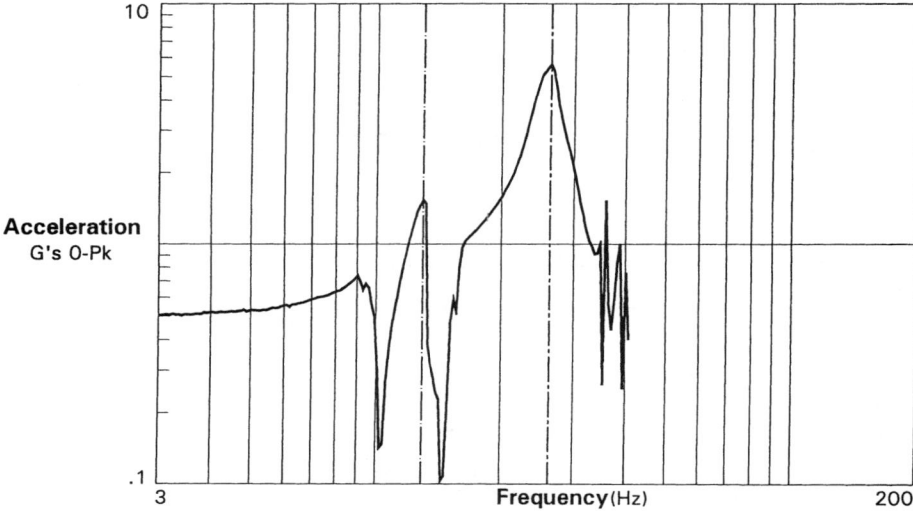

FIGURE 9.49 Swept sine wave test.

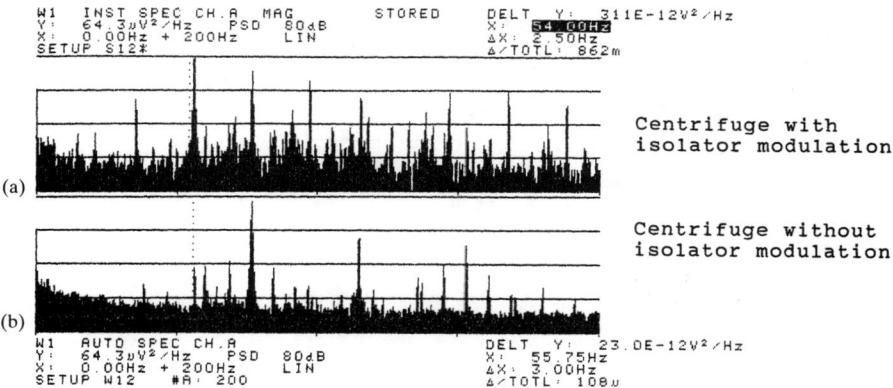

FIGURE 9.50 Spectra of centrifuge: (*a*) centrifuge with isolator modulation, $x = 54.00$ Hz, and (*b*) centrifuge without isolator modulation, $x = 55.75$ Hz.

9.6 FATIGUE AND LUBRICATION TESTS

A typical shaft fatigue test consists of motors mounted to shelves with weights hung on steel shafts to produce a bending fatigue condition similar to a rotating beam machine. In Fig. 9.51, the right motor has a hanger bearing and a weight posi-

FIGURE 9.51 Fatigue and lubrication tests.

tioned at a calculated distance to provide the necessary shaft stress without exceeding bearing fatigue limits during the test. Different distances and weights will provide a bearing fatigue or grease test without overstressing the shaft. The recorded data for shaft fatigue includes the number of cycles at each bending stress level and the measured shaft geometry. Shafts which do not fail, after a sufficient number of cycles, to reach the plateau of the S/N curve (usually between 10^8 and 10^9 cycles) will repeat the test at a load increased by a uniform increment of about one to two standard deviations until a failure occurs. Following statistical analysis, the S/N diagram in Fig. 9.52 derived from the data collected from this type of test. For convenience, this chart uses inch-pounds of bending rather than actual stress.

Linear regression and exponential distribution show poor correlation because of data scatter. However, the mean and three-sigma limits for points in the region of the plateau yield a reasonable endurance limit in agreement with linear regression. After adjusting individual points for geometry variation, allowing for differences between test specimen strength and minimum specified strength, compensating for data scatter, and applying an appropriate factor of safety, these data provided a safe design limit for this material. Given this correlation, fatigue calculations can adjust for design variations and still provide good confidence in the results. Major advantages of this test procedure over photoelastic or normal rotating beam tests include empirical confirmation of notch sensitivity factors, surface factors, size factors, and various process effects such as residual stresses and plating. Furthermore, this test method serves as much to validate the manufacturing process as the design and material. Seemingly minor process changes can produce significant changes in endurance limits.

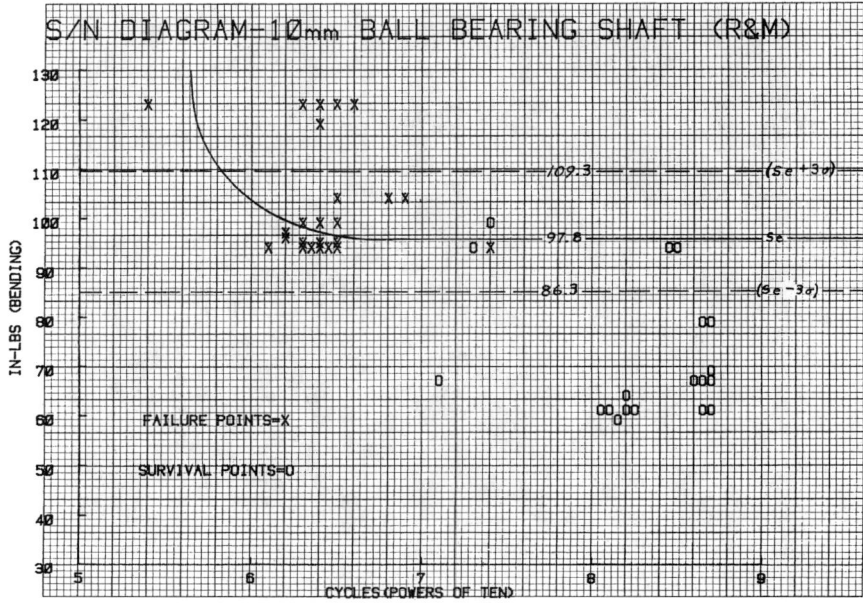

FIGURE 9.52 Signal-to-noise diagram.

9.6.1 Gear Dynamic Load Tests

The cycling inertia test carts shown in Fig. 9.53 maintain the same gear motor orientation as in the customer unit. Each pair of gear motors connects to a controller with a joystick and an interface box. The gearbox shaft supports an inertia wheel that simulates the worst-case inertia for the customer unit. A programmed controller sets the forward, reverse, acceleration, and braking levels to simulate the dynamic loads for severe field operation.

A programmable logic controller (PLC) provides a cycling relay closure between the common and the forward pin of the interface box for 7.5 s, and between the common and the reverse pin for the next 7.5 s. The PLC sequencing maintains 1.25 s of separation between each acceleration current spike to avoid affecting other motor pairs from spike overlap.

Two 100-A supplies feed each cart (at least three times the required capacity) to avoid limiting the controllers. A current probe and digital oscilloscope provide plots of current spikes for each motor at the beginning of the tests. A surveillance camera and time-elapse VCR record the failure time and date for each gear motor.

9.6.2 Constant Load Gear Tests

The gear motor in Fig. 9.54 drives a magnetic brake at the rated load for most of a day and automatically shuts down. After a cooling period, the motor restarts at an

FIGURE 9.53 Cycling inertia test charts.

intermittent-duty load and reduces to rated load after reaching the thermal duty limit. This setup will test for gear fatigue and lubrication life as well as for gearbox bearing and shaft fatigue. If necessary, the brake will apply multiple torque profiles for specific applications. This kind of testing proves especially useful in finding unexpected design or process weaknesses, while also helping to select lubricants and check fatigue limits.

Plastic materials, in general, do not have plateaus on their fatigue curves (as do most steels), but instead the safe fatigue limit falls continuously as their expected life increases (see Fig. 9.55). For this reason, plastic gears will eventually fail from fatigue if run long enough. Steel gears, however, will not fail from fatigue in a clean environment if operated below a given endurance limit. Consequently, properly designed and operated steel gears will generally fail from wear, while properly designed plastic gears could fail either from wear or from fatigue when run beyond the design life. The possibility of continuous usage during the normal warranty therefore places more severe design requirements upon plastic gears than upon steel gears. In addition, fatigue tests require much longer test times for plastic than for steel.

Significant drops in fatigue strength often occur for molded plastic gears as compared to hobbed plastic gears. Consequently, molded gears usually need more face width than equivalent hobbed plastic gears. In many cases, assembly must also utilize ultrasonic techniques or double-D joints to overcome added stresses from post mold shrinkage and from reduced part stiffness caused by uniform wall thickness webbing.

FIGURE 9.54 Gear motor.

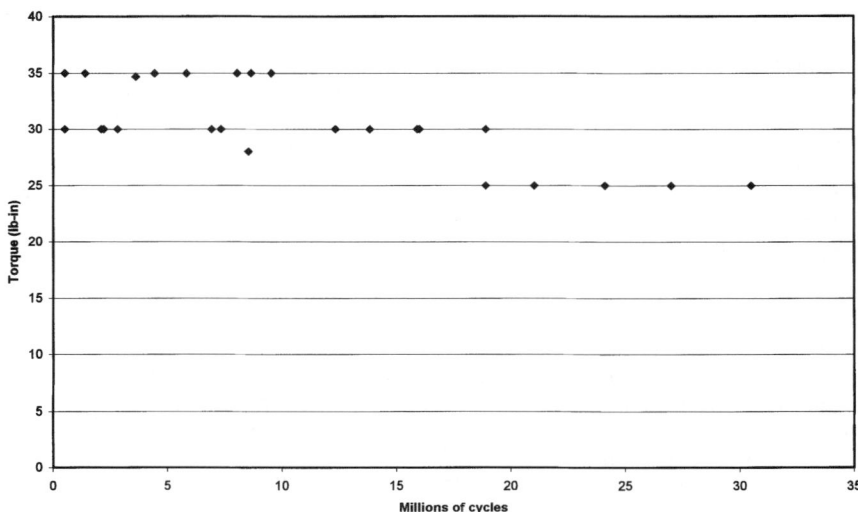

FIGURE 9.55 Molded plastic gear fatigue.

9.6.3 Bearing Grease Tests

In most cases, grease and bearing manufacturers do not provide damping curves for particular greases. After measuring the bearing drag at different speeds and plotting the data, the damping constant equals the slope (usually in ounce-inches per 1000 rpm) and the friction equals the intercept point at zero speed (see Fig. 9.56). This test especially affects grease selection for servo drives.

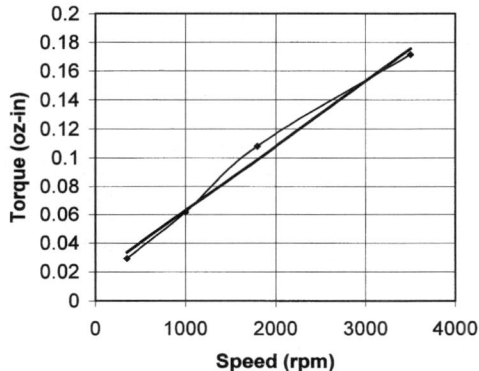

FIGURE 9.56 Bearing grease torque test for bearing 0002-1695, 0.0002 to 0.0005 in clearance. Line with diamonds indicates test 2, bearing 2; solid line indicates linear performance. $y = 5\text{E-}05x + 0.0177$; $R^2 = 0.9899$.

Continuous and cycling eccentric load tests will also compare greases for the ability to withstand vibration during operation, while also checking bearing support structures. This accelerated test often finds problems within days that may not reach failure in normal life tests for months or years. Although eccentric load tests will not replace long-term life tests, they significantly reduce product development time by quickly wringing out problems prior to lengthier tests.

Grease life tests at elevated temperatures provide a quick means of predicting application life based on the thermal stability of the grease. Although considerable industry data exists for older greases, the newer greases that offer major life advantages frequently do not have enough independent tests to overcome the variability common to this type of testing. Furthermore, the failure criteria for different tests may range from a 10°C increase in bearing temperature at the outer race to actual seizure, leading to widely different results. In many cases, a two- to three-month test can provide the high-temperature data necessary to proceed with confidence in grease selection. However, thermal performance of grease alone will not qualify a candidate for a specific application and serves strictly as a minimum requirement. Resistance to corrosion, operating and nonoperating vibration, and other application-specific conditions can easily outweigh thermal stability in importance. Unfortunately, many customers do not know the exact conditions of their applications.

9.7 QUALIFICATION TESTS FOR ADHESIVES AND PLASTIC ASSEMBLIES

Since the bonding process can affect strength more than the adhesive material itself, the qualification of production assemblies applies just as much to process control as to the design itself. For this reason, test parts must always reflect exact production procedures rather than ideal lab conditions. In particular, any adhesive process change should first undergo the appropriate tests to avoid serious production problems and potential field failures.

Regardless of the type of test, documentation of failed parts should include the percentage of adhesive or substrate failure and the percentage of adhesive coverage at the joints. At a minimum, comparisons of test results to calculations must consider the effects of temperatures, surface finish, air gaps, cleanliness, and especially adhesive coverage in the joint.

Short-term tests for plastic parts should at least include impact, thermal shock, environmental resistance, and deformation under application load and temperature. Long-term creep tests should occur under the application load and temperature for at least 10 percent of the expected product life to permit good extrapolation of results. Fatigue tests for significant variable loads should not exceed 10 times the application cycling rate to provide reliable data.

9.7.1 Thermal Shock Tests

Static thermal shock tests expose bonded or molded items to application, shipping, or storage temperature extremes. Subsequent tests should use the same thermal shock test parts to incorporate any degradation from hidden flaws and to more closely simulate the application. Furthermore, thermal cycles will more quickly initiate the thermoplastic postmold shrinkage that generally occurs over a long period of time in use. When added to prior assembly stresses, these shrinkage stresses often will initiate cracks during the static thermal tests and will affect the results of later tests. In addition, thermal stresses from materials with different coefficients of thermal expansion could combine with these other stresses to induce failures.

In most cases, the test usually begins with the lowest temperature to avoid potential corrosion at the end of the test from frost formation and to expedite hot shear tests by reducing oven time. The test specimens stay in the cold chamber until reaching equilibrium and then quickly rotate to the heated chamber for a similar period. After six cycles, any signs of cracks or other failures generally disqualify the assembly from further evaluation.

9.7.2 Impact Tests

A typical impact test setup may include a pendulum of known mass dropped from incrementally higher measured heights until the assembly fails. The sample must securely fasten to the base of the fixture to absorb the full impact at the most desirable point. For a large part mass, dropping the sample vertically to the point of impact from various heights may suffice for failure to occur. With either method, the impact force equals mass times vertical distance. For a small sample mass, a larger known mass may vertically impact a test specimen resting against a hard surface.

Given low enough friction, guides can ensure a consistent and uniform impact for an evenly distributed stress with any of these methods.

The impact method should remain constant for the same assembly to obtain good comparisons of different adhesives or different joint types. However, construction variations between assemblies often force the use of different impact techniques to fully test the joint, while avoiding extraneous deformations or brittle failures that would introduce premature failures unrelated to the joint under test. The same principles apply to plastic coatings or to assemblies that include plastic parts. Ideally, the substrate should fail before the adhesive in bonded assemblies. Otherwise, normal variations in adhesive strength require very large factors of safety to avoid field failures. Nevertheless, applications with shock requirements would also require testing the complete assembly after component qualification.

9.7.3 Stress Tests

Shear tests on parts at the maximum operating temperature and at room temperature provide a means of checking temperature factors used in stress calculations. Since plastics and adhesives often do not provide the same strength in all three axes, spin tests for rotating parts at operating temperatures will also permit a check of axial versus radial strength. Production specifications for nondestructive, room-temperature shear tests will then provide confidence that conforming parts will perform in the application.

9.8 TRENDS IN TEST AUTOMATION*

The conflicting pressures to reduce test time and generate more test data force an increasing level of computerization. In particular, extremely time-consuming UL thermal tests often require up to a week to test for all conditions specified by UL. Fully automated 24-h test setups can easily reduce this time to less than a third of single-shift test duration.

Manual spectral analysis can require hours or days of detailed analysis on a limited number of motors, especially for brush dc gear motors. In contrast, automated systems can perform most types of standardized spectral analysis and diagnostics in seconds or minutes, while accepting or rejecting large numbers of motors against established limits. Although a certain amount of manual analysis will remain necessary over the next few years, the growing sophistication of automated systems will increasingly obsolete the manual techniques. Any motor company that fails to stay on top of this technology will experience considerable difficulty in overcoming the required learning curve.

The best engineering networks provide seamless data transfer among design, analysis, test, and manufacturing. This new technology opens the way to automatic comparison of calculated and empirical results, and even automated adjustments of FEA models. Furthermore, the cost of these systems has dropped enough that

---

* Portions of this section adapted from Robert Band, *Twenty-four Hour Test* procedure, courtesy of Electro-Craft Motion Control, Gallipolis, Ohio—a Rockwell Automation business.

smaller companies can now afford to obtain the greater engineering productivity and reduced development time offered by this software. However, test programs must change to take full advantage of these new capabilities.

9.8.1 Twenty-four-Hour Test Bench

The greater precision of computer control reduces thermal test leveling times while nearly eliminating transition times in running-overload, safe operating area, and reduced-voltage tests. In addition, virtual charts eliminate manual transcription of data while improving accuracy.

When combined with automatic voltage control, frequency changes, and switching of the cooling fan and test motor, computerized sequencing of tests permits 24-h unattended testing (see Fig. 9.57). The number of test days can easily drop to a third or less of the time required for single-shift testing, especially when considering the lost thermal test time often found at the beginning and end of shifts. These major

FIGURE 9.57 24-h computerized test bench.

reductions in test time can result in an order-of-magnitude higher productivity with far more data, thereby freeing test personnel for other types of engineering work such as life testing and various kinds of engineering analysis. The following ac motor tests increasingly utilize this type of automation with 24-h test sequencing. (See Figs. 9.58 through 9.61.)

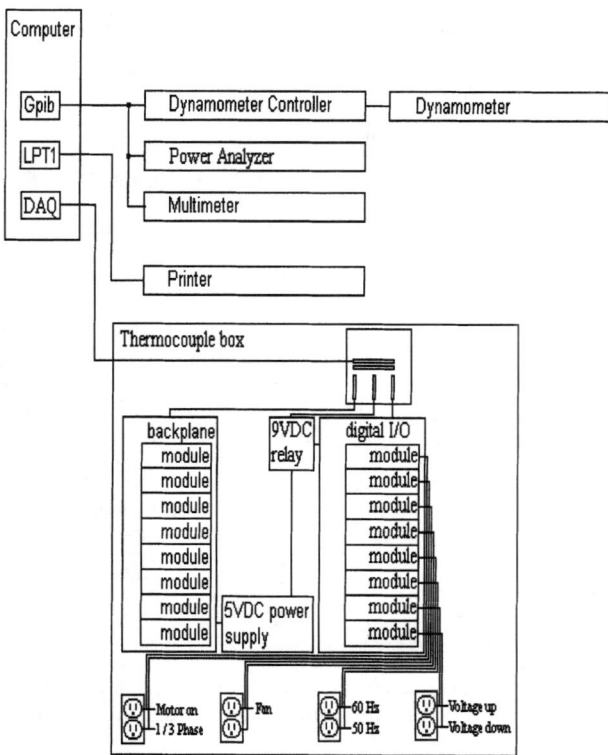

FIGURE 9.58 Block diagram of 24-h test station.

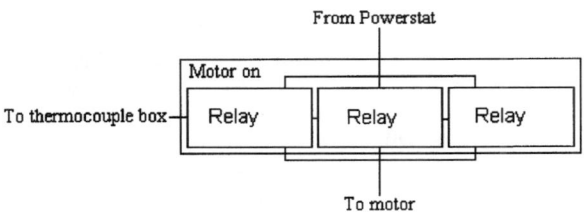

FIGURE 9.59 Meter control block diagram.

9.8.2 Locked-Rotor Test

This test determines the trip and reset temperatures at stall for a motor-protector combination. Before beginning additional locked tests at other conditions or intermittent-full-load or idle tests, the computer will start a fan to cool the motor to

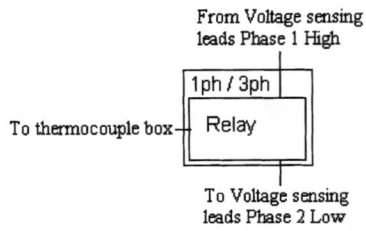

FIGURE 9.60 Voltage control block diagram.

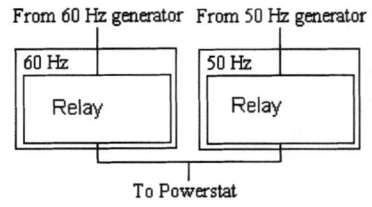

FIGURE 9.61 Frequency control block diagram.

within 5° of ambient temperature. For continuous duty, the motor will restart and the fan will stop after the protector resets.

9.8.3 Full-Load Heat Run

Full-load tests include three common variations. For a given full-load torque, the computer maintains a constant torque until the winding temperature levels. For a given full-load current, the load torque changes during the test to hold a constant current until the winding temperature levels. For a given winding temperature, the load torque changes during the test to level the winding temperature as specified. The computer automatically acquires data, sends it to a file, and follows with a hard copy at the end of the test.

9.8.4 Running-Overload Test

Following a full-load-by-current test, the computer adjusts the torque to increase the current by 0.2 A. After the winding temperature levels, the computer increments

TABLE 9.2 Test Methods for Electric Motors and Generators

Test specification number	Test or description
112-1996	IEEE Standard Test for Polyphase Induction Motors and Generators
115-1995	IEEE Guide: Test Procedure for Synchronous Machines, Part 1—Acceptance and Performance Testing, Part II—Test Procedures and Parameter Determination for Dynamic Analysis
304-1977	IEEE Standard Test Procedure for Evaluation and Classification of Insulation Systems for Direct-Current Machines
522-1992	IEEE Guide for Testing Turn-to-Turn Insulation of Form-Wound Stator Coils for Alternating-Current Rotating Electric Machines
620-1996	IEEE Guide for the Presentation of Thermal Limit Curves for Squirrel Cage Induction Machines
1107-1996	IEEE Recommended Practice for Thermal Evaluation of Sealed Insulation Systems for AC Electric Machinery Employing Random Wound Stator Coils

another 0.2 A and repeats the test until the protector trips. The running-overload test may also follow the full-load-by-torque test.

9.8.5 Full- and Reduced-Voltage Idle Test

The motor runs without load, uncoupled from the dynamometer, until the winding temperature levels. If the protector trips, the computer reduces the voltage by 10 V, cools the motor to reset the protector, and continues the test until the winding temperature levels or the protector trips to initiate the next increment.

9.8.6 IEEE Tests

The Institute of Electrical and Electronics Engineers (IEEE) has, over the years, developed test methods for electric motors and generators. They are listed in Table 9.2 for reference. They may be obtained through the IEEE.

CHAPTER 10
DRIVES AND CONTROLS

Chapter Contributors

Birch L. Devault
Duane C. Hanselman
Daniel P. Heckenkamp
Dan Jones
Douglas W. Jones
Ramani Kalpathi
Todd L. King
Robert M. Setbacken

10.1 MEASUREMENT SYSTEMS TERMINOLOGY*

10.1.1 Measurement Units

The linear unit of length is the meter (m). It is the distance light travels in approximately 1/300,000,000 (1/299,792,458) s. Linear measurement systems commonly define design parameters in units of the micron (μ), or 0.000001 m. 1 μ is equivalent to approximately 0.000040 in. 0.0001 in = 2.54 μ. The angular unit is the radian (rad), which is the angle subtended by an arc whose length is equal to the radius of a circle. This unit of measurement is most commonly used in military applications. The degree (°) is used mostly in commercial applications. Fine angles are represented as both fractions of degrees and as minutes (') and seconds ("). $1' = 1/60°$; $1'' = 1/60'$.

10.1.2 Accuracy and Resolution Defined

Accuracy is the ability to repeatably indicate an exact location, while *resolution* is the ability to detect motion in finer and finer increments. For a rotary encoder, this is in cycles per revolution (cpr) or pulses per revolution (ppr). For a linear system,

---

* Sections 10.1 to 10.5 contributed by Robert M. Setbacken, Renco Encoders.

this is counts per inch, or it is defined in terms of the graduation pitch in microns. Accuracy and resolution are not directly related. Although it is generally true that high accuracy systems usually resolve smaller increments, a measuring device could in principle have very coarse resolution and still be very accurate.

10.1.3 Quadrature

In Fig. 10.1, the 90° electrical separation (one-quarter period) between the two signals is referred to as *quadrature*. Quadrature signals allow the user to know what direction the system is turning, and provide additional resolution by allowing edge counting.

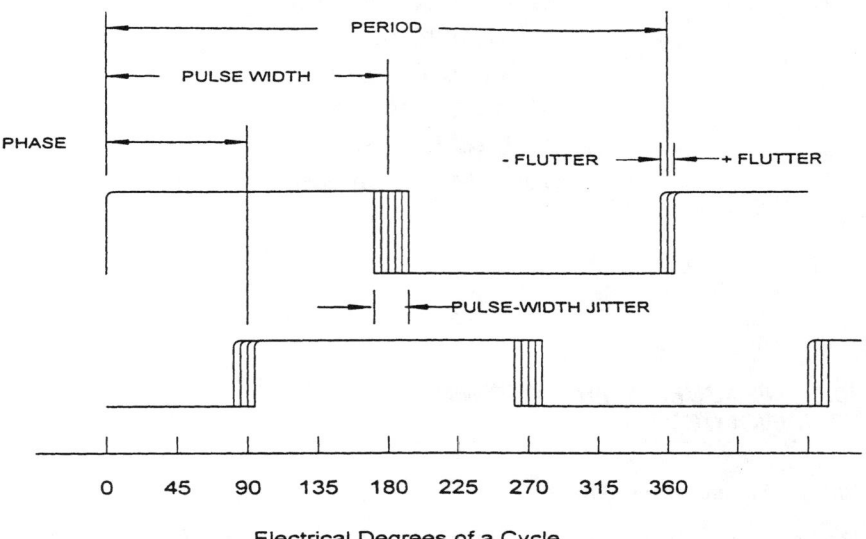

FIGURE 10.1 Output waveform definitions.

10.1.4 Edge Counting

Again referring to Fig. 10.1, it can be seen that within one cycle, there are four edge transitions between the two output signals. This can effectively be used to provide a resolution of 4 times the base resolution.

10.1.5 Direction Sensing

Referring to Fig. 10.2, one can see that when B makes a low transition, the value of A is locked into Q. When the system is moving clockwise (CW), Q will be low. When the system is moving counterclockwise (CCW), Q will be high. This scheme can be used within the sensor to provide a pulse output with a high/low direction indicator.

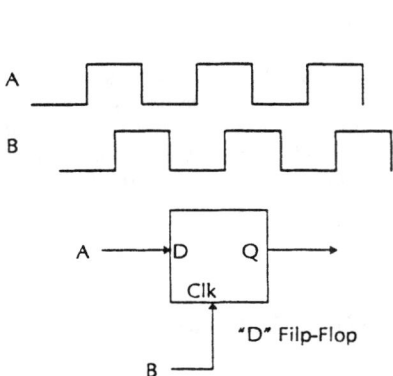

FIGURE 10.2 Quadrature direction encoding.

10.1.6 Interpolation or Multiplication

Interpolation is the process of dividing an analog signal into phase-shifted copies, which are then recombined to give a higher effective resolution. When the output of a sensor is sinusoidal and there are two outputs in quadrature, the signals can be interpolated. Transistor-transistor logic (TTL) signals can not be interpolated. As a result, interpolation can be used to improve overall accuracy by reducing the error component due to quantization.

10.1.7 Contacting Systems

There are various interpretations of what this term means. Linear encoders that use bearings to control the gap between the read head and the scale are called *noncontacting*. Linear encoders that use low-friction coatings on the glass surfaces to float the read head over the scale are *contacting*. A more explicit definition of contacting sensors includes potentiometers and pin-contact encoders. Although contact methods are still used, and some companies have developed very robust examples, long-term reliability is favoring noncontacting designs. Some applications still find uses for contacting sensors, especially pin-contact encoders. One major example is in the nuclear industry, where pin-contact encoders generally last as long as the measured system itself. Magnetic systems loose magnetization, and optical systems using plastic are fogged due to the radiation in these environments, so pin-contact encoders work very well.

10.1.8 Non-contacting Systems

These systems generally offer higher reliability, and are typified by the following

- Optical, capacitive, and magnetic encoders
- Brushless resolvers

- Most modular or kit encoders
- Open-frame linear scales

Note that the use of incorporation of bearings into a feedback device does not exclude it from being described as a noncontacting sensor. Make sure you are fully aware of the manufacturing principles when specifying a noncontacting sensor. Truly noncontacting sensors, like modular rotary encoders or brushless resolvers, can still become partially contacting devices if seals are incorporated in the final installation to the application.

10.2 ENVIRONMENTAL STANDARDS

10.2.1 Specification of the Application Environment

The end user needs to have some idea of the environment in which the sensor is to be placed. Many times the final installed environment cannot be known. This is generally the case for motor manufacturers that ship to original equipment manufacturers (OEMs), which then ship products of various types all over the world. In order to address such situations, various standards organizations have developed guidelines which can be used to characterize the applications a device should be able to withstand. In Europe, the International Electrotechnical Commission (IEC) has developed a large suite of specifications covering every imaginable detail. The United States has been relying on military standards (MIL-STDs) when such guidance is required. Finally, the various industries themselves develop de-facto standards through the published specifications for their products. In the United States, the two most widely referenced standards for feedback elements are the following:

MIL-STD-810	Environmental test methods
MIL-STD-202	Test methods for electronic and electrical component parts

Similar standards from the IEC are the following:

IEC 68-1	Part 1: General and guidance
IEC 68-2-1	Test A: Cold
IEC 68-2-2	Test B: Dry heat
IEC 68-2-3	Test Ca: Damp heat, steady state
IEC 68-2-6	Test Fc and guidance: Vibration (sinusoidal)
IEC 68-2-27 IEC 68-2-47	Test Ea and guidance: Shock mounting of components, equipment, and other articles for dynamic tests including shock (Test Ea), bump (Test Eb), vibration (Tests Fc and Fd), steady-state acceleration (Test Ga), and guidance
IEC 68-2-48	Guidance on the application of the tests in IEC Publication 68 to simulate the effects of storage
IEC 529	Degrees of protection provided by enclosures (IP code)

IEC 34-5   Classification of degrees of protection provided by enclosures of rotating electrical machines

When possible, the user should request a test program report for the device being considered. Even if a device is tested, it is important to know what passing the test entails. The IEC uses the following definitions:

A   No degradation during or after
B   Degradation during, not after
C   Loss of function but undamaged; operation restored by reset

Standard Atmospheric Conditions

IEC specifications for ambient atmospheric conditions are as follows:

Temperature	Relative humidity	Air pressure
23 ± 2°C	45 to 55%	86 to 106 kPa

Test Programs. A reasonable test program for sensor design verification should consist of both environmental and mechanical testing.

Environmental Testing. This testing should consist of climatic sequencing. The IEC guidelines suggest the following order:

1. Dry heat
2. Damp heat
3. Cold
4. Low air pressure
5. Damp heat, cyclic

Not all test programs must include all tests, but the tests included should run in this order. An interval of not more than 3 days is permitted between any of these conditionings, except for the interval between the first cycle of the damp heat cyclic conditioning and the cold conditioning. For this period, the interval shall not be more than 2 h, including recovery.

Suggested severity levels for environmental testing of feedback devices are as follows:

- *Dry heat.* 1000 h of dry heat at 110 ± 2°C, with relative humidity during the testing not exceeding 50 percent.
- *Damp heat steady state.* 500 h of damp heat at 85 ± 2°C, with relative humidity during the testing at 85 ± 10 percent.
- *Cold.* 500 h of cold at −30 ± 2°C, with relative humidity during the testing not specified.

Mechanical Testing. This testing must provide assurance that the sensor can withstand the effects of storage, transportation, and the final application environment.

IEC guidelines provide model environments, such as would be found in ground, air, or space applications. Suggested severity levels for mechanical testing of feedback devices are as follows:

- *Vibration testing.* Between 10 and 2000 Hz, with an amplitude of gs above 57 Hz. Below this frequency, the motion will be amplitude limited to approximately 0.030 in maximum, with frequency sweep from low to high and back 10 times at a sweep rate of 1 octave/min. This test should be conducted in the vertical and horizontal axes.
- *Shock testing.* Using a half-sine wave form at 100 g for 6 m, 3 shocks in the positive and negative direction for each axis, for a total of 6 shocks.

Responsibility for Test Certifications. If you are involved with the shipment of motion-control products to Europe, the CE mark is now the means through which the European Community will check to see if you have done your homework. Suppliers of products which require the CE mark must not only have designed the units using safe practices, used proper design rules, and validated the designs with proper testing, they must also make the design process records available to anyone who needs them within 3 days of a request.

10.2.2 Environmental Protection

Sealed. Although there are National Electrical Manufacturers Association (NEMA) specifications for many types of devices and enclosures, the IEC specifications seem to be the most common. Tables 10.1 and 10.2 summarize the International Protection (IP) codes. For a complete discussion of these ratings, the specifications IEC 529 and/or IEC 34-5 should be examined.

Exposed. Open-frame tachometers, resolvers, and encoders must be protected by the application equipment from environmental concerns. In the servo industry today, three basic technologies are used in the majority of applications. These consist of sensors using either magnetic, inductive, or optical methods.

Magnetic sensors are of two types: those using ac technology, such as synchros, inductosyns, and resolvers; and those using permanent-magnet (PM) technology, such as magnetic encoders, Hall devices, and the like. They tend to be used in very low cost, low-accuracy applications, or when the sensor must be run exposed to the elements (e.g., in submerged or high-particulate environments).

Inductive transducers, particularly resolvers, are used in extremely rugged environments where accuracy is not of first importance.

Optical encoders are chosen for applications in which accuracy and stability are of primary importance.

The cost of an inductive transducer is generally lower than that of an optical one, but the costs equalize or begin to favor the encoder when interface electronics and overall performance issues are directly compared. Today, integrated circuit (IC) technology and application-specific IC (ASIC) integration capabilities are making the inductive interface circuits more simple, robust, and cost-effective, while manufacturers of optical sensors are using the same methods to lower product part count and overall costs.

The Institute for Applied Microelectronics has developed a two-chip set that will implement the entire drive electronics for a brushless dc (BLDC) motor. The chip set will accept sinusoidal commutation signals and incremental encoder and resolver inputs, and has a small-scale integration (SSI) interface for communication with absolute encoders. When components of this capability become available, system cost will depend exclusively on performance requirements.

TABLE 10.1 IP Nomenclature—Degrees of Protection Indicated by the First Characteristic Numeral

First characteristic numeral	Degree of protection of equipment	
	Brief description*	Definition
0	Machine nonprotected	No special protection.
1[†]	Machine protected against solid objects >50 mm	Accidental or inadvertent contact with or approach to live and moving parts inside the enclosure by a large surface of the human body, such as a hand (but no protection against deliberate access). Ingress of solid objects exceeding 50 mm in diameter.
2[†]	Machine protected against solid objects >12.5 mm	Contact with or approach to live or moving parts inside the enclosure by fingers or similar objects not exceeding 80 mm in length. Ingress of solid objects exceeding 12 mm in diameter.
3[†]	Machine protected against solid objects >2.5 mm	Contact with or approach to live or moving parts inside the enclosure by tools or wires exceeding 2.5 mm in diameter. Ingress of solid objects exceeding 2.5 mm in diameter.
4[†]	Machine protected against solid objects >1 mm	Contact with or approach with live or moving parts inside the enclosure by wires or strips of thickness greater than 1 mm. Ingress of solid objects exceeding 1 mm in diameter.
5[‡]	Machine dust-protected	Contact with or approach to live or moving parts inside the enclosure. Ingress of dust not totally prevented, but dust does not enter in sufficient quantity to interfere with satisfactory operation of the machine.
6[§]	Machine dust-tight	No ingress of dust.

* This description should not be used to specify the form of protection.
[†] Machines assigned a first characteristic numeral of 1,2,3, or 4 will exclude both regularly or irregularly shaped solid objects provided that three normally perpendicular dimensions of the object exceed the appropriate description in the Definition column.
[‡] The degree of protection against dust defined by this standard is a general one. When the nature of the dust (dimensions of particles and, their nature; for instance, fibrous particles) is specified, test conditions should be determined by agreement between the manufacturer and the user.
[§] Not specified under IEC 34-5 for rotating machines.

The sensor configuration of the motor and sensor package chosen depends ultimately on the intended application. Cost is always an important issue, and for BLDC motors, there appear to be five categories of applications.

1. *Low-cost motors for basically constant-speed operation.* Typical examples are fan motors, fuel pumps, and disk drives. These are very high volume, low-cost applications where tooling of molded magnets and Hall structures can be justified. Alternatively, many are doing away with Hall sensors and going to smart IC controls. Control chips made by Allegro Microsystems, Inc., Hitachi America

TABLE 10.2 IP Nomenclature—Degrees of Protection Indicated by the Second Characteristic Numeral

Degree of protection of equipment

Second characteristic numeral	Degree of protection of equipment	
	Brief description*	Definition
0	nonprotected	No special protection.
1	Machine protected against dripping water	Dripping water (vertically falling drops) shall have no harmful effect.
2	Machine protected against dripping water when tilted up to 15°	Vertically dripping water shall have no harmful effect when the machine is tilted at any angle up to 15° from its normal position.
3	Machine protected against spraying water	Water falling as a spray at an angle up to 60° from the vertical shall have no harmful effect.
4	Machine protected against splashing water	Water splashing against the machine from any direction shall have no harmful effect.
5	Machine protected against water jets	Water projected by a nozzle against the machine from any direction shall have no harmful effect.
6	Machine protected against powerful water jets	Water from heavy seas or water projected in powerful jets shall not enter the machine in harmful quantities.
7	Machine protected against the effects of temporary immersion in water	Ingress of water in the machine in a harmful quantity shall not be possible when the machine is immersed in water under stated conditions of pressure and time.
8	Machine protected against continuous submersion	The machine is suitable for continuous submersion in water under conditions which shall be specified by the manufacturer.†

* This brief description should not be used to specify the form of protection.
† Normally, this will mean that the machine is hermetically sealed. However, with certain types of machines it can mean that water can enter but only in such a manner that it produces no harmful effect.

Ltd., Micro Linear Corporation, Signetics Company, Silicon Systems, Inc., and SGS-Thomson Microelectronics, Inc., can provide complete commutation of BLDC motors. Some of these controllers even provide braking and speed control as part of the package, so an external sensor like an encoder or a resolver is not needed for this type of servo application.

2. *Traditional BLDC motors with resolver or encoder feedback.* These are motors which contain an encoder or a resolver for position feedback and possibly a tachometer as well, depending on the control system being implemented. Encoder-based systems also require Hall sensors for commutation. Resolver systems used with a rectangular drive could use Hall sensors as well, but this is usually all that is needed. These types of motors have been the backbone of the BLDC motor industry for the past decade and are found in a wide variety of applications.

3. *Integrated-sensor motors.* These use optical encoders which generate rotor-position as well as incremental-position signals. The rotor-position signals are electrically the same as can be obtained from Hall switches, and they can be used for commutation of two-, three-, or four-pole-pair motors. Integrated-sensor BLDC motors are being used in Japan and the United States to provide high-performance servodrive solutions to cost-critical applications. The encoders are built-in hollow-shaft encoders, and generally come in resolutions up to 13 bits ($2^{13} = 8192$ cpr).

4. *High-performance integrated-sensor motors.* These are used in systems requiring large dynamic range in the speed control (such as z-axis control in a machine tool), very high resolution, or very low speed operation. These are being developed primarily in Europe and are distinguished by sinusoidal rather than TTL output signals.

5. *Smart motors.* These are high-performance integrated-sensor motors requiring additional capabilities such as absolute positioning, bus interfaces, storage for motor data, temperature monitoring, etc. This is currently a very small portion of the market, but it is definitely growing. The sensors for these motors provide commutation outputs, incremental outputs, and up to 25 bits of absolute-position data, 13 bits per turn with 12-bit turn counting.

10.3 FEEDBACK ELEMENTS

10.3.1 Rotary and Linear Incremental Optical Encoders

Optical encoders (Fig. 10.3) can be characterized by the physical measurement principle they use (diffraction or directed light), their design features, and the protection requirements to which they are built. They range from completely enclosed and sealed units to open-frame kit units. They are typically used in velocity- or position-feedback systems such as those found in tape transport equipment, machine-tool spindle controls, bed positioning equipment, woodworking machines, robots, material-handling equipment, textile machines, plotters, printers, tape drives, and a variety of measuring and testing devices. Commercial encoders are generally defined as being capable of measuring angles of up to 30″. For higher resolutions, an *angular measurement device* must be used. These devices are capable of measuring angles as fine as 0.000010° (0.036″).

There are three categories of encoders from an environmental protection viewpoint. *Sealed encoders* are generally protected to the levels of IP 64 or better. These are stand-alone units that have internal bearings and seals and are not intended to allow user access to internal workings. *Self-contained* encoders are not necessarily dust proof. These have internal bearings and are stand-alone units, but some customer access may be possible or may even be necessary during installation. *Modular encoders* are completely open units which rely entirely on the application for protection. These units do not contain bearings. They are sometimes referred to as *kit encoders* or *tach kits.*

Sealed units are the most expensive, and generally are not well suited for high-speed operation because of the seals. However, these can be very high accuracy, high-resolution devices, capable of resolutions ranging up to 10,000 cpr.

Modular encoders are the lowest in cost. These units generally have the best price-to-performance rating, but they require some care on the part of the user as

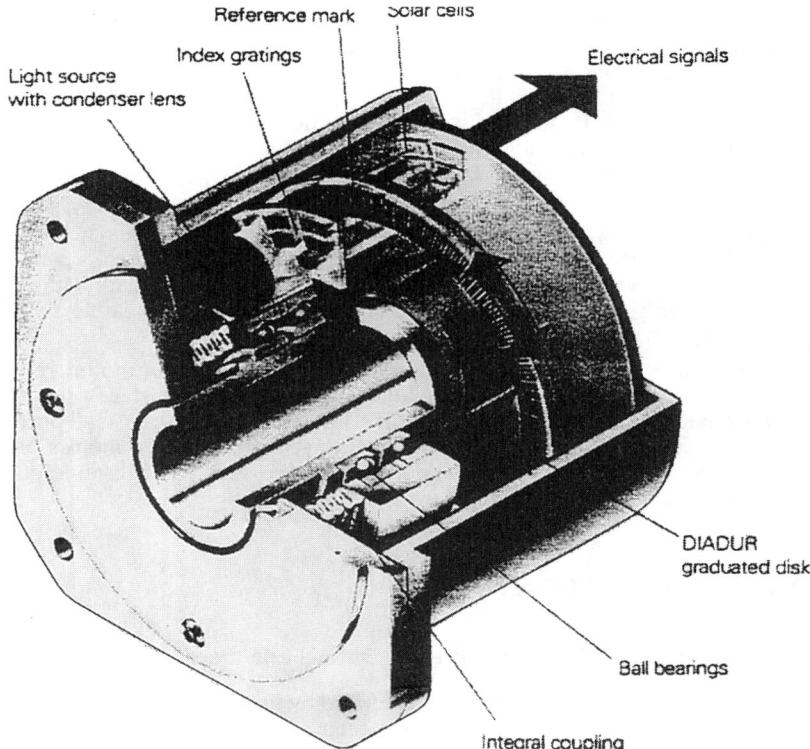

FIGURE 10.3 Rotary optical encoder.

they can be damaged if not installed properly. Modular units are available with resolutions up to 2500 cpr.

Self-contained encoders span the entire performance envelope, at a slightly higher cost than modular devices. The self-contained hollow-shaft encoders are widely used in the drive industry, as they eliminate coupling resonance.

Hollow-shaft encoders are also widely used with integrated commutation electronics. This provides a simplified assembly process to the manufacturer by allowing elimination of the Hall board. This approach also simplifies overall alignment.

Terms
amplitude modulation Using the code wheel and mask as an optical shutter, or to create Moiré patterns to modulate the intensity of light impinging on the photodetectors.
code wheel A circular disk of transparent material with patterns of transmissive and opaque regions equally spaced about the perimeter. Light shining through the clear regions is passed onto the mask. The spacing on the code wheel defines the line count of the encoder. 1024 opaque regions separated by 1024 clear spaces will create a 1024-cpr encoder.
disk Another term for **code wheel**.

grating A pattern of closely spaced lines which is used to shutter light passing through the code wheel.
index See **reference mark**.
mask A glass plate mounted on the encoder housing so as to remain stationary with respect to the rotating code disk. The mask supports the optical gratings or patterns.
Moiré patterns When light is transmitted through a set of gratings that are equally spaced but at a slight angle to each other, patterns of brightness and dark are created. These patterns are called Moiré patterns. As the gratings are moved relative to each other, periodic brightness fluctuations can be seen.
phase modulation Using a reflective mask with a stepped grating pattern to modulate light impinging on the photo-detector via constructive and destructive interference.
phase plate Another term for **mask**. More appropriately used when referring to encoders using phase modulation of light rather than amplitude modulation.
reference mark A once-per-revolution output that is one period wide.
reticle Another term for **mask**.

Principles of Operation. The basic components of rotary optical encoders are as follows (see Fig. 10.4):

- Light source, which can be a lamp or a light-emitting diode (LED)
- Collimating (condenser) lens to improve light power density and reduce diffraction effects
- Code wheel
- Mask
- Signal detectors
- Output-conditioning circuitry

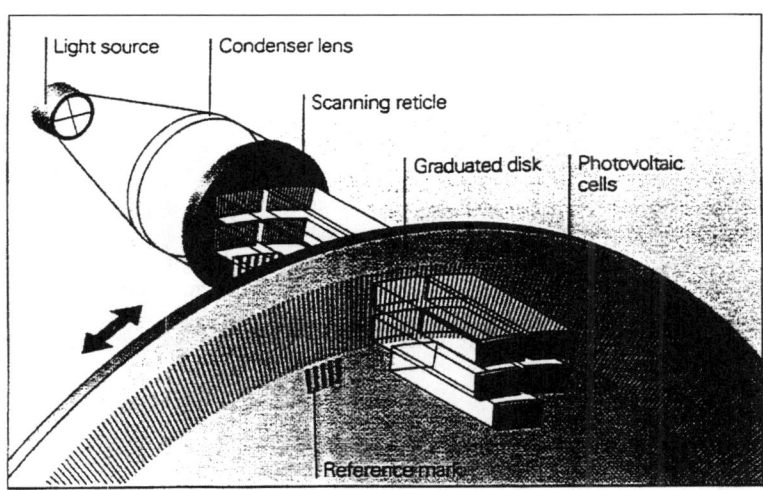

FIGURE 10.4 Directed light scanning.

The sensor operation results from photoelectrically scanning very fine gratings on the disk. A disk with a radial grating of lines and gaps serves as the measuring standard. The opaque lines can be made using a number of methods, such as plating chromium onto the glass. The lines are placed so that the spacing between lines and gaps is equal, and the lines are spaced uniformly around the circumference of the disk, so as to make a circular graduation.

In close proximity to the rotating disk is a scanning reticle, with grating fields for the data channels and one or more fields for the reference mark. The data-channel windows are placed onto the scanning reticle such that they are phase-shifted in relation to each other and the graduation pitch by one-quarter of the grating period. All of these fields are simultaneously illuminated by a beam of collimated light. As the graduation rotates, the light is modulated onto the sensors, and the sensors then output two sinusoidal signals with a 90° phase shift between them.

Reference Mark. The reference mark is created by a peripheral set of gratings in tandem with or adjacent to the main data windows. Sometimes the reference is made by a constant light source outside the code wheel and a single window in the mask area. The reference mark produces a single pulse that is one period wide. The reference can also be digitally combined with the main quadrature signals so that it is active only during a specific portion of the quadrature cycle. This is called *gating* the reference mark.

Typical Resolution. Commercial encoder products are available with resolutions up to 10,000 cpr. Above this value, different techniques must be used in the design and manufacture, increasing overall cost significantly.

Methods of Fabrication. Rotary optical encoders can be constructed using either amplitude modulation (AM) or phase modulation (PM) techniques, but AM is far more common due to the lower cost of manufacture.

PM methods are used for very high resolution devices which might be found on the z axis of a machine tool or, more commonly, in a linear optical encoder used for measurement equipment. For a discussion of PM methods, refer to subsec. 10.3.3, Linear Optical Encoders.

Graduations. Three major materials are used for manufacturing the mask or disk graduations:

- Chrome on glass
- Estar-based film or photoplastic
- Metal

The disk graduations can be made by either expose-and-etch processes or plate-up processes. Expose and etch is very similar to processes used by the printed-circuit board industry. The plate-up approach was developed by Dr. Johannes Heidenhain, GmbH, and is called the *Diadur process.*

Plate-up processes yield much better edge quality, but require extensive investment by the manufacturer. Etch processes utilize the same materials and techniques developed for the semiconductor industry, and so require very little investment on the part of the manufacturer to implement. Etch processes are used exclusively for graduations on photoplastics and metal.

The expose-and-etch printing method is as follows:

1. The graduation is produced by placing a master plate against a blank plate that has been coated first with chrome and then a photoresist.
2. The blank is exposed using a high-intensity ultraviolet (UV) light.

3. The exposed blank is then developed and etched. The etching process produces a duplicate of the master image.
4. The disks are cut from the blank, cleaned, and are then ready to be installed.

Centering Process. The centering process for placing the disk onto the encoder shaft is crucial to the performance and accuracy of the encoder. The graduation must be placed as precisely as possible with respect to the rotational axis of the shaft or hub. Typically, the concentricity of the disk pattern to the rotational axis must be better than 0.0004 in (10° μm).

Light Sources. Light sources can be incandescent or solid state (LED), depending on the environmental constraints and cost targets.

Solid-state light sources are used more predominantly due to their long service life (in excess of 100,000 h). They also have excellent resistance to shock and vibration. However, because they are silicon devices, they are limited to junction temperatures of approximately 150°C. This results in limitations on their use at high ambient temperatures. The output of LEDs also drops about 1 percent per degree Celsius, so use at higher temperatures must be evaluated carefully.

Incandescent illumination sources are used when environmental temperatures are extreme, 125°C or higher, due to their ability to withstand a higher ambient temperature (up to 200°C). They also have about twice the output of LEDs.

Most sources provide light in all directions, most of which will not fall on the detectors. To improve this situation, a collimating lens is used. Collimation gathers the light and focuses it at a point at infinity. The result is a parallel beam which can be precisely directed at the photoelements. This provides three improvements:

- It serves to combat the intensity loss due to the inverse-square law.
- The reduction in scattered light reduces crosstalk and noise at the detector.
- When parallel light passes through the disk-mask "shutter", there is less leakage due to stray light, which results in better modulation and more useable signal from the detectors.

PhotoDetectors. There are three primary types of photodetectors used in optical encoders. These are the *solar cell* or *photovoltaic device,* the *photodiode,* and the *phototransistor.*

- *Photovoltaic devices.* These are solar cells, or photodiodes being used in a photovoltaic mode. These devices generate electricity when light impinges on the detector surface. They do not require external power. When connected to a load, the voltage potential created by the illumination results in the generation of a current. These devices have a very broad spectral response, and are particularly sensitive to the infrared region. They have excellent frequency response and are resistant to most environmental contaminants.

- *Photodiodes.* By connecting a photodiode anode to a power supply, and the cathode to a load resistor, the photodiode operates in the photoconductive mode. In this mode, the device acts like a valve which controls the amount of current flowing through the resistor from the voltage source, depending on how much incident light is present. These devices retain the excellent frequency response characteristics of the photovoltaic devices, and generally need less detector surface area to develop equivalent output signals.

- *Phototransistors.* These devices trade off the frequency response of photodiodes for increased output levels. Phototransistors can generate significant output voltages (>1 V), which makes them superior for use in noisy environments. However, they are significantly slower to respond than photodiodes, which results in reduced frequency response of the encoder. Phototransistors can also be implemented in less area than can photovoltaic devices. Significant signals can be generated for a device as small as 0.021 in^2 (0.5 mm^2).

Signal Conditioning. Figure 10.5 diagrams a typical single-sided-supply photodiode sensor arrangement. This circuit uses comparators to create square-wave quadrature output signals with 50 percent duty cycles.

The selection of values for R_2 and R_3 controls the amount of hysteresis in the circuit, while the values for R_1 and R_4 control the photodiode output-signal levels.

Balance Adjustment. To develop a 50 percent duty cycle at the output of the comparator, the input offset levels must be identical. This will never occur naturally for a number of reasons, but primarily because the amount of light shining on the two detectors will never be exactly the same, and the photodetectors will not have exactly the same characteristics. Most encoders therefore require adjustment as part of the final manufacturing process. This can be accomplished in the following ways.

- *Shading screws.* By physically blocking light to one or both of the complementary sensors, their outputs can be matched. This is a robust process, but somewhat slow and difficult to automate.
- *Analog and digital pots.* By replacing one of the load resistors with a potentiometer, the input voltages to the comparator can be made equal. This process is very easily accomplished, but the different resistance values can cause problems over temperature, and potentiometers decrease the overall reliability. Digital potentiometers can be adjusted via a computer interface, which makes this approach very amenable to automation.
- *Test-Select.* This process selects fixed-load resistor values at the final test of the encoder. This is very time consuming, but the fixed values are more stable than a potentiometer. This process is moderately difficult to automate.

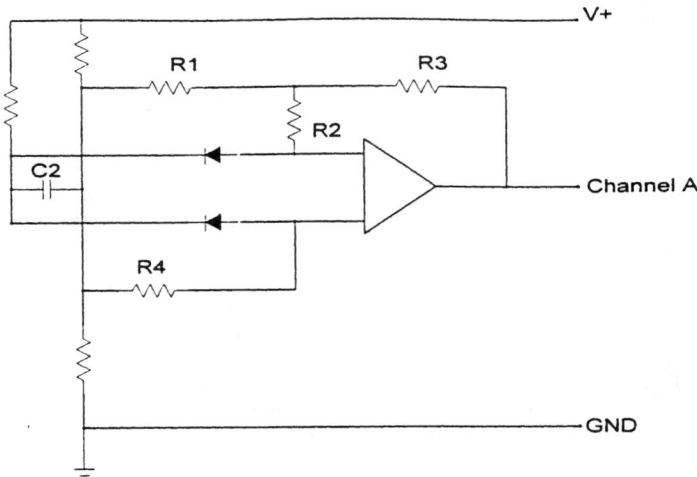

FIGURE 10.5 Optical encoder signal conditioning.

- *Balanced sensor array.* This process uses an interdigitated pattern of detectors to eliminate the need for balance adjustment. The distribution of sensors throughout the area of illumination results in an overall averaging which balances the outputs. This process is very efficient to manufacture as it eliminates adjustment entirely. However, each resolution must be tooled uniquely, which makes the capital investment in this approach very high and causes long lead times when new resolutions are needed.

Output Signal Qualities. Signal processing of the sinusoidal outputs is handled in two ways, either as analog information or as digital information.

Analog outputs come in a number of flavors, the most basic form of which is to supply the raw sensor signals. These are low-level signals in the microamp range, which must be carefully shielded and cannot be sent over long distances. The next most common analog output is to provide simple amplified signals. These can be implemented as dc-biased ac signals, which can be driven by a single-sided power supply, or as amplified zero-referenced signals using a dual power supply. Amplitudes of either approach are somewhat user driven, but 100 mV for the single supply and 2.5 V peak to peak for the amplified approach would be common values. The third form is very common in Europe, and is termed the 1-V peak-to-peak output. This output is guaranteed to hold this level (+0/−3 dB) over the rated frequency range, which can be as high as 200 Hz. These units are also capable of driving significant lengths of cable, and the constant level sinusoidal signal is excellent for use with interpolation electronics.

Digital signals also come in some variety. When the encoder produces quadrature outputs, the signals can be formatted as TTL, HTL, line-driver, high-voltage line-driver, complementary metal-oxide semiconductor (CMOS) line-driver, buffered, and open-collector variations. In any case, the signals are digital in nature, switching between ground and the supply or high-voltage value determined by the application. The goal of these outputs is to retain pulse width and symmetry over all frequencies and temperatures. These signals cannot be interpolated, although edge counting of the quadrature signals is common. Another version is direction sensing. These signals are usually either TTL or HTL, but anything is possible and has probably been sold at one time or another.

Accuracy and Resolution. The resolution, or measuring step, of an encoder is the angle corresponding to the distance between two edges of the square-wave pulse-train output. Basically, this is one-quarter of the grating period. Accuracy can usually be approximated as 5 percent of the grating period for resolutions up to 5000 cpr. Between 5000 and 10,000 cpr, accuracy is basically constant at approximately ± 12 arc sec. (Dr. Johannes Heidenhain, GmbH). There are many texts discussing this issue (*Electro-Craft Handbook,* 1980; Ernst, 1989). Error consists of intrinsic instrument errors in the encoder, plus system errors.

System errors are due to the following causes:

- *Hysteresis effects.* The amount of hysteresis used to control noise will effect overall accuracy, as this changes the switching point of the output in a TTL system and introduces phase lag in an analog system.
- *Runout due to eccentricity of the disk and hub assembly with respect to center of rotation.* Eccentricity errors are created by manufacturing process accuracies associated with putting the disk on the hub, tolerance between the hub and the motor shaft, bearing runout, and the accuracy of the pattern itself. This type of error will result in amplitude modulation of the output A, approximated by

$$R = 1 \text{ in}$$

$$\Delta R = 0.0005 \text{ in}$$

$$\Delta A = \frac{\Delta R}{R + \Delta R}$$

$$\Delta A = 0.05\%$$

- *Surface runout.* This is either due to poor mounting of the disk to the hub flange or, in a modular encoder, due to tolerances between the hub and motor shaft of the motor shaft runout. All of these can result in variations in the gap between the disk and the mask. The angular error due to shaft runout (arc minutes) can be approximated as follows:

$$60 \times \sin^{-1} \frac{\text{TIR}}{R_t}$$

where
TIR = motor shaft runout
R_t = nominal data track radius

For a 0.75-in track radius, this is 0.458′/0.0001 in.

- *Pattern errors which cause both amplitude and frequency variation errors.* Frequency errors appear as "flutter" on an oscilloscope. This is caused by irregular spacing of the opaque patterns on the code wheel. These errors can result from errors in master generation or from printing errors. Many times, these errors will be cyclic, occurring every 45 or 90° mechanical. These errors result from certain types of master generation processes in which a section of the disk pattern is stepped and repeated to make the entire 360° pattern. Other errors occur with pattern-generation equipment, called *closing errors*. These occur when a small error results over the 360° printing cycle, so that the last line generated is slightly larger or smaller than all the rest.
- *Jitter.* This can occur when the alignment of the elements in the optical path is incorrect, the illumination source is poorly collimated, or contamination is present on the disk or mask surface.
- *Sensor output drift.* Most encoders use a push-pull configuration to minimize the effects of detector changes, light variation, and voltage variation. When the sensors drift out of balance with each other, symmetry in the quadrature output will change.

Interpolation. There are many methods of developing higher-resolution TTL outputs by processing the analog sinusoidal signals developed in the measurement system. One consists of developing phase-shifted copies of the original signal using resistor networks. Taking advantage of the relationship

$$\sin(\alpha + \phi) = \cos\alpha \sin\phi + \sin\alpha \cos\phi$$

the base sinusoidal signals $\sin\alpha$ and $\cos\alpha$ are multiplied by phase-shifted copies. For example, a 5× interpolator would use 5 sets of signals, each shifted 18°. The results are converted into square waves via comparators, and all of the outputs are routed through an exclusive OR gate. The result is a set of square waves in quadrature at a frequency equal to 5 times the original. Figure 10.6 shows how interpolation of 5× would compare with the original output.

Interpolation of this type can be used for multiplication up to 25× with reasonable success. Higher subdivisions are obtained using digital methods. One such

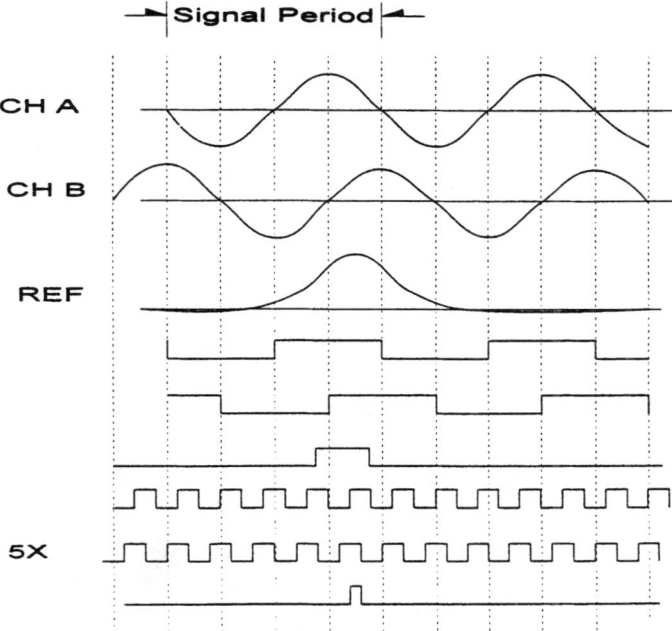

FIGURE 10.6 Interpolation of 5× compared to original output.

method computes the arctangent using the values of the two analog quadrature signals as the sine and cosine values, then uses table look-up methods to determine the corresponding angle. Quadrant detectors complete the calculation. Another method makes readings of the analog values at two discrete times, and then creates an artificial pulse train to get between the two at the desired resolution. Similar methods are used for resolver-to-digital converters, and are discussed in that subsection.

Application Considerations

Environment. Encoders are very robust sensors, but they need to be selected for the intended environment. The main limitation in the application of encoders is temperature. Most commercial encoders are rated at 85°C or lower. Industrial ratings increase this to −10 to 100°C. Severe-environment encoders operate up to 125°C. Shock and vibration are rarely a problem. Even though many encoders utilize glass disks, these assemblies are very robust and can withstand most military levels of shock and vibration. In fact, it is very difficult to damage an encoder mechanically and not damage the motor it is mounted on.

Interface Requirements. In any encoder application, it must be decided what signal levels are needed for interface with the controls, what type of circuitry the encoder will be connected to, what frequency response is needed, and what type of signal will be sent through the cable, as well as mounting and coupling requirements

Slew Rate. The encoder slew rate is limited by either mechanical or electrical considerations. Mechanical limits are encountered when bearing limits are exceeded, or when testing has shown that the assembly is not capable of remaining intact under the rotational stresses. Electrical limits are encountered when the input frequency from the sensors to the signal conditioning circuit exceeds the response capabilities of that circuit. This relationship is stated as follows:

$$n_{\max} = \frac{f_{\max}}{z} \times 10^3 \times 60 \text{ rpm}$$

where f = scanning frequency, Hz
z = encoder line count, cpr

Figure 10.7 shows how frequency response, encoder resolution and input rpm are related.

Interconnection. Applications in very noisy environments, or which must drive long cables, should use differential line drivers. Shielding and grounding are also

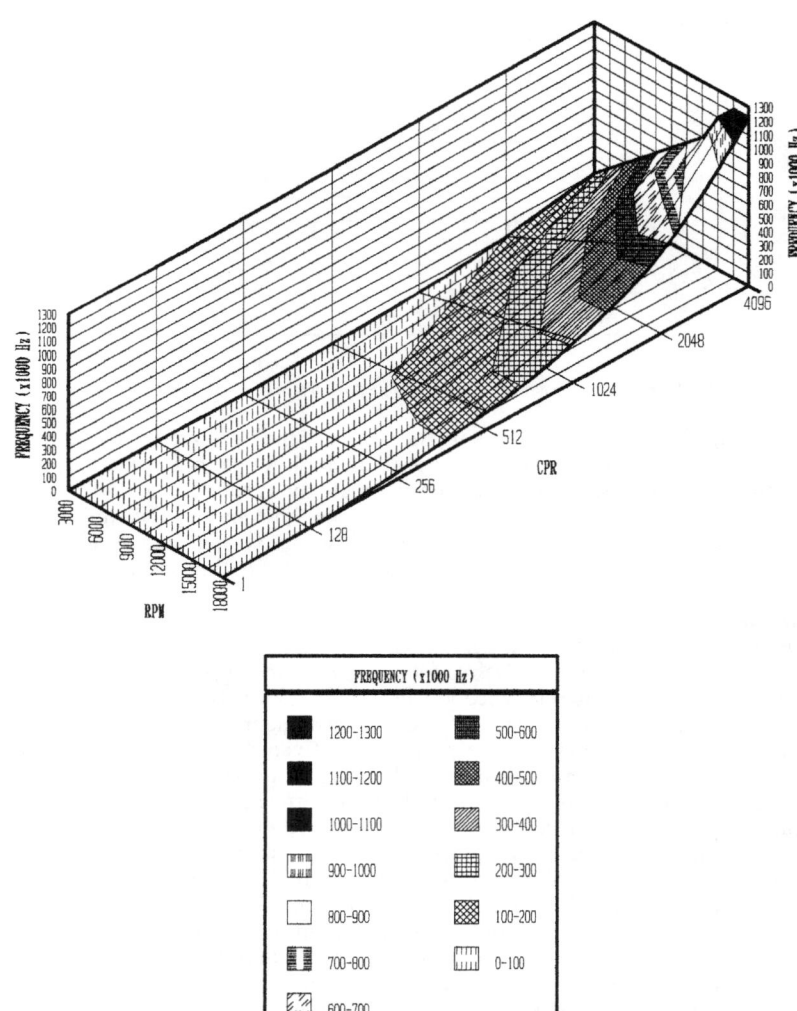

FIGURE 10.7 Frequency response capability.

important, but this can also drive sensor cost dramatically. Cable length, environmental protection, and signal types all play together. Long cables in a high-noise environment are best dealt with using amplified sinusoidal signals with current drivers, feeding into a line receiver. For cables less than 100 m in length, TTL signals can get by, but care should be used at this distance. For best control, cable shields should be tied to the control, and the control to ground. In Europe, it is also desired that the encoder case be tied to the cable shield and that the power ground remain isolated. Some examples of suggested interface circuits are shown in Fig. 10.8.

Mounting Requirements. There are several standard mounting patterns. For hollow-shaft encoders, there are also various styles of spring-plate adapters, which are very important to the performance of the installed device. These couplings must be designed to allow for high torsional rigidity, while being compliant in the axial direction. It is a design goal for these couplings to have a natural frequency exceeding the application bandwidth by a factor of 10. For example, a system with a planned servo bandwidth of 100 Hz should have an encoder flex-coupling mount with natural frequency of >1000 Hz.

Motor End-Play. A modular unit will require approximately ±0.010 in of motor shaft end-play to maintain disk integrity. Hollow-shaft encoders with flexible mounting plates can usually accommodate as much as ±0.040 in.

Power Supply Constraints. Because encoders utilize LED or incandescent illumination, they can draw significant amounts of power. It is not uncommon for an encoder to require 250 mA or more in a high-temperature brushless servo application. The designer should check to be sure that the drive system has sufficient power to support an encoder application. This is especially important in commutation encoder applications, in which a system that was designed to support Hall sensors is now being connected to a commutation encoder. Most power supplies for Hall sensors are very low wattage units, and they may not be able to support the encoder requirements.

10.3.2 Single and Multiturn Absolute Rotary Encoders

Absolute encoders are manufactured in exactly the same manner as incremental encoders. The main difference is that more sensors are used, and so they are more complex than incremental encoders. The overall complexity depends on the number of bits, or word size of the encoder—the more bits, the more complex and expensive. They are used where motion can occur when power is removed, such as to provide level control or fail-safe operation. Machine-tool and robotics applications are the primary users of these devices.

Principles of Operation. An absolute encoder uses one track of the code disk for each bit in the output. Therefore, an 8-bit absolute encoder has 8 tracks on the disk and requires at least 8 sensors to detect light passing through these tracks. Depending on the size of the encoder, the sensors, and the tracks, it may be necessary to use multiple sources of illumination to assure adequate signal levels.

The data tracks can be encoded to provide position information in a number of ways. One method is to encode the data as pure binary information. In this approach, each track is equal to a power of 2. One disadvantage of this approach is that it requires many simultaneous bit transitions. For example, when counting from 15 to 16, 4 bit transitions are required simultaneously.

$$15 \Rightarrow 01111 \text{ binary}$$

$$16 \Rightarrow 10000 \text{ binary}$$

10.20 CHAPTER TEN

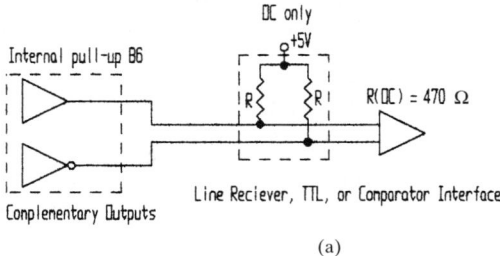

(a)

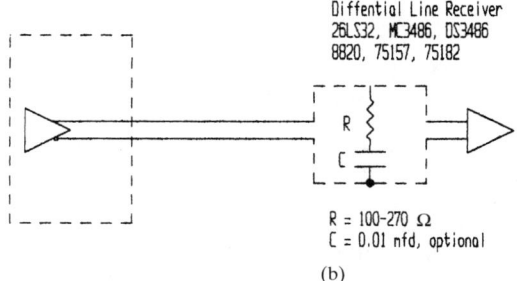

(b)

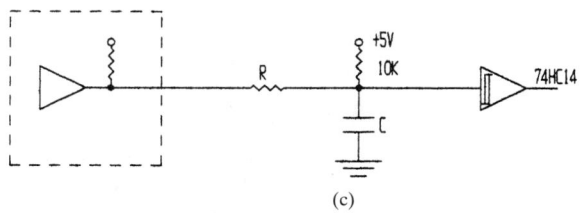

(c)

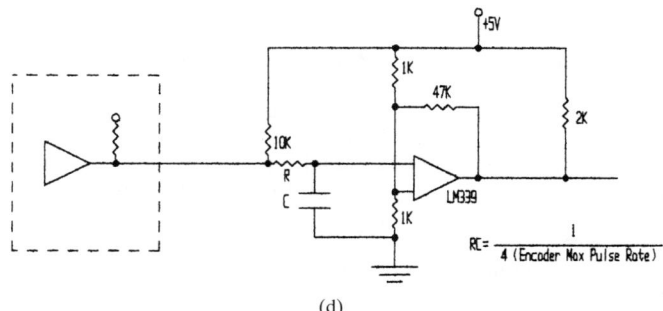

(d)

FIGURE 10.8 Interconnection schematics: (*a*) TTL buffered output (7404/7406), (*b*) RS-442 line driver, (*c*) voltage comparator, and (*d*) voltage comparator with improved noise immunity.

This situation is distinctly unique to the absolute encoder, as it cannot occur with an incremental device. The binary code is termed *polystrophic* because of this characteristic of multiple bit changes.

Polystrophism is a problem because in a real-world situation, all these bits will not change simultaneously. There will be some slight ambiguity, for however small a time, which will result in the possibility of the encoder generating incorrect outputs. All of the problems associated with the manufacture of an accurate incremental encoder apply here, compounded by the number of data tracks being implemented. Hysteresis, eccentricities, noise, and so forth can all add up to slight variations. Were the bit error to occur in the most significant bit (MSB), the user could receive a feedback signal that is in error by 180°. Encoder manufacturers have developed specialized scanning methods, called *U-scan* and *V-scan,* to orchestrate the transitions of the many bits simultaneously. V-scan uses the least significant bit (LSB) to determine which direction the scale is moving—that is, is the bit transition from high to low or from low to high. The sensors are arranged in two banks, in a V shape, the distribution allowing for tolerances in the system (Fig. 10.9). Once the direction is determined, logic selects the correct side of the V to obtain the reading without transition error.

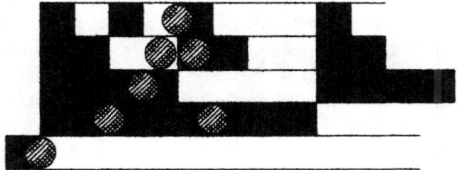

FIGURE 10.9 V scan.

Although proper design can result in the successful implementation of an absolute encoder using binary encoding, the problem is real enough that many other codes have been developed. The Gray code is a monostrophic code. This is a very popular code which allows only one bit change between any two monotonic values. Table 10.3 shows the difference between decimal, binary, and Gray coding. Once the values are read by the computer, they can be readily translated into whatever form is most appropriate.

TABLE 10.3 Differences in Monotonic Values

Decimal	Gray code	Binary
0	0000	0000
1	0001	0001
2	0011	0010
3	0010	0011
4	0110	0100
5	0111	0101
6	0101	0110
7	0100	0111
8	1100	1000
9	1101	1001
10	1111	1010

Note. strophe (from Greek, act of turning; to turn; to twist; action of whirling): The movement of the classical Greek chorus while turning from one side to the other of the orchestra (*Webster's Seventh New Collegiate Dictionary,* 1971).

Methods of Fabrication. A single-turn absolute encoder can generally be produced with up to 14 bits of position information. 14 bits results in 16,384 unique positions per revolution of the encoder. In many cases, this is not enough. For a machine tool, where the bed must traverse several feet and the absolute encoder is connected to the lead screw, each rotation is unique, as well as the angle within each rotation. To accommodate these requirements, multiturn encoders have been developed. Typical multiturn absolute encoders provide 13 or 14 bits per turn, and up to 12 more bits for turn counting. The combination provides up to 26 bits of absolute position data. Even if the resolution per bit were 0.000007 in, this would allow for over 39 ft of absolute position control.

The manner in which turn-counting is implemented determines the cost of the device. The least expensive approach is to use a battery backup for the encoder. The disadvantage of this approach is that, during power loss, the battery must also energize the encoder so that information will not be lost if movement occurs during this event. Because the LED can be a significant drain on the battery, an encoder like this can usually not last more than a few days before power must be restored or information is lost. Many companies have developed ingenious methods to improve the battery life for these devices, and they are widely used throughout the industry.

The most robust multiturn absolutes are built using gearboxes driving additional code wheels for turn counting. By continually gearing down the output shaft, and using this gearing to drive smaller encoders, an additional 12 bits of information can be obtained. The multiple encoder outputs must be carefully combined, using overlap bits, to ensure that transition errors will not occur. Of course, these devices are complex and require that precision mechanical components work properly. They are available from a number of manufacturers.

Application Considerations. Because of large output word sizes (up to 26 bits), absolute encoder interfaces have developed many interface methods. For word lengths up to 10 bits, parallel interfaces are used. All 10 bits, and sometimes an additional quadrature channel, are provided via direct wiring. For larger word sizes, this is not practical. For these encoders, there are typically two forms of interface. Since the encoder is used as the primary feedback device during operation, a standard incremental encoder is provided with standard wiring. When used as an absolute reference at power-up, some of the databus system is used to pass the longer digital value over to the main controller. This eliminates the need to handle long cables with many wires. Once the drive has been initialized and begins operation, the incremental encoder interface is used exclusively.

10.3.3 Linear Optical Encoders

Linear optical encoders are no different from rotary optical encoders. However, their form factor and the way they are used result in some differences in the typical manufacturing processes and end-user handling.

Linear optical encoders are available in lengths from several centimeters to hundreds of meters. In their most basic form, they are comprised of a graduated scale, a read head, and mounting hardware. The read head contains the illumination source, the scanning reticle, and the signal-conditioning electronics. The scale can be made of glass, steel, or plastic. Linear encoders are found in a wide variety of applications, and because of this, there is a need for various types of environmental protection, just as is the case for rotary encoders. However, because the linear encoder must of necessity include a large opening over its entire length for the read head to exit, sealing and protection methods are quite different and not as robust as for rotary encoders. Like rotary encoders, linear encoders have frequency response limitations. However, these are defined as meters per minute or feet per minute rather than revolutions per minute. Unlike rotary encoders, linear systems are usually found on machine-tool beds and measuring systems, neither of which are normally subjected to ambient temperature extremes. For this reason, they are generally limited to operation over lower temperature ranges. Linear optical encoders are capable of very high resolution, in some cases rivaling that of laser interferometers. They are far more accurate than similar devices using magnetic or inductive systems, as their grating periods can be much smaller and they have superior interpolation accuracy.

Terms
Abbe error Measuring error caused by guideway imperfections and the distance between the tool point and the scale. This results from deviations between the linear scale straight axis and curvatures in the machine tool.
carriage The framework which connects the read head to the scale.
read head The movable portion of the scale containing the signal-conditioning electronics, illumination source, and scanning reticle.
response threshold Error which results from hysteresis and backlash as a result of a directional change.

Principles of Operation. Linear encoders can be manufactured to use either the directed-light principle or the diffracted-light principle. When grating periods of less than 8 μm are employed, the diffracted-light method must be utilized. Figure 10.10 depicts a linear encoder scanning mechanism which senses movement by diffraction and interference techniques. Note that only three photodetectors are rewired because of the use of the interference mechanism. As the plane wave of light generated by the collimating lens passes through the transparent scanning reticle, it is diffracted into three directions. At the phase grating of the scale, the light is reflected and diffracted again. The diffracted light returns back through the scanning grating and is diffracted a third time, resulting in three interfering unidirectional light beams. These are collected through a lens and projected onto the photodetectors.

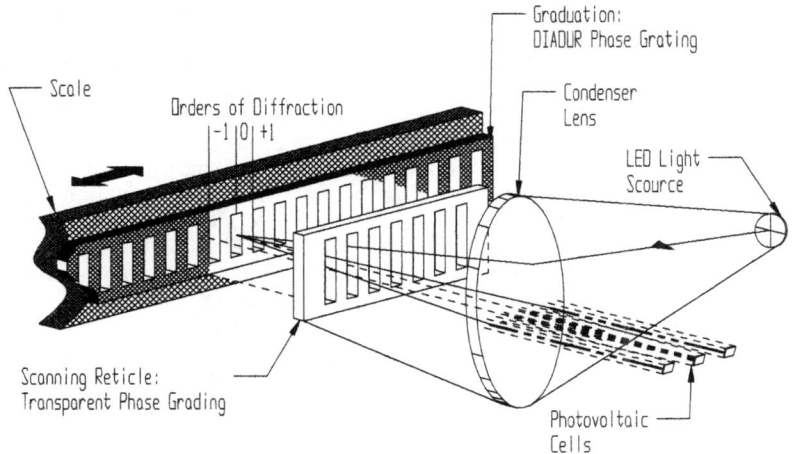

FIGURE 10.10 Scanning using diffraction and interference.

Scales using this technique can achieve measuring steps down to a few nanometers and can be as accurate as a laser interferometer when temperature and atmospheric errors are accounted for. It is of interest that although these devices are quite sensitive to angular alignment of the scanning reticle to the scale, they can be relatively insensitive to gap. This is not true for scales using the directed-light principle, in which gap must be very tightly controlled at small grating pitches or diffraction effects will destroy the signal.

Reference Mark. Most linear encoders contain at least one reference mark. Since some linear scales are quite long, it can be awkward to attempt to find this mark when the system is started or when the power has been lost. To minimize this problem, linear scales sometimes use *distance-coded* reference marks. In this approach, many reference marks are used. The distance between every other mark is constant, but the distance between any two will vary by a line width. In this way it can be known what section of the scale is in use and how far it is to the last mark, so the position is absolutely determined. This method can reduce the seek motion to 100 mm or less, instead of the entire scale length.

Methods of Fabrication. Linear scales can be glass, metal, metal tape, or Mylar, and can measure over inches or feet.

Scales. Typical grating pitches are 10 and 20 µm for encoders using the transmitted-light principle. For protection against contamination, the glass graduation is mounted within an aluminum extrusion, which is sealed to the environment with lip seals. At these grating pitches, the diffraction of light is significant, so it is important to maintain a precise gap and alignment between the carriage and the scale. The mounting of the glass scale to the housing is done with an elastic compound so that thermal differences can be accommodated. The housing is mounted firmly to the machine at its midpoint, with elastic blocks at the ends. This also is done to allow for thermal differences between the scale and the machine it is mounted on. The maximum glass scale length is 3 m in a single piece.

Steel scales can be manufactured in any length, and they are designed to use reflective techniques. Highly reflective gold is plated onto the steel scale, with opaque etched spaces defining the grating pattern. The typical pitch for steel scales is 40 µm. Resolution using interpolation of the 40 µm pitch can be as good as 0.2 µm (200×). For long sections, the scale is supplied in sections which are assembled at the site.

Interferential measuring systems can be made of steel, and a reflective steel phase grating is used to define the graduation. An 8-µm grating results in a 4-µm signal period, which can be interpolated up to 400 times to produce a 0.01-µm measuring step. This is in the same range as a laser interferometer. In some cases this is a better system, because the steel scale is thermally matched to the steel workpiece, so it will better track the machine tool than would a laser interferometer.

Read head. There are two basic methods for controlling gap. One method is mechanically simple, and involves coating the glass scale with a friction-resistant coating. The carriage is then allowed to ride in contact with this coating, allowing very close gaps with good consistency. Because the gap is based on the thickness of this coating, it is very important to apply it in a manner which promotes a constant film thickness. Encoders of this design are termed *contacting encoders*.

Another method is to use ball-bearing rollers to support the carriage of the read head above the glass scale surface. Although the rollers contact the scale, there is no interaction in the region where light is being transmitted or reflected. For this reason, these are termed *noncontacting encoders*.

In both cases, angular alignment is provided by ball-bearing rollers riding on the outer edge of the scale.

Light source. The light source is designed to illuminate as large an area as possible so that the photodetectors will average the light and eliminate any problem resulting from contamination or scale imperfections. With a 10-μm pitch, several hundred lines can be averaged to develop the detected signal. Typically, light sources are LED type.

Output Signal Qualities. Output signals are equivalent to what is available for rotary optical encoders, but TTL outputs tend to be only of the line-drive type (RS-422). Analog outputs are either amplified sine-wave or current outputs of the 11-μA peak-to-peak type.

Accuracy and Resolution. Linear encoders are capable of accurately measuring 1 μm/m, [1 part per million (ppm)]. An absolute accuracy of 0.5 μm is readily available as well, but not as common. The major source of error in a system using linear encoders is the Abbe error. Abbe error can be compensated for by calibration of the machine after the scale has been installed.

Typical resolution for scales using a 10-μm pitch is 0.25 μm, using a 10× interpolation and edge transitions as the measuring step.

Application Considerations. Although there are no real standards, linear encoders have consumer-oriented requirements that become industry standards. Most scales limit traverse rates to 30 m/min due to frequency-response limitations. Traverse rate also has an impact on the distance design life, depending on the type of scale. Contacting scales have design lifetimes of >1 million ft. Bearing systems can exceed this, but bearings have lifetimes as well. The user should consult with the manufacturer for this information should this issue need to be addressed.

Flatness of the mounting is very important to preserving accuracy. Because of this, many scales can be significantly more troublesome to use than others. The user should evaluate the bracketry and adjustments available in the scale mounting hardware for ease of use and practicality. It will do no good to have a scale capable of 0.5 μm accuracy if the installation is good to only 5 μm.

10.3.4 Magnetic Encoders

Magnetic encoders have many useful features. They have lower power requirements than optical encoders, a simple and robust structure, good performance characteristics, excellent resistance to humid and dirty environments, and they are well suited to large-volume manufacturing techniques. Some of the disadvantages are sensitivity to temperature effects, and generally lower resolution capabilities.

Terms
gauss (G)   Unit of magnetic flux density. 0.05 T = 500 G.
tesla (T)   Unit of magnetic flux density in the SI system of units. 1 T = 1 Wb/m^2.

Principles of Operation. Magnetic encoders utilize a magnetoresistive (MR) sensing element and a magnetic code wheel (Fig. 10.11). The MR sensor changes resistance in the presence of a magnetic field. MR sensors are slightly more sensitive than Hall devices. MR elements can sense fields of approximately 0.005 T (50 G). This is at the low end for Hall devices. This is necessary because flux density is proportional to the pole width, so as line count goes up, density goes down. High-resolution magnetic encoders result in very modest magnetic flux densities.

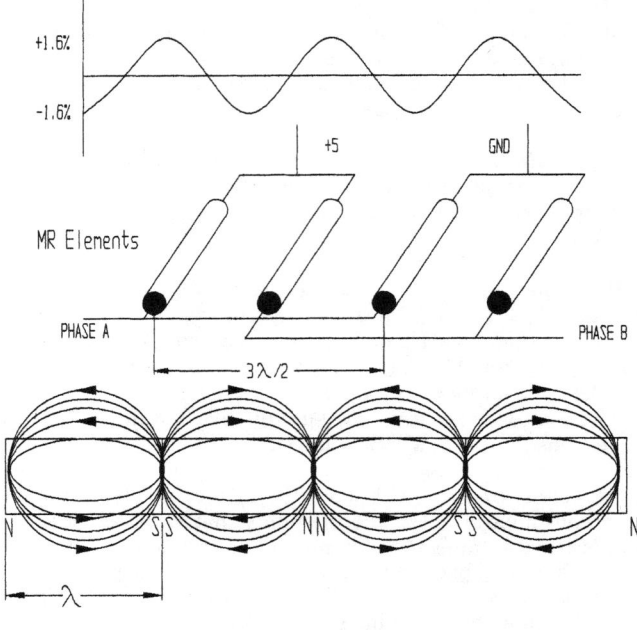

FIGURE 10.11 Magnetoresistive (MR) sensor operational principles.

As the magnetoresistive elements pass through the magnetic field of the code wheel, the material resistance changes by approximately ±1.6 percent, depending on the polarity of the field. The sensor is constructed with an aspect ratio that is much wider than it is high. A change in resistance is observed when flux passes across the width, but when the MR element is over a magnetic field that is normal to the conductor, no change in resistance is observed. When connected as shown in Fig. 10.11, if a 5-V potential is applied to the element, the output at phase A or phase B will vary by 0.040 V peak to peak. Because both magnetic poles affect the sensor in the same way but with opposite polarity, when a code wheel of radius r is rotated through an angle $\tan^{-1} r\lambda$, the output will complete one electrical cycle.

Methods of Fabrication. Magnetic encoders consist of a sensor element and a code wheel. In many cases, these are provided to the end user as separate components, and the application must provide suitable mounting structures.

The code wheel is often constructed as a cylinder, with magnetic poles recorded on the outer circumference. This allows the sensor to be placed along the outside surface of the drum. This is a major advantage for this type of sensor in that it provides for substantial motor end-play allowance. This is a benefit to motor manufacturers, and will sometimes allow lower-cost motors to be used. The code wheel is generally made by injection molding of an isotropic magnetic material which has been mixed with a plastic carrier, such as nylon or polycarbonate. It can be made in arbitrarily large or small sizes, depending only on the mechanical structure required to support it. For this reason it is well suited for large through-shaft requirements, such as in spindle motors.

The molded code wheel is magnetized in the same way as a recording tape. The wheel position is carefully controlled while a series of magnetic pole pairs are recorded onto the magnetic media. This provides a very flexible manufacturing approach. Any line count can be created by simply recording a different "song." The minimum width that can be recorded is proportional to the coercivity of the magnetic material, so small pole pitches require low-coercivity magnets. Isotropic ferrite is most often used for high resolution.

The MR sensor is made by sputtering a magnetoresistive substance, such as Ni-Fe permalloy, onto an insulating substrate, usually glass. The permalloy is etched to produce a grating structure with spacing appropriate to the resolution of the encoder being fabricated. The film thickness of the MR sensors is on the order of 200 nm (0.2 µm). Usually, the grating structure is created so that more than one resolution can be achieved. This is done by connecting the appropriate grating fingers at final assembly. Figure 10.12 shows how one such sensor was implemented. Various resolutions could be supported by soldering connections to the appropriate "fingers" on the etched pattern.

FIGURE 10.12 MR sensor.

Output Signal Qualities. Frequency-response capabilities into the megahertz range have been documented (Hunt, 1987) but there are few if any manufacturers printing this on a product data sheet. A magnetic encoder with this performance would have to have a very high resolution to be practical, hence a small gap. In this situation, many of the advantages of the device would be lost. Most manufacturers specify output frequency capabilities of 100 to 200 kHz.

Signals are usually provided in TTL square-wave format. This is because the interaction between the magnetic flux and the MR sensors does not easily yield sinusoidal signals. As a result, they are not easily interpolated, nor used as sinusoidal devices. Typically, the MR outputs are digitized via comparators, as shown in Fig. 10.13.

Accuracy and Resolution. Many manufacturers produce magnetic encoders, Sony being one of the more prominent in the United States. The Sony RE20, RE21, and RE30 encoders support resolutions up to 2048 cpr, and will also provide commutation signals. Special versions are also being fabricated with resolutions of 8192 cpr.

VPL of Topanga, California, has developed an incremental magnetic encoder with resolutions of up to 26,112 cpr, with 402,124 cpr being possible.

Accuracy information is not readily available. Accuracy can be affected by uneven heating of the MR elements. As current is supplied, the resistive elements heat up. If the heat loading is uneven, the resistance values change unequally, causing the sensor to deviate from its balanced condition. Careful design of the sensor to provide heat-sinking capacity can minimize this issue.

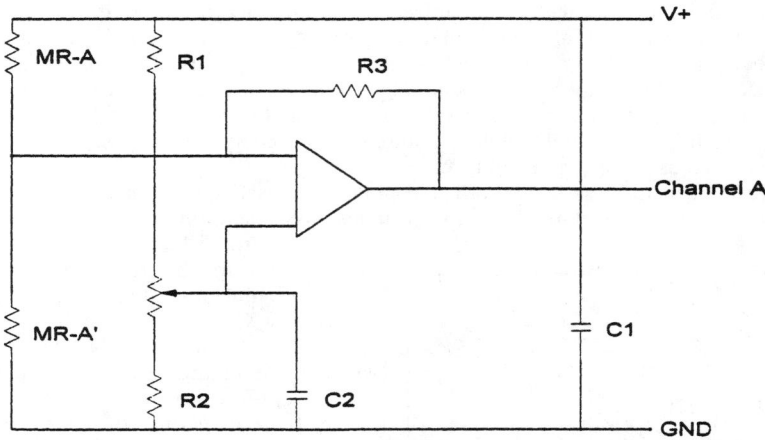

FIGURE 10.13 Typical MR sensor signal-conditioning circuit.

Interpolation. It is quite easy to develop multiplied outputs for this type of encoder. Figure 10.14 shows a proposed method for developing a 4× output for a magnetic encoder. In this method, 16 MR sensors are equally spaced over distance λ. Although the output frequency is multiplied by a factor of four, the input circuit frequency requirement has not changed. In this application, the author was able to create an encoder capable of developing 10,000 ppr (2500 cpr) and operating up to 700 kHz (Campbell, 1990).

Application Considerations. Air gaps must be approximately 80 percent of the pole pitch. For a 40-mm code wheel, 500 ppr, pole pitch is 0.25 mm. A gap of 0.2 mm would suffice. Successful operation with variations of as much as 50 percent has been documented (Campbell, 1990).

Adequate shielding must be provided, as these devices can be affected by a strong magnetic field. This is particularly important in machine-tool environments where magnetic chucks can be placed in close proximity to the ways where the sensor might be located. It is sometimes more difficult than expected to completely shield these devices from magnetic effects in motor applications. Improper handling of this issue can result in partial or total demagnetization.

One environment in which magnetic encoders work well is one with substantial contamination. Dust, chalk, oil, and other substances do not affect them (Micro Switch, 1982).

10.3.5 Hall Devices

Hall devices offer the following characteristics (Micro Switch, 1982):

- True solid-state construction
- Long service life (20 billion operations)
- Reasonable frequency response (100 kHz)
- Work at zero speed

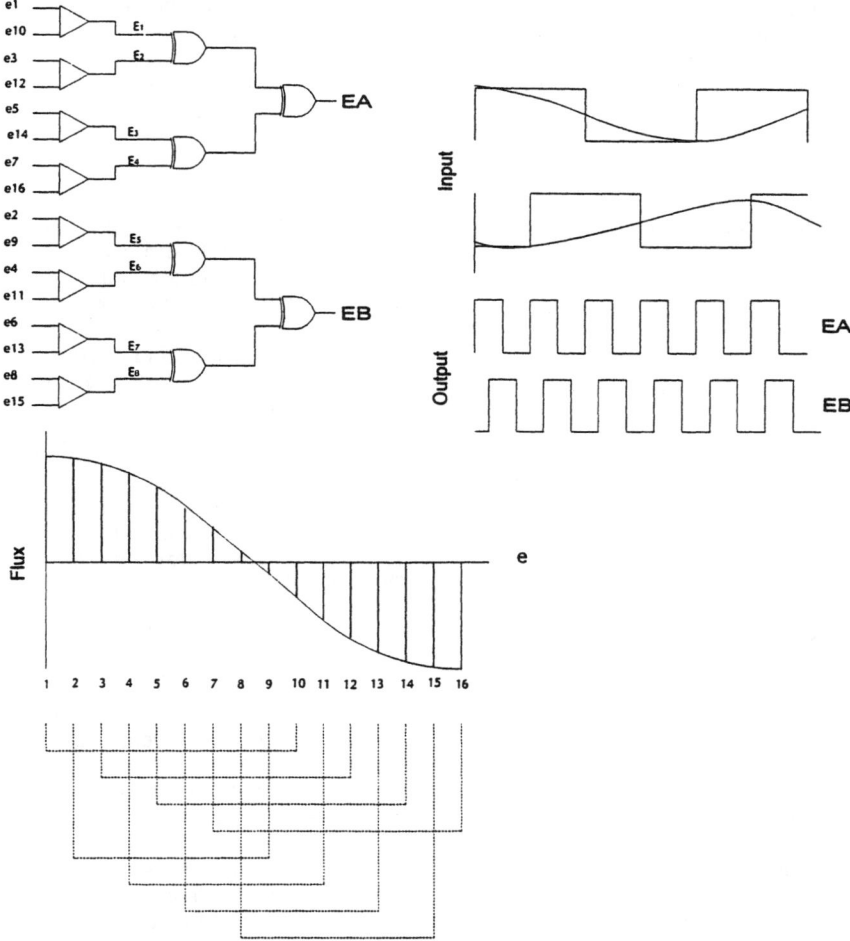

FIGURE 10.14 Interpolation scheme for MR sensor elements.

- Noncontacting system
- Good output interface characteristics
- Good operating temperature limits (–40 to 150°C)

Principles of Operation. Dr. Edwin Hall discovered the Hall effect in 1879. He was using a foil of gold to study electron flow. When he placed a magnet so that its field was perpendicular to the current flow, a voltage potential was developed across the two edges of the foil.

By placing output connections perpendicular to the direction of current flow, no potential difference exists in the absence of a magnetic field. When a magnetic field is present, current flow is acted on by Lorentz forces, disturbing the current distribution and causing a potential difference across the foil. This is called the *Hall voltage*. A balanced field will produce no output.

Linear sensors have an output voltage characteristic like that shown in Fig. 10.15. Output varies from 1.5 to 4.5 V as magnetic flux varies, and can be approximated as follows:

$$V_{out} = (6.25 \times 10^{-4})V_{supply})B + 0.5V_{supply}$$

where $-400 \leq B \leq 400$

Digital output devices have on/off characteristics. The on values are related to supply voltages, and can be anything up to 15 V. These devices basically consist of a linear sensor driving a Schmitt trigger, which is simply a comparator with hysteresis. The effect of hysteresis on the sensor output is shown in Fig. 10.16.

Both unipolar and bipolar digital devices are available. Unipolar devices switch at increasing or decreasing levels of the same pole. Bipolar devices can be selected to have the hysteresis require positive and negative polarities to switch. The response curve for a unipolar Hall element is shown in Fig. 10.17. The sensor must be brought up to D_1 before it will switch on. The sensor must be moved away to point D_2 before it will switch off.

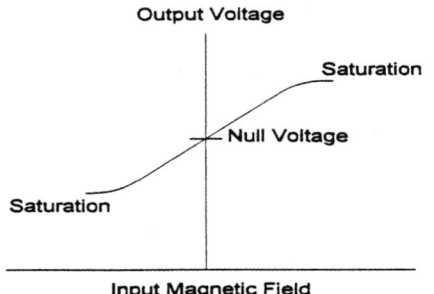

FIGURE 10.15 Linear Hall sensor output voltage characteristics.

Methods of Fabrication. Hall sensors are entirely solid-state devices. They are mass produced by large manufacturers such as Seimens, Honeywell, and others. Each of these suppliers has its own method of manufacture.

Magnets, on the other hand, can be purchased, and many times can be optimized, when specified by the end user. The user can implement them in a system in many ways. Unipolar slide-by Hall sensors (Fig. 10.18) are used in low-precision applications and are not suitable for linear sensors. Bipolar slide-by Hall sensors (Fig. 10.19) are more accurate and provide crisp response. Unipolar Hall sensors with pole pieces added (Fig. 10.20) allow gain adjustment. The addition of a pole piece allows actuation at a greater distance (D_4 instead of D_3; D_2 instead of D_1). Bias magnets can

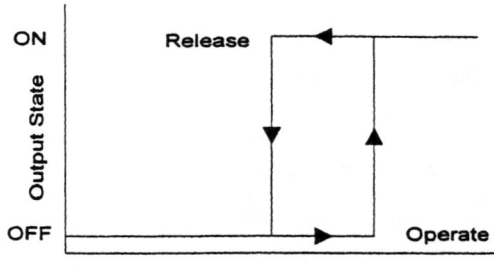

FIGURE 10.16 Sensor output hysteresis curve.

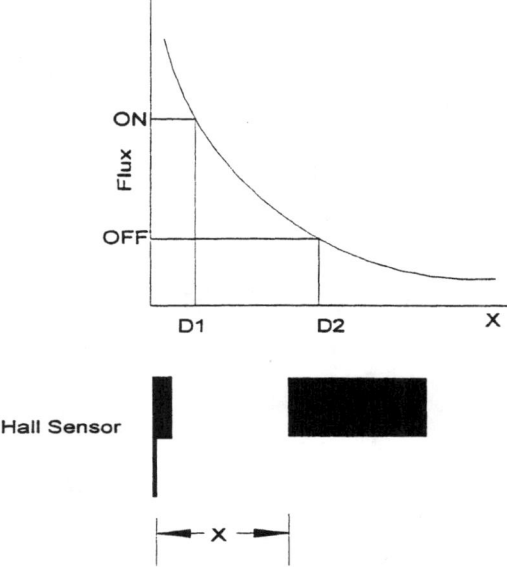

FIGURE 10.17 Digital unipolar Hall sensor operational characteristics.

allow the addition of an offset to the flux curve, allowing the user to fine-tune it. Care should be taken to avoid demagnetization if opposing fields are used.

Output Signal Qualities. Both linear and digital devices are available. Either may contain internal voltage regulators, or may require regulated voltage inputs. Both linear and digital devices are reasonably slow, with switching times on the order of 1 μs.

For linear units, the output loading must not drop below the minimum resistance. For microswitching this is around 2200 Ω. The output source can be up to 10 mA, and be used with supply voltages from 6 to 16 V dc. The sensitivity for linear sensors is typically on the order of 7 mV/G nominal, with a range of ±400 G.

Digital devices are basically the same, but there are *sourcing* and *sinking* types. Sinking types are active low, so when operational, the output drops to 0.4 V maximum. Sourcing typos are active high, with output voltage V_{supply} of 1.5 V typical. They can source or sink about 10 mA.

Accuracy and Resolution. These systems are generally used for coarse angular measurement. A typical accuracy value for rotation is between 1 and 5° mechanical. Resolution to 10° mechanical is less common but possible. This is basically due to difficulties in getting ring magnets with adequate pole densities, and the reduction of flux density as pole density increases. Accuracy is obtained by operating the Hall device in a steep portion of the flux curve, so that the distance required to cause switching is minimized, and the transitions are crisp. Most applications of this type require a bipolar magnet design. Most devices use internal circuitry to provide temperature stabilization, but the overall accuracy of these devices can be expected to average about 5 percent (Pippenger and Tobaben, 1982).

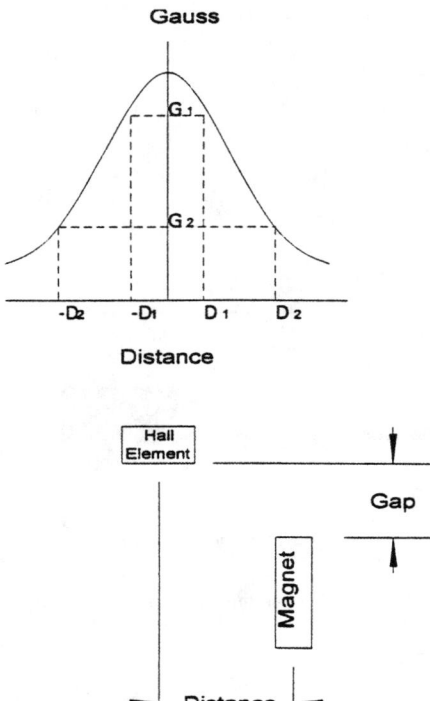

FIGURE 10.18 Unipolar slide-by Hall sensor.

Application Considerations. The Hall voltage is defined as follows:

$$V_H = kI_c (B \sin \theta)$$

where I_c is the applied current; $B \sin \theta$ is the component of magnetic field perpendicular to the current path; and k is a function of the geometry of the Hall element, ambient temperature, and the strain placed upon the Hall element. Compensation for temperature can be provided in the device. Typical compensated outputs are ±3 percent null shift and 0.077 percent gain shift over a −25 to 85°C range.

In general, systems using Hall sensors can be designed readily using trial and error. If the signal is not big enough, get a bigger magnet or add a pole piece. However, sometimes the user must specify the magnet and application mechanics, and there are a few rules which can help to support this.

If the magnetic properties are not known, they can be measured using a calibrated Hall sensor. These are devices which are the same size as the part planned for use, but which have calibrated linear outputs that can be used to determine magnetic field strength in the desired application configuration. Once this is known, operating margins can be determined. If a specific magnet needs to be specified, it is best to work closely with a magnet supplier. Here are a few issues which need to be addressed in any magnet design consideration. Magnet selection uses the *BH* curve

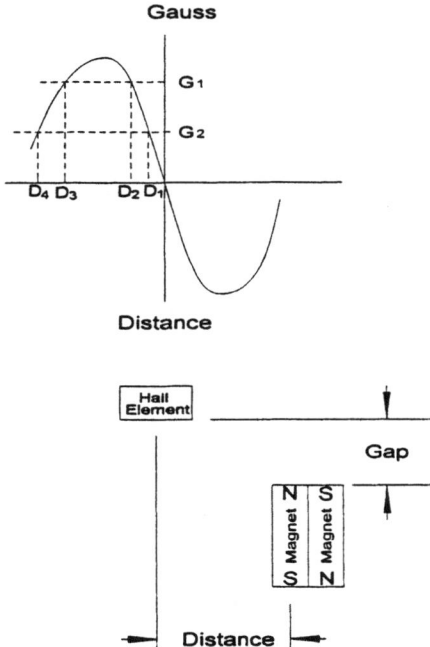

FIGURE 10.19 Bipolar slide-by Hall sensor.

(Fig. 10.21) to select between materials. The peak flux available from a magnet, and which must be used to trigger the Hall sensor, depends on the magnetic geometry and the material.

When a magnetic material is magnetized by magnetizing force H, the material moves from the origin to the maximum flux density B_{max}. At this point, the material is saturated. When H is released, flux density reduces to the residual value B_R. The amount of magnetizing force required to then demagnetize the material is the coercive force H_c. The shaded region in Fig. 10.21 is the demagnetization curve. Plotting the product of BH in this region against B produces a curve known as the *energy product curve,* which has an extremum value known as the *peak energy product.* This value is used to compare magnetic materials. Although the designer can perform the magnetic calculations, and sometimes must, many manufacturers supply charts showing flux density versus distance from the detector for various standard magnet materials and shapes. In most cases, this will be enough to allow determination of design parameters. Handling of magnet structures can also be an issue.

If the magnetic material is purchased already magnetized, handling requirements are more stringent, and the handling of bulk magnets can be quite an issue for the stock room. Accidental demagnetization can result in a number of ways:

- Dropping magnetic materials can result in "knock-down" demagnetization.
- Placement near uncontrolled strong ac fields can result in demagnetization.
- Bulk handling sometimes places magnets into temperature extremes, which may result in loss of magnet integrity.

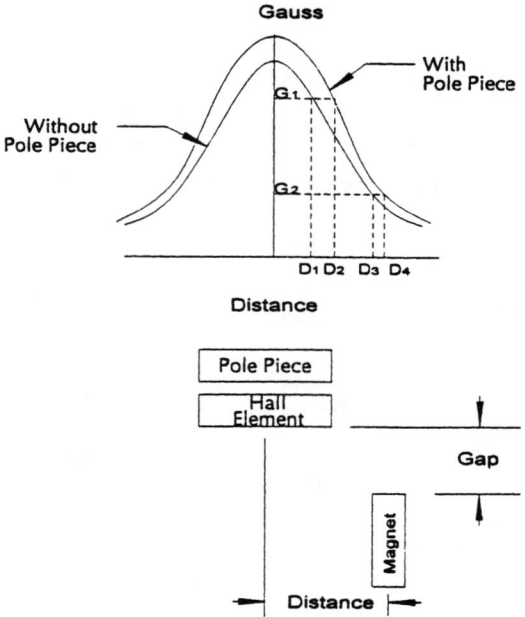

FIGURE 10.20 Unipolar Hall sensor with pole piece.

- Contact of the magnetic surfaces by ferromagnetic materials (screwdrivers, etc.) can affect flux levels.

Lodex is a material which can be affected by temperatures as low as 100°C. However, for most magnets the threshold is 250°C or above. Still, temperature-related loss of magnetization can account for as much as 5 percent flux change. These changes are nonreversible, requiring remagnetization to correct. Although this may

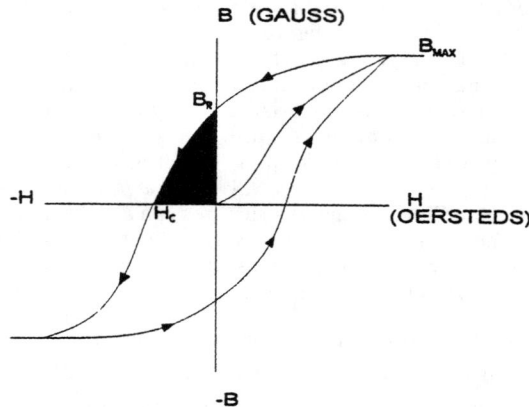

FIGURE 10.21 Magnetization curve.

have no impact on a digital sensor application, linear circuits are more susceptible to this problem. Keepers are highly recommended to minimize flux interactions, improve handling, and provide overall structural protection. If large volumes of magnetic material are to be handled, it is recommended that the magnetization be done as part of the manufacturing process. When properly designed, handled, and applied, both linear and digital devices are well suited to high temperature operation, with −40 to 150°C operational limits.

10.3.6 Linear and Rotary Inductosyns

Inductosyn is a trademark of Farrand Industries, Inc. Both rotary and linear versions are manufactured. Rotary Inductosyns consist of two disks, each carrying radial meander-shaped circuits. Linear versions have a long portion carrying the measurement standard and a slider which moves across it, acting as the output device. They can be manufactured using the same materials as the machine they are to be used on, so thermal mismatch effects can be minimized. This ability to use steel and other engineering materials also makes them mechanically robust and insensitive to contamination. Although the intrinsic resolution of these devices is not high, they are readily interpolated to a high factor. Rotary devices generate 2048 poles, and linear products have cycle lengths of 0.01, 0.02, 0.1, and 0.2 in and 2 mm. Interpolation electronics allowing the base resolution to be multiplied by 1000, 2000, or 2048 are readily available. The use of 2048-pole rotary device with a 2048 converter yields >4 million cpr.

Principles of Operation. Figure 10.22 provides a schematic representation of the manner in which the Indoctosyn operates. Electromagnetic coupling between conductors on the slider and the scale causes signal output. The excitation frequency can range from 200 Hz to 200 kHz, but the recommended input is 2 V peak to peak at 10 kHz. With this excitation, an output signal as large as 100 mV can be obtained. The sensor output is sinusoidal and depends on the relative motion between two sets of conductors. Inductive coupling between the two sets of windings is used to measure displacement. An alternating current supplied to the excited conductor induces voltage in the other which changes as the relative position varies.

Coupling is at a nominal maximum at the position shown. After displacement of one-fourth of the conductor pitch λ, the coupling is nulled out. Displacement of another ¼ λ produces full output again, but with reversed polarity to the original position.

Rotary devices obtain an additional benefit by the nature of their design. In this case, the sensor and the excited conductor are the same size. As a result, the sensor

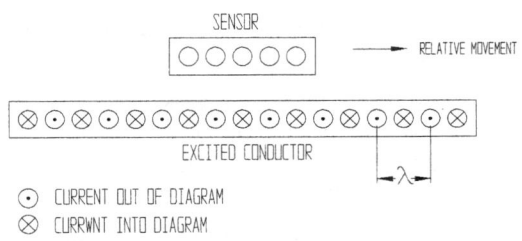

FIGURE 10.22 Inductosyn operating principles.

is able to simultaneously scan the entire circular pattern of the code disk. This helps to minimize both eccentricity and graduation errors.

Methods of Fabrication. Generally, these devices are manufactured using hot-rolled steel or aluminum. Special designs can use stainless steel or beryllium and can vary in size and thickness. Rotary devices consist of two plates, which must be mounted on a surface prepared by the user. This eliminates coupling errors, but places the burden of assembly and final device accuracy on the skill of the user. Standard devices are rather large, 3 in being the smallest. Both rotary and linear devices are produced using plate-and-etch techniques, similar to those used in the printed-circuit board industry. Once the base pattern is created on the substrate, the entire device is coated with a shielding material to minimize the effects of stray EMI.

Output Signal Qualities. Inductosyn digital interfaces typically provide excitation for the device and a user-selectable interpolation value, such as 1000 or 2000. These devices are closed-loop servos using null-seeking techniques, and have tracking rate limits from 840 to 3600 in/min at 0.0001 in resolution. The Inductosyn elements behave as transformer windings, with coupling ratio k that is at maximum when the windings are directly adjacent to each other. The sine and cosine windings can be excited by constant amplitude sine and cosine signals, or both can be excited with a common carrier. When sine and cosine excitation is used, the output is a constant amplitude signal that is phase shifted 360° for each movement of the slider by a distance λ. When common excitation is used, the output amplitude varies according to the following relationship:

$$V_{out} = kV_e \cos\left(2\pi \frac{x}{\lambda}\right)$$

where
 X = relative displacement ($0 \le x \le \lambda$)
 V_e = excitation voltage
 V_{out} = output voltage

These devices typically consume approximately 3.75 W of power at 5 V dc.

Accuracy and Resolution. Standard linear products are capable of 0.0002 in (0.005 mm) accuracy, rotary products to 2″ (0.00056°). Selected units can reduce these numbers by half. Repeatability is actually better than the rated accuracy by a factor of 10. Bigger is better in terms of accuracy for rotary devices, because error terms are reduced as diameter increases.

Application Considerations. Abbe errors and misalignment errors are probably the biggest error contributors in Inductosyn applications. The next is electrical noise.

Direct attachment to the machine tool provides a major advantage in that it eliminates the possibility of backlash. However, the alignment must be done properly and is nontrivial. For rotary devices, the recommended pattern concentricity requires a TIR of 0.0002 in. The smaller the disk, the larger the impact of eccentricity becomes. Eccentricity error is inversely proportional to the diameter of the disk. As an example, if the same amount of eccentricity were to be measured on two rotors, one 12 in in diameter and the other 3 in in diameter, the error would increase by a factor.

The allowable air gap is generous at 0.0005 to 0.015 in, and the devices will operate well at any value in this range. However, for rotary products, TIR of the surface must be held to less than 10 percent. This can be a challenge for a 12-in disk. Linear products must be mounted on ground surfaces such that the overall gap variation is less than 0.002 in.

To control noise on the output signal, twisted-pair cables with good separation and magnetic shielding sleeves are required. Cable shields must be insulated from each other and carried independently through the harness to be grounded at one common point in the system (Farrand Controls). EMI shielding is normally built into the device, but it must be properly grounded to the machine. Errors can be generated if the sine/cosine excitation is unbalanced or if there is cross-coupling. Shielding, ground loops, stray coupling at terminations, power-supply regulation, or phase-shift problems can all contribute to errors in the output.

10.3.7 Synchros and Resolvers

The synchro and the resolver are probably the earliest applications of inductive measuring techniques. Synchros resemble three-phase motors, and in fact are in some ways used like them. The torque transmitter (TX) is a synchro that is capable of driving enough current to actually do work. When the output of a TX is supplied to the stator windings of a torque receiver (TR), the rotor of the TR will rotate. The TR is simply a low-impedance synchro being run backward. The rotor of the TR will take a position closely approximating that of the TX in order to balance the current flowing in the loops. The control transmitter (CX) and control transformer (CT) are high-impedance versions of the TX and TR. Now real power is developed in the loop, and the CT simply generates an error voltage.

Resolvers are similar, but they have two stator coils at a 90° orientation rather than the 120° configuration of the synchro. The output signal amplitudes vary with the sine and cosine of the rotation angle, as the inductive coupling between rotor and stator coils varies with the angular position of the rotor. Resolvers are absolute-position devices, because the outputs of a simple two-pole resolver complete a full electrical cycle for each mechanical revolution. Additional poles can be added to increase resolution, with 16- or 36-pole configurations being fairly common, but when this is done the device is no longer absolute. The two-speed resolver contains a two-pole and a multipole device on a common housing. This allows the coarse sensor to be used to determine the basic angle, and the fine-speed device to give additional accuracy to the measurement.

The rotary differential transformer is another variation. The stator has three windings, and a magnetic iron core is rotated within it. One winding is excited with alternating current, and the voltage on the secondary windings varies depending on the position of the core.

Resolvers are not available as torque components.

Terms
CDX   Control differential transmitter.
CT   Control transformer.
CX   Control transmitter.
RC   Resolver control transformer.
RDC   Resolver-to-digital converter.
SDC   Synchro-to-digital converter.
TDX   Torque differential transmitter.
TR   Torque receiver.
TX   Torque transmitter or resolver transmitter.

Principles of Operation. Synchros and resolvers both have a rotor which rotates inside a fixed stator. Schematically, they are constructed as shown in Fig. 10.23.

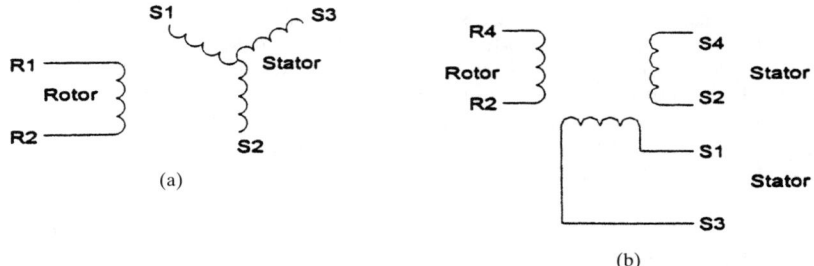

FIGURE 10.23 Schematic drawings of (a) synchros and (b) resolvers.

The three stator windings of the synchro are physically wound 120° apart and are connected in a star fashion. The three terminals are brought directly out. The rotor can be accessed via slip rings and brushes, or it can be excited by an additional circular transformer wound internally at the end of the unit. When the rotor is excited with an ac voltage, there are induced voltages in the stator that are proportional to the sine of the rotor coil axis and the stator coil axis. For example, if the voltage across R_1 and R_2 is $A \sin \omega t$, the voltages at S_1, S_2, and S_3 will be

$$S_1 \text{ to } S_3 = A \sin \omega t \sin \theta$$

$$S_3 \text{ to } S_2 = A \sin \omega t \sin (\theta + 120°)$$

$$S_2 \text{ to } S_1 = A \sin \omega t \sin (\theta + 240°)$$

For a resolver, this relationship would be

$$S_1 \text{ to } S_3 = A \sin \omega t \sin \theta$$

$$S_4 \text{ to } S_2 = A \sin \omega t \cos \theta$$

Actually, since more and more servo systems are going digital, understanding these synchros and resolvers requires knowledge of the digital converter, as well. These devices are little servos in and of themselves, and the selection and setup of these devices is crucial to the performance of the overall feedback system.

Resolver-to-digital converters (RDCs) interpolate the resolver output signals and provide 10-, 12-, 14-, and 16-bit results, depending on the converter used. (Some even have programmable word sizes.) When a single-speed resolver is coupled to a 12-bit RDC, a position measurement resolution of 360°/4096 = 0.0879° (5.27′) is obtained. The frequency at which this conversion must be obtained depends on the speed in rpm of the motor, and Table 10.4 shows this relationship. For example, a 12-bit RDC must be able to update a conversion result at the rate of 200 kHz, or 5 μs, when the motor is turning at 3000 rpm to avoid data latency. Although most converters can present data without a problem at this rate, usually needing only 1 μs to transfer data, the data may not be completely accurate. Most RDCs are tracking converters, which are implemented using a type 2 servo. The type 2 servo is a closed-loop control system which is characterized as having zero error for constant velocity

TABLE 10.4 RDC Update Rates—Output Frequency, kHz

Motor speed, rpm	Resolution, cpr						
	1	128	256	512	1024	2048	4096
3,000	0.1	6.4	12.8	25.6	51.2	102.4	204.8
4,500	0.1	9.6	19.2	38.4	76.8	153.6	307.2
6,000	0.1	12.8	25.6	51.2	102.4	204.8	409.6
7,500	0.1	16.0	32.0	64.0	128.0	256.0	512.0
9,000	0.2	19.2	38.4	76.8	153.6	307.2	614.4
10,500	0.2	22.4	44.8	89.6	179.2	358.4	716.8
12,000	0.2	25.6	51.2	102.4	204.8	409.6	819.2
13,500	0.2	28.8	57.6	115.2	230.4	460.8	921.6
15,000	0.3	32.0	64.0	128.0	256.0	512.0	1024.0
17,500	0.3	37.3	74.7	149.3	298.7	597.3	1194.7
18,000	0.3	38.4	76.8	153.6	307.2	614.4	1228.8

or stationary inputs. Conversely, this type of system will demonstrate errors in all other situations, and the magnitude of these errors must be controlled through optimized tuning of the converter. The frequency response of the converter will play an important part in the overall loop stabilization.

The tracking converter works by multiplying a "guess" angle ϕ by the input voltages. If the resolver is at an angle θ, then the resultant will be

$$V \sin \omega t \sin \theta \cos \phi \quad \text{and} \quad V \sin \omega t \cos \theta \sin \phi$$

These can be subtracted from each other, and when the reference voltage is factored out, the remainder is reduced by the trigonometric difference relationship to

$$V \sin \omega t \sin (\theta - \phi)$$

This signal is demodulated and sent to a phase-sensitive detector, which results in the generation of an error signal that is proportional to the difference between θ and ϕ. This difference is then used to modify the guess, and this is fed back to close the loop. It is the value of ϕ that is output as the converter result. Because it is a type 2 system, which means that the error is integrated until it goes away, the guess ϕ will reach the exact value quickly and with no error for a constant position or velocity situation. There will be errors during acceleration, but one is not normally trying to control this variable anyway.

Methods of Fabrication. Synchros and resolvers are constructed much like a motor. Iron laminations are created for the stator, and the windings are inserted. The rotor is usually made from solid iron, and one or more windings are applied, depending on whether it is to be a CX, CT, CDX, or whatever.

Output Signal Qualities. Amplitude or phase evaluation can be used for interpolation. By driving the stators with ac signals 90° phase shifted to each other, an ac signal is developed in the rotor which has a phase relationship to the supply voltage that depends on the rotor position.

Accuracy and Resolution. Resolvers have accuracy ratings of ±2 to ±20′. The corresponding RDC adds an uncertainty of ±2 to ±8′ + 1 LSB (Analog Devices, 1992). Resolver errors also have both static and dynamic contributions, which result from the acceleration error in the RDC tracking loop, offset voltages that are uncompensated, phase shift between the signals and the reference voltage, and capacitive or inductive crosstalk between the resolver signals and the reference cabling. Noise in the interconnection or on the reference will generate speed-dependent errors proportional to the phase shift in reference and inversely proportional to the reference frequency. Errors can develop if sine/cosine gain is unbalanced, or if there is cross-coupling. Shielding, ground loop, stray coupling at terminations, power-supply regulation, or phase-shift problems are also resulting errors.

Application Considerations. In addition to the problems just discussed, high-speed operation of resolver systems generally requires higher reference frequencies. Whether this is an acceptable configuration depends on the system and the application. For a retrofit or refurbishment situation, it may be that the reference frequency and voltage is already set, thus constraining the upgrade possibilities. In the past, 400- or 60-Hz at 26- or 115-V RMS references have been used extensively, with occasional 1200-Hz applications. Today, it seems that users are moving toward higher frequencies and lower voltages in order to obtain higher tracking rates. The result is that there are many opportunities for low-level modifications in most refurbishment applications, as references are not standard and probably never will be. New drive systems must be able to accommodate this as well, resulting in added complexity.

The fact that the converter itself has dynamics becomes an important part of the system design. Being a type 2 device, the converter can introduce up to 180° of phase lag into the system. For a 12-bit converter using a 400-Hz reference (*Analog Devices AD2S80A*), the RDC bandwidth (−3 dB point) will be less than 100 Hz. Using the same reference, a 14-bit converter will have a bandwidth of 66 Hz, and a 16-bit converter will have a bandwidth of 53 Hz. A 100-Hz −3-dB bandwidth means that there will be approximately 3 dB of peaking and 45° of phase shift at 40 Hz. As many servos attempt to close position loops near these frequencies, and an added 45° phase shift would be undesirable, it should also be noted that although the RDC tracking rate may not be exceeded, a system with difficult load dynamics could well prove unstable when the RDC dynamics are introduced. The situation only worsens when 14- or 16-bit converters are utilized.

10.3.8 DC Tachometers

Tachometers are used as velocity feedback sensors on speed-control systems. These devices generate an electrical signal which is proportional to the angular velocity of the motor shaft. They are used for monitoring open-loop systems and as the primary feedback element in velocity control systems. As shown earlier, they can also be used for inner-loop stabilization in position-control systems, of which there are three basic types, iron core, moving coil, and brushless, the most predominant type being the iron core.

Terms
dc generator   Used interchangeably with dc tachometer.
K_g Tachometer voltage sensitivity, V/krpm
Ripple   Noise voltage which can be assumed to be superimposed upon a linear output.

Principles of Operation. A tachometer is the opposite of a motor. A motor converts electrical energy into motion, and a tachometer converts mechanical motion into electrical energy. The output of the tachometer can be modeled as a main signal and the ripple component. The main signal is directly proportional to the rotor angular velocity, much like the counter-emf generated by a motor. The constant defining this proportionality is termed the gain K_g. Typical values for K_g range from 0.3 to 25 V/krpm.

Methods of Fabrication. Iron-core tachometers are made using rotor laminations, much like a motor. The rotor for a tachometer has more slots than a rotor for a motor would. Because of this use of iron, these devices have significant inertia. The moving-coil tachometer is a wound-coil rotor with magnets on the stator. These have very low inertia. For both iron-core and moving-coil tachometers, the current generated in the individual windings is routed to the output terminals via a commutator and brushes. Once again, this is similar in a fashion to a motor, but the current flow is in the opposite direction. The current flows out of the tachometer, and into the motor. Brushless tachometers are like iron-core devices, but instead of brushes they use optical or magnetic commutation circuits for current steering.

Output Signal Qualities. Tachometer outputs are bipolar, and are positive for one direction and negative for the other. A perfect device would have a completely linear relationship between rpm and output voltage, with zero ripple. This is never the case, however, and voltage ripple defines the quality of the output. The output voltage sensitivity is affected by the load the tachometer sees.

The tachometer output is dependent upon the load resistor value by

$$K_g \omega = R_G I = R_L I$$

$$V_{out} = \frac{R_L K_G \omega}{R_L + R_G}$$

R_L also affects the current in the circuit, shown in Fig. 10.24.

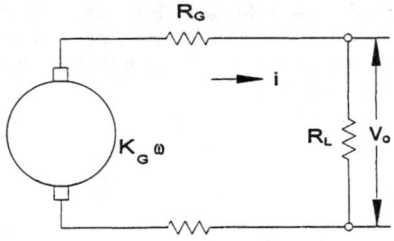

FIGURE 10.24 Tachometer circuit.

Accuracy and Resolution. The accuracy of a tachometer is determined by its linearity, ripple voltage, and temperature stability. Accuracy values needed vary depending upon the application. For the paper and pulp industry, accuracy values of ±0.03 percent are standard. Linearity of 0.5 percent is normal up to 3000 rpm. At higher velocities, nonlinearities become more apparent. This can result from brush bounce, commutator eccentricity, and brush skew due to directional changes at high speeds. Linearity is also affected by eddy current and hysteresis losses in the armature due to shorting during switching.

A good value for voltage ripple is 1 percent. With ripple values this small, small-order effects like shaft eccentricity can be seen if present. Levels this low are generally achieved only by moving-coil types because of their very low inductance. Ripple is composed mainly of noise created by brush transition between commutator segments. The signal is periodic and is related to the number of commutation segments and the shaft velocity. The frequency of this second-order contributor to ripple is armature eccentricity. It causes low-frequency amplitude variations of the same frequency as the rotor rpm. Finally, inductive effects in the windings can also affect the tachometer output. This contribution is generally very high frequency, however, and can be easily filtered out.

Temperature stability can also have some effect on tachometer performance. A thermal stability of 0.01 percent per degree Celsius is the best available, while the low-end commercial grade can be as poor as 0.2 percent. For a particular application, look at the temperature differential expected during operation. If the servo needs to maintain 2 percent of set point over this range, then 2 percent divided by range equals the percent stability required. Choose a tachometer with a temperature stability rating that is better than this value.

Application Considerations. In order to eliminate backlash, tachometers are generally mounted directly onto the motor shaft. In some cases, the tachometer is actually built right into the motor. This manufacturing approach is cost effective, but leads to coupling between the motor and the tachometer due to interaction between their magnetic fields. The electromagnetic coupling between the motor and the tachometer will be stronger at higher frequencies, so it is a sort of high-pass filter. The phase relationship of the electromagnetic coupling with respect to the motor voltage depends on the angular orientation of the tachometer to the motor. If the tachometer is improperly aligned to the motor, the combination of the high-pass filter characteristics and an improper phase relationship can result in instabilities in the motor/tachometer outputs even when used in an open-loop system. Many manufacturers provide motors with integral tachometers, and when properly built, they can be used for systems with servo bandwidths in the 15- to 30-Hz region. However, above this frequency, magnetic coupling between the tachometer and the motor will not be manageable, and the two devices should be separated.

Tachometers are not designed to provide any significant output power. In order to maintain commutation quality, they should be terminated into a load resistor which will maintain current levels on the order of 1 mA. If linearity is the prime consideration, the R_L should be chosen to be at least 100 times the dc resistance of the tachometer. If the intended application will run at very low speeds, such that this will be difficult to maintain, a silver commutator should be used.

10.4 COMPARISONS BETWEEN THE VARIOUS TECHNOLOGIES

The following discussion is basically a comparison between magnetic/inductive technologies and optical technologies. In many cases, the cost and performance issues make the technology selection self-evident. For example, it would probably not be a good idea to use an Inductosyn for an application that could get by with a Hall device. In the area of servo applications for resolvers and optical encoders, the issues are more gray. Cost and performance issues overlap significantly for these devices, so this comparison focuses more on this issue than on others. Finally, since resolvers, synchros, and Inductosyns all operate in much the same way, they are generically discussed as *resolvers* unless specific details warrant.

10.4.1 Power

Magnetic sensors enjoy the most favorable position in this respect. The magnetic field is free and lasts forever. As a result, these devices consume as little as 125 mW maximum. Encoders come in second place. Encoders with commutation included consume approximately 950 to 1030 mW, depending on the manufacturer and the temperature rating of the encoder. Encoders without commutation run less than this, averaging around 650 mW. A resolver with its associated converter comes in last place. A typical RDC chip alone typically consumes 300 mW, and the resolver consumes more, depending on its design, reference frequency, etc. Typically, resolvers require approximately 1000 mW (Clifton Precision).

10.4.2 Noise Control

For all the sensors discussed, there are methods that can or should be used to improve noise immunity in any given application. Noise can be minimized by the following methods:

- Increasing signal amplitude or signal to noise (S/N) ratio
- Using signal filtering
- Using hysteresis
- Using shielding

With the exception of shielding, all of these methods have performance implications.

Signal Amplitude or Signal to Noise (S/N) Ratio. In magnetic encoders, this can be optimized through gap, magnetic material, and sensor geometry tradeoffs. However, stronger magnetic materials cannot produce short pole pairs, and so resolution is lost.

In optical encoders, signal strength can be increased through using phototransistors, increasing illumination output, increasing sensor area, and reducing optical gap. Using of phototransistors reduces frequency response and causes problems at higher temperatures. Increasing illumination by raising LED current or using an incandescent light source reduces overall life and increases power requirements. Increasing sensor area works only if illumination is available, and reducing gap can result in mechanical stability problems or increased cost.

In resolvers, not much can be done to modify the output signal without modifying the power supply. This may be feasible in an analog system, within limits, but it is constrained by the digital converter in all other cases.

Signal Filtering. Adding filters in the signal-conditioning circuit can help to eliminate transducer errors, but it can add phase lag into the system transfer function, and can reduce the slew capability of the device. Resolvers include significant filtering by virtue of the converter electronics. For encoders with TTL outputs, digital filtering methods must be used at the receiver.

Hysteresis. Using of hysteresis in the signal conditioning circuitry can help to eliminate errors caused by low-speed operation in the region of a transition point, or when the transducer is stopped at a transition. Hysteresis is created in the input signal-conditioning circuitry (Fig. 10.25) in the following way.

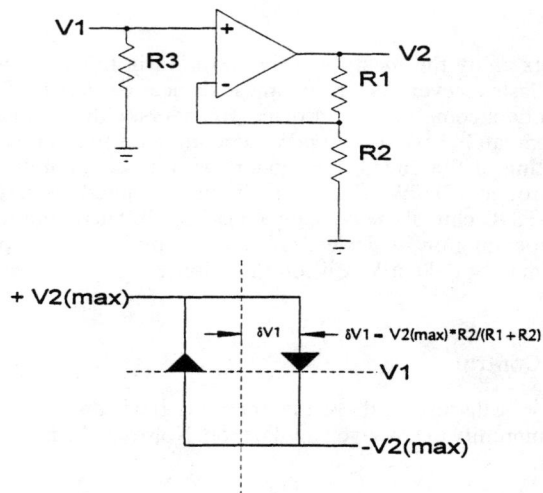

FIGURE 10.25 Hysteresis in the signal-conditioning circuitry.

The signal input current at V_1 can be from a voltage or current source, and must result in a voltage across R_3 large enough to exceed $V_{2,\max} R_2/(R_1 + R_2)$. Switching will not occur again unless the voltage across R_3 goes lower than $-\delta V_1$. This eliminates the possibility of small oscillations about a switching point causing spurious counts. A disadvantage of hysteresis is that it induces phase shift in the output signal, and it affects absolute accuracy. All position-sensing sensor technologies make use of hysteresis to condition their switching circuitry. Optical encoders design this into the amplifier, Hall devices utilize the Schmitt trigger, and RDC circuits implement it in the up/down count logic.

Shielding. No one technology has intrinsically better noise immunity for all environments than another, although for a specific application one technology may be preferred. Magnetic encoders must be shielded from magnetic fields in order to minimize the possibility of demagnetization. Optical encoders must be shielded from EMI in order to minimize switching errors. Resolvers and Inductosyns rely heavily upon proper shielding and grounding techniques to ensure accuracy.

10.4.3 Performance

For lower-accuracy systems, performance is based upon cost. In this area the Hall sensor is currently the low-cost leader, followed closely by simple optical products, and then MR sensors. As accuracy requirements increase, resolvers and optical encoders compete. The next performance level has optical products competing with Inductosyns. At these levels of performance, system requirements are so specialized that costs become secondary, and performance is everything. The application-specific needs will determine the appropriate technology.

Performance must also be derived through proper design, and the difficulties of successful implementation must be considered. For example, resolvers are infinitely tunable devices. The user must carefully consider a number of parameters and select fixed components in order to successfully utilize a resolver in any given application. The minimum resolver application must consider reference frequency, bandwidth, maximum tracking rate, number of bits in the conversion result, input filtering, ac coupling of the reference, phase compensation of the signal and reference, and offset adjustment. All of these issues affect the overall accuracy and performance of the installed device. In contrast, the designer choosing to utilize an incremental or absolute encoder in his system is faced with only two issues, line count and the maximum rpm to frequency-response rating. Many encoder designs can now provide data at up to 1 MHz. This would allow a 4000-line encoder, equivalent to a 12-bit RDC, to turn at 15,000 rpm. There are also no encoder dynamics to deal with, and the digitization process in an encoder eliminates any analog consideration on the part of the user. Encoders provide data with guaranteed signal separation and symmetry at up to the rated speed, and that is all there is. The added flexibility of interpolation at low speeds is similar to what is done via a multispeed resolver, but it can be implemented at a much lower cost.

Inductosyns must consider similar issues, but they achieve better overall performance than resolves, due to the greater number of poles and the good signal qualities they obtain. This is especially true for rotary Inductosyns.

10.4.4 Accuracy

In general, optical encoders have an advantage in this area, followed closely by Inductosyns. Angle measurement systems can be as good as 0.18″ (0.00005°). Inductosyns can reach 0.32″ for minimum resolution. High-resolution self-contained encoders have a minimum resolution of about 12″ or ½₀ cycle, which for a 2000-count encoder can be as low as 32″. Modular incremental rotary encoders usually have an accuracy rating of 2′ or better.

Although specially fabricated resolvers can be obtained which are accurate to ±2′, in general, resolvers have accuracy ratings of ±2 to ±20′.

Hall switches can achieve a resolution of about 5°, and are typically very difficult to align. This can be a real problem in a brushless motor commutation application. In these applications, the switching point should be optimized in order to ensure minimum torque ripple and to maximize motor performance. For these applications, resolver and optical encoder commutation can be good to ±10′ with some careful planning, and they are much simpler to align.

10.4.5 Resolution

The resolution, or measuring step, of a sensor is the amount of motion the device must experience before a change in output will occur. Figure 10.26 shows the range of resolutions each of the major position sensor technologies can accommodate.

10.4.6 Slew Rate

Position sensors generate data based on the resolution of the sensor, multiplied by the interpolation factor of the interface electronics. The maximum slew, or tracking rate, of an RDC is limited to ⅟₁₆ of the resolver reference frequency. For example, the Analog Devices AD2S80A RDC, using a 400-Hz reference, has a tracking limit of

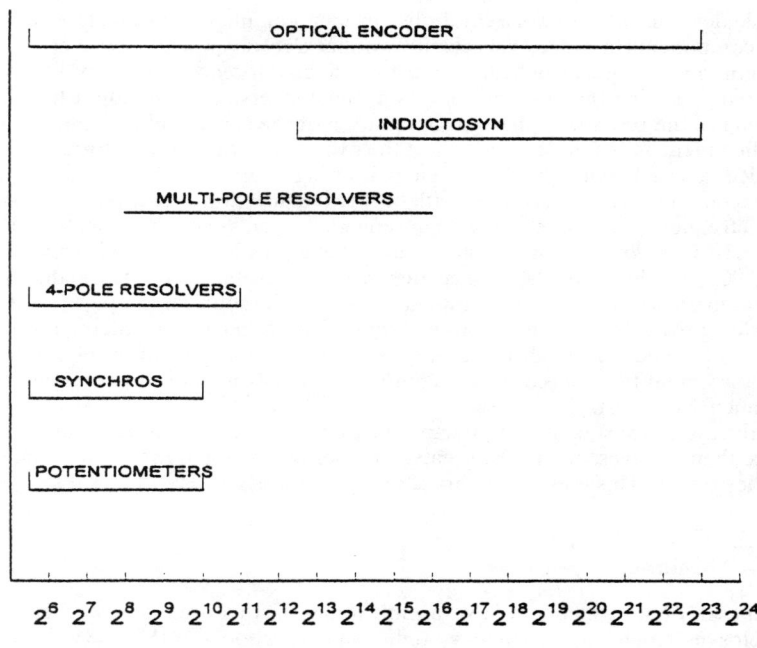

FIGURE 10.26 Comparative resolution of various technologies.

1500 rpm. This value can be increased to 18,750 rpm by using a 5-kHz reference. When the tracking rate of the converter is exceeded, the position value cannot be computed fast enough to keep up with the input, and the digital output becomes totally unpredictable (Boyes, 1980). A similar condition exists if the maximum acceleration rate is exceeded.

For encoders, the tracking rate is limited primarily by the frequency response of the sensor. Low- to intermediate-cost encoders generally have a maximum frequency response capability of 200 kHz, and there is no reference frequency dependency. Newer product offerings are beginning to push the envelope up to approximately 400 kHz, which allows a 4096-cpr encoder to turn at 6000 rpm without missing counts. Higher performance is available at a higher cost, with a maximum frequency response capability of approximately 1 MHz.

10.4.7 Dynamic Range

Low-speed operation requires high resolution and accuracy from the sensor. Encoder systems have dealt with this in a number of ways in the past, with new ideas and approaches being limited by available technology until recently. Many encoder-based BLDC systems have simply compromised by using a resolution based on the maximum rpm and frequency-response capabilities of the encoder, then letting the system designer deal with low-speed operation via drivetrain design. For example, many drives use a 2000-cycle encoder with a 200-kHz frequency-response capability. This allows the motor to run at 6000 rpm without missing counts. However, for a digital drive system with a 400-µs sampling interval, the one increment per sample speed would be 18.75 rpm. It is obvious that the system could run slower, but control accuracy degrades from this point. Improved low-speed operation is usually obtained by using interpolation electronics, which use encoder signals that are sinusoidal, as in resolvers. There are many types of interpolation electronics, and interpolation values of up to 4096 times the base resolution can be obtained, allowing encoder users to develop extremely high count-per-revolution values. Table 10.5 shows how this can be obtained. For example, a 2048-cpr encoder paired with 512× interpolation electronics yields >1 million cpr, and the minimum controllable speed now becomes 0.0006 rpm.

Resolver users have similar capabilities, but not to the same degree. Many RDC circuits allow for programmable resolutions, and tracking rates are resolution dependent. As an example, the AD2S80A has programmable resolutions of 10, 12, 14, and 16 bits. A 16-bit position resolution used with a 400-µs sample interval would allow a minimum speed control of 0.038 rpm.

From the preceding discussion, it can be seen that the two sensor types can utilize system software to realize a substantial dynamic range. For typical resolver systems, a maximum dynamic range of 18,750 to 0.038 rpm (113 dB) can be achieved. For encoder systems, using a 4096-cycle baseline device, a maximum dynamic range of 15,000 to 0.0006 rpm (144 dB) can be obtained. In order to obtain these dynamic ranges, both systems require some gymnastics on the part of the designer. The RDC system requires switching of gain resistors in order to switch resolution, and the encoder must switch to the interpolation outputs for low-speed operation. Both situations require thought on the part of the system designer to ensure that the transfer is "bumpless." As a point of reference, the typical range available from a dc tachometer is approximately 30,000 to 1 rpm (~90 dB).

10.4.8 Reliability and Durability

Reliability depends on the application environment, the overall temperature of operation, the shock, vibration, humidity, and so forth. It is common knowledge that resolvers are capable of operating at high temperatures, and in many cases can be operated at up to 150°C. The question is, how useful is this? In many high-performance applications, it is the motor winding that is the limiting factor. The feedback device is usually mounted at the end of the motor, and in many applications is thermally isolated to some extent.

TABLE 10.5 Effective Resolution of Interpolated Outputs

Interpolation factor	Line count					
	256	512	1024	2048	4096	5120
256	65,536	131,072	262,144	524,288	1,048,576	1,310,720
512	131,072	262,144	524,288	1,048,576	2,097,152	2,621,440
1024	262,144	524,288	1,048,576	2,097,152	4,194,304	5,242,880
2048	524,288	1,048,576	2,097,152	4,194,304	8,388,608	10,485,760
4096	1,048,576	2,097,152	4,194,304	8,388,608	16,777,216	20,971,520

As a result, it is usually not necessary to require a feedback device to be operated above 100°C, and this temperature can be met by many modern encoders as well as resolvers. No one can discount the sturdiness of the resolver. It is a simple device with a similar makeup to that of the motor, consisting of windings, bearings, and sometimes even brushes. However, for the majority of the environments encountered, an encoder is completely adequate. Typical specifications for vibration, shock, and temperature quoted by a variety of manufacturers can be summarized as shown in Table 10.6.

TABLE 10.6 Environmental Specifications

Factor	Encoder	Resolver
Vibration	15g, 10 to 2000 Hz, 3 axes	Same
Shock	50g, 11 ms duration	Same
Temperature	−10 to 100°C	−55 to 125°C

Because both encoders and resolvers can be purchased with or without bearings, neither can claim an advantage in this respect. However, frameless resolvers may have slightly less sensitivity to axial play than a modular encoder would. With respect to electronics, it is true that the resolver electronics can be mounted remotely from the motor in a less extreme environment. However, they are much more complex than those of the encoder. Typical encoder designs use a small number of very basic components, and can be implemented using commercial-, industrial-, or military-rated devices. It is possible that for extremely high impact shock environments, a resolver could claim some advantage over an encoder. However, if resolutions of less than 1000 counts per revolution are desired, than an encoder with a metal code wheel can compete favorably with resolver designs.

10.5 FUTURE TRENDS IN SENSOR TECHNOLOGY

10.5.1 Smart Sensors

Smart sensors may be used in temperature logging, diagnostics, failure detection, and fault isolation. Also, they may be used in device parameter storage, allowing a drive to interrogate the motor, determine what type it is, and set itself up automatically in plug-and-play drives, databus interfaces, and local interloop control systems, where a localized velocity loop could be implemented using a very high sample rate, allowing the position loop to be closed over a slower network bus. Full local control allows some companies to put the entire servo inside the motor. Databus interfaces need only provide the set point.

10.5.2 Redundant Sensors

Some technologies may benefit from massively redundant sensors and activators, such as robotics, teleoperated devices, and so forth. There are Advanced Research Project Administration (ARPA) programs, such as the microelectromechanical systems (MEMS) project, where the development of distributed arrays of miniature low-cost sensor elements is being promoted. One example of this program is a 13-bit single-turn absolute encoder the size of a dime.

10.5.3 Distributed Sensor Networks—Sensor Databus Systems

One less fanciful area with what appear to be near-term opportunities is in higher-end sensors capable of interfacing with factory automation networks. The only problem is that there are so many candidates that a manufacturer will never build the right one. The problem is that every one has a better idea. The military has developed standards: MIL-STD-1553, MIL-STD-1760, and ARINC 429. In the commercial sector, there are many standards that are in wide use at the hardware level, such as RS-485, but no real application interface standards. As an example of current and evolving approaches, the following are examples of three bus systems that are in use today, and one that is in development.

Synchronous Serial Interface (SSI). This is a simple, dedicated bus structure which allows a single device to be connected to a remote processor using only four wires. SSI is used as a device-level bus to connect multiturn encoders to controls. It is a single-drop clocked serial bus using RS-422 or RS-485 media for data transmission. The controller sources the clock, and the encoder returns data at the appropriate bit times. The data format consists of 25-bit words. The first 12 are dedicated to turn counting, the next 12 to the position within the turn or measuring steps. The last bit is always 0. This allows a simple serial-in/parallel-out shift-register implementation to be used.

Some companies are putting the entire servo inside the motor for full local control, and the interface need only provide the set point. This is an efficient, low-overhead system allowing economical interconnection between two devices to be made. It does not allow for expansion or multidrop applications.

SERCOS. This is a network that has grown out of the machine-tool industry but is being used more and more for general-purpose applications. Started in Germany around 1986, it became a European standard in 1995 (IEC 1491). SERCOS is now being championed in the United States by the SERCOS North America group. [Membership information is available by calling (800) 5-SERCOS.] Worldwide, there are more than 25 manufacturers supplying SERCOS-compatible controls and drives, with Indramat being one of the large companies involved. In the United States, Pacific Scientific has committed to SERCOS and has developed its product line and software products around SERCOS device interfaces.

The SERCOS architecture is flexible and allows the combination of many types of devices on the same backbone. The topology and technology are sound and make good use of current technology. A master/slave architecture is used with a ring topology and time-division multiplexing for data transmission. The bus media is fiberoptic, and can transmit at up to 4Mbit/s. The similarity to MIL-STD-1553 is strong, and the reliability of that bus has been proven in thousands of applications. The SERCOS technology exceeds MIL-STD-1553 in a few important areas, such as datarate and the number of slaves allowed. The fiberoptic media is also a bonus for noise immunity, and it reduces cost over the MIL-STD-1553 hardware.

SERCOS can address up to 254 slaves per ring. A novel feature is the ability to exactly synchronize all devices on the network through what is called the *master synchronization telegram*. This is very advantageous in control applications where harmonics could result from slightly out-of-phase actions by distributed actuators.

FieldBus. FieldBus is almost a generic name for a factory-floor databus. The FieldBus Foundation has been developing the specifications for what was to have been the definitive bus implementation. Called the Foundation FieldBus, its only problem is that the specification has been in development for years, and it is not completed yet. As a result, several manufacturers have taken the lead and developed their own implementations. This has caused many versions of "FieldBus" systems to come into existence. The parameters that vary include the number of connected devices, data packet size, connectors, certification requirements, datarate, bus length, and implementation cost. The specification of a FieldBus usually follows the definitions set forth by the Open Systems Interconnection (OSI) terminology, and will generally apply to the following layers:

Layer 7	Application layer	Management functions
Layer 2	Data-link layer	Definition of dataframe format
Layer 1	Physical layer	Circuit for bus electrical protocol
Layer 0	Media	Cable (not a real OSI layer)

Layer 0 varies significantly between bus systems. It can be as simple as two wires in flat cable, as used in AS-1, or it can be a fiberoptic cable, as in SERCOS. As an example of how FieldBus networks are implemented, the following gives the details of three fairly popular examples.

Process FieldBus (PROFIBUS) DP. There are actually three variations of this bus:

- PROFIBUS FMS, for general-purpose automation
- PROFIBUS DP, for high-speed data transfer to decentralized peripherals
- PROFIBUS PA, for process control with support for intrinsically safe operation

PROFIBUS is used by approximately 40 percent of the German FieldBus market, and certification to market it is required. It is used by Siemens, AEG, and Mitsubishi Electric, among others. This bus differs from SERCOS in both topology and architecture. The PROFIBUS physical architecture allows for multimaster/multislave systems, and is a linear bus topology. A multimaster system can be of benefit in that it will allow the system to withstand single-point failures more readily. The physical bus can be twisted-pair based upon the RS-485 standard or fiberoptic cable, allowing for up to 12Mbit/s datarates with 32 terminals. The physical-layer implementation has been developed by Siemens in a single-chip ASIC.

INTERBUS-S. This bus was developed by Phoenix Contact as a sensor/actuator bus, and it is used in the automotive industry. This is a single-master/multislave ring bus, similar to that of SERCOS. The bus is implemented as a twisted-pair wire using the RS-485 protocol, and it transmits at up to 2Mbit/s with a bus length of up to 400 m. INTERBUS-S can accommodate more slaves than PROFIBUS, and it allows up to 64 stations. The physical-layer implementation has been developed by and is available from Phoenix.

Controller Area Network (CAN). This bus was developed by Bosch and Intel for automotive real-time applications. It is a very low cost and reliable system, with many features built in to support redundancy and failure detection or isolation. It is currently used and supported by Honeywell (ADA Smart Distributed System), Allen-Bradley (DeviceNet), and Selectron (SELECAN).

This is an example of a multimaster bus in which the masters all arbitrate for control based on priorities assigned in their object lists. The physical media uses a twisted-pair cable which will transmit data at up to 1Mbit/s over a 40-m cable. Up to 110 stations can be on the bus at one time. It is a linear bus with terminations at both ends. Interface hardware is marketed by Intel, Motorola, Phillips, National, Siemens, and NEC. This bus may become the dominant embodiment of FieldBus as time goes on. Many manufacturers are producing products with interfaces to the DeviceNet protocol in particular. Sensor manufacturers are getting onto the bus, too. Leine & Linde, a Swedish encoder manufacturer, is providing multiturn absolute encoders which interface to the CAN bus.

Smart Transducer Interface for Sensors and Actuators. The National Institute of Standards and Technology (NIST) is actively developing a standard for smart transducer interfaces via its TC-9 committee. Using the fact that more and more sensors are utilizing built-in intelligence via micro-processors, NIST wants to leverage this capability into provision of a unified bus interface system.

The standard for this new interface system is being encapsulated in IEEE Standards 1451.1 and 1451.2, which are intended to define an interface that transducer manufacturers can build. The interface system is built upon three components: the smart transducer interface module (STIM), the network-capable application processor (NCAP), and the transducer independent interface (TII). The STIM may handle up to 255 transducers of seven defined types. The different transducer types are managed using transducer electronic data sheets (TEDS), which appear to be similar to the INTERBUS-S device profiles. The basic idea is that the user network communicates to the NCAP, which routes data to the STIM via the TII.

The network communications and NCAP are defined in IEEE 1451.1. The STIM, TEDS, and TII are defined in IEEE 1451.2. The network aspects of these systems are being heavily emphasized, and go beyond the single factory implementation. In May 1997, NIST demonstrated real-time control of remote sensors over the Internet at Sensors Expo. It may soon be possible for engineers to monitor and control plant processes anywhere in the world from any location having Internet access.

10.6 SELECTION OF SHORT-CIRCUIT PROTECTION AND CONTROL FOR DESIGN E MOTORS*

The 1996 National Electrical Code [Article 430-7 (9)] introduced the Design E motor to the electrical industry (National Fire Prevention Association, 1996). A new motor designation was necessary because the Design E motor has higher efficiency, higher locked-rotor current, and reduced torque requirements compared to existing NEMA motor designs. Design E motors provide increased energy savings that are important to many users. But the possible additional costs of protective and control equipment as well as more horsepower must be part of a cost-savings calculation.

This section is intended to give users of Design E motors electrical application information concerning the selection of the appropriate motor disconnection means; motor branch-circuit short-circuit protection; motor circuit conductor, motor controller, and motor overload protection; and guidance concerning torque characteristics. It describes how motor efficiency is increased and how the design

---

* Section contributed by Birch L. DeVault and Daniel P. Heckenkamp, Cutler-Hammer, and Todd L. King, Eaton Corporation.

changes related to increased efficiency affect the mechanical and electrical characteristics of motors. The section reviews the requirements of Article 430 of the 1996 National Electrical Code and provides generic application tables that compare the protection and control requirements of Design E motors with those of Designs B, C, and D.

10.6.1 Definitions

Breakdown torque   The maximum torque the motor will develop under rated voltage and frequency which will not result in an abrupt decrease in speed. This torque level typically represents a significant overload situation, and the motor should be operated at this level for only short periods of time and very low duty cycles to prevent overheating.

Full-load torque   The torque necessary to produce rated power at full-load speed.

Locked-rotor torque   The minimum torque the motor can produce at locked rotor (zero speed) at any angular position with rated voltage and frequency applied.

Pull-up torque   The minimum torque developed by the motor during acceleration from rest to the speed corresponding to breakdown torque. The load must require less than this value to obtain reliable starting.

Slip   The difference between the synchronous speed of the motor and the actual motor speed. It is usually designated as a percent of synchronous speed. In an induction motor, the slip increases with increasing load.

X/R ratio   The ratio of inductive reactance to resistance measured at the motor terminals during starting. This ratio is proportional to the electrical time constant during starting.

10.6.2   Origins of the Design E Motor

Motor Characteristics and Performance.   The Comprehensive Energy Policy Act (Public Law 102-486) of 1992 specifies that motor manufacturers meet an efficiency standard for "energy efficient" polyphase squirrel-cage induction motors as defined by NEMA Standard MG 1-1993, Table 10-9. Motors designed to the previous NEMA Standard (MG 1-1987, Table 12-6B) were also labeled "energy efficient." However, the motors that meet the old standard are generally referred to as "standard efficiency" motors in today's nomenclature. Therefore, in the context of this section, *standard efficiency* refers to motors that meet the old (1987) NEMA requirements, whereas *energy efficient* or *premium efficiency* refers to motors that meet the new (1993) NEMA requirements. Design E is a special category of energy-efficient motors whose efficiency design requirements exceed those of other NEMA-design motors.

Premium-Efficiency Motors.   Higher efficiencies are typically obtained by using more material, better materials, and better manufacturing processes, which results in a higher-cost motor. Energy-efficient motors have lower operating costs than standard-efficiency motors because they use less power; therefore, the higher initial cost can be offset by the lower operating costs over the life of the motor. However, the payback period is very much dependent on the application. For example, a pump motor that is run at full load for weeks without stopping will have a shorter payback period than an underloaded pump motor that is run occasionally.

How Efficiency Is Increased. The efficiency of the induction motor is defined as the mechanical output power measured at the motor shaft divided by the electrical input power measured at the motor terminals. The difference is due to losses in the motor. Motor efficiency is increased by decreasing these losses. There are five components of losses in the induction motor, typically referred to as (1) stator I^2R copper losses, (2) rotor I^2R copper losses, (3) iron (core) losses, (4) windage and friction losses, and (5) stray load losses. The premium-efficiency motor addresses each one of these loss components.

The *stator copper losses* are caused by the resistance of the stator windings. These losses are reduced by increasing the cross section of the stator conductors and therefore reducing the stator resistance. Adding cross section to the stator conductors adds more copper and cost to the motor.

The *rotor copper losses* are caused by the resistance of the conductors in the rotor—the rotor cage and end rings. Reducing the resistance of these conductors could also reduce losses, but at the expense of the starting torque and locked-rotor current. The starting torque is also proportional to the rotor resistance, and the locked-rotor current is inversely proportional to the rotor resistance. Since the motor usually has to meet NEMA requirements for the minimum starting torque and maximum locked-rotor current, there is a limit to the amount the rotor resistance can be reduced. But motor designers can optimize the shape of the rotor bars to minimize locked-rotor current and maximize efficiency.

The *core losses* are caused primarily by hysteresis and eddy current losses in the stator iron. The hysteresis is improved by reducing the flux density in the stator iron by increasing the stator stack length. It is also improved by using a higher grade of stator lamination material. The eddy current losses are reduced by optimization of the quality and thickness of the stator lamination material. These changes all increase the cost of the motor.

The *windage and friction losses* are due to the rotation of the motor. In fan-cooled motors, the fan is connected directly to the rotor shaft and provides circulating air to limit the motor temperature due to motor losses via convection cooling. An energy-efficient motor requires less cooling than a standard-efficiency motor because it has lower losses and therefore requires a smaller fan and as such requires less power to operate. The friction losses are improved by incorporating better bearings.

The *stray losses* are more difficult to identify but are very important in terms of efficiency. These losses are minimized by careful motor design and manufacturing practices.

Effect of Efficiency Improvements on Motor Performance. The motor design modifications done to increase efficiency have significant effects on other motor performance characteristics, as shown in Table 10.7. Table 10.7 shows the typical performance characteristics of the Design E motor compared to an energy-efficient Design B motor. In any particular case, the actual comparison depends on how the motor efficiency is obtained. However, NEMA provides limits for some of the values listed in Table 10.7, and the modifications necessary for obtaining increased efficiency tend to push the performance in a direction that may exceed these limits. For example, an effort to increase efficiency by reducing the resistance of the stator windings is limited by restrictions on locked-rotor amperage (LRA). Therefore, the high-efficiency design becomes a compromise between efficiency and performance with respect to the NEMA requirements.

TABLE 10.7 Design E Motor Performance Characteristics

	Relative to energy-efficient design B motor	
Characteristic*	Average change	Range
Nominal efficiency, 1%	1 to 4%	
Maximum locked-rotor current	0.42%	15 to 54%
Minimum locked-rotor torque	−1%	−16 to 13%
Minimum pull-up torque	−3%	−20 to 0%
Minimum breakdown torque	−6%	−20 to 5%
Full-load current	None	
Full-load torque	None	

* In addition, the power factor and slip of Design E motors tend to decrease relative to Design B motors; the X/R ratio tends to increase relative to that of Design B motors.

10.6.3 NEMA Motor Designs

Designs A, B, C, D, and E. NEMA designates letter classifications for induction motors according to the performance characteristics of locked-rotor torque, breakdown torque, locked-rotor current, and full-load slip. Table 10.8 shows the typical values for the different classifications. Note that the Design A motor is not specifically defined in the table but has characteristics similar to the Design B except for higher locked-rotor starting current. NEMA designates limits for minimum locked-rotor torque, minimum breakdown torque, minimum pull-up torque, and maximum locked-rotor current for each of these motor classifications. NEMA also requires minimum efficiencies for Design B and Design E motors.

Design B Energy-Efficient Motors. Many Design B motors had been redesigned to meet the 1993 NEMA efficiency standards given in NEMA MG 1, Table 12-10. Efficiency gains come at the expense of other performance characteristics. Whereas the standard-efficiency motors typically fall within the NEMA limits with room to spare, the energy-efficient motors tend to push the NEMA limits. These motors are typically physically longer and heavier and are more expensive than the standard-efficiency motors. It is not clear what will happen to the standard-efficiency motor products.

Design E. The Design E motor is intended to provide a satisfactory solution to most of the general-purpose applications provided by the Design B motor but at a higher efficiency. Because of the modifications needed to obtain this higher efficiency, the Design E motor possibly cannot meet the Design B NEMA performance requirements. Therefore, NEMA has defined new requirements for the Design E motor. The efficiency requirements are included in Table 10.8, and Table 10.9 compares the NEMA LRA, efficiency, and torque requirements of Design E and Design B motors at 460 V.

Comparison of Designs E and B. Figure 10.27 compares the nominal efficiency of Designs E and B. Figure 10.28 compares the locked-rotor amperage (LRA) to full-load amperage (FLA) ratios of Designs E and B. Figure 10.29 compares the LRAs of Designs E and B. Figure 10.30 compares the locked rotor torque, breakdown torque, and pull-up torque of Design E and B motors.

TABLE 10.8 Motor Designs

Classification	Locked-rotor torque (% rated-load torque)	Breakdown torque (% rated-load torque)	Locked-rotor current (% rated-load current)	Slip, %	Typical applications	Relative efficiency
Design B—normal locked-rotor torque and normal locked-rotor current	70–275*	175–300*	600–700	0.5–5	Fans, blowers, centrifugal pumps and compressors, motor generator sets, etc., where starting torque requirements are relatively low	Medium or high
Design C—high locked-rotor torque and normal locked-rotor current	200–250*	190–225*	600–700	1–5	Conveyors, crushers, stirring machines, agitators, reciprocating pumps and compressors, etc, where starting torque under load is required	Medium
Design D—high locked-rotor torque and high slip	275	275	600–700	5–8	High peak loads with or without flywheels such as punch presses, shears, elevators, extractors, winches, hoists, oil-well pumping and wire-drawing machines	Medium
Design E—IEC 34-12 Design N locked-rotor torques and currents	75–190*	160–200*	800–1000	0.5–3	Fans, blowers, centrifugal pumps and compressors, motor generator sets, etc., where starting torque requirements are relatively low	High

Note. Design A motor performance characteristics are similar to those for Design B, except that the locked-rotor starting current is higher than the values shown.
* Higher values are for motors having lower horsepower ratings.

TABLE 10.9 460-V Four-Pole, Open-Frame Design B and E Motors—Comparison of NEMA-Defined Performance

HP	Full-load amperes (FLA) per NEC Table 430-150 Design B and E	Maximum locked-rotor amperes (LRA) per NEMA MG 1, Tables 12.35 and 12.35A		LRA/FLA ratio		Nominal full-load efficiency per NEMA MG 1, Tables 12-10 and 12-11 (open 4-pole), %		Efficiency ratio Design E/B	Minimum locked-rotor torque per NEMA MG 1, Tables 12-2 and 12.38.4, %		Minimum breakdown torque per NEMA MG 1, Tables 12.39.1 and 12.39.3, %		Minimum pull-up torque per NEMA MG 1, Tables 12.40.1 and 12.40.3, %	
		Design B	Design E	Design B	Design E	Design B*	Design E		Design B	Design E	Design B	Design E	Design B	Design E
3	4.8	32	37	6.7	7.7	86.5	89.5	1.04	215	180	250	200	150	120
5	7.6	46	61	6.1	8.0	87.5	90.2	1.03	185	170	225	200	130	120
7²	11	64	92	5.8	8.4	88.5	91.0	1.03	175	160	215	200	120	110
10	14	81	113	5.8	8.1	89.5	91.7	1.03	165	160	200	200	115	110
15	21	116	169	5.5	8.0	91.0	92.4	1.02	160	150	200	200	110	110
20	27	145	225	5.4	8.3	91.0	93.0	1.02	150	150	200	200	105	110
25	34	183	281	5.4	8.3	91.7	93.6	1.02	150	140	200	190	105	100
30	40	218	337	5.5	8.4	92.4	94.1	1.02	150	140	200	190	105	100
40	52	290	412	5.6	7.9	93.0	94.5	1.02	140	130	200	190	100	100
50	65	363	515	5.6	8.0	93.0	95.4	1.03	140	130	200	190	100	100
60	77	435	618	5.6	7.5	93.6	95.4	1.02	140	120	200	180	100	90
75	96	543	723	5.7	7.6	94.1	95.4	1.01	140	120	200	180	100	90
100	124	725	937	5.8	7.5	94.1	95.4	1.01	125	110	200	180	100	80
125	156	908	1171	5.8	7.8	94.5	95.4	1.01	110	110	200	180	100	80
150	180	1085	1405	6.0	7.8	95.0	95.8	1.01	110	100	200	170	100	80
200	240	1450	1873	6.0	7.8	95.0	95.8	1.01	100	100	200	170	100	80
250	302	1825	2344	6.0	7.8	95.4	96.2	1.01	80	90	175	170	90	70
300	361	2200	2809	6.1	7.8	95.4	96.2	1.01	80	90	175	170	75	70
350	414	2550	3277	6.2	7.9	95.4	96.5	1.01	80	75	175	160	75	60
400	477	2900	3745	6.1	7.9	95.4	96.5	1.01	80	75	175	160	75	60
450	515	3250	4214	6.3	8.2	95.8	96.8	1.01	80	75	175	160	75	60
500	590	3625	4682	6.1	7.9	95.8	96.8	1.01	80	75	175	160	75	60

* Applies to induction motors labeled "premium efficiency" or "energy efficient."

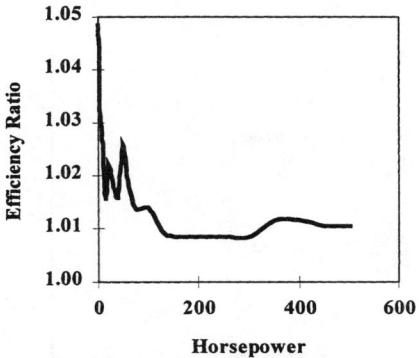

FIGURE 10.27 Design E to Design B nominal efficiency ratio.

10.6.4 Electrical and Mechanical Considerations Regarding the Application of Design E Motors

Modifications were made to the 1996 National Electrical Code to accommodate Design E motor applications. Article 430-7 (9) acknowledges the Design E marking on motors. Table 430-151B includes maximum locked-rotor current values for Design E motors. Table 430-152 includes branch-circuit short-circuit protective devices for Design E motors.

Motor Disconnection Means. Article 430-109 Exception 1 specifies a derating factor on motor-circuit switches that are not marked "rated for use with a Design E motor." The marking may appear on the motor-switch nameplate or in the instruction literature. If there is no marking, the motor-circuit switch "shall have a horsepower rating not less than 1.4 times the rating of a motor rated 3 through 100 horsepower, or not less than 1.3 times the rating of a motor rated over 100 horsepower." In certain cases, as shown in Tables 10.10 through 10.14, a larger motor-circuit switch is required with Design E motors. The horsepower ratings of the motor-circuit switches are based on the NEMA KS 1-1990 Standard, Table 4-3.

Motor Branch-Circuit Short-Circuit Protection. Article 430-52 (3) Exception 1 allows instantaneous-trip circuit-

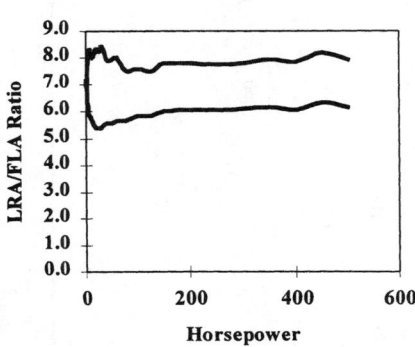

FIGURE 10.28 LRA/FLA ratios of Design E (upper curve) and Design B (lower curve).

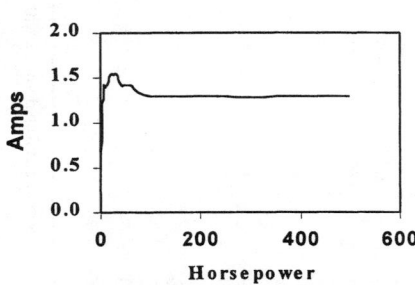

FIGURE 10.29 Design E to Design B LRA ratio.

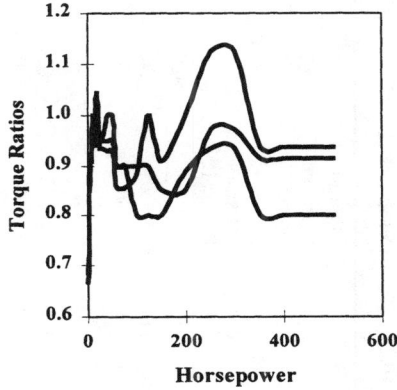

FIGURE 10.30 Design E to Design B locked rotor torque (upper curve), breakdown torque (middle curve) and pull-up torque (lower curve) ratios.

TABLE 10.10 200-V Design B, C, D or E Motors—Branch-Circuit Short-Circuit Protection and Controller Selection

HP	Typical FLA per NEC	Typical magnetic circuit-breaker rating, A Design B, C, or D	Design E	Typical thermal-magnetic breaker rating, A Design B, C, or D	Design E	Typical time-delay fuse, A Design B, C, or D	Design E	Motor switch rating, A Design B, C, or D	Design E*	Motor starter size Design B, C, or D	Design E*
3	11.0	15	30	40	40	15	25	30	30	0	1
5	17.5	30	30	50	50	25	40	30	30	1	1
7.5	25.3	50	50	70	70	35	50	30	60	1	3
10	32.2	50	70	90	90	45	70	60	60	2	3
15	48.3	70	100	125	125	70	100	60	100	3	3
20	62.1	100	150	125	125	80	125	100	100	3	4
25	78.2	150	150	125	150	100	175	100	200	3	4
30	92	150	150	125	175	125	200	100	200	4	5
40	120	150	150	175	225	150	250	200	200	4	5
50	150	250	250	200	300	200	300	200	400	5	5
60	177	250	400	250	350	225	350	200	400	5	6
75	221	400	600	300	400	300	450	400	400	5	6
100	285	400	600	400	500	400	600	400	600	6	6
125	359	600		600	900	450	600	400		6	
150	414	600		1000	1000	600	800	600		6	
200	552										
250											
300											
350											
400											
450											
500											

* Unless marked "Rated for use with a Design E motor."

TABLE 10.11 208-V Design B, C, D, or E Motors—Branch-Circuit Short-Circuit Protection and Controller Selection

HP	Typical FLA per NEC	Typical magnetic circuit-breaker rating, A Design B, C, or D	Design E	Typical thermal-magnetic breaker rating, A Design B, C, or D	Design E	Typical time-delay fuse, A Design B, C, or D	Design E	Switch rating, A Design B, C, or D	Design E*	Starter size Design B, C, or D	Design E*
3	10.6	15	30	40	40	15	20	30	30	0	1
5	16.7	30	50	50	50	25	35	30	30	1	1
7.5	24.2	50	50	70	70	35	50	30	60	1	3
10	30.8	50	70	90	90	45	60	60	60	2	3
15	46.2	70	100	125	125	70	100	60	100	3	3
20	59.4	100	150	125	125	80	125	100	100	3	4
25	74.8	100	150	125	150	100	150	100	200	3	4
30	88	150	150	125	175	125	175	100	200	4	5
40	114	150	150	175	225	150	250	200	200	4	5
50	143	250	250	200	300	200	300	200	400	5	5
60	169	250	400	250	350	225	350	200	400	5	6
75	211	400	600	300	400	300	450	400	400	5	6
100	273	600	600	400	500	400	600	400	600	6	6
125	343	600		600		450		400		6	
150	396	600		1000		500		600		6	
200	528										
250											
300											
350											
400											
450											
500											

* Unless marked "Rated for use with a Design E motor."

TABLE 10.12 230-V Design B, C, D, or E Motors—Branch-Circuit Short-Circuit Protection and Controller Selection

HP	Typical FLA per NEC	Typical magnetic circuit-breaker rating, A Design B, C, or D	Design E	Typical thermal-magnetic breaker rating, A Design B, C, or D	Design E	Typical time-delay fuse, A Design B, C, or D	Design E	Switch rating, A Design B, C, or D	Design E*	Starter size Design B, C, or D	Design E*
3	9.6	15	30	30	30	12	20	30	30	0	1
5	15.2	30	50	40	40	20	30	30	30	1	1
7.5	22	30	50	60	60	30	50	30	60	1	2
10	28	50	70	70	70	35	60	60	60	2	2
15	42	70	100	100	100	60	90	60	100	2	3
20	54	100	150	125	100	70	110	100	100	3	3
25	68	100	150	125	125	90	150	100	200	3	4
30	80	150	150	125	150	100	175	100	200	3	4
40	104	150	150	150	175	150	225	200	200	4	5
50	130	150	250	175	225	175	250	200	400	4	5
60	154	250	400	225	300	200	300	200	400	5	5
75	192	400	400	250	350	250	400	400	400	5	6
100	248	400	600	350	500	350	500	400	600	5	6
125	312	600		400	600	400	600	400	600	6	6
150	360	600		600	600	450	600	600	600	6	6
200	480			900	1200	600	800	600		6	7
250										7	8
300										7	8
350										8	9
400										8	9
450										8	9
500										9	9

* Unless marked "Rated for use with a Design E motor."

TABLE 10.13 460-V Design B, C, D, or E Motors—Branch-Circuit Short-Circuit Protection and Controller Selection

HP	Typical FLA per NEC	Typical magnetic circuit-breaker rating, A Design B, C, or D	Typical thermal-magnetic breaker rating, A Design B, C, or D		Typical time-delay fuse, A Design B, C, or D	Design E	Switch rating, A Design B, C, or D	Design E*	Starter size Design B, C, or D	Design E*
			Design E							
3	4.8	7	15	15	6	10	30	30	0	1
5	7.6	15	15	15	10	15	30	30	0	1
7.5	11	15	30	30	15	25	30	30	1	2
10	14	30	40	40	17½	30	30	30	1	2
15	21	30	50	50	30	45	30	60	2	2
20	27	50	70	70	35	60	60	60	2	3
25	34	50	70	70	45	70	60	60	2	3
30	40	70	90	90	50	90	60	100	3	3
40	52	70	100	100	70	110	100	100	3	4
50	65	100	125	125	90	125	100	200	3	4
60	77	150	125	125	100	150	100	200	4	4
75	96	150	150	175	125	200	200	200	4	5
100	124	150	175	225	175	250	200	400	4	5
125	156	250	225	300	200	350	400	400	5	5
150	180	250	250	350	225	400	400	400	5	5
200	240	400	350	400	300	500	400	600	5	6
250	302	400	400	600	400	600	400	600	6	6
300	361	600	600	900	500	800	600	600	6	7
350	414	600	800	1000	600	800	600	800	6	7
400	477		900	1200	600				6	7
450	515		1000	1200					7	7
500	590		1200	1400					7	8

* Unless marked "Rated for use with a Design E motor."

TABLE 10.14 575-V Design B, C, D, or E Motors—Branch-Circuit Short-Circuit Protection and Controller Selection

HP	Typical FLA per NEC	Typical magnetic circuit-breaker rating, A Design B, C, or D	Design E	Typical thermal-magnetic breaker rating, A Design B, C, or D	Design E	Typical time-delay fuse, A Design B, C, or D	Design E	Switch rating, A Design B, C, or D	Design E*	Starter size Design B, C, or D	Design E*
3	3.9	7	15	15	15	5	6	30	30	0	1
5	6.1	15	15	15	15	8	10	30	30	0	1
7.5	9	15	30	20	20	12	20	30	30	1	2
10	11	15	30	25	25	15	25	30	30	1	2
15	17	30	50	40	40	25	35	30	60	2	2
20	22	30	50	50	50	30	50	30	60	2	3
25	27	50	70	70	70	35	60	60	60	2	3
30	32	50	70	70	70	40	70	60	60	3	3
40	41	70	100	100	100	60	90	60	100	3	4
50	52	70	150	125	125	70	110	60	100	3	4
60	62	100	150	125	125	80	125	100	200	4	5
75	77	150	150	125	150	100	150	100	200	4	5
100	99	150	150	150	175	125	200	200	200	4	5
125	125	150	150	175	225	175	250	200	400	5	5
150	144	150	400	200	250	200	300	200	400	5	6
200	192	150	400	250	350	250	400	400	400	5	6
250	242	400	600	350	500	350	500	400	400	6	6
300	289	400	600	400	500	400	600	400	600	6	7
350	336	600		600	600	450	600	400	600	6	7
400	382	600		600	900	500		600		6	7
450	412	600		900	1000					7	7
500	472			900	1200					7	8

* Unless marked "Rated for use with a Design E motor."

breaker settings for Design E motors to be in the range of 1100 to 1700 percent of full-load current. Settings for Design B motors are limited to a range of 800 to 1300 percent of full-load current. This higher rating for Design E motors avoids nuisance tripping during start-up due to higher transient and steady-state LRA. In certain cases, as shown in Tables 10.10 through 10.14, a larger instantaneous-trip circuit breaker will be required to obtain the necessary current range to start Design E motors.

In the case of thermal-magnetic circuit breakers, the nonadjustable magnetic trip is typically set by the manufacturer at 1600 percent of full-load current. So to avoid nuisance tripping during start-up, thermal-magnetic circuit breakers should be chosen near the maximum (400 percent of FLA for FLA $\geq$ 100 A, 300 percent of FLA for FLA > 100 A) allowed by Article 430-52 (c)(1)(c) of the National Electrical Code. In certain cases, as shown in Tables 10.10 through 10.14, a larger thermal-magnetic trip circuit breaker than the one normally recommended by the circuit-breaker manufacturer is required with Design E motors.

In the case of time-delay fuses, there is a potential danger of nuisance fuse blowing on start-up due to the higher locked-rotor currents of Design E motors. So, time-delay fuses used with Design E motors should be chosen near the maximum (225 percent of FLA) allowed by Article 430-52 (c)(1)(b) of the National Electrical Code. In certain cases, as shown in Tables 10.10 through 10.14, a larger fuse rating than the one normally recommended by the fuse manufacturer is required with Design E motors.

Motor Circuit Conductor. Article 430-22 (a) specifies that "Branch-circuit conductors supplying a single motor shall have an ampacity not less than 125 percent of the motor full-load current rating." Article 430-6 (a) specifies that Table 430-150 be used to determine the motor full-load current. In Table 430-150, motor full-load currents for Designs B, C, D, and E are equal. Therefore, no conductor derating is required when Design E motors are used.

Motor Controller. Article 430-83 (a) Exception 1 specifies that a controller for a Design E motor must either be marked "Rated for use with a Design E motor," or the controller horsepower rating must be derated. The marking may be on the controller nameplate or in the instruction literature. If there is no marking, the controller "shall have a horsepower rating not less than 1.4 times the rating of a motor rated 3 through 100 horsepower, or not less than 1.3 times the rating of a motor rated over 100 horsepower." In certain cases, as shown on Tables 10.10 through 10.14, a larger-size controller is required with Design E motors. The horsepower rating of the controllers is based on the NEMA ICS 12-1993 Standard, Table 2-4-1.

Motor Overload Protection. Article 430-32 specifies that overload devices are selected based on motor nameplate full-load current. Full-load currents for Design E and B motors are equal, but service factor is a consideration when selecting overload heater elements. While open-type Design A, B, and C motors are required by NEMA to have service factors of 1.15 (NEMA MG 1 Standard, Table 12-4), Design E may have a service factor of 1.0. Normal practice is to select heaters one size lower for 1.0 service factor applications. Class 20 (standard-trip) overload relays provide ample motor protection for general-purpose applications of Design B, C, D, or E motors unless specified otherwise by the motor control manufacturer. The locked-rotor current of Design E motors may necessitate an increase in overload class from Class 20 to Class 30 or may require that the heater size be increased as allowed by NEC Article 430-34 in order to start the motor. Consult with the motor control manufacturer regarding applications such as hermetic refrigerant motor compressors that may require Class 10 (fast-trip) or applications such as high-inertia ball mills, reciprocating pumps, or loaded conveyors that may require Class 30 (slow-trip) protection.

Application Example. Using the retrofit of a four-pole Design B 460-V 50-hp motor having a minimum efficiency of 93 percent with a four-pole Design E 460-V 50-hp motor having a minimum efficiency of 95.4 percent as an example, each of the protection and control requirements must be considered.

1. *Switch and fuse combination.* The Design B motor requires a 100-A disconnect switch. A Design E motor requires a 200-A disconnect switch unless the existing disconnect switch is marked "Rated for use with a Design E motor." A 90-A time-delay fuse is typically recommended for use with a 50-hp Design B motor. This fuse is at 138 percent of the full-load current. To avoid nuisance fuse blowing, a 125-A time-delay fuse at 192 percent of full-load current can be recommended. If the 125-A fuse is used, 200-A fuse clips are required in place of the original 100-A fuse clips used on this application.

2. *Thermal-magnetic breaker.* The Design B motor requires a 125-A circuit breaker. The same circuit breaker can be used with a Design E motor.

3. *Magnetic circuit breaker.* The Design B motor uses a 100-A magnetic circuit breaker with a typical setting range of 462 to 1538 percent of full-load current. A 150-A magnetic circuit breaker with a typical setting range of 692 to 2308 percent of full-load current is required with the 50-hp Design E motor.

4. *Motor starter.* The Design B motor requires a NEMA size 3 motor starter. A size 4 motor starter is required with a Design E motor unless the original size 3 motor starter is marked "Rated for use with a Design E motor."

5. *Overload relay.* A Class 20 (standard-trip) overload relay is typically used with the normal 1.15 service factor Design B motor. The same Class 20 overload may be used with the Design E motor, but one lower heater rating is recommended due to the 1.0 service factor.

6. *Locked-rotor torque.* The minimum locked-rotor torque of the 50-hp Design B motor is 140 percent of the full-load torque. The minimum locked-rotor torque of the Design E motor is 130 percent of the full-load torque. If this 7 percent reduction in torque is indicated by the motor manufacturer, it must be considered in the mechanical application of the motor.

7. *Breakdown torque.* The minimum breakdown torque of the 50-hp Design B motor is 200 percent of the full-load torque. The minimum breakdown torque of the Design E motor is 190 percent of the full-load torque. Such a 5 percent reduction in torque must be considered in the mechanical application of the motor.

8. *Pull-up torque.* The minimum pull-up torque of the 50-hp Design B motor is 100 percent of the full-load torque. The minimum pull-up torque of the Design E motor is also 100 percent of the full-load torque.

Summary. As the preceding example shows, increased locked-rotor current, and reduced locked-rotor torque, breakdown torque, and pull-up torque of Design E motors can result in costly errors in retrofit or replacement situations if all of the parameters are not thoroughly studied. If a Design E motor is being used to replace an existing general-purpose Design B motor, torque characteristics, existing protective device rating, and controller size must be considered.

10.6.5 Availability of Design E Motors

Motor manufacturers are not marketing Design E motors at this time. Among the reasons for the lack of market demand are market satisfaction with the efficiency levels of today's energy-efficient Design B motors, failure of manufacturers to offer switches and controllers rated for use with Design E motors, application issues related to the adjustment requirements of magnetic circuit breakers used with energy-efficient motors, and application issues related to the torque provided by Design E motors.

10.6.6 Conclusion

The engineering solution used to calculate the viability of Design E motors to an application must balance the energy savings with the possible increased cost of more horsepower and increased cost of protection and control equipment. Tables 10.10 through 10.14 give a guide as to the relative protection and control equipment requirements of NEMA Design B and E motors. The user of this information must contact the control equipment manufacturer to determine specific product requirements. The user of this information must also verify that the suggested combination of equipment meets all of the applicable safety codes. Applications using IEC (Design N) metric motors and controls have already demonstrated that the technology for control and protection of Design E motors has been proven in European applications where reduced-voltage starting is more common than in the United States. New equipment using Design E motors will comply with the protection and control standards referenced in this section. But care must be taken that the proper protection and control be applied in motor replacement situations.

10.7 SWITCHED-RELUCTANCE MOTOR CONTROLS*

Switched-reluctance motors have been available for over 100 years, but the means to control them have made significant strides in the last decade. The availability of high-power high-frequency semiconductor switches combined with the power of microcontrollers has made these motors a strong contender in the race to provide elegant motion solutions. Switched-reluctance motors (SRMs) are known for their unique capability to work at high ambient temperature conditions.

10.7.1 Basic Principles

Figure 10.31 shows the cross-sectional view of a four-phase SRM. When the coils wound around diametrically opposing poles are energized by a dc source, the rotor poles have a tendency to align with the corresponding stator poles. Continuous torque production is achieved by energizing the phase coils in sequence. As the rotor moves, the phase coil inductance changes from a maximum value when the stator and rotor poles are aligned to a minimum value when the corresponding poles are unaligned.

---

* Section contributed by Ramani Kalpathi, Dana Corporation.

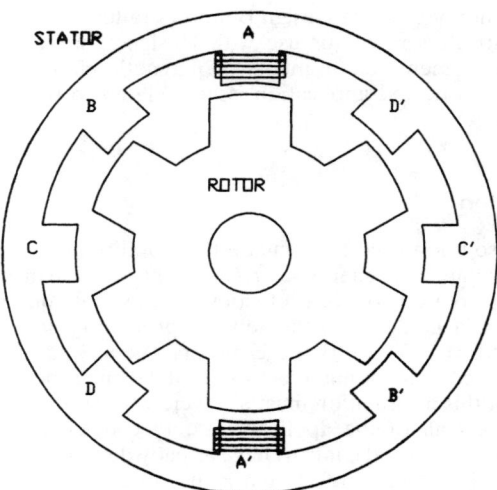

FIGURE 10.31 Cross-sectional view of an 8/6 four-phase switched-reluctance motor (SRM).

Figure 10.32 shows the idealized inductance profile of one phase of an SRM. The placement of coil current in the rising slope of inductance produces positive torque, while the placement of current in the falling slope of inductance produces opposing or negative torque. For optimal torque production, the coil current should be initiated prior to the rising slope of inductance and must be commutated before the onset of the falling slope of inductance. Such a control is conventionally accomplished by the use of a rotor-position sensor that indirectly indicates the occurrence of the inductance slope.

10.7.2 Control Methods

Power Inverter. Figure 10.33 shows the power inverter circuit used to energize the phase coils of a four-phase SRM. A full-wave diode rectifier is used to provide a dc source for the SRM. The inverter is comprised of an upper switch and a lower switch between which one phase coil is connected. The semiconductor switches are implemented by metal-oxide semiconductor field-effect transistors (MOSFETs) in the case of low-voltage systems (under 100 V) and insulated-gate bibolar transistors (IGBTs) in the case of high-voltage, high-power systems. The two-switch-per-phase inverter is used to energize the phase coils such that when both switches are closed, the current in the phase coil starts rising. When the current in the phase coil reaches a predetermined limit, the upper switch is controlled in such a manner that the coil current can be made to remain within a predetermined band. The magnitude of the band-limited current is proportional to the torque produced by the motor. When the rotor reaches a position of alignment with the stator poles, the switches in series with the phase coil are commutated, disconnecting the dc source from the phase coils. The stored electrical energy in the phase coil is circulated back to the dc source capacitor through the diodes. The capacitor across the dc bus voltage assists in recovering the energy after a cycle of commutation.

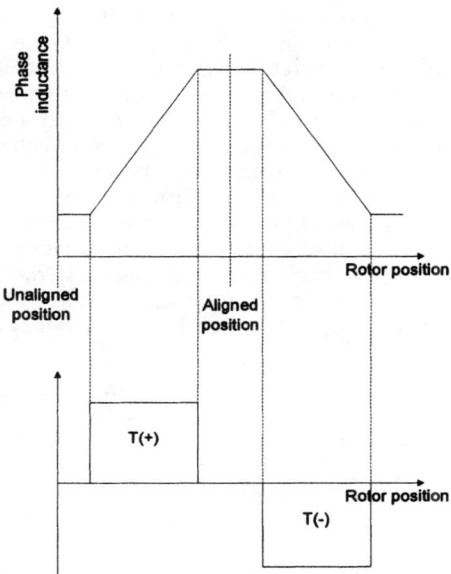

FIGURE 10.32 Idealized inductance and torque profiles.

Current Control. The method of controlling the current within a band using high-frequency switching of the semiconductor devices is called *hysteresis control* or *current control*. In this method, a current sensor provides a feedback of the motor current, which is compared to the commanded current, and the controlling signals to the power switches are realized. Torque control is realized by varying the commanded level of current. When both the switches in series with the phase coil are turned on, the current starts rising to the commanded level. The voltage to the phase coil is equal to the dc bus voltage. When the motor current reaches the commanded current, both the switches are turned off, causing the diodes to freewheel the motor current back to the dc source. As a result, the voltage across the phase coil is reversed in polarity due to the action of the diodes, and the current starts falling

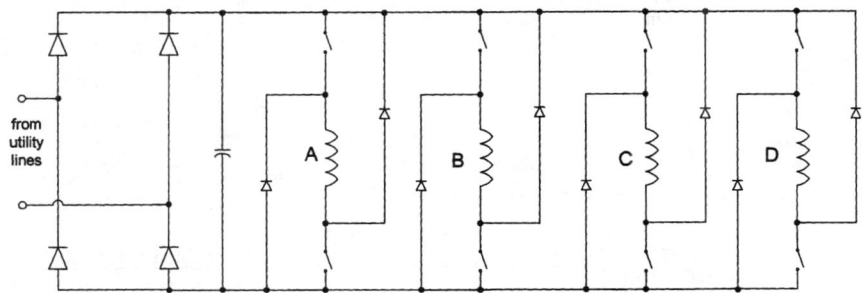

FIGURE 10.33 Configuration of the classic converter.

down. When the current reaches the lower level of the hysteresis band, the switches are turned on again, and the switching process continues. Figure 10.34 shows the corresponding waveforms during current control. Current control is typically used in variable-speed automotive applications, such as radiator cooling fans and cooling pumps. A typical control block diagram for a current-controlled system appears in Fig. 10.35. The angle calculator, commutator, and speed calculator are implemented using a microcontroller. The angle calculator typically calculates the turn-on and turn-off angles for the coil current after measuring the motor speed and knowing the commanded current. As an example, for high motor speeds, the turn-on angle is advanced with respect to the onset of the rising slope of inductance in order to permit the current to rise to the commanded value. Similarly, the turn-off angle is calculated such that there is no negative torque when positive torque is commanded, which would occur due to the presence of current in the falling slope of inductance.

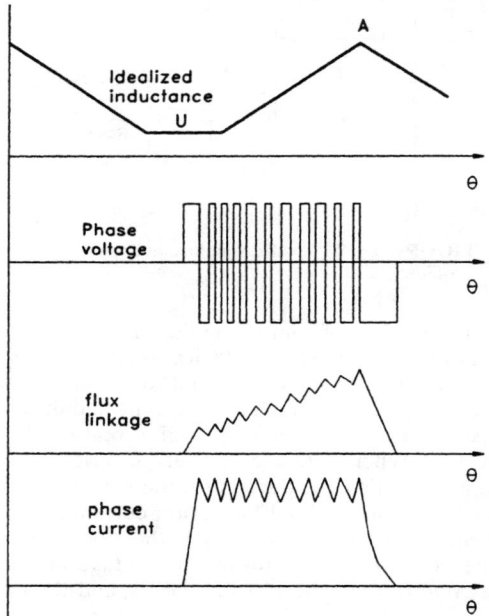

FIGURE 10.34 Low-speed operation phase inductance, phase voltage, flux linkage, and phase current.

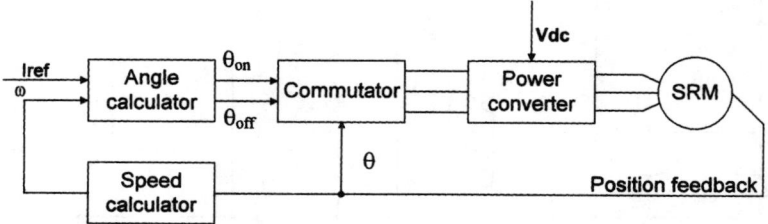

FIGURE 10.35 Typical SRM drive system with position feedback.

The commutator sequences the phase coil currents in the correct sequence of commanded direction. For a three-phase motor, if the energization sequence of ABC produces a clockwise direction, then the energization sequence of ACB would rotate the motor in a counterclockwise direction. Position feedback is achieved by using a shaft encoder or a position estimator.

Voltage Control. Voltage control is the method of switching semiconductors by providing a high-frequency gate signal to control the voltage input to the phase coil. Torque control is achieved by varying the duty cycle of the control signal while keeping the frequency constant. This scheme is typically implemented by generating the duty cycle signal using a PWM channel from a microcontroller. This method is commonly used to vary the duty cycle within one commutation cycle such as to profile the current waveform. Such control leads to minimizing the torque ripple that is often common in SRMs. Voltage control is used in servopositioning automotive applications, such as electric power steering and electronic throttle control.

The torque-speed characteristics of a typical SRM drive show a flat torque capability from zero speed to base speed, after which the curve follows a constant power curve. The high-speed operation follows a natural characteristic, as indicated in Fig. 10.36.

10.7.3 Open-Loop and Closed-Loop Control

The cost of position sensors and the complexity associated with position estimators have resulted in economical open-loop controllers that sequence the phase coil commutation without any position feedback at low operation frequencies, resulting in low speeds of operation. The greatest advantage of such systems is that a motor with gear reduction can achieve extremely accurate positioning capability with high torque output. An example of such a system is found in automotive cruise control,

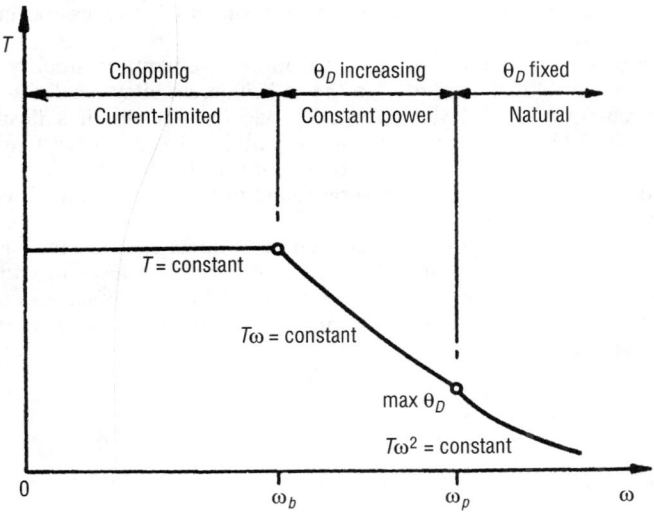

FIGURE 10.36 Torque-speed characteristics of the SRM drive.

where the throttle plate is positioned in real time to match the commanded speed set by the driver. The motor operating speed required to achieve this positioning is relatively low; hence, an open-loop system can be used.

In variable-speed drives, the operational speed is high; hence, the use of closed-loop feedback for commutation is mandatory. Position feedback mechanisms using optointerrupters with slotted disks, resolvers, and encoders are commonly employed at present. However, research in position estimators has led to the development of sensorless commutation methods that use the variation in the coil inductance to estimate commutation instants.

10.7.4 Conclusion

Advanced control methods for elimination of position sensors, minimization of torque ripple, and reduction of acoustic noise have made SRMs suitable for various applications. However, control cost and complexity have prevented them from replacing conventional dc motors in several other applications. The capability of withstanding high temperatures and the ability to perform servo applications without any brushes or magnets is seen as a significant advantage as far as automotive applications are concerned. As reliability and performance become crucial over the next few years, the use of SRMs will start to dominate in the industry.

10.8 BASIC STEPPING-MOTOR CONTROL CIRCUITS*

This section deals with the basic final-stage drive circuitry for stepping motors. This circuitry is centered on a single issue, switching the current in each motor winding on and off and controlling its direction. The circuitry discussed in this section is connected directly to the motor windings and the motor power supply, and this circuitry is controlled by a digital system that determines when the switches are turned on or off.

This section covers all types of motors, from the elementary circuitry needed to control a variable-reluctance motor to the H-bridge circuitry needed to control a bipolar permanent magnet (PM) motor. Each class of drive circuit is illustrated with practical examples, but these examples are not intended as an exhaustive catalog of the commercially available control circuits, nor is the information given here intended to substitute for the information found in the manufacturer's component data sheets for the parts mentioned.

This section covers only the most elementary control circuitry for each class of motor. All of these circuits assume that the motor power supply provides a drive voltage no greater than the motor's rated voltage, and this significantly limits motor performance. Section 10.9, on current-limited drive circuitry, covers practical high-performance drive circuits.

---

* Sections 10.8 to 10.10 contributed by Douglas W. Jones, University of Iowa.

10.8.1 Variable-Reluctance Motors

Typical controllers for variable-reluctance stepping motors are variations on the outline shown in Fig. 10.37. Boxes are used to represent switches; a control unit, not shown, is responsible for providing the control signals to open and close the switches at the appropriate times in order to spin the motors. In many cases, the control unit will be a computer or programmable interface controller, with software directly generating the outputs needed to control the switches, but in other cases, additional control circuitry is introduced, sometimes gratuitously.

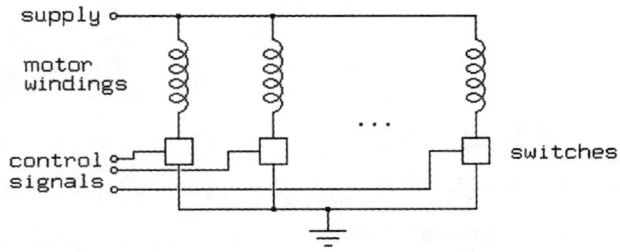

FIGURE 10.37 Typical controller for variable-reluctance stepping motors.

Motor windings, solenoids, and similar devices are all inductive loads. As such, the current through the motor winding cannot be turned on or off instantaneously without involving infinite voltages. When the switch controlling a motor winding is closed, allowing current to flow, the result of this is a slow rise in current. When the switch controlling a motor winding is opened, the result of this is a voltage spike that can seriously damage the switch unless care is taken to deal with it appropriately.

There are two basic ways of dealing with this voltage spike. One is to bridge the motor winding with a diode, and the other is to bridge the motor winding with a capacitor. Figure 10.38 illustrates both approaches.

The diode shown in Fig. 10.38 must be able to conduct the full current through the motor winding, but it will conduct only briefly each time the switch is turned off, as the current through the winding decays. If relatively slow diodes, such as the common 1N400X family, are used together with a fast switch, it may be necessary to add a small capacitor in parallel with the diode.

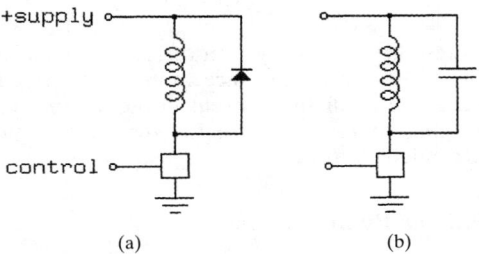

FIGURE 10.38 Voltage-spike control with (a) a diode and (b) a capacitor.

The capacitor shown in Fig. 10.38 poses more complex design problems. When the switch is closed, the capacitor will discharge through the switch to ground, and the switch must be able to handle this brief spike of discharge current. A resistor in series with the capacitor or in series with the power supply will limit this current. When the switch is opened, the stored energy in the motor winding will charge the capacitor up to a voltage significantly above the supply voltage, and the switch must be able to tolerate this voltage. To solve for the size of the capacitor, we equate the two formulas for the stored energy in a resonant circuit.

$$P = \frac{CV^2}{2}$$

$$P = \frac{LI^2}{2}$$

where P = stored energy, W · s or C · V
C = capacity, F
V = voltage across capacitor
L = inductance of motor winding, H
I = current through motor winding

Solving for the minimum size of capacitor required to prevent overvoltage on the switch is fairly easy.

$$C > \frac{LI^2}{(V_b - V_s)^2}$$

where:
V_b = breakdown voltage of the switch
V_s = supply voltage

Variable-reluctance motors have variable inductance that depends on the shaft angle. Therefore, worst-case design must be used to select the capacitor. Furthermore, motor inductances are frequently poorly documented, if at all.

The capacitor and motor winding in combination form a resonant circuit. If the control system drives the motor at frequencies near the resonant frequency of this circuit, the motor current through the motor windings, and therefore the torque exerted by the motor, will be quite different from the steady-state torque at the nominal operating voltage. The resonant frequency is

$$f = \frac{1}{2\pi \, (L \, C)^{0.5}}$$

Again, the electrical resonant frequency for a variable-reluctance motor will depend on shaft angle. When a variable-reluctance motor is operated with the exciting pulses near resonance, the oscillating current in the motor winding will lead to a magnetic field that goes to zero at twice the resonant frequency, and this can severely reduce the available torque.

10.8.2 Unipolar PM and Hybrid Motors

Typical controllers for unipolar stepping motors are variations on the outline shown in Fig. 10.39. As in Fig. 10.37, boxes are used to represent switches; a control unit, not shown, is responsible for providing the control signals to open and close the switches at the appropriate times in order to spin the motors. The control unit is commonly a computer or programmable interface controller, with software directly generating the outputs needed to control the switches.

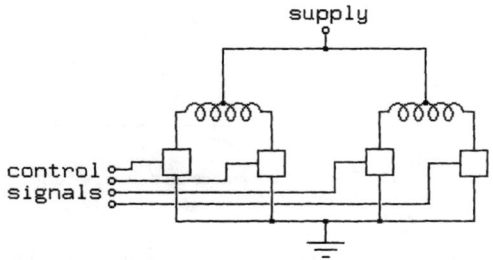

FIGURE 10.39 Typical controller for unipolar stepping motors.

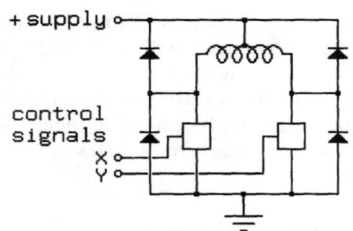

FIGURE 10.40 Hybrid drive with diode shunt suppression.

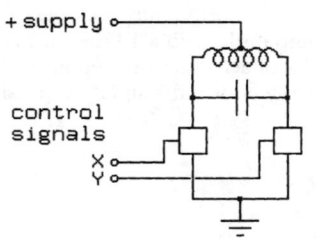

FIGURE 10.41 Hybrid drive with capacitive suppression.

As with drive circuitry for variable-reluctance motors, we must deal with the inductive kick produced when each of these switches is turned off. Again, we may shunt the inductive kick using diodes, but now four diodes are required, as shown in Fig. 10.40.

The extra diodes are required because the motor winding is not two independent inductors; it is a single center-tapped inductor with the center tap at a fixed voltage. This acts as an autotransformer. When one end of the motor winding is pulled down, the other end will fly up, and vice versa. When a switch opens, the inductive kickback will drive that end of the motor winding to the positive supply, where it is clamped by the diode. The opposite end will fly downward, and if it was not floating at the supply voltage at the time, it will fall below ground, reversing the voltage across the switch at that end. Some switches are immune to such reversals, but others can be seriously damaged.

A capacitor may also be used to limit the kickback voltage, as shown in Fig. 10.41. The rules for sizing this capacitor are the same as the rules for sizing the capacitor shown in Fig. 10.38, but the effect of resonance is quite different. With a PM motor, if the capacitor is driven at or near the resonant frequency, the torque will increase to as much as twice the low-speed torque. The resulting torque-speed curve may be quite complex, as illustrated in Fig. 10.42.

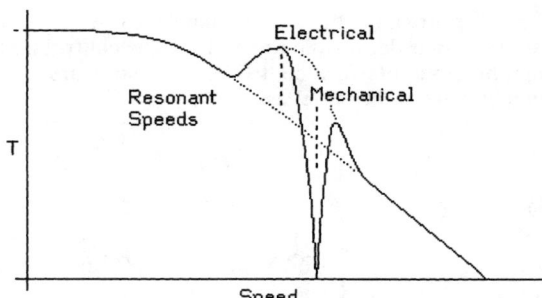

FIGURE 10.42 Effect of resonance on motor performance.

Figure 10.42 shows a peak in the available torque at the electrical resonant frequency and a valley at the mechanical resonant frequency. If the electrical resonant frequency is placed appropriately above what would have been the cutoff speed for the motor using a diode-based driver, the effect can be a considerable increase in the effective cutoff speed.

The mechanical resonant frequency depends on the torque, so if the mechanical resonant frequency is anywhere near the electrical resonance, it will be shifted by the electrical resonance. Furthermore, the width of the mechanical resonance depends on the local slope of the torque-speed curve. If the torque drops with speed, the mechanical resonance will be sharper, while if the torque climbs with speed, the mechanical resonance will be broader or even split into multiple resonant frequencies.

10.8.3 Practical Unipolar and Variable-Reluctance Drivers

In the preceding circuits, the details of the necessary switches are deliberately ignored. Any switching technology, from toggle switches to power MOSFETS, will work. Figure 10.43 contains some suggestions for implementing each switch, with a motor winding and protection diode included for orientation purposes.

Each of the switches shown in Fig. 10.43 is compatible with a TTL input. The 5-V supply used for the logic, including the 7407 open-collector driver used in the figure, should be well regulated. The motor power, typically between 5 and 24 V, needs only

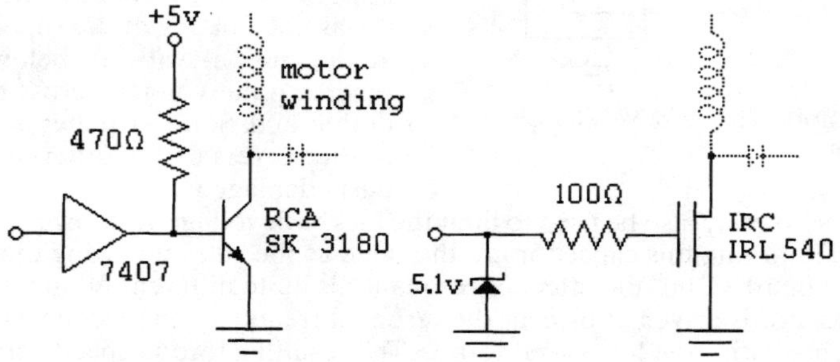

FIGURE 10.43 Possible switching schemes.

minimal regulation. It is worth noting that these power-switching circuits are appropriate for driving solenoids, dc motors, and other inductive loads as well as for driving stepping motors.

The SK3180 transistor shown in Fig. 10.43 is a power Darlington with a current gain >1000; thus, the 10 mA flowing through the 470-Ω bias resistor is more than enough to allow the transistor to switch a few amps of current through the motor winding. The 7407 buffer used to drive the Darlington may be replaced with any high-voltage open-collector chip that can sink at least 10 mA. In the event that the transistor fails, the high-voltage open-collector driver serves to protects the rest of the logic circuitry from the motor power supply.

The IRC IRL540 shown in Fig. 10.43 is a power field-effect transistor. This can handle currents of up to about 20 A, and it breaks down nondestructively at 100 V; as a result, this chip can absorb inductive spikes without protection diodes if it is attached to a large enough heat sink. This transistor has a very fast switching time, so the protection diodes must be comparably fast or bypassed by small capacitors. This is particularly essential with the diodes used to protect the transistor against reverse bias. In the event that the transistor fails, the zener diode and 100-Ω resistor protect the TTL circuitry. The 100-Ω resistor also acts to somewhat slow the switching times on the transistor.

For applications where each motor winding draws under 500 mA, the ULN200x family of Darlington arrays from Allegro Microsystems, also available as the DS200x from National Semiconductor and as the Motorola MC1413 Darlington array, will drive multiple motor windings or other inductive loads directly from logic inputs. Figure 10.44 shows the pinout of the widely available ULN2003 chip, an array of seven Darlington transistors with TTL compatible inputs.

The base resistor on each Darlington transistor is matched to standard bipolar TTL outputs. Each NPN Darlington is wired with its emitter connected to pin 8, intended as a ground pin. Each transistor in this package is protected by two diodes, one shorting the emitter to the collector, protecting against reverse voltages across the transistor, and one connecting the collector to pin 9; if pin 9 is wired to the positive motor supply, this diode will protect the transistor against inductive spikes.

The ULN2803 chip is essentially the same as the ULN2003 chip just described, except that it is in an 18-pin package and contains eight Darlingtons, allowing one chip to be used to drive a pair of common unipolar permanent-magnet or variable-reluctance motors.

For motors drawing under 600 mA per winding, the UDN2547B quad power driver made by Allegro Microsystems will handle all four windings of common unipolar stepping motors. For motors drawing under 300 mA per winding, Texas Instruments SN7541, 7542 and 7543 dual power drivers are a good choice. Both of these alternatives include some logic with the power drivers.

10.8.4 Bipolar Motors and H Bridges

Things are more complex for bipolar PM stepping motors, because these have no center taps on their windings. There-

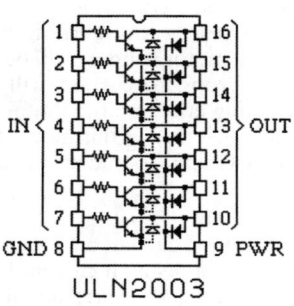

FIGURE 10.44 Pinout diagram of the ULN2003 chip.

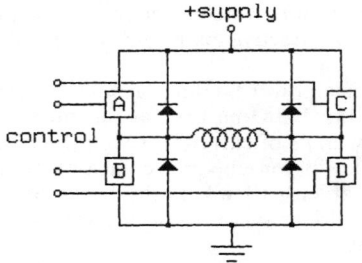

FIGURE 10.45 H-bridge driver.

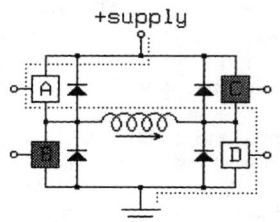

FIGURE 10.46 H-bridge driver in forward mode.

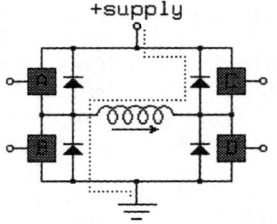

FIGURE 10.47 H-bridge driver in fast-decay mode.

fore, to reverse the direction of the field produced by a motor winding, we need to reverse the current through the winding. We could use a double-pole double-throw switch to do this electromechanically; the electronic equivalent of such a switch is called an *H bridge* and is outlined in Fig. 10.45.

As with the unipolar drive circuits discussed previously, the switches used in the H bridge must be protected from the voltage spikes caused by turning the power off in a motor winding. This is usually done with diodes, as shown in Fig. 10.45.

It is worth noting that H bridges are applicable not only to the control of bipolar stepping motors, but also to the control of dc motors, push-pull solenoids (those with PM plungers) and many other applications.

With four switches, the basic H bridge offers 16 possible operating modes, 7 of which short out the power supply. The following operating modes are of interest:

- Forward mode, switches A and D closed.
- Reverse mode, switches B and C closed.

These are the usual operating modes, allowing current to flow from the supply, through the motor winding, and onward to ground. Figure 10.46 illustrates forward mode.

Fast-Decay or Coasting Mode, All Switches Open. Any current flowing through the motor winding will be working against the full supply voltage, plus two diode drops, so current will decay quickly. This mode provides little or no dynamic-braking effect on the motor rotor, so the rotor will coast freely if all motor windings are powered in this mode. Figure 10.47 illustrates the current flow immediately after switching from forward running mode to fast-decay mode.

Slow Decay or Dynamic Braking Modes. In these modes, current may recirculate through the motor winding with minimum resistance. As a result, if current is flowing in a motor winding when one of these modes is entered, the current will decay slowly, and if the motor rotor is turning, it will induce a current that will act as a brake on the rotor. Figure 10.48 illustrates one of the many useful slow-decay modes,

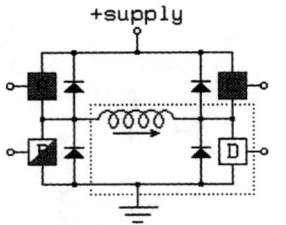

FIGURE 10.48 H-bridge driver in slow-decay or dynamic-braking mode.

with switch D closed; if the motor winding has recently been in forward running mode, the state of switch B may be either open or closed.

Most H bridges are designed so that the logic necessary to prevent a short circuit is included at a very low level in the design. Figure 10.49 illustrates what is probably the best arrangement. Here, the operating modes shown in Table 10.15 are available.

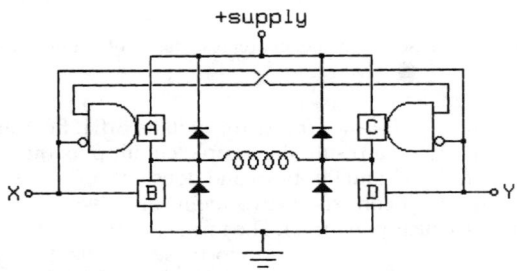

FIGURE 10.49 H-bridge driver with short-circuit protection.

The advantage of this arrangement is that all of the useful operating modes are preserved, and they are encoded with a minimum number of bits; the latter is important when using a microcontroller or computer system to drive the H bridge, because many such systems have only limited numbers of bits available for parallel output. Sadly, few of the integrated H-bridge chips on the market have such a simple control scheme.

10.8.5 Practical Bipolar Drive Circuits

There are a number of integrated H-bridge drivers on the market, but it is still useful to look at discrete component implementations for an understanding of how an H bridge works. Antonio Raposo suggested the H-bridge circuit shown in Fig. 10.50.

The X and Y inputs to this circuit can be driven by open-collector TTL outputs as in the Darlington-based unipolar drive circuit in Fig. 10.43. The motor winding will be energized if exactly one of the X and Y inputs is high and exactly one of them is

TABLE 10.15 Operating Modes of Driver Circuit in Fig. 10.49

XY	ABCD	Mode
00	0000	Fast decay
01	1001	Forward
10	0110	Reverse
11	0101	Slow decay

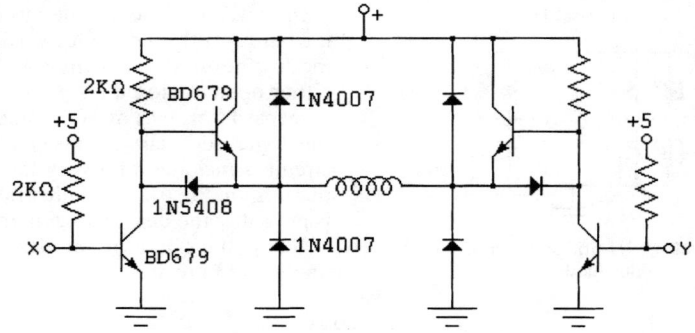

FIGURE 10.50 Discrete component representation of H-bridge driver.

low. If both are low, both pull-down transistors will be off. If both are high, both pull-up transistors will be off. As a result, this simple circuit puts the motor in dynamic-braking mode in both the 11 and 00 states and does not offer a coasting mode.

The circuit in Fig. 10.50 consists of two identical halves, each of which may be properly described as a push-pull driver. The term *half-H bridge* is sometimes applied to these circuits. It is also worth noting that a half-H bridge has a circuit quite similar to the output drive circuit used in TTL logic. In fact, TTL tristate line drivers such as the 74LS125A and the 74LS244 can be used as half-H bridges for small loads, as illustrated in Fig. 10.51.

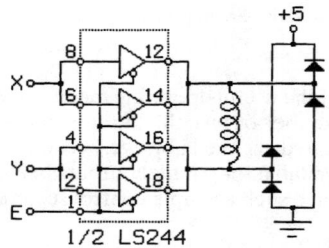

FIGURE 10.51 Half-bridge driver.

This circuit is effective for driving motors with up to about 50 Ω per winding at voltages up to about 4.5 V using a 5-V supply. Each tristate buffer in the LS244 can sink about twice the current it can source, and the internal resistance of the buffers is sufficient, when sourcing current, to evenly divide the current between the drivers that are run in parallel. This motor drive allows for all of the useful states achieved by the driver in Fig. 10.49 (see Table 10.16), but these states are not encoded as efficiently. The second dynamic-braking mode, $XYE = 110$, provides a slightly weaker braking effect than the first because of the fact that the LS244 drivers can sink more current than they can source.

TABLE 10.16 Operating Modes of Driver Circuit in Fig. 10.51

XYE	Mode
001	Fast decay
000	Slower decay
010	Forward
100	Reverse
110	Slow decay

One of the problems with commercially available stepping-motor control chips is that many of them have relatively short market lifetimes. For example, the Seagate IPxMxx series of dual H-bridge chips (IP1M10 through IP3M12) were very well thought out; unfortunately, it appears that Seagate made these only while the company used stepping motors for head positioning in Seagate disk drives. The Toshiba TA7279 dual H-bridge driver would be another excellent choice for motors under 1 A, but again, it appears to have been made for internal use only.

The SGS-Thompson (and others) L293 dual H bridge is a close competitor for the preceding chips, but unlike them, it does not include protection diodes. As a result, each motor winding must be set across a bridge rectifier (1N4001 equivalent). Despite this drawback, the L293 is an excellent choice for driving small bipolar steppers drawing up to 1 A per motor winding at up to 36 V. Figure 10.52 shows the pinout of this chip. This chip may be viewed as four independent half-H bridges, enabled in pairs, or as two full H bridges. This is a power dual in-line package (DIP), with pins 4, 5, 12, and 13 designed to conduct heat to the PC board or to an external heat sink.

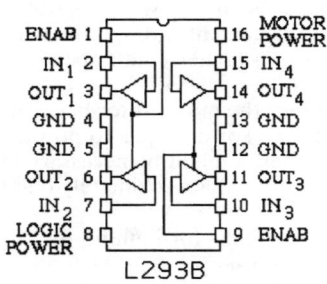

FIGURE 10.52 Pinout of the L293B chip.

The SGS-Thompson (and others) L298 dual H bridge is quite similar to the preceding, but is able to handle up to 2 A per channel and is packaged as a power component; as with the LS244, it is safe to wire the two H bridges in the L298 package into one 4-A H bridge (the data sheet for this chip provides specific advice on how to do this). One warning is appropriate concerning the L298—this chip has very fast switches, fast enough that commonplace protection diodes (1N400X equivalent) do not work. Instead, use a diode such as the BYV27. The National Semiconductor LMD18200 H bridge is another good example; this handles up to 3 A and has integral protection diodes.

While integrated H bridges are not available for very high currents or very high voltages, there are well-designed components on the market to simplify the construction of H bridges from discrete switches. For example, International Rectifier sells a line of half-H-bridge drivers; two of these chips plus four MOSFET switching transistors suffice to build an H bridge. The IR2101, IR2102, and IR2103 are basic half-H-bridge drivers. Each of these chips has two logic inputs to directly control the two switching transistors on one leg of an H bridge. The IR2104 and IR2111 have similar output-side logic for controlling the switches of an H bridge, but they also include input-side logic that, in some applications, may reduce the need for external logic. In particular, the 2104 includes an enable input, so that four 2104 chips plus eight switching transistors can replace an L293 with no need for additional logic.

A number of manufacturers make complex H-bridge chips that include current-limiting circuitry. It is also worth noting that there are a number of three-phase bridge drivers on the market, appropriate for driving Y- or delta-configured three-phase pm steppers. Few such motors are available, and these chips were not developed with steppers in mind. Nonetheless, the Toshiba TA7288P, GL7438, TA8400 and TA8405 are clean designs, and two such chips, with one of the six half-bridges ignored, will cleanly control a five-winding 10-step-per-revolution motor.

10.9 CURRENT LIMITING FOR STEPPING MOTORS

Small stepping motors, such as those used for head positioning on floppy disk drives, are usually driven at a low dc voltage, and the current through the motor windings is usually limited by the internal resistance of the winding. High-torque motors, on the other hand, are frequently built with very low resistance windings; when driven by any reasonable supply voltage, these motors typically require external current-limiting circuitry.

There is good reason to run a stepping motor at a supply voltage above that needed to push the maximum rated current through the motor windings. Running a motor at higher voltages leads to a faster rise in the current through the windings when they are turned on, and this, in turn, leads to a higher cutoff speed for the motor and higher torques at speeds above the cutoff.

Microstepping, where the control system positions the motor rotor between half steps, also requires external current-limiting circuitry. For example, to position the rotor one-fourth of the way from one step to another, it might be necessary to run one motor winding at full current while the other is run at approximately one-third of that current.

The remainder of this section discusses various circuits for limiting the current through the windings of a stepping motor, starting with simple resistive limiters and moving up to choppers and other switching regulators. Most of these current limiters are appropriate for many other applications, including limiting the current through conventional dc motors and other inductive loads.

10.9.1 Resistive Current Limiters

The easiest current limiter to understand is a series resistor. Most motor manufacturers recommended this approach in their literature up until the early 1980s, and most motor data sheets still give performance curves for motors driven by such circuits. The typical circuits used to control the current through one winding of a pm or hybrid motor are shown in Fig. 10.53.

R_1 in Fig. 10.53 limits the current through the motor winding. Given a rated current of I and a motor winding with a resistance R_W, Ohm's law sets the maximum supply voltage as $I(R_W + R_1)$. Given that the inductance of the motor motor winding is L_W, the time constant for the motor winding will be $L_W/(R_W + R_1)$. Figure 10.54

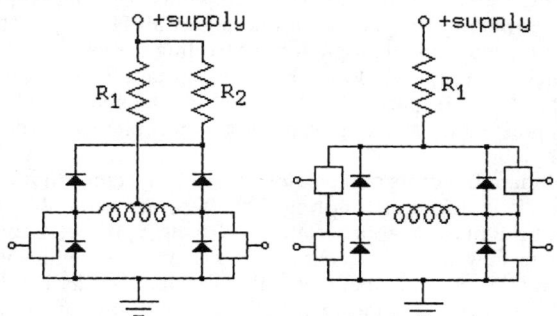

FIGURE 10.53 Typical hybrid current-control circuit.

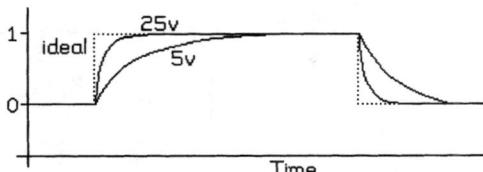

FIGURE 10.54 Effect of increasing resistance on rise and fall times of stepping-motor circuit.

illustrates the effect of increasing the resistance and the operating voltage on the rise and fall times of the current through one winding of a stepping motor.

R_2 is shown only in the unipolar example in Fig. 10.53 because it is particularly useful there. For a bipolar H-bridge drive, when all switches are turned off, current flows from ground to the motor supply through R_1, so the current through the motor winding will decay quite quickly. In the unipolar case, R_2 is necessary to equal this performance.

Note. When the switches in the H-bridge circuit shown in Fig. 10.53 are opened, the direction of current flow through R_1 will reverse almost instantaneously. If R_1 has any inductance—for example, if it is wire-wound—either it must be bypassed with a capacitor to handle the voltage kick caused by this current reversal, or R_2 must be added to the H bridge.

Given the rated maximum current through each winding and the supply voltage, the resistance and wattage of R_1 are easy to compute. If it is included, R_2 poses more interesting problems. The resistance of R_2 depends on the maximum voltage the switches can handle. For example, if the supply voltage is 24 V, and the switches are rated at 75 V, the drop across R_2 can be as much as 51 V without harming the transistors. Given an operating current of 1.5 A, R_2 can be a 34-Ω resistor. Note that an interesting alternative is to use a zener diode in place of R_2.

Figuring the peak average power R_2 must dissipate is a wonderful exercise in dynamics; the inductance of the motor windings is frequently undocumented and may vary with the rotor position. The power dissipated by R_2 also depends on the control system. The worst case occurs when the control system chops the power to one winding at a high enough frequency that the current through the motor winding is effectively constant; the maximum power is then a function of the duty cycle of the chopper and the ratios of the resistances in the circuit during the on and off phases of the chopper. Under normal operating conditions, the peak power dissipation will be significantly lower.

10.9.2 Linear Current Limiters

A pair of high-wattage power resistors can cost more than a pair of power transistors plus a heat sink, particularly if forced-air cooling is available. Furthermore, a transistorized constant-current source, as shown in Fig. 10.55, will give faster rise times through the motor windings than the current-limiting resistor shown in Fig. 10.53. This is because a current source will deliver the full supply voltage across the motor winding until the current reaches the rated current; only then will the current source drop the voltage.

In Fig. 10.55, a transistorized current source ($T_1 + R_1$) has been substituted for the current-limiting resistor R_1 used in the examples in Fig. 10.53. The regulated voltage

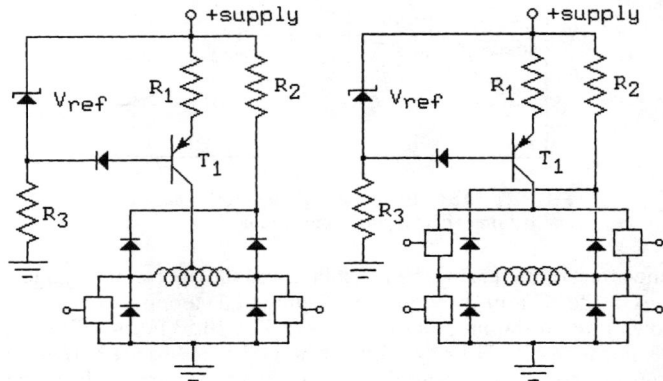

FIGURE 10.55 Constant-current circuits.

supplied to the base of T_1 serves to regulate the voltage across the sense resistor R_1, and this, in turn, maintains a constant current through R_1 so long as any current is allowed to flow through the motor winding. Typically, R_1 will have as low a resistance as possible, in order to avoid the high cost of a power resistor. For example, if the forward voltage drops across the diode in series with the base T_1 and V_{BE} for T_1 are both 0.65 V, and if a 3.3-V zener diode is used for a reference, the voltage across R_1 will be maintained at about 2.0 V, so if R_1 is 2 Ω, this circuit will limit the current to 1 A, and R_1 must be able to handle 2 W.

R_3 in Fig. 10.55 must be sized in terms of the current gain of T_1 so that sufficient current flows through R_1 and R_3 to allow T_1 to conduct the full rated motor current.

The transistor T_1 used as a current regulator in Fig. 10.55 is run in linear mode; therefore, it must dissipate quite a bit of power. For example, if the motor windings have a resistance of 5 Ω and a rated current of 1 A, and a 25-V power supply is used, $T_1 + R_1$ will dissipate, between them, 20 W. The circuits discussed in the following sections avoid this waste of power while retaining the performance advantages of the circuit given here.

When an H-bridge bipolar drive is used with a resistive current limiter, as shown in Fig. 10.53, the resistor R_2 is not needed because current can flow backward through R_1. When a transistorized current limiter is used, current cannot flow backward through T_1, so a separate current path back to the positive supply must be provided to handle the decaying current through the motor windings when the switches are opened. R_2 serves this purpose here, but a zener diode may be substituted to provide even faster turn-off.

The performance of a motor run with a current-limited power supply is noticeably better than the performance of the same motor run with a resistively limited supply, as illustrated in Fig. 10.56.

With either a current-limited supply or a resistive current limiter, the initial rate of increase of the current through the inductive motor winding when the power is turned on depends only on the inductance of the winding and the supply voltage. As the current increases, the voltage drop across a resistive current limiter will increase, dropping the voltage applied to the motor winding, and therefore dropping the rate of increase of the current through the winding. As a result, the current will approach the rated current of the motor winding only asymptotically.

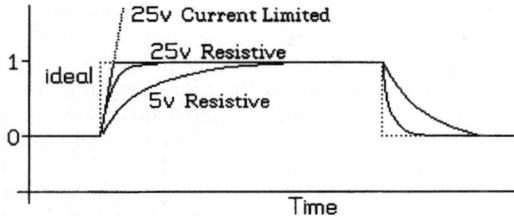

FIGURE 10.56 Performance of resistively limited supply.

In contrast, with a pure current limiter, the current through the motor winding will increase almost linearly until the current limiter cuts in, allowing the current to reach the limit value quite quickly. In fact, the current rise is not linear; rather, the current rises asymptotically toward a limit established by the resistance of the motor winding and the resistance of the sense resistor in the current limiter. This maximum is usually well above the rated current for the motor winding.

10.9.3 Open-Loop Current Limiters

Both the resistive and the linear transistorized current limiters just discussed automatically limit the current through the motor winding, but at a considerable cost in terms of wasted heat. There are two schemes that eliminate this expense, although at some risk because of the lack of feedback about the current through the motor.

Use of a Voltage Boost. If we plot the voltage across the motor winding as a function of time, assuming the use of a transistorized current limiter such as is illustrated in Fig. 10.54, and assuming a 1-A 5-Ω motor winding, the result will be something like that illustrated in Fig. 10.57. As long as the current is below the current limiter's set point, almost the full supply voltage is applied across the motor winding. Once the current reaches the set point, the voltage across the motor winding falls to that needed to sustain the current at the set point, and when the switches open, the voltage reverses briefly as current flows through the diode network and R_2. An alternative way to get this voltage profile is to use a dual-voltage power supply, turning on the high voltage for as long as it takes to bring the current in the motor winding up to the rated current, and then turning off the high voltage and turning on the sus-

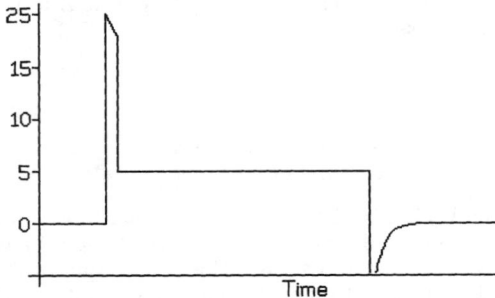

FIGURE 10.57 Plot of voltage versus time with a transistorized current limiter.

taining voltage. Some motor controllers do this directly, without monitoring the current through the motor windings. This provides excellent performance and minimizes power losses in the regulator, but it offers a dangerous temptation.

If the motor does not deliver enough torque, it is tempting to simply lengthen the high-voltage pulse at the time the motor winding is turned on. This will usually provide more torque, although saturation of the magnetic circuits frequently leads to less torque than might be expected, but the cost is high. The risk of burning out the motor is quite real, as is the risk of demagnetizing the motor rotor if it is turned against the imposed field while running hot. Therefore, if a dual-voltage supply is used, the temptation to raise the torque in this way should be avoided.

The problems with dual-voltage supplies are particularly serious when the time intervals are under software control, because in this case, it is common for the software to be written by a programmer who is insufficiently aware of the physical and electrical characteristics of the control system.

Use of Pulse-Width Modulation (PWM). Another alternative approach to controlling the current through the motor winding is to use a simple power supply controlled by PWM or by a chopper. During the time the current through the motor winding is increasing, the control system leaves the supply attached with a 100 percent duty cycle. Once the current is up to the full rated current, the control system changes the duty cycle to that required to maintain the current. Figure 10.58 illustrates this scheme.

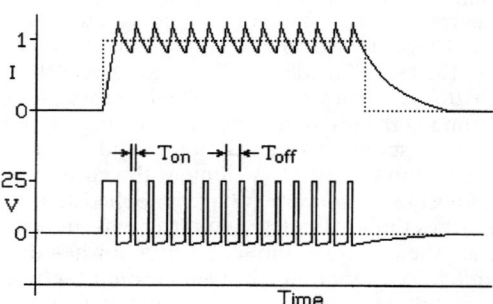

FIGURE 10.58 Pulse-width modulation current control.

For any chopper or PWM, we can define the duty cycle D as the fraction of each cycle that the switch is closed.

$$D = \frac{T_{on}}{T_{on} + T_{off}}$$

where T_{on} = time switch is closed during each cycle
T_{off} = time switch is open during each cycle

The voltage curve shown in Fig. 10.58 indicates the full supply voltage being applied to the motor winding during the on phase of every chopper cycle, while when the chopper is off, a negative voltage is shown. This is the result of the forward voltage drop in the diodes that are used to shunt the current when the switches turn off, plus the external resistance used to speed the decay of the current through the motor winding.

For large values of T_{on} or T_{off}, the exponential nature of the rise and fall of the current through the motor winding is significant, but for sufficiently small values, we can approximate these as linear. Assuming that the chopper is working to maintain a current I and that the amplitude is small, we can approximate the rates of rise and fall in the current in terms of the voltage across the motor winding when the switch is closed and when it is open.

$$V_{on} = V_{supply} - I(R_{winding} + R_{on})$$

$$V_{off} = V_{diode} + I(R_{winding} + R_{off})$$

Here, we lump together all resistances in series with the winding and power supply in the on state as R_{on}, and we lump together all resistances in the current recirculation path when the switches are open as R_{off}. The forward voltage drops of any diodes in the current recirculation path are lumped as V_{diode}; if the off-state recirculation path runs from ground to the power supply (H-bridge fast-decay mode), the supply voltage must also be included in V_{diode}. Forward voltage drops of any switches in the on-state and off-state paths should also be incorporated into these voltages. To solve for the duty cycle, we first note that:

$$\frac{dI}{dt} = \frac{V}{L}$$

where I = current through motor winding
V = voltage across winding
L = inductance of winding

We then substitute the specific voltages for each phase of operation:

$$\frac{I_{ripple}}{T_{off}} = \frac{V_{off}}{L}$$

$$\frac{I_{ripple}}{T_{on}} = \frac{V_{on}}{L}$$

where I_{ripple} = peak-to-peak ripple in the current

Solving for T_{off} and T_{on} and then substituting these into the definition of the duty cycle of the chopper, we get

$$D = \frac{T_{on}}{T_{on} + T_{off}} = \frac{V_{off}}{V_{on} + V_{off}}$$

If the forward voltage drops in diodes and switches are negligible, and if the only significant resistance is that of the motor winding itself, this simplifies to

$$D = \frac{IR_{winding}}{V_{supply}} = \frac{V_{running}}{V_{supply}}$$

This special case is particularly desirable because it delivers all of the power to the motor winding, with no losses in the regulation system, without regard for the difference between the supply voltage and the running voltage.

The ac ripple I_{ripple} superimposed on the running current by a chopper can be a source of minor problems; at high frequencies, it can be a source of radio frequency

(RF) emissions, and at audio frequencies, it can be a source of annoying noise. For example, with audio-frequency chopping, most stepper-controlled systems will squeal, sometimes loudly, when the rotor is displaced from the equilibrium position. To find the ripple amplitude, first recall that:

$$\frac{I_{\text{ripple}}}{T_{\text{off}}} = \frac{V_{\text{off}}}{L}$$

Then solve for I_{ripple}.

$$I_{\text{ripple}} = \frac{T_{\text{off}} V_{\text{off}}}{L}$$

Thus, to reduce the ripple amplitude at any particular duty cycle, it is necessary to increase the chopper frequency. This cannot be done without limit because switching losses increase with frequency. Note that this change has no significant effect on ac losses; the decrease in such losses due to decreased amplitude in the ripple is generally offset by the effect of increasing frequency.

The primary problem with use of a simple chopping or PWM control scheme is that it is completely open loop. Design of good chopper-based control systems requires knowledge of motor characteristics such as inductance that are frequently poorly documented, and as with dual-voltage supplies, when motor performance is marginal, it is very tempting to increase the duty cycle without attention to the long-term effects of this on the motor. In the designs that follow, this weakness is addressed by introducing feedback loops into the low-level drive system to directly monitor the current and determine the duty cycle.

10.9.4 One-Shot Feedback Current Limiting

The most common approach to automatically adjusting the duty cycle of the switches in the stepper driver involves monitoring the current to the motor windings; when it rises too high, the winding is turned off for a fixed interval. This requires a current-sensing system and a one-shot, as illustrated in Fig. 10.59.

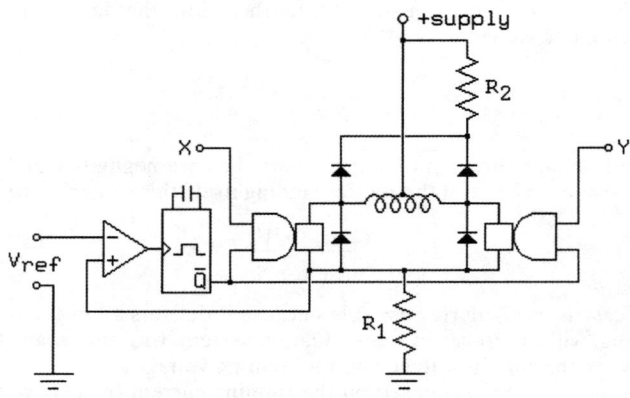

FIGURE 10.59 One-shot current-sensing circuit.

Figure 10.59 illustrates a unipolar drive system. As with the circuit given in Fig. 10.55, R_1 should be as small as possible, limited only by the requirement that the sense voltage provided to the comparator must be high enough to be within its operating range. Note that when the one-shot output $\overline{Q}$ is low, the voltage across R_1 no longer reflects the current through the motor winding. Therefore, the one-shot must be insensitive to the output of the comparator between the time it fires and the time it resets. Practical circuit designs using this approach involve some complexity to meet this constraint.

Selecting the value of R_2 for the circuit shown in Fig. 10.59 poses problems. If R_2 is large, the current through the motor windings will decay quickly when the higher level control system turns off this motor winding, but when the winding is turned on, the current ripple will be large, and the power lost in R_2 will be significant. If R_2 is small, this circuit will be very energy efficient, but the current through the motor winding will decay only slowly when this winding is turned off, and this will reduce the cutoff speed of the motor.

The peak power dissipated in R_2 will be $I^2 R_2$ during T_{off} and 0 during T_{on}; thus, the average power dissipated in R_2 when the motor winding is on will be

$$P_2 = \frac{I^2 R\, T_{\text{off}}}{T_{\text{on}} + T_{\text{off}}}$$

Recall that the duty cycle D is defined as $T_{\text{on}}/(T_{\text{on}} + T_{\text{off}})$ and may be approximated as $V_{\text{running}}/V_{\text{supply}}$. As a result, we can approximate the power dissipation as

$$P_2 = I^2 R_2 \left(1 - \frac{V_{\text{running}}}{V_{\text{supply}}}\right)$$

Given the usual safety margins used in selecting power resistor wattages, a better approximation is not necessary.

When designing a control system based on PWM, note that the cutoff time for the one-shot determines T_{off} and that this is fixed, determined by the timing network attached to the one-shot. Ideally, this should be set as follows:

$$T_{\text{off}} = \frac{L I_{\text{ripple}}}{V_{\text{off}}}$$

This presumes that the inductance L of the motor winding is known, that the acceptable magnitude of I_{ripple} is known, and that V_{off}, the total reverse voltage in the current recirculation path, is known and fixed.

Note that this scheme leads to a variable chopping rate. As with the linear current limiters shown in Fig. 10.55, the full supply voltage will be applied during the turn-on phase, and the chopping action begins only when the motor winding reaches the current limit set by V_{ref}. This circuit will vary the chopping rate to compensate for changes in the counter-emf of the motor winding, for example, those caused by rotor motion; in this regard, it offers the same quality of regulation as the linear current limiter.

The one-shot current regulator shown in Fig. 10.59 can also be applied to an H-bridge regulator. The encoded H bridge shown in Fig. 10.49 is an excellent candidate for this application, as shown in Fig. 10.60.

Unlike the circuit in Fig. 10.59, this circuit does not provide design tradeoffs in the selection of the resistance in the current decay path; instead, it offers the same selection of decay paths as is available in the original circuit from Fig. 10.49. If the X and Y control inputs are held in a running mode (01 or 10), the current limiter will alter-

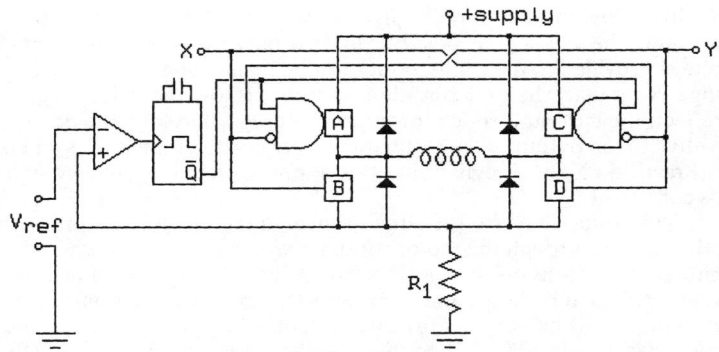

FIGURE 10.60 Current-regulated H bridge.

nate between that running mode and slow-decay modes, maximizing energy efficiency. When the time comes to turn off the current through the motor winding, the X and Y inputs may be set to 00, using fast-decay mode to maximize the cutoff speed, while if the damping effect of dynamic braking is needed to control resonance, X and Y may be set to 11.

Note that the current recirculation path during dynamic braking does not pass through R_1; as a result, if the motor generates a large amount of power, burnt-out components in the motor or controller are likely. This is unlikely to cause problems with stepping motors, but when dynamic braking is used with dc motors, the current limiter should be arranged to remain engaged while in braking mode.

Practical Examples. SGS-Thompson (and others) L293 (1 A) and L298 (2 A) dual H bridges are designed for easy use with partial-feedback current limiters. These chips have enable inputs for each H bridge that can be directly connected to the output of the one-shot, and they have ground connections for motor power that are isolated from their logic ground connections; this allows sense resistors to be easily incorporated into the circuit.

The 3952 H bridge from Allegro Microsystems can handle up to 2 A at 50 V and incorporates all of the logic necessary for current control, including comparators and one-shot.

This chip is available in many package styles; Fig. 10.61 illustrates the DIP configuration wired for a constant-current limit.

If R_t is 20 Ω and C_t is 1000 pF, T_{off} for the PWM will be fixed at 20 (±2) μs. The 3952 chip incorporates a 10-to-1 voltage divider on the V_{ref} input, so attaching V_{ref} to the 5-V logic supply sets the actual reference voltage to 0.5 V. Thus, if the sense resistor R_S is 0.5 Ω, this arrangement will attempt to maintain a regulated current through the load of 1 A.

Note that all power-switching chips are potentially serious sources of electromagnetic interference. The 47 μF capacitor shown between the motor power and ground should be as close to the chip as possible, and the path from the sense pin through R_S to ground and back to a ground pin of the chip should be very short and have a very low resistance.

On the 5-V side, because V_{ref} is taken from V_{cc}, a small decoupling capacitor should be placed very close to the chip. It may even be appropriate to isolate the V_{ref} input from V_{cc} with a small series resistor and a separate decoupling capacitor. If this

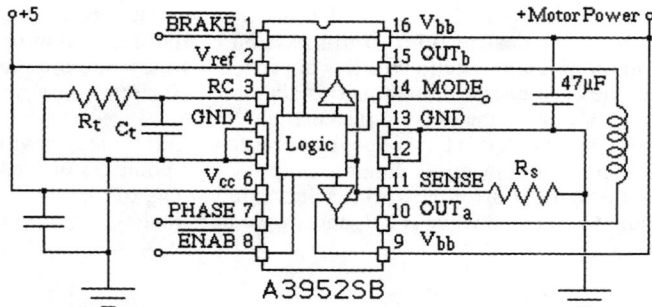

FIGURE 10.61 H-bridge DIP configuration wired for a constant current limit.

is done, note that the resistance from the V_{ref} pin to ground through the chip's internal voltage divider is around 50 kΩ.

One of the more dismaying features of the 3952 chip, as well as many of its competitors, is the large number of control inputs. These are summarized in Table 10.17.

In the forward and reverse running modes, the mode input determines whether fast- or slow-decay modes are used during T_{off}. In the dynamic-braking modes, the mode input determines whether the current limiter is enabled. This is of limited value with stepping motors, but use of dynamic braking without a current limiter can be dangerous with dc motors.

In sleep mode, the power consumption of the chip is minimized. From the perspective of the load, sleep and standby modes put the load into fast-decay mode (all switches off) but in sleep mode, the chip draws considerably less power, both from the logic supply and the motor supply.

10.9.5 Hysteresis Feedback Current Limiting

In many cases, motor-control systems are expected to operate acceptably with a number of different stepping motors. The one-shot-based current regulators illustrated in Figs. 10.59 and 10.60 have an accuracy that depends on the inductance of the motor windings. Therefore, if fixed accuracy is required, any motor substitution must be balanced by changes to the RC network that determines the off time of the one-shot.

TABLE 10.17 Control Inputs of 3952 Chip

		Pin				
Brake	Enable	Phase	Mode	Out_a	Out_b	Notes
0	—	—	0	0	0	Brake
0	—	—	1	0	0	Limited brake
1	1	—	0	—	—	Standby
1	1	—	1	—	—	Sleep
1	0	0	0	0	1	Reverse, slow
1	0	0	1	0	1	Reverse, fast
1	0	1	0	1	0	Forward, slow
1	0	1	1	1	0	Forward, fast

This subsection deals with alternative designs that eliminate the need for this tuning. These alternative designs offer fixed-precision current regulation over a wide range of load inductances. The key to this approach is to arrange the recirculation paths so that the current-sense resistor R_1 is always in the circuit, and then turn the switches on or off depending only on the current.

The usual way to build this type of controller is to use a comparator with a degree of hysteresis—for example, by feeding the output of the comparator back into one of its inputs through a resistor network, as illustrated in Fig. 10.62.

To compute the desired values of R_2 and R_3, we note that:

$$V_{\text{ripple}} \geq V_{\text{hysteresis}}$$

where
$V_{\text{ripple}} = I_{\text{ripple}} R_1$
I_{ripple} = maximum ripple allowed in the current
$V_{\text{hysteresis}} = V_{\text{swing}} R_2 / (R_2 + R_3)$
V_{swing} = voltage swing at comparator output

We can solve this for the ratio of the resistances as follows:

$$\frac{R_2}{R_2 + R_3} \leq \frac{I_{\text{ripple}} R_1}{V_{\text{swing}}}$$

For example, if $R_1 = 0.5\ \Omega$ and we wish to regulate the current to within 10 mA, using a comparator with TTL compatible outputs and a voltage swing of 4 V, the ratio must be no greater than 0.00125. Note that the sum $R_2 + R_3$ determines the loading on V_{ref}, assuming that the input resistance of the comparator is effectively infinite. Typically, therefore, this sum is made quite large. One problem with the circuit given in Fig. 10.62 is that it does not limit the current through the motor in dynamic-braking or slow-decay modes. Even if the current through the sense resistor vastly exceeds the desired current, switches B and D will remain closed in dynamic-braking mode, and if the reference voltage is variable, rapid drops in the reference voltage will not be enforced by this control system.

The designers of the Allegro 3952 chip faced this problem and passed the solution back to the user, providing a mode input to determine whether the chopper alternates between running and fast-decay mode or running and slow-decay mode.

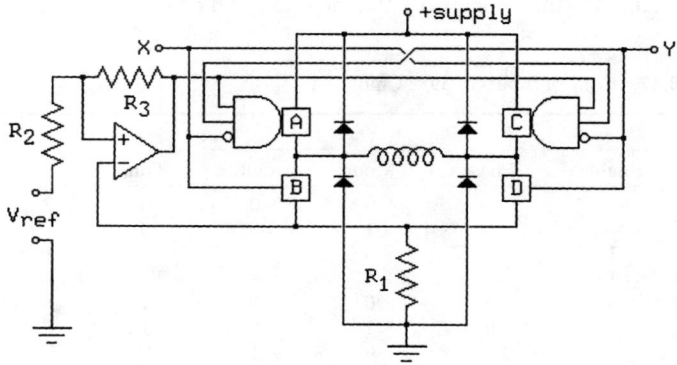

FIGURE 10.62 Resistive-feedback controller.

Note that this chip uses a fixed off-time set by a one-shot; therefore, switching between the two decay modes will change the precision of the current regulator. Given that such a change in precision is acceptable, we can modify the circuit from Fig. 10.62 to automatically throw the system into fast-decay mode if the running or dynamic-braking current exceeds the set point of the comparator by too great a margin. Figure 10.63 illustrates how this can be done using a second comparator.

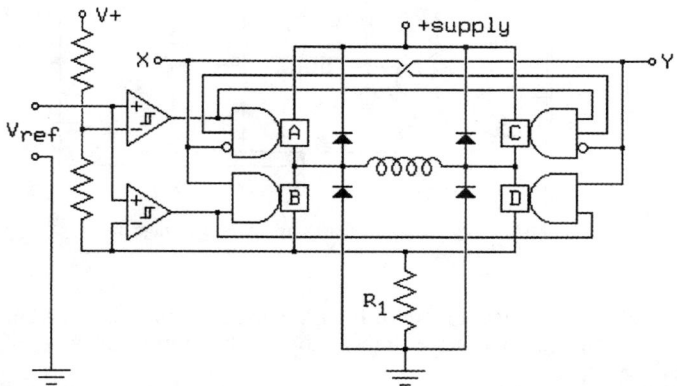

FIGURE 10.63 Fast-decay restrictive-feedback controller.

As shown in Fig. 10.63, the lower comparator directly senses the voltage across R_1, while the upper comparator senses a higher voltage, determined by a resistor network. This network should hold the negative inputs of the two comparators just far enough apart to guarantee that, as the voltage across R_1 rises, the top comparator will always open the top switches before the bottom comparator opens the bottom switches, and as the voltage across R_1 falls, the bottom comparator will always close the bottom switches before the top comparator closes the top switches.

As a result, this system has two basic steady-state running modes. If the motor winding is drawing power, one of the bottom switches will remain closed while the opposite switch on the top is used to chop the power to the motor winding, alternating the state of the system between running and slow-decay mode.

If the motor winding is generating power, the top switches will remain closed, and the bottom switches will do the chopping, alternating between fast-decay and slow-decay modes as needed to keep the current within limits.

If the two comparators have accuracies on the order of 1 mV with hysteresis on the order of 5 mV, it is reasonable to use a 5-mV difference between the top and bottom comparators. If we use the 5-V logic supply as the pull-up supply for the resistor network, and we assume a nominal operating threshold of around 0.5 V, the resistor network should have a ratio of 900 to 1; for example, a 90-kΩ resistor from +5 and a 100-Ω resistor between the two comparator inputs.

Practical Examples. The basic idea described in this section is also applicable to unipolar stepping-motor controllers, although in this context, it is somewhat easier to apply if the reference voltage is measured with respect to the unregulated motor power supply. Figure 10.64 illustrates a practical example, using the forward voltage drop across an ordinary silicon diode as the reference voltage. This circuit uses a 2.4-

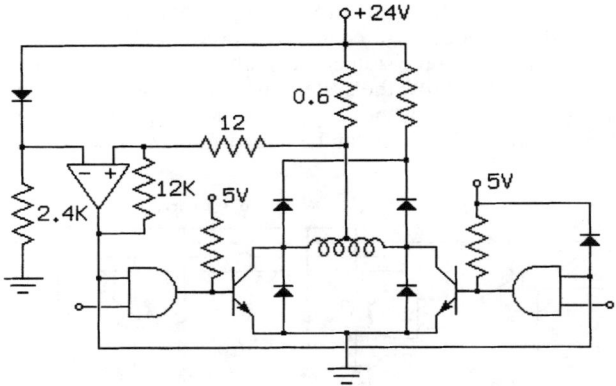

FIGURE 10.64 Unipolar controller using forward voltage drop.

kΩ resistor to provide a bias current of 10 mA to the reference diode. A small capacitor should be added across the reference diode if the motor power supply is minimally regulated.

The 0.6-Ω value used for the current-sensing resistor sets the regulator to 1 A, assuming that the reference voltage is 0.6 V. The 1000-to-1 ratio on the feedback network around the comparator sets the allowed ripple in the regulated current to around 8 mA. The comparator shown in Fig. 10.64 can be powered from the minimally regulated motor power supply, but only if it is able to operate with the inputs very close to its positive supply voltage. The Mitsubishi M5249L comparator appears to be ideally suited to this job; it can work from a positive supply of up to 40 V, and the input voltages are allowed to slightly exceed the positive supply voltage. The output of this comparator is open collector, so the hysteresis network shown in Fig. 10.64 also acts as a pull-up network, providing a pull-up current of a few milliamps. The diode to +5 shown in the figure clamps the comparator output to the logic supply voltage, protecting the AND gate inputs from overvoltage.

10.9.6 Other Current-Sensing Technologies

The feedback loops of all of the current limiters given in the preceding subsections use the voltage drop across a small resistor to measure the current. This is an excellent choice for small motors, but it poses difficulties for large high-current motors. There are other current-sensing technologies appropriate for such settings, most notably those that deliver only a fraction of the motor current to the sensing resistor and those that measure the current by sensing the magnetic field around the conductor.

National Semiconductor had incorporated a very clever current sensor into a number of its H bridges. This sensor delivers a current to the sense resistor that is proportional to the current through the motor winding, but far lower. For example, on the LMD18200 H-bridge, the sense resistor receives exactly 377 mA per ampere flowing through the motor winding. The key to the current sensing technology used in the National Semiconductor line of H bridges is found in the internal structure of the diffusion metal-oxide semiconductor (DMOS) power-switching transistors they use. These transistors are composed of thousands of small MOSFET transistor cells wired in parallel. A small but representative fraction of these cells, typically 1 in 4000, is used to extract the sense current while the remainder of the cells control the motor current. The data sheet for the National LMD18245 LMD18245 H bridge contains an excellent writeup on how this is done.

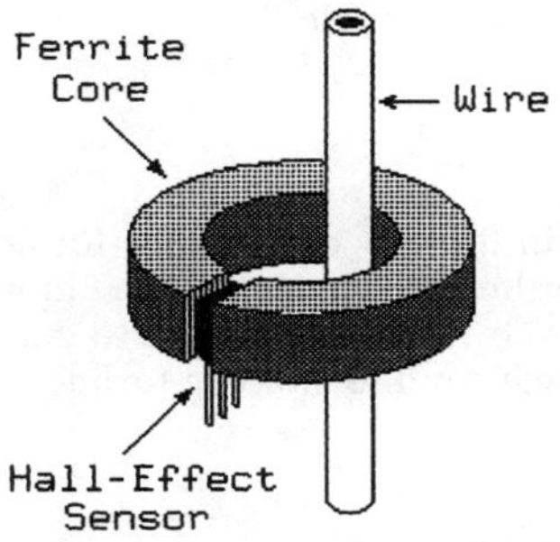

When very high currents are involved, precluding use of an integrated H bridge, an appealing and well-established current-sensing technology involves the use of a split ferrite core and a Hall-effect sensor, as illustrated in Fig. 10.65.

Simple linear Hall-effect sensors require a small regulated bias current between two of their terminals, and they generate a dc voltage proportional to the magnetic field on a third terminal. The magnetic field across the gap sawed in the ferrite core is proportional to the current through the wire; therefore, the voltage reported by the Hall-effect sensor is proportional to the current.

FIGURE 10.65 Current sensor using a split ferrite core and a Hall-effect sensor.

Allegro Microsystems and others make full lines of Hall-effect sensors, but precalibrated Hall-effect current sensors are available; these include the split core, the Hall-effect sensor, and auxiliary components, all mounted on a small PC board or potted as a unit. Newark Electronics (1997) lists a few sources of these, including Honeywell, F. W. Bell, and LEM Instruments.

An intriguing new current sensor became available as of 1998, based on a thin-film magnetoresistive sensor; the sensitivity of this technology eliminates the need for the ferrite core and the result is a very compact current sensor. The NT series sensors made by F. W. Bell use this technology.

10.10 MICROSTEPPING

Microstepping serves two purposes. First, it allows a stepping motor to stop and hold a position between the full or half-step positions; second, it largely eliminates the jerky character of low-speed stepping-motor operation and the noise at intermediate speeds; and third, it reduces problems with resonance.

Although some microstepping controllers offer hundreds of intermediate positions between steps, it is worth noting that microstepping does not generally offer great precision, both because of linearity problems and because of the effects of static friction.

10.10.1 Sine-Cosine Microstepping

For an ideal two-winding variable-reluctance or permanent-magnet motor, the torque–shaft angle curve is determined by the following formulas:

$$h = (a^2 + b^2)^{0.5}$$

$$x = \frac{S}{\pi/2} \arctan \frac{b}{a}$$

where a = torque applied by winding with equilibrium at angle 0
b = torque applied by winding with equilibrium at angle S
h = holding torque of composite
x = equilibrium position
S = step angle

This formula is quite general, but it offers little in the way of guidance for how to select appropriate values of the current through the two windings of the motor. A common solution is to arrange the torques applied by the two windings so that their sum h has a constant magnitude equal to the single-winding holding torque. This is referred to as *sine-cosine microstepping*.

$$a = h_1 \sin\left(\frac{\pi/2}{S} \theta\right)$$

$$b = h_1 \cos\left(\frac{\pi/2}{S} \theta\right)$$

where

h_1 = single-winding holding torque

$\frac{\pi/2}{S} \theta$ = electrical shaft angle

Given that none of the magnetic circuits are saturated, the torque and the current are linearly related. As a result, to hold the motor rotor to angle 0, we set the currents through the two windings as

$$I_a = I_{max} \sin\left(\frac{\pi/2}{S} \theta\right)$$

$$I_b = I_{max} \cos\left(\frac{\pi/2}{S} \theta\right)$$

where I_a = current through winding with equilibrium at angle 0
I_b = current through winding with equilibrium at angle S
I_{max} = maximum allowed current through any motor winding

Keep in mind that these formulas apply to two-winding pm or hybrid stepping motors. Three-pole or five-pole motors have more complex behavior, and the magnetic fields in variable-reluctance motors do not add according to the simple rules that apply to the other motor types.

10.10.2 Limits of Microstepping

The utility of microstepping is limited by at least three considerations. First, if there is any static friction in the system, the angular precision achievable with microstepping will be limited. This effect is discussed in more detail in Sec. 5.2.10, in the discussion of friction and the dead zone.

Detent Effects. The second problem involves the nonsinusoidal character of the torque–shaft angle curves on real motors. Sometimes this is attributed to the detent torque on pm and hybrid motors, but in fact, both detent torque and the shape of the torque-angle curves are products of poorly understood aspects of motor geometry—specifically, the shapes of the teeth on the rotor and stator. These teeth are almost always rectangular, and this author is aware of no detailed study of the impact of different tooth profiles on the shapes of these curves.

Most commercially available microstepping controllers provide a fair approximation of the sine-cosine drive current that would drive an ideal stepping motor to uniformly spaced steps. Ideal motors are rare, and when such a controller is used with a real motor, a plot of the actual motor position as a function of the expected position will generally look something like the plot shown in Fig. 10.66.

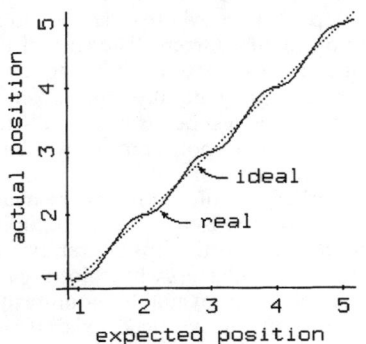

FIGURE 10.66 Actual position versus expected position of a microcontrolled stepping motor.

Note that the motor is at its expected position at every full step and at every half step, but that there is significant positioning error in the intermediate positions. The curve shown is the curve that would result from a perfect sine-cosine microstepping controller used with a motor that had a torque-position curve that included a significant fourth harmonic component, usually attributed to the detent torque.

Quantization. The third problem arises because most applications of microstepping involve digital control systems; thus, the current through each motor winding is quantized, controlled by a digital-to-analog converter. Furthermore, if typical PWM current-limiting circuitry is used, the current through each motor winding is not held perfectly constant, but rather oscillates around the current-control circuit's set point. As a result, the best a typical microstepping controller can do is approximate the desired currents through each motor winding.

The effect of this quantization is easily seen if the available current through one motor winding is plotted on the X axis and the available current through the other motor winding is plotted on the Y axis. Figure 10.67 shows such a plot for a motor controller offering only four uniformly spaced current settings for each motor winding. Of the 16 available combinations of currents through the motor windings, 6 combinations lead to roughly equally spaced microsteps. There is a clear trade-off between minimizing the variation in torque and minimizing the error in motor position, and the best available motor positions are hardly uniformly spaced. Use of higher-precision digital-to-analog conversion in the current-control system reduces the severity of this problem, but it cannot eliminate it.

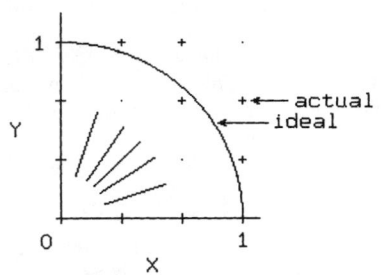

FIGURE 10.67 Plot of actual versus ideal winding currents.

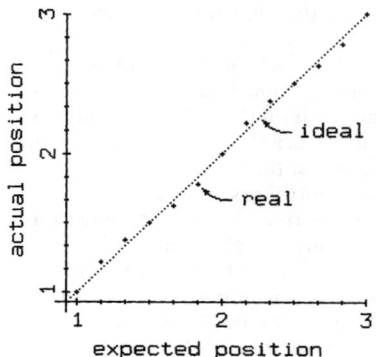

FIGURE 10.68 Plot of actual versus expected rotor position of a microcontrolled stepping motor.

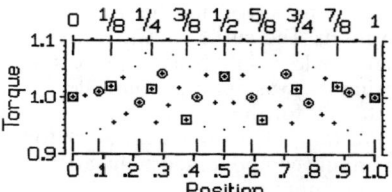

FIGURE 10.69 Holding-torque position.

Plotting the actual rotor position of a motor using the microstep plan outlined in Fig. 10.67 versus the expected position gives the curve shown in Fig. 10.68.

It is very common for the initial microsteps taken away from any full-step position to be larger than the intended microstep size, and this tends to give the curve a staircase shape, with the downward steps aligned with the full-step positions where only one motor winding carries current. The sign of the error at intermediate positions tends to fluctuate, but generally, the position errors are smallest between the full-step positions, when both motor windings carry significant current.

Another way of looking at the available microsteps is to plot the equilibrium position on the horizontal axis, in fractions of a full step, while plotting the torque at each available equilibrium position on the vertical axis. If we assume a 4-bit analog-to-digital converter, giving 16 current levels for each motor winding, there are 256 equilibrium positions. Of these, 52 offer holding torques within 10 percent of the desired value, and only 33 are within 5 percent; these 33 points are shown in bold in Fig. 10.69.

If torque variations are to be held within 10 percent, it is fairly easy to select eight almost uniformly spaced microsteps from among those shown in Fig. 10.69; these are boxed in the figure. The maximum errors occur at the ¼-step points; the maximum error is 0.008 step or 0.06 microstep. This error will be irrelevant if the dead zone is wider than this.

If 10 microsteps are desired, the situation is worse. The best choices, still holding the maximum torque variation to 10 percent, gives a maximum position error of 0.026 step or 0.26 microstep. Doubling the allowable variation in torque approximately halves the positioning error for the 10-microstep example, but does nothing to improve the 8-microstep example.

One option which some motor control system designers have explored involves the use of nonlinear digital-to-analog converters (DACs). This is an excellent solution for small numbers of microsteps, but building converters with essentially sinusoidal transfer functions is difficult if high precision is desired.

10.10.3 Typical Control Circuits

As typically used, a microstepping controller for one motor winding involves a current-limited H-bridge or unipolar drive circuit, where the current is set by a reference voltage. The reference voltage is then determined by an analog-to-digital converter, as shown in Fig. 10.70.

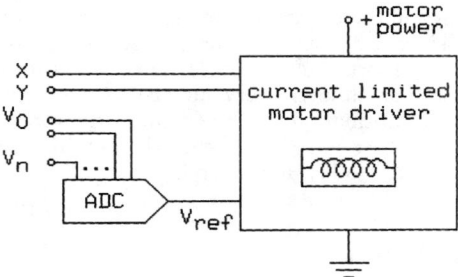

FIGURE 10.70 Microstepping controller circuit.

Figure 10.70 assumes a current-limited motor controller such as is shown in Fig. 10.59, 10.60, 10.62, and 10.63. For all of these drivers, the state of the X and Y inputs determines whether the motor winding is on or off and, if on, the direction of the current through the winding. The V_0 through V_n inputs determine the reference voltage and thus the current through the motor winding.

Practical Examples. There are a fair number of nicely designed integrated circuits combining a current-limited H-bridge with a small DAC to allow microstepping control of motors drawing under 2 A per winding. The PBL3717 and PBL3770 from Ericsson Microelectronics are excellent examples; the latter is also available as the UC3770 from Unitrode. These chips integrate a 2-bit DAC with a PWM-controlled H bridge, packaged in either 16-pin power-DIP format or in surface-mountable form. The 3717 is a slightly cleaner design, good for 1.2 A, while the 3770 is good for up to 1.8 or 2 A, depending on how the chip is cooled. The 3955 from Allegro Microsystems incorporates a 3-bit nonlinear DAC and handles up to 1.5 A; this is available in 16-pin power DIP or small-outline integrated-circuit (SOIC) formats. The nonlinear DAC in this chip is specifically designed to minimize step-angle errors and torque variations using 8 microsteps per full step.

The LMD18245 from National Semiconductor is a good choice for microstepped control of motors drawing up to 3 A. This chip incorporates a 4-bit linear DAC, and an external DAC can be used if higher precision is required. As indicated by the data shown in Fig. 10.69, a 4-bit linear DAC can produce 8 reasonably uniformly spaced microsteps, so this chip is a good choice for applications that exceed the power levels supported by the Allegro 3955.

10.11 BRUSHLESS DC MOTOR DRIVE SCHEMES*

While elaborate and expensive drive schemes are possible, in many applications simplifying assumptions are made that lead to readily implemented drive schemes that perform reasonably well. This section illustrates these simple drive schemes for two- and three-phase motors. The fundamental task for a motor drive is to apply current

---

* Section contributed by Duane C. Hanselman, University of Maine.

to the correct windings, in the correct direction, at the correct time. This process is called *commutation*, since it describes the task performed by the commutator (and brushes) in a conventional brush dc motor. The goal here is to develop an intuitive understanding, rather than discuss every nuance of every possible motor drive scheme. More detailed information can be found in references such as Leonhard (1985) and Murphy and Turnbull (1988). With this intuitive understanding, more complex drive schemes are readily understood.

10.11.1 Two-Phase Motors

Previously, torque and back-emf expressions are developed considering just one motor phase. When there is more than one phase, each individual phase acts independently to produce torque. Consider the two-phase motor illustrated in Fig. 10.71. Power dissipated in the phase resistances produces heat; the phase inductances store energy but dissipate no power; and power absorbed by the back-emf sources E_A and E_B is converted to mechanical power T_ω. (Think about it: where else could it go?) Writing this last relationship mathematically gives

$$E_A i_A + E_B i_B = T_\omega \qquad (10.1)$$

Here the back-emf sources are determined by the motor design, and the currents are determined by the motor drive. Because of the BLv law, the back-emf sources are linear functions of speed; that is, $E = k\omega$, where k, the emf waveform, is a function of motor parameters and position. Substituting this relationship into Eq. (10.1) gives

$$k_A i_A + k_B i_B = T \qquad (10.2)$$

Thus, the mutual torque produced is a function of the back-emf waveforms and the applied currents. Most important, Eq. (10.2) applies instantaneously. Any instantaneous variation in the back-emf waveforms, or the phase currents will produce an instantaneous torque variation.

Equation (10.2) provides all the information necessary to design drive schemes for the two-phase motor. Since the back-emf waveforms are a function of position, it

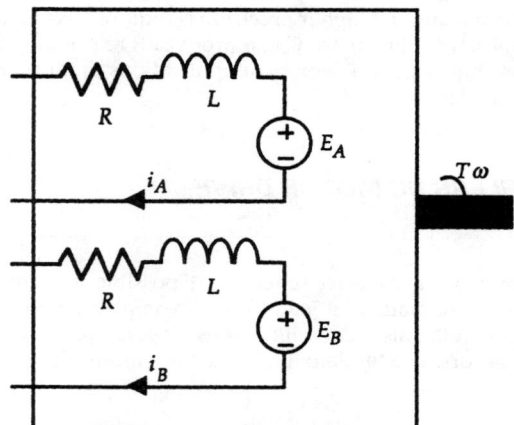

FIGURE 10.71 Two-phase motor schematic.

is convenient to consider Eq. (10.2) graphically. Making the simplifying assumption that the emf is an ideal square wave, Fig. 10.72 shows the back-emf waveforms, with that from phase B delayed by $\pi/2$ rad electrical with respect to phase A.

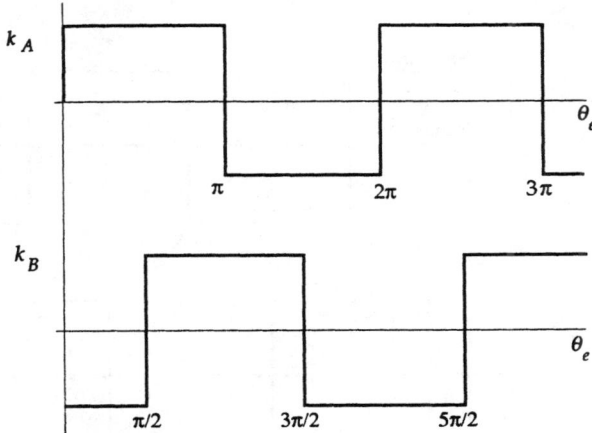

FIGURE 10.72 Square-wave back-emf waveforms for a two-phase motor.

One-Phase-On Operation. Given the waveforms shown in Fig. 10.72, several drive schemes become apparent. The first, shown in Fig. 10.73, is one-phase-on operation where only one phase is conducting current at any one time. In this figure, the phase currents are superimposed over the back-emf waveforms and Eq. (10.2) is applied instantaneously to show the resulting motor torque on the lower axes. The overbars are used to signify current flowing in the reverse direction. Some important aspects of this drive scheme include the following:

- Ideally, constant ripple-free torque is produced.
- The shape of the back-emf of the phase not conducting at any given time (e.g., phase A over $3\pi/4 \leq \theta \leq 5\pi/4$) has no influence on the torque production since the associated current is zero. The back-emf need only be flat when the current is applied. The smoothing of the transitions in the back-emf that exist in a real motor do not add torque ripple.
- Neither phase is required to produce torque in regions where its associated back-emf is changing sign.
- Each phase contributes an equal amount to the total torque produced. Thus each phase experiences equal losses and the drive electronics are identical for each phase.
- Copper utilization is said to be 50 percent, since at any time only half of the windings are being used to produce torque; the other half have no current flowing in them.
- The amount of torque produced can be varied by changing the amplitude of the current flowing in them.

- Square pulses of current are required but not achievable in the real world, since the inductive phase windings limit the current slope to $di/dt = v/L$, where v is the applied voltage and L is the inductance. Using $\theta = \omega t$, this relationship can be stated in terms of position as $di/d\theta = v/(\omega L)$. With either interpretation, the rate of change in current is finite, whereas Fig. 10.73 assumes that it is periodically infinite.

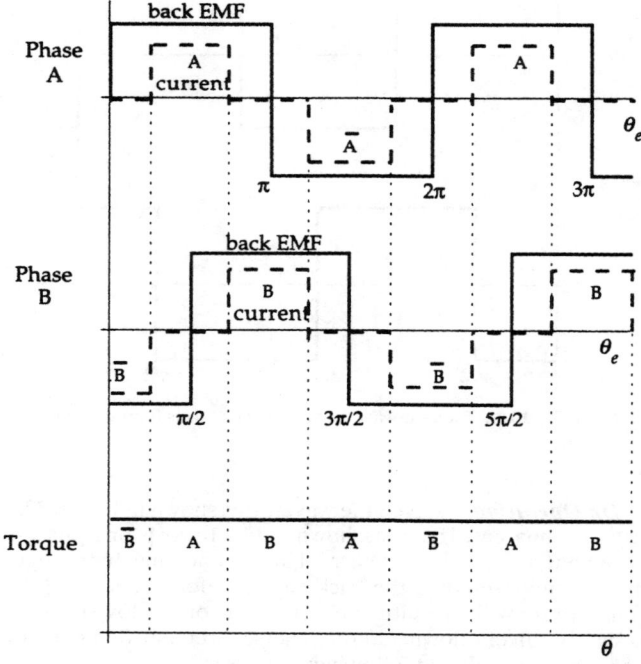

FIGURE 10.73 One-phase-on torque production.

Two-Phase-On Operation. Following the same procedure used to construct Figs. 10.73 and 10.74 shows two-phase-on operation, where both phases are conducting at all times. Important aspects of this drive scheme include the following:

- Ideally, constant ripple-free torque is produced.
- The shape of the back-emf is critical at all times, since torque is produced in each phase at all times.
- If either current does not change sign at exactly the same point that the emf does, negative phase torque is produced, which leads to torque ripple.
- Both phases are required to produce torque in regions where their associated back-emf is changing sign.
- Each phase contributes an equal amount to the total torque produced. Thus, each phase experiences equal losses, and the drive electronics are identical for each phase.
- Copper utilization is 100 percent.

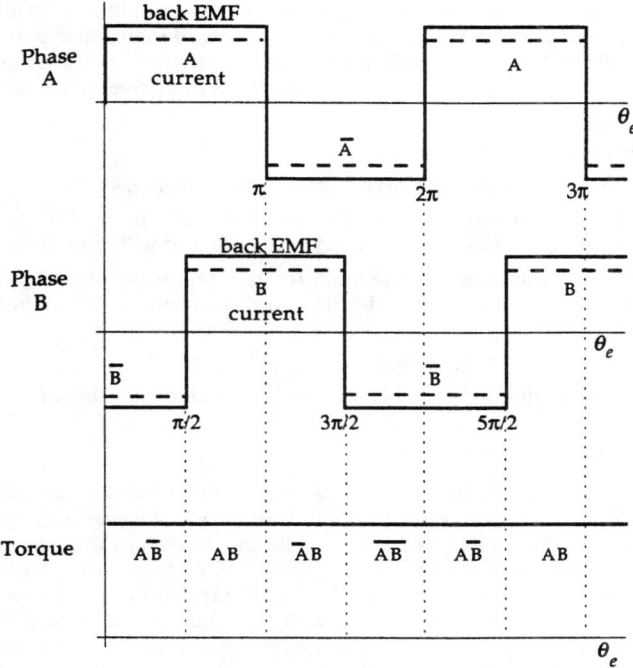

FIGURE 10.74 Two-phase-on torque production.

- The amount of torque produced can be varied by changing the amplitude of the square-wave currents.
- Impossible-to-produce square-wave currents are required.
- For a constant torque output, the peak phase current is reduced by half compared with the one-phase-on scheme.

The Sine-Wave Motor. A square-wave back-emf motor driven by square current pulses in either one- or two-phase-on operation as previously described represents what is usually called a *brushless dc* (BLDC) *motor*. On the other hand, if the back-emf is sinusoidal, the motor is commonly called a *synchronous motor*. Operation of this motor follows Eq. (10.2) also. However, in this case it is easier to illustrate torque production analytically. The key to understanding the two-phase synchronous motor is by recalling the trigonometric identity $\sin^2 \theta + \cos^2 \theta = 1$.

Let phase A have a back-emf shape of $k_A = K \cos \theta$ and be driven by a current $i_A = I \cos \theta$. If as before the back-emf of phase B is delayed by $\pi/2$ rad electrical from phase A, $k_B = K \sin \theta$, and the associated phase current is $i_B = I \sin \theta$. Applying these expressions to Eq. (10.2) gives

$$k_A i_A + k_B i_B = T \qquad (10.3)$$
$$KI(\cos^2 \theta + \sin^2 \theta) = KT = T$$

Thus, once again the torque produced is constant and ripple-free. In addition, the currents are continuous, and only finite $di/d\theta$ is required to produce them. Just as in the square-wave case considered earlier, the currents must be synchronized with the motor back-emf. To summarize, important aspects of this driven scheme include the following:

- Ideally, constant ripple-free torque is produced.
- The shape of the back-emf and drive currents must be sinusoidal.
- If both phase currents are out of phase an equal amount with their respective back-emf the torque will have a reduced amplitude but will remain ripple-free.
- Each phase contributes an equal amount to the total torque produced. Thus, each phase experiences equal losses, and the drive electronics are identical for each phase.
- Copper utilization is 100 percent.
- The amount of torque produced can be varied by changing the amplitude of the sinusoidal currents.
- The phase currents have finite $di/d\theta$.

Based on the three examples just considered, it is clear that there are an infinite number of ways to produce constant ripple-free torque. All that is required is that the left-hand side of Eq. (10.2) instantaneously sum to a constant. The trouble with the square-wave back-emf schemes is that infinite $di/d\theta$ is required. The torque ripple that results from not being able to generate the required square pulses is called *commutation torque ripple*. The trouble with the sinusoidal emf case is that pure sinusoidal currents must be generated. In all cases, the back-emf and currents must be very precise whenever the current is nonzero; any deviation from ideal produces torque ripple. For the square-wave back-emf schemes, position information is required only at the commutation points (i.e., 4 points per electrical period). On the other hand, for the sinusoidal back-emf case, much higher resolution is required if the phase currents are to closely follow the back-emf waveforms. Thus, simple and inexpensive Hall-effect sensors are sufficient for the brushless dc motor, whereas an absolute position sensor (e.g., an absolute encoder or resolver) is required in the sinusoidal current drive case.

Despite the fact that the square-wave back-emf schemes inevitably produce torque ripple, they are commonly implemented because they are simple and inexpensive. In many applications, the cost of higher performance cannot be justified.

H-Bridge Circuitry. Based on Figs. 10.73 and 10.74, it is necessary to spend positive and negative current pulses through each motor winding. The most common circuit topology used to accomplish this is the full-bridge or H-bridge circuit, as shown in Fig. 10.75. In the figure, V_{cc} is a dc supply, switches S_1 through S_4 are commonly implemented with MOSFETs or IGBTs (though some still use bipolar transistors

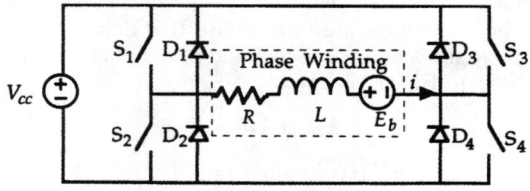

FIGURE 10.75 H-bridge circuit.

because they are cheap), diodes D_1 through D_4, called *freewheeling diodes*, protect the switches by providing a reverse current path for the inductive phase current, and R, L, and E_b represent one motor phase winding.

Basic operation of the H bridge is fairly straightforward. As shown in Fig. 10.76a, if switches S_1 and S_4 are closed, current flows in the positive direction through the phase winding. On the other hand, when switches S_2 and S_3 are closed, current flows in the negative direction through the phase winding, as shown in Fig. 10.76b. In either case, the current climbs exponentially according to the L/R time constant and reaches the value of $(\pm V_{cc} - E_b)/R$ if the switches are left closed long enough.

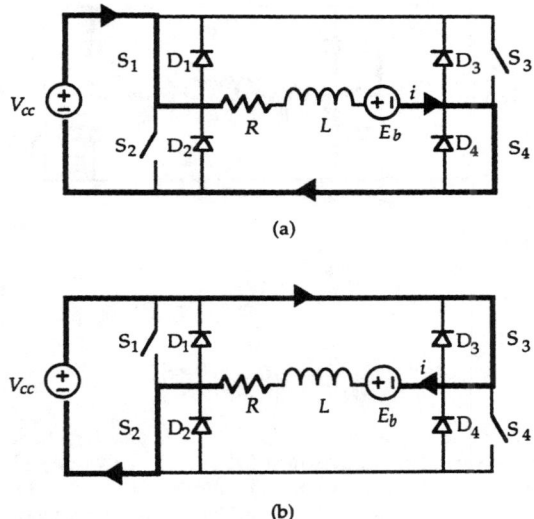

FIGURE 10.76 H-bridge circuit: (*a*) positive current conduction, and (*b*) negative current conduction.

Turn-Off Behavior. What takes more work to understand is the turn-off behavior of the H bridge and how phase current is controlled to limit its magnitude. Current control is accomplished by chopping, that is, by employing PWM techniques. Because of its fundamental nature, PWM is discussed at length later. For the time being, consider the turn-off behavior of the H bridge. This behavior is guided by the fundamental behavior of inductors. That is, that current cannot change instantaneously but must be continuous, and the larger the voltage across an inductor, the faster the current through it will change.

To start, let the phase current be a constant I_m with switches S_1 and S_4 closed, as shown in Fig. 10.76a. Given these initial conditions, consider what happens when both switches are opened to bring the current back to zero. Now, since current no longer flows through S_1 and S_4, a negative voltage appears across the inductor because di/dt is negative. At the same time, the phase current continues to flow in the same direction because it cannot change instantaneously. The only path for current flow is through diodes D_2 and D_3, as shown in Fig. 10.77a. No current can flow through diodes D_1 or D_4. During this time, the voltage across the phase inductance is

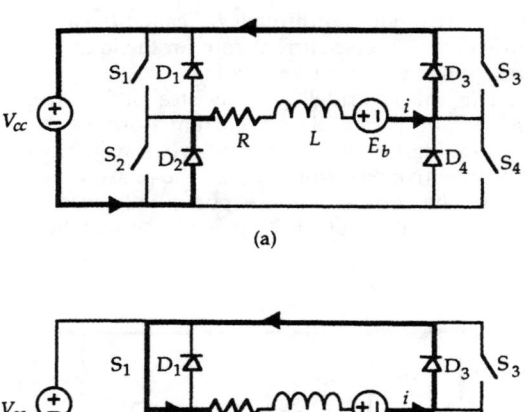

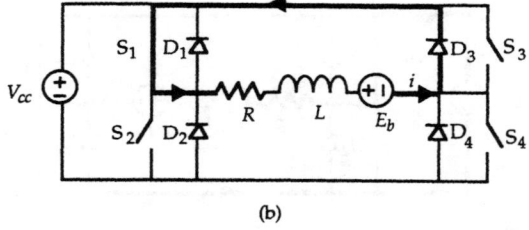

FIGURE 10.77 Current decay in an H-bridge circuit: (a) switches S_1 and S_4 open, and (b) only switch S_4 open.

$$L \frac{di}{dt} = -Ri - V_{cc} - E_b \tag{10.4}$$

which is clearly large and negative when $i > 0$, $V_{cc} < 0$, and $E_b > 0$. As time progresses, the current decreases exponentially toward the negative value $-(V_{cc} + E_b)/R$. Upon reaching zero current, the diodes turn off, the energy in the inductor $0.5Li^2$ is returned to the supply, and the circuit rests. If the circuit lacks freewheeling diodes, the switches are destroyed in an attempt to provide a current path for the inductor current.

In some situations, just one of the two switches is opened. To illustrate this action, assume the conditions shown in Fig. 10.76a and open only switch S_4; let S_1 remain closed. The path for decaying current flow in this case is through D_3 and S_1, as shown in Fig. 10.77b, giving an inductor voltage of

$$L \frac{di}{dt} = -Ri - E_b \tag{10.5}$$

which is much smaller in magnitude than that given in Eq. (10.4) because $-V_{cc}$ is missing. Hence, the inductor current decays much more slowly in this situation. Later, this turn-off mode will prove helpful in implementing PWM current control.

Switch Current. A major task in drive-circuit design is to size the switches, that is, to determine their rms currents. In the H bridge, switches S_1 and S_4 carry the positive portion of the phase current, whereas switches S_2 and S_3 carry the negative portion of the phase current. Because of this division, the rms switch current is less than the RMS phase current. As illustrated for the two-phase-on scheme in Fig. 10.78, the RMS value of the switch current is easily shown to be $100/\sqrt{2} = 70.7$ percent of the RMS phase current. Though not shown, the same ratio applies to the one-phase-on scheme.

Summary. Important aspects of the H-bridge circuit include the following:

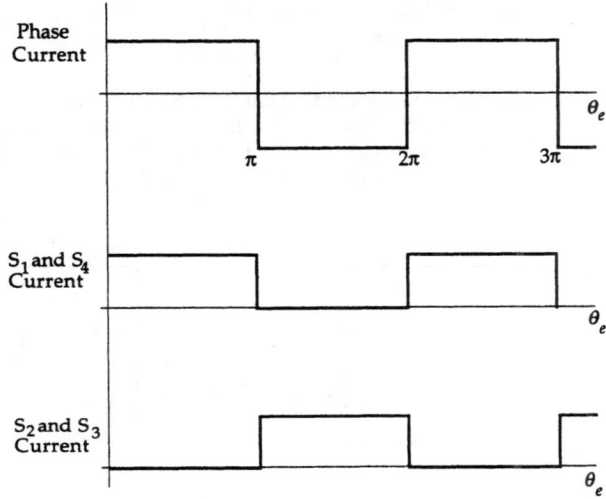

FIGURE 10.78 Phase and switch currents for two-phase-on operation.

- Bidirectional current flow is easily achieved.
- Given that the back-emf and current have the same sign in Figs. 10.73 and 10.74, the back-emf acts to fight the increase in phase current amplitude during turn-on.
- In the one-phase-on drive scheme in Fig. 10.73, the back-emf and current have the same sign at the turn-off points. Thus, by Eq. (10.4), the back-emf acts to assist the decrease in phase current during turn-off.
- In the two-phase-on drive scheme in Fig. 10.74, the back-emf and current have opposite signs immediately after the turn-off points. Thus, the back-emf acts to fight the decrease in phase current during turn-off. Thus, the back-emf hinders commutation at both turn-on and turn-off in the two-phase-on drive scheme.
- At no time can vertical pairs of switches (i.e., S_1 and S_2 or S_3 and S_4) be closed simultaneously. If this happens, a shoot-through fault occurs where the motor supply is shorted. In implementation, a short delay is often added between commutations to guarantee that no shoot-through condition occurs.
- For the square-wave back-emf schemes, the rms switch current is equal to 70.7 percent of the RMS phase current.
- For two-phase motors, H bridges are required, giving a total of eight switches to be implemented by power electronic devices.

10.11.2 Three-Phase Motors

Three-phase motors overwhelmingly dominate all others. The exact reasons for this dominance are not known, but the historical dominance of three-phase induction and synchronous motors and the minimal number of power electronic devices required are likely contributing factors. The addition of a third phase provides an additional degree of freedom over the two-phase motor, which manifests itself in more drive schemes and terminology. For example, wye (Y) and delta (Δ) connections are possible.

10.106 CHAPTER TEN

In three-phase motors, the power balance equation leads to

$$k_A i_A + k_B i_B + k_C i_C = T \tag{10.6}$$

where k_C and i_C are the back-emf shape and current, respectively, of the third phase. By construction, the back-emfs of each phase have the same shape but are offset from each other by $2\pi/3$ rad electrical, or $120°$ electrical. The back-emf shapes for the ideal square-wave back-emf motor are shown in Fig. 10.79.

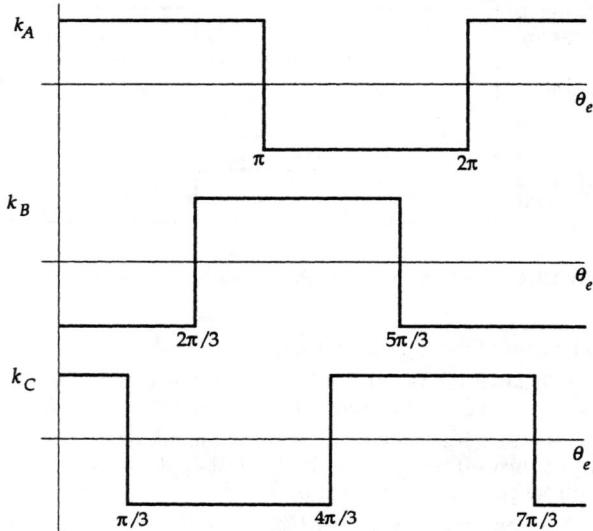

FIGURE 10.79 Square-wave back-emf waveforms for a three-phase motor.

Three-Phase-On Operation. The most obvious drive scheme for the three-phase motor is to extend the two-phase-on operation of the two-phase motor, as shown in Fig. 10.80. Here each phase conducts current at all times and contributes equally to the torque at all times. At each commutation point, one phase current changes sign and the others remain unchanged. The important aspects previously listed for the two-phase-on two-phase motor apply here as well.

Despite the conceptual simplicity of this drive scheme, it is hardly ever implemented in practice because three H bridges as shown in Fig. 10.75 are required, one for each phase winding. The resulting 12 power electronic devices make the drive expensive compared with other drive schemes.

Y Connection. Just as the Y connection is a popular configuration in three-phase power systems, it is also the most common configuration in three-phase brushless PM motors. As shown in Fig. 10.81, the center or neutral of the Y is not brought out, each external terminal or line is connected to a half-bridge circuit, and the collection of three half bridges is called a *three-phase bridge*. In this way, an H bridge appears

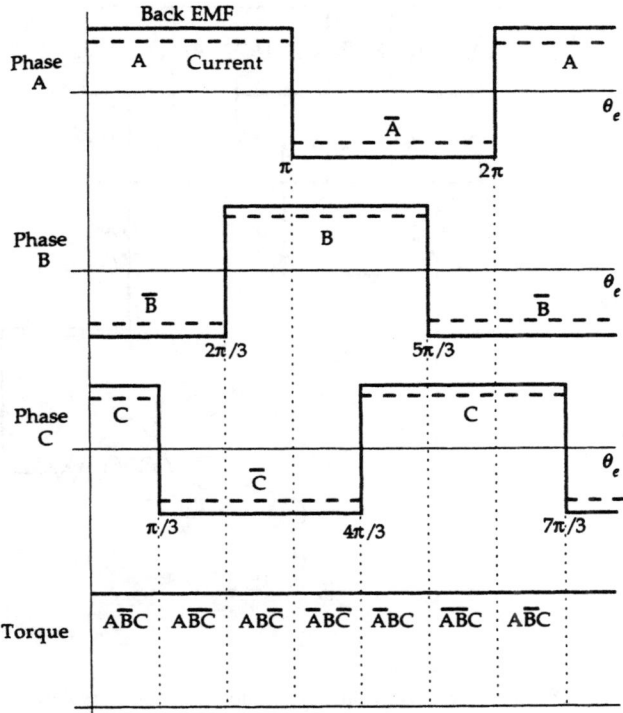

FIGURE 10.80 Three-phase-on operation.

between each set of terminals. Only six power electronic devices are needed for the switches in the three-phase bridge, as opposed to eight for a two-phase motor. The supply voltage is applied from line to line through the switches rather than from line to neutral. Compared with the three-phase-on case, the supply voltage works against two back-emf sources to force current into the motor. Furthermore, independent control of the phase currents is not possible since Kirchhoff's current law, $I_A + I_B + I_C = 0$, must be satisfied.

Torque production follows the idea that current should flow in only two of the three phases at a time, and that there should be no torque production near the back-emf sign crossings. Figure 10.82 shows the phase currents superimposed on the back-emfs. Each phase conducts currents over the central $2\pi/3$ rad electrical of each half cycle. The resulting torque is shown at the bottom of the figure with the letter designating the current polarities contributing to the torque. At each commutation point, one switch remains closed, one opens, another closes, and the rest remain open. There are six commutations per electrical period, and thus this drive scheme is often called a *six-step drive* (Murphy and Turnbull, 1988). The six numbered arrows shown in Fig. 10.81 illustrate these steps, as do the respective circled step numbers in Fig. 10.82.

Because only two phases are conducting current and contributing to torque production at any one time, the amplitude of the current must be 50 percent larger here

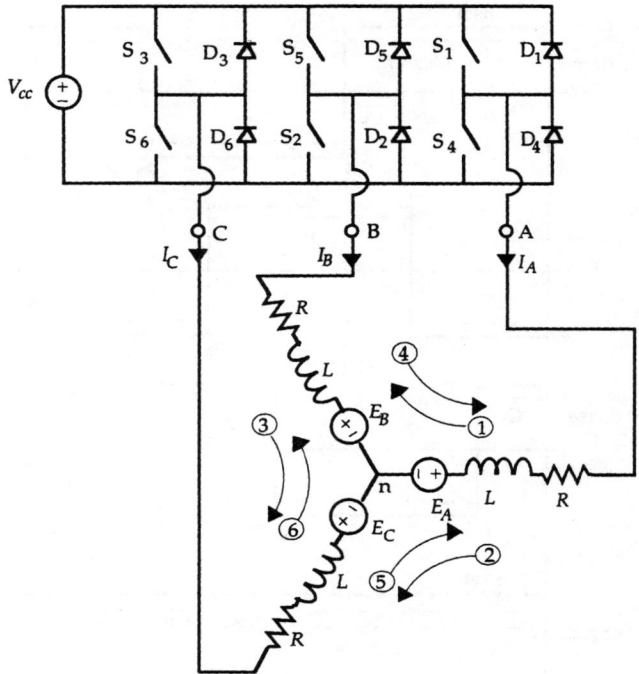

FIGURE 10.81 Y-connected three-phase motor and drive circuitry.

than in the three-phase-on case, where all three phases contribute simultaneously. When two phases are called upon to produce the same torque that three phases do, the current in each phase must be 3/2 as large, since (3/2)(2 phases) = (1)(3 phases). As a result, if this drive scheme is implemented, the current equations must be modified to reflect the current waveforms shown in Fig. 10.82.

The RMS phase currents are required to produce a specified rated torque. Based on the preceding discussion, these currents must be increased in amplitude by a factor of 3/2. Moreover, the equations must reflect the RMS value of the phase currents, which is $\sqrt{2/3}I_{peak}$, based on the waveforms shown in Fig. 10.82. Combining these factors, the phase current I_{ph} becomes:

$$I_{ph} = \frac{3}{2}\sqrt{\frac{2}{3}}\frac{I_s}{3n_s} = \sqrt{\frac{3}{2}}\frac{I_2}{3n_s} \tag{10.7}$$

where I_s = slot current
n_s = number of turns per slot

for the six-step driven three-phase motor. Compared with the three-phase-on case, the RMS phase current is approximately 22 percent larger and the ohmic motor loss is 50 percent greater. Thus, while the Y connection minimizes the number of power electronic devices used, it does not minimize losses.

To summarize, important aspects of this drive scheme include the following:

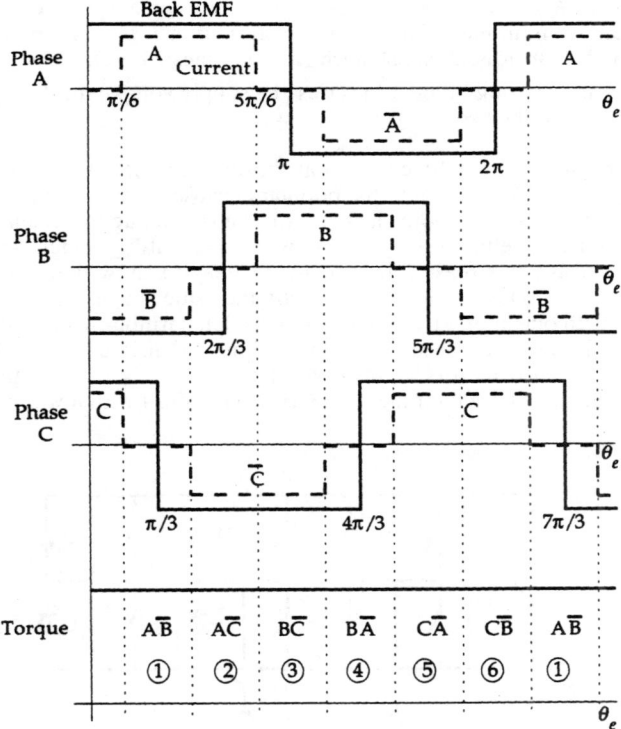

FIGURE 10.82 Torque production in a Y-connected three-phase motor.

- Ideally, constant ripple-free torque is produced.
- Only six switches are required, which is a minimum number.
- Phases are not required to produce torque in regions where their associated back-emf is changing sign. Thus, the back-emf can be more trapezoidal than square.
- Each phase contributes an equal amount to the total torque produced. Thus, each phase experiences equal losses, and the drive electronics are identical for each phase.
- Copper utilization is 67 percent, since at any one time only two of the three phases are conducting current.
- For the same output, ohmic motor losses are 50 percent greater than those in the three-phase-on drive scheme.
- The amount of torque produced can be varied by changing the amplitude of the square-wave currents.
- Impossible-to-produce 120°-wide square-wave currents are required. The inherent finite rise and fall time of the current creates torque ripple, commonly called *commutation torque ripple*.
- Independent control of phase currents is not possible.

- From $I_A + I_B + I_C = 0$, it can be shown that the phase currents cannot have any harmonics that are multiples of three, that is, triple-n or *triplen* harmonics (Murphy and Turnbull, 1988; Kassakian, Schlecht, and Verghese, 1991).
- Because phase windings appear in series, the supply voltage must be greater than the vector sum of the back-emfs at rated speed.

Delta Connection. The delta connection shown in Fig. 10.83 is the dual of the Y connection. This connection is not that popular because it has a major weakness, that being the additional ohmic motor loss and torque ripple due to circulating currents flowing around the delta. Three-phase power-system utility generators are never delta-connected for this reason. It is relatively easy to show that if the back-emf waveforms of each phase do not have exactly the same shape, are not exactly 120° out of phase with one another, or contain any triplen harmonics, circulating currents will flow around the delta. Because of this weakness, connected motors appear only in lower-performance motors at low-output power levels (e.g., in the fractional horsepower range), where their higher losses can be offset with lower material costs (Miller, 1989).

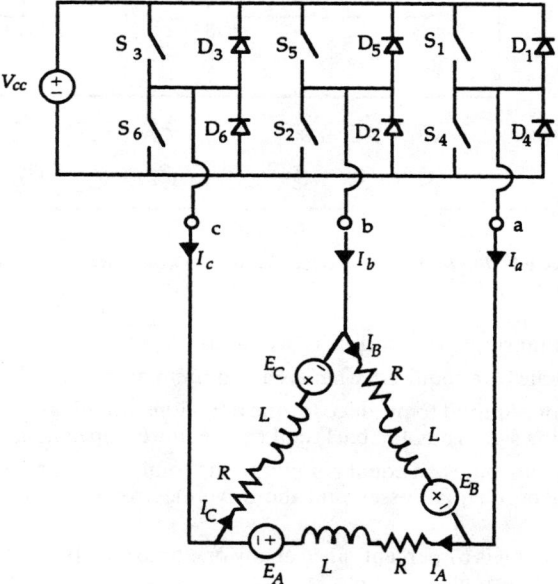

FIGURE 10.83 Delta-connected three-phase motor and drive circuitry.

Based on the preceding discussion, a motor having the ideal square-wave back-emf shape as shown in Fig. 10.79 cannot be connected in the delta connection because a square-wave back-emf motor has very high triplen harmonic content. Given the nature of dual circuits, it is not surprising that swapping the current and back-emf waveforms of the Y connection gives a workable solution for the delta connection, as shown in Fig. 10.84. Creating a motor with 120°-wide square-wave

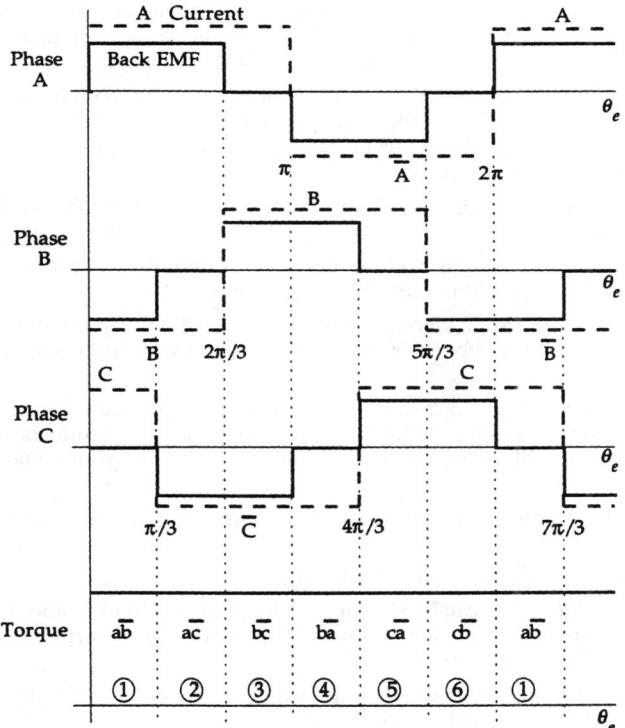

FIGURE 10.84 Torque production in a delta-connected three-phase motor.

back-emf waveforms is not difficult. Simply making the magnet arc narrower works, which results in the use of less magnet material.

To ease the explanation of the delta connection, the rising edges of the back-emfs and currents are aligned in Fig. 10.84. As shown, the back-emf of one phase is zero at all times. Each takes a turn at being zero for 60°. Because of this zero back-emf, the line current splits approximately equally through the remaining two phases, which conduct current in opposite directions. As before, the torque produced is given by applying Eq. (10.6). The lowercase letters under the torque curve signify the line currents during the respective commutation intervals. The line not given in each commutation interval is left floating electrically and is associated with the phase having zero back-emf. A comparison of these states with those of the Y connection in Fig. 10.82 show that the three-phase bridge circuit switches identically for both configurations. It is for this reason that the commutation logic in commercial driver ICs for small brushless motors works with either Y- or delta-connected motors.

To summarize, important aspects of this drive scheme include the following:

- Ideally, constant ripple-free torque is produced.
- Only six switches are required, which is a minimum number.

- Each phase contributes an equal amount to the total torque produced. Thus, each phase experiences equal losses, and the drive electronics are identical for each phase.
- Copper utilization remains 67 percent, even though all three phases conduct current simultaneously. At all times, one phase is conducting current and adding to the ohmic motor loss but is not producing torque, since the back-emf is zero in each phase one-third of the time.
- The amount of torque produced can be varied by changing the amplitude of the square-wave currents.
- Impossible-to-produce square-wave currents are required. The inherent finite rise and fall time of the current creates torque ripple.
- With all else being equal, ohmic motor losses are 50 percent greater than those in the Y connection, but the motor requires only two-thirds of the magnetic material (Miller, 1989).
- Just as in the Y-connected case, the phase current amplitude must be increased by 50 percent to make up for the fact that only two phases are producing the required torque. Since the phase currents are square waves, the current equation becomes $I_{ph} = I_s/2n_s$.
- Compared with the three-phase-on case, ohmic motor losses are 125 percent greater.
- Independent control of phase currents is not possible.
- From $I_A + I_B + I_C = 0$, it can be shown that the phase currents cannot have any harmonics that are multiples of three, that is, triple-n or *triplen* harmonics (Kassakian, Schlecht, and Verghese, 1991).
- Because phases appear in parallel, the supply voltage need only be greater than the peak phase back-emfs at rated speed.
- The delta connection is traditionally found in low-power, lower-performance motors.

The Sine-Wave Motor. The sine-wave back-emf motor completes the discussion of three-phase motors. A three-phase motor with sinusoidal back-emf can be Y or delta connected because there are by definition no triplen harmonics. Excitation of a sinusoidal motor with sinusoidal current gives constant ripple-free torque just as the two-phase sinusoidal motor does. In this case, the back-emfs and currents are all offset from each other by 120° electrical. Following the notion used earlier, the torque is found by substitution into Eq. (10.6) and is given by

$$k_A i_A + k_B i_B + k_C i_C = T$$

$$KI \sin^2 \theta + KI \sin^2 (\theta - 120°) + KI \sin^2 (\theta - 240°) = T \qquad (10.8)$$

$$\frac{3}{2} KI = T$$

The simple elegance of Eqs. (10.3) and (10.8) is due to the pure sinusoidal content of back-emf and phase currents. Because of this elegance, a greater deal of work goes into the design of some motors to minimize the higher harmonics in the back-emf so that a sinusoidal drive can be implemented. The sinusoidal motor commonly appears in high-performance applications where high accuracy and minimal torque ripple are required.

As shown by Eq. (10.8), each phase produces torque proportional to half the peak value of the current and back-emf as compared with a unity ratio for the square-wave back-emf motor driven three-phase-on. Therefore, in a sinusoidal motor driven by sinusoidal currents, correction of the phse current equation is necessary to establish the rms phase current required to produce a specified torque. The factor of ½ is taken into account by increasing the current amplitude by a factor of 2. Combining this information with the factor of $1/\sqrt{2}$ for the RMS values of a sinusoid, the phase current becomes

$$I_{ph} = \sqrt{2}\frac{I_s}{3n_s} \qquad (10.9)$$

10.11.3 PWM Methods

Specific current waveforms are assumed in each of the motor drive schemes discussed previously. To produce these waveforms from a voltage source requires current control. For maximum efficiency, this current control cannot require sustained operation of the power electronic device in its linear operating region. Rather, devices should act as switches having two states: off, where power dissipation is zero because there is no current flow, and on, where power dissipation is low because the voltage across the device is minimized. As a result, current control is implemented as a switching strategy in which the switch duty cycle is varied according to some error criterion, and the current maintains the correct shape in an average sense only. If switching action occurs at a much higher rate than any variation in the desired current waveform, the deviation between the actual and desired current can be made small. As a whole, these switching strategies are called *pulse-width modulation* (PWM).

Because PWM is applied in countless applications in addition to motor drives, there are hundreds of articles on PWM in the literature. Many of these articles pertain to voltage PWM, where one seeks to control voltage rather than current (Holtz, 1992). A smaller number pertain to current-control PWM, which is of interest here. As before, the goal is to develop an intuitive understanding, rather than discuss every nuance of every PWM scheme. More detailed information can be found in references such as Holtz (1992), Anunciada and Silva (1991), Brod and Novotny (1985), and Murphy and Turnbull (1988).

In motor drive applications, PWM is almost always implemented by controlling the bridge switches themselves. However, switching can also be implemented external to the bridge. Moreover, because motor windings have inductance, PWM action causes the phase inductance to charge and discharge, giving a continuous current despite the presence of a discontinuous applied voltage. As discussed earlier and shown in Figure 10.77, inductor discharge can be fast or slow depending on which switch or switches are controlled by PWM. Intelligent use of this capability can lead to improved performance (Freimanis, 1992).

Hysteresis PWM. Hysteresis PWM, conceptually the simplest PWM scheme, controls the on-off state of switches to keep the current within a band around the desired value, as shown in Fig. 10.85. In the figure, I^* is the reference current waveform (i.e., the desired current), $2\Delta I$ is the tolerance band, $I = I^* - \Delta I$ is the lower bound, and $I = I^* + I$ is the upper bound. Whenever the current crosses the upper bound, a switch is opened, allowing current to decay or discharge. Likewise, when-

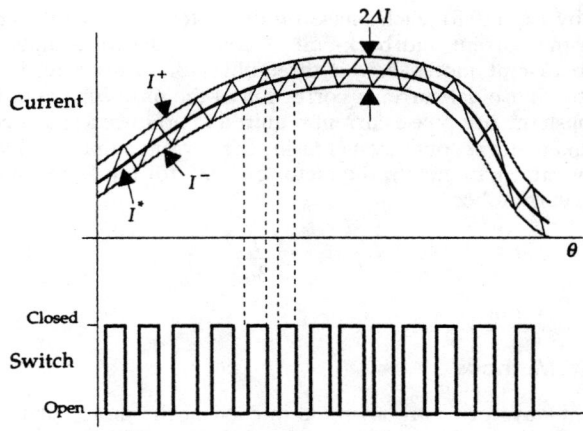

FIGURE 10.85 Hysteresis PWM waveforms.

ever the current crosses the lower bound, a switch is closed, forcing current to climb in amplitude or change. Clearly, the rate at which the inductance involved charges and discharges influences the rate at which switching occurs. In a motor drive, where the voltage across the inductance is a function of the difference between a supply voltage and the back-emf, the switching frequency will be high at low speeds and low at high speeds. The switching frequency at low speeds can be decreased by increasing the tolerance band. However, this increases the percentage ripple in the current.

Important aspects of this PWM scheme include the following:

- Precise current control is possible, as the tolerance band width is a design parameter.
- The frequency at which switches change state is not a design parameter. As a result, the switching frequency can vary by an order of magnitude or more.
- Acoustic and electromagnetic noise are difficult to filter, because their respective spectral components vary with the switching frequency.
- This PWM method is more commonly implemented in motor drives where motor speed and load are constant. Under these circumstances, the variation in switching frequency is small.

Clocked Turn-On PWM. This PWM method is the most commonly implemented scheme. Rather than control the peak-to-peak error as the hysteresis controller does, here the switching frequency is held constant. Clocked turn-on PWM is shown in Fig. 10.86, where the top trace is a synchronizing clock. Whenever this clock pulse appears, a switch is closed, causing the inductance to charge. At some point later when the current reaches I^+, a switch opens, initiating inductor discharge, which continues until the next clock pulse appears.

Important aspects of this PWM scheme include the following:

- Current control is not as precise here, since there is no fixed tolerance band that bounds the current.
- The frequency at which switches change state is a fixed design parameter.

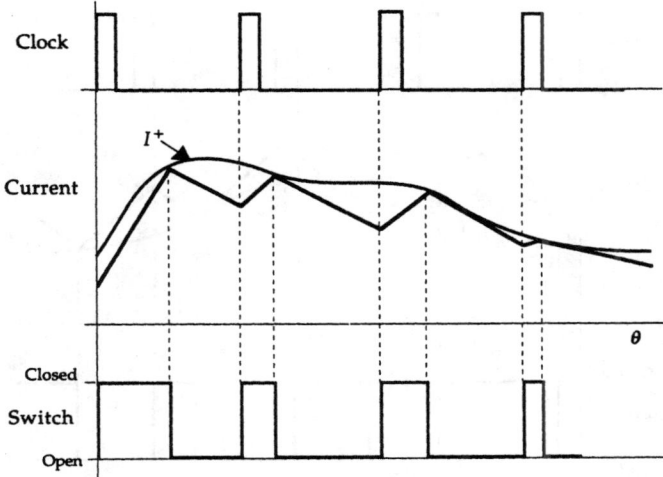

FIGURE 10.86 Clocked turn-on PWM waveforms.

- Acoustic and electromagnetic noise are relatively easy to filter, because the switching frequency is fixed.
- This PWM method has ripple instability that produces subharmonic ripple components for duty cycles *above* 50 percent (Anunciada and Silva, 1991; Kassakian, Schlecht, and Verghese, 1991). While this instability does not lead to any destructive operating mode, it is a chaotic behavior that reduces performance. The predominant current ripple occurs at half the switching frequency.
- Ripple instability can be eliminated by adding a *stabilizing ramp* to the reference current (Kassakian, Schlecht, and Verghese, 1991).

Clock Turn-Off PWM. Clocked turn-off PWM is the complement of clocked turn-on PWM. In this method, shown in Fig. 10.87, the clock pulse initiates inductor discharge. Later, when the current decays to I^-, a switch closes and the inductance charges until the next clock pulse appears. Once again, the switching frequency is fixed by the clock frequency.

Important aspects of this PWM scheme include the following:

- Current control is not as precise here, since there is no fixed tolerance band that bounds the current.
- The frequency at which switches change state is a fixed design parameter.
- Acoustic and electromagnetic noise are relatively easy to filter, because the switching frequency is fixed.
- This PWM method has ripple instability that produces subharmonic ripple components for duty cycles *below* 50 percent (Anunciada and Silva, 1991; Kassakian, Schlecht, and Verghese, 1991). While this instability does not lead to any destructive operating mode, it is a chaotic behavior that reduces performance. The predominant current ripple occurs at half the switching frequency.

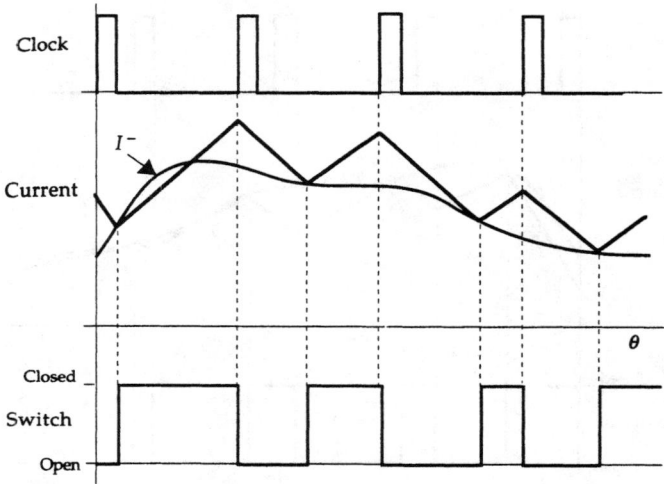

FIGURE 10.87 Clocked turn-off PWM waveforms.

Dual Current-Mode PWM. This PWM method was developed by Anunciada and Silva (1991) to eliminate the ripple instability present in the previous two methods. Their scheme combines the clocked turn-on and clocked turn-off methods in a clever way. For duty cycles below 50 percent, the method implements stable clocked turn-on PWM, whereas for duty cycles above 50 percent, the method implements stable clocked turn-off PWM.

As illustrated in Fig. 10.88, this method has two clock signals, where the turn-off clock is delayed one-half period with respect to the turn-on clock. Operation is determined by logic that initiates inductor charging when the turn-on clock pulse appears or the current reaches I^-, and initiates inductor discharge when the turn-off clock appears or the current reaches I^+. As shown in Fig. 10.88, the method smoothly moves from one mode to the other. This scheme has all the attributes of the two previous PWM schemes, except for the ripple instability. Furthermore, this scheme reduces to hysteresis PWM if the clock frequency is low compared with the rate at which the inductance charges and discharges.

Triangle PWM. Triangle PWM is a popular voltage PWM scheme that is commonly used to produce a sinusoidal PWM voltage. When used in this way, it is called *sinusoidal PWM* (Kassakian, Schlecht, and Verghese, 1991).

Application of this scheme to current control is accomplished by letting the PWM input be a function of the difference between the desired current and the actual current. As shown in Fig. 10.89, both the turn-on and turn-off of the switch are determined by the intersections of the triangle waveform and the processed current error. As the processed current error increases, so does the switch duty cycle. Typically, the processed current error is equal to a linear combination of the current error and the integral of the current error (i.e., *PI* control is used). As a result, as the steady-state error goes to zero, the switch duty cycle will go to the correct value to maintain it there. Though Fig. 10.89 shows a unipolar triangle waveform and error signal, both signals can also be bipolar, in which case zero current error produces a 50 percent duty cycle PWM signal (Murphy and Turnbull, 1988).

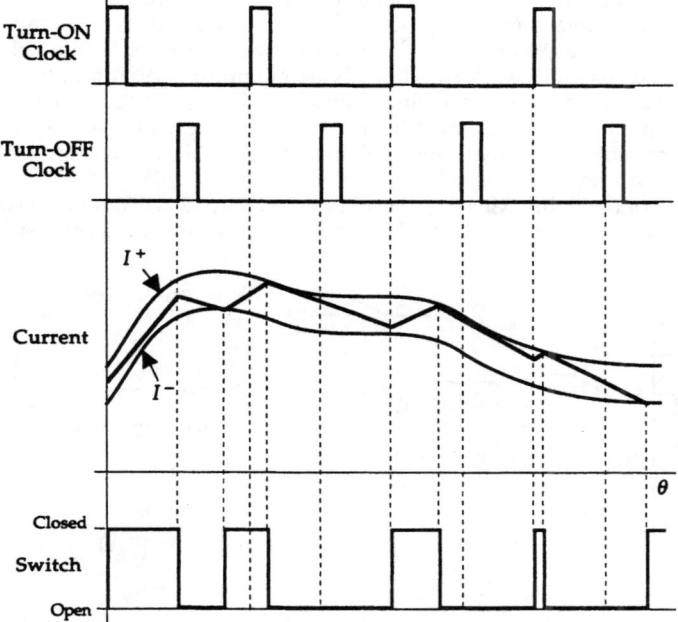

FIGURE 10.88 Dual-current-mode PWM waveforms.

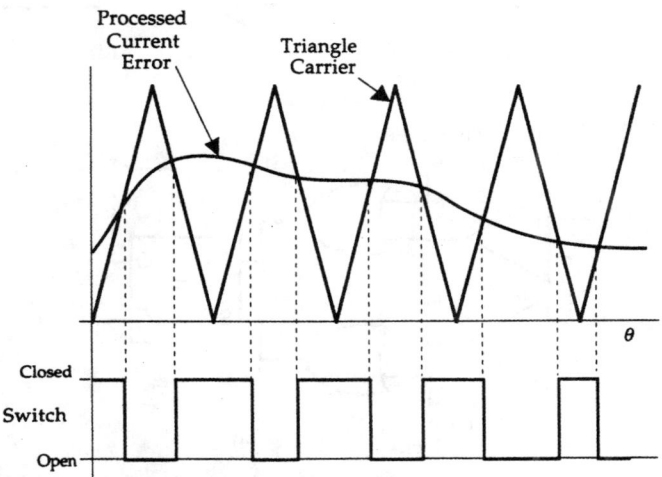

FIGURE 10.89 Triangle PWM waveforms.

Summary. The PWM methods discussed in this section represent the most common methods implemented in practice. Each method has its own strengths and weaknesses; no one PWM scheme is the best choice for every motor drive. Implementation details for these PWM methods are not presented in order to focus attention on the fundamental switching concepts. For reference, conceptual logic diagrams for all the methods are shown in Fig. 10.90. These diagrams apply for positive currents only. When the reference current is bipolar, more complex logic diagrams are required.

Because of the finite switching time of power electronic devices, duty cycles near 0 and near 1 must be avoided in all PWM methods. Switching devices must remain on

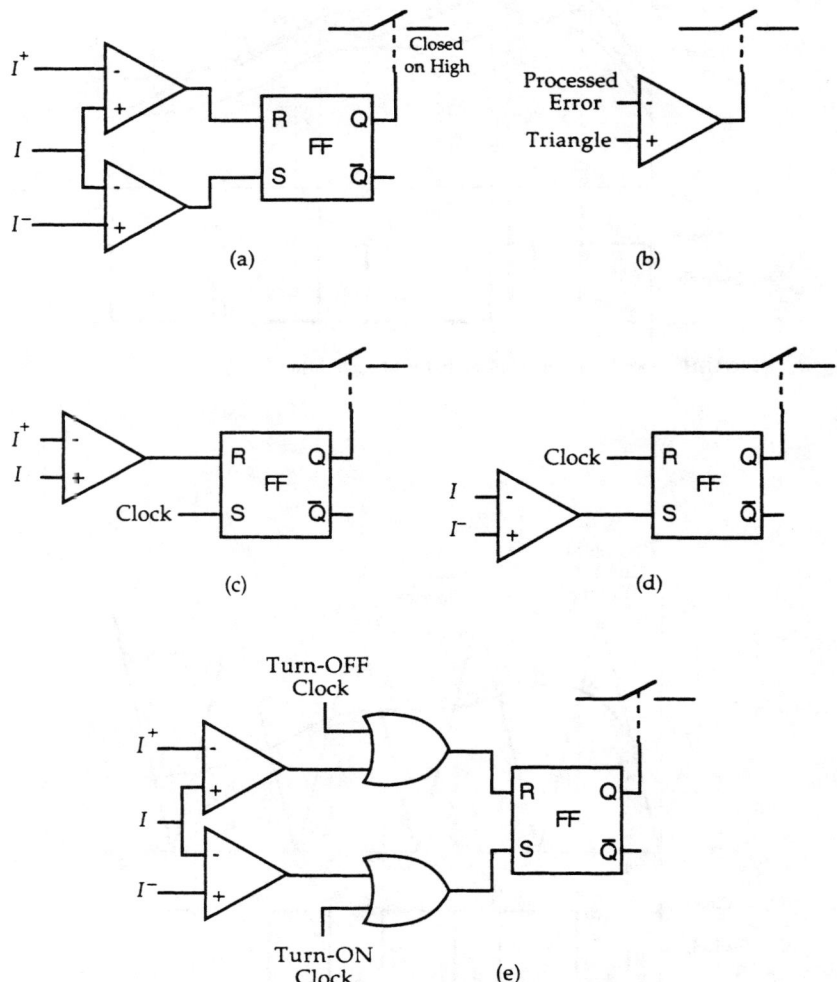

FIGURE 10.90 Conceptual logic PWM implementation: (a) hysteresis PWM, (b) triangle PWM, (c) clocked turn-on PWM, (d) clocked turn-off PWM, and (e) dual-current-mode PWM.

and off for sufficient time to reach equilibrium before being switched back to the opposite state. Sustained operation at either duty cycle extreme increases the loss experienced by the switching devices and can lead to destructive device heating. The choice of PWM frequency is a tradeoff. Generally, the higher the switching frequency, the smaller the current error will be. On the other hand, the higher the switching frequency, the greater the switching loss incurred by the switches. Furthermore, PWM schemes are only as accurate as the current sensors used. Sensor type, placement, shielding, and signal processing are all critical to accurate operation of a current-control PWM method.

10.12 MOTOR DRIVE ELECTRONIC COMMUTATION PATTERNS*

The understanding of BLDC motor commutation in BLDC motors is as important as brush placement in brush dc motors. The Hall device is the single component that made brushless dc motors possible with their product announcement in the mid-1970s. They are still necessary in positioning applications, but recently many suppliers have utilized the PM motor's K_e signals to commutate or switch to the appropriate windings at the appropriate time.

The Hall device is an electronic switch that can be used to turn on and turn off the appropriate stator windings sequentially. Positioning of the Hall device is an important condition, with higher rotor pole counts being more sensitive to its location based on the multiple effect of electrical to mechanical degrees. For example, the eight-pole rotor is 4 times more sensitive than a two-pole rotor to positioning the Hall device for optimum commutation.

One straightforward method of locating the Hall sensors is to measure the generated voltage from the PM brushless motor and align the switching pattern of the Hall device for that winding by using an oscilloscope. Fig. 10.91 shows the generated

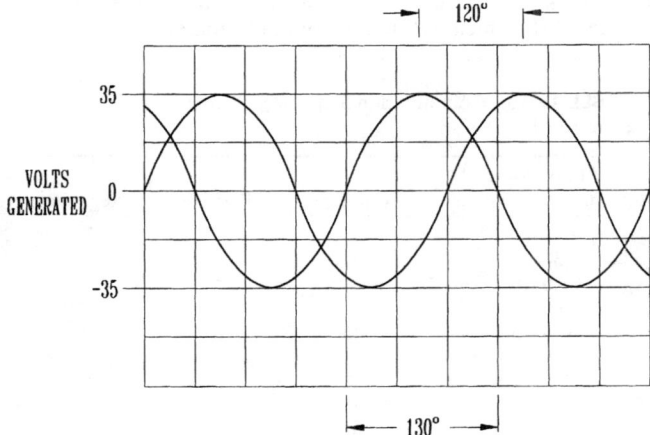

FIGURE 10.91 Generated voltage signals from a BLDC motor: voltage output versus position of two adjacent stator phases. Rotor velocity = 1000 rpm.

---

* Sections 10.12 and 10.13 contributed by Dan Jones, Incremotion Associates.

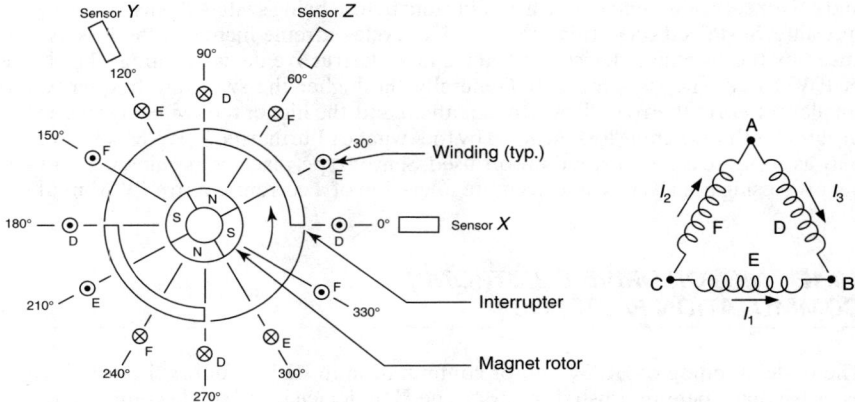

FIGURE 10.92 Location of Hall sensors at the beginning of the C^+B^- commutation cycle.

voltage signals from two of three phases of a BLDC motor. The angular positions are measured, and the Hall devices are placed at specific positions shown in Fig. 10.92. Just substitute a four-pole magnet pattern in place of the photo interrupter in Fig. 10.92, and the positions for the coils are shown for a 12-slot stator with elements of the 12 coils located every 30° mechanical. A delta winding is the line configuration. Figure 10.92 displays the sensor signals X_1, Y_1, and Z_1 for two turn-on events and two turn-off events per 360° mechanical or 720° electrical. The excitation pattern for the six power devices is shown in Fig. 10.93a. The commutation pattern is shown in Fig. 10.93b and Table 10.18, which define the commutation sequence for a three-phase

TABLE 10.18 Commutation Sequence—Three-Phase Four-Pole

Stator-rotor mechanical degrees	Drive windings	Positive	Negative
0–30	E	C	B
30–60	D	A	B
60–90	F	A	C
90–120	E	B	C
120–150	D	B	A
150–180	F	C	A
180–210	E	C	B
210–240	D	A	B
240–270	F	A	C
270–300	E	B	C
300–330	D	B	A
330–360	F	C	A

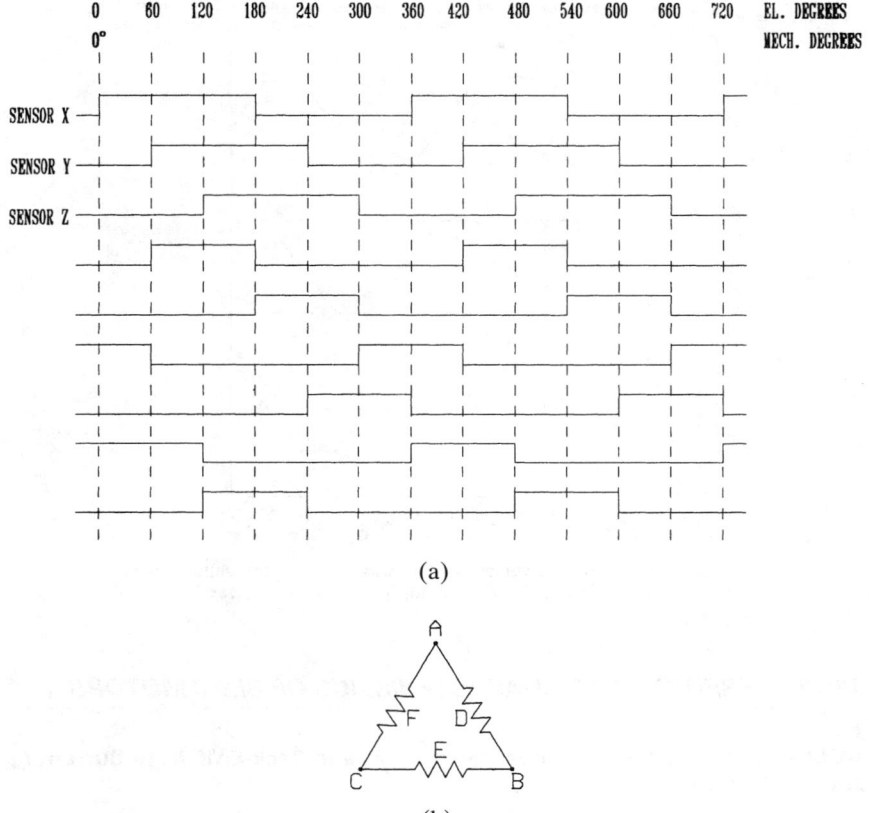

FIGURE 10.93 Delta configuration: (*a*) excitation pattern, and (*b*) commutation pattern.

four-pole delta line configuration. Figure 10.94 displays the torque versus position waveform for the delta line configuration for the 12 commutation points per full mechanical rotation.

Figures 10.95 and 10.96 and Table 10.19 illustrate equivalent patterns for a full-wave Y line configuration. Many other patterns are possible, particularly with Hall devices that turn on and off for only N or S rotor magnet poles. Using a separate magnet for commutation will also accomplish the same task. Most BLDC motor designs use the rotor magnets' leakage flux to energize the electronic switch. Although relatively expensive, the Hall device does possess temperature limitations (above 125°C). Many servomotor suppliers use resolvers for position signals; their ability to generate absolute position signals allow the resolver and its companion R and D circuit to commutate the motor as well.

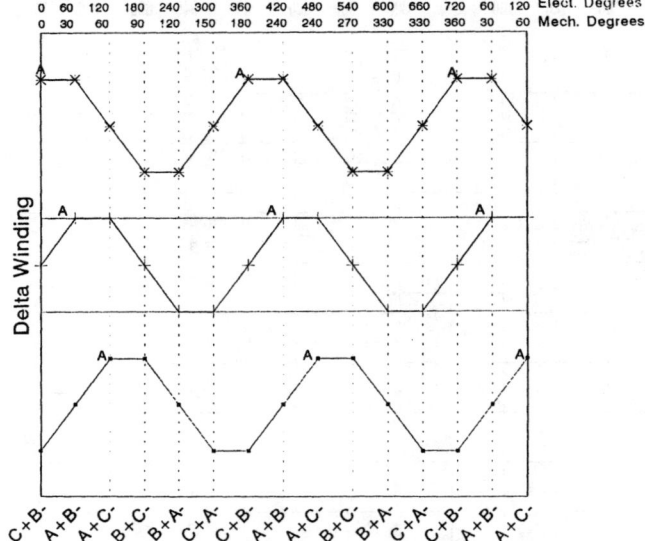

FIGURE 10.94 Torque versus position waveform for the delta line configuration: (top) winding F ~ T_{AC}, (middle) winding D ~ T_{AB}, and (bottom) winding E ~ T_{CB}.

10.13 PERFORMANCE CHARACTERISTICS OF BLDC MOTORS

10.13.1 Relationship of Torque Constant K_1 and Back-EMF K_e to Current I_{pk} and I_{rms} Profiles

There are two major approaches to driving a BLDC motor: square-wave current drives, also known as trapezoidal drives, and sine wave current drives also known

TABLE 10.19 Y Commutation Sequence—Three-Phase Four-Pole

Stator-rotor mechanical degrees	Drive windings	Positive	Negative
0–30	DF	C	B
30–60	EF	A	B
60–90	ED	A	C
90–120	FD	B	C
120–150	FE	B	A
150–180	DE	C	A
180–210	DF	C	B
210–240	EF	A	B
240–270	ED	A	C
270–300	FD	B	C
300–330	FD	B	A
330–360	DE	C	A

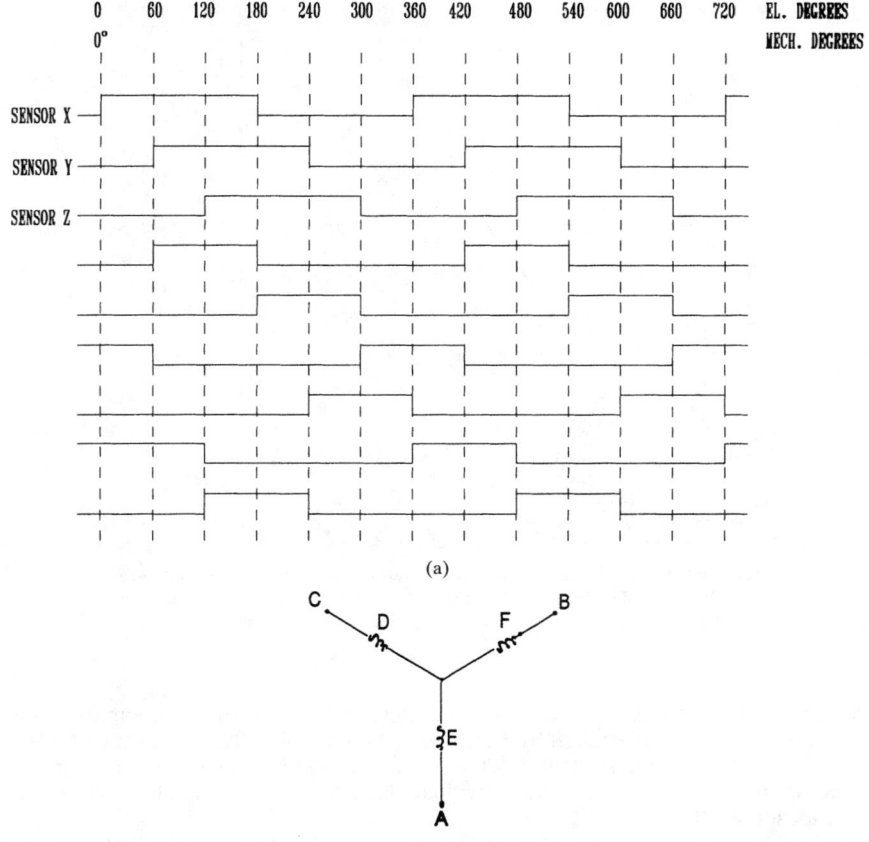

FIGURE 10.95 Y configuration: (*a*) excitation pattern, and (*b*) commutation pattern.

as sinusoidal drives. Figure 10.97 describes the torque and current profiles for the trapezoidally driven BLDC motor. The current winding excitation requires that the individual phase currents be excited for 120° electrical to achieve a 60° electrical torque commutation zone, as shown in Fig. 10.97. The sinusoidally driven BLDC motor's current and torque waveforms are shown in Fig. 10.98. A series of sinusoidally generated current waveforms create sinusoidally shaped torque waveforms.

In reviewing the available information concerning back-emf K_e, torque constant K_t, and torque and phase currents for both sinusoidal and trapezoidal motor drive combinations, there appear to be conflicting and confusing data. The trapezoidal motor drive produces more peak torque for the same peak current inputs, while the sinusoidal drive produces slightly more peak torque for the same RMS current. The major difference between the two drive schemes is unit price, with the sinusoidal

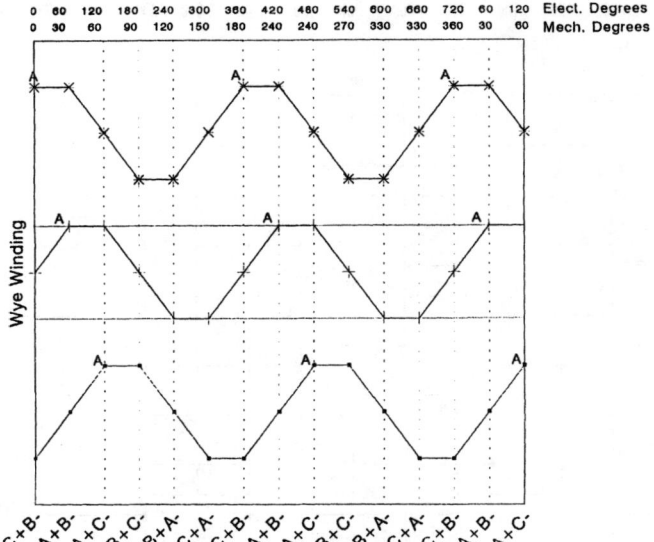

FIGURE 10.96 Torque versus position waveform for the Y line configuration: (top) winding ED ~ T_{AC}, (middle) winding EF ~ T_{AB}, and (bottom) winding DF ~ T_{CB}.

drive being the more expensive. Table 10.20 details the performance comparisons of both drive systems undertaken by Tomasek (1986), Welsh (1994), Comstock (1986), and Miller (1992). Table 10.20 also details the K_e, K_t, and R_{LL} values from each motor drive when one has computed the individual phase K_t and R_{ph} values for a sinusoidal or trapezoidal BLDC motor.

How does one design a sinusoidal BLDC motor? The answer to that question may have been provided by Erland Persson (1989). Persson focused on varying the PM arc width or length for three commercially available laminations, a 12-slot, a 24-slot, and a 36-slot lamination with a four-pole PM arc segment rotor structure. The stator lamination ODs were 49 mm (1.929 in) OD for the 12- and 24-slot laminations

TABLE 10.20 Ratios of Torque and Voltage Constants for Brushless Motors in Three-Phase Sine-Wave Systems, Motor Speed in krpm

K_t [N · m/A_{RMS}] = $1.654 \times 10^{-2} K_e$	[$V_{RMS,LL}$/krpm]
K_t [lb · ft/A_{RMS}] = $1.22 \times 10^{-2} K_e$	[$V_{RMS,LL}$/krpm]
K_t [lb · ft/A_{RMS}] = $0.1464 K_e$	[$V_{RMS,LL}$/krpm]
K_t [oz · in/A_{RMS}] = $2.342 K_e$	[$V_{RMS,LL}$/krpm]
K_e [$V_{RMS,LL}$/krpm] = $60.46 K_t$	[N · m/A_{RMS}]
K_e [$V_{RMS,LL}$/krpm] = $81.97 K_t$	[lb · ft/A_{RMS}]
K_e [$V_{RMS,LL}$/krpm] = $6.831 K_t$	[lb · in/A_{RMS}]
K_e [$V_{RMS,LL}$/krpm] = $0.4269 K_t$	[oz · in/A_{RMS}]

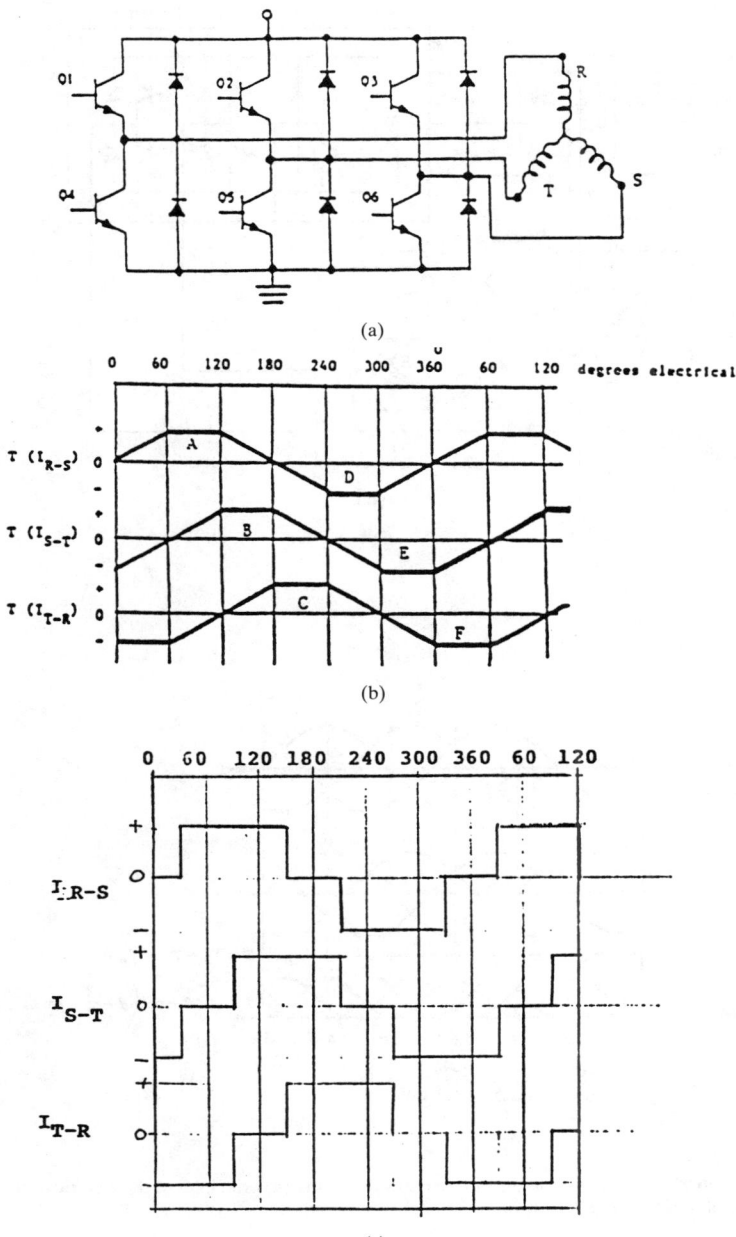

FIGURE 10.97 Trapezoidal torque and current waveforms: (*a*) typical 3/0 brushless motor inverter, (*b*) composite torque function for a 3/0 BLDC motor, and (*c*) square-wave current waveforms for a 3/0 BLDC motor.

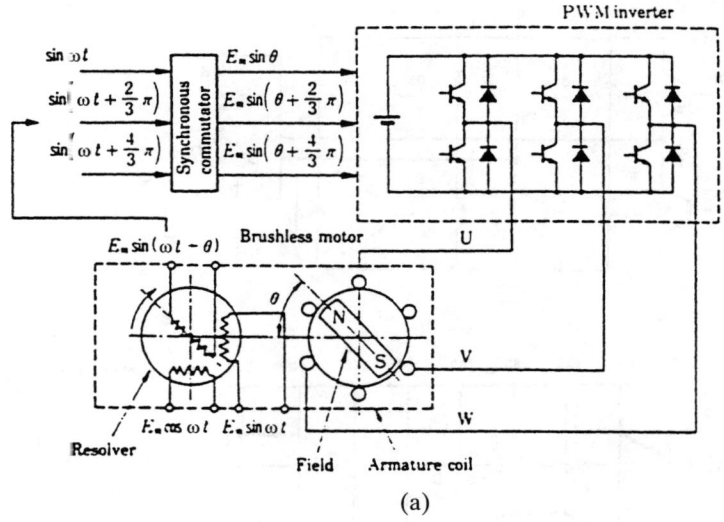

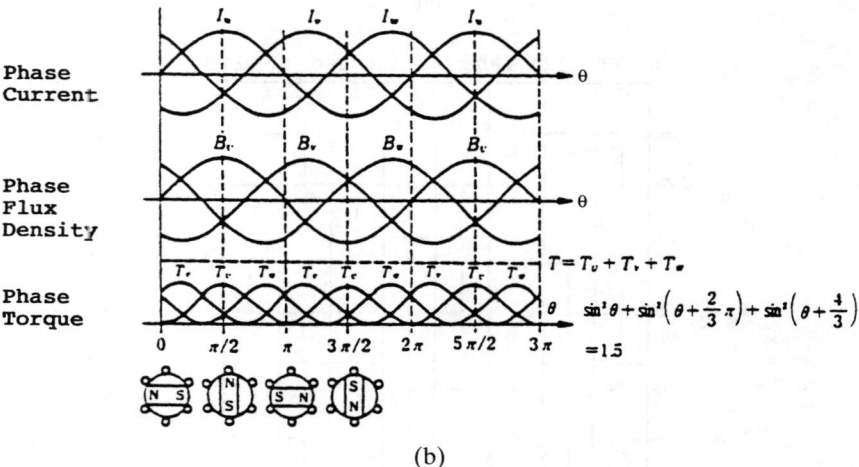

FIGURE 10.98 Sinusoidal torque and current waveforms: (*a*) principles of the electrical circuit, and (*b*) principles of torque generation.

TABLE 10.21 Ratios of Torque and Voltage Constants for Brushless Motors in Three-Phase Sine-Wave Systems, Motor Speed in rad/s.

K_t [N·m/A_{RMS}] = 1.732 K_e	[$V_{RMS,LL}$/rad·s^{-1}]
K_t [lb·ft/A_{RMS}] = 1.2775 K_e	[$V_{RMS,LL}$/rad·s^{-1}]
K_t [lb·in/A_{RMS}] = 15.33 K_e	[$V_{RMS,LL}$/rad·s^{-1}]
K_t [oz·in/A_{RMS}] = 245.28 K_e	[$V_{RMS,LL}$/rad·s^{-1}]
K_e[$V_{RMS,LL}$/rad·s^{-1}] = 0.5774 K_t	[N·m/A_{RMS}]
K_e[$V_{RMS,LL}$/rad·s^{-1}] = 0.7828 K_t	[lb·ft/A_{RMS}]
K_e[$V_{RMS,LL}$/rad·s^{-1}] = 6.523 × 10^{-2} K_t	[lb·in/A_{RMS}]
K_e[$V_{RMS,LL}$/rad·s^{-1}] = 4.007 × 10^{-3} K_t	[oz·in/A_{RMS}]

and 54 mm (2.126 in) OD for the 36-slot. The magnet arc widths were varied from 70 to 85° mechanical, and the actual measured torque versus position waveforms were recorded by using a plotter and mounted torque transducer. The stator stack axial lengths (SALs) were all 3.0 in long. Persson also tested stators with no slot skew and with either a ¾-slot skew or a 1-slot skew.

Table 10.21 shows Persson's data summary. Persson was interested in the torque ripple measured for the 18 stator slot and skew combinations. The impact of skew on reducing torque ripple is also demonstrated. The impact of more teeth per pole in reducing torque ripple was not demonstrated by the experimental data. There was no attempt to alter tooth shapes, air gaps, or magnet material, just to supply the reader with a reference point in terms of solid experimental data. Magnet material used was a low grade of ferrite magnet material, so there was no magnetic saturation in any of the motor soft iron (steel) members.

What is more astounding is the shape of the motor T versus θ waveform. The

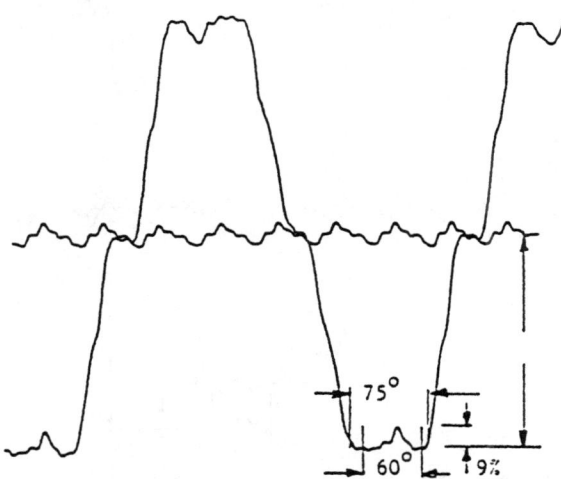

FIGURE 10.99 Static torque function: 12 slots, 80° arc, no skew.

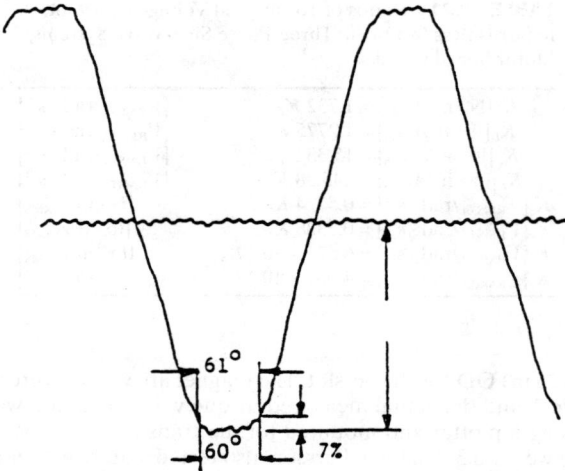

FIGURE 10.100 Static torque function: 12 slots, 80° arc, ¾-slot skew.

commutation interval (angle) is 60° electrical as in the previous theoretical examples. Figures 10.99 and 10.100 show the static torque function for the 12-slot stator in both unskewed and ¾-skewed stator configurations, respectively. The overall shape is flat for trapezoidal drive conditions with a maximum drop of 9 percent in torque over the 60° commutation interval. Most of the torque drop in Fig. 10.100 was caused

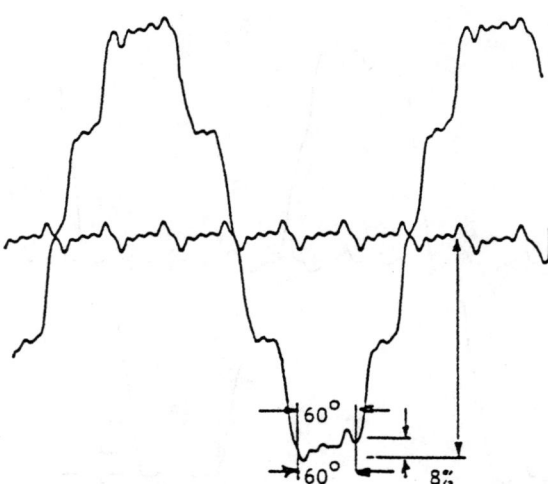

FIGURE 10.101 Static torque function: 12 slots, 70° arc, no skew.

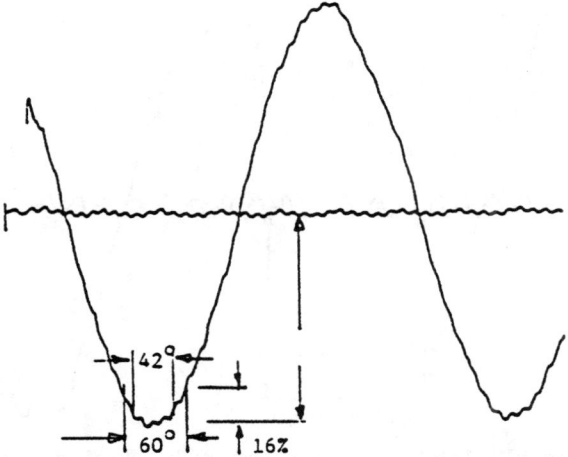

FIGURE 10.102 Static torque function: 12 slots, 70° arc, ¾-slot skew.

by the motor's cogging torque. The ¾-slot skewed version (Fig. 10.100) reduces the torque drop to 7 percent of total (peak) value over the 60° electrical interval. The overall torque versus position waveform shape for both skewed and unskewed stators (with an 80° arc width) is trapezoidal in shape.

Figures 10.101 and 10.102 show the torque versus position plot for a 12-slot stator with a 70° magnet arc width. Figure 10.101 displays the torque vs. position plot with no stator slot skew. The cogging torque signature is the primary cause of the 8 percent torque variation over the 60° commutation interval in this setup. Skewing the stator stack by ¾ slot changes the torque profile into a quasi-sinusoidal torque profile with lower instantaneous torque variation (see Table 10.22).

Figure 10.103 illustrates the torque profile of the 24-slot BLDC stator, no skew, with an 85° rotor magnet arc width. There is a 9 percent torque drop-off, but the waveform has a near-trapezoidal torque profile. Adding a 1-slot skew (Fig. 10.104) smoothes the torque profile, and the waveform over the 60° electrical interval is closer to a trapezoidal waveform. Figures 10.105 and 10.106 show the effects on waveform of a 70° arc with a 24-slot stator, both skewed and unskewed, respectively. The waveform becomes more sinusoidal. The 36-slot lamination performance is shown in Figures 10.107 and 10.108. Figure 10.107 displays the 80° rotor magnet arc width with no stator skew. The overall torque profile is decidedly sinusoidal. There is

TABLE 10.22 Subjective Evaluation of Expected Torque Ripple

Magnet arc	12 slot		24 slot		36 slot	
	Straight	Slot skew	Straight	Slot skew	Straight	Slot skew
85°	8	5	8	7	14	5
80°	9	7	9	7	13	10
70°	8	16	16	14	26	20

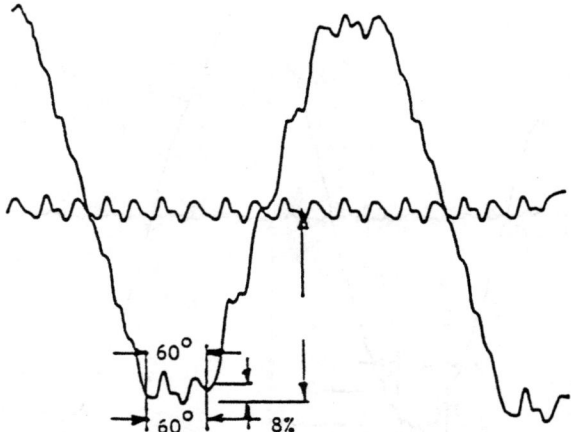

FIGURE 10.103 Static torque function: 24 slots, 85° arc, no skew.

a 13 percent drop-off in torque magnitude over the 60° electrical commutation interval, and the torque waveform shows the cogging torque "bumps" quite prominently. The 1-slot skew version of the 80° rotor magnet and 36-slot lamination displays a quasi-trapezoidal waveform with a 10 percent torque drop-off over the 60° electrical commutation interval.

The final torque-position waveform shows the last torque profile, which describes a 36-slot lamination with a 70° rotor magnet arc width. The overall shape is definitely sinusoidal in both unskewed and skewed examples. The torque drop-off has increased to 26 percent (Fig. 10.109) and 20 percent (Fig. 10.110). Figures 10.111 and 10.112 display copies of the actual 12-slot and 24-slot laminations used by Persson.

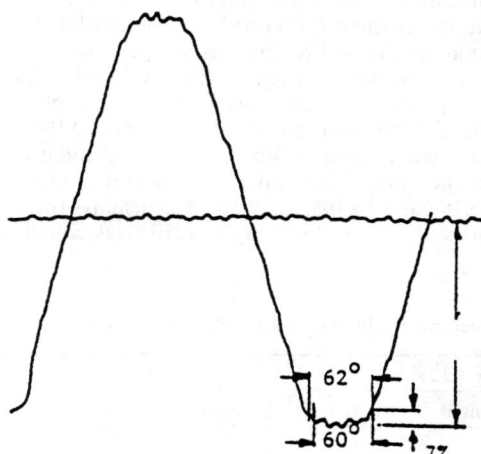

FIGURE 10.104 Static torque function: 24 slots, 85° arc, 1-slot skew.

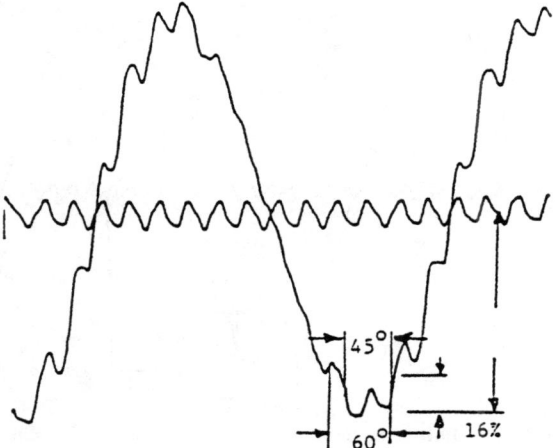

FIGURE 10.105 Static torque function: 24 slots, 70° arc, no skew.

Summarizing Persson's results:

- Decreasing the magnet arc width will lead to sinusoidal torque waveforms.
- Skewing the rotor will reduce cogging and increase the tendency toward quasi-sinusoidal torque waveforms.
- Increasing the number of stator slots per phase per pole ($n \geq 3$) will increase the tendency toward sinusoidal torque waveforms.

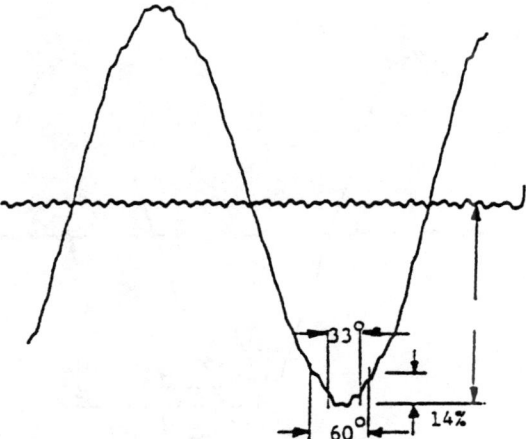

FIGURE 10.106 Static torque function: 24 slots, 70° arc, 1-slot skew.

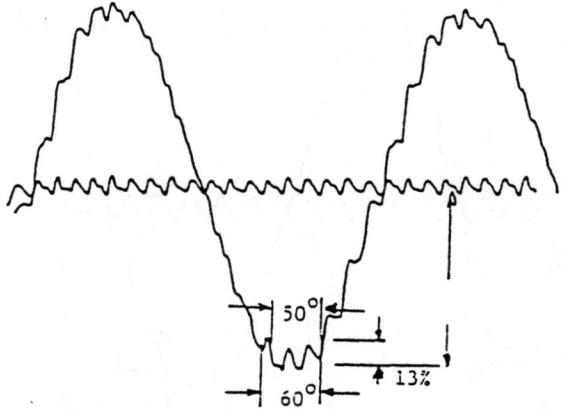

FIGURE 10.107 Static torque function: 36 slots, 80° arc, no skew.

- Use a wide pole arc magnet and a minimum number of stator slots per phase per pole ($n = 1$) to achieve a trapezoidal torque profile.

10.13.2 Phase Resistance

This section develops a basic comparison chart for the theoretically anticipated values for line-to-line resistance for the four major line-winding configurations. Using the phase resistance X values, the line-to-line resistance for a given configuration is as shown in Table 10.23.

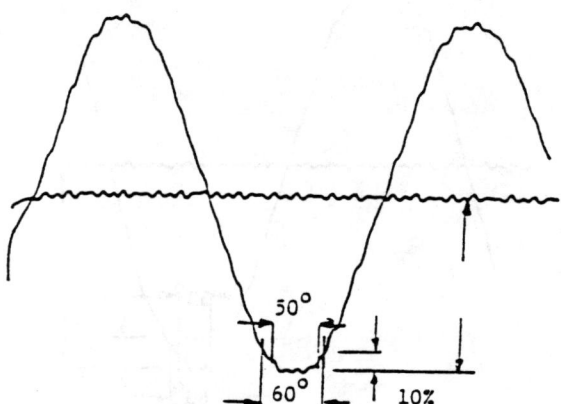

FIGURE 10.108 Static torque function: 36 slots, 80° arc, 1-slot skew.

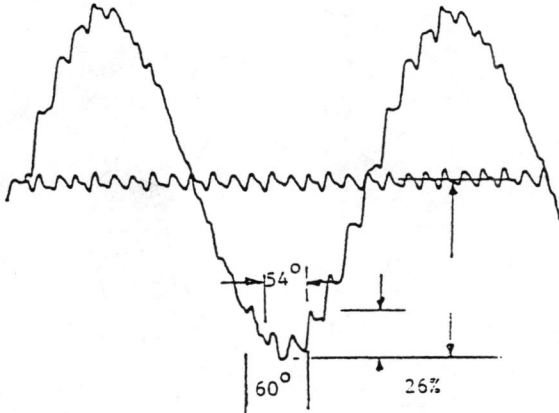

FIGURE 10.109 Static torque function: 36 slots, 70° arc, no skew.

10.13.3 Establish Speed-Torque Curve—No-Load, Load, and Peak Torque Values

The brushless dc motor has one major difference from its brush dc counterpart. All testing including the "simple" back-emf K_e measurement requires that the brushless dc motor be tested with its electronic drive package. The brushless dc motor that uses rare earth magnets can usually outperform its companion drive in the region of maximum torque capability, and there are commutation limits based on motor speed

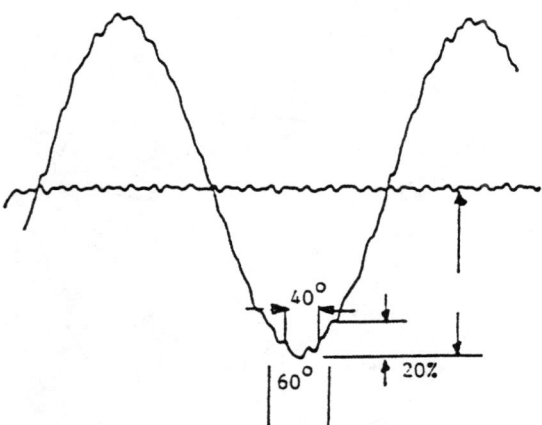

FIGURE 10.110 Static torque function: 36 slots, 70° arc, 1-slot skew.

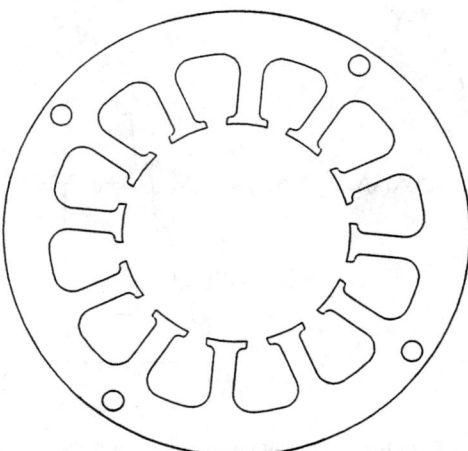

FIGURE 10.111 12-slot lamination.

times the number of commutation switching events per revolution. These two limits shape the torque-speed profile. Figure 10.113 shows a representative torque-speed limit curve for a sinusoidal drive showing these limitations.

The calculations for establishing the various torque-speed performance points does proceed in the same manner used for a PM brush dc motor, except that the torque and speed limits established for the specific motor-drive combination must be used to limit the calculated values at both ends of the motor torque-speed curve.

The BLDC motor's torque and speed performance is linear over a major portion of the torque-speed curve as described.

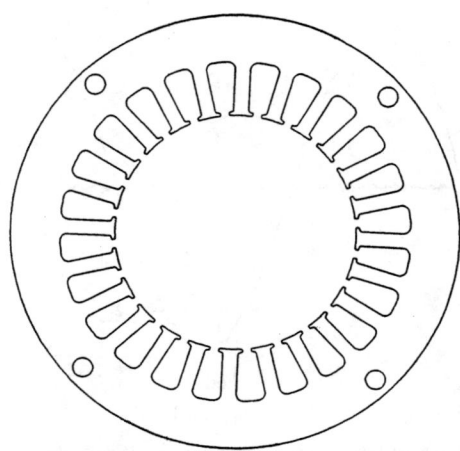

FIGURE 10.112 24-slot lamination.

TABLE 10.23 Basic Comparison Chart

Phase resistance	Hookup configuration	Line-to-line resistance
X	Y (half)	X
X	Y (full)	2X
X	Delta	⅔X
X	Independent	X

Once the torque constant K_t and the phase resistance R_{ph} have been established for a representative design, one can use simple algebraic formulas to compute the motor's basic torque versus speed performance. The basic performance parameters from the design example are shown in the Fig. 10.114 computer simulation. The computed values for $K_t = 21.046$ (oz · in)/A or 0.149 (N · m)/A and $R_{LL} = 0.275\ \Omega$ for a full-wave Y-connected BLDC motor are used to establish the load torque T_{load}, peak torque T_{pk}, and other operating parameters.

$$T_{load} = I_{load}\, K_t \tag{10.10}$$

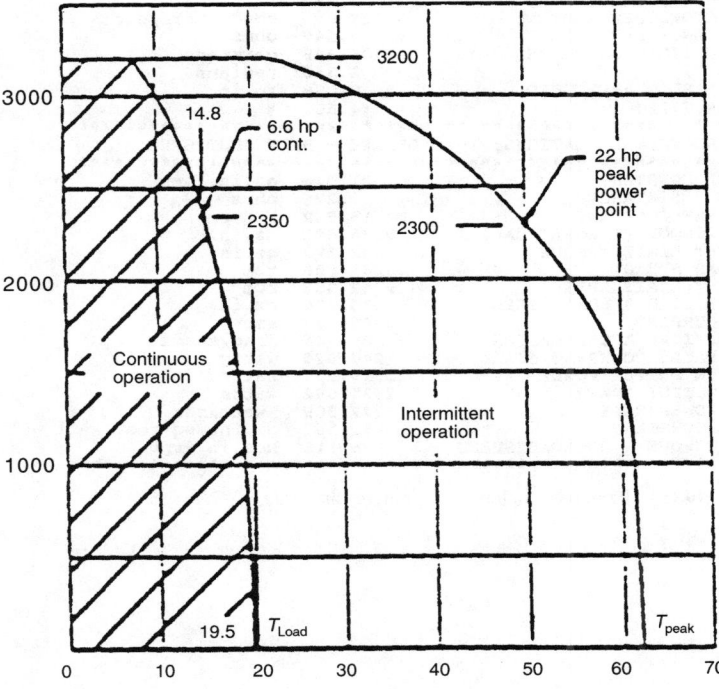

FIGURE 10.113 Typical BLDC motor torque-speed limit curve.

```
CUSTOMER:SMMA 95                          DATE:03-02-1995
MAGNET CONFIGURATION:ARC SEGMENT          SOLUTION CODE NUMBER:
PARALLEL WINDING HOOKUP            PART II
SATURATION CHECK, MOTOR PARAMETER PERFORMANCE AND UNIT PHASE RESISTANCE:
1.MATERIAL USED FOR BH CURVE:    NeI 30
2.Br:                            9351.942   gausses
3.Hc:                            2148.057   oersteds
4.TOTAL CONDUCTORS:              1656.
5.NUMBER OF PHASES:              3.
6.SHAFT O.D.:                    0.500      in.
7.VOLTAGE:                       160.000    volts
8.CURRENT LIMIT:                 30.000     amps
9.LOAD SPEED:                    10000.000  rpm
AWG VALUES FOR CALCULATIONS:
1.WIRE SIZE:                     22.0       single coated
2.AWG VALUE No. 1:               16.2000    ohms/Kfeet
3.AWG VALUE No. 2:               1410.000   turns/sq.in.

MAGNET CIRCUIT CALCULATIONS:
*************************************************************
   FLUX MAGNET PER POLE              128.308  kilomaxwells
   FLUX GAP PER POLE W/O OVERHANG    109.294  kilomaxwells
   FLUX GAP PER POLE WITH OVERHANG   118.801  kilomaxwells
   FLUX GAP PER POLE WITH LEAKAGE    103.305  kilomaxwells
   CROSSECTIONAL HOUSING AREA          0.516  sq.in.
   HOUSING FLUX DENSITY               15.512  kilogausses
   CROSSECTIONAL TOOTH AREA            0.338  sq.in.
   TOOTH FLUX DENSITY                 15.819  kilogausses
   HUB AREA                            0.625  sq.in.
   HUB FLUX DENSITY                   15.914  kilogausses
*************************************************************
MOTOR WINDING CALCULATIONS:
*************************************************************
   TURNS PER COIL                       69.   turns/coil
   MEAN LENGTH OF TURN                 5.894  in./turn
   LENGTH OF COIL                     33.892  ft.
   CONDUCTORS PER SLOT                 138.   conductors/slot
   RESISTANCE PER COIL                 0.549  ohms
   DEMAGNETIZATION FORCE            1706.849  oersteds
   GAMMA                               1.345  radians
   ACTUAL SLOT AREA USED               0.098  sq.in.
   COPPER FILL                        44.566  %
*************************************************************
WYE WINDING CALCULATIONS: 15.0 DEGREE - HALF SLOT SKEW
*************************************************************
   TORQUE CONSTANT                    21.046  oz.in./amp
   PHASE RESISTANCE                    0.275  ohms
   BACK EMF CONSTANT                  15.562  V/Krpm
   PEAK TORQUE (THEORETICAL)       12265.872  oz.in.
   CURRENT LIMIT TORQUE              631.373  oz.in.
   NO LOAD SPEED                   10281.586  rpm
   CURRENT LIMIT SPEED              9752.352  rpm
   TORQUE LOAD @ LOAD SPEED          335.930  oz.in.
   PEAK CURRENT                      582.819  amps
   THEORETICAL ACCELERATION           84.646  Krad/sec^2
   DISSIPATED POWER AT STALL         247.075  watts
   MAXIMUM OUTPUT POWER            23319.641  watts
   LOAD OUTPUT POWER                2484.692  watts
   PEAK POWER RATE                   377.309  Kwatt/sec
   MOTOR CONSTANT                     40.167  oz.in./sq root watt
   PEAK TORQUE - NO LOAD SPEED       126.113  Moz.in.-rpm
*************************************************************
```

FIGURE 10.114 Example simulation program results.

where I_{load} is the value established by the motor's thermal constants

$$T_{pk} = \frac{E_t}{R_{LL}} K_t \qquad (10.11)$$

where

$$E_T = \text{terminal voltage} \qquad (10.12)$$

$$I_{pk} = \frac{E_T}{R_{LL}} \qquad (10.13)$$

The load torque can be computed by using Eq. (10.10) if the load current has been established. The inner line on the Fig. 10.113 representative torque-speed curve is the motor's safe operating ambient conditions (SOAC) curve. All load points (torque and speed) to the left of this boundary line are rated and continuous values. The I_{load} is located somewhere along this curve. The actual motor terminal voltage would create a straight-line torque-speed curve based on constant voltage which would intersect the SOAC curve at a specific point.

For purposes of illustration, let us assume that this BLDC motor can dissipate 74 W at a specific point on the SOAC line. Using the estimated hot resistance value for R_{LL} which is assumed to be at 100°C above ambient, the actual load current can be computed using the formula in Eq. (10.14):

$$I_{LD} = \frac{\sqrt{P_{diss}}}{R_{LL,hot}} = 13.8 \text{ A} \qquad (10.14)$$

where $R_{H,hot} = (0.275)(1.4) = 0.385 \; \Omega$.

Therefore, using Eqs. (10.10), (10.11), and (10.12), the load torque T_{load} is computed at 290 oz · in. The T_{pk} is calculated as $(160/0.385)(21.046) = 8746$ oz · in or 61.94 N · m. The peak current is calculated as 415.8 A. The peak values are theoretical, because one can expect the drive limits to engage well before these calculated peak values would be reached. The establishment of the theoretical no-load speed $N_{NL,th}$ can be done by using the K_e value computed in equation 10.15.

$$N_{NL,th} = \frac{E_T}{K_e} \qquad (10.15)$$

where: E_T = motor terminal voltage
K_e = motor's Y-configuration back-emf

$$N_{NL,th} = \frac{160}{15.562} = 10.28 \text{ krpm}$$

where $N_{NL,th}$ is the motor's no-load speed.

Therefore, 10,280 rpm is the peak theoretical no-load speed. One can estimate or calculate the actual no-load speed if no load current is available. In any case, since the BLDC no-load rotating losses (windage, bearing system, friction, etc.) are very low, conservatively in the region of 2 to 3 percent, an adjusted no-load speed value of 9972 rpm (97 percent of theoretical value) can be established. Load speed N_L can be computed as displayed in Eq. (10.16).

$$N_L = N_{NL}(1 - \alpha) \qquad (10.16)$$

where α is the ratio of $T_{load}/T_{PK,th}$

$$N_L = (9972)\left(1 - \frac{290}{8746}\right) = (9972)(0.967) = 9641 \text{ rpm}$$

Any speed point along the constant-voltage derived torque-speed curve can be computed in a similar manner until the boundary for peak performance (intermittent duty limit line in Fig. 10.113) is reached.

The values used in the Fig. 10.114 simulation are the original cold (20°C) value for $R_u = 0.275 \, \Omega$, so the calculated values in the simulation in Fig. 10.114 are 40 percent higher than the ones computed in this section. Still, the procedures are the same.

10.13.4 Optimizing Motor Constant K_m

The thousands of applications that use a BLDC motor make it impossible to determine a simple criteria in selecting the best BLDC motor for a given application. For motor designers, there is a single figure of merit that can determine if the motor design engineer has optimized the design in terms of magnetics and stator winding conditions. The figure of merit (motor constant K_m), along with the motor's dimensions does provide a relatively simple but effective method of evaluating a BLDC motor design. This also assumes that the proper magnet material has been selected, based on cost as well as dimensional requirements.

$$K_m = \frac{T}{\sqrt{W_{in}}} = \frac{K_t}{\sqrt{R_{LL}}} \tag{10.17}$$

where T = any torque value
W_{in} = input power at the chosen torque value
K_t = motor torque constant
R_{LL} = motor line-to-line resistance

This constant, when maximized for a given motor volume, ensures the design engineer of the best possible design. Many motor mechanical, magnetic, and electrical parameters are integrated into motor constant K_m. Maximizing K_m is maximizing the motor's copper to iron ratio for that specific motor size.

If one reviews the design process, the K_t value is primarily tied to magnet flux, number of turns, pole count, unit path reluctance, and so forth. R_{LL} is a function of available slot space, winding pattern, pole count, and so forth. Optimizing both performance parameters optimizes the motor design in terms of the motor's magnetic circuit and winding selection.

Remember that other user requirements, such as the lowest possible winding inductance, would bias the motor constant K_m to very low numbers of turns (conductors). Minimizing rotor inertia for certain incremental motion applications would also bias the K_m optimization process as described.

The K_m value listed in Fig. 10.114 is 40.17 (oz · in)/$\sqrt{W}$. One could significantly increase K_m by the following means:

- Increasing the number of poles
- Increasing the rotor OD up to a certain OD dimension; then the K_m value would decrease beyond this dimensional limit

- Changing to high-energy magnets
- Increasing motor OD or AL

Usually the motor's available input voltage and current levels will act as a limit to optimizing K_m beyond a certain limit. The challenge for the motor design engineer is to maximize K_m within specified electric inputs or friction and inertia requirements which will yield the best overall magnetic design.

10.13.5 Power, Losses, Efficiency, and Others

There are two groups of motor performance parameters, the conventional and the special performance parameters. A list of the conventional motor performance parameters includes those parameters or figures of merit that relate to motor power. For example, Eq. (10.18) determines dissipated power I^2R, Eq. (10.19) computes input power, Eq. (10.20) calculates output power, and Eq. (10.21) establishes the power efficiency at that specific load point P_{LD}. These parameters are designated as the conventional performance parameters and are computed as follows:

$$P_{diss} = I_{LD,RMS}\, R_{LL} \tag{10.18}$$

$$P_{in} = E_T\, I_{LD,RMS} \tag{10.19}$$

$$P_{out} = \frac{T_{LD} N_{LD}}{K} \tag{10.20}$$

where $K = 1352$ for oz · in and rpm values
$K = 1$ for N · m and rad/s values

$$\text{Efficiency} = \frac{P_{out}}{P_{in}} \tag{10.21}$$

The measuring of motor voltage and currents has become much more difficult with the inverter and the power amplifier unable to be separated from the motor. The V_{RMS} and I_{RMS} current values are the most important ones to measure and require special equipment.

In most design computations, a 100°C temperature rise value of R_{LL} is used. Computations of these values are shown in Fig. 10.114, where a more optimistic room ambient R_{LL} value is employed.

The more specialized figures of merit include motor characteristics such as theoretical acceleration T/J, peak power rate (PPR), rated power rate (RPR), peak power density (PPD), and rated power density (RPD). The first three figures of merit are used in incremental motion applications. The other two are more often used in automotive, aerospace, and portable instrument applications. Theoretical acceleration can be determined by Eq. (10.22).

$$T_{acc} = J_m\, \alpha \quad \text{or} \quad \frac{T_{acc}}{J_T} = \alpha \tag{10.22}$$

where α = acceleration, rad/s^2
$T_{acc} = T_{pk}$
J_m = the motor rotor inertia

This equation can be modified in the inertia term (J_m) to include the load inertia J_L for load matching conditions.

Power rate is defined as the rate of change of power with respect to time and is important in incremental motion applications. Equations (10.23) and (10.24) detail the method of computation of peak and rated power rates (PPR and RPR).

$$\text{PPR} = \frac{T_{pk}^2}{J_m} = \frac{I_{pk}^2 R_{LL}}{\Gamma_m} \tag{10.23}$$

where Γ_m is the machine's mechanical time constant,

$$\text{RPR} = \frac{T_R^2}{J_m} = \frac{I_R^2 R_{LL}}{\Gamma_m} \tag{10.24}$$

Power rate can be expressed in kilowatts per second. The BLDC motor with the highest power rate will produce the shortest time to move from one point to another.

Power density or power per unit volume has become popular in recent years with the advent of battery operated motors and actuators. These figures of merit relate output power to the unit volume and, indirectly, unit weight. Equations (10.25) and (10.26) show the simple formulas for computing these figures of merit.

$$\text{PPD} = \frac{\text{Peak output power}}{\text{unit volume}} \tag{10.25}$$

$$\text{RPD} = \frac{\text{Rated output power}}{\text{unit volume}} \tag{10.26}$$

The units are watts per cubic inch or watts per cubic meter. The higher the value of power density, the more power per unit volume available.

REFERENCES

Analog Devices: *Analog Devices AD2S80A: Variable Resolution Monolithic Resolver-to-Digital Converter* (product specifications), Halifax, Nova Scotia, pp. 11, 12.

———: "S/D Converters, Rev. A," *Data Converter Reference Manual*, vol. 1, Analog Devices, Halifax, Nova Scotia, 1992, pp. 3–96.

Anunciada, V., and M. M. Silva: "A New Current Mode Control Process and Application," *IEEE Transactions on Power Electronics*, vol. 6, no. 4, 1991, pp. 601–610.

ASM: *Metals Handbook, Desk Edition*, 2d ed., vol. 7, ASM International, Materials Park, Ohio, 1998.

Band, Robert: *Twenty-four Hour Test*, Electro-Craft Motion Control, Gallipolis, Ohio.

Boyes, Geoffrey S. (ed.): *Synchro and Resolver Conversion*, Analog Devices, Halifax, Nova Scotia, 1980.

Bozorth, R. M.: *Ferromagnetism*, IEEE Press, Piscataway, N.J., 1993.

Brod, D. M., and D. W. Novotny, "Current Control of VSI-PWM Inverters," *IEEE Transactions on Industry Applications*, vol. 21, no. 4, 1985, pp. 562–570.

Bularzik, J. H., R. F. Krause, and H. R. Kokal: "Accucore Extends PM's Magnetic Reach," *Metal Powder Report*, vol. 53, no. 7, July/August 1988, pp. 38–42.

Campbell, Peter: *IEEE Transactions on Magnetics*, vol. 26, no. 5, September 1990.

Chai, H. P.: *Motion Devices and Transformers*, San Jose State University, Department of Electrical Engineering, San Jose, Calif., 1995.

Chang, S. S. L, and J. H. Karr: "Armature Iron Losses in Series Motors," *AIEE Transactions*, vol. 68, 1949.

——— and ———: "A-C Series, A Design Method," *AIEE Transactions*, vol. 50, 1950.

Clifton Precision: *Synchro and Resolver Engineering Handbook*, Clifton Precision, Clifton Heights, Pa.

Comstock, Robert: "Trends in Brushless PM Drive and Motor Technology Overview," 86 Motor-Con Conference, Boston, 1986.

Cullity, B. D.: *Introduction to Magnetic Materials*, Addison-Wesley, Reading, Mass., 1972, p. 502.

Dr. Johannes Heidenhain, GmbH: *Incremental Rotary Encoders Catalog and General Information*, p. 14.

Electro-Craft Handbook: DC Motors, Speed Controls, Servo Systems, 5th ed., Electro-Craft Motion Control, Gallipolis, Ohio, 1980.

Ernst, Alfons: *Digital Linear and Angular Metrology: Position Feedback Systems for Machines and Devices*, Verlag Moderne Industrie AG & Co.

Eshelman, Ronald: *Vibration Control*, Vibration Institute.

Evershed, "PM Magnets in Theory and Practice," *Journal of the Institute of Electrical Engineers*, vol. 58, 1920, pp. 780–837.

Farrand Controls: *Inductosyn Precision Linear and Rotary Position Transducers* (catalog).

Figel, M., and D. Labahn: "Forts chritte bei der Konstruktion von Universal Motoren," *Siemens-Oeitschrift*, H9, 1972, pp. 761–766.

Fortesque, C. L.: "Methods of Symmetrical Components Applied to the Solution of Polyphase Networks," Transactions of AIEE, vol. 37, 1918, pp. 1027–1115.

Frayman, L., D. R. Ryan, and J. B. Ryan: "The Role of the Secondary Operations in the Manufacturing of P/M Automotive Components for Soft Magnetic Applications," *Advances in Powder Metallurgy and Particulate Materials*, vol. 20, Metal Powder Industries Federation, Princeton, N.J., 1996, pp. 25–37.

Freimanis, M.: "Hybrid Microstepping Chopper Can Reduce Iron Losses," *Motion Control*, April 1992, pp. 36–39.

Gray, Alexander: *Electrical Machine Design*, 2d ed., McGraw-Hill, New York, 1926.

Hendershot, Jim: "Brushless DC Motor Phase, Pole, and Slot Configurations," IMCSS 90 Symposium, San Jose, Calif., 1990.

Holtz, J.: "Pulsewidth Modulation—A Survey," *IEEE Transactions on Industrial Electronics*, vol. 39, no. 5, 1992, pp. 410–420.

Hunt, Robert P.: "A Magnetoresistive Readout Transducer," *IEEE Transactions on Magnetics*, vol. 23, no. 5, September 1987.

International Electrotechnical Commission: "Dimensions of Brushes and Brush Holders for Electrical Machinery—Part 3," *IEC Technical Questionnaire for Users of Carbon Brushes*, IEC 136, IEC, Germany, 1980.

Ireland, James R.: *Ceramic Permanent-Magnet Motors*, McGraw-Hill, New York, 1968.

Kassakian, J. G., M. F. Schlecht, and G. C. Verghese: *Principles of Power Electronics*, Addison-Wesley, Reading, Mass., 1991.

Kokal, H. R., J. H. Bularzik, and R. F. Krause: "High Density Pressed Soft Magnetic Components for AC and DC Applications," *Advances in Powder Metallurgy and Particulate Materials*, pt. 1, Metal Powder Industries Federation, Princeton, N.J., 1997, pp. 61–75.

Kuo, B. C.: *Incremental Motion Control Step Motors and Control Systems*, SRC, 1979.

Leonhard, W.: *Control of Electrical Devices*, Springer-Verlag, 1985.

Micro Switch: *Hall Effect Transducers: How to Apply Them as Sensors*, Micro Switch, Freeport, Ill., 1982.

Miller, Tim: "Definition of K_T and K_E for Brushless DC Motors," IMCSS 92 Symposium, San Jose, Calif., 1992.

Miller, T. J. E.: *Brushless Permanent-Magnet and Reluctance Motor Drives*, Oxford University Press, New York, 1989.

———: *Switched Reluctance Motors and Their Control*, Oxford University Press, New York, 1993.

Murphy, J. M. D., and F. G. Turnbull: *Power Electronic Control of AC Motors*, Pergamon Press, Oxford, U.K., 1988.

MWS Wire Industries: *Technical Data Catalog*, MWS, Westlake Village, Calif., 1985.

National Fire Prevention Association: *National Electrical Code, 1996 Edition*, NFPA, Battery March Park, Mich., 1996.

National Electrical Manufacturers Association: Standard MG 1-1987, NEMA, Washington, D.C., 1987.

———: Standard MG 1-1993, NEMA, Washington, D.C., 1993.

Newark Electronics: *Newark Electronics Components Catalog*, Newark Electronics, Chicago, 1997.

Packer, L. C.: "Commutation in Universal Type Motors," *AIEE Transactions*, vol. 70, 1951, pp. 1–9.

Persson, Erland: "Influence of Magnet and Winding Configuration on the Torque Function of Brushless Motors," PCIM 89 Conference, Long Beach, Calif., 1989.

Pippenger, D. E., and E. J. Tobaben, *Linear and Interface Circuits Applications*, vol. 3: *Peripheral Drivers, Data Acquisition Systems, Hall Effect Devices*, Texas Instruments, Niskayuna, N.Y., 1982.

Plonsey, R., and R. E. Collin: *Principles and Applications of Electromagnetic Fields*, McGraw-Hill, New York, 1961.

Puchstein, A. F.: *The Design of Small Direct Current Motors,* John Wiley & Sons, New York, 1961.

Roters, H. C.: *Electromagnetic Devices,* John Wiley & Sons, New York, 1941.

Roye, Daniel: "Communication in Small Uncompensated Motors," *IEEE T-PAS,* vol. 97, no. 1, January/February, 1978, pp. 242–250 (3e04).

Tomasek, Jaroslav: "Structural and Performance Fundamentals of Sine Wave Controlled Brushless Servo Drives," IEEE Applied Power Electronics Conference, Baltimore, Md., 1986.

Trickey, P. H.: *Electric Motor and Generator Design,* vols. 1–17, Duke University, Durham, N.C./Matt Gavin, Sacramento, Calif., 1971.

Veinott, Cyril G.: *Theory and Design of Small Induction Motors,* McGraw-Hill, New York, 1959.

——— and Joseph Martin: *Fractional and Subfractional Horsepower Electric Motors,* 4th ed., McGraw-Hill, New York, 1987.

Webster's Seventh New Collegiate Dictionary, G. & C. Merriam, Springfield, Mass., 1971.

Welsh, Richard, Jr.: "Fundamental Operation of a Brushless DC Motor," 94 Motion Expo, San Jose, Calif., 1994.

BIBLIOGRAPHY

Adams, L. F.: "Alternating Current Single Phase Commutator Motors," *General Electric Rev.*, vol. 20, 1917.
Anderson, Edward, and Rex Miller: *Electric Motors*, Theodore Audel, Indianapolis, Ind., 1968.
Arnold, Frank: (1988), "Power Rate Simplifies Drive Selection," *PTD*, June 1988.
Automatic Armature Winding Pioneering—Theory and Practice, Globe Corporation, Dayton, Ohio, March 1970.
Britani, Giampiero: "The Effects of the Torque Ripple and Cogging Torque on ISV of a Brushless DC Motor," 85 Motor-Con Conference, Boston, 1985.
Carbon Brushes and Electrical Machines, Morganite Electrical Carbon, Ltd., Dunn, N.C., 1988.
Carter, F. W.: *Electrical World* (New York), vol. 38, November 30, 1901.
———: *JAIEE*, vol. 46, 1927, p. 431.
Clayton, Albert: *Performance and Design of Direct Current Machines*, 3d ed., 1959.
Cooper, W.: "Regeneration of Power with Single Phase Electric Railway Motors," *AIEE*, vol. 26, pt. 2, 1907.
Creedy, F.: "Alternating Current Commutator Motors," *JIEE* (British), vol. 33, 1903–1904.
———: "The Alternating Current Series Motor," *JIEE* (British), vol. 3S, 1904–1905.
Dijken, R. N.: "Optimization of Small AC Series Commutator Motors," *Philips Res. Repts. Suppl.*, No. 6, 1971, pp. 1–176.
Dote, Yasuhiko, and Sakan Kinoshita: *Brushless Servo Motors Fundamentals and Applications*, Claredon Press, Oxford, U.K., 1990.
Duplessis, John: "An Attractive Proposal," *Machine Design*, June 11, 1993.
Elsey, H. M. and C. Lynn: "Effects of Commutator Surface Film Conditions on Commutation," *AIEE Tr.*, vol. 68, pt. 1, 1949, pp. 106–112.
Fejes, William, Jr.: "Computer Simulations of Torque Speed Characteristics of a Brushless Servo System," 85 Motor-Con Conference, Boston, 1985.
Fitzgerald, A. E., Charles Kingsley Jr., and Stephen D. Umans: *Electric Machinery* 5th ed., McGraw-Hill, New York, 1990.
Fogiel, M., and staff: *The Electrical Machines Problem Solver*, Research and Education Association, New York, 1983.
Franklin, W. S., and S. S. Seyfert: "On the Space Economy of the Single Phase Series Motor," *AIEE*, vol. 29, pt. 1, 1910.
Fuchs, E. F.: "Aging of Electrical Appliances due to Power System Harmonics: Transformers, Universal Motors, and Induction Motors," *T-PWR D*, July 1986, pp. 301–307.
Fujii, I., and Hanazana, T.: "Commutation of Universal Motors," IEEE Industry Applications Society, 1989.
Fujii, T.: "Improving Design Procedure of Universal Motor by Vector Diagram Performance Analysis." *Tr. IEEE 96-B*, December 1976, pp. 623–630 (Japanese).
———: "Discussion." *IEEE Tr. PAS-97*, January/February 1978, p. 249.
———: "Study of Universal Motors," *Bulletin of the Kvushi-Sangyo University*, November 1978, pp. 157–170.
———: "With Lag Angle Brushes: Performance Analysis," *IEEE Tr. PAS 1982*, June 1982, pp. 1288–1296.

Fynn, V. A.: "A New Single Phase Commutator Motor," *JIEE* (British), vol. 36, 1905–1906.

Granborg, Bertil S. M.: "Transient Analysis Using Digital Simulation," *IEEE 75*, July/August 1975.

Gray, C. B.: (1989), *Electrical Machines and Drive Systems,* Longman Scientific and Technical Publishing Ltd., Harlow, U.K., 1989.

Hague, F. T., and G. W. Penney: "Influence of Temperature on Large Commutator Operations," *AIEE Journal,* vol. 48, June 1929, pp. 473–476.

Hellmund, R. E.: "Single Phase Commutator Motors," *Elec. Journal,* vol. 14, 1917.

——— and J. V. Dobson: "Single Phase Commutator Motors," *Elec. Journal,* March 1916.

———, and L. R. Ludwig: "Sparking Under Brushes of Commutator Machines," *IEEE Tr.,* vol. 54, March 1935, pp. 315–321.

Hendershot, Jim: "Brushless DC Motor Phase, Pole and Slot Configurations," IMCSS 90 Symposium, San Jose, Calif., 1990.

Hibbard, L. J.: "Systems of Single Phase Regeneration for Use with Series Type Commutator Motors," *AIEE,* vol. 42, 1923.

Jiles, David: *Introduction to Magnetism and Magnetic Materials,* Chapman and Hall, New York, 1991.

Jones, Dan: "Motor Magnetics Design Changes in Higher Energy Rare Earth PM DC Motors," PCIM Conference 1985, Chicago, 1985.

———: "Brushless DC Motor Magnetics in Both Tooth and Toothless Configurations," IMCSS 86 Symposium, Champaign, Ill., 1986.

———: "Comparison of BDC and BLDC Motors in High Power Density Applications," PCIM 89 Conference, Long Beach, Calif., 1989.

——— and George Gogue: "Magnetics Design of Brush and Brushless DC Motors through Lumped Reluctance and FEM Simulation Programs," IMCSS 88 Symposium, Champaign, Ill., 1988.

Jufer, Marcel: "Brushless DC Motors, Gap Permeance and PM-MMF Distribution Analysis," IMCSS 88 Symposium, Champaign, Ill., 1988.

Jungk, H. G.: "Progress in the Design of the Single Phase Series Motor," *AIEE,* vol. 50, March 1931.

Kenjo, Takashi, and Nagamori Shingenobu: *Permanent Magnet and Brushless DC Motors,* Oxford University Press, Oxford, U.K., 1985.

Kolb, W. C.: "Maintenance of Good Brush Performance," *IEEE Tr.,* vol. 64, December 1945, pp. 819–825.

Konn, F.: "The Single Phase Commutator Type Traction Motor," pts. 1–3: Basic Principles and Analysis of Commutation Factors," *General Electric Rev.,* vol. 3S, 1932.

Lamme, B. G.: "The Single Phase Commutator Type Motor," *AIEE,* vol. 27, pt. 1, 1908.

———: "A Theory of Commutation and its Application to Commutating Pole Machines," *IEEE Tr.,* vol. 30, pt. 2, 1911, pp. 2359–2431.

———: "Physical Limitations in D-C Commutating Machinery," *IEEE Tr.,* vol. 34, pt. 2, 1915, pp. 1739–1800.

La Tour, Marius: "Commutation in Single-Phase Commutator Motors at Starting," *Electrical World,* December 3, 1904, January 14, 1905, March 10, 1906, September 8, 1906.

———: "Commutation in Alternating Current Machinery," *AIEE,* vol. 37, pt. 1, 1918.

Lawrence, R. R.: *Principles of Alternating Current Machinery,* McGraw-Hill, New York, 1940.

Liwschitz-Garik, Michail, and Whipple, Clyde C.: *Direct-Current Machines,* D. Van Nostrand, New York, 1956.

Lynn, C.: "Effects of Temperature on Mechanical Parts of Rotating Electrical Machinery," *IEEE Tr.,* vol. 58, October 1939, pp. 514–518.

Marcos, Anthony: "Brushless DC Motor Magnetics in Both Tooth and Toothless Configurations," IMCSS 86 Symposium.

Matsch, Leander, and J. Derald Morgan: *Electromagnetic and Electromechanical Machines,* 3d ed., John Wiley & Sons, New York, 1987.

McAllister, A. S.: "Improvement of Power Factor and Commutation Conditions in Single Phase Series Motors," *Journal Franklin Institute,* vol. 168, 1909.

McPherson, George: *An Introduction to Electrical Machines and Transformers,* John Wiley & Sons, New York, 1981.

——— and R. Laramore, *Electrical Machines and Transformers,* John Wiley & Sons, New York, 1990.

Nasar, S. A., and L. E. Unnewehr: *Electromechanics and Electric Machines,* John Wiley & Sons, New York, 1979.

Niethammer, F.: "Iron Losses in Single-Phase Commutator Motors," *Electrical World,* March 24, 1906.

O'Kelly, Dennis: *Performance and Control of Electrical Machines,* McGraw-Hill, Maidenhead, U.K., 1991.

Ostovic, Vlado: *Dynamics of Saturated Electric Machines,* Springer-Verlag, Berlin, 1989.

Packer, L. C.: "Universal Type Motors," *AIEE Journal,* vol. 44, December 1925, pp. 1319–1323.

———, W. E. Wier, and R. P. Perey: "Universal Motors," *Electrical Manufacturing* (New York), September 1948.

Page, G. W. P., and G. J. Scott: "Single Phase Commutator Motors, Especially the Latour-Winter-Eichberg Type," *JIEE* (British), vol. 47, 1911.

Patten, F. J.: "Alternating Current Motors: The Evolution of a New Type," *AIEE,* vol. 6, November 1888 to November 1889.

Pessina, G., C. Zinaglia, et al.: "Fractional Power Universal Motor; Equivalent Circuit and Test Settlements," IEEE Second International Conference on Electric Machines, London, England, 1985.

Poritsky, "Graphical Field Plotting Methods in Engineering," *AIEE,* vol. 57, 1938.

Prina, Steven: "Considerations in the Design of Brushless DC Motors," IMCSS 92 Symposium, San Jose, Calif., 1992.

Pritchard, F. H., and F. Icons: "The Modern Single Phase Motor for Railroad Electrification," *AIEE,* vol. 50, March 1931.

——— and Campbell: *Voltage Relations and Losses in Small Universal Motors,* Ohio State Engineering Experiment Station Bulletin No. 58, 1931.

Puchstein, A. F., and Kimberly: *Universal Electric Motors,* Ohio State Engineering Experiment Station Bulletin No. 53, 1930.

Punga, F.: "Das Funken von Kommutator Motoren," Gebrüder Jänecke, Hanover, 1905. English translation by R. F. Looser, *A.M.I.E.E.,* Whittaker & Co., New York, 1906.

Richards, E. A., and D. Dunham: "Comparative Tests on Single Phase Alternating Current Commutator Motors," *L* (British), vol. 52, 1914.

Roye, D., and M. Poloujadoff: "Contributions to the Study of Commutation in Small Uncompensated Universal Motors," *IEEE Tr. PAS-97,* January/February 1978, pp. 242–250.

Roys, C. S., J. Nader, and M. Spotts: "A Variable Speed Constant Frequency Generator," *AIEE,* vol. 66, 1947.

Say, M. G.: *Electrical Engineering Design Manual,* Chapman and Hall, New York, 1962.

———: *Performance and Design of AC Machines,* Pitman.

Schroeder, T. W., and J. C. Aydelott: "The Black Band Method of Commutation Observation," *IEEE Tr.,* vol. 60, 1941, pp. 446–451.

Sen, P. C.: *Principles of Electric Machines and Power Electronics,* John Wiley & Sons, New York, 1989.

Slemon, Gordon: *Electric Machines and Drivers,* Addison-Wesley, Reading, Mass., 1992.

Slichter, W. I.: "The Vector Diagram of the Compensated Single Phase Alternating Current Motor," *AIEE,* vol. 26, pt. 2, 1907.

Smith, S.: "Single and Three Phase Alternating Current Commutator Motors with Series and Shunt Characteris," *JIEE* (British), vol. 60, 1922.

Sokolowski, W.: *Series-Wound Motors,* Bodine Electric Company, Chicago, 1988.

Stnat, Karl: "The Outlook for Permanent Magnet Materials," *Motion,* January and February 1988.

Stone, E. C.: "Circle Diagram of Compensated Series Single-Phase Motor," *Electrical World,* March 24, 1906.

Taft, Charles: "Brushless Motor Torque per Square Root of Watt and Dissipation as a Function of Motor Size," IMCSS 92 Symposium, San Jose, Calif., 1992.

Taylor, E. O.: "Performance and Design of AC Commutation Motors," Pitman, 1958.

Tomasek, Jaraslov: "Basic Performance Specifications and Ratings for Sine Wave Brushless Servo Systems," IMCSS 86 Symposium, Champaign, Ill., 1986.

Trickey, P. H.: *Electric Motor and Generator Design,* Duke University, Durham, N.C., PTS 19, 1932; PTS 40, 1934; PTS 4B, 1960.

Veinott, C. G.: *Fractional Horsepower Motors,* McGraw-Hill, New York, 1939.

Weinert, H.: "Kommutierung & Bürstenverschlei B als Optimierungsproblem bei Universalmotoron," *Ringsdorf-Werke GmbH Technishe Mitteilungen,* N9, 1978, pp. 1–24.

Wier, W. E.: "Iron Loss in Universal Motors," *AIEE Transactions,* vol. 73, pt. 3B, 1954, pp. 1546–1552.

Yohe, Wilford E.: *Carbon Brush Engineering Handbook,* Stackpole Corporation, Saint Marys, Pa., 1982.

INDEX

Abbe error:
  defined, 10.23
  and Inductosyns, 10.36
  and linear encoders, 10.25
Absolute count stacking, 3.38–3.39
Absolute encoders, 10.19–10.22
AC. *See* AC induction motors; AC series motors; Alternating current; Synchronous motors
Acceleration:
  determining, 8.4–8.9
  testing, 9.2–9.5
  theoretical, 10.139
  unit conversions, 1.5, 1.6
Accucore, 2.52
Accuracy, 10.1–10.2
  absolute encoder, 10.25
  incremental encoder, 10.15–10.16
AC induction motors:
  overview, 6.1–6.5
  permanent-split-capacitor (PSC), performance, 6.61–6.72
  polyphase, performance, 6.93–6.102
  shaded-pole, performance, 6.72–6.93
  single-phase, performance, 6.45–6.61
  single-phase, theory, 6.5–6.38
  stator and coil assemblies, 6.1–6.2
  testing, 9.5–9.9, 9.44–9.47
  three-phase, theory, 6.38–6.44
  typical applications, 8.34
  windings, 6.2–6.5
Acrylics, as motor adhesives, 3.116, 3.122
AC series motors. *See also* AC induction motors
  performance, 4.116–4.130
  theory, 4.17–4.23

Actuators:
  applying force and energy equations to, 1.15–1.20
  magnetizing sources for, 1.33–1.35
  moving-coil, 1.40–1.41, 1.83–1.88
  reluctance, 1.35–1.43, 1.78–1.82
ADA Smart Distributed System, 10.50
Adhesives, used in motor manufacturing:
  applications for, 3.117–3.123
  chemistries of, 3.115–3.117
  dispensing equipment, 3.123
  joint design, 3.114–3.115
  for magnets, 2.87, 3.122–3.123
  overview, 3.109–3.114
  safety factors, 3.124
  testing of, 9.42–9.43
Advanced Research Project Administration (ARPA), 10.48–10.49
AEG, and PROFIBUS, 10.50
Aging (carbon-induced), defined, 2.44
Air flux, defined, 1.44
Air gap:
  in BLDC motors, 5.40
  controlling, 3.21–3.22
  geometry, 4.75–4.84
  linear equations, 1.33–1.43
  magnetic coenergy in, 1.18–1.20
  and mmf in dc series motors, 4.5–4.10, 4.117–4.119, 4.122–4.126
  and mmf in PMDC motors, 4.27–4.32
  and mmf in universal ac motors, 4.122–4.126
  in permanent-magnet versus induction motors, 3.25
  permeance, 1.21–1.32
  predicting, 4.97–4.106

Air gap (*Cont.*):
  in step motors, 5.55, 5.69
  in switched-reluctance motors, 5.100
  in VR step motors, 5.55
AISI, steel grade standards, 2.4
Allegheny-Teledyne Company:
  on magnetic test methods, 2.71n
  representative magnetic curves, 2.25–2.30
Allegro Microsystems, Inc.:
  control circuits, 10.97
  and current regulation, 10.90
  Darlington arrays, 10.75
  Hall devices, 10.93
  H bridges, 10.88
  smart IC controls, 10.7
Allen-Bradley, 10.50
Alnico, 2.90
  demagnetization curves for, 2.103–2.107
  density of, 5.21, 5.69
  and eddy current effect, 3.143
  in hybrid step motors, 5.76
  leakage flux paths and, 1.30–1.31
  and PMDC motors, 4.36–4.37
  properties of, 1.49, 2.84, 2.87–2.89
  typical magnetizing forces for, 3.125, 3.142
Alpha pattern winding, 4.162, 4.169
Alternating current. *See also* AC induction motors; AC series motors; Synchronous motors
  coil design, 1.72–1.73, 1.76–1.77
  dynamic analysis, 1.80–1.81, 1.85–1.88
  magnetic properties, 2.53–2.56
  magnetic test methods, 2.74–2.80
  motor manufacturing process flow, 3.1–3.2
  in powder metallurgy applications, 2.65–2.71
Aluminum:
  alloys, 2.87
  coating, 2.86
  coefficient of thermal expansion, 3.20
  in end frame construction, 3.4
  in housing construction, 3.10–3.12
  in stator assembly processing, 3.21

Aluminum (*Cont.*):
  thermal conductivity value, 8.16
  thermal properties of, 8.19
American Bearing Manufacturers Association (ABMA):
  load ratings for bearings, 3.66–3.67
  reliability ratings for bearings, 3.65–3.66
American Hoffman Corporation, 3.87n
American Society for Testing and Materials (ASTM):
  ac test methods, 2.74–2.76
  dc test methods, 2.72, 2.73
  magnetic test methods, 2.71–2.78
  P/M-related standards, 2.61
  sample B-H magnetization loops, 2.6–2.7
  steel grade specs, 2.4
American Wire Gauge (AWG):
  and coil design, 1.69–1.77
  and lamination design, 3.28
  in stator winding design, 5.33–5.35
  and wire properties, 2.179–2.183
  and wire sizes, 2.176–2.177
Ampere's law:
  in determining armature mmf, 4.5, 4.28
  and PMDC motors, 4.38
Amplitude modulation (AM), defined, 10.10
Anaerobic adhesives, defined, 3.115–3.116
Analog Devices, 10.40, 10.45–10.46
Anchored lead loops, 4.157
Ancorsteel:
  magnetic characteristics of, 2.66–2.67
  permeability of, 2.63–2.64
  saturation induction for, 2.62
Anderson, Larry C., 3.87n
Angle of misalignment, defined, 3.58
Angular measurement device, 10.9
Anisotropic material, defined, 2.87
Annealing, lamination, 2.6–2.46
Antiferromagnetism, defined, 1.63
Antilock brake wheel sensors, materials used for, 2.62

INDEX

Anunciada, V.:
  and clocked PWM, 10.115
  and dual-current-mode PWM, 10.116
  and voltage PWM, 10.113
Applications, motor:
  for ac motors, 8.34
  current density and, 8.9–8.10
  for dc motors, 8.33
  electromagnetic interference (EMI), controlling, 8.32–8.38
  and environmental standards, 10.4–10.7
  requirements for, 8.1–8.4
  thermal analysis and, 8.10–8.32
  velocity profiles, determining, 8.4–8.9
Arcing, minimizing, 4.9–4.10, 4.31–4.32
Armature:
  balancing, 3.87–3.98, 4.61
  conductors, 4.57–4.61
  force, determining, 1.15–1.20
  four-pole, 4.165–4.181
  geometry, in PMCD motors, 4.106–4.108
  geometry, in universal motors, 4.48–4.54
  magnetic circuits, 4.54–4.57
  manufacturing/assembly, 3.22
  mmf in dc series motors, 4.5–4.10
  mmf in PMDC motors, 4.27–4.42
  reactances, 4.82–4.84
  varnishing of, 3.106
  winding, theory and practice, 4.140–4.181
  yoke, defined, 4.54
Armco:
  product properties, 1.46
  and steel grade designations, 2.4
Arnold Engineering Company, typical magnetic curves, 2.102–2.133
Arnox, demagnetization curves for, 2.108–2.112
ARPA, 10.48–10.49
ASIC technology, and environmental concerns, 10.6
ASTM. *See* American Society for Testing and Materials (ASTM)
Automated testing, 9.43–9.47

Avogadro's number, 1.63
AWG. *See* American Wire Gauge (AWG)
Axial air gap motor, 5.9–5.11
Axial field, 1.65–1.66
Axial play, defined, 3.56–3.57
Axis SPA, 3.79n

Babbit, as bearing material, 3.72
Back electromotive force (emf):
  constant, 4.115
  defined, 1.80–1.81, 1.86
  in three-phase motors, 10.112–10.113
  trapezoidal versus sinusoidal drives, 10.122–10.124
  in two-phase motors, 10.98–10.99, 10.101–10.102
Back iron, 5.13, 5.40
Back-iron thickness, defined, 3.16
Balancing:
  PSC motor, 6.69–6.72
  rotor assembly, 3.19, 3.87–3.98
Ball bearing analysis, 9.17–9.23
Ball bearings. *See* Bearings
Band, Robert, 9.43n
Bank, John S., 3.99n
Barium ferrite, as magnet material, 2.90
Bearings:
  assembly and fitting of, 3.69–3.72
  ball-type, 3.47–3.48, 3.51–3.55
  components, 3.51–3.55
  geometry of, 3.55–3.58
  grease tests, 9.41
  lubricants for, 3.51, 3.61–3.66, 3.73–3.79
  materials used for, 3.58–3.61
  overview, 3.46–3.48
  preloading of, 3.67–3.69
  selection of, 3.49–3.51
  sleeve-type, 3.72–3.79
  static capacity of, 3.66–3.67
Bell, F. W. (Hall device manufacturer), 10.93
B-H curve, 1.8–1.9, 2.81–2.84
  and Hall devices, 10.32–10.34
  and PMDC motors, 4.38, 4.42, 4.110–4.112

B-H curve (*Cont.*):
  and series dc and ac motors, 4.122
  and shaded-pole motor, 6.81–6.82
Bifilar winding, 5.49, 5.104
Biot-Savart law, 1.65–1.66
Bipolar motors, 5.49, 5.86–5.87, 10.75–10.79
Bipolar slide-by Hall sensor, 10.30, 10.33
Bitter, Francis, 2.98
Bitter coil, 2.98
Blade gap, in lamination design, 3.30–3.31
BLDC motors. *See* Brushless dc (BLDC) motors
Blue coating (bluing), 2.45
Bode plot, 9.31, 9.35
Bohr magneton, 1.5, 1.62–1.63
Bonded cores, 3.26
Bonding agents, 2.87, 3.121–3.123. *See also* Adhesives
Boron, properties of, 1.49
Bosch (as developer of CAN), 10.50
Boundary element analysis, 5.65
Boyes, Geoffrey S., 10.46
Bozorth, R. M., 1.56
Brackets. *See* End frames
Breakdown torque, defined, 10.52
Brinell hardness values, 1.43
Brod, D. M., 10.113
Bronze, in sleeve bearings, 3.72
Brown, Warren, 3.42n
Brushes:
  arcing, 4.9–4.10, 4.31–4.32
  contact loss, 4.16–4.17, 4.44–4.46
  current density calculations, 4.115
  effect on polarity, 4.158
  in four-pole armatures, 4.170
  holders, 3.99–3.103
  selection, 4.84–4.87, 4.163–4.165
  as source of heat loss, 8.12, 8.14
Brushless dc (BLDC) motors:
  applications, 10.7–10.9
  configuration, 5.4–5.13
  design, 5.36–5.46
  drive schemes, 10.97–10.119

Brushless dc (BLDC) motors (*Cont.*):
  overview, 5.1–5.3
  performance characteristics, 10.122–10.140
  sizing and shaping, 5.11–5.22
  stator winding design, 5.22–5.36
Bularzik, Joseph H., 2.51n
Bureau of Standards, and dc test methods, 2.74

Cages, bearing, 3.49, 3.51, 3.59–3.61
Caine, Peter, 3.103n
Campbell, Peter, 10.28
Canadian Standards Association, and insulation requirements, 2.167
Capacitance. *See also* Capacitive-discharge magnetizers
  during magnetization, 2.95, 3.144
  thermal, 8.13–8.16, 8.31–8.32
Capacitive-discharge magnetizers, 2.93–2.97, 3.131–3.133
  CDM systems, 3.138–3.142, 3.145–3.146
  fixture design for, 3.133–3.137
  process of magnetizing, 3.142–3.147
Capacitor-start motors:
  performance calculations, 6.45–6.59, 6.60–6.61
  typical applications, 8.34
Carbon contamination:
  annealing as antidote to, 2.44
  from brush dust, 4.87
Carbonitriding, defined, 3.14
Carburizing process, 3.14
Carpenter, soft magnetic material properties, 1.46
Carriage, defined, 10.23
Carter's coefficient:
  calculating, 4.75
  and magnetic air gap length, 4.99–4.100, 6.78
Cartridge-style brush holders, 3.100
Case hardening, explained, 3.14–3.15
Cast iron:
  coefficient of thermal expansion, 3.20
  in end frame construction, 3.4

Cast iron (*Cont.*):
  in housing construction, 3.10–3.11
  in stator assembly processing, 3.21
CDM systems. *See* Capacitive-discharge magnetizers
CDX, defined, 10.37
CE mark, European Community, 10.6
CEN, and EMI regulation, 8.32–8.35
Centimeter, gram, second. *See* CGS system of units
Centripetal force, and rotor balancing, 3.91
Ceramic materials, 2.90
  B-H curves and, 2.83–2.85
  and PMDC motors, 4.36–4.37
  properties of, 1.49, 2.87–2.89
  thermal conductivity value, 8.16
  typical magnetizing forces for, 3.125
C-frame shaded-pole motors, 6.73, 6.74, 6.78, 6.80, 6.85–6.86
CGS system of units, 1.1–1.3, 2.5, 5.88
Chai, H. P., 5.77
Chang, S. S. L., 6.73, 6.83
Chillers, for magnetizing process, 2.98–2.99
Choppers, in controlling current, 10.84–10.86
Chord factor, 5.32–5.33
Chrome steel, in bearing manufacture, 3.49, 3.58–3.59, 3.65
CIP gaskets, 3.120–3.121
Circular-mil slot-fill percentages, 3.32
Clean-sheet methodology, and BLDC motor design, 5.45–5.46
Cleated cores, 3.27
Cleavage stress, defined, 3.115
Clifton Precision, 10.43
Clocked PWM, 10.114–10.116
Closed-loop control, 10.69–10.70
Closed-pocket retainers, 3.59
CNC machines:
  in end frame manufacturing, 3.4, 3.10
  in housing manufacturing, 3.11
  in rotor manufacturing, 3.16
  in shaft manufacturing, 3.13
Coasting mode, 10.76

Coatings, magnetic:
  and lamination annealing, 2.44–2.45
  role in suppressing oxidization, 2.86–2.87
Cobalt. *See also* Samarium-cobalt
  intrinsic saturation flux density, 1.63–1.64
  properties of, 1.49
Cocco, John, 3.109n
Code wheel, defined, 10.10
Coefficient of thermal expansion, 3.19–3.20. *See also* Thermal analysis
Coenergy. *See also* Energy-coenergy; Magnetic coenergy
  equations, 1.12, 1.14
  as force and torque determinant, 1.17–1.20, 1.91–1.96
Coercivity, magnetic, 2.80, 3.142
C of F, 2.184
Cogging:
  and BLDC motors, 5.16
  and step motors, 5.89
Coil:
  in ac induction motors, 6.1–6.3
  actuator, permeance value, 1.31
  during commutation, 4.90–4.93
  design, 1.67–1.77
  magnetizing, 2.92–2.93, 3.141–3.142–3.147
Cold-rolled motor lamination steel. *See* CRML steel
Cold-rolled steel. *See* CRS
Collin, R. E., 1.66–1.67
Commutation:
  and brush selection, 4.84–4.87
  in dc series motors, 4.10–4.14
  defined, 1.91, 4.145, 10.98
  flashover and ring fire, 4.95–4.96
  patterns, 10.119–10.122
  in PMDC motors, 4.32–4.37
  system design, 4.87–4.95
  torque ripple, 10.109
Commutator fusing, 3.79–3.87
Compaction process, slot fill, 3.34
Component slot milling, 3.93–3.94

Compound-wound dc motor calculations, 4.134–4.137
Comprehensive Energy Policy Act, 10.52
Compressive stress, defined, 3.115
Computer-aided design (CAD), in armature calculations, 4.108
Computerized testing, 9.1, 9.43–9.47
Computer numerically controlled (CNC) machines. *See* CNC machines
Comstock, Robert, 10.124
Concentricies, maintaining, 3.21–3.22
Conductance, thermal. *See* Thermal analysis
Conduction, defined, 8.11
Conductor resistance tests:
  for armatures, 4.57–4.61
  for stators, 4.73–4.84
  for wire, 2.186
Conformal coating, defined, 2.86
Conservation of energy, 1.10, 1.12–1.15, 1.91–1.96
Constant-current flux resetting test, 2.78
Constant load gear tests, 9.38–9.40
Constant-pitch winding, 5.24–5.32
Constants:
  for BLDC motor calculations, 10.122–10.123, 10.138–10.139
  friction and windage, 4.128–4.129
  for PMDC motor calculations, 4.115–4.116
  for stator reactance calculations, 4.68–4.69
  for synchronous motors, 7.17–7.19
  thermal resistance, 9.12
  Trickey's B, 6.91–6.93
  used in magnetics, 1.5
  voltage, 9.10
Contact angle, ball-bearing, 3.57, 3.58
Contacting encoders, defined, 10.3, 10.24
Contour milling, 3.93
Controller Area Network (CAN), 10.50–10.51
Controls and drives. *See* Drives and controls
Convection. *See* Thermal analysis
Conversions, measuring unit, 1.3–1.6, 2.5
Cooling, during magnetizing process, 2.98–2.99
Copper:
  alloys, 2.87
  in armature, 4.60
  and coil design, 1.67–1.69, 1.73–1.77
  in the commutation process, 4.14
  losses, and Design E motors, 10.53
  losses, and thermal analysis, 8.12–8.13, 8.14, 8.30
  losses, in dc series motors, 4.16–4.17
  losses, in PMDC motors, 4.44–4.46, 4.114–4.115
  losses, in single-phase induction motors, 6.27, 6.32, 6.71
  losses, in switched-reluctance motors, 5.103
  and powder metallurgy processing, 2.60
  properties of, 1.46
  in stator, 4.74
  thermal conductivity value, 8.16
  thermal properties of, 8.19
Core losses, 2.46–2.50, 2.53–2.59, 10.53. *See also* Eddy current loss; Hysteresis loss; Magnetic cores
  in dc series motors, 4.16–4.17
  defined, 2.4
  and iron powder composites, 2.67–2.70
  measuring at ultrasonic frequencies, 2.79–2.80
  in PMDC motors, 4.45–4.46
  role of coating in reducing, 2.44–2.45, 3.37–3.38
  in silicon-iron steels, 2.38–2.43
  and thermal analysis, 8.12, 8.14
  and variable-speed motors, 2.51–2.52
Cores, magnetic. *See* Core losses; Magnetic cores
Coulomb friction, defined, 3.76

Coulomb's law, and units of measure, 1.2
Counter-emf (cemf):
  and BLDC motors, 5.41–5.42
  defined, 4.4, 4.14, 4.26, 4.42
  and step motors, 5.82, 5.98
Couple unbalance, 3.89
Critical speed, defined, 3.92
CRML steel, versus pressed material, 2.53–2.56
Cross-field theory, 6.29
  and shaded-pole motor calculations, 6.74–6.93
  and single-phase motor calculations, 6.45–6.46
Cross-magnetizing armature reaction, defined, 4.7, 4.30
Crown retainers, 3.59
CRS, in shaft manufacture, 3.12
CT, defined, 10.37
Cumulative compound motor, 4.135–4.137
Cunico, 2.87
Cunife, 2.90
  as conductor, 2.84
  naming convention, 2.87
  properties of, 2.88, 2.90
Curie temperature:
  defined, 1.63, 2.85
  samarium-cobalt and, 2.91
  and stepping motors, 5.98–5.99
Current:
  and BLDC motors, 5.42–5.43
  density guidelines, 8.9–8.10
  limiting, for stepping motors, 10.80–10.93
  and switched-reluctance motors, 10.67–10.69
Current loop:
  magnetic field for, 1.65–1.67
  magnetic moment for, 1.59–1.60
Current-torque performance curve, defined, 4.96–4.97
Cutler-Hammer, 10.51n
Cutoff speed, defined, 5.97
CX, defined, 10.37
Cyaniding process, 3.15

Cyanoacrylate, as motor adhesive, 3.116, 3.121

Damping:
  coefficient, 5.82, 5.83
  to control resonance, 5.95–5.96
  defined, 3.92
  equations, 2.96–2.97
Dana Corporation, 10.65n
Darlington transistors, 10.75
DC. *See* DC motors; Direct current
DC motors:
  automatic armature winding, 4.140–4.181
  commutation, 4.84–4.96
  compared to ac series motors, 4.17–4.23
  compound-wound dc motor calculations, 4.134–4.137
  lamination, field, and housing geometry, 4.46–4.84
  permanent-magnet, 4.23–4.46
  PMDC motor performance, 4.96–4.116
  series dc motor performance, 4.116–4.130
  shunt-connected dc motor performance, 4.130–4.134
  testing of, 9.9–9.15
  theory, 4.1–4.17
  typical applications, 8.33
  winding patterns, 4.138–4.140
DC tachometers. *See* Tachometers
Dead zone, in stepping-motor physics, 5.90–5.91
Deceleration, determining, 8.4–8.9
Degrees of springback, 2.178
Delta connections, 5.23–5.24, 10.110–10.112
Demagnetization:
  Curie temperature and, 2.85
  effect of armature reaction, 4.9, 4.31
  percentage, 3.127
  in PMDC motors, 4.36–4.42
  representative curves, 2.102–2.163
  testing, 9.11–9.12

Design E motors:
  availability, 10.64–10.65
  background, 10.51–10.54
  performance, 10.54–10.64
Detent torque, defined, 5.89, 10.95
DeVault, Birch L., 10.51n
DeviceNet, 10.50, 10.51
Diadur process, 10.12
Dielectric breakdown test, 2.184–2.185
Dies. *See* Stamping dies
Diesters, as bearing lubricants, 3.61, 3.62
Differential compound motor, 4.135–4.137
Differential scanning calorimetry, 2.187
Diffusion metal-oxide semiconductor (DMOS), 10.922
Dipole moment, 1.60–1.61, 1.64–1.65
Dipping systems (for applying varnish), 3.104–3.105
Direct current. *See also* DC motors
  coil design, 1.70–1.72, 1.74–1.76
  magnetic test methods, 2.71–2.74
  magnetizers, 3.130
  motor manufacturing process flow, 3.3
  in powder metallurgy applications, 2.61–2.71
  steady-state analysis, 1.78–1.80, 1.83–1.85
Direction sensing, defined, 10.2
Disk. *See* Code wheel
Dissipated power equation, 10.139
Distribution factor, 5.32–5.33
Dolan, Phil, 3.38n
Domain boundaries, defined, 1.43–1.44
Donart ac tester, 2.53
Double-end lacers, 3.43
Double-revolving-field theory. *See* Revolving-field theory
Drawn-over-mandrel (DOM) tube, in housing manufacture, 3.11
Drives and controls:
  for BLDC motors, 10.7–10.9, 10.97–10.119
  commutation patterns, 10.119–10.122
  comparative technologies, 10.42–10.48

Drives and controls (*Cont.*):
  Design E motors, 10.51–10.65
  environmental standards, 10.4–10.7
  microstepping, 10.93–10.97
  performance characteristics, 10.122–10.140
  redundant sensors, 10.48–10.49
  sensor databus systems, 10.49–10.51
  smart sensors, 10.48
  stepping-motor control circuits, 10.70–10.79
  stepping-motor current limiting, 10.80–10.93
  for switched-reluctance motors, 10.65–10.70
  terminology, 10.1–10.4
  units of measure, 10.1
D-type bearing seals, 3.50, 3.60
Dual-current-mode PWM, 10.116, 10.117
DuPont, insulating materials chart, 2.174
Dust core technology, defined, 2.65–2.66
Dynamic analysis:
  moving-coil actuator, 1.85–1.88
  stepping motor, 5.92
Dynamic braking mode, 10.76–10.77
Dynamic coefficient of friction test, 2.184–2.185
Dynamic friction, versus static friction, 5.90
Dynamic radial load rating, defined, 3.63
Dynamic unbalance, 3.89
Dynamic viscosity, 3.77
Dynamometer, 5.80

Eaton Corporation, 1.1n, 10.51n
Eccentricity errors, encoder, 10.15–10.16
Eccentric load tests, 9.41
E-coat, defined, 2.86
Eddy current loss, 1.52–1.57, 2.46–2.50
  in bonded cores, 3.26
  in cleated cores, 3.27

Eddy current loss (*Cont.*):
  defined, 2.3–2.4
  during magnetizing process, 3.142–3.143
  in welded cores, 3.25
Edge counting, defined, 10.2
Efficiency:
  defined, 10.53
  improving via rotor assembly, 3.16–3.17
Elastomeric material, to control resonance, 5.95–5.96
Electrical energy. *See* Energy
Electrical Steels (AISI), products manual, 2.4
Electro-Craft Handbook (Rockwell Automation), 9.13n, 10.15
Electro-Craft Motion Control (Rockwell Automation), 9.43n
Electrolytic versus oil-filled capacitors, 3.140–3.141
Electromagnetic fields, and radiation, 8.35–8.38
Electromagnetic forces, 1.89–1.91
Electromagnetic interference (EMI):
  controlling, 8.35–8.38
  emissions standards, 8.32–8.35
Electromechanical forces and torques, 1.32–1.43
Electronically commutated motors:
  brushless dc (BLCD) motors, 5.1–5.46
  step motors, 5.47–5.99
  switched-reluctance motors, 5.99–5.107
Electrons:
  magnetic moment of orbit, 1.62
  magnetic moment of spin, 1.5, 1.62–1.63
Electro-press process (General Electric), 3.34
Elongation wire testing, 2.178
EMF, defined, 4.2
EMI. *See* Electromagnetic interference (EMI)
EMU, defined, 1.2, 1.3

Encoders:
  absolute, 10.19–10.22
  comparisons between, 10.42–10.48
  incremental, 10.9–10.19
  linear, 10.22–10.25
  magnetic, 10.25–10.28
End bells. *See* End frames
End frames, manufacturing, 3.4–3.10
End leakage reactance, 4.83
End play, defined, 3.56–3.57
End shield. *See* End frames
End-turn factor, 5.32
Energy, electromechanical:
  conservation of, 1.10, 1.12–1.15, 1.91–1.96
  equations for, 1.10–1.15
  unit conversions, 1.4
Energy-coenergy applications, 1.15–1.21, 1.91–1.96
Energy-efficient motors, defined, 10.52
Energy product curve, defined, 10.33
Environmental standards:
  application-related, 10.4–10.6
  safety-related, 10.6–10.7
Epoxy:
  for insulation coating, 2.163–2.164, 2.175, 2.176
  as motor adhesive, 3.116, 3.122
  for suppressing oxidization, 2.86
Epstein tests:
  ac-related data, 2.74–2.76, 2.79
  core losses and, 2.47, 2.53–2.54
  dc-related data, 2.72
  and properties of magnetic motor steels, 2.8–2.18
Equipment, testing:
  adhesives and plastic assemblies, 9.42–9.43
  computerized, 9.43–9.47
  dc motors, 9.9–9.15
  fatigue and lubrication, 9.36–9.41
  resonance, 9.27–9.36
  spectral analysis, 9.15–9.27
  speed-torque curve, 9.1–9.5
  thermal analysis, 9.8

Equivalent circuits:
  combined-winding, 6.28–6.32
  dynamic equations, 6.13–6.17
  Fortesque's symmetrical component approach, 6.32–6.37
  rotor transformation, 6.18–6.25
  for shaded-pole motors, 6.74–6.89
  single-winding, 6.25–6.28
  stator transformation, 6.17–6.18
  for three-phase motors, 6.38–6.44
Ericsson Microelectronics, 10.97
Ernst, Alfons, 10.15
ESU, defined, 1.2
Etching processes:
  in Inductosyns, 10.36
  in optical encoders, 10.12
European Commission (EC), and EMI regulation, 8.32–8.35
European Community CE mark, 10.6
European pole design, 3.32
Evershed, 2.83
Exchange force, impact on magnetic moment, 1.63
Excitation frequency, defined, 3.92

Faraday's law:
  in ac coil design, 1.72, 1.80, 1.85
  in ac series motor analysis, 4.19–4.20
  in dc series motor analysis, 4.15
  and eddy currents, 1.53–1.54
  and energy-coenergy approach, 1.91
  equations for, 1.8, 1.11, 1.20, 2.100
  and magnetic flux changes, 2.84, 2.100
  in PMCD motor analysis, 4.43
Far field, magnetic, 1.64–1.65
Farrand Industries, Inc., 10.35, 10.37
Fast-decay mode, 10.76, 10.90–10.91
Fast Fourier transform (FFT), 9.15
Fatigue:
  of bearings, 3.63–3.66
  tests for, 9.36–9.41
Federal Communications Commission (FCC), and EMI regulation, 8.32
Ferrite:
  density of, 5.21, 5.69
  and eddy current effect, 3.143
  hard versus soft, 2.90

Ferrite (*Cont.*):
  leakage flux paths and, 1.30–1.31
  properties of, 1.46
  typical magnetizing forces for, 3.142
Ferromagnetism, defined, 1.63
FieldBus, 10.50–10.51
Field intensity, magnetic:
  defined, 1.8–1.9, 1.20
  as magnetic property, 2.1–2.2
  unit conversions, 1.4
Field resistance, defined, 4.73–4.74
Finite element analysis (FEA):
  and computerized testing, 9.43
  and step motors, 5.65
  and universal motors, 4.55–4.57
FIP gaskets, 3.120
First law of thermodynamics, 1.10
Fixed-source unbalance, defined, 3.87
Fixtures, magnetizing, 2.97–2.99, 3.133–3.137
Flange sealing, 3.120–3.121
Flashover, 4.95–4.96. *See also* Arcing
Flemming's law, 6.45
Flux:
  and B-H curve geometry, 4.77–4.80, 4.110–4.112
  calculations, polyphase motors, 6.98–6.99
  calculations, single-phase motors, 6.54–6.56
  defined, 1.7–1.21
  permeance, 1.21–1.32
  predicting air gap, 4.97–4.106
Flux density:
  armature-related, 4.5–4.10
  and BLCD motors, 5.13–5.14, 5.17, 5.40–5.41
  and core loss, 2.46–2.50
  defined, 1.8, 1.20
  equations, 1.2, 1.3
  as magnetic property, 2.2–2.4, 2.80
  unit conversions, 1.3–1.4
Fluxmeters, 2.100–2.101, 3.130, 3.142
Flux path permeance equations, 1.21–1.32

Force, electromechanical:
  and dc series motors, 4.1–4.4
  energy-coenergy approach, 1.91–1.96
  equations, 1.12–1.21, 1.32–1.43
  explained, 1.89–1.91
  magnetizing, 3.125, 3.142
  unit conversions, 1.5
Fortesque, C. L., 6.32–6.34
Foundation FieldBus, 10.50
Fourier analysis:
  of air gap flux harmonics, 4.6–4.8, 4.28–4.30
  and motor efficiency, 2.58
  and spectral analysis, 9.15
Four-pole armature. See Armature
Fractional-pitch winding, 5.24–5.32
Frayman, L., 2.63
Free angle, ball-bearing, 3.58
Free space, permeability of:
  and the B-H curve, 2.80–2.83
  and cgs system, 1.3
  and intrinsic saturation flux density calculations, 1.63
  and mks system, 1.2
  unit conversions, 1.5
Freewheeling diodes, defined, 10.103
Freimanis, M., 10.113
Freon, role in cooling, 2.98
Frequency response functions (FRFs), 9.30–9.36
FRFs, 9.30–9.36
Friction, in bearing systems, 3.76–3.79
Friction and windage losses:
  and Design E motors
  in PMDC motors, 4.45–4.46, 4.115
  in polyphase motors, 6.100
  in series dc and ac motors, 4.128–4.130
  in single-phase motors, 6.56
  in switched-reluctance motors, 5.102
  in synchronous motors, 7.23
  in thermal analysis, 8.12, 8.14
Fringing flux paths, 1.29
Frustaglio, Brad, 6.45n
Full-load heat run test, 9.5–9.6, 9.46
Full-load torque, defined, 10.52

Full-voltage idle test, 9.47
Full-wave wye, defined, 5.23–5.24
Furnaces, annealing, 2.45–2.46
Fusing, 3.79–3.87

Gasketing, 3.120–3.121
Gauss, defined, 10.25. See also CGS system of units
Gaussian system of units, 1.2–1.3
Gaussmeters, 2.100, 3.130, 3.142
Gear dynamic load tests, 9.38
General Electric Electro-press process of slot fill, 3.34
Globe Products, 4.140n, 4.162, 4.163
Gradient pole design, 3.32
Grain boundaries, defined, 1.43–1.44
Grain growth, annealing process and, 2.44, 2.46
Graphite:
  and brush contact loss, 4.16
  in the commutation process, 4.14
  and powder metallurgy processing, 2.60
Grating, defined, 10.11
Gray, Alexander, 4.96
Gray coding, 10.21
Grease, bearing. See Lubricants
Green paper, explained, 2.101–2.102
Gross slot area (GSA), calculating, 4.53, 4.67–4.68, 5.62–5.63
GSA, defined, 5.62
Gun-wound salient pole motors, 3.35–3.36

H. A. Holden Co., lead wire application chart, 2.188
Half-cycle magnetizers:
  advantages of, 3.130–3.131
  fixture design for, 3.133–3.137
  power surges and, 2.93
  selection of, 3.132–3.133
Half-cylinder flux paths. See Flux path permeance equations
Half-H bridge, 10.78–10.79
Half-pitch winding, 5.24–5.32
Half-stepping, defined, 5.89–5.90
Half-wave wye, defined, 5.23–5.24

Hall, Edwin Herbert, 2.100, 10.29
Hall devices, 10.28–10.35
   and BLDC motors, 5.1–5.2, 5.5, 5.9
   and commutation, 10.119–10.121
   comparative technologies,
      10.42–10.48
   and current limiting, 10.93
   and gaussmeters, 2.100
Hanejko, Francis, 2.59n
Hanselman, Duane:
   on design equations for BLDC
      motors, 5.36n
   on drive schemes for BLDC motors,
      10.97n
Hard-bearing balancers, 3.19,
   3.94–3.97
Hardening process, shaft, 3.14–3.15
Hard magnetic materials, 1.43–1.52
Harmonics:
   in armature mmf, 4.6–4.8, 4.28–4.30
   in shaded-pole motor calculations,
      6.83
Hay bridge test method, 2.76
H bridge:
   in bipolar motors, 5.86, 10.75–10.79
   circuitry, 10.102–10.105
   in multiphase motors, 5.87
Heat aging tests, bond strength, 3.114
Heat-shrinking:
   in rotor assembly process, 3.17–3.18
   in wound stator assembly process,
      3.21
Heat transfer. *See* Thermal analysis
Heckenkamp, Daniel P., 10.51n
Heidenhain, Dr. Johannes, GmbH,
   10.12, 10.15
Helmholtz coil, 1.67, 1.68, 2.101
Hendershot, Jim, 5.40
High-voltage continuity test,
   2.184–2.186
Hilsch tube, 2.98
H insulators, 2.176, 3.34–3.35
Hitachi America Ltd., 10.7–10.8
Hoeganes Corporation, 2.59n
Holding tests, 9.13
Hollow-shaft encoders, 10.10
Holtz, J., 10.113

Honeywell:
   Hall devices, 10.30, 10.93
   as user of CAN, 10.50
Hooke's law, 5.93
Hot-rolled steel. *See* HRS
Hot-staking method, 4.159
Hot strength tests, bond strength, 3.114
Housing manufacture, 3.10–3.12. *See
   also* Lamination, field, and housing
   geometry
H paper, 3.34–3.35
HRS, in shaft manufacture, 3.12
H-type bearing shields, 3.59, 3.60
Hunt, Robert P., 10.27
Hybrid step motors:
   controllers for, 10.72–10.74
   design, 5.74–5.79
   operation, 5.50–5.51
Hydrocarbons, as bearing lubricant,
   3.61
Hysteresis:
   and current limiting, 10.89–10.92
   effect on encoder accuracy, 10.15
   pulse-width modulation (PWM),
      10.113–10.114
Hysteresis loss:
   defined, 2.2–2.3
   equations for, 1.52–1.53, 2.46–2.50
   loop tracer data, 2.77–2.78
   and noise control, 10.43–10.44
   in soft versus hard magnetic materi-
      als, 1.44–1.45
Hysteresis synchronous motors, 7.5–7.9

IBRAs. *See* Internal brush ring assemblies (IBRAs)
Idle heat run tests, 9.7, 9.46
ID machining, rotor assembly,
   3.15–3.16
IEC:
   and brush holder design, 3.102
   and Design E motors, 10.65
   and EMI regulation, 8.32
   and environmental standards,
      10.4–10.7
   and SERCOS, 10.49
   and thermal testing conditions, 9.8

IEEE:
 on brush contact loss, 4.16
 and smart transducer interfaces, 10.51
 test methods, 9.46, 9.47
IGBTs:
 and H-bridge circuitry, 10.102
 in SRMs, 10.66
Ignitron, role in magnetizers, 2.93–2.95, 3.131, 3.141
Imaginary permeability, 1.57–1.58
Imbalanced-phase winding, 3.32
Impact tests, 9.42–9.43
Impregnation resins. *See* Sealants
Impulse magnetizer. *See* Capacitive-discharge magnetizers
Incremental encoders, 10.9–10.19
Incremotion Associates, 5.22n, 10.119n
Independent winding connection, 5.23–5.24
Index. *See* Reference mark
Indramat (SERCOS-compatible controls), 10.49
Inductance:
 and BLDC motors, 5.43–5.44
 current loop, 1.66–1.67
 defined, 1.9–1.10, 1.20
 during magnetization, 2.95, 3.144
 single-phase motor, calculations, 6.13–6.17
 and step motors, 5.80
 three-phase motor, calculations, 6.38–6.44
 unit conversions, 1.4
Induction synchronous motors, 7.1–7.4
Inductive transducers, and environmental protection, 10.6
Inductosyns (Farrand Industries, Inc.), 10.35–10.37, 10.42–10.48
Inertia, armature, 4.60–4.61, 4.106
Inertia, rotor:
 in BLDC motors, 5.20–5.22
 in step motors, 5.52–5.55, 5.68–5.69, 5.74–5.76, 5.81
Inner-rotor motors, 5.8–5.9, 5.12–5.22
Input power equation, 10.139

Inside-diameter (ID) rotor machining, 3.15–3.16
Institute for Applied Microelectronics, 10.6
Institute of Electrical and Electronics Engineers (IEEE). *See* IEEE
Insulated-gate bipolar transistors (IGBTs). *See* IGBTs
Insulation material, 2.163–2.176
Integrated circuit (IC) technology, and environmental concerns, 10.6
Intel (as developer of CAN), 10.50, 10.51
INTERBUS-S, 10.50
Internal brush ring assemblies (IBRAs), 3.100–3.101
International Electrotechnical Commission (IEC). *See* IEC
International Protection (IP) codes, 10.6–10.8
International Rectifier, 10.79
Interpolar leakage reactance, 4.83
Interpolation:
 defined, 10.3
 optical encoder, 10.16–10.17
Intrinsic curve, defined, 2.82
Intrinsic saturation flux density, 1.63–1.64
Intrinsic value, defined, 1.44
Ireland, James R., 4.99
Iron:
 and armature force, 1.15–1.20
 intrinsic saturation flux density, 1.63
 losses, polyphase motors, 6.100
 losses, single-phase motors, 6.56, 6.57
 and powder metallurgy applications, 2.51, 2.59–2.71
ISO Standards, for acceptable armature balance, 4.61
Isotropic material, defined, 2.87

Jackson, Leon, 3.138n
Jitter, effect on encoder accuracy, 10.16
John C. Dolph Company, resins, 2.171–2.173
Joint design, 3.114–3.115

Jones, Dan:
 on commutation patterns in BLDC motors, 10.119n
 on stator winding design for BLDC motors, 5.22n
Jones, Douglas:
 on operation of stepping motors, 5.83n
 on stepping-motor control circuits, 10.70n
Judd, Robert R., 2.4n
Judd Consulting, 2.4n
Juds, Mark A., 1.1n

Kalpathi, Ramani, 10.65n
Karr, J. H., 6.73
Kassakian, J. G.:
 and clocked PWM, 10.115
 and delta connection, 10.112
 and triangle PWM, 10.116
 and Y connection, 10.110
Kinematic viscosity, 3.77
King, Todd L., 10.51n
Kirchhoff's law, 1.16, 1.91
 in analysis of dc series motors, 4.14
 in analysis of PMDC motors, 4.42
 in analysis of single-phase motors, 6.13, 6.74
 in analysis of three-phase motors, 10.107
Kit encoders, defined, 10.9
Knee of the curve, defined, 2.82
Kokal, Harold R., 2.51n
Krause, Robert F., 2.51n
Kuo, B. C., 5.73, 5.76

Lacing, stator, 3.42–3.46
Lamination:
 design factors, 3.27–3.38, 5.58, 5.100
 geometry (*see* Lamination, field, and housing geometry)
 in motor manufacturing process, 3.1–3.4
 steel, 2.4–2.46

Lamination, field, and housing geometry:
 armature conductors, 4.57–4.61
 field conductors, 4.73–4.74
 magnetic paths, 4.54–4.57, 4.69–4.73, 4.75–4.84
 stator, 4.61–4.69
 universal motor construction, 4.46–4.54
Laplace operator, thermal analysis and, 8.15
Lap winding:
 in dc motors, 4.138–4.140.4.143–4.156
 versus wave winding, 4.165–4.168, 4.181
LaValley, Roger O., 3.124n
Lawrence, Bill, 3.103n
LDJ Electronics, 3.138n
Lead wire, 2.88–2.189
Leakage factor calculations:
 for polyphase motors, 6.97
 for series dc and ac motors, 4.119–4.128
 for single-phase motors, 6.52–6.53, 6.64–6.67
 for synchronous motors, 7.19–7.21
Leakage flux paths:
 defined, 1.15
 permeance of, 1.29–1.32
Least significant bit (LSB), 10.21
Left flier top coming (LFTC), defined, 4.145
Left-hand rule, 1.6
Leine & Linde, 10.51
LEM Instruments (Hall device manufacturer), 10.93
Length unit conversions, 1.5
Lenz's law:
 explained, 3.142–3.143
 reactance voltage and, 4.10
Leonhard, W., 10.98
Level wind area, explained, 3.31
Linear commutation, defined, 4.10–4.12, 4.33–4.35, 4.84
Linear current limiters, 10.81–10.83
Linear encoders, 10.9–10.19, 10.22–10.25

Link Engineering Company:
  lacing machines, 3.43, 3.45
  on stator lacing, 3.42n
Load line, 1.48–1.51
  approximation of, 4.39
  plotting, 4.99, 5.73–5.74
Locked-rotor test, 9.44–9.45
Locked-rotor torque:
  defined, 10.52
  in PMDC motors, 4.113–4.114
  in PSC motors, 6.70
Locking-wire condition, explained, 3.30–3.31
Loctite Corporation, 3.109n
Lodestone, history of, 1.6, 2.90
Lodex, 2.87, 10.34
Long-shunt connection, defined, 4.134
Lorentz force:
  defined, 1.33
  and Hall devices, 2.100, 10.29–10.30
  in moving-coil motors, 1.35, 1.40–1.41, 1.85
LRA/FLA ratios, 10.54–10.63
L-type bearing seals, 3.50, 3.60
Lubricants:
  for bearings, 3.51, 3.61–3.66, 3.73–3.79
  testing, 9.36–9.41

Machining:
  rotor, 3.15–3.16
  shaft, 3.12–3.14
Macroporosity, defined, 3.119
Magnequench Company:
  demagnetization curves, 2.134–2.163
  properties of products, 1.49
Magnesia, history of, 1.6, 2.90
Magnetic coenergy, 1.12, 1.14, 1.17–1.21
Magnetic cores, manufacture of. *See also* Core losses
  bonded, 3.26
  cleated, 3.27
  lamination design factors, 3.27–3.38
  stack-in-die techniques, 3.37–3.42
  stator lacing, 3.42–3.46
  welded, 3.25

Magnetic curves, representative:
  Allegheny-Teledyne Company, 2.25–2.30
  Arnold Engineering Company, 2.102–2.133
  Magnequench Company, 2.134–2.163
  for nonoriented silicon steels
  Temple Steel Company, 2.6–2.24
Magnetic encoders, 10.25–10.28, 10.42–10.48
Magnetic field indicating sheet, 2.101–2.102
Magnetic field intensity unit conversions, 1.4
Magnetic field lines, explained, 1.6
Magnetic flux unit conversions, 1.3–1.4
Magnetic Instrumentation, 3.124n
Magnetic Material Producers Association (MMPA), 2.87
Magnetic materials. *See* Materials
Magnetic moment, 1.58–1.65
  in atoms, 1.62–1.63
  for current loop, 1.59
  equations, 1.2, 1.3
  far field, 1.64–1.65
  intrinsic saturation flux density, 1.63
  for magnetic material, 1.59
  overview, 1.58–1.59
  torque on, 1.61–1.62
  unit conversions, 1.5
Magnetic noise analysis, 9.23–9.27
Magnetic paths:
  armature, 4.54–4.57
  stator, 4.69–4.73
Magnetics:
  coil design, 1.67–1.77
  electromagnetic forces, 1.89–1.91
  electromechanical forces and torques, 1.32–1.43
  energy approach, 1.91–1.96
  flux path permeance, 1.21–1.32
  Helmholtz coil, 1.67
  losses, 1.52–1.58
  magnetic field, 1.65–1.67
  magnetic moment, 1.58–1.65
  materials, 1.43–1.52

Magnetics (*Cont.*):
  moving-coil actuator analysis,
    1.83–1.88
  reluctance actuator analysis,
    1.78–1.82
  terminology, 1.6–1.21
  units used for, 1.1–1.6
Magnetic sensors, and environmental
    protection, 10.6
Magnetics International, 2.51n
Magnetizers:
  capacitive-discharge, 2.93–2.97,
    3.138–3.147
  fixtures for, 2.97–2.99, 3.133–3.137
  overview, 2.91–2.93, 3.124–3.130
  safety factors, 2.99–2.100, 3.141,
    3.146–3.147
  types of, 2.93–2.97, 3.130–3.133
Magnetizing process. *See* Magnetizers
Magnetomotive force (mmf):
  defined, 4.4
  in PMDC motors, 4.27–4.42
  in series motors, 4.5–4.10,
    4.117–4.119, 4.122–4.126
  in shaded-pole motors, 6.81–6.82
  in two-phase motors, 6.6–6.13
Magnetomotive unit conversions, 1.4
Magnetoresistive (MR) sensors,
    10.25–10.28
Magnets:
  history of, 1.6, 2.80
  permanent (*see* Permanent magnets)
  safety issues of, 2.91
Magnet wire, 2.176–2.188
Major loop, defined, 2.82
Mandrel flexibility wire testing,
    2.178–2.187
Manufacturing of motors:
  adhesives, 3.109–3.124
  armatures, 3.22, 3.87–3.98
  assembly and testing, 3.23–3.24
  bearings, 3.46–3.79
  brush holders, 3.99–3.103
  commutator fusing, 3.79–3.87
  end frames, 3.4–3.10
  housing, 3.10–3.12
  magnetic cores, 3.25–3.46

Manufacturing of motors (*Cont.*):
  magnetizers, 3.124–3.147
  painting and packing, 3.25
  process flow, 3.1–3.3
  rotors, 3.15–3.203.87–3.98
  shafts, 2.12–3.15
  stators, 3.21–3.22
  testing, 3.129–3.130
  varnish impregnation, 3.103–3.109
Martin, Joseph, on winding patterns,
    5.24, 5.30, 5.32
Mask, defined, 10.11
Mass, defined, 3.91–3.92
Master synchronization telegram, 10.49
Materials:
  for bearings, 3.58–3.61
  in BLDC motors, 5.19–5.20
  characteristics of, 1.43–1.52, 2.1–2.4,
    2.80–2.163
  and core loss, 2.46–2.50
  end-frame, 3.4
  housing, 3.10–3.11
  insulation and, 2.163–2.173
  lamination steels, 2.4–2.46
  lead wire, 2.189
  magnet, 3.125
  magnet wire, 2.176–2.188
  and powder metallurgy, 2.59–2.71
  pressed core, 2.51–2.59
  shaft, 3.12
  test methods for, 2.71–2.80
  thermal analysis values for selected,
    8.16, 8.17, 8.19
Maximum energy product, 2.83–2.84
Maxwell bridge test method, 2.76
Maxwell's equation, 1.2
Mean length of turn (MLT), and BLDC
  motors, 5.32, 5.35
Mean turn length (MTL), and step
  motors, 5.63–5.64, 5.72
Measurement, electromagnetic,
  2.100–2.102. *See also* Drives and
  Controls; Units of measure
Mechanical energy. *See* Energy
Metal-oxide semiconductor field-effect
  transistors (MOSFETs).
  See MOSFETs

Metal Powder Industries Federation
  (MPIF), 2.61
Metals Handbook (ASM), 2.60
Meter, kilogram, second. *See* MKS system of units
Meter-mix pumping systems,
  3.106–3.107
Metglas, product properties, 1.46
Microelectromechanical systems
  (MEMS) project, 10.48–10.49
Micro Linear Corporation, 10.8
Microporosity, defined, 3.119
Microstepping, 10.93–10.97
  and current limiting, 10.80
  physics, 5.89–5.90
Micro Switch, 10.28
Midfrequency resonance, defined,
  5.83
Military standards. *See* MIL-STDs
Miller, Andrew E.:
  on ac induction motors, 6.72n
  on dc motors, 4.116n
Miller, T. J. E. (Tim), 10.124
  on delta connection, 10.110, 10.112
  on switched-reluctance motors,
    5.100
Milling, unbalance-correction,
  3.92–3.94
MIL-STDs:
  and application environment guidelines, 10.4
  and SERCOS technology, 10.49
Minor loop, defined, 2.82
Mitsubishi Electric:
  and current regulation, 10.92
  and PROFIBUS, 10.50
MKS system of units, 1.1–1.2, 2.5, 5.88
MMF drops, calculating:
  for series motors, 4.122–4.126
  for shaded-pole motors, 6.81–6.82
MMPA brief designation, defined,
  2.87
Modal analysis, 9.28–9.36
Modular encoders, defined, 10.9
Moiré patterns, defined, 10.11
Moment unbalance, 3.89
Monofilar coil windings, 5.49

MOSFETs:
  in H bridges, 10.92–10.93, 10.102
  in SRMs, 10.66
Most significant bit (MSB), 10.21
Motorola, 10.51, 10.75
Motors, electric. *See also* AC induction
    motors; DC motors; Electronically
    commutated motors; PMDC
    motors
  brief history of, 4.140–4.141
  defined, 1.1
  efficiency of pressed core versus laminated core, 2.56–2.59
  magnetizing sources for, 1.33
  manufacturing of (*see* Manufacturing
    of motors)
  materials used in (*see* Materials)
  versus tachometers, 10.40–10.41
Moving-coil actuator:
  Lorentz force, 1.40–1.41
  static and dynamic analysis, 1.83–1.88
MPIF, 2.61
Multiphase motors, 5.87
Multiplication. *See* Interpolation
Murphy, J. M. D.:
  and motor drive schemes, 10.98
  and PWM methods, 10.113, 10.116
  and Y connection, 10.107, 10.110
MWS Wire Industries, 1.69

NACAP, defined, 10.51
National Electrical Code (NEC). *See*
  NEC
National Electrical Manufacturers
  Association (NEMA). *See* NEMA
National Fire Prevention Association,
  10.51
National Institute of Standards and
  Technology (NIST). *See* NIST
National Semiconductor:
  Darlington arrays, 10.75
  H bridges, 10.79, 10.92, 10.97
  as interface hardware provider, 10.51
Natural frequency, explained, 2.96
NEC:
  and Design E motors, 10.51–10.52,
    10.57, 10.63

NEC (*Cont.*):
  exceeding guidelines during testing, 9.1
  as interface hardware provider, 10.51
Needle-wound salient pole motors, 3.35–3.37
Negative sequence network, defined, 6.32–6.33
NEMA:
  and brush holder design, 3.102
  environmental protection standards, 10.6
  magnet wire rating, 2.176
  motor designs, 10.51–10.65
  standards for motor applications, 8.1–8.2
Neodymium-iron, 2.91
  density of, 5.21, 5.69
  and eddy current effect, 3.143
  properties of, 2.87–2.89
  typical magnetizing forces for, 3.125, 3.142
Net armature resistance, calculating, 4.59
Net slot area (NSA), calculating, 4.51, 4.53, 4.68, 5.63
Newark Electronics, 10.93
Newton's law, 5.93, 5.94
Nickel:
  alloys, 2.87
  as coating, 2.86
  intrinsic saturation flux density, 1.63–1.64
  and powder metallurgy processing, 2.60
NIST, and smart transducer interfaces, 10.51
Nitriding process, 3.15
NMB Corporation, on bearings, 3.46n, 3.49n
Noise:
  analysis of magnetic, 9.23–9.27
  minimizing, 10.43–10.44
  in switched-reluctance motors, 5.105–5.107
Noncontacting encoders, defined, 10.3–10.4, 10.24

Nonlinear commutation, defined, 4.12
Nonoriented material, defined, 2.87
Normal curve, defined, 2.82
Normal flux, defined, 1.44
North magnetic pole, defined, 1.6, 2.80
Novotny, D. W., 10.113
NSA, defined, 5.63
Nuclear spin, impact on magnetic moment, 1.63
Nye Lubricants, 3.46n

Oberg Industries, Inc.:
  on manufacturing rotor/stator stacks, 3.38n
  stamping dies, 3.40
OD machining, rotor assembly, 3.16
Oersted, Hans, 1.6
Oersted Technologies, 2.80n
Oersted units, cgs system and, 1.2–1.3
Ohm's law:
  and ac coil current, 1.70, 1.72, 1.80, 1.85
  and current limiters, 10.80
  equation for, 1.8, 1.20
  and PMDC motors, 4.38
  and stepping motors, 5.98
Oil-filled versus electrolytic capacitors, 3.140–3.141
Oils, bearing. *See* Lubricants
One-phase-on operation, 10.99–10.100
One-shot feedback current limiting, 10.86–10.89
Open-circuiting, defined, 3.134
Open-loop controllers, 10.69–10.70
Open-loop current limiters, 10.83–10.86
Open Systems Interconnection (OSI). *See* OSI
Optical encoders:
  absolute, 10.19–10.22
  comparative technologies, 10.42–10.48
  and environmental issues, 10.6
  incremental, 10.9–10.19
  linear, 10.22–10.25
Orbital magnetic field, defined, 1.43
Oriented material, defined, 2.87
OSI, and FieldBus specs, 10.50

Outer-rotor motors, 5.4–5.9, 5.11–5.21
Output power equation, 10.139
Outside-diameter (OD) rotor machining, 3.16
Oven Systems, Inc., 3.103n
Overcommutation, 4.12, 4.33–4.35, 4.84
Owen bridge test method, 2.75
Oxidization, suppressing, 2.86

Pacific Scientific (SERCOS-compatible controls), 10.49
Packing and shipping issues, 3.23–3.25, 9.17, 9.35
Painting of motors, 3.25
Parasitic mass, as limitation of soft-bearing balancers, 3.96–3.97
Pattern errors, and encoder accuracy, 10.16
Payne, Stanley D., 3.27n
Peak energy product, defined, 10.33
Peak power density (PPD), 10.139–10.140
Peak power rate (PPR), 10.139–10.140
Peel stress, defined, 3.115
Perfluoropolyether (PFPE) oils, as bearing lubricants, 3.61, 3.76–3.78
Performance:
 optimizing motor constant, 10.138–10.139
 phase resistance, 10.132, 10.135
 power efficiency, 10.139–10.140
 speed-torque curve, 10.133–10.138
 torque and current profiles, 10.122–10.132
Peristaltic pumps, 3.107
Permanent-magnet (PM) dc motors. See PMDC motors
Permanent-magnet (PM) step motors:
 characteristics, 5.53
 controllers for, 10.72–10.74
 design, 5.68–5.74
 operation, 5.48–5.50
Permanent-magnet (PM) synchronous motors, 7.9–7.15
Permanent magnets:
 characteristics of, 2.80–2.163
 defined, 1.43–1.52

Permanent magnets (*Cont.*):
 magnetic moment for, 1.59–1.61
 magnetizing of, 2.91–2.100, 3.130
Permanent-split-capacitor (PSC) motors:
 and hysteresis synchronous motors, 7.5
 performance calculations, 6.45–6.59, 6.61–6.72
 and permanent-magnet synchronous motors, 7.9
 typical applications, 8.34
Permeability:
 ac-related test methods, 2.76–2.77, 2.79
 defined, 1.8–1.9, 1.21
 of free space, 1.5, 2.80–2.83
 unit conversions, 1.4
Permeameter, explained, 2.81
Permeance:
 coefficient, 1.48, 4.98–4.105, 5.73–5.74
 defined, 1.9–1.10, 1.20
 of probable flux paths, 1.21–1.32
Persson, Erland, 10.124–10.132
Peterman, Derrick, 3.138n
Petroleum, as bearing lubricant, 3.61, 3.62
Phase insulation, 2.176, 3.34–3.35
Phase modulation (PM), defined, 10.11
Phase plate, defined, 10.11
Phasor diagram, in performance calculations, 4.126–4.127
Phelps Dodge Company:
 on magnet wire, 2.176n
 wire properties charts, 2.180–2.183
Phillips (interface hardware provider), 10.51
Phoenix Electric Manufacturing:
 on brush holders, 3.99n
 and INTERBUS-S, 10.50
Phosphorus, in powder metallurgy processing, 2.62, 2.64–2.65
Photodetectors, 10.13
Photodiodes, 10.13
Phototransistors, 10.13–10.14
Photovoltaic devices, 10.13
Phytron (motor manufacturers), 5.89

Pin-contact encoders, 10.3
Pippenger, D. E., 10.31
Planck's constant, 1.5, 1.62
Plastic assemblies, testing of, 9.42–9.43
Plastic versus steel gears, 9.39–9.40
Plastiform, demagnetization curves for, 2.113–2.133
Plate-up processes, in optical encoders, 10.12
PLC. *See* Programmable logic controller (PLC)
Plonsey, R., 1.66–1.67
PMDC motors:
    comparison to BLDC motors, 5.1
    performance, 4.96–4.116
    synchronous motors as, 7.9
    theory, 4.23–4.46
    thermal analysis for, 8.10–8.32
Polarity, armature, 4.145–4.161
Pole arc distribution factor, 4.112
Pole configuration:
    and BLDC motors, 5.13, 5.15–5.16
    design factors, 3.31–3.32, 3.35–3.36
    effect on magnetizing, 3.126–3.127, 3.128
    effect on performance, 1.38, 1.39
    and step motors, 5.55–5.58, 5.60–5.62
    and switched-reluctance motors, 5.99–5.100
Polyester:
    as insulating material, 2.174, 3.35
    as lacing-cord material, 3.43–3.44
Polymer and iron powder composites, 2.65–2.71
Polyphase induction motors, 6.47, 6.93–6.102. *See also* Three-phase induction motors; Two-phase induction motors
Polystrophism, 10.21
Porosity, defined, 3.119
Positive sequence network, defined, 6.32–6.33
Potentiometers, 10.3
Powder metallurgy (P/M), 2.59–2.71
Power efficiency calculations, 10.139–10.140

Power losses, 1.52–1.58
Preaesodymium-iron, 2.87
Precision wind area, explained, 3.30–3.31
Premium-efficiency motors, defined, 10.52
Pressed core materials, 2.51–2.59
Press-fitting process, 3.17–3.18
Process FieldBus, 10.50
PROFIBUS, 10.50
Programmable logic controller (PLC):
    in magnetizing process, 3.130
    in varnishing process, 3.108
Progressive winding, defined, 4.145, 4.165
PSC motors. *See* Permanent-split-capacitor (PSC) motors
Puchstein, A. F.:
    on calculating field distortion, 4.133
    on determining air gap flux, 4.99
    on field leakage reactance, 4.81
    on flashover, 4.96
    on optimum commutation, 4.86
Pull-in torque, 5.81, 5.92
Pull-out torque, 5.81, 5.92
Pull-up torque, defined, 10.52
Pulse-width modulation (PWM), 10.113–10.119
    and current limiting, 10.84–10.86, 10.95
    and SRM voltage control, 10.69
    and turn-off behavior, 10.103

q–d transformation, 6.16–6.22, 6.25
Quadrature, defined, 10.2
Quantization, as limiting factor in microstepping, 10.95–10.96
Quarter-cylinder flux paths, 1.23–1.24, 1.25, 1.28–1.29
Quasi-static unbalance, 3.89–3.90

Race and ball defects, 9.17–9.23
Raceway:
    curvature, 3.57–3.58
    defined, 3.47, 3.55
Radial play, defined, 3.56–3.57

Radiation:
    as cause of EMI, 8.35–8.38
    as method of heat transfer, 8.11
Radio-frequency interference (RFI).
    See Electromagnetic interference
    (EMI)
Raposo, Antonio, 10.77
Rare-earth magnets:
    and eddy current effect, 3.143–3.144
    in hybrid step motors, 5.76
    leakage flux paths and, 1.30–1.31
    and PMDC motors, 4.36–4.37
Rated power density (RPD),
    10.139–10.140
Rated power rate (RPR),
    10.139–10.140
Rating life, defined, 3.64–3.65
RC, defined, 10.37
RDC, defined, 10.37, 10.38–10.39
Reactance:
    armature, 4.82–4.84
    of commutation, 4.94–4.95
    in dc series motors, 4.10–4.14
    field leakage, 4.81–4.82
    in PMDC motors, 4.32–4.37
    in shaded-pole motors, 6.82–6.86
    and stator geometry, 4.68–4.69
    in synchronous motors, 7.22, 7.23
Read head, defined, 10.23
Recoil permeability, defined, 2.82
Reduced-voltage idle test, 9.7, 9.47
Redundant sensors, 10.48–10.49
Reference mark, defined, 10.11, 10.12,
    10.24
Reflected core loss resistance, 1.57
Refrigerator magnets, composition of,
    2.89
Reluctance:
    actuator analysis, 1.15–1.20,
        1.78–1.82
    air gaps and saturation, 1.35–1.43,
        4.109–4.112
    defined, 1.7–1.10, 1.20
    motor, defined, 1.33
    permanent magnet, 1.51–1.52
Remanance *B,* defined, 2.81
Renco Encoders, 10.1n

Resins. *See also* Varnish
    John C. Dolph Company, 2.171–2.173
    3M Company, 2.175
Resistance:
    armature, 4.57–4.61
    and BLDC motors, 5.43
    defined, 1.7–1.10
    during magnetization, 2.95, 3.144
    in polyphase motors, 6.100
    in shaded-pole motors, 6.86–6.88
    in single-phase motors, 6.53–6.54
    and step motors, 5.63–5.64, 5.79
    in synchronous motors, 7.23
    testing, 9.10, 9.12
    thermal (*see* Thermal analysis)
    wire, 2.186
Resistive current limiters, 10.80–10.81
Resolution:
    and absolute encoders, 10.25
    defined, 10.1–10.2
    and incremental encoders,
        10.15–10.16
    and sensor technology, 10.45, 10.46
Resolvers, 10.37–10.40, 10.42–10.48
Resonance:
    analysis and control, 9.27–9.36
    and rotor balancing, 3.91–3.92
    and stepping-motor physics, 5.92–5.98
Response threshold, defined, 10.23
Retainers, bearing. *See* Cages
Retaining compounds, as motor adhe-
    sives, 3.121
Reticle. *See* Mask
Retrogressive winding, defined, 4.145,
    4.165
Revolving-field theory, 6.13–6.25, 6.29,
    6.33, 6.37, 6.45–6.46
Reynolds number, 8.11
Ribbon retainer, bearing, 3.59
Richards, Earl F.:
    on ac induction motors, 6.5n
    on application of motors, 8.10n
    on dc motor theory, 4.1n
    on energy-coenergy, 1.91n
Richter, R. C., 2.3–2.4
Right-hand rule, 1.6
Ring fire, 4.95–4.96

Ripple, defined, 10.40
RMS value:
  and ac series motor voltages, 4.19
  and eddy current density, 1.55
  and H-bridge circuitry, 10.104
  and sine-wave three-phase motors, 10.113
  and torque, 5.35
  and Y-connected three-phase motors, 10.108
Rockwell Automation, 9.43n
Rockwell C scale, and shaft hardness, 3.14, 3.72
Rolled steel, in housing manufacture, 3.11
Root mean square (RMS) value. *See* RMS value
Rose pattern winding, 4.165, 4.169
Rotary optical encoders, 10.9–10.19
Rotor:
  assembly, 3.15–3.20
  balancing of, 3.87–3.98
  copper losses, 10.53
  lamination process, 2.45, 3.1–3.4, 3.38–3.42
  transformation, 6.16–6.25, 6.41–6.44
  in VR step motors, 5.52–5.55
Rotors, H. C.:
  and flux path permeance, 1.22–1.28
  on predicting air gap flux, 4.100
  and skin effect, 1.56
Round-frame shaded-pole motors, 6.73, 6.79–6.81, 6.85–6.86
RTV gaskets, 3.120
Run capacitor, defined, 6.71
Runge-Kutta technique, 1.82, 1.87
Running-overload test, 9.6–9.7, 9.46–9.47
Running torque, defined, 3.50, 5.92
Runout errors, and encoder accuracy, 10.16
Ryan, D. R., 2.63
Ryan, J. B., 2.63

Safe operating area curve test (SOAC), 9.12–9.13

Safety issues:
  chemical-related, 3.124
  insulation-related, 2.167
  during magnetizing process, 2.99–2.100, 3.141, 3.146–3.147
  neodymium-iron, 2.91
Salient pole motors, 3.35–3.37
Samarium-cobalt, 2.90–2.91
  density of, 5.21, 5.69
  and eddy current effect, 3.143
  properties, 2.87–2.89
  typical magnetizing forces for, 3.125, 3.142
Saturation factor, in synchronous motors, 7.21–7.22
Saturation induction:
  and armature-induced flux distortion, 4.8–4.9
  as property of P/M materials, 2.62, 2.63
Saturation point, defined, 2.2
Schlecht, M. F.:
  and clocked PWM, 10.115
  and delta connection, 10.112
  and triangle PWM, 10.116
  and Y connection, 10.110
Schmitt trigger, 10.30, 10.44
Schultz, Karl H., 3.1n, 3.21n
Schultz Associates, 3.1n, 3.21n
SDC, defined, 10.37
Seagate, control chips, 10.79
Sealants, 3.119–3.121
Sealed encoders, defined, 10.9
Seals, bearing, 3.47, 3.50, 3.59–3.61
Seamless tubing, in housing manufacture, 3.11
Selectron, 10.50
Self-contained encoders, defined, 10.9–10.10
Sensing devices. *See* Drives and controls
Sensor databus systems, 10.49–10.51
Sensors Expo, 10.51
Separators, bearing. *See* Cages
SERCOS, 10.49

Series motors, theory. *See also* AC induction motors; DC motors
  alternating current, 4.17–4.23
  direct current, 4.1–4.17
Series-wound motor construction, 4.46–4.54
Servomotor lacing machines, 3.45
Setbacken, Robert M., 10.1n
SGS-Thomson Microelectronics, Inc., 10.8, 10.79, 10.88
Shaded-pole motors:
  calculating performance, 6.72–6.93
  induction synchronous, 7.1–7.4
  typical applications, 8.34
Shafts:
  attaching rotors to, 3.17–3.20
  manufacture of, 3.12–3.15
Shaker tests, 9.34–9.35
Shear stress:
  adhesive-related, 3.115, 9.43
  bearing-related, 3.74
Sheet-metal step motor. *See* Permanent-magnet (PM) step motors
Shields, bearing, 3.47, 3.50, 3.59–3.61
Shipping issues, 3.23–3.25, 9.17, 9.35
Short-shunt connection, defined, 4.134
Shrink-fitting process, 3.17–3.20
Shunt-connected dc motor performance, 4.130–4.134
Shunt PM field motors, 4.23–4.46
Siemens:
  and CAN, 10.51
  Hall devices, 10.30
  and PROFIBUS, 10.50
Signal conditioning, optical encoder, 10.14
Signal-to-noise ratio, optimizing, 10.43
Signetics Company, 10.8
Silicon, and powder metallurgy processing, 2.62, 2.64–2.65
Silicon-controlled rectifier (SCR), 2.93–2.95, 3.131, 3.141
Silicone:
  as bearing lubricant, 3.61, 3.62
  as motor adhesive, 3.117, 3.120–3.121
Silicon Systems, Inc., 10.8

Silva, M. M.:
  and clocked PWM, 10.115
  and dual-current-mode PWM, 10.116
  and voltage PWM, 10.113
Sine-cosine microstepping, 10.93–10.94
Sine dwell tests, 9.17
Sine-wave current drives, 10.122–10.132
Sine-wave motor, 10.101–10.102, 10.112–10.113
Single-end lacers, 3.43
Single-phase induction motors:
  balancing networks, 6.32–6.37
  circuit model, 6.13–6.32
  design procedure, 6.37–6.38
  overview, 6.1–6.5
  performance calculations, 6.45–6.61
  permanent-split-capacitor (PSC), 6.61–6.72
  rotor transformation, 6.18–6.25
  shaded-pole, 6.72–6.93
  stator transformation, 6.17–6.18
  theory, 6.5–6.13
Sintering, 2.61–2.65
Sinusoidal distribution:
  determining, 6.2–6.5
  for single-phase motors, 6.50–6.51
Sinusoidal drives, 10.122–10.132
Sinusoidal PWM, 10.116
SI units. *See* Système International (SI) system
Six-step drive scheme, defined, 10.107
Skeleton-frame shaded-pole motors, 6.73
Skin effect:
  and eddy current power loss, 1.53–1.57
  minimizing, 5.98–5.99
Sleeve bearings. *See* Bearings
Slew rate:
  optical encoder, 10.17–10.18
  and sensor technology, 10.45–10.46
Sliding friction. *See* Static friction
Slip, defined, 10.52
Slip-fitting process, 3.17–3.19
Slot configuration:
  in BLDC motors, 5.33–5.36, 5.37–5.40
  in lamination design, 3.28–3.30
  in step motors, 5.62–5.63

Slot constants, calculating:
    polyphase motors, 6.95–6.96
    single-phase motors, 6.51–6.52, 6.63–6.64
    synchronous motors, 7.18–7.19
Slot-fill percentages, 3.32–3.34
Slot leakage reactance, 4.82–4.83
Slotless brushless inner-rotor motors, 5.9–5.13
Slow-decay mode, 10.76–10.77, 10.90–10.91
Smart sensors, 10.48
Smearing, effect on motor efficiency, 3.16–3.17
Snap flex test, 2.184–2.185
SOAC. *See* Safe operating area curve test (SOAC)
Soft-bearing balancers, 3.19, 3.94–3.97
Soft magnetic materials, 1.43–1.44
Solderability test, 2.187
Solenoid-type magnetizing fixtures, 3.133–3.134
Sony, 10.27
South magnetic pole, defined, 1.6, 2.80
Spectral analysis, 9.15–9.27
Speed, and rotor balancing, 3.90–3.91
Speed-torque curve:
    acquiring data, 9.11
    calculating, 10.133–10.138
    defined, 4.96–4.97
    testing stations, 9.1–9.5
Spherical flux paths. *See* Flux path permeance equations
Spin magnetic field, defined, 1.43
Spin tests, for plastics and adhesives, 9.43
Split-phase motors:
    performance calculations, 6.45–6.60
    typical applications, 8.34
Springback, degrees of, 2.178
Square-wave current drives, 10.122–10.132
Square-wire slot-fill percentages, 3.32
Squirrel-cage rotors:
    in single-phase motors, 6.13, 6.21–6.22, 6.72
    in three-phase motors, 6.44

SSD21-type bearing shields, 3.60–3.61
Stack-in-die techniques, 3.37–3.38
Stainless steel:
    applications for, 2.62
    in bearing manufacture, 3.49, 3.59, 3.65
Stamped-construction step motor. *See* Permanent-magnet (PM) step motors
Stamping dies, and rotor/stator stacking, 3.38–3.42
Standard-efficiency motors, defined, 10.52
Standardization:
    sensor-related, 10.4–10.7
    of steel grade specs, 2.4–2.6
Starting torque, defined, 3.50
Static analysis:
    moving-coil actuator, 1.83–1.85
    stepping motor, 5.88–5.89
Static friction, defined, 3.76, 5.90–5.91
Static radial load rating, 3.66–3.67
Static unbalance, 3.88–3.89
Stator:
    assembly, 3.21–3.22, 6.1–6.3
    copper losses, 10.53
    cost factors, 2.51
    geometry of, 4.61–4.69
    lacing process, 3.42–3.46
    lamination process, 2.44–2.45, 3.1–3.3, 3.38–3.42
    magnetic paths in, 4.69–4.73
    pressed metal versus laminated, 2.59
    in step motors, 5.70–5.71
    transformation, 6.16–6.18, 6.41–6.44
    varnishing of, 3.106
    winding, in BLDC motors, 5.22–5.36
Steady-state analysis, 1.83–1.85
Steady-state equivalent circuit:
    in single-phase induction motors, 6.13–6.32
    in three-phase induction motors, 6.38–6.44
Steady-state model of heat transfer, 8.11

Steel. *See also* Materials
   in armature, calculating weight of, 4.53–4.54
   Brinell hardness of, 1.43–1.44
   coefficient of thermal expansion, 3.20
   flux path permeance equations, 1.21–1.29
   and gear fatigue, 9.39
   lamination, 2.4–2.46
   magnetizing forces in, 1.33, 1.35–1.40
   in motor construction (*see* Manufacturing of motors)
   properties of, 1.46
   selection of, 2.1–2.4
   thermal conductivity value, 8.16
   thermal properties, 8.19
Steinmetz, C. P.
   defining hysteresis loss, 2.2–2.3
   and skin effect, 1.56
Step angle, defined, 5.55–5.59
Step motors:
   control circuits, 10.70–10.79
   current limiting, 10.80–10.93
   hybrid step motors, 5.50–5.51, 5.74–5.79
   measurement of characteristics, 5.79–5.83
   operation, 5.83–5.99
   overview, 5.1
   performance, 5.51
   permanent-magnet (PM) step motors, 5.48–5.50, 5.68–5.74
   sizing, 5.51–5.52
   variable-reluctance (VR) step motors, 5.47–5.48, 5.52–5.67
Stiffness, defined, 3.92
STIM, defined, 10.51
Stored magnetic field energy, defined, 1.11–1.14
Straddle loop, defined, 4.162
Stray load loss:
   in dc series motors, 4.16
   and Design E motors, 10.53
   in PMDC motors, 4.45–4.46, 4.115
Stress tests, 9.43
Strontium ferrite, as magnet material, 2.90

Stupak, Joseph J., 2.80n
S-type bearing seals, 3.60
Superglue, as motor adhesive, 3.116
Swenski, Chris:
   on ac induction motors, 6.45n
   and shrink-fitting calculations, 3.19
   on synchronous motors, 7.16n
Swept sine wave test, 9.17, 9.31, 9.36
Switched-reluctance motors (SRMs):
   controls for, 10.65–10.70
   design, 5.103–5.107
   lamination properties, 5.100
   overview, 5.1, 5.99
   performance, 5.100–5.103
   selection of poles and phases, 5.99–5.100
   sizing the air gap, 5.100
Synchronous motors:
   drive schemes, 10.101–10.102
   hysteresis, 7.5–7.9
   induction, 7.1–7.4
   performance, 7.16–7.24
   permanent-magnet, 7.9–7.15
Synchronous serial interface (SSI), 10.49
Synchros, 10.37–10.40
Synthetic hydrocarbons, as bearing lubricants, 3.61
Système International (SI) system, 1.1–1.2, 2.5

Tach kits, 10.9
Tachometers, DC, 10.40–10.42
Tang-type commutators:
   fusing process, 3.79–3.87
   hot-staking and, 4.159
TDX, defined, 10.37
Teeth configuration:
   and BLDC motors, 5.13, 5.15, 5.16–5.19
   and step motors, 5.60–5.62, 5.70–5.71
Teflon:
   in bearing seals, 3.50, 3.60
   to control resonance, 5.96
Temperature:
   and BLDC motor performance, 5.44–4.45
   encoders versus resolvers, 10.47

Temperature (*Cont.*):
  and Hall devices, 10.34–10.35
  impact on magnetic materials, 2.85
  and motor lubricant selection, 3.62, 3.76–3.79
  and optical encoders, 10.17
  in shrink-fitting rotors to shafts, 3.19–3.20
  and switched-reluctance motors, 10.65
  and tachometer performance, 10.42
  thermal analysis, 8.10–8.32, 9.5–9.9
  in varnishing process, 3.107
Temple Steel Company:
  on core loss, 2.46n
  representative magnetic curves, 2.6–2.24
  steel material specs, 2.4–2.6
Tensile stress, defined, 3.114
Terminal resistance test, 9.10
Tesla units, defined, 10.25. *See also* Système International (SI) system
Testing:
  ac motor thermal tests, 9.5–9.9
  of adhesives and plastic assemblies, 9.42–9.43
  alternating-current tests, 2.74–2.80
  automation of, 9.43–9.47
  bond strength, 3.114
  commutator fuse, 3.80–3.82
  dc motor testing, 9.9–9.15
  direct-current tests, 2.71–2.74
  fatigue and lubrication tests, 9.36–9.41
  magnetizer, 3.129–3.130
  of motors prior to shipping, 3.23–3.25, 9.17, 9.35
  particle evaluation, 2.52–2.53
  resonance control, 9.27–9.36
  sensor-related, 10.5–10.6
  spectral analysis, 9.15–9.27
  speed-torque curve, 9.1–9.5
  wire, 2.178–2.187
Texas Instruments, 10.75
Thermal analysis, 8.10–8.32, 9.5–9.9
Thermal resistance tests, 9.12
Thermal shock tests, 9.42
Threadlocking adhesives, 3.117–3.118

3M Company, resins, 2.175
Three-phase induction motors:
  circuit model, 6.38–6.44
  drive schemes for, 10.105–10.113
  typical applications, 8.34
TII, defined, 10.51
Time unit conversions, 1.5
Tobaben, E. J., 10.31
Toluene/alcohol boil test, 2.187
Tomasek, Jaroslav, 10.124
Tooth pitch angle, calculating, 4.50, 4.100, 4.107–4.108, 5.61, 7.19
Tooth tip leakage reactance, 4.84
Torque:
  of bearings, 3.50
  and BLDC motors, 5.41
  constant, 4.115
  and dc series motors, 4.14–4.16
  electromagnetic, 1.89–1.91
  energy-coenergy approach, 1.91–1.96
  equations, 1.12–1.21, 1.32–1.43, 1.90, 1.96
  loss, 4.128–4.130
  on magnetic moment, 1.61–1.62
  performance curves, 4.96–4.97
  and PMDC motors, 4.42–4.44
  profiles, 10.122–10.132
  ripple, 9.13–9.15
  and shaded-pole motors, 6.90—6.91
  and single-phase motors, 6.27–6.28, 6.32, 6.36–6.37, 6.59–6.61, 6.68–6.69
  versus speed, 5.97–5.98
  and step motors, 5.64–5.65, 5.64–5.67, 5.73–5.74, 5.76–5.77, 5.80–5.81, 5.80–5.82, 5.82
  and switched-reluctance motors, 5.100–5.102
  and universal motors, 4.128–4.130
Torque ripple test, 9.13–9.15
Toshiba, 10.79
TR, defined, 10.37
Track dimensions, raceway, 3.55–3.56
Transient model of heat transfer, 8.11
Trapezoidal drives, 10.122–10.132
Triangle PWM, 10.116–10.117
Tri-arc magnet configuration, 3.122–3.123

Trickey, P. H.:
  and cross-field theory, 6.45, 6.74
  equations for B constants, 6.91–6.93
  on field reactance, 4.81–4.82
  on plotting B-H curve, 4.80, 4.110
  on shaded-pole motors, 6.72–6.89
  on synchronous motors, 7.16
Trickey factor, and mmf drop, 4.76, 4.79–4.82
Trickle system (for applying varnish), 3.105–3.107
Triplen harmonics, 10.110–10.112
Tube steel, in housing manufacture, 3.11
Turnbull, F. G.:
  and motor drive schemes, 10.98
  and PWM methods, 10.113, 10.116
  and Y connection, 10.107, 10.110
Turn-off behavior, and H-bridge circuitry, 10.103–10.104
Turns ratio:
  of auxiliary winding to main winding, 6.29
  field-to-armature, defined, 4.74
  of stator and rotor windings, 6.19
Twenty-four-Hour Test (Band), 9.43n
Twenty-four-hour test bench, 9.44–9.45
Two-phase induction motors:
  drive schemes for, 10.98–10.105
  theory, 6.6–6.13
Two-phase-on operation, 10.100–10.101
TX, defined, 10.37

UL requirements. *See* Underwriters Laboratories (UL)
Unbalance:
  defined, 3.87
  effects of, 4.61
Undercommutation, 4.12, 4.33–4.35, 4.84
Underwriters Laboratories (UL):
  and good test practices, 9.5
  and insulation requirements, 2.167
  and thermal tests, 9.6, 9.8, 9.43
Unipolar motors, 5.49, 5.84–5.85, 10.72–10.75
Unipolar slide-by Hall sensor, 10.30–10.31, 10.32

Unitrode, control circuits, 10.97
Units of measure:
  drives and controls, 10.1
  electromagnetics, 1.1–1.6, 2.5, 5.88
Universal motor:
  construction, 4.46–4.54
  defined, 4.142
  performance, 4.116–4.130
University of Iowa, 5.83n, 10.70n
University of Maine, 5.36n, 10.97n
Urethane, as motor adhesive, 3.116
U.S. Steel, 1.52
U-scan method, 10.21
UV light, as curing mechanism for adhesives, 3.117

Vanadium permendur:
  core loss in, 2.26–2.30
  dc hysteresis loop for, 2.30
  induction and permeability of, 2.25
Variable-pitch winding, 5.24–5.32
Variable-reluctance (VR) step motors:
  controllers for, 10.71–10.72
  design, 5.52–5.67
  operation, 5.47–5.48, 5.83–5.87
Variable-source unbalance, defined, 3.87
Varnish:
  impregnation process, 3.103–3.109
  properties of, 2.166, 2.168–2.173
Vectors, defined, 2.80
Veinott, C.:
  on single-phase motor design, 6.37
  on winding patterns, 5.24, 5.30, 5.32
Velocity:
  as component of centripetal force, 3.91
  of light, 1.5
  profiles, determining, 8.4–8.9
  unit conversions, 1.5–1.6
Verghese, G. C.:
  and clocked PWM, 10.115
  and delta connection, 10.112
  and triangle PWM, 10.116
  and Y connection, 10.110
Vibration Institute, 9.15
Vibration testing. *See* Spectral analysis

Vically (obsolete material), 2.87
Virgin curve, defined, 2.81
Viscosity:
　defined, 3.74
　as factor in choosing adhesive dispensing system, 3.123
Viscous friction, defined, 3.76, 5.90
Visible light, as curing mechanism for adhesives, 3.117
V-mill cutter, 3.93–3.94
Voltage boosting, and current limiting, 10.83–10.84
Voltage constant test, 9.10
Vortex tube, role in cooling, 2.98
VPL (as encoder manufacturer), 10.27
VR step motors. *See* Variable-reluctance (VR) step motors
V-scan method, 10.21

Walker dc tester, 2.53
Walters, Harry J., 3.38n
Waveform analysis, motor efficiency, 2.56–2.58
Wave winding:
　in dc motors, 4.138–4.140
　versus lap winding, 4.165–4.168, 4.181
WCI, steel grade designations, 2.4
Welded cores, 3.25
Welding process, rotor-to-shaft, 3.17–3.19
Welsh, Richard, Jr., 10.124
Westinghouse, magnet wire chart, 5.33–5.34
Wheatstone bridge, 5.79
White noise, and bearing contamination, 9.19
Windage losses. *See* Friction and windage losses
Windamatic Systems, Inc., 3.27n
Winding:
　in ac induction motors, 6.2–6.5
　in BLDC motors, 5.22–5.36
　distribution factor in PMDC motors, 4.112
　double-flier, 4.161–4.165, 4.175–1.181
　four-pole armature, 4.165–4.181
　history of, 4.140–4.141

Winding (*Cont.*):
　patterns in dc motors, 4.138–4.140, 5.24–5.33
　polarity, 4.145–4.161
　specs, 3.27, 3.30–3.31, 4.145
　in step motors, 5.72, 5.78–5.79
　theory, 4.142–4.145
Wire:
　in the fusing process, 3.84–3.85
　lead, 2.189
　magnet, 2.176–2.188
Wound stator assembly processing, 3.1, 3.21–3.25
Wrapped steel, in housing manufacture, 3.11
Wye connections, 10.105–10.110

X/R ratio, defined, 10.52

Y connections. *See* Wye connections
Yeadon, Alan W.:
　on bearing systems, 3.46n
　on dc motors, 4.46n, 4.116n
　on magnetic cores, 3.25n
Yeadon, William H.:
　on ac induction motors, 6.1n
　on bearing systems, 3.46n, 3.72n
　on BLDC motors, 5.1n, 5.45n
　on dc motors, 4.130n
　on insulation, 2.163n
　on lead wire and terminations, 2.189n
　on magnetic cores, 3.25n
　on magnetic materials, 2.1n
　on motor applications, 8.1n
　on step motors, 5.47n
　on switched-reluctance motors, 5.99n
　on synchronous motors, 7.1n
Yeadon Energy Systems, Inc., 3.19n
Yeadon Engineering Services, PC. *See* Yeadon, Alan W.; Yeadon, William H.

Zener diode, 10.82
Zero-impedance damping coefficient, 4.116
Zinc, in end frame construction, 3.4
Z-type bearing shields, 3.59

ABOUT THE CONTRIBUTORS

LARRY C. ANDERSON (Sec. 3.14) is an applications consultant with American Hofmann Corporation, one of the world's leading manufacturers of precision balancing machines. He has been with the company since 1990 and performs unbalance analysis on rotating assemblies for manufacturers worldwide. He holds a BS degree in electrical engineering technology and has over 20 years experience, with the past 8 focused on the electric motor industry.

JOHN S. BANK (Sec. 3.15) is the executive vice president of Phoenix Electric Manufacturing Company and is responsible for coordinating new product development and developing advanced strategies. He received his bachelor's degree in business administration (magna cum laude) from the University of Michigan in 1981 and his JD from UCLA in 1984. He is also a Certified Public Accountant in the state of Illinois (1981) and a licensed real estate broker in the state of Illinois (1981). Mr. Bank currently serves on the board of directors of SMMA (1995–present) and EMERF (1997–present). He is the Company Representative and Voting Member of NEMA (1992–present), EMCWA (1992–present) and NAM (1992–present).

WARREN C. BROWN (Sec. 3.10.6) graduated with a BSME from Michigan State University in 1966 and with an MBA from Michigan State University in 1968. He was the Manager/Director MIS of Burroughs Corporation in Detroit, Michigan, from 1968 to 1982. He directed sales and marketing at Link Engineering Company from 1982 to 1990. Since 1990, he has been vice president for motor products of Link Engineering Company. He has been a member of SAE, ESD, and SMMA.

JOSEPH H. BULARZIK (Sec. 2.5) is a staff engineer. He received a BS in chemistry from Arizona State University, Tempe, in 1982. He received a PhD in chemistry from the University of California, Berkeley, in 1987. He conducted postdoctoral research in the field of superconducting oxides at Princeton University, Princeton, New Jersey, in 1989. He was an assistant professor of chemistry at Lycoming College, Williamsport, Pennsylvania, from 1987 to 1989. He has seven years of experience in magnetic materials research. He is a member of ASM. He has worked in research for Magnets International, Inc., East Chicago, Indiana, since 1994, and he worked in research at Inland Steel Company, East Chicago, Indiana, from 1990 to 1994.

PETER CAINE (Sec. 3.16) graduated from the University of Wisconsin in Platteville with a BS in industrial engineering. His career at Oven Systems, Inc., has included applications engineering, custom product sales, and management. For the past three years, he has managed the electric motor equipment division.

DAVID CARPENTER (Chap. 4 finite-element plots) received a first-class honors BSc in electrical engineering from the University of Southampton, England, in 1979 and joined GEC Ltd. as an induction motor design engineer. In 1986 to 1987 he was appointed as visiting professor at Lakehead University, Canada, and in the following year he received an MSc from Coventry University, England. After joining Vector Fields Ltd. as an application engineer in 1991, he received a PhD from the University of Bath, England, in 1993. He was appointed to the position of vice president of Vector Fields, Inc., United States, in 1995. He is a Charter Engineer and a member of the IEEE.

JOHN COCCO (Sec. 3.17) is the director of Loctite Corporation's North American Application Engineering Center. For the past 10 years, he has been working with Loctite Corporation's customer base, developing adhesive and sealant applications for use in small motors. In the past two years, he has conducted several design seminars at original equipment manufacturers focusing on this topic. He holds a bachelor's degree in chemical engineering and is a licensed Professional Engineer.

PHILIP DOLAN (Sec. 3.10.5) graduated from Marquette University with a BA. He was vice president of Marketing for Oberg Industries and had previous experience in plant management and strategic planning.

BIRCH L. DEVAULT (Sec. 10.6) was born in Pittsburgh, Pennsylvania, in 1946. He received a BS in electrical engineering from the University of Pittsburgh in 1967. He joined the Westinghouse Electrical Graduate Student Course in 1967. In 1968, he joined the Westinghouse Standard Control Division, Beaver, Pennsylvania, as an associate design engineer. In 1981, he joined the Control Division in Asheville, North Carolina. Since February of 1994, he has been a senior development engineer with Cutler-Hammer, Milwaukee, Wisconsin, responsible for the design and application of magnetic motor control. He is a Registered Professional Engineer in the state of Pennsylvania. He has eight patents in the area of motor control. He is a member of IEEE. He has published papers related to motor control in TAPPI and IEEE publications.

BRAD FRUSTAGLIO (Sec. 6.4) has a BSME from Michigan Technological University and is a design engineer for Yeadon Energy Systems, Inc.

FRANCIS HANEJKO (Sec. 2.6) is a metallurgical engineer and received his BS and MS degrees from Drexel University. He has been employed by the Hoeganaes Corporation for 22 of the last 25 years. During that time, he has held numerous positions in the sales and marketing and research and development departments. His current position is manager of electromagnetics and customer applications in the research and development department, with responsibilities for customer service and product development. He is a past chairman of the Philadelphia Section of the APMI.

DUANE C. HANSELMAN (Secs. 5.1.4 and 10.11) is an associate professor in electrical engineering at the University of Maine. He holds PhD and MS degrees from the University of Illinois. He is a senior member of IEEE and an associate editor of the *IEEE Transactions of Industrial Electronics*. He is the author of numerous articles on motors and motion control. He has published several textbooks, including *Brushless Permanent-Magnet Motor Design and MATLAB Tools for Control System Analysis and Design* (McGraw-Hill, 1994).

DANIEL P. HECKENKAMP (Sec. 10.6) received his BS in mechanical engineering from the University of Wisconsin, Milwaukee, in 1983. In 1981, he joined the Square D Company in Milwaukee, where he was responsible for the design of industrial lifting magnets and their applications. In 1983, he transferred to the Square D Controls Division, where he was responsible for contactor development. He joined Cutler-Hammer's controls division in 1988 as a product development engineer, where he has been responsible for the design and maintenance of contactors and overload relays. His current position is principal engineer.

LEON JACKSON (Sec. 3.19) received an AS from Port Huron Junior College in 1957. He also received a BS in electrical engineering from Wayne State University in 1960. He also attended the University of Loyola for business administration. He received honors from the Tau Betta Pi educational honor society and the Etta Kappa Nu engineering honor society for academic achievement. He has worked for General Magnetic Corporation and LDJ Electronics, Inc., where he is currently president. He is a member of the IEEE Magnetics Society.

DAN JONES (Secs. 5.1.3, 10.12, and 10.13) has a BSEE from Hofstra University and a MS in mathematics from Adelphi University. He is a member of ASME, IEEE, ISA, and AIME. He has 38 years experience in the motor business. He founded Incremotion Associates in 1982 and has previously worked for such companies as Vernitron, Printed Motors, Inc., Singer-Kearfott, Electro-Craft Corporation, Data Products Corporation and IMC Magnets Corporation.

ABOUT THE CONTRIBUTORS

DOUGLAS W. JONES (Secs. 5.2.10 and 10.8 to 10.10) is an associate professor of computer science at the University of Iowa. He received his PhD in computer science from the University of Illinois, Urbana, in 1980. He completed his BS in physics at Carnegie-Mellon University in 1973. His research interests are discrete event simulation, resource protection in architecture, operating systems, system Programming Languages, and the history of computing.

MARK A. JUDS (Secs. 1.1 to 1.12) has BS and MS degrees in mechanical engineering from the University of Wisconsin. He is currently a senior principal engineer for Eaton Corporation's Innovations Center, where he designs electromagnetic devices. He also has expertise in heat transfer and mechanical dynamics.

ROBERT R. JUDD (Secs. 2.2 and 2.3) is currently president of Judd Consulting Associates, Inc., a general ferrous metallurgy and electrical-sheet consulting firm. He acquired his doctorate in materials science from Carnegie-Mellon University and holds a bachelor's degree in mechanical engineering from the University of Rochester. He spent 30 years in principal research positions for U.S. Steel and Ispat-Inland. For three years he served as director of research and development for Johnstown Corporation, a large ferrous foundry and fabrication firm. He has also taught general metallurgy at Carnegie-Mellon University. His professional activities include ASM, AIME, MPIF and the ASTM A-6 subcommittee on magnetics. He is also the treasurer and organizing committee member of the annual Conference on the Properties and Application of Magnetic Materials. He holds patents in the powder metallurgy and soft magnetic material fields.

RAMANI KALPATHI (Sec. 10.7) was a senior project engineer with Dana Corporation. He completed his PhD in electrical engineering at Texas A&M University in 1994 and has been with Dana for the past five years. Recently he has returned home to start his own consulting firm in Madras, India. His interests are in the areas of power electronics and control of switched-reluctance motors.

JOHN KAUFFMAN (Sec. 2.10) graduated from Purdue University with a BA in industrial economics in 1963. He has worked for Phelps Dodge Magnet Wire Company for 35 years. He holds four patents for magnet wire and cable products and equipment.

TODD L. KING (Sec. 10.6) received BS and MS degrees in electrical engineering from the University of Wisconsin-Madison in 1978 and 1980, respectively. He joined Borg Warner Corporate Research Center, Des Plaines, Illinois, in 1980, where he worked in analysis of motors and actuators and the design of automotive controls, actuators, and sensors. He joined Eaton Corporate Research and Development Center, Milwaukee, Wisconsin, in 1988 as a senior engineer specialist, where he worked in the design of actuators for appliance, automotive, aerospace, hydraulic, and truck products. He also worked in the design and analysis of commercial and industrial motor controls. He became the engineering manager for the Design Analysis Technology Group in 1990 and added systems technology in the Eaton Innovation Center, where he has responsibility for defining the strategic direction of systems technology for the corporation.

HAROLD R. KOKAL (Sec. 2.5) is a senior staff engineer. He received his BS and MS degrees in metallurgical engineering from the University of Minnesota, Minneapolis, in 1964 and 1970, respectively. He has 30 years experience in process and product research. He is a member of APMI and AIME. He has worked in research at Magnetics International, Inc., East Chicago, Indiana, since 1992. He worked in research at Inland Steel Company, East Chicago, Indiana, from 1985 to 1992, and at U.S. Steel Corporation, Coleraine, Minnesota, and Monroeville, Pennsylvania, from 1968 to 1985. He was an MRRC Research Fellow at the University of Minnesota, Minneapolis, from 1965 to 1966.

ROBERT F. KRAUSE (Sec. 2.5) is a technical director. He received his BS and PhD degrees in material science from Notre Dame University, South Bend, Indiana, in 1962 and 1966, respectively. He has 31 years experience in metallurgy and magnetic materials. He is a member of the ASM and IEEE. He has worked in research at Magnetics International, Inc., Burns Harbor, Indiana, since 1991. He worked in research at Inland Steel Company, East Chicago, Indiana, from 1987 to 1991; at Crucible Steel Company, Pittsburgh, Pennsylvania, from 1986 to 1987; at Westinghouse Electric Corporation, Churchill, Pennsylvania, from 1972 to 1986; and at U.S. Steel Corporation, Monroeville, Pennsylvania, from 1966 to 1972.

BARRY LANDERS (Chap. 9) has 24 years of experience in the design and testing of ac and dc motors, including writing electrical and mechanical design and testing software for fractional-horsepower ac, brush DC, and brushless dc motors, as well as for fine-pitch custom gearing. In addition, he has 17 years of experience in spectral analysis of sound, vibration, and current on these motor types and on ball bearings as received, as well as in failure analysis of field problems. As a senior project engineer and registered Professional Engineer, he currently has responsibility for an engineering development, analysis, and test group for ac and dc products at Electro-Craft Motion Control, Gallipolis, Ohio (a Rockwell Automation business).

ROGER O. LAVALLEY (Sec. 3.18) is a senior application engineer with Magnetic Instrumentation, Inc. He has 25 years experience in the area of magnetic applications. In his present position he is responsible for reviewing customer requirements for the magnetizing, demagnetizing, and measuring of permanent magnets and magnet assemblies and for proposing the appropriate equipment and complete systems.

BILL LAWRENCE (Sec. 3.16) has a BSME and an MBA from Marquette University. He has worked in sales of servo electric motors at Moog, Inc., and in sales of specialty motors at Doerr Electric. He is currently the vice president of Oven Systems, Inc.

ANDREW E. MILLER (Secs. 4.5, 6.4.3, and 6.4.4) has a BS in chemical engineering from Michigan Technological University. He has several years of experience in software design and three years of experience in the motor design industry.

STANELY D. PAYNE (Sec. 3.10.4) is the vice president engineer at Windamatics Systems, Inc., Fort Wayne, Indiana.

DERRICK PETERMAN (Sec. 3.19) has over eight years experience with magnetics research and instrumentation. He completed a BA in physics at Washington University, St. Louis, Missouri, in 1989 and a PhD in physics at Ohio State University in 1996. He currently holds the position of magnetic measurement specialist at LDJ Electronics.

CURTIS REBIZANT (Figs. 5.58 to 5.61 boundary element plots) is an engineer at Integrated Engineering Software, which produces and markets software for electromagnetic, thermal, and structural system simulation. He has a BS in electrical engineering from the University of Manitoba and has extensive experience with electromagnetic CAE software.

EARL F. RICHARDS (Secs. 1.14, 4.1, 6.2, 6.3, and 8.4 to 8.8) is Professor Emeritus of Electrical Engineering in the School of Engineering, University of Missouri, Rolla. He received his PhD from the University of Missouri. He has 16 years of field experience in motor design and over 36 years of experience in the instruction of motor technology. His professional emphasis is on electromechanical, power, and control systems. He currently instructs graduate-level engineering courses and is frequently sought as an industrial and legal consultant.

ROBERT M. SETBACKEN (Secs. 10.1 to 10.5) is vice president of engineering at Renco Encoders, Inc. He received his MSME degree in 1979. He has developed and tested analog and digital electromechanical and hydraulic servosystems for the military and commercial interests. Since joining Renco in 1990, he has been involved with the design and manufacture of incremental rotary optical encoders for the industrial and office automation industries.

KARL H. SCHULTZ (Secs. 3.1 to 3.9) holds a BSME from Western Michigan University. He is a senior member of SME and a member of SMMA. He has 25 years experience in manufacturing and management with such companies as General Signal, General Electric, Emerson Electric, Clark Equipment, Chrysler, and Cincinnati Milacron, and his own consulting firm.

ABOUT THE CONTRIBUTORS

JOSEPH J. STUPAK JR. (Sec. 2.8) received BSME and MSME degrees from the California Institute of Technology, Pasadena, California, in 1965 and 1969, respectively. He is a licensed Professional Engineer in the state of California, in the field of control. He has been a senior engineer; chief scientist with Synektron Corporation, a manufacturer of brushless dc motors; and a professor at California Polytechnic Institute. He worked as an independent consultant in the fields of magnetics and electromagnetics for 10 years, and included the U.S. Naval Undersea Warfare Center, Newport, Rhode Island, among his major clients. He is now the president of Oersted Technology Corporation, Portland, Oregon, a manufacturer of magnetizing equipment and instruments for the magnetics industry. He has 16 issued patents, with 3 more applied for, and has published 20 papers. He speaks Danish and German, as well as native U.S. English. He is a member of the IEEE. He is an amateur magician and is a licensed commercial pilot with instrument rating.

CHRIS A. SWENSKI (Secs. 3.6.5, 6.4, and 7.4) is an engineering technician for Yeadon Energy Systems, Inc.

HARRY J. WALTERS (Sec. 3.10.5) is a graduate of the Johns Hopkins University in mechanical engineering. He has patents in press transfers, stamping die mechanisms, and die sensing. He has a background in plastics extrusion and injection molding, stamping die and mold design, automation, and machine design. He is currently employed by Oberg Industries.

ALAN W. YEADON, P.E. (Secs. 3.10, 3.11, and 4.2 to 4.5) is vice president of Yeadon Engineering Services, PC, and Yeadon Energy Systems Inc. He holds a BSEE degree from the University of Illinois. He has 12 years experience in product design, consulting for the motor industry, and development of software for electric motor design and analysis.

LUCI YEADON is the owner of Luci's Photography, Stambaugh, Michigan. She contributed most of the photographs for the book.

WILLIAM H. YEADON, P.E. (Secs. 1.13, 2.1, 2.9, 2.11, 3.10 to 3.12, 4.3, 4.6 to 4.8, 5.1 to 5.3, 6.1, 6.4, 7.1 to 7.4, and 8.1 to 8.3) is president of Yeadon Engineering Services, PC, and Yeadon Energy Systems, Inc. He is a graduate of the University of Dayton and has 33 years experience in the motor industry. He has expertise in the areas of design and development, production, quality control, and management. He has worked for such companies as Redmond Motors, A. O. Smith, Warner Electric, and Barber Colman. He is an instructor with the SMMA Motor College.

ABOUT THE EDITORS

WILLIAM H. YEADON, P.E. is the president of Yeadon Engineering Services, PC, and Yeadon Energy Systems, Inc. He helped to establish the motor college for the Small Motor and Motion Association (SMMA). He is a member of SMMA, Electrical Manufacturing and Coil Winding Association (EMCWA), National Society of Professional Engineers (NSPE), and the Institute of Electrical and Electronics Engineers, Inc. (IEEE). He currently writes and teaches courses for the SMMA and EMCWA, designs motors, and is a consultant. He has more than 30 years of experience in electric motors, holding positions in design and development, management, production, and quality control with companies that include Redmond Motors, A. O. Smith, Warner Electric, and the motor division of Barber-Colman Company. He has design and development experience in electric motors and generators including ac induction motors, dc permanent-magnet and wound-field motors, and electronically commutated, brushless dc, stepper, and switched-reluctance motors. He has done failure analysis and served as a manufacturing and cost-reduction consultant. He also has served as an expert witness. He is a graduate of the University of Dayton and is a registered professional engineer in Michigan, Ohio, Illinois, and Wisconsin.

ALAN W. YEADON, P.E. holds a BSEE degree from the University of Illinois. He assisted in the establishment of the SMMA motor college and has taught PMDC motor design classes. He has design experience in ac induction motors, dc permanent-magnet and wound-field motors, electronically commutated bushless dc, and switched-reluctance motors. He has 12 years experience in product design, consulting, and development of software for electric motor of design and analysis. He is a registered professional engineer in Michigan and Illinois.